Useful constants extracted from P. J. Mow and B. N. Taylor, *Reviews of Modern Physics*, **72**, 351 (2000).

$$\mathbb{C} \text{ (speed of light)} = 2.99792458 \times 10^{10} \text{ cm/sec}$$

$$\mathbf{N_A} \text{ (Avagadro's number)} = 6.0221420 \times 10^{23} \text{ molecules/mol}$$

$$h_P \text{ (Planck's constant)} = 6.62606876 \times 10^{-27} \text{ erg-sec}$$
$$= 6.62606876 \times 10^{-34} \text{ kg-m}^2\text{/sec}$$
$$= 4.13566727 \times 10^{-15} \text{ eV-sec}$$

$$h_P\mathbf{N_A} = 9.530698 \times 10^{-14} \text{ (kcal/mol)-sec}$$
$$= 3.990312689 \times 10^{-13} \text{ (kJ/mol)-sec}$$

$$\mathbf{N_A}h_P\mathbb{C} = 2.857229 \times 10^{-3} \text{ kcal/molecule/(cm}^{-1})$$
$$= 1.1962656492 \times 10^{-2} \text{ kJ/molecole/(cm}^{-1})$$

$$k_B \text{ (Boltzman's constant)} = 1.3806503 \times 10^{-16} \text{ erg/molecule/}^\circ\text{K}$$
$$= 1.3806503 \times 10^{-23} \text{ kg-m}^2\text{/sec/molecule/}^\circ\text{K}$$

$$R = k_B/\mathbf{N_A} = 1.9865 \times 10^{-3} \text{ kcal/mol}$$
$$= 0.0820575 \text{ lit-atm/mol/}^\circ\text{K}$$
$$= 8.314472 \text{ kg-m}^2\text{/sec}^2/^\circ\text{K/mol}$$

Conversion factors

$$1 \text{ kcal/mol} = 4.186800 \text{ kJ/mol} = 4.33931 \times 10^{-2} \text{ eV/molecule}$$

$$1 \text{ kcal/mol corresponds to } 349.8 \text{ cm}^{-1}$$

$$1 \text{ amu} = 1.6605387 \times 10^{-24} \text{ gm}$$

Other useful formulas

$$\frac{k_B T}{h_P} = 6.2510 \times 10^{12}/\text{sec} \left(\frac{T}{300 \text{ K}}\right)$$

$$\bar{v} = \left(\frac{8k_B T}{\pi m_g}\right)^{1/2} = 2.52028 \times 10^{13} \frac{\text{Å}}{\text{sec}} \left(\frac{T}{300 \text{ K}}\right)^{1/2} \left(\frac{1 \text{ amu}}{m_g}\right)$$

$$\frac{h_P v}{k_B T} = \left(\frac{v}{209.2 \text{ cm}^{-1}}\right)\left(\frac{300 \text{ K}}{T}\right)$$

CHEMICAL KINETICS AND CATALYSIS

CHEMICAL KINETICS AND CATALYSIS

Richard I. Masel
University of Illinois at Urbana-Champaign

WILEY-
INTERSCIENCE

A JOHN WILEY & SONS, INC., PUBLICATION

New York • Chichester • Weinheim • Brisbane • Singapore • Toronto

For ordering and customer service, call 1-800-CALL-WILEY.

Library of Congress Cataloging-in-Publication Data:

Masel, Richard I., 1951-
   Chemical kinetics and catalysis / by Richard I. Masel.
     p. cm.
   Includes index.
   ISBN 0-471-24197-0 (cloth : alk. paper)
    1. Chemical kinetics. 2. Catalysis. I. Title.

QD502 .M35 2002
541.3'94 — dc21   00-043303

Printed in the United States of America.

10 9 8 7 6 5 4 3 2 1

CONTENTS

PREFACE

The idea for this book arose out of an undergraduate/first-year graduate course that we teach at the University of Illinois called "kinetics and catalysis." When I first started teaching this course over 20 years ago, there were several good books in the area. The second edition of Laidler's classic book, *Chemical Kinetics*, was about 10 years old at that point. Moore and Pearson had just produced an updated version of Frost and Pearson's old standby, *Kinetics and Mechanism*. Also available were the excellent books by Wilkinson, Weston and Schwartz, Espenson, Kondat'ev, and Benson.

As with any field, even the most wonderful books become dated after many years. The initial versions of Laidler and Frost and Pearson were composed almost 50 years ago. At that time, scientists predicted preexponentials for reactions from first principles to within a factor of 100. Activation barriers and mechanisms could not be predicted at all. In comparison, today scientists routinely predict preexponentials for gas-phase reactions to within a factor of 2, calculate activation barriers within a few kilocalories per mol, and predict mechanisms of gas-phase reactions. Additionally, chemical kinetics continues to gain ground in a large number of disciplines. Scientists now use the concepts of chemical kinetics to understand the behavior of biological systems, to predict rates of semiconductor growth, and to aid in the design of new materials and processes. Such broad and far-reaching examples did not exist when Laidler and Frost and Pearson were conceived. The revolution in chemical kinetics, along with the ongoing need for current examples, provided my motivation to conceive and author a modern chemical kinetics textbook.

The purpose of this book is to update the standard texts on chemical kinetics and to provide an overview of the kinetics concepts and examples since 1960. Although I have maintained a traditional introduction to kinetics via rate equations and mechanisms, the material is presented with a fresh perspective. For example, we can now predict mechanisms of many reactions, and, thus, obtain an a priori rate equation. With the advent of computers and graphing programs, it is no longer necessary to utilize complex methods to fit data to appropriate rate equations. Let the computer do this portion of the work!

The next part of the book discusses the theory of reaction rates. I maintain an introduction to our old friends, collision theory, transition state theory, the Rice–Ramsperger–Kassel–Marcus (RRKM) model, again, however, with a new perspective. For example, I demonstrate the use of trajectory calculations to simulate reactions and to develop improved models to describe the reaction rate.

The third section discusses the prediction of potential energy surfaces. I begin with a description of the use of quantum-mechanical methods to estimate the potential surfaces for reactions, and in particular, discuss how codes have advanced to the stage where most scientists, regardless of their training, are able to generate accurate and useful predictions of potential surfaces. I also focus on the recent advances in potential energy surface science, and how scientists can change the potential energy surface for a reaction in a directed fashion to modify the rate.

Finally, I close with a brief discussion of catalysis. In particular, I discuss how solvents, metals, and oxides can be used to modify the potential energy surface for reactions, and thereby modify reaction rates.

As with any book, it was necessary to compromise on some issues to provide a reasonably sized book for comprehension purposes. I decided to emphasize the chemistry of kinetics, rather than physics or dynamics. Therefore, I believe that this book gives a thorough description of why reactions occur in the way they do and how one influences rates. Still, I provided limited coverage of certain traditional issues, including the mathematics of integrating rate equations, photochemistry, state-to-state chemistry, and reactions in condensed phases. In my book entitled *Principles of Adsorption and Reaction on Solid Surfaces,* I focused in great detail on solid surfaces. With respect to the integrating rate equations, the widespread use of computers and graphics programs have made those issues trivial. Finally, I did not include a detailed discussion of experimental methods, which I find a very interesting topic. Again, because of my desire to create a useful yet manageable textbook, those issues were not addressed.

In order to create a useful textbook for students, professors, and researchers alike, I included many solved examples, useful homework problems, and illustrative computer programs. I have personally found a need for good problems in my kinetics courses. I also provide a good deal of background material, such as a brief introduction to quantum mechanics and statistical mechanics. Although not essential to a kinetics course, I have found that they provide a useful background for students to apply in a number of learning situations. Finally, I provide detailed derivations of all key equations.

In the text I have used boldface to highlight key terms or concepts and for special emphasis. I have used italics for ordinary emphasis and for definition or introduction of general terms. I have used block capital letters (e.g., HOT MOLECULE) to alert readers to points that are frequently missed by students or that often generate errors. Also, I have enclosed certain equations and portions of text in boxes to indicate items that warrant closer attention.

I routinely teach the material in this text to both undergraduate students and graduate students. When I teach my undergraduate course, I usually focus on Chapters 1–7, use the examples in Chapters 8 and 9, and then skip to Chapters 11 and 12. If time permits, I also use Chapters 13 and 14. For the graduate course, I alternate between two different methods. Some years, I teach a two-semester course, with kinetics during the fall semester and catalysis and surfaces in the spring. For the two-semester scenario, I teach the material in Chapters 1 to 12 in the first semester, and include the material from the later chapters in the second semester. Other years, I teach kinetics and catalysis as a one-semester course. In the one-semester course scenario, I use a course outline similar to the undergraduate course, but spend less time on Chapter 3 and focus more on Chapters 10 and 11. The book is written so that professors can select chapters to tailor the course to their particular needs. I have found, however, that Chapters 2, 5, and 7 are required for students prior to more advanced material.

During my years of writing, it was always my goal to create a book that would serve professors, researchers, and students alike. I have tried to make this book a useful tool for all those interested in chemical kinetics. I hope that you enjoy it.

In writing the book, I very much appreciate the helpful comments of Scott Fogler, Andy Gellman, Ed Seebauer, Ali Shah, Rajeev Kilkarni, Paul Blowers, Laura Ford, and Ivan Lee. Scott Fogler's careful review of the manuscript is much appreciated. If there are errors in the book, they are all mine.

RICHARD I. MASEL

CHEMICAL KINETICS AND CATALYSIS

INTRODUCTION

1.1 INTRODUCTION

Chemical kinetics is the study of reaction rates. The field got its start more than 200 years ago. At that time people were just developing the atomic view of matter, and were just starting to understand what chemical reactions were like. One of the early discoveries was that chemical reactions did not occur instantaneously. For example, in 1771 Wenzel noted that the dissolution of zinc or copper in acid was not instantaneous. Rather, it took a finite amount of time. Further, the rate of dissolution depended on the concentration of the acid. In 1778, Priestly found that the amount of time it took to convert mercury oxide into elemental mercury depended on the amount of oxygen present. Lavosier made similar measurements with tin calx. These early measurements showed that chemical reactions had finite rates that could be measured and, in principle, understood.

Chemical kinetics has changed considerably since Priestly and Lavosier's time; however, the fundamental goals have remained the same: finding ways to understand and predict rates of important chemical processes. Prior to 1900, the main focus of kinetics was on establishing the nature of rate laws. In a series of papers published between 1860 and 1870, Harcourt and Essen showed that the rate of reaction varied with the concentration of the reactants. They also showed that one could fit the variation an equation. Van't Hoff [1886] expanded this idea to show that rates of reactions were a function of the concentrations in the reactor and the temperature. Arrhenius quantified the temperature behavior. Menschutkin showed that the rate also varied with the structure of the molecules and the nature of the solvent. These workers showed that one could develop mathematical expressions for rates of reaction as a function of the conditions of the reaction. The idea that rates can be quantified with equations is still important today.

The first theories of reaction rates were proposed during the period from 1889–1930. In 1889 Arrhenius wrote a famous paper where he showed that reactions are activated because only hot molecules can actually react. That lead to the idea that rates of reactions were determined by the rate of collisions of hot molecules. In 1918, Trautz and Lewis quantified the ideas by showing that the rate of reaction was equal to the rate of collisions times the probability that the collision leads to reaction. The resultant model, called *collision theory*, is still used today. Trautz and Lewis did not have an expression for the reaction probability. However, in 1928, Tolman showed how to formulate the probability of reaction in terms of the partition function of the reacting molecules. Polanyi and Eyring then expanded the idea to produce what is now called *transition state theory*. These works showed that one can estimate fairly accurate preexponentials for reactions from simple thermodynamic ideas and a knowledge of the nature of molecular collisions. Those concepts are still used today.

People could not directly examine molecular collisions in the 1930s and 1940s. However, in about 1960 Lee and Hershbach found that they could directly measure rates of molecular collisions in a device called a *molecular beam*. They directed two streams of molecules together, and measured the reaction probability directly. They discovered, for example, that the collision of two atoms can never lead to reaction. Instead one needs a collision partner for reaction to occur. In a larger way they provided the data that one needed to check the models of molecular collisions. Lee and Hershbach changed collision theory from a computational to an experimental science. That drove kinetics for 30 years.

In the late 1980s laser spectroscopy lead to many major advances. Laser spectroscopy allows one to probe molecular events on a femtosecond (FS) timescale. That allows one to directly probe reactions on the timescale of a collision and to examine the molecular details. Since 1980, another major driving force for improvements in our understanding of reaction rates has been computational. People can now calculate rates about as well as they can measure them. Ab initio calculations are starting to be almost as accurate as experiments in predicting rates. I expect computations to grow over the next 50 years and to become the predominant technique in chemical kinetics.

1.2 OUTLINE OF THE PRESENTATION

The objective of this book is to provide a general overview of chemical kinetics. Our goal is to

- Provide students with a working knowledge of the kinetics of chemical processes.
- Show how rates of reactions are related to the microscopic processes
- Show how rate constants can be estimated or calculated accurately
- Explain how activation barriers arise
- Show how a solvent or a catalyst affects rates

Chapter 2 provides a general overview of chemical kinetics and kinetic rate laws. Themes in the chapter include definitions of some key quantities, introduction to rate equations and rates of reaction, and temperature dependence. The chapter compares catalytic and noncatalytic reactions and shows how the rates are the same and different. The chapter also provides many examples from outside of chemistry to show how the principles are applicable to them.

Chapter 3 considers the analysis of kinetic data. The chapter briefly discusses how rates are measured, with a particular emphasis on distinguishing between direct or indirect measurements or rate data. The chapter then describes a variety of different ways to analyze rate data, and discusses the strengths and weaknesses of each.

Chapters 4 and 5 consider mechanisms of reactions. Chapter 4 reviews the idea of a mechanism and describes how mechanisms relate to rate laws. It shows how one can integrate a series of rate equations analytically or numerically, to simulate behavior. It then discusses a common approximation called the *quasi-steady-state approximation* and describes when and why it works.

Chapter 5 discusses mechanisms in more detail. In particular, the chapter describes what mechanisms are like in the gas phase and on surfaces, and how one can predict mechanisms from information about the thermodynamics of elementary steps, and the knowledge of bond energies in key species.

Chapter 6 is a review of thermodynamics and statistical mechanics. It includes a short section on the empirical prediction of physical properties. Then the chapter reviews the key concepts from statistical mechanics that students need to do reaction rate calculations.

Chapters 7–9 reviews reaction rate theory. Chapter 7 gives an overview of collision theory, transition state theory, and related ideas to give students a qualitative idea of how one can estimate preexponentials of reactions.

Chapter 8 examines molecular collisions in more detail. In particular, it discusses how reactions happen on a molecular level and shows how one can model the process using techniques from molecular dynamics. The chapter then describes the collisions that lead to reaction, and mentions the forces that control reaction behavior.

Chapter 9 concentrates on the accurate calculation of preexponentials for reactions. First we derive a general expression for rates of bimolecular reactions. We then simplify the expression to get to transition state theory. We briefly discuss some extensions of transition state theory, including variational transition state theory and multiconfiguration transition state theory. We then move on to discuss the theory of unimolecular reactions: principally RRKM theory and phase space theory.

Chapter 10 discusses why reactions are activated. We derive the Polanyi equation, the Marcus equation, and the Blowers–Masel equation and study their similarities and differences. We then discuss the strengths and limitations of the various methods, and describe why they fail in the quantum limit. We then describe the configuration mixing model, as a way of bridging the gap between the semiempirical methods and the exact quantum results.

Chapter 11 describes how one can estimate an activation barrier for reaction. We show how to use the semiempirical methods from Chapter 10 to estimate activation barriers. We then go on and discuss calculation of activation barriers using ab initio methods, and provide many examples for the reader to use. At the end of Chapter 11 we demonstrate that one can accurately calculate activation barriers for reactions using ab initio methods.

Chapter 12–14 provide a detailed discussion of catalysis. Chapter 12 provides an overview of catalytic action. We describe what catalysts are like, and what they do. We then have a detailed discussion of how catalysts work, and the modes of catalytic action. We also review the kinetics of catalytic action and the role of mass transfer on catalytic rates.

Chapter 13 shows how one can use the results from Chapter 12 to understand how solvents affect rates of reaction. In the literature, people do not think of solvents as catalysts. However, the chapter will show that solvents work just like catalysts. Solvents stabilize intermediates and thereby enhance rates.

Chapter 14 covers catalysis by metals. We will review the modes of catalytic action and see how the ideas apply to reaction on metals. We will then review how the activity of metals varies over the periodic table. Finally we provide a quantitative basis for catalyst selection.

1.3   A GUIDE FOR STUDENTS

When you read the book you should pay attention to a few key things.

- Look for equations in boxes. They are the most important equations in the chapter, and you should probably memorize them.
- Be sure to look at the solved examples. I put the solved examples at the end of the chapter so that you can find them easily when you are doing the homework. Be sure to look at them.
- Read the chapter twice. I usually recommend that students (1) read the chapter through once, skipping the derivations, (2) go through the solved examples, and then (3) read the chapter a second time to concentrate on the details.
- Download the computer programs. I have posted computer programs for many of the examples on my Website, *http://www.uiuc.edu/~r-masel/*. Be sure to download them and try them out.
- Remember to read the words as well as the equations. I know that students often use a book to help figure out how to solve their homework. Still, in my experience a qualitative understanding of kinetics is more important than knowing how to put numbers into the equations. Be sure to read the words as well as the equations.
- Do not trust everything in the textbook. I know that the statement "Do not trust everything in the textbook" sounds strange to a student, but remember that kinetics is an evolving field. If I look back to 10 years ago, things that I was sure then were right are now known to have serious limitations. I expect that 10 years from now, some of the ideas that I discuss here will be known to not work as well as I thought when I wrote this book. Evolution of knowledge is part of science; and in 10 years something in this book is bound to be wrong.

I have tried to make this book as useful of a teaching tool as possible given the length limitations. I hope that you enjoy it.

In Chapters 3–14, I will list a series of bibliographic works for further information. Reviews of the history of kinetics include:

V. A. Kritsman, G. E. Zaikov, and M. M. Emanuel, *Chemical Kinetics and Chain Reactions: Historical Aspects*, Nova Science Publishers, 1995.

K. J. Laidler, *The World of Physical Chemistry*, Oxford University Press, New York, 1996.

M. C. King, Experiments with time, progress and problems in the development of chemical kinetics. *Ambix* **28**, 70 (1981).

M. C. King, *Ambix* **29**, 49 (1982).

Specific references mentioned in this chapter are given at the end of this book.

2

REVIEW OF SOME
ELEMENTARY CONCEPTS

PRÉCIS

The objective of this chapter is to provide some background material that will be used throughout the rest of the book. Specifically, we review the definition of the reaction rate, the stoichiometric coefficient, the rate equation, the order of reaction, the preexponential, and the activation energy. We will then discuss empirical rate laws, and the effect of temperature and catalysts on rates. Discussions of these topics are necessarily brief, since most of the readers of this book will already be familiar with these topics. However, the objective of this chapter is to bring the student's understanding of these topics to the next level. We will provide precise definitions of terms that were defined loosely in freshman chemistry. We will point out the historical basis for ideas, and describe how the ideas can be applied to nonstandard problems. We will also give real examples. There is necessarily some review material in the chapter. However, we directed the chapter toward someone who already has seen the material and who wants to understand it better, rather than to someone learning the material for the first time.

2.1 DEFINITIONS

To start off, it is useful to define some key terms. In this section we will define

- The stoichiometric coefficient
- The rate of reaction
- Homogeneous reactions
- Heterogeneous reactions

First, it is useful to start with some history. Chemical reactions were first studied in detail at the end of the eighteenth century. The initial work was done before people knew that there were atoms or molecules and the idea that there were chemical compounds was still in dispute. It was clear that chemical reactions did occur. However, the nature of the reactions was still unclear.

One of the difficulties in the early work was that when one did an experiment, everything seemed to change. For example, Lavoisier [1789] did a series of experiments where he oxidized tin to tin–calx (SnO_2). The volume of the tin changed during the reaction and the mass changed during reaction. All of the properties of the tin seemed to change during the reaction. Lavoisier initially had trouble finding anything that was constant. However, in a series of famous experiments, Lavoisier oxidized the tin in a sealed flask and weighed the flask as the reaction proceeded. The experiment required the most sensitive balance that had ever been constructed, and the work was possible only because the French Academy of Science decided to finance the work. Lavoisier found that the weight of the sealed flask did not change during reaction. Instead, the weight change occurred only when air was able to rush into the vessel. This work showed, for the first time, that mass was conserved during chemical reactions.

Lavoisier also found that a fixed amount of air reacted with a fixed amount of tin. This work lead to the law of fixed proportions, which states that chemicals react in fixed proportions during chemical reactions.

Finally, Lavoisier found that only part of the air was reactive. This finding lead Lavoisier to propose that air was made up of different elements: a reactive element and an unreactive element.

Lavoisier called the reactive element in the air the "atmospheric principle". Later, Joseph Priestley visited France at Lavoisier's invitation. Lavoisier and Priestley discussed Priestley's discovery [1790] of '*dephlosgenated air*', which was liberated when mercury oxide was heated. Lavoisier quickly recognized that the dephlosgenated air was the reactive element in the air. Lavoisier renamed the reactive element *oxygen*. This work lead to the idea that air, and all mixtures and chemical compounds are made up of fundamental components called elements. Lavoisier also showed that reactions happen when an element moves from one chemical compound to another chemical compound.

Lavoisier then did a series of experiments where he reacted oxygen with a series of metals, and found that in most cases, a fixed amount of oxygen reacted with a fixed amount of metal to yield a fixed amount of oxide. Soon thereafter, Lavoisier proposed that **molecules react in fixed proportions**. That is, if one has a given reaction between A and B, one can define coefficients α_1 and α_2 such that during a given reaction, α_1 molecules of A always react with α_2 molecules of B to yield α_3 molecules of C and α_4 molecules of D:

$$\alpha_1 A + \alpha_2 B \Longrightarrow \alpha_3 C + \alpha_4 D \qquad (2.1)$$

In Lavoisier's case, he said that 1 kilogram (kg) of air would always react with 0.78 kg of tin to yield 0.99 kg of tin–calx and leave 0.79 kg of inerts. Lavoisier decided to call the study of these fundamental proportions **stoichiometry**, from the Greek word stoichio-, which means fundamental. Lavoisier also proposed that one could write an equation for each chemical reaction by balancing the production of all of the essential elements in the reaction. In other words, during the reaction between oxygen and tin, the number of kilograms of oxygen used up during the reaction would be equal to the number of kilograms of oxygen in the product calx.

Lavoisier also defined the **stoichiometric coefficient** β_n for molecule n participating in a given reaction, where Lavoisier defined the stoichiometric coefficient as the number of molecules of n *produced* when the given reaction goes once. Lavoisier's definition is different from the one you are used to in freshman chemistry. The stoichiometric coefficient is negative for the reactants and positive for the products. For example, in the reaction

$$2CO + O_2 \Longrightarrow 2CO_2 \qquad (2.2)$$

$\beta_{CO} = -2$, $\beta_{O_2} = -1$, $\beta_{CO_2} = 2$. The advantage of this definition is that it is much easier to use in a computer program and a spreadsheet than a definition where all stoichiometric coefficients are positive. Later in this book, we will find that this definition also allows you to simplify the equations for a complex reaction pathway. The disadvantage of the definition is that you have to watch out for the negative numbers.

One thing to be careful about is that the value of the stoichiometric coefficient changes according to how one writes the equation. For example, reaction (2.2) can be rewritten

$$CO + \tfrac{1}{2}O_2 \Longrightarrow CO_2 \qquad (2.3)$$

In reaction (2.3), $\beta_{CO} = -1$, $\beta_{O_2} = -\tfrac{1}{2}$, and $\beta_{CO_2} = 1$.

Dalton [1805] extended Lavoisier's idea with his atomic theory. Dalton proposed that each of Lavoisier's elements were composed of tiny particles called *atoms*. Dalton showed that the number of each type of atom was conserved during chemical reactions. Each particle had a fixed weight. Although Dalton could not weigh each atom, he could define a relative scale by arbitrarily defining the relative weight of hydrogen to be one atomic mass unit (1 amu), and then calculating the weights of all of the elements accordingly. Dalton's calculations were not particularly accurate, but they did allow Lavosier to calculate stoichiometric coefficients for many reactions.

Inspired by Lavoisier and Dalton's findings, many gentlemen-scientists set out to explore the nature of molecules. A particularly famous person at that time was Louis Thénard, [1818] who discovered hydrogen peroxide. Thénard thought that hydrogen peroxide was a wondrous molecule. It was composed of hydrogen and oxygen — the same components as ordinary water. Yet hydrogen peroxide behaved much differently than ordinary water. It could change the colors of dyes and would convert alcohol to vinegar. Thénard, being a wine connoisseur, was particularly interested in how hydrogen peroxide could convert alcohol into vinegar. He studied the reaction and measured the rate. Thénard found that when he diluted the alcohol, the rate went down. Additional hydrogen peroxide made the reaction go faster. Thénard then showed that by eliminating peroxides, he could make wine last longer. Thénard never quantified his findings, but his work showed for the first time that rates of chemical reactions varied with the concentrations of the reactants.

Many people expanded on Thénard's findings. In 1869, Guldberg and Waage proposed that rates of reaction were proportional to the "active masses" of the reactants, where in most cases the active mass roughly corresponded to the concentration of the species. Many other rate laws were proposed between 1860 and 1880. Still, it was not until Van't Hoff published his famous book, *Etudes de Dynamique Chemie* in 1884 that rates were put into a quantitative framework.

Van't Hoff was considered to be the greatest chemist of his day. Van't Hoff started the field *physical chemistry* and was the first person to really push the idea that quantitative reasoning was useful to a chemist. Van't Hoff wrote several influential books, including *Theoretische Chemie*, which set out the field of physical chemistry for the first time, and *Etudes de Dynamique Chemie*, which established chemical kinetics as a separate subdiscipline.

In *Etudes de Dynamique Chemie*, Van't Hoff quantified his ideas about rates of chemical reactions. Van't Hoff (1878) noted is that if one wants to compare rates of different reactions, one has to do so on a consistent basis. At that time, people had used all different kinds of reactors to examine rates of reactions. Some people had worked in tiny reactors, while other people had worked on big reactors. Generally, big reactors produced products faster than did small reactors, independent of what reaction was occurring in the reactor. Van't Hoff asserted that if one wanted to understand reactions, one would have to remove the effects of the size of the reactor from the analysis. For example, Van't Hoff noted that if a reaction is occurring uniformly over the volume of a vessel, then if one doubles the size of the vessel, one will produce twice as many moles of products in a given time. Consequently, Van't Hoff noted that it was not very useful to compare the total number of moles of product produced in one reactor to the number of moles of a different product produced in a different-sized reactor. Rather, it was important to consider how the rate per unit volume varied from one reaction to the next. Van't Hoff defined r_A, the rate that a given molecule A reacts as the rate of *production* of A in moles per volume per time. With this definition, the reaction rate, r_A, for a species A is negative if A is a reactant and positive if A is a product. Van't Hoff showed that if a reaction is run in a closed vessel with perfect mixing, one can determine r_A by measuring C_A, the concentration of A, as a function of time, t, and then substituting into the following equation:

$$r_A = \frac{dC_A}{dt} \tag{2.4}$$

Van't Hoff also noted that equation (2.4) is useful only when a reaction is occurring uniformly over the volume of a reactor. Van't Hoff considered several reactions that were instead occurring on the surface of his reactor. In that case, it is useful to define R_A, a rate per unit surface (i.e., $mol/(cm^2 \cdot second)$).

In the work that follows, we will use a small r to indicate a rate per unit volume and a capital R to indicate a rate per unit area. We will also define **heterogeneous reactions** as reactions that take place on the walls of the reactor or on other surfaces within the reactor. We will also define **homogeneous** reactions as reactions that occur in the gas phase, in liquids, or throughout the bulk at a solid.

Again, these definitions are slightly different from the definitions you learned in freshman chemistry. In freshman chemistry, one usually talked about homogeneous reactions occurring in the same phase as the reactants and heterogeneous reactions as occurring in the interface between two phases. Note, however, that the freshman chemistry definition is not accurate. For example, consider the hydrogenation of ethylene:

$$H_2 + C_2H_4 \Longrightarrow C_2H_6 \tag{2.5}$$

One can run ethylene hydrogenation by bubbling ethylene and hydrogen through a liquid containing rhodium atoms or by bubbling the same mixture through a reactor containing a solid platinum catalyst. In the former case, the ethylene and hydrogen absorb (i.e., dissolve) into the liquid and react while they are in solution. In the latter case, the ethylene and hydrogen adsorb onto (i.e., bind to) the surface of the catalyst and react on the surface. Therefore, both processes are very similar. Still, we call the former case a *homogeneous* reaction because the reaction is occurring in a liquid. We call the latter case a *heterogeneous* reaction because the reaction is occurring on a surface. Nevertheless, in both cases the reaction is occurring in the same phase as the absorbed or adsorbed

Table 2.1 Summary of the key definitions

Stoichiometric coefficient	The amount of product produced when the reaction goes once; the stoichiometric coefficient is positive for a product and negative for a reactant
r_{A1}	The net rate of production of a species A; r_A is positive for a product and negative for a reactant
Rate of reaction 1, r_1	$r_1 = \dfrac{1}{\beta_A} r_A$ for any species A participating in reaction 1.
Homogeneous reaction	A reaction that happens throughout the reacting mixture.
Heterogeneous reaction	A reaction that happens near the boundary of a reacting phase

reactants, so one cannot distinguish between homogeneous and heterogeneous reactions by looking to see if the reaction occurs in the same phase as do the reactants.

For the material later in this book, we will have to distinguish between r_A, the rate of production of species A; and, r_1, the rate of reaction 1. Van't Hoff defined r_1, the rate of reaction 1, as

$$r_1 = \frac{1}{\beta_A} r_A \tag{2.6}$$

where r_1 is the rate of reaction 1, r_A is the rate of formation of A during reaction 1, and β_A is the stoichiometric coefficient of A. Note in general that $r_A \neq r_1$. Again this definition is slightly different from the one you learned in freshman chemistry. In Example 2.A at the end of this chapter, we show how to use measurements on a simple reactor to calculate r_A and r_1. The reader may want to look at this solved example before proceeding.

Table 2.1 summarizes the key definitions so far in this chapter. The reader should memorize these definitions before proceeding.

2.2 VARIATION IN RATE WITH CONDITIONS

Next we want to discuss how rates of reaction vary with conditions. Van't Hoff showed that generally rates vary with

- Concentrations of all of the reactants, products, and other species in the system
- Temperature
- The presence of solvents
- The presence of catalysts

In the next several sections, we will discuss the effects or concentration and temperature at length. The role of solvents and catalysts will be discussed in Chapters 12–14.

2.3 EMPIRICAL RATE LAWS

We will first consider the role of concentration on the rate of reaction. Our objective is to define the rate equation and the order of a reaction.

Studies of the role of concentration on rates of chemical reactions started in the middle part of the nineteenth century. Van't Hoff summarized the effects in his 1896 book.

Van't Hoff reported data on many different reactions, and showed how to correlate the measurements to the concentrations of the various species. Van't Hoff proposed that the rate was a function of only the temperature of the reactor, the concentration of all of the species, the concentration of solvents or catalysts, and the total pressure. Further, Van't Hoff asserted that if one fixed the temperature and solvent, one could formulate an equation for the rate as a function of concentration. An equation that describes the relationship between the rate of reaction and the concentration of all of the species in the reactor is called a **rate equation**.

Van't Hoff published the findings described above in a series of books that were classics for the day. The interest generated by Van't Hoff's books and the fact that he won the first Nobel Prize in chemistry stimulated many other scientists to examine the effect of concentration on rates of reactions.

The most famous person to follow in Van't Hoff's footsteps was Friedrich Wilhelm Ostwald. Ostwald had a very strong character. On one hand, he argued vehemently that while atoms were useful theoretical constructions, atoms do not really exist. On the other hand, Ostwald was the first person to propose that ions formed when species dissolve in solution. Ostwald created electrochemistry and did much of the early work on thermodynamics of solutions.

Ostwald was fascinated with rates of reaction in solution. A particularly interesting example was the hydration of (+)sucrose ($C_{12}H_{22}O_{11}$) to glucose ($C_6H_{12}O_6$), and fructose, ($C_5H_9O_5CH_2OH$):

$$C_{12}H_{22}O_{11} + H_2O \xrightarrow{H^+} C_6H_{12}O_6 + C_5H_9O_5CH_2OH \qquad (2.7)$$

Reaction (2.7) is a common natural process called *sugar inversion*. Honeybees use sugar inversion to convert sugar from plants into honey. Humans invert sucrose as part of digestion.

Experimentally, sucrose is stable in water for months. Still, if one adds a small amount of either an acid such as HCl, or an enzyme called *invertase*, the sucrose is quickly converted into glucose and fructose.

Wilhelmy [1850] and Berthelot (1862) examined the reaction in some detail and found that it obeyed a simple rate law as follows:

$$r_G = -r_S = k_3 C_S C_A \qquad (2.8)$$

where r_G is the rate of formation of glucose, r_S is the rate of production of sucrose, C_S is the sucrose concentration, C_A is the acid concentration, and k_3 is a constant.

Ostwald reexamined reaction (2.7) and found that he could catalyze the reaction with a wide range of acids. Equation (2.8) did not explain his data very well. Still, he could fit the data with the following rate form:

$$r_G = k_6 C_S C_{H^+} \qquad (2.9)$$

where C_{H^+} is the concentration of hydrogen ions in solution.

Ostwald also generalized his results by noting that if one takes an arbitrary reaction of the form $A \Rightarrow B$, one might be able to fit the data to a form like equation (2.8). Thus r_A, the rate of production of A, is given by

$$r_A = -k_1 C_A \qquad (2.10)$$

where C_A is the concentration of the species A that is being reacted away; r_A is the rate of production of A, and k_1 is a constant. Note that r_A is negative.

Ostwald did extensive measurements of rates of many reactions in solution. He also reanalyzed all of the rate data that had appeared prior to 1890 and found that the rate equation given by equation (2.10) sometimes works great. However, more often than not equation (2.10) is a poor representation of the rate of reaction. Ostwald also noted that a slightly modified equation could represent a wide range of data:

$$r_A = -k_n(C_A)^n \qquad (2.11)$$

where n is a fitting constant.

Table 2.2 illustrates some rate data that one might fit to equation (2.11). In the data in Table 2.2 the rate doubles when the concentration doubles. Consequently, n = 1. One can also plug the data in Table 2.2 back into equation (2.11) to calculate k_n. k_n works out to be 0.5/min.

Ostwald called n the **order** of the reaction, and k_n the **rate constant**. The relationship between the rate and the concentrations of the reactants is called the **rate equation**. The definition of a rate equation as *the rate as a function of concentration* is quite important. The reader should memorize this definition before proceeding.

Ostwald also proposed that heterogeneous reactions could also be fitted by a similar rate form

$$R_A = k_n(C_A^S)^n \qquad (2.12)$$

where C_A^S is the surface concentration of A in mol/cm^2.

For future reference, we note that reactions that obey equation (2.11) with n = 1 are called **first-order** reactions, reactions that obey equation (2.11) with n = 2 are called **second-order** reactions, while reactions that obey equation (2.11) with n = 3 are called **third-order** reactions. The terms *rate constant*, *rate equation*, *first-order*, and *second-order* are quite important. The reader should memorize the definitions before proceeding.

As an exercise, the reader might want to show that the rate constant for a homogeneous or heterogeneous first-order reaction has the dimension $(time)^{-1}$ (e.g., $hour^{-1}$), the rate constant for a homogeneous second-order reaction has the dimensions volume/(molecule or mole)/time [e.g., liter/(mol·hour)], while the rate constant for a heterogeneous second-order reaction has dimensions of area/(molecule or mole)/time [e.g., $cm^2/(mol·hour)$]. Similarly, the rate constant for a homogeneous third-order reaction has the dimensions $volume^2$/(molecule or mole)^2/time [e.g., $liters^2/(mol^2·hour)$]. While the rate constant for a third-order heterogeneous catalyst has dimensions of $area^2$/time/(moles or molecule)^2 [e.g., $cm^{-4}/(mol^2·minute)$].

The idea that there was a rate equation was a sizable advance. At the time this work was going on, people did not know whether chemical processes would be amenable to quantitative analysis. The fact that the rate could be written as a function (i.e., equation)

Table 2.2 Sample rate data to illustrate equation (2.11)

C_A, mol/liter	Rate, (mol·liter/hour)	C_A, (mol/liter)	Rate, (mol·liter)/hour
0.25	0.13	1	0.5
0.5	0.25	2	1.0

of the concentrations of all of the species in the reactor meant that many chemical processes would be amenable to quantitative analysis. Most students do not realize that the observation that the rate can be written as a rate equation is important. In fact, however, this observation forms the basis of all studies of kinetics.

In 1896, it was not obvious that the rate was a *function* of the concentration where we use the word *function* in a mathematical sense. Recall from freshman calculus, that certain criteria must be satisfied before it is mathematically correct to say that one variable is a function of another. However, Van't Hoff (1884, 1896) examined all of the reactions that had been studied prior to 1896, and found that in all cases, he could express the rate as a function of the concentration. Therefore, Van't Hoff asserted that one could write a rate equation for any reaction.

More recent data show that Van't Hoff is usually correct; one can usually fit rate data to a rate equation. However, there are some exceptions where the rate is not a mathematical function of just the temperature, pressure, and composition (see Section 2.8). Still, those are rare exceptions. In most cases, the rate is a function of the temperature, pressure, and composition.

Rate equations, of the form in equation (2.11), apply only to cases where there is only one reactant. When there are two reactants, A and B, one often finds rate equations such as:

$$r = k_m (C_A)^n (C_B)^m \qquad (2.13)$$

In the case of reaction (2.13), one defines the **overall order of the reaction** to be $n + m$. If $n = 1$ and $m = 2$, the reaction will be third-order. One might also say that the reaction is first-order in A and second-order in B.

Table 2.3 illustrates some data that one might fit to equation (2.13). In the data in the table, the rate goes up a factor of 2 when C_A doubles, so $n = 1$. However, the rate goes up by a factor of approximately $2^2 = 4$ when C_B doubles. Therefore, $m = 2$. If one plugs back into equation (2.13), one finds k_m equals 0.5 liter/(mol·minute).

Table 2.4 summarizes the definitions in this section; one should memorize these definitions before proceeding.

Table 2.3 Sample data to illustrate equation (2.12)

C_A, mol/liter	C_B, mol/liter	Rate, (mol·liter)/minute	C_A, mol/liter	C_B, mol/liter	Rate, (mol·liter)/minute
1	0.25	0.031	0.25	1	0.13
1	0.5	0.13	0.5	1	0.25
1	1	0.5	1	1	0.5
1	2	2.0	2	1	1.0

Table 2.4 The key definitions from Section 2.3

Rate equation	The rate as a function of the concentration of the reactants
Order	The exponent n is the expression
First-order reaction	A reaction whose rate is preparation to the reactant concentration to the first power [e.g., $n = 1$ in equation (2.11)]
Second-order reaction	A reaction whose rate is proportional to the reactant concentration to the second order
Overall order of reaction	The sum of the orders for all of the reactants

2.3.1 Notation

In the next section, we will be discussing the rate equations for a variety of reactions. Before we do so, we will need to define some notation. We will define small k with a subscript as being the rate constant for a reaction; so k_1, k_2, k_3, and k_{-1} will be the rate constants for reactions 1, 2, 3, and -1, respectively. Similarly, large K with a numerical subscript will be an equilibrium constant for a reaction; so K_1, K_2, and K_3 will be the equilibrium constants for reactions 1, 2, and 3. k_B, where the k has a tail, will be Boltzmann's constant where Boltzmann's constant equals the gas law constant (R) divided by Avogadro's number. We will use two different notations for a concentration of a species A: C_A, and [A] (concentration of A), where:

$$[A] \equiv C_A \tag{2.14}$$

We will use the [A] notation mainly where we have a molecule reacting. In the later notation, equation (2.9) will be rewritten

$$r_G = k_6[\text{sucrose}][H^+] \tag{2.15}$$

Again, we note that we define r_A as the rate per unit volume of a homogeneous reaction while R_A is the rate per unit surface area of a heterogeneous reaction. The reader should memorize this notation before proceeding.

2.4 THE EXPERIMENTAL SITUATION

Now that we have the notation out of the way, it is interesting to go back and compare equation (2.13) to the data. Our key point will be that equation (2.13) works for a wide variety of reactions provided the data are being taken over a limited range of concentration. To start, it is useful to recall that Ostwald first proposed equation (2.13) in 1890. At the time there were reliable data for three gas-phase reactions and six reactions in solution. Table 2.5 shows the rate equations for these reactions as cited by Van't Hoff (1896). Notice that all of these rate equations are of the general form of equation (2.13). Since 1896, thousands of reactions have been examined. Most follow simple first- or second-order rate laws over moderate concentration ranges. There are some exceptions. Generally, the rate equation is different or there are concentrations of 0.001 and 10 mol/liter. However, equation (2.13) works for 95% of the reactions that have been examined so far. The main exceptions are catalytic reactions. They will be discussed later in this chapter.

   Equation (2.13) has also proved to be useful in a wide range of systems outside of chemistry. For example, Figure 2.1 shows data for the reproduction rate of *E. coli* (a common bacterium) in a sugar solution. Notice that the reproduction rate of *E. coli* follows

$$r_{CO1} = k_{CO1}[E.\ coli][\text{sugar}] \tag{2.16}$$

where r_{CO1} is the rate of production of *E. coli*, [*E. coli*] is the *E. coli* concentration, [sugar] is the sugar concentration, and k_{CO1} is a constant. The plot is linear over most of the range, indicating that the reproduction of *E. coli* is first-order in the sugar concentration.

   Similarly, Figure 2.1 shows how the reproduction rate of paramecium varies with the population of the paramecium. Notice that at paramecium concentrations below

Table 2.5   Some of the rate equations that were discovered before 1886^a

Reaction		Rate equation	
$4PH_3 \rightleftharpoons P_4 + 6H_2$	(2.T.1)	$r_{PH_3} = -k_3[PH_3]$	(2.T.2)
$2AsH_3 \rightleftharpoons As_2 + 3H_2$	(2.T.3)	$r_{AsH_3} = -k_4[AsH_3]$	(2.T.4)
$2PH_3 + 4O_2 \rightleftharpoons P_2O_5 + 3H_2O$	(2.T.5)	$r_{PH_3} = -k_5[PH_3][O_2]^{1/2}$	(2.T.6)
$C_{12}H_{22}O_{11} + H_2O \xrightarrow{H^+}$	(2.T.7)	$r_S = -k_6[\text{sucrose}][H^+]$	(2.T.8)
$\quad C_6H_{12}O_6 + C_5H_9O_5CH_2OH$			
$CH_3COOR + H_2O \xrightarrow{H^+} CH_3COOH + ROH$	(2.T.9)	$r_{Ac} = +k_7[CH_3COOR][H^+]$	(2.T.10)
$CH_3COOH + ROH \xrightarrow{H^+} CH_3COOR + H_2O$	(2.T.11)	$r_{Ac} = -k_8[CH_3COOH][ROH][H^+]$	(2.T.12)
$ClCH_2COOH + H_2O$	(2.T.13)	$r_{C_2H_3ClO_2} = -k_9[C_2H_3ClO_2]$	(2.T.14)
$\quad \rightleftharpoons HOCH_2COOH + HCl$			
$2FeCl_3 + SnCl_2 \rightleftharpoons 2FeCl_2 + SnCl_4$	(2.T.15)	$r_{Fe^{3+}} = -k_{10}[Fe^{3+}]^2[Sn^{2+}]$	(2.T.16)
$KClO_3 + 6FeO \rightleftharpoons KCl + 3Fe_2O_3$	(2.T.17)	$r_{Fe^{3+}} = k_{11}[Fe^{2+}][ClO_3^-]$	(2.T.18)

^{a}In these equations, r_{PH_3}, r_{AH_3}, r_S, r_{Ac}, $r_{C_2H_3ClO_2}$, and r_{Fe}^{3+} are the rates of formation of phosphine, arsine, sucrose, acetic acid, chloroacetic acid, and Fe^{3+}, respectively; $[PH_3]$, $[AsH_3]$, $[O_2]$, [sucrose], $[H^+]$, $[CH_3COOR]$, $[CH_3COOH]$, $[ROH]$, $[Sn^{2+}]$, $[ClO_3^-]$, and $[C_2H_3ClO_2]$ are the concentrations of phosphine, arsine, oxygen, sucrose, hydrogen ion, acetate, acetic acid, alcohol, Sn^{2+}, ClO_3^-, and chloroacetic acid, respectively; and k_3, k_4, k_5, k_6, k_7, k_8, k_9, k_{10}, and k_{11} are constants.

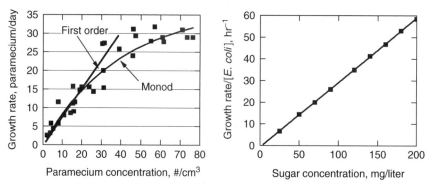

Figure 2.1   The reproduction rate of paramecium as a function of the paramecium concentration and the rate of *Escherichia coli* growth in sugar solutions as a function of the sugar concentration. [Paramecium data of Meyers (1927) 1; *E. coli* data from Monod (1942).]

$20/cm^3$ the reproduction rate of the paramecium increases linearly with the paramecium population:

$$r_{par} = k_{par}[par] \tag{2.17}$$

where r_{par} is the rate of paramecium reproduction, [par] is the concentration of paramecium, and k_{par} is a constant. Consequently, the reproduction of paramecium is a first-order process. Equation (2.13) fits data in a wide variety of systems, which is why it has proved to be so useful.

Of course, data taken over a wider range of concentrations show that reactions do not follow equation (2.13). For example, the growth of *E. coli* follows so-called **Monod** (after Jacques L. Monod, biochemist) kinetics:

$$r_{ecoli} = \frac{k_1 K_2 [E.\ coli][\text{sugar}]}{(1 + K_2[\text{sugar}])} \tag{2.18}$$

where k_1 and K_2 are constants.

According to equation (2.18), the rate goes up and then saturates as shown in Figure 2.2. If one plots the log of the rate versus concentration, one gets a characteristic S-shaped curve. The one special feature of reaction (2.16) is that as the reaction proceeds, more bacteria are born, so the rate increases. In the usual case, the rate decreases as the reaction proceeds because the reactants are used up. One calls reactions where the rate increases as products build up **autocatalytic reactions**. In autocatalytic reactions the product catalyses (i.e., speeds up) the reaction. Many biological processes are autocatalytic.

Rate data for most biological growth processes can fit to equation (2.13) when concentrations are small. Such a result shows the power of equation (2.13). Still, equation (2.13) seldom fits at high concentrations. Instead, one needs a more complex rate equation like that in equation (2.18).

In the previous paragraph, we noted that the *E. coli* growth curve does not follow equation (2.13) at high concentrations. Similar effects are seen in most chemical reactions. Figure 2.3 shows a plot of the rate of CH_3NC isomerization as a function of pressure. To keep the plot in perspective, if we take the log of equation (2.11), we obtain

$$\ln(-r_A) = \ln(K_n) + n \ln(C_a) \tag{2.19}$$

Therefore, the plot of the log of the rate versus the log of the pressure should be linear with a slope of 1 for a first-order reaction and linear with a slope of 2 for a second-order reaction. Notice that the slope is a 1.0 at high pressure, implying a first-order reaction. However, the reaction changes to second-order at low pressure. Experimentally, changes in the order of a reaction with pressure are quite common. In fact, it is unusual to see a change in the kinetics of a gas-phase reaction when one works over a wide range of pressures.

It is also common to find that the rate data for a given reaction *cannot* be fit to a rate equation of the form in equation (2.13). For example, the rate of the reaction

$$H_2 + Br_2 \Longrightarrow 2HBr \tag{2.20}$$

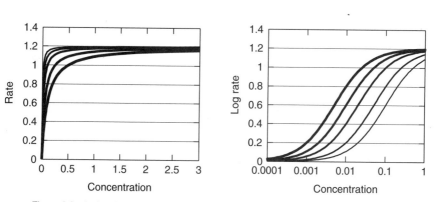

Figure 2.2 A plot of the rate for Monod kinetics, for $k_1 = 1,2$ and $K_2 = 10,20,50,100,200$.

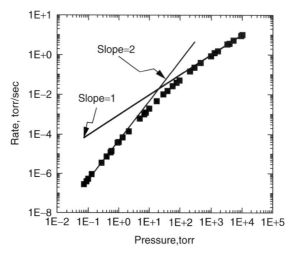

Figure 2.3 The rate of CH_3NC isomerization to CH_3CN as a function of the CH_3NC pressure. [Data of Schneider and Rabinovitz (1962).]

is thought to obey a complex rate equation, as follows:

$$r_{HBr} = \frac{k_1[H_2][Br_2]}{1 + K_2 \dfrac{[HBr]}{[Br_2]}} \tag{2.21}$$

where r_{HBr} is the rate of formation of HBr, $[H_2]$ is the H_2 concentration, $[Br_2]$ is the Br_2 concentration, $[HBr]$ is the HBr concentration, and, k_1 and K_2 are constants. One cannot fit HBr rate data to equation (2.13).

Note also that one cannot easily define a reaction order when one has a complex rate equation such as that in equation (2.21). In my experience, equation (2.13) works great most of the time. However, it can fail, especially when the kinetic data are taken over a wide range of concentration.

2.4.1 The Relationship between Stoichiometry and Rate

Next we want to consider whether there is any relationship between the rate equation for a reaction and the stoichiometry of the reaction. The experimental data shown in Table 2.3 say no. Notice that in reaction (2.T.5), (in Table 2.5), two oxygens react with each phosphine. Yet reaction (2.T.5) is only half-order in oxygen. Similarly, the rate equation for reaction (2.T.7) is first-order in the H^+ concentration, even though no H^+ is produced or consumed in the reaction. Reaction (2.T.1) seems to be an exception to the rule. If one rewrites reaction (2.T.1) as

$$PH_3 \Longrightarrow \tfrac{1}{4}P_4 + \tfrac{3}{2}H_2 \tag{2.22}$$

One might say that the kinetics of the reaction bear some relationship to the stoichiometry. However, the data in Figure 2.3 show that the first-order behavior does not persist over a wide range of conditions. *Experimentally, there is seldom any relationship between the rate of a reaction and the stoichiometry of the reaction.* Consequently, stoichiometry does not seem to have an important influence on the form of the rate equation.

We will discuss this point further in Chapter 4. However, the key thing to remember for now is that rate data can often be fit to a rate equation but that the rate equation may not be related to the stoichiometry of the reaction in a simple way.

2.4.2 Summary of the Effect of Composition on Rates

In summary, then, so far in this chapter we have considered how rates of reaction vary with concentration. We found that in most cases the rate of reaction varies in a simple way with composition. We often observe first- or second-order data. There are some exceptions, especially when we take data over a wide range of conditions. However, we can usually find a rate equation that works even over a wide temperature range.

2.5 TEMPERATURE EFFECTS

At this point, we will be changing topics. So far, we have been talking about the dependence of rate on composition. Now, we want to change topics and discuss how reaction rates depend on temperature. Our objective will be to present Arrhenius' law, and the Harcourt–Essen equation.

Studies of the influence of temperature effects in reactions date back to the medieval alchemists. The alchemists mixed substances together and saw what happened. One of the things that was discovered was that many reactions turn on suddenly over a modest temperature range. For example, the alchemists discovered that oil will suddenly ignite when the oil is heated to 320°C. Similarly, a hydrogen–oxygen mixture is stable at 400°C, but ignites suddenly at about 440°C. In a series of important experiments, Meyer and Raum (1895) showed that if they held a mixture of hydrogen and oxygen at 300°C for 65 days, 1–5% of the hydrogen and oxygen is converted into water. Therefore, it was concluded that reactions do not really turn on suddenly. Instead, rates of reaction change with temperature.

People explored the effects of temperature on rates of reaction in the latter part of the nineteenth century. The studies were usually limited to a modest temperature range (50°C). The variations in rate were so small that no one was able to accurately test models for the temperature dependence of rates. Still, considerable data were generated.

In 1893, Kooij and Van't Hoff examined the decomposition of phosphine between 310 and 510°C. This was the first time that the kinetics of a single reaction was measured over a wide temperature range. Figure 2.4 shows a plot of the rate of phosphine decomposition via reaction (2.T.1) (in Table 2.2) as measured by Kooij and Van't Hoff. Notice that the rate increases rapidly with temperature, doubling once every 35 K.

Prior to Kooij and Van't Hoff's work, there were two competing models in the literature on the effects of temperature on rates: Perrin's model and Arrhenius' model. Perrin (1919) had proposed that reactions were activated because the reactants had to accumulate enough energy to break bonds before reaction could occur. Perrin noted that the energy transfer could come from radiation or convection. If one calculates the rate of energy transfer to a cold molecule, the rate of energy transfer from the walls of the vessel can be approximated by

$$E_T = E_T^0 T^n \tag{2.23}$$

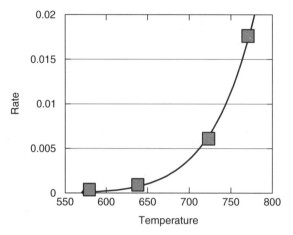

Figure 2.4 The rate of PH_3 decomposition as a function of temperature. [Data of Kooij (1893).]

where E_T is the rate of energy transfer, T is the absolute temperature, E_T^0 is a constant, and n is a constant that varies from 1 (when forced convection dominates) to 4 (when radiation dominates). Consequently, Perrin suggested that if the rate of energy transfer from the walls of a vessel controls the rate of reaction, the rate constant for a given reaction k_1 can be written as

$$k_1 = k_1^T T^n \qquad (2.24)$$

where k_1^T is a constant. Equation (2.24) was also proposed by Harcourt and Essen on empirical grounds (the model worked).

An alternative model was proposed by Arrhenius. Arrhenius suggested molecules needed to go over a barrier before reaction could occur. Wigner [1932] and Polanyi [1931] later represented the barrier as a hill between reactants and products as seen in Figure 2.5. Only hot molecules (i.e., molecules with a total energy greater than E_a) can react, and there are more hot molecules at higher temperatures. Consequently, rates increase with temperature.

Arrhenius quantified his ideas by assuming that the hot molecules were in equilibrium with the reactants. At equilibrium, F_{E_A}, the fraction of the molecules that are hotter than

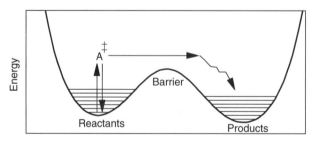

Figure 2.5 Wigner and Polanyi's representation of Arrhenius' model of activation barriers to reactions.

E_A, is given by

$$F_{E_A} = e^{-E_A/k_B T} \qquad (2.25)$$

where E_A is called the *activation energy*, k_B is Boltzmann's constant, and, T is the absolute temperature. Note that Boltzmann's constant is just the ideal-gas law constant (R) divided by Avogadro's number. Also note the notation k_B. We use the tail on the k to indicate that k_B is not a rate constant.

The rate constant becomes

$$k_1 = k_1^0 e^{-E_a/k_B T} \qquad (2.26)$$

where k_1^0 is called the *preexponential*. Equation (2.26) is called *Arrhenius' law*. It is a key equation for the remainder of this book.

Equation (2.28) is correct for the case when the activation energy is measured in kilocalories per/molecule. If you measure activation barriers in kilocalories per *mole*, then the following equation should be used:

$$k_1 = k_1^0 \exp\left(-\frac{E_a}{RT}\right) \qquad (2.27)$$

where again R is the gas law constant.

Van't Hoff (1896) compared equations (2.24) and (2.26) to his phosphine data and found that at low pressure, equation (2.24) worked best. However, when he examined a wider data set, equations (2.24) and (2.26) worked equally well much of the time. Still, Van't Hoff found it hard to believe that radiation played an important role in chemical reactions. Later, Langmuir [1920] showed that the emissivity of the walls did not affect chemical reaction rates, so energy transfer from the walls of the vessel does not affect reactions. As a result, Perrin's equation was discarded in the literature. Arrhenius' equation on the other hand is seen in most textbooks.

Still, Dunbar and McMahon (1998) have shown that radiation from the walls can have an important influence on the rate of unimolecular reactions. Later in this book, we will find that energy transfer barriers to reactions are just as important as activation barriers. The energy transfer barriers are not barriers to energy transfer to the walls. Rather, they are barriers to energy transfer within a molecule. Reactions with energy transfer barriers do not follow Arrhenius' law [equation (2.26)]. Instead, they follow

$$k_1 = k_m^0 (T)^m e^{-E_A/k_B T} \qquad (2.28)$$

where k_1 is the rate constant, E_A is the activation barrier, k_B is Boltzmann's constant, T is temperature, and m and k_m^0 are constants. According to Arrhenius' law, a plot of the log of the rate versus $1/T$ should be linear. However, Figure 2.6 shows some data for a typical reaction, and one observes significant curvature. Equation (2.28) fits the data in Figure 2.6 much better than does Arrhenius' law. Nevertheless, Arrhenius' law equation (2.26) is cited much more often than equation (2.24) or (2.28). Therefore, the reader should memorize equation (2.26) before proceeding.

In my experience, Arrhenius' law works great when one is working over perhaps a 50–100 K temperature range. It only fails when one measures data over a much wider temperature range.

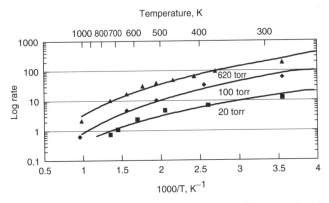

Figure 2.6 The rate of the reaction $CH + N_2 \rightarrow HCN + N$ as a function of the temperature. [Data of Becker, et al. (1996).]

2.5.1 Implications of Arrhenius' Law

In freshman chemistry, you learned that Arrhenius' law can be used to calculate activation barriers. The idea is to measure the rate of reaction as a function of temperature, and then make a plot of the log of the rate versus one over the temperature. The slope of the plot is proportional to the activation barrier. Figure 2.7 shows an example of that. Arrhenius' law equation (2.26) fits the data okay, although not as well as does equation (2.28).

In my view, if the only thing you could do with Arrhenius' law were to fit data, Arrhenius' law would not be as important. After all, equation (2.28) fits data much better. However, the real strength of Arrhenius' law is that it allows one to make predictions without doing very many experiments. In this section, we will show how Arrhenius' law can be used to make useful predictions from very little data.

According to Arrhenius' law, the rate constant for a reaction, k_1, will be given by

$$k_1 = k_1^0 e^{-E_a/k_B T} \qquad (2.29)$$

where k_1^0 is the preexponential, E_a is the activation energy, k_B is Boltzmann's constant, and T is temperature.

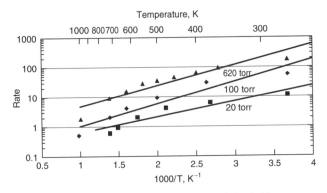

Figure 2.7 A fit of the data in Figure 2.6 to Arrhenius' law.

In Chapter 7, we will find that for an elementary reaction, k_1^0 is usually about $10^{14\pm3}$/second for first-order reactions, $10^{13\pm2}$ Å^3/(molecule·second) for second-order reactions; and $10^{13\pm2}$ Å^6/(molecule^2·second) for third-order reactions. The reader should memorize these typical values of the preexponentials before proceeding.

Now, let's consider a first-order reaction A $\Rightarrow$ B. In Chapter 3, we will find that if we load A into a beaker and let it react, then 50% of the A will be reacted after a time $\tau_{1/2}$, given by

$$\tau_{1/2} = \frac{\ln 2}{k_1} \tag{2.30}$$

Figure 2.8 shows a plot of $\tau_{1/2}$ versus E_A calculated from equation (2.30) for $k_1^0 = 10^{13}$/second. Notice that at 300 K, a reaction with an activation barrier of 14 kcal/mol takes about one millisecond (1 ms) to go to 50% completion. A reaction with an activation barrier of 18 kcal/mol takes about 1 second to go to 50% completion. A reaction with an activation barrier of 20.4 kcal/mol takes about 1 minute to go to 50% completion. A reaction with an activation barrier of 23 kcal/mol takes about an hour to go to 50% completion. A reaction with an activation barrier of 24.8 kcal/mol will take a day to go to 50% completion. A reaction with an activation barrier of 28 kcal/mol will take about a year to go to 50% completion. A reaction with an activation barrier of 31 kcal/mol will take a century to go to 50% completion. A reaction with an activation barrier of 36 kcal/mol will take about a million years to go to 50% completion.

Physically, this is very important because it means that if you run a reaction at room temperature and the rate is easily measurable (i.e., the reaction takes between 1 second and 1 hour), the activation barrier for the reaction will usually be 20 ± 6 kcal/mol.

In Professor Masel's lab, students often do experiments where they slowly heat a system and watch for reaction. Generally, if one observes a reaction taking a minute at some temperature T_{minute}, then the activation energy for the reaction will be approximately

$$E_a = (1/15 \text{ kcal/mol-}°K)T_{minute} \tag{2.31}$$

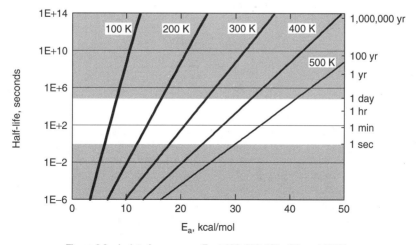

Figure 2.8 A plot of $\tau_{1/2}$ versus. E_a at 100, 200, 300, 400, and 500 K.

Note that T_{minute} in equation (2.31) is temperature in degrees Kelvin. Similarly, if the reaction takes a second at some temperature, T_{second}, then the activation barrier is approximately

$$E_a = [(0.06 \text{ kcal·mol})K]T_{second} \qquad (2.32)$$

We use the expression "the reaction takes a second" very loosely in equations (2.31) and (2.32). The equations were derived using the time where 50% of the product is used up. However, if you would instead consider the time it takes for 10% to be used up, or the time it takes for 90% to be used up, you could derive almost the same expression, except that the constant would change by perhaps 10%. Equations (2.31) and (2.32) are very important because they allow one to get a rough estimate of the activation energy from very little data. If you see a reaction at room temperature, you know that the reaction *probably* has a barrier of 20 ± 6 kcal/mol, independent of the details of the reaction. As an example, most of the chemical reactions occurring in the metabolic processes in living cells have activation barriers of 15–25 kcal/mol. So do the reactions involved with digesting your food or bleaching your clothes in a clothes washer. One can use equations (2.31) and (2.32) to get a reasonable value of the activation barrier for reaction for a wide variety of reactions without knowing much about the reaction. That makes equations (2.31) and (2.32) very powerful (although sometimes inaccurate.)

Still, equations (2.31) and (2.32) are approximations that assume that there is only one reaction occurring. If you have two competing reactions, or other complicated effects, then equations (2.31) and (2.32) will not work.

In my experience, equations (2.31) and (2.32) work in about 90% of the cases. In most of the other cases, there is something complicated going on, either there are competing reactions, or some other complexities. If you measure an activation barrier and it is way off from that predicted from equations (2.31) and (2.32), it is safe to assume that some complex process is happening during the reaction. You need to be very careful with your data in such a case.

Another application of equations (2.31) and (2.32) is to solve the equations to see at what temperature a given reaction will have a measurable rate. For example, one can solve equation (2.31) for T_{minute} to show

$$T_{minute} = \frac{15 \text{ K·mol}}{kcal}E_a \qquad (2.33)$$

where T_{minute} is the temperature you would have to go to get a half-life of about a minute. According to equation (2.33), if a reaction has an activation barrier of 30 kcal/mol, you need to run the reaction at 450 K to get a reasonable reaction rate. Similarly, a reaction with an activation energy of 50 kcal/mol needs to be run at 750 K to get a reasonable reaction rate. Consequently, equation (2.33) allows one to plan experiments for a wide variety of systems.

Equation (2.33) can also be used to estimate ignition temperatures. If you have a gas burner in your kitchen, the gas usually stays in the flame for about a tenth of a second. If the ignition process has an activation barrier of 50 kcal/mol, then, according to equation (2.33), you need to heat the system to about 800 K before ignition will occur.

Equations (2.31)–(2.33) are very useful in estimating activation barriers and planning experiments. The reader should memorize these equations before proceeding.

2.5.2   Changes in Rate with Temperature

Next, it is useful to quantify the extent to which rates go up with increasing temperature. Consider a reaction occurring at a rate r_1, at temperature T_1. According to equation (2.26), for an nth-order reaction

$$r_1 = k_1^0 \exp\left(\frac{-E_a}{k_B T_1}\right) (C_A)^n \tag{2.34}$$

Now consider changing the temperature to T_2. At that temperature, the rate is given by

$$r_2 = k_2^0 \exp\left(\frac{-E_a}{k_B T_2}\right) (C_A)^n \tag{2.35}$$

dividing equation (2.34) by equation (2.35) and rearranging yields

$$r_2 = r_1 \exp\left(\frac{E_a}{k_B}\left(\frac{1}{T_1} - \frac{1}{T_2}\right)\right) \tag{2.36}$$

Equation (2.36) allows one to calculate how the rate of a given reaction changes with changing temperature.

Figure 2.9 shows the fractional change in the rate of an nth-order reaction when the temperature is changed from 25 to 35°C as a function of the activation energy of the reaction. The increase varies from a factor of 1.1 to a factor of 10 for reasonable values of the activation barrier. To put the plot in perspective, according to equation (2.32), a reaction that takes about a second at room temperature should have an activation energy of about 20 kcal/mol. Plugging into equation (2.36) shows that the rate of a reaction with a 20-kcal/mol barrier should go up about $2\frac{1}{2}$ times for every 10°C.

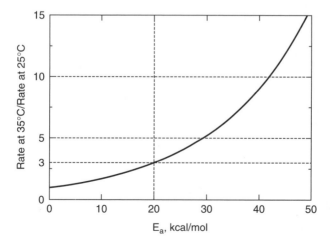

Figure 2.9   The fractional change in the rate of an nth-order reaction when the temperature is changed from 25 to 35°C.

Table 2.6 The variation in rate of a series of reactions with a 10-K change in temperature

Reaction	Temperature range, °C	Rate Change with a 10-K Temperature Change
$CH_3COOCH_2CH_3 + H_2O \xrightarrow{H^+} CH_3COOH + CH_3CH_2OH$	3.6–30.4	2.03
$CH_3CH_2Cl + NaOH \Longrightarrow H_2C{=}CH_2 + NaCl + H_2O$	23.5–43.6	2.87
$CH_3CH_2CH_2Cl + NaOH \Longrightarrow CH_3CH{=}CH_2 + NaCl$	24.5–43.6	2.68
$HPO_3 + H_2O \Longrightarrow H_3PO_4$	0–61	3.0

Source: Data from Van't Hoff (1884).

Table 2.6 shows how the rates of several room-temperature reactions actually vary with temperature. Notice that the rate of room-temperature reactions increase by a factor of 2 or 3 for every 10 K increase in temperature, just as you learned in freshman chemistry. This happens because all room-temperature reactions have activation barriers of 20 ± 6 kcal/mol, and when the activation barrier is 20 ± 6 kcal/mol, the rate doubles or triples every 10°C.

You probably did not learn this in freshman chemistry, but the fact that rates double every 10°C is a very universal phenomenon. Plant growth follows Arrhenius' law. Growth of yeast or bacteria follows Arrhenius' law. Many other processes in nature also follow Arrhenius' law.

For example, Clausen (1890) examined the respiration rate of plants. He found that the respiration rate of wheat went up 2.47 times every 10°, lilac went up 2.48 times; and lupine (a blue flower) went up 2.46 times (see Table 2.7). Bailey and Ollis (1977) report the rate of *E. coli* (bacteria) growth as a function of temperature. Figure 2.10 shows their results. Notice that again, the rate of *E. coli* growth goes up about 2.5 times every 10 K until the yeast begins to die at 40°C.

A different example is the chirping of crickets. Male crickets chirp to attract mates, and the chirping rate changes as their temperature changes. Figure 2.11 shows how the rate of cricket chirping varies with temperature. Again the rate doubles every 10°C.

There are similar data on the growth of insects, lizards, and other cold-blooded animals. Even the speed at which ants walk doubles every 10°C. Another example includes the fact that bleach cleans clothes twice as fast when one raises the water temperature by 10°C. These examples show the idea that the rate of room temperature reactions goes up by about a factor of 2–3 for every 10°C increase in temperature; this rate change applies to a wide range of systems, and is not limited to things that happen in a chemical reactor.

Another application of Arrhenius' law is to learn how to control the rate of a chemical reaction. If you have a reaction that is too slow, the easiest way to speed it up is to increase the reaction temperature. Heating up a room temperature reaction usually doubles

Table 2.7 The variation in the respiration rate of plants with a 10° change in temperature

Wheat	2.47
Lilac	2.48
Lupine	2.46

Source: Data from Clausen (1890).

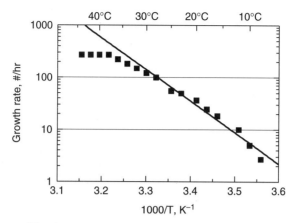

Figure 2.10 The rate of *E. coli* growth as a function of temperature. [Adapted from Bailey and Ollis (1977).]

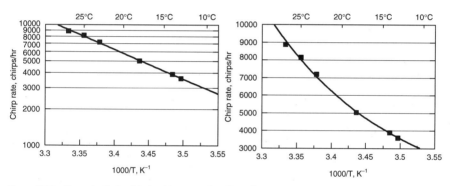

Figure 2.11 The rate that crickets chirp as a function of temperature. Data for field crickets (*Gryllys pennsylvanicus*). From Heinrich (1993).

or triples the rate of reaction. When you are running a reactor in industry, you often adjust the temperature to control the rate. Equation (2.33) will allow you to estimate the required temperature.

2.5.3 Exceptions to Arrhenius' Law

There is one key subtlety in Arrhenius' law, in that it applies only to elementary reactions, not overall reactions. The terms *elementary reactions* and *overall reactions* will be defined in Chapter 3.) If you consider a complex reaction, Arrhenius' law will work for each of the elementary reactions. However, the overall reaction might not show Arrhenius behavior. In this section, we want to work out a simple example, to illustrate the fact that overall rates might not show Arrhenius behavior even though elementary rates do.

Consider a reversible reaction A $\rightleftharpoons$ B. In Chapter 3 we will find

$$r_B = k_2[A] - k_2[B] \tag{2.37}$$

where k_1 is the rate constant for the forward reaction and k_2 is the rate constant for the reverse reaction. Both k_1 and k_2 follow Arrhenius' law:

$$k_1 = k_1^0 \exp\left(\frac{-E_a^1}{k_B T}\right)$$

$$k_2 = k_2^0 \exp\left(\frac{-E_a^2}{k_B T}\right)$$

where k_1^0 and k_2^0 respectively are the preexponentials for the forward and reverse reactions and E_a^1 and E_a^2 are the activation barriers for each reaction.

In Chapter 3 we will show that for a reversible reaction

$$E_a^2 = E_a^1 + \Delta H_r$$

where ΔH_r is the heat at reaction.

Figure 2.12 shows a plot of the rate calculated from equation (2.37) as a function of temperature. Notice that the overall rate of the endothermic reaction shows significant deviations from Arrhenius' law even though k_1 and k_2 follow Arrhenius' law. This example illustrates that Arrhenius' law applies only to elementary reactions.

2.5.4 Important Exceptions to Arrhenius' Law

The example in the previous section was not a typical example, but in fact all reactions that follow complex rate laws do not also follow Arrhenius' law. Instead, a plot of the log of the overall rate versus one over temperature shows nonlinear behavior.

For example, earlier in this chapter we noted that the growth of bacteria follows Monod kinetics:

$$r_{ecoli} = \frac{k_1 K_2 [E.\ coli]}{(1 + K_2 [E.\ coli])} \tag{2.38}$$

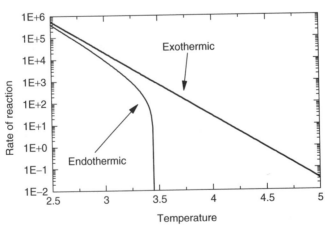

Figure 2.12 The rate of the reaction $A \rightleftharpoons B$ with $k_1^0 = k_2^0 = 10^{13}$/second, $E_a = 15$ kcal/mol, $\Delta H = +3$ kcal/mol, and $[A] = 1$ mol/liter $[B] = 0.01$ mol/liter.

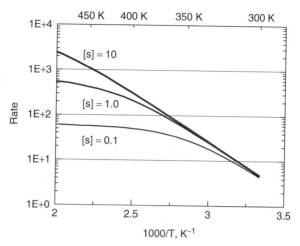

Figure 2.13 The temperature dependence of a reaction that follows Monod's law $r_S = k_1 K_2[S]/(1 + K_2[S])$ with $k_1 = 10^{13}$/second -20 (kcal/mol)/$k_B T$ and $K_2 = 6 \times 10^{-6}$ (10 kcal/mol)/$k_B T$.

In equation (2.38), k_1 and k_2 follow Arrhenius' law. However, *E. coli* shows nonlinear behavior.

Figure 2.13 shows a plot of the rate versus temperature calculated from Monod's law with some typical values of the parameters. Notice that the rate shows no linear behavior with temperature. Our rules of thumb from Section 2.14 do not work in this example.

In my experience, most simple reactions follow Arrhenius' law. However, the overall rate of most industrially important reactions cannot be accurately fit with Arrhenius' law. Again, we note that each elementary reaction will follow Arrhenius' law. It is just the rate at overall reaction that shows derivations from Arrhenius' law.

2.6 CATALYSTS

Industrially, the deviations from Arrhenius' law are quite important. Recall that most industrial reactions are run with catalysts. Well, most catalytic reactions do not follow Arrhenius' law. These exceptions are so important that I wanted to highlight them early in this book. In the remainder of this chapter, I want to discuss catalytic reactions, and show how the kinetics at catalytic reactions are different from the kinetics of simple reactions in solution.

To start, it is useful to review some background material on catalysis. Ostwald defined a catalyst as "a substance one adds to a chemical reaction to speed up the reaction without the catalyst undergoing a chemical change itself." Later in this book, we will find that this definition is not quite accurate. Catalysts do undergo chemical changes during the course of reaction. It is just that the changes are reversible, so that the catalyst is not consumed as the reaction proceeds. Examples of catalysts include the acids in your stomach that you use to break down food, and the enzymes that people put in detergents to make the detergent work better. Most chemical processes use catalysts at some stage in the production process.

There are two kinds of catalysts: **homogeneous catalysts** and **heterogeneous catalysts**. Homogeneous catalysts are things that one dissolves in a liquid or a gas to promote a reaction. Homogeneous catalysts include acids and bases, enzymes, transition metal ions, and alkyls. In some cases, solvents can act as homogeneous catalysts.

A **heterogeneous catalyst** is a solid that one adds to a reaction mixture. The solid does not dissolve in the mixture, but still is able to promote a desired reaction or a series of desired reactions. Typical heterogeneous catalysts include transition metals (e.g., platinum on alumina) or solid acids such as silica/alumina.

Heterogeneous and homogeneous catalysts are fundamentally different. A homogeneous catalyst dissolves into the gas phase or solution and acts uniformly throughout the liquid or gas mixture. On the other hand, heterogeneous catalysts do not dissolve. Instead, the reaction occurs on the surface of the catalyst.

Table 2.8 shows some typical homogeneous catalysts. Generally, homogeneous catalysts consist of transition metal atoms, peroxoradicals, and acids or bases. Industrially, one uses homogeneous catalysts in most polymerization and carbonylation reactions, and in some partial oxidations.

Homogeneous catalysts are also used in the home. For example, most detergents contain enzymes to allow the oxidation and depolymerization of stains (i.e., dissolved proteins). Cleansers contain metal atoms to speed up the action of bleach. Most biological reactions are also controlled by homogeneous catalysts, such as enzymes or RNA.

Heterogeneous catalysts are more important industrially. Figure 2.14 shows some typical heterogeneous catalysts. Heterogeneous catalysts are generally powders or pellets that one can add to a reacting mixture to speed up the reaction. Heterogeneous catalysts are used extensively in chemical processing because heterogeneous catalysts are easier to separate from the products of a reaction mixture than a homogeneous catalyst. Table 2.9 shows some typical heterogeneous catalysts. Over 90% of all bulk chemicals and petroleum products are made via catalytic processes.

Heterogeneous catalysts are also used in people's homes. For example, people used to sell catalytic igniters that one could use to start a gas range or fireplace. Now the most common heterogeneous catalyst in your possession is the catalytic converter in your car. The catalytic converter reduces the pollution and toxic emissions produced by your car. People also make catalytic converters to destroy pollutants in homes or in woodstove chimneys. Heterogeneous catalysts are not as common as homogeneous catalysts in the home but they are equally important.

Catalysts are so important because they make tremendous differences in reaction rate. For example, a hydrogen/oxygen mixture may be stable for years at 25°C. However, if one adds a platinum wire to the mixture, the mixture explodes. Table 2.10 shows some

Table 2.8 Some Common Homogeneous Catalysts

Reaction	Catalyst
Ethylene $\rightarrow$ polyethylene (polymerization)	$TiCl_4/Al(C_2H_5)_3$ (Ziegler–Natta catalyst)
Ethylene $\rightarrow$ polyethylene, styrene $\rightarrow$ polystyrene	Peroxides
$C_2H_4 + \frac{1}{2}O_2 \rightarrow CH_3(C=O)H$ (Wacker process)	$PdEt_3$
Olefins $+ CO + H_2 \rightarrow$ aldehydes (hydroformylation)	$Co(CO)_6$
$CH_3OH + CO \rightarrow CH_3COOH$	$RhCl_3$
$SO_2 \rightarrow SO_3$ (lead chamber process)	NO/NO_2
$CH_3COOH + CH_3OH \rightarrow CH_3COOCH_3 + H_2O$	Acids or bases; invertase
(sucrose $\rightarrow$ glucose + fructose)	

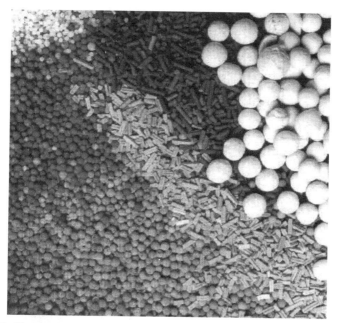

Figure 2.14 Photographs of some heterogenous catalysts. (From Wijngaarden and Westerterp, 1998).

Table 2.9   Common Heterogeneous Catalysts

Catalyst	Reaction
Platinum on alumina, nickel on alumina	Hydrogeneration/dehydrogenation
Platinum/tin on acidic alumina	Reforming
Solid acids (zeolites)	Hydrocarbon isomerization, cracking
Silver	$C_2H_4 + \frac{1}{2}O_2 \rightarrow$ ethylene oxide
$(B_2O_3)_x(MO_3)_y$	$CH_2{=}CHCH_3 + \frac{3}{2}O_2 + NH_3 \rightarrow CH_2CH{-}CHO + H_2O$
V_2O_5	$SO_2 + \frac{1}{2}O_2 \rightarrow SO_3$
Platinum gauze	$2NH_3 + 4O_2 \rightarrow N_2O_5 + 3H_2O$

other examples. Notice that in the examples shown, the catalyst lowers the activation barrier for the reaction by 19–30 kcal/mol. It is not unusual for a catalyst to lower the barrier for a reaction by 19–30 kcal/mol, and Masel (1996) gives some special examples where the catalyst lowers the activation barrier for a reaction by more than 50 kcal/mol. A 19–30-kcal/mol change in the barrier for a reaction changes the rate tremendously. The examples in Table 2.10 show rate enhancements of 10^8–10^{13}. Similar changes in rate are quite common. Masel (1996) reports some special cases where the rate enhancements are as large as 10^{21}.

From a practical standpoint, one of the key roles of a catalyst is to lower the temperature where a reaction takes place. According to Table 2.10, the reaction

$$(C_2H_5)_2O \Longrightarrow 2C_2H_4 + H_2O \tag{2.39}$$

Table 2.10   The change in rate of some typical reactions seen when a catalyst is added to the reaction mixture

Reaction	Catalyst	E_a Uncatalyzed, kcal/mol	E_a, Catalyzed, kcal/mol	Rate of Enhancement Calculated at 500 K
$H_2 + I_2 \rightarrow 2HI$	Pt	44	14	10^{13}
$2N_2O \rightarrow 2N_2 + O_2$	Au	58	29	10^{13}
$(C_2H_5)_2O \rightarrow 2C_2H_4 + H_2O$	I_2	53	34	10^8

Source: Table adapted from data in Bond (1987).

has an activation barrier of 53 kcal/mol in the absence of a catalyst and 38 kcal/mol in the presence of a catalyst. According to equation (2.33), you need to run a reaction with an activation barrier of 53 kcal/mol at about 800 K. While you can run a reaction with an activation barrier of 38 kcal/mol at 600 K, that temperature difference is huge if you need to pay the heating costs. Besides, there is an explosion hazard at the higher temperature. In industry, one hardly ever runs a reaction without a catalyst, which is why catalysis is so important. In 1995, almost 700 billion dollars of products were made with catalysts in the United States.

2.7   CATALYTIC KINETICS

One of the key features of catalysts is that they change the form of the rate equation. Over the years, there have been many attempts to determine how rates of reactions on heterogeneous catalysts vary with the partial pressure of the reactants. One of the things that has been discovered is that rates of catalytic reactions do not bear any simple relationship to the stoichiometry. In the gas phase, rates of reactions are often proportional to the reactant concentrations to some simple powers. However, catalytic reactions follow much more complex rate equations. It is common for the rate of a catalytic reaction to be constant or even go down as the concentration of one of the reactants increases. This is quite different from gas-phase reactions, where rates generally increase with increasing reactant pressure.

For example, Figure 2.15 shows data on the rate of CO oxidation on a catalyst called Rh(111). Notice that the rate increases linearly with the CO concentration up to a CO partial pressure of 10^{-7} torr. However, then the rate decreases again. One cannot fit data with a rate equation like that in equation (2.11). However, one can fit them with a more complex rate form:

$$r_{CO} = \frac{k_1 P_{CO} P_{O_2}}{(1 + K_2 P_{CO} + K_3 P_{O_2})^2} \qquad (2.40)$$

where r_{CO} is the rate of reaction; P_{CO} is the CO pressure; P_O is the oxygen pressure; and, k_1, K_2, and K_3 are constants.

In the discussion that follows, we will provide examples of this effect for two separate cases: unimolecular surface reactions, where a single reactant rearranges or decomposes to yield products; and bimolecular surface reactions, where two or more reactants combine and rearrange to form products.

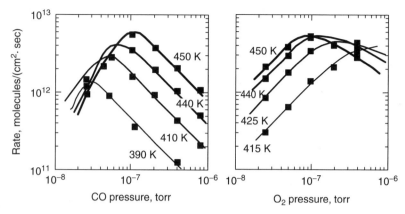

Figure 2.15 The influence of the CO pressure on the rate of CO oxidation on Rh(111). [Data of Schwartz, et al. (1986).]

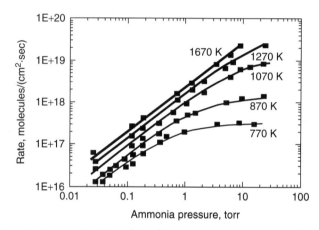

Figure 2.16 The rate of the reaction $NH_3 \Rightarrow \frac{1}{2}N_2 + \frac{3}{2}H_2$ over a platinum wire catalyst. [Data of Loffler and Schmidt (1976a,b).]

Figure 2.16 shows some data for the rate of a simple unimolecular surface reaction: the decomposition of ammonia over a polycrystalline platinum wire catalyst.

$$NH_3 \Longrightarrow \tfrac{1}{2}N_2 + \tfrac{3}{2}H_2 \qquad (2.41)$$

Notice that at moderate temperatures, the rate of reaction increases with increasing ammonia pressure and then levels off. At higher temperatures the rate continues to increase over the pressure range shown. However, other data show that the rate will eventually level off at sufficiently high pressures.

Figure 2.17 shows how the reaction rate varies with temperature at a series of fixed pressures. Notice that the rate increases with increasing temperature and then begins to level off.

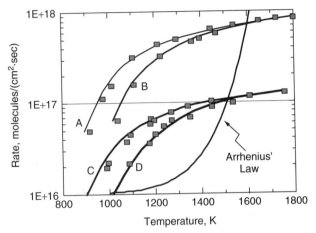

Figure 2.17 The variation in the rate of the reaction in Figure 2.16 with temperature: (a) $P_{NH_3} = 0.3$, $P_{H_2} = 0.15$; (b) $P_{NH_3} = 0.3$, $P_{H_2} = 0.44$; (c) $P_{NH_3} = 0.05$, $P_{H_2} = 0.15$; (d) $P_{NH_3} = 0.05$, $P_{H_2} = 0.45$. [Data of Loffler and Schmidt (1976a,b).]

The fact that the data do not follow Arrhenius' law shows that there is some complication during the reaction. Later in this book, we will find that the ammonia sticks onto the platinum, and then there are two competing pathways: (1) reaction and (2) desorption back into the gas-phase. The presence of two competing pathways causes the curvature in Figure 2.17.

The results in Figures 2.16 and 2.17 are typical of those for unimolecular reactions on heterogeneous catalysts. Generally, one observes rates that increase with increasing temperature and pressure. However, the rate seldom increases monotonically with pressure. Rather, the rate increase slows with increasing pressure, and eventually plateaus at high pressures. The rate also increases with temperature. However, the slope of the curve decreases with increasing temperature.

Note that gas-phase reactions show behavior very different from that seen in Figures 2.16 and 2.17. In the gas-phase, the rate of a simple unimolecular reaction increases continuously with increasing pressure and never levels off. The rate also increases continuously with increasing temperature. According to Arrhenius' law

$$\text{Rate} = r_0 \exp\left(-\frac{E_a}{k_B T}\right) \tag{2.42}$$

According to equation (2.42), the slope of the rate–temperature curve should increase with increasing temperature. However, Figure 2.17 shows that the slope actually decreases. Therefore, it seems that the kinetics of unimolecular catalytic reactions are quite different from the kinetics of unimolecular reactions in the gas-phase.

Even larger differences are seen for bimolecular reactions. For example, Figure 2.18 shows how the rate of the reaction

$$CO + \tfrac{1}{2}O_2 \Longrightarrow CO_2 \tag{2.43}$$

varies with temperature over a Rh(111) catalyst. Notice that at fixed reactant pressure, the rate reaches a maximum with increasing temperature and then declines. Figure 2.15 shows the pressure dependence of the reaction rate. Notice that the rate first increases with

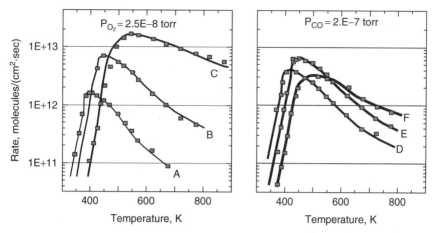

Figure 2.18 The rate of the reaction $CO + \frac{1}{2}O_2 \Rightarrow CO_2$ on Rh(111): (a) $P_{CO} = 2.5 \times 10^{-8}$ torr, $P_{O_2} = 2.5 \times 10^{-8}$ torr; (b) $P_{CO} = 1 \times 10^{-7}$ torr, $P_{O_2} = 2.5 \times 10^{-8}$ torr; (c) $P_{CO} = 8 \times 10^{-7}$ torr, $P_{O_2} = 2.5 \times 10^{-8}$ torr; (d) $P_{CO} = 2 \times 10^{-7}$ torr, $P_{O_2} = 4 \times 10^{-7}$ torr; (e) $P_{CO} = 2 \times 10^{-7}$ torr, $P_{O_2} = 2.5 \times 10^{-8}$ torr; (f) $P_{CO} = 2.5 \times 10^{-8}$ torr, $P_{O_2} = 2.5 \times 10^{-8}$ torr. Data of Schwartz et al. (1986).

increasing CO pressure and then declines. At lower temperatures, the rate continuously decreases with increasing CO pressure, while at higher temperatures the rate increases with increasing CO pressure.

The behavior in Figures 2.15 and 2.18 is typical of that for a bimolecular reaction on a heterogeneous catalyst. The reaction rate generally shows complex behavior with temperature and concentration. The behavior is usually quite different than that of bimolecular gas-phase reactions.

Typically, one fits data for the rate of a surface reaction with a complex rate equation. Table 2.11 shows a selection of the rate equations that have been fit to data on supported catalysts. The table is extracted from a longer compilation of Mezaki and Inoue (1991). Note that the rate equations are often quite complex and the rate can go up or down with increasing reactant pressure.

For example, the rate of the reaction $CO + 2H_2 \Rightarrow CH_3OH$ increases with increasing CO pressure and then reaches a plateau. However, the rate continuously increases with a continuous increase in H_2 pressure. In contrast, the rate of the reaction $SO_2 + \frac{1}{2}O_2 \Rightarrow SO_3$ plateaus as the partial pressure of either reactant increases while the rate of the reaction $4NH_3 + 6NO \Rightarrow 5N_2 + 6H_2O$ reaches a maximum with increasing reactant pressure and then declines. Generally, the kinetics of reactions on surfaces are quite different from the kinetics of reactions in the gas-phase or in solution, so much different rate equations are generally used.

Table 2.11 shows several other examples. Notice that the rate equations for surface reactions are generally more complex than the rate equations for typical gas-phase reactions. Generally, the rate equation has a term for each reactant in both the numerator and the denominator. The term in the numerator can cause the rate to decrease with increasing reactant concentration. By comparison, the rate of gas-phase reactions always increases with increasing concentration.

This unusual behavior arises because catalytic reactions are generally surface reactions. The reactants adsorb on the surface of the catalyst. Then there is a reaction between the reactants to form products.

Table 2.11 A selection of some of the rate equations for some common catalytic reactions extracted from the compilation of Mezaki and Inoue (1991)

Reaction	Catalyst	Rate Equation
$SO_2 + \frac{1}{2}O_2 \rightarrow SO_3$	V_2O_5	$\dfrac{k_1 K_3 K_4 P_{SO_2} P_{O_2}^{1/2} - k_2 K_5 P_{SO_3}}{1 + K_3 P_{O_2}^{1/2} + K_4 P_{SO_2} + K_5 P_{SO_3}}$
$N_2 + \frac{3}{2}H_2 \rightarrow NH_3$	Fe/Al_2O_3	$k_1 P_{N_2} \left(\dfrac{P_{H_2}^3}{P_{NH_3}^2}\right)^a - k_2 \left(\dfrac{P_{NH_3}^2}{P_{H_2}^3}\right)^{1-a}$
$CO + 2H_2 \rightarrow CH_3OH$	$CuO/ZuO/Al_2O_3$	$\dfrac{k_1 P_{CO} P_{H_2}^2 - k_2 P_{CH_3OH}}{1 + k_3 P_{H_2} + k_4 P_{CO} + k_5 P_{CO} P_{H_2}^{3/2}}$
$C_2H_4 + H_2 \rightarrow C_2H_6$	Ni/Al_2O_3	$\dfrac{k_1 K_2 K_3 P_{H_2} P_{C_2H_4}}{(1 + K_2 P_{H_2} + K_3 P_{C_2H_4})^2}$
$C_2H_4 + \frac{1}{2}O_2 \rightarrow C_2H_4O$	Ag/Al_2O_3	$\dfrac{k_1 K_2 P_{O_2} P_{C_2H_4}}{1 + K_2 P_{O_2} P_{C_2H_4}}$
$CO + H_2O \rightarrow H_2 + CO_2$	Fe_2O_3/Cr_2O_3	$\dfrac{k_1 K_3 K_4 P_{CO} P_{H_2O} - k_2 K_5 K_6 P_{H_2} P_{CO_2}}{(1 + K_3 P_{CO} + K_4 P_{H_2O} + K_5 P_{CO_2} + K_6 P_{H_2})^2}$
$4NH_3 + 6NO \rightarrow 5N_6 + 6H_2O$	Pt	$\dfrac{k_1 K_2 K_3 P_{NO} P_{NH_3}^{1/2}}{(1 + K_2 P_{NO} + K_3 P_{NH_3}^{1/2})^2}$

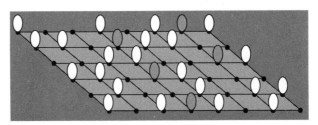

Figure 2.19 Langmuir's model for the adsorption of gas on a solid catalyst. The light gray area represents the surface of the catalyst. The black dots are the sites on the catalyst are sites that are available to adsorb gas. The white ovals are the adsorbed A molecules. The dark ovals are adsorbed B molecules.

Figure 2.19 is a snapshot of what the surface during the reaction $A + B \Rightarrow C$. The light gray area represents the surface of the catalyst. The black dots are the sites on the catalyst are sites that are available to adsorb gas. The white ovals are the adsorbed A molecules. The dark ovals are adsorbed B molecules. During the reaction, an adsorbed A molecule reacts with an adsorbed B molecule to produce products. The rate of the reaction is proportional to the surface concentration of A and B.

An important detail is that the surface has a finite capacity to adsorb A and B. When you put more A on the surface, there is not as much room to hold B. Consequently, the B concentration will decrease.

Now consider a simple reaction $A + B \Rightarrow C$ that follows

$$R_C = 1 \text{ Å}^2/(\text{molecule·second}) \; [A_{(ad)}][B_{(ad)}] \qquad (2.44)$$

In this equation R_c is the rate of formation of C, and $[A_{(ad)}]$ and $[B_{(ad)}]$ are the surface concentrations of A and B, in molecules/per square angstrom.

Table 2.12 The changes in the rate of production of C as [A$_{(ad)}$] varies, assuming [A$_{(ad)}$] + [B$_{(ad)}$] = 1 molecule/Å^2, R$_C$ = 1 Å^2/(molecule·second) [A$_{(ad)}$][B$_{(ad)}$]

[A$_{(ad)}$], molecule/Å^2	[B$_{(ad)}$], molecule/Å^2 Calculated from [A$_{(ad)}$] + [B$_{(ad)}$] = 1	R$_C$, (molecule·Å^2)/second Calculated from Equation (2.44)
0.1	0.9	0.09
0.2	0.8	0.16
0.5	0.5	0.25
0.8	0.2	0.16
0.9	0.1	0.09

Table 2.12 shows how the rate of formation of C changes as [A$_{(ad)}$] varies, assuming [A$_{(ad)}$] + [B$_{(ad)}$] = 1 molecule/Å^2. Notice that initially the rate goes up as the A concentration is increased. But then the rate reaches a maximum and decreases again with further increases in A. The decrease in rate causes the unusual behavior in Figure 2.18.

We will derive a qualitative equation for this effect in Section 12.17.1. The thing to remember for now is that rates of catalytic reactions show complex behavior because whenever the concentration of one reactant goes up, the surface concentrations of the other reactants go down.

The other thing that is special about a heterogeneously catalyzed reaction is that the reaction rate scales as the surface area of the catalyst, not the volume of the reactor. The rate is often measured as R, a rate in (mol·cm^2)/min. That is different than the rate of a reaction in a fluid, where r is the rate in (mol·cm^3)/min or (mol·liter)/hour.

Experimentally, the rate of a heterogeneously catalyzed reaction is proportional to the surface area of the catalyst. A catalyst with a surface area of 1000 m^2 will produce twice as much product as a catalyst with a surface area of 500 m^2.

People often call reactions on heterogeneous catalysts **surface reactions** since the reaction occurs on the surface of the catalyst.

People have developed a number of special materials to squeeze as much surface area as possible in as little volume as possible. Figure 2.20 shows some of the materials. Generally, it is possible to squeeze the surface area of 5000 m^2 (a football field) into 100 cm^3 or less of material!

Figure 2.20 The amount of Linde molecular sieve, activated carbon, and γ-alumina needed to get a surface area of about 5000 m^2 (i.e., about the area of a football field). [From Masel (1996).]

For a typical catalytic reaction, one might have a catalyst with an effective surface area of 10^4 m^2/liter of catalyst and a reaction rate of 10^{-6} (mol·m²)/second.

In the literature, people rarely report a reaction rate in (mol·m²)/second. Instead, they use a unit called a **turnover number**, T_N, or turnover frequency. The turnover number is defined as the rate that molecules are converted per atom in the surface of the catalyst per second. One can calculate the turnover number from

$$T_N = \frac{R_A}{N_S} \tag{2.45}$$

where R_A is the rate that molecules are converted in (molecule·cm²)/second and N_S is the number of surface atoms/cm². One usually sees turnover numbers expressed in units of reciprocal seconds (second^{-1}).

Figure 2.21 shows some typical values of the turnover numbers for some typical catalytic reactions. Note that typically, the turnover numbers are on the order of 1/second (i.e., 1 second^{-1}). Rates of 1/sec are also seen in semiconductor growth.

Students usually have trouble relating to a turnover number of 1/second. Just to give you an idea, a liter of a typical catalyst might have an effective surface area of $10^4 m^2$. If you run a reaction over that catalyst, then, at a turnover rate of 1/second, you would produce about 10 mol/minute of product. If you had 2 liters of catalyst, you would produce 20 mol/minute of product. In general, we can write the reaction rate as 1 (mol·minute)/liter of catalyst.

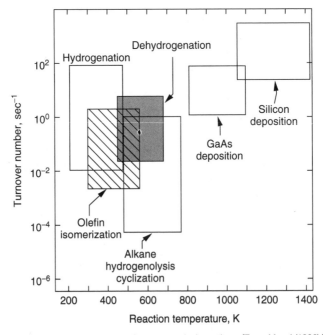

Figure 2.21 Turnover numbers for some typical reactions. [From Masel (1996).]

Now think back to your organic chemistry days and recall the reactions you ran in organic chemistry lab. It might have taken 30 minutes to 0.1 mol of product in a 200-cm flask. A production of 0.1 mol in 30 minutes in a 200-cm vessel corresponds to a reaction rate of

$$r_A = \frac{0.1 \text{ mol}}{(30 \text{ minutes})(0.2 \text{ liter})} = 0.0167 \frac{\text{mol}}{\text{liter·min}} \qquad (2.46)$$

Catalytic reactions generally go 1000–10,000 times faster than the reactions you ran in organic chemistry lab. Further, they will speed up reactions that are too slow to run in a chemistry lab. As a result, catalysts are very useful.

We will be discussing catalysts in detail in Chapter 12 in this book. For now, the key thing to remember is that catalysts tremendously change rates and also change the rate equation for a reaction.

2.8 MULTIPLE STEADY STATES, OSCILLATIONS, AND OTHER COMPLEXITIES

There is one other subtler with catalytic reactions: sometimes, one cannot write a rate equation for a catalytic reaction. All of the examples so far in this chapter had relatively simple rate equations. The rate could be written as a function of easily measured variables such as the concentrations of the reactants and products and perhaps a catalyst concentration (e.g., $[H^+]$). There were no other variables in the rate equation. In most reactions, one can write the rate of reaction as a function of a series of easily measurable concentrations. There are some exceptions. In this section, we will discuss the exceptions.

There are two kinds of systems that do not follow simple rate equations: those that show something called *multiple steady states*, and those that show something called *oscillations*. We will discuss multiple steady states first.

Figure 2.22 shows some data for the rate of CO oxidation:

$$CO + \tfrac{1}{2}O_2 \Longrightarrow CO_2 \qquad (2.47)$$

over a rhodium catalyst. In the experiment, Schwartz et al. (1986) fed a fixed amount of CO and O_2 over a rhodium catalyst in a well-stirred reactor. Then the rate of CO_2 production was measured as a function of temperature. The conversion of the system was low, so the concentrations of CO and O_2 were constant within the reactor. Schwartz et al. found that when they heated the surface from a low temperature, the reaction rate followed the lower curve in the figure. In contrast, when they cooled from a high temperature, the rate followed the higher curve in the figure. Notice that at temperatures between 500 and 600 K, the rate is about an order of magnitude lower when the system is being heated up than when the system is being cooled down. Yet the concentration of all of the species and the temperature is constant.

Now, think about how to write a rate equation for the data in Figure 2.22. One can imagine expressing the rate as a function of the reactant concentration, the product concentration, and temperature. However, that will not work for the data in Figure 2.22. After all, one observes two different rates at what are nominally the same conditions. As a result, one cannot write a simple rate equation for the data in Figure 2.22.

Gray and Scott (1990) review several examples showing two different rates at what are nominally the same conditions. The feature that all of these systems have in common is that while the compositions are constant, some other internal variable is different when

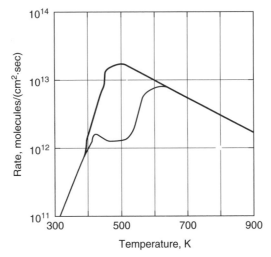

Figure 2.22 Rate data for CO oxidation on Rh(100) catalyst. [Data of Schwartz et al. (1986).]

the rate is high than when the rate is low. The fact that you have an extra variable complicates the rate equation. For example, the rhodium catalyst in Figure 2.22 shows two different crystal structures. One crystal structure is more reactive than the other. The crystal structure then becomes another variable that one needs to consider in the rate equation. A surface with 30% of the first crystal structure and 70% of the second will have a reactivity quite different from that of a sample, which has 70% of the first crystal structure and 30% of the second crystal structure.

One can define a new internal variable, the percentage of the catalyst in each crystal structure. The different crystal structure is an internal variable which alters the rate. The presence of the extra internal variable makes the analysis of rate data difficult.

Some other examples have other complications. For example, Figure 2.23 shows data for the rate of the Belousov–Zhabotinskii (BZ) reaction:

$$HOOCCH_2COOH + HBrO \xrightarrow{Ce^{4+}} products \qquad (2.48)$$

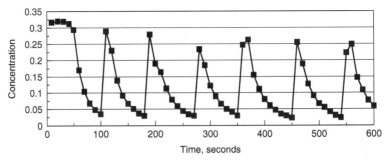

Figure 2.23 The concentration versus time measured during the Belousov–Zhabotinskii reaction. (Unpublished data of G. E. Poisson, D. A. Tuchman, and R. I. Masel.)

as a function of time. This reaction is special because if one loads a fixed amount of HBrO and maleic acid into the reactor, the reactor does not come to steady state. Instead, the composition in the reactor oscillates with time. The rate never reaches steady state.

The oscillations in rate seen with the BZ reaction are more complex versions of the same physics that is causing the CO oxidation reaction to exhibit multiple steady states. The rate depends on the oxidation state of the cerium catalyst. The cerium can be +3 or +4 state, and the reactivity is different for the +3 and +4 states. In addition, there are at least three reaction pathways that occur simultaneously.

One can write a rate equation for reaction (2.48). However, the rate is a function of the Ce^{3+}, Ce^{4+} and malate concentration. One cannot express the rate as a simple function of the concentrations of the reactants and products. The result is a very complex rate equation. See Gray and Scott (1990) for details.

The key point to remember for the discussion in this book is that in a complex case, the rate of reaction can depend on variables other than the concentration of the reactants and products and the temperature. The additional variables make any kinetic analysis difficult.

Generally, Van't Hoff's assumption that the rate is a function of the concentration works for 99.9% of the reactions which have been studied so far. However, there are a few exceptions.

2.9 SUMMARY

In summary, then, in this chapter, we reviewed some of the elementary concepts in kinetics that most of our readers had seen before. We defined the rate of reaction, the rate equation and the rate constant. We saw what rate equations are like, and we defined first- and second-order reaction. We described how temperature affects rates and how to use that information. We briefly discussed catalysis. Most of the material was qualitative. We will quantify the material in the next several chapters.

2.10 SOLVED EXAMPLES

Example 2.A   Illustration of Some of the Concepts from Section 2.1   Ethane emissions from cars are one of the major sources of air pollution in the United States. Most cars today are equipped with a catalytic converter, which, among other things, oxidizes ethane to CO_2 and water. Consider the oxidation of ethane in the 10 liter catalytic converter shown in Figure 2.A.1. The overall reaction is

$$7O_2 + 2C_2H_6 \longrightarrow 4CO_2 + 6H_2O \tag{2.A.1}$$

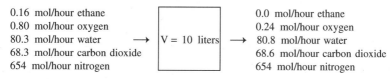

0.16 mol/hour ethane		0.0 mol/hour ethane
0.80 mol/hour oxygen		0.24 mol/hour oxygen
80.3 mol/hour water	$\longrightarrow$ V = 10 liters $\longrightarrow$	80.8 mol/hour water
68.3 mol/hour carbon dioxide		68.6 mol/hour carbon dioxide
654 mol/hour nitrogen		654 mol/hour nitrogen

Figure 2.A.1   The flow rates into and out of a catalytic converter for the complete combustion of ethane.

(a) Use a mass balance to show that the average rate of production of any species A in the reactant r_A is

$$r_A = \frac{\text{(mol/hour of A out of reactor)} - \text{(mol/hour of A into the reaction)}}{\text{volume of the reactor}}$$

(b) Calculate r_A for all of the species in the reactor.

(c) Calculate all of the stoichiometric coefficients (i.e., β_A values).

(d) Calculate the rate of the reaction from the equation

$$r_{2,A} = \frac{1}{\beta_A} r_A$$

Do a separate calculation for each species.

(e) How do the calculations compare?

Solution

(a) A mass balance on species A in the reactor is

$$\text{Moles of A out per hour} - \text{moles of A in per hour}$$

$$= \text{moles of A produced per hour} \qquad (2.A.2)$$

The definition of r_A the rate of production of species A is

$$r_A = \frac{\text{moles of A produced per hour}}{\text{volume of reactor}} \qquad (2.A.3)$$

Combining (2.A.2) and (2.A.3) yields

$$r_A = \frac{\text{moles of A out per hour} - \text{moles of A in per hour}}{\text{volume of reactor}} \qquad (2.A.4)$$

(b) Equation (2.A.4) is the key equation for this problem plugging in the numbers.

$$r_{\text{ethane}} = \frac{0 \text{ mol/hr} - 16 \text{ mol/hr}}{10 \text{ liters}} = -0.016 \text{ mol/(liter·hour)}$$

$$r_{O_2} = \frac{0.24 \text{ mol/hour} - 0.80 \text{ mol/hour}}{10 \text{ liters}} = -0.056 \text{ mol/(liter·hour)}$$

$$r_{H_2O} = \frac{80.8 \text{ mol/hour} - 80.3 \text{ mol/hour}}{10 \text{ liters}} = 0.05 \text{ mol/(liter·hour)}$$

$$r_{CO_2} = \frac{68.6 \text{ mol/hour} - 68.3 \text{ mol/hour}}{10 \text{ liters}} = 0.03 \text{ mol/(liter·hour)}$$

$$r_{N_2} = \frac{654 \text{ mol/hour} - 654 \text{ mol/hour}}{10 \text{ liters}} = 0 \text{ mol/(liter·hour)}$$

(c) $\beta_{O_2} = -7$, $\beta_{C_2H_6} = -2$, $\beta_{CO_2} = +4$, $\beta_{H_2O} = +6$.

(d) Next calculate $r_{2.A.1}$ the rate of reaction 2.A.1. According to equation (2.6)

$$r_A = \frac{1}{\beta_A} r_A \tag{2.A.5}$$

$$r_{2.A.1} = \frac{1}{\beta_{O_2}} r_{O_2} = \frac{1}{(-7)}[-0.056 \text{ mol/(liter·hour)}] = 0.008 \text{ mol/(liter·hour)}$$

$$r_{2.A.1} = \frac{1}{\beta_{C_2H_6}} r_{C_2H_6} = \frac{1}{(-2)}[-0.016 \text{ mol/(liter·hour)}] = 0.008 \text{ mol/(liter·hour)}$$

$$r_{2.A.1} = \frac{1}{\beta_{CO_2}} r_{CO_2} = \frac{1}{4}[0.03 \text{ mol/(liter·hour)}] = 0.0075 \text{ mol/(liter·hour)}$$

$$r_{2.A.1} = \frac{1}{\beta_{H_2O}} r_{H_2O} = \frac{1}{6}[0.05 \text{ mol/(liter·hour)}] = 0.0083 \text{ mol/(liter·hour)}$$

(e) Notice that all of the values of $r_{2.A.1}$ are the same to within the measurement error.

Example 2.B   Comparing First- and Second-Order Reactions   Consider a first-order reaction and a second-order reaction: How much would the rate change if you diluted the reactants by a factor of 3?

Solution   For a first-order reaction

$$r = k_A C_A \tag{2.B.1}$$

Consider two different concentrations C_A and C_A^1. The rates are

$$r_A = k_A C_A \tag{2.B.2}$$
$$r_A^1 = k_A C_A^1 \tag{2.B.3}$$

Dividing (2.B.2) by (2.B.3) yields

$$\frac{r_A^1}{r_A} = \frac{C_A^1}{C_A} \tag{2.B.4}$$

Therefore, if we dilute the reactants by a factor of 3, the rate will go down by a factor of 3:

$$\frac{r_A^1}{r_A} = \left(\frac{\frac{1}{3}}{1}\right) = \frac{1}{3}$$

If instead we have a second-order reaction, then

$$r_A = k_2 (C_A)^2 \tag{2.B.5}$$
$$r_A^1 = k_2 (C_A^1)^2 \tag{2.B.6}$$

$$\frac{r_A^1}{r_A} = \left(\frac{C_A^1}{C_A}\right)^2 \tag{2.B.7}$$

Therefore, if we dilute the reactants by a factor of 3, the rate will go down by a factor of 9:

$$\frac{r_A^1}{r_A} = \left(\frac{\frac{1}{3}}{1}\right)^2 = \frac{1}{9}$$

The difference between these results gives one a quick way to determine the order of a reaction.

Example 2.C   Using Equation (2.32) to Estimate Ignition Temperatures   How hot would you need to heat a methane–oxygen mixture before the mixture ignites? Assume that the activation barrier for the reaction is half the bond energy of the weakest bond in the methane molecule (bond energy = 104 kcal/mol).

Solution   Ignition is a complex problem to solve exactly because one needs to consider the increases in temperature during ignition, how much heat is lost to the walls, and other effects. However, one can approximate the autoignition temperature from equation (2.34) assuming an ignition time of about 1 second.

$$T_{sec} = \frac{E_a}{0.06 \ (kcal/(mol \cdot K)}$$

If we assume $E_a = 52$ kcal/mol (i.e., half the C–H bond energy), we calculatate

$$T_{sec} \left(\frac{(52 \ kcal/mol)}{0.06 \frac{kcal}{mol \cdot K}} \right) = 866 \ K$$

By comparison, the experimental value is 810 K.

Example 2.D   Fitting Data to Equation 2.28   In Section 2.6, we noted that if one measures rate data over a wide temperature range, it is often necessary to fit the data to

$$k_1 = k_1^0 T^N \exp\left(\frac{-E_a}{k_B T}\right) \tag{2.D.1}$$

rather than Arrhenius' law. The objective here is to do an example of the fitting procedure. Yang et al. (1995) examined the rate of the reaction

$$NCO + NO \Longrightarrow CO_2 + N_2 \tag{2.D.2}$$

As a function of temperature. Their data are given in Table 2.D.1. In the table, the apparent rate constant is defined as rate/([NCO][NO]) and is measured in units of $10^{-11} cm^3/(molecule \cdot second)$.
    Fit the data in Table 2.D.1 to equation (2.D.1).

Solution   The easiest way to solve this problem is to use the linear regression tool in Microsoft Excel or Lotus 1-2-3. I used Excel.

Table 2.D.1   The rate of reaction 2.C.2 as a function of temperature

Temperature, K	Apparent Rate Constant	Temperature, K	Apparent Rate Constant	Temperature, K	Apparent Rate Constant
293	3.8	432	2.92	632	1.72
344	3.36	538	2.28	836	1.28

I rearranged equation (2.D.1) as follows:

$$\ln(k_1) = \ln(k_1^0) + N \ln(T) - \frac{E_a}{k_B T} \qquad (2.D.3)$$

Equation (2.D.3) is a linear equation, with two variables, $\ln(T)$ and $1/T$. Therefore, a linear regression package can be used to fit the unknown parameters $\ln(k_1^0)$, N, and E_a/k_B. I decided to solve the equation in Microsoft Excel.

I set up my spreadsheet as shown in Table 2.D.2. The temperature is listed in column A; the rate constant K is listed in column B. I want to do a linear fit of $\ln(k)$ versus $\ln(T)$ and $1/T$. Column C gives $\ln(k)$, column D gives $\ln(T)$, and column E gives $1/T$.

I then used the regression analysis tool (under analysis tools) in Excel. I called c2 through c7 Y, and d2 through e7 X. I then used the analysis tool to solve for the coefficients. The analysis tool gives lots of output. The key part of the output is

	Coefficients	Standard Error
Intercept	14.362	1.07
X variable 1	−2.0136	0.1486
X variable 2	−469.29	70.0

Therefore the best fit of the line is

$$\ln(k) = (14.4 \pm 1.0) - (2 \pm 0.15)\ln T + \frac{(-469 \pm 70)}{T}$$

Therefore

$$k = \exp(14.4)T^{-2}\exp\left(\frac{-469}{T}\right)$$

Table 2.D.2   The formulas used to solve Example 2.D.1

	A	B	C	D	E
1	temp	k	ln k	ln(T)	1/T
2	293	3.8	=LN($B2)	=LN(A2)	=1/$A2
3	344	3.36	=LN($B3)	=LN(A3)	=1/$A3
4	432	2.92	=LN($B4)	=LN(A4)	=1/$A4
5	538	2.28	=LN($B5)	=LN(A5)	=1/$A5
6	682	1.72	=LN($B6)	=LN(A6)	=1/$A6
7	836	1.28	=LN($B7)	=LN(A7)	=1/$A7

There is one other detail of note. Different spreadsheet programs give different results. For example, Quatrapro gives

	Coefficient	Error
Intercept	13.634	0.0467
X variable 2	−1.92033	0.3562
X variable 2	−408.09	166.479

This difference arises because of a bug in Excel.

Example 2.E   Determining Reaction Order   Under some conditions the platinum-catalyzed reaction $CO + \frac{1}{2}O_2 \rightleftharpoons CO_2$ follows the rate equation

$$r_{CO_2} = \frac{k_1[CO][O_2]^{1/2}}{(1 + K_2[CO])^2} \tag{2.E.1}$$

(a) What is the order of the reaction?
(b) If $K_2[CO] \gg 1$, what is the order of the reaction?
(c) If $K_2[CO] \ll 1$, what is the order of the reaction?

Solution

(a) No general order.
(b) If $K_2[CO] \gg 1$, the rate equation becomes

$$r_{CO_2} = \frac{k_1[CO][O_2]^{1/2}}{(\cancel{1} + K_2[CO])^2} = \frac{k_1[CO][O_2]^{1/2}}{(K_2)^2[CO]^2} = \frac{k_1}{(K_2)^2}\frac{[O_2]^{1/2}}{[CO]}$$

Therefore, half order in O_2; negative one order in CO; negative half-order overall.
(c) If $K_2[CO] \ll 1$, then

$$r_{CO_2} = \frac{k_1[CO][O_2]^{1/2}}{(1 + \cancel{K_2[CO]})^2} = k_1[CO][O_2]^{1/2}$$

Thus, first-order in [CO]; half-order in O_2; $\frac{3}{2}$-order overall.
Examples of fitting catalytic rate constants will be given in Chapter 3.

Example 2.F   Numerical Integration of a Rate Equation   Assume that you are running a reaction $A \Rightarrow B$ that follows

$$r_A = -(1 \times 10^{13}/\text{second}) \exp\left(\frac{-20 \text{ kcal/mol}}{k_B T}\right)[A] \tag{2.F.1}$$

where r_A is the rate of reaction, T is temperature, k_B is Boltzmann's constant, and [A] is the A concentration. The temperature varies during the course of the reaction according to:

$$T = 300 \text{ K} + 10 \text{ K} \sin\left(\frac{t}{10 \text{ seconds}}\right) \qquad (2.F.2)$$

where t is time. How long will it take to reduce the A concentration from 1 mol/liter to 0.1 mol/liter?

Solution   According to equation (2.4)

$$\frac{d[A]}{dt} = r_A \qquad (2.F.3)$$

Combining equations (2.F.1)–(2.F.3) yields

$$\frac{d[A]}{dt} = -(1 \times 10^{13}/\text{second}) \exp\left[-\frac{(20 \text{ kcal/mol})}{k_B(300 \text{ K} + 10 \text{ K} \sin(t/10 \text{ seconds}))}\right] [A] \quad (2.F.4)$$

Rearranging and integrating

$$-\int_{C_A^0}^{C_A} \frac{d[A]}{[A]} = (1 \times 10^{13}/\text{second}) \int_0^t \exp\left[-\frac{20 \text{ kcal/mol}}{k_B(300 \text{ K} + 10 \text{ K} \sin(t/10 \text{ seconds}))}\right] dt$$
$$(2.F.5)$$

where C_A^0 is the initial concentration of A and C_A is the concentration at time t. Performing the integral on the left-hand side yields

$$\ln\left(\frac{C_A^0}{C_A}\right) = (1 \times 10^{13}/\text{second}) \int_0^t \exp\left[-\frac{(20 \text{ kcal/mol})}{k_B(300 \text{ K} + 10 \text{ K} \sin(t/10 \text{ seconds}))}\right] dt$$
$$(2.F.6)$$

The right-hand side can be integrated with the trapezoidal rule. In the trapezoidal rule, you divide time into small segments Δt, then approximate the integral as a sum. For example, if you want to integrate a function F(t), you say

$$\int_0^t F(t) \, dt = \sum_{i=0}^{n-1} F(t_i)\Delta t - \frac{1}{2}(F(t_0) + F(t_n))\Delta t \qquad (2.F.7)$$

Consequently, according to the trapezoidal rule

$$\int_0^t \exp\left[-\frac{(20 \text{ kcal/mol})}{k_B(300 \text{ K} + 10 \text{ K} \sin(t/10 \text{ seconds}))}\right] dt$$
$$= \sum_i \exp\left[-\frac{(20 \text{ kcal/mol})}{k_B(300 \text{ K} + 10 \text{ K} \sin(t_i/10 \text{ seconds}))}\right] \Delta t$$
$$- 0.5\left(\exp\left[-\frac{(20 \text{ kcal/mol})}{k_B(300 \text{ K} + 10 \text{ K} \sin(t_0/10 \text{ seconds}))}\right] \Delta t\right.$$
$$\left. + \exp\left[-\frac{(20 \text{ kcal/mol})}{k_B(300 \text{ K} + 10 \text{ K} \sin(t_n/10 \text{ seconds}))}\right] \Delta t\right) \qquad (2.F.8)$$

where t_i is the time at time increment i, t_0 is the initial time, and t_n is the final time. Combining (2.F.6) and (2.F.8) yields

$$\ln\left(\frac{C_A^0}{C_A}\right) = \sum_i (1 \times 10^{13}/\text{second}) \exp\left[-\frac{20 \text{ kcal/mol}}{k_B(300 \text{ K} + 10 \text{ K} \sin(t_i/10 \text{ seconds}))}\right]\Delta t$$

$$- 0.5 \times 10^{13}\left(\exp\left[-\frac{(20 \text{ kcal/mol})}{k_B(300 \text{ K} + 10 \text{ K} \sin(t_0/10 \text{ seconds}))}\right]\Delta t\right.$$

$$\left.+ \exp\left[-\frac{(20 \text{ kcal/mol})}{k_B(300 \text{ K} + 10 \text{ K} \sin(t_n/10 \text{ seconds}))}\right]\Delta t\right) \qquad (2.F.9)$$

It is useful to simplify this expression by defining the temperature at time t_i by

$$T_i = 300 \text{ K} + 10 \text{ K} \sin(t_i/10 \text{ seconds}) \qquad (2.F.10)$$

Combining equations (2.F.9) and (2.F.10) yields

$$\ln\left(\frac{C_A^0}{C_A}\right) = \sum_i (1 \times 10^{13}/\text{second}) \exp\left(-\frac{20 \text{ kcal/mol}}{k_B T_i}\right)\Delta t - 0.5$$

$$\times 10^{13}\left[\exp\left(-\frac{20 \text{ kcal/mol}}{k_B T_0}\right)\Delta t + \exp\left(-\frac{20 \text{ kcal/mol}}{k_B T_n}\right)\Delta t\right] \qquad (2.F.11)$$

Rearranging

$$C_A = C_A^0 \exp\left\{-\sum_i (1 \times 10^{13}/\text{second}) \exp\left(-\frac{20 \text{ kcal/mol}}{k_B T_i}\right)\Delta t\right.$$

$$\left.+ 0.5 \times 10^{13}\left[\exp\left(-\frac{20 \text{ kcal/mol}}{k_B T_0}\right)\Delta t + \exp\left(-\frac{20 \text{ kcal/mol}}{k_B T_n}\right)\Delta t\right]\right\} \qquad (2.F.12)$$

Therefore, I can integrate equation (2.F.12) numerically to calculate the concentration versus time. I used a Microsoft Excel worksheet to solve the problem. Table 2.F.1 shows the formulas in my worksheet; Table 2.F.2 shows the numerical values. I named cell b1 to be dt so I could set delta time. I also named cell D1 Ca0. I made column A time and column B, temperature. I defined column C as the term I needed to sum

$$\text{Term} = (1 \times 10^{13}/\text{second}) \exp\left[-\frac{(20 \text{ kcal/mol})}{k_B T_i}\right]\Delta t \qquad (2.F.13)$$

Next I want to compute the integral from 0 to t_n where n varies. I defined column E to be the approximation to the integral

$$\text{Integral} = \left\{-\sum_i (1 \times 10^{13}/\text{second}) \exp\left(-\frac{20 \text{ kcal/mol}}{k_B T_i}\right)\Delta t\right.$$

$$\left.+ 0.5 \times 10^{13}\left[\exp\left(-\frac{(20 \text{ kcal/mol})}{k_B T_0}\right)\Delta t + \exp\left(-\frac{20 \text{ kcal/mol}}{k_B T_n}\right)\Delta t\right]\right\}$$

$$(2.F.14)$$

Table 2.F.1   The formulas in the spreadsheet to calculate C_A

	A	B	C	D	E	F
1	dt=	2	Ca0=	1		
2	time	temp	term	sum	integral	Ca
3	0	=300+10*SIN(A3/10)	=EXP(−20/0.00198/B3)*1.e13*dt	=SUM(C$3:C3)	=0.5*(C$3+C3)−D3	=Ca0*EXP(E3)
4	=A3+dt	=300+10*SIN(A4/10)	=EXP(−20/0.00198/B4)*1.e13*dt	=SUM(C$3:C4)	=0.5*(C$3+C4)−D4	=Ca0*EXP(E4)
5	=A4+dt	=300+10*SIN(A5/10)	=EXP(−20/0.00198/B5)*1.e13*dt	=SUM(C$3:C5)	=0.5*(C$3+C5)−D5	=Ca0*EXP(E5)
6	=A5+dt	=300+10*SIN(A6/10)	=EXP(−20/0.00198/B6)*1.e13*dt	=SUM(C$3:C6)	=0.5*(C$3+C6)−D6	=Ca0*EXP(E6)
7	=A6+dt	=300+10*SIN(A7/10)	=EXP(−20/0.00198/B7)*1.e13*dt	=SUM(C$3:C7)	=0.5*(C$3+C7)−D7	=Ca0*EXP(E7)
8	=A7+dt	=300+10*SIN(A8/10)	=EXP(−20/0.00198/B8)*1.e13*dt	=SUM(C$3:C8)	=0.5*(C$3+C8)−D8	=Ca0*EXP(E8)
9	=A8+dt	=300+10*SIN(A9/10)	=EXP(−20/0.00198/B9)*1.e13*dt	=SUM(C$3:C9)	=0.5*(C$3+C9)−D9	=Ca0*EXP(E9)
10	=A9+dt	=300+10*SIN(A10/10)	=EXP(−20/0.00198/B10)*1.e13*dt	=SUM(C$3:C10)	=0.5*(C$3+C10)−D10	=Ca0*EXP(E10)
11	=A10+dt	=300+10*SIN(A11/10)	=EXP(−20/0.00198/B11)*1.e13*dt	=SUM(C$3:C11)	=0.5*(C$3+C11)−D11	=Ca0*EXP(E11)
12	=A11+dt	=300+10*SIN(A12/10)	=EXP(−20/0.00198/B12)*1.e13*dt	=SUM(C$3:C12)	=0.5*(C$3+C12)−D12	=Ca0*EXP(E12)
13	=A12+dt	=300+10*SIN(A13/10)	=EXP(−20/0.00198/B13)*1.e13*dt	=SUM(C$3:C13)	=0.5*(C$3+C13)−D13	=Ca0*EXP(E13)
14	=A13+dt	=300+10*SIN(A14/10)	=EXP(−20/0.00198/B14)*1.e13*dt	=SUM(C$3:C14)	=0.5*(C$3+C14)−D14	=Ca0*EXP(E14)

Table 2.F.2 The numerical values in the spreadsheet to calculate C$_A$

	A	B	C	D	E	F
1	dt=	2	CaO=	1		
2	time	temp	term	sum	integral	Ca
3	0	300	0.047678	0.047678	0	1
4	2	301.9867	0.0595	0.107179	−0.05359	0.947821
5	4	303.8942	0.073401	0.180579	−0.12004	0.886885
6	6	305.6464	0.088809	0.269388	−0.20114	0.817794
7	8	307.1736	0.104667	0.374055	−0.29788	0.742389
8	10	308.4147	0.119477	0.493532	−0.40995	0.663681
9	12	309.3204	0.131501	0.625033	−0.53544	0.58541
10	14	309.8545	0.139115	0.764148	−0.67075	0.511324
11	16	309.9957	0.141197	0.905345	−0.81091	0.444454
12	18	309.7385	0.137427	1.042772	−0.95022	0.386656
13	20	309.093	0.128379	1.171152	−1.08312	0.338537
14	22	308.085	0.115361	1.286513	−1.20499	0.299694
			spreadsheet continues			
38	70	306.5699	0.098104	2.295655	−2.22276	0.108309
39	72	307.9367	0.113554	2.409208	−2.32859	0.097433

I summed all of the terms from 0 to t and subtracted that sum from the second term in equation (2.F.14). I defined column F to be the concentration at time t. I used equation (2.F.12) to calculate it.

Therefore it takes between 70 and 72 seconds to get to 90% conversion.

Additional information about Arrhenius' law is given in

K. J. Laidler, The development of Arrhenius's equation, *J. Chem. Educ.* **61**, 494 (1984).

B. Heinrich, *Hot Blooded Insects*, Harvard University Press, Cambridge, MA, (1993).

J. R. Heulett, Deviations from the Arrhenius equation, *Q. Rev., Chem. Soc.* **18**, 277 (1964).

2.11 PROBLEMS

2.1 Define the following terms in three sentences or less:

(a) Rate equation	**(i)** Multiple steady state
(b) Order	**(j)** Catalyst
(c) Rate constant	**(k)** Stoichiometric coefficient
(d) Preexponential	**(l)** First order
(e) Activation barrier	**(m)** Second order
(f) Arrhenius' law	**(n)** Half order
(g) Catalysis	**(o)** Heterogeneous reaction
(h) Oscillating reaction	**(p)** Homogeneous reaction

2.2 What did you learn that was new to you in this chapter? Were you aware of all of the definitions? Was the historical information new? Were you aware of the limitations of the Arrhenius equation and of representing data by a rate law? Were you aware that Arrhenius' law applies to so many practical situations? Did you know that you could estimate an activation barrier without doing many experiments? What else did you learn that was new?

2.3 Find 20 examples of kinetic processes in your home, such as cooking different kinds of meals, washing clothes, you digesting different types of food, or plants growing on your window sill.

2.4 Identify the units for

(a) r_A, the rate per unit volume

(b) R_A, the rate per unit surface area

(c) A first-order gas-phase reaction

(d) A first-order surface reaction

(e) A second-order gas-phase reaction

(f) A second-order surface reaction

2.5 A first-order reaction has a rate constant of 0.15/minute.

(a) Calculate the rate of reaction at a reactant concentration of 0.57 mol/liter.

(b) Repeat for a second-order reaction with a rate constant of 0.73 liter/(mol·minute).

2.6 Calculate the activation barrier for a reaction whose rate exactly doubles from

(a) 100 to 110 K

(b) 200 to 210 K

(c) 300 to 310 K

(d) 500 to 510 K

(e) 750 to 760 K

(f) 1000 to 1010 K

2.7 According to Table 2.3, the oxidation of phosphine (PH_3) obeys

$$r_{PH_3} = k_5[PH_3][O_2]^{1/2} \qquad\qquad (P2.7.1)$$

(a) What is the overall order of the reaction?

(b) What is the order with respect to phosphine?

(c) What is the order with respect to oxygen?

(d) How much will the rate increase if you double the oxygen pressure keeping the phosphine pressure constant?

(e) How much will the rate increase if you double the phosphine pressure?

(f) How much will the rate increase if you double the total pressure constant?

Hint: First define $x_{PH_3} = P_{PH_3}/P_{Total}$. Show $P_{O_2} = P_{Total} - P_{PH_3} = P_{Total}(1 - x_{PH_3})$. Then substitute into equation (P2.7.1).

2.8 Later in this book, we will provide some methods to estimate the preexponentials and activation barriers for reactions. Assume that we solve a problem and make a mistake, and calculate a 25 kcal/mol barrier for a reaction that actually has a barrier of 30 kcal/mol. The preexponential was calculated accurately.

(a) If you need to calculate the rate constant at 500 K, how large an error in the rate will you make?

(b) How much would you have to change the temperature to compensate for the error (i.e., to increase the rate to the rate you calculate)?

2.9 Milk goes sour because bacteria in the milk converts the lactose in the milk into lactic acid.

(a) The reaction goes 40 times faster at 25°C than at 4°C. Estimate the activation barrier.

(b) How does your estimate compare to that from Figure 2.9? Assume a half-life of 12 hours.

(c) How do you account for the differences?

2.10 Your taste buds work by a complex pathway. First the receptors in your tongue bind to some of the components in food. That sets off a complex chain reaction leading to a change in polarity of the cell, and eventually the sensation of taste. The object of this problem is to use the material in this chapter to guess how changes in the temperature of your tongue will affect the perception of taste.

(a) How much would you expect the sensitivity of your tongue to change when you change the temperature of your taste buds by 10°C?

(b) Guess or measure how much the temperature of your tongue changes when you drink a glass of soda pop (carbonated beverage).

(c) Guess or measure how much the temperature of your tongue changes when you eat a dish of ice cream.

(d) Ice cream manufacturers say that they need about twice as much sugar to sweeten ice cream as to sweeten soda pop to the same perceived level of sweetness. Are your expectations consistent with these findings?

(e) Try an experiment to verify that this works. Take a sample of ice cream and store it in a sealed container in your refrigerator overnight. Let the ice cream warm up to 10°C. Now do a taste test. Taste the frozen ice cream. Taste the melted ice cream. Which tastes sweeter?

2.11 When you cook rice, you hydrolyze the cellulose in the rice to starch via the following chemical reaction:

$$\text{Cellulose} + H_2O \Longrightarrow \text{starch}$$

Brown rice takes twice as long to cook as white rice because there is twice as much cellulose to hydrolyze. What does that tell you about the kinetics of the hydrolysis reaction?

(a) If you had a first-order reaction, how would the cooking time vary with the cellulose concentration?

(b) What does the fact that the cooking time doubles as the cellulose concentration doubles tell you about the kinetics of the reaction?

(c) How would the cooking time vary if you were working at an altitude of 10,000 feet (ft)? (*Hint*: The boiling point of water at 10,000 feet is 92°C.)

2.12 The reaction $4PH_3 \xrightarrow{1} P_4 + 6H_2$ follows:

$$r_1 = 0.08/\text{second [PH}_3]$$

(a) Calculate r_{PH_3}, r_{P_4}, and r_{H_2} at 500°C, and a PH_3 partial pressure of 5 torr. [*Note*: 760 torr = 1 atm. (atmosphere of pressure).]

(b) How many milligrams of phosphorous do you deposit in 100 seconds?

2.13 In Section 2.1, we reported that Lavoisier found that 1 kg of air always reacted with 0.78 kg of tin to yield 0.99 kg of tin calx (SnO) and leave 0.79 kg of inerts.

(a) Did Lavoisier measure things correctly? Does 1 kg of air react with 0.79 kg of tin to yield 0.99 kg of tin calx?

(b) How many kilograms of the following substances will react with 1 kg of air: aluminum forming Al_2O_3, iron forming Fe_2O, Ni forming NiO_2, silicon forming SiO_2?

(c) What are the stoichiometric coefficients in the reaction to form tin calx? Assume a basis of one mole of tin.

2.14 In Section 2.1, we noted that Lavoisier showed that mass is conserved during chemical reactions by sealing a flask filled with tin and air and measuring the weight change as the oxygen in the air reacted with the tin.

(a) If you have a sealed 1 liter flask filled with air and 10 grams of tin, how much SnO_2 can you produce if you use up all of the oxygen in the air?

(b) By much would the weight of the sealed flask change?

(c) By much would the weight change if you opened the flask?

(d) Now think about doing this in the year 1796. How would you measure the weight change?

2.15 Consider the reaction $4PH_3 \rightarrow P_4 + 6H_2$. (P2.15)

(a) What are the stoichiometric coefficients of each of the species in the reaction (P2.15)?

(b) If we load 2 mol of PH_3 into a reactor, and convert 50% of the PH_3 into products, how much P_4 and H_2 would be produced?

(c) Assume that we are using reaction (P2.15) to deposit phosphorous onto a silicon wafer. Assume that we deposit 10^{-4} mol/(hour·cm^2) of P_4 on our wafer. Calculate

(1) The rate of reaction (P2.15)

(2) The rate of hydrogen production

(3) The rate of phosphine (PH_3) production

(d) If we write the reaction $PH_3 \xrightarrow{2.15a} \frac{1}{4}P_4 + \frac{3}{2}H_2$, how will the stoichiometric coefficients change?

2.16 Are the following reactions homogeneous or heterogeneous?

(a) Cooking a steak (i.e., converting the collagen to glycogen and oxidizing the myoglobin)

(b) Bleaching your clothes (i.e., reactions of bleach with dirt)

(c) Buildup of rust on you car (i.e., reacting oxygen with iron)

(d) Combustion in your car's engine (i.e., burning gasoline)

(e) Burning a candle

(f) Digesting the sugar in a glass of beer (i.e., turning sucrose to glucose in your stomach)

2.17 In Example 2.A we considered the oxidation of ethane in a catalytic converter. However, another important reaction is the oxidation of ethanol:

$$CH_3CH_2OH + 3O_2 \Longrightarrow 2CO_2 + 3H_2O$$

Assume that you measure the oxidation of ethanol in a catalytic converter and the data in Table P2.17 are obtained.

(a) Calculate the rate of formation of each species.

(b) Calculate the overall rate of reaction.

(c) Notice that the nitrogen and oxygen do not quite balance. This indicates that another reaction is occurring. What is the stoichiometry of other reaction? (*Hint*: How much excess oxygen and nitrogen are in the exit? What other species could decompose or react to form the excess oxygen and nitrogen?)

2.18 In Problem 2.D we fit the data in Table 2.D.1 to equation (2.D.1).

(a) Set up the problem yourself and see if you can reproduce the findings. The spreadsheet is available from Professor Masel's website. http://www.uiuc.edu/~rl-masel or Wiley's website.

(b) What is the activation energy (with error bars)?

(c) By much would the activation barrier change if the point at 293 K were misrecorded so k(273) = 3.08?

(d) What does the result in (c) tell you about the importance of experimental error in the measurement of activation barriers?

2.19 In Example 2.F we used a numerical procedure to integrate a rate expression.

(a) Set up the problem yourself and see if you can reproduce the findings. The spreadsheet is available from Professor Masel's website.

(b) Assume that the temperature instead varies as $T = 300\ K + 5\ K\sin(t/10\ seconds)$. How will your answer change?

Table P2.17 The inlet and outlet flow rates (in mol/hour) from an experimental catalytic converter with a volume of 10 liters

Species	Inlet Flow Rate	Exit Flow Rate
Ethanol	0.21	0.01
Oxygen	0.80	0.30
Water	80.5	81.1
Carbon dioxide	60.9	61.3
Nitrogen	637	637.1

(c) Assume that the temperature instead varies as $T = 300 \text{ K} + 1 \text{ K} \sin(t/10 \text{ seconds})$. How will your answer change?

(d) Assume that the temperature instead varies as $T = 300 \text{ K} + 0.1 \text{ K} \sin(t/10 \text{ seconds})$. How will your answer change?

(e) What do these results tell you about the importance of temperature control in kinetics measurements?

2.20 Fit the data in Figures 2.7 and 2.10 to Arrhenius' law. How well do the data fit in each case? (*Hint*: Draw a line on the graph. Do not use a spreadsheet.)

2.21 Figure P2.21 shows V. Van Spaendonk and R. I. Masel's unpublished data for the pyrolysis of a mixture of ethylene and hydrogen over a platinum catalyst. During the experiment, Van Spaendonk and Masel adsorbed ethylene and hydrogen onto the catalyst and then heated the catalyst and looked for reaction. Each of the peaks in Figure P2.21 corresponds to a different reaction pathway; For instance, the ethane peak at 310 K corresponds to a reaction where two adsorbed species react to form ethane.

(a) From the data shown, how many chemical reactions do you see? (*Hint*: Each chemical reaction will produce one peak.)

(b) Use the data in Figure P2.21 to estimate the activation barrier for each of these chemical reactions. Assume that the peak temperature corresponds to the temperature where the half-life is one second.

2.22 The objective of this problem is to use a spreadsheet (e.g., Lotus or Excel) to examine the implications of Arrhenius' law.

(a) Assume that you are running a first-order reaction in a batch reactor, and that the reaction has a half-life of 1 minute at some temperature T. Calculate

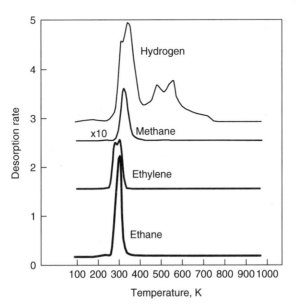

Figure P2.21   Van Spaendonk and Masel's data.

the activation energy of the reaction as a function of T, assuming that the reaction follows Arrhenius' law with a preexponential of 10^{13}/second. For the purposes of this problem, consider T to range between 100 and 1000 K.

(b) Vary the preexponential from 10^{11} to 10^{15}/second. To what extent do your estimates of the activation barrier change?

(c) Change your estimate of the half-life from a minute to an hour. By how much do your estimates of the activation barrier change?

(d) Now go back to your estimate of the activation barrier in part (a). At each temperature, calculate how much you would have to increase the temperature in order to double the rate. (*Hint*: At 500 K you would need to increase the temperature to perhaps 514 to double the rate.)

(e) Use your results to see if the idea that rates double every 10 K works for this example.

2.23 Yang, Lee and Wang, *International J. Of Chemical Kinetics*, **27**, (1995), 1111 examined the rate of the reaction

$$NCO + NO_2 \Longrightarrow CO + 2NO \qquad (P2.23)$$

as a function of temperature. Their data are given in Table P2.23. In the table, the rate constant is defined as rate/([NCO][NO$_2$]) and is measured in units of 10^{-11} cm^3/(molecule·second).

(a) Are the units of the rate constant correct?

(b) Fit the data to Arrhenius' law. How well do they fit?

(c) Is the fit improved with equation (2.28)? Use the methods in Example 2.D.

(d) How do you interpret the sign of the activation barrier?

(e) On the basis of your fits and the error bars in Excel, how much confidence do you have in the value of the activation barrier?

2.24 Eberhard and Howard, *International J. Of Chemical Kinetics*, **28**, (1996) 731 examined the rate of the reaction

$$CH_3CH_2OO + NO \Longrightarrow CH_3CH_2O + NO_2 \qquad (P2.24.1)$$

as a function of temperature. Their data are given in Table P2.24. In the table, the rate constant is defined as rate/([CH$_3$CH$_2$OO][NO]) and is measured in units of 10^{-12} cm^3/(molecule·second).

Table P2.23   The rate of reaction (P2.23) as a function of temperature

Temperature, K	Apparent Rate Constant	Temperature, K	Apparent Rate Constant	Temperature, K	Apparent Rate Constant
294	2.68	384	2.08	579	1.14
320	2.43	431	2.02	665	1.12
351	2.57	493	1.58	774	1.05

Table P2.24   The rate of reaction (P2.24) as a function of temperature

Temperature, K	Apparent Rate Constant	Temperature, K	Apparent Rate Constant	Temperature, K	Apparent Rate Constant
206	16.9	252	11.9	325	9.2
208	17.9	261	11.6	355	8.02
223	14.0	298	9.06	402	6.76
244	14.1	298	9.67	403	6.58

(a) Are the units of the rate constant correct?

(b) Fit the data to Arrhenius' law. How well do they fit?

(c) Is the fit improved with equation (2.28)? Use the methods in Example 2.D.

(d) How do you interpret the sign of the activation barrier?

(e) On the basis of your fits, how much confidence do you have in the value of the activation barrier?

2.25 As noted earlier in this chapter, crickets chirp to attract mates; the cricket's metabolic rate varies with the cricket's body temperature, so the chirping rate also varies with the cricket's temperature. [Walker, *Evolution* **16**, (1962) 407.] measured the rate that a group of crickets chirped and obtained the following data:

Temperature, °F	Chirp rate, chirps/minute	Temperature, °F	Chirp rate, chirps/minute	Temperature, °F	Chirp rate, chirps/minute
55	60	64	122	77	136
58	65	73	136	81	148

(a) Estimate the activation energy for the process leading to chirping.

(b) A cricket often chirps from dusk to dawn. Assume that during the course of the night, the temperature varied as follows:

Time	Temperature, °F	Time	Temperature, °F	Time	Temperature, °F
9 P.M.	80	12 P.M.	73	3 A.M.	67
10 P.M.	77	1 A.M.	68	4 A.M.	67
11 P.M.	75	2 A.M.	67	5 A.M.	67

How many chirps would a cricket make from 9 P.M. to 5 A.M.? (*Hint:* Set up a problem as an integral that needs to be solved, then use the methods in Example 2.F to do the integration.)

(c) Next time you are walking at night, listen for the crickets. How fast are they chirping? Does the chirp rate agree with the data above? Can you use the rate at which crickets chirp to estimate the air temperature? How accurate is your measurement?

2.26 A.M. Smith, *Environ, Entomol.* **21**, (1992), 314 examined the speed that pea weevil (*Brochus pisorum*) eggs matured as a function of temperature. The following data were obtained:

Temperature, °C	Days for Weevils to Hatch	Temperature, °C	Days for Weevils to Hatch	Temperature, °C	Days for Weevils to Hatch
10.7	38	18.1	15.6	23.7	7.3
14.4	19.5	18.1	9.6	24.7	4.5
16.2	15.6	21.4	9.5	28.6	7.1

(a) Fit these data to Arrhenius' law and estimate the activation energy for the process. For the purposes of this problem, assume that the time to hatch is the time where the weevils are 90% developed (i.e., $X_A = 0.9$).

(b) Assume that a weevil lays her eggs on your garden on a Monday. How long will it take before the weevil larvae hatch and start to eat your plants? Temperature data follow:

Day	Temperature, High/Low, °C	Day	Temperature, High/Low, °C	Day	Temperature, High/Low, °C
Mon.	18/24	Fri.	18/21	Tues.	17/27
Tues.	14/21	Sat.	18/25	Wed.	19/21
Wed.	17/23	Sun.	16/21	Thurs.	18/28
Thurs.	22/28	Mon.	19/28	Fri.	17/25

For the purposes of discussion, assume that the daily temperature varies linearly from a high at 2 P.M. to a low at 5 A.M. Use the Methods in Example 2.F to do the integration. A more detailed discussion of the temperature effect can be found in P.G. Allsopp, *Agricultural Forecasting and Meteorology*, **41**, (1987).

2.27 Harlow Shapley was a famous astronomer who discovered the Milky Way galaxy. As a sideline, on cloudy days, he measured the walking speed of ants outside the Mount Wilson Observatory. Shapley, *Proc. National Academy of Science*, **6**, (1920, 1924) 436 reported the following data:

Temperature, °C	Running Speed, cm/second	Temperature, °C	Running Speed, cm/second	Temperature, °C	Running Speed, cm/second
9.0	0.44	21	2.23	33	4.32
10.3	0.64	22	2.16	33.5	4.77
12.5	0.77	23.5	2.29	34	4.35
14.5	1.10	24.5	2.65	35	5.08
17	1.28	25.5	2.94	36	5.57
18.6	1.48	26	2.56	37	5.67
19	1.62	30	3.05	38	6.06
20	1.90	31	3.06	38.5	6.60

(a) Fit these data to Arrhenius' law and estimate the activation energy for the process. Assume that the running speed corresponds to the rate of a metabolic process in the ants.

(b) Try an experiment to see if Shapley is correct. Put some ants on a skillet and watch how quickly they move as you heat the skillet. Do the ants follow our expectations?

(c) Assume that you and your boyfriend or girlfriend go for a picnic in the park on a warm (28°C) day. You lay out the picnic basket on a blanket and start to eat. Unbeknownst to you, there is an anthill 10 meters away. How long will it take before the ants discover your picnic basket and begin to raid your basket?

(d) Heinrich's book, the *Hot-Blooded Insects*, Harvard University Press, (1993) has other examples of changes in insect behavior with temperature. Find another example to suggest a problem like this one. Be sure to solve the problem to make sure that it works.

2.28 Explain when you can represent rate data with a rate expression (i.e., determine which criterion needs to be satisfied for the rate to be a function of only the concentrations of all the species in the reactor, pressure and temperature). (*Hint*: Think back to the phase rule from your thermodynamics class. When are temperature, pressure, and composition insufficient to uniquely define all of the properties of the system such as density?)

2.29 The growth of tumors follows kinetic expressions similar to those discussed in this chapter. Look at the following papers and describe the findings: Hasenclever et al. *Annals of Oncology*, **7**, (1996); Chaplain et al. *Wounds* **8** (1996) 42; Groebe, *International Journal of Radiation Oncology*, **34** (1996) 395; Wassermann et al. *Mathematical Biosciences*, **136** (1996) 111; Costa et al. *Mathematical Bioscience*, **125** (1995) 191. Skip any papers not available in your university's library.

2.30 Sponge et al. *Mathematical Bioscience*, **138** (1996) 1, described a kinetic model for the spread of an HIV (human immunodeficiency virus) infection. Look up the paper and describe the findings.

2.31 What did you learn new doing this homework set?

3

ANALYSIS OF RATE DATA

PRÉCIS

At this point, we will be changing our focus. In Chapter 2, we discussed general principles that can be applied to a variety of reactions. In this chapter, we will be discussing some of the techniques one uses to fit rate data to a rate equation. First, we will briefly discuss experimental techniques. Then we will discuss how to analyze rate data.

When this book was written, it was assumed that our readers had seen this material before. Maybe they had measured a rate in freshman chemistry or integrated a rate equation in their physical chemistry or reactor design course. However, people seldom know where to start when presented with a new problem. What needs to be measured, and how does one analyze the data? The objective of the material that follows is to summarize the key ideas that one needs to understand in order to measure rate data for a new system.

3.1 INTRODUCTION

Studies of rates of chemical reactions date back to Wilhelmy's (1850) measurement of the rate of sugar conversion in the grapejuice used to make wine. They have been improved ever since.

In this chapter we will discuss how one actually determines a rate equation. First we will briefly discuss some of the experimental techniques that one uses to determine a rate equation. In particular, we will distinguish between direct and indirect methods, and describe the properties of each. Then we will discuss the data analysis. Our approach is a little nonstandard, in that we point out that some common data analysis schemes are subject to some uncertainty. Understanding the uncertainty is as important as understanding the methods.

The outline of the material is as follows:

- In Sections 3.2–3.4 we will briefly review the experimental techniques. The experimental techniques are evolving all of the time, and we had to keep up with the latest advances. Still, we wanted to give a picture of how the data are taken, and in particular to distinguish between direct and indirect methods.
- In Sections 3.5–3.8 we will examine the analysis of the data. We will derive the key equations and explain how they are used.
- Then we will include solved examples. The solved examples are a key part of this chapter, and one should be sure to examine them carefully.

I have focused the discussion to make it most understandable to a senior undergraduate. Please refer to your physical chemistry or reaction engineering text for a less advanced treatment.

3.2 BACKGROUND

Studies of the kinetics of chemical reactions began in earnest in the middle part of the nineteenth century. At the time, winemaking was one of the key chemical industries, so many of the studies used wine, and winemaking equipment.

One of the key issues in a winery is controlling how sweet the wine tastes. Generally, sweeter grapes will produce a sweeter wine, so it is very important to control the sugar content of the grapes used to produce the wine.

In the 1600s, it was discovered that the refractive index of sugar solutions varied linearly with the sugar concentration. Consequently, one can measure the refractive index of a grape squeezing and use that measurement to estimate the sugar content of the grapes. If you have been to a winery, you will know that when the grapes are delivered, people use a refractometer to measure the sugar concentration of the grape juice.

There is a problem with that measurement, however. Sometimes when you let grape juice sit, the sucrose in the grape juice is converted to fructose and lactose via reaction (2.7). The refractive index of a fructose/lactose solution is quite different from the refractive index of a sucrose solution. Consequently, if some of the sucrose in grape juice has been converted to fructose and lactose, one will not get an accurate sugar measurement with a refractometer.

In 1850, Wilhelmy did some measurements to try to understand how quickly the sucrose was converted to lactose and fructose. He started with a sucrose solution and used a polarimeter to measure the sucrose concentration as a function of time. Wilhelmy also added various acids because it had previously been found that acids would speed the spoilage of wine or grape juice.

Figure 3.1 shows some of Wilhelmy's results. Notice that the sucrose concentration decays with time. Later in this chapter, we will show that for a first-order reaction, the sucrose concentration, C_S, should follow

$$C_S = C_S^0 \exp(-k_S \tau) \tag{3.1}$$

where C_S^0 is the initial sucrose concentration, τ is time, and k_S is the rate constant. Ostwald (1885) fit Wilhelmy's data to equation (3.1) and then inferred a rate equation for the conversion of sucrose to lactose and fructose.

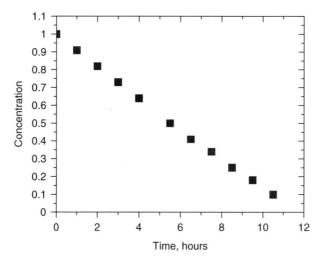

Figure 3.1 Wilhelmy's (1850) measurements of the changes in sucrose concentration in grape juice after acid is added.

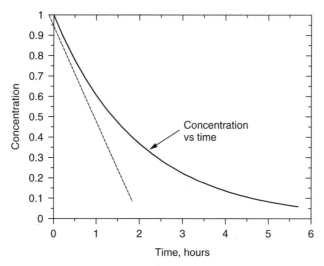

Figure 3.2 Concentration versus time for a simple reaction.

Most simple kinetic measurements are done using a technique that is very similar to those pioneered by Wilhelmy. One finds a way to initiate the reaction and then measures the concentration of some key species as a function of time. Typical data are shown in Figure 3.2. One then uses the data to infer the order of the reaction and the rate constant. Over the years, the details have changed. However, one still induces a change in the system and then uses spectroscopy to measure the concentration as a function of time.

In the next several sections we briefly describe the techniques, and mention how they are used.

3.3   BRIEF OVERVIEW OF THE EXPERIMENTAL TECHNIQUES

Most kinetics books talk about the various methods one uses to measure rates of reaction. Generally, the methods discussed in most books are similar. One finds a way to initiate a reaction and then uses some sort of spectroscopic technique to measure the concentration of a key species as a function of time. One then fits the data to an equation to infer a rate law.

There are lots of schemes to initiate the reaction. If the reaction is slow, you can simply mix the reactants together. If the reaction is fast, you might start the reaction by mixing the reactants at low temperature and then quickly heating the mixture. People also use spark plugs and lasers to initiate reactions. For a really fast reaction, one uses a molecular beam system to measure rates.

Table 3.1 summarizes the methods discussed in most textbooks. The simplest method, called the *conventional method*, is to mix the reactants in a beaker, and then measure concentrations as a function of time. Generally, this is a very easy technique. However, it usually, takes several seconds to mix the reactants, and possibly several seconds to make a concentration measurement. As a result, the conventional methods are useful only for reactions that are slow enough that the reactants can be mixed and measurements made before there is significant conversion of reactants to products. In Table 3.1, we say that the reaction must take 10 seconds or more. That is because it usually takes several seconds to pour the reactants into a beaker and thoroughly mix them.

The next method is the *stopped-flow method*. In the stopped flow, one runs the reaction in a flow cell in a spectrophotometer. You start the experiment by flowing the reactants into the flow cell, and increase the velocity of the reactants until the conversion of the reactants is negligible. You wait for everything to be well mixed, and then you stop the flow and measure the concentrations of the reactants as a function of time. The stopped-flow method avoids the difficulties of the conventional method, since the reactants are already mixed at the start of the reaction. Still, it takes time to stop the flow. As a result, the stopped-flow method is limited to reactions that take 10^{-3} seconds or more.

The third method is the *temperature-jump method*. In this method, one mixes the reactants at a low enough temperature that the reaction rate is negligible and then zaps the mixture with a CO_2 laser to suddenly heat the reactants to the desired reaction temperature. Again, mixing problems are avoided, but there is the difficulty that one can heat things only so fast. In practice, the temperature-jump method is usually limited to reactions that take 10^{-6} seconds or more, although people are developing techniques to heat liquids as fast as 10^{-9} seconds.

The *shock tube* is a way to examine fast reactions in gases. One constructs a long tube, with a thin membrane in the middle. One then puts one reactant on one side of the membrane, and a second reactant on the other side of the membrane. One adds an atmosphere or two of an inert gas, to ensure that there is a large pressure difference between the two sides of the membrane. One then ruptures the membrane, and allows the gases to suddenly mix. Generally, a shock wave forms at the interface between the high-pressure and low-pressure gas. The shock wave heats the reactants and allows them to mix. One can generally observe reactions that take between 10^{-3} and 10^{-5} seconds in this way. One cannot study faster reactions with the technique because of the finite mixing times. One cannot study slower reactions because the gas cools off after the shock wave has passed.

Flash photolysis is the fastest method. One zaps a mixture with a laser, and uses a variety of spectroscopic techniques to observe the resultant reactants. Generally, one

Table 3.1   Some techniques used to measure rates of reaction

Method	Description	Timescale, seconds
	Batch Methods	
Conventional	1) Mix reactants together in a batch reactor 2) Measure concentration versus time	≥ 10
Stopped flow	1) Set of continuous-flow systems where reactants are fed into the reactor, and flow out again so quickly that there is negligible reaction 2) Stop the flow so that the reactants can react 3) Measure conversion versus time	$\geq 10^{-1}$
Temperature jump	1) Mix reactants at such a low temperature that the reaction rate is negligible 2) Use CO_2 laser to suddenly heat reaction 3) Measure concentration vs. time	$\geq 10^{-6}$
Shock tube	1) Put 10^{-1} atm of one reactant and 10 atm at helium on one side of a diaphragm 2) Put 10^{-3} atm of the other reactant on the other side of the diaphragm 3) Suddenly break the diaphragm so that the gas flows from the high-pressure side to the low-pressure side Measure the reactant concentration vs. time	$10^{-3}-10^{-5}$
Flash photolysis	1) Put the reactants into a vessel under conditions where reaction is negligible 2) Pulse a laser or flash lamp to start reaction 3) Measure the reactant concentration vs. time	$10^{-9}-10^{-1}$
NMR[a]	1) Initiate a change with a magnetic pulse 2) Measure the decay of spins by NMR	$10^{-2}-10^{-9}$
	Flow Methods	
Conventional flow system	1) Continuously feed reactants into a reactor (CSTR[b] or plug flow) 1) Measure the steady-state reaction rate	$\geq 10^{-3}$
Molecular beam	1) Direct beams of reactants toward each together in a vacuum system 2) Measure the steady-state reaction rate	$10^{-13}-10^{-9}$

[a] Nuclear magnetic resonance.
[b] Continuously stirred tank reactor.

can use the laser to examine the reactions of radicals or other reactive intermediates by creating the intermediates with a laser and watching how they react. Generally, laser methods are limited to reactions that take 10^{-9} seconds or more because laser pulses usually last perhaps 10^{-10} seconds. However, people are working on femtosecond lasers to get around those difficulties.

The methods described in the preceding paragraphs are called *batch experiments*. In a batch experiment, one mixes a batch of reactants in a reactor and then allows the reactants to react. One can also run reactions in a flow reactor where one continuously feeds reactants into the reactor. There are two kinds of flow reactors, continuously stirred tank reactors (CSTRs) and plug-flow reactors. In either case, one flows reactants into the reactor and measures the concentration of the products as a function of the residence time in the reactor τ, where τ is given by

$$\tau = \frac{(\text{reactor volume, liters})}{(\text{volumetric flow rate of reactants, liters/hour})} \quad (3.2)$$

The CSTR is the simplest reactor. The CSTR consists of a mixing tank with flow in and out, as indicated in Figure 3.3. Later in this chapter, we will show that when the general reaction A $\Rightarrow$ B occurs in a CSTR, the reaction rate, r_{A1}, in the CSTR is related to the residence time in the CSTR by

$$r_A = \frac{C_A^{in} - C_A^{out}}{\tau} \quad (3.3)$$

where C_A^{in} is the concentration of A in the inlet to the reactor, C_A^{out} is the concentration in the outlet of the pipe, and τ is the residence time. Plug-flow reactors are basically pipes with baffels to prevent backward mixing. One feeds reactants into the pipe and takes products at the end. One can show that the residence time for a plug-flow reactor is related to the reaction rate by

$$\tau = \int_{C_A^0}^{C_A^F} \frac{dC_A}{-r_A} \quad (3.4)$$

Plug-flow reactors (Figure 3.4) can be used to measure kinetics, although they are used less often than CSTRs.

One last experimental technique, that is especially important, is called *molecular beam*. Molecular beam measurements are fundamentally different than all of the other measurements in Table 3.1. Figure 3.5 shows a schematic of the molecular beam system

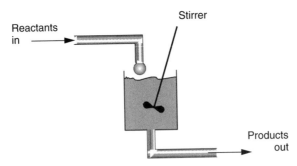

Figure 3.3 A continuously stirred tank reactor.

Figure 3.4 A plug-flow reactor.

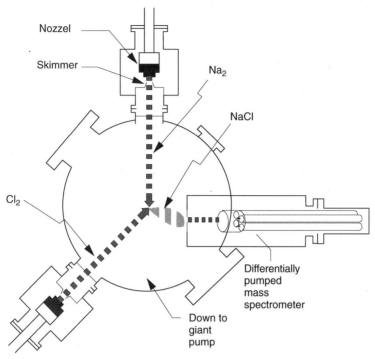

Figure 3.5   A molecular beam system used to measure the rate of the reaction Na + Cl₂ ⇒ NaCl + Cl. After Goudan et al (1968), and Lee et al (1968).

being used to examine the rate of the reaction:

$$Na + Cl_2 \Longrightarrow NaCl + Cl \qquad (3.5)$$

Sodium atoms are generated in the sodium source and flow into a vacuum system. The sodium atoms collide with the stream of chlorine, and the NaCl and Cl products are detected with a mass spectrometer. A laser can also be used to measure the properties of the reactant and product molecules. The system is designed so that the unreacted sodium atoms and chlorine molecules are pumped away.

The advantage of a molecular beam system is that the sodium atoms are in close proximity to the Cl_2 molecules for perhaps only 10^{-11} seconds. Consequently, very fast processes can be studied via molecular beam techniques. Further, one can catch the product NaCl molecules before the products collide with the walls or other molecules. One can directly measure the properties of the molecules that have reacted, and simultaneously determine all of the properties of the product molecules that form. Therefore, one can get more information about what happens during the reaction than with conventional techniques.

In practice, the beam systems have allowed people to examine much faster processes than could be examined with conventional techniques. Also, many of the details of a reaction can be probed only by using molecular beam techniques. Consequently, molecular beam techniques have been popular in the literature.

Unfortunately, there is insufficient room for us to discuss molecular beam techniques here. One should refer to Herghbauh (1966) Levine and Bernstein (1987) or Scoles (1988) for a discussion of molecular beam techniques and how they contribute to kinetics. We also do not have room to discuss all of the experimental techniques in Table 3.1 in detail. Several older books, including the books by Moore and Pearson (1981) and Laidler (1987), discuss the experimental techniques.

Still, there is one other important point to recognize in the table — one needs to use a different experimental method for a fast reaction than for a slow reaction. If a reaction takes 10^{-2} seconds, one needs to initiate the reaction within 10^{-3} seconds. A fast measurement system is needed. A much slower measurement technique can be used if the reaction takes several minutes. Table 3.1 indicates which methods can be used to measure rates for fast reactions and which ones are limited to slow reactions. Consequently, the reader should memorize the information in Table 3.1 before proceeding.

3.4 DIRECT AND INDIRECT METHODS

The listing in Table 3.1 separates the experimental methods according to the timescale of the measurement. However, there is another key distinction: whether the technique produces a **direct** or **indirect** measurement of the rate equation. I am assuming that the reader probably is not familiar with the terms *direct* and *indirect methods*. Therefore, I thought that I would define them before proceeding.

Recall that, by definition, the rate equation is an expression for the rate as a function of the concentration of the reactants. One can measure the rate as a function of concentration directly or indirectly. The experimental methods in Section 3.2 were mainly indirect measurements. If one loads a reactant into a reactor and measures the reactant concentration as a function of time, one is not measuring the rate as a function of concentration. Instead, one is measuring the concentration as a function of time, and fitting that data to infer a rate law. This is an indirect measurement of the rate law, so I will refer to it as an *indirect method*.

On the other hand, it is possible to directly measure the rate as a function of concentration. In that case, one can fit the rate equation directly. I will call such a measurement a *direct rate measurement*. More precisely, I will define a direct method as any experimental method where one actually measures the rate of reaction as a function of the concentrations in the reactor. I will define an indirect method as a method where one does not actually measure the rate. Instead, one measures some other property, for example, a concentration as a function of time, and fits that data to infer a rate law.

In the literature, it has become common to refer to a direct method as a **differential method,** while an indirect method is referred to as an **integral method**.

It is useful to consider an example, the decomposition of arsine (AsH_3), to illustrate the difference between direct and indirect measurements. The decomposition of arsine on silicon or gallium arsenic is quite important to semiconductor device manufacture. During the reaction, the arsine decomposes to yield an arsenic film and liberate hydrogen:

$$2AsH_3 \Longrightarrow 2As + 3H_2 \tag{3.6}$$

Reaction (3.6) is used to deposit arsenic as a dopant for silicon in the manufacture of integrated circuits. Reaction (3.6) is also used as an arsenic source in gallium arsenide production for light-emitting diodes (LEDs), flat-panel displays, and compact-disk (CD) players.

Figure 3.6 shows a typical reactor used for this reaction. The reactor consists of a quartz tube in a tube furnace. You load silicon wafers into the reactor, evacuate, turn on the oven, and let the wafers heat to 1000°C. You then feed arsine onto the hot wafers. The arsine decomposes, depositing arsenic onto the wafer.

Now, let's consider trying to measure the rate of reaction (3.6). One can measure the rate of reaction (3.6) using the apparatus in Figure 3.7. The apparatus consists of a reactor containing a very sensitive balance called a '*microbalance*.' During a kinetic measurement, one loads a microchip onto the microbalance, runs the reaction, and then weighs the chip as a function of time. The change in weight of the chip is equal to the weight of the arsenic that is deposited. Therefore, one can use the change in weight of the chip to determine how much arsenic is deposited.

There are two ways to run the reactor. First, one could continuously feed arsine into the reactor, ensure a constant concentration of the arsine in the reactor and then determine the weight of the chip as a function of time. In that case, one would get a steady-state reaction rate. Second, one could load a fixed amount of arsine into the reactor and determine how much arsine is deposited as a function of time. In that case, one would determine a transient reaction rate.

In the first case, arsenic would be continuously deposited onto the chip. If one measured the weight of the chip as a function of time, one could calculate R_D, the arsenic deposition

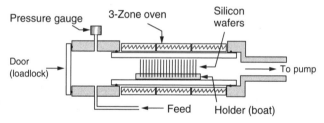

Figure 3.6   A typical arsine decomposition reactor.

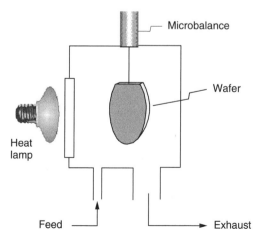

Figure 3.7   A possible apparatus to examine the decomposition of arsine (AsH_3) on silicon.

rate in mol/(hour·cm^2), from

$$R_D = \frac{1}{(A_w)} \frac{1}{MW_{As}} \frac{dWeight_w}{dt} \qquad (3.7)$$

where A_w is the area of the chip, MW_{AS} is the molecular weight of arsenic, and $Weight_w$ is the weight of the chip. In this way, one could measure the rate of arsenic deposition at any arsine concentration directly.

If one wanted to use this method to determine a rate law, one would repeat the measurements dozens of times, varying the arsine and hydrogen concentration within the reactor. Eventually, one could get a plot of the rate of reaction as a function of the arsine and hydrogen concentration within the reactor. Notice that during such an experiment, one is directly measuring a rate of reaction as a function of the concentration in the reactor. Therefore, I like to call such a measurement a **direct** measurement of the rate equation. In the literature, people also call it a **differential method** because one generally has to differentiate data to calculate a rate.

An alternate experiment is to load a fixed amount of arsine into the reactor and measure the arsine pressure as a function of time while the arsine is deposited. Figure 3.8 shows typical data. In this case, the rate varies with time because the arsine is being used up. One could still differentiate the data to get the rate as a function of time. However, an alternative is to derive a theoretical equation for the pressure versus time, and compare the theoretical equation to the data.

Later in this chapter, we will integrate the rate equation to show that, for a first-order reaction, the weight of the chip will vary as follows:

$$P_{AsH_3} = P^0_{AsH_3} \times e^{-k_1 t} \qquad (3.8)$$

where P_{AsH_3} is the measured pressure of arsine, which varies with time; $P^0_{AsH_3}$ is the initial arsine pressure; k_1 is the rate constant for arsine decomposition; and t is time. One could measure the weight as a function of time and fit the data to equation (3.8) to calculate

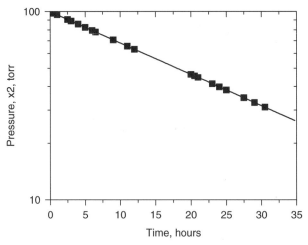

Figure 3.8   Typical batch data for reaction (3.7). [Data of Tamaru et al. (1955).]

the rate constant. In the literature, people call such an analysis an **integral method** since one derived equation (3.8) by integrating the rate equation.

I like to instead call it an **indirect method** because one is not measuring the rate as a function of concentration directly. Instead, one is determining how the rate varies with concentration indirectly by fitting the measurements to a theoretical expression.

Now one might ask, "How are the direct method and the indirect method fundamentally different?" Notice that when one uses the direct method, one is actually measuring the rate law directly. In other words, when one differentiates the weight versus time data, one gets a direct measurement of the rate for each arsine concentration. One does not have to make any assumptions about the form of the rate equation to calculate a rate; one just has to fit data. Alternatively, with the indirect method, the concentration is varying, so one has to make an assumption about the form of the rate equation to derive equation (3.8). That is less accurate. However, the advantage is that one does not need to differentiate data to get useful information.

In the literature, it has become common to refer to any method where one gets a direct measurement of the rate at each concentration a **direct method** or **differential method**, even if one does not have to actually differentiate data to calculate a rate. Alternatively, one refers to any method where one gets an indirect measurement of the rate as an **indirect method** or **integral method**, even though one would not necessarily need to integrate a rate equation to analyze the data.

3.4.1   Advantages of Direct and Indirect Methods

When one plans an experiment to determine the rate equation for a given reaction, one first needs to do is to decide whether to use a direct or indirect method.

Table 3.2 compares the advantages and disadvantages of direct and indirect methods. Generally, indirect measurements require easier experiments than do direct measurements, but the resulting rate equations are less accurate. With an indirect method, one simply loads reactants into a reactor and measures concentration versus time. Those are usually relatively easy measurements. In contrast, direct measurements generally require a flow system and a very sensitive measurement device to directly measure the rate. Consequently, the actual experiments are harder with a direct method than with an indirect method. The experiments are impossible with very fast or very slow reactions. Also, one doing a direct measurement gets only one point on the rate versus concentration curve

Table 3.2   Comparison of the advantages and disadvantages of direct and indirect methods

Direct Method	Indirect Method
Advantages	Disadvantages
Get rate equation directly	Must infer rate equation
Easy to fit data to a rate law	Hard to analyze rate data
High confidence on final rate equation	Low confidence on final rate equation
Disadvantages	Advantages
Difficult experiment	Easier experiment
Need many runs	Can do a few runs and get important information
Not suitable for very fast or very slow reactions	Suitable for all reactions including very fast or very slow ones

at a time, so it takes many measurements to determine a rate equation (i.e., the rate as a function of concentration). One can get useful data from a single run with an indirect measurement.

Still, the advantage of a direct method is that one measures the rate equation (i.e., the rate as a function of the reactant concentration) directly. One does not have to make any assumptions about the form of the rate equation to get an answer.

Generally direct measurements are much easier to fit to a rate equation than indirect measurements because, in a direct measurement, one determines the rate equation directly, while in an indirect measurement, one needs to infer the rate equation by fitting a curve to the data. The latter process can introduce some degree of error.

In my experience, when you are trying to determine kinetics for a new system, it is usually better to start with an indirect method. The indirect method gives you an approximate rate equation with a quick and easy experiment. That is often good enough. A direct method is required only when you need a precise rate equation. If you are designing a process where a 10% change in rate matters, you need to do direct rate measurements. If you can accept a 10% error and adjust the process accordingly, an indirect method will suffice. In my experience, direct measurements take 10–100 times longer than indirect measurements, so direct methods are useful only when a high degree of accuracy is needed.

3.5   EXAMPLES OF DIRECT AND INDIRECT METHODS

Indirect methods are the most common kinetic measurements in the older literature. You measure a concentration as a function of time and then fit the data to a rate equation. Direct measurements are harder. One has to find a way to measure the rate of reaction directly.

Most direct methods involve differentiating a rate equation. However, that is not a necessity. For example, in the reactor in Figure 3.8, one could measure how much arsine flows into the reactor and how much flows out. If one knows how many moles per hour of arsine flow into the reactor and how many moles per hour flow out of the reactor, one can calculate the rate from a mass balance:

$$R_{As} = \frac{1}{A_W}(F_{in} - F_{out}) \tag{3.9}$$

where R_{As} is the rate of arsenic deposition per unit area, A_W is the area of the wafer, F_{in} is the flow rate of arsine into the reactor in mol/hour, and, F_{out} is the flow rate of arsine out of the reactor in mole/hour.

Notice that one is still directly measuring the rate. Therefore, one would call the measurement a *direct method* or a *differential method*, even though you are not differentiating anything to get a direct measurement of the rate.

Note, however, that equation (3.9) applies only if the rate is constant across the wafer and there is no reaction anywhere else in the reactor. If there were a reaction somewhere else in the reactor, one would have to do analysis to eliminate those effects. One would call the measurement *integral methods* or *indirect methods* since one needs to do an analysis to determine the rate. Generally, one would only call a measurement a *direct determination* of the rate equation when one can directly measure the rate as a function of concentration. If one has to do some analysis, it will be an indirect method.

There are many variations of this idea. Today, most direct kinetic measurements are made in a continuously stirred tank reactor (CSTR). Figure 3.4 shows a diagram of a CSTR. Basically, you continuously feed reactants into the reactor. Some of the reactant molecules react, while other reactant molecules just flow through the reactor. You then measure the reactant concentration in the inlet and outlet of the reactor. The average reaction rate $\bar{r}_A$ in the reactor is given by

$$ r_A = \frac{C_A^{in} - C_A^{out}}{\tau} \tag{3.10} $$

One then runs the reactor so that the mixture stays well mixed. In that case, the rate is constant throughout the reactor, so one can calculate the rate directly from equation (3.10).

Another important direct method is the **method of initial rates**. In the initial rate method, one runs the reaction in a batch reactor, as in an indirect measurement. However, one analyzes the rate differently. Consider the data in Figure 3.2. Figure 3.2 shows concentration–time data. Notice that according to equation (2.4), the rate at any time is the slope of the target to the line. Therefore, one can use the slope to get a rate.

In the initial rate method, one fits a (dashed) line to the initial part of the concentration versus the data. The rate is the slope of the line. One then changes the initial concentration, and generates the rate versus concentration data. The advantage of this approach is that one can operate direct data quickly, by running several small reactors at once.

3.6 EXAMPLES OF INDIRECT MEASUREMENTS

Indirect measurements are generally made in a batch system. Most of the rate measurements you did in your chemistry lab were indirect measurements. For example, you might have loaded some species into a beaker, measured the concentration versus time, and fit the data to a first-order or second-order rate law. That is an indirect measurement. The flash photolysis experiments that students sometimes do are also indirect measurements of the rate law. Direct measurements require that you actually measure the rate. The measurements are harder. They are seldom done in undergraduate labs. In Table 3.1, we briefly mentioned several techniques that one uses to measure rates of reaction. As an exercise, the reader should go back and decide whether each method is a direct or indirect one.

3.7 FITTING DATA TO EMPIRICAL RATE LAWS: SINGLE REACTANTS

At this point, we will be changing topics. We will assume that you have used either a direct or an indirect method to measure the rate data for a given reaction. We will now discuss how one fits the rate data to a rate equation.

The general scheme will be to

1. Determine the order of the reaction. (In a complicated case, one also has to determine the form of the rate equation.)
2. Fit the constants.

The hard part is to determine the order of the reaction.

3.8 ANALYSIS OF DATA FROM A DIFFERENTIAL REACTOR

The easiest data to analyze are data from a differential reactor. As noted above, in a differential reactor, one can measure the rate directly at a fixed concentration of reactants. One then varies the concentration of the reactants and measures the rate again. The data can then be fit to a rate equation using a regression technique.

If the data are simple, a log–log plot of the rate versus concentration can be used to infer the order of the reaction. In a more complex case, one needs to assume a form for the rate equation and then fit the data to constants. In Chapters 4 and 5, we will discuss how one can guess a suitable rate form for the rate equation. However, the analysis is easy once the data are obtained.

For example, copper and platinum have been suggested as components in memory chips. Steger and Masel (1994, 1998) have examined the rate that copper is etched in a mixture of oxygen and hexafluoropentanedione ($CF_3COCH_2COCF_3$). Steger and Masel found that the main reaction is

$$\tfrac{1}{2}O_2 + 2CF_3COCH_2COCF_3 + Cu \Longrightarrow Cu(CF_3COCHCOCF_3)_2 + H_2O \qquad (3.11)$$

Steger and Masel loaded a copper disk into a flow reactor similar to the one shown in Figure 3.7. They then turned on the feed and measured the weight of the copper disk as a function of time to yield the data in Figure 3.9. The data were then plugged into equation (3.10) to calculate a rate. Steger and Masel then changed the O_2 concentration in the reactor and measured the rate again. After several runs, they were able to generate the data shown in Figure 3.10. The data are a little unusual in that Steger and Masel observed two different rate laws: one rate law when the copper is oxidized and a different rate law when the copper is reduced. Further, there is a change from one rate law to another as the oxygen pressure is increased in the reactor and the copper is oxidized. Still, the analysis of this data are simple. For example, Steger and Masel used a regression technique to fit the data on the unoxidized surface to the following rate form:

$$R_{Cu} = \frac{k_5 P_{O_2}^{1/2}}{1 + K_2 P_{O_2}^{1/2}} \qquad (3.12)$$

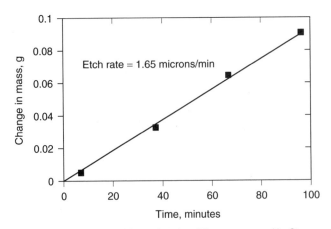

Figure 3.9 The weight of a copper disk as a function of time as measured by Steger and Masel.

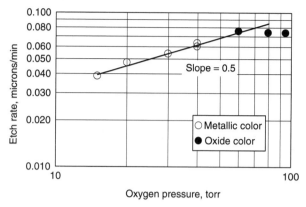

Figure 3.10 The rate of copper etching as a function of the oxygen concentration. [Data of Steger and Masel (1998).]

Notice how easy it is to do the analysis even though this is a very complex reaction. The technique can easily be generated. If one can measure a reaction in a differential reactor, one can get data for the rate as a function of the reactant concentration. One can then fit the rate data using a regression method. Detailed analysis of the data are given in Example 3.E. The result is a straightforward analysis of rate data.

3.8.1 Pitfalls in the Analysis of Rate Data from a Differential Reactor

There are a few pitfalls in analyzing the rate data from a differential reactor.

- It is not uncommon for more than one rate equation to fit the measured kinetics within the experimental uncertainties. One should not infer that the rate equation is correct just because the data fits.
- The quality of kinetic data varies with the equipment used and the method of temperature measurement and control. Data taken on one apparatus are seldom directly comparable to data taken on a different apparatus.
- It is not uncommon to observe 10–30% variations in rate taken in the same apparatus on different days. Usually, these variations can be traced to variations in the temperature, pressure, or flow rate in the reactor.
- The procedure used to fit the data can have a major effect on the values of the parameters obtained in the data analysis.
- The quality of the regression coefficient (r^2) does not tell you how well a model fits your data.

Examples 3.A to 3.C illustrate this effect. In the examples, we fit the paramecium data in Figure 2.1 to two different rate equations:

$$\text{Rate} = \frac{k_1 K_2 [\text{par}]}{1 + K_2 [\text{par}]} \qquad (3.13)$$

$$\text{Rate} = \frac{k_1 K_2 [\text{par}]}{1 + K [\text{par}]^{3/2}} \qquad (3.14)$$

and three different fitting procedures.

Table 3.3 Results of four different ways to fit the rate data

Rate Equation	Fitting Method	k_1	K_2	r^2	Total Error
(3.13)	Lineweaver–Burke	140	0.0370	0.901	9454
(3.13)	Eadie–Hofstee	246	0.0156	0.344	5647
(3.13)	Nonlinear least squares	204	0.0221	0.905	4919
(3.14)	Nonlinear least squares	—	—	0.908	4576

Table 3.3 illustrates the result of the analysis. Notice that the rate constants vary by a factor of 2 according to which fitting procedure is used. The Eadie–Hofstee fitting method gives a terrible r^2 but a reasonable error. The Lineweaver–Burke method gives a great r^2, but a large error. Equations (3.13) and (3.14) fit the data almost equally well. The data show a small preference for equation (3.14). However, the analysis in Example 3.C shows that the differences are not statistically significant.

These examples illustrate the pitfalls in analyzing rate data. A rate equation can fit data poorly and still give a reasonable value of r^2. A rate equation can fit well and give a poor value of r^2. Different models can fit the data within the statistical noise in the data.

There is one other difficulty with direct measurements — the measurements are rather tedious. One needs to make a separate measurement for each set of conditions. Generally, each point in Figure 3.10 took about a day to take and a second day to reproduce. The entire plot took over a month. Further, we needed several plots: one for each hexafluoropentamedione concentration to determine a rate law. These measurements took months to complete. Also, differential reactors tend to be hard to operate. There were many bad runs. One does have to think about whether it is necessary to take such difficult data when one is planning the experiment.

3.9 ANALYSIS OF RATE DATA FROM AN INTEGRAL REACTOR

The alternative is to use an indirect method. The advantage of an indirect method is that one can generate a significant amount of data in a short time. The disadvantage is that the data are harder to analyze than data from a differential method and the uncertainty in the data analysis is large with differential data.

In the next several sections, we will discuss how one analyzes data from what is called an **integral reactor**, which is a reactor you use to do both indirect and direct rate measurements. The simplest integral reactor is a **batch reactor**. A batch reactor is a closed vessel like a flask with a stirrer. One loads the reactants into the batch reactor and measures the concentration of the reactants as a function of time. Typical data are shown in Figure 3.2. One then fits the data to a universal curve to infer the order of the reaction and the rate constant. We will derive an equation for the performance of the batch reaction in Section 3.9.1 and use the equation in Section 3.10. The actual derivation is not important. One could skip to Section 3.10 without loss of continuity.

3.9.1 Derivation of the Concentration versus Time for Batch Reactions

Generally, if one wants to use a batch reactor to determine kinetics, one loads the reactants into the reactor and runs the reaction while measuring the concentration versus time. One then fits the resulting data to a universal curve to infer a rate equation. In this section, we will derive expressions for the universal curves. We will assume perfect mixing and kinetics of the form:

$$r_A = -k_n[A]^n \tag{3.15}$$

We will then derive a universal curve for the concentration versus time. We include the derivation for completeness. However, one can skip to equation (3.42) without loss of continuity.

Consider a reaction $A \Rightarrow B$ in the batch reactor shown in Figure 3.11. For the purpose of derivation, we will assume that the concentration of A is the same everywhere in the reactor, and that the reactor is sealed so no A evaporates, enters the reactor, or leaves the reactor.

If there is no flow in or out of the reactor, then all of the A that is generated in the reactor must accumulate in the reactor. Therefore

$$(\text{Rate of accumulation of A}) = \begin{pmatrix} \text{Rate of} \\ \text{generation} \\ \text{of A} \end{pmatrix} \tag{3.16}$$

where

$$\begin{pmatrix} \text{Rate of} \\ \text{generation} \\ \text{of A} \end{pmatrix} = \begin{pmatrix} \text{moles of} \\ \text{A reacted} \\ \text{per unit time} \end{pmatrix} \tag{3.17}$$

Substituting the definition of r_A into the right side of equation (3.17) yields

$$\begin{pmatrix} \text{Rate of} \\ \text{generation} \\ \text{of A} \end{pmatrix} = r_A V \tag{3.18}$$

where V is the volume of liquid in the batch reactor. Similarly

$$(\text{Rate of accumulation of A}) = \frac{dN_A}{dt} \tag{3.19}$$

where N_A is the number of moles of A in the batch reactor.

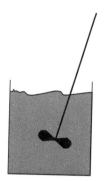

Figure 3.11   A batch reactor.

Substituting equations (3.18) and (3.19) into equation (3.16) shows

$$r_A V = \frac{dN_A}{dt} \tag{3.20}$$

Integrating equation (3.20) yields

$$\int_{N_A^0}^{N_A^F} \frac{dN_A}{V \, r_A} = \int_0^\tau dt = \tau \tag{3.21}$$

where N_A^0 is the initial concentration of A, N_A^F is the final concentration of A, and τ is the reaction time. Inverting the limits in equation (3.21) yields

$$\int_{N_A^F}^{N_A^0} \frac{dN_A}{V(-r_A)} = \tau \tag{3.22}$$

It is often useful to work in terms of quantity called the *conversion*, X_A, where X_A is defined by

$$X_A = \frac{N_A^0 - N_A}{N_A^0} \tag{3.23}$$

Physically, X_A is the fraction of the reactant A that has been converted into products. One should memorize this definition before proceeding.

Differentiating equation (3.23) and rearranging yields

$$dN_A = N_A^0(-dX_A) \tag{3.24}$$

Changing the integral in equation (3.22) from moles to conversion yields

$$\int_0^{X_A} \frac{N_A^0(-dX_A)}{V(r_A)} = \tau \tag{3.25}$$

rearranging equation (3.25) yields

$$N_A^0 \int_0^{X_A} \frac{dX_A}{V(-r_A)} = \tau \tag{3.26}$$

Note that r_A is negative in equation (3.26), that is, $(-r_A)$ is positive. Consequently τ works out to be positive in equation (3.26). Equations (3.22) and (3.26) are the key equations for batch reactors.

Note that when a reaction proceeds, $(-r_A)$ decreases. $1/V(-r_A)$ increases as indicated in Figure 3.12. The time is proportional to the area under the curve in Figure 3.12. Initially, it takes little time to increase the conversion. However, as time proceeds, the curve in Figure 3.13 increases so it takes proportionally more time to get to higher conversion.

Another important idea is that the time to get to a certain conversion is proportional to the area under the curve in Figure 3.12. This idea is often used in reactor design.

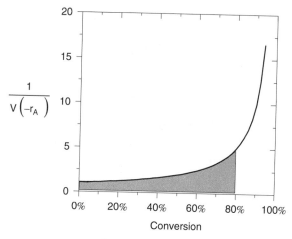

Figure 3.12 A plot of $\dfrac{1}{V(-r_A)}$ for a first-order reaction in a batch reactor.

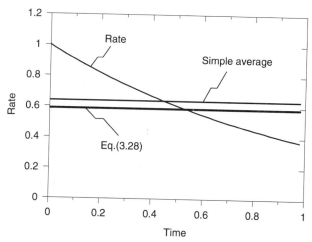

Figure 3.13 $\left(\dfrac{1}{r_A}\right)$ versus $\tilde{C}$ for a first-order reaction in a batch reactor. Notice that the rate decreases with time.

Next we will derive an approximation for constant density systems. Examples of constant density systems include

1. Liquid reactions in solution
2. Gas-phase reactions with no net change in moles, such as

$$CO + H_2O \Longrightarrow CO_2 + H_2$$

3. Reactions in constant-volume closed-batch reactors

For constant density, we can remove V from the integral in equation (3.26):

$$\frac{N_A^0}{V} \int_0^{X_A} \frac{dX_A}{-r_A} = \tau \tag{3.27}$$

$$C_A^0 \int_0^{X_A} \frac{dX_A}{-r_A} = \tau \tag{3.28}$$

where C_A^0 is the initial concentration of the reactant A in the reactor. It is also useful to transform equation (3.28) as follows. Note that if we divide equation (3.23) by V, and rearrange, we find that for a constant-density system

$$C_A = C_A^0 (1 - X_A) \tag{3.29}$$

where C_A is the concentration of A and, C_A^0 is the initial concentration of A. Substituting equation (3.29) into equation (3.28) yields

$$\int_{C_A^0}^{C_A} \frac{-dC_A}{-r_A} = \tau \tag{3.30}$$

where C_A in the upper limit of the integral is the concentration after a time, τ. Switching the limits in equation (3.30) yields

$$\int_{C_A^F}^{C_A^0} \frac{dC_A}{-r_A} = \tau \tag{3.31}$$

Equations (3.28) and (3.31) form the basis of batch reactor design for constant-density systems.

Notice that according to equation (3.31), the time is given as an integral of the rate equation, which is why this type of analysis is called an *integral method.*

Equation (3.31) is different from equation (3.26) in that it considers a plot of $1/(-r_A)$. $1/(-r_A)$ decreases with increasing concentration as indicated in Figure 3.13. Note, however, that during the reaction you start with a high concentration of the reactants, and end up with a low concentration of reactions. Consequently, $1/(-r_A)$ increases as the reaction proceeds.

The time needed to obtain to a given is proportional to the area under the curve in Figure 3.14. The idea that time is proportional to an area under the curve is quite important to reactor design.

Next, it is useful to discuss how equation (3.31) relates to the definition of a reaction rate in Chapter 2. Recall that in Chapter 2 we defined the reaction rate to be a rate in molecules per hour per reactor volume. Next, we want to demonstrate that equation (3.31) implies that the production rate of species in the reactor is proportional to the average rate in the reactor times the reactor volume, where the average reaction rate, $\bar{r}_A$ is given by

$$\frac{1}{\bar{r}_A} = \frac{\int_{C_A}^{C_A^0} \dfrac{dC_A}{r_A}}{\int_{C_A}^{C_A^0} dC_A} \tag{3.32}$$

Note that during a reaction in a batch reactor, r_A is changing with time as indicated in Figure 3.13. The function $\bar{r}_A$ is an average value of r_A over the course of the reaction. Performing the integral in the denominator of equation (3.32) and rearranging yields

$$\frac{(C_A^0 - C_A)}{\bar{r}_A} = \int_{C_A^0}^{C_A} \frac{dC_A}{r_A} \tag{3.33}$$

Substituting equation (3.31) into equation (3.32) yields

$$-\frac{C_A^0 - C_A}{\bar{r}_A} = \tau \tag{3.34}$$

Rearranging equation (3.34) yields

$$\frac{C_A - C_A^0}{\tau - \tau_0} = \bar{r}_A \tag{3.35}$$

where $\tau_0 = 0$. One can rewrite equation (3.35) as

$$\frac{\Delta C_A}{\Delta t} = \bar{r}_A \tag{3.36}$$

This compares to the corresponding equation for a differential reactor.

$$\frac{dC_A}{d\tau} = r_A \tag{3.37}$$

Notice that equations (3.36) and (3.37) are very similar. In a differential reactor, one measures an instantaneous rate. In contrast in an integral reactor, one measures an average rate where the average is defined by equation (3.32). When one averages, one washes out some of the details. Consequently, integral methods tend to be less accurate than differential methods.

3.9.2 Derivation of the Performance Equations for nth-Order Kinetics

Next, we will derive the performance equations for a first-order reaction and an nth-order reaction in a batch reactor. The derivation is included because many of our students need to see it again. However, I am assuming that most of our readers have already seen the derivation. Therefore, one could skip this section without loss of continuity.

Let's begin by considering first-order reactions. For a first-order reaction A $\Rightarrow$ B

$$r_A = -k_1 C_A \tag{3.38}$$

Substituting equation (3.38) into equation (3.31) and integrating yields

$$\frac{1}{k_1} \ln \left(\frac{C_A^0}{C_A} \right) = \tau \tag{3.39}$$

Substituting equation (3.29) into equation (3.39) yields

$$\frac{1}{k_1} \ln \left(\frac{1}{1 - X_A} \right) = \tau \tag{3.40}$$

Similarly, for an nth-order reaction

$$r_A = -k_n (C_A)^n \tag{3.41}$$

Substituting equation (3.41) into equation (3.31), integrating, and rearranging yields

$$\frac{1}{(n-1)k_n(C_A^0)^{n-1}} \left[\left(\frac{C_A^0}{C_A} \right)^{n-1} - 1 \right] = \tau \tag{3.42}$$

where C_A^0 is the initial A concentration; C_A is the concentration at time, τ; n is the order of the reaction; and, k_n is the rate constant. As an exercise, the reader may want to show that equation (3.42) goes to equation (3.39) as n goes to 1.0.

3.9.3 Reversible Reactions

The derivations in Section 3.9.1 were for irreversible reactions. However, many reactions are reversible. Consider the reaction

$$A \underset{2}{\overset{1}{\rightleftharpoons}} B \tag{3.43}$$

which obeys

$$(r_A) = -k_1[A] + k_2[B] \tag{3.44}$$

From a mass balance

$$[B] = [A]_0 - [A] \tag{3.45}$$

Substituting (3.45) into (3.43) yields

$$\frac{d[A]}{dt} = (r_A) = -k_1[A] + k_2([A]_0 - [A]) \tag{3.46}$$

Table 3.4 Integrated rate laws for a number of reactions

Reaction	Rate Law	Differential Equation	Integral Equation	See Section
A → products A + B → products	$r_A = k_A$	$\dfrac{dX_A}{d\tau} = k_A$	$k_A = \dfrac{X_A}{\tau}$	3.9.1
A → products A + B → products	$r_A = k_A[A]$	$\dfrac{dX_A}{d\tau} = k_A X_A$	$k_A = \dfrac{1}{\tau}\ln\left(\dfrac{1}{1-X_A}\right)$	3.9.1
A → products A + B → products	$r_A = k_A[A]^n$	$\dfrac{dX_A}{d\tau} = k_A(C_A^0)^{n-1}X_A^n$	$k_A = \dfrac{1}{\tau(n-1)(C_A^0)^{n-1}}\left[\left(\dfrac{1}{1-X_A}\right)^{n-1} - 1\right]$	3.9.1
A + B → products	$r_A = k_A[A][B]$	$\dfrac{dX_A}{d\tau} = k_A(1-X_A)$ $\times (C_B^0 - C_A^0 X_A)$	$k_A = \dfrac{1}{\tau(C_A^0 - C_B^0)}\ln\left(\dfrac{C_B^0(1-X_A)}{C_B^0 - X_A C_A^0}\right)$	3.9.2
A + 2B → products	$r_A = k_A[A][B]$	$\dfrac{dX_A}{d\tau} = k_A(1-X_A)$ $\times (C_B^0 - 2C_A^0 X_A)$	$k_A = \dfrac{1}{\tau(2C_A^0 - C_B^0)}\ln\left(\dfrac{C_B^0(1-X_A)}{C_B^0 - 2X_A C_A^0}\right)$	3.9.2
A → B (reversible)	$r_A = k_1[A] - k_2[B]$	$\dfrac{dX_A}{d\tau} = k_A(1-X_A) - k_2 X_A$	$(k_1 + k_2) = \dfrac{1}{\tau}\ln\left(\dfrac{1}{X_e - X_A}\right)$ where X_e is the equilibrium conversion	3.9.2

The reaction rate goes to zero when the rate reaches equilibrium. Let's define $[A]_e$ as the concentration of A when the system reaches equilibrium. By definition

$$0 = -k_1[A]_e + k_2([A]_0 - [A]_e) \tag{3.47}$$

Subtracting equation (3.47) from (3.46) yields

$$\frac{d[A]}{dt} = (k_1 + k_2)([A] - [A]_e) \tag{3.48}$$

The solution of equation (3.48) is

$$\tau = \frac{1}{(k_1 + k_2)} \ln\left(\frac{[A]_0 - [A]_e}{[A] - [A]_e}\right) \tag{3.49}$$

which looks just like a first-order approach to equilibrium. Table 3.4 shows several other examples. Derivations are given in later sections.

3.10 QUALITATIVE BEHAVIOR

Equations (3.39), (3.42), and (3.49) are the key performance equations for nth-order kinetics in a batch reactor.

It is interesting to plot these results as a function of time. Figure 3.14 shows a plot of the concentration as a function of time calculated from equations (3.39) and (3.42) for $n = \frac{1}{2}, 1, 2, 3$. All of the curves are qualitatively the same. The concentration starts out varying linearly with time and then levels off. Generally, the curve levels off more quickly with larger values of n, namely, higher-order reactions.

We plotted the results in two ways in Figure 3.14: first with $k(C_A) = 1$, and then with $k = 1$ and $C_A^0 = 1.5$. Notice that all of the plots are similar. The concentration of

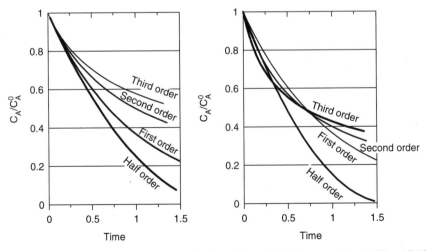

Figure 3.14 A plot of the concentration as a function of time calculated from equations (3.39) and (3.42).

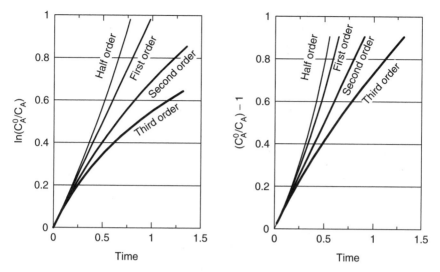

Figure 3.15 A replot of the data from Figure 3.14 as a function of $\ln(C_A/C_A^0)$ and $(C_A^0/C_A) - 1$.

the reactant, A, initially decays quickly with time. However, the rate of decay slowly decreases, giving an exponential-looking curve in all cases.

The key point that one needs to draw from Figure 3.14 is that the concentration–time plots for a batch reactor look qualitatively the same, independent of the order of the reaction. The concentration–time profiles are the same at low conversion independent of the order of the reaction. There are some quantitative differences, especially at high conversion. However, these differences are often subtle, especially when there is some error in the data.

There are two ways to look at this result. On one hand, the performance of a batch reactor does not vary that much with the order of the reaction, except at high conversion. Consequently, for many purposes, you are going to get the right answer even if you assume the wrong order for the reaction. On the other hand, it will be hard to use a batch reactor to determine the order of the reaction because the results are not going to vary that much with the reaction order. Consequently, there are some uncertainties in using batch reactor data to determine kinetics.

3.11   FITTING RATE DATA TO IDEAL BEHAVIOR

3.11.1   Essen's Method

In the literature, people say that one can get some useful information by replotting the data in Figure 3.14. For example, Figure 3.15 shows a plot of $\log (C_A/C_A^0)$ versus time and $[(C_A^0/C_A) - 1]$ versus time. Notice that a plot of $\log (C_A/C_A^0)$ is linear for a first-order reaction, while a plot of $(C_A^0/C_A) - 1$ is lightly curved. In contrast, the plot of $[(C_A^0/C_A) - 1]$ is linear for a second-order reaction and slightly curved for a first-order reaction. Therefore, in principle, one can distinguish between first- and second-order kinetics by preparing curves like those in Figure 3.15 and seeing which plot is the most linear.

Essen proposed that one could generalize these ideas to determine the order of the reaction. The idea is to

1. Measure the concentration versus time in a batch reactor.
2. Fit the data to the batch reactor equations [equation (3.39) for first-order, equation (3.42) for second-order].
3. Whichever fits best is assumed to be the correct rate equation for the reaction.

Note that according to equation (3.42), a plot of $(C_A)^{1-n}$ versus time should be linear for $n \neq 1$ while according to equation (3.39), a plot of $\ln(C_A^0/C_A)$ versus time should be linear for $n = 1$ (i.e., a first-order reaction).

Essen proposed that one can determine the order of a reaction by constructing plots of $\ln(C_A^0/C_A)$ and $[(C_A^0/C_A)^{n-1} - 1]$ versus time for various values of n, and then doing some analysis to check the results. See Harcourt and Essen (1865, 1866, 1867).

For example, in our undergraduate labs, we examine the reaction between Red Dye 4 and hydrogen peroxide to yield a yellow dye. Table 3.5 shows the concentration of Red Dye 4 as a function of time as measured in a batch reactor starting with equal dye and peroxide concentration. According to Essen, one should analyze this data by making a plot of $\ln(C_A^0/C_A)$ (for first-order reactions), $(C_A^0/C_A) - 1$ (for second-order reactions), and $(C_A^0/C_A)^2 - 1$ (for third-order reactions), and see which plot is the most linear.

The easiest way to do this is to put the data into a spreadsheet on a computer. One then uses the spreadsheet to plot $\ln(C_A^0/C_A)$, $(C_A^0/C_A) - 1$, $(C_A^0/C_A)^2 - 1$ versus time to see which plot is the most linear. Sample spreadsheets are given in Example 3.D. A linear regression process can be used to fit the rate constants.

Figure 3.16 shows the plots. The data are actually second-order, but you could not tell that from the figure. Notice that all of the plots look linear. If we exclude the first two points, then a plot of $\ln(C_A^0/C_A)$ is linear with a regression coefficient of 0.981, a plot of $(C_A^0/C_A) - 1$ is linear with a regression coefficient of 0.999, while a plot of $(C_A^0/C_A)^2 - 1$ is linear with a regression coefficient of 0.984. A regression coefficient of 0.981 is as good as the measurements.

There are two ways to look at this result. First, all the regression coefficients are reasonable, which means that you cannot accurately determine the order of the reaction from the data. On the other hand, the plot of (C_A^0/C_A) is more linear than the other plots. According to Figure 3.15, this is the expected behavior for a second-order reaction. Therefore, one might say that the data in Table 3.5 are as expected for a second-order reaction (i.e., $n = 2$).

My own opinion is that such an assertion is not valid. In the supplementary material we note that a regression coefficient measures the uncertainty in the parameters obtained

Table 3.5 The concentration of dye as a function of time

C_A, mmol/liter	τ, minutes	C_A, mmol/liter	τ, minutes	C_A, mmol/liter	τ, minutes
1	0	0.63	6	0.45	12
0.91	1	0.59	7	0.43	13
0.83	2	0.56	8	0.42	14
0.77	3	0.53	9	0.40	15
0.71	4	0.50	10	0.38	16
0.67	5	0.48	11	0.37	17

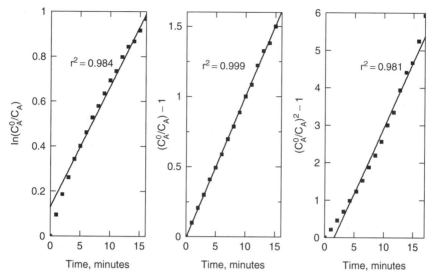

Figure 3.16   An Essen plot of the data in Table 3.5.

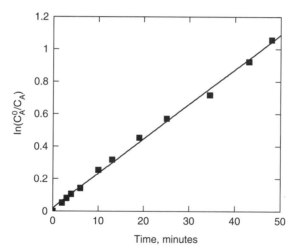

Figure 3.17   Essen Plot of the Data in Table 3.6.

from a linear fit to the data assuming that the model is correct. It does not tell you how well a model works. In Example 3.A, we show that the regression coefficient can vary by a factor of 2 according to how it is calculated. As a result, it is not useful to compare regression coefficients calculated for different models; each regression coefficient is calculated using a different model (i.e., a first-order model vs. a second-order model). As a result, comparisons of small differences in the regression coefficients are rarely meaningful. Still, that subtlety is often lost in the literature. You often see people making the mistake of saying that a rate equation fits better because the regression coefficient is closer to 1.0. That assertion is incorrect.

I do not want to imply that the results in Figure 3.16 are useless. The one advantage of Figure 3.16 is that one can calculate the rate constant for the reaction from the slope of the line in Figure 3.16. According to equation (3.42), the slope of the curve is given by

$$\text{Slope} = (n - 1)k_n(C_A^0)^{n-1} \qquad (3.50)$$

For the data in Figure 3.17, $n = 2$. Therefore, if we know the slope and C_A^0, we can calculate k_n.

Essen and Harcourt (1865) and Van't Hoff (1884) showed that one can use this method to analyze data for a wide variety of simple reactions. The advantage of this approach is that it is conceptually simple and it is easy to assess the goodness of the fit. The disadvantage of the method is that it requires many plots, although that is not a problem if one has a spreadsheet available.

3.11.2 Van't Hoff's Method

Van't Hoff (1884) used Essen's method to analyze a variety of rate data. He found that it usually worked. However, there were some examples where the method did not work well.

Table 3.6 shows data for the reaction of chloroacetic acid and water at 100 K. Figure 3.17 shows a plot of $\ln(C_A^0/C_A)$ versus τ. The plot is very linear, which suggests that the reaction is first-order. However, Van't Hoff examined the reaction in more detail and noted that if one rearranges equation (3.39), one can show that a first-order reaction should obey

$$k_1 = \frac{1}{\tau} \ln \left(\frac{C_A^0}{C_A} \right) \qquad (3.51)$$

Table 3.6 shows values of k_1 computed in this way. Notice that k decreases slightly as the reaction proceeds. Schwab (1883) examined this effect in some detail, and proved that the rate constant varied with the product HC1 concentration. Therefore, the reaction

Table 3.6 Buchanan's (1871) data for the reaction $ClCH_2COOH + H_2O \Rightarrow HOCH_2COOH + HCl$ at $100°C$

Time, hours	[ClCH$_2$COOH], grams/liter	$k_1 = \ln \left(\dfrac{[ClCH_2COOH]^0}{[ClCH_2COOH]} \right)$, hour^{-1}
0	4	—
2	3.80	0.026
3	3.69	0.026
4	3.60	0.023
6	3.47	0.025
10	3.10	0.024
13	2.91	0.023
19	2.54	0.022
25	2.26	0.021
34.5	1.95	0.020
43	1.59	0.021
48	1.39	0.021

is not really first-order even though a plot of $\ln(C_A^0/C_A)$ versus τ is linear. Rather, the reaction follows a complicated rate equation.

Van't Hoff noted that one has to be very careful in using the Essen plots to analyze data because one can easily be fooled into thinking that a reaction with a complex rate equation follows simple first- or second-order kinetics. Van't Hoff asserted that it was important to check the results from Essen's analysis. To do the check, one computes k_1 from equation (3.51) for $n = 1$, or k_n from

$$k_n = \frac{1}{(n-1)\tau(C_A^0)^{n-1}} \left[\left(\frac{C_A^0}{C_A} \right)^{n-1} - 1 \right] \tag{3.52}$$

for $n \neq 1$. One then makes a plot to check that k_n is constant. k_n should always work out to be constant within experimental error. If k_n varies in any sort of systematic way, then the reaction is not really first- or second-order. Instead, a complex rate equation will be needed to fit the rate data.

In my experience, Van't Hoff's method is much more accurate than Essen's. I find it useful to plot what I call a *Van't Hoff plot*, which is a plot of k_n versus τ for various values of n, and see if it is constant. For example, Figure 3.18 is a Van't Hoff plot of the dye data in Table 3.5. Notice that k_2 is constant, which means that the reaction really is second-order. On the other hand, k_1 is not constant for the chloracetic acid data in Table 3.6, which means that the reaction is not first- or second-order. We give further examples in the solved problems. In my experience, Van't Hoff plots are much more revealing than Essen plots. Surprisingly, though, you usually see Essen plots, not Van't Hoff plots, in most textbooks. Example 3.D illustrates this method more carefully. One should carefully examine this example before proceeding.

3.11.3 Powell's Method

There are many other ways to analyze kinetic data. A particularly powerful method that was used mainly before computers were available is called **Powell's method**.

Powell's method is derived from equations (3.39) and (3.42).

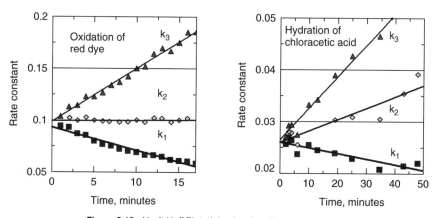

Figure 3.18 Van't Hoff Plot of the data from Tables 3.5 and 3.6.

Rearranging equation (3.42) and taking the logarithm of both sides yields

$$\log_{10}\left\{ \frac{1}{n-1}\left[\left(\frac{C_A^0}{C_A}\right)^{n-1} - 1 \right] \right\} = \log_{10}[k(C_A^0)^{n-1}] + \log_{10}(\tau) \qquad (3.53)$$

Similarly, taking the logarithms of both sides of equation (3.39) yields

$$\ln\left(\ln\left(\frac{C_A^0}{C_A}\right) \right) - \ln(k) = \ln(\tau) \qquad (3.54)$$

Figure 3.19 shows a plot of (C_A/C_A^0) versus $\log(\tau)$ calculated from equation (3.53) for various values of $k(C_A^0)^{n-1}$ and n. Notice that all of the curves calculated for a first-order reaction have an identical shape. They are just shifted to the right or left of each other. In contrast, the curves for second-order reactions have a decidedly different shape. Consequently, one should be able to distinguish first-order reactions from second-order reactions by plotting C_A/C_A^0 versus $\log(\tau)$ and looking at the general shape of the plot.

The easiest way to use Powell's method is to employ a spreadsheet on a computer. The general procedure is to

1. Make a plot of C_A/C_A^0 versus $\log 10\ (\tau)$.
2. Program the spreadsheet to calculate ideal curves for C_A^0/C_A versus $\log 10\ (\tau)$ for $n = \frac{1}{2}, 1, 2, 3$.
   (a) Use equations (3.53) and (3.54) to calculate the values of $\log(\tau)$ for each value of C_A/C_A^0.
   (b) Treat $\log_{10}(k(C_A^0)^{n-1})$ as a variable (i.e., $B$1). Set the variable initially to zero.
3. Vary $\log_{10}(k(C_A^0)^{n-1})$ to see which curve fits.

Example calculations are given in solved Example 3.D.

Powell actually proposed his method before computers were available, and so he did not use a spreadsheet. Instead, Powell used a set of universal curves for zero-, first-, second-, and third-order reactions to see how data fit. Figure 3.20 is a suitable set of

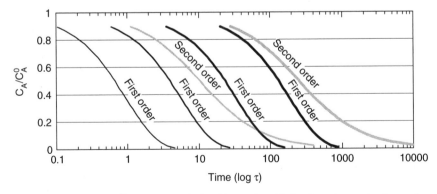

Figure 3.19   A plot of (C_A/C_A^0) versus $\log \tau$ calculated for a series of first-order and second-order reactions.

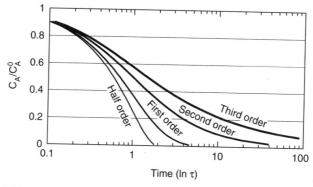

Figure 3.20 Powell's universal curves for half-, first-, second-, and third-order reactions.

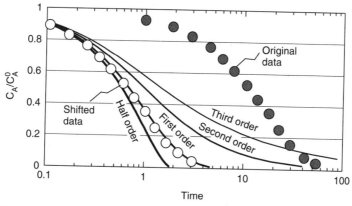

Figure 3.21 An illustration of a Powell plot for data taken for a second-order reaction.

universal curves. One can use these curves to calculate the order of a reaction. The general scheme is to

1. Make a plot of C_A/C_A^0 versus ln (τ) as shown in Figure 3.20, making sure that the scales on each of the axes are the same as in Figure 3.20.
2. Slide the data to the left and right to see which curve in Figure 3.20 fits best.

Figure 3.21 illustrates this method. We took data and plotted it on the Powell plot. We then shifted the data to the right to see which curve fits best. In this case, the second-order curve fits better than the rest, so we conclude that the reaction is second-order. We do need to go back and make a table of k_1 versus τ as described in Section 3.12 to calculate the rate constant. Example 3.A gives additional examples of analysis of data with Powell's method.

3.12 THE HALF LIFE METHOD

Another indirect method to analyze rate data is called the **half-life method**. The half-life method was quite popular before the days of computers. We include it because the ideas are useful even if the method is not used very much anymore.

The half-life, $\tau_{1/2}$, is defined as the time to get to 50% conversion. In the half-life method, one makes a log–log plot of the half-life versus time and calculates the order from the slope of the plot.

Next, we will derive an expression for the half-life as a function of C_A^0 for a first-order reaction and an nth-order reaction. Rearranging equation (3.39) shows that for a first-order reaction

$$\frac{1}{k_1} \ln\left(\frac{C_A^0}{C_A}\right) = \tau \tag{3.55}$$

at 50% conversion $(C_A^0/C_A) = 2$. Therefore, $\tau_{1/2}$, the half-life, is given by

$$\tau_{1/2} = \frac{\ln(2)}{k_1} \tag{3.56}$$

Similarly, rearranging equation (3.42) shows that for an nth-order reaction:

$$\frac{1}{(n-1)k_n(C_A^0)^{n-1}}\left[\left(\frac{1}{1-X_A}\right)^{n-1} - 1\right] = \tau \tag{3.57}$$

$X_A = 0.5$ at $\tau_{1/2}$. Therefore, for an nth-order reaction:

$$\tau_{1/2} = \frac{1}{(n-1)k_n(C_A^0)^{n-1}}(2^{n-1} - 1) \tag{3.58}$$

Figure 3.22 shows a plot of the half-life versus concentration calculated for half-order, a first-order, second-order, and third-order reactions. Notice that in all cases, one should observe a linear plot on a log–log scale where the slope of the plot is proportional to

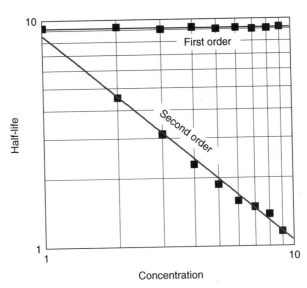

Figure 3.22   A plot of the half-life versus C_A^0 for a first-order reaction and a second-order reaction.

$(n - 1)$. The fact that we get linear plots on a log–log scale is a great advantage because it allows one to calculate the order directly even for fractional order systems. Consequently, the half-life method was quite popular before computers were readily available. It is used less now because Van't Hoff's method and Essen's method are so easy to put in a spreadsheet that they have replaced most other methods.

Still, the half-life method is very useful in getting a very rough idea about the kinetics of a reaction. One loads reactants into the reactor and measures how long it takes to convert about half the reactants into products. One then dilutes the reactants by a factor of 10 and runs the reaction again.

If the time is about constant in the two experiments, the reaction is probably first-order. If the time increases markedly with each data run, the reaction is most likely second-order. One can then get a very rough value of the rate constant from equation (3.57) or (3.58).

If one needs to know only an approximate rate constant, one does not have to actually run to exactly 50% conversion. Rather, one can run to any fixed conversion and get useful information.

Now, if one needs to get an exact rate constant, one needs to add some extra work. One will have to run the reaction versus time and calculate an exact time to 50% conversion. Note that one can calculate $\tau_{1/2}$ versus C_A^0 from data for a single run. Figures 3.23 and 3.24 illustrate the process.

Let's start the experiment at $\tau = 0$. If we start at $\tau = 0$, $C_A^0 = 1$ mol/liter, then $\tau_{1/2}$ is the time to get to $C_A = 0.5$ as shown in the left part of Figure 3.23. Now imagine starting the experiment at $\tau = 0.3$. At $\tau = 0.3$, $C_A^0 = 0.9$. Therefore, the half-life is the time it takes to go from $C_A = 0.9$ to $C_A = 0.45$. Similarly, we can imagine starting the experiment at any value of C_A in Figure 3.23 and calculate the half-life for that value of C_A. The result is that one can construct a half-life plot from a single run.

Figure 3.24 shows data generated in this way. Notice that the half-life decreases linearly with $\ln(C_A^0)$. The slope of the plot is -1.0, which implies that $1 - n = -1.0$ or $n = 2$.

In my experience, half-life plots were the easiest way to analyze batch reactor data before computers were readily available. The biggest advantage of the half-life method is

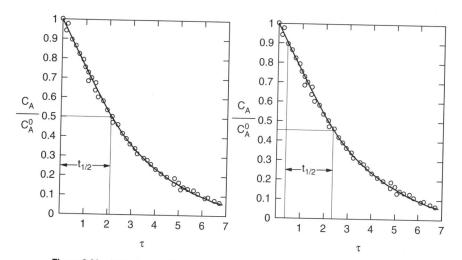

Figure 3.23   An illustration of how one can calculate the half-life from a single data set.

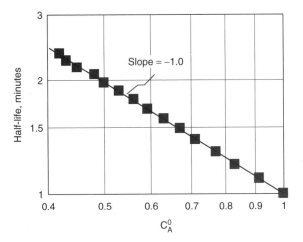

Figure 3.24 The half-life as a function of C_A for the data in Table 3.5 (with some extra points added).

that there are large differences between, for example, first- and second-order data. Further, if one takes data on a system that does not follow a simple rate equation, one knows it because the half-life plot is curved on a log–log scale. Still, the half-life method is difficult to automate. Consequently, half-life plots are now rarely seen in the literature.

Equations (3.56) and (3.58) are very useful, however. One can use these equations to decide how to run experiments, or to estimate activation barriers from very little data. See Section 2.5.1 for more information.

There are other methods in the literature, including the Gugenheim method. These methods have largely disappeared as computers made their appearance. For an older review, see Roseveare (1931).

3.13 FITTING DATA TO EMPIRICAL RATE LAWS: MULTIPLE REACTANTS

The results in Section 3.7 are useful only in the case of a reaction where a single reactant is converted into products. In a more typical reaction, two reactants, for example, C and B, react to form products. One needs a more complex analysis to consider those cases. In the work that follows, we will derive an expression for the conversion versus time of a system with multiple reactions. However, we need some further information first.

For example, consider the Diels–Alder reaction of benzoquinone (B) and cyclopentadiene (C) to yield an adduct (A):

$$(3.59)$$

(C) (B) (A)

It is useful to consider what data we need to analyze rate data for reaction (3.59). Next, we will derive an expression for the conversion of benzoquinone as a function of time assuming that we run the reaction in a constant-volume batch reactor and that the rate equation for the reaction is of the form

$$r_B = -k_B C_B C_C \tag{3.60}$$

where C_B is the benzoquinone concentration, C_C is the cyclopentadiene concentration, r_B is the rate of formation of benzoquinone and k_B is a constant.

According to equation (3.31)

$$\tau = C_B^0 \int_0^{X_B} \frac{dX_B}{(-r_B)} \tag{3.61}$$

where X_B is the conversion, and τ is the reaction time.

Combining equations (3.60) and (3.61) yields

$$\tau = C_B^0 \int_0^{X_B} \frac{dX_B}{k_B C_B C_C} \tag{3.62}$$

In order to use equation (3.62), we need an expression for C_B and C_C in terms of X_B. In the next section, we will describe how to get it.

3.13.1 The Stoichiometric Table

Consider a general reaction

$$aA + bB \longrightarrow cC + dD \tag{3.63}$$

where $-a$ $-b$, c, and d are the stoichiometric coefficients for species A, B, C, and D, respectively. Assume that N_A^0 moles of A and N_B^0 moles of B are loaded into a closed isothermal vessel. Next, we will derive a relation between the concentrations of all of the species as a function of the concentration of a single species, A. To simplify the algebra, it is useful to work in terms of a new variable, X_A, the fractional **conversion** of A, where X_A is defined as

$$X_A = \frac{N_A^0 - N_A}{N_A^0} \tag{3.64}$$

Physically X_A is the fraction of the reactant A that has been converted into products. One should memorize this definition before proceeding.

Next, it is useful to calculate the concentrations of each of the species in the reactor as a function of X_A. Fogler [1998] has a neat trick that makes it easier to keep track of this. The trick is called a **stoichiometric table**.

The stoichiometric table is a table where the following information is presented (for a batch system):

Column 1:   The particular species
Column 2:   The number of moles of each species initially present
Column 3:   The change in the number of moles due to reaction
Column 4:   The number of moles remaining in the reactor after conversion X_A

Table 3.7 is a stoichiometric table for reaction (3.63) with some of the information missing.

Next, let's calculate the changes in moles of each of the species as a function of X_A. Assume that we load N_A^0 moles of A and N_B^0 moles of B into the reactor. Then we run

Table 3.7 The start of a stoichiometric table for reaction (3.63)

Species	Initial Moles	Change in Moles	Final Moles
A	N_A^0		N_A^f
B	N_B^0		N_B^f
C	N_C^0		N_C^f
D	N_d^0		N_D^f
Inerts	N_I^0		N_I^f
Total	$N_T^0 = N_A^0 + N_A^0 + N_B^0 + N_D^0 + N_I^0$		N_T^f

the reaction for a time, t, so that the conversion of species A is X_A. From the definition of the conversion we obtain

$$\text{Moles of A converted} = N_A^0 X_A \tag{3.65}$$

From the stoichiometry of the reaction, the number of moles of B converted is given by

$$\text{Moles of B converted} = \left(\frac{b}{a}\right) N_A^0 X_A \tag{3.66}$$

Similarly:

$$\text{Moles of C formed} = \left(\frac{c}{a}\right) N_A^0 X_A \tag{3.67}$$

$$\text{Moles of D formed} = \left(\frac{d}{a}\right) N_A^0 X_A \tag{3.68}$$

It is important to keep track of the sign. During the reaction, the concentrations of A and B decrease while the concentrations of C and D rise. Consequently, the change in moles of B is:

$$\Delta N_B = -\left(\frac{b}{a}\right) X_A N_A \tag{3.69}$$

while the change in moles of C is

$$\Delta N_C = +\left(\frac{c}{a}\right) X_A N_A \tag{3.70}$$

Therefore, the stoichiometric table is as given in Table 3.8:
  One can slightly simplify Table 3.8 by noting

$$-a - b + c + d = \sum_n \beta_v = \Delta \text{mol} \tag{3.71}$$

where Δmol is the change in moles when reaction (3.63) goes one time. Substituting equation (3.67) into Table 3.8 yields Table 3.9.

3.13.2 Derivation of the Performance Equation for Reaction (3.59)

Now it is useful to go back and derive a performance equation for reaction (3.59). Let's go back to equation (3.62). Equation (3.62) was a performance equation for reaction (3.59).

Table 3.8   The stoichiometric table for reaction (3.63)

Species	Initial Moles	Change in Moles	Final Moles
A	N_A^0	$-N_A^0 X_A$	$N_A^0(1 - X_A)$
B	N_B^0	$-\left(\dfrac{b}{a}\right) N_A^0 X_A$	$N_B^0 - \left(\dfrac{b}{a}\right) N_A^0 X_A$
C	N_C^0	$+\left(\dfrac{c}{a}\right) N_A^0 X_A$	$N_C^0 + \dfrac{c}{a} N_A^0 X_A$
D	N_D^0	$+\left(\dfrac{d}{a}\right) N_A^0 X_A$	$N_D^0 + \dfrac{d}{a} N_A^0 X_A$
Inerts	N_I^0	0	N_I^0
Total	N_T^0	$\left(\dfrac{-a - b + c + d}{a}\right) N_A^0 X_A$	$N_T^0 + N_A^0\left(\dfrac{-a - b + c + d}{a}\right) X_A$

Table 3.9   An alternative stoichiometric table for reaction (3.63)

Species	Initial Moles	Change in Moles	Final Moles
A	N_A^0	$-N_A^0 X_A$	$N_A^0(1 - X_A)$
B	N_B^0	$-\left(\dfrac{b}{a}\right) N_A^0 X_A$	$N_B^0 - \left(\dfrac{b}{a}\right) N_A^0 X_A$
C	N_C^0	$+\left(\dfrac{c}{a}\right) N_A^0 X_A$	$N_C^0 + \dfrac{c}{a} N_A^0 X_A$
D	N_D^0	$+\left(\dfrac{d}{a}\right) N_A^0 X_A$	$N_D^0 + \dfrac{d}{a} N_A^0 X_A$
Inerts	N_I^0	0	N_I^0
Total	N_T^0	$+N_A^0\left(\dfrac{\Delta\text{mol}}{a}\right) X_A$	$N_T^0 + N_A^0\left(\dfrac{\Delta\text{mol}}{a}\right) X_A$

However, we needed an expression for C_C as a function of C_B so that we can integrate the equation. We can get the needed expression from a stoichiometric table. Table 3.10 shows a general stoichiometric table, while Table 3.11 shows the specific stoichiometric table for the case considered here.

Substituting results from Table 3.11 into equation (3.60) yields

$$(-r_B) = K_B\{C_B\}\{C_C\} = K_B\{C_B^0(1 - X_B)\}\{(C_C^0 - X_B C_B^0)\} \tag{3.72}$$

According to equation (3.30)

$$\tau = C_B^0 \int_0^{X_B} \frac{dX_B}{(-r_B)} \tag{3.73}$$

Table 3.10   The stoichiometric table for the reaction bB + cC ⇒ aA

Species	Initial Concentration	Change	Final Concentration
B	C_B^0	$-C_B^0 X_B$	$C_B^0(1 - X_B)$
C	C_C^0	$-(c/b)C_B^0 X_B$	$C_C^0 - (c/b)C_B^0 X_B$
A	C_A^0	$(a/b)C_B^0 X_B$	$C_A^0 + (a/b)C_B^0 X_B$

Table 3.11 The stoichiometric table for the reaction C + B ⇒ A

Species	Initial Concentration	Change	Final Concentration
B	C_B^0	$-C_B^0 X_B$	$C_B^0(1 - X_B)$
C	C_C^0	$-C_B^0 X_B$	$C_C^0 - C_B^0 X_B$
A	C_A^0	$C_B^0 X_B$	$C_A^0 + C_B^0 X_B$

Substituting equation (3.72) into equation (3.73) yields

$$\tau = C_B^0 \int_0^{X_B} \frac{dX_B}{(k_B C_B^0 (1 - X_B)(C_C^0 - X_B C_B^0))} \tag{3.74}$$

The integral equation (3.74) looks imposing. Fortunately, I was able to look it up in Gradshteyn and Ryzhik (1965).

$$\tau = \frac{\ln \left[\dfrac{C_B^0}{C_C^0} \left(\dfrac{\frac{C_C^0}{C_B^0} - X_B}{1 - X_B} \right) \right]}{k_B (C_C^0 - C_B^0)} \tag{3.75}$$

Equation (3.75) has two interesting limits: the limit where $C_C^0 \gg C_B^0$ and the limit where $C_C^0 = C_B^0$. The first limit, called the "swamping limit," corresponds to running the reaction with a huge excess of cyclopentadiene (C). If C_C^0 is in large excess, $C_C^0/C_B^0 \gg 1 \geq X_B$. Therefore, the X_B term in the numerator of the log term in equation (3.75) will be negligible. Similarly, C_B^0 in the denominator of equation (3.75) will be negligible. Consequently, when there is a large excess of cyclopentadiene, equation (3.75) reduces to

$$\tau = \frac{1}{k_B C_C^0} \ln \left(\frac{1}{1 - X_B} \right) \tag{3.76}$$

A comparison of equations (3.76) and (3.39) shows that equation (3.76) is a first-order rate equation with

$$k_1 = k_B C_C^0 \tag{3.77}$$

Physically, what is happening is that when $C_C^0 \gg C_B^0$, the concentration of cyclopentadiene does not change significantly during the reaction. The C_C is constant in the rate equation. Therefore, the reaction appears as if it were first-order.

The other key limit is when $C_B^0 = C_C^0$. Note that during reaction (3.59), benzoquinone (B) and cyclopentadiene (C) are used up at the same rate. Consequently, if initially $C_B^0 = C_C^0$, then

$$C_B = C_C \tag{3.78}$$

everywhere during the reaction. Consequently, the rate equation for reaction (3.59) will reduce to the equation for a second-order reaction.

One can derive the same result by letting C_B^0 approach C_C^0 in equation (3.84), and using l'Hôspital's rule to do the limit. The result is

$$\tau = \frac{1}{k_B C_B^0} \left(\frac{C_B^0}{C_B} - 1 \right) \tag{3.79}$$

Equation (3.79) is equivalent to equation (3.42), with n = 2. Equations (3.75) and (3.79) are used to plan experiments. First, one swamps the reactor with one species, say, C, and measures the concentration of B as a function of time, and uses Essen's method or Powell's method to see if the reaction follows equation (3.76). One then runs the reaction with equal B and C concentrations, and uses Powell's method or Essen's method to see if equation (3.79) works. If both equations fit the data, then one can be assured that the rate data follows equation (3.60). If either equation (3.76) or (3.79) fails, one needs a more complex procedure to fit the data. Such procedures are beyond the scope of the discussion here.

3.14 SEQUENTIAL REACTIONS

There is one other example that we will need to consider later in this book, which is the limit where a reactant A is first converted into an intermediate, I, and then into a product, P:

$$A \xrightarrow{\ 1\ } I \xrightarrow{\ 2\ } P \tag{3.80}$$

Next, we will derive equations for the concentrations of A, B, and C versus τ assuming that reactions 1 and 2 are both first-order.

If reaction 1 is first-order, then

$$\frac{dC_A}{d\tau} = r_A = -k_1 C_A \tag{3.81}$$

where C_A is the concentration of A. Solving equation (3.81) yields

$$C_A = C_A^0 e^{-k_1 \tau} \tag{3.82}$$

If reactions 1 and 2 are first-order, then

$$\frac{dC_I}{dt} = k_1 C_A - k_2 C_I \tag{3.83}$$

where C_I is the concentration of I.

Combining equations (3.82) and (3.83) and integrating yields

$$C_I = \frac{k_1 C_A^0}{k_2 - k_1} (e^{-k_1 \tau} - e^{-k_2 \tau}) + C_I^0 e^{-k_2 \tau} \tag{3.84}$$

where C_I^0 is the initial concentration of B. We could integrate and calculate the concentration of P. However, it is much easier to calculate the concentration of P from a mass balance.

$$C_P = C_A^0 + C_I^0 + C_P^0 - C_A C_I \qquad (3.85)$$

Equations (3.82)–(3.84) are the key equations for sequential reactions.

Figure 3.25 shows a plot of equation C_A, C_I, and C_0 versus $k_1\tau$ for various values of k_2/k_1 calculated from equations (3.82), (3.84), and (3.85). We have plotted the data as a function of $k_1\tau$, to eliminate the k_1 dependence.

The concentration of the reactant A decreases exponentially with time as was shown in Figure 3.14. However, in Figure 3.26, we plot the data as a function of $k_1\tau$. All of the C_A data fall on a universal curve, independent of k_1 and k_2.

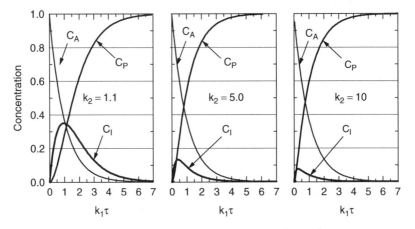

Figure 3.25   A plot of equations (3.82)–(3.85) with $C_A^0 = 1$, $C_I^0 = 0$.

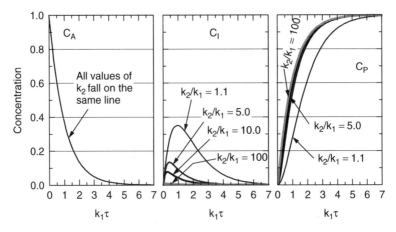

Figure 3.26   A plot of C_I and C_A versus k_1 t for $k_2/k_1 = 1.1$, 5.0, 10, 100.

The concentration of the intermediate shows quite different behavior. The intermediate concentration increases, reaches a maximum, and then declines again. The height of the maximum decreases as k_2/k_1 increases. Physically, what is happening is that in the initial part of the reaction, the reactant A is being converted into the intermediate, so initially the concentration of the intermediate rises. However, as the concentration of the intermediate builds up, the rate of conversion of the intermediate into the product P begins to become important. Eventually, the rate of destruction of the intermediate through reaction 2 gets to be larger than the rate of production of the intermediate through reaction 1. At that stage, the concentration of the intermediate falls again.

For future reference, it is useful to plot two other quantities:

$$C_P^s = C_A^0 - C_A \tag{3.86}$$

and

$$C_I^s = \frac{k_1}{k_2} C_A \tag{3.87}$$

Notice that equation (3.85) reduces to equation (3.86) when $C_I = C_I^0 = 0$. Therefore, C_P^s in equation (3.86) will be the product concentration if the intermediate concentration is negligible. C_I^s is more complicated. C_I^s is the concentration of the intermediate that one would calculate if the derivative on the left side of equation (3.82) were zero.

Figure 3.27 shows a plot of the C_P^s and C_I^s versus $k_1\tau$. We have multiplied C_P^s and C_I by k_2/k_1 to make the figure easier to see. Notice that when $k_2/k_1 > 10$, C_I^s is barely distinguishable from C_I, except at the very start of the reaction. C_P^s is barely distinguishable from C_P. That will be very important for the discussion in the next

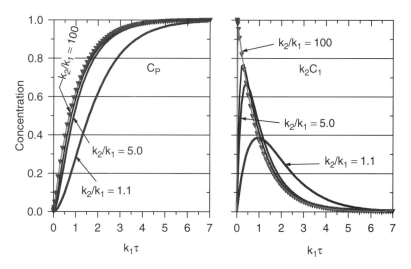

Figure 3.27 Left — a plot of C_P (lines) and C_P^s (triangles) versus $k_1\tau$. Right — a plot of k_2C_I (lines) and $k_2C_I^s$ (points) versus $k_1\tau$.

Table 3.12   Summary of key concepts

Two methods to measure rates: direct and indirect	Methods to analyze direct data
Direct: high accuracy; many runs needed	Least squares
Indirect: lower accuracy; fewer experiments	Nonlinear least-squares
Methods to analyze indirect data	Nonlinear least-squares easier and more accurate
	Key equations for indirect data

Two methods to measure rates: direct and
   indirect
  Direct: high accuracy; many runs needed
  Indirect: lower accuracy; fewer
    experiments
Methods to analyze indirect data
  Essen
    Construct plots of
     $\ln(C_A^0/C_A)$, $(C_A^0/C_A)^n - 1$
    See if linear
  Van't Hoff
    Calculate k_1, k_2, k_3
    See if constant
  Van't Hoff—easier and more accurate

Methods to analyze direct data
  Least squares
  Nonlinear least-squares
  Nonlinear least-squares easier and more accurate
  Key equations for indirect data

$$N_A^0 \int_0^{X_A} \frac{dX_A}{V(-r_A)} = \tau \tag{3.26}$$

$$C_A^0 \int_0^{X_A} \frac{dX_A}{-r_A} = \tau \tag{3.28}$$

$$\int_{C_A^F}^{C_A^0} \frac{dC_A}{-r_A} = \tau \tag{3.31}$$

$$\frac{1}{k_1} \ln \left(\frac{C_A^0}{C_A} \right) = \tau \tag{3.39}$$

$$\frac{1}{(n-1)k_n(C_A^0)^{n-1}} \left[\left(\frac{C_A^0}{C_A} \right)^{n-1} - 1 \right] = \tau \tag{3.42}$$

chapter, because it means that one can calculate an accurate value of the concentration of the intermediate by assuming the derivative in equation (3.82) to be zero whenever $k_2/k_1 > 10$.

3.15 SUMMARY

Table 3.12 summarizes the key results from this chapter. In this chapter, we reviewed some basic kinetic analysis. We discussed the order of a reaction and the activation energy, and showed how those quantities could be measured. We then briefly discussed planning experiments and noted that a key decision in planning an experiment is to decide between direct and indirect methods. Direct methods require harder experiments, but the data are easier to analyze. Indirect methods require easy experiments, but the analysis of the rate data are less certain. We also discussed how one analyzes data from batch reactors. We expect that most of these concepts are already familiar to our readers. Still, we recommend that the reader review the solved examples at the end of this chapter before going on to Chapter 4.

3.16 SUPPLEMENTAL MATERIAL: MEASURING RATE DATA FROM VOLUME AND PRESSURE CHANGES

There is one other topic that I wanted to include in the text, but it did not really fit, which is to discuss how one uses pressure and volume changes to infer rate data. As noted previously, kinetic measurements first became popular during the latter half of the nineteenth century. At the time, it was hard to make kinetic measurements. None of the spectroscopic techniques described in Chapter 1 had been invented. Accurate microbalances were not generally available. Analytic techniques were primitive. As a result, people needed to find a series of tricks to measure the concentration changes. A common experiment was to run a gas-phase reaction in a closed vessel and measure the change in the pressure of the gas as a function of time.

For example, in 1883, Van't Hoff examined the decomposition of arsine, specifically, reaction (3.5), on a glass vessel. Today, one would examine that reaction with a microbalance as described in Section 3.5. In 1883, however, no one had a balance that was sensitive enough to measure how much arsenic was deposited. Van't Hoff had to find another method to make the measurements. He loaded the arsine into a closed flask and measured the pressure as a function of time. Note that reaction (3.5) converts 2 mol of arsine gas to 3 mol of hydrogen gas, plus 1 mol of solid arsenic. If 2 mol are converted into 3 mol in a fixed-size vessel, the pressure in the vessel will go up.

Van't Hoff noted that if one measures the pressure as a function of time, one can back-calculate the amount of arsine that was converted and therefore calculate a rate. It is unusual to see this type of measurement today. However, there are many examples of this type in the literature. Consequently, it is useful to derive a series of equations to see how one can use the pressure in a system to calculate the concentrations of all of the species and the amount of gas reactant that is converted.

The derivation will start with the stoichiometric table we derived previously for the reaction

$$aA + bB \longrightarrow cC + dD \tag{3.88}$$

where $-a$ $-b$, c, and d are the stoichiometric coefficients for species A, B, C, and D, respectively. Assume that $N_A^0 N_B^0$ moles of B are loaded into a closed isothermal vessel. Next, we will derive a relation between the number of moles of A in the vessel and the pressure in the vessel.

Recall from the discussion in Section 3.13.1 that the stoichiometric table for this reaction is given by a table such as Table 3.13 (see also Table 3.7).

Next, we will calculate the pressure in the system as a function of the initial pressure. According to the ideal-gas law:

$$P_T V = N_t RT \tag{3.89}$$

Table 3.13   A stoichiometric table for reaction (3.87)

Species	Initial Moles	Change in Moles	Final Moles
A	N_A^0	$-N_A^0 X_A$	$N_A^0(1 - X_A)$
B	N_b^0	$-\left(\dfrac{b}{a}\right) N_A^0 X_A$	$N_B^0 - \left(\dfrac{b}{a}\right) N_A^0 X_A$
C	N_c^0	$+\left(\dfrac{c}{a}\right) N_A^0 X_A$	$N_C^0 + \dfrac{c}{a} N_A^0 X_A$
D	N_d^0	$+\left(\dfrac{d}{a}\right) N_A^0 X_A$	$N_D^0 + \dfrac{d}{a} N_A^0 X_A$
Inerts	N_I^0	0	N_I^0
Total	N_T^0	$+N_A^0 \left(\dfrac{\Delta mol}{a}\right) X_A$	$N_T^0 + N_A^0 \left(\dfrac{\Delta mol}{a}\right) X_A$

where P_T is the total pressure of the reactor, V is the volume in the reactor, N_T is the number of moles in the reactor, R is the gas law constant, and T is the temperature. At the start of the reaction

$$P_T^0 V = N_T^0 RT \tag{3.90}$$

where P_T^0 is the initial pressure of the reactor and N_T^0 is the total number of moles in the reactor at the start of the reactions. Dividing equation (3.89) by equation (3.90) and rearranging yields

$$P_T = P_T^0 \left(\frac{N_T}{N_T^0} \right) \tag{3.91}$$

Substituting N_T and N_T^0 from the stoichiometric table into equation (3.91) and rearranging yields

$$\frac{P_T}{P_T^0} = 1 + \left(\frac{N_A^0}{N_T^0} \right) \left(\frac{\Delta mol}{a} \right) X_A \tag{3.92}$$

Therefore, if one knows the pressure in a vessel as a function of time, one can calculate the conversion as a function of time.

Figure 3.28 shows a plot of the pressure calculated from equation (3.92) as a function or conversion for various values of the Δmol. Notice that the pressure always varies linearly with conversion. The pressure goes up when Δmol is positive, while the pressure goes down when Δmol is negative. Consequently, whenever Δmol is nonzero, one can use the pressure to estimate the conversion.

Examples 3.B and 3.C illustrate the use of the stoichiometric table to calculate the conversion as a function of time.

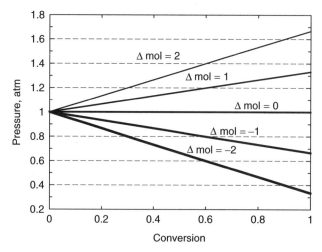

Figure 3.28 A plot of the pressure versus conversion calculated from equation (3.92) with $N_A^0 = 1$ mol, $N_T^0 = 3$ mol, and a = 1.0.

3.17 SOLVED EXAMPLES

Example 3.A   Fitting Data to Monod's Law   Table 3.A.1 shows some data for the growth rate of paramecium as a function of the paramecium concentration. Fit the data to Monod's law (Monod [1942]).

$$r_p = \frac{k_1 K_2 [par]}{1 + K_2 [par]} \tag{3.A.1}$$

where [par] is the paramecium concentration and k_1 and K_2 are constants.

Solution   There are two methods that people use to solve problems like this:

- Rearranging the equations to get a linear fit and using least-squares
- Doing nonlinear least-squares

I prefer the latter, but I wanted to give a picture of the former.
    There are two versions of the linear plots:

- Lineweaver–Burke plots
- Eadie–Hofstee plots

In the Lineweaver–Burke method, one plots 1/rate against 1/concentration. Rearranging equation (3.A.1) shows

$$\frac{1}{r_p} = \frac{1}{k_1 K_2 [par]} + \frac{1}{k_1} \tag{3.A.2}$$

Therefore, a plot of $1/r_P$ versus $1/[par]$ should be a straight line. The intercept should be $\dfrac{1}{k_1}$. The slope should be $\dfrac{1}{k_1 K_2}$. Once k_1 is determined from the intercept, K_2 can be

Table 3.A.1   The rate of paramecium reproduction as a function of the paramecium concentration

Paramecium concentration, N/cm^3	Rate, $N/cm^3 \cdot hour$	Paramecium concentration, N/cm^3	Rate, $N/(cm^3 \cdot hour)$	Paramecium concentration, N/cm^3	Rate, $N/(cm^3 \cdot hour)$
2	10.4	16	36	46	96
3.6	12.8	16.6	46.4	46.2	124.8
4	23.2	19	59.2	47.4	117.6
5.2	17.6	20	62.4	55	112
7.8	46.4	23.8	62.4	57	127.2
8	23.2	26	57.6	61	116
8	46.4	30.4	108.8	61.6	111.2
11	32	31	80	71	124
14.4	34.4	31.2	61.6	74	116
15.6	44.8	31.6	109.6	76.4	116
15.6	63.2	39.2	103.2		

Source: Data of Meyers (1927).

determined from the slope. Figure 3.A.1 shows the plot. This is not a wonderful linear relationship, but such a result is typical.

Next, I want to calculate the rate constants using a least-squares procedure. Table 3.A.2 shows the formulas used to do the least-squares fit. I listed the concentration in column A and the rate data in column B. Column C is one over the concentration; column D is one over the rate. I then used the SLOPE, INTERCEPT and RSQ Functions in Excel to calculate the slope of the line.

Table 3.A.3 shows the numerical values in the plot. From the least-squares fit

$$\frac{1}{r_p} = \frac{0.194}{[par]} + 0.00711 \qquad (3.A.3)$$

Comparison of equations (3.A.2) and (3.A.3) shows

$$k_1 = \frac{1}{0.00711} = 140.5, \qquad K_2 = \frac{1}{(0.194 * k_1)} = 0.037.$$

Figure 3.A.2 is a plot taking $k_1 = 140.4$ and $K_2 = 0.0366$. The curve does not do a bad job of fitting the data, although there is a systematic error at high concentrations. This is typical. One can even get cases where the Lineweaver–Burke plot misses the trends in the data.

We got the systematic error because we fit to $1/r_p$. A plot of $1/r_p$ gives greater weight to data taken at small concentrations, and that is usually where the data are the least accurate.

The Eadie–Hofstee plot avoids the difficulty at low concentrations by instead finding a way to linearize the data without calculating $1/r_p$.

Rearranging equation (3.A.1), we have

$$r_p(1 + K_2[par]) = k_1 K_2[par] \qquad (3.A.4)$$

Further rearrangement yields

$$\frac{r_p}{[par]} = k_1 K_2 - K_2 r_p \qquad (3.A.5)$$

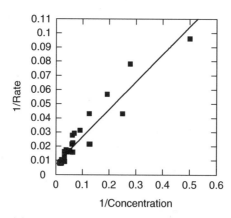

Figure 3.A.1   A Lineweaver–Burke plot of the data in Table 3.A.1.

Table 3.A.2 The formulas in the spreadsheet for the Lineweaver–Burke plot

	A	B	C	D	E	F
01		k_1	=1/D2	=SLOPE(D6:D37,C6:C37)		2
02		K_2	=1/C1/D1	=INTERCEPT(D6:D37,C6:C37)		=SUM(F5:F37)
03				=RSQ(D6:D37,C7:C37)		
04	conc	rate	1/conc	1/rate	rate calculated from rate equation	error
05	0	0			=C$1*C$2*$A5/(1+C$2*$A5)	=ABS(E5–$B5)^$F$1
06	2	10.4	=1/A6	=1/B6	=C$*1C$2*$A6/(1+C$2*$A6)	=ABS(E6–$B6)^$F$1
07	3.6	12.8	=1/A7	=1/B7	=C$1*C$2*$A7/(1+C$2*$A7)	=ABS(E7–$B7)^$F$1
08	4	23.2	=1/A8	=1/B8	=C$1*C$2*$A8/(1+C$2*$A8)	=ABS(E8–$B8)^$F$1
09	5.2	17.6	=1/A9	=1/B9	=C$1*C$2 $A9/(1+C$2*$A9)	=ABS(E9–B9)^$F$1
10	7.8	46.4	=1/A10	=1/B10	=C$1*C$2*$A10/(1+C$2*$A10)	=ABS(E10–B10)^$F$1
11	8	23.2	=1/A11	=1/B11	=C$1*C$2*$A11/(1+C$2*$A11)	=ABS(E11–B11)^$F$1
12	8	46.4	=1/A12	=1/B12	=C$1*C$2*$A12/(1+C$2*$A12)	=ABS(E12–$B12)^$F$1
13	11	32	=1/A13	=1/B13	=C$1*C$2*$A13/(1+C$2*$A13)	=ABS(E13–$B13)^$F$1

Table 3.A.3 The numerical values in the spreadsheet for the Lineweaver–Burke plot

	A	B	C	D	E	F
01		k 1	140.5	0.194		2.000
02		K 2	0.037	0.00711		9454.2
03				0.901		
04	conc	rate	1/conc	1/rate	rate	error
05	0.000	0.000			0.000	0.000
06	2.000	10.400	0.500	0.096	9.58	0.664
07	3.600	12.800	0.278	0.078	16.35	12.67
08	4.000	23.200	0.250	0.043	17.94	27.62
09	5.200	17.600	0.192	0.057	22.47	23.69
10	7.800	46.400	0.128	0.022	31.20	230.87
11	8.000	23.200	0.125	0.043	31.82	74.37
12	8.000	46.400	0.125	0.022	31.82	212.45
13	11.000	32.000	0.091	0.031	40.33	69.42

Note: These numbers are from Excel 98. Excel 95 gives slightly different results.

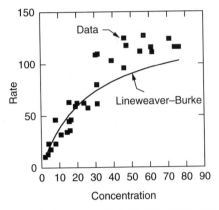

Figure 3.A.2   The Lineweaver–Burke fit of the data in Table 3.A.1.

Therefore a plot of $r_P/[par]$ versus r_p should yield a straight line. Figure 3.A.3 shows the plot. In this case, the line does not fit the data very well. I did a least-squares fit through the data using a spreadsheet like that in Table 3.A.3. The results showed

$$\frac{r_p}{[par]} = 4.21 - 0.0171 r_p \tag{3.A.6}$$

Comparing equation (3.A.6) and (3.A.5) shows $K_2 = 0.0171$, $k_1 = 4.20/0.0171 = 246$. Figure 3.A.4 shows how well the data actually fit's the line. Notice that there is still an error at high concentration. In this case the predicted rate is slightly too high. When you divide by [par], you give lower weight to the high concentration points. The result is that there are still errors. Still, the Eadie–Hofstee method fits better than the Lineweaver–Burke fit, even though R^2 is 0.34 in Figure 3.A.3 while R^2 is 0.90 in Figure 3.A.1. The last way to fit the data are with a nonlinear least-squares.

The idea in nonlinear least-squares is to use the solver function of a spreadsheet to calculate the best values of the coefficients based on some criterion. A common criterion

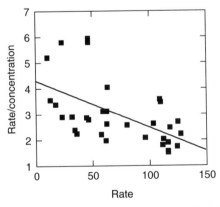

Figure 3.A.3   An Eadie–Hofstee plot of the data in Table 3.A.1.

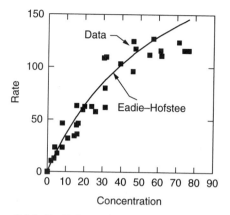

Figure 3.A.4 The Eadie–Hofstee fit of the data in Table 3.A.1.

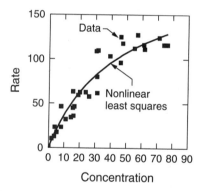

Figure 3.A.5 A nonlinear least-squares fit to the data in Table 3.A.1.

is to minimize the total error, where the total error is defined by

$$\text{Total error} = \sum_{\text{Data}} \{(\text{abs}[(\text{measured rate}) - (\text{calculated rate})]\}^2$$

where abs is absolute value. For our example, this becomes

$$\text{Total error} = \sum_{\text{data}} \left[\text{abs} \left(r_p - \frac{k_1 K_2 [\text{par}]}{1 + K_2 [\text{par}]} \right) \right]^2 \tag{3.A.7}$$

One often uses powers other than 2 to do the fitting. Table 3.A.4 shows a spreadsheet for the fitting. The spreadsheet calculates the rate in column C. The error is calculated in column D. Element $D$2 is the total error. I then used the solver function in Microsoft

Table 3.A.4   Part of the spreadsheet used to calculate values of k_1 and K_2 to minimize the total error

	A	B	C	D
01		k_1=	204.3 (Calculated by solver)	2
02		K_2=	0.0221 (Calculated by solver)	=SUM(D4:D50)
03	conc	rate	monod	error
04	0	0	=k_1*K_2*A4/(1+K_2*A4)	=ABS(C4 − $B$4)^$D$1
05	2	10.4	=k_1*K_2*A5/(1+K_2*A5)	=ABS(C5 − $B$5)^$D$1
06	3.6	12.8	=k_1*K_2*A6/(1+K_2*A6)	=ABS(C6 − $B$6)^$D$1
07	4	23.2	=k_1*K_2*A7/(1+K_2*A7)	=ABS(C7 − $B$7)^$D$1
08	5.2	17.6	=k_1*K_2*A8/(1+K_2*A8)	=ABS(C8 − $B$8)^$D$1
09	7.8	46.4	=k_1*K_2*A9/(1+K_2*A9)	=ABS(C9 − $B$9)^$D$1
10	8	23.2	=k_1*K_2*A10/(1+K_2*A10)	=ABS(C10 − $B$10)^$D$1
11	8	46.4	=k_1*K_2*A11/(1+K_2*A11)	=ABS(C11 − $B$11)^$D$1
12	11	32	=k_1*K_2*A12/(1+K_2*A12)	=ABS(C12 − $B$12)^$D$1
13	14.4	34.4	=k_1*K_2*A13/(1+K_2*A13)	=ABS(C13 − $B$13)^$D$1
14	15.6	44.8	=k_1*K_2*A14/(1+K_2*A14)	=ABS(C14 − $B$14)^$D$1
15	15.6	63.2	=k_1*K_2*A15/(1+K_2*A15)	=ABS(C15 − $B$15)^$D$1
16	16	36	=k_1*K_2*A16/(1+K_2*A16)	=ABS(C16 − $B$16)^$D$1
17	16.6	46.4	=k_1*K_2*A17/(1+K_2*A17)	=ABS(C17 − $B$17)^$D$1

Excel[1] to minimize $D$2 by varying k_1 and K_2. The result is that $k_1=204.3$ and $K_2=0.0221$.

Next, it is useful to compare the numerical values of the parameters calculated via the various methods. Figure 3.A.6 shows the fits to the data, and all three lines are reasonable. Still, the Lineweaver–Burke method is too low at high concentrations while the Eadie–Hofstee method is too high. The nonlinear least-squares method gives the best balance in the fit.

Next, it is useful to compare the parameters calculated via the various methods. Notice that the three methods give k_1 values between 140 and 246 and K_2 values between 0.0171 and 0.0366. Yet all three curves fit the data. This example illustrates the difficulty in fitting kinetic data in that when you have lots of parameters, you can use different values of the parameters and still fit the data pretty well.

Well, which fits best? That is not an easy question. In the literature, people often assess the "goodness of fit" by looking at the R^2 values. They are given in Table 3.A.5. Table 3.A.5 also shows the errors created by each method. Where the total error is defined by equation (3.A.7). The Lineweaver–Burke plot has an excellent value of R^2 as measured

---

[1] One needs to be very careful with the solver programs in Excel. In my experience, the solver programs often goes to a local minimum in the error rather than a global minimum. Therefore it is important to start the algorithm at several sets of initial guesses for k_1 and K_2 minimize and then take the best answer. There are also global minimizers, such as Sahinidi's BARON program. They are the best chance when you want to be sure you minimize the error.

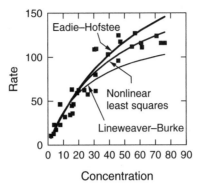

Figure 3.A.6 A comparison of the three fits to the data.

Table 3.A.5 A comparison of the various fits to the data in Table 3.A.1

Method	k_1	k_2	Total error	R^2
Lineweaver–Burke	140	0.0366	9454	0.901 (linear plot)
Eadie–Hofstee	246	0.0171	5648	0.344 (linear plot)
Nonlinear least-squares	204	0.0221	4919	0.905 (nonlinear)

by the linear regression, but the highest total error. The Eadie–Hofstee method has the worst R^2, but a lesser total error. The nonlinear least-squares method has the lowest total error but a value of R^2 similar to that for Lineweaver–Burke.

My own view is that one has to be very careful in using R^2 as a measure of the goodness of fit. In linear regression, R^2 measures the uncertainty in the slope of the line. However, it does not give you any uncertainty in the intercept, or the relative importance of the slope and intercept in fitting your data. If you fit some data Y to a function f(x), then R^2 is defined by

$$R^2 = 1 - \frac{\sum_{\text{data}}[\text{abs}(Y - f(x))]^2}{\sum_{\text{data}}[\text{abs}(Y - \overline{Y})]^2} \qquad (3.A.8)$$

where $\overline{Y}$ is the average value of Y. When you use a least-squares technique, you define a function that you are fitting to a variable Y. According to how you define Y, you can get wildly different values of R^2. The Lineweaver–Burke, Eadie–Hofstee, and nonlinear least-squares methods define Y differently, as indicated in Table 3.A.6. The net result is that the values of R^2 are not comparable.

In the homework set, we have an example where we fit two different models to a single data set. The first model fits the data to two significant figures even though R^2 is 10^{-5}. The second model does not fit the data at all. (It sometimes gets the wrong sign.) Yet, R^2 is 0.75. This example clearly illustrates the ideas that R^2 does not tell you how well a given model fits your data.

Table 3.A.6   The values of Y and f(x) used to calculate R^2 using the different methods

Method	Y	f(x)
Lineweaver–Burke	$\dfrac{1}{r_p}$	$\dfrac{1 + K_2[\text{par}]}{k_1 K_2[\text{par}]}$
Eadie–Hofstee	$\dfrac{r_p}{[\text{par}]}$	$k_1 K_2 - K_2 r_p$
Nonlinear least-squares	r_p	$\dfrac{k_1 K_2[\text{par}]}{1 + K_2[\text{par}]}$

One way around the difficulty is to define a uniform value of R^2. For example, two definitions that we can use are

$$R^2 = 1 - \frac{\sum\limits_{\text{data}}\left[\text{abs}\left(r_p - \dfrac{k_1 K_2[\text{par}]}{1 + K_2[\text{par}]}\right)\right]^2}{\sum\limits_{\text{data}}[\text{abs}(r_p - \bar{r}_p)]^2} \tag{3.A.9}$$

where $\bar{r}_p$ is the average value of the rate.

$$R^2 = 1 - \frac{\sum\limits_{\text{data}}\left[\text{abs}\left(\dfrac{r_p}{[\text{par}]} - \dfrac{k_1 K_2}{1 + K_2[\text{par}]}\right)\right]^2}{\sum\limits_{\text{data}}\left[\text{abs}\left(\dfrac{r_p}{[\text{par}]} - \left(\dfrac{r_p}{[\text{par}]}\right)_{\text{av}}\right)\right]^2} \tag{3.A.10}$$

where $(r_p/[\text{par}])_{\text{av}}$ is the average value of $(r_p/[\text{par}])$. Table 3.A.7 shows the values. Notice that the different definitions of R^2 give wildly different values of R^2.

In the literature, people often use R^2 as a measure of the goodness of fit and to distinguish between various models. One has to be very careful when doing that. After all, Table 3.A.7 shows that you can calculate wildly different values of R^2 according to how you do the calculations. Still, in the literature people often ignore these complexities and naively use the R^2 values that come out of their linear regression to assess how well their model fits the data. Readers can judge for themselves whether that is a valid approach.

Example 3.B   Tests of Statistical Significance: Analysis of Variance   The next question is how can one really tell if one model fits a given data set better than another.

Table 3.A.7   The values of R^2 calculated using the different methods

Method	R^2 from linear regression	R^2 from equation (2.B.9)	R^2 from equation (2.B.10)
Lineweaver–Burke	0.901	0.818	0.552
Eadie–Hofstee	0.344	0.890	0.525
Nonlinear least-squares	0.905	0.905	0.558

The objective of this example is to provide an objective test. Let's consider the example in Example 3.A (see Table 3.B.1).

Which model fits best? Is the difference statistically significant?

Solution   First, let us see which model fits best. We do that by calculating the variance of the data and seeing which model has the lowest variance. The variance V_i is defined by

$$V_i = \frac{\sum_{\text{points}} ((\text{experimental rate}) - (\text{calculated rate}))^2}{(\text{number of samples}) - (\text{number of independent parameters in model})} \quad (3.B.1)$$

Substituting in equation (3.A.7) yields

$$V_i = \frac{\text{total error from equation 3.A.7}}{\text{number of samples} - \text{number of parameters}} \quad (3.B.2)$$

It is important to calculate the variance as shown in (3.B.1) and not, for example, the variance of one over the rate. In order to use the statistical tests below, one will have to assume that the error in the data follows what statisticians call a *chi-square* (χ^2) *distribution*. If you calculate the errors in the rate, the errors usually do follow a χ^2 distribution. However, the errors in one over rate *do not* follow a χ^2 distribution. As a result, although the statistical measures below are meaningful for variances calculated via equation (3.B.1), they are not meaningful for variances in one over the rate.

The variances are easily calculated from the total errors in Table 3.A.5. For example, in the nonlinear least-squares case

$$V_i = \frac{4919}{(32 \text{ points} - 2 \text{ parameters})} = 164 \quad (3.B.3)$$

Similarly, for Eadie–Hofstee

$$V_i = \frac{5648}{32 - 2} = 188 \quad (3.B.4)$$

Table 3.B.1   Fits to the data in Example 3.A

Concentration	Experimental Rate	Calculated Rate		
		Nonlinear Least Squares	Lineweaver–Burke	Eadie–Hofstee
0	0	0	0	0
2	10.4	8.65	9.6	8.14
3.6	12.8	15.06	16.36	14.26
4	23.2	16.60	17.94	15.75
5.2	17.6	21.07	22.46	20.08
7.8	46.4	30.05	31.20	28.95
8	32	30.71	31.82	29.60
8	34.4	30.71	31.82	29.60
Variance		164	315	188

while for the Lineweaver–Burke plot

$$V_i = \frac{9454}{32 - 2} = 315 \tag{3.B.5}$$

So the nonlinear least-squares method fits the data best.

The next question is whether the difference is statistically significant. This is important, because one model could fit better, but the difference between the models could be within the noise in the data. Statisticians are still developing methods to test whether one model is better than another. It is easy to test what are called "nested models": two models that are the same except that one has one extra parameter. However, independent models are much harder to test. The Cox algorithm can do the testing rigorously. One can also do a test using a Bayesian maximum-likelihood or minimum-entropy algorithm. Both are beyond the scope of this book.

In this problem we will provide an approximation *that is not mathematically rigorous* but has the advantage that it can be used for practical calculations. The method is based on a statistical test called the *F test*. The idea in the F test is to compute $F_{inverse}$ given by

$$F_{inverse} = \frac{\text{variance in weaker model}}{\text{variance in better model}} \tag{3.B.6}$$

So, if we want to compare the Lineweaver–Burke and nonlinear least-squares methods, we calculate

$$F_{inverse} = \frac{315}{164} = 1.91 \tag{3.B.7}$$

Statistically, if the two variances are independent and the value of $F_{inverse}$ is large enough, we can say that the difference is statistically significant. Table 3.B.2 gives values of $F_{inverse}$. In the table, $F_{inverse}$ is listed as a function of nf given by

$$nf = \text{number of data points} - \text{parameters in the model} \tag{3.B.8}$$

If you want to see if one model fits rate data better than another does, you should always do an F test to see if the difference between two models is statistically significant.

I want to say clearly that the F test is an approximation. It assumes that the two variances are independent, which is clearly not true. Still, it does give useful information even though the F test cannot be rigorously applied to this case.

To read Table 3.B.2, if nf is 30, then $F_{inverse}$ must be 1.84 to have 95% confidence that one model is better than another and 2.39 to have 99% confidence that one model is better than another. We are between 95 and 99%, so we can say that we are between 95 and 99% and are certain that the nonlinear least-squares model fits the data better than does the Lineweaver–Burke model.

One can calculate a more accurate value for the confidence using the FDIST function in Excel. FDIST calculates the probability that a given value of $F_{inverse}$ occurs by chance. So, 1-FDIST is the probability that it occurred by other than chance.

$$\%\text{confidence} = 1 - \text{FDIST}(F_{inverse}, \text{nf for better model, nf for worse model}) \tag{3.B.9}$$

I used Excel to calculate

$$1 - \text{FDIST}(1.91, 30, 30) = 0.96$$

Table 3.B.2   Values of $F_{inverse}$ as a function of nf when both models have the same value of nf

nf	\multicolumn{4}{c}{Significance Level}
	90%	95%	99%	99.5%
1	39.86	161.5	4052	16212
2	9.0	19	99	199
3	5.39	9.28	29.46	47
4	4.11	6.39	15.98	23
5	3.45	5.05	10.97	14.94
6	3.05	4.28	8.42	11.07
7	2.78	3.79	6.99	8.89
8	2.59	3.44	6.03	7.50
9	2.44	3.18	5.35	6.54
10	2.32	2.98	4.85	5.85
20	1.79	2.12	2.84	3.32
30	1.61	1.84	2.39	2.63
40	1.51	1.69	2.11	2.3
50	1.44	1.60	1.95	2.1

so I am 96% sure that the nonlinear least-squares fit better than the Lineweaver–Burke plot. Excel also has a FINV function that calculates $F_{inverse}$ via

$$F_{inverse} = FINV(1 - \%significance, nf \text{ for better model, } nf \text{ for worse model})$$

So you do not have to use Table 3.B.2 to calculate $F_{inverse}$.

In the literature, people often use R^2 values to see which model is better. I do not believe that is meaningful. There are other methods to compare models (see Bevinson and Robinson, 1992).

Example 3.C   Fitting the Data in Example 3.A with Another Kinetic Model   The objective of this problem is to illustrate the difficulty in using kinetic data to distinguish between rate data. In order to illustrate the ideas, we will fit the data in Example 3.A with the model:

$$r_p = \frac{k_1 K_2 [par]}{1 + K_2 [par]^{1.5}} \qquad (3.C.1)$$

and see if the difference is statistically significant.

Solution   In Table 3.C.1, column A is the concentration, column B is the experimental rate data. Column C is the rate calculated from equation (3.C.1), whereas column D is the square of the difference between the calculated rate and the measured rate. I then used the solver function to minimize cell D2, by varying cells C1 and C2.

The spreadsheet is the same as in Example 3.A (see Table 3.C.1).

When you solve the equation, you find that the total error is 4576, compared to 4919 with equation (3.A.1). Therefore, equation (3.C.1) fits the data slightly better than does equation (3.A.1).

Next, we will consider whether the difference is statistically significant. Again we can do an F test.

Table 3.C.1 Part of the spreadsheet used to calculate values of k_1 and K_2 to minimize the total error

	A	B	C	D
01		k_1=	1944 (Calculated by solver)	2
02		K_2=	0.00188	=SUM(D4:D50)
			(Calculated by solver)	
03	conc	rate	equation 3.C.1^1.5	error
04	0	0	=k_1*K_2*A4/(1+K_2*A4^1.5)	=ABS(C4 – $B4)^$D$1
05	2	10.4	=k_1*K_2*A5/(1+K_2*A5^1.5)	=ABS(C5 – $B5)^$D$1
06	3.6	12.8	=k_1*K_2*A6/(1+K_2*A6^1.5)	=ABS(C6 – $B6)^$D$1
07	4	23.2	=k_1*K_2*A7/(1+K_2*A7^1.5)	=ABS(C7 – $B7)^$D$1
08	5.2	17.6	=k_1*K_2*A8/(1+K_2*A8^1.5)	=ABS(C8 – $B8)^$D$1
09	7.8	46.4	=k_1*K_2*A9/(1+K_2*A9^1.5)	=ABS(C9 – $B9)^$D$1
10	8	23.2	=k_1*K_2*A10/(1+K_2*A10^1.5)	=ABS(C10 – $B10)^$D$1
11	8	46.4	=k_1*K_2*A11/(1+K_2*A11^1.5)	=ABS(C11 – $B11)^$D$1
12	11	32	=k_1*K_2*A12/(1+K_2*A12^1.5)	=ABS(C12 – $B12)^$D$1
13	14.4	34.4	=k_1*K_2*A13/(1+K_2*A13^1.5)	=ABS(C13 – $B13)^$D$1
14	15.6	44.8	=k_1*K_2*A14/(1+K_2*A14^1.5)	=ABS(C14 – $B14)^$D$1
15	15.6	63.2	=k_1*K_2*A15/(1+K_2*A15^1.5)	=ABS(C15 – $B15)^$D$1
16	16	36	=k_1*K_2*A16/(1+K_2*A16^1.5)	=ABS(C16 – $B16)^$D$1
17	16.6	46.4	=k_1*K_2*A17/(1+K_2*A17^1.5)	=ABS(C17 – $B17)^$D$1

$V_{3.A.1}$, the variance of equation (3.A.1), is

$$V_{3.A.1} = \frac{4919}{32 \text{ points} - 2 \text{ parameters}} = 164$$

$V_{3.C.1}$, the variance of equation (3.C.1), is

$$V_{3.C.1} = \frac{4576}{32 \text{ points} - 2 \text{ parameters}} = 152$$

The ratio of the variances is

$$F_{inverse} = \frac{164}{152} = 1.07$$

The probability that equation (3.C.1) is better than (3.A.1) is given by equation (3.B.8). Plugging in the numbers yields

$$\text{Probability} = 1 - \text{FDIST}(1.07, 30, 30) = 0.58$$

So there is a 58% chance that equation (3.C.1) fits better than does (3.A.1). Conversely, there is a 42% chance that equation (3.A.1) fits better than does equation (3.C.1).

Statisticians usually say that one needs a 95–99% chance to say that a difference is statistically significant. Here we see only a 58 : 42 chance. Consequently, we can say that equations (3.A.1) and (3.C.1) both fit the data to within the error bars. This is important because people often say that one model fits data better than another. Often, however, multiple models are within the error bars in the data.

Example 3.D Analysis of the data in Table 3.5 Use Essen's method, Van't Hoff's method, and Powell's method to analyze the data in Table 3.5.

Solution

Essen's Method According to equations (3.38) and (3.40)

$$\frac{1}{k_1} \ln \left(\frac{C_A^0}{C_A} \right) = \tau \qquad (3.38)$$

$$\frac{1}{(n-1)k_n(C_A^0)^{n-1}} \left[\left(\frac{C_A^0}{C_A} \right)^{n-1} - 1 \right] = \tau \qquad (3.40)$$

First, let's start with Essen's method. According to Essen's method, one should make plots of $\ln(C_A^0/C_A)$, $(C_A^0/C_A) - 1$, $(C_A^0/C_A)^2 - 1$ and see which plot is the most linear. I programmed an Excel spreadsheet to do the calculations.

Table 3.D.1 shows the formulas in the spreadsheet to do the calculations. Column A in the spreadsheet is the time, column B is concentration, column C is $\ln(C_A^0/C_A)$, column D is $(C_A^0/C_A) - 1$, and column E is $(C_A^0/C_A)^2 - 1$. Cell B6 is the concentration at time $= 1$. =LN(1/B6) computes $\ln(C_A^0/C_A)$ at time $= 1$ with $C_A^0 = 1$. The regression statements at the bottom of the table compute the slope, the intercept, and ρ for a least-squares fit to the data.

Table 3.D.2 shows the numerical values for the plots. The regression coefficients are particularly interesting. Notice that the regression coefficients are always above 0.98, independent of what order is assumed.

Plots of the data are given in Figure 3.18. Notice that all three lines fit the data reasonably well, which means that we could not distinguish between the various orders with an Essen plot. If we had much more data, going to above 90% conversion, we would be able to distinguish between the curves. However, there is no reason to do additional experiments since Van't Hoff's method allows one to distinguish between zero-, first-, and second-order kinetics using the existing data.

Van't Hoff's Method In Van't Hoff's method, one calculates k_1, k_2, and k_3 from

$$k_1 = \frac{1}{\tau} \ln \left(\frac{C_A^0}{C_A} \right) \qquad (3.51)$$

$$k_n = \frac{1}{n^{(n-1)}\tau(C_A^0)^{n-1}} \left[\left(\frac{C_A^0}{C_A} \right)^{n-1} - 1 \right] \qquad (3.52)$$

When $n \neq 1$, I find it easier to plot a new function KN defined by

$$KN = k_n n^{n-1} (C_A^0)^{n-1} \qquad (3.D.2)$$

Table 3.D.1 Formulas for the spreadsheet for Essen's method

	A	B	C	D	E
1					
2			Essen's Method		
3	time	conc	first	second	third
4			ln(Ca0/Ca)	(Ca0/Ca)−1	=(CA0/CA)^2−1
5	0	1	=LN(1/B5)	=(1/B5−1)	=((1/B5)^2−1)
6	1	0.91	=LN(1/B6)	=(1/B6−1)	=((1/B6)^2−1)
7	2	0.83	=LN(1/B7)	=(1/B7−1)	=((1/B7)^2−1)
8	3	0.77	=LN(1/B8)	=(1/B8−1)	=((1/B8)^2−1)
9	4	0.71	=LN(1/B9)	=(1/B9−1)	=((1/B9)^2−1)
10	5	0.67	=LN(1/B10)	=(1/B10−1)	=((1/B10)^2−1)
11	6	0.63	=LN(1/B11)	=(1/B11−1)	=((1/B11)^2−1)
12	7	0.59	=LN(1/B12)	=(1/B12−1)	=((1/B12)^2−1)
13	8	0.56	=LN(1/B13)	=(1/B13−1)	=((1/B13)^2−1)
14	9	0.53	=LN(1/B14)	=(1/B14−1)	=((1/B14)^2−1)
15	10	0.5	=LN(1/B15)	=(1/B15−1)	=((1/B15)^2−1)
16	11	0.48	=LN(1/B16)	=(1/B16−1)	=((1/B16)^2−1)
17	12	0.45	=LN(1/B17)	=(1/B17−1)	=((1/B17)^2−1)
18	13	0.43	=LN(1/B18)	=(1/B18−1)	=((1/B18)^2−1)
19	14	0.42	=LN(1/B19)	=(1/B19−1)	=((1/B19)^2−1)
20	14	0.4	=LN(1/B20)	=(1/B20−1)	=((1/B20)^2−1)
21	16	0.38	=LN(1/B21)	=(1/B21−1)	=((1/B21)^2−1)
22	17	0.37	=LN(1/B22)	=(1/B22−1)	=((1/B22)^2−1)
23		Intercept	=Intercept ($C$5..$C$22, A$7..A$22)	=Intercept ($D$5..$D$22, A$5..A$22)	=Intercept ($E$5..$E$22, A$5..A$22)
24		Slope	=slope ($C$5..$C$22, A$5..A$22)	=slope ($D$5..$D$22, A$7..A$22,110)	=slope ($E$7..$E$22, A$5..A$22)
25		Rho	=RSQ ($C$5..$C$22, A$5..A$22)	=RSQ ($D$5..$D$22, A$5..A$22)	=RSQ ($E$5..$E$22, A$5..A$22)

Substituting

$$KN = \left[\left(\frac{C_A^0}{C_A} \right)^{n-1} - 1 \right] \tag{3.D.3}$$

KN should be a constant!

Again, I used an Excel spreadsheet to do the calculations. When I solve this, I find it useful to use the module macro in Excel to set up all of the equations. The module macro capability allows you to define your own functions, in Visual BASIC.

Table 3.D.2   Numerical values for the Essen plots

	A	B	C	D	E
1					
2			Essen's Method		
3	time	conc	first	second	third
4			$\ln(1/Ca)$	$(Ca0/Ca)-1$	$(CA0/CA)^2-1$
5	0	1.00	0.000	0.000	0.000
6	1	0.91	0.094	0.099	0.208
7	2	0.83	0.186	0.205	0.452
8	3	0.77	0.261	0.299	0.687
9	4	0.71	0.342	0.408	0.984
10	5	0.67	0.400	0.493	1.228
11	6	0.63	0.462	0.587	1.520
12	7	0.59	0.528	0.695	1.873
13	8	0.56	0.580	0.786	2.189
14	9	0.53	0.635	0.887	2.560
15	10	0.50	0.693	1.000	3.000
16	11	0.48	0.734	1.083	3.340
17	12	0.45	0.799	1.222	3.938
18	13	0.43	0.844	1.326	4.408
19	14	0.42	0.868	1.381	4.669
20	15	0.40	0.916	1.500	5.250
21	16	0.38	0.968	1.632	5.925
22	17	0.37	0.994	1.703	6.305
23		Intercept	0.087	−0.005	−0.477
24		Slope	0.057	0.100	0.373
25		Rho	0.984	0.999	0.981

You can define your own functions by first inserting a module page into your workbook. In my version of Excel you do that by using the INSERT-MACRO-MODULE tab. Once you create a module, you can create your own functions by typing them onto the module page.

Table 3.D.3 shows the module to calculate k_1, k_2 and k_3, where k_1 is given by equation (3.51) and k_2 and k_3 are given by equation (3.D.3). The first three lines define a function "kone" (k_1). The first line defines kone as a public function. I called the function "kone", not k_1 because of a limitation of Excel. The word "public" in the function definition makes the function available to Excel. The term "as variant" says that the return type is variant. "Variant" is a general return type that can be used for anything. Table 3.D.4 gives a list of other return types.

The second line of the definition gives a value to "kone" according to equation 3.51. The third line tells the function to return. We use log in the function. In modules, log returns the natural log of a number while log10 returns a base 10 log

Table 3.D.3 Module used to calculate k_1, k_2, k_3, where k_1, k_2, and k_3 are defined by equation (3.D.2)

```
Public Function kone(ca0, ca, tau) As Variant
kone = Log(ca0/ca)/tau
End Function

Public Function ktwo(ca0, ca, tau) As Variant
ktwo = ((1#/ca)-(1#/ca0))/tau
End Function

Public Function kthree(ca0, ca, tau) As Variant
kthree = ((1#/ca)2-(1#/ca0)2)/tau/2
End Function
```

Table 3.D.4 Some Microsoft Excel/Visual BASIC return types

Type	Meaning	Type	Meaning
As variant	General return type (can be an integer, real, vector, matrix logical or text)	As Double	Double-precision real
As single	Single precision real	As Integer	Integer

Table 3.D.3 also shows the function definitions to calculate k_2 and k_3 according to equation (3.50). The only thing that is weird in the definition is that the "1#" makes 1 a floating-point number.

Once you type in the module, you can then use the function kone like any other function in Excel. So if you want to calculate kone with Ca0 = 1.0 Ca = 0.6 and tau = 3, you enter

```
=kone(1, 0.6, 3)
```

in a cell in your spreadsheet.

I use these functions to calculate k_1, k_2, and k_3 in my spreadsheet. Table 3.D.5 shows my spreadsheet. I listed time in column B and concentration in column C, and I wanted to calculate k_1 in column D, k_2 in column E, and k_3 in column F.

I used the following steps to get the answer:

1. I named cell e1 CaO, and set it equal to 1.
2. I used the kone, ktwo, and kthree functions to calculate k_1, k_2, and k_3.

For example, row 6 is for time=1. Ca0=1, Ca=c6, tau=b6, so kone calculates k_1 for CaO=ca0, ca=c6, tau=b6.

I also included a spreadsheet (Table 3.D.6) that does not use the user-defined functions.

Table 3.D.6 shows the formulas used in the spreadsheet. Column A is the time, column B is the concentration, column C is k_1, column D is k_2, and column E is k_3. At time = 2, $C_A = B33$, $\ln(C_A^0/C_A) = \ln(1/B33)$. Also, $\ln(C_A^0/C_A)/\tau = \ln(1/B33)/A33$.

The actual calculations are listed in Table 3.D.7.

Table 3.D.5 The formulas in the spreadsheet for Van't Hoff's method

	B	C	D	E	F
1			CaO=1.0		
2			Essen's Method		
3	time	conc	first	second	third
4			ln(CaO/Ca)	(CaO/Ca)−1	(CAO/CA)^2−1
5	0	1	=kone(cao,C5,B5)	=ktwo(cao,C5,B5)	=kthree(cao,C5,B5)
6	1	0.91	=kone(cao,C6,B6)	=ktwo(cao,C6,B6)	=kthree(cao,C6,B6)
7	2	0.83	=kone(cao,C7,B7)	=ktwo(cao,C7,B7)	=kthree(cao,C7,B7)
8	3	0.77	=kone(cao,C8,B8)	=ktwo(cao,C8,B8)	=kthree(cao,C8,B8)
9	4	0.71	=kone(cao,C9,B9)	=ktwo(cao,C9,B9)	=kthree(cao,C9,B9)
10	5	0.67	=kone(cao,C10,B10)	=ktwo(cao,C10,B10)	=kthree(cao,C10,B10)
11	6	0.63	=kone(cao,C11,B11)	=ktwo(cao,C11,B11)	=kthree(cao,C11,B11)
12	7	0.59	=kone(cao,C12,B12)	=ktwo(cao,C12,B12)	=kthree(cao,C12,B12)
13	8	0.56	=kone(cao,C13,B13)	=ktwo(cao,C13,B13)	=kthree(cao,C13,B13)
14	9	0.53	=kone(cao,C14,B14)	=ktwo(cao,C14,B14)	=kthree(cao,C14,B14)
15	10	0.5	=kone(cao,C15,B15)	=ktwo(cao,C15,B15)	=kthree(cao,C15,B15)
16	11	0.48	=kone(cao,C16,B16)	=ktwo(cao,C16,B16)	=kthree(cao,C16,B16)
17	12	0.45	=kone(cao,C17,B17)	=ktwo(cao,C17,B17)	=kthree(cao,C17,B17)
18	13	0.43	=kone(cao,C18,B18)	=ktwo(cao,C18,B18)	=kthree(cao,C18,B18)
19	14	0.42	=kone(cao,C19,B19)	=ktwo(cao,C19,B19)	=kthree(cao,C19,B19)
20	15	0.4	=kone(cao,C20,B20)	=ktwo(cao,C20,B20)	=kthree(cao,C20,B20)
21	16	0.38	=kone(cao,C21,B21)	=ktwo(cao,C21,B21)	=kthree(cao,C21,B21)
22	17	0.37	=kone(cao,C22,B22)	=ktwo(cao,C22,B22)	=kthree(cao,C22,B22)
23					

Notice that k_2 is constant while k_1 and k_3 vary. Figure 3.18 also shows a plot of these data. According to Van't Hoff's analysis, if k_2 is constant, the reaction is second-order; therefore, we conclude that the reaction is second-order.

Powell's Method In Powell's method (see formulas listed in Table 3.D.8), one uses plots of log of τ versus ln C_A. One then shifts the curves left and right until things fit. In my spreadsheet, I defined a new variable shift to do the calculations. I then varied shift by hand. I put the top two points in the middles of the curves and then saw which curve fit best.

The values for Powell's method are shown in Table 3.D.9.

A plot of the data in Table 3.D.9 shows that these data follow second-order kinetics (see Figure 3.D.1).

Example 3.E Using the Stoichiometric Table to Solve Stoichiometric Problems
Let's use the stoichiometric table to do a problem from page 118 in Felder and Rousseau, (1978). Ethylene is being made via the dehydrogenation of ethane.

$$C_2H_6 \Longrightarrow C_2H_4 + H_2 \qquad (3.E.1)$$

Assume that 100 mol/minute are fed into a flow reactor. Analysis of the exit stream indicates that 40 mol/minute of hydrogen leave the reactor. Calculate the exit flow rate.

Table 3.D.6   The formulas used for Van't Hoff's method

	A	B	C	D	E
30	time	conc	k1	k2	k3
31	0	1	$\ln(1/Ca)/t$	$((Ca0/Ca)-1)/t$	$((CA0/CA)^2-1)/t/2$
32	1	0.91	=LN(1/B32)/A32	=(1/B32-1)/A32	=((1/B32)^2-1)/A32/2
33	2	0.83	=LN(1/B33)/A33	=(1/B33-1)/A33	=((1/B33)^2-1)/A33/2
34	3	0.77	=LN(1/B34)/A34	=(1/B34-1)/A34	=((1/B34)^2-1)/A34/2
35	4	0.71	=LN(1/B35)/A35	=(1/B35-1)/A35	=((1/B35)^2-1)/A35/2
36	5	0.67	=LN(1/B36)/A36	=(1/B36-1)/A36	=((1/B36)^2-1)/A36/2
37	6	0.63	=LN(1/B37)/A37	=(1/B37-1)/A37	=((1/B37)^2-1)/A37/2
38	7	0.59	=LN(1/B38)/A38	=(1/B38-1)/A38	=((1/B38)^2-1)/A38/2
39	8	0.56	=LN(1/B39)/A39	=(1/B39-1)/A39	=((1/B39)^2-1)/A39/2
40	9	0.53	=LN(1/B40)/A40	=(1/B40-1)/A40	=((1/B40)^2-1)/A40/2
41	10	0.5	=LN(1/B41)/A41	=(1/B41-1)/A41	=((1/B41)^2-1)/A41/2
42	11	0.48	=LN(1/B42)/A42	=(1/B42-1)/A42	=((1/B42)^2-1)/A42/2
43	12	0.45	=LN(1/B43)/A43	=(1/B43-1)/A43	=((1/B43)^2-1)/A43/2
44	13	0.43	=LN(1/B44)/A44	=(1/B44-1)/A44	=((1/B44)^2-1)/A44/2
45	14	0.42	=LN(1/B45)/A45	=(1/B45-1)/A45	=((1/B45)^2-1)/A45/2
46	15	0.4	=LN(1/B46)/A46	=(1/B46-1)/A46	=((1/B46)^2-1)/A46/2
47	16	0.38	=LN(1/B47)/A47	=(1/B47-1)/A47	=((1/B47)^2-1)/A47/2
48	17	0.37	=LN(1/B48)/A48	=(1/B48-1)/A48	=((1/B48)^2-1)/A48/2

Table 3.E.1   Stoichiometric table for Example 3.E

Species	Flow-rate in Feed	Change Moles	Moles out of Reactor
Ethane	$F_{C_2H_6}^0$	$-F_{C_2H_6}^0 X_{C_2H_6}$	$F_{C_2H_6}^0(1-X_{C_2H_6})$
Ethylene	$F_{C_2H_4}^0$	$(b/a)F_{C_2H_6}^0 X_{C_2H_6}$	$F_{C_2H_4}+(b/a)F_{C_2H_6}^0 X_{C_2H_6}$
Hydrogen	$F_{H_2}^0$	$(c/a)F_{C_2H_6}^0 X_{C_2H_6}$	$F_{H_2}+(c/a)F_{C_2H_6}^0 X_{C_2H_6}$

Table 3.E.2   The results after substituting data into Table 3.E.1

Species	Flow Rate in Feed	Change Moles	Moles out of Reactor
Ethane	100 mol/minute	-100 mol/minute $X_{C_2H_6}$	100 mol/minute $(1-X_{C_2H_6})$
Ethylene	0	100 mol/minute $X_{C_2H_6}$	100 mol/minute $X_{C_2H_6}$
Hydrogen	0	100 mol/minute $X_{C_2H_6}$	100 mol/minute $X_{C_2H_6}$

Solution   Table 3.7 is a general stoichiometric table for the reaction. There are one reactant, two products, and no inerts in reaction (3.E.1). Therefore, the stoichiometric table will be in the format of Table 3.E.1.

In Table 3.7, a is the minus stoichiometric coefficient of C_2H_6, b is the stoichiometric coefficient of C_2H_4 and c is the stoichiometric coefficient of H_2. Substituting a = b = c = 1, $F_{C_2H_6}^0 = 100$ mol/minute, $F_{H_2}^0 = F_{C_2H_4}^0 = 0$ yields Table 3.E.2.

Table 3.D.7   The numerical values for Van't Hoff's method

	B	C	D	E	F
3	time	conc	k1	k2	k3
4	0	1	ln(1/Ca)/t	((Ca0/Ca)−1)/t	((CA0/CA)^2−1)/t/2
5	1	0.91	0.094	0.099	0.104
6	2	0.83	0.093	0.102	0.113
7	3	0.77	0.087	0.100	0.114
8	4	0.71	0.086	0.102	0.123
9	5	0.67	0.080	0.099	0.123
10	6	0.63	0.077	0.098	0.127
11	7	0.59	0.075	0.099	0.134
12	8	0.56	0.072	0.098	0.137
13	9	0.53	0.071	0.099	0.142
14	10	0.5	0.069	0.100	0.150
15	11	0.48	0.067	0.098	0.152
16	12	0.45	0.067	0.102	0.164
17	13	0.43	0.065	0.102	0.170
18	14	0.42	0.062	0.099	0.167
19	15	0.4	0.061	0.100	0.175
20	16	0.38	0.060	0.102	0.185
21	17	0.37	0.058	0.100	0.185

Table 3.E.3   The results (in mol/minute) after substituting $X_{C_2H_6} = 0.4$ into Table 3.E.2

Species	Flow Rate Feed	Change Moles	Moles out of Reactor
Ethane	100	−40	60
Ethylene	0	+40	40
Hydrogen	0	+40	40

But 40 mol/minute of hydrogen leave the reactor. Therefore

$$[100 \text{ mol/hour}]X_{C_2H_6} = 40 \text{ mol/minute} \tag{3.E.2}$$

or

$$X_{C_2H_6} = 0.4 \tag{3.E.3}$$

plugging $X_{C_2H_6}$ into Table 3.E.2 yields Table 3.E.3.

Example 3.F   Chuchani and Martin (1997) examined the pyrolysis of maleic acid:

$$C_6H_5C(H)(OH)COOH \longrightarrow C_6H_5CHO + CO + H_2O$$

by loading the maleic acid into a batch reactor and measuring the pressure as a function of time. The data in Table 3.F.1 were obtained at 320.1°C, after correcting for a side reaction.

(a) Calculate the concentration of maleic acid as a function of time.

(b) Fit the data to a simple rate equation.

Table 3.D.8 Formulas used to analyze the data using Powell's method

	A	B	C	D	E	F	G	H
52			shift=	−1				
53	time	conc	log time	In time +shift	Ca/Ca0	Log Tau First order	Log Tau Second Order	Log Tau Third Order
54	0	1						
55	1	0.91	=LOG10(A55)	=C55+SHIFT	+B55/1	=LOG(LN(1/B55))	=LOG(1/B55−1)	=LOG((1/B55^2−1)/2)
56	2	0.83	=LOG10(A56)	=C56+SHIFT	+B56/1	=LOG(LN(1/B56))	=LOG(1/B56−1)	=LOG((1/B56^2−1)/2)
57	3	0.77	=LOG10(A57)	=C57+SHIFT	+B57/1	=LOG(LN(1/B57))	=LOG(1/B57−1)	=LOG((1/B57^2−1)/2)
58	4	0.71	=LOG10(A58)	=C58+SHIFT	+B58/1	=LOG(LN(1/B58))	=LOG(1/B58−1)	=LOG((1/B58^2−1)/2)
59	5	0.67	=LOG10(A59)	=C59+SHIFT	+B59/1	=LOG(LN(1/B59))	=LOG(1/B59−1)	=LOG((1/B59^2−1)/2)
60	6	0.63	=LOG10(A60)	=C60+SHIFT	+B60/1	=LOG(LN(1/B60))	=LOG(1/B60−1)	=LOG((1/B60^2−1)/2)
61	7	0.59	=LOG10(A61)	=C61+SHIFT	+B61/1	=LOG(LN(1/B61))	=LOG(1/B61−1)	=LOG((1/B61^2−1)/2)
62	8	0.56	=LOG10(A62)	=C62+SHIFT	+B62/1	=LOG(LN(1/B62))	=LOG(1/B62−1)	=LOG((1/B62^2−1)/2)
63	9	0.53	=LOG10(A63)	=C63+SHIFT	+B63/1	=LOG(LN(1/B63))	=LOG(1/B63−1)	=LOG((1/B63^2−1)/2)
64	10	0.5	=LOG10(A64)	=C64+SHIFT	+B64/1	=LOG(LN(1/B64))	=LOG(1/B64−1)	=LOG((1/B64^2−1)/2)
65	11	0.48	=LOG10(A65)	=C65+SHIFT	+B65/1	=LOG(LN(1/B65))	=LOG(1/B65−1)	=LOG((1/B65^2−1)/2)
66	12	0.45	=LOG10(A66)	=C66+SHIFT	+B66/1	=LOG(LN(1/B66))	=LOG(1/B66−1)	=LOG((1/B66^2−1)/2)
67	13	0.43	=LOG10(A67)	=C67+SHIFT	+B67/1	=LOG(LN(1/B67))	=LOG(1/B67−1)	=LOG((1/B67^2−1)/2)
68	14	0.42	=LOG10(A68)	=C68+SHIFT	+B68/1	=LOG(LN(1/B68))	=LOG(1/B68−1)	=LOG((1/B68^2−1)/2)
69	15	0.4	=LOG10(A69)	=C69+SHIFT	+B69/1	=LOG(LN(1/B69))	=LOG(1/B69−1)	=LOG((1/B69^2−1)/2)
70	16	0.38	=LOG10(A70)	=C70+SHIFT	+B70/1	=LOG(LN(1/B70))	=LOG(1/B70−1)	=LOG((1/B70^2−1)/2)
71	17	0.37	=LOG10(A71)	=C71+SHIFT	+B71/1	=LOG(LN(1/B71))	=LOG(1/B71−1)	=LOG((1/B71^2−1)/2)

Table 3.D.9 **Numerical values of the data for the analysis of Powell's method**

	A	B	C	D	E	F	G	H
52			shift=	−1.000				
53	time	conc	In time	In time + shift	Ca/CaO	Log Tau First order	Log Tau Second Order	Log Tau Third Order
54	0.000	1.000						
55	1.000	0.910	0.000	−1.000	0.910	−1.025	−1.005	−0.984
56	2.000	0.830	0.301	−0.699	0.830	−0.730	−0.689	−0.646
57	3.000	0.770	0.477	−0.523	0.770	−0.583	−0.525	−0.464
58	4.000	0.710	0.602	−0.398	0.710	−0.465	−0.389	−0.308
59	5.000	0.670	0.699	−0.301	0.670	−0.397	−0.308	−0.212
60	6.000	0.630	0.778	−0.222	0.630	−0.335	−0.231	−0.119
61	7.000	0.590	0.845	−0.155	0.590	−0.278	−0.158	−0.029
62	8.000	0.560	0.903	−0.097	0.560	−0.237	−0.105	0.039
63	9.000	0.530	0.954	−0.046	0.530	−0.197	−0.052	0.107
64	10.000	0.500	1.000	0.000	0.500	−0.159	0.000	0.176
65	11.000	0.480	1.041	0.041	0.480	−0.134	0.035	0.223
66	12.000	0.450	1.079	0.079	0.450	−0.098	0.087	0.294
67	13.000	0.430	1.114	0.114	0.430	−0.074	0.122	0.343
68	14.000	0.420	1.146	0.146	0.420	−0.062	0.140	0.368
69	15.000	0.400	1.176	0.176	0.400	−0.038	0.176	0.419
70	16.000	0.380	1.204	0.204	0.380	−0.014	0.213	0.472
71	17.000	0.370	1.230	0.230	0.370	−0.003	0.231	0.499

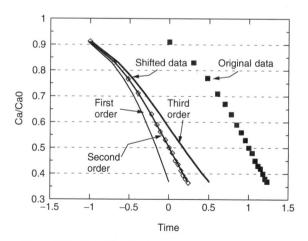

Figure 3.D.1 Plot of the data in the spreadsheet in Table 3.D.9.

Solution   According to equation (3.88)

$$\frac{P_T}{P_T^0} = 1 + \left(\frac{N_A^0}{N_T^0}\right)\left(\frac{\Delta mol}{a}\right) X_A \tag{3.F.1}$$

Solving

$$X = \left(\frac{P_T}{P_T^0} - 1\right)\left(\frac{N_T^0}{N_A^0}\right)\left(\frac{a}{\Delta mol}\right)$$

But $N_T^0 = N_A^0$, $a = 1$, $\Delta mol = 2$.
Therefore

$$X = \frac{1}{2}\left(\frac{P_T}{P_T^0} - 1\right)$$

$$C_M = \frac{N}{V} = \frac{P_A^0}{RT}(1 - X_A) = \frac{52.1 \text{ torr}}{\left(\frac{0.052 \text{ liter·atm}}{\text{mol·K}}\right)(593.1 \text{ K})}\left(\frac{1 \text{ atm}}{760 \text{ torr}}\right)(1 - X_A)$$

or

$$C_M = 1.4 \times 10^{-3}(1 - X_A)$$

Plugging in the numbers yields Table 3.F.2.

I analyzed these data using the spreadsheet in Table 3.C.3. The results are given in Table 3.F.3.

Notice that k_1 is constant up until about 70% conversion, which implies that the reaction is first-order in that conversion range (the side reaction starts at about 60% conversion).

Table 3.F.1   Data for Example 3.F

Pressure, atm	τ, minutes	Pressure, atm	τ, minutes	Pressure, atm	τ, minutes
52.1	0	110.9	5.2	124.7	9
72.8	1.5	114.4	6.0	134.7	40
92.5	3.0	121.6	7.0	139.8	51

Table 3.F.2   Concentration versus time for the data in Table 3.F.1

Time, minutes	X_A	C_M, mol/liter	Time, minutes	X_A	C_M, mol/liter
0	0	1.40×10^{-3}	9	0.667	4.66×10^{-4}
1.5	0.199	1.12×10^{-3}	10	0.697	4.27×10^{-4}
3	0.388	8.57×10^{-4}	40	0.763	2.90×10^{-4}
5.2	0.564	6.10×10^{-4}	51	0.842	2.21×10^{-4}
6	0.598	5.63×10^{-4}	—	—	—

Example 3.G   Develop a stoichiometric table for the reaction

$$CO + H_2O \Longleftrightarrow CO_2 + H_2$$

and show that the density is constant.

Solution   Following the analysis in Section 3.12.1, Table 3.G.1 the stoichiometric table for this system becomes.

Since the number of moles is constant, the density is constant (for an isothermal ideal gas).

Example 3.H   Limiting Reactants

(a) Develop a stoichiometric table for the homogeneous gas-phase reaction:

$$2NO_2(g) + \tfrac{1}{2}O_2(g) \Longleftrightarrow N_2O_5$$

(b) Develop an expression for the concentration of the oxygen as a function of the conversion of the nitrogen dioxide in a constant pressure reactor.

Table 3.F.3   Numerical Values in Table 3.F.2

	A	B	C	D	E
30	time	$1-X_A$	k1	k2	k3
31	0	1	$\ln(1/Ca)/t$	$((Ca0/Ca)-1)/t$	$((Ca0/Ca)^2-1)/t/2$
32	1.5	0.80	0.148	0.166	0.186
33	3.9	0.61	0.164	0.211	0.278
34	5.2	0.44	0.160	0.246	0.410
35	6	0.40	0.152	0.248	0.432
36	7	0.33	0.157	0.286	0.573
37	9	0.30	0.133	0.256	0.550
38	40	0.21	0.039	0.096	0.279
39	51	0.16	0.036	0.104	0.383

Table 3.G.1   Stoichiometric table for Example 3.G

Species	Initial Moles	Change Moles	Moles Unreacted
CO	N_{CO}^0	$-N_{CO}^0 X_{CO}$	$N_{CO}^0(1 - X_{CO})$
H_2	$N_{H_2}^0$	$-N_{CO}^0 X_{CO}$	$N_{H_2}^0 - N_{CO}^0 X_{CO}$
CO_2	$N_{CO_2}^0$	$+N_{CO}^0 X_{CO}$	$N_{CO_2}^0 + N_{CO}^0 X_{CO}$
H_2	$N_{H_2}^0$	$+N_{CO}^0 X_{CO}$	$N_{H_2}^0 + N_{CO}^0 X_{CO}$
Total	$N_{CO}^0 + N_{H_2}^0 + N_{CO_2}^0 + N_{H_2}^0$	0	$N_{CO}^0 + N_{H_2}^0 + N_{CO_2}^0 + N_{H_2}^0$

(c) If we put 2 mol of nitrogen dioxide and 1 mol of oxygen into a closed reactor, will oxygen or nitrogen dioxide be used up first? We call the reactant that is used up first the **limiting reactant**.

Solution

(a) Refer to Table 3.H.1.

(b) The oxygen concentration: by definition

$$C_{O_2} = \frac{N_{O_2}}{V} \tag{3.H.1}$$

where V is the volume. From the ideal-gas law

$$PV = N_{total}RT \tag{3.H.2}$$

Substituting V from equation (3.H.2) into equation (3.H.1) and substituting expressions for N_{total} and the concentration of NO_2 from the stoichiometric table yields

$$C_{O_2} = \left(\frac{N_{O_2}^0 - \frac{1}{4}X_{NO_2}N_{NO_2}^0}{N_{O_2}^0 + N_{NO_2}\left(1 - \frac{3}{4}X_{NO_2}\right)} \right) \frac{P}{RT}$$

(c) The NO_2 is used up when $X_{NO_2} = 1$. At that point

$$N_{O_2} = N_{O_2}^0 - \frac{1}{4}X_{NO_2}N_{NO_2}^0 = 1 - \left(\frac{1}{4}\right)(1)(2) = 0.5 \text{ mol}$$

NO_2 is used up first. Therefore NO_2 is the limiting reactant.

Example 3.I   Using the Equations in Section 3.13.2 for Design   Ethylbenzoate (E) is made by reacting benzoyl chloride (B) with ethyl alcohol (A) (see Figure 3.I.1).

Table 3.H.1   Stoichiometric table for Example 3.H

Species	Initial Moles	Change Moles	Moles Unreacted
NO_2	$N_{NO_2}^0$	$-X_{NO_2}N_{NO_2}^0$	$N_{NO_2}^0(1 - X_{NO_2})$
O_2	$N_{O_2}^0$	$-\frac{1}{4}X_{NO_2}N_{NO_2}^0$	$N_{O_2}^0 - \frac{1}{4}X_{NO_2}N_{NO_2}^0$
N_2O_5	0	$+\frac{1}{2}X_{NO_2}N_{NO_2}^0$	$\frac{1}{2}X_{NO_2}N_{NO_2}^0$
Total	$N_{NO_2}^0 + N_{O_2}^0$	$-\frac{3}{4}X_{NO_2}N_{NO_2}^0$	$N_{O_2}^0 + N_{NO_2}^0(1 - \frac{3}{4}X_{NO_2})$

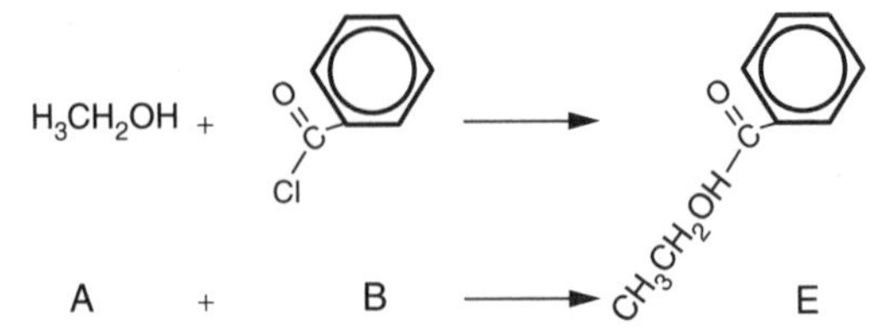

Figure 3.I.1   The reaction to form ethylbenzoate.

Assume that you mix 1 mol of benzoyl chloride into 1 liter of ethyl alcohol and find that 30% of the benzoyl chloride is converted after 30 minutes. How long will it take for 95% of the benzoyl chloride to be converted?

Solution

Note that $C_A \gg C_B$. Therefore, equation 3.77 applies. Rearranging equation 3.77 yields

$$\frac{1}{k_1} \ln \left(\frac{C_B^0}{C_B} \right) = \tau \tag{3.I.1}$$

with $k_1 = k_B C_A^0$. Rearranging yields

$$k_1 = \frac{1}{\tau} \ln \left(\frac{C_B^0}{C_B} \right) = \frac{1}{30 \text{ minutes}} \ln \left(\frac{1}{1 - 0.3} \right) = 1.1 \times 10^{-2} / \text{minute} \tag{3.I.2}$$

when $X_B = 0.95$, $C_B = (1 - X_B)$ 1 mol/liter. Substituting into equation (3.I.1) yields

$$\tau = \frac{1}{k} \ln \left(\frac{1}{1 - 0.95} \right) = 250 \text{ minutes} \tag{3.I.3}$$

Example 3.J   Multiple Reactants In Section 3.13, we did a problem on the Diels–Alder reaction of benzoquinone (B) and cyclopentadiene (C) to yield an adduct. Assume that you now want to use dimethyl cyclopentadiene (C′) for the reaction. Adding the methyl groups increases k. Assume a 20% increase (I made up a number).

First, 0.1 mol/liter of C and 0.08 mol/liter of B are loaded into a well-stirred batch reactor.

(a) Determine the residence time needed to convert 95% of the benzoquinone to adduct.

(b) Reminder, the reaction took 2.2 hours with cyclopentadiene.

Solution   From equation (3.79)

$$\tau = \frac{1}{k_B} \left(\int_{C_B^F}^{C_B^0} \frac{dC_B}{C_B C_C} \right)$$

The term in brackets is constant. If k_B goes up 20%, τ goes down by 20%:

$$\tau = \frac{2.2 \text{ hours}}{1.2} = 1.8 \text{ hours}$$

Example 3.K   Van't Hoff Plots for More Complex Reactions Bodenstein and Lund (1904) examined the rate of the reaction:

$$H_2 + Br_2 \Longrightarrow 2HBr \tag{3.K.1}$$

in a batch reactor. They did runs where they mixed known amounts of H_2 and Br_2, and measured the HBr concentration as a function of time. The data listed in Table 3.K.1 were obtained at $301.3°C$.

How well do the data in Table 3.K.1 fit the rate equations

$$r_{HB_r} = k_2[H_2][Br_2] \tag{3.K.2}$$

$$r_{HB_r} = k_{1.5}[H_2][Br_2]^{1/2} \tag{3.K.3}$$

Solution   I will use equation (3.30) to solve this problem.

$$\int_{C_A^0}^{C_A} \frac{-dC_A}{-r_A} = \tau \tag{3.30}$$

I am going to work in terms of the HBr concentration. According to equation (3.30)

$$\tau = \int_{C_{HBr}^F}^{C_{HBr}^O} \frac{d[HBr]}{(-r_{HBr})} \tag{3.K.4}$$

Inverting the limits yields

$$\tau = \int_{C_{HBr}^O}^{C_{HBr}^F} \frac{d[HBr]}{(r_{HBr})} \tag{3.K.5}$$

Next, I need an expression for r_{HBr}. Equations (3.K.2) and (3.K.3) give me expressions for r_{HBr} in terms of the concentrations $[H_2]$, $[Br_2]$ in the reactor. In order to use the equations, I will need to know how $[H_2]$ and $[Br_2]$ are changing as the reaction proceeds.

Next, I will use the stoichiometric table (Table 3.K.2) to develop an expression for $[H_2]$ and $[Br_2]$ in terms of the initial concentration in the reactor and $[HBr]$.

If we call $[HBr]$ the amount of HBr that forms, then the change in the Br_2 concentration is

$$\Delta[Br_2] = \frac{(-1)}{(2)}[HBr] \tag{3.K.6}$$

Table 3.K.1   Bodenstein and Lund's [1904, 1907] data for HBr production in a batch reactor

Run 1		Run 2		Run 3	
$[H_2]^0 = 0.5637$ $[Br_2]^0 = 0.2947$		$[H_2]^0 = 0.2281$ $[Br_2]^0 = 0.1517$		$[H_2]^0 = 0.3103$ $[Br_2]^0 = 0.5069$	
Time, minutes	[HBr]	Time, minutes	[HBr]	Time, minutes	[HBr]
0	0	0	0	0	0
14.5	0.0669	19.5	0.0322	15	0.0492
24.5	0.0985	34.5	0.0527	35	0.1031
34.5	0.1262	54.5	0.0713	55	0.1406
49.5	0.1644	79.5	0.0912	80	0.1752
79.5	0.2093	99.5	0.1040	102	0.1963
99.5	0.2306	124.5	0.1142	125	0.2179
124.5	0.2502	149.5	0.1217	155	0.2360
149.5	0.2619	174.5	0.1295	196	0.2533

Table 3.K.2   Stoichiometric table for Example 3.K

Species	Initial Concentration	Change	Final Calculation
HBr	0	[HBr]	[HBr]
Br_2	$[Br_2]$	$+\dfrac{(-1)}{(2)}[HBr]$	$[Br_2]^0 - \dfrac{1}{2}[HBr]$
H_2	$[H_2^0]$	$\dfrac{(-1)}{(2)}[HBr]$	$[H_2^0] - \dfrac{1}{2}[HBr]$

where (-1) is the stoichiometric coefficient of Br_2 and (2) is the stoichiometric coefficient of HBr. The rest of the stoichiometric table is given above.

From the stoichiometric table

$$[Br_2] = [Br_2]^0 - 0.5[HBr] \tag{3.K.7}$$

$$[H_2] = [H_2]^0 - 0.5[HBr] \tag{3.K.8}$$

where $[H_2]^0$ and $[Br_2]^0$ are the initial concentrations of H_2 and Br_2, respectively. If equation (3.K.2) works, then

$$r_{HBr} = k_2([H_2]^0 - 0.5[HBr])([Br_2]^0 - 0.5[HBr]) \tag{3.K.9}$$

According to equation (3.K.5)

$$\tau = \int_{C_{HBr}^0}^{C_{HBr}^F} \frac{d[HBr]}{(r_{HBr})} \tag{3.K.10}$$

Substituting equation (3.K.9) into equation (3.K.10) and looking up the integral in an integral table yields

$$\tau = \frac{2}{k_2([H_2]^0 - [Br_2]^0)} \ln\left(\frac{[Br_2]^0([H_2]^0 - 0.5[HBr])}{[H_2]^0([Br_2]^0 - 0.5[HBr])}\right) \tag{3.K.11}$$

Solving for k_2 yields

$$k_2 = \frac{2}{\tau([H_2]^0 - [Br_2]^0)} \ln\left(\frac{[Br_2]^0([H_2]^0 - 0.5[HBr])}{[H_2]^0([Br_2]^0 - 0.5[HBr])}\right) \tag{3.K.12}$$

Here is an Excel spreadsheet to do the calculations. Column A is the initial hydrogen concentration, column B is the initial bromine concentration, column C is the final HBr concentration, column D is τ, and column E is k_2 calculated from equation (3.K.11).

	A	B	C	D	E
3	$[H_2]°$	$[Br_2]°$	[HBr]	tau	k_2
4	0.5637	0.2947	0.0699	14.5	=2/D4/(A4−B4)*LN(B4/A4* (A4−0.5*C4)/(B4−0.5*C4))
5	0.5637	0.2947	0.0985	24.5	=2/D5/(A5−B5)*LN(B5/A5* (A5−0.5*C5)/(B5−0.5*C5))
6	0.5637	0.2947	0.1262	34.5	=2/D6/(A6−B6)*LN(B6/A6* (A6−0.5*C6)/(B6−0.5*C6))
7	0.5637	0.2947	0.1644	49.5	=2/D7/(A7−B7)*LN(B7/A7* (A7−0.5*C7)/(B7−0.5*C7))
8	0.5637	0.2947	0.2093	79.5	=2/D8/(A8−B8)*LN(B8/A8* (A8−0.5*C8)/(B8−0.5*C8))
9	0.5637	0.2947	0.2306	99.5	=2/D9/(A9−B9)*LN(B9/A9* (A9−0.5*C9)/(B9−0.5*C9))
10	0.5637	0.2947	0.2502	124.5	=2/D10/(A10−B10)*LN(B10/A10* (A10−0.5*C10)/(B10−0.5*C10))
11	0.5637	0.2947	0.2619	149.5	=2/D11/(A11−B11)*LN(B11/A11* (A11−0.5*C11)/(B11−0.5*C11))
12					
13	0.2281	0.1517	0.0322	19.5	=2/D13/(A13−B13)*LN(B13/A13* (A13−0.5*C13)/(B13−0.5*C13))
14	0.2281	0.1517	0.0527	34.5	=2/D14/(A14−B14)*LN(B14/A14* (A14−0.5*C14)/(B14−0.5*C14))
15	0.2281	0.1517	0.0713	54.5	=2/D15/(A15−B15)*LN(B15/A15* (A15−0.5*C15)/(B15−0.5*C15))
16	0.2281	0.1517	0.0912	79.5	=2/D16/(A16−B16)*LN(B16/A16* (A16−0.5*C16)/(B16−0.5*C16))
17	0.2281	0.1517	0.104	99.5	=2/D17/(A17−B17)*LN(B17/A17* (A17−0.5*C17)/(B17−0.5*C17))
18	0.2281	0.1517	0.1142	124.5	=2/D18/(A18−B18)*LN(B18/A18* (A18−0.5*C18)/(B18−0.5*C18))
19	0.2281	0.1517	0.1217	149.5	=2/D19/(A19−B19)*LN(B19/A19* (A19−0.5*C19)/(B19−0.5*C19))
20	0.2281	0.1517	0.1295	174.5	=2/D20/(A20−B20)*LN(B20/A20* (A20−0.5*C20)/(B20−0.5*C20))
21					
22	0.3103	0.5069	0.0492	15	=2/D22/(A22−B22)*LN(B22/A22* (A22−0.5*C22)/(B22−0.5*C22))
23	0.3103	0.5069	0.1031	35	=2/D23/(A23−B23)*LN(B23/A23* (A23−0.5*C23)/(B23−0.5*C23))

(continued overleaf)

	A	B	C	D	E
24	0.3103	0.5069	0.1406	55	=2/D24/(A24−B24)*LN(B24/A24*(A24−0.5*C24)/(B24−0.5*C24))
25	0.3103	0.5069	0.1752	80	=2/D25/(A25−B25)*LN(B25/A25*(A25−0.5*C25)/(B25−0.5*C25))
26	0.3103	0.5069	0.1963	102	=2/D26/(A26−B26)*LN(B26/A26*(A26−0.5*C26)/(B26−0.5*C26))
27	0.3103	0.5069	0.2179	125	=2/D27/(A27−B27)*LN(B27/A27*(A27−0.5*C27)/(B27−0.5*C27))
28	0.3103	0.5069	0.236	155	=2/D28/(A28−B28)*LN(B28/A28*(A28−0.5*C28)/(B28−0.5*C28))
29	0.3103	0.5069	0.2533	196	=2/D29/(A29−B29)*LN(B29/A29*(A29−0.5*C29)/(B29−0.5*C29))

Here are the results:

	A	B	C	D	E
3	$[H_2]°$	$[Br_2]°$	[HBr]	tau	k_2
4	0.5637	0.2947	0.0699	14.5	0.03191
5	0.5637	0.2947	0.0985	24.5	0.027749
6	0.5637	0.2947	0.1262	34.5	0.026342
7	0.5637	0.2947	0.1644	49.5	0.025444
8	0.5637	0.2947	0.2093	79.5	0.021819
9	0.5637	0.2947	0.2306	99.5	0.019989
10	0.5637	0.2947	0.2502	124.5	0.01801
11	0.5637	0.2947	0.2619	149.5	0.016076
12					
13	0.2281	0.1517	0.0322	19.5	0.052353
14	0.2281	0.1517	0.0527	34.5	0.051628
15	0.2281	0.1517	0.0713	54.5	0.047042
16	0.2281	0.1517	0.0912	79.5	0.044285
17	0.2281	0.1517	0.104	99.5	0.04236
18	0.2281	0.1517	0.1142	124.5	0.038715
19	0.2281	0.1517	0.1217	149.5	0.035441
20	0.2281	0.1517	0.1295	174.5	0.033406
21					
22	0.3103	0.5069	0.0492	15	0.022279
23	0.3103	0.5069	0.1031	35	0.021633
24	0.3103	0.5069	0.1406	55	0.019903
25	0.3103	0.5069	0.1752	80	0.018055
26	0.3103	0.5069	0.1963	102	0.01646
27	0.3103	0.5069	0.2179	125	0.015504
28	0.3103	0.5069	0.236	155	0.014012
29	0.3103	0.5069	0.2533	196	0.012302

Similarly, let's assume that r_{HBr} is given by

$$r_{HBr} = k_{1.5}[H_2][Br_2]^{1/2} \tag{3.K.13}$$

Substituting equations (3.K.7) and (3.K.8) into (3.K.13) yields

$$r_{HBr} = k_{1.5}([H_2]^0 - 0.5[HBr])([Br_2]^0 - 0.5[HBr])^{1/2} \tag{3.K.14}$$

Substituting equation (3.K.14) into (3.K.10), looking up the integral in Gradsteyn and Ryzhik or Mathematica, and rearranging yields for $[H_2]^0 = [Br_2]^0$:

$$k_{1.5} = \frac{4}{\tau}\left(\frac{1}{([Br_2]^0 - 0.5[HBr])^{0.5}} - \frac{1}{([Br_2]^0)^{0.5}}\right) \tag{3.K.15}$$

for $[H_2]^0 > [Br_2]^0$:

$$k_{1.5} = \frac{4}{\tau\sqrt{[H_2^0] - [Br_2]^0}}$$

$$\times \left(\arctan\left(\frac{\sqrt{[Br_2]^0}}{\sqrt{[H_2^0] - [Br_2]^0}}\right) - \arctan\left(\frac{\sqrt{[Br_2]^0 - 0.5[HBr]}}{\sqrt{[H_2^0] - [Br_2]^0}}\right)\right) \tag{3.K.16}$$

for $[Br_2]^0 > [H_2]^0$:

$$k_{1.5} = \frac{2}{\tau\sqrt{([Br_2]^0 - [H_2]^0)^{0.5}}}$$

Table 3.K.3   The Excel module used to calculate $k_{1.5}$

```
Public Function kk(h20, br20, hbr, tau) As Variant
If (br20 = h20) Then
  kk = 4#/tau*(1#/Sqr(br20 - 0.5*hbr) - 1#/Sqr(br20))
  Exit Function
End If

If (h20 > br20) Then
  kk = Atn(Sqr(br20)/Sqr(h20 - br20))
  kk = kk - Atn(Sqr(br20 - 0.5*hbr)/Sqr(h20 - br20))
  kk = kk*4#/tau/Sqr(h20 - br20)
  Exit Function
End If

x1 = Sqr(br20 - 0.5*hbr) + Sqr(br20 - h20)
x2 = Sqr(br20) - Sqr(br20 - h20)
x3 = Sqr(br20 - 0.5*hbr) - Sqr(br20 - h20)
x4 = Sqr(br20) + Sqr(br20 - h20)
kk = 2/tau/Sqr(br20 - h20)*Log((x1*x2)/(x3*x4))

End Function
```

$$\times \ln \left(\frac{ \{([Br_2]^0 - 0.5[HBr])^{0.5} + ([Br_2]^0 - [H_2]^0)^{0.5} \} }{ \dfrac{ \{([Br_2]^0)^{0.5} - ([Br_2]^0 - [H_2]^0)^{0.5} \} }{ \{([Br_2]^0 - 0.5[HBr])^{0.5} - ([Br_2]^0 - [H_2]^0)^{0.5} \} } } \{([Br_2]^0)^{0.5} + ([Br_2]^0 - [H_2]^0)^{0.5} \} \right) \quad (3.K.17)$$

Again, I used a spreadsheet to do the calculations:

This function is so complicated that I found it easier to write an Excel module to calculate $k_{1.5}$. Table 3.K.3 is a printout of my module.

Table 3.K.4 is the spreadsheet using the Excel module. I listed the values of $[H_2]^0$ in column A, $[Br_2]^0$ in column B, $[HBr]$ in column C, and τ in column D. I then used my function to calculate the values of $k_{1.5}$ in column E.

Next, I show a speadsheet that calculates the formulas by hand.

Table 3.K.4 The formulas in the spreadsheet used to calculate $k_{1.5}$

	A	B	C	D	E
3	H_2^0	Br_2^0	[HBr]	tau	k_1.5
4	0.5637	0.2947	0.0699	14.5	=kk(A4,B4,C4,D4)
5	0.5637	0.2947	0.0985	24.5	=kk(A5,B5,C5,D5)
6	0.5637	0.2947	0.1262	34.5	=kk(A6,B6,C6,D6)
7	0.5637	0.2947	0.1644	49.5	=kk(A7,B7,C7,D7)
8	0.5637	0.2947	0.2093	79.5	=kk(A8,B8,C8,D8)
9	0.5637	0.2947	0.2306	99.5	=kk(A9,B9,C9,D9)
10	0.5637	0.2947	0.2502	124.5	=kk(A10,B10,C10,D10)
11	0.5637	0.2947	0.2619	149.5	=kk(A11,B11,C11,D11)
12					
13	0.2281	0.1517	0.0322	19.5	=kk(A13,B13,C13,D13)
14	0.2281	0.1517	0.0527	34.5	=kk(A14,B14,C14,D14)
15	0.2281	0.1517	0.0713	54.5	=kk(A15,B15,C15,D15)
16	0.2281	0.1517	0.0912	79.5	=kk(A16,B16,C16,D16)
17	0.2281	0.1517	0.104	99.5	=kk(A17,B17,C17,D17)
18	0.2281	0.1517	0.1142	124.5	=kk(A18,B18,C18,D18)
19	0.2281	0.1517	0.1217	149.5	=kk(A19,B19,C19,D19)
20	0.2281	0.1517	0.1295	174.5	=kk(A20,B20,C20,D20)
21					
22	0.3103	0.5069	0.0492	15	=kk(A22,B22,C22,D22)
23	0.3103	0.5069	0.1031	35	=kk(A23,B23,C23,D23)
24	0.3103	0.5069	0.1406	55	=kk(A24,B24,C24,D24)
25	0.3103	0.5069	0.1752	80	=kk(A25,B25,C25,D25)
26	0.3103	0.5069	0.1963	102	=kk(A26,B26,C26,D26)
27	0.3103	0.5069	0.2179	125	=kk(A27,B27,C27,D27)
28	0.3103	0.5069	0.236	155	=kk(A28,B28,C28,D28)
29	0.3103	0.5069	0.2533	196	=kk(A29,B29,C29,D29)

	A	B	C	D	E
3	tau	$[H_2]°$	$[Br_2]°$	[HBr]	$k_{1.5}$
4	14.5	0.5637	0.2947	0.0699	=4/D4/SQRT(A4−B4)*(ATAN(SQRT(B4)/SQRT(A4−B4))−ATAN(SQRT(B4−0.5*C4)/SQRT(A4−B4)))
5	24.5	0.5637	0.2947	0.0985	=4/D5/SQRT(A5−B5)*(ATAN(SQRT(B5)/SQRT(A5−B5))−ATAN(SQRT(B5−0.5*C5)/SQRT(A5−B5)))
6	34.5	0.5637	0.2947	0.1262	=4/D6/SQRT(A6−B6)*(ATAN(SQRT(B6)/SQRT(A6−B6))−ATAN(SQRT(B6−0.5*C6)/SQRT(A6−B6)))
7	49.5	0.5637	0.2947	0.1644	=4/D7/SQRT(A7−B7)*(ATAN(SQRT(B7)/SQRT(A7−B7))−ATAN(SQRT(B7−0.5*C7)/SQRT(A7−B7)))
8	79.5	0.5637	0.2947	0.2093	=4/D8/SQRT(A8−B8)*(ATAN(SQRT(B8)/SQRT(A8−B8))−ATAN(SQRT(B8−0.5*C8)/SQRT(A8−B8)))
9	99.5	0.5637	0.2947	0.2306	=4/D9/SQRT(A9−B9)*(ATAN(SQRT(B9)/SQRT(A9−B9))−ATAN(SQRT(B9−0.5*C9)/SQRT(A9−B9)))
10	124.5	0.5637	0.2947	0.2502	=4/D10/SQRT(A10−B10)*(ATAN(SQRT(B10)/SQRT(A10−B10))−ATAN(SQRT(B10−0.5*C10)/SQRT(A10−B10)))
11	149.5	0.5637	0.2947	0.2619	=4/D11/SQRT(A11−B11)*(ATAN(SQRT(B11)/SQRT(A11−B11))−ATAN(SQRT(B11−0.5*C11)/SQRT(A11−B11)))
12					

	A	B	C	D	E
13	19.5	0.2281	0.1517	0.0322	=4/D13/SQRT(A13−B13)*(ATAN(SQRT(B13)/SQRT(A13−B13))−ATAN(SQRT(B13−0.5*C13)/SQRT(A13−B13)))
14	34.5	0.2281	0.1517	0.0527	=4/D14/SQRT(A14−B14)*(ATAN(SQRT(B14)/SQRT(A14−B14))−ATAN(SQRT(B14−0.5*C14)/SQRT(A14−B14)))
15	54.5	0.2281	0.1517	0.0713	=4/D15/SQRT(A15−B15)*(ATAN(SQRT(B15)/SQRT(A15−B15))−ATAN(SQRT(B15−0.5*C15)/SQRT(A15−B15)))
16	79.5	0.2281	0.1517	0.0912	=4/D16/SQRT(A16−B16)*(ATAN(SQRT(B16)/SQRT(A16−B16))−ATAN(SQRT(B16−0.5*C16)/SQRT(A16−B16)))
17	99.5	0.2281	0.1517	0.104	=4/D17/SQRT(A17−B17)*(ATAN(SQRT(B17)/SQRT(A17−B17))−ATAN(SQRT(B17−0.5*C17)/SQRT(A17−B17)))
18	124.5	0.2281	0.1517	0.1142	=4/D18/SQRT(A18−B18)*(ATAN(SQRT(B18)/SQRT(A18−B18))−ATAN(SQRT(B18−0.5*C18)/SQRT(A18−B18)))
19	149.5	0.2281	0.1517	0.1217	=4/D19/SQRT(A19−B19)*(ATAN(SQRT(B19)/SQRT(A19−B19))−ATAN(SQRT(B19−0.5*C19)/SQRT(A19−B19)))
20	174.5	0.2281	0.1517	0.1295	=4/D20/SQRT(A20−B20)*(ATAN(SQRT(B20)/SQRT(A20−B20))−ATAN(SQRT(B20−0.5*C20)/SQRT(A20−B20)))
21					
22	15	0.3103	0.5069	0.0492	=2/D22/SQRT(B22−A22)*LN((SQRT(B22−0.5*C22)+SQRT(B22−A22))*(SQRT(B22)−SQRT(B22−A22))/(SQRT(B22−A22)−SQRT(B22−0.5*C22))/(SQRT(B22)+SQRT(B22−A22)))

	A	B	C	D	E
23	35	0.3103	0.5069	0.1031	`=2/D23/SQRT(B23-A23)*LN((SQRT(B23-0.5*C23)+SQRT(B23)-SQRT(B23-0.5*C23))/(SQRT(B23-A23))-SQRT(B23-A23))/(SQRT(B23-A23))+SQRT(B23-A23)))`
24	55	0.3103	0.5069	0.1406	`=2/D24/SQRT(B24-A24)*LN((SQRT(B24-0.5*C24)+SQRT(B24)-SQRT(B24-0.5*C24))/(SQRT(B24-A24))-SQRT(B24-A24))/(SQRT(B24-A24))+SQRT(B24-A24)))`
25	80	0.3103	0.5069	0.1752	`=2/D25/SQRT(B25-A25)*LN((SQRT(B25-0.5*C25)+SQRT(B25)-SQRT(B25-0.5*C25))/(SQRT(B25-A25))-SQRT(B25-A25))/(SQRT(B25-A25))+SQRT(B25-A25)))`
26	102	0.3103	0.5069	0.1963	`=2/D26/SQRT(B26-A26)*LN((SQRT(B26-0.5*C26)+SQRT(B26)-SQRT(B26-0.5*C26))/(SQRT(B26-A26))-SQRT(B26-A26))/(SQRT(B26-A26))+SQRT(B26-A26)))`
27	125	0.3103	0.5069	0.2179	`=2/D27/SQRT(B27-A27)*LN((SQRT(B27-0.5*C27)+SQRT(B27)-SQRT(B27-0.5*C27))/(SQRT(B27-A27))-SQRT(B27-A27))/(SQRT(B27-A27))+SQRT(B27-A27)))`
28	155	0.3103	0.5069	0.236	`=2/D28/SQRT(B28-A28)*LN((SQRT(B28-0.5*C28)+SQRT(B28)-SQRT(B28-0.5*C28))/(SQRT(B28-A28))-SQRT(B28-A28))/(SQRT(B28-A28))+SQRT(B28-A28)))`
29	196	0.3103	0.5069	0.2533	`=2/D29/SQRT(B29-A29)*LN((SQRT(B29-0.5*C29)+SQRT(B29)-SQRT(B29-0.5*C29))/(SQRT(B29-A29))-SQRT(B29-A29))/(SQRT(B29-A29))+SQRT(B29-A29)))`

Here are the results:

	A	B	C	D	E
3	$[H_2]°$	$[Br_2]°$	$[HBr]$	tau	$k_{1.5}$
4	0.5637	0.2947	0.0699	14.5	0.016782
5	0.5637	0.2947	0.0985	24.5	0.014386
6	0.5637	0.2947	0.1262	34.5	0.013456
7	0.5637	0.2947	0.1644	49.5	0.012715
8	0.5637	0.2947	0.2093	79.5	0.010596
9	0.5637	0.2947	0.2306	99.5	0.009564
10	0.5637	0.2947	0.2502	124.5	0.008494
11	0.5637	0.2947	0.2619	149.5	0.007513
12					
13	0.2281	0.1517	0.0322	19.5	0.019823
14	0.2281	0.1517	0.0527	34.5	0.01916
15	0.2281	0.1517	0.0713	54.5	0.017116
16	0.2281	0.1517	0.0912	79.5	0.015742
17	0.2281	0.1517	0.104	99.5	0.014815
18	0.2281	0.1517	0.1142	124.5	0.013355
19	0.2281	0.1517	0.1217	149.5	0.012096
20	0.2281	0.1517	0.1295	174.5	0.01127
21					
22	0.3103	0.5069	0.0492	15	0.015663
23	0.3103	0.5069	0.1031	35	0.014984
24	0.3103	0.5069	0.1406	55	0.013632
25	0.3103	0.5069	0.1752	80	0.012232
26	0.3103	0.5069	0.1963	102	0.011073
27	0.3103	0.5069	0.2179	125	0.010352
28	0.3103	0.5069	0.236	155	0.009294
29	0.3103	0.5069	0.2533	196	0.008107

Neither fit perfectly, but equation (3.K.3) fits better. One does need to do an analysis of variance (ANOVA) to see if the difference is statistically significant.

3.18 SUGGESTIONS FOR FURTHER READING

Techniques for analysis of kinetic data are discussed in:

J. H. Van' Hoff, *Studies in Chemical Dynamics*, Edward Arnold, London, 1896.

K. J. Laidler, *Chemical Kinetics*, Harper & Row, New York, 1987.

J. I. Steinfeld, J. S. Fransisco, and W. L. Hase, *Chemical Kinetics and Dynamics*, 2nd ed., Prentice Hall, Upper Saddle River, ND, 1998.

S. Fogler, *Elements of Chemical Reaction Engineering*, 3rd ed., Prentice-Hall, Upper Saddle River ND, 1998.

S. R. Logan, *Fundamentals of Chemical Kinetics*, Longman, Essex (1996).

J. H. Espenson, *Chemical Kinetics and Reaction Mechanisms*, McGraw-Hill, New York, 1995.

A treatment of statistical methods and their pitfalls in kinetics is given in:

A Cornish-Bowden, *Analysis of Enzyme Kinetic Data*, Oxford University Press, Oxford, UK, 1995.

R. de Levie, When, why and how to use weighted least-squares, *J. Chem. Educ.*, **63** 10 (1986). *Grit. Rev. Analytic Chem.*, **30**, 59.

P. R. Bevington and D. K. Robinson, *Data Reduction and Error Analysis for the Physical Sciences*, McGraw-Hill, New York, 1992.

3.19 PROBLEMS

3.1 Define the following terms:

(a) Conversion	**(k)** CSTR
(b) Integral method	**(l)** Plug-flow reactor
(c) Differential method	**(m)** Residence time
(d) Direct method	**(n)** Batch method
(e) Indirect method	**(o)** Flow method
(f) Powell's method	**(p)** Stopped-flow method
(g) Essen's method	**(q)** Temperature-Jump method
(h) Van't Hoff's method	**(r)** Shock tube
(i) Half-life	**(s)** Flash photolysis
(j) Batch reactor	**(t)** Molecular beam

3.2 What did you learn in this chapter that was new? Had you heard of direct methods before? How about indirect methods? Were you aware of the inaccuracies in Essen's method? What else did you learn that was new?

3.3 Compare integral methods and differential methods for the analysis of rate data. What are the advantages and disadvantages of each method?

3.4 Compare the various experimental methods in Table 3.1. How does each method work? What determines how long it takes to initiate the reaction?

3.5 Find 10 examples of kinetic processes in your home, such as cooking different kinds of meals, washing clothes, you digesting different types of food, or plants growing on your windowsill.

   (a) What experiments would you do to determine a rate equation for each of them? What variables would you think were important?

   (b) Find an approximation to the rate equation for each of the reactions from your everyday observations.

   (c) Pick one of the examples and explain how you would go about determining a *direct* measurement of the rate equation. Be sure to say specifically what you would do.

   (d) For the same example, explain how you would go about determining an *indirect* measurement of the rate equation. Be sure to say specifically what you would do.

3.6 The half-life of tritium is 13.6 years.

    (a) Calculate a rate constant for the decomposition of tritium. Assume a first-order reaction.

    (b) How long will it take for 99.99% of the tritium to disappear?

3.7 Assume that you are working in the semiconductor industry. IBM just announced that they have a new process to replace aluminum with copper in their chips. Your boss tells you, "We need a copper process, too. Get me one." You look in the literature, and find that you can deposit copper via the chemical vapor deposition (CVD) reaction:

$$Cu(hfac)_2 + H_2 \Longrightarrow Cu + H(hfac)$$

where hfac is a hexafluoroacetylacetonate ligand. Your company already makes CVD reactors, so this seems like a good process for you to bring back to your boss. What would you do to measure the kinetics of the process to enable your company to sell a copper deposition process, too? Be sure to say what you would do during the experiment, what you would measure, and how you would analyze your data.

3.8 Assume that you are working in the pharmaceutical industry. Your company just started selling Interluken-II and noticed that it degrades when it sits in a bottle for about 4 weeks. How would you measure the kinetics of the process? Be sure to consider what criteria you would use to decide whether to make direct or indirect measurements, qualitatively how you would make the measurements, and how you will analyze your data.

3.9 A first-order polymerization reaction is being run in a batch reactor. A concentration of 0.007 mol/liter of monomer is loaded into the reactor, and then a catalyst is added to initiate the reaction. Experiments show that the reaction is 30% complete in 10 minutes.

    (a) Calculate the rate constant.

    (b) Calculate the half-life.

    (c) How long will it take for the reaction to be 90% complete?

    (d) How would the time in (c) change if you increased the concentration in the reactor to 0.16 mol/liter?

    (e) Repeat for a second-order reaction.

3.10 N_2O_5 can be made via oxidation of ammonia over a platinum gauze. You do an experiment and find that you get 50% conversion of the ammonia with a 0.1-second residence time (τ) in the reactor at 1000 K.

    (a) Estimate the rate constant for the reaction assuming that the reaction is first-order in the ammonia pressure and zero-order in the oxygen pressure.

    (b) How long of a residence time will you need to get to 90% conversion at 1000 K?

    (c) Now assume that the reaction is instead second-order in the ammonia pressure.

    (d) Estimate the rate constant for the reaction assuming 50% conversion in 0.1 second. Assume a stoichiometric feed at 1 atm pressure.

Table P3.12   Additional data for Example 3.12

Time, minutes	Concentration, mol/liter	Time, minutes	Concentration, mol/liter	Time, minutes	Concentration, mol/liter
30	0.25	60	0.14	90	0.10

(e) What would your conversion be if you used the residence time you calculated in part (b)?

(f) Calculate a rate constant that would give you 90% conversion for part (e).

(g) The results in (F) have a lot of industrial significance. People often design their reactors assuming that they have a first-order reaction, and then adjust the temperature of the reactor to get the conversion that they want. Explain how you could change the temperature to increase the rate constant for the reaction.

(h) Assume that you used your results in (a) to design your reactor, but in fact the reaction is second-order, so the actual conversion is the value you calculate in problem (e). How much would you have to increase the temperature to get 90% conversion?

3.11 In Example 3.A we fit some data for the growth of paramecium.

(a) Reproduce the results yourself. A suitable spreadsheet is available in the instructions materials.

(b) Change the first point. Assume that the measured rate at a paramecium concentration of 2 is 5.4, not 10.4. How will your results change?

(c) Next, compare the fits obtained with the various methods. How do the r^2 values compare? How do the variances compare?

(d) What do your results in (c) tell you about the influence of errors in data on the various methods to analyze data?

(e) Do an F test as in Example 3.B. Are the differences between the two models significant?

(f) Try the model in Example 3.C. How well does it work?

3.12 In Example 3.D we used a number of methods to analyze the rate data in Table 3.5.

(a) Reproduce the results yourself. A suitable spreadsheet is in the instructions materials.

(b) Assume that we have three more points as given in Table P3.12. How will that change your results?

(c) How does r^2 change?

(d) Now assume that you mixed up the point for time $= 90$ and recorded a concentration of 0.05 mol/liter. How will that change your results?

(e) According to the Essen plot, which model has the lowest value of r^2 with the one bad point?

(f) What do you conclude about the utility of r^2 as a way of assessing the reliability of kinetic data?

3.13 In Example 3.F, we used Van't Hoff's method to analyze the data in Table 3.F.1.

    (a) Set up your own spreadsheet to calculate the conversions from the data.

    (b) Verify the numbers in Table 3.F.2.

    (c) Verify the numbers in Table 3.F.3.

    (d) Analyze the data using Essen's method.

    (e) Analyze the data using Powell's method.

    (f) How do your results differ?

3.14 In Example 3.K, we used Van't Hoff's method to analyze the data in Table 3.K.1.

    (a) Set up your own spreadsheets and verify the results.

    (b) Analyze the data using Essen's method.

    (c) How do your results in (b) differ from those in Example 3.K?

3.15 Steger and Masel examined the etching of copper in a reactor used to produce electronic materials. The main reaction is $Cu + 2hfacH + 0.5O_2 \rightarrow H_2O + Cu(hfac)_2$ where hfac is a hexafluoroacetylacetonate ligand. The following data were obtained:

HfacH Pressure, torr	Etch Rate, μm/minute	HfacH Pressure, torr	Etch Rate, μm/minute	HfacH Pressure, torr	Etch Rate, μm/minute
0.25	0.031	0.35	0.081	0.45	0.113
0.30	0.055	0.40	0.099	0.50	0.126

    (a) Fit these data to equation (2.12) to determine the order of the reaction.

    (b) Steger and Masel also measured the temperature dependence of the rate, and obtained the following data:

Temperature, K	Etch rate, μm/minute	Temperature K	Etch rate, μm/minute	Temperature K	Etch rate, μm/minute
548	0.101	573	0.132	598	0.189
563	0.123	583	0.162	613	0.214

    Estimate the activation barrier for the reaction.

    (c) How well do these data fit Perrin's equation? Which fits better, Arrhenius' law or Perrin's equation?

    (d) Does the activation barrier agree with equation (2.31)? What is the significance of this result?

3.16 Assume that you have modeled a reaction, $A + B \Rightarrow$ products, and find that it follows the rate equation

$$-r_B = \frac{k_1 K_2[A]}{1 + K_2[A]}(k_4 + K_3[B])$$

with known values of k_1 and K_2. You do not know K_3 and k_4 so you decide to go into the lab and measure it. Your data are given in Table P3.16.

(a) Use linear regression to estimate a value of K_3. (*Hint*: Plot $r_B/[(k_1K_2[A]/1 + K_2[A])]$ versus [B].)

(b) How good is your regression coefficient?

(c) Make a plot of the calculated rate versus the predicted rate. How well does the model actually fit the data? Now assume that the reaction follows the rate equation:

$$-r_B = \left(\frac{k_1 K_2[A]}{1 + K_2[A]} \right)^2 (k_4 + K_3[B])$$

(d) Use linear regression to estimate a value of K_3. Hint: Plot $r_b/\left(\dfrac{k_1 K_2[A]}{1 + K_2[A]} \right)^2$ versus [B].

(e) How good is your regression coefficient?

(f) Make a plot of the calculated rate versus the predicted rate. How well does the model actually fit the data?

(g) Notice that the first model fits the data to two significant figures, even though the regression coefficient is 4×10^{-5}. In contrast, the second model has a much better regression coefficient but does not fit the data at all. What does this result tell you about the utility of using regression coefficients to distinguish between kinetic models?

(h) Use the variances to see which model works best.

(i) Do an F test to see if the difference is statistically significant.

3.17 Table P3.17 gives Schneider and Rabinovitz' data for the isomerization of CH_3NC to CH_3CN.

(a) Try to fit the data with a simple first- or second-order rate law. How well does it work?

(b) Try fitting the data to rate $= k_1[CH_3NC]^2/(1 + K_2[CH_3NC])$. How well does the equation fit? (*Hint*: You could plot $[CH_3NC]/$rate vs. $1/[CH_3NC]$. However, I find it more accurate to simply program the rate equation in a spreadsheet and use the solver function to find k_1 and K_2 until the rate equation fits all the data.)

(c) Are the differences statistically significant? Do an F test on the error in the natural logarithm of the rate.

Table P3.16   Rate data for Example 3.16

[B], mol/liter	$k_1K_2[A]/1 + K_2[A]$, mol/(liter·hour)	$-r_B$ mol/(liter·hour)	[B], mol/liter	$k_1K_2[A]/1 + K_2[A]$, mol/(liter·hour)	$-r_B$ mol/(liter·hour)
0.25	1.001	1.0	2.8	5.002	5.0
1.5	2.001	2.0	3.5	6.001	6.0
2	3.000	3.0	4.6	7.003	7.0
2.3	4.001	4.0	5	8.008	8.0

Table P3.17   The rate of methyl isocyanide isomerization

Methylisocinide Pressure (mol/liter)	Rate (mol/liter)	Methyliscocinide Pressure (mol/liter)	Rate (mol/liter)
10,520	9.8	18.1	0.0047
10,250	9.4	10.1	0.0019
9,880	9.1	8	0.0012
5,580	5.1	7.14	0.0010
4,020	3.5	5.1	0.00062
3,850	3.5	2.2	0.00014
3,610	3.3	1.39	0.000067
3,580	3.2	1.05	0.000039
1,757	1.5	0.95	0.000036
1,349	1.2	0.59	0.000014
1,050	0.85	0.56	0.000012
486	0.39	0.41	0.0000073
309	0.23	0.286	0.0000036
222	0.15	0.272	0.0000035
100	0.05	0.13	0.00000092
80.6	0.04	0.101	0.00000054
59.6	0.027	0.0876	0.00000040
40.8	0.015	0.0725	0.00000029
29.8	0.010		

Source: Data of Schneider and Rabinovitz (1962).

3.18   In our undergraduate labs, we measure the rate of oxidation of Red Dye 40 with bleach. The main reaction is

$$\text{Red Dye} + ClO^- \longrightarrow Cl^- + \text{Yellow Dye} + H_2O$$

Over the years, we have done many different measurements, and the data in Table P3.18 were obtained:

The objective of this problem is to fit the data to equation (2.13) and determine the order of the reaction in bleach and dye. The easiest way to solve this problem is to use the regression capabilities of your spreadsheet.

Table P3.18   Rate data for Example 3.18

Dye Concentration, mol/liter	Bleach Concentration, mol/liter	Rate, mol/(liter·minute)	Dye Concentration, mol/liter	Bleach Concentration, mol/(liter·minute)	Rate, mol/liter
0.011	0.031	0.018	0.033	0.030	0.053
0.015	0.0315	0.023	0.034	0.039	0.073
0.018	0.0270	0.024	0.039	0.044	0.092
0.022	0.039	0.041	0.041	0.051	0.115
0.023	0.036	0.032	0.045	0.024	0.053
0.025	0.009	0.009	0.044	0.010	0.028
0.028	0.0189	0.023	0.052	0.052	0.145

(a) Convert equation (2.13) so that you can use linear regression.

(b) Set up your spreadsheet to do the regression using the Data-Analysis/Regression tool in Microsoft Excel.

(c) Try nonlinear regression as in Table 3.A.4 to see how that changes your answers.

3.19 Commercial sterilizers work by heating bacteria to high temperatures where the bacteria die. The FDA (U.S. Food and Drug Administration) requires all sterilizers to meet a standard of an overkill of 10^{12}; specifically, that each bacterial or bacteria spore has one chance in 10^{12} of surviving. Generally people test sterilizers with a thermobacteria spore that is particularly able to survive high temperatures. It is hard to detect a 10^{12} overkill, so people measure the time to a 10^{6} overkill and assume that if they double the sterilization time, a 10^{12} overkill will be achieved.

(a) Show that if the death of bacteria follow a first-order rate law, the time to achieve a 10^{12} overkill is twice the time to achieve a 10^{6} overkill.

(b) What will the overkill be if the reaction is instead second-order? (*Hint*: Assume an initial concentration of $10^{8}/cm^{3}$. At a 10^{6} overkill, you need to get to a final concentration of $10^{2}/cm^{3}$. At a 10^{12} overkill, you need to get to a final concentration of $10^{-4}/cm^{3}$. Calculate the time in each case.)

(c) You can increase the overkill by increasing the temperature. How much would you have to increase the temperature to get the overkill up to 10^{12} in the case in (b)?

(d) Assume that you are a canned milk manufacturer who uses a sterilizer to kill the bacteria in the cans before the cans leave your plant. The cans start out with 10,000 thermobacteria each. If the reaction is first-order, what fraction of the cans will have at least one bacterium left after sterilization?

(e) If you produce 50,000,000 cans/per year, how many will go bad?

(f) How would your results in (e) change if the reaction were second-order?

3.20 Ammonium dinitramide (ADN), $NH_4N(NO_2)_2$, is an oxidant used in solid fuel rockets and plastic explosives. The ADN is difficult to process because it can blow up. Oxley et al. *J. Phys Chem A*, **101** (1997) 5646, examined the decomposition of ADN to try to understand the kinetics of the explosion process. At 160°C they obtained the data in Table P3.20.

(a) Is this a direct or indirect measurement of the rate?

(b) Use Essen's method to fit these data to a rate equation. Assume an initial concentration of 10^{-3} molar.

Table P3.20   Oxley's measurements of the decomposition of dinitramide at 160°C

Time, seconds	Fraction of the ADN Remaining	Time, seconds	Fraction of the ADN Remaining	Time, seconds	Fraction of the ADN Remaining
0	1.0	900	0.58	2400	0.24
300	0.84	1200	0.49	—	—
600	0.70	1500	0.41	—	—

(c) Use Van't Hoff's method to fit these data to a rate equation.

(d) Use Powell's method to fit these data to a rate equation.

(e) If you had to process ADN at 160°C, how long could you run the process without blowing anything up? Assume that there is an explosion hazard once 5% of the ADN has reacted to form unstable intermediates.

(f) If you wanted to process for 5 minutes, what temperature would you choose? Assume that the reaction follows Arrhenius' law with a preexponential of 10^{13}/second. (*Hint:* First, estimate the activation energy from your value of the rate constant and the known preexponential.)

3.21 Chlebicki, et al. *Int J. Chem. Kinetics*, **29** (1997) 73, examined the sodium cresolate (S) catalyzed decomposition of epichlorohydrin (E). At 71°C they obtained the results in Table P3.21.

(a) Is this a direct or indirect measurement of the rate?

(b) Fit these data to a rate equation. (*Hint:* Assume that C_S is constant during each run. First, fit the rate data at each C_S to a rate equation, and then determine how the rate constant varies with C_S. Assume an initial concentration of 0.1 mol/liter.)

3.22 In Problem 3.21 we noted that Chlebicki, et al. *Int J. Chem. Kinetics*, examined the sodium cresolate (S)–catalyzed decomposition of epichlorohydrin (E) in a batch reactor. However, they could have instead run the reaction in a CSTR.

(a) Explain what they would have needed to do to measure the rate in a CSTR.

(b) What value of the residence time, τ, will give a conversion of 0.45 at $C_S = 0.76$ mol/dm^3 (mole per cubic decimeter)?

3.23 Bodenstein and Lund *Z. Physik Chem*, **57**, (1907)(168), examined the kinetics of the reaction

$$H_2 + Br_2 \Longrightarrow 2HBr \qquad\qquad (P3.23.1)$$

by loading equal amounts of bromine an hydrogen into a reactor and measuring the concentration as a function of time. Table P3.23 shows some of their data.

Table P3.21   The decomposition of epichlorhydrin in the presence of sodium cresolate

$C_S = 1.2$ mol/dm^3		$C_S = 0.88$ mol/dm^3		$C_S = 0.76$ mol/dm^3		$C_S = 0.65$ mol/dm^3	
Time, minutes	Fraction of the E Remaining	Time, minutes	Fraction of the E Remaining	Time, minutes	Fraction of the E Remaining	Time, minutes	Fraction of the E Remaining
0	1.0	0	1	0	1.0	0	1.0
5	0.90	5	0.93	5	0.94	5	0.94
15	0.74	15	0.80	15	0.82	15	0.84
25	0.61	25	0.69	25	0.72	25	0.75
35	0.50	35	0.59	35	0.63	35	0.66
45	0.41	45	0.51	45	0.55	45	0.59
55	0.33	55	0.44	55	0.48	55	0.53

(a) Is this a direct or indirect measurement of the rate?

(b) Use Essen's method to fit these data to a simple rate equation.

(c) Use Van't Hoff's method to fit these data to a simple rate equation.

(d) Use Powell's method to fit these data to a simple rate equation.

(e) What do you conclude from the nonlinearity of your plots?

3.24 In Problem 3.23 we noted that Bodenstein and Lund Z. *Physik Chem*, **57**, (1907) 168, examined the kinetics of the reaction

$$H_2 + Br_2 \Longrightarrow 2HBr \tag{P3.24.1}$$

by loading equal amounts of bromine an hydrogen into a reactor and measuring the concentration as a function of time. Table P3.23 shows some of their data. Bodenstein and Lund fit their data to the expression

$$r_{HBr} = \frac{k_1[H_2][Br_2]^{1/2}}{1 + K_2\dfrac{[HBr]}{[Br_2]}} \tag{P3.24.2}$$

(a) Use the stoichometric table to derive an expression for $[H_2]$ and $[HBr]$ as a function of the Br_2 conversion.

(b) Plug into equation (P3.24.2) to prove

$$\frac{dX_{Br_2}}{dt} = \frac{K_1(1 - X_{Br_2})^{1/2}(C_{H_2}^0 - X_{Br}C_{Br_2}^0)}{1 + 2K_2\dfrac{X_{Br_2}}{(1 - X_{Br_2})}} \tag{P3.24.3}$$

where X_{Br_2} is the conversion of Br_2 and $C_{H_2}^0$ and $C_{Br_2}^0$ are the initial H_2 and Br_2 concentrations.

(c) Show that the solution of equation (P3.24.3) is

$$\frac{C_{Br_2}^0 k_1}{4K_2}\tau = \left(\frac{1 - 2K_2}{2K_2\sqrt{1 - X_{Br_2}}} + \frac{1}{3(1 - X_{Br_2})^{3/2}} - \frac{1}{2K_2} + \frac{2}{3} \right)$$

when $C_{Br_2}^0 = C_{H_2}^0$

$$\frac{C_{H_2}^0 k_1}{4K_2}\tau = \frac{\sqrt{C_{Br_2}^0}}{\sqrt{C_{H_2}^0 - C_{Br_2}^0}} \left(\frac{1 - 2K_2}{2K_2} - \frac{C_{Br_2}^0}{C_{H_2}^0 - C_{Br_2}^0} \right)$$

Table P3.23 Bodenstein and Lund's data for the reaction $H_2 + Br_2 \Rightarrow 2HBr$

Time, minutes	$[H_2]=[Br_2]$, mol/liter	Time, minutes	$[H_2]=[Br_2]$, mol/liter	Time, minutes	$[H_2]=[Br_2]$, mol/liter
0	0.2250	90	0.1158	300	0.0478
20	0.1898	128	0.0967	420	0.0305
60	0.1323	180	0.0752		

$$\times \arctan\left(\frac{(1 - \sqrt{1 - X_{Br_2}})(\sqrt{C_{H_2}^0 - C_{Br_2}^0}\sqrt{C_{Br_2}^0})}{(C_{H_2}^0 - C_{Br_2}^0 + C_{Br_2}^0\sqrt{1 + X_{Br_2}})}\right)$$

$$+ \frac{C_{Br_2}^0}{C_{H_2}^0 - C_{Br_2}^0}\left(\frac{1}{\sqrt{1 - X_{Br_2}}} - 1\right)$$

when $C_{H_2}^0 > C_{Br_2}^0$

$$\frac{C_{H_2}^0 k_1}{2K_2}\tau = \frac{\left(\left(\dfrac{1 - 2K_2}{K_2}\right) + \left(\dfrac{C_{Br_2}^0}{C_{Br_2}^0 - C_{H_2}^0}\right)\right)\sqrt{C_{Br_2}^0}}{\sqrt{C_{Br_2}^0 - C_{H_2}^0}}$$

$$\times \ln\left(\frac{(\sqrt{C_{Br_2}^0}\sqrt{1 - X_{Br_2}} + \sqrt{C_{Br_2}^0 - C_{H_2}^0})(C_{Br_2}^0 - \sqrt{C_{Br_2}^0 - C_{H_2}^0})}{(\sqrt{C_{Br_2}^0}\sqrt{1 - X_{Br_2}} - \sqrt{C_{Br_2}^0 - C_{H_2}^0})(C_{Br_2}^0 - \sqrt{C_{Br_2}^0 - C_{H_2}^0})}\right)$$

$$- \frac{2C_{Br_2}^0}{C_{Br_2}^0 - C_{Hr_2}^0}\left(\frac{1}{\sqrt{1 - X_{Br_2}}} - 1\right)$$

(d) Use your results in (c) to devise new Van't Hoff and Essen plots for the reaction.

(e) Construct the Van't Hoff and Essen plots and see if they work.

3.25 Ranley, Rust and Vaughn, *JACS* **70** (1948) 88 examined the decomposition of di-tertiary butyl peroxide. The main reaction is

$$(CH_3)_3COOC(CH_3)_3 \Longrightarrow 2CH_3COCH_3 + C_2H_6 \qquad (P3.25.1)$$

Ranley et al. loaded the ditertiarybutyl peroxide into a batch reactor at 154.6°C and measured the pressure as a function of time. They obtained the data listed in Table P3.25.

(a) Is this a direct or indirect measurement of the rate?

(b) Develop a stoichiometric table for the reaction.

(c) Calculate the conversion as a function of time from the data in Table (P3.25).

(d) Use Essen's method to fit these data to a simple rate equation.

(e) Use Van't Hoff's method to fit these data to a simple rate equation.

(f) Use Powell's method to fit these data to a simple rate equation.

(g) How long of a residence time would you need to decompose 99.9% of the ditertiarybutyl peroxide in a CSTR?

3.26 Silicon dioxide (SiO_2) films are used as dielectrics in electronic devices. SiO_2 films are made by decomposing TEOS [tetraethylorthosilicate, $(Si(OC_2H_5)_4)$] on a silicon wafer. Kim and Gill *J. Electrochemical Society* **142** (1995) 676, examined the thermal decomposition of TEOS in a microbalance. The data shown in Table P3.2.6 were obtained.

Table P3.25   Raney, et al.'s data for reaction (P3.25)

Time, minutes	Pressure, atm	Time, minutes	Pressure, atm	Time, minutes	Pressure, atm
0	0.223	9	0.295	18	0.355
3	0.249	12	0.316	21	0.372
6	0.273	15	0.336	—	—

Table P3.26   The rate of SiO$_2$ deposition from TEOS at 1070 K

TEOS Pressure, torr	Deposition Rate, µg/hour	TEOS Pressure, torr	Deposition Rate, µg/hour	TEOS Pressure, torr	Deposition Rate, µg/hour
0.15	148	0.29	175	0.42	190
0.55	200	0.68	209	0.81	215

(a) Is this a direct or indirect measurement of the rate?

(b) What is the order of the reaction?

(c) Kim and Gill fit the data to

$$r_{SiO_2} = \frac{k_1 P_{TEOS}^{1/2}}{1 + k_2 P_{TEOS}^{1/2}} \qquad (P3.26.1)$$

How well does equation (P3.26) fit the data?

(d) Assume that the reaction follows equation (P3.26.1). Derive an equation for the TEOS pressure as a function of time when TEOS is loaded in a batch reactor and the reactor is heated to 1070°C. Assume the following overall reaction:

$$Si(OC_2H_5)_4 \Longrightarrow SiO_2 + 2(C_2H_5)_2O$$

where $Si(OC_2H_5)$ is TEOS. Calculate the TEOS pressure as a function of time, starting with an initial TEOS pressure of 1 torr. Calculate for a long enough period that 60% of the TEOS is used up.

(e) Use Essen's method to fit your results in (d) to a zero-order, half-order, first-order, or second-order rate equation. How well do your calculated results fit zero-order, half-order, first-order or second-order rate expressions?

(f) Do an F test to see which model fits best.

3.27 Chung and Lu, *J. Polymer Sci A*, **36** (1998) 1017, studied the production of a polyethylene-styrene copolymer. Chung and Lu loaded ethylene into a reactor, added a small amount of styrene, and then initiated the reaction. Chung and Lu then measured the conversion of styrene as a function of time. They repeated the same experiments substituting methylstyrene for styrene. (The styrene and methylstyrene runs were done separately.) Some of Chung and Lu's results are given in Table P3.27.

(a) Is this a direct or indirect measurement of the rate?

(b) Use Essen's method to fit these data to a rate equation.

Table P3.27 The conversion of styrene
as a function of time and the conversion
of methylstyrene as a function of time,
and reported by Chung and Lu

Time, minutes	Styrene Conversion	Methylstyrene Conversion
5	—	13
15	19	27
30	38	55
45	52	70
60	60	85

(c) Use Van't Hoff's method to fit these data to a rate equation.

(d) Use Powell's method to fit these data to a rate equation.

(e) How long would you have to run to get 90% conversion of styrene?

3.28 The growth of bacteria is often thought to follow Monod kinetics, where the growth rate of bacteria, r_B, in bacteria/(liter·hour) is related to the bacteria concentration by

$$r_B = k_B[B]\frac{K_F[F]}{1 + K_F[F]} \tag{P3.28.1}$$

where [B] is the bacteria concentration in bacteria/liter and [F] is the food concentration in mol/liter.

(a) What are the units of k_B and K_F?

(b) Assume that the bacteria are growing under conditions where there is a large excess of food ($K_F[F] \gg 1$). Develop an equation expressing how quickly the population of bacteria doubles.

(c) In the literature, it is common to report values at a constant k_G, where k_G is given by

$$k_G = \frac{r_B}{B}$$

Chang and Hong, *J. Biotechnology*, **42** (1995) 189, examined the growth of a potentially toxic bacteria, *Pseudomonas aeruginosa*, PU21 in a glucose solution, and obtained the data in Table P3.28. How well do these data fit Monod kinetics?

3.29 The adsorption and destruction of alcohol in a human body can be modeled as two first-order reactions in series. When you drink an alcoholic beverage, the alcohol in the beverage reacts with the blood in your stomach walls to yield an alcohol/blood complex. The alcohol/blood complex is then quickly transported throughout your entire body, including your liver. In a second process, the enzymes in your liver break down the alcohol into other products.

(a) Assume that the adsorption and destruction of alcohol are first-order processes. Use the equations in this chapter to obtain an expression for your blood alcohol level as a function of time.

Table P3.28 k_G for the growth of *Pseudomonas aeruginosa* PU21 in glucose solution

Glucose Concentration grams/ml	k_G, hour^{-1}	Glucose Concentration grams/ml	k_G, hour^{-1}	Glucose Concentration grams/ml	k_G, hour^{-1}
0.0	0.0	0.7	0.80	2.25	0.93
0.01	0.4	1.0	0.82	4	0.98
0.05	0.65	1.25	0.85	8	1.0
0.3	0.74	2	0.91		

(b) How would you get values for the rate constants in your model? Assume that you had a Breathalyzer available. What experiments would you do? Would you do experiments only on yourself, or would you include friends?

(c) Welling et al. *J. Clinical Pharmacology* **17** (1977) 199, measured the alcohol level in subjects and the data in Table P3.29 were obtained. How well does your model fit Welling's data?

(d) What is the maximum amount of alcohol you can drink and still keep your blood alcohol level below 0.1 gram/liter? (*Hint*: I am looking for you to calculate the answer, NOT do an experiment.)

3.30 Aranda et al. *Int J. Chemical kineties* **30** (1998) 249, used the swamping method to measure the kinetics of the reaction

$$CH_3O + Br \longrightarrow CH_2O + HBr \qquad (P3.30.1)$$

They produced the bromine atoms via a reaction with chlorine, while the CH_3O was produced via photolysis of methanol. They ran their experiments so that there was always a large excess of bromine in the reactor.

(a) Is this a direct or an indirect measurement of the rate equation?

(b) Set up a stoichiometric table for the reaction.

(c) Assume that the kinetics follow

$$r_{CH_2O} = k_2[CH_3O][Br] \qquad (P3.30.2)$$

Table P3.29 Blood alcohol levels measured on a 75-kg test subject after fasting and then drinking 15 ml of 95% alcohol

Time, minutes	Blood Alcohol Concentration mg/liter	Time, minutes	Blood Alcohol Concentration, mg/liter	Time, minutes	Blood Alcohol Concentration, mg/liter
0	0	30	160	90	60
10	150	45	130	110	40
20	200	80	70	170	20

Source: Data of Welling et al. (1977).

Show that when there is a large excess of bromine atoms in the reactor, the CH_3O concentration will follow

$$\ln\left(\frac{[CH_3O]_0}{[CH_3O]}\right) = k_2[Br]t \qquad (P3.30.3)$$

where $[CH_3O]_0$ is the initial CH_3O concentration and t is time.

(d) Table P3.30.1 shows some data for the reaction. Use Essen's method to determine the order of the reaction.

(e) Use Van't Hoff's method to determine the order of the reaction.

(f) Use Powell's method to determine the order of the reaction.

(g) Aranda et al. also report values of $\ln\{([CH_3O]_0/[CH_3O])/t\}$ for various bromine concentrations. Table P3.30.2 shows the data. Use an Essen plot (i.e., $\ln\{([CH_3O]_0/[CH_3O])/t\}$ vs. [Br]) to see how well these data follow equation (P3.30.2).

(h) Repeat part (g) using a Van't Hoff plot (i.e., k_2 vs. [Br]).

(i) Can you find another rate law that fits the data better?

3.31 In Section 3.9.1, we derived an expression for the average rate of a reaction as a function of the conversion. The objective of this problem is to see how the average rate compares to the rate at the average concentration.

(a) Derive an expression for the average rate of a first-order reaction and a second-order reaction as a function of the conversion in a batch reactor.

Table P3.30.1   The CH_3O conversion versus time reported by Aranda et al

Time, ms	CH_3O Conversion [Br] = 2.22 $\times 10^{11}$ molecules/cm^3	CH_3O Conversion [Br] = 4.48 $\times 10^{11}$ molecules/cm^3	CH_3O Conversion [Br] = 6.52 $\times 10^{11}$ molecules/cm^3
0	0	0	0
0.5	0.08	0.15	0.21
1	0.15	0.28	0.38
1.5	0.21	0.39	0.51
2	0.27	0.48	0.61
2.5	0.33	0.56	0.7
3	0.38	0.63	0.76
5	0.55	0.81	0.91

Table P3.30.2   Values of $\{\ln([CH_3O]_0/[CH_3O])\}/t$ reported by Aranda et al

[Br] $\times 10^{11}$ molecules/cm^3	$\{\ln([CH_3O]_0/[CH_3O])\}/t$	[Br] $\times 10^{11}$ molecules/cm^3	$\{\ln([CH_3O]_0/[CH_3O])\}/t$
1.59	116	6.96	503
2.94	210	7.26	506
3.38	250	8.71	578
4.78	334	10.8	755
5.07	427	11.84	913
6.23	505	12.46	823

(b) Derive an expression for the rate at the average concentration.

(c) How do the two compare?

(d) Find a concentration where the average rate equals that rate at that concentration.

3.32 In Section 3.13 we derived a number of equations for the behavior of a reaction

$$A + B \Longrightarrow C + D \qquad \text{(P3.32.1)}$$

(a) Show that equation (3.60) goes to equation (3.61) in the limit that $[B] \gg [A]$.

(b) Show that equation (3.60) goes to equation (3.64) in the limit that $[A] = [B]$.

(c) How can you use the results in (a) and (b) to determine the kinetics of a reaction?

(d) Assume that you try to run reaction (P3.32.1) with $[A] = [B]$, but make a mistake so $[A] = 0.30$ mol/liter, and $[B] = 0.32$ mol/liter. Calculate the concentration as a function of time with $k_2 = 0.45$ liter/(mol·hour). Assume that your final A concentration is 0.01 mol/liter.

(e) Make an Essen plot of your results in (d) assuming that the reaction follows equation (3.64).

(f) Make a Van't Hoff plot of your results.

(g) Repeat for $[A] = 0.30$ $[B] = 0.62$.

(h) What do the results in (e)–(g) tell you about the utility of running the reaction with $[A] = [B]$?

3.33 Estenfelder, Lintz, Stein Gaube, *Chemical Engineering & Processing*, **37**, (1998) 109, compared the use of an integral and differential reactor to measure the partial oxidation of an unsaturated aldehyde.

(a) Describe the integral reactor use in these studies.

(b) Describe the differential reactor used in these studies.

(c) How do the data obtained by the two methods compare?

(d) Are there any unexpected findings in the paper?

(e) When do the authors say that each method should be used?

(f) How do the findings compare to your expectations from this chapter?

3.34 The hydrolysis of ethylacetate is a reversible reaction, which is catalyzed by acids. The main reaction is

$$CH_3COOCH_2CH_3 + H_2O \overset{H^+}{\Longrightarrow} CH_3COOH + HOCH_2CH_3 \qquad \text{(P3.34.1)}$$

The reaction obeys

$$r_{EA} = -k_1[H^+][CH_3COOCH_2CH_3] + k_2[H^+][CH_3COOH][HOCH_2CH_3] \qquad \text{(P3.34.2)}$$

(a) Develop a stoichiometric table for the reaction.

(b) Rearrange equation (P3.30.2) to prove that if there is no ethanol or acetic acid in the reactor at the beginning of the reaction, then

$$\frac{dX_{EA}}{dt} = -k_1[H^+](1 - X_{EA}) + k_2 C_{EA}^O[H^+](X_{EA})^2 \qquad (P3.34.3)$$

where X_{EA} is the conversion of ethylacetate and C_{EA}^O is the initial ethylacetate concentration.

(c) Show that the solution of equation (P3.34.3) is

$$k_1[H^+]\tau = \left(\frac{1}{1 + 2X_{EA}^{eq}}\right) \ln \left(\frac{(X_{EA}^{eq} - X_{EA})(1 + X_{EA}^{eq})}{(X_{EA}^{eq})(X_{EA} + 1 + X_{EA}^{eq})}\right) \qquad (P3.34.4)$$

with

$$2X_{EA}^{eq} = \left(\sqrt{4\left(\frac{k_2 C_{EA}^O}{k_1}\right) + 1}\right) - 1$$

(d) Make a plot of the rate with various values of the parameters. How does the rate of reaction vary as you vary k_1 and X_{EA}^{eq}?

More Advanced Problems

3.35 People often use bacteria to digest hazardous materials in wastestreams. The rate usually follows Monod kinetics:

$$r_B = \frac{d[B]}{dt} = k_B[B]\frac{K_F[W]}{1 + K_F[W]} \qquad (P3.35.1)$$

$$r_W = \frac{d[W]}{dt} = -k_W[B]\frac{K_F[W]}{1 + K_F[W]} \qquad (P3.35.2)$$

where [B] is the bacteria concentration in bacteria/liter and [W] is the waste concentration in mol/liter. Assume $K_F = 220$ liters/mol, $k_B = 0.35$/hour, and $k_W = 2.5 \times 10^{-6}$ (mol·hour)/bacteria.

(a) Derive an expression for $\frac{d[B]}{d[W]}$.

(b) Integrate your expression in (a) to derive an expression for [B] as a function of [W], the waste concentration at any time, t, and $[W]_0$ and $[B]_0$, the initial bacteria and waste concentrations.

(c) Rearrange your expression in (b) to derive an expression for [B] as a function of the X_W, the fractional conversion of the waste.

(d) Compare your results to those in the stoichiometric table. Can you see that you are converting waste into bacteria?

(e) Substitute your expression into equation (P3.35.2) to calculate the rate of waste reduction as a function of X_W.

(f) Integrate your expression to obtain an expression for the time to get a conversion X_W.

(g) Assume that you start with 10^6 bacteria/liter and 1 mol/liter of waste. You have a choice of two bacteria: one with a $K_F = 2.2 \times 10^2$ liters/mol, $k_B = 0.35$/hour, $k_W = 2.5 \times 10^{-6}$ (mol·hour)/bacteria a second with a $K_F = 2.2 \times 10^5$ liters/mol, $k_B = 0.35$/hour, $k_W = 2.5 \times 10^{-7}$ (mol·hour)/bacteria. Which bacteria will get to 99% conversion first?

(h) Repeat (f) for 99.999% conversion.

(i) How would your results change with a CSTR?

(j) On the basis of your results in (f) and (g), could you design a system that starts with one bacteria, then adds a second bacteria to finish the job?

3.36 Read the following papers and write a one-page report on the kinetics described in each paper. Why were kinetics measured? What techniques were used to do the kinetic measurements? How were the kinetic data analyzed? What were the key results?

(a) Koch, R., Palm, Wu., and Setzsch, C. The first rate constants for the reactants of OH radicals with amides. *Int. J. Chem. Kinet.*, **29**, 81 (1997).

(b) Crivello, J. V., and Liu, S. S. Synthesis and cationic polymerization of glycidyl ether. *Poly. Sci. A*, **36**, 1017 (1998).

(c) Simakov, P. A., Martinez, F. N., Horner, J. H., and Newcomb, M. Absolute rate constants for alkoxycarbonyl radical reactions. *J. Org. Chem.*, **63**, 1226 (1998).

(d) Musa, O. M., Choi, S. Y., Horner, J. H., and Newcomb, M. Absolute rate constants for alpha-amide radical reactions. *J. Org. Chem.*, **63**, 786 (1998).

(e) Kettling, U., Koltermann, A., Schwille, P., and Eigen, M. Real-time enzyme kinetics monitored by dual-color fluorescence cross-correlation spectroscopy. *Proc. Nat. Acad. Sci. U.S.*, **95**, 1416 (1998).

(f) Wallington, T. J., Guschin, A., Steinn, T. N. N., Platz, J., Sehested, J., Christensen, L. K., and Nielsen, O. J. Atmospheric chemistry of $CF_3CH_2OCH_2CF_3$-UV spectra and kinetic data for $CF_3CH \cdot ICH_2CF_3$ and $CF_3CH \cdot OCH_2CF_3$ radicals and atmospheric fate of $CF_3CH \cdot OCH_2CF_3$ radicals. *J. Phy. Chem.*, **102**, 1152 (1998).

(g) Tolti, N. P., and Leigh, W. J. Direct detection of 1,1-diphenylgermene in solution and absolute rate constants for germene trapping reactions. *J. Amer. Chem. Soc.*, **120**, 1172 (1998).

(h) Lepicard, S. D., and Canosa, A. Measurement of the rate constant for the association reaction $CH+N_{-2}$ at 53 K and its relevance to tritons atmosphere. *Geophys. Res. Let.*, **25**, 485 (1998).

(i) Johnson, K. A. Advances in transient-state kinetics (review). *Curr. Opin. Biotechnol.*, **9**, 87 (1998).

(j) Campbell, M. L. Gas-phase kinetics of ground-state platinum with O_{-2}, NO, N_2O and CH_4. *J. Chem. Soc. Faraday Trans.*, **94**, 353 (1998).

(k) Decker, C. The use of UV irradiation in polymerization (review). *Polym. Int.*, **45**, 133 (1998).

(l) Wolter, S. D., Mohney, S. E., Venugopalan, H., Wickenden, A. E., and Koleske, D. D. Kinetic study of the oxidation of gallium nitride in dry air. *J. Electrochem. Soc.*, **145**, 629 (1998).

(m) Bedjanian, Y., Laverdet, G., and Lebras, G. Low-pressure study of the reaction of CL atoms with isoprene. *J. Phys. Chem.*, **102**, 953 (1998).

(n) Blaser, H. U., Jalett, H. P., Garland, M., Studer, M., Thies, H., and Wirthtijani, A. Kinetic studies of the enantioselective hydrogenation of ethyl pyruvate catalyzed by a cinchona modified Pt/Al203 catalyst. *J. Catal.*, **173**, 282 (1998).

(o) Bradford, M. C. J. CO_2 reforming of CH_4 over supported PT catalysts. *J. Catal*, **173**, 157 (1998).

(p) Madras, G., Smith, J. M., and McCoy, B. J. Thermal degradation kinetics of polystyrene in solution. *Polym. Degradation Stab.*, **58**, 131 (1997).

(q) Deters, R., Otting, M., Wagner, H. G., Temps, F., Laszlo, B., Dobe, S., and Berces, T. A direct investigation of the reaction CH_3OH — Overall rate constant and CH_2 formation at T = 298 K. *Ber. Bunsenges. Phys. Chem.*, **102**, 58 (1998).

(r) Kunz, A., and Roth, P. A high temperature study of the reaction $SIH_4 + H– \rightarrow H_{-2}$. *Ber. Bunsenges. Phys. Chem.*, **102,** 73 (1998).

(s) Sehested, J., Christensen, L. K., Mogelberg, T., Nielsen, O. J., Wallington, T. J., Guschin, A., Orlando, J. J., and Tyndall, G. S. Absolute and relative rate constants for the reaction $CH_3C(0))_{-2} + NO$ and $CH_3C(0)O_{-2} + NO_2$ and thermal stability of $CH_3C(O)O_2NO_2$. *J. Phys. Chem.*, **102**, 1779 (1998).

(t) Harwood, M. H., Rowley, D. M., Cox, R. A., and Jones, R. L. Kinetics and mechanism of the bro self-reaction-temperature-and pressure-dependent studies. *J. Phys. Chem.*, **102**, 1790 (1998).

(u) Pereira, R. D., Baulch, D. L., Pilling, M. J., Robertson, S. H., and Zeng, G. Temperataure and pressure dependence of the multichannel rate coefficients for the $CH_3 + OH$ system. *J. Phys. Chem.*, **101**, 9681 (1997).

(v) Stutz, J., Ezell, M. J., and Finlaysonpitts, B. J. Inverse kinetic isotope effect in the reaction of atomic chlorine with C_2H_4 and C_2D_4. *J. Phys. Chem.*, **101**, 9187 (1997).

(w) Bokenkamp, D., Desai, A., Yang, X., Tai, Y. C., Marzluff, E. M., and Mayo, S. L. Microfabricated silicon mixers for submillisecond quench-flow analysis. *Anal. Chem.*, **70**, 232 (1998).

(x) Rotaru, P., Blejoiu, S. L., Constantinescu, R., Pometescu, N., Uliu, F., and Bunescu, O. Perfectly stirred catalytic reactor. *Appl. Catal. A*, **166**, 363 (1998).

(y) Manke, G. C., and Setser, D. W. Measuring gas-phase chlorine atom concentrations — Rate constants for $CL + HN_3$, CF_{31}, and C_2F_{51}. *J. Phys. Chem.*, **102**, 153 (1998).

3.37 Go to the **(a)** *International Journal of Chemical Kinetics*, or if this journal is not available in your library, try **(b)** the kinetics section of *Physical Chemistry A*, **(c)** *J. Physical Organic Chemistry*, **(d)** *Biotechnology and Bioengineering*, **(e)** *Reaction Kinetics and Catalysis Letters*, **(f)** *J. Polymer Science A*. Find an article where someone measures the kinetics of a reaction. Write a one-page report on the findings in the article to describe:

(a) Why the study was undertaken

(b) What techniques were used

(c) How the data were analyzed

(d) What the key results were

4

RELATIONSHIP BETWEEN RATES AND MECHANISMS

PRÉCIS

The objective of this chapter is to examine the relationship between rates and mechanisms. So far in this book we have treated rate equations as empirical relationships that are fit to rate data. However, in fact, rate equations have a more fundamental basis. The form of the rate equation is determined by the elementary chemical reactions that occur as a reaction proceeds. The rate constants are related to the rate of the elementary reactions.

The objective of this chapter is to explore the relationship between rates and mechanisms. We will introduce the concept of a mechanism, and show how one can use the mechanism and the pseudo-steady-state approximation to calculate a rate equation for a given reaction. Most of the students who take our course already know that reactions obey distinct mechanisms, and many have seen the pseudo-steady-state approximation previously. However, most students do not know where the pseudo-steady-state approximation comes from, when it works, and when it fails. The objective of this chapter is to extend students' knowledge so that they will know how the pseudo-steady-state approximation arises, and where it fails. We will also define some more key terms that we will use later in the book.

4.1 INTRODUCTION

Studies of the relationship between rates and mechanisms have had a long history. Work started with Dobereiner (1829) and Wilhelmy (1850), who supposed that rates of reaction would be simply related to the stoichiometry of the reaction. However, in 1878, Van't Hoff showed that the rate of reaction had little correlation to stoichiometry. For example, Table 2.5 (page 14) shows the rate equation for several reactions discussed by Van't Hoff. Notice that the rate of phosphine oxidation is first-order in the phosphine concentration and half-order in the oxygen concentration. Yet during the reaction, each phosphine reacts with two oxygens to yield products. Further, one needs two phosphines to yield each P_2O_5.

This example shows that there is no relationship between the rate equation for a reaction and the stoichiometry of the reaction.

The purpose of this chapter is to try to understand the molecular basis of the rate equation, and in particular why rates do not correlate with stoichiometry. Section 4.2 will discuss the idea that reactions actually occur by a complex series of chemical reactions. Sections 4.3 and 4.4 will review the rates of elementary reactions and how the rates are related to the stoichiometry of the reactions.

Next, we will examine rate equations. In Section 4.5 we will show that if one knows the mechanism one will be able to derive a set of differential equations for the change in concentration of all of the species in the reactor. The differential equations are the fundamental rate equations for a chemical reaction, and they fully determine the behavior. We will briefly review the behavior to give the reader a qualitative picture of the results. Next, we describe the pseudo-steady-state approximation and see how it can be used to determine a rate equation. We will also discuss the rate-determining-step approximation. Finally, we will close by discussing the failure of the methods, and what one does in such a case. The discussion will assume that the reader has had an introduction to these topics before. If you have not had an introduction, please look back over the kinetics section of your physical chemistry text before proceeding with this chapter.

4.2 HISTORICAL OVERVIEW

To start, it is useful to review some history. In the 1896 edition of his book, *l'Etudes Dynamique Chemie*, Van't Hoff speculated why the kinetics of a reaction did not correspond to the stoichiometry of the reaction. Van't Hoff supposed that when only one molecule participated in a critical step of reaction, the reaction would be first-order. In contrast, if two molecules participated in some critical step of a reaction, the reaction would be second-order. Consequently, Van't Hoff proposed that the kinetics of a reaction were related to what Van't Hoff termed the **molecularity** of a reaction, where the molecularity was loosely defined as the number of molecules participating in some critical step in the reaction.

For further reference, we will need to know that there are **unimolecular** reactions, where only one reactive molecule participates; **bimolecular** reactions, where two molecules participate; and **termolecular** reactions, where three molecules participate. An example of a unimolecular reaction is

$$\text{Cyclopropane} \Longrightarrow \text{propylene} \tag{4.1}$$

where one cyclopropane molecule rearranges to yield propylene. An example of a bimolecular reaction is

$$\text{OH} + \text{C}_2\text{H}_6 \Longrightarrow \text{H}_2\text{O} + \text{C}_2\text{H}_5 \tag{4.2}$$

where a hydroxyl grabs a hydrogen from an ethane molecule. An example of a termolecular reaction is

$$\text{CH}_3 + \text{CH}_3 + \text{N}_2 \Longrightarrow \text{C}_2\text{H}_6 + \text{N}_2 \tag{4.3}$$

This example is more complex; two methyl radicals are reacting to form ethane. However, no reaction occurs unless a nitrogen collides with the two methyls. The nitrogen facilitates the reaction.

Van't Hoff proposed that all first-order reactions are unimolecular, all second-order reactions are bimolecular, and all third-order reactions are termolecular.

If one looks at the stoichiometry of the reaction, though, one finds that complicated reactions have simple rate equations. For example, according to the information in Table 2.1 (page 14) the reaction

$$4PH_3 \Longrightarrow P_4 + 6H_2 \tag{4.4}$$

is first-order in the phosphine pressure. Yet, it is hard to imagine how reaction (4.4) could be a unimolecular reaction. After all, one needs four phosphines to produce a single P_4. Van't Hoff tried to get around this difficulty by assuming that only one molecule was participating in some critical step in reaction (4.4). However, when Van't Hoff's book was written (1896), it was not obvious how it was possible to have only a single phosphine molecule participate in the reaction when four phosphines are needed to produce a P_4.

In the time between 1890 and 1919, a number of papers were written that looked at the details of many different reactions. An attempt was made to try to understand how the kinetics of a reaction was related to the stoichiometry of the reaction. David Chapman (1913) Muriel Chapman, and Max Bodenstein (1907) examined the kinetics of HCl formation. They showed that the reaction

$$H_2 + Cl_2 \Longrightarrow 2HCl \tag{4.5}$$

did not occur via a direct reaction between molecular chlorine and molecular hydrogen. Instead, the chlorine needed to dissociate into atoms before the reaction proceeded. Slowly a consensus emerged that the reactants are not directly converted into products during a chemical reaction. Rather, the reactants are converted into species called **reactive intermediates**, and then the reactive intermediates are converted into products. For example, during the acid-catalyzed isomerization of 1-butene to 2-butene in solution

$$CH_3CH_2HC{=}CH_2 \xrightarrow{\ H^+\ } CH_3HC{=}CHCH_3 \tag{4.6}$$

The H^+ reacts with the 1-butene to produce a $\left[CH_3CH_2HC\overset{\overset{\displaystyle H}{/\backslash}}{-}CH_2 \right]^+$ intermediate where

the hydrogen is held in a three-center two-electron bond. The $\left[CH_3CH_2HC\overset{\overset{\displaystyle H}{/\backslash}}{-}CH_2 \right]^+$

intermediate then loses a 3-proton to form 2-butene. The sequence of reactions is

$$CH_3CH_2HC{=}CH_2 + H^+ \longrightarrow \left[CH_3CH_2HC\overset{\overset{\displaystyle H}{/\backslash}}{-}CH_2 \right]^+ \tag{4.7a}$$

$$\left[CH_3CH_2HC\overset{\overset{\displaystyle H}{/\backslash}}{-}CH_2 \right]^+ \longrightarrow CH_3HC{=}CHCH_3 + H^+ \tag{4.7b}$$

There are two reactions in mechanism (4.7):1-butene is the reactant, $\left[\text{CH}_3\text{CH}_2\text{HC} \overset{\text{H}}{\underset{\diagup \diagdown}{-}} \text{CH}_2\right]^+$

is the intermediate, and 2-butene is the product. Each of the steps in mechanism 4.7 are called **elementary reactions**, while reaction (4.6) is called the **overall reaction**, or the **stoichiometric reaction**.

> More precisely, an elementary reaction is defined as a chemical reaction going from reactants to products without going through any *stable* intermediates. In this context, a species is said to be stable if it has a lifetime longer than $\sim 10^{-11}$ seconds (i.e., much longer than vibrational or collisional times).

There are some important notations. We will use a single arrow $\rightarrow$ to designate an elementary reaction but a double arrow $\Rightarrow$ to designate an overall reaction. In this notation, the reaction

$$\text{H}_2 + \text{Br}_2 \Longrightarrow 2\text{HBr} \qquad (4.8)$$

will be an overall reaction, while the reaction

$$\text{H} + \text{Br}_2 \longrightarrow \text{HBr} + \text{Br} \qquad (4.9)$$

will be an elementary reaction.

In the gas phase, most elementary reactions are of the type

$$\text{A} + \text{B} \longrightarrow \text{C} + \text{D} \qquad (4.10)$$

where at least two reactants come together to produce at least two products. Examples include

$$\text{H} + \text{HBr} \longrightarrow \text{H}_2 + \text{Br} \qquad (4.11)$$

and

$$\text{X} + \text{H}_2 \longrightarrow 2\text{H} + \text{X} \qquad (4.12)$$

where X is a something called a **collision partner**. The collision partner is a species that collides with the H_2 and initiates the reaction.

Next, we want to define a concept called the **mechanism** of a reaction. The mechanism of a reaction is defined as the sequence of elementary reactions that occur at appreciable rates when the reactants come together and react to form products. For example, if one actually runs reaction (4.6) in solution, one finds that four elementary reactions can occur:

$$\text{CH}_3\text{CH}_2\text{HC}=\text{CH}_2 + \text{H}^+ \longrightarrow \left[\text{CH}_3\text{CH}_2\text{HC} \overset{\text{H}}{\underset{\diagup \diagdown}{-}} \text{CH}_2\right]^+ \qquad (4.13a)$$

$$\left[\text{CH}_3\text{CH}_2\text{HC} \overset{\text{H}}{\underset{\diagup \diagdown}{-}} \text{CH}_2\right]^+ \longrightarrow \text{CH}_3\text{HC}=\text{CHCH}_3 + \text{H}^+ \qquad (4.13b)$$

$$\left[CH_3CH_2HC \overset{\overset{\displaystyle H}{\diagup \diagdown}}{-} CH_2 \right]^+ \longrightarrow CH_3CH_2HC{=}CH_2 + H^+ \qquad (4.13c)$$

$$\left[CH_3CH_2HC \overset{\overset{\displaystyle H}{\diagup \diagdown}}{-} CH_2 \right]^+ \longrightarrow H_2C{=}CHCH_2CH_3 + H^+ \qquad (4.13d)$$

The first two steps in (4.13) are the steps listed in reaction (4.7). The last two steps are processes where the $\left[CH_3CH_2HC \overset{\overset{\displaystyle H}{\diagup \diagdown}}{-} CH_2 \right]^+$ intermediate loses a proton to produce either 1-butene or the equivalent 3-butene. Note that the proton that leaves the $\left[CH_3CH_2HC \overset{\overset{\displaystyle H}{\diagup \diagdown}}{-} CH_2 \right]^+$ intermediate in steps (4.13c) and (4.13d) may be different from the proton which reacts in step (4.13a).[1]

The sequence of elementary reactions listed in (4.13) is called the *mechanism of acid-catalyzed isomerization* of 1-butene and 2-butene.

There are two subtle points in the definition of a mechanism that students often miss. First, notice that there are four elementary reactions in equation (4.13). However, only the first two of the four lead from reactants to products. The other two steps lead back to the reactants. Therefore, it is incorrect to think of the mechanism as consisting of just the steps leading from reactants and products. Rather, the mechanism includes the steps that lead from reactions to products, plus other steps that either lead to new products or convert the intermediates back to the reactants.

A second subtlety is that when someone reports a mechanism in the literature, they do not necessarily list all of the reactions that would occur. They list only reactions that occur at appreciable rates. For example, during reaction (4.6), one also gets the reaction

$$\left[CH_3CH_2HC \overset{\overset{\displaystyle H}{\diagup \diagdown}}{-} CH_2 \right]^+ \longrightarrow \left[CH_3 \right]^+ + CH_2{=}CH{-}CH_3 \qquad (4.14)$$

Reaction (4.14) is much slower than reactions (4.13a)–(4.13d). Consequently, people often ignore reaction (4.14) even though it occurs at a nonnegligible rate. The mechanism of a reaction, then, does not necessarily include all reactions that occur. Rather, they include only reactions that occur at nonnegligible rates.

An important point is that

> Every overall chemical reaction can be divided into a sequence of elementary reaction. Every reaction has a mechanism.

Sometimes reactions have more than one mechanism. This occurs when there are, for example, competing reaction pathways.

---

[1] If reaction (4.13) would occur in the gas phase, one would need collision partners. For the purposes here, we will ignore the collision partners, since in solution, the solvent can act as a collision partner.

The mechanism in (4.13) is relatively simple. Here are some other examples. The reaction

$$H_2 + Br_2 \Longrightarrow 2HBr \tag{4.15}$$

follows the following mechanism:

$$Br_2 \xrightarrow{\ 1\ } 2Br$$

$$Br + H_2 \xrightarrow{\ 2\ } HBr + H$$

$$H + Br_2 \xrightarrow{\ 3\ } HBr + Br \tag{4.16}$$

$$2Br \xrightarrow{\ 4\ } Br_2$$

$$H + HBr \xrightarrow{\ 5\ } H_2 + Br$$

The reaction

$$2H_2 + O_2 \Longrightarrow 2H_2O \tag{4.17}$$

follows a complex mechanism with over 30 steps. However, one can get the structure of the solution by assuming that the reaction obeys the following mechanism:

$$H_2 + O_2 \xrightarrow{\ 1\ } 2OH\bullet$$

$$OH\bullet + H_2 \xrightarrow{\ 2\ } H_2O + H\bullet$$

$$H\bullet + O_2 \xrightarrow{\ 3\ } OH\bullet + O:$$

$$O: + H_2 \xrightarrow{\ 4\ } OH\bullet + H\bullet \tag{4.18}$$

$$H\bullet \xrightarrow{\ 5\ } wall$$

$$H\bullet + O_2 + X \xrightarrow{\ 6\ } HO_2\bullet + X$$

$$HO_2\bullet + X \xrightarrow{\ 7\ } wall$$

$$2HO + X \xrightarrow{\ 8\ } H_2O_2 + X$$

Notice that even through the overall reactions in (4.15) and (4.17) appear to be simple, the mechanisms are not simple. This is typical. Physically, what is happening is that reactive intermediates are, by definition, reactive. They undergo many reactions. As a result, there are usually many reactions in an overall mechanism.

Let's repeat that again because it is really important.

> Reactive intermediates are by definition reactive. They undergo many reactions.

We will describe the mechanisms of many different reactions in Chapters 5 and 12. However, the thing to remember for now is that a typical chemical reaction occurs via a number of elementary steps. One needs to understand the kinetics of those individual steps to determine the mechanism of a reaction.

4.3 KINETICS OF ELEMENTARY REACTIONS

Next, we want to discuss how the mechanism of a reaction is related to the kinetics of the reaction. We will assume for the moment that you already know the mechanism of a reaction, and ask how one can determine the rate equation of reaction from the mechanism.

We will start with some material that we will eventually derive in Chapter 9. Consider an elementary reaction between an atom or molecule, A, and another atom or molecule, B, to form two product atoms or molecules, P and Q:

$$A + B \longrightarrow P + Q \tag{4.19}$$

In Chapter 9, we will show that r_2, the rate of reaction (4.19), is given by

$$r_2 = k_2[A][B] \tag{4.20}$$

Similarly, the rate of formation of A is given by

$$r_A = -k_2[A][B] \tag{4.21}$$

where [A] and [B] are the concentrations of A and B, respectively, and k_2 is a constant. Equation (4.20) is a key result. It says that if we know the stoichiometry of an elementary reaction, we also know the rate equation for the reaction. Notice that the rate equation for an elementary reaction is proportional to the product of the concentrations of the reactants of the elementary reaction. However, the rate does not depend on the concentrations of the products of the elementary reactions.

One can also consider elementary reactions of the form

$$2A \xrightarrow{\;\;4\;\;} P + Q \tag{4.22}$$

In this case the rate of reaction, r_4, obeys

$$r_4 = k_4[A]^2 \tag{4.23}$$

Similarly, the rate of formation of A, r_A is given by

$$r_A = -2k_4[A]^2 \tag{4.24}$$

where k_4 is the rate constant for reaction (4.22).

The factor of 2 in equation (4.24) is very important. The factor of 2 arises because two molecules of A are destroyed every time reaction (4.22) goes once. Students often forget the factors of 2 when they derive rate equations. The reader should be very careful to avoid that error.

One way to avoid this error is to use the definitions from Chapter 2 to quantify the rate of formation of any species. For example, consider species A participating in a group of reactions numbered $1, 2, \ldots, i$. One can show that the net rate of formation of the species, r_A, is given by

$$r_A = \beta_{A,1}r_1 + \beta_{A,2}r_2 + \cdots + \beta_{A,i}r_i \qquad (4.25)$$

where r_i, $r_2, \ldots, r_i$ are the rates at reactions $1, 2, \ldots, i$, respectively, and $\beta_{A,1}$, $\beta_{A,2}, \ldots, \beta_{A,i}$ are the stoichiometric coefficients of species A in reaction $1, 2, \ldots, i$. Again, we remind you that in our definition, the stoichiometric coefficient is negative for a reactant and positive for a product. In reaction (4.22), $\beta_{A,4.26}$ is -2, which is why there is a -2 in equation (4.24). When I am deriving rate equations, I find it easier to remember that the coefficient in the rate equation is ALWAYS the stoichiometric coefficient instead of having to work out the coefficient every time.

Equation (4.20) applies only to elementary reaction of the form in (4.19). However, people often generalize equations (4.20) to first- and third-order reactions. For example, with an elementary reaction of the form

$$A \longrightarrow P \qquad (4.26)$$

it is often *incorrectly* assumed that the reaction obeys

$$r_A = k_1[A] \qquad (4.27)$$

where [A] is the concentration of A and k_1 is a constant. Similarly, with an elementary reaction of the form

$$A + B + C \longrightarrow \text{products} \qquad (4.28)$$

it is often assumed that the reaction obeys:

$$r_A = -k_3[A][B][C] \qquad (4.29)$$

where [A], [B] and [C] are the concentration of A, B, and C, respectively, and k_3 is a constant.

In actual practice, elementary reactions of the form in equation (4.26) are never seen experimentally and reactions of the form in equation (4.28) are rarely seen. In Chapter 8, we will show that the rate constant for any elementary reaction of the form in (4.26) (i.e., where one reactant is converted into one product without colliding with any other molecule in the system) is ZERO, which means that elementary reactions of the form in equation (4.26) are impossible. Similarly, we will show that the rate constant for reactions of the form in equation (4.28) is very small. Therefore, in practice, one rarely sees reactions of the form in equations (4.26) and (4.28). One can, however, see reactions of the form

$$A + X \longrightarrow P + X \qquad (4.30)$$

where X is any other molecule in the system that can collide with A. We call X the *collision partner*. Reaction (4.30) obeys

$$r_A = -k_2[A][X] \tag{4.31}$$

where [A] and [X] are the concentrations of A and X, and k_2 is a constant. Equation (4.31) is very similar to equation (4.27) in cases where [X] does not change during the reaction. As a result, people often say that the kinetics of a reaction like that in reaction (4.30) obey (4.27) approximately, even though they do not fit exactly.

There is an important point I want to repeat here that for emphasis. One can never have an elementary reaction with only one reactant or only one product. It looks as though you only have one reactant or one product in some elementary reaction. There will always be some other species, X, contributing to the reaction. Therefore, the elementary reactions are

$$A + X \longrightarrow B + C + X$$
$$A + B + X \longrightarrow P + X$$

and NOT

$$A \longrightarrow B + C$$
$$A + B \longrightarrow C$$

4.4   THE RELATIONSHIP BETWEEN KINETICS AND EQUILIBRIUM (MICROSCOPIC REVERSIBILITY)

Before we proceed, it is useful to point out that the results in Section 4.3 imply that there is a simple relationship between the kinetics of a reaction and the equilibrium constant for the reaction. Consider the simple reversible elementary reaction:

$$A + B \underset{2}{\overset{1}{\rightleftharpoons}} C + D \tag{4.32}$$

The equilibrium constant for the reaction, $K_{1,2}$, is given by

$$K_{1,2} = \frac{[C][D]}{[A][B]} \tag{4.33}$$

where [A], [B], [C], and [D] are the concentrations of A, B, C, and D. There is a principle called **microscopic reversibility**, which states

> At equilibrium, the rate of any forward chemical reaction (elementary or not) must equal the rate of the reverse chemical reaction. (4.34)

Principle (4.34) implies that at equilibrium, the rate of all processes must equal the rate of the backward processes.

If we apply principle (4.34) to our example, we conclude that at equilibrium, the rate of the forward reaction 1 should be equal to the rate of the reverse reaction 2. Therefore

$$k_1[A][B] = k_2[C][D] \tag{4.35}$$

where k_1 and k_2 are the rate constants for reactions 1 and 2. Rearranging equation (4.35) yields

$$\frac{[C][D]}{[A][B]} = \frac{k_1}{k_2} \tag{4.36}$$

Substituting equation (4.36) into equation (4.33) yields

$$K_{1,2} = \frac{k_1}{k_2} \tag{4.37}$$

Consequently, the equilibrium constant for a reversible elementary reaction is equal to the ratio of the forward and reverse rate constants for the reaction.

One can generalize this result to a more complex reaction:

$$A + B \underset{-1}{\overset{1}{\rightleftharpoons}} C + D \underset{-2}{\overset{2}{\rightleftharpoons}} E + F \underset{-3}{\overset{3}{\rightleftharpoons}} G + H \underset{-4}{\overset{4}{\rightleftharpoons}} I + J \underset{-5}{\overset{5}{\rightleftharpoons}} L + M \tag{4.38}$$

At equilibrium

$$\frac{[E][F]}{[A][B]} = \frac{k_1 k_2}{k_{-1} k_{-2}} \tag{4.39}$$

$$\frac{[L][M]}{[A][B]} = \frac{k_1 k_2 k_3 k_4 k_5}{k_{-1} k_{-2} k_{-3} k_{-4} k_{-5}} \tag{4.40}$$

$$\frac{[L][M]}{[G][H]} = \frac{k_4 k_5}{k_{-4} k_{-5}} \tag{4.41}$$

where [A], [B], [E], [F], [G], [H], [L], and [M] are the concentrations of A, B, E, F, G, H, L, and M, respectively, and k_1, k_{-1}, and so on are the rate constants for reactions 1, −1, and so forth. I find that a good mnemonic to remember these results is that if I want to calculate the equilibrium concentration of two species G and H and two other species C and D, I say that the product of the concentration of G and H divided by the concentrations of C and D is equal to the product of all the rate constants leading from C and D to G and H divided by the rate constants for the reverse reactions. There are two rate constants leading from C and D to G and H, k_2 and k_3. Therefore

$$\frac{[G][H]}{[C][D]} = \frac{k_2 k_3}{k_{-2} k_{-3}} \tag{4.42}$$

Similarly, I say that the product of the concentration of C and D divided by the concentrations of G and H is equal to the product of all of the rate constants leading from G and H to C and D divided by the rate constants for the reverse reactions. There are two rate constants leading from G and H to C and D, k_{-2} and k_{-3}. Therefore

$$\frac{[C][D]}{[G][H]} = \frac{k_{-2} k_{-3}}{k_2 k_3} \tag{4.43}$$

4.5 RATES OF OVERALL REACTIONS

Next, we will change topics slightly and use the results in Section 4.2 to derive a rate equation for an overall reaction. Our approach will be to consider the mechanism in equation (4.13) and derive an equation. We will then generalize to a mechanism.

Consider the mechanism in equation (4.13). During the mechanism in reaction (4.13), 1-butene reacts with a proton to yield an intermediate. The intermediate then decomposes to either form 2-butene or decomposes back to form 1-butene. If we call 1-butene, "A", 2-butene "P", and the intermediate "I", then one can view the reaction as follows:

$$A + H^+ \underset{2}{\overset{1}{\rightleftharpoons}} I \overset{3}{\longrightarrow} P + H^+ \qquad (4.44)$$

where we have ignored the collision partners since we are running the reaction in solution.

In equation (4.44), we call the reaction that converts A into I "reaction 1". The reaction that converts I back to A is called "reaction 2". The reaction that converts I to P is called "reaction 3".

Next, we will to derive a general equation for the change in the concentration of all of the species in the system. Our approach will be to use equation (4.25) to derive a differential equation for the production rate of all of the species. We will then integrate the differential equation to compute all of the rates.

4.5.1 Derivation of a Differential Equation for the Concentration

First, we derive a general differential equation for the rate of production of each of the species. The general approach will be to plug into equation (4.25) to obtain a differential equation for each species.

Consider the intermediate I. The intermediate I is formed in reaction 1 and destroyed in reactions 2 and 3. Therefore, according to equation (4.25), one can write the change in the concentration of the intermediate I as

$$\frac{d[I]}{dt} = r_1 - r_2 - r_3 \qquad (4.45)$$

where r_1, r_2, and r_3 are the rates of reactions 1, 2, and 3 respectively; [I] is the concentration of the intermediate I; and t is time. Similarly, [A], the concentration of A, obeys:

$$\frac{d[A]}{dt} = r_2 - r_1 \qquad (4.46)$$

where r_1, and r_2 are the rates of reactions 1 and 2, respectively.
According to equation (4.21), r_1 is given by

$$r_1 = k_1[A][H^+] \qquad (4.47)$$

where k_1 is the rate constant for reaction 1, [A] is the concentration of A, and [H^+] is the concentration of protons. Similarly, r_2 and r_3 are given by

$$r_2 = k_2[I] \qquad (4.48)$$

$$r_3 = k_3[I] \qquad (4.49)$$

Substituting equations (4.47) and (4.48) into equation (4.46) yields

$$\frac{d[A]}{dt} = k_2[I] - k_1[A][H^+] \tag{4.50}$$

Similarly, substituting equations (4.47)–(4.49) into equation (4.45) yields

$$\frac{d[I]}{dt} = k_1[A][H^+] - (k_2 + k_3)[I] \tag{4.51}$$

Equations (4.50) and (4.51) are the fundamental differential equations for the behavior of the system. They are the key results in this section.

4.5.2 Integration of the Rate Equation

Next, we want to integrate the equations to calculate the overall rate of reaction. There are three approaches:

- Analytic integration of the differential equations
- Numerical integration of the differential equations
- Approximate integration of the rate equation

In this section we will concentrate on the analytic treatment. Numerical integration is discussed in the supplementary material. An approximation scheme is discussed in Section 4.7. In order to simplify the algebra, we will assume that $[H^+]$ is constant [equation (4.50)] and that we start with pure A (no intermediate or product) so that

$$[A] = [A]^0 \qquad [I] = 0 \qquad [P] = 0 \qquad \text{at } t = 0 \tag{4.52}$$

One can then solve equations (4.50) and (4.51) simultaneously to calculate the concentration of A and I as a function of time. The algebra is complex, but the answer is

$$[A] = [A]^0 \left\{ \frac{(k_4 - k_6)\exp(-k_5 t) - (k_5 - k_6)\exp(-k_4 t)}{(k_4 - k_5)} \right\} \tag{4.53}$$

$$[I] = [A]^0 \left\{ \frac{(k_4 - k_6)(k_6 - k_5)}{k_2(k_4 - k_5)} \right\} \{\exp(-k_5 t) - \exp(-k_4 t)\} \tag{4.54}$$

where $[A]^0$ is the initial concentration of A and

$$k_4 = \tfrac{1}{2}(k_6 + k_2 + k_3 + \sqrt{(k_6 + k_2 + k_3)^2 - 4k_6 k_3}) \tag{4.55}$$

$$k_5 = \tfrac{1}{2}(k_6 + k_2 + k_3 - \sqrt{(k_6 + k_2 + k_3)^2 - 4k_6 k_3}) \tag{4.56}$$

$$k_6 = k_1[H^+] \tag{4.57}$$

Note that every mole of A that is used up must be converted into P or I. The number of moles of A used up is the initial concentration of A minus the current concentration of A. Therefore, the sum of the number of moles of I and P is given by

$$[P] + [I] = [A]^0 - [A] \tag{4.58}$$

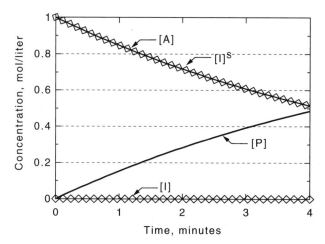

Figure 4.1 [A], [I], and [P] as a function of time calculated from equations (4.53), (4.54), and (4.58), respectively, with $k_1[H^+] = 0.2/$minute, $k_2 = 5.7 \times 10^6/$minute, and $k_3 = 3.8 \times 10^7/$minute.

where [A], [I], and [P] are the concentrations of A, I, and P, respectively, at any time during the reaction, and $[A]^0$ is the initial concentration of A. Equation (4.58) can be used to calculate [P]. One should verify equations (4.53) and (4.54) before proceeding with this chapter.

Figure 4.1 shows a plot of [A], [I], and [P] as a function of time calculated from equations (4.53), (4.54), and (4.58), respectively, for some typical values of the parameters. Typically $k_1[H^+]$, the rate constant for the formation of the reactive intermediate, will be much less than k_2 and k_3, the rate constants for the destruction of the reactive intermediate. If one substitutes values into equation (4.54), one finds that the concentration of the intermediates is small.

4.5.3 An Approximation

For future reference, Figure 4.1 also provides a plot of another quantity, $[I]^S$, where $[I]^S$ is defined by

$$[I]^S = [I] \left(\frac{k_2 + k_3}{k_1[H^+]} \right) \tag{4.59}$$

Notice that $[I]^S$ and [A] are almost equal, except near $t = 0$.

Next, we will compute a new quantity, where $[I]^X$ is an approximation to [I], by substituting [A] for $[I]^S$ in equation (4.59), substituting $[I]^X$ for [I] and rearranging. The result is

$$[I]^X = [A] \left(\frac{k_1[H^+]}{k_2 + k_3} \right) \tag{4.60}$$

Figure 4.2 compares [I] calculated from equation (4.54) to $[I]^X$ computed from equation (4.60). The left curve shows the behavior from $t = 0$ to $t \leq 10^{-7}$ minutes; the right curve shows the behavior from 2×10^{-2} minutes on. Notice that there are some deviations for $t < 10^{-7}$ minutes. However, the two curves are virtually identical when $t > 10^{-7}$ minutes.

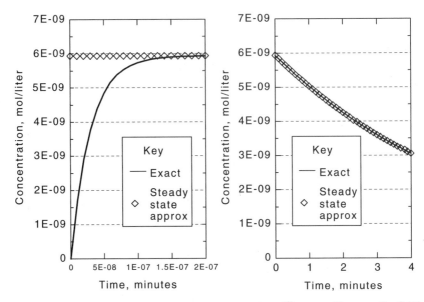

Figure 4.2 A comparison of [I] calculated from equation (4.54) and $[I]^X$ computed from equation (4.60).

We rarely care about the first 10^{-7} minutes in a process that takes several minutes. Consequently, the approximation in (4.60) works quite well in most practical cases.

4.6 GENERALIZATION TO OTHER REACTIONS

Next, we want to generalize these results to other reactions. First, it is important to note that there is nothing special about the example we considered in Section 4.4. If we start with any reaction, we can always set up a series of differential equations similar to equations (4.50) and (4.51), and then solve the equations to predict the behavior of the system. In a general case, the differential equations can be complex. However, one can always numerically integrate the differential equations to calculate all of the concentrations. There are some details that one needs to worry about when one does the numerical integrations because initially the intermediate concentration changes much more quickly than does the reactant concentration. Still, the equations can be solved numerically, using standard commercial software as shown in the solved examples at the end of this chapter. Consequently, if one knows the rate equations for a mechanism, one can always solve the rate equations to calculate the behavior of the reacting system. Algorithms are given in Example 4.B.

The algorithms are very general, and can be used for any rate equation. Generally, the numerical solution is the most accurate way to understand the behavior of a reaction.

4.7 THE PSEUDO-STEADY-STATE APPROXIMATION

Often however, one needs an analytical expression for the rate equation that one then uses in, for example, a reactor design. In order to get the approximation, we will generalize the approximation in equation (4.60) and use that approximation to estimate the concentrations of all the intermediates. In particular, we will derive what is called the *pseudo-steady-state approximation* for the intermediate concentrations and see how this approximation works.

I know that most of our readers have already seen the pseudo-steady-state approximation. However, I was not sure whether people know where this approximation comes from. Therefore, I decided to "derive" this approximation rather than just present it.

We will start by computing the size of the various terms in equation (4.51). Figure 4.3 is a plot of the various terms in equation (4.51) calculated by substituting equations (4.53) and (4.54) into equation (4.51). Notice that the first term on the right side of equation (4.51) is between 0.1 and 0.2 over the entire range at times shown. The second term on the right side of equation (4.51) is between 0.1 and 0.2 after 2×10^{-8} minutes. In contrast, while the derivative on the left side of equation (4.51) is significant for the first 10^{-7} minutes, the derivative decreases rapidly so that after the first 2×10^{-7} minutes, it is much smaller than the other two terms in the equation. In fact, on the scale shown, after the first 10^{-7} minutes, the derivative appears to be virtually zero.

One can use the fact that the derivative in equation (4.51) is almost zero to compute useful quantities. Let's assume, for the moment, that we can set the derivative in equation (4.51) to zero. In that case, equation (4.51) becomes

$$0 = k_1[A][H^+] - (k_2 + k_3)[I] \tag{4.61}$$

Rearranging equation (4.61) yields

$$[I] = [A]\left(\frac{k_1[H^+]}{k_2 + k_3}\right) \tag{4.62}$$

Notice that equations (4.60) and (4.62) are virtually the same. According to Figure 4.2, [I] calculated from equation (4.62) will be virtually identical to [I] computed from the exact

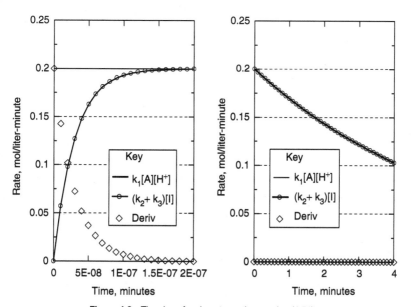

Figure 4.3 The size of various terms in equation (4.51).

result, equation (4.54), except at t = 0. Therefore, one can compute useful properties by assuming that the derivative in equation (4.51) is zero.

Physically, the derivative is the *net* rate of formation of the intermediate. If we set the derivative equal to zero, we are in effect assuming that the intermediates react as soon as they form. Physically, even though the intermediates form relatively quickly, they react very quickly, too, so that the net rate of formation of the intermediates is small.

One can easily generalize this result using something called the **pseudo-steady-state approximation**, discussed above. Most of our readers have already seen the pseudo-steady-state approximation. However, I want to state it a little differently than you have seen before

---

According to the pseudo-steady-state approximation, one can compute accurate values of the concentrations of all of the intermediates in a reaction by assuming that the net rate of formation of the intermediates is negligible (i.e., the derivatives with respect to time of the concentrations of all intermediates are negligible).

Note that there is a subtle point here. One is not actually assuming that the derivatives are exactly zero. Rather, one is assuming that the derivatives are much smaller than the other terms in the rate equations, so that the intermediates are consumed as quickly as the intermediates are formed. Consequently, the net rate of formation of the intermediates (i.e., the difference between the rate that the intermediates are formed and destroyed) can be neglected when one uses the rate equation to compute the concentration of all of the intermediates.

Physically, what is happening is that the concentrations of the intermediates are usually tiny, so the derivatives of the intermediate concentrations are tiny, too. Consequently, the derivatives of the intermediate concentration can be ignored in equation (4.51).

One needs to be careful because one **cannot** just ignore all of the terms in the rate equation containing the intermediate concentration; one can ignore only the derivatives. For example, in equation (4.51), there is a term:

$$r_3 = k_3[I] \tag{4.63}$$

If one proceeded naively, one might guess that according to equation (4.62), the intermediate concentration in equation (4.63) is tiny, so r_3 is tiny, too. However, note that we plotted r_3 in Figure 4.3, and r_3 is not tiny. The reason is that even though the intermediate concentration is tiny (6×10^{-9}), k_3 is huge ($2.8 \times 10^{+7}$). Consequently, r_3 is not negligible.

This is a general result. While the concentrations of all of the reactive intermediates are generally small, the intermediates are by definition very reactive; the rate constants for reactions involving the intermediates are generally huge. As a result, one cannot simply eliminate the terms involving the intermediate concentrations in the rate equation. Rather, one does have to consider the intermediate concentrations explicitly. One can, however, often eliminate terms that have to do with the derivatives of the intermediate concentration since the derivatives are usually small, except near time = 0.

There is another subtle point — the pseudo-steady-state approximation works only for reactive intermediates. If you consider a stable species such as a reactant, a product, or even a side product, you cannot assume that the species will be consumed as soon as it is formed. In most, but not all cases, radicals, hot molecules, highly strained species, or species with some unusual bonding [e.g., the three-center bonded intermediate in reaction (4.13)] will obey the pseudo-steady-state approximation. There are exceptions, of course. For example, in the upper atmosphere, OH radicals and HO_2 species are stable. More specific criteria for when the pseudo-steady-state approximation works will be given in Section 4.9. The key point, however, is that the pseudo-steady-state approximation works for most highly reactive species. Consequently, this approximation is quite useful.

4.8   APPLICATIONS OF THE PSEUDO-STEADY-STATE APPROXIMATION TO DETERMINE THE KINETICS OF A REACTION

Next, we will use the pseudo-steady-state approximation to derive a rate law for a number of example reactions. I realize that most of the readers have seen these types of derivations before. However, we needed some of the results for the discussion later in this chapter. Also, the material is a useful review.

We will start with the simple example that was discussed in Section 3.14. Consider the reaction

$$A \xrightarrow{\;1\;} I \xrightarrow{\;2\;} P \tag{4.64}$$

According to the analysis in Section 3.9, [I], the concentration of the intermediate, can be computed exactly from equation (3.84). In this section, we will use the pseudo-steady-state approximation to calculate an approximation for [I].

According to equation (4.31), r_I, the net rate of formation of the intermediate, is given by

$$r_I = k_1[A] - k_2[I] \tag{4.65}$$

where [A] and [I] are the concentrations of A and I, respectively, and k_1 and k_2 are the rate constants for reactions 1 and 2, respectively. According to the pseudo-steady-state approximation, r_I will be much smaller than the other terms in equation (4.65). Consequently, r_I can be ignored in equation (4.65). Setting r_I to zero in equation (4.65) and solving for [I] yields

$$[I] = \frac{k_1}{k_2}[A] \tag{4.66}$$

It is useful to compare equation (4.66) to equation (3.87). The notation is different in the two equations, but otherwise the two equations are identical. Figure 3.28 shows that equation (3.87) is an excellent approximation whenever k_2/k_1 is greater than about 10. With a typical intermediate, k_2/k_1 is in the order of 10^6. Clearly, the pseudo-steady-state approximation works in such a case.

Next, let's consider a more complex example, the reaction between hydrogen and bromine to yield HBr:

$$H_2 + Br_2 \Longrightarrow 2HBr \tag{4.67}$$

The mechanism of the reaction is

$$X + Br_2 \xrightarrow{\;1\;} 2Br + X$$

$$Br + H_2 \xrightarrow{\;2\;} HBr + H$$

$$H + Br_2 \xrightarrow{3} HBr + Br \tag{4.68}$$

$$X + 2Br \xrightarrow{4} Br_2 + X$$

$$H + HBr \xrightarrow{5} H_2 + Br$$

where X is a collision partner (i.e., any other species that can collide with Br_2 as described in Section 4.2). Next, let's derive an expression for the rate that Br_2 is used up during the reaction. Note that Br_2 is formed in reaction 4 and destroyed in reactions 1 and 3. Therefore

$$r_{Br_2} = -r_1 - r_3 + r_4 \tag{4.69}$$

where r_1, r_3, and r_4 are the rates of reactions 1, 3, and 4, respectively. One can compute r_1, r_3, and r_4 from equation (4.28):

$$r_1 = k_1[X][Br_2] \tag{4.70}$$

$$r_3 = k_3[H][Br_2] \tag{4.71}$$

$$r_4 = k_1[X][Br]^2 \tag{4.72}$$

where $[Br_2]$, $[Br]$, $[H]$, and $[X]$ are respectively the concentrations of bromine molecules, bromine atoms, hydrogen atoms, and collision partners. Substituting equations (4.70)–(4.72) into equation (4.69) yields

$$r_{Br_2} = -k_1[X][Br_2] - k_3[H][Br_2] + k_4[X][Br]^2 \tag{4.73}$$

Equation (4.73) is an exact expression for the rate of consumption of bromide during the reaction. In principle, therefore, equation (4.73) can be used to calculate the concentration of all of the species in the system. In practice, however, the concentration's of bromine atoms and hydrogen atoms are tiny. Therefore, equation (4.73) is not very useful in practice.

Now, one might think that one could just ignore the second and third terms on the right side of (4.73). After all, if one runs the reaction at 1 atm and 800 K, one finds that the concentrations of hydrogen atoms and bromine atoms are less than 10^{-8} mol/liter. By comparison, the Br_2 concentration is 0.01 mol/liter under the same conditions. However, it is incorrect to ignore the second and third terms in equation (4.73) because the rate constants for the second the third terms in that equation (4.73) are large. According to the data in Benson (1976), k_3 is about 1×10^{10} $Å^3$/(molecule·second) at 800 K, while k_1 is only 1×10^2 $Å^3$/(molecule·second). Plugging in numbers shows that the second term on the right side of equation (4.73) is about 100 times bigger than the first term in the equation. A similar argument also applies to the third term on the right side of equation (4.73). As a result, it would be certainly incorrect to ignore the second and third terms on the right side of equation (4.73). Instead, one must include these two terms in the analysis.

With three terms, we have a very complicated equation. Fortunately, the pseudo-steady-state approximation can be used to get simple expressions for the bromine and hydrogen atom concentrations in equation (4.73).

Bromine atoms are formed in reactions 1, 3, and 5, and lost in reactions 2 and 4. Therefore

$$\frac{d[Br]}{dt} = 2r_1 - r_2 + r_3 - 2r_4 + r_5 \qquad (4.74)$$

The factors of 2 in equation (4.74) arise because two bromine atoms are formed in reaction 1 and two atoms of bromine are lost in reaction 4. Substituting expressions similar to equations (4.70)–(4.72) into equation (4.74) yields

$$\frac{d[Br]}{dt} = r_{Br} = 2k_1[X][Br_2] - k_2[H_2][Br] + k_3[H][Br_2]$$

$$- 2k_4[Br]^2[X] + k_5[H][HBr] \qquad (4.75)$$

A similar analysis shows

$$\frac{d[H]}{dt} = r_H = k_2[H_2][Br] - k_3[H][HBr_2] - k_5[H][HBr] \qquad (4.76)$$

(*A note for students*: Notice that I first wrote out the equations in terms of the rates of the individual reactions, and then substituted in the rate laws. In my experience, students make fewer errors if they first write down the rates in terms of $r_1, r_2, r_3, \ldots$ and then substitute the rate in terms of laws into the subsequent equations.)

Next, we will assume that the concentrations of the bromine atoms and hydrogen atoms follow the pseudo-steady-state approximation, that is, that the derivatives in equations (4.75) and (4.76) are negligible. Physically, we can make that assumption because the concentrations of the reactive intermediates are so small that the derivatives of the intermediate concentrations will be small, too. We note again, however, that k_2, k_3, k_4, and k_5 are huge, so one cannot assume that the other terms in equations (4.75) and (4.76) are negligible.

If one assumes that the derivatives in equations (4.75) and (4.76) are negligible, one finds

$$0 = 2k_1[X][Br_2] - k_2[H_2][Br] + k_3[H][Br_2]$$

$$- 2k_4[Br]^2[X] + k_5[H][HBr] \qquad (4.77)$$

$$0 = +k_2[H_2][Br] - k_3[H][Br_2] - k_5[H][HBr] \qquad (4.78)$$

We now have two equations — equations (4.77) and (4.78) — and two unknowns — [H] and [Br] — so we can solve the equations simultaneously to calculate [H] and [Br].

Adding equations (4.77) and (4.78) yields

$$0 = 2k_1[X][Br_2] - 2k_4[Br]^2[X] \qquad (4.79)$$

Solving equation (4.79) for [Br] yields

$$[Br] = \left(\frac{k_1}{k_4}\right)^{1/2} [Br_2]^{1/2} \qquad (4.80)$$

Equation (4.80) is the pseudo-steady-state expression for the concentration of bromine atoms produced during reaction (4.67).

Next, we will obtain an expression for the concentration of hydrogen atoms. Substituting equation (4.80) into equation (4.78) yields

$$k_2[H_2]\left(\frac{k_1}{k_4}\right)^{1/2}[Br_2]^{1/2} = k_3[H][Br_2] + k_5[H][HBr] \tag{4.81}$$

Solving equation (4.81) for [H] yields

$$[H] = \frac{k_2\left(\dfrac{k_1}{k_4}\right)^{1/2}[H_2][Br_2]^{1/2}}{k_3[Br_2] + k_5[HBr]} \tag{4.82}$$

Equation (4.82) is the pseudo-steady-state approximation for the concentration of hydrogen atoms produced during reaction (4.67).

At this point, we have expressions for all of the unknown quantities in equation (4.73), so we can calculate an expression for the rate of destruction of bromine atoms. Substituting equation (4.82) and (4.80) into equation (4.73) and rearranging yields

$$r_{Br_2} = -\frac{k_2\left(\dfrac{k_1}{k_4}\right)^{1/2}[H_2][Br_2]^{1/2}}{1 + \dfrac{k_5}{k_3}\dfrac{[HBr]}{[Br_2]}} \tag{4.83}$$

Equation (4.83) is the pseudo-steady-state approximation to the rate of reaction (4.67).

The analysis above was for reaction (4.67). However, one can use an analysis similar to that above to calculate a rate equation for any reaction when one knows the mechanism of the reaction. For example, we can use the steady approximation to calculate a rate equation for the reaction between hydrogen and oxygen by assuming that the reaction follows the mechanism in equation (4.17). Generally, one has to do some algebra to solve all of the equations. However, the process is straightforward.

For the work that follows, we will need to consider one other example — the reaction between hydrogen and oxygen:

$$2H_2 + O_2 \Longrightarrow 2H_2O \tag{4.84}$$

Reaction (4.84) follows a complex mechanism with over 40 important steps. Still, one can get a reasonable approximation to the rate if one assumes that the reaction obeys the following *approximate* mechanism:

$$H_2 + O_2 \xrightarrow{\ 1\ } 2OH\bullet$$

$$OH\bullet + H_2 \xrightarrow{\ 2\ } H_2O + H\bullet$$

$$H\bullet + O_2 \xrightarrow{\ 3\ } OH\bullet + O:$$

$$O: + H_2 \xrightarrow{\ 4\ } OH\bullet + H\bullet \tag{4.85}$$

$$H\bullet \xrightarrow{\ 5\ } wall$$

$$H\bullet + O_2 + X \xrightarrow{\ 6\ } HO_2\bullet + X$$

$$HO_2\bullet + X \xrightarrow{\ 7\ } wall$$

In equation (4.85), the term "wall" in reactions 5 and 7 refers to processes where radicals absorb onto the walls of the vessel. We want to emphasize that reaction (4.85) is not the correct mechanism of the reaction, but one gets a reasonable rate equation anyway. In the work in the next section, we will need an expression for [H•], the concentration of the hydrogen atoms during this reaction. We will use the pseudo-steady-state approximation to get it. Let's start by deriving an equation for the rate of formation of hydrogen atoms. Hydrogen atoms are formed in reactions 2 and 4 while hydrogen atoms are lost in reactions 3, 5, and 6. Consequently, r_H, the rate of formation of hydrogen atoms is given by

$$r_H = r_2 - r_3 + r_4 - r_5 - r_6 \tag{4.86}$$

where r_2, r_3, r_4, r_5, and r_6 are the rate of reactions 2–6. Substituting the rate reactions for reactions 2–6 into equation (4.86) yields

$$\frac{d[H\bullet]}{dt} = r_{H\bullet} = k_2[OH\bullet][H_2] - k_3[H\bullet][O_2]$$
$$+ k_4[O\bullet][H_2] - k_5[H\bullet] - k_6[H\bullet][O_2][X] \tag{4.87}$$

where [H$_2$], [H•], [OH•], [O$_2$] [O:], and [X] are the concentration of hydrogen molecules, hydrogen atoms, hydroxyl radicals, oxygen molecules, oxygen atoms, and collision partners, respectively. Similarly, the rate of formation of hydroxyl radicals and oxygen atoms, r_{OH} and r_O, is given by

$$\frac{d[OH\bullet]}{dt} = r_{OH} = 2k_1[H_2][O_2] - k_2[OH\bullet][H_2]$$
$$+ k_3[H\bullet][O_2] + k_4[O:][H_2] \tag{4.88}$$

$$\frac{d[O:]}{dt} = r_O = k_3[H\bullet][O_2] - k_4[O:][H_2] \tag{4.89}$$

If we assume that r_{OH}, r_H and r_O are negligible, we can solve equations (4.87), (4.88), and (4.89) simultaneously to calculate expressions for [H], [OH], and [O]. Adding equation (4.87), (4.88) and 2 times equation (4.89), and neglecting r_{OH}, r_H, and r_O yields

$$0 = 2k_1[H_2][O_2] + 2k_3[H\bullet][O_2] - k_5[H\bullet] - k_6[H][O_2][X] \tag{4.90}$$

Solving equation (4.90) for [H•] yields

$$[H\bullet] = \frac{2k_1[H_2][O_2]}{k_5 + (k_6[X] - 2k_3)O_2} \tag{4.91}$$

Equation (4.91) is the pseudo-steady-state approximation for the hydrogen atom concentration.

We give several other examples of the use of the pseudo-steady-state approximation in the problems at the end of this chapter. The pseudo-steady-state approximation is very important, so it is important that reader's solve some of these examples themselves to verify that they know how to use the pseudo-steady-state approximation to calculate a rate equation.

4.8.1   Tricks to Simplify the Rate Equation

The one difficulty with the pseudo-steady-state approximation is that the algebra is complicated. I find that students often get lost in the algebra and never get to an answer.

Just to keep track of everything, in the example at the end of Section 4.7, we are trying to get an expression for the hydrogen atom concentration in terms of the concentration of all of the other stable species. The general approach we take is to

1. Set up the differential equation for the species of interest in terms of rate of all of the elementary reactions using equation (4.25) to keep track of the coefficients.
2. Substitute the expression for the rate of each of the elementary reactions using equations from Section 4.3.
3. Set the derivatives of the intermediate concentrations to zero.
4. Eliminate terms in the expression in (1) that contain the concentrations of unstable intermediates other than the species of interest.
5. Solve the resultant expression for the concentration of the species of interest.

For example, if one wanted to calculate the concentration of hydrogen atoms during reaction (4.85), one would start with the expression for change in the concentration of hydrogen atoms:

$$\frac{d[H\bullet]}{dt} = r_2 - r_3 + r_4 - r_5 - r_6 \tag{4.92}$$

One substitutes in the appropriate rate terms, and sets the derivative of the intermediate concentration to zero.

$$0 \approx \frac{d[H\bullet]}{dt} = k_2[\bullet OH][H_2] - k_3[H\bullet][O_2]$$
$$+ k_4[\bullet O][H_2] - k_5[H\bullet] - k_6[H\bullet][O_2][X] \tag{4.93}$$

Next, one would have to determine what terms in this expression need to be eliminated. Notice that the k_3, k_5, and k_6 terms contain $[H\bullet]$. We are trying to find $[H\bullet]$, so those terms are okay in the expression. In contrast, the k_2 and k_4 terms contain $[O\bullet]$ or $[OH\bullet]$. These terms need to be eliminated.

The k_2 term contains the hydroxyl concentration, so one can eliminate it by looking at the steady-state approximation for hydroxyls. Setting the derivative in equation (4.88) to zero and solving for $k_2[OH][H_2]$ yields

$$k_2[OH\bullet][H_2] = 2k_1[H_2][O_2] + k_3 - [H\bullet][O_2] + k_4[O:][H_2] \tag{4.94}$$

Substituting equation (4.94) into equation (4.93) yields

$$0 = 2k_1[H_2][O_2] + 2k_4[O:][H_2] - k_5[H\bullet] - k_6[H\bullet][O_2][X] \tag{4.95}$$

Equation (4.95) has only one term to eliminate the k_4 term. The k_4 term contains the concentration of oxygen atoms. One can eliminate the term from the steady-state approximation on oxygen atoms in equation (4.89) by setting the derivative in equation (4.89) equal to zero; then, solving for k_4 [O][H$_2$] yields

$$k_4[O:][H_2] = k_3[H\bullet][O_2] \qquad (4.96)$$

Substituting equation (4.96) into equation (4.95) yields

$$0 = 2k_1[H_2][O_2] + 2k_3[H\bullet][O_2] - k_5[H\bullet] - k_6[H][O_2][X] \qquad (4.97)$$

Solving equation (4.97) for [H•] yields

$$[H\bullet] = \frac{2k_1[H_2][O_2]}{k_5 + (k_6[X] - 2k_3)O_2} \qquad (4.98)$$

The point of this exercise is that you need to focus on where you are going when you are simplifying the rate equation. Look at the terms that you need to eliminate, and find a way to eliminate them. It usually works out that other terms cancel as well. The algebra is usually simple unless you have a reaction where two other reactive intermediates collide to form products. In that case, only brute-force algebra works.

The key thing to remember when simplifying rate expressions are

- Write down the differential equations in terms of the rates and then substitute in the rate equations.
- Keep track of what terms you want to eliminate, and eliminate them.

The solutions are normally straightforward and the methods work when you keep track of where you are going.

4.9 RATE-DETERMINING STEPS

The pseudo-steady-state approximation is the key approximation that people use to derive rate equations. Consequently, this approximation is very important. There are a series of other approximations that are occasionally used. I do not think they are important. Still, I wanted to mention them so that you know what people are doing when they use another approximation.

One key simplification occurs if one can assume that one step, called the **rate-determining step** or **rate-limiting step**, is much slower than all of the other steps in the mechanism. The objective of this section is to work out the details. One can skip this section without loss of continuity.

Let us start by considering a simple reaction mechanism where the idea of a rate-determining step might be important. Consider a reaction where molecule A collides with a collision partner to yield an unstable excited species, $A^\ddagger$. Then the excited species reacts to form a product B:

$$A + X \underset{2}{\overset{1}{\rightleftharpoons}} A^\ddagger + X$$

$$A^\ddagger \xrightarrow{\ 3\ } B \qquad (4.99)$$

The rate of formation of B, r_B is given by

$$r_B = k_3[A^{\ddagger}] \tag{4.100}$$

where $[A^{\ddagger}]$ is the concentration of excited A molecules and k_3 is the rate constant for reaction 3. One can derive an equation for $[A^{\ddagger}]$ using the pseudo-steady-state approximation:

$$r_A^{\ddagger} = k_1[A][X] - k_2[A^{\ddagger}][X] - k_3[A^{\ddagger}] \approx 0 \tag{4.101}$$

where $r_A^{\ddagger}$ is the rate of formation of excited A molecules; $[A]$, $[A^{\ddagger}]$, and $[X]$ are respectively the concentrations of A molecules, excited A molecules, and collision partners; and, k_1, k_2, and k_3 are the rate constants for reactions 1, 2, and 3, respectively. Solving equation (4.101) for $[A^{\ddagger}]$ and then substituting that into equation (4.100) yields

$$r_B = \frac{k_3 k_1[A][X]}{k_2[X] + k_3} \tag{4.102}$$

Now consider the case where $k_2[X] \gg k_3$ so that reaction 3 is much slower than the rest. In that limit, equation (4.102) simplifies to

$$r_B = \frac{k_3 k_1}{k_2}[A] \tag{4.103}$$

This is a key result because it shows that the reaction looks first order when $k_2[X] \gg k_3$.

Next, we will show that one could have gotten to equation (4.103) more simply. When $k_2[X] \gg k_3$, reactions 1 and 2 are just about in equilibrium. At equilibrium, the ratio of the concentration of excited A molecules to total A molecules is given by the equilibrium constant $K_{1,2}$:

$$\frac{[A^{\ddagger}]}{[A]} = K_{1,2} = \frac{k_1}{k_2} \tag{4.104}$$

In equation (4.104), we noted that the equilibrium constant for a reaction is given by the ratio of the equilibrium constants for a reaction, as noted in Section 4.3. Substituting equation (4.104) into equation (4.100) yields

$$r_B = \frac{k_3 k_1}{k_2}[A] \tag{4.105}$$

Notice that equations (4.103) and (4.105) are identical. Therefore, in the limit that $k_2[X] \gg k_3$, one can calculate an accurate rate equation by assuming that reactions 1 and 2 are in equilibrium.

One can generalize this process to any reaction. The idea is to assume that one step, called the **rate-determining step** or **rate-limiting step**, is much slower than the rest of the steps in the mechanism. One can show that if one step is extremely slow, then all of the steps before the slow step can be assumed to be in equilibrium with the reactants, and all of the steps after the rate-determining step can be considered to be in equilibrium with the products. If so, one can often derive a suitable rate equation for the reaction using somewhat less algebra than in the case where none of the steps are slow.

For example, consider a reversible reaction:

$$A + B \underset{-1}{\overset{1}{\rightleftharpoons}} C + D \underset{-2}{\overset{2}{\rightleftharpoons}} E + F \underset{-3}{\overset{3}{\rightleftharpoons}} G + H \underset{-4}{\overset{4}{\rightleftharpoons}} I + J \underset{-5}{\overset{5}{\rightleftharpoons}} L + M \qquad (4.106)$$

One can write, r_A, the rate formation of the reactant A by

$$r_A = r_{-1} - r_1 \qquad (4.107)$$

where r_1 and r_{-1} are the rates of reactions 1 and -1, respectively. The steady-state approximation on C shows

$$\frac{d[C]}{dt} = r_1 - r_{-1} - r_2 + r_{-2} \approx 0 \qquad (4.108)$$

where r_1, r_{-1}, r_2, and r_{-2} are respectively the rates of reaction 1, -1, 2, and -2. Similarly, the steady-state approximation on E shows

$$\frac{d[E]}{dt} = r_2 - r_{-2} - r_3 + r_{-3} \approx 0 \qquad (4.109)$$

where r_2, r_{-2}, r_3, and r_{-3} are the rates of reaction 2, -2, 3 and -3, respectively. Adding equations (4.107)–(4.109) yields

$$r_A = r_{-3} - r_3 \qquad (4.110)$$

Substituting the expressions for r_3 and r_{-3} into equation (4.110) yields

$$r_A = k_{-3}[G][H] - k_3[E][F] \qquad (4.111)$$

Equation (4.110) is a key result. It says that if one has a reaction consisting of several steps, then at steady state, the net rates of each of the steps are equal. Consequently, one can calculate the rate of reaction by considering the net rate of reaction at any point during the reaction sequence.

Now, let's assume that reaction 3 and -3 are much slower than the rest. Then it will be permissible to assume that the concentrations of E and F are in equilibrium with the reactants. The analysis in Section 4.4 shows:

$$[E][F] = \frac{k_1 k_2}{k_{-1} k_{-2}}[A][B] \qquad (4.112)$$

Similarly, the concentration of G and H will be in equilibrium with the products:

$$\frac{[L][M]}{[G][H]} = \frac{k_4 k_5}{k_{-4} k_{-5}} \qquad (4.113)$$

Substituting equation (4.112) and (4.113) into equation (4.111) yields

$$r_A = \frac{k_{-3} k_{-4} k_{-5}}{k_4 k_5}[L][M] - \frac{k_1 k_2 k_3}{k_{-1} k_{-2}}[A][B] \qquad (4.114)$$

Equation (4.114) is the equivalent of the pseudo-steady-state approximation when k_3 and k_{-3} are small.

In my experience, there is not that much simplification in assuming that one of the steps in a mechanism is at equilibrium, and one often misses some key information. Therefore, there is some uncertainty in using the idea of a rate-determining step to derive a rate equation.

Still, the idea of a rate-determining step is a useful construct to understand reactions. Typically, there is one step in a reaction mechanism, called the *rate-determining* step, which is much slower than the other reactions in the mechanism. This slow step is the bottleneck. Anything one can do to eliminate the bottleneck will greatly enhance the rate of reaction.

4.10   TESTING THE ACCURACY OF THE PSEUDO-STEADY-STATE APPROXIMATION

Next, I want to change topics and consider when the pseudo-steady-state approximation is accurate. In practice, this approximation is used to calculate rate equations for a wide range of reactions. It usually works but unfortunately, there are a few exceptions. The exceptions come in systems that can explode, or in systems with other unusual behavior, such as oscillations. These are rare examples, but there are still a few cases where the pseudo-steady-state approximation does not work.

The fact that the pseudo-steady-state approximation may not work is pretty significant. For example, let's assume that you are trying to determine the kinetics for a reaction that has never been studied before. In principle, you can derive a mechanism and then use the pseudo-steady-state approximation to derive a rate equation. The difficulty, though, is that at that point you will not know whether the pseudo-steady-state approximation is accurate. As a result, you will not know whether to believe the results derived using the pseudo-steady-state approximation.

In this section, we will discuss how one can determine whether the pseudo-steady-state approximation is accurate for a given reaction network. Let's start with the case discussed in Section 4.4:

$$A + H^+ \underset{2}{\overset{1}{\rightleftharpoons}} I \overset{3}{\longrightarrow} P + H_2 \qquad (4.115)$$

Figure 4.4 compares the intermediate concentration calculated via the pseudo-steady-state approximation [equation (4.62)] to the concentration calculated via the exact result [equation (4.54)] for several values of the rate constants. Notice that in the exact case ($k_2 = 5.7 \times 10^6$/minute), the pseudo-steady-state approximation works extremely well, and in fact, this approximation is still accurate when k_2 is reduced by a factor of 10^5. Significant deviations are seen only when k_2 is reduced to about 1.

Notice that at the point where the pseudo-steady-state approximation fails, the intermediate concentration is already a few percent of the reactant concentration. This is a general finding. The pseudo-steady-state approximation is essentially exact when the intermediate concentration is much smaller than the reactant and product concentrations. One can, however, start to see deviations when the intermediate concentration begins to be significant compared to the reactant (or product) concentration. Later in this chapter, we will note that there are some cases where the pseudo-steady-state approximation works even though the intermediate concentrations are significant. However, the key

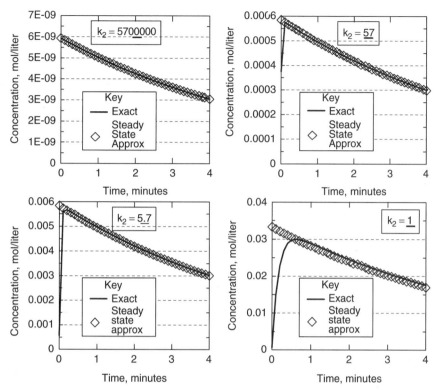

Figure 4.4 A comparison of the intermediate concentration calculated via the steady-state approximation to the concentration calculated via the exact result: equations (4.54) to (4.57) with $k_6 = 0.2$, $k_3 = 5k_2$, and various values of k_2.

point to remember now is that the pseudo-steady-state approximation always works when the intermediate concentrations are much less than the reactant or product concentrations.

Physically, when we set the net rate of formation of the intermediate to zero, we are saying that the amount of the intermediate that we are forming is negligible. That is a good approximation if the intermediate concentration is tiny so that the amount we are forming is negligible. However, it is a bad approximation if we form a nonnegligible amount of impurity.

One can use the idea that the concentration of the intermediates is negligible to check whether the pseudo-steady-state approximation works for a specific example. The idea is to use the pseudo-steady-state approximation to calculate the concentrations of all of the intermediates and see whether the concentration of any of the intermediates is significant compared to the reactants or products.

For example, there are two intermediates in reaction (4.67): [H] and [Br]. One can estimate the intermediate concentrations from equation (4.80) and (4.82). Rate constants are given in Westley (1980). When we plug numbers into equations (4.80) and (4.82), we find that at 300 K the intermediate concentrations are a factor of 10^{-8} lower than the reactant concentrations. The intermediate concentrations are low, which implies that the steady-state approximation will work.

4.11 FAILURE OF THE PSEUDO-STEADY-STATE APPROXIMATION

4.11.1 Explosions

The reaction between hydrogen and oxygen [reaction (4.84)] is different. In Section 4.7, we derived an expression for the pseudo-steady-state approximation to [H], the hydrogen atom concentration. Rearranging the expression yields

$$[H] = \frac{2k_1[H_2][O_2]}{k_5 + k_6[X][O_2] - 2k_3[O_2]} \qquad (4.116)$$

Figure 4.5 is a plot of the radical concentration as a function of the oxygen pressure calculated from equation (4.116) for some typical values of the parameters. Notice that the radical concentration goes first to infinity and then to negative infinity and then comes back again.

This weird behavior occurs because there is a negative term in the denominator of equation (4.116). That term increases as the concentration of molecular oxygen in the reactor increases. When the denominator of equation (4.116) decreases, the calculated value of the hydrogen atom concentration increases. If one plugs in the rate constants, one finds that at temperatures above 450 K, the denominator of equation (4.116) goes to zero at moderate oxygen concentrations. Consequently, according to the pseudo-steady-state approximation, under some conditions, the radical concentration should approach infinity. One can even get to the situation where the denominator is negative, which means that according to the pseudo-steady-state approximation, the radical concentration is negative.

What is going on is that the pseudo-steady-state approximation fails in cases where the hydrogen atom concentration is comparable to or larger than the H_2 concentration, or when the calculated concentrations are negative. Physically, it is impossible for the hydrogen atom concentration to be negative. Large hydrogen atom concentrations are impossible as well since the only hydrogen source is the H_2 initially fed into the reactor.

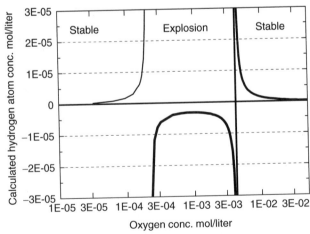

Figure 4.5 A plot of the hydrogen atom concentration calculated from equation (4.116) as a function of the oxygen concentration.

Consequently, the hydrogen atom concentration can never exceed twice the initial H_2 concentration.

The pseudo-steady-state approximation fails whenever the intermediate concentration is comparable to the H_2 concentration, which one can test by plugging numbers into equation (4.116). The weird behavior in Figure 4.5 is associated with the failure of the steady-state approximation.

That does not imply that the pseudo-steady-state approximation was useless in this example. If one runs the hydrogen/oxygen reaction under conditions where the denominator in equation (4.116) is small or negative, one finds that the concentration of radicals does increase rapidly when one loads the reactants into the reactor. The radical concentration does not actually increase to infinity; instead the radical concentration is finite. However, there is a rapid production of radicals, which means that the reaction rate gets to be very large. One calls a rapid increase in the radical concentration an **explosion**. You already know that a hydrogen/oxygen mixture can explode. Now you can see mathematically how the explosion happens.

If one actually runs the hydrogen/oxygen reaction experimentally, one finds that there is some complex behavior because the rate of the reactions with the walls [i.e., reactions 5 and 7 in equation (4.85)] depends on the rate at which molecules diffuse through the gas mixture. One can also find circumstances where the heat generated during a reaction heats up the reaction mixture leading to an explosion. However, the thing to remember is that an explosion happens when the concentration of reactive intermediates grows quickly during the course of the reaction. Once you start to build up radicals, the radicals are so reactive that the reaction rate quickly grows toward a large value. In the $H_2 + O_2$ case, one observes an upper and lower explosion limit. The lower explosion occurs because, as you raise the oxygen pressure, the $-k_3[O_2]$ term in the denominator of equation (4.116) grows until the denominator becomes zero. The upper explosion limit occurs because at high pressure, the $k_6[X][O_2]$ term becomes important, and so the denominator becomes positive again. One can also get a thermal explosion, where the heat generated by the reaction raises the temperature, which, in turn, raises the radical concentration.

There are many other examples where the pseudo-steady-state approximation fails. For example, OH and HO_2 radicals are stable in the upper atmosphere, so one cannot apply the pseudo-steady-state approximation to them. The pseudo-steady-state approximation works only when the intermediate concentrations are much smaller than the reactant and product concentrations. The pseudo-steady-state approximation does not work well in cases where the intermediate concentrations are comparable to the reactant concentrations. In those cases, one can solve the rate equations numerically as described in Section 4.5. The pseudo-steady-state approximation is quite useful, but it does not work all the time.

4.11.2 Oscillations

There is one major class of reactions where the pseudo-steady-state approximation fails dismally: systems where oscillations in rate are observed. We briefly mentioned oscillating reactions in Chapter 2. The classic example is a modified Belousov–Zhabotinskii (BZ) reaction:

$$HOOCCH_2COOH + HBrO_3 \xrightarrow{Fe^{3+}} products \qquad (4.117)$$

If you run reaction (4.117) in a beaker with phenanthroline to complex the iron, the color of the mixture oscillates from red to blue. This makes a great demonstration during class. In a batch reactor, the oscillations eventually stop because one runs out of reactants.

However, one can also run the reaction in a flow reactor if one continuously feeds reactants into the flow reactor, the oscillations can be sustained forever (see Figure 4.6).

Clearly, if the reaction rate oscillates, the pseudo-steady-state approximation is not going to work.

Unfortunately, the BZ reaction is very complicated, so it does not make a good example for the discussion here. Therefore, I would like to instead consider a simpler example: the oxidation of a molecule called *nicotinamide adenine dineuclotide hydride* (NADH), shown in Figure 4.7. The oxidation reaction can be written

$$\text{NADH} + \text{H}^+ + \tfrac{1}{2}\text{O}_2 \longrightarrow \text{NAD}^+ + \text{H}_2\text{O} \tag{4.118}$$

The oxidation is catalyzed by an enzyme called *peroxidase*. Reaction (4.118) is one of the key steps in the energy cycle of cells. It is also used to produce cell walls in plants.

Benson and Scheeline (1996) examined the chemistry in detail using the mechanism in Figure 4.6. They find that the key reactions are

$$\text{NADH} + \text{O}_2 + \text{H}^+ \xrightarrow{\;1\;} \text{NAD}^+ + \text{H}_2\text{O}_2 \tag{4.119}$$

$$\text{Per}^{3+} + \text{H}_2\text{O}_2 + 2\text{H}^+ \xrightarrow{\;2\;} \text{Per}^{5+} + 2\text{H}_2\text{O} \tag{4.120}$$

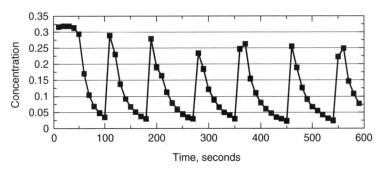

Figure 4.6 Rate oscillations during reaction (4.117). (Unpublished data of G. E. Poisson, D. A. Tuchmann, and R. I. Masel).

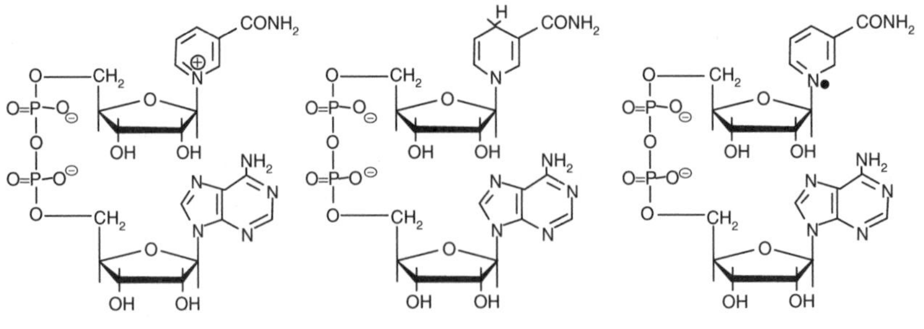

Figure 4.7 NAD$^+$, NADH and NAD$^{\bullet}$.

$$Per^{5+} + NADH \xrightarrow{\ 3\ } Per^{4+} + NAD\bullet + H^+ \tag{4.121}$$

$$Per^{4+} + NADH \xrightarrow{\ 4\ } Per^{3+} + NAD\bullet + H^+ \tag{4.122}$$

$$NAD\bullet + O_2 \xrightarrow{\ 5\ } NAD^+ + O_2^- \bullet \tag{4.123}$$

$$Per^{3+} + O_2^- \bullet + 4H^+ \xrightarrow{\ 6\ } Per^{6+} + 2H_2O \tag{4.124}$$

$$2O_2^- \bullet + 2H^+ \xrightarrow{\ 7\ } H_2O_2 + O_2 \tag{4.125}$$

$$Per^{6+} + NAD\bullet \xrightarrow{\ 8\ } Per^{5+} + NAD^+ \tag{4.126}$$

$$NADH_R \xrightarrow{\ 9\ } NADH \tag{4.127}$$

$$NADH \xrightarrow{\ 10\ } NADH_R \tag{4.128}$$

$$O_{2(g)} \longrightarrow O_{2(aq)} \tag{4.129}$$

$$O_{2(aq)} \longrightarrow O_{2(g)} \tag{4.130}$$

$$2NAD\bullet \xrightarrow{\ 11\ } (NAD)_2 \tag{4.131}$$

where $NAD\bullet$ is the NAD radical shown in Figure 4.7; Per^{3+}, Per^{4+}, Per^{5+}, and Per^{6+} are the different oxidation states of the peroxidase enzyme; and $NADH_R$ is a different isomer of NADH.

Figure 4.8 shows a pictorial representation of the mechanism. The mechanism contains a classic catalytic cycle with five important isomers of horseradish peroxidase, and two forms of O_2 (i.e., O_2 and O_2^- with different properties).

The NAD radicals are classic unstable intermediates. As a result, one would expect their concentrations to be low. However, the radicals can form dimers. The dimers are stable and can accumulate in the system. Consequently, one cannot use the steady-state approximation to calculate the dimer concentration.

What one needs to do, instead, is to follow the procedures in Section 4.4 and set up the differential equations for all of the species in the system. For example, one would say that the concentration of NAD radicals follows

$$\frac{d[NAD\bullet]}{dt} = k_3[NADH][Per^{5+}] + k_4[NAD\bullet][Per^{4+}] - k_5[NAD\bullet][O_2]$$

$$- k_8[NAD^+][Per^{6+}] - k_{11}[NAD\bullet]^2 \tag{4.132}$$

Benson and Scheeline (1996) list the other differential equations to describe the behavior of the species produced during NADH oxidation.

One can numerically integrate these differential equations to calculate the behavior of the system using a modification of the computer program at the end of Example 4.B.6.C. Benson and Scheeline did so, and the key result is shown in Figure 4.9. Notice that the concentrations of many species oscillate with time. Therefore, the system never reaches steady state.

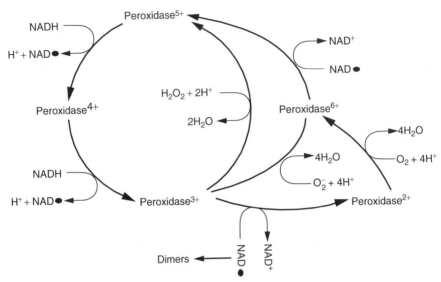

Figure 4.8 The mechanism of NADH oxidation as suggested by Benson and Scheeline (1996).

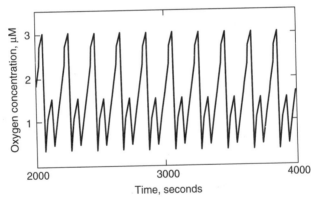

Figure 4.9 Solution of the differential equations of Bensen and Scheeline (1996).

There are many other systems that oscillate. Generally, one can determine whether a given reaction shows oscillations by numerically integrating the rate equations for all of the species in the system. If the system can oscillate, then one will observe oscillations in the numerical solutions. One can also do analysis on the differential equations for each of the species to see when oscillations can occur. The details are beyond the scope of this book. See Gray and Scott (1990) for more information.

The key point to remember is that the steady-state approximation fails when the concentrations of the reactive intermediates become significant. One must integrate the rate equations for the system in cases where the steady-state approximation fails. In such a case, one can get unusual behavior, including, possibly, oscillations or explosions.

4.12 USING THE PSEUDO-STEADY-STATE APPROXIMATION TO INFER MECHANISMS

In Sections 4.6 and 4.10, we discussed cases where the mechanism of a reaction was known and showed how one can use the pseudo-steady-state approximation to derive a rate equation for the reaction. I have to tell you that in practice I have rarely used the pseudo-steady-state approximation in that way. It is usually hard to determine the mechanism of a reaction from experiments. Kinetic measurements are much easier. Consequently, if all that one could do with the pseudo-steady-state approximation were to calculate something that is easy to measure, namely, a rate equation, from something that is hard to measure, namely, a mechanism, the pseudo-steady-state approximation would not be very useful.

In Chapter 5, we will show that one can often predict mechanisms of reactions from first principles. In that case, the pseudo-steady-state approximation is useful. However, in a more general case, one will not know the mechanism very well. In that case, one can apply the pseudo-steady-state approximation in reverse and use kinetic measurements to infer something about the mechanism of a reaction.

Generally, I find that it is best to try to infer the rate law and the mechanism simultaneously. The general procedure is to first do some kinetic measurements. You need to know if the reaction is closer to first-order or second-order in each reactant or something in between. Once you have some kinetic information, you

1. Guess at a mechanism.
2. Use the pseudo-steady-state approximation to derive a rate equation for that mechanism.
3. Take rate data.
4. See if the rate data fits the mechanism. Generally, once you have a mechanism, you need to take more rate data to verify that the rate equation fits the kinetic data.
5. Modify the proposed mechanism and iterate.

Unfortunately, there is a lot of black art in trying to find a mechanism. For years, people mainly guessed, using their best judgment, and then iterated. Such a procedure allows you to fit kinetic data. However, often more than one mechanism will fit the data. Consequently, it is incorrect to assume that just because a mechanism fits the rate data, the mechanism is correct. On the other hand, if the rate equation does not fit the data, the mechanism is incorrect. Consequently, one can use the pseudo-steady-state approximation to eliminate proposed mechanisms.

Let's solve an equation to show how one can use these ideas to fit rate data. Assume that you were trying to determine the mechanism of the following reaction:

$$H_2 + Br_2 \Longrightarrow 2HBr \tag{4.133}$$

As a first guess, let's assume that you could write the rate equation for the reaction as follows:

$$r_{HBr} = k_{HBr}[Br_2]^n[H_2]^m \tag{4.134}$$

where r_{HBr} is the rate of HBr formation; $[H_2]$ and $[Br_2]$ are the concentrations of hydrogen and bromine in the reactor; and, k_{HBr}, m and n are constants. Now the question is how to fit the constants k_{HBr}, m, and n.

Well, the procedures in Chapter 3 give you a clue. For example, one could use the swamping method. You could load Br_2 and a large excess of hydrogen into a glass vessel, as described in Section 3.13.2. If there is a large excess of hydrogen in the reactor, then the $[H_2]$ term in equation (4.118) will be constant. Consequently, one can use Van't Hoff's method or some equivalent described in Section 3.11 to determine n and k.

When one actually does that procedure, one finds that the reaction order varies during the course of the reaction, from $n = 0.5$ at low conversions to $n = 1.5$, when there is excess HBr in the reactor. Other measurements show that the reaction is first-order in the hydrogen pressure.

The next question is where to go from there. One could try to empirically pick a rate equation, but it is difficult to know where to start. Therefore, the next step is to guess a mechanism, use the pseudo-steady-state approximation to derive a rate equation for the mechanism, and then see if the mechanism fits.

In Chapter 5, we will provide some rules to guess the mechanisms of gas-phase reactions. One key finding is that most gas-phase reactions go by what is called an *initiation–propagation mechanism*. Consider the simple reaction

$$A + B \Longrightarrow P_1 + P_2 \tag{4.135}$$

where a molecule of A reacts with a molecule of B to produce two products, $P_1 + P_2$. According to the initiation–propagation mechanism, the reaction starts when a molecule, X, collides with one of the reactant molecules and breaks a bond in the reactants. The result is that two radicals are formed, which I will call R_1 and R_2:

$$X + B \longrightarrow R_1 + R_2 + X \tag{4.136}$$

Then the radicals react with the reactants or the products, to yield a series of new radicals R_3 and R_4:

$$R_1 + A \longrightarrow P_1 + R_3$$
$$R_2 + B \longrightarrow P_2 + R_4 \tag{4.137}$$

R_3 and R_4 can also react to regenerate the original radicals:

$$R_4 + A \longrightarrow P_1 + R_2$$
$$R_3 + B \longrightarrow P_2 + R_1 \tag{4.138}$$

Finally, the radicals can recombine:

$$R_1 + R_2 \longrightarrow B \tag{4.139}$$

There are many different combinations. Let's assume that reaction (4.122) goes via an initiation–propagation mechanism. If the Br_2 bond breaks in the initiation step, then the simplest initiation–propagation reaction leading to HBr is

Initiation

$$X + Br_2 \xrightarrow{\ 1\ } 2Br + X$$

Propagation

$$Br + H_2 \xrightarrow{\ 2\ } HBr + H \tag{4.140}$$

$$H + Br_2 \xrightarrow{\;3\;} HBr + Br$$

Termination

$$X + 2Br \xrightarrow{\;4\;} Br_2 + X$$

Alternatively, if one assumes that the H_2 bond breaks during the initiation step, the simplest reaction leading to products is

Initiation

$$X + H_2 \xrightarrow{\;1\;} 2H + X$$

Propagation

$$Br + H_2 \xrightarrow{\;2\;} HBr + H \tag{4.141}$$

$$H + Br_2 \xrightarrow{\;3\;} HBr + Br$$

Termination

$$X + 2H \xrightarrow{\;4\;} H_2 + X$$

In Section 4.7, we used the pseudo-steady-state approximation to derive a rate equation for mechanism (4.68). The result was equation (4.83). Mechanism (4.140) is equivalent to mechanism (4.68) with $k_5 = 0$. Therefore, if the reaction followed mechanism (4.140), the rate would be equation (4.83) with $k_5 = 0$, or

$$r_{HBr} = 2k_{140} \left(\frac{k_1}{k_4} \right)^{1/2} [H_2][Br_2]^{1/2} \tag{4.142}$$

A similar analysis shows that if the reaction follows mechanism (4.141), the reaction would obey

$$r_{HBr} = 2k_{141} \left(\frac{k_1}{k_4} \right)^{1/2} [H_2]^{1/2}[Br_2] \tag{4.143}$$

Now we have two possible rate equations that we could compare to data. There are many different experiments that we could do. For example, if equation (4.142) worked, the rate would be first-order in the hydrogen pressure. If reaction (4.143) worked, the reaction would be half-order in the hydrogen pressure. One can measure the order in hydrogen by swamping the system with excess hydrogen or bromine as discussed in Section 3.13.2, and then use Van't Hoff's method to determine the order of the reaction in the hydrogen pressure. In actual practice, Bodenstein and Lund did not use the swamping method. Still they were able to get a useful Van't Hoff plot. The analysis appears in Example 3.K.

Figure 4.10 shows a plot of k_{140} versus k_{141} as a function of τ obtained in this way. The k_{141} data shows some scatter at small τ and a drop in k_{140} at high τ. The k_{141} curve shows a similar trend except that the scatter is much larger. To put Figure 4.10 in perspective, if equation (4.142) fit the data, then k_{140} should have been constant. In contrast, if equation (4.143) had fit the data, then k_{141} should have been constant. Neither

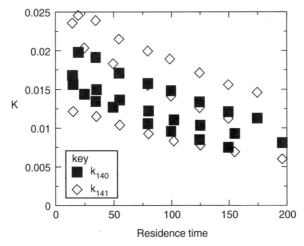

Figure 4.10 The rate of HBr production plotted as suggested by Van't Hoff. [Data of Bodenstein and Lund (1904).]

was constant. However, k_{140} is more constant than k_{141}. Therefore, equation (4.142) is closer to the truth than equation (4.143) is.

Equation (4.142) is appropriate for reaction (4.141), while equation (4.143) is appropriate for reaction (4.141). Consequently, the observation that k_{140} is more constant than k_{141} implies that mechanism (4.140) is closer to the true mechanism than is mechanism (4.141).

Bodenstein and Lund did a series of experiments starting with various initial concentrations of H_2 and Br_2, and then used Van't Hoff's method to estimate k_{140} and k_{141}. The procedure follows the procedure in Example 3.K.

One would then repeat the experiment with excess bromine in the reactor. When one did that, one would find that in the early part of the reaction (4.142) would work OK, but at the latter part of the reaction neither equation (4.142) nor equation (4.143) fit the bromine dependence very well. Therefore, one would conclude that neither mechanism (4.140) nor mechanism (4.141) is quite correct for this system, although mechanism (4.140) is closer to reality.

At that point, you would have to modify mechanism (4.140) in some way to make it fit better. Up to 1990, there was no good way to do that except by trial and error. The idea was to modify the proposed mechanism and iterate to the exact mechanism, equation (4.68). In Chapter 5, we will provide some empirical rules that let you predict a mechanism without having to resort to a trial and error procedure. However, the trial-and-error procedures do work, and most of the mechanisms that we know about were determined in this way.

4.12.1 Mechanisms to Get Different Reaction Orders

The hardest part of this procedure is to find the first guess of a mechanism. Let's consider a simple reaction:

$$CH_3COH \Longrightarrow CH_4 + CO + \text{other products} \qquad (4.144)$$

If we presume that the reaction follows

$$
\begin{gathered}
\text{Initiation}\\
X + CH_3COH \xrightarrow[2]{1} CH_3 + COH + X\\
\text{Propagation}\\
CH_3 + CH_3COH \xrightarrow{2} CH_4 + CH_3CO\\
CH_3CO + X \xrightarrow{3} CH_3 + CO + X\\
\text{Termination}\\
X + 2CH_3CO \xrightarrow{4} CH_3CO - COCH_3 + X
\end{gathered}
\tag{4.145}
$$

One can show that the kinetics will follow:

$$
r_{CO} = k_3[X]\left(\frac{k_1}{2k_4}\right)^{1/2}[CH_3COH]^{1/2} \tag{4.146}
$$

If one changes the mechanism to

$$
\begin{gathered}
\text{Initiation}\\
X + CH_3COH \xrightarrow{1} CH_3 + COH + X\\
\text{Propagation}\\
CH_3 + CH_3COH \xrightarrow{2} CH_4 + CH_3CO\\
CH_3CO + X \xrightarrow{3} CH_3 + CO + X\\
\text{Termination}\\
X + 2CH_3 + X \xrightarrow{4} C_2H_6 + X
\end{gathered}
\tag{4.147}
$$

The kinetics will follow

$$
r_{CH_4} = k_2\left(\frac{k_1}{2k_4}\right)^{1/2}[CH_3COH]^{3/2} \tag{4.148}
$$

One can also consider a third mechanism:

$$
\begin{gathered}
\text{Initiation}\\
X + CH_3COH \xrightarrow{1} CH_3 + COH + X\\
\text{Propagation}\\
CH_3 + CH_3COH \xrightarrow{2} CH_4 + CH_3CO\\
CH_3CO + X \xrightarrow{3} CH_3 + CO + X
\end{gathered}
\tag{4.149}
$$

Termination

$$X + CH_3 + CH_3CO \xrightarrow{\;4\;} CH_3COH$$

The algebra is more difficult for this case. Still, one finds that the rate follows

$$r_{CH_4} = \frac{k_2 k_3}{2k_4}[CH_3COH] + \text{higher-order terms} \qquad (4.150)$$

The implication of these results is that the form of the rate equation depends in a critical way on the nature of the termination reaction.

Now a key issue is what is different about the termination steps in the various reaction schemes above to cause the kinetics of reaction (4.140) to be different from the kinetics of reaction (4.142).

Goldfinger et al. (1945) noted that there are two different intermediates, CH_3 and CH_3CO, in the propagation steps in mechanisms (4.140), (4.142), and (4.144). CH_3 reacts via a bimolecular reaction, while CH_3CO reacts via a pseudounimolecular pathway. Goldfinger et al. called any intermediate that reacts via a bimolecular reaction a β intermediate, while any intermediate that reacts via a pseudo-first-order reaction is an μ intermediate. They then showed that for first-order initiation reactions (reactants $+ X \rightarrow$ radicals $+ X$), if the termination step is

$$\beta + \beta + X \xrightarrow{\hspace{1cm}} \text{products} + X \qquad (4.151)$$

The reaction will always be $\frac{3}{2}$-order. Table 4.1 gives several other examples.

One uses Table 4.1 to quickly get the kinetics from a mechanism. For example, in 1935, Rootsaert and Farkas discovered that H_2 can exist in two forms: parahydrogen (p-H_2), where the nuclei on the two hydrogens are in the same spin state, and orthohydrogen (o-H_2), where the nuclei have different spin states. At low temperature, the para state is most stable, but as the temperature is raised, the p-H_2 is converted to o-H_2. Let's assume that the reaction follows the following mechanism:

$$p\text{-}H_2 + X \xrightarrow{\;1\;} 2H + X$$

$$H + p\text{-}H_2 \xrightarrow{\;2\;} o\text{-}H_2 + H \qquad (4.152)$$

$$2H \xrightarrow{\;3\;} p\text{-}H_2$$

Table 4.1   The kinetics of reactions of the form A $\Rightarrow$ products following Goldfinger, Letort and Niclause (1945)

Initiation Reaction	Termination Reaction	Order
$A + X \rightarrow$ radicals $+ X$	$\beta + \beta + X \rightarrow$ products $+ X$	3/2
$A + X \rightarrow$ radicals $+ X$	$\beta + \mu + X \rightarrow$ products $+ X$	1
$A + X \rightarrow$ radicals $+ X$	$\mu + \mu + X \rightarrow$ products $+ X$	1/2
$2A + X \rightarrow$ radicals $+ X$	$\beta + \beta + X \rightarrow$ products $+ X$	2
$2A + X \rightarrow$ radicals $+ X$	$\beta + \mu + X \rightarrow$ products $+ X$	3/2
$2A + X \rightarrow$ radicals $+ X$	$\mu + \mu + X \rightarrow$ products $+ X$	1

According to mechanism (4.152), hydrogen atoms are β radicals. Mechanism (4.152) shows $\beta-\beta$ termination. Therefore, one would expect the rate to be $\frac{3}{2}$-order in H_2. A detailed derivation shows that the rate is given by

$$\frac{d[o\text{-}H_2]}{dt} = k_2 \left(\frac{k_1}{k_3}\right)^{1/2} [p\text{-}H_2]^{3/2} \tag{4.153}$$

in agreement with the Goldfinger, Letort and Niclause rules. I am not aware of similar rules for bimolecular reactions. Most bimolecular reactions follow rather complex reaction mechanisms. However, a few mechanisms do yield simple rate laws. Rate laws for these mechanisms are given in Table 4.2.

Again, we see that the form of the rate equation depends in a critical way on the form of the termination reaction.

There are two ways to look at that result. First, from a practical standpoint, termination reactions are exceedingly difficult to study because the termination rates are so small.

Table 4.2   Rate equations for a few reactions of the form $A_2 + B_2 \Rightarrow 2AB$

Mechanism	Rate Equation
$X + A_2 \xrightarrow{\ 1\ } 2A + X$ $A + B_2 \xrightarrow{\ 2\ } AB + B$ $B + A_2 \xrightarrow{\ 3\ } AB + A$ $X + 2A \xrightarrow{\ 4\ } A_2 + X$	$r_{AB} = 2k_2 \left(\dfrac{k_1}{k_4}\right)^{1/2} [A_2]^{1/2}[B_2]$
$X + A_2 \xrightarrow{\ 1\ } 2A + X$ $A + B_2 + X \xrightarrow{\ 2\ } AB_2 + X$ $AB_2 + A_2 \xrightarrow{\ 3\ } 2AB + A$ $2AB_2 \xrightarrow{\ 4\ } A_2 + 2B_2$	$r_{AB} = 2\left(\dfrac{k_3}{k_4}\right)[A]_2$ $\times(\sqrt{(k_3[A_2])^2 + 4k_4(k_1k_2)^{1/2}[A_2]^{1/2}[B_2][X]} - k_3[A_2])$
$X + A_2 \xrightarrow{\ 1\ } 2A + X$ $A + B_2 \xrightarrow{\ 2\ } AB + B$ $B + A_2 \xrightarrow{\ 3\ } AB + A$ $X + 2B \xrightarrow{\ 4\ } B_2 + X$	$r_{AB} = 2k_3 \left(\dfrac{k_1}{k_4}\right)^{1/2} [A_2]^{3/2}$
$X + A_2 \xrightarrow{\ 1\ } 2A + X$ $A + B_2 \xrightarrow{\ 2\ } AB + B$ $B + A_2 \xrightarrow{\ 3\ } AB + A$ $A \xrightarrow{\ 4\ } \text{wall}$	$r_{AB} = 4\left(\dfrac{k_1k_2}{k_4}\right)[A_2][B_2][X]$

Consequently, if one tries to determine the mechanism by doing an experiment, one can often miss some key termination. In such a case, one will get the kinetics wrong.

The other way to look at this, though, is that the kinetics are very sensitive to the form of the termination reaction. Consequently, if one gets the rate equation right, one will often have the termination right. Unfortunately, one could have the wrong propagation reactions.

Note that the details of the propagation reactions do not have a large influence on the form of the rate equation. Consequently, it is often the case that more than one mechanism will fit a given set of rate data.

To make things worse, real data have inaccuracies. When there are inaccuracies in the data, it is often hard to distinguish between different rate laws. Further, real kinetic models have five or six parameters. When there are that many parameters, it is often possible to do a reasonable job of fitting rate data even with a wrong mechanism. In my experience, it is useful to use kinetics to suggest the mechanism of a reaction. However, I usually like to see some spectroscopic evidence for the presence of the key intermediates or some calculations to see if the rate laws are correct before I believe that the mechanism is correct.

4.13 SUMMARY

Table 4.3 summarizes the key findings in this chapter. The main topic of the chapter was the relationship between rates of reactions and mechanisms of reactions. We introduced

Table 4.3 Summary of key concepts

Commercially important reactions rarely go directly from reactants to products; instead, the reaction goes by a complete mechanism where an unstable intermediate forms and then reacts
  Elementary reaction—reaction where reactants converted to products without intermediates
  Overall reaction—a reaction going from stable reactants to stable products
  Mechanism—the sequence of reactions that occur when the reactants are converted to products
Key equations
  For an *elementary* reaction $A + B \xrightarrow{\;1\;} C + D$

$$r_1 = k_1[A][BC] \tag{4.20}$$

  For a species A participating in elementary reactions 1, 2, 3,..., i

$$r_A = \beta_{A,1}r_1 + \beta_{A,2}r_2 + \cdots + \beta_{A,i}r_i \tag{4.25}$$

Equation (4.25) can be used to set up differential equations for the concentrations of all of the species; equations can be solved using
  Analytic solution
  Numerical solution
  Pseudo-steady-state formation
Pseudo-steady-state approximation
  The net rate of formation of key reactive intermediates is much smaller than the rate of formation or distinction of the intermediates; consequently, the net rate can be set to zero when solving the rate equation
  Does not always work
    Stable by-products
    Explosions

the concept of a mechanism and showed how one can use the mechanism to derive a series of differential equations to analyze the behavior of a reacting mixture. We then derived the pseudo-steady-state approximation and showed how this approximation can be used to derive a rate equation for a given reaction or to infer a mechanism of a reaction from kinetic data. We also discussed the limitations of the pseudo-steady-state approximation — in particular, cases where the pseudo-steady-state approximation fails because the system explodes. Most of our discussion up to this point was relatively brief because we have assumed that most of our readers have seen this material before. However, the derivation and analysis of the pseudo-steady-state approximation is difficult to find in the literature. Therefore, we wanted to include it for our readers.

4.14   SOLVED EXAMPLES AND SUPPLEMENTAL MATERIAL

Example 4.A   Computations using Equations (4.49), (4.50), and (4.54)   In the homework set, I ask students to verify the steady-state approximation themselves by plugging into equations (4.53), (4.54), and (4.58) for various values of the rate constants. That is not as easy as it seems, because equation (4.56) is what is called "ill-behaved." Equation (4.56) computes k_5 as a difference between two big numbers, and so unless one computes things to many significant figures, one can get the wrong answer. The following derivation shows how to get the right numbers.

Solution   According to equation (4.56)

$$k_5 = \tfrac{1}{2}\left(k_6 + k_2 + k_3 - \sqrt{(k_6 + k_2 + k_3)^2 - 4k_6k_3}\right) \qquad (4.A.1)$$

Rearranging equation (4.A.1) yields

$$k_5 = \frac{(k_6 + k_2 + k_3)}{2}\left(1 - \sqrt{1 + \frac{4k_6k_3}{(k_6 + k_2 + k_3)^2}}\right) \qquad (4.A.2)$$

Recall from calculus that one can expand $\sqrt{1-x}$ as a Taylor series as follows:

$$\sqrt{1-x} = 1 - \frac{x}{2} - \frac{x^2}{8} - \frac{x^3}{16} + \cdots \qquad (4.A.3)$$

Plugging into equation (4.A.3) yields

$$k_5 = \frac{(k_6 + k_2 + k_3)}{2}$$
$$\times \left(1 - 1 + \frac{2k_6k_3}{(k_6 + k_2 + k_3)^2} + \frac{2k_6^2k_3^2}{(k_6 + k_2 + k_3)^4} + \frac{4k_6^3k_3^3}{(k_6 + k_2 + k_3)^6}\right) + \cdots \quad (4.A.4)$$

Performing the algebra yields

$$k_5 = \frac{k_6k_3}{(k_6 + k_2 + k_3)} + \frac{k_6^2k_3^2}{(k_6 + k_2 + k_3)^3} + \frac{2k_6^3k_3^3}{(k_6 + k_2 + k_3)^5} + \cdots \qquad (4.A.5)$$

In practice, equation (4.A.5) gives a much more accurate value of k_5 than does equation (4.A.1) when $(k_2 + k_3) > 100 k_6$.

Table 4.A.1 shows the spreadsheet I actually used to do the calculations, while Table 4.A.2 shows the numerical values. I defined column H as time, column I as [A], column J as [I], column K as [P], column L as [I]^s, and column M as [I]^x. Cells K1 to K6 were k_1 to k_6. I was given k_1, k_2, and k_3 and [H$^+$], so I entered them. First, I needed to calculate k_4, k_5, and k_6. I calculated k_4 from equation (4.55). I calculated k_5 from equation (4.A.5). I calculated k_6 from equation (4.57). Column I (i.e., [A]) was calculated from equation (4.53), column J (i.e., [I]) was calculated from equation (4.54), column K (i.e., [P]) was calculated from equation (4.58), column L (i.e., [I]^s) was calculated from equation (4.59), and column M was calculated from equation (4.60).

The numerical values are given in Table 4.A.2. Notice that [I] = [I] after the first 2×10^{-7} seconds.

Example 4.B   Numerical Integration   The alternative way to solve Example 4.A is to do it numerically. Just by way of introduction, the reaction $A + H^+ \underset{2}{\overset{1}{\rightleftharpoons}} I \overset{3}{\longrightarrow} P$ follows the following differential equations:

$$\frac{d[A]}{dt} = -k_6[A] + k_2[I] \tag{4.B.1}$$

$$\frac{d[I]}{dt} = +k_6[A] - (k_2 + k_3)[I] \tag{4.B.2}$$

$$\frac{d[P]}{dt} = k_3[I] \tag{4.B.3}$$

In the next several sections I will show how these differential equations can be solved numerically.

4.B.1 Numerical Methods

When I teach this course, I usually need to give a short aside about numerical solution of differential equations. Our discussion will be brief. See Lorenzini and Passoni (1999) or Press et al. 1996 for a more comprehensive discussion of numerical simulation of the differential equations used in kinetics.

Let's define two new vectors, $\vec{C}$ and $\vec{R}$, by $C(1) = [A]$, $C(2) = [I]$, $C(3) = [P]$:

$$R(1) = \frac{dC(1)}{dt}$$

$$R(2) = \frac{dC(2)}{dt} \tag{4.B.4}$$

$$R(3) = \frac{dC(3)}{dt}$$

Physically, $\vec{C}$ is a vector containing the concentrations of all of the species, while $\vec{R}$ is a vector containing all of the rates of reaction. For our example

$$R(1) = -k_6[A] + k_2[I]$$

$$R(2) = +k_6[A] - (k_2 + k_3)[I] \tag{4.B.5}$$

$$R(3) = k_3[I]$$

Table 4.A.1   The formulas in the spreadsheet used for the calculations in Example 4.A.1

	H	I	J	K	L	M
1	H+=	0.2	k1=	1		
2			k2=	5700000		
3	1E-008	=time increment 1	k3=	28000000		
4	0.1	=time increment 2	k4=	(@SQRT((K2+K3+K6)^2- 4*K6*K3) +K2+K3+K6)/2		
5			k5=	@SUM(I8..L8)		
6			k6=	+$I$1*K1		
7						
8	k5= sum(	+K6*K3/ (K2+K3+K6)	+I8*$K$6*$K$3/ ($K$2+$K$3+ $K$6)^2	+2*J8 *$K$6*$K$3/ $K$2+$K$3+$K$6)^2	+5/16*K8 *$K$6*$K$3/ ($K$2+$K$3+$K$6) ^2	
9						
10	time	[A]	[I]	[P]	[I]s	[I]x
11	0	(($K$4-$K$6)* @EXP(- $K$5*$H11) -($K$5-$K$6)* @EXP(-$K$4*$H11))/ ($K$4- $K$5)	(@EXP(-$K$5*$H11))γ* @EXP(-$K$4*$H11))/ ($K$4- $K$6)*($K$6- $K$5)/$K$2/($K$4-$K$5)γ	1-I11-J11	($K$2+$K$3)/$K$6 *J11	+I11*$K$6/($K$2 +$K$3) +$K$3
12	+H11+$H$3	(($K$4-$K$6)*@EXP(- $K$5*$H12)-($K$5- $K$6)*@EXP(- $K$4*$H12))/($K$4-$K$5)	(@EXP(-$K$5*$H12)- @EXP(-$K$4 $H12)) ($K$4- $K$6)*($K$6- $K$5)/$K$2/($K$4-$K$5)	1-I12-J12	($K$2+$K$3)/$K$6 *J12	+I12*$K$6/($K$2 +$K$3)
13	+H12+$H$3	(($K$4-$K$6)*@EXP(- $K$5*$H13)-($K$5- $K$6)*@EXP(- $K$4*$H13))/($K$4-$K$5)	(@EXP(-$K$5*$H13)- @EXP(-$K$4*$H13))*($K$4- $K$6)*($K$6- $K$5)/$K$2/($K$4-$K$5)	1-I13-J13	($K$2+$K$3)/$K$6 *J13	+I13*$K$6/($K$2 +$K$3)

Table 4.A.1 *(continued)*

	H	I	J	K	L	M
14	+H13+$H$3	(($K$4-$K$6)*@EXP(-$K$5*$H14)-($K$5-$K$6)*@EXP(-$K$4*$H14))/($K$4-$K$5)	(@EXP(-$K$5*$H14)-@EXP(-$K$4*$H14))*($K$4-$K$6)*$K$6)/$K$5)/$K$2/($K$4-$K$5)	1-I14-J14	($K$2+$K$3)/$K$6 *J14	+I14*$K$6/($K$2 +$K$3)
15	+H14+$H$3	(($K$4-$K$6)*@EXP(-$K$5*$H15)-($K$5-$K$6)*@EXP(-$K$4*$H15))/($K$4-$K$5)	(@EXP(-$K$5*$H15)-@EXP(-$K$4*$H15))*($K$4-$K$6)*$K$6)/$K$5)/$K$2/($K$4-$K$5)	1-I15-J15	($K$2+$K$3)/$K$6 *J15	+I15*$K$6/($K$2 +$K$3)
16	+H15+$H$3	(($K$4-$K$6)*@EXP(-$K$5*$H16)-($K$5-$K$6)*@EXP(-$K$4*$H16))/($K$4-$K$5)	(@EXP(-$K$5*$H16)-@EXP(-$K$4*$H16))*($K$4-P$K$6)*$K$6)/$K$5)/$K$2/($K$4-$K$5)	1-I16-J16	($K$2+$K$3)/$K$6 *J16	+I16*$K$6/($K$2 +$K$3)
17	+H16+$H$3	(($K$4-$K$6)*@EXP(-A$K$5*$H17)-($K$5-$K$6)*@EXP(-$K$4*$H17))/($K$4-$K$5)	(@EXP(-$K$5*$H17)-@EXP(-$K$4*$H17))*($K$4-$K$6)*$K$6)/$K$5)/$K$2/($K$4-$K$5)	1-I17-J17	($K$2+$K$3)/$K$6 *J17	+I17*$K$6/($K$2 +$K$3)
18	+H17+$H$3	(($K$4-$K$6)*@EXP(-$K$5*$H18)-($K$5-$K$6)*@EXP(-$K$4*$H18))/($K$4-$K$5)	(@EXP(-$K$5*$H18)-@EXP(-$K$4*$H18))*($K$4-$K$6)*$K$6)/$K$5)/$K$2/($K$4-$K$5)	1-I18-J18	($K$2+$K$3)/$K$6 *J18	+I18*$K$6/($K$2 +$K$3)
19	+H18+$H$3	(($K$4-$K$6)*@EXP(-$K$5*$H19)-($K$5-$K$6)*@EXP(-$K$4*$H19))/($K$4-$K$5)	(@EXP(-$K$5*$H19)-@EXP(-$K$4*$H19))*($K$4-$K$6)*$K$6)/$K$5)/$K$2/($K$4-$K$5)	1-I19-J19	($K$2+$K$3)/$K$6 *J19	+I19*$K$6/($K$2 +$K$3)

Note: The table was cut short to save space. You need to copy the formulas down and change h4 in column H to h5 to duplicate the results in the figures.

Table 4.A.2   Numerical values in the spreadsheet

	H	I	J	K	L	M
1	H+ =	0.2	k1=	1		
2			k2=	5.7E+ 006		
3	1.00E−008	=time incrmnt 1	k3=	2.8E+ 007		
4	0.1	=time incrmnt 2	k4=	3.37E+ 007		
5			k5=	0.166		
6			k6=	0.2		
7						
8	k5= sum(	0.166	8.19E−010	8.08E−018	1.24E−026	
9						
10	time	[A]	[I]	[P]	[I]s	[I]x
11	0.000	1.000	0.0	2.22E−016	0.000	5.93E−009
12	1.00E−008	1.000	1.70E−009	2.51E−010	0.286	5.93E−009
13	2.00E−008	1.000	2.91E−009	9.06E−010	0.490	5.93E−009
14	3.00E−008	1.000	3.78E−009	1.85E−009	0.636	5.93E−009
15	4.00E−008	1.000	4.39E−009	3.00E−009	0.740	5.93E−009
16	5.00E−008	1.000	4.83E−009	4.29E−009	0.815	5.93E−009
17	6.00E−008	1.000	5.15E−009	5.69E−009	0.868	5.93E−009
18	7.00E−008	1.000	5.37E−009	7.17E−009	0.905	5.93E−009
19	8.00E−008	1.000	5.53E−009	8.70E−009	0.933	5.93E−009
20	9.00E−008	1.000	5.65E−009	1.03E−008	0.952	5.93E−009
21	1.00E−007	1.000	5.73E−009	1.19E−008	0.966	5.93E−009
22	1.10E−007	1.000	5.79E−009	1.35E−008	0.975	5.93E−009
23	1.20E−007	1.000	5.83E−009	1.51E−008	0.982	5.93E−009
24	1.30E−007	1.000	5.86E−009	1.67E−008	0.987	5.93E−009
25	1.40E−007	1.000	5.88E−009	1.84E−008	0.991	5.93E−009
26	1.50E−007	1.000	5.90E−009	2.00E−008	0.994	5.93E−009
27	1.60E−007	1.000	5.91E−009	2.17E−008	0.995	5.93E−009
28	1.70E−007	1.000	5.92E−009	2.33E−008	0.997	5.93E−009
29	1.80E−007	1.000	5.92E−009	2.50E−008	0.998	5.93E−009
30	1.90E−007	1.000	5.92E−009	2.66E−008	0.998	5.93E−009
31	2.00E−007	1.000	5.93E−009	2.83E−008	0.999	5.93E−009
32	2.10E−007	1.000	5.93E−009	3.00E−008	0.999	5.93E−009
33	2.20E−007	1.000	5.93E−009	3.16E−008	0.999	5.93E−009
34	2.30E−007	1.000	5.93E−009	3.33E−008	1.000	5.93E−009
35	2.40E−007	1.000	5.93E−009	3.50E−008	1.000	5.93E−009
36	2.50E−007	1.000	5.93E−009	3.66E−008	1.000	5.93E−009
37	2.60E−007	1.000	5.93E−009	3.83E−008	1.000	5.93E−009
38	2.70E−007	1.000	5.93E−009	3.99E−008	1.000	5.93E−009
39	2.80E−007	1.000	5.93E−009	4.16E−008	1.000	5.93E−009
40	2.90E−007	1.000	5.93E−009	4.33E−008	1.000	5.93E−009
41	3.00E−007	1.000	5.93E−009	4.49E−008	1.000	5.93E−009
42	0.1	0.984	5.84E−009	1.65E−002	0.984	5.84E−009
43	0.2	0.967	5.74E−009	3.27E−002	0.967	5.74E−009
44	0.3	0.951	5.65E−009	4.86E−002	0.951	5.65E−009
45	0.4	0.936	5.55E−009	6.43E−002	0.936	5.55E−009
46	0.5	0.920	5.46E−009	7.97E−002	0.920	5.46E−009
47	0.6	0.905	5.37E−009	9.49E−002	0.905	5.37E−009
48	0.7	0.890	5.28E−009	1.10E−001	0.890	5.28E−009
49	0.8	0.876	5.20E−009	1.24E−001	0.876	5.20E−009
50	0.9	0.861	5.11E−009	1.39E−001	0.861	5.11E−009
51	1.0	0.847	5.03E−009	1.53E−001	0.847	5.03E−009

(*continued overleaf*)

Table 4.A.2 *(continued)*

	H	I	J	K	L	M
52	1.1	0.833	4.94E–009	1.67E–001	0.833	4.94E–009
53	1.2	0.819	4.86E–009	1.81E–001	0.819	4.86E–009
54	1.3	0.806	4.78E–009	1.94E–001	0.806	4.78E–009
55	1.4	0.792	4.70E–009	2.08E–001	0.792	4.70E–009
56	1.5	0.779	4.63E–009	2.21E–001	0.779	4.63E–009
57	1.6	0.767	4.55E–009	2.33E–001	0.767	4.55E–009
58	1.7	0.754	4.47E–009	2.46E–001	0.754	4.47E–009
59	1.8	0.741	4.40E–009	2.59E–001	0.741	4.40E–009
60	1.9	0.729	4.33E–009	2.71E–001	0.729	4.33E–009
61	2.0	0.717	4.26E–009	2.83E–001	0.717	4.26E–009
62	2.1	0.705	4.19E–009	2.95E–001	0.705	4.19E–009
63	2.2	0.694	4.12E–009	3.06E–001	0.694	4.12E–009
64	2.3	0.682	4.05E–009	3.18E–001	0.682	4.05E–009
65	2.4	0.671	3.98E–009	3.29E–001	0.671	3.98E–009
66	2.5	0.660	3.92E–009	3.40E–001	0.660	3.92E–009
67	2.6	0.649	3.85E–009	3.51E–001	0.649	3.85E–009
68	2.7	0.638	3.79E–009	3.62E–001	0.638	3.79E–009
69	2.8	0.628	3.73E–009	3.72E–001	0.628	3.73E–009
70	2.9	0.618	3.67E–009	3.82E–001	0.618	3.67E–009
71	3.0	0.607	3.60E–009	3.93E–001	0.607	3.60E–009
72	3.1	0.597	3.55E–009	4.03E–001	0.597	3.55E–009
73	3.2	0.588	3.49E–009	4.12E–001	0.588	3.49E–009
74	3.3	0.578	3.43E–009	4.22E–001	0.578	3.43E–009
75	3.4	0.568	3.37E–009	4.32E–001	0.568	3.37E–009
76	3.5	0.559	3.32E–009	4.41E–001	0.559	3.32E–009
77	3.6	0.550	3.26E–009	4.50E–001	0.550	3.26E–009
78	3.7	0.541	3.21E–009	4.59E–001	0.541	3.21E–009
79	3.8	0.532	3.16E–009	4.68E–001	0.532	3.16E–009
80	3.9	0.523	3.10E–009	4.77E–001	0.523	3.10E–009
81	4.0	0.514	3.05E–009	4.86E–001	0.514	3.05E–009
82	4.1	0.506	3.00E–009	4.94E–001	0.506	3.00E–009
83	4.2	0.498	2.95E–009	5.02E–001	0.498	2.95E–009
84	4.3	0.489	2.90E–009	5.11E–001	0.489	2.90E–009
85	4.4	0.481	2.86E–009	5.19E–001	0.481	2.86E–009
86	4.5	0.473	2.81E–009	5.27E–001	0.473	2.81E–009
87						

Equations (4.B.1)–(4.B.3) become

$$\frac{d\vec{C}}{dt} = \vec{R}(\vec{C}, t) \tag{4.B.6}$$

We indicated that the rate vector, $\vec{R}$, can be a function of $\vec{C}$ and t in equation (4.B.6). The concentration dependence is obvious. The time dependence arises because the temperature can change with time, which, in turn, changes the rate constants.

Numerical methods start by looking at some small time increment Δt and approximating $d\vec{C}/dt$ by $\Delta\vec{C}/\Delta t$. You probably learned the mean value theorem in freshman calculus. According to the mean value theorem

$$\frac{\Delta \vec{C}}{\Delta t} = \vec{R}(\zeta, \xi) \tag{4.B.7}$$

where $\vec{R}(\zeta, \xi)$ is an average value of $\vec{R}$ between t and $t + \Delta t$.

Rearranging equation (4.B.6) yields

$$\vec{C}(t + \Delta t) = \vec{C}(t) + \Delta t \vec{R}(\zeta, \xi) \tag{4.B.8}$$

where $\vec{C}(t)$ is the value of the concentrations at time $= t$, and $\vec{C}(t + \Delta t)$ is the value at $t + \Delta t$.

The numerical methods used to solve ordinary differential equations all use an approximation for $\vec{R}(\zeta, \xi)$ in equation (4.B.8). Recall that the concentrations are changing during every timestep. So the rate will change too. People approximate the change in rate so that they can integrate equation (4.3.8) numerically. For example, consider the simplest approximation, called *Euler's method*. In Euler's method one assumes that the rate is constant during each timestep so that $\vec{R}(\zeta, \xi)$ can be replaced by $\vec{R}(\vec{C}(t), t)$, the rate at the beginning of the timestep:

$$\vec{C}(t + \Delta t) = \vec{C}(t) + \Delta \vec{R}(\vec{C}(t), t) \tag{4.B.9}$$

Students are sometimes confused by the notation in equation (4.B.9). The last term is the value of $\vec{R}$ calculated at a concentration equal to the concentration at time t.

Equation (4.B.9) says that if you know the concentration of any time t, and the rate equation you can calculate the concentration at some time $t + \Delta t$, by plugging into equation (4.B.9).

In practice, you use equation (4.B.9) iteratively. You start with $t = 0$, and use equation (4.B.9) to calculate the behavior some small time increment later, for instance, 0.1 second. You then substitute the concentration at 0.1 second in to equation (4.B.9) to calculate the concentration at 0.2 second. By repeating the process many times, one can calculate the concentration at any time τ.

Example 4.B.1.a   Solving a Differential Equation   Consider solving the differential equation

$$\frac{dC_A}{dt} = (-k_1)C_A = r_A \tag{4.B.10}$$

with an initial concentration of 1 mol/liter, and $k_1 = -2/\text{second}$.

Solution   In equation (4.B.10)

$$\vec{R} = r_A(C_A(t)) \tag{4.B.11}$$

Combining equations (4.B.9) and (4.B.11) yields

$$C_A(t + \Delta t) = C_A(t) + \Delta t \, r_A(C_A(t)) = C(t) - k_1(\Delta t)C(t) \tag{4.B.12}$$

Equation (4.B.12) gives us a way to solve for $C_A(t + \Delta t)$ given $C_A(t)$.

Next, we will use a spreadsheet to solve this problem. Table 4.B.1 shows my spreadsheet. I set up the spreadsheet so column T is time, column CA is the concentration

Table 4.B.1 The formulas used to integrate equation (4.B.10)

	T	CA	DA
01	dt=	0.1	
02	k_1=	2	
03	0	1	=−k_1*CA3
04	=T3+dt	=CA3+dt*DA3	=−k_1*CA4
05	=T4+dt	=CA4+dt*DA4	=−k_1*CA5
06	=T5+dt	=CA5+dt*DA5	=−k_1*CA6
07	=T6+dt	=CA6+dt*DA6	=−k_1*CA7
08	=T7+dt	=CA7+dt*DA7	=−k_1*CA8
09	=T8+dt	=CA8+dt*DA8	=−k_1*CA9
10	=T9+dt	=CA9+dt*DA9	=−k_1*CA10
11	=T10+dt	=CA10+dt*DA10	=−k_1*CA11
12	=T11+dt	=CA11+dt*DA11	=−k_1*CA12
13	=T12+dt	=CA12+dt*DA12	=−k_1*CA13
14	=T13+dt	=CA13+dt*DA13	=−k_1*CA14
15	=T14+dt	=CA14+dt*DA14	=−k_1*CA15
16	=T15+dt	=CA15+dt*DA15	=−k_1*CA16
17	=T16+dt	=CA16+dt*DA16	=−k_1*CA17
18	=T17+dt	=CA17+dt*DA17	=−k_1*CA18

of A, and column DA is dC_A/dt. I named cell CA1 dt, and CA2 k_-1. I then hid all of the rest of the columns so that the spreadsheet was easy to see.

Let us go through the formulas in the spreadsheet. Start with column T. Recall that column T is time, and I decided to start time at T3. Therefore I set cell T3 as time = 0. Cells T4–T18 were then calculated by adding dt to the cell above (i.e. I incremented time by dt each time). As a result, column T gives the time at the end of each time increment.

Now focus on column DA. DA is (dC_A/dt), which I wanted to calculate from equation (4.B.10). I did that by taking the concentration in column CA and multiplying by $(-k_-1)$.

Now let's focus on column CA. CA is the concentration at time T. Cell CA3 was set to the initial concentration, 1 mol/liter. Now let's try to try to understand cell CA4. I calculated cell CA4 by plugging into equation (4.B.2); that is, I increment C_A by DA Δt each timestep.

I used the same formulas all the way down the spreadsheet.

Table 4.B.2 shows the numerical values in the spreadsheet. I have also included the exact concentration, calculated by integrating (4.B.10) analytically, [i.e., the solution of equation (4.B.10) is $C_A = C_A^0 \exp(-k_-1t)$].

Notice that Euler's method gives a fair approximation to the concentration, but the results are by no means exact.

Euler's method has failed because it assumes that the rate is constant over the entire time increment. For example, during the first time increment, the concentration of A

Table 4.B.2 The numerical values in the spreadsheet in equation (4.B.1)

	T	CA	DA	
01	dt=	0.1		Exact
02	k_1=	2		concentration
03	0	1	−2	1
04	0.1	0.8	−1.6	0.818731
05	0.2	0.64	−1.28	0.67032
06	0.3	0.512	−1.024	0.548812
07	0.4	0.4096	−0.8192	0.449329
08	0.5	0.32768	−0.65536	0.367879
09	0.6	0.262144	−0.52429	0.301194
10	0.7	0.209715	−0.41943	0.246597
11	0.8	0.167772	−0.33554	0.201897
12	0.9	0.134218	−0.26844	0.165299
13	1	0.107374	−0.21475	0.135335
14	1.1	0.085899	−0.1718	0.110803
15	1.2	0.068719	−0.13744	0.090718
16	1.3	0.054976	−0.10995	0.074274
17	1.4	0.04398	−0.08796	0.06081
18	1.5	0.035184	−0.07037	0.049787

drops from 1.0 to 0.8. The rate of reaction changes from -2.0 to -1.6. However, Euler's method ignores the change in rate. Instead, it assumes that the rate stays at -2.0 for the entire time increment. This leads to errors.

4.B.2 *Roundoff Error* Years ago, people tried to avoid these difficulties by choosing very small timesteps. The idea is that if the timestep is small enough, the rate will be constant over the timestep. Consequently, the numerical procedure could have small errors.

Unfortunately, it does not work out that way. In order to illustrate the difficulty, let us set $\Delta t = 10^{-6}$ seconds. According to equation (4.B.12):

$$C_A(1 \times 10^{-6} \text{ seconds}) = C_A(0) - (2/\text{seconds})(10^{-6} \text{ seconds})(C_A(0)) \qquad (4.B.13)$$

where $C_A(0)$ is the initial concentration and $C_A(1 \times 10^{-6}$ seconds) is the concentration after running the reaction for 1×10^{-6} seconds.

Plugging in the numbers, we get

$$C_A(1 \times 10^{-6} \text{ seconds}) = 1 \text{ mol/liter} - (0.2/\text{second})(10^{-6} \text{ seconds})(1 \text{ mol/liter})$$

$$= 0.999998 \text{ mol/liter} \qquad (4.B.14)$$

Similarly, we can repeat the process to calculate the concentration after running the reaction for 2×10^{-6} seconds:

$$C_A(2 \times 10^{-6} \text{ seconds}) = C_A(1 \times 10^{-6} \text{ seconds}) - (0.2/\text{second})(10^{-6} \text{ seconds})$$
$$\times C_A(1 \times 10^{-6} \text{ seconds}) \qquad (4.B.15)$$

Substituting in the numbers, we obtain

$$C_A(2 \times 10^{-6} \text{ seconds}) = 0.999998 \text{ mol/liter} - (0.02/\text{ second})(10^{-6} \text{ seconds})$$
$$\times 0.999998 \text{ mol/liter}$$
$$= 0.999998000004 \text{ mol/liter} \qquad (4.B.16)$$

where I have carried the calculation to too many significant figures because I want to make a point below.

If we repeated this process 1,000,000 times, we could calculate how the concentration changed after one second.

One does have to worry about the accuracy of the calculations. If we could do the calculation exactly, then the smaller the time increment we choose, the more accurate the calculation. In actual practice, computers do not calculate numbers exactly. For example, the computer might round off 0.999998000004 to 0.999998. In that case, one creates some error at each Δt step. After 10^7 steps, that error can add up to be something significant. In practice, Euler's method is not particularly accurate because you need tiny steps (i.e., Δt values) and round off error builds up too quickly. Consequently Eulers method does not work much of the time.

4.B.3 Implicit Methods   Fortunately, there are a series of other methods. The basic idea in the other methods is that one uses an approximation to get a good value of $R(\varepsilon, \xi)$ in equation (4.B.8). In this chapter, we will consider an implicit Euler's method, which assumes that $\vec{R}(\zeta, \xi) = \vec{F}(\vec{C}(t + \Delta t), t + \Delta t)$:

$$\vec{C}_A(t + \Delta t) = \frac{\Delta t(\vec{R}_A(t) + \vec{R}_A(t + \Delta t))}{2} \qquad (4.B.17)$$

Equation (4.B.17) replaces the exact value of the rate, with an average of the initial and final values of the rate so it gives much more accurate values than does Euler's method.

Equation (4.B.17) is more complex than it initially appears because $\vec{R}_A(t + \Delta t)$ is a function of $\vec{C}(t + \Delta t)$, so one has to generally solve equation (4.B.9) using some iterative numerical procedure to calculate $\vec{C}(t + \Delta t)$. Methods where one has to solve for $\vec{C}(t + \Delta t)$ iteratively are called **implicit methods**. Methods where one can compute $\vec{C}(t + \Delta t)$ directly are called **explicit methods**.

Example 4.B.3.a Implicit Solution of Differential Equations   Solve equation (4.B.10) using equation (4.B.17).

Solution   I will set up a spreadsheet to calculate the rate as before. The spreadsheet is given in Table 4.B.3. The spreadsheet is almost the same as above except that we used an average value of DA to calculate CA. For example, consider cell CA4. In the spreadsheet

Table 4.B.3 The formulas used to integrate equation (4.B.10) according to equation (4.B.17)

	T	CA	DA
01	dt=	0.1	
02	k_1=	2	
03	0	1	
			=-k_1*CA3
04	=T3+dt	=CA3+dt*(DA3+DA4)/2	=-k_1*CA4
05	=T4+dt	=CA4+dt*(DA4+DA5)/2	=-k_1*CA5
06	=T5+dt	=CA5+dt*(DA5+DA6)/2	=-k_1*CA6
07	=T6+dt	=CA6+dt*(DA6+DA7)/2	=-k_1*CA7
08	=T7+dt	=CA7+dt*(DA7+DA8)/2	=-k_1*CA8
09	=T8+dt	=CA8+dt*(DA8+DA9)/2	=-k_1*CA9
10	=T9+dt	=CA9+dt*(DA9+DA10)/2	=-k_1*CA10
11	=T10+dt	=CA10+dt*(DA10+DA11)/2	=-k_1*CA11
12	=T11+dt	=CA11+dt*(DA11+DA12)/2	=-k_1*CA12
13	=T12+dt	=CA12+dt*(DA12+DA13)/2	=-k_1*CA13
14	=T13+dt	=CA13+dt*(DA13+DA14)/2	=-k_1*CA14
15	=T14+dt	=CA14+dt*(DA14+DA15)/2	=-k_1*CA15
16	=T15+dt	=CA15+dt*(DA15+DA16)/2	=-k_1*CA16
17	=T16+dt	=CA16+dt*(DA16+DA17)/2	=-k_1*CA17
18	=T17+dt	=CA17+dt*(DA17+DA18)/2	=-k_1*CA18

in Table 4.B.3, cell CA4 uses an average of DA3 and DA4 to estimate the derivative, while only the initial value is used in the spreadsheet in Table 4.B.1.

In order to use spreadsheet Table (4.B.3) in Excel, one needs to change the calculation options to allow circular references. I set the options so Excel would iterate, with a maximum of 1000 iterations and an maximum error of 0.0001. The result is the numbers in Table 4.B.4. Notice that the concentrations are much more accurate than before. All of the predicted concentrations are within 0.005 of the exact results.

The spreadsheet in Table 4.B.3 did so much better than one in Table 4.B.1 because it considered the fact that the rate is dropping during the timestep. That gives much more accurate results.

4.B.4 Runge–Kutta Methods

Another important method is called the *Runge–Kutta method*. The Runge-Kutta method approximates $\vec{R}(\zeta, \xi)$ in equation (4.B.8) with some intermediate values of $\vec{R}(\vec{C}, t)$. That gives you more accuracy. For example, a second-order Runge–Kutta routine approximates $\vec{C}(t + \Delta t)$ by

$$\vec{C}(t + \Delta t) = \vec{C}(t) + \frac{\Delta t}{2}(\vec{R}(\vec{C}(t), t) + \vec{R}(\vec{G}_A, t + \Delta t)) \qquad (4.B.18)$$

where $\vec{G}_A$ is given by

$$\vec{G}_A(t + \Delta t) = \vec{C}(t) + \Delta t \vec{R}(t) \qquad (4.B.19)$$

Table 4.B.4 The numerical values in the spreadsheet in Table 4.B.3[a]

	T	CA	DA	Exact concentration
01	dt=	0.1		
02	k_1=	2		
03	0	1	−2	1
04	0.1	0.818182	−1.63636	0.818731
05	0.2	0.669421	−1.33884	0.67032
06	0.3	0.547708	−1.09542	0.548812
07	0.4	0.448125	−0.89625	0.449329
08	0.5	0.366648	−0.7333	0.367879
09	0.6	0.299985	−0.59997	0.301194
10	0.7	0.245442	−0.49088	0.246597
11	0.8	0.200816	−0.40163	0.201897
12	0.9	0.164304	−0.32861	0.165299
13	1	0.134431	−0.26886	0.135335
14	1.1	0.109989	−0.21998	0.110803
15	1.2	0.089991	−0.17998	0.090718
16	1.3	0.073629	−0.14726	0.074274
17	1.4	0.060242	−0.12048	0.06081
18	1.5	0.049289	−0.09858	0.049787

[a]I also added column for the exact concentration.

You use equations (4.B.18) and (4.B.19) in the same way as you use Euler's method. You start at time = 0, and calculate G_A at some small time increment later using equation (4.B.19). You then plug into equation (4.B.18) to calculate C at the end of the time increment. You repeat the step through the time increment. Table 4.B.7 lists two fourth-order Runge–Kutta methods. In a fourth-order method, one calculates four intermediate values of C, and uses those intermediate values to get better estimates of the average rate over the time increment $\vec{R}(\zeta, \xi)$.

Example 4.B.4.a Solve equation (4.B.10) using the Runge–Kutta method.

Solution I used a spreadsheet to do the calculations. Table 4.B.5 shows the spreadsheet. In this spreadsheet column GA is the concentration calculated according to equation (4.B.19), and HA is the rate at that concentration. As above, I used an average rate to calculate the change in concentration, where the average rate is the average of the rate at the start of the time increment and the rate at a concentration of G_A.

Table 4.B.6 shows the numerical values in the spreadsheet. I also inserted the exact concentration for comparison purposes. Notice that the results are better than those in Table 4.B.2, but not as good as those in Table 4.B.4. That is because the derivative approximation used in 4.B.4 is better than the other derivative approximations.

Table 4.B.5 The formulas used to integrate equation (4.B.10) according to the second-order Runge–Kutta method

	T	CA	DA	GA	HA
01	dt=	0.1			
02	k_1=	2			
03	0	1	= -k_1*CA3		
04	=T3+dt	=CA3+dt*(DA3+HA4)/2	= -k_1*CA4	=CA3+dt*DA3	=-k_1*GA4
05	=T4+dt	=CA4+dt*(DA4+HA5)/2	= -k_1*CA5	=CA4+dt*DA4	=-k_1*GA5
06	=T5+dt	=CA5+dt*(DA5+HA6)/2	=-k_1*CA6	=CA5+dt*DA5	=-k_1*GA6
07	=T6+dt	=CA6+dt*(DA6+HA7)/2	= -k_1*CA7	=CA6+dt*DA6	=-k_1*GA7
08	=T7+dt	=CA7+dt*(DA7+HA8)/2	= -k_1*CA8	=CA7+dt*DA7	=-k_1*GA8
09	=T8+dt	=CA8+dt*(DA8+HA9)/2	= -k_1*CA9	=CA8+dt*DA8	=-k_1*GA9
10	=T9+dt	=CA9+dt*(DA9+HA10)/2	= -k_1*CA10	=CA9+dt*DA9	=-k_1*GA10
11	=T10+dt	=CA10+dt*(DA10+HA11)/2	= -k_1*CA11	=CA10+dt*DA10	=-k_1*GA11
12	=T11+dt	=CA11+dt*(DA11+HA12)/2	= -k_1*CA12	=CA11+dt*DA11	=-k_1*GA12
13	=T12+dt	=CA12+dt*(DA12+HA13)/2	= -k_1*CA13	=CA12+dt*DA12	=-k_1*GA13
14	=T13+dt	=CA13+dt*(DA13+HA14)/2	= -k_1*CA14	=CA13+dt*DA13	=-k_1*GA14
15	=T14+dt	=CA14+dt*(DA14+HA15)/2	= -k_1*CA15	=CA14+dt*DA14	=-k_1*GA15
16	=T15+dt	=CA15+dt*(DA15+HA16)/2	= -k_1*CA16	=CA15+dt*DA15	=-k_1*GA16
17	=T16+dt	=CA16+dt*(DA16+HA17)/2	= -k_1*CA17	=CA16+dt*DA16	=-k_1*GA17
18	=T17+dt	=CA17+dt*(DA17+HA18)/2	= -k_1*CA18	=CA17+dt*DA17	=-k_1*GA18

4.B.5   Other   Methods   Table 4.B.7 lists some other methods. The first is the Bilirush–Stoer method. The Bilirush-Stoer method extrapolates the second-order Runge–Kutta method to zero step size for higher accuracy.

The second method is the Rosenbrock routine. Rosenbrock's method is an implicit version of Runge–Kutta. Finally, there are so-called predictor-corrector methods, which use previous values of C(t) to get better values of $\bar{C}(t + \Delta t)$, which are then plugged into equation (4.B.9) to get better values of $\bar{R}(C_1, t + \Delta t)$. Generally, Runge–Kutta is the real workhorse for solving simple differential equations. Bilirush–Stoer is used for high accuracy. Rosenbrock is used for what are called stiff differential equations described in the next section. Predictor–corrector methods are not used that much anymore, although they were used extensively years ago.

Table 4.B.7 lists several of these methods. The details are not important for the purposes here. What is important is that the methods work well for simple differential equations, and the algorithms are readily available. Suitable algorithms are included with Microsoft FORTRAN, Microsoft Visual C, or any book on numerical recipes.

4.B.6 Solving Stiff Equations Using Euler's Method, and Numerical Solution

Sections 4.B.1–4.B.5 listed all of the methods one usually sees during numerical simulations. However, there are some special issues that arise in kinetics that do not arise in other systems—in kinetics, the equations are hard to solve numerically. Recall that equation (4.B.2) computes d[I]/dt as a difference between two large numbers. If you

Table 4.B.6   The numerical values in Table 4.B.5

	T	CA		DA	GA	HA
01	dt=	0.1	Exact			
02	k_1=	2	concentration			
03	0	1	1	-2		
04	0.1	0.82	0.818731	-1.64	0.8	-1.6
05	0.2	0.6724	0.67032	-1.3448	0.656	-1.312
06	0.3	0.551368	0.548812	-1.10274	0.53792	-1.07584
07	0.4	0.452122	0.449329	-0.90424	0.441094	-0.88219
08	0.5	0.37074	0.367879	-0.74148	0.361697	-0.72339
09	0.6	0.304007	0.301194	-0.60801	0.296592	-0.59318
10	0.7	0.249285	0.246597	-0.49857	0.243205	-0.48641
11	0.8	0.204414	0.201897	-0.40883	0.199428	-0.39886
12	0.9	0.16762	0.165299	-0.33524	0.163531	-0.32706
13	1	0.137448	0.135335	-0.2749	0.134096	-0.26819
14	1.1	0.112707	0.110803	-0.22541	0.109958	-0.21992
15	1.2	0.09242	0.090718	-0.18484	0.090166	-0.18033
16	1.3	0.075784	0.074274	-0.15157	0.073936	-0.14787
17	1.4	0.062143	0.06081	-0.12429	0.060628	-0.12126
18	1.5	0.050957	0.049787	-0.10191	0.049715	-0.09943

make an error in the calculations, that error is amplified. This makes the calculations difficult.

Another difficulty is that there are two different timescales. Figure 4.7 shows that [I] changes rapidly in the first 10^{-7} seconds, and then slowly decays. Systems where there are large differences in timescales, and things are calculated as a difference between two big numbers, are called "**stiff differential equations.**" One needs to be very careful doing calculations with stiff equations.

When I start to discuss stiff equations, students often have difficulty. Therefore, it is useful to consider and example illustrating the difficulties.

Example 4.B.6.a   Stiff Equations   Solve equation (4.B.2) using Euler's method.

Solution   If we solve the equation with Euler's method, equation (4.B.9), we would have to solve:

$$C_A(t + \Delta t) = C_A(t) + \Delta t(k_2 C_I(t) - k_6 C_A(t)) \tag{4.B.20}$$

$$C_I(t + \Delta t) = C_I(t) + \Delta t\{(k_6[C_A(t + \Delta t)] - (k_2 + k_3)C_I(t + \Delta t)\}$$

Notice that equation (4.B.20) expresses C_I as a small difference of two big numbers: $k_6 C_A(t)$ and $(k_2 + k_3)C_I(t)$. Whenever you compute a quantity as a small difference of two big numbers, errors can arise.

Table 4.B.7 Some of the common numerical methods used to solve differential equations

Method	Formula	Speed	Accuracy	Useful for Stiff Equations?
	Explicit Methods			
Eulers	$\vec{C}(t + \Delta t) = \vec{C}(t) + \Delta \vec{R}(t)$	Slow	Poor	No
Runge–Kutta Second-order	$\vec{C}(t + \Delta t) = \vec{C}(t) + \Delta \vec{R}(t)$	Fair	Good	No
Fourth-order	$\vec{C}(t + \Delta t) = \vec{C}(t) + \dfrac{\Delta t}{6}(\vec{R}_1 + 2\vec{R}_2 + 2\vec{R}_3 + \vec{R}_4)$	Excellent	Good	No
	$\vec{R}_1 = \vec{R}(\vec{C}(t), t)\vec{C}_1 = C(t) + \dfrac{\Delta t}{2}\vec{R}_1$			
	$\vec{R}_2 = \vec{R}\left(\vec{C}_1(t) + \dfrac{\Delta t}{2}\right), \quad \vec{C}_2 = \vec{C}(t) + \dfrac{\Delta t}{2}\vec{R}_2$			
	$\vec{R}_3 = \vec{R}\left(\vec{C}_2, t + \dfrac{\Delta t}{2}\right), \quad \vec{C}_3 = \vec{C}(t) + \vec{R}_3\Delta t$			
	$\vec{R}_4 = \vec{R}\left(\vec{C}_3, t + \dfrac{\Delta t}{2}\right)$			
Runge–Kutta/Gill	$\vec{C}(t + \Delta t) = \vec{C}(t) + \dfrac{\Delta t}{C}(\vec{R}_1 + 0.586\vec{R}_2 + 3.414\vec{R}_3 + \vec{R}_4)$	Outstanding	Excellent	No
	$\vec{R}_1 = \vec{R}(\vec{C}(t), t), \vec{C}_1 = \vec{C}(t) + \dfrac{\Delta t}{2}\vec{R}_2$			
	$\vec{R}_2 = \vec{R}\left(\vec{C}_1, t + \dfrac{\Delta t}{2}\right)\vec{C}_2 = \vec{C}(t) + (0.207\vec{R}_1 + 0.293\vec{R}_2)\Delta t$			
	$\vec{R}_3 = \vec{R}\left(\vec{C}_2(t) + \dfrac{\Delta t}{2}\right)\vec{C}_3 = \vec{C}(t) + \Delta t(1.707\vec{R}_3 - .707\vec{R}_2)$			
	$\vec{R}_4 = \vec{R}\left(\vec{C}_3, t + \dfrac{\Delta t}{2}\right)$			

(continued overleaf)

Table 4.B.7 *(continued)*

Method	Formula	Speed	Accuracy	Useful for Stiff Equations?
Bilirush-Stoer	Extrapolate second-order Runge–Kutta to 0 step size	Good	Outstanding	Maybe
Adams–Milne predictor–correcter	Iteratively solve equation (4.B.10)	Good	Fair	No
	Implicit Methods			
Gear	Iterative solution of: $$C(t + \Delta t) = C(t) + \Delta t \eta \vec{R}(C(t + \Delta t), t + \Delta t)$$	Slow	Fair	Yes
Rosenbrock	$$\vec{C}(t + \Delta t) = \vec{C}(t) + \frac{\Delta t}{4}\left(\frac{19}{9}\vec{R}_1 + \frac{1}{2}\vec{R}_2 + \frac{25}{108}\vec{R}_3 + \frac{125}{108}\vec{R}_4\right)$$ $$\vec{R}_1 = \vec{R}\left(C(t)t\right) + \vec{R}\left(C(t) + \vec{R}_1 C\left(t + \frac{\Delta t}{2}\right), t + \Delta t\right), \text{etc.}$$	Fair	Excellent	Yes

In the first iteration, with $\Delta t = 10^{-6}$ seconds, we obtain

$$C_A(1 \times 10^{-6} \text{ seconds}) = C_A(0) - 10^{-6} \text{ seconds } (0.2C_A(0) - 5.8 \times 10^6 \ C_I(0))$$
$$(4.B.21)$$

or

$$C_A = 1 \text{ mol/liter} - (10^{-6})(0.2(1 \text{ mol/liter}) - 5.8 \times 10^6 \times 0)$$
$$C_A(1 \times 10^{-6} \text{ seconds}) = 0.9999998 \text{ mol/liter}$$

Similarly

$$C_I(1 \times 10^{-6} \text{ seconds}) = C_I(0) + 10^{-6}(0.2C_A^0 - 3.38 \times 10^7 C_I(0))$$
$$C_I(1 \times 10^{-6} \text{ seconds}) = 0 + 10^{-6}(0.21 - 3.38 \times 10^7 \times 0) \qquad (4.B.22)$$
$$C_I(1 \times 10^{-6} \text{ seconds}) = 2 \times 10^{-7} \text{ mol/liter}$$

On the second iteration we find

$$
\begin{aligned}
C_A(10^{-6} \text{ seconds}) &= C_A(10^{-6} \text{ seconds}) - (10^{-6} \text{ seconds})(0.2 \times C_A(10^{-6}) \\
&\quad - 5.8 \times 10^6 C_I(10^{-6})) \\
&= 0.9999998 - 10^{-6}(0.2 \times 0.9999998 - (5.8 \times 10^{-6})(2 \times 10^{-2})) \\
&= 1.0000076 \qquad\qquad\qquad\qquad\qquad\qquad\qquad (4.B.23)
\end{aligned}
$$

Similarly

$$
\begin{aligned}
C_I(2 \times 10^{-6}) &= C_I(1 \times 10^{-6} \text{ seconds}) + 10^{-6}(0.2 \times C_A(1 \times 10^{-6} \text{ seconds}) \\
&\quad - 3.38 \times 10^7 * C_I(1 \times 10^{-6} \text{ seconds})) \\
&= 2 \times 10^7 + 10^{-6}(0.02 * 0.999998 - 3.38 \times 10^7 * 2 \times 10^{-7}) \\
&= -6.36 \times 10^{-6} \qquad\qquad\qquad\qquad\qquad\qquad (4.B.24)
\end{aligned}
$$

Notice that according to the calculation, C_A is greater than the initial A concentration while C_I is negative! Concentrations can never be negative, so the method has failed.

Table 4.B.8 shows values of the $C_I(0)$, $C_I(\Delta t)$, $C_I(2\Delta t)$ and $C_I(3\Delta t)$ for various values of Δt. Notice that with $\Delta T = 10^{-6}$ seconds, C_2 is negative, and C_3 is about 10^{-4}. By comparison, the exact solution is given in Table 4.A.2. and C_I is never greater than

Table 4.B.8 Values of C_1, C_2, and C_3 calculated by Euler's method [equation (4.B.8)] and the backward difference Euler method [Equation (4.B.17)]

	Euler's Method (Table 4.B.1)			Implicit Euler's (Table 4.B.3)		
—	$\Delta t = 10^{-6}$	$\Delta t = 10^{-7}$	$\Delta t = 3 \times 10^{-8}$	$\Delta t = 2 \times 10^{-7}$	$\Delta t = 10^{-6}$	$\Delta t = 10^{-3}$
$C_I(0)$	0	0	0	0	0	0
$C_I(1\Delta t)$	2×10^{-7}	2×10^{-8}	6×10^{-9}	5.17×10^{-9}	5.76×10^{-9}	5.93×10^{-9}
$C_I(2\Delta t)$	-6.3×10^{-6}	-2.8×10^{-7}	5.92×10^{-9}	5.84×10^{-9}	5.93×10^{-9}	5.93×10^{-9}
$C_I(3\Delta t)$	2.1×10^{-4}	8.6×10^{-7}	5.92×10^{-9}	5.92×10^{-9}	5.93×10^{-9}	5.93×10^{-9}

6×10^{-9}. C_I is always positive. Consequently, it appears that Euler's method is giving qualitatively incorrect results. One does get a reasonable answer if one uses a step size of 3×10^{-8} minutes. In that case, one would need 1.7×10^8 steps to compute the concentrations after the reaction has run for 5 minutes.

Example 4.B.6.b Using Euler's Implicit method   Solve the differential equation (4.B.9) using Euler's implicit method, equation (4.B.25).

$$C_I(t + \Delta t) = C_I(t) + \Delta t \vec{R}_A(t + \Delta t) \qquad (4.B.25)$$

where $R_A(t + \Delta t)$ is the value of $\vec{R}_A$ at the end of the timestep.

Solution   One cannot solve equation 4.B.25 with a spreadsheet. However, one can solve it analytically. Substituting the rate equation into equation (4.B.25) yields

$$C_I(t + \Delta t) = C_I(t) + \Delta t\{(k_6[C_A(t + \Delta t)] - (k_2 + k_3)C_I(t + \Delta t)\} \qquad (4.B.26)$$

$$C_A(t + \Delta t) = C_A(t) + \Delta t\{k_2 C_I(t + \Delta t) - k\ C_A(t + \Delta t)\} \qquad (4.B.27)$$

Solving equations (4.B.26) and (4.B.27) simultaneously for $C_A(t + \Delta t)$ and $C_I(t + \Delta t)$ yields

$$C_I(t + \Delta t) = \frac{(1 + k_6\Delta t)C_I(t) + k_6\Delta t C_A(t)}{1 + k_6\Delta t + k_2 k_6(\Delta t)^2 + k_2\Delta t + k_3\Delta t} \qquad (4.B.28)$$

$$C_A(t + \Delta t) = \frac{C_A(t) + k\Delta t C_I(t + \Delta t)}{1 + k_6\Delta t} \qquad (4.B.29)$$

Notice that one no longer needs to take a difference between two big numbers, so equation (4.B.29) is much easier to use than equation (4.B.20). Further, when $\Delta t(k_2 + k_3) \gg 1$, the first three terms in the denominator cancel, so that

$$C_I(t + \Delta t) = \frac{k_6[C_A(t)]}{k_2 + k_3} \qquad (4.B.30)$$

which is the steady-state approximation to C_I. Table 4.B.2 shows concentrations calculated from equation (4.B.28) for several values of Δt. Notice that equation (4.B.29) always gives values similar to the exact results in Table 4.A.2.

This example shows that we have to be careful with stiff differential equations because the numerical methods can give weird results. Generally, implicit methods work better than do explicit methods in stiff problems, but the cost is that one generally has to solve a nonlinear equation at each step in the process.

Example 4.B.6.c Numerical Solution   Use a standard computer package to solve equation (4.B.9).

Solution   Algorithms for implicit methods are readily available. For example, Microsoft FORTRAN comes with a program called STIFF, which is designed to solve stiff ordinary differential equations using Rosenbrock's method.

In order to use STIFF, you need to provide two subroutines: DERIV, which computes $\bar{R}$, and a subroutine JACOBIAN, which computes what is called the Jacobian of $\bar{R}$, where the Jacobian is defined by

$$df\,dc(i,\,j) = \frac{dR(i)}{dC(j)} \qquad (4.B.31)$$

Here are the subroutines needed to do the calculations:

```
      PROGRAM intrat
C     Solve diffequ for reaction problem
      INTEGER KMAXX,NMAX,NCONCS
      PARAMETER (KMAXX=200,NMAX=50,NCONCS=3)
      INTEGER kmax,kount,nbad,nok,i
      REAL dxsav,eps,hstart,time1,time2,conc(NCONCS),xp,yp,
     totcon
      COMMON /path/kmax,kount,dxsav,xp(KMAXX),yp(NMAX,KMAXX)
      REAL k6,k2,k3
      COMMON /rates/k6,k2,k3
      EXTERNAL stiff,derivs,rkqs


c
c The k's are the rate constants
      k6=0.2
      k2=5.7e6
      k3=2.8e7

      eps=1.e-6
      hstart=1.e-3
      kmax=0

      time2=0.
c
c the concs are the concentrations
c conc(1) is the reactant concentration
c conc(2) is the intermediate concentration
c conc(3) is the product concentration
c
      conc(1)=1.
      conc(2)=0.
      conc(3)=0.
c next open data file and write out initial conditions
      open(unit=4,file='difdat.txt',action='write',status=
     'replace')
      totcon=conc(1)
      do i=2,NCONCS
      totcon=totcon+conc(i)
      end do
```

```
      write(4,'(f5.2,4(1h,,2x,e9.3))')time2,conc(1),conc(2),
       conc(3),totcon

c next start solving the equation

      do while (time2.lt. (4.99))
      time1=time2
      time2=time1+0.1
c
c odeint integrates the rate equation from time1 to time2
c
c note: change "stiff" to "rkqs" in the call odeint
      statement
c if you want to use runge kutta rather than a stiff solver
c
      call odeint(conc,NCONCS,time1,time2,eps,hstart,0,nok,
       Snbad,derivs, stiff)
c

**NOTE:  *'s designates continuation lines

c total concentration is the total concentration
c
      totcon=conc(1)
      do i=2,NCONCS
      totcon=totcon+conc(i)
      end do
      write(*,*)'conc(', time2,')=',conc(1),conc(2),conc(3),
       ' total conc = ',totcon
      write(4,'(f5.2,4(1h,,2x,e9.3))')time2,conc(1),conc(2),
       conc(3),totconc
10    end do

      write(*,'(/1x,a,t30,i3)') 'Successful steps:',nok
      write(*,'(1x,a,t30,i3)') 'Bad steps:',nbad
      write(*,*) 'conc(END)=',conc(1),conc(2),conc(3)
999   write(*,*) 'NORMAL COMPLETION'
      STOP
      END


      SUBROUTINE derivs(time,conc,f)
c
c  subroutine to calculate the rates of formation of each of
      the species
c
      REAL time,conc(*),f(*)
      REAL k6,k2,k3
```

```
        COMMON /rates/k6,k2,k3

c
c  F(1) is the rate of formation of species one
c

        f(1)= -k6*conc(1)+k2*conc(2)
        f(2)=k6*conc(1)-k2*conc(2)-k3*conc(2)
        f(3)=k3*conc(2)
        return
        END

        SUBROUTINE jacobn(time,conc,dfdt,dfdc,n,nmax)
        INTEGER n,nmax,i

        REAL time,conc(*),dfdt(*),dfdc(nmax,nmax)
        REAL k6,k2,k3
        COMMON /rates/k6,k2,k3

        do 11 i=1,3
          dfdt(i)=0.
c
c   dfdt is the df(i)/dt
c
11        continue

c
c  dfdc(i,j) is df(i)/dc(j)
c  so dfdc(1,2) is the derivative of df(1)/dc(2)
c

        dfdc(1,1)=-k6
         dfdc(1,2)=k2
         dfdc(1,3)=0.0
         dfdc(2,1)=k6
         dfdc(2,2)=-(k2+k3)
         dfdc(2,3)=0.0
         dfdc(3,1)=0.0
         dfdc(3,2)=k3
         dfdc(3,3)=0.0
         return
         END
```

Here is the output:

```
time,    [A]          [I]              [C]
Total conc
```

```
 .00,  .100E+01,  .000E+00,  .000E+00,  .100E+01
 .10,  .984E+00,  .584E-08,  .165E-01,  .100E+01
 .20,  .967E+00,  .574E-08,  .327E-01,  .100E+01
 .30,  .951E+00,  .565E-08,  .486E-01,  .100E+01
 .40,  .936E+00,  .555E-08,  .643E-01,  .100E+01
 .50,  .920E+00,  .546E-08,  .797E-01,  .100E+01
 .60,  .905E+00,  .537E-08,  .949E-01,  .100E+01
 .70,  .890E+00,  .528E-08,  .110E+00,  .100E+01
 .80,  .876E+00,  .520E-08,  .124E+00,  .100E+01
 .90,  .861E+00,  .511E-08,  .139E+00,  .100E+01
1.00,  .847E+00,  .503E-08,  .153E+00,  .100E+01
1.10,  .833E+00,  .494E-08,  .167E+00,  .100E+01
1.20,  .819E+00,  .486E-08,  .181E+00,  .100E+01
1.30,  .806E+00,  .478E-08,  .194E+00,  .100E+01
1.40,  .792E+00,  .470E-08,  .208E+00,  .100E+01
1.50,  .779E+00,  .463E-08,  .221E+00,  .100E+01
1.60,  .767E+00,  .455E-08,  .233E+00,  .100E+01
1.70,  .754E+00,  .447E-08,  .246E+00,  .100E+01
1.80,  .741E+00,  .440E-08,  .259E+00,  .100E+01
1.90,  .729E+00,  .433E-08,  .271E+00,  .100E+01
2.00,  .717E+00,  .426E-08,  .283E+00,  .100E+01
2.10,  .705E+00,  .419E-08,  .295E+00,  .100E+01
2.20,  .694E+00,  .412E-08,  .306E+00,  .100E+01
2.30,  .682E+00,  .405E-08,  .318E+00,  .100E+01
2.40,  .671E+00,  .398E-08,  .329E+00,  .100E+01
2.50,  .660E+00,  .392E-08,  .340E+00,  .100E+01
2.60,  .649E+00,  .385E-08,  .351E+00,  .100E+01
2.70,  .638E+00,  .379E-08,  .362E+00,  .100E+01
2.80,  .628E+00,  .373E-08,  .372E+00,  .100E+01
2.90,  .618E+00,  .367E-08,  .382E+00,  .100E+01
3.00,  .607E+00,  .360E-08,  .393E+00,  .100E+01
3.10,  .597E+00,  .355E-08,  .403E+00,  .100E+01
3.20,  .588E+00,  .349E-08,  .412E+00,  .100E+01
3.30,  .578E+00,  .343E-08,  .422E+00,  .100E+01
3.40,  .568E+00,  .337E-08,  .432E+00,  .100E+01
3.50,  .559E+00,  .332E-08,  .441E+00,  .100E+01
3.60,  .550E+00,  .326E-08,  .450E+00,  .100E+01
3.70,  .541E+00,  .321E-08,  .459E+00,  .100E+01
3.80,  .532E+00,  .316E-08,  .468E+00,  .100E+01
3.90,  .523E+00,  .310E-08,  .477E+00,  .100E+01
4.00,  .514E+00,  .305E-08,  .486E+00,  .100E+01
4.10,  .506E+00,  .300E-08,  .494E+00,  .100E+01
4.20,  .498E+00,  .295E-08,  .502E+00,  .100E+01
4.30,  .489E+00,  .290E-08,  .511E+00,  .100E+01
4.40,  .481E+00,  .286E-08,  .519E+00,  .100E+01
4.50,  .473E+00,  .281E-08,  .527E+00,  .100E+01
4.60,  .466E+00,  .276E-08,  .534E+00,  .100E+01
4.70,  .458E+00,  .272E-08,  .542E+00,  .100E+01
4.80,  .450E+00,  .267E-08,  .550E+00,  .100E+01
```

```
4.90, .443E+00, .263E-08, .557E+00, .100E+01
5.00, .436E+00, .259E-08, .564E+00, .100E+01
```

Example 4.C Deriving the Langmuir Rate Law Schwab and Schwab-Agallidis (1943) measured the kinetics of the metal-catalyzed decomposition of formic acid:

$$HCOOH \Longrightarrow CO_2 + H_2 \qquad (4.C.1)$$

Schwab et al. (1943) and Fukuda et al. (1968) proposed that, on nickel, the reaction follows

$$HCOOH + S \rightleftharpoons HCOOH_{(ad)} \qquad (4.C.2)$$

$$HCOOH_{(ad)} \xrightarrow{3} H_2 + CO_2 + S \qquad (4.C.3)$$

where S is a bare surface site and $HCOOH_{(ad)}$ is an adsorbed formic acid complex. Use the steady-state approximation to calculate a value for the rate.

Solution Let's formulate an expression for the rate of CO_2 production. If reaction (4.C.3) is elementary, then

$$R_3 = k_3[HCOOH_{(ad)}] \qquad (4.C.4)$$

where R_3 is the rate of reaction (4.C.3) in molecules/($cm^2 \cdot$second), k_3 is the rate constant for reaction 3, and $[HCOOH_{(ad)}]$ is the concentration of adsorbed formic acid complex.

We want an expression of $[CO_2]$ in molecules/($cm^3 \cdot$second). The correct expression is

$$r_{CO_2} = R_3 \left(\frac{S}{V} \right) \qquad (4.C.5)$$

where (S/V) is the surface to volume ratio of the system.

Combining equation (4.C.4) and (4.C.5) yields

$$r_{CO_2} = k_3[HCOOH_{(ad)}](S/V) \qquad (4.C.6)$$

One can calculate $[HCOOH_{(ad)}]$ from the steady-state approximation:

$$\phi = \frac{d[HCOOH_{ad}]}{dt} = k_1[HCOOH][S] - (k_2 + k_3)[COOH_{ad}] \qquad (4.C.7)$$

where [S] is the concentration of bare surface sites and [HCOOH] is the concentration of formic acid in the gas phase. Rearranging equation (4.C.7) yields

$$[HCOOH_{(ad)}] = \frac{k_1[HCOOH][S]}{(k_2 + k_3)} \qquad (4.C.8)$$

Combining equation (4.C.6) and (4.C.8) yields

$$r_{CO_2} = \left(\frac{S}{V} \right) \left(\frac{k_1 k_3}{k_2 k_3} \right) [HCOOH][S] \qquad (4.C.9)$$

Equation (4.C.9) is an exact expression for the rate of CO_2 formation. However, the expression contains the quantity [S], which Schwab was not able to measure. Consequently, Schwab used an approximation first derived by Langmuir (1915). Langmuir's model will be discussed in detail in Chapter 12. However, the idea is that when gases bind to metal surfaces, the gases attach to specific sites on the metal surface. Generally, there are a finite number of sites to hold gas. Each of the sites can be empty or covered with a formic acid. If there are S_o total site, then a site balance shows

$$S_o = [S] + [HCOOH_{(ad)}] \tag{4.C.10}$$

Substituting equation (4.C.8) into equation (4.C.10) yields

$$S_o = [S] + \left(\frac{k_1}{k_2 + k_3} \right) [HCOOH_{(ad)}][S] \tag{4.C.11}$$

Rearranging equation (4.C.11) yields

$$[S] = \frac{S_o}{1 + \left(\dfrac{k_1}{k_2 + k_3} \right) [HCOOH]} \tag{4.C.12}$$

Combining equations (4.C.9) and (4.C.12) yields

$$r_{CO_2} = \frac{(k_1 k_3 S_0)}{(k_2 + k_3)} \frac{[HCOOH]}{1 + \left(\dfrac{k_1}{k_2 k_3} \right) [HCOOH]} \tag{4.C.13}$$

Equation (4.C.13) is called the *Langmuir–Hinshelwood expression for the rate*. Fukuda et al. (1968) showed that it fits their data for formic acid decomposition.

Additional examples of derivation of Langmuir–Hinshelwood rate laws are given in the solved examples for Chapter 12.

4.15 SUGGESTION FOR FURTHER READING

General references to the quasi-steady-state approximation include:

A. N. Yannacopoulos, A. S. Tomlin, J. Brindley, J. H. Merkin, and M. J. Pilling, The error of the quasi steady-state approximation in spatially distributed systems. *Chem. Phys, Lett.* **248**, 63–70 (1996).

R. A. B. Bond, B. S. Martincigh, J. R. Mika, R. H. Simoyi, The quasi-steady-state approximation — numerical validation. *J. Chem. Educ.* **75**(9), 1158–1165 (1998).

V. Viossat and R. I. Ben-Aim, Validity of the quasi-stationary-state approximation in the case of two successive first-order reactions. *J. Chem. Educ.* **75**, 1165 (1998).

The numerical methods are discussed in:

R. Lorenzini and L. Passoni, Test of numerical methods for the integration of kinetic equations in tropospheric chemistry, *Comput. Phys. Commun.* **117**(3), 241–249 (1999).

W. H. Press, S. A. Teukolsky, and M. Metcalf, eds., *Numerical Recipes in Fortran 90*, Oxford University Press, Oxford, UK, 1996.

S. C. Chapra, R. P. Canale, S. Chapra, and R. Canale, *Numerical Methods for Engineers: With Programming and Software Applications*, McGraw-Hill, New York, 1998.

4.16 PROBLEMS

4.1 Define the following terms in three sentences or less:

(a) Mechanism

(b) Rate-determining step

(c) Pseudo-steady-state approximation

(d) Reactive intermediate

(e) Overall reaction

(f) Stoichiometric reaction

(g) Elementary reaction

(h) Molecularity

(i) Unimolecularity

(j) Bimolecularity

(k) Termolecularity

(l) Explosion

(m) Collision partner

(n) Stiff differential equation

(o) Implicit method to solve ordinary differential equations (ODEs)

(p) Explicit method to solve ODEs

(q) Runge–Kutta method

(r) Microscopic reversibility

4.2 Describe the similarities and differences between reaction order and molecularity:

(a) Do both terms apply to elementary reactions?

(b) Do both terms apply to overall reactions?

(c) When are they the same?

(d) When are they different?

4.3 What did you learn new in this chapter? Had you seen the pseudo-steady-state approximation before? Were you aware of its limitations? Did you know why explosions arise? Did you know what to do when the steady-state approximation fails?

4.4 Describe the pseudo-steady-state approximation in your own words. Explain when the approximation works and when it fails. Your answer should be 10 sentences or less.

4.5 Table 4.2 gives rate expressions for a number of reactions.

(a) Derive all of the rate equations to make sure that the table is correct.

(b) Notice that you get wildly different rate equations on what are demonstrably similar mechanisms. What are the implications of that result?

(c) Let's assume that you measure the rate of a reaction in what is called a "seasoned glass vessel," namely, a glass vessel that has been treated so that wall reactions are negligible. How would the glass vessel change the kinetics of the reaction?

(d) Would you expect the kinetics to change if you instead ran the reaction in a stainless-steel pressure vessel?

(e) What are the implications of this result for measuring kinetics in a lab that you are going to use for reactor design?

4.6   Rice and Herzfeld, *JACS*, **56** (1934) 284, proposed the following mechanism for the decomposition of acetaldehyde:

$$CH_3CHO + X \longrightarrow CH_3 + CHO + X$$
$$CH_3 + CH_3CHO \longrightarrow CH_3CO + CH_4$$
$$CH_3CO + X \longrightarrow CH_3 + CO + X$$
$$CHO + X \longrightarrow H + CO$$
$$H + CH_3CHO \longrightarrow CH_3CO + H_2$$
$$2CH_3 + X \longrightarrow C_2H_6 + X$$

In Chapter 5 we will find that this mechanism is not quite correct. However, we will use it here anyway.

(a) Derive an equation for the rate of formation of CH_3 radicals in terms of the concentrations of all of the other species in the system.

(b) Use the pseudo-steady-state approximation to eliminate the concentrations of the CH_3, H, and CH_3CO species from your expression.

(c) Also obtain an approximation to the rate of CO formation during the reaction.

4.7   Carbonic anhydrase is an enzyme that catalyzes the reversable reaction $CO_2 + H_2O \Rightarrow HCO_3^- + H^+$ in the human body. The main reaction is thought to be

$$H_2O + [\text{free anhydrase}] \longrightarrow [\text{anhydrase–OH}]^- + H^+ \qquad \text{(P4.7.1)}$$
$$[\text{anhydrase–OH}]^- \longrightarrow H_2O + [\text{free anhydrase}] \qquad \text{(P4.7.2)}$$
$$[\text{anhydrase–OH}]^- + CO_2 \longrightarrow [\text{anhydrase–HCO}_3]^- \qquad \text{(P4.7.3)}$$
$$[\text{anhydrase–HCO}_3]^- \longrightarrow HCO_3^- + [\text{free anhydrase}] \qquad \text{(P4.7.4)}$$

whereas $[\text{anhydrase–OH}]^-$ is an anhydrase molecule that is bound to a hydroxyl, $[\text{anhydrase–HCO}_3]^-$ is an anhydrase molecule that is bound to a bicarbonate, and [free anhydrase] is an anhydrase molecule that is not bound to either a hydroxyl or a bicarbonate.

(a) Use the steady-state approximation to calculate the rate of bicarbonate (HCO_3^-) production as a function of the CO_2 and [free anhydrase] concentration in the aqueous solution. Assume that water is available in large excess so that you can ignore the water in your rate equations.

(b) Also derive expressions for the concentrations of $[\text{anhydrase–OH}]^-$ and $[\text{anhydrase–HCO}_3]^-$.

(c) In actual practice one does not normally know the concentration of the [free anhydrase]. Instead, one knows the total anhydrase concentration, where

$$[\text{Total anhydrase}] = [\text{free anhydrase}] + [\text{anhydrase–OH}]^-$$
$$+ [\text{anhydrase–HCO}_3]^-. \qquad \text{(P4.7.5)}$$

Use equation (P4.7.5) and the steady-state approximation to eliminate the concentration of the free anhydrase from your expression in part (a). Rearrange your expression to obtain what is called the *Michaelis–Menten* expression for the rate of an enzyme-catalyzed reaction:

$$r_{CO_2} = -k[\text{total} - \text{anhydrase}]\frac{[CO_2]}{k_{MM} + [CO_2]}$$

4.8 Ozone is produced in the upper atmosphere via the reactions

$$O_2 + h_p\nu_1 \longrightarrow 2O \qquad\qquad (P4.8.1)$$

$$O + O_2 + X \longrightarrow O_3 + X \qquad\qquad (P4.8.2)$$

$$O_3 + h_p\nu_2 \longrightarrow O_2 + O \qquad\qquad (P4.8.3)$$

$$O_3 + 2O \longrightarrow O_2 + X \qquad\qquad (P4.8.4)$$

where $h_p\nu_1$ is a UV photon with a wavelength of less than 242 nm and $h_p\nu_2$ is a photon with a wavelength of 240–320 nm.

(a) Use the steady-state approximation to derive an expression for the steady-state ozone concentration as a function of the O_2 concentration, the rate constant for the dissociation of O_2 and O_3, and the density of photons.

(b) Chlorine atoms in the upper atmosphere can decompose the ozone. Assume that the chlorine reacts via

$$Cl + O_3 \longrightarrow ClO + O_2 \qquad\qquad (P4.8.5)$$

$$ClO + O_3 \longrightarrow Cl + 2O_2 \qquad\qquad (P4.8.6)$$

Use a steady-state approximation on ClO to derive an equation for the effect of Cl on the ozone concentration. Assume that ozone reacts via reactions (P4.8.1)–(P4.8.6).

(c) What is the significance of this result? Will chlorine increase or decrease the ozone concentration?

4.9 The objective of this problem is to test the pseudo-steady-state approximation for the mechanism discussed in Section 4.4 and Example 4.A

(a) Set up a spreadsheet similar to that in Example 4.A and reproduce the results in Table 4.A.2. The original spread sheet is available from Wiley's website.

(b) Make a plot of your results similar to that in Figure 4.2.

(c) Set $k_3 = 5.0*k_2$, and vary k_2 and see when the steady-state approximation to [I] is within 5% of the exact results for times of 0.1 sec or more.

(d) Check your values of [A] and [P]. Are they also within 5% of the exact results?

(e) Change to $k_3 = 2.5*k_2$. How does that change your conclusions in (c) and (d)?

4.10   The objective of this problem is to put together the results in Chapters 3 and 4 to solve the rate equations for the following mechanisms:

$$Br_2 + X \longrightarrow 2Br + X \qquad (P4.10.1)$$

$$Br + H_2 \longrightarrow HBr + H \qquad (P4.10.2)$$

$$H + Br_2 \longrightarrow HBr + Br \qquad (P4.10.3)$$

$$X + 2Br \longrightarrow Br_2 + X \qquad (P4.10.4)$$

(a) First derive a steady-state approximation for [Br] and [H] and r_{Br_2}.

(b) Assume that you load 1 atm of H_2, 1 atm of Br_2, and 8 atm of inerts into an isothermal batch reactor at 900 K. Substitute your rate equation into equation (3.61) and integrate as shown in Example 3.K to derive an expression for the concentrations of H_2 and Br_2 as a function of time. Also use the steady-state approximation to calculate [H] and [Br] as a function of time.

(c) Make a plot of your data assuming $k_1 = 0.067$ liter^2/(mol^2·second), $k_2 = 8.2 \times 10^5$ liter/(mol·second), $k_3 = 6.2 \times 10^9$ liter/(mol·second), and $k_4 = 4 \times 10^5$ liter^2/(mol^2·second).

4.11   In Section 4.8 we provided mechanisms for the formation of HBr and the oxidation of H_2. Mixtures of H_2 and O_2 can explode, but mixtures of H_2 and Br_2 will not explode.

(a) Examine the two mechanisms carefully. What are the similarities and differences between the two mechanisms?

(b) Count the radicals in the initiation steps, the propagation steps, and the termination steps. Are there any key differences in the net production of radicals during the reactions?

(c) On the basis of your findings, can a mixture of H_2 and F_2 undergo a free-radical explosion? (*Note*: There can also be a thermal explosion where the heat generated by the reaction causes a thermal runaway).

4.12   The objective of this problem is to use the computer program in Example 4.B to simulate a series of chemical reactions for the mechanisms

$$Br_2 + X \longrightarrow 2Br + X \qquad (P4.12.1)$$

$$Br + H_2 \longrightarrow HBr + H \qquad (P4.12.2)$$

$$H + Br_2 \longrightarrow HBr + Br \qquad (P4.12.3)$$

$$X + 2Br \longrightarrow Br_2 + X \qquad (P4.12.4)$$

(a) Set up a computer program similar to that in Example 4.B.6.c and reproduce the results given at the end of Example 4.B.6.c. The original computer program is available from Wiley's website.

(b) Derive the differential equations needed to simulate the mechanism P4.12.1, P4.12.2, P4.12.3, P4.12.4.

(c) Develop an expression for the Jacobians.

(d) Change your computer program so that it uses the differential equations that you derived and the rate constants from Problem 4.10.

(e) Run the program and calculate the concentrations of all of the species as a function of time.

(f) Make a plot of your results similar to that in Figure 4.4 comparing to your analytic solution in Problem 4.10.

(g) Set $k_3 = 7.5 \times 10^3 * k_2$, and vary k_2 to see when the steady-state approximation to [Br] is within 5% of the exact results.

(h) Check your values of $[Br_2]$ and $[H_2]$. Are they also within 5% of the exact results?

(i) Change to $k_3 = k_2$. How does that change your conclusions in (c) and (d)?

(j) Now try changing from a stiff differential equation solver to a Runge–Kutta equation solver (i.e., change the stiff keyword in the CALL ODTINT statement to rkqs). First try the original rate constants, and see if the solution converges. Then try $k_3 = k_2$. Keep reducing k_2 until the Runge–Kutta method gives an answer. Is the answer accurate?

4.13 Rootsaert and Schwab (1943) examined the nickel catalyzed decomposition of formic acid:

$$HCOOH \Longrightarrow H_2 + CO_2 \qquad\qquad (P4.13.1)$$

In Example 4.C we used the steady-state approximation to derive an expression for the rate of the reaction. Schwab worked at pressures between 0.01 and 1 atm. The steady-state approximation was excellent under those conditions. However, more recently McCarthy et al. (1973) examined the reaction at 10^{-11} atm pressure. The steady-state approximation no longer worked. The objective of this problem is to use the computer program in Example 4.B to see when the steady-state approximation works for reaction (P4.13.1).

(a) Set up a computer program similar to that in Example 4.B and reproduce the results in Table 4.B.7.

(b) Derive the differential equations needed to simulate the mechanism Assume that you start with a one-liter vessel containing 10^{-11} atm of formic acid. *Hint*: Derive all of your equations in terms of the rate per unit surface area, then multiply by the surface to volume ratio to get the rate per unit volume.

(c) Develop an expression for the Jacobians.

(d) Change your computer program so that it uses the differential equations you derived in (b).

(e) Run the program and calculate the concentrations of all of the species as a function of time. Assume $S_0 = 5 \times 10^{14}$ molecules/cm^2, $k_1 = 2 \times 10^{-8}$ cm^3/(molecule·second), $k_2 = 10$/second, and $k_3 = 1$/second. (S/V) = 0.01/cm. (*Hint*: Measure your concentration of adsorbed species in molecules/cm^2.)

(f) How long does it take for the steady-state approximation to work?

4.14 One of the steps in the production of a memory circuit is deposition of *polysilicon*. One can make polysilicon by decomposing silane (SiH_4) on a silicon wafer. The main reaction is

$$SiH_4 \Longrightarrow Si + 2H_2$$

Assume that the reaction occurs via the following mechanisms:

$$SiH_4 + 2S \longrightarrow SiH_{3(ad)} + H_{(ad)} \qquad (P4.14.1)$$

$$SiH_{3(ad)} \longrightarrow SiH + H_2 \qquad (P4.14.2)$$

$$SiH + H_{ad} \longrightarrow Si + H_2 \qquad (P4.14.3)$$

$$2SiH \longrightarrow 2Si + H_2 \qquad (P4.14.4)$$

$$2H_{ad} \longrightarrow 2S + H_2 \qquad (P4.14.5)$$

(a) Use the steady-state approximation for the deposition rate of silane. For the purposes of derivation assume that each time you deposit a silicon atom, the silicon atom can be a site for more reaction.

(b) How would your rate equation change if SiH_4 could also undergo the following reactions:

$$SiH_4 \longrightarrow SiH_2 + H_2 \qquad (P4.14.6)$$

$$SiH_2 + S \longrightarrow SiH_{2(ad)} \qquad (P4.14.7)$$

$$SiH_{2(ad)} \longrightarrow Si + H_2 \qquad (P4.14.8)$$

4.15   The objective of this problem is to work out the implications of microscopic reversibility for the reaction

$$H_2 + Br_2 \Longrightarrow 2HBr$$

In 1904 Bodenstein and Lund examined the reaction at temperatures below $1000°C$ and proposed that the reaction occurred via the following mechanisms:

$$Br_2 + X \longrightarrow 2Br + X \qquad (P4.15.1)$$

$$Br + H_2 \longrightarrow HBr + H \qquad (P4.15.2)$$

$$H + Br_2 \longrightarrow HBr + Br \qquad (P4.15.3)$$

$$X + 2Br \longrightarrow Br_2 + X \qquad (P4.15.4)$$

$$H + HBr \longrightarrow H_2 + Br \qquad (P4.15.5)$$

The back reaction rate was negligible under the conditions used by Bodenstein and Lund. However, the back reaction will occur at $2000°C$. The main mechanism is thought to be

$$HBr + X \longrightarrow Br + H + X \qquad (P4.15.6)$$

$$Br + HBr \longrightarrow Br_2 + H \qquad (P4.15.7)$$

$$H + HBr \longrightarrow H_2 + Br \qquad (P4.15.8)$$

$$X + H + Br \longrightarrow HBr + X \qquad (P4.15.9)$$

$$H_2 + Br \longrightarrow H + HBr \qquad (P4.15.10)$$

The objective of this problem is to see if these two mechanisms are consistent with microscopic reversibility.

(a) Derive an equation for the rate of the forward reaction.

(b) Derive an equation for the rate of the back reaction.

(c) Are the two rates the same at equilibrium? At equilibrium $[HBr]^2 = K_e[H_2][Br_2]$. If you plug $[HBr]^2 = K_e[H_2][Br_2]$ into the two expressions, are the two expressions the same?

(d) If you did everything right in (c), you will find that the results are NOT consistent with microscopic reversibility. What does that tell you about the mechanisms above?

(e) One of the implications of microscopic reversibility is that if a given elementary reaction is part of the forward mechanism then at equilibrium, then the reverse of the given elementary reaction must be part of the mechanism of the back reaction. What reactions do you need to add to the mechanisms above to satisfy microscopic reversibility?

(f) One could in principle derive a new rate equation for the forward and reverse reactions given the elementary reactions that you added in part (d). The algebra is pretty difficult. Still, one can get useful information without deriving a rate equation. Derive an equation for the overall rates of the forward and reverse reactions in terms of the rates of all of the elementary reactions.

(g) Use the steady-state approximation to show that at equilibrium the two rates are equal. *Hint*: Use the expressions for the rates of the elementary reactions first, then when you have simplified your expression, plug in the individual rate equations.

4.16 The free-radical-initiated polymerization of styrene with a benzoyl peroxide initiator can be modeled by the following mechanism:

$$C_6H_5(CO)OO(CO)C_6H_5 \longrightarrow 2C_6H_5O \tag{P4.16.1}$$

$$C_6H_5O + [styrene] \longrightarrow [styrene \ radical] \tag{P4.16.2}$$

$$[styrene \ radical] + [styrene] \longrightarrow [polystyrene \ (2) \ radical \ 4]$$

$$[polystyrene \ (n) \ radical] + [styrene] \longrightarrow [polystyrene \ (n+1) \ radical]$$

$$\tag{P4.16.4}$$

$$[polystyrene \ (m) \ radical] + [polystyrene \ (n) \ radical] \longrightarrow [polystyrene]$$

$$\tag{P4.16.5}$$

where a styrene radical is a radical containing one benzyloxy (C_6H_5O) unit and one styrene unit. A [styrene (2)—radical] is a species containing one benzyloxy unit and two styrene units, a [styrene (n) radical] is a species containing one benzyloxy units and n styrene units, and a [styrene (n + 1) radical] is a species containing one benzyloxy unit and n + 1 styrene units. For the purposes of this problem, assume that all of the styrene (n) + radicals are kinetically equivalent to the styrene (2) radicals, the styrene (n + 1) radicals, and the styrene (m) radicals so that all can be treated as a single species in the rate equation.

(a) Use the steady-state approximation to show that at steady state the rate of polystyrene production is equal to the rate of reaction 1 (i.e., you get one polystyrene molecule for every benzyl peroxide molecule that you use up).

(b) Does your answer in (a) make sense? Are you using one benzyl peroxide molecule to produce one polystyrene molecule?

(c) Use the steady-state approximation to develop an expression for the rate of consumption of styrene.

(d) Compare the rate of reaction (P4.16.1) and the rate of styrene consumption. If the rate of reaction (P4.16.1) is equal to the number of polystyrenes you produce and the rate of styrene consumption, then the ratio of the two rates should be the average number of polystyrenes that are incorporated into each polystyrene. Derive an expression for the average number of styrenes incorporated in each polymer chain.

(e) Is the steady-state approximation reasonable for this example? (*Hint*: Do the criteria from this chapter hold?)

(f) Try numerically simulating the rate equations above using an adaptation of the computer code for Example 4.B. Be sure to test your code for Example 4.B.6.c before proceeding to use the code for the new problem. (*Note*: The source code is available in the instructions materials).

(g) Derive an expression for the molecular weight for polymer as a function of time. [*Hint*: Each time reaction (P4.16.4) goes once, the molecular weight goes up by the weight of one styrene.]

4.17 Van't Hoff [1896] proposed that the oxidation of phosphine follows the rate law, $r = k[PH_3][O_2]^{1/2}$.

$$PH_3 + 4\,S \rightleftharpoons P_{(ad)} + 3\,H_{(ad)} \qquad (P4.17.1)$$

$$O_2 + 2\,S \rightleftharpoons 2\,O_{(ad)} \qquad (P4.17.2)$$

$$P_{(ad)} + O_{(ad)} \longrightarrow PO_{(ad)} + S \qquad (P4.17.3)$$

$$2PO_{(ad)} + 3\,O_{(ad)} \rightleftharpoons P_2O_5 + 5\,S \qquad (P4.17.4)$$

(a) Use the methods in Example 4.C to derive a rate equation for the reaction. Assume that reaction P4.17.3 is rate-determining.

(b) How well does your rate equation explain Van't Hoff's observations?

4.18 So far in this chapter we have been using the pseudo-steady-state approximation to derive an expression for rates of reaction. However, the pseudo-steady-state approximation is used in a wide variety of other contexts. Consider using a thermocouple to measure the temperature of a water bath.

(a) Use a heat balance on the thermocouple to obtain an expression for the derivative of the thermocouple temperature with respect to time as a function of the thermocouple temperature, the water bath temperature, the heat transfer coefficient between the water bath and the thermocouple, and any other variables that you need. Assume that the thermocouple has a mass of 30 grams, a specific heat of 0.6 cal/(K·cm²), a surface area of 5 cm² and a heat transfer coefficient of 5 cal/(second·K).

(b) Set up an equation for the change in the bath temperature with respect to time. Assume that there are 10 liters of water in the bath. Assume that there is a heater in the bath that supplies 2 kcal/minute. Be sure to consider the heat transferred to the thermocouple in your heat balance.

(c) You should end up with equations that look very similar to the equations in Section 4.3. Solve the equations analytically to calculate the bath temperature and the thermocouple temperature as a function of time.

(d) Also try to solve the equations numerically using an algorithm like that in problem 4.8; try using both a Runge–Kutta integrator (RKQS) and a stiff integrator and see which works best.

(e) Finally, use the steady-state approximation for the thermocouple temperature (i.e., set the derivative of the thermocouple temperature respect to time equal to zero and solve for the thermocouple temperature).

(f) How well does the steady-state approximation work in these examples?

4.19 The purpose of this problem is to verify the results in Sections 4.5.2 and 4.5.3 analytically.

(a) Show that equations (4.53) and (4.54) satisfy the differential equations, equations (4.50) and (4.51).

(b) Show that equations (4.53) and (4.54) satisfy the boundary conditions equation in equation (4.52).

(c) Let $k_2 = k$, $k_3 = \alpha k$, substitute into equations (4.53)–(4.56).

(d) Take the limit of your expression k approaches ∞. Show that it approaches equation (4.62).

4.20 In Table 4.B.1 we used Euler's method to solve the differential equation

$$\frac{dC_A}{dt} = -k_1 C_A \qquad (P4.20.1)$$

with $k_1 = 2$ liter/(mol·minute) and $C_A^0 = 1$ mol/liter.

(a) Set up the spreadsheet yourself and reproduce the findings. The original spread sheet is available in the instructions materials.

(b) Try varying the timestep. How small of a timestep do you need to get within 1% of the exact results at t = 2 minutes.

(c) Now try decreasing k_1 to 1 liter/(mol·minute). How does the needed timestep change.

(d) Now try doubling k_1. How does that affect the needed timestep?

4.21 In Table 4.B.3 we used a modified Euler method to solve equation (P4.20.1) with $k_1 = 2$ liter/(mol·minute) and $C_A^0 = 1$ mol/liter.

(a) Set up the spreadsheet yourself and reproduce the findings.

(b) Try varying the timestep. How small of a timestep do you need to get within 1% of the exact results at t = 2 minutes?

(c) Now try decreasing k_1 to 1 liter/mol·minute. How does the needed timestep change?

(d) Now try doubling k_1. How does that affect the needed timestep?

    (e) Why did the modified Euler method need so many fewer timesteps than the simple Euler method?

4.22 In Table 4.B.5 we used a second-order Runge–Kutta method to solve equation (P4.20.1) with $k_1 = 2$ liters/(mol·minute) and $C_A^0 = 1$ mol/liter.

    (a) Set up the spreadsheet yourself and reproduce the findings.

    (b) Try varying the timestep. How small of a timestep do you need to get within 1% of the exact results at $t = 2$ minutes?

    (c) Now try decreasing k_1 to 1 liter/(mol·minute). How does the needed timestep change?

    (d) Now try doubling k_1. How does that affect the needed timestep?

    (e) Why did the modified Euler method need so many fewer timesteps than the simple Euler method?

4.23 The objective of this problem is to use Euler's method to solve the set of differential equations

$$\frac{dC_A}{dt} = -k_1 C_A + k_2 C_I \tag{P4.23.1}$$

$$\frac{dC_I}{dt} = k_1 C_A - (k_2 + k_3)C_I \tag{P4.23.2}$$

with $k_1 = 2$ liters/(mol·minute), $k_2 = 10$ liters/(mol·minute), $k_3 = 20$ liters/(mol·minute), $C_A^0 = 1$ mol/liter, and $C_I^0 = 0.1$ mol/liter. The method will be to modify the spreadsheet in Table 4.B.1.

    (a) Set up the spreadsheet in Table 4.B.1 and make sure that you can reproduce the findings. (Use the original differential equation for this part of the problem.)

    (b) Add extra columns for C_I and D_I and substitute in the correct rate equation. Does the calculation converge?

    (c) Compare your results to the exact solution to the differential equations given by equations (4.53) and (4.54).

    (d) How small of a step size do you need to be within 1% of the exact result at $t = 2$ minutes?

    (e) Change $k_1 = 0.2$ liters/(mol·minute), $k_2 = 100$ liters/(mol·minute), and $k_3 = 200$ liters/(mol·minute), and see how well things work. Can you get within 1% of the exact result?

4.24 The objective of this problem is to use the modified Euler method to solve equations (P4.23.1) and (P4.23.2) with $k_1 = 2$ liters/(mol·minute), $k_2 = 10$ liters/(mol·minute), $k_3 = 20$ liters/(mol·minute), $C_A^0 = 1$ mol/liter, and $C_I^0 = 0.1$ mol/liter. The method will be to modify the spreadsheet in Table 4.B.3.

    (a) Set up the spreadsheet in Table 4.B.3 and make sure that you can reproduce the findings. (Use the original differential equation for this part of the problem.)

    (b) Add extra columns for C_I and D_I and substitute in the correct rate equation. Does the calculation converge?

    (c) Compare your results to the exact solution to the differential equations given by equations (4.53) and (4.54).

(d) How small of a step size do you need to be within 1% of the exact result at $t = 2$ minutes?

(e) Change $k_1 = 0.2$ liters/(mol·minute), and $k_2 = 100$ liters/(mol·minute), $k_3 = 200$ liters/(mol·minute), and see how well things work. Can you get within 1% of the exact result?

4.25 The objective of this problem is to use the modified Euler method to solve equations (P4.23.1) and (P4.23.2). with $k_1 = 2$ liters/(mol·minute), $k_2 = 10$ liters/(mol·minute), $k_3 = 20$ liters/(mol·minute), $C_A^0 = 1$ mol/liter, and $C_I^0 = 0.1$ mol/liter. The method will be to modify the spreadsheet in Table 4.B.5.

(a) Set up the spreadsheet in Table 3.B.5 and make sure that you can reproduce the findings. (Use the original differential equation for this part of the problem.)

(b) Add extra columns for C_I and D_I and substitute in the correct rate equation. Does the calculation converge?

(c) Compare your results to the exact solution to the differential equations given by equations (4.53) and (4.54).

(d) How small of a step size do you need to be within 1% of the exact result at $t = 2$ minutes?

(e) Change $k_1 = 0.2$ liters/(mol·minute), $k_2 = 100$ liters/(mol·minute), $k_3 = 200$ liters/(mol·minute), and see how well things work. Can you get within 1% of the exact result?

More Advanced Problems

4.26 The computer programs in this chapter used the numerical recipes distributed by Microsoft. However, you could instead use the IMSL (International Mathematics and Statistics Library) software packages. Rewrite the FORTRAN programs in Example 4.B to work with the IMSL routines. How do your results compare to those in Example 4.B?

4.27 Look up one of the following papers and write a one-page report describing how the findings in this class apply to the results in the paper.

(a) Zeng, G. Pilling, M. J., and Saunders, S. M. Mechanism reduction for tropospheric chemistry — butane oxidation. *J. Chem. Soc., Faraday Trans.* **93**(16), 2937–2946 (1997).

(b) Buxton, G. V., Malone, T. N., and Salmon, G. A. Reaction of SO4.- with Fe2+, Mn2+ and Cu2+ in aqueous solution. *J. Chem. Soc. Faraday Trans.* **93**(16), 2893–2897 (1997).

(c) Juang, E., and Peacocklopez, E. Steady-state approximation in the minimal model of the alternative pathway of complement. *Biophys. Chem.* **65**(2–3), 143–156 (1997).

(d) Bikrani. M., Fidalgo, L., Garralda M. A., and Ubide, C. Catalytic hydrogen transfer activity of cationic iridium(I) complexes containing alpha-diimines. *J. Mol. Catal. A.* **118**(1) 47–53 (1997).

(e) Mathew, J., Ghadage, R. S., Lodha, A., Ponrathnam, S., and Prasad, S. D. Copolyesters of poly(ethylene terephthalate), hydroquinone diacetate, and terephthalic acid — a simple rate model for catalyzed synthesis in the melt. *Macromolecules* **30**(6), 1601–1610 (1997).

(f) Jay, L. O., Sandu, A., Potra, F. A., and Carmichael, G. R. Improved quasi-steady-state-approximation methods for atmospheric chemistry integration. *Siam J. Sci. Comput.* **18**(1), 182–202 (1997).

(g) Johnson, B. R., Scott, S. K., and Thompson, B. W. Modeling complex transient oscillations for the BZ reaction in a batch reactor. *Chaos* **7**(2), 350–358 (1997).

(h) Yannacopoulos, A. N. Tomlin. A. S., Brindley, J., Merkin, J. H., and Pilling M. J. Error propagation in approximations to reaction-diffusion-advection equations. *Phys. Lett. A* **223**(1–2), 82–90 (1996).

(i) Goussis, D. A. On the construction and use of reduced chemical kinetic mechanisms produced on the basis of given algebraic relations. *J. Comput. Phys.* **128**(2), 261–273 (1996).

(j) Hasegawa, T., Yasuda, N., and Yoshioka, M. Photocyclization of 2-(dialkylamino)ethyl acetoacetates — remote proton transfer and stern-volmer quenching kinetics in the system involving two reactive excited states. *J. Phys. Org. Chem.* **9**(4), 221–226 (1996).

(k) Segal, E., Urbanovici, E., and Popescu, C. On the validity of the steady-state approximation in non-isothermal kinetics .2. *Thermochim. Acta* **274**, 173–177 (1996).

(l) Harris, S. First-layer island growth during epitaxy. *Phys. Rev. B* **53**(11), 7500–7503 (1996).

(m) Roberts, J. R., Xiao, J., Schleisman, B., Parsons. D. J., and Shaw C. F. Kinetics and mechanism of the reaction between serum albumin and auranofin (and its isopropyl analogue) in vitro. *Inorg. Chem.* **35**(2), 424–433 (1996).

(n) Yannacopoulos, A. N., Tomlin, A. S., Brindley, J., Merkin, J. H., and Pilling, M. J. The error of the quasi steady-state approximation in spatially distributed systems. *Chem. Phys. Lett.* **248**(1–2), 63–70 (1996).

(o) Olson, T., and Hamill, P. A time-dependent approach to the kinetics of homogeneous nucleation. *J. Chem. Phys.* **104**(1), 210–224 (1996).

(p) Juang, E., and Peacocklopez, E. Steady-state approximation in the minimal model of the alternative pathway of complement. *Biophys. Chem.* **65**(2–3), 143–156 (1997).

(q) Wu, J. F., and Luther, G. W. Complexation of Fe(Iii) by natural organic ligands in the northwest atlantic ocean by a competitive ligand equilibration method and a kinetic approach. *Mar. Chem.* **50**(1–4) 159–177 (1995).

(r) Knoth, O., and Wolke, R. Numerical methods for the solution of large kinetic systems. *Appl. Numer. Mathe.* **18**(1–3), 211–221 (1995).

(s) Uchibo, A., Tamura, H., and Furuichi, R. Modeling of the kinetics of lithium ion incorporation into a spinel type manganese oxide, a template ion exchanger. *Bunseki Kagaku* **44**(6), 449–455 (1995).

(t) Schnell, S., and Mendoza, C. Closed form solution for time-dependent enzyme kinetics. *J. Theor. Biol.* **187**(2), 207–212 (1997).

(u) Borghans, J. A. M., Deboer, R. J., and Segel, L. A. Extending the quasi-steady-state approximation by changing variables. *Bull. Math. Biol.* **58**(1), 43–63 (1996).

(v) Chappell, M. J. Structural identifiability of models characterizing saturable binding — comparison of pseudo-steady-state and non-pseudo-steady-state model formulations. *Math. Biosci.* **133**(1), 1–20 (1996).

(w) Rigalleau, V., Beylot, M., Laville, M., Guillot, C., Deleris, G., Aubertin, J., and Gin, H. Measurement of post-absorptive glucose kinetics in non-insulin-dependent diabetic patients — methodological aspects. *Eur. J. Clin. Invest.* **26**(3) 231–236 (1996).

(x) Aivazyan, R. G. Critical conditions of monosilane self-ignition with oxygen and chain branching reactions. *Kinet. Catal.* **38**(2), 174–184 (1997).

(y) Schnell, S., and Mendoza, C. Closed form solution for time-dependent enzyme kinetics. *J. Theor. Biol.* **187**(2), 207–212 (1997).

(z) Winiarski, D. W., and Oneal, D. L. A quasi-steady-state model of attic heat transfer with radiant barriers. *Energy Buildings* **24**(3), 183–194 (1996).

The reader might notice that all of these references are from 1996 or 1997. One can obtain additional references by searching *Current Contents* with the keywords "steady-state approximation" or "quasi–steady state".

5

PREDICTION OF THE MECHANISMS OF REACTIONS

PRÉCIS

In Chapter 4, we showed that if one knows the mechanism of a reaction and the rate constants for all of the elementary steps in the mechanism, one can calculate a rate equation for the reaction.

The objective of this chapter is to give some insights into the mechanisms of reaction. We will start with gas-phase reactions and present some empirical rules to predict reaction mechanisms. We will then move on to reactions in liquids and on solid surfaces, and discuss how mechanisms can be predicted in those cases. We will find that we can often get accurate predictions with relatively simple methods.

5.1 INTRODUCTION

The study of the mechanisms of simple reactions had its start in the latter part of the nineteenth century. At the time, people were still unsure why the kinetics of simple reactions were so different from what one would expect from the stoichiometry of a reaction. In 1898, people first realized that reactions followed complex mechanisms. In the 1920s and 1930s, there were several attempts to predict mechanisms. However, the work was largely unsuccessful. That is as far as the theory went up until 1950.

In fact, up until 1985, no one seriously discussed the idea that one might be able to predict the mechanisms of simple reactions. However, as computers became faster, people started to talk about the idea that one might be able to predict mechanisms, and not just measure mechanisms. Unfortunately, much of the information has been discussed only at meetings and has not made it into the general literature. Still, I believe that the ideas that are coming out now are important enough, and simple enough, that students and others ought to know about them. In this chapter, we will discuss how one can predict reactions mechanisms.

First, we will discuss why reaction follow complex mechanisms and then we will work out some rules for prediction of the mechanisms of gas-phase reactions, surface reactions, and reactions in solution. If you can predict mechanisms, you know all of chemistry, so clearly we cannot cover everything. However, we wanted to provide some principles that readers could use to understand mechanisms.

5.2   WHY DO REACTIONS FOLLOW COMPLEX MECHANISMS?

The first question we want to ask is why reactions follow complex mechanisms; why don't the reactants instead combine in one step to produce products? The answer is pretty simple. In most reactions in industrial practice, the reactants are stable molecules. A stable molecule is by definition stable, that is, relatively unreactive. As a result, a direct reaction between the reactants is usually slow. There are exceptions, of course. If you react sodium with chlorine, you get a direct reaction between the reactants. However, most organic or inorganic reactions start with stable, unreactive species. If you start with a stable unreactive species, you need to do something to the species before that species becomes reactive. Generally, most reactions follow the general mechanism

$$\text{Reactants} \rightleftharpoons \text{reactive species}$$

$$\text{Reactants} + \text{reactive species} \longrightarrow \text{products} \qquad (5.1)$$

As seen in reaction (5.1), you first produce a reactive species, and then the reactive species reacts with more reactants to form products. You can form the reactive species by breaking a bond in the species or by exciting the species with, for example, a laser. The excited species can be a vibrationally excited molecule or something like a radical or an ion. In the gas phase, the excited species is most often a radical. However, if you start with an unreactive molecule, you almost always need to create a reactive molecule before significant reaction can occur.

For example, let's consider the case that was discussed in Chapter 4: the reaction between hydrogen and bromine to yield HBr:

$$H_2 + Br_2 \Longrightarrow 2HBr \qquad (5.2)$$

In Chapter 4, we noted that the mechanism of the reaction is

$$X + Br_2 \xrightarrow{\ 1\ } 2Br + X$$

$$Br + H_2 \xrightarrow{\ 2\ } HBr + H$$

$$H + Br_2 \xrightarrow{\ 3\ } HBr + Br \qquad (5.3)$$

$$X + 2Br \xrightarrow{\ 4\ } Br_2 + X$$

$$H + HBr \xrightarrow{\ 5\ } H_2 + Br$$

where X is a collision partner (i.e., any other species that can collide with Br_2 as described in Sections 4.3 and 8.11). Notice that we formed a reactive species, namely, a radical in step 1. The radical then reacts to form products. The formation of a reactive species is a common feature of most chemical reaction mechanisms.

5.3 PREDICTION OF MECHANISMS OF REACTIONS OF GAS-PHASE SPECIES

In the remainder of this chapter, we will discuss how one can predict the mechanism of a variety of different reactions. We will first do gas-phase reactions, then move on to reactions in liquids and on surfaces.

Let's start with gas-phase reactions. Most gas-phase reactions follow what are called *initiation–propagation mechanisms*. Initiation–propagation mechanisms were mentioned in Chapter 4. Consider the idealized reaction

$$A + B \Longrightarrow P_1 + P_2 \tag{5.4}$$

where a molecule of A reacts with a molecule of B to produce two products, P_1 and P_2. This reaction is said to follow an initiation–propagation mechanism when the reaction goes as follows:

First, there is an **initiation step** where a collision partner X collides with one of the reactant molecules, and breaks a bond in the reactant to produce two radicals, which I will call R_1 and R_2:

$$X + B \longrightarrow R_1 + R_2 \tag{5.5}$$

Then there may be a **transfer step** where one of the radicals reacts to form a more reactive species. R_3 and R_4, and two side products C and D:

$$R_1 + B \longrightarrow R_3 + C$$
$$R_2 + A \longrightarrow R_4 + D \tag{5.6}$$

Then the radicals react with the reactants or the products in a series of **propagation steps** to yield a series of new radicals, R_5 and R_6:

$$R_3 + A \longrightarrow P_1 + R_5$$
$$R_4 + B \longrightarrow P_2 + R_6 \tag{5.7}$$

Additional propagation steps occur where R_5 and R_6 react to regenerate R_3 and R_4:

$$R_6 + A \longrightarrow P_1 + R_4$$
$$R_5 + B \longrightarrow P_2 + R_3 \tag{5.8}$$

Finally, the radicals can recombine in what is called a **termination step**:

$$X + R_1 + R_2 \longrightarrow B + X \tag{5.9}$$

There are many different combinations. Still, the key feature of an initiation–propagation mechanism is that first there is an initiation step where radicals are formed. Then, there are a series of steps where radicals react to form products. Finally, there are a series of termination steps where radicals are destroyed.

Most gas-phase reactions go via initiation–propagation mechanisms. For example, Rice (1932) examined the decomposition of a wide variety of organic compounds, including aliphatic and olefinic hydrocarbons, aldehydes, ketones, esters, amines, and carbolic acids, and found that in all cases, the reaction went by an initiation–propagation mechanism. Therefore, if one wanted to guess at a mechanism of a gas-phase reaction, one would guess an initiation–propagation mechanism.

Let's look back at mechanism (5.3) and show that it follows an initiation–propagation mechanism. Reaction 1 in mechanism (5.3) is an initiation step. Reactions 2 and 3 are propagation steps. Reaction 4 is a termination step. Reaction 5 is the reverse of reaction 2. Clearly, mechanism (5.3) is an initiation–propagation mechanism.

Another important initiation–propagation reaction is the reaction that happens in the cylinders of your car. During the power cycle of your car, a mixture of air and gasoline is put into the cylinders of the car. Nothing happens until the spark plugs fire, producing radicals. Then the propagation reactions occur, producing power. Termination reactions include reactions with tetraethyl lead (TEL) or methyl *tert*-butyl ether (MTBE), and reactions that occur on the walls of the cylinders.

Table 5.1   Examples of initiation–propagation mechanisms

Reaction	Example Mechanism
Combustion, e.g., $CH_4 + O_2 \Rightarrow CO_2 + 2H_2O$ + other products	$O_2 \rightleftharpoons 2O$ $O + CH_4 \rightarrow CH_3 + OH$ $OH + CH_4 \rightarrow H_2O + CH_3$ $CH_3 + O_2 \rightarrow CH_3 + O + O$ $CH_3 + OH \rightarrow CH_2 + H_2O$ + other products $OH \rightarrow$ walls $CH_3 \rightarrow$ walls
Free-radical polymerization, e.g., ethylene $\Rightarrow$ polyethylene with a free-radical catalyst, R_2	$R_2 \rightarrow 2R\bullet$ $R\bullet + C_2H_4 \rightarrow R(C_2H_4)\bullet$ $RC_2H_4\bullet + C_2H_4 \rightarrow R(C_2H_4)_2\bullet$ $R(C_2H_4)_n\bullet + C_2H_4 \rightarrow R(C_2H_4)_{n+1}\bullet$ $R(C_2H_4)\bullet_m + R(C_2H_4)\bullet_n \rightarrow R(C_2H_4)_{m+n}R$
Ozone depletion	$O_2 + h\nu_1 \rightarrow 2O$ $O + O_2 + X \rightarrow O_3$ $O_3 + h\nu_2 \rightarrow O_2 + O$ $Cl + O_3 \rightarrow O_2 + ClO$ $ClO + O \rightarrow O_2 + Cl$
Hydrocarbon pyrolysis	$X + CH_3COH \rightarrow CH_3 + COH + X$ $CH_3 + CH_3OH \rightarrow CH_3CO + CH_4$ $CH_3CO + CH_3OH \rightarrow CH_4 + CH_3CO$ $COH + X \rightarrow CO + H + X$ $H + CH_3COH \rightarrow CH_4 + COH$ $H + CH_3COH \rightarrow CH_3 + CO + H_2$ $2CH_3 + X \rightarrow C_2H_6 + X$ $H + CH_3 + X \rightarrow CH_4 + X$ $H + CH_3CO + X \rightarrow CH_3COH + X$

Table 5.1 gives several other examples of initiation–propagation reactions. Readers should convince themselves that each of these reactions is an initiation–propagation mechanism.

> Most gas-phase reactions of neutral species go via initiation–propagation reactions.

In the literature, initiation–propagation reactions of gas-phase species are also called *radical reactions*, since the reactive species are radicals. They are also called *Rice–Herzfeld mechanisms* since Rice and Herzfeld were early proponents of the mechanisms.

5.4 PREDICTION OF THE MECHANISM OF INITIATION–PROPAGATION MECHANISMS

Next, we will discuss the prediction of the mechanism of simple gas-phase reactions. Our objective will be to develop some quantitative tools that one can use to predict the mechanisms of reactions.

To start, let's again consider the following reaction:

$$H_2 + Br_2 \Longrightarrow 2HBr \tag{5.10}$$

Our general scheme to predict the mechanism of these reactions will be to

- Guess or predict all of the species that are likely to form during the reaction.
- Write down all of the possible reactions of those species.
- Use various rules to pare down the list to manageable number of steps.

Generally the rules are that

- There must be at least one initiation reaction.
- The propagation reactions must occur in a cycle where radicals react to form new radicals and then the new radicals react to form the original radicals again.
- All of the steps in the catalytic cycle must have low activation barriers.
- There should be at least one termination reaction where two radicals react to form stable species.

Next, let us apply this scheme to predict a mechanism for reaction (5.10). For the sake of discussion, we will assume that HBr is the only product of the reaction, and that hydrogen and bromine atoms (i.e., radicals) are the only intermediates in the system. Under such circumstances, there are two possible initiation reactions:

$$H_2 + X \longrightarrow 2H + X \tag{5.11}$$

$$Br_2 + X \longrightarrow 2Br + X \tag{5.12}$$

The hydrogen atoms can react with each of the stable species in the system:

$$H + Br_2 \longrightarrow HBr + Br \tag{5.13}$$

$$H + HBr \longrightarrow H_2 + Br \tag{5.14}$$

$$H + H_2 \longrightarrow H_2 + H \tag{5.15}$$

The bromine atoms can react with each of the stable species in the system:

$$Br + H_2 \longrightarrow HBr + H \tag{5.16}$$

$$Br + HBr \longrightarrow Br_2 + H \tag{5.17}$$

$$Br + Br_2 \longrightarrow Br_2 + Br \tag{5.18}$$

The radicals can react with each other:

$$2Br + X \longrightarrow Br_2 + X \tag{5.19}$$

$$2H + X \longrightarrow H_2 + X \tag{5.20}$$

$$H + Br + X \longrightarrow HBr + X \tag{5.21}$$

I also want to mention that years ago, people thought about the possibility of the reaction occurring via a four-centered complex:

$$Br_2 + H_2 \longrightarrow \begin{matrix} H-H \\ | \quad | \\ Br-Br \end{matrix} \longrightarrow 2HBr \tag{5.22}$$

They also considered reactions without collision partners, such as

$$Br_2 \longrightarrow 2Br \tag{5.23}$$

$$2Br \longrightarrow Br_2 \tag{5.24}$$

Notice that reactions (5.11)–(5.24) are all of the possible reactions between H_2, Br_2, HBr, H, and Br. Therefore, the mechanism must involve *only* these elementary reactions.

Next, let's try to eliminate reactions. In Chapter 4, we noted that

Every elementary reaction must have at least two reactants and at least two products.

$$(5.25)$$

Consequently, reactions (5.23) and (5.24) cannot occur.

In contrast, reactions (5.11)–(5.22) are perfectly good elementary reactions, There are two or more reactants and two or more products in each step, which means that in principle all of these reactions could occur. A more precise analysis shows that they *all* do occur at some rate. Therefore, in a sense, reactions (5.11)–(5.22) represent the complete mechanism at the reaction between H_2 and Br_2.

The question is: Why did we say that reaction (5.2) follows mechanism (5.3) and not include reactions (5.11), (5.15), (5.17), (5.18), and (5.20)–(5.22)? This brings up a subtle point. When we say that reaction (5.2) follows mechanism (5.3), we are not saying that reactions (5.12)–(5.14), (5.16), and (5.19) are the only reactions that occur. Rather, we are saying that reactions (5.11), (5.15), (5.17), (5.18), and (5.20)–(5.22) occur so slowly that they have a negligible effect on the overall rate. More precisely, we can state that

The mechanism represents the principal route from reactants to products, but other reactions can, and do occur.

Now, the next question we want to address is: Could one have guessed the mechanism of reaction (5.2) without doing any experiments? The answer is "Yes". There are a series of empirical rules that one can use to tell you about the mechanisms of chemical reactions. One can often use those empirical rules to guess the preferred mechanism of a chemical reaction.

5.4.1 Empirical Rules to Predict Mechanisms of Gas-Phase Radical Reactions: Reactions to Include

There are two kinds of empirical rules: empirical rules to *exclude* reactions, and empirical rules to *include* reactions.

Let's start with the empirical rules to include reactions. Note that when one has an initiation–propagation reaction, there are certain requirements that one needs to satisfy to make the reaction happen. The first empirical rule is that

> There must be at least one initiation reaction. (5.26)

I include this rule because students often forget that there must be an initiation step in any initiation–propagation reaction. Typically, the initiation step has a small rate. Consequently, students sometimes assume that the initiation step will not be important to the reaction. However, in practice, no reaction happens until the initiation step occurs. Earlier in this chapter, we noted that the reactions in the engine of your car follow an initiation–propagation reaction in which the initiation process is activated by the spark plugs. If you never fire the spark plugs and start the reaction, no reaction ever occurs. The same is true for any initiation–propagation reaction. Typically, the initiation steps have a much lower rate than do any other steps in the reaction mechanism. However, if one ignores the initiation step, one will never derive a correct rate equation.

You need to include at least one initiation step in every mechanism independent of whether the rate of the initiation reaction is small.

The next key requirement is

> The propagation reactions must occur in a cycle where radicals react with the reactants to form new radicals and then the new radicals react to form the original radicals again. (5.27)

Requirement (5.27) is more complicated to understand, so let's go back to mechanism (5.3) and see what we mean. Notice that there are two propagation reactions in mechanism (5.3), which we have relabeled reactions (5.13) and (5.16). In reaction (5.16), a bromine atom reacts with H_2 to yield HBr plus a hydrogen atom. In reaction (5.13), a hydrogen atom reacts with a Br_2 to yield a second HBr and regenerate the original bromine atom. Consequently, one can view reactions (5.13) and (5.16) as a cyclic process where hydrogen atoms are created via reaction (5.16) and lost via reaction (5.13). Bromine atoms are created in reaction (5.13) and then lost via reaction (5.16).

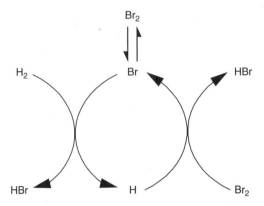

Figure 5.1 A cycle for HBr formation via reaction (5.3).

Figure 5.1 shows a pictorial representation of this mechanism. Figure 5.1 is a different representation of the mechanism than you are used to. In the figure, we label the key radicals and show how those radicals are formed and destroyed. Reactants and products are indicated by the ends of the arrows. Two arrows come together when a given pair of species react. The products of the reaction are indicated by arrows leading away from the encounter. For example, the curved arrows on the left of the figure indicate that H_2 is reacting with a bromine atom to yield a HBr and a hydrogen atom. This should be convinced that Figure 5.1 is a valid representation of the propagation steps for mechanism (5.3) before proceeding with this chapter.

Notice the cycle in Figure 5.1. Hydrogen atoms are formed and destroyed in a cyclic process. The cyclic nature of the propagation reaction is a key feature of initiation–propagation reactions. Radicals are very reactive species. Consequently, if one had a significant concentration of radicals in the system, one could produce a considerable amount of product by going around the cycle many times. Later in this book, we will refer to the cycle as a **catalytic cycle**. Catalytic cycles are key features of most reactions.

All kinds of cycles have been observed experimentally. One can have a simple cycle with only two radicals as shown in Figure 5.1 or a more complex cycle with eight intermediate species as shown in Figure 5.2. Still the key feature is that the cycle needs to loop back on itself (i.e., form the original intermediates) in order to produce a significant amount of product. For example, if one produced a chain of radicals and never got back to the original radicals, the reaction would go around the cycle once and stop. Relatively little product would be produced. In contrast, if the cycle loops back on itself, you can go around and around the cycle and produce lots of product. Consequently, cycles that loop back on themselves have a much higher net rate than do cycles that do not loop back on themselves. In actual practice, usually only cycles that loop back on themselves show appreciable steady-state rates. Noncyclic reactions are seen in laser photolysis, however.

A related requirement is that

All of the steps in the catalytic cycle must have low barriers.	(5.28)

Again, in order to get much product, each of the steps in the catalytic cycle must have low barriers or else the cycle will be too slow. As a general rule of thumb, if you are

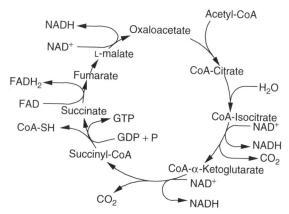

Figure 5.2 The tricarboxylic acid cycle. This cycle is the principal way that cells produce NADH. The NADH is subsequently used to convert ADP to ATP (adenosine diphosphate to adenosine triphosphate).

running the reaction at a temperature T, then the E_A, the activation barrier for each of the steps should satisfy

$$E_A < [0.07 \ (kcal \cdot mol)/K]T \qquad (5.29)$$

to get a reasonable amount of product. Equation (5.29) tells you that if the propagation steps have high barriers, you need to run the reaction at high temperature to get a reasonable amount of product.

The last key requirement for an initiation–propagation mechanism is that:

> There *should* be at least one termination reaction where two radicals combine to yield stable species. $\qquad (5.30)$

Requirement (5.30) is different from requirements (5.26) and (5.27) because it says that there *should* be a termination reaction, not that there *must* be a termination reaction. There are some reactions where the termination reactions have extremely tiny rates. Still, if there were no termination reactions, radicals would build up in the system and could never be destroyed. Consequently, the system would have to get to steady state. Instead, the reactive mixture would explode. One can show that one can never get a steady-state reaction unless there is at least one termination reaction. Therefore, if one observes a steady-state reaction, there must be at least one termination reaction. One often sees several termination reactions.

Explosive reactions are exceptions. One can design a high explosive so that the substance explodes in the absence of a significant termination reaction. In actual practice, such systems are dangerous. However, they can be designed.

Industrially, however, one often adds something to a reacting mixture to enhance the termination reactions and thereby prevent unnecessary explosions. For example, in your car, people add octane enhancers [e.g., methyl t-butyl ether (MTBE) and tetraethyl-lead (TEL)] to better control the combustion process. The MTBE and TEL participate in termination reactions, so the fuel burns more smoothly. If you do not add MTBE, you get little explosions (knock), which tears your engine apart. In practice, one needs termination

Table 5.2 Key requirements for the mechanism for reaction $H_2 + Br_2 \Rightarrow 2HBr$

Reaction	Requirement	Result
(5.26)	There must be an initiation reaction	Must include reaction (5.11) or (5.12)
(5.27)	Must have an catalytic cycle	Must include reaction (5.1) and (5.16)
(5.30)	There should be a termination reaction	Should include either (5.19), (5.20) or (5.21).

reactions to prevent explosions. That is why people add MTBE to gasoline. In practice, most reactions have termination steps, which is why we can state (5.28) as a requirement, even if it is not an absolute requirement.

Before we proceed, it is useful for the student to look back to Table 5.1 and verify that each of the conditions in Table 5.2 is obeyed for the reactions in Table 5.1.

Let's apply these rules to the HBr reaction. Requirement (5.26) (see Table 5.2) says that there must be an initiation reaction. If we apply requirement (5.26) to the HBr case, then we conclude that either reaction (5.11) or reaction (5.12) must occur during the reaction. Requirement (5.27) says that we must have a catalytic cycle to have an appreciable rate. If we apply requirement (5.27) to the HBr case, then we conclude that reactions (5.13) and (5.16) must be included in the mechanism. Requirement (5.30) says that there should be a termination step. If we apply requirement (5.30) to the HBr case, we conclude that either reaction (5.19), (5.20), or (5.21) should be included in the mechanism. Therefore, we know that the mechanism of HBr production must include reactions (5.13) and (5.16), and either reaction (5.11) or reaction (5.12), and either reaction (5.19) or reaction (5.20) or reaction (5.21). We do not know anything about reactions (5.14), (5.15), (5.17), (5.18), or (5.22).

5.4.2 Empirical Rules to Predict Mechanisms of Gas-Phase Radical Reactions: Reactions to Exclude

Next, we will provide some rules to exclude reactions from a mechanism. Generally, these rules are based on the fact that if the rate of a reaction is small enough, the reaction will have a negligible effect on the overall rate.

These rules are harder to apply because you must include at least one initiation reaction even though the initiation reaction has a low rate, and you must include at least one termination reaction even though the termination reaction has a low rate. Still, if there is more than one possible initiation reaction or termination reaction, one can use the rules in this section to choose the one that will be the most important.

The overall plan will be to estimate rate constants for reactions, and if a reaction has a small enough rate, exclude it. Generally, the methods in this chapter will let us determine rate constants to within only a few orders or magnitude. That is not good enough for detailed kinetics. However, it is good enough to eliminate reactions from mechanisms.

Our general approach will be to use various methods to estimate the activation barriers for reactions. We will eliminate all reactions that do not satisfy equation 5.29.

For the purpose of discussion we will assume that all of the reactions follow Arrhenius' law:

$$k_u = k_u^0 \exp\left(-\frac{E_a}{k_B T}\right) \tag{5.31}$$

where k_u is the rate constant for the reaction, k_u^0 is the preexponential, E_a is the activation energy; k_B is Boltzmann's constant, and T is the absolute temperature.

Next, consider the initiation steps. The initiation steps, reactions (5.11) and (5.12), are called *unimolecular reactions*. We have a single reactant colliding with a collision partner to yield products. In Chapter 8, we will find that unimolecular reactions usually have a preexponential of about $10^{15\pm2}$ Å^3/(molecule·second) independent of the details. Reactions (5.11) and (5.12) are both unimolecular reactions, and their preexponentials are very similar. Consequently, one could not distinguish between these two reactions on the basis of the preexponentials. The two reactions do, however, have quite different activation energies. To see that, we have to introduce our next empirical rule.

The activation energy for an initiation reaction, E_a^i can be approximated via

$$E_a^i = \Delta H_{A-A} + 1 \text{ kcal/mol} \qquad (5.32)$$

where ΔH_{A-A} is the enthalpy change during the initiation reaction.

Let's use equation (5.32) to estimate the activation barriers for reactions (5.11) and (5.12). According to data in the Chemical Rubber Company Handbook (CRC) hydrogen has a bond energy of 103 kcal/mol while bromine has a bond energy of 45 kcal/mol. Therefore, according to equation (5.32), reaction (5.11) will have an activation barrier of about 104 kcal/mol while reaction (5.12) will have an activation energy of about 46 kcal/mol. If we plug back into equation (5.31), we find that at 800 K, reaction (5.11) will be a factor of 10^{15} slower than reaction (5.12). Reaction (5.11) is much slower than reaction (5.12). Therefore, little error will be created if we ignore reaction (5.11).

One can generalize the ideas in the previous paragraph to show

> The most important initiation reaction will be the reaction that breaks the weakest bond in the reactants. $\qquad (5.33)$

Physically, the weakest bond in the molecule is the easiest to break. Consequently, the weakest bond breaks first. There are some special cases where there are two bonds in the reactants that have almost the same energy. In those cases, both bonds break with reasonable rates. Still, such cases are exceptions. There is usually only one initiation step. During that step the weakest bond in the molecule breaks.

The implication of the discussion in the last paragraph is that while we have to include reaction (5.12) in the mechanism of HBr formation, we can ignore reaction (5.11).

Now, let's consider reactions (5.15) and (5.18), which are called **exchange reactions**. In an exchange reaction, one atom in a molecule is replaced by another identical atom in the same molecule. Exchange reactions often occur at a reasonably high rate. However, exchange reactions do not consume any reactants or radicals or produce any products. Consequently, they do not affect the overall rate of reaction. Therefore

> Exchange reactions can be ignored when one is deriving a rate equation. $\qquad (5.34)$

As a result, there is no reason to include reactions (5.15) and (5.18) in the mechanism of HBr formation. Physically, the reactions do occur at reasonably high rates. However, the reactions do not contribute to the formation of HBr. Therefore, they are not part of the mechanism of HBr formation.

Next, let's consider reactions (5.14) and (5.17), which are called **inhibition reactions**. In inhibition reactions, the products of the reaction react with a radical to produce a reactant molecule. In the case of reactions (5.14) and (5.17), HBr reacts with a radical to produce either H_2 or Br_2. If one starts with the product and ends up with the reactants, one will definitely slow down the reaction, which is why reactions of this type are called "inhibition" reactions.

Now the question is whether we need to include reactions (5.14) and (5.17) in the mechanism. The answer is "Yes", provided that the reactions satisfy equation (5.29). According to data in the CRC, reaction (5.14) is 16 kcal/mol exothermic. The heat of reaction is similar to that of reaction (5.13). Consequently, there is no reason to exclude reaction (5.14) on the basis of energetics, and we will see in the next paragraph that there is no reason to exclude reaction (5.14) on the basis of kinetics, either. Reaction (5.17), however, is 40 kcal/mol endothermic. Consequently, if everything else is equal, one would expect reaction (5.17) to be much less important than reaction (5.14).

We can quantify the effect by using another empirical rule, called the **Polanyi relationship**, to estimate the difference in the rates of the two reactions. The Polanyi relationship was proposed by Polanyi and Evans to try to correlate activation barriers to reactions. According to the Polanyi relationship, E_a, the activation barrier for a given reaction, is related to ΔH_r, the heat of reaction, by

$$E_a = E_a^0 + \gamma_P \Delta H_r \qquad (5.35)$$

where E_a^0 is the intrinsic activation barrier and γ_P is the transfer coefficient. We will discuss the Polanyi relationship in detail in Chapters 10 and 11. Generally, E_a^0 and γ_P vary with the reaction types and the shape of the bonds which form. One needs to consider these effects in detail, if one wants to predict an accurate value of the activation barrier. Fortunately, in Section 5.4.5, we will find it possible to get a rough approximation to the activation barrier if we assume that E_a^0 is 12–16 kcal/mol for atom transfer reactions while γ_P is 0.2–0.5 for exothermic reactions and 0.5–0.8 for endothermic reactions. If we substitute into equation (5.35) with $\gamma_P = 0.7$, we find that reaction (5.17) should have an activation barrier of about 34 kcal/mol. By comparison, the same calculation shows that reaction (5.14) will have an activation barrier of only 6 kcal/mol. Both reactions have similar preexponentials. Clearly, at 400 K, reaction (5.14) will be much more important than reaction (5.17).

Still, in the homework set, we will ask the reader to show that if one loads HBr into the reactor and runs reaction (5.2) in reverse, one will need to include reaction (5.17). Actually, the HBr formation reaction is not very reversible. One needs to heat the system to almost 4000 K to get a reasonable conversion. Still, reaction (5.17) is important for the reverse reaction at 4000 K, even though it is not so important for the forward reaction at 400 K.

One can quantify this idea more carefully, using the material from Section 2.6. Generally, in order for an elementary reaction to have a significant rate, the activation energy for the reaction needs to be less than a critical value, E_a^{crit}, where

$$E_a < E_a^{crit} = \begin{cases} 0.07\dfrac{kcal}{mol \cdot K}T & \text{for propagation reactions} \\[2mm] 0.15\dfrac{kcal}{mol \cdot K}T & \text{for initiation reactions} \\[2mm] 0.05\dfrac{kcal}{mol \cdot K}T & \text{for reactions in the catalytic cycle} \end{cases} \qquad (5.36)$$

There are different values of the critical temperature for initiation and propagation reactions because initiation reactions have much smaller rates than do propagation reactions. Equation (5.36) allows you to tell whether a given reaction will be important at a given temperature.

According to equation (5.36), at 400 K we should include all reactions where $E_a <$ (0.07 kcal/mol-$^\circ$K)(400 K) = 28 kcal/mol. On the previous page we found that Reaction (5.14) has an activation barrier of 6 kcal/mol while reaction (5.17) has a barrier of 34 kcal/mol. Reaction (5.14) satisfies equation (5.36) so it should be included at 400 K. Reaction (5.17) does not satisfy equation (5.36) at 400 K, so it should not be included at 400 K. Reaction (5.17) should be included at temperatures above 500 K, however.

At this point, we have eliminated reactions (5.11), (5.15), (5.17), and (5.18) from the mechanism of HBr formation and shown that we must include reactions (5.12)–(5.14) and (5.16).

Next, let's consider reactions (5.19)–(5.21), which are all termination reactions. According to rule, we know that we must have at least one termination reaction. The question is which one, and whether we need to have only one termination reaction or to include all three termination reactions. The question of which one is easy. There is another empirical rule:

> One should include all termination reactions where the species produced in the initiation reaction combine to yield stable products. If there are transfer steps (defined later), one should include the species produced in the transfer steps, too. (5.37)

In reaction (5.12), we produce bromine atoms, so we must include reaction (5.19) in the mechanism. It is not obvious whether to include reaction (5.20) and (5.21). Another empirical rule is that

> Radical recombination reactions will usually have an activation energy of less than 2 kcal/mol (5.38)

Thus, the rate constant for reactions (5.20) and (5.21) should be significant. One can, in fact, find conditions where reactions (5.20) and (5.21) are quite significant. Consequently, if I were guessing the mechanism of reaction (5.2), I would include reactions (5.20) and (5.21). However, when Bodenstein and Lund did experiments on HBr in 1906, they picked conditions where the hydrogen atom concentration was much less than the bromine atom concentration. In those special circumstances, the rate of reaction (5.19) is much larger than the rate of reactions (5.20) and (5.21). Consequently, it was acceptable to ignore reaction (5.20) and (5.21).

I want to note, however, that if you are predicting a mechanism, you are not going to know a priori that people are going to only want to take data under conditions where the hydrogen atom concentration is much less than the bromine atom concentration. Consequently, if one is asked to predict a mechanism, one should include reactions (5.20) and (5.21) even though under some conditions reactions (5.20) and (5.21) will not be important.

At this point, we have considered all of the reactions above except reaction (5.22). Reaction (5.22) is different from all of the rest in that it is a four-centered reaction.

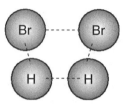

Figure 5.3 A stick diagram of a proposed transition state for reaction (5.22).

Many years ago, people used to talk about four-centered reactions. They would assume a transition state such as in Figure 5.3 and calculate properties. However 1965, Woodword Hoffman pointed out that a symmetric four-centered transition state is impossible with neutral molecules. We provide a detailed derivation of the rule in Chapter 10. However, the thing to remember for now is that in a symmetric four-centered transition state, one ends up trying to put four electrons into a single molecular orbital. That is quantum-mechanically forbidden. One can get sort of a four-centered transition state by first breaking two of the bonds. However, that is an energetically unfavorable process. Consequently

> Four-centered transition states can be ignored when the reactants are neutral nonpolar molecules. (5.39)

In summary, according to the rules in the last two sections, the mechanism of HBr formation must include reaction (5.12)–(5.14), (5.16), and (5.19) and probably should include reaction (5.20) and (5.21). If we look back to Bodenstein's mechanism of the reaction in equation (5.3), we see that we got the answer almost right. We predicted that of all the reactions which Bodenstein said occurs do occur. The only issue was that we included reactions (5.20) and (5.21), while in Bodenstein's original work [1906] these two reactions were ignored. Now it is hard to say that Bodenstein was wrong, since reaction (5.20) and (5.21) were not important under the specific conditions he considered. Still, I would include those two reactions in the mechanism since they are important under many conditions.

5.4.3 Additional Rules for More Complex Reactions

Next, we want to extend the ideas in the last two sections to more complex reactions. To keep this effort in perspective, note that we have spent several pages analyzing reaction (5.2), the formation of HBr, and believe it or not, reaction (5.2) was a particularly easy case. There were only two reactants. They were both diatomics, and all of the radicals were atoms. Anything else is going to be more difficult. Still, one can make useful predictions.

There are two general difficulties with extending these ideas to more complex reactions: (1) there are additional intermediates to consider and (2) there are many more reactions to consider. Let's talk about the issues of the number of intermediates first.

One of the problems with calculating the mechanism for a more complex reaction is that there are many more intermediates to consider. For instance, I have found that my chemical intuition is usually not able to predict all of the intermediates that could form. For example, let's consider the mechanism of HI formation:

$$H_2 + I_2 \Longrightarrow 2HI \qquad (5.40)$$

HI formation is very similar to HBr formation, and so one might expect the mechanism of reaction (5.40) to be very similar to the mechanism of HBr formation [mechanism (5.3)]. In fact, however, an H_2I intermediate forms during reaction (5.40) while no similar intermediate is seen during HBr formation.

I cannot tell you a simple way to predict that the H_2I intermediate will be important during HI formation, even though the H_2Br intermediate is not important to HBr formation. One can calculate the stability of the intermediates using the quantum-mechanical methods discussed in Chapter 11. However, that is complicated. For now, my best advice is that if you are unsure whether an intermediate is stable, include it in the mechanism. The worst that will happen is that you will have some extra reactions to consider. However, if you are wrong, you will miss some key chemistry.

5.4.4 Intrinsic Barriers for Propagation Reactions

Once you know which intermediates are stable, you can start to predict mechanisms. The procedure is the same as in Section 5.4:

- Start with a guess of all of the intermediates in the system. Use quantum mechanics (Chapter 11) to verify that you are not missing any intermediates.
- Write down those intermediates.
- Use various rules to pare down the list of reactions.

The hard part is the last step. The general scheme is to use an empirical rule to estimate the activation barriers for each of the reactions in a mechanism. Earlier in this chapter, we presented three empirical rules to determine barriers to reactions: equation (5.32) for initiation reactions; equation (5.38) for termination reactions; and, a more general relationship, equation (5.35) for propagation reactions. According to equation (5.35), E_a, the activation barrier for a given reaction, is related to ΔH_r, the heat of reaction, by

$$E_a = E_a^0 + \gamma_P \Delta H_r \qquad (5.41)$$

where E_a^0 is the intrinsic activation barrier and γ_P is the transfer coefficient. Equation (5.41) is the most important relationship in this chapter because it allows you to estimate an activation barrier for a given reaction from the heat of reaction. In this section, we will examine equation (5.41) in more detail to see how it can be used to produce mechanisms of reaction.

There are two terms in equation (5.41): an intrinsic barrier, and the transfer coefficient times the heat of reaction. The second term on the right side of equation (5.41), $\gamma_P \Delta H_r$ is easy to understand. This term is associated with the energy you gain or lose in converting the reactants into products. If you have an exothermic reaction, you gain energy as the reaction proceeds. The heat of reaction is negative, which means that, according to equation (5.41), the reaction is a little easier. In contrast, if you have an endothermic reaction, you lose energy as the reaction proceeds. The heat of reaction is positive, which means that, according to equation (5.41), the reaction is a little harder. There is a coefficient, γ_P, in propagation reactions. γ_P is usually between 0.3 and 0.7. Therefore, one does not get the full effect of the heat of reaction in equation (5.41). Still, there is the qualitative feature that exothermic reactions tend to be favored over endothermic reactions.

The intrinsic barrier in equation (5.41) is harder to understand. Physically, the intrinsic barrier provides a measure of the energy it takes to make the reaction happen if the heat of reaction were zero.

In order to get a physical picture of why intrinsic barriers arise, consider an example, the reaction

$$H + CH_3CH_3 \longrightarrow CH_4 + CH_3 \tag{5.42}$$

Reaction (5.42) is a simple reaction where a hydrogen atom approaches an ethane molecule and induces the carbon–carbon bond in the ethane to break.

If one looks at reaction (5.42) in detail, it is not at all obvious that the reaction would be activated. After all, one can imagine the reaction proceeding as in Figure 5.4. One starts off with a hydrogen and an ethane with a 88-kcal/mol carbon–carbon bond. During the reaction, the hydrogen comes up to the ethane and forms a partial carbon–hydrogen bond. Simultaneously, the carbon–carbon bond breaks. If one imagines that one has a partial carbon–hydrogen and a partial carbon–carbon bond, then one might imagine writing the bond energy of the system, E_{bond}, at any time during the reaction it:

$$E_{bond}(t) = E_{H-C}(t) + E_{C-H_3}(t) \tag{5.43}$$

where $E_{M-C}(t)$ is the energy of the H–C bond at time t and $E_{C-CH_3}(t)$ is the energy of the carbon methyl bond at time t.

If you have half of a carbon–carbon bond and half of a carbon–hydrogen bond, the E_{bond} comes out to be

$$E_{bond} = (0.5)(102 \text{ kcal/mol}) + 0.5(88 \text{ kcal}) = 95 \text{ kcal/mol} \tag{5.44}$$

According to this simple model, it seems that you have more total bond energy in the transition state than you do in the reactants, and this conclusion is verified by detailed quantum-mechanical calculations. Therefore, it is not obvious that reaction (5.42) is activated.

In fact, however, reaction (5.42) has an activation barrier of 44 kcal/mol. Lee and Masel [1996, 1997] examined the barriers to reaction (5.42) in some detail, and found that the main contribution to the barrier comes from the orbital distortions that occur during the reaction. During reaction (5.42), the 1s orbital in the hydrogen interacts with what is called the $3A_{1g}$ orbital on the ethane to induce the ethane's carbon–carbon bond to break. Diagrams of both orbitals are shown in Figures 5.5 and 5.6. In the figure, lightly colored lobes are used to designate positive orbitals, while darkly colored lobes are used to designate negative orbitals. When the hydrogen atom is far away from the ethane, the 1s orbital in hydrogen looks spherical, while the $3A_{1g}$ orbital has a central balloon-shaped orbital plus two kidney-shaped non-bonding lobes.

The $3A_{1g}$ orbital is constructed from a p orbital on each carbon plus some s orbitals on the hydrogens as indicated in Figure 5.5. The diagram on the left of Figure 5.5 shows

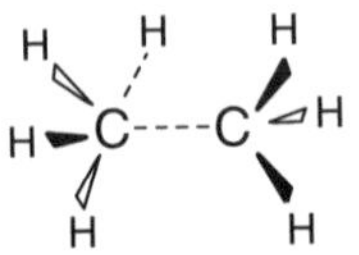

Figure 5.4 A stick diagram of the bonds during reaction (5.42).

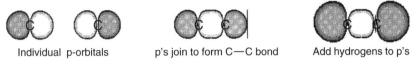

Individual p-orbitals	p's join to form C—C bond	Add hydrogens to p's

Figure 5.5   Construction of the $3A_{1g}$ orbital in ethane.

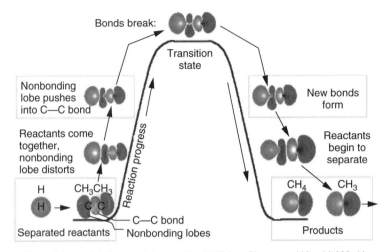

Figure 5.6   Orbital diagram during reaction (5.46) from Blowers and Masel (1999a,b).

what the p orbitals are like. The p orbitals have two lobes in a figure-eight configuration. One lobe of the p orbital has a positive sign. The other p orbital has a negative sign. I have shaded the positive orbital white in the figure while the negative orbital is shaded gray.

When the two carbon atoms come together to form a carbon–carbon bond, the two positive orbitals join to form a bond. The negative lobes stick out away from the carbon–carbon bond. The result is that the p orbital forms a carbon–carbon bond, plus two extra negative lobes. If one were forming C_2, the orbitals would like the center diagram in Figure 5.5, with a C–C bond and two small negative lobes. However, in ethane, the hydrogens contribute electrons to the negative parts of the p orbital, causing the lobes to enlarge. The result is that the outer lobes of the $3A_{1g}$ orbital are larger than the bonding lobes. The outer lobe is kidney shaped, because we are showing a cross section where one of the hydrogens is in the plane while the other two hydrogens are out of the plane.

Now consider what happens when the hydrogen approaches the ethane. Figure 5.6 shows how the 1s and $3A_{1g}$ orbitals change when the hydrogen approaches the ethane. The 1s orbital on the hydrogen flattens out, while the kidney-shaped orbital on the ethane is pushed aside. I like to think about the orbitals as being two balloons. You know that when you push two balloons together, the balloons push apart. In the same way, the orbitals on the hydrogen atom and on the ethane push apart. This is really an electron–electron repulsion, moderated by bond distortion. Physicists call this a **Pauli repulsion**. It costs energy to overcome that repulsion. As a result, reaction (5.42) is activated.

The intrinsic barrier in equation (5.41) gives a measure of how much one needs to distort the orbitals to get reaction to happen. Generally, the intrinsic barrier is small (1–2 kcal/mol) for initiation reactions because one does not have to distort orbitals to get the reaction to start. Similarly, in termination reactions, the lone pairs on the radicals form a bond during reaction. In those cases, the intrinsic barriers are also small. γ_P is approximately 1.0 for both initiation and termination reactions. In contrast, during a propagation reaction, one bond breaks and another forms. Figure 5.6 shows that one has to push the bonding orbitals out of the way before new bonds can form. It costs energy to push the orbitals out of the way. Consequently, propagation reactions generally have intrinsic barriers between 8 and 70 kcal/mol.

Next, we want to discuss how intrinsic barriers change as one changes the reaction. First, we want to note that there is a fundamental difference between transferring an individual hydrogen atom and transferring a molecular ligand. Let's go back to the ethane hydrogenolysis case mentioned earlier. If we ignore exchange, then there are two possible reactions which can occur when ethane reacts with a hydrogen atom. The carbon–carbon bond can break to yield methane:

$$H + CH_3CH_3 \longrightarrow CH_4 + CH_3 \tag{5.45}$$

or the incoming hydrogen can pick off a hydrogen atom from the ethane:

$$H + CH_3CH_3 \longrightarrow H_2 + CH_2CH_3 \tag{5.46}$$

Lee and Masel [1996] found that reaction (5.45) has an intrinsic barrier of 45 kcal/mol while reaction (5.46) has an intrinsic barrier of 12 kcal/mol. Experimentally, only reaction (5.46) is seen.

It is important to consider why reaction (5.46) has a much lower intrinsic barrier than does reaction (5.45). When reaction (5.46) occurs, the hydrogen atom comes up along with the CH bond axis in the ethane. One has to distort the orbitals in the break in the C–H bond. However, the remaining orbitals in the ethane are largely unperturbed.

In contrast, when reaction (5.45) occurs, the hydrogen approaches the ethane along its C–C axis. The incoming hydrogen needs to push the C–H bonds in the ethane out of the way before any reaction can occur. That costs extra energy, so reaction (5.45) has a much higher intrinsic barrier than does reaction (5.46).

Lee and Masel [1996] showed that this extra barrier is associated with the energy to distort orbitals and not just the energy to move the C–H bonds. For example, the reaction

$$CH_3OH + H \longrightarrow CH_3 + H_2O \tag{5.47}$$

has almost the same intrinsic barrier as reaction (5.45) even though the hydrogen does not have to displace any atoms during reaction (5.47). The hydrogen does need to displace the lone-pair orbitals on the OH, however. This example shows that the intrinsic barriers are associated with orbital displacements and not just motions of individual bonds.

A key point is that it does not matter what molecular ligand one is transferring, one always sees intrinsic barriers of a similar magnitude. So for example, if we look at a series of radical transfer reactions such as

$$R_1 - R + H \longrightarrow R_1 + RH \tag{5.48}$$

where R and R_1 are two hydrocarbon ligands one usually finds that the intrinsic barrier is between 40 and 60 kcal/mol independent of R and R_1. In contrast, hydrogen transfer reactions of the form

$$R_1\text{–}H + R \longrightarrow R_1 + HR \tag{5.49}$$

Table 5.3 The intrinsic barriers for the exchange reaction

$$X^- + CH_3X \rightarrow XCH_3 + X^-$$

X	Intrinsic Barrier, kcal/mol
F	11
Cl	12.8
Br	10.8
I	9.7

Source: Results of Glukhoutsev, Pross and Radom [1995].

have intrinsic barriers between 8 and 15 kcal/mol, independent of R and R_1. If one looks in detail, one finds that the intrinsic barriers do depend slightly on R and R_1. Typical variations in activation barriers are ±5 kcal/mol. A 5-kcal/mol change in the barrier is sufficient to change the rate by a factor of 10^3 at 300 K. However, such a change is unimportant to the prediction of the reaction mechanism. Therefore, we will ignore those variations for the discussion in this chapter. These variations will be discussed in Chapter 11.

The other thing that one finds is that if one goes to stiffer orbitals, the intrinsic barriers rise. For example, Table 5.3 give the intrinsic barriers for a number of isotopic exchange reactions as reported by Glukhoutsev et al. (1995). Notice that as one goes to stiffer atoms, the intrinsic barriers rise and then fall. Consequently, it is harder to transfer a stiff atom than a spongier one. The differences are only a couple of kilocalories, though.

Table 5.4 gives a selection of other intrinsic barriers and transfer coefficients for different types of reaction. The reader should refer to this table when predicting mechanisms of reactions.

Examples 5.A and 5.B show several examples to indicate how Table 5.4 is used to determine whether a certain mechanism is feasible. One should study those examples before proceeding with this chapter.

5.4.5 Extra Reactions with Larger Molecules

So far, we have been discussing reactions of simple diatomic molecules. Next, we want to discuss reactions of organic molecules containing several atoms.

Hyser (1970), Fossey, et al. (1995), Perkins (1994), and Alfassi (1999) offer good reviews of the reactions of organic molecules in the gas phase. Generally, most organic molecules follow mechanisms such as those we have already discussed, but, there are some extra reactions that can occur:

- Chain transfers
- Addition reactions
- Isomerizations
- Disproportionations
- Fragmentations
- β Scissions

We will review each of these types of reaction next.

Table 5.4 Intrinsic barriers and transfer (coefficients for different types of reaction of neutral species).

Reaction	Example	Actual E_a^0, kcal/mol	E_a^0 to Assume When Predicting Mechanisms, kcal/mol	Actual γ_P	γ_P to Assume when Predicting Mechanisms
Simple bond scission	$AB + X \rightarrow A + B + X$ (X = a collision partner)	0–1	1	1.0	1.0
Recombination	$A + B + X \rightarrow AB + X$ (X = a collision partner)	0–3	1	0.0	0.0
Atom transfer reaction $0 > \Delta Hr > -40$	$Rx + R_1 \rightarrow R + x - R_1$ (x = an atom)	8–16	12	0.2–0.6	0.3
Atom transfer reaction $0 \leq \Delta Hr \leq 40$	$R - x + R \rightarrow R + x - R_1$ (x = an atom)	8–16	12	0.4–0.8	0.7
Atom transfer reaction $\Delta Hr < -40$	$R - x + R \rightarrow R + x - R_1$ (x = an atom)	0–2	0	0	0
Atom transfer reaction $\Delta Hr > 40$	$R - x + R \rightarrow R + x - R_1$ (x = an atom)	0–5	2	0.9–1.0	1.0
Ligand transfer reaction to hydrogen	$H + R - R_1 \rightarrow HR + R_1$	40–50	45	0.4–0.6	0.5
Other ligand transfer reactions	$x + R - R_1 \rightarrow xR + R_1$ (x = an atom)	≥ 50	50	0.3–0.7	0.5
Endothermic β-hydrogen elimination	$RCH_2CH_2\bullet + X \rightarrow RCH = CH_2 + H\bullet + H$	12–25	15	0.5–1	0.7

A *chain transfer reaction* is one where the initial step produces a simple radical, and then that radical reacts to form a new species before the catalytic cycle begins. Chain transfer is most important in a case where you produce one radical but another radical is more reactive.

For example, consider the reaction

$$C_2H_6 \Longrightarrow C_2H_4 + H_2 \tag{5.50}$$

In Section 5.4.2, we noted that the weakest bond in the reactants usually breaks during the initiation step. Well, in ethane ($H_3C–CH_3$) the carbon–carbon bond is 11 kcal/mol weaker than any of the C–H bonds. Consequently, the carbon–carbon bond breaks during the initiation step.

$$X + C_2H_6 \longrightarrow X + 2CH_3 \tag{5.51}$$

In principle, the methyl radicals could react through a catalytic cycle to produce product. However, in the Examples 5.A and 5.B show that it is more likely that the methyl group reacts with ethane, to form an ethyl radical:

$$CH_3 + C_2H_6 \longrightarrow CH_4 + C_2H_5 \tag{5.52}$$

The ethyl radical is much more easily converted into the ethylene product.

The reaction mechanism becomes

Initiation

$$X + C_2H_6 \longrightarrow 2CH_3 + X \tag{5.53}$$

Transfer

$$CH_3 + C_2H_6 \longrightarrow CH_4 + C_2H_5 \tag{5.54}$$

Propagation

$$C_2H_5 + X \longrightarrow C_2H_4 + H + X \tag{5.55}$$

$$H + C_2H_6 \longrightarrow C_2H_5 + H_2 \tag{5.56}$$

Termination

$$X + 2H \longrightarrow H_2 + X \tag{5.57}$$

$$X + 2CH_3 \longrightarrow C_2H_6 + X \tag{5.58}$$

$$X + CH_3 + C_2H_5 \longrightarrow C_3H_8 + X \tag{5.59}$$

$$X + 2C_2H_5 \longrightarrow C_4H_{10} + X \tag{5.60}$$

Reaction (5.52) is called a **chain transfer process**, since you convert one radical into another.

Another common reaction is called a β **scission**. Reactions (5.61) and (5.62) are examples of β-scission reactions.

$$R - CH_2CH_2\bullet + X \longrightarrow R\bullet + C_2H_4 + X \tag{5.61}$$

$$R - \overset{\displaystyle O}{\underset{\displaystyle}{C}} - O\bullet + X \longrightarrow R\bullet + CO_2 + X \tag{5.62}$$

In both cases, one breaks the σ bond at the β position from the radical, and forms a double bond with the β carbon. All β-scission reactions break σ bonds at the β position. β-Scission reactions are very common in the decomposition or combustion reactions of modest-sized molecules.

β Scissions are one example of a larger class of reaction called **fragmentation reactions**. Fragmentation reactions are reactions where a radical breaks apart to form a stable species and a new radical. For example

$$(CH_3)_3CO\bullet + X \longrightarrow (CH_3)_2C{=}O + CH_3\bullet + X \tag{5.63}$$

is a fragmentation reaction. Again, one forms a double bond that drives the reaction.

The fragmentation reactions can also occur in reverse:

$$CH_3\bullet + H_2C{=}CH_2 \longrightarrow CH_3CH_2CH_2\bullet \tag{5.64}$$

In this case, the radical adds across the double bond. Reaction (5.64) and the reactions where you add species to radicals are called **addition reactions**.

There is also the possibility of a **disproportion reaction**. In a disproportion reaction, two radicals react to form a stable species. For example, reaction (5.65) is a disproportion reaction:

$$CH_3CH_2\bullet + CH_3CH_2\bullet \longrightarrow CH_3CH_3 + CH_2{=}CH_2 \tag{5.65}$$

Reaction (5.65) is similar to the termination reactions discussed earlier in this chapter, except that we get two species as products instead of one.

Finally, if you have radicals with double bonds, there is the possibility of the radicals isomerizing as shown in equation (5.66):

$$\tag{5.66}$$

In practice, reaction (5.66) does not occur with small molecules because the ring strain is too high. However, it can occur when there is a chain of five or more carbons. Therefore, it is called *five-centered isomerization*. This behavior gives a lot more reactions to consider when predicting a mechanism.

5.4.6 Application to the Predication of the Mechanisms of Reactions

Now, let's apply the findings in the last two sections to predict mechanisms of reactions. Lee and Masel (1997), and Broadbelt et al. (1994) have separately shown that one can use equation (5.41) to predict mechanisms of reactions. The idea is to follow the method described in Section 5.2. First, we postulate a series of possible intermediates, then look at all of the possible reactions of the intermediates, and use equation (5.41) to see which reactions are favored.

As noted previously

> The most important initiation reaction will be the reaction that breaks the weakest bond in the reactants.   (5.67)

One can derive equation (5.67) from equation (5.41) by noting that the intrinsic barrier for an initiation reaction is usually $1-2$ kcal/mol and γ_P is about 1.0. ΔH_r for the scission of a weak bond is smaller than ΔH_r for scission of a strong bond. Consequently, according to equation (5.41), a weak bond will always be easier to break than a strong bond during the initiation step.

In contrast, during a propagation reaction, a weak bond may or may not be the easiest to break. For example, reactions (5.45) and (5.46) are both propagation reactions. Reaction (5.45) is 15 kcal/mol exothermic, while reaction (5.46) is only 8 kcal/mol exothermic. Consequently, on the basis of thermodynamics alone, one would expect reaction (5.45) to be favored over reaction (5.46). Note, however, that the intrinsic barrier to reaction (5.46) is 30 kcal/mol less than the intrinsic barrier reaction (5.45). That difference in intrinsic barriers is more than enough to overcome the difference in ΔH_r for the two reactions. Consequently, according to equation (5.41), one would expect reaction (5.46) to be favored over reaction (5.45).

Experimentally, reaction (5.46) has an activation barrier of 10 kcal/mol, while reaction (5.45) has never been observed. Lee and Masel calculate an activation barrier of 40 kcal/mol for reaction (5.45).

One can generalize this result to say that

> During propagation reactions, if everything else is equal, it is usually easier to transfer atoms than to transfer molecular ligands.   (5.68)

There are exceptions to equation (5.68) in cases where the differences in ΔH_r are large. However, equation (5.68) works about 98% of the time.

5.4.7 Example: Acetaldehyde Decomposition

Next, let's use the methods we have discussed so far to predict a mechanism for a simple gas reaction: the decomposition of acetaldehyde:

$$CH_3CHO \Longrightarrow CH_4 + CO \qquad (5.69)$$

First, we note that this reaction is likely to go by an initiation–propagation mechanism. One cannot be 100% sure of that, but 99.9% of gas-phase reactions go by initiation–propagation reactions.

Next, let's try to predict the mechanism. According to rule (5.33), the first step in the mechanism will be a step where the weakest bond in the acetaldehyde breaks. According to the data in the CRC, the carbon–carbon bond is the weakest bond in acetaldehyde. Therefore, one would expect the initial step in acetaldehyde decomposition to be scission of the carbon–carbon bond

$$CH_3COH + X \longrightarrow CH_3 + HCO + X \tag{5.70}$$

to yield a methyl group and a formyl group. Next, the methyl and formyls need to react via a catalytic cycle to produce the CO and CH_4 products, and to regenerate each of the radicals. For example, we might postulate that the methyl radical reacts with the acetaldehyde to produce methane. There are two possible reactions:

$$CH_3 + CH_3COH \longrightarrow CH_4 + CH_3CO \tag{5.71}$$

$$CH_3 + CH_3COH \longrightarrow CH_4 + CH_2COH \tag{5.72}$$

However, the H–C bond in the methyl group has a bond energy of 9 kcal/mol while the C–H bond in the formyl group has a bond energy of 96 kcal/mol. Reactions (5.71) and (5.72) have similar intrinsic barriers because both are atom transfer reactions. Therefore, one would expect reaction (5.71) to be slighthy faster than reaction (5.72).

One can quantify these ideas as follows. First note that according to the CRC, the C–H bond in CH_4 has a bond strength of 104 kcal/mol.

Consequently, reaction (5.71) has a heat of reaction of $96 - 104 = -8$ kcal/mol, while reaction (5.72) has a heat of reaction of $99 - 104 = -5$ kcal/mol. One can calculate the activation barriers for both reactions from equation (5.35).

$$E_a[5.71] = 12 \text{ kcal/mol} + 0.3^*(-8 \text{ kcal/mol}) = 9.6 \text{ kcal/mol}.$$

$$E_a[5.72] = 12 \text{ kcal/mol} + 0.3^*(-5 \text{ kcal/mol}) = 10.5 \text{ kcal/mol}.$$

Therefore reaction (5.71) should be slightly faster than reaction (5.72), although both would likely occur at room temperature.

One might also consider the reaction

$$CH_3 + CH_3OCH \longrightarrow CH_3CH_3 + OCH \tag{5.73}$$

We analyze reaction (5.73) in problem 5.C. It has a high barrier so it is unlikely to occur.

In order to complete the cycle, we need the CH_3CO fragment to decompose to yield CO and regenerate the methyl group:

$$CH_3CO + X \longrightarrow CH_3 + CO + X \tag{5.74}$$

Similarly, the formyl (HCO) radical can react via a chain transfer process:

$$HCO + HCOCH_3 \longrightarrow H_2CO + CH_3CO \tag{5.75}$$

A second chain transfer process is

$$HCO + X \longrightarrow H + CO + X \tag{5.76}$$

$$H + CH_3COH \longrightarrow H_2 + CH_3CO \tag{5.77}$$

Analysis is given in Example 5.C.

One could also consider a closed cycle:

$$H + CH_3COH \longrightarrow CH_4 + COH \qquad (5.78)$$

Reaction (5.78) has a high intrinsic barrier, but it is 19 kcal exothermic. Consequently, we need to do a calculation to see when to include it [we include it if equation (5.36) is satisfied].

The calculation is given in Example 5.C.

One also needs termination reactions. The following reactions are possible:

$$CH_3 + COH + X \longrightarrow CH_3COH + X \qquad (5.79)$$

$$CH_3 + HCO \longrightarrow CH_4 + CO \qquad (5.80)$$

$$X + 2CH_3 \longrightarrow C_2H_6 + X \qquad (5.81)$$

$$2COH \longrightarrow HCOCOH \qquad (5.82)$$

$$2COH \longrightarrow CO + H_2CO \qquad (5.83)$$

One then has to go back and use equation (5.41) to see which of these reactions are feasible. Reaction (5.80) is thermodynamically favored by 3.5 kcal/mol over reaction (5.79). Reaction (5.79) is a standard termination reaction, so it has an intrinsic barrier of 1–2 kcal/mol. Reaction (5.80), on the other hand, is a hydrogen atom transfer reaction. Hydrogen transfer reactions have intrinsic barriers of 10–14 kcal/mol. If one substitutes into equation (5.46) one finds that reaction (5.80) is unactivated. Therefore reactions (5.79) and (5.80) should both occur. Note however, that reaction (5.79) is required according to (5.37) while (5.80) is optional. Consequently, one would expect reaction (5.79) to predominate over reaction (5.80). A similar argument applies to reaction (5.83). At this point, we have considered reactions (5.70)–(5.83). There are no other alternatives or any other reactions. Consequently, reactions (5.70), (5.71), (5.76)–(5.79), (5.81), and (5.82) constitute the mechanism of acetaldehyde decomposition.

Benson analyzes this case in more detail and notes that at 700 K and 1 atm:

$$[CH_3] = 10^{-11.3} \text{ mol/liter}$$

$$[CH_3CO] = 10^{-13.6} \text{ mol/liter}$$

$$[CHO] = 10^{-14.6} \text{ mol/liter}$$

$$[H] = 10^{-17} \text{ mol/liter}$$

Therefore, Benson concluded under these conditions, reaction (5.81) is much faster than reaction (5.79), (5.80), (5.82), or (5.83). Reaction (5.79) is still required, however.

Benson's analysis can be generalized to any reaction. If one radical has a much higher concentration than the rest, then only that radical will matter to the termination reactions. In the literature, it has been proposed that one way to identify the most important termination reaction is to look at the catalytic cycle. If one reaction is much slower than all the rest, then the intermediates involved in that reaction will have the highest concentrations. Consequently, the most important termination reactions will be the reactions involving the intermediates participating in the slowest step in the catalytic cycle.

Unfortunately, that criterion is not easy to use. For example, in the acetaldehyde case, reaction (5.71) is exothermic while reaction (5.74) is endothermic. Reaction (5.71) is an atom transfer reaction, so according to Table 5.4, reaction (5.71) should have a barrier of $12 + 0.3 \times (-14.8) = 7.6$ kcal/mol. In contrast, reaction (5.74) should have a barrier of $1 + 1.0 \times 11.3 = 12.3$ kcal.mol. Consequently, if everything else were equal, reaction (5.71) would be quicker than reaction (5.74).

In fact, however, reaction (5.74) has an unusually large preexponential, 10^{17} Å^3/ (molecule·second). As a result, reaction (5.71) is slower than reaction (5.74) at high temperature. Under such conditions, CH_3 is the key intermediate so only reaction (5.81) matters. Still, I am not sure how you would know that the preexponentials were so different. (We discuss the preexponentials in Chapter 9.) Therefore, my advice is that if you have a case where the barriers are close, include all the termination reactions. That way you will be sure that you are right.

One can extend this analysis to many different reactions. More than 95% of the mechanisms of gas-phase reactions can be predicted by these procedures. I find it pretty easy to guess a mechanism of a gas phase reaction, if I know what kinds of intermediates are stable. In my experience, the hard part is deciding what intermediates and what initial group of reactions to consider. For example, is a species such as CH_2COH stable enough to form? If one can find a way to determine which species are stable, and what types of reactions to consider, one can predict the mechanism of the reaction using equation (5.46), assuming that the reaction occurs via an initiation–propagation mechanism.

5.5   GAS-PHASE REACTIONS THAT DO NOT FOLLOW INITIATION–PROPAGATION MECHANISMS

Next, we will discuss the gas-phase reactions that do not follow initiation–propagation mechanism. As noted previously, in 1932, Rice showed that all of the mechanisms of gas-phase reactions of hydrocarbons that had been studied before 1930 follow initiation–propagation mechanisms. However, since 1930, people have discovered a few gas-phase reactions that do not follow initiation–propagation mechanisms. In the next few sections, we will discuss gas-phase reactions that do not follow initiation–propagation mechanisms.

There are three groups of exceptions:

- Reactions where there is no feasible initiation–propagation mechanism from reactants to products
- Reactions where one of the steps in the catalytic cycle has a high barrier so that no reaction is feasible
- Reactions where concerted processes are possible

In the next several sections, we will discuss these exceptions.

5.5.1   Reactions with No Feasible Initiation–Propagation Mechanism

First, let's consider reactions where no initiation–propagation reactions is feasible. A key feature of the initiation–propagation reactions is that there is a catalytic cycle where radicals are formed and destroyed. If there is no catalytic cycle, one cannot have an initiation–propagation mechanism.

For example, Polanyi (1932) examined the reaction of sodium with chlorine in the gas phase:

$$2Na + Cl_2 \Longrightarrow 2NaCl \tag{5.84}$$

Reaction (5.84) is special in that there is no initiation–propagation mechanism leading from reactants. For example, one can imagine a first step such as:

$$Na + Cl_2 \longrightarrow NaCl + Cl \qquad (5.85)$$

However, then what happens? If the chlorine atom reacted with the sodium atom, one would get stable NaCl and no radicals. There would be no catalytic cycle. As a result, this reaction cannot follow an initiation propagation mechanism. In fact, no one has found a catalytic cycle for reaction (5.84).

Polanyi (1932) showed that if you react Cl_2 with sodium dimers, the reaction does go via a catalytic cycle:

$$X + Na_2 \longrightarrow 2Na + X \qquad (5.86)$$

$$Na + Cl_2 \longrightarrow NaCl + Cl \qquad (5.87)$$

$$Cl + Na_2 \longrightarrow NaCl + Na \qquad (5.88)$$

$$X + 2Na \longrightarrow Na_2 + X \qquad (5.89)$$

However, reaction (5.88) will not occur unless there are sodium dimers in the system. Polanyi showed that if there are no sodium dimers, the reaction goes by

$$Na + Cl_2 \longrightarrow NaCl + Cl \qquad (5.90)$$

$$Na_{wall} + Cl \xrightarrow{\text{wall}} NaCl \qquad (5.91)$$

In reaction (5.91), the chlorine atoms react with sodium atoms that are on the walls of the vessel.

This brings up an important point: if there is no catalytic cycle leading from reactants to products, then there will be no way for an initiation–propagation mechanism to occur. In such a case, the products must be formed by some other route: either a direct, concerted reaction between the reactants, as in step (5.90); or a wall reaction, as in reaction (5.91).

5.5.2 Association Reactions

One key class of reactions where no direct conversion is feasible are called association reactions. In an association reaction, two stable species come together to form products.

For example, the reaction

$$NH_3 + HCl \rightleftharpoons NH_4Cl \qquad (5.92)$$

is a simple association reaction, because two stable species come together to form products. It is easy to show that with an association reaction, there is no simple initiation–propagation mechanism leading from reactants to products. Consequently, the reaction must follow some other mechanism.

Usually, an association reaction will occur via a direct concerted process. For example, reaction (5.92) goes via

$$NH_3 + HCl + X \longrightarrow NH_4Cl + X \qquad (5.93)$$

No radicals are needed. In general, simple association reactions occur via concerted reactions. Initiation–propagation reactions seldom apply.

Another kind of association reaction is a Diels–Alder reaction:

$$(5.94)$$

Again, there is no initiation–propagation reaction leading from reactants to products, and so the reaction must occur via some other pathway such as a concerted addition. That is what occurs.

Generally, association reactions occur via concerted processes. Initiation–propagation reactions do not apply.

5.5.3 Reactions with High Barriers to Propagation Steps

In all of the reactions in Sections 5.4.1 and 5.4.2, one does not observe an initiation–propagation reaction because there is no feasible initiation–propagation mechanism leading from reactants to products. There are other groups of reactions, where there is a feasible initiation–propagation mechanism. However, one of the steps in the catalytic cycle has a high barrier. As a result, the catalytic cycle is slow; and most of the product molecules are produced by some other route. A key example is the *methane coupling reaction*:

$$2CH_4 \Longrightarrow C_2H_6 + H_2 \tag{5.95}$$

Reaction (5.95) was first studied in 1919, but in 2000 people were still trying to determine a conclusive mechanism. In 1932, Kassel considered whether the reaction could occur via a simple initiation–propagation mechanism with an initiation step

$$CH_4 \longrightarrow CH_3 + H \tag{5.96}$$

with simple propagation cycle

$$H + CH_4 \longrightarrow CH_3 + H_2 \tag{5.97}$$
$$CH_3 + CH_4 \longrightarrow C_2H_6 + H \tag{5.98}$$

and three termination steps:

$$H + CH_3 \longrightarrow CH_4 \tag{5.99}$$
$$2H \longrightarrow H_2 \tag{5.100}$$
$$2CH_3 \longrightarrow C_2H_6 \tag{5.101}$$

However, Kassel found that this mechanism did not fit his data. There have been many arguments on this point. Hinshelwood showed that methane coupling could occur via this sequence of reactions. However, in practice, one does not produce much product via this route.

The cycle is so slow because there is a large barrier to reaction (5.98). Reaction (5.98) is the reverse of reaction (5.42), which was illustrated in Figure 5.4. Figure 5.4 shows that if you start with methane and methyls (i.e., the products in Figure 5.4), you need to distort the orbitals in the methane before the reaction can happen. Consequently, reaction (5.98) would be expected to have a large barrier. Lee and Masel [1996] calculate an activation barrier of 54 kcal/mol. The analysis in Example 5C gives 52.5 kcal/mol. Therefore, the catalytic cycle violates requirement (5.28). The reaction is never going to be fast enough to produce much product at a reasonable temperature.

What happens in practice is that the overall reaction is just really slow. You cannot get the catalytic cycle to occur at a reasonable rate. No other set of gas-phase reactions is feasible. Consequently, the reaction is slow unless you heat to 800 K.

5.5.4 Unimolecular Isomerization Reactions

The complications discussed in this section are particularly important to what I will call *unimolecular isomerization reactions*. A unimolecular isomerization reaction is a reaction where a single molecule rearranges to form a new product. Examples include

$$\text{Cyclopropane} \Longrightarrow \text{propene} \qquad (5.102)$$

$$CH_3NC \Longrightarrow CH_3CN \qquad (5.103)$$

Many isomerization reactions proceed via initiation–propagation reactions. However, in reaction (5.102) there is no feasible initiation–propagation reaction leading from reactants to products. There is a possible mechanism for reaction (5.103), with an initiation step

$$CH_3NC \longrightarrow CH_3 + NC \qquad (5.104)$$

a propagation step

$$CH_3NC + CH_3 \longrightarrow CH_3 + NCCH_3 \qquad (5.105)$$

and a termination step:

$$CH_3 + NC \longrightarrow CH_3CN \qquad (5.106)$$

However, reaction (5.105) includes C–N bond scission, so it has a large intrinsic barrier. The reaction violates requirement (5.28). Consequently, the catalytic cycle for reaction (5.105) is not feasible.

Most isomerization reactions follow what is called the **Lindemann mechanism**. In the Lindemann mechanism, the reactants first collide with a collision partner to yield an excited species, and then the excited species reacts to form products. One can write the mechanism in a simple form as

$$\text{Methylisocyanide} + X \longrightarrow \text{excited methylisocyanide} + X$$
$$\text{excited isocyanide} \longrightarrow \text{methylcyanide} \qquad (5.107)$$

Qualitatively, the mechanism in (5.107) can explain much of the data on unimolecular isomerization reactions. Experimentally, collision partners are needed to get unimolecular reactions to happen. Further, it has been shown that excited species are involved in the reaction process.

Still, mechanism (5.107) is too simple. If you look at reaction (5.102) in detail, you find that several excited states of cyclopropane contribute to the reaction. When the cyclopropane collides with a collision partner, the cyclopropane goes into one of the excited states. Then that excited state reacts to form a second excited state. Eventually, the excited state decomposes into products. It is not unusual for 20 excited states to be important to a unimolecular isomerization reaction. The result is a complex reaction network, even though the reaction appears simple.

5.5.5 Concerted Eliminations

A larger molecule also shows extra complications in the reaction path; there may be some extra pathways in addition to the standard initiation–propagation steps. For example, during butane pyrolysis, the initiation step is the scission of a carbon–carbon bond:

$$X + CH_3CH_2CH_2CH_3 \longrightarrow 2CH_3CH_2 + X \qquad (5.108)$$

However, a competing process is a concerted hydrogen elimination.

$$X + C_4H_{10} \longrightarrow CH_3CHCHCH_3 + H_2 + X \qquad (5.109)$$

Concerted reactions, where you break two or more bonds simultaneously, have high intrinsic barriers. For example, reaction (5.109) has an intrinsic barrier of about 60 kcal/mol. By comparison, reaction (5.108) has an intrinsic barrier of less than 2 kcal/mol. At first, one might think that it is obvious that reaction (5.108) will predominate. However, note that reaction (5.108) is 80.6 kcal/mol exothermic while reaction (5.109) is only 22.4 kcal/mol endothermic. Therefore, one does have to consider whether the difference in the heats of reactions is sufficient to overcome the difference in intrinsic barriers. If you plug into equation (5.41), you find that the answer is "No". Besides, reaction (5.109) does not produce any radicals, so you will not get a catalytic cycle. Still, in the literature, people discuss the concerted elimination processes in great detail, and say that there are cases where the concerted eliminations dominate.

In my experience in the gas phase, it is usually okay to ignore the concerted eliminations if one is considering only small molecules with no fluorines or chlorines. Under such circumstances, one can predict mechanisms easily. Large molecules are harder to consider because there are so many reactions and concerted eliminations occur. The result is a very complex reaction pathway. You need a computer to keep track of it.

The one place where there is some reasonable evidence for concerned eliminations is the reactions of chlorinated or fluorinated molecules. For example, Swihart and Carr (1998) examined decomposition of dichlorosilane and suggested that the primary initiation step is

$$X + SiH_2Cl_2 \longrightarrow HCl + SiHCl + X \qquad (5.110)$$

At first sight, this reaction would not be expected to occur, because of the large barriers. However, the hydrogen in HCl is ionic. Later in this chapter, we will find that the reactions of ionic species have much lower intrinsic barriers than do the reactions of neutral species. The ionic transition state allows reaction (5.110) to occur.

Generally, one observes concerted eliminations only with chlorine and fluorine, and even then the concerted elimination process occurs only in the initiation step of the mechanisms.

Table 5.5   Summary of the initiation–propagation mechanisms of radicals

Initiation step—weakest bond in reactants break to yield radicals: Must run at a temperature T

satisfying $E_a \leq 0.15 \dfrac{kcal}{mol \cdot K} T$

Radical reacts via a catalytic cycle

   Atoms transferred one atom at a time

   Must be cycle

   Include all reactions satisfying $E_a \leq 0.07 \dfrac{kcal}{mol \cdot K} T$

   Should have a cycle satisfying $E_a \leq 0.05 \dfrac{kcal}{mol \cdot K} T$

Termination Step—radicals recombine

Exceptions: no catalytic cycle where atoms are transferred one atom at a time

5.6   SUMMARY OF INITIATION–PROPAGATION MECHANISMS

Table 5.5 summarizes the key requirements for initiation–propagation mechanisms. First, there is an initiation step where the weakest bonds in the reactants break to yield radicals. Then the radicals react via a catalytic cycle where atoms are transferred one at a time, and the cycle leads back to the original radicals. Then there are termination steps. There are a few reactions that do not follow these general rules. Still, one can use the rules in Table 5.5 to predict mechanisms of a wide variety of gas-phase reactions, and be right 99% of the time.

I find it pretty easy to guess a mechanism of a gas-phase reaction, if I know what kinds of intermediates are stable. In my experience, the hard part is deciding what intermediates and what initial group of reactions to consider. For example, is a species such as CH_2COH stable enough to form? If one can find a way to determine which species are stable, and what types of reactions to consider, one can predict the mechanism of the reaction using the methods in this chapter.

Still, I have to admit that students find the prediction of mechanisms rather difficult. I think that the issue is partly information overload. In order to predict a mechanism, you need to know what the relative strengths of bonds are like, and what types of radicals can form. There is the trial-and-error procedure, where one guesses at a mechanism and then plugs into equation (5.46) to see if the mechanism is feasible. The only thing that I have to say about this is to stick with it. First, examine the solved examples, and convince yourself that the rules work for these reactions. Next, go back to all of the reactions mentioned so far in this chapter. First consider all of the reactions in Table 2.1. Label each step. Is it an initiation step or a propagation step? Is the step an association reaction, a β-hydride elimination...? Estimate the barrier for each step to see if the reaction follows equation 5.36. Then consider possible side reactions, and ask which side reactions are feasible. Finally try to predict the main reaction pathway. In my experience, most students learn to predict mechanisms, if they work at it.

5.7   REACTIONS OF IONS

At this point, I want to change topics and start to look at other classes of reactions. So far, the discussion in this chapter has focused on the reactions of neutral species. However, next, we will discuss the reactions of ions.

Table 5.6 Some examples of ion reactions

Reaction	Simplified Mechanism	Typical Application
Isomerization $CH_3CH_2CH{=}H_2$ $\xrightarrow{H^+} CH_3CH{=}CHCH_3$	$CH_3CH_2CH{=}CH_2 + H^+ \rightarrow$ $[CH_3CH_2CH \overset{H}{-\!\!-} CH_2]^+ \rightarrow$ $CH_3CHC{=}CHCH_3 + H^+$	Octane enhancement, monomer production
Cracking $C_{10}H_{20} \xrightarrow{H^+} 2C_5H_{10}$	$C_{10}H_{20} + H^+ \rightarrow [C_{10}H_{21}]^+ \rightarrow$ $C_5H_{10} + C_5H_{11}^+$ $C_5H_{11}^+ \rightarrow C_5H_{10} + H^+$	Crude-oil conversion, biological conversions
Alkylation $CH_3OH + C_6H_6 \xrightarrow{H^+}$ $CH_3C_6H_5 + H_2O$	$CH_3OH + H^+ \rightarrow CH_3^+ + H_2O$ $CH_3^+ + C_6H_6 \rightarrow CH_3C_6H_6^+$ $CH_3C_6H_6^+ \rightarrow CH_3C_6H_5 + H^+$	Pharmaceutical production, monomer production, fine chemicals
Esterfication $CH_3COOH + CH_3OH \xrightarrow{H^+}$ $CH_3COOCH_3 + H_2O$	$CH_3COOH \rightarrow CH_3COO^- + H^+$ $CH_3OH + H^+ \rightarrow CH_3^+ + H_2O$ $CH_3^+ + CH_3COO^- \rightarrow$ CH_3COOCH_3	Soap production, fragrance production

Table 5.6 gives some examples of important ionic reactions, including isomerization, cracking, alkylation, and esterfication. *Isomerization* is a process where one takes a molecule and rearranges it. *Cracking* is a process where you take a big molecule and break it into two smaller molecules. *Alkylation* is the opposite of cracking, where you take two molecules and combine them into one bigger molecule. *Esterfication* is a special case of alkylation where you combine acids and alcohols to form soaps or detergents. These four reactions represent more than $400 billion of yearly production, so they are certainly quite important.

Most ionic reactions occur in solution or on solid catalysts. However, ionic reactions are also seen in certain other systems such as plasma reactors and mass spectrometers. In this section, we will discuss the reactions of ions in the gas phase. Reactions in solution will be briefly mentioned in Section 5.9.

First, it is important to note that the reactions of ions are not fundamentally different from reactions of neutral species. The reactions often proceed via an initiation–propagation mechanism, where first the ions are formed, then the ions react through a catalytic cycle, and finally ions are neutralized. However, there are some important differences, as outlined in Table 5.7.

In practice, many ionic reactions are run in solution, and in solution reactions do not necessarily follow a catalytic cycle. For example, in an S_N1 reaction, you form an ion and it reacts, but there is no catalytic cycle reforming the initial ion. Instead you feed the ions into the solution.

Industrially, most ionic reactions do follow catalytic cycles, as do most ionic reactions in biology. Thus, although there are some exceptions, the majority of important ionic reactions follow catalytic cycles.

For example, the dehydration of ethanol

$$CH_3CH_2OH \xrightarrow{H^+} H_2C = CH_2 + H_2O \qquad (5.111)$$

Table 5.7   Key differences between ionic reactions and radical reactions

Radical Reaction	Ionic Reaction
Most follow initiation–propagation reactions	Many follow initiation–propagation reactions, but several do not
Bonding like that in stable molecules; intermediates have structures similar to those in stable molecules	Species often have three-centered two-electron bonds
Atom transfer reactions predominate	Isomerization and cracking can go with modest barrier
Intrinsic barriers determined by Pauli repulsions between the reactants	Pauli repulsions small for bare ions; intrinsic barriers determined by interactions in the solvent cage (hard to predict)
Preferred mechanisms easy to predict	Preferred mechanisms hard to predict; determined by solvents and other protecting groups

goes by the following mechanism in the presence of HCl. First, the acid dissociates, yielding protons (i.e., H^+) and chloride ions:

$$HCl \xrightarrow{e^-} H^+ + Cl^- \qquad (5.112)$$

Then the protons react with the ethanol to yield a charged complex:

$$H^+ + C_2H_5OH \longrightarrow [C_2H_5OH_2]^+ \qquad (5.113)$$

Then the charged complex decomposes, yielding a water and an ethyl cation:

$$[C_2H_5OH_2]^+ \longrightarrow [C_2H_5]^+ + H_2O \qquad (5.114)$$

Then the ethyl cation decomposes, yielding ethylene and the proton:

$$[C_2H_5]^+ \longrightarrow C_2H_4 + H^+ \qquad (5.115)$$

Finally, the proton can recombine with chlorine:

$$H^+ + Cl^- \longrightarrow HCl \qquad (5.116)$$

There is one subtle point—it is hard to form ions in the gas phase. One often uses electrons to produce ions. So, for example, one could replace reaction (5.112) by

$$H_2 + e^- \longrightarrow H^+ + H + 2e^- \qquad (5.117)$$

where e^- is an electron.

Reaction (5.112) is a classic initiation reaction. Reactions (5.113)–(5.115) are classic propagation reactions. Reaction (5.116) is a classic recombination reaction. Therefore, there are some similarities between reactions of ions and reactions of radicals. Still, one finds that the actual individual steps are quite different with ions than with radicals. Consequently, even though both radicals and ions follow initiation–propagation reactions, the products of ion–molecule reactions are quite different than the products of radical–molecule reaction.

For example, in Section 5.4.4 we noted that there are two key reactions when a deuterium reacts with ethane, an exchange reaction:

$$D + CH_3CH_3 \longrightarrow H + DCH_2CH_3 \qquad (5.118)$$

and a dehydrogenation:

$$D + CH_3CH_3 \longrightarrow DH + CH_2CH_3 \qquad (5.119)$$

In principle, one might also see bond scission:

$$D + CH_3CH_3 \longrightarrow DCH_3 + CH_3 \qquad (5.120)$$

However, reaction (5.120) has never been observed experimentally.

If one instead runs the reaction with D^+, the main reactions are

$$D^+ + CH_3CH_3 \longrightarrow \left[H_3C \overset{D}{-} CH_3 \right]^+ \qquad (5.121)$$

$$\left[H_3C \overset{D}{-} CH_3 \right]^+ \longrightarrow DCH_3 + CH_3^+ \qquad (5.122)$$

In reaction (5.121), one forms a complex with a three-centered two-electron bond, while in reaction (5.122), the species decomposes to yield deuterated methane and methyl radicals. Li et al. (1989) examined these reactions with a molecular beam and found that reaction (5.121) had an activation barrier of zero while reaction (5.122) had an activation barrier of less than 2 kcal/mol. Computations by Carnero (1994) suggest that both reactions are unactivated as is loss of H_2 from the complex. By comparison, Lee and Masel (1996) calculated that reaction (5.120) has an activation barrier of 45 kcal/mol.

This example illustrates two key differences between the reactions of ions and those of neutral species: (1) species with unusual bonding configurations are stable and (2) the activation barriers for reactions of ions are quite different from the activation barriers of neutral species.

The fact that the activation barriers change greatly can be easily understood in light of ideas presented in Section 5.4.5. In Section 5.4.5, we found that the activation barrier for reaction (5.120) was associated with a repulsion between the electrons in the incoming deuterium and the electrons in the ethane, causing the balloonlike orbitals to flatten out. In fact, if one replaces the deuterium with a D^+, there will be no electrons on the D^+, so there will be no repulsions. As a result, the reaction is not activated. Consequently, positive ions are much more reactive than neutral radicals.

The absence of electrons also allows the unusual bonding in the intermediates to occur. As noted above, when a D^+ reacts with an ethane, the deuterium ends up in a three-centered two-electron bond. At first, this may seem to be a very strange bonding configuration. However, if you think about it a little bit, you will realize that there is no place else that the deuterium can go. A bare deuterium has a huge electron affinity, 13 electronvolts (eV) (325 kcal/mol), so there is a huge driving force for the deuterium to grab onto an electron. An ethane has all of its normal bonding orbitals filled. However, with a 13-eV driving force, the incoming hydrogen will form a bond somewhere. The incoming deuterium craves electrons, and the richest source of electrons in ethane is the C–C bond. One could imagine the deuterium instead bonding to a single methyl group. However, that is a much less electron-rich environment than the carbon–carbon bond. Therefore, the most likely place for the D^+ to react is along the carbon–carbon bond. Consequently, when a deuterium reacts with an ethane, one would expect the D^+ to form a three-centered two-electron bond as is observed.

One can extend these arguments to the reaction of olefins. For example, consider the reaction of a D^+ with a substituted ethylene. The carbon–carbon double bond on the ethylene is a very rich course of electrons, so that initially the D^+ will react to form a three-centered two-electron complex:

$$D^+ + R_2C{=}CR_2 \longrightarrow \left[R_2C\overset{\displaystyle D}{\diagup\diagdown}CR_2 \right]^+ \tag{5.123}$$

In other words

$$\left[R_2C\overset{\displaystyle D}{\diagup\diagdown}CR_2 \right]^+ \tag{5.124}$$

It is not obvious that the species in equation (5.124) is stable. Other possibilities include an ion with charge on one center:

$$\left[R_2\overset{+}{C}{-}\overset{\displaystyle D}{}CR_2 \right]^+ \tag{5.125}$$

or a species where the R groups have migrated:

$$\left[R_2C\overset{\displaystyle R}{\diagup\diagdown}CRD \right]^+ \tag{5.126}$$

Experimentally, when $R = H$, species (5.124) is the most stable; when $R = CH_3$, species (5.125) is the most stable; however, when $R = C_5H_6$ the species in (5.126) is the most stable. For the analysis here, it is not really important to know which of these species is the most stable. However, the key thing is that the ions can exist in several isomers, including an isomer where one species bridges between two carbon centers. Further, there are hardly any barriers to convert one isomer into another because there are no Pauli repulsions.

Now those two factors lead to interesting chemistry. There are two key types of reaction that usually happen only with ions: cracking and isomerization. **Cracking** is a reaction where one starts with a heavy hydrocarbon and breaks up the hydrocarbon into smaller fragments. The actual reaction involves a series of carbon–carbon bond scission processes like those in equations (5.120) and (5.122).

Notice that such reactions will occur with low barriers if the species are positively charged. In contrast, high barriers will be seen with neutral species. Therefore, if one wanted to crack a hydrocarbon, one would run the reaction under conditions where a positive ion would form.

Similarly, **isomerization** is a process where one takes a ligand (i.e., an R group) and moves it around a molecule. With an ion, one can form a bridging intermediate like that in equation (5.126), so isomerization is easy. On the other hand, with a radical, the bridging species is unstable. Consequently, isomerization reactions have much lower barriers with ions than with radicals.

5.7.1 The Mechanisms of Reactions of Carbocations

Next, I want to talk about the detailed mechanism of the reactions of carbocations. A carbocation is a positively charged hydrocarbon ion. Examples include $C_3H_7^+$ and $C_3H_9^+$. Carbocation chemistry has been studied in great detail in solutions and in solid catalysts. The reactions have been less heavily studied in the gas phase. Generally, the reactions of carbocations are fairly complex. There are many intermediates and very many interconnecting reactions. Many of the species have unusual bonding. I generally find it difficult to predict the stability of the various species. Consequently, it is slightly harder to make detailed predictions of the reaction pathways of the ions than to predict reaction pathways of radicals. Still, as we will see, the analysis is easy provided you know what intermediates can form.

To start off, it is useful to review some of the literature. Martens and Jacobs (1990) and Vogel (1985) point out that there are two key types of carbocations that one can consider: carbonium ions and carbenium ions. **Carbonium ions** are formed by adding a H^+ or a R^+ to a paraffin. Examples of carbonium ions include

$$\left[H_3C \overset{\overset{\displaystyle R}{\diagup\ \diagdown}}{-} CH_3 \right]^+ \tag{5.127}$$

$$\left[H_3C \overset{\overset{\displaystyle R}{|}}{-} CH_3 \right]^+ \tag{5.128}$$

$$\left[H_3C \overset{\overset{\displaystyle R}{|}}{-} OH \right]^+ \tag{5.129}$$

$$\left[H_3C \overset{\overset{\displaystyle R}{\diagup\ \diagdown}}{-} OH \right]^+ \tag{5.130}$$

Years ago, people perceived carbonium ions as having five coordinated carbon atoms, three coordinated oxygen atoms, and fairly localized charges. However, people now know that there are bridging species and that the charge is distributed throughout the molecule.

Carbenium ions are formed by adding a H^+ or a R^+ to a double bond or the removal of a proton from a paraffin. Examples of carbenium ions include

$$H_3C^+ \tag{5.131}$$

$$RH_2C^+ \tag{5.132}$$

$$R_2HC^+ \tag{5.133}$$

$$R_3C^+ \tag{5.134}$$

$$\left[R_2C \overset{\displaystyle \overset{H}{\diagup\diagdown}}{} CH_2 \right]^+ \tag{5.135}$$

Carbenium ions often have three coordinated carbon atoms. However, there are also bridging species and cyclic structures.

Generally, carbonium ions react via two pathways: cracking and isomerization. During cracking of a carbonium ion, a proton reacts with a paraffin to form a bridging species:

$$H^+ + RCH_2{-}CH_2R \longrightarrow \left[RH_2C \overset{\displaystyle \overset{H}{\diagup\diagdown}}{} CH_2R \right]^+ \tag{5.136}$$

Then the bridging species decomposes to yield a smaller hydrocarbon fragment and a carbenium ion:

$$\left[RH_2C \overset{\displaystyle \overset{H}{\diagup\diagdown}}{} CH_2R \right]^+ \longrightarrow RCH_3 + CH_2R^+ \tag{5.137}$$

Isomerization of carbonium ions occurs when a linear carbonium ion rearranges to form a bridged species and then goes back to a linear species:

$$\left[H_3\overset{\displaystyle \overset{R}{|}}{C}{-}CH_3 \right]^+ \longrightarrow \left[H_3C \overset{\displaystyle \overset{R}{\diagup\diagdown}}{} CH_3 \right]^+ \longrightarrow \left[H_3C{-}\overset{\displaystyle \overset{R}{|}}{C}H_3 \right]^+ \tag{5.138}$$

Carbenium ions can also crack and isomerize. In the gas phase, reaction (5.138) has an intrinsic barrier of less than 4 kcal/mol, which makes the reaction very rapid. Cracking occurs via β-scission, where the bond between the α carbon and the β carbon breaks to yield an olefin and a smaller carbenium ion:

$$R'{-}\overset{\overset{H}{|}}{\underset{\underset{H}{|}}{C}}{-}\overset{\overset{H}{|}}{\underset{\underset{H}{|}}{C}}{-}\overset{\overset{R}{|}}{\underset{\underset{H}{|}}{C}}{-}\overset{\overset{+}{}}{\underset{\underset{H}{|}}{C}}{-}H \longrightarrow R'{-}\overset{\overset{H}{|}}{\underset{\underset{H}{|}}{C}}{-}\overset{\overset{H}{|}}{\underset{\underset{H}{|}}{C}}+ + \overset{\overset{R}{|}}{\underset{\underset{H}{|}}{C}}{=}\overset{\overset{H}{|}}{\underset{\underset{H}{|}}{C}} \tag{5.139}$$

There are two pathways for isomerization: a standard 1,2 shift, where an R group migrates to an adjacent site; and a reaction where a cyclic intermediate forms. Reaction (5.140) is an example of a 1,2 shift:

$$R'{-}\overset{\overset{H}{|}}{\underset{\underset{H}{|}}{C}}{-}\overset{\overset{H}{|}}{\underset{\underset{H}{|}}{C}}{-}\overset{\overset{R}{|}}{\underset{\underset{H}{|}}{C}}{-}\overset{\overset{+}{}}{\underset{\underset{H}{|}}{C}}{-}R'' \longrightarrow R'{-}\overset{\overset{H}{|}}{\underset{\underset{H}{|}}{C}}{-}\overset{\overset{H}{|}}{\underset{\underset{H}{|}}{C}}{-}\overset{\overset{+}{}}{\underset{\underset{H}{|}}{C}}{-}\overset{\overset{R}{|}}{\underset{\underset{H}{|}}{C}}{-}R'' \tag{5.140}$$

Notice that the R group shifts to the right. These 1,2 shifts are the primary isomerization route with a small hydrocarbon. With a modest-sized hydrocarbon, one also sees isomerization via cyclic intermediates. First the molecule reacts to form a cyclic intermediate as shown in equation (5.141):

$$(5.141)$$

Then the cyclic intermediate undergoes bond scission as shown in reaction (5.142):

$$(5.142)$$

One can also stop reaction (5.142) at the first step to yield a cyclic product. In gas-phase reactions (5.140)–(5.142) are extremely rapid and often reach equilibrium in a very short time. In a solvent, the reactions are a little slower because you need to move solvent molecules out of the way. In the latter case, the intrinsic barriers to reaction are associated with motions of the solvent molecules, and not the intrinsic barriers to rearrangements of the reactants.

Carbenium ions can also participate in alkylation reactions, where a carbenium ion B^+ reacts with a paraffin or olefin to yield a carbonium or carbenium ion:

$$B^+ + R_3C{-}CR_2H \longrightarrow \left[R_3C\overset{B}{\underset{\diagup\diagdown}{-}}CR_2H \right]^+ \qquad (5.143)$$

The carbonium or carbenium ion can then lose a proton to yield a higher-molecular-weight species:

$$\left[R_3C\overset{B}{\underset{\diagup\diagdown}{-}}CR_2H \right]^+ \longrightarrow R_3C{-}CR_2B + H^+ \qquad (5.144)$$

There are many combinations here. One can form a cyclic product, and then isomerize the product to form a larger ring. One can then undergo bond scission to yield a myriad of products.

For example, Figure 5.7 shows some of the reactions of tridecylcation ($C_{13}H_{25}^+$). Only sixteen reactions are shown in the figure. However, Martens and Jacobs actually found 70 total reactions. In the top reaction in the figure, the molecule starts out with the charge centered on the carbon atom labeled zero in the figure. During the reaction, the molecule pivots around the 2 carbon, bringing the 0 carbon and the 3 carbon in close proximity. Then a bond forms between the 0 carbon and the 3 carbon, forming a four centered ring. A proton migrates, moving the charge.

Then the ring opens again. The other reactions on the page are similar, except that different size rings form. This example illustrates the complexity of ion reactions. Table 5.8 shows the products of the reaction. Note that there are 12 main products and at least 70 reactions in the reaction network. Consequently, the mechanism is rather complex.

One thing that people do know is that the stability of the ions varies with the degree of branching. For example, Table 5.9 shows the heat of formation of several $C_{10}H_{21}^+$

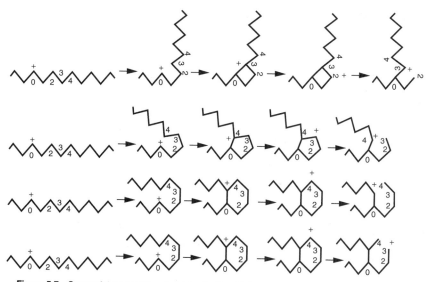

Figure 5.7 Some of the reactions of a tridecylcation. [Adapted from Martens and Jacobs (1990)].

Table 5.8 The products of tridecylcation ($C_{13}H_{25}^+$) isomerization in different environments

Species	Production in the Gas Phase	Production in Platinum/CAY Zeolite Catalyst	Production in Platinum/USY Zeolite Catalyst
2 Methyldodecane	9.3	7.0	11
3-Methyldodecane	17.7	15.1	17.9
4-Methyldodecane	19.0	18.4	18.0
5-Methyldodecane	19.1	22.8	19.6
6-Methyldodecane	19.1	23.4	19.4
3 Ethylundecane	2.7	1.9	2.7
4 Ethylundecane	3.4	3.4	3.4
5 Ethylundecane	3.9	3.6	3.9
6 Ethylundecane	2.2	2.0	2.2
4 Propyldecane	1.0	0.7	1.0
5 Propyldecane	1.8	1.4	1.8
5 Propyldecane	0.7	0.3	0.7

Source: Results of Martens and Jacobs [1990].

ions. Notice that the branched ions are more stable than the linear ions. Consequently, at equilibrium, one would expect to form highly branched species. You remember from organic chemistry that in small molecules, a primary ion is less stable than a secondary ion, which is less stable than a tertiary ion. That rule also works from a simple ion. With a more complex ion, it is easier to think about the ion stability in terms of Brauman and Blair's (1971) model, where you view the stabilization of the ion, as an interaction between the charged site and the polar stable alkyl groups. The more alkyl groups you get near the charged site, the more the charge will be stabilized. One can get alkyl groups next to the

Table 5.9   The relative stability of some $C_{10}H_{21}{}^+$ intermediates

Intermediate	Heat of Formation, kcal/mol	Intermediate	Heat of Formation, kcal/mol
1-Butyl cation	+138	2-Methyl-2-butyl cation	−12
2-Methyl-1-propyl cation	+130	2-Methyl-2-pentyl cation	−19
2-Butyl-cation	+67	3-Methyl-3-pentyl cation	−19
2-Methyl-2-propyl cation	0	2,3-Dimethyl-2-pentyl cation	−21

Source: After Martens and Jacobs (1990).

charged site by moving from a primary to a secondary to a carbocation. However, one can also stabilize charge by making the alkyl groups more polarizable, or by adding long chains that can coil up to bring the ends of the chains in close proximity to the charged center. In my experience, it is not so easy to predict the relative stability of carbocations because of chain coiling. However, Martens and Jacobs (1990) show that if one knows the stability of the ion, one can sometimes predict the products of the carbocation reaction from equilibrium. See Martens and Jacobs for details.

5.8   PREDICTION OF THE MECHANISM OF IONIC REACTIONS

Next, we will be discussing the prediction of the mechanism of gas-phase ionic reactions. Generally, it is pretty easy to guess the mechanism of a ionic reaction in the gas phase, provided the key intermediates are known. For example, let's start with the reaction

$$CH_3-CH=CH-CH_3 + H^+ \Longrightarrow \text{products} \tag{5.145}$$

We would expect that the first step would be the formation of a carbenium ion:

$$CH_3-CH=CH-CH_3 + H \longrightarrow [CH_3-CH-CH_2-CH_3]^+ \tag{5.146}$$

Then the species could undergo a series of isomerization and cracking reactions as indicated in Figure 5.8. As with radicals, the first step in predicting a mechanism is to write down all of the reactions that could occur. One then does calculation using equation (5.41) to see which reactions can occur. Generally, with a carbenium ion, the intrinsic barriers are all small; usually 1–5 kcal/mol. This means that in the gas phase all of the reactions in Figure 5.8 will occur. As a result, the system will quickly reach equilibrium. The cracking reaction [reaction (5.137)] is slow because it is thermodynamically unfavorable. However, the situation in a solvent is different because there are significant barriers to rearrange the solvent molecules. Unfortunately, we cannot predict the barriers in a solvent, so mechanisms are difficult to guess.

As the butene pressure in the system goes up, one also gets alkylation reactions

$$C_4H_9{}^+ + C_4H_8 + X \longrightarrow C_8H_{17}{}^+ + X \tag{5.147}$$

and transfer reactions

$$C_4H_9{}^+ + C_4{}^*H_8 \longrightarrow C_4H_8 + C_4{}^*H_9{}^+ \tag{5.148}$$

$$C_4H_9{}^* + C_4{}^*H_8 \longrightarrow C_3H_6{}^+ (CH_3C_4H_8) \tag{5.149}$$

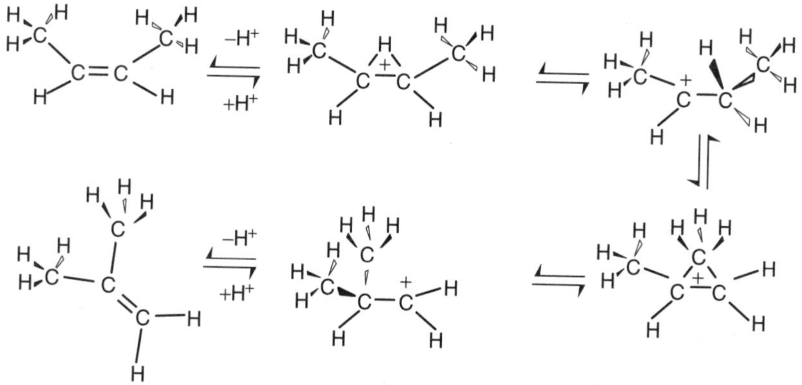

Figure 5.8 The mechanism of the reaction $CH_3CH{=}CH_2CH_3 + H^+ \rightarrow$ products at low pressure.

In reaction (5.147), the $C_8H_{17}^+$ is a transient species, not a true intermediate. The result is some very complex chemistry.

Now, with a carbenium ion, all of the reactions occur with low barriers. However, if you do the same reactions with a carbonium ion by replacing the butene in reaction (5.147) with butane, there are significant barriers. Unfortunately, at present, such reactions have not been studied in enough detail to permit useful predictions.

5.9   REACTIONS IN POLAR SOLUTION

So far in this chapter we have been discussing the mechanisms of reactions in the gas phase. However, now we are going to change topics and start to consider reactions in solution. From a practical standpoint, reactions in solution are quite important. Virtually all organic and inorganic syntheses are run in solution. All biological reactions are run in solution. Most of the reactions you ran in freshman chemistry and organic chemistry occurred in solution. Most of the chemical reactions in the human body and in other living things occur in solution.

Unfortunately, there is insufficient room in this book to discuss solution chemistry of molecules in any detail. You should consult your organic or biochemistry books for discussions of these reaction systems.

Still, we will need a few concepts for the discussion later in this book. You probably remember S_N1 and S_N2 reactions from your organic chemistry class. During an S_N1 reaction, a molecule containing a neucleophile breaks apart in solution to yield a carbenium ion and a neucleophile:

$$R'{-}\underset{\underset{H}{|}}{\overset{\overset{H}{|}}{C}}{-}\underset{\underset{H}{|}}{\overset{\overset{H}{|}}{C}}{-}\underset{\underset{H}{|}}{\overset{\overset{R}{|}}{C}}{-}\underset{\underset{H}{|}}{\overset{\overset{N}{|}}{C}}{-}H \longrightarrow R'{-}\underset{\underset{H}{|}}{\overset{\overset{H}{|}}{C}}{-}\underset{\underset{H}{|}}{\overset{\overset{H}{|}}{C}}{-}\underset{\underset{H}{|}}{\overset{\overset{R}{|}}{C}}{-}\overset{+}{\underset{\underset{H}{|}}{\overset{\overset{}{|}}{C}}}{-}H + N^- \qquad (5.150)$$

where N^- is a neucleophile. The carbenium ion can then rearrange or crack to yield a new ion:

$$R'{-}\underset{\underset{H}{|}}{\overset{\overset{H}{|}}{C}}{-}\underset{\underset{H}{|}}{\overset{\overset{H}{|}}{C}}{-}\underset{\underset{H}{|}}{\overset{\overset{R}{|}}{C}}{-}\overset{+}{\underset{\underset{H}{|}}{\overset{\overset{}{|}}{C}}}{-}H \longrightarrow R'{-}\underset{\underset{H}{|}}{\overset{\overset{H}{|}}{C}}{-}\underset{\underset{H}{|}}{\overset{\overset{H}{|}}{C}}{-}\overset{+}{\underset{\underset{H}{|}}{\overset{\overset{}{|}}{C}}}{-}\underset{\underset{H}{|}}{\overset{\overset{R}{|}}{C}}{-}H \qquad (5.151)$$

Then the species can react with a second neucleophile to yield a stable product:

$$N^- + R'-\underset{\underset{H}{|}}{\overset{\overset{H}{|}}{C}}-\underset{\underset{H}{|}}{\overset{\overset{H}{|}}{C}}-\underset{\underset{H}{|}}{\overset{\overset{+}{|}}{C}}-\underset{\underset{H}{|}}{\overset{\overset{R}{|}}{C}}-H \longrightarrow R'-\underset{\underset{H}{|}}{\overset{\overset{H}{|}}{C}}-\underset{\underset{H}{|}}{\overset{\overset{H}{|}}{C}}-\underset{\underset{H}{|}}{\overset{\overset{N}{|}}{C}}-\underset{\underset{H}{|}}{\overset{\overset{R}{|}}{C}}-H \qquad (5.152)$$

In S_N2 reactions the neucleophile reacts with a stable molecule and exchanges:

$$[N']^- + R_3 C N \longrightarrow N' C R_3 + N^- \qquad (5.153)$$

One neucleophile replaces another without the appendant groups undergoing significant rearrangements.

The S_N2 reactions are very similar to the reactions of radicals in the gas phase. One observes the same basic chemistry — single atom transfers are fast; ligand transfers are slower. Everything is more complex because all of the species are charged.

The similarities are not as close with S_N1 reactions. S_N1 reactions look something like ion reactions in the gas phase. First, there is a step where an ion forms. Then the ion rearranges. One observes isomerization cyclizations and cracking. Then a stable species forms again. The reaction is not a complete cyclic process, so the kinetics are different. However, there are some qualitative similarities between an S_N1 reaction in solution and an ion reaction in the gas phase.

The one major difference between reactions in the gas phase and in solution is that ionic species are stabilized by the presence of the solution. For example, as noted above in the gas phase, carbonium ions are hard to form. Well, in water or other polar solvents, carbonium and carbenium ions are much more stable than in the gas phase. Consequently, one can form carbocations without using electrons to ionize the species.

The presence of the solvent also changes the intrinsic barriers to ionic reactions. Generally, intrinsic barriers are higher in solution than in the gas phase. Recall that in the gas phase, ion reactions like those in equation (5.121) are often virtually unactivated, because there are no Pauli repulsions between the reactants. When the reactants dissolve in solution, everything changes. It is unusual to get a bare ion in solution. Instead, the ion is generally surrounded by a solvent cage as shown in Figure 5.9. In order for a reaction to occur, one needs to first distort the ion cage so the reactants can get close enough to react. The distortion of the ion cage generally produces a repulsion similar to the Pauli repulsions seen in radical reactions.

Unfortunately, at present, it is very hard to predict how much the intrinsic barrier changes as a result of the solvent. For example, if one runs a reaction in water, the water

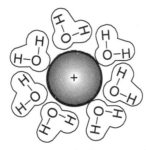

Figure 5.9   A schematic of the solvent cage around an ion in solution.

forms a hydrogen-bonded cage around each ion. The cage needs to be distorted when a reaction occurs, but the exact nature of the distortion is unclear. Consequently, it is difficult to make any predictions.

Fortunately, the changes in the intrinsic barriers do not really change the mechanism of a reaction. One sees all the same elementary reactions in solution that one sees in the gas phase. It is just that the relative rates of reactions change in the solution. The changes in rate will be discussed in Chapter 13.

Olah et al. (1997) show that one can often, but not always, predict mechanisms of reaction without considering the changes in the intrinsic barriers because of the solvent. It is not obvious, but if one species can be in two different reactions, then the intrinsic barriers to both reactions will change in the presence of the solvent. Often, the intrinsic barriers to all of the reactions of a species are changed by a similar amount. In such as case, one can still predict accurate mechanisms by ignoring the changes that occur in the intrinsic barriers in the presence of the solvent.

In practice, however, you want the solvent to affect the rate. After all, in organic synthesis, you want to only make a few products and not make others. People use the solvent to direct reactions. For example, if you wanted to get a specific reaction, you would use a solvent that surrounds the part of the molecule that you want to protect, so no chemistry could happen there. You can also use a counterion or an enzyme for a similar purpose. In practice, the solvent should be an active participant in reactions because that gives you opportunities to optimize selectivity.

One substantial difference between carbocation reactions in the gas phase and in solution is that in solution one can get direct reactions with the solvent. For example, water can take up protons. The reactions with the solvent limit the stability of the carbocations. That can have important implications for mechanisms. This will be discussed in Chapter 13.

In my experience, it is much harder to predict the reaction mechanisms in a solvent than in the gas phase. One needs much more background. Consequently, the discussion will be deferred to Chapter 13.

5.10   REACTIONS ON METAL SURFACES

Next we want to discuss the mechanisms of another important class of reactions: reactions on metal surfaces. Table 5.10 lists some important reactions that occur on the surfaces of metal. Metal surfaces are used extensively as catalysts to promote hydrogenation, dehydration, and oxidations. Reactions on metal surfaces are also important to corrosion and the production of integrated circuits.

Reactions on metal surfaces look similar to radical reactions except that the reactants are not radicals. Instead, the reactants are species that have a bond to the surface. The surface acts like a solvent to stabilize radicals in the same way that water acts as a solvent to stabilize ions.

For example, the reaction $O_2 + 2H_2 \longrightarrow 2H_2O$ can occur in the gas phase. According to Hinshelwood (1946), some of the key steps in the mechanism are

$$X + H_2 \longrightarrow 2H + X$$
$$X + O_2 \longrightarrow 2O + X$$
$$O + H_2 \longrightarrow OH + H$$

Table 5.10   Some examples of reactions on metal surfaces

Hydrogenation	Chemical production
$N_2 + 3H_2 \Longrightarrow 2NH_3$	Crude–oil upgrade
$C_2H_4 + H_2 \Longrightarrow C_2H_6$	Essential oil upgrade
Dehydrogenation	Octane enhancement
$C_2H_4 \xrightarrow{\text{Pt}} C_2H_2 + H_2$	Monomer production
Oxidation	Catalytic converters
$2CO + CO \xrightarrow{\text{Pt}} 2CO_2 + N_2$	Monomer production
$C_2H_4 + \frac{1}{2}O_2 \xrightarrow{\text{Ag}} C_2H_4O$	Chemicals production
$2NH_3 + 4O_2 \xrightarrow{\text{Pt}} N_2O_5 + 3H_2O$	
Chemical vapor deposition	Connections on integrated circuits
$Al(C_2H_5)_3 \xrightarrow{\text{Al}} Al + 3C_2H_4 + \frac{3}{2}H_2$	
$Fe_{(s)} \Longrightarrow Fe^{3+}_{(Aq)}$	Corrosion

$$H_2 + OH \longrightarrow H_2O + H$$

$$H + O_2 \longrightarrow OH + O$$

$$X + 2H \longrightarrow H_2 + X \tag{5.154}$$

Anton and Cadogan (1991) examined the mechanism of hydrogen oxidation on a Pt(111) surface. They found that the main reactions were

$$H_2 + 2S \longrightarrow H_{(ad)}$$

$$O_2 + 2S \longrightarrow 2O_{(ad)}$$

$$O_{(ad)} + H_2 \longrightarrow OH_{(ad)} + H_{(ad)}$$

$$2O_{(ad)} + H_2 \longrightarrow 2OH_{(ad)}$$

$$O_{(ad)} + H_{(ad)} \longrightarrow OH_{(ad)} \tag{5.155}$$

$$2OH_{(ad)} \longrightarrow H_2O + O_{(ad)}$$

$$H_{(ad)} + OH_{(ad)} \longrightarrow H_2O$$

$$H_2 + 2S \longrightarrow 2H_{(ad)}$$

where the notation (ad) is used to designate that the species are attached to the platinum surface. Notice there is a close similarity between the mechanism of hydrogen oxidation in the gas phase [reaction (5.154)] and the mechanism on the surface [reaction (5.155)]. The elementary steps and intermediates are almost the same. Consequently, one can often think about a reaction on a surface as being a gas-phase reaction except that the intermediates of the reaction are bound to the surface.

The fact that the reacting species are bound to the surface during the surface reaction has important implications. The surface acts like a solvent to stabilize the radicals. Unpaired electrons in the radicals pair up with the free electrons in the metal to produce stable species. The result is that the intermediates of the reaction are stable species rather than radicals. Still, the species adsorbed on metal surfaces often follow reaction pathways that are quite similar to reaction pathways of radicals in the gas phase, although they do not always do so.

People often picture surface reactions via a catalytic cycle, where surface intermediates are created and then destroyed. For example, Figures 5.10a and 5.10b show the catalytic cycles for the production of water via the mechanism in equation (5.155). Water is produced in two reactions in equation (5.155): (1) the disproportion of OH and (2) the direct reaction between H_{ad} and $OH_{(ad)}$. The first reaction dominates when the oxygen concentration is low. The second pathway dominates when the oxygen concentration is high.

In the first pathway, the surface starts out covered by oxygen. Hydrogen adsorbs on the oxygen to yield OH groups. Two OHs then couple to yield H_2O and a bare site. Oxygen then adsorbs on the bare site to get back to the initial conditions.

In the second pathway, the surface starts out empty. Hydrogen and oxygen adsorb. The adsorbed species then reacts to yield water and regenerate the clean surface.

All steady-state catalytic reactions can be viewed as occurring via a catalytic cycle where surface species are formed and destroyed. The cycles are similar to those discussed previously for gas-phase reactions. However, there is an important difference between reactions in the gas phase and on a surface. In the gas phase, radicals are rather unstable species. As a result, one has to go to rather high temperatures before one gets a high enough concentration of radicals to get a reasonable rate. In contrast, when a radical binds to a surface, the dangling bond in the radical forms a bond to the surface, producing a species that is stable. As a result, one can get a reasonable concentration of adsorbed radicals at low temperatures. For example, at 300 K, less than one hydrogen molecule in 10^{35} dissociates in the gas phase, while on a platinum surface, more than 99.9% of the hydrogen dissociates. The result is that hydrogen on platinum is much more reactive than gas phase H_2.

The hydrogen on platinum is not as reactive as a hydrogen radical in the gas phase, however. The hydrogen atoms on the surface are bound to the platinum and are no longer radicals. As a result, the hydrogen atoms on the surface are less reactive than free hydrogen radicals in the gas phase. However, the effect of the reduced reactivity is smaller than the effects of the enhanced concentration of hydrogen atoms. As a result, at moderate temperatures, the hydrogen/oxygen reaction occurs much faster over a platinum catalyst

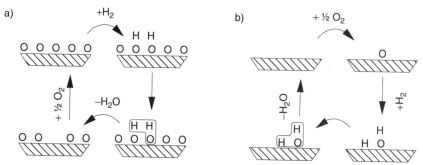

Figure 5.10 Catalytic cycles for the production of water (a) via disproportion of OH groups and (b) via the reaction $OH_{(ad)} + H_{(ad)} \rightarrow H_2O$.

than in the gas phase. At room temperature, a hydrogen oxygen mixture remains stable for years. However, if one puts a platinum wire in the mixture, the mixture will explode.

In general, metal surfaces stabilize reactive intermediates. That greatly speeds up the initiation steps in the mechanism. The speedup of the initiation reaction is why catalytic reactions have so much higher rates than do gas-phase reactions.

Masel (1996) also notes that surfaces can stabilize di- and triradicals. That allows new chemistry to occur on a surface that would not occur in the gas phase. Details are given in Chapter 12.

There is one key difference in the nature of the initiation reaction between the mechanism of the reaction in the gas phase and the mechanism of a reaction on the surface. In the gas phase, the initiation step always involves the scission of a bond. On a surface, the key initiation is the **adsorption** of one or more of the reactants onto the surface. During adsorption, a molecule attaches itself to a surface that changes the properties at the molecule so that the reactions change as well. We call the species bond to surfaces **adsorbed** molecules or **adsorbates**. Masel (1996) has presented a detailed discussion of adsorption.

There is some important notation. We will use a subscript (ad) to designate an adsorbed species. That way we can distinguish between adsorbed species, and species in the gas phase.

5.11 GENERAL CONCEPTS ABOUT ADSORPTION AND REACTION ON SURFACES

Before we talk about specific mechanisms of surface reactions, we need to discuss some general concepts for surface reactions. There is a lot of information here, and students often find it overwhelming. My best advice is to stick with it. The material will get easier after you have read it a couple of times.

The first general concept is that surface reactions occur in what is called an *adsorbed phase*. The adsorbed phase is a layer of gas that is bound to the surface. Figure 5.11 shows what the adsorbed phase is like when ethylene adsorbs on platinum. The surface consists of an ordered layer of platinum atoms indicated as circles in the figure. Platinum, is a metal, and metals have many *lots of* free electrons. When ethylene adsorbs on platinum, some of the ethylene adsorbs intact. Other ethylene molecules react to form a variety of hydrocarbon fragments. Some of the hydrocarbon fragments form strong bonds to the platinum atoms, while some of the other fragments are free to move around. The result is a complex mixture that has properties somewhere between those of a liquid and those of a solid.

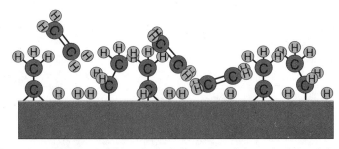

Figure 5.11 An illustration of the adsorbed phase when ethylene adsorbs on platinum.

People spend their entire lifetimes studying adsorption. There is a tremendous amount of information about the topic. For now, the key thing to remember is that a surface reaction usually occurs in a thin layer of molecules attached to the surface. The layer can be disordered like a liquid or ordered like a solid. Still, the key feature is that there is a thin layer of adsorbed molecules and reaction occurs in that layer.

The next key point is that there are two types of adsorption: **molecular adsorption** and **dissociative adsorption**. In molecular adsorption, the molecule adsorbs intact, while in dissociative adsorption, a bond breaks in the adsorbed molecule. One finds that bonds break easily during a surface reaction, so it is not unusual for a molecule to dissociate on adsorption.

For example, when O_2 adsorbs onto platinum at 300 K, the oxygen dissociates, yielding two oxygen atoms. H_2 dissociates into two hydrogen atoms. C_2H_4 dissociates to yield a $CH_3C\equiv$group and an adsorbed hydrogen. At 100 K, H_2 still dissociates. However, O_2 and C_2H_4 adsorb molecularly.

The fact that bonds can break during adsorption is quite important, because it means that one can produce atoms or other very reactive species in the adsorbed layer. Consequently, surface reactions are often quite rapid.

The high reaction rates on metal surfaces make metal surfaces very active catalysts. Metals are good at producing radicals, so they tend to catalyze radical reactions. Ionic reactions are rarely catalyzed by metals, although there are a few exceptions.

In order to know whether a surface will catalyze a given reaction, it is important to know whether a given molecule dissociates on a given surface. Figure 5.12 gives a chart that one can use to see if a given molecule dissociates. Note that metals in the middle part of the periodic table are the most active for dissociation.

Another key concept for surface reactions is that the reactions occur on distinct **surface sites**. Surface sites are defined as the place on the surface where reactions can occur.

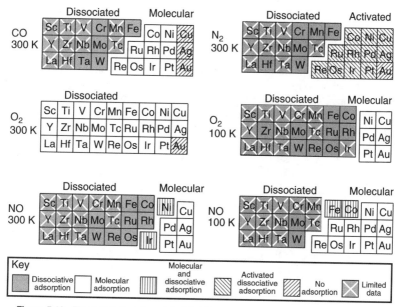

Figure 5.12 The metals that dissociate CO, NO, N_2, O_2, and CO at various temperatures.

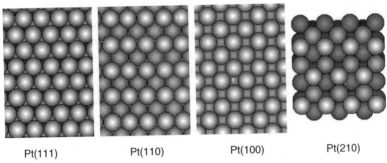

Pt(111)   Pt(110)   Pt(100)   Pt(210)

Figure 5.13   Pictures of some common surface structures.

Sometimes, the surface site consists of a specific arrangement of metal atoms. Other times, only a single bare metal atom is sufficient. Experimentally, one often finds that the rates of surface reactions are greatly enhanced if a specific arrangement of atoms is available on the surface of the catalyst. However, that is never an absolute requirement.

Often, there is an ordered arrangement of sites in the surface of a catalyst. Figure 5.13 shows a picture of some surface structures. The surface atoms are indicated as spheres in the photo. Notice that one has an ordered arrangement of surface atoms. If each surface atom is a reactive site, then one will also have an ordered arrangement of surface sites.

5.12   PREDICTION OF THE MECHANISMS OF SURFACE REACTIONS

Next, we want to discuss how one can predict the mechanisms of reactions on metal surfaces. The material will build on the analysis in Sections 5.4.1–5.4.3, 5.4.5, and 5.4.6, where we showed that one can combine a series of empirical rules and the Polanyi relationship to predict the mechanism of reactions.

First, I want to note that all reactions on metal surfaces are propagation reactions. In Section 5.2, we noted that when we run a reaction in the gas phase, the first step is always an initiation step. Reactions on metal surfaces are a little different, in that there are always dangling bonds and free electrons in a metal surface, so that the surface is, in effect, a radical. Consequently, one does not have to form radicals at the start of the surface reaction. Instead, the bare surface is a radical. Consequently, one does not need an initiation step.

In Section 5.4.1, we noted that all propagation reactions go in catalytic cycles where radicals are formed and destroyed. In the same way

> All surface reactions occur in cycles
> where bare surface sites are formed and destroyed.   (5.156)

Figure 5.10 shows a typical catalytic cycle. The surface starts out empty. Hydrogen and oxygen adsorb. The adsorbed species then react to yield water and regenerate the clean surface. Notice that we start with a bare surface site at the beginning of the reaction, and end with a bare surface site at the end of the reaction in agreement with requirement (5.156).

According to rule (5.68), when radicals react in the gas phase, the mechanism generally consists of a series of elementary reactions where a single atom is transferred from one species to the next as described in Section 5.4.6.

Reactions on metal surfaces are very similar to radical reactions in the gas phase. Generally, molecules first adsorb. Then, they transfer one atom at a time to yield products.

For example, Figure 5.14 shows the mechanism of methanol decomposition

$$CH_3OH \Longrightarrow CO + 2H_2 \qquad (5.157)$$

on a platinum surface called Pt(110). The methanol adsorbs molecularly. Then the methanol loses one hydrogen at a time to yield products. This mechanism differs from the mechanism of methanol decomposition in the gas phase because there is no initiation reaction. Still, the methanol follows the general rule that atoms are transferred one atom at a time into products.

Before we proceed, we need to define some notation. We will designate a surface site by the symbol S. With this notation, the second reaction in Figure 5.14, where an adsorbed methanol reacts with the surface site to yield an adsorbed methoxy (CH_3O) plus an adsorbed hydrogen atom, can be written

$$CH_3OH_{(ad)} + S \longrightarrow CH_3O_{(ad)} + H_{(ad)} \qquad (5.158)$$

where the subscript (ad) is used to indicate that we have an adsorbed species.

Next, we need to note that, just as in the gas phase, one does not always simply transfer one atom at a time during a surface reaction. For example, Figure 5.15 shows the mechanism of ethanol decomposition on platinum. The ethanol first sequentially dehydrogenates to yield an acetyl (CHCO) intermediate. Then the C–C bond breaks to yield CO and an adsorbed methyl group.

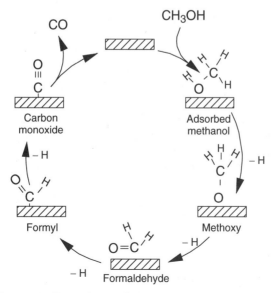

Figure 5.14   The mechanism of methanol decomposition on Pt(110).

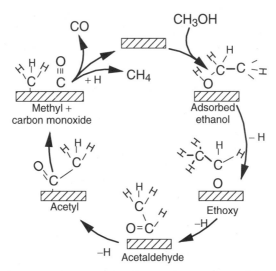

Figure 5.15 The mechanism of ethanol decomposition on Pt(111).

Notice the acetyl group could still lose hydrogens, so the fact that the C–C bond breaks is an exception to the rule that molecules gain or lose one atom at a time during reactions on solid surfaces.

5.12.1 Intrinsic Barriers to Surface Reactions

Next we will discuss how one can predict mechanisms of reactions on solid surfaces. In Section 5.4.6, we noted that one can use the Polanyi relationship to predict the mechanism of the reactions in the gas phase. Lee and Masel (1996) showed that one can use the same ideas to predict the mechanism of reactions of simple hydrocarbons, oxygenates, and amines on metal surfaces. The idea is that when you run a reaction on a surface, the reactants and products of the reaction are stabilized. However, in many cases, the intrinsic barriers to the reaction do not change. In that case, one can use equation (5.41) to predict the barriers to surface reactions. One then follows the procedures in Section 5.4.6 to predict the mechanism of the reaction.

There are two major changes when we have reactions on metal surfaces. There is a proximity effect that bonds only in close proximity to the surface can break, and there is a role of d electrons, which can weaken bonds and thereby make them easier to break.

The proximity effect is discussed in Section 10.11 of Masel (1996). The basic idea is that bonds can break only when the bonds are in close proximity to the surface. For example, Figure 5.16 shows the bond energies and geometry of methanol adsorbed on platinum. Notice that the CH bond is weaker than the OH bond. Therefore, on the basis of thermodynamics, one would expect the C–H bond to be easier to break than the O–H bond. However, experimentally, the OH bond is easier to break than the CH bond. Lee and Masel [1996] analyzed this case and found that the issue is what Masel [1996] called a proximity effect. Figure 5.17 shows a diagram of the transition state for C–H scission bond. During reaction, the carbon and hydrogen both need to form a bond to the platinum. There is no problem forming a platinum–hydrogen bond, but it is hard to form a platinum–carbon bond because the hydrogen gets in the way. Masel [1996] calls

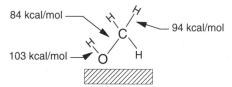

84 kcal/mol —

94 kcal/mol

103 kcal/mol →

Figure 5.16   Bond energies in methanol.

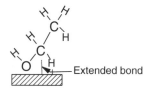

—Extended bond

Figure 5.17   The transition state for C–H scission in adsorbed ethanol.

this a proximity effect because only bonds close to the surface can break. Generally, the proximity effect adds about 10–20 kcal/mol to intrinsic barriers, so it is quite important.

5.12.2   A Specific Example: Ethanol Decomposition of Platinum

As an example of the analysis, in this section, we will try to predict the mechanism of ethanol decomposition on platinum. We will assume that the intrinsic barrier for the scission of a O–H or C–H bond is about 15 kcal/mol. The intrinsic barrier for the scission of an C–C or C–O bond is about 45 kcal/mol as described in Section 5.4.6.

In principle, ethanol could decompose by a host of different reaction pathways. The ethanol can sequentially dehydrogenate. The C–O bond could break before any other reaction occurs. Now, let's ask which reaction is most favored if the intrinsic barrier for the scission of a C–C or O–C bond is 45 kcal/mol while the intrinsic barrier to the scission of a C–H or O–H bond is 15 kcal/mol.

Figure 5.18 shows the bond energies in ethanol. Notice that the C–C bond is the weakest bond in the molecule. Therefore, if all of the intrinsic barriers to bond scission were equal, one would expect the C–C bond in ethanol to be easier to break than the C–H, O–H and C–O bonds. However, if one considers the difference in intrinsic barrier, one would come to a very different conclusion. Notice that the intrinsic barrier for the scission of a C–C or O–C bond is 30 kcal/mol higher than the intrinsic barrier of the scission of a C–H or O–H bond. Consequently, the C–O bond would be harder to break

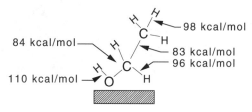

98 kcal/mol

84 kcal/mol—

83 kcal/mol
96 kcal/mol

110 kcal/mol →

Figure 5.18   Bond energy in ethanol.

than either the O–H or C–H bond. Therefore, both the C–H and O–H bond should break at lower temperatures than the C–C bond.

Now consider whether the O–H or C–H bond should be easier to break. Notice that the C–H bond is weaker than the O–H bond. Therefore, on the basis of bond energies alone, the C–H bond should break more easily than the O–H bond. However, if one looks at the reaction in detail, one finds that there is a proximity effect that substantially raises the intrinsic barrier for C–H bond scission. During reaction, carbon–surface and hydrogen–surface bonds form. Figure 5.17 shows a possible transition state for the reaction. Notice that it is hard to form a carbon-surface bond because the carbon-hydrogen bonds get in the way. One does not know how much energy this costs without doing detailed calculations. However, Masel and Lee [1996] did quantum-mechanical calculations that suggest that these repulsions raise the intrinsic barriers by 24 kcal/mol.

One can do the same analysis on the O–H bond scission step. In the case of O–H bond scission, there are some extra intrinsic barriers due to the repulsions of the empty dangling bonds on the ethanol. However, those effects would be expected to be much smaller than the extra 24 kcal/mol repulsion if the C–H bond breaks. Therefore, on the basis of an analysis of the intrinsic barriers, one would expect O–H bond scission to have a lower activation barrier than C–H, C–C, or C–O bond scission, even though the O–H bond is the strongest bond in ethanol. Therefore, we conclude that, per an analysis of the intrinsic barriers, the first step in ethanol decomposition should be O–H bond scission to yield an ethoxy intermediate, as shown in Figure 5.15. That is what has been observed experimentally on all of the transition face metals that have been examined previously except Pt(110)(1 × 2) and Pt(331), where the notation Pt(110) (2 × 1) and (311) refers to platinum catalysts with surface arrangements called (110) (2 × 1) or (311) as shown in Figure 5.19.

One can continue the analysis to predict a mechanism of ethoxy decomposition on metals. Again, we note that the C–C bond in ethoxy is 10 kcal/mol weaker than the C–H bond. Therefore, if one considered bond energies only, one would expect the C–C bond in ethoxy to break before the C–H bond. However, the intrinsic barriers go in the opposite directions. The intrinsic barriers for C–H bond scission are much lower than the intrinsic barriers for C–C bond scission. If one plugs numbers into equation (5.41), one finds that

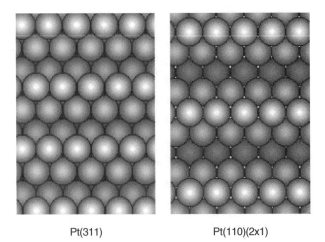

Pt(311)   Pt(110)(2x1)

Figure 5.19 The arrangement of the platinum atoms in Pt(311) and Pt(110) (1 × 2).

the overall activation barrier for ethoxy dehydration should be 10–20 kcal/mol lower than the activation barrier for C–C bond scission in ethoxy. Consequently, the ethoxy should sequentially dehydrogenate to produce an acetaldehyde and an acetyl intermediate as indicated in Figure 5.15.

The analysis changes after the acetyl intermediate forms. In Chapter 5, we will provide some methods to estimate heats of reactions. If one uses the methods from Chapter 6, one finds that on platinum, the reaction

$$CH_3CO_{ad} \longrightarrow CH_{3ad} + CO_{ad} \tag{5.159}$$

is favored by about 44 kcal/mol over the reaction

$$CH_3CO_{ad} \longrightarrow H_{ad} + CH_2CO_{ad} \tag{5.160}$$

Reaction (5.160) has a 20–40 kcal/mol lower intrinsic barrier than does reaction (5.159), but that lower intrinsic barrier is overcome by the 44-kcal/mol difference in ΔH_r. Therefore, according to equation (5.41), reaction (5.159) should have a lower activation barrier than reaction (5.160). Consequently, one would expect the acetyl intermediate to decompose via reaction (5.159) rather than reaction (5.160). Experimentally, one finds that ethoxy decomposition follows the mechanism in Figure 5.15 in all of the cases examined so far except in the case of ethoxy decomposition on Rh(111).

The point of this exercise is that one can make useful predictions with very few numbers: rough estimates of the heats of adsorption and intrinsic barriers. In other work, Lee, and Masel (1997) have suggested that one can successfully predict the mechanism of ethylene, methanol, ethanol, acetaldehyde, methyl iodide, methylamine, and ethyl iodide decomposition on most of the loose-packed faces of transition metals examined to date using the same simple analysis!

Now, I would like to say that this procedure will always work, but unfortunately, I cannot. For example, Cong et al. (1998) found that on a stepped surface, such as that of Pt(110) (1 × 2) shown in Figure 5.19, ethanol decomposes by one of two pathways: the pathway shown in Figure 5.15 or, a pathway where the C–C bond breaks at low temperature. This second pathway is not expected from the materials in this section. Brown and Barteau [1996] also found an exception during ethanol decomposition on Rh(100). These exceptions are currently being actively investigated but, are not well understood at present.

In my experience, one can guess at a mechanism of a surface reaction and be right 95% of the time. However, there are occasional exceptions and the exceptions are not understood.

5.13 GENERIC TYPES OF SURFACE REACTIONS

There is one other detail of importance to the prediction of mechanism of surface reactants. Some of the species that participate in the reaction might not be strongly attached to the surface. People in the literature often discuss the participation of gas-phase species, and weakly bound complexes on surface reactions.

There are three generic types of surface reaction: **Langmuir–Hinshelwood** mechanisms, **Rideal–Eley** mechanisms, and **precursor** mechanisms. Figure 5.20 shows a schematic of these three mechanisms for a hypothetical reaction A + B → AB. In the

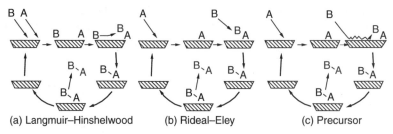

(a) Langmuir–Hinshelwood   (b) Rideal–Eley   (c) Precursor

Figure 5.20 Schematic of (a) Langmuir–Hinshelwood, (b) Rideal–Eley, and (c) precursor mechanisms for the reaction $A + B \Rightarrow AB$ and $AB \Rightarrow A + B$.

Langmuir–Hinshelwood mechanism, A and B first adsorb onto the surface of the catalyst. Next, the adsorbed A and B react to form an adsorbed A–B complex. Finally, the A–B complex then desorbs. In what's now called the *Rideal–Eley* mechanism,[1] the reactant A chemisorbs. The A then reacts with an incoming B molecule to form an A–B complex. The A–B complex then desorbs. In the precursor mechanism, A adsorbs. Next, B collides with the surface, and enters a mobile precursor state. The precursor rebounds along the surface until it encounters an adsorbed A molecule. The precursor then reacts with the A to form an A–B complex which desorbs.

Each of these reactions in Figure 5.20 can also occur in reverse as shown in Figure 5.21. For example, one can run the Langmuir–Hinshelwood reaction in reverse by adsorbing an A–B molecule, heating to allow the A–B to decompose into adsorbed A and B, and then desorbing the A and B. Alternatively, if the adsorbed A–B molecule

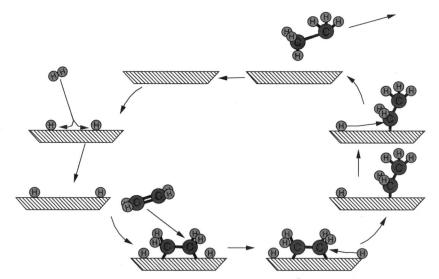

Figure 5.21 A Langmuir–Hinshelwood mechanism for the reaction $C_2H_4 + H_2 \Rightarrow C_2H_6$.

[1] In their original papers, Rideal and Eley [1940, 1941] did not distinguish between what is now called a *Rideal–Eley* mechanism and a precursor mechanism. However, more recent workers now make a distinction. See Weinberg [1992] for details.

decomposes to yield an adsorbed A molecule and a gas-phase species B, one would have the reverse of the Rideal–Eley mechanism. If a precursor forms on the way to the products, then one would have the precursor mechanism.

In the literature, the precursor mechanism is sometimes referred to as the **trapping-mediated** mechanism. Similarly, the Rideal–Eley mechanism is sometimes referred to as the **Eley–Rideal** mechanism.

All of the surface reactions that have been studied so far can be viewed as following a Langmuir–Hinshelwood mechanism, a Rideal–Eley mechanism, a precursor mechanism, or some combination of these three mechanisms.

For example, Figure 5.21 shows the mechanism of ethylene hydrogenation seen at very low pressure on a platinum catalyst. First, the hydrogen adsorbs and dissociates. Then, the ethylene adsorbs. Then, one adds one hydrogen at a time to the adsorbed ethylene to yield products. Notice that in the mechanism in Figure 5.21, the reactants adsorb and then react. Therefore, the mechanism in Figure 5.21 is a classic Langmuir–Hinshelwood mechanism. Langmuir–Hinshelwood mechanisms are quite common in the literature. Most catalytic reactions go via Langmuir–Hinshelwood mechanisms.

Reactions used to deposit films on surfaces, on the other hand, are thought to often go (proceed) via Rideal–Eley mechanism. For example, one grows CVD (chemical vapor deposition) diamonds by heating methane to produce CH_3 groups in the gas phase. The CH_3 groups then react with a cool surface to produce a diamond film. Harris (1990, 1993) examined the mechanism theoretically and proposed the mechanism shown in Figure 5.22. First, the methyl groups adsorb on the surface, forming a hydrocarbon layer. The CH_3 groups from the gas phase react with the hydrocarbon layer to form a hydrogen-terminated diamond surface. Finally, other CH_3 groups react with the layer to remove excess hydrogen (hydrogen also desorbs). Notice that each of these reactions follow the general reaction scheme

$$CH_3 + A_s \longrightarrow \text{products} \tag{5.161}$$

where a gas-phase species, $CH_{3,g}$, reacts with a surface intermediate, A_s, to form products. Hence, classically, this is a Rideal–Eley mechanism.

In my experience, the key difference between catalytic reactions and film growth reactions is that in catalysis, one usually runs the reaction at low temperatures. When the temperatures are low, the reactants do not dissociate until the reactants adsorb onto the surface. In that case, reactions happen only after all of the species adsorb. Only Langmuir–Hinshelwood reactions are seen. On the other hand, film growth reactions often

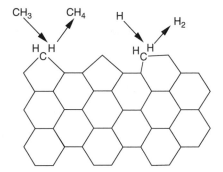

Figure 5.22 A Rideal–Eley mechanism for diamond deposition.

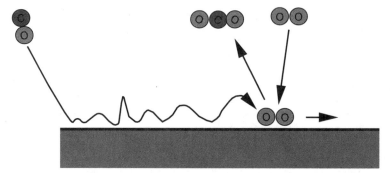

Figure 5.23   A precursor mechanism for the reaction $CO + O_2 \rightarrow CO_2 + O_{ad}$.

run at much higher temperatures. Radicals can form in the gas phase. When a radical hits a surface, the radical can react before it adsorbs. One can observe Rideal–Eley mechanisms of radicals.

I do not have a good example of a precursor mechanism. People cannot actually observe precursors because the precursors remain on the surface for perhaps only 10^{-6} seconds. Consequently, one can never tell for sure whether a reaction goes via a precursor. There is some experimental evidence that the reaction

$$2CO + O_2 \Longrightarrow 2CO_2 \tag{5.162}$$

can go via the precursor mechanism shown in Figure 5.23. First the oxygen adsorbs and dissociates. Then a CO collides with the surface and wanders around until the CO finds an adsorbed oxygen atom. Then the two species react to form a CO_2 molecule, which desorbs.

If one glances at Figure 5.23, it is not immediately obvious that this is a precursor mechanism and not a Langmuir–Hinshelwood mechanism. After all, the CO and O_2 both enter the surface phase and then react. However, people say that this is a precursor mechanism because the CO reacts with the oxygen before the CO has had time to form a strong bond with the surface. Of course, at this point, no one has directly observed the precursor, so no one knows for sure whether the CO forms a bond to the surface before it reacts. In my view, the best available evidence is that the CO bonds before the CO reacts. However, there are others who disagree.

In my experience, most reactions that were originally thought to proceed via a precursor mechanism are now thought to go by other mechanisms. Still, people discuss precursor mechanisms at length in the literature. Therefore, I thought that it would be important to mention them.

5.14   REACTIONS ON ACID SURFACES

So far we have been talking about reactions on metal surfaces. Next, we want to move on and briefly mention reactions on solid acids. Students usually think about acids as being liquids, but there is a whole class of solid acids. For example, one can put protons into an ion exchange resin to create a solid acid. In the analysis and detergent industries, one uses special silica aluminates, specifically mixtures of SiO_2 and alumina

(Al$_2$O$_3$) called *zeolites*. The zeolites are highly porous structures whose surfaces have been specially treated to create Lewis or Brønsted acid sites. It is possible to make a zeolite with a pK$_a$ of -10. Recall that by comparison, 1 N H$_2$SO$_4$ has a pK$_a$ of 1 and water has a pK$_a$ of 7.

Reactions on solid acids are a cross between reactions on metal surfaces and reactions in acid solution. As in a metal, the reactions occur at distinct sites on the acid's surface and as in a metal, the reaction usually follows a Langmuir–Hinshelwood rate law. Also as in a metal, the reactions go via a catalytic cycle where protons are consumed and regenerated. Still, on a solid acid, most of the key intermediates are ions. The reaction pathways look more like reactions in solution than reactions on metal surfaces.

In particular, reactions on solid acids usually follow the mechanisms outlined in Sections 5.4 and 5.4.1. Olefins, paraffins, and alcohols react with protons in the acid to yield carbenium or carbonium ions. The carbenium and carbonium ions react as described in Sections 5.4 and 5.4.1. One observes the same cracking, isomerization, and alkylation reactions in a solid acid that one observes in the gas phase or in a superacid solution. The only difference is that there are mass transfer limitations that limit the products produced when the reacting molecules are big.

Solid acids are widely used as catalysts. They are also used as ion traps, for example, in detergents and other cleaners.

The analysis of mechanisms of reactions in solid acids is very similar to the analysis of reactions of ions in the gas phase. Martens and Jacobs (1990) suggest that all of the ideas in Sections 5.4 and 5.4.1 apply to solid acids, and indeed the results in Table 5.8 show that reactions in solid acids behave very similarly to ionic reactions in the gas phase. Generally, one can predict the qualitative behavior of solid acid catalysts using the material in Sections 5.4 and 5.4.1 (see Martens and Jacobs (1990) for details).

5.15 SUMMARY

In summary, then, in this chapter we reviewed the mechanisms of simple reactions in the gas phase and in various environments. We found that there are basically two types of reaction pathways: radical pathways and ionic pathways. The radical pathways usually consist of an initiation step where radicals are formed, a series of propagation steps where products are produced in a catalytic cycle, and a series of termination steps where radicals recombine. We found that one can often predict the reaction pathway using the Polanyi relationship, some knowledge of intrinsic barriers, and a few empirical rules. In particular, the intrinsic barriers for atom transfer reactions are usually 30 kcal/mol lower than the intrinsic barriers for transfer of molecular ligands. As a result, atomic transfer reactions usually dominate in the gas phase.

Ionic reactions are more complicated in that there are no empirical rules for prediction of reaction mechanisms. Generally, ionic pathways consist of a series of additions, isomerizations, and bond scissions. In the gas phase all of the processes have low intrinsic barriers, which means that the reactions quickly reach equilibrium. However, if one runs the reaction in a solvent, the rearrangements of the solvent cage determines the intrinsic barrier. The forces due to the rearrangements of the solvent cage are still not well understood, which makes prediction of reaction pathways more difficult.

5.16 SOLVED EXAMPLES

Example 5.A The reaction $CH_3CH_3 \Rightarrow H_2C = CH_2 + H_2$ obeys the following mechanism:

$$CH_3CH_3 + X \xrightarrow{1} 2 \cdot CH_3 \cdot + X$$

$$\cdot CH_3 + CH_3CH_3 \xrightarrow{2} CH_4 + \cdot CH_2CH_3$$

$$\cdot CH_2 + CH_3X \xrightarrow{3} CH_2CH_2 + H \cdot + H$$

$$2CH_3 \cdot + X \xrightarrow{5} CH_3CH_3$$

(+ other reactions)

a) Make a diagram of the reaction similar to that in Figure 5.3
b) Identify the initiation step, the transfer step, the propagation steps, the termination steps. Also indicate whether the propagation steps are associations, β-hydrogen transfer, etc.
c) Does the mechanism obey the rules in the Section 5.4.7
d) Estimate the activation barrier for each step.
e) Calculate the temperature where equation 5.35 is satisfied.

Solution

a)

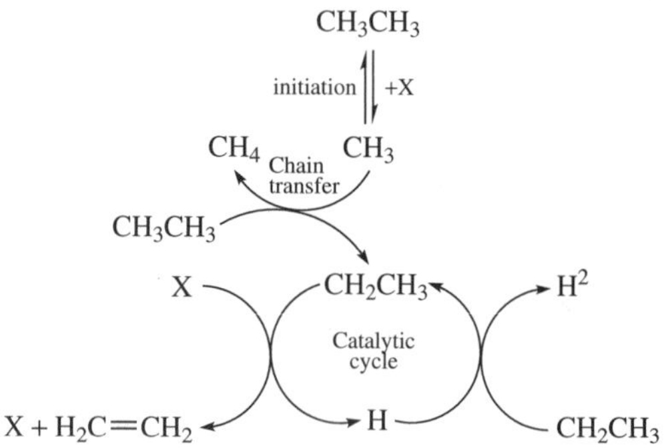

b) Step 1 — initiation
   Step 2 — chain transfer
   Step 3 — propagation (β-hydrogen elimination)
   Step 4 — propagation (hydrogen transfer)
   Step 5 — termination
c) Does follow rules
   There is an initiation step (step 1)
   There is a catalytic cycle (step 3 and 4)
   There is a termination step (step 5)
   We will verify the activation barriers below

d) Estimate the activation barriers

First consider: $CH_3CH_3 + X \rightarrow 2CH_3\bullet + H$

Step 1: estimate ΔH_r

    From NIST Web book (*http://webbook.nist.gov*)

$$\Delta H_r(CH_3CH_3) = -20.0$$

Therefore

$$\Delta H_r(CH_3\bullet) = +34.8 \text{ kcal/mol}$$

$$\Delta H_r = 2(34.8) - (-20.0) = 89.6 \text{ kcal/mol}$$

Step 2: estimate E_A using Table 5.4. This is a simple bond scission reaction. From Table 5.4

$$E_A = 1 + \Delta H_r = 90.6 \text{ kcal/mol}$$

$$CH_3\bullet + CH_3CH_3 \longrightarrow CH_4 + \bullet CH_2CH_3$$

Step 1: estimate ΔH_r

    From the NIST web book

$$\Delta H_f(CH_3CH_3) = -20.0 \text{ kcal/mol}$$

$$\Delta H_f(CH_3\bullet) = +34.8 \text{ kcal/mol}$$

$$\Delta H_f(CH_4) = -17.9 \text{ kcal/mol}$$

$$\Delta H_f(\bullet CH_2CH_3) = +28.4 \text{ kcal/mol}$$

Therefore

$$\Delta H_r = 17.9 + 28.4 - 34.8 - (-20.0) = -4.3 \text{ kcal/mol}$$

Step 2: estimate E_A using Table 5.4. This is an atom transfer reaction. From Table 5.4

$$E_A = 10 \text{ kcal/mol} + 0.3(-4.3) = 8.7 \text{ kcal/mol}$$

$$CH_2CH_3 + X \longrightarrow CH_3CH_2 + H + X$$

Step 1: estimate ΔH_r

    From the NIST web book

$$\Delta H_f(CH_2CH_3) = +28.4 \text{ kcal/mol}$$

$$\Delta H_f(CH_2CH_2) = +12.5 \text{ kcal/mol}$$

$$\Delta H_f(H) = +52.1 \text{ kcal/mol}$$

$$\Delta H = 52.1 + 12.5 - 28.4 = +36.2 \text{ kcal/mol}$$

Step 2: estimate E_A using Table 5.4. This is as endothermic β-scission reaction. From Table 5.4

$$E_A = 15 + 0.7(36.2) = 40.3 \text{ kcal/mol}$$

$$H\bullet + CH_3CH_3 \longrightarrow H_2 + \bullet CH_2CH_3$$

Step 1: estimate ΔH_r

From the NIST web book

$$\Delta H_f(H) = +52.1 \text{ kcal/mol}$$
$$\Delta H_f(CH_3CH_3) = -20.0 \text{ kcal/mol}$$
$$\Delta H_f(CH_2CH_3) = +28.4 \text{ kcal/mol}$$
$$\Delta H_2 = 0 \text{ kcal/mol}$$

Therefore

$$\Delta H_r = 0 + 28.4 - 52.1(-20.0) = -3.7 \text{ kcal/mol}$$

Step 2: estimate H_r using Table 5.4. This is a hydrogen transfer reaction. From Table 5.4

$$E_A = 10 + 0.3(-3.7) = 5.9 \text{ kcal/mol}$$

$$\bullet CH_3 + \bullet CH_3 + X \longrightarrow CH_3CH_3 + X$$

Step 1: estimate ΔH_r

$$\Delta H_r = -89.6 \text{ kcal/mol (reverse reaction 1)}$$

Step 2: estimate H_r using Table 5.4. This is a recombination reaction. From Table 5.4

$$E_A = 1 \text{ kcal/mol}$$

e) Calculate temperature such that

$$E_A \leq 0.15 \frac{\text{kcal}}{\text{mol}^\circ \text{ K}} \text{ for initiation}$$

$$E_A \leq 0.07 \frac{\text{kcal}}{\text{mol}^\circ \text{ K}} \text{ for all propagation}$$

$$E_A \leq 0.05 \frac{\text{kcal}}{\text{mol}^\circ \text{ K}} \text{ for catalytic cycle}$$

For initiation, $E_A = 90.6 \text{ kcal/mol}$

$$T \geq 90.6/0.15 = 604 \text{ K}$$

For propagation.

$$T \geq 40.3/0.05 = 806 \text{ K}$$

Therefore any temperature above 806 K will satisfy all constraints.

Example 5.B   Consider the following alternate mechanism for ethylene production from ethane

$$CH_3CH_3 + X \xrightarrow{\;1\;} 2CH_3\bullet$$

$$CH_3\bullet + CH_3CH_3 \xrightarrow{\;2\;} CH_4 + \bullet CH_2CH_3$$

$$\bullet CH_2 + CH_3X \xrightarrow{\;3\;} H_2C = CH_2 + H\bullet + X$$

$$H\bullet + CH_3CH_3 \xrightarrow{\;6\;} CH_4 + \bullet CH_3$$

$$2CH_3\bullet + X \xrightarrow{\;5\;} CH_3CH_3 + X$$

a) Does this mechanism follow all of the rules at 810 K?

b) Is this mechanism more or less likely than the mechanism in Example 5.A?

Solution

a) This does follow the rules!
   1) There is an initiation step (step 1)
   2) There is a catalytic cycle (steps 2,3,6)
   3) There is a termination step (step 5)
   Check all steps obey constraint in equation 5.36 Steps 1,2,3,5 do (see Example 5.A)
   Check step 6 $H\bullet + CH_3CH_3 \rightarrow CH_4 + \bullet CH_3$

Step 1: estimate ΔH_r
   From the Nist webbook

$$\Delta H_f(H\bullet) = +52.1 \text{ kcal/mol}$$
$$\Delta H_f(CH_3CH_3) = -20.0 \text{ kcal/mol}$$
$$\Delta H_f(CH_4) = -17.9 \text{ kcal/mol}$$
$$\Delta H_f(\bullet CH_3) = +34.8$$
$$\Delta H_r = 34.8 + (-17.9) - 52.1 - (20.0) = -15.2 \text{ kcal/mol}$$

Step 2: estimate E_A using Table 5.4. This is a ligand transfer reaction to hydrogen. From Table 5.4
$$E_A = 45.0 + 0.5 + (-15.2) = 37.4 \text{ kcal/mol}$$

This reaction is in the catalytic cycle

$$E_A \le \left(0.05 \frac{\text{kcal}}{\text{mol}} \right) T = (0.05)(810 \text{ K}) = 40.5 \frac{\text{kcal}}{\text{mol}}$$

   Therefore all constraints are satisfied

b) Which mechanism is better?

The difference between the two mechanisms is that mechanism 5.A assume H reacts like reaction 4, while mechanism 5.B assumes H reacts like reaction 6.

$$H\bullet + CH_3CH_3 \xrightarrow{\ 4\ } H_2 + \bullet CH_2CH_3$$

$$H\bullet + CH_3CH_3 \xrightarrow{\ 6\ } CH_4 + \bullet CH_3$$

According to the analysis above, reaction 4 has an activation barrier of 8.9 kcal/mol while reaction 6 has an activation barrier of 37.4 kcal/mol, so clearly reaction 4 will dominate.

Note however, that both reactions satisfy equation 5.36 at 810 K. Consequently both reactions will occur at 810 K. Reaction 4 will just be much faster than reaction 6.

Example 5.C: E_a for other reactions in this chapter Estimate the activation barriers for reactions 5.73 to 5.83 and 5.98

Solution

Reaction 5.73
Step 1: estimate ΔH_r
  From the NIST webbook

$$\Delta H_f(OCH\bullet) = +10.4 \text{ kcal/mol}$$

$$\Delta H_f(CH_3CH_3) = -20.0 \text{ kcal/mol}$$

$$\Delta H_f(CH_3CHO) = -40.8 \text{ kcal/mol}$$

$$\Delta H_f(\bullet CH_3) = +34.8 \text{ kcal/mol}$$

$$\Delta H_r = 10.4 + (-20) - (-40.8) - (+34.8) = -3.6 \text{ kcal/mol}$$

Step 2: estimate E_A using Table 5.4. This is a ligand transfer reaction. From Table 5.4

$$E_a = 50 \text{ kcal/mol} + 0.5^*(-3.6 \text{ kcal/mol}) = 48.2 \text{ kcal/mol}$$

Reaction 5.74
Step 1: estimate ΔH_r
  From the NIST webbook

$$\Delta H_f(CO) = -26.4 \text{ kcal/mol}$$

$$\Delta H_f(CH_3CO) = +2.9 \text{ kcal/mol}$$

$$\Delta H_f(\bullet CH_3) = +34.8 \text{ kcal/mol}$$

$$\Delta H_r = 34.8 + (-26.4) - (+2.9) = +5.5 \text{ kcal/mol}$$

Step 2: estimate E_A using Table 5.4. This is an initiation reaction. From Table 5.4

$$E_a = 1 \text{ kcal/mol} + (5.5 \text{ kcal/mol}) = 6.5 \text{ kcal/mol}$$

Reaction 5.75

Step 1: estimate ΔH_r

From the NIST webbook

$$\Delta H_f(CH_2CO) = -27.7 \text{ kcal/mol}$$
$$\Delta H_f(CH_3CO) = +2.9 \text{ kcal/mol}$$
$$\Delta H_f(OCH\bullet) = +10.4 \text{ kcal/mol}$$
$$\Delta H_f(CH_3COH) = -40.8 \text{ kcal/mol}$$
$$\Delta H_r = (-27.7) + (2.9) - 10.4 - (-40.8) = 5.6 \text{ kcal/mol}$$

Step 2: estimate E_A using Table 5.4. This is an initiation reaction. From Table 5.4

$$E_a = 12 \text{ kcal/mol} + 0.7^*(5.6 \text{ kcal/mol}) = 15.9 \text{ kcal/mol}$$

Reaction 5.76

Step 1: estimate ΔH_r

From the NIST webbook

$$\Delta H_f(OCH\bullet) = +10.4 \text{ kcal/mol}$$
$$\Delta H_f(CO) = -26.4 \text{ kcal/mol}$$
$$\Delta H_f(\bullet H) = +52.1 \text{ kcal/mol}$$
$$\Delta H_r = (52.1) + (26.4) - 10.4 = 15.3 \text{ kcal/mol}$$

Step 2: estimate E_A using Table 5.4. This is an initiation reaction. From Table 5.4

$$E_a = 1 \text{ kcal/mol} + 1.0^*(15.3 \text{ kcal/mol}) = 16.3 \text{ kcal/mol}$$

Reaction 5.77

Step 1: estimate ΔH_r

From the NIST webbook

$$\Delta H_f(CH_3CO) = +2.9 \text{ kcal/mol}$$
$$\Delta H_f(H_2) = 0 \text{ kcal/mol}$$
$$\Delta H_f(\bullet H) = +52.1 \text{ kcal/mol}$$
$$\Delta H_f(CH_3CHO) = -40.8 \text{ kcal/mol}$$
$$\Delta H_r = (2.9) + (0) - 52.1 - (-40.8) = -8.4 \text{ kcal/mol}$$

Step 2: estimate E_A using Table 5.4. This is a hydrogen transfer reaction. From Table 5.4

$$E_a = 12 \text{ kcal/mol} + 0.3^*(-8.4 \text{ kcal/mol}) = 9.5 \text{ kcal/mol}$$

Reaction 5.78

Step 1: estimate ΔH_r

From the NIST webbook

$$\Delta H_f(HCO\bullet) = +10.4 \text{ kcal/mol}$$
$$\Delta H_f(CH_4) = -17.9 \text{ kcal/mol}$$
$$\Delta H_f(\bullet H) = +52.1 \text{ kcal/mol}$$
$$\Delta H_f(CH_3CHO) = -40.8 \text{ kcal/mol}$$
$$\Delta H_r = (10.4) + (-17.9) - 52.1 - (-40.8) = -18.8 \text{ kcal/mol}$$

Step 2: estimate E_A using Table 5.4. This is a ligand transfer reaction. From Table 5.4

$$E_a = 45 \text{ kcal/mol} + 0.5^*(-18.8 \text{ kcal/mol})$$
$$= 35.6 \text{ kcal/mol} \ (\text{importantabove} 35.6/0.07 = 510 \text{ K})$$

Reaction 5.79

Step 1: estimate ΔH_r

From the NIST webbook

$$\Delta H_f(CH_3CHO) = -40.8 \text{ kcal/mol}$$
$$\Delta H_f(HCO\bullet) = +10.4 \text{ kcal/mol}$$
$$\Delta H_f(\bullet CH_3) = +35.1 \text{ kcal/mol}$$
$$\Delta H_r = (-40.8) - (+10.4) - 35.1 = -86.0 \text{ kcal/mol}$$

Step 2: estimate E_a using Table 5.4. This is a recombination reaction. From Table 5.4

$$E_a = 1 \text{ kcal/mol}$$

Reaction 5.80

Step 1: estimate ΔH_r

From the NIST webbook

$$\Delta H_f(CH_4) = -17.9 \text{ kcal/mol}$$
$$\Delta H_f(CO) = -26.4 \text{ kcal/mol}$$
$$\Delta H_f(HCO\bullet) = +10.4 \text{ kcal/mol}$$
$$\Delta H_f(\bullet CH_3) = +35.1 \text{ kcal/mol}$$
$$\Delta H_r = (-40.8) - (+10.4) - 35.1 = -86.0 \text{ kcal/mol}$$

Step 2: estimate E_a using Table 5.4. This is a hydrogen reaction. From Table 5.4

$$E_a = 0 \text{ kcal/mol}$$

Reaction 5.81 and 5.82

Both are simple recombination reactions. Therefore $E_a = 1$ kcal/mol

Reaction 5.83

Step 1: estimate ΔH_r

From the NIST webbook

$$\Delta H_r = (-27.7) + (-26.4) - 2(+10.4) = -84.9 \text{ kcal/mol}$$

Step 2: estimate E_A using Table 5.4. This is a hydrogen reaction. From Table 5.4

$$E_a = 0 \text{ kcal/mol}$$

Reaction 5.98

Step 1: estimate ΔH_r

From the NIST webbook

$$\Delta H_f(CH_3CH_3) = -20.0 \text{ kcal/mol}$$

$$\Delta H_f(\bullet H) = +52.1 \text{ kcal/mol}$$

$$\Delta H_f(CH_4) = -17.9 \text{ kcal/mol}$$

$$\Delta H_f(\bullet CH_3) = +35.1 \text{ kcal/mol}$$

$$\Delta H_r = (-20.0) + (52.1) - (+35.1) - (-17.9) = 15.1 \text{ kcal/mol}$$

Step 2: estimate E_a using Table 5.4. This is a ligand transfer reaction from hydrogen. From Table 5.4

$$E_a = 45 \text{ kcal/mol} + 0.5^*15.1 \text{ kcal/mol} = 52.5 \text{ kcal/mol}.$$

5.17 SUGGESTIONS FOR FURTHER READING

Good general references for organic reaction mechanisms include:

J. March, *Advanced Organic Chemistry: Reactions, Mechanisms, and Structure*, 4th ed., Wiley, New York, 1992.

R. B. Grossman, *The Art of Writing Reasonable Organic Reaction Mechanisms*, Springer-Verlag, Berlin, 1998.

A. Miller, *Writing Reaction Mechanisms in Organic Chemistry*, Academic Press, San Diego CA, 1999.

Many of the concepts in this chapter are extensions of the ideas discussed in:

N. N Semenov, *Chemical Kinetics and Chain Reactions*. Oxford, University Press, Oxford, UK, (1935).

Good discussions of radical reactions are found in:

Z. B. Alfassi, ed., *General Aspects of the Chemistry of Radicals*, Wiley, Chichester, 1999.

J. M. Perkins, *Radical Chemistry*, Ellis Horwood, New York, 1994.

J. Fossey, D. Lefort, and J Sorba, *Free Radicals in Organic Chemistry*, Wiley, New York, 1995.

J. E. Leffler, *An Introduction to Free Radicals*, Wiley, New York, 1993.

Ionic reactions are reviewed in:

H. Pines, *The Chemistry of Catalytic Hydrocarbon Conversions*, Academic Press, San Diego, 1991.

The quantitative prediction of mechanisms is discussed in:

L. J Broadbelt, S. M. Stark, and M. T. Klein, Computer generated pyrolysis modeling-on-the-fly generation of species, reactions, and rates. *Ind. Eng. Chem. Res.* **33**; 790–799 (1994).

Applications on surfaces are discussed in:

R. I. Masel, *Principles of Adsorption and Reaction on Solid Surfaces*, Chapter 10, Wiley, New York, 1996.

E. Shustorovich, ed., *Metal-Surface Reaction Energetics*: *Theory and Applications to Heterogeneous Catalysis, Chemisorption, and Surface Diffusion*, VCH, New York, 1991.

5.18 PROBLEMS

5.1 Define the following terms in three sentences or less:

(a) Mechanism	**(l)** Pauli repulsion
(b) Initiation–propagation mechanism	**(m)** Isomerization
(c) Initiation step	**(n)** Cracking
(d) Propagation step	**(o)** Alkylation
(e) Termination step	**(p)** Three-centered bond
(f) Catalytic cycle	**(q)** Carbenium ion
(g) Polanyi relationship	**(r)** Carbonium ion
(h) Intrinsic activation barrier	**(s)** Carbocation
(i) Transfer coefficient	**(t)** S_N1 reaction
(j) Four-centered transition state	**(u)** S_N2 reaction
(k) Lindemann mechanism	

5.2 What did you learn new in this chapter? Had you known about initiation–propagation mechanisms? Had you realized that the mechanism of a reaction does not include all of the reactions that occur? Had you known that you could predict mechanisms before?

5.3 Describe in your own words the key differences between the reactions of radicals and the reactions of ions. What types of reaction occur via radical mechanisms? What types of reaction occur via ionic mechanisms? Why is there such a difference between the reactions of radicals and the reactions of ions? If you were designing a catalyst, when would you want the catalyst to promote radical reactions, and when would you want the catalyst to promote ionic reactions?

5.4 Describe in your own words why gas-phase molecules follow such complex reaction pathways.

(a) Why do initiation–propagation mechanisms dominate over direct reactions between the reactants?

(b) Why do you need catalytic cycles?

(c) Why do transfer reactions occur?

(d) Why do you need an initiation step?

(e) Why are termination steps important?

5.5 Look at each of the mechanisms in Table 5.1. The objective of this problem is to show that each mechanism is an initiation–propagation mechanism.

(a) Find the initiation reaction for each mechanism.

(b) Identify the propagation steps.

(c) Identify the termination steps.

(d) Diagram the catalytic cycle.

(e) Are there any transfer steps?

(f) Is the initiation step the one that you would expect?

(g) Would you expect any other reactions? If so, identify them.

5.6 The objective of this problem is to examine the ion reactions in Table 5.6.

(a) Diagram the catalytic cycle for each reaction.

(b) Identify the initiation steps. (*Hint*: Not all of the initiation processes are included in the mechanisms in the table.)

(c) Make a rough diagram of the orbitals for each step.

(d) Which reactions would be expected to have significant activation barriers in the gas phase?

(e) How would the barriers change in water solution?

5.7 Look carefully at the catalytic cycle in Figure 5.2. Which reactions are ionic reactions? Which reactions are radical reactions? Are there any reactions that do not fall into either general category?

5.8 In Chapter 4, we described a mechanism for the oxidation of hydrogen. The objective of this problem is to show that the mechanism is an initiation–propagation mechanism.

(a) Find the initiation reaction for each mechanism.

(b) Identify the propagation steps.

(c) Identify the termination steps.

(d) Diagram the catalytic cycle.

(e) Are there any transfer steps?

(f) Is the initiation step the one that you would expect?

(g) Would you expect any other reactions? If so, identify them.

5.9 Westley, *et al. Tables of Recommended Rate Constants for Reactions in Combustion*, NBS (1980), lists the measured rate constants for a number of combustion reactions. Table P5.9 lists some of his data:

(a) Calculate the intrinsic barriers for each of the reactions, assuming that $\gamma_P = 0.5$.

(b) Calculate the intrinsic barriers for each of the reactions, assuming that $\gamma_P = 0.3$ for exothermic reactions and $\gamma_P = 0.7$ for endothermic reactions.

(c) How do your results compare to the expectations from Table 5.3?

5.10 Thon, (1926), examined the reaction between H_2 and Cl_2 in the presence of O_2.

(a) Predict a mechanism for the reaction assuming that the only stable intermediates are Cl, H, HO_2, and ClO_2. (*Hint*: HO_2 and ClO_2 react with H_2 to yield H_2O_2.)

(b) Derive a rate equation for your reaction mechanism.

(c) How well do your results agree with the findings of Thon?

5.11 Christiansen, (1922) examined the following reaction:

$$Cl_2 + CO \Longrightarrow Cl_2CO$$

(a) Predict a mechanism for the reaction.

(b) Diagram the catalytic cycle.

(c) Use your predicted mechanism to derive a rate equation for the reaction.

(d) Christiansen actually found that the rate was given by rate $= k[CO][Cl_2]^{3/2}$. Is your mechanism consistent with Christiansen's data?

5.12 Pease (1928) examined the reaction

$$C_2H_6 \Longrightarrow C_2H_4 + H_2$$

(a) Predict a mechanism for the reaction.

(b) Use your predicted mechanism to derive a rate equation for the reaction. Assume that ethylene, hydrogen, and methane are the only products.

(c) Pease actually found that the rate was given by the rate $= k_{5.12}[C_2H_6]$. Is your mechanism consistent with Pease' data?

(d) Laidler and Wojciechowski, (1961) examined the reaction at higher pressures, where a small amount of C_4H_{10} is formed. They found that the reaction changed

Table P5.9 Rate data for Problem 5.9

Reaction	Activation Barrier, kcal/mol	Reaction	Activation Barrier, kcal/mol
$H + OH \rightarrow H_2 + O$	7.0	$H_2 + CH_3 \rightarrow H + CH_4$	11
$H + CH_3O \rightarrow H_2 + H_2CO$	0	$O_2 + CO \rightarrow CO_2 + O$	48
$H + CH_3CH_2CH_3 \rightarrow H_2 + C_3H_7$	6.3	$O_2 + CH_3 \rightarrow O + CH_3O$	30
$H + (CH_3)_2CO \rightarrow H_2 + CH_3COCH_2$	8.4	$O_2 + N_2 \rightarrow O + N_2O$	110
$H + (CH_3)_2C{=}CH_2 \rightarrow (CH_3)_2CH + CH_2^0$	0	$H + CH_4 \rightarrow H_2 + CH_3$	12
$H_2 + CO_2 \rightarrow H_2) + CO$	15	$H + CH_3CH_3 \rightarrow H_2 + CH_3CH_2$	9.7

to $\frac{3}{2}$ order. Suggest a mechanism which forms C_4H_{10} as a small product. Is your mechanism consistent with Laidler's data?

5.13 Chlorocarbons have been implicated in the destruction of ozone in the atmosphere. The basic reaction is for the chlorocarbons to be photolyzed, yielding chlorine atoms. The chlorine atoms subsequently decompose the ozone.

(a) Propose a mechanism for ozone decomposition in the absence of chlorine.

(b) Notice that the reaction does not go via a catalytic cycle. What are the implications of that result?

(c) Now add chlorine atoms. Propose a mechanism assuming that the chlorine does participate in a catalytic cycle, where chlorine atoms react with ozone and then the chlorine atoms are regenerated.

(d) Use the steady-state approximation to derive a rate equation for the reaction in (a) and the reaction in (c).

(e) Plug in reasonable values for the rate constants, to calculate the rate of ozone decomposition in the troposphere. Assume that the preexponentials for all of the reactions are 10^{13} Å^3/(molecule·second), that there is 10^{-3} atm of ozone in a total pressure of 0.1 atm, and that the temperature is 100°C.

(f) According to your estimates, by how much do chlorine atoms increase the rate of ozone depletion from the atmosphere?

(g) Would chlorine also catalyze the production of ozone?

(h) Now put in the hydroxyls that are naturally in the atmosphere. The hydroxyls can react to form HO_2, which then would participate in a catalytic cycle. How would hydroxyls affect your conclusions in (f)?

5.14 Frey (1934) examined the reaction

$$\text{Cyclohexane} \Longrightarrow \text{ethylene} + \text{propylene} + \text{other products}$$

(a) Find a feasible mechanism for the reaction.

(b) What other products would you expect?

(c) How do your predictions compare to the results of Frey (1934).

5.15 The polymerization of styrene can proceed via an initiation–propagation mechanism. One adds a free-radical initiator such as benzoyl peroxide, $C_6H_5(CO)OO(CO)C_6H_5$. The benzoyl peroxide dissociates at the O–O bond to yield radicals. The radicals then catalyze the polymerization of the styrene.

(a) Suggest a mechanism for the reaction, assuming that radicals add across the double bonds in the styrene.

(b) What will the termination reactions be like? Will the termination reactions result in loss of active centers (i.e., radicals), or in transfer of the radicals to a new species?

(c) The reaction can also occur via what is called *cationic polymerization*, where H_2SO_4, or a mixture of BF_3 and water, produces H^+, which acts as an initiator. Suggest a mechanism for this reaction.

(d) What kinds of termination processes would you expect in (c)? Would the main termination be loss of the proton, or transfer of the proton (and the counterion) to another styrene? Will the termination reactions result in loss of active centers (i.e., radicals), or transfer of the radicals to a new species?

(e) Now try initiating the reaction with an anion, such as $(CH)_3C^-$. Will the reaction be harder to start? What kinds of termination process do you expect?

(f) How will the polystyrene differ when it is produced by the three different mechanisms? For a discussion, see Young and Lovel (1991).

5.16 *The US Chemical Industry Statistical Handbook for 1997* gives a chart showing the feedstocks used to produce all of the major chemicals in the United States. Figure P5.16 shows part of the ethylene chain, specifically, the chemicals produced starting with ethylene. The chart includes the major organic chemicals used in the process, but not hydrogen, oxygen, CO, or water.

(a) Look at each step (i.e., arrow) and write a balanced reaction for the step, adding hydrogen, oxygen, water, or CO as needed.

(b) Suggest a possible mechanism for each reaction.

(c) Look in the chemical processes handbook. How do your mechanisms compare to the ones actually observed?

5.17 In Section 5.5.3 we noted that the methane coupling reaction $2CH_4 \Rightarrow CH_3CH_3 + H_2$ has a high activation barrier.

(a) Go through the arguments and explain in your own words why there is a large barrier.

(b) Use the results in Section 5.3 to determine what temperature you would need to run the reaction via the mechanism in Section 5.5.3.

(c) How would the mechanism change if you had a source of protons in the system?

(d) Could the reaction be run at a reasonable temperature in case (c)?

(e) Would it be better to try to couple the methane to ethylene via the reaction $CH_4 + C_2H_4 \Rightarrow CH_3CH_2CH_3$?

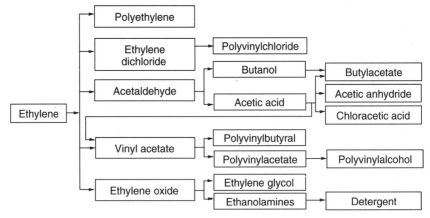

Figure P5.16   A part of the ethylene chain for the production of industrial chemicals. [Adapted from the U.S. Chemical Industry (1997)].

5.18 People make microelectronic devices by depositing doped silicon layers onto a silicon substrate. When people first started the process, they made the silicon layers via deposition of silane:

$$SiH_4 \Longrightarrow Si + 2H_2$$

However, now it is more common to make the layers via decomposition of dichlorosilane:

$$SiH_2Cl_2 \Longrightarrow Si + 2HCl$$

(a) Use the knowledge that you have acquired in this chapter to devise a mechanism for the decomposition of silane and dichlorosilane. Assume that reactions occur both in the gas phase and on the silicon surface.

(b) Which mechanism would be more likely to produce particles in the gas phase?

5.19 The isomerization of n-octane to isooctane is important in upgrading the octane number of gasoline. Generally, you run the reaction in three steps. First you dehydrogenate the n-octane to octane over a platinum catalyst, then you isomerize on an acid catalyst, before you rehydrogenate over platinum.

(a) Predict a feasible mechanism for each step.

(b) What side products do you expect?

(c) Why is isooctane important? Use the material in this chapter to predict why isooctane is better able to prevent knock than n-octane.

5.20 Huang et al. (1987) examined the dehydrogenation of methylamine (CH_3NH_2) over a platinum catalyst.

(a) Predict a mechanism for the reaction.

(b) What products do you expect?

(c) Use the steady-state approximation to derive a rate equation for the mechanism.

(d) How do your results compare to those of Huang et al.?

5.21 In Section 5.3 we stated that in order to get a reasonable amount of product out of a catalytic cycle, all of the steps in the cycle must satisfy

$$E_A < [0.5 \text{ kcal}/(\text{mol·K})]T. \qquad \text{(P5.21.1)}$$

The objective of this problem is to derive equation (P5.21.1). Consider the rate-determining step (i.e the slowest step) in a catalytic cycle. Assume that the rate determining step is of the form

$$Re + In \longrightarrow Pr + In^a \qquad \text{(P5.21.2)}$$

where Re is a reactant, In and In^a are intermediates, and Pr is a product. Further assume that there are three other reactions in the catalytic cycle:

$$Re + In^a \longrightarrow Pr + In^b \qquad \text{(P5.21.3)}$$

$$Re + In^b \longrightarrow Pr + In^c \qquad \text{(P5.21.4)}$$

$$Re + In^c \longrightarrow Pr + In \qquad \text{(P5.21.5)}$$

(a) Use the steady-state approximation to show that the rate of production of product, r_{Pr}, is given by

$$r_{Pr} = 4r_{5.21.2} \qquad (P5.21.6)$$

where $r_{5.21.2}$ is the rate of reaction (P5.21.2).

(b) Generalize equation (P5.21.6) to show that for an arbitrary catalytic cycle

$$r_{Pr} = n_{cy}r_{5.21.2} \qquad (P5.21.7)$$

where n_{cy} is the number of moles of product produced in going around the cycle once.

(c) Next, provide an expression for the rate of reaction (P5.21.2) in terms of the concentration of the intermediate, In, and the concentration of the reactant, Re.

(d) Substitute Arrhenius' law into your equation from (C). Assume a preexponential of 10^{13} $\mathring{A}^3$/(molecule·second).

(e) Next, use your expression in (d) to derive an expression for the number of cycles per/minute that the reaction undergoes.

(f) Calculate the activation barrier needed to get one cycle per second, one cycle per minute, one cycle per hour, or one cycle per day at 300, 500, and 1000 K.

(g) How do your activation barriers in (f) compare to the predictions of equation (P5.21.1)?

(h) Use your results to estimate the minimal temperature for the reaction
$CH_3CH_3 \Rightarrow CH_2=CH_2 + H_2$

5.22 In Section 5.4.4 we noted that the isomerization of methylisocyanide (CH_3NC) goes via the Lindemann mechanism:

$$X + CH_3NC \rightleftharpoons [CH_3NC]^* + X \qquad (P5.22.1)$$

$$[CH_3NC]^* \longrightarrow CH_3CN \qquad (P5.22.2)$$

where $[CH_3NC]^*$ is an excited methylisocyanide. Theoretically, the reaction could also go via an initiation–propagation mechanism:

$$X + CH_3(CH_3NC \rightleftharpoons CH_3 + CN + X \qquad (P5.22.3)$$

$$CH_3NC + CH_3 \longrightarrow CH_3 + CH_3CN \qquad (P5.22.4)$$

(a) Derive a rate equation for the Lindemann mechanism.

(b) Assume that you have 1 atm of methylisocyanide in a reactor, that reaction (P5.22.1) has an free energy two-thirds that of the CH_3–CN bond energy in methylisocyanide, and that reaction (P5.22.2) has a rate constant of 10^{17}/second. Calculate the rate of methylisocyanide isomerization via the Lindemann mechanism.

(c) Derive a rate equation for the initiation–propagation mechanism.

(d) Repeat part (b) for the initiation–propagation mechanism. Assume that the methylisocyanide is also a collision partner and that all of the preexponentials are 10^{13} $\mathring{A}^3$/(molecule·second).

(e) Compare the rates. Why does the Lindemann mechanism predominate?

More Advanced Problems

5.23 Boni and Penner, (1976) examined the formation of intermediates during the combustion of methane. Figure P5.23 shows a redraft of their data at 1 atm pressure and 2000 K.

(a) Postulate a mechanism for methane combustion. What initiation reactions do you expect? (*Hint*: Be sure to consider the fact that O_2 is paramagnetic, so that it can do radical chemistry.)

(b) What propagation reactions do you expect?

(c) How do your expected intermediates compare to those seen experimentally?

(d) Are there any expected intermediates missing?

(e) Boni and Penner actually took transient data, where they measured the intermediate concentration after starting the flame. Initially they saw the intermediates produced only in the initiation step. However, after a while, they also saw intermediates produced during propagation steps. How does the order of appearance of the intermediates compare to your expectations?

(f) Are any intermediates formed earlier than you might expect? Are any intermediates formed later than you might expect?

5.24 Look up the following paper: Watson, B.A., Klein, M.T. and Harding, R.H. Catalytic cracking of alkylcyclohexanes: Modeling the reaction pathways and mechanisms. *Int. J. Chem. Kinet.*, **29**, 545–560, (1997).

(a) How do the findings in this paper compare to your expectations from this chapter?

(b) Are there any reactions that you expect to happen that do not happen?

(c) Are there any reactions that you do not expect?

(d) Are the intrinsic barriers within expectations?

(e) Are the transfer coefficients within expectations?

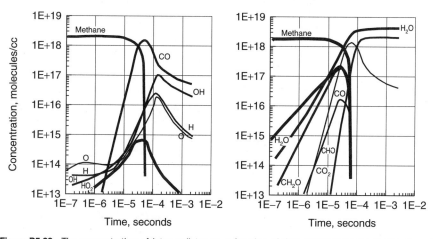

Figure P5.23 The concentration of intermediates as a function of time during methane combustion at 2000 K. [Results of Boni and Penner, *Combust. Sci. and Technol.*, 15, (1976) 99.

5.25 Look up one of the following papers. Write a one-page paper describing how the findings in that paper and how the material in this course apply to the results in the paper. Also suggest a follow-up study that you could do to expand on the results in the paper.

(a) Zummallen. M. P., and Schmidt. L. D., Oxidation of methanol over polycrystalline Rh and Pt — Rates, OH desorption, and model. *J. Catal.* **161**, (1), 230–246 (1996).

(b) Susnow, R. G., Dean, A. M., Green, W. H., Peczak P., and Broadbelt, L. J. Rate-based construction of kinetic models for complex systems. *J. Phys. Chem.* **101**, (20), 3731–3740 (1997).

(c) Pecullan, M., Brezinsky, K., and Glassman, I. Pyrolysis and oxidation of anisole near 1000 K. *J. Physi. Chem. A* **101**, (18), 3305–3316 (1997).

(d) Hausmann, M., and Homann, K. H., Scavenging of hydrocarbon radicals from flames with dimethyl-disulfide.2. Hydrocarbon radicals in fuel-rich low-pressure flames of acetylene, ethylene, 1,3-butadiene and methane with oxygen *Ber. Bunsenges Physi. Chem. Chem. Phys.* **101**, (4), 651–667 (1997).

(e) Alawadi, N. A., and Eldusouqui, O. M. E. Pyrolysis of beta-hydroxyketones and alpha-ketoesters — Gas-phase elimination kinetics of 3-hydroxy-3-methyl-2-butanone and methyl benzoylformate *Int. J. Chem. Kinet.* **29**, (4), 295–298 (1997).

(f) Martin, G., Ascanio, J., Rodriguez, J. Gas-phase thermolysis of N-substituted diallyl- and allylpropargylamines. *J. Phys. Org. Chem.* **10** (1), 49–54. (1997).

(g) Kruse T., and Roth, P. Kinetics of C-2 reactions during high-temperature pyrolysis of acetylene *J. Phy. Chem. A* **101**, (11), 2138–2146 (1997).

(h) Safarik I., and Strausz, O. P. The thermal decomposition of hydrocarbons. Polycyclic N-alkylaromatic compounds. *Res. Chem. Intermed.* **23**, (2), 179–195 (1997).

(i) Lattimer, R. P. Pyrolysis field ionization mass spectrometry of hydrocarbon polymers. *J. Anal. Appl. Pyrolysis* **39**, (2), 115–127 (1997).

(j) Madden. L. K., Mebel, A. M., Lin, M. C., and Melius, C. F. Theoretical study of the thermal isomerization of fulvene to benzene. *J. Phys. Org. Chem.* **9**, (12), 801–810 (1996).

(k) Blazso, M. Recent trends in analytical and applied pyrolysis of polymers. *J. Anal. Appl. Pyrolysis* **39**, 1, 1–25 (1997).

(l) Fan, G. H., Maung N., Ng, T. L., Heelis, P. F., Williams, J. O., Wright, A. C., Foster, D. F., and Colehamilton, D. J. Thermal decomposition of di-tertiary-butyl selenide and dimethylzine in a metalorganic vapour phase epitaxy reactor. *J. Cryst. Growth* **170**, (1-4), 485–490 (1997).

(m) Rettner, C. T., Auerbach, D. J., and Lee, J. Dynamics of the formation of CD4 from the direct reaction of incident D atoms with CD3/Cu(111). *J. Chem. Phys.* **105**, (22), 10115–10122 (1996).

(n) Jorand, F., Heiss, A., Sahetchian, K., Kerhoas, L., and Einhorn, J. Identification of an unexpected peroxide formed by successive isomerization reactions of the N-butoxy radical in oxygen. *J. Chem. Soc. Faraday Trans.* **92**, (21), 4167–4171 (1996).

(o) Thomas, S. D., Bhargava, A., Westmoreland, P. R., Lindstedt, R. P., and Skevis, G. Propene oxidation chemistry in laminar premixed flames. *Bull. Soc. Chim. Belg.* **105**, (9), 501–512 (1996).

(p) Ortego, J. D., Richardson, J. T., and Twigg, M. V. Catalytic steam reforming of chlorocarbons — Methyl chloride. *Appl. Catal.B-* **12**(4), 339–355 (1997).

(q) Ranzi, E. Dente, M., Faravelli, T., Bozzano, G., Fabini, S., Nava, R., Cozzani, V., and Tognotti, L. Kinetic modeling of polyethylene and polypropylene thermal degradation. *J. Anal. Appl. Pyrolysis* **40**(1), 305–319 (1997).

6

REVIEW OF SOME THERMODYNAMICS AND STATISTICAL MECHANICS

PRÉCIS

This chapter will go back to the main theme of the book: to provide a series of principles that can be applied to a wide range of reactions. In Chapter 5, we found that if we know the heat of reaction for a given elementary reaction, we can estimate the activation barrier for the reaction, and thereby tell if the elementary reaction is going to be an important part of the mechanism. In this chapter, we will review some of the ideas that appear in the literature to relate the heats of reaction and other thermodynamic properties of a molecule to the structure of the molecule. In particular, we will first consider group contribution methods and describe how these methods can be used to estimate heats and entropies of reactions. We will then review statistical mechanics and show how one can calculate the thermodynamic properties of a molecule from the energy levels and vibrational properties of the molecule. Most of our discussion will be brief because we want to cover only the things that we will be using later in this book. One can get a more detailed treatment of group contribution methods in Benson (1971, 1976). Introductory statistical mechanics is covered well in McQuarrie (2000), Hecht (1998), Chandler (1987), or Davis (1995). There is also Hills' old classic, *Statistical Thermodynamics* (1986).

6.1 GROUP CONTRIBUTION METHODS

The objective of this chapter is to provide the reader with an overview of how one can estimate heats of reaction and other thermodynamic properties of a molecule from a knowledge of the structure, composition, and other properties of the molecule. Group contribution methods will be covered first. Then we will review statistical mechanics.

To start off, we will consider group contribution methods. The idea of a group contribution method is to estimate thermodynamic properties of a molecule from the structure of the molecule. For example, if one wanted to calculate heats of formation of a molecule, one might add up the heats of formation of all of the bonds in the molecule and then correct for resonance energies and other factors.

The earliest work on group contribution methods was G. N. Lewis' (1923a,b) paper on valence. M. Born and W. Nernst also apparently did some early work on the subject, although their work was not published. There are some very nice measurements done by Von Steger (1920) and Fajans (1920). Still, it was not until Pauling's work (1931a,b, 1939, 1960) that all the work was put together. In 1931, Pauling wrote two key articles where he showed that heats of formation of molecules could be estimated by knowing the bond energies and electronegativity of all of the atoms in the molecule.

At first, Pauling tried to calculate absolute energies of formation with limited success. However, Pauling realized that it was much easier to calculate relative energies, that is, energies relative to some standard state. As in other thermodynamic quantities, Pauling started with a standard state of the pure elements and tried to work from there.

Benson (1955, 1958, 1971, 1976) was the person who did the most to expand on Pauling's findings. Benson took all of the thermodynamic data in the literature, up until 1976, and correlated it to the properties of the molecule. Generally, Benson's methods work to within a few tenths of a kcal/mol, which is about as good as the experiment. Consequently, Benson's methods allow you to estimate thermodynamic properties for many systems. In the next section, we will summarize Benson's findings. Further details are given by Benson (1976).

6.1.1 Groups Contribution Methods for Gas-Phase Species

Benson (1976) notes that there is a hierarchy of group contribution methods that one can employ. If one wants an answer to within 5 kcal/mol, one can simply assume that the standard heat of formation of a molecule is equal to the energy of all of the atoms in the system (i.e., zero if the standard state is pure elements) plus a correction for all of the bonds in the molecule. On the other hand, if one wants an answer to within 0.5 kcal/mol, one also has to make corrections for all of the appendant groups on the molecule. Further corrections are needed to get to within 0.1 kcal/mol. In this book, we will review the former methods. Additional details are given in Benson (1976).

Let's start with the simplest method: adding up the bonds of a molecule to calculate heats of formation. Table 6.1 shows Benson's values for the contribution of various bonds to the standard heat of formation, entropy of formation, and specific heat of various molecules. In the table, C–H refers to an sp_3-hybridized carbon bonded to a

Table 6.1   The contribution of various bonds to key thermodynamic properties

Bond	C_p, cal/(mol·K)	S, cal/(mol·K)	ΔH_f, kcal/mol	Bond	C_p, cal/(mol·K)	S, cal/(mol·K)	ΔH_f, kcal/mol
C–H	1.74	12.90	−3.83	C_D–C	2.6	−14.3	6.7
C–C	1.98	−16.40	2.73	C_D–H	2.6	13.8	3.2
C–F	3.34	16.9	−52.5	C_D–F	4.6	18.6	−3.9
C–O	2.7	−4.0	−12.0	C_D–C_D	—	—	7.5
O–H	2.7	24.0	−27.0	CO–H	4.2	26.8	−13.9
C–N	2.1	−12.8	9.3	CO–N	3.7	−0.6	−14.4
N–H	2.3	17.7	−2.6	C_B–H	3.0	11.7	3.25
C_B–C_B	—	—	10.0	C_B–C	4.5	−17.4	7.25

Source: Data of Benson (1976).

hydrogen, C_D–H refers to a sp_2-hybridized carbon (i.e., a double-bonded carbon) bonded to a hydrogen, C_T–H refers to a sp-hybridized carbon (i.e., a triple-bonded carbon) bonded to a hydrogen, and C_B–H refers to a carbon in a benzene ring that is bonded to a hydrogen.

Notice that there are values for many of the groups found in most hydrocarbons. One uses Table 6.1 to calculate standard heats of formation by simply adding up the contributions from all of the bonds in the molecule. For example, if one wanted to calculate the standard heat of formation of methane, CH_4, one would add up the contributions from the four C–H bonds:

$$\Delta H_f = 4 \times H_f(C\text{–}H) = 4 \times (-3.83) = -15.3 \text{ kcal/mol} \qquad (6.1)$$

By comparison, the experimental value is -17.9 kcal/mol. If one wanted to calculate the standard heat of formation of ethane, C_2H_6, one would add up the contributions from the six C–H bonds and the one C–C bond:

$$\Delta H_f = 6 \times H_f(C\text{–}H) + H_f(C\text{–}C) = 6 \times (-3.83) + 2.73 = -20.3 \text{ kcal/mol} \qquad (6.2)$$

By comparison, the experimental value is -20.2 kcal/mol. Generally, bond additivity methods get within 1–5 kcal/mol, which is often good enough to predict mechanisms of reactions.

Benson (1971) also noted that one can even make better predictions if one considers individual functional groups in a molecule instead of all of the bonds. In Benson's notation, a functional group is a carbon, oxygen, or similar, but not a hydrogen. Table 6.2 shows Benson's values for the contribution of various functional groups to the standard heat of formation, entropy of formation, and specific heat of various molecules. In the table, C–$(H_3)(C)$ refers to a carbon atom bound to another carbon atom and three hydrogens.

One uses Table 6.2 to calculate standard heats of formation by simply adding up the contributions from all of the functional groups in the molecule. For example, if one wanted to calculate the standard heat of formation of ethane, C_2H_6, one would add up the contributions from the two CH_3 groups:

$$\Delta H_f = 2 \times H_f(C - (C)(H)_3) = 2 \times (-10.0) = -20.0 \text{ kcal/mol} \qquad (6.3)$$

By comparison, the experimental value is -20.0 kcal/mol. If one wanted to calculate the standard heat of formation of ethylene, C_2H_4, one would add up the contribution from the two CH_2 groups:

$$\Delta H_f = 2 \times H_f(C_D - (H)_2) = 2 \times (+6.26) = +12.52 \text{ kcal/mol} \qquad (6.4)$$

By comparison, the experimental value is $+12.54$ kcal/mol. Benson shows that there are corrections to the group activity rules when there are ring structures or when there are two large groups on either side of a single bond. Generally, simple group additivity methods get within 0.5 kcal/mol using the results in Table 6.2. Corrections described in Benson (1971) allow corrections to 0.1 kcal/mol. That is as good as the data. All of Benson's work is for gas-phase species. Cohen (1996) has values for liquids and solids.

Benson also has a table of the functional group contributions of radicals, and Table 6.3 shows the results. One uses Table 6.3 just like Table 6.2. For example, if one wanted to calculate the standard heat of formation of an ethyl radical, $CH_3CH_2\bullet$, one would add up

Table 6.2 The contribution of various functional to key thermodynamic properties in the gas

Ligand	C_p, cal/(mol·K)	S, cal/(mol·K)	ΔH_f, kcal/mol	Ligand	C_p, cal/(mol·K)	S, cal/(mol·K)	ΔH_f, kcal/mol
$C-(H)_3C$	6.19	30.41	−10.00	$O-(H)_2$	8.0	45.1	−57.8
$C-(H)_2(C)_2$	5.50	9.42	−5.00	$O-(H)(C)$	4.3	29.07	−37.9
$C-(H)(C)_3$	4.54	−12.07	−2.4	$O-(H)(C_B)$	4.3	29.1	−37.9
$C-(C)_4$	4.37	−35.10	−0.1	$O-(H)(C_D)$	3.8	24.5	−58.1
$C_D-(H)_2$	5.10	27.61	6.26	$O-(C)_2$	−3.4	8.68	−23.2
$C_D-(H)(C)$	4.16	7.97	8.6	$CO-(H)_2$	8.5	52.3	26.0
$C_D-(C)_2$	4.10	−12.70	10.34	$CO-(H)(C)$	7.0	34.9	29.1
$C_D-(C_D)(H)$	4.46	6.38	6.8	$CO-(C)_2$	7.0	—	29.1
$C_D-(C_D)(C)$	4.40	−14.6	8.8	$CO-(H)(C_B)$	5.6	15.0	−31.4
$C_D-(C_B)(H)$	4.46	6.38	6.8	$CO-(C)_2$	6.0	14.8	−35.1
$C_D-(C_B)C$	4.40	−14.6	8.64	$CO-(C)(O)$	6.19	30.41	−10.8
$C_D-(C_T)(H)$	4.46	6.38	6.78	$C-(H)_3(O)$	4.99	9.8	−8.1
$C-(C_D)(C)(H)_2$	5.2	9.80	−4.76	$C-(H)_2(O)(C)$	6.19	30.41	−10.08
$C-(C_D)(H)_2$	4.7	10.2	−4.29	$C-(H)_3(CO)$	6.2	9.6	−5.2
$C-(C_D)(C_B)(H)_2$	4.7	10.2	−4.29	$C-(H)_2(CO)(C)$	3.9	−10.2	−0.9
$C-(C_T)(C)(H)_2$	4.95	10.3	−4.73	C_B-O	6.19	30.41	−10.08
$C_T-(H)$	5.27	24.7	26.93	$C-(N)(H)_3$	5.25	9.8	−6.6
$C_T-(C)$	3.13	6.35	27.55	$C-(N)(C)(H)_2$	5.72	29.71	4.8
$C_T-(C_D)$	2.57	6.43	29.20	$N-(C)(H)_2$	4.20	8.94	15.4
$C_T-(C_B)$	2.57	6.43	29.20	$N-(C)_2(H)$	12.7	42.5	−161
$C_B-(H)$	3.25	11.53	3.30	$C-(F)_2(H)(C)$	9.9	39.1	−102.3
C_B-C	2.07	−7.69	5.51	$C-(F)(H)_2(C)$	8.1	35.4	−51.5
C_B-C_D	3.59	−7.80	5.68	$C-(F)_2(C)_2$	9.9	17.4	−99

Source: Data of Benson *Thermochemical Kinetics* (1976), with revision due to Cohen, *J. Phys. Chem. Ref. Data,* **25** (1996) 141.

Table 6.3 The contribution of various functional to key thermodynamic properties of radicals[a]

Ligand	C_p, cal/(mol·K)	ΔS, cal/(mol·K)	ΔH_f, kcal/mol	Ligand	C_p, cal/(mol·K)	ΔS, cal/(mol·K)	ΔH_f, kcal/mol
•C–(C)(H)$_2$	5.99	30.7	38.47	C–(O•)(C)$_2$(H)	7.7	14.7	7.8
•C–(C)$_2$(H)	5.16	10.74	41.58	C–(O•)(C)$_3$	7.2	–7.5	8.6
•C–(C)$_3$	4.06	–10.77	41.50	•C–(H)$_2$(C$_D$)	5.39	27.65	23.7
C–(C•)(H)$_3$	6.19	30.41	–10.07	•C–(H)(C)(C$_D$)	4.58	7.02	25.2
C–(C•)(C)(H)$_2$	5.50	9.42	–4.95	•C–(C)$_2$(C$_D$)	4.00	–15.0	24.8
C–(C•)(C)$_2$(H)	4.54	–12.07	–1.90	•C–(C$_B$)(H)$_2$	4.10	26.85	23.0
C–(C•)(C)$_3$	4.37	–35.10	1.50	•C–(C$_B$)(C)(H)	5.30	6.38	24.7
C–(O•)(C)(H)$_2$	7.9	36.4	6.1	•C–(C$_B$)(C)$_2$	4.72	–15.46	25.5
H•	4.97	27.42	52.1	O•	5.23	38.44	59.56
CH$_3$	9.25	46.4	34.82	•CH$_2$	8.27	46.35	92.35

[a]The original source is Benson (1976), but we have corrected some of the values based on data from the NIST (National Institute of Standards and Technology) Web book. The values in the last (lowest) two lines of the table are data for specific radicals.

Table 6.4   The heat of reaction of $H^+ + NH_2R \Rightarrow [NH_3R]^+$ as a function of the R group

R	ΔH, kcal/mol	R	ΔH, kcal/mol	R	ΔH, kcal/mol
H	−205	CH_3	−214.1	C_2H_5	−217.1
$n\text{-}C_3H_7$	−218.5	$n\text{-}C_4H_9$	−219.0	$n\text{-}C_6H_{13}$	−220.1
$n\text{-}C_8H_{17}$	−220.4	$n\text{-}C_{10}H_{17}$	−220.7	$i\text{-}C_4H_9$	−219.5
$s\text{-}C_4H_9$	−220.5	$t\text{-}C_4H_9$	−221.3	—	—

Source: Data of Bowers (1977).

the contributions from the CH_3 group and a CH_2 radical:

$$\Delta H_f = H_f(C - (C\bullet)(H_3)) + H_f(\bullet C - (C)(H_2)) = (-10.07) + 38.47 = 28.4 \text{ kcal/mol} \tag{6.5}$$

By comparison, the experimental value is +28.4 kcal/mol. In general, Benson's numbers are less accurate for radicals than for neutral species because they are based on older data. Nevertheless, Benson's methods are still accurate enough for most purposes.

One is not as lucky with ions. At present, there are no similar correlations for ions. There are lots of data on ions. However, Brauman and Blair (1971) showed that there is long-range effect, where the energy varies with the length of the chain attached to the ion. For example, Table 6.4 shows some of the data for substituted amine. Notice that the heat of formation of the C_{10}-substituted amine is 0.6 kcal/mol larger than the C_6-substituted amine. Consequently, if one wanted to form a table like Table 6.3, one would not just have to know about the first nearest neighbors as in Table 6.3. Instead, one would need to know about the tenth nearest neighbors. That makes a simple method much more difficult.

At present, Benson's methods are accurate enough for simple molecules and radicals. However, similar methods do not work for ions.

6.2   APPLICATIONS TO METAL SURFACES

Group contribution methods have also been applied to the prediction of heats of adsorption of radicals on metal surfaces. Generally, the approach is to start with the energy of a radical, and then add on energy because the radical is stabilized with a metal–radical bond. In the literature, people report absolute bond energies. However, we have converted the values to a contribution to ΔH_f, and the results are shown in Table 6.5. One uses Table 6.5 just like Table 6.2. For example, if one wanted to calculate the standard heat of formation of an ethyl radical, CH_3CH_2, adsorbed on platinum, one would add up the contributions from the CH_3 group, a CH_2 radical, and a platinum–carbon bond:

$$\Delta H_f = H_f(C–(C)(H_3)) + H_f(\bullet C–(C)(H_2)) + H_f(Pt–C)$$

$$= (-10.2) + 38.47 + (-40) = -11.7 \text{ kcal/mol} \tag{6.6}$$

There is no experimental value for this quantity, but higher-level calculations indicate energies between 10 and 20 kcal/mol.

In Chapter 5, we noted that molecules can be adsorbed either molecularly or dissociatively. Table 6.5 can be used only in cases where the molecule dissociates (i.e., a bond breaks) on adsorption. Presently, there is nothing in the literature that would allow

Table 6.5 **Approximate contributions of metal surface bond to ΔH_f kcal/mol[a]**

	Group							
	IVA	VA	VIA	VIIA	VIII	VIII	VIII	IB
Element	Ti	V	Cr	Mn	Fe	Co	Ni	Cu
H(M–C)	[−62]	[−56]	−53	−50	−49	−48	−50	−41
H(M–N)	[−77]	[−61]	[−44]	[−36]	−14	[−1]	−10	−3
H(M–O)	−68	−55	−58	−44	−45	−40	−38	−30
H(M–H)	−19	−15	−14	−12	−11	−12	−12	−5
Element	Zr	Nb	Mo	Tc	Ru	Rh	Pd	Ag
H(M–C)	−62	−59	−53.1	−49	−43	−40	−40	−25
H(M–N)	−34	−23	−19	[−12]	[−10]	[−9]	[−7]	+10
H(M–O)	−78	−58	[−41]	[−37]	[−36]	−28	−24	−22
H(M–H)	−20	−13	−13	−12	−11	−10	−10	0
Element	Hf	Ta	W	Re	Os	Ir	Pt	Au
H(M–C)	−65	−81	−72.5	−52.5	−43	−40	−40	−20
H(M–N)	−34	−26	−13	−15	[−11]	[−8]	[−5]	+10
H(M–O)	−80	−61	−47	−33	−24	−29	−24.5	−19.5
H(M–H)	?	−19	−16	−12	−11	−8	−6	+10

[a] The data in the table are calculated from results in Benziger (1991) and results in Masel (1996). Most of the numbers are ±5–10 kcal/mol. The numbers in brackets are based on extrapolations. Consequently, those numbers may have larger errors.

one to simply calculate heats of molecular adsorption. One has to look up data for the various quantities.

6.3 EXTENSION TO SEMICONDUCTORS AND INSULATORS

Unfortunately, one cannot simply extend these ideas to adsorption on semiconductors and insulators. With semiconductors, one would expect the ideas to carry over directly. However, there is a complication due to what is called a *surface reconstruction*. This is a change in the arrangement of the surface atoms when gases adsorb. For example, when you form a silicon surface, the dangling bonds on the silicon atoms pair up, forming highly strained bonds. Those bonds distort or break when a gas adsorbs. One needs to consider all of the changes in strain energy during adsorption to make useful predictions of heats of adsorption, but the changes in strain are difficult to consider with group additivity methods. As a result, group contribution methods do not work on semiconductor surfaces. One can make useful predictions by considering all of the strain energies explicitly.

Insulators are even harder. The problem with insulators is that all of the key species are ions. No one has published a series of group additivity values for ions, so one cannot make useful predictions. If one knows the energy of an ion, there is a technique called *electronegativity equalization* that can be used to calculate how the ion changes in the presence of a solid. See Masel (1996) or Mortier (1991) for details. However, one still has to calculate the energy of the ion to begin with, and so one usually uses data to estimate thermodynamic properties.

6.4 CALCULATION OF THERMODYNAMIC PROPERTIES FROM STATISTICAL MECHANICS

The other way to estimate thermodynamic properties is from statistical mechanics. Statistical mechanics provides a way to, in principle, calculate thermodynamics properties to any degree of accuracy. In practice, statistical methods are still less accurate than Benson's method. However, statistical methods can be applied to a wide variety of systems where Benson's methods do not apply. If one has a radical, an ion, or a transition state, statistical methods are much more accurate than Benson's method. We will be using statistical mechanics later in this book, both to estimate heats of reactions and to estimate rate constants. When I teach my course, I find that while many of the students have seen statistical mechanics before, they are not comfortable enough with the subject to understand the results presented later in this book. Therefore, I find it useful to review some of the key ideas. The remainder of this chapter contains the notes from my review.

First, it is useful to give a brief overview of statistical mechanics:

- The key concept in statistical mechanics is that one can calculate all thermodynamic properties as an average. For example, the internal energy of molecules in a box is an average of the internal energies of each molecule, which is then also averaged over time. The entropy is an average of all of the entropies of the molecules averaged over time.
- There are alternative ways to compute the averages. For example, instead of computing a time average, one can compute an *ensemble average*. The ensemble average will be defined later in this chapter. If you do everything right, all of the averages should come out to be the same value, which is why statistical mechanics is so valuable.
- When you do statistical mechanics, you use all of the normal state variables that you learned about in thermodynamics: pressure, temperature, volume, free energy, enthalpy, and so on. In addition, there are some special state variables called the *partition functions*.
- The partition functions are like any other state variable. The partition functions are completely defined if you know the state of the system. You can also work backward, so if you know the partition functions, you can calculate any other state variable of the system.
- The partition functions are defined by equations (6.15) and (6.16). These equations allow the partition functions to be calculated from the properties of the molecules in the system (energy levels, atomic masses, etc). The fact that the partition functions can be calculated easily makes them particularly convenient thermodynamic variables. If you know the properties of all of the molecules, you can calculate the partition functions. You can then work backward and calculate any thermodynamic property of the system.
- The key variable for the work later in this book is the equilibrium constant for a reaction. K, the equilibrium constant for the reaction $A + B \leftrightharpoons C + D$ is given by

$$K = \frac{q_C q_D}{q_A q_B}$$

(6.7)

In equation (6.7), q_A, q_B, q_C, and q_D are the partition functions per unit volume for molecules A, B, C, and D, respectively. Equation (6.7) is the key result for the rest of the book.

- One can calculate the molecular partition functions as a product of atomic partition functions as described in Table 6.6 and Example 6.B.

We will describe all of these results later in this chapter. When I teach this material, I cover it rather quickly. There are a few key concepts that a student needs to know:

- How can properties be calculated statistically? (See Section 6.4.1 and Example 6.B.)
- What is a statistical ensemble? (See Section 6.4.1.)
- What is a partition function? [See equations (6.15) and (6.16).]
- How are partition functions related to equilibrium constants? [See equation (6.7).]
- How do we evaluate partition functions? (See Table 6.6 and Example 6.C.)

I have included detailed derivations, because some students need to see them to understand the results. However, one could learn enough to understand the rest of the book, if one carefully reviews Section 6.4.1 and Examples 6.B and 6.C.

Table 6.6   Equations for the partition function for translational, rotational, vibrational modes, and electronic levels

Type of Mode	Partition Function	Approximate Value of the Partition Function for Simple Molecules
Translation of a molecule of an ideal gas in a one-dimensional box of length a_x	$q_t = \dfrac{(2\pi m_g k_B T)^{1/2} a_x}{h_p}$	$q_t \approx 1 - 10/\text{Å } a_x$
Translation of a molecule of an ideal gas at a pressure P_A and a temperature T	$q_T^3 = \dfrac{(2\pi m_g k_B T)^{3/2}}{h_p^3}\left(\dfrac{k_B T}{P_A}\right)$	$q_T^3 \approx 10^6 - 10^7$
Rotation of a linear molecule with moment of inertia I	$q_r^2 = \dfrac{8\pi^2 I k_B T}{S_n h_p^2}$ where S_n is the symmetry number	$q_r^2 \approx 10^2 - 10^4$
Rotation of a nonlinear molecule with a moment of inertia of I_a, I_b, I_c, about three orthogonal axes	$q_r^3 = \dfrac{(I_a I_b I_c)^{1/2}(8\pi k_B T)^{3/2}}{S_n h_p^3}$	$q_r^3 \approx 10^4 - 10^5$
Vibration of a harmonic oscillator when energy levels are measured relative to the harmonic oscillator's zero-point energy	$q_v = \dfrac{1}{1 - \exp(-h_p \nu / k_B T)}$ where ν = the vibrational frequency	$q_v \approx 1 - 3$
Electronic level (assuming that the levels are widely spaced)	$q_e = \exp\left(-\dfrac{\Delta E}{k_B T}\right)$ ΔE = level spacing	$q_e = \exp(-\beta \Delta E)$

6.4.1   Historical Introduction

To start our discussion, it is useful to review some history. Statistical mechanics was invented by Clausius (1857), Maxwell (1860), Boltzmann (1872), and Gibbs (1902). To put the work in historical perspective, throughout the eighteenth century and the early part of the nineteenth century, it was thought that gases and liquids bonded similarly to solids. Boyle (1660,1663), Newton (1687), Charles (1787), and Gay-Lussac (1802) showed that one could calculate the pressure–volume relationships of a number of ideal and nonideal gases by assuming that what we today would call *gas molecules* were held in a lattice connected by springs. Bernoilli (1738) showed that one could calculate the pressure in a flow system by assuming that there were long-range interactions between the gas molecules. Hence, the idea that there are long-range interactions between the atoms in an ideal gas explained many of the measurable properties of the gas.

In 1787, however, Lavoisier and co-workers proposed that gases were composed of individual isolated particles that did not interact strongly. Dalton (1804) identified the particles as atoms and molecules. Similar work was done by Avogadro (1811), Priestley (1790), Faraday (1819), Dobereiner (1829), Scheele (1777), Kelvin (1870), and others. By 1840, the view that gases were composed of isolated molecules was accepted by most scientists. There was a problem, however, in that it was not obvious how the continuum properties of gases arose from the atomic/molecular picture of matter.

Clausius (1858), Maxwell (1860), Boltzmann (1872), and Gibbs (1902) proposed statistical mechanics as a way to bridge the gap between the molecular model of matter, and the continuum properties of the matter. Their idea was that one can add up all of the microscopic properties of the matter in a given system, and by taking a suitable average, one can calculate any macroscopic (continuum) property of the system. In Maxwell's (1860) and Boltzmann's (1872) formation of statistical mechanics, one computed a macroscopic property of the system, $\langle F \rangle$, as the time average of the instantaneous values of F at a series of snapshots of the system. Later Gibbs showed that one could replace the time average with an "ensemble" average of F. The idea is to consider each of the snapshots of the system to be a separate system, and compute the average by summing over all of the individual systems, s, in the ensemble as follows:

$$\langle F \rangle = \frac{1}{n_{sy}} \sum_{s} F_s \qquad (6.8)$$

where n_{sy} is the number of systems in the ensemble.

It is useful to discuss a simple example that illustrates the equivalence of the time average and the ensemble average. Consider using a thermocouple (i.e., a digital thermometer) to measure the average temperature of a small beaker of water. In principle, a thermocouple can be used to measure the water temperature to arbitrary accuracy. However, when one does the measurement, one often finds that the thermocouple reading is noisy because of eddies in the beaker. In order to get very accurate readings, therefore, one has to find a way to average over all of the noise. One can use a simple beaker and measure the time average. However, an alternative idea is to take a large number of beakers that are identical to the first, place all of them in identical conditions, put a thermocouple in each beaker, and average over the readings of all of the thermocouples as illustrated in Figure 6.1.

In the same way, one can calculate the thermodynamic properties of a system by creating a series of systems that are identical to the first and then averaging over all of

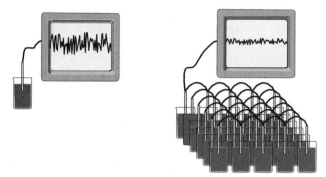

Figure 6.1 A diagram showing how the use of multiple thermocouples can be used to lower the noise in a temperature measurement.

the systems. More formally, we will choose a set of systems that are representative of the first to do the average. We call the representative set of systems the **statistical ensemble**.

It is useful to comment on what we mean by systems that are representative of the first. In general, people use the ensembles to do thermodynamic calculations. For example, one might want to calculate the pressure of a fixed number of molecules of gas in a box at a given volume and a given temperature. In such a case, the only systems that are equivalent to the first are those that have the same volume, temperature, and number of molecules as the original system. Hence, one would want to choose systems that have the same volume, temperature, and number of molecules as the first system to do the ensemble average. One can generalize this result to show that one does not want to include all possible systems in the statistical ensemble. Rather, one wants to include only those systems that satisfy the same temperature, pressure, volume, and other constraints as the original system.

Statistical ensembles are very useful in the calculation of thermodynamic properties. Generally, one can step through the ensemble and estimate properties. An algorithm is given in Example 6.G. One should read that example, and run the associated computer program if it is available before proceeding with this chapter.

We are going to use two different statistical ensembles for the work in this book: the **canonical ensemble** and the **grand canonical ensemble**. The *canonical ensemble* is defined as an ensemble of systems that have a fixed value of N, V, and T, where N is the number of molecules in a given system, V is the volume of the system, and T is the temperature. Physically, the canonical ensemble contains all of the states and internal arrangements of a set of N molecules confined to a closed box at constant-volume and temperature that is in thermal equilibrium with a heat bath.

The *grand canonical* ensemble is defined as an ensemble of systems that has a fixed value of μ, V, and T, where μ is the chemical potential and V is the volume of the system, and T is the temperature rates. The chemical potential is defined as the derivative of the Helmholtz free energy, A, with respect to the number of molecules in the system at constant entropy, S, and constant-volume, V:

$$\mu = \left(\frac{dA}{dN} \right)_{S,V} \tag{6.9}$$

Physically, the grand canonical ensemble contains all of the states and internal arrangements of a system consisting of an open box in equilibrium with a heat bath. In the grand canonical ensemble, the average number of molecules in the system is allowed to vary with the constraint that the chemical potential of the system be constant. Physically, this constraint is equivalent to saying that the number of molecules in the system must remain in equilibrium with all of the other systems in the ensemble.

Next, we will derive some of the key results in statistical mechanics. The results will follow the derivations of Gibbs (1902). Gibbs (1902) rearranged equation (6.8) as follows. First, he noted that rather than keeping track of all of the systems in the ensemble, one can instead keep track of all of the configurations of the system, where a *configuration* is defined as a set of all of the arrangements of molecules in the system plus all of the vibrational, rotational, and translational states of the molecules in the system. If we define p_n as the probability that the system is in configuration n, then equation (6.8) becomes

$$\langle F \rangle = \sum_n F_n p_n \tag{6.10}$$

where F_n is the value of the property F when the system is in configuration n.

In Maxwell and Boltzmann's original formulation of statistical mechanics, only the configurations that would be reached by a system in a finite time were considered. However, in Gibbs' analysis, all of the possible configurations of the system were included. The assumption that all possible configurations should be included in the sum is called the **ergodic hypothesis**.

The implication of the ergodic hypothesis is that the index n should go over all of the possible configurations of the system, namely, all possible arrangements of the molecules of the system, and all possible vibrational, rotational, and translation states of each molecule in the system.

Gibbs (1902) proposed that for the canonical ensemble, p_n is given by

$$p_n = \frac{g_n e^{-\beta U_n}}{Q_{canon}^N} \tag{6.11}$$

where U_n is the internal energy of the system when it is in configuration n; g_n is the degeneracy of the state n (i.e., the number of states that have the same energy as a given state n); $\beta = 1/k_B T$, where k_B is Boltzmann's constant; and T is the temperature.

For the grand canonical ensemble:

$$p_n = \frac{g_n e^{-\beta E_n}}{Q_{grand}} \tag{6.12}$$

with

$$E_n = U_n - \mu N_n \tag{6.13}$$

where N_n is the number of molecules in the system when the system is in configuration n. The Q values in equations (6.11) and (6.12) are normalization constants that we call **partition functions**. A detailed derivation of equations (6.11) and (6.12) can be found in any statistical mechanics text.

We require

$$\sum_n p_n = 1 \tag{6.14}$$

Combining equations (6.11), (6.12), and (6.14) and rearranging yields

$$Q_{canon}^N = \sum_n g_n e^{-\beta U_n} \qquad (6.15)$$

$$Q_{grand} = \sum_n g_n e^{-\beta E_n} \qquad (6.16)$$

Example 6.E illustrates the use of equation (6.15). One should read that example before proceeding with this chapter.

Next, we will show that Q_{canon}^N is proportional to the average occupancy number for the system. Let's define a function, F, as the occupancy number for the system where the occupancy number is defined as

$$F_0 = 1 \quad \text{for the ground state}$$

$$F_n = \frac{g_n}{g_0} \exp(-\beta(U_n - U_0)) \qquad (6.17)$$

where g_0 is the degeneracy of the ground state, g_n is the degeneracy of state n, U_0 is the energy of the ground state, and U_n is the energy of state n. Physically, F_n is the probability that a given state is occupied relative to the ground state.

F_{total}, the total occupancy, becomes

$$F_{total} = \sum_n F_n = \frac{\exp(\beta U_0)}{g_0} \sum_n g_n \exp(-\beta U_n) \qquad (6.18)$$

Comparing equations (6.15) and (6.18) shows

$$Q_{canon}^N = g_0 \exp(-\beta U_0) F_{total} \qquad (6.19)$$

The implication of this result is that the canonical partition is proportional to the occupancy number of the system. Physically, the total occupancy number can be thought of as the number of states of the system that are actively contributing to the dynamic properties. The more states that contribute, the larger Q_{canon}^N will be. People talk about Q_{canon}^N as though it were a volume. It is not a real volume. Instead, it is proportional to the volume of states that are contributing to reaction. Still, the analogy to volume is an important concept because increasing the real volume increases the number of states.

6.4.2 Partition Functions are State Variables

Next, I want to make an important point. The partition functions are thermodynamic state variables. Notice that if one knows the state of the system (e.g., the temperature) and all of the energy levels, one can calculate the partition functions directly from equations (6.15) and (6.16). In thermodynamics, any variable that you can calculate exactly once you know the state of the system is called a *thermodynamic state variable*. Consequently, the partition function is a state variable. That is very important because one can use the partition function like any other thermodynamic state variable. In particular, one can use the partition functions to calculate an equation of state for a system. Recall from thermodynamics that if one knows P and T for a closed system, one can calculate all of the other properties of the system via the Maxwell relations. In the same way

> If one knows Q_{canon}^N and T for a system, one can calculate all of the other thermodynamic properties of the system via some modified Maxwell relationships.

It happens that one can easily calculate partition functions using a computer as illustrated in Example 6.G. As a result, the partition functions are quite useful in practice.

In the next several sections, we will derive equations for various thermodynamic properties as a function of the partition function for the molecules. We include complete derivations in case the reader is interested. However, the reader could skip Section 6.4.3 without loss of continuity.

6.4.3 Derivation of the Maxwell Relationships for the Partition Function

First, we will calculate an expectation value for the internal energy of the system, $\langle U \rangle$, in terms of the canonical partition function.

From equation (6.10)

$$\langle U \rangle = \sum_n U_n p_n \tag{6.20}$$

Substituting p_n from equation (6.11) into equation (6.20) yields

$$\langle U \rangle = \frac{1}{Q_{canon}^N} \sum_n U_n g_n e^{-\beta U_n} \tag{6.21}$$

In Example 6.4G, we will use the Monte Carlo method to directly calculate the expectation value of the energy from equation (6.21). However, here we will use a mathematical trick to solve equation (6.8) for the expectation value of the internal energy in terms of the partition function and β. Consider the quantity $\left(-\dfrac{\partial \ln Q_{canon}^N}{\partial \beta} \right)$. One can show

$$-\frac{\partial \ln Q_{canon}^N}{\partial \beta} = -\frac{1}{Q_{canon}^N} \frac{\partial Q_{canon}^N}{\partial \beta} \tag{6.22}$$

Equation (6.15) shows

$$\frac{\partial Q_{canon}^N}{\partial \beta} = \sum_n g_n \frac{\partial}{\partial \beta} e^{-\beta U_n} \tag{6.23}$$

Performing the algebra, we obtain

$$\frac{\partial Q_{canon}^N}{\partial \beta} = -\sum_n g_n U_n e^{-\beta U_n} \tag{6.24}$$

Substituting equation (6.24) into (6.22) yields

$$-\frac{\partial \ln Q_{canon}^N}{\partial \beta} = \frac{1}{Q_{canon}^N} \sum_n g_n U_n e^{-\beta U_n} \tag{6.25}$$

Comparing equations (6.21) and (6.25) shows

$$\boxed{-\frac{\partial (\ln Q_{canon}^N)}{\partial \beta} = \langle U \rangle} \tag{6.26}$$

Equation (6.26) is a key result because it allows the internal energy to be calculated form the partition function.

Next, we will derive some relationships for other thermodynamic properties. Assume that we have a system of N molecules confined to a box of volume, V. Consider doing a reversible process on the box. From thermodynamics, for any reversible process

$$d\langle U \rangle = T\,dS - P\,dV \tag{6.27}$$

where P is the pressure and S is the entropy system. Now consider breaking up the reversible process into a thermodynamic path where we first add heat at constant-volume and then expand or compress the system adiabatically to the correct final volume.

The energy change in the constant-volume part of the thermodynamic path is

$$d\langle U \rangle_V = T\,dS \tag{6.28}$$

where we have used the subscript V to indicate that equation (6.28) only applies to the constant-volume part of our thermodynamic path.

From equation (6.20)

$$d\langle U \rangle = \sum_n U_n\,dp_n + \sum_n p_n\,dU_n \tag{6.29}$$

Note that U_n is the sum of the vibrational, rotational, and translational energy of all of the molecules in the box when the system is in state n. If we have N molecules in a box, then the rotational and vibrational energy levels are dependent on neither the size of the box nor the temperature. From the solution of a particle in a box, we know that translational energy levels of the molecules are dependent on the dimensions of the box but not on temperature. Therefore, when the system is in a fixed state n, U_n is a function of the dimensions of the box but not the temperature. For a constant-volume process, the dimensions of the box are constant; therefore, U_n is constant for any state of the system n. Hence, $dU_n = 0$ in equation (6.29). Therefore, for the constant-volume part of our thermodynamic path:

$$T\,dS = d\langle U \rangle_V = \sum_n U_n\,dp_n \tag{6.30}$$

Substituting equation (6.30) into equation (6.29), and substituting the resultant expression for $d\langle U \rangle$ into equation (6.27) and rearranging yields

$$P \, dV = - \sum_n P_n \, dU_n \tag{6.31}$$

As an exercise, one might want to derive equation (6.31) by noting that from thermodynamics:

$$\left(\frac{\partial \langle U \rangle}{\partial V} \right)_{N,S} = -P \tag{6.32}$$

Next, we will do some algebra to provide an expression for S in terms of Q^N_{canon}. Solving equation (6.11) for U_n yields

$$U_n = -\frac{1}{\beta} (\ln P_n + \ln Q^N_{canon} - \ln g_n) \tag{6.33}$$

Substituting equation (6.33) into equation (6.30) yields

$$T \, dS = -\frac{1}{\beta} \sum_n \left[\ln \left[\frac{P_n}{g_n} \right] dp_n \right] - \frac{1}{\beta} (\ln Q^N_{canon}) \sum_n dp_n \tag{6.34}$$

but since $\sum_n P_n = 1$, $\sum_n dp_n = 0$. Therefore

$$T \, dS = -\frac{1}{\beta} \sum_n \ln \left[\frac{P_n}{g_n} \right] dp_n \tag{6.35}$$

Next, we will use another mathematical trick to simplify equation (6.35). Let's consider $d\left(\sum_n P_n \ln \left[\frac{P_n}{g_n} \right] \right)$. From the chain rule

$$d\left(\sum_n P_n \ln \left[\frac{P_n}{g_n} \right] \right) = \sum_n \ln \left[\frac{P_n}{g_n} \right] dp_n + \sum_n P_n \, d(\ln p_n) - \sum_n P_n \, d(\ln(g_n)) \tag{6.36}$$

Note that since $g_n = $ constant, $dg_n = 0$ for any thermodynamic path. Further $\sum_n dp_n = 0$. Therefore

$$d\left(\sum_n P_n \ln \left[\frac{P_n}{g_n} \right] \right) = \sum_n \ln \left[\frac{P_n}{g_n} \right] dp_n + \sum_n \frac{P_n}{P_n} \, dp_n$$

$$= \sum_n \ln \left[\frac{P_n}{g_n} \right] dp_n \tag{6.37}$$

Combining equations (6.35) and (6.37) yields

$$T \, dS = -\frac{1}{\beta} d\left(\sum_n P_n \ln \left[\frac{P_n}{g_n} \right] \right) \tag{6.38}$$

Substituting $\beta = 1/k_B T$ into equation (6.37) and dividing both sides of the equation by T yields

$$dS = -d\left(k_B \sum_n p_n \ln\left[\frac{p_n}{g_n}\right]\right) \tag{6.39}$$

Integrating yields

$$S = -k_B \sum_n p_n \ln\left[\frac{p_n}{g_n}\right] \tag{6.40}$$

Note that there could be an integration constant in equation (6.40). However, it is taken to be zero.

Equation (6.40) is the statistical mechanics formulation of the entropy. We illustrate how equation (6.40) can be used in practice in Example 6.F. One should read that example before proceeding with this chapter.

Substituting equation (6.11) into the ln term in equation (6.40) yields

$$S = k_B \sum_n p_n (U_n \beta + \ln Q_{canon}^N) \tag{6.41}$$

Rearranging equation (6.41) yields

$$S = k_B \beta \sum_n p_n U_n + k_B \ln(Q_{canon}^N) \sum_n p_n \tag{6.42}$$

According to equation (6.20), the first sum in equation (6.42) is the internal energy. According to equation (6.16), the second sum in equation (6.42) is unity. Therefore

$$S = \frac{1}{T}\langle U \rangle + k_B \ln Q_{canon}^N \tag{6.43}$$

From thermodynamics

$$\langle U \rangle - TS = A \tag{6.44}$$

where A is the Helmholtz free energy. Substituting S from equation (6.43) into equation (6.44) yields

$$\boxed{A = -k_B T \ln(Q_{canon}^N)} \tag{6.45}$$

Equation (6.45) is a key result.

We can use equation (6.45) to calculate other thermodynamic properties. Note that for any thermodynamic path

$$dA = -S\,dT - P\,dV + \mu\,dN \tag{6.46}$$

Therefore

$$S = -\left(\frac{\partial A}{\partial T}\right)_{V,N} = k_B T\left(\frac{\partial \ln Q_{canon}^N}{\partial T}\right)_{V,N} + k_B \ln Q_{canon}^N \tag{6.47}$$

$$P = -\left(\frac{\partial A}{\partial V}\right)_{T,N} = k_B T \left(\frac{\partial \ln Q_{canon}^N}{\partial V}\right)_{T,N} \tag{6.48}$$

$$\mu = \left(\frac{\partial A}{\partial N}\right)_{T,V} = k_B T \left(\frac{\partial \ln Q_{canon}^N}{\partial N}\right)_{T,V} \tag{6.49}$$

One has to be careful with the units in equations (6.40), (6.45), (6.47) to (6.49). Energies are measured in kcal/molecule in equations (6.40) to (6.49). If one wants energies in kcal/mol, or entropies in cal/mol-K one needs to replace k_B by R in each equation where R is the ideal gas law constant.

6.4.4 Results for the Grand Canonical Ensemble

These arguments can also be extended to the grand canonical ensemble. Comparing equation (6.15) and (6.16) shows

$$Q_{grand} = \sum_N Q_{canon}^N e^{\beta N \mu} \tag{6.50}$$

The derivation of equation (6.40) was independent of whether the box was open or closed. Therefore, for the grand canonical ensemble

$$S = -k_B \sum_n P_n \ln \left(\frac{P_n}{g_n}\right) \tag{6.51}$$

Substituting equation (6.12) into equation (6.51) yields

$$S = \beta k_B \sum_n P_n U_n - \beta k_B \mu \sum_n N P_n + k_B \ln(Q_{canon}^N) \sum_n P_n \tag{6.52}$$

However, $k_B \beta = 1/T$, the first sum, is the average internal energy; the second sum is the average number of atoms in the box; and the third sum is unity. Therefore

$$S = \frac{\langle U \rangle}{T} - \frac{\mu \langle N \rangle}{T} + k_B \ln(Q_{grand}) \tag{6.53}$$

However, from thermodynamics

$$S = \frac{U}{T} - \frac{N\mu}{T} + \frac{PV}{T} \tag{6.54}$$

Substituting S from equation (6.54) into equation (6.53) and rearranging yields

$$PV = k_B T \ln(Q_{grand}) \tag{6.55}$$

Equation (6.55) is a key result for open systems.

One can use equation (6.55) to deduce other thermodynamic properties. Again from thermodynamics

$$d(PV) = S\,dT + N\,d\mu + P\,dV \tag{6.56}$$

Therefore

$$S = \left(\frac{\partial PV}{\partial T}\right)_{V,\mu} = k_B T \left(\frac{\partial \ln Q_{grand}}{dT}\right)_{V,\mu} + k_B \ln(Q_{grand}) \tag{6.57}$$

$$N = \left(\frac{\partial PV}{\partial \mu} \right)_{T,V} = k_B T \left(\frac{\partial \ln Q_{grand}}{\partial \mu} \right)_{V,T} \tag{6.58}$$

6.4.5   Maxwell Equations for Thermodynamic Properties

In summary, the results in Sections 6.4.1 and 6.4.2 show that if one knows the partition function and the temperature, one can calculate all of the other thermodynamic properties of the system. The key equations for closed systems are

$$A = -k_B T \ln(Q_{canon}^N) \tag{6.59}$$

$$\langle U \rangle = \frac{\partial (\ln Q_{canon}^N)}{\partial \beta} \tag{6.60}$$

$$S = -\left(\frac{\partial A}{\partial T} \right)_{V,N} = k_B T \left(\frac{\partial \ln Q_{canon}^N}{\partial T} \right)_{V,N} + k_B \ln Q_{canon}^N \tag{6.61}$$

$$P = -\left(\frac{\partial A}{\partial V} \right)_{T,N} = k_B T \left(\frac{\partial \ln Q_{canon}^N}{\partial V} \right)_{T,N} \tag{6.62}$$

$$\mu = \left(\frac{\partial A}{\partial N} \right)_{T,V} = k_B T \left(\frac{\partial \ln Q_{canon}^N}{\partial N} \right)_{T,V} \tag{6.63}$$

The key equations for open systems are

$$S = \left(\frac{\partial PV}{\partial T} \right)_{V,\mu} = k_B T \left(\frac{\partial \ln Q_{grand}}{dT} \right)_{V,\mu} + k_B \ln(Q_{grand}) \tag{6.64}$$

$$N = \left(\frac{\partial PV}{\partial \mu} \right)_{T,V} = k_B T \left(\frac{\partial \ln Q_{grand}}{\partial \mu} \right)_{V,T} \tag{6.65}$$

Equations (6.59) to (6.65) are the Maxwell relationships for the canonical and grand canonical partition functions. One can use these equations to calculate all of the thermodynamic properties of any system. Examples are given in Examples 6.D and 6.F. One should read those examples before proceeding.

6.4.6   Simplifications for Ensembles of Ideal Gases

Next, we will simplify the equations for an ideal gas containing noninteracting molecules. Consider N molecules in a box. If the molecules are isolated, then in Example 6.E we show we can divide the sum over n (i.e., all of the configurations of the system) in equation (6.15) into a sum over n_i, the configurations of each of the individual molecules in the system; and a sum over n_a, all of the arrangements of the atoms in the system. Details are given in Example 6.E. One should review that example before proceeding with this chapter. Equation (6.15) becomes

$$Q_{canon}^N = \sum_{n_a} \sum_{n_1} \sum_{n_2} \cdots \sum_{n_N} g_n e^{-\beta U_n} \tag{6.66}$$

where n_a is an index of all of the arrangements of the system and $n_1, n_2, n_3, \ldots, n_N$ are indices of all of the configurations of molecules $1, 2, 3, \ldots, N$:

$$g_n = g_{n_a} g_{n_1} g_{n_2} \cdots g_{n_N} \tag{6.67}$$

$$U_n = U_{n_a} + U_{n_1} + U_{n_2} + \cdots + U_{n_N} \tag{6.68}$$

where g_{n_a} is the degeneracy of the arrangement of molecules n_a, and $g_{n_1}, g_{n_2}, \ldots, g_{N_n}$ are the degeneracies of the internal state of molecules $1, 2, \ldots, N$. $U_{n_1}, U_{n_2}, \ldots, U_{n_N}$ are the internal energies of molecules $1, 2, \ldots, N$, and U_{n_a} is the interaction energy of the molecules in the box when the system is in arrangement n_a. If the molecules are nearly isolated, then U_{n_a} will not depend on the internal states of the individual molecules. Combining equations (6.66)–(6.68) yields

$$Q_{canon}^N = \left(\sum_{n_a} g_{n_a} e^{-\beta U_{n_a}} \right) \left(\sum_{n_1} g_{n_1} e^{-\beta U_{n_1}} \right) \left(\sum_{n_2} g_{n_2} e^{-\beta U_{n_2}} \right) \cdots \left(\sum_{n_N} g_{n_N} e^{-\beta U_{n_N}} \right)$$

(6.69)

It is useful to define q_i, the partition function for an isolated molecule, by

$$q_i = \sum_{n_i} g_{n_i} e^{-\beta U_{n_i}}$$

(6.70)

Substituting equation (6.70) into equation (6.69) yields

$$Q_{canon}^N = \sum_{n_a} g_{n_a} e^{-\beta U_{n_a}} \prod_{i=1}^{N} q_i$$

(6.71)

For identical, noninteracting particles $U_{n_a} = 0$, $q_i = q_j = \cdots = q$. Equation (6.71) becomes

$$Q_{canon}^N = g_a q^N$$

(6.72)

where g_a is the number of equivalent arrangements of the molecules.

A similar derivation shows

$$Q_{grand} = \sum_{n_a} g_{n_a} e^{-\beta(U_a + \mu N)} q^N$$

(6.73)

Rearranging, we obtain

$$Q_{grand} = \sum_{n_a} g_{n_a} (e^{-\beta(U_a + N(\mu - k_B T \ln q))})$$

(6.74)

6.4.7 The Molecular Partition Function

In the next section, we will consider some of the properties of q, the partition function for an individual molecule. From equation (6.70)

$$q = \sum_{n_i} g_{n_i} e^{-\beta U_{n_i}}$$

(6.75)

It is useful to divide the internal energy of molecule i into the rotational, translational, vibrational, electronic, and zero-point energy for molecule i. A derivation similar to that in Section 6.4.4 shows that for any molecule

$$q = \Pi_i q_i$$

(6.76)

where q_i is a partition function for an individual vibrational, translational, rotational, or electronic mode of the molecule. Equation (6.76) is quite important, because it allows one to estimate the partition function for a molecule.

In order to use equation (6.76), one has to know how many modes a molecule has. According to quantum mechanics, a molecule with n_a atoms will have $3n_a$ total vibrational, translational, and rotational modes; thus, a molecule with three atoms will have nine total translational, vibrational, and rotational modes. A molecule with nine atoms will have 27 total translational, vibrational, and rotational modes. Every molecule can translate in three dimensions, so it will have three translational modes. A nonlinear molecule can rotate around three axes, so it will have three rotational modes. A linear molecule can rotate in only two dimensions, because rotation around the molecular axis does not move any of the atoms. Consequently, a linear molecule has only two rotational modes. Everything else must be vibration, so a nonlinear molecule will have $3n - 6$ vibrational modes while a linear molecule will have $3n - 5$ vibrational modes.

In order to calculate the partition function exactly for a molecule, one has to calculate the partition function for each of the modes of the molecule and multiply, as in equation (6.76). Often, however, one needs only the orders of magnitude of the partition function. In that case, one can calculate the partition function using average values of all of the parameters, as given in Table 6.5. For a nonlinear molecule, one obtains

$$q = (q_T)^3 (q_r)^3 (q_V)^{3n_a-6} g_e e^{-\beta U_0} \qquad (6.77)$$

where U_0 is the zero point energy of the molecule and q_T, q_r, q_V are the partition functions for the individual translational, rotational, and vibrational modes of the molecule. In deriving equation (6.77), we made explicit use of the observation that a nonlinear molecule has three translational modes, three rotational modes, and $3n_a - 6$ vibrational modes, where n_a is the number of atoms in the molecule. If the individual rotational and vibrational modes are different, one should use a product of the partition functions for the individual modes rather than some average partition function to some power in equation (6.77).

We do not have room in this book to derive equations for q_V, q_r, and q_t. Derivations appear in most introductory physical chemistry texts and will not be repeated here. The final equations are given in Table 6.6.

Table 6.7 shows the same table, except that we have inserted numerical values of k_B and h_p into the equations, and converted the results into convenient units. Derivations of these results are given in Examples 6.B and 6.C.

There is one other important distinction. Later in this book we will use two different translational partition functions, $(q_T)^3$ and $(q_t)^3$. $(q_T)^3$ is the normal translational partition function defined by equation 6.75. It is cubed because there are three translational modes. It happens that the translational partition function is proportional to the molecular volume, and so is has been common to define another translational partition function $(q_t)^3$. $(q_t)^3$ is given by

$$(q_t)^3 = \frac{(q_T)^3}{V_M}$$

where V_M is the volume per molecule. One should be careful with these two definitions.

6.4.8 The Classical Partition Function

Equations (6.10)–(6.12) were derived for a quantum system with distinct energy levels. They are most useful when we know all of the energy levels of the system. However,

Table 6.7 Simplified expressions for the average velocity and the translational, rotational, and vibrational partition function (derivations are given in Examples 6.B and 6.C)

Type of Mode	Partition Function	Partition Function after Substituting in Values of k_B, h_p
Average velocity of a molecule	$\bar{v} = \left(\dfrac{8k_B T}{\pi m_g}\right)^{1/2}$	$\bar{v} = 2.52 \times 10^{13}\, \dfrac{\text{Å}}{\text{second}} \left(\dfrac{T}{300\ \text{K}}\right)^{1/2} \left(\dfrac{1\ \text{amu}}{m_g}\right)$
Translation of a molecule in three dimensions (partition function per unit volume)	$q_t^3 = \dfrac{(2\pi m_g k_B T)^{3/2}}{(h_p)^3}$	$q_t^3 = \dfrac{0.977}{\text{Å}^3} \left(\dfrac{T}{300\ \text{K}}\right)^{3/2} \left(\dfrac{m_g}{1\ \text{amu}}\right)^{3/2}$
Rotation of a linear molecule	$q_r^2 = \dfrac{8\pi I k_B T}{S_n (h_p)^2}$	$q_r^2 = \left(\dfrac{12.4}{S_n}\right) \left(\dfrac{T}{300\ \text{K}}\right) \left(\dfrac{I}{1\ \text{Å·amu}}\right)$
Rotation of a nonlinear molecule	$q_r^2 = \dfrac{(8\pi k_B T)^{3/2}(I_a I_b I_c)^{1/2}}{S_n (h_p)^3}$	$q_r^3 = \left(\dfrac{43.7}{S_n}\right) \left(\dfrac{T}{300\ \text{K}}\right)^{3/2} \left(\dfrac{I_a I_b I_c}{1\ \text{Å}^3 \cdot \text{amu}^3}\right)^{3/2}$
Vibration of a harmonic oscillator	$q_v = \dfrac{1}{1 - \exp(-h_p \nu / k_B T)}$	$q_v = \dfrac{1}{1 - \exp\left(-\left(\dfrac{\nu}{209.2\ \text{cm}^{-1}}\right)\left(\dfrac{300\ \text{K}}{T}\right)\right)}$

with a complex system, we often know only the force laws between the molecules, and therefore, the classical dynamics of the molecules and not the energy levels. In such systems, it is easier to do calculations based on classical mechanics rather than quantum mechanics. Therefore, it is useful to review the connection between classical partition functions and quantum partition functions.

In a quantum system, we usually define the configuration of a molecule in terms of the position and momentum of the center of mass of the molecule plus the vibrational, rotational, and electronic quantum numbers (i.e., energy levels) for the molecule. Classically, however, a molecule does not have distinct rotational or vibrational levels. Instead, all of the atoms in the molecule have distinct positions and momentums. As a result, Boltzmann (1872) defined a classical partition function, $Q'_{classical}$, via

$$Q'_{classical} = \int \int \int \cdots \int e^{-\beta U}\, d\vec{r}_1\, d\vec{r}_2 \cdots d\vec{r}_m\, d\vec{p}_1\, d\vec{p}_2 \cdots d\vec{p}_m \qquad (6.78)$$

where the $\vec{r}_i$ and $\vec{p}_i$ terms are respectively the position and momentum of all of the atoms in the system, and m is the number of atoms. Note that $Q'_{classical}$ is not dimensionless, while Q^N_{canon} and Q_{grand} are. However, one can define a different classical partition function:

$$Q_{classical} = \dfrac{Q'_{classical}}{h_p^{3m}} = \int \int \int \cdots \int e^{-\beta U}\, d\vec{r}_1\, d\vec{r}_2 \cdots d\vec{r}_m\, d\vec{p}_1\, d\vec{p}_2 \cdots d\vec{p}_m / h_p^{3m} \qquad (6.79)$$

where $Q_{classical}$ is dimensionless and h_p is Planck's constant. It happens for a system with very closely spaced energy levels:

$$Q_{classical} \approx Q_{canon}^N \tag{6.80}$$

As a result, one can calculate approximate values for the internal energy, entropy, and free energy by substituting $Q_{classical}$ for Q_{canon}^N in equations (6.26), (6.47), and (6.45), respectively. Classically one can calculate $\langle f \rangle$, the average value of any function f, from

$$\langle f \rangle = \frac{1}{Q_{classical}\,(h_p)^{3m}} \int\int \cdots \int f e^{-\beta U}\, d\vec{r}_1\, d\vec{r}_2 \cdots d\vec{r}_m \quad d\vec{p}_1\, d\vec{p}_2 \cdots d\vec{p}_m \tag{6.81}$$

6.5 SUPPLEMENTAL MATERIAL: MOLECULAR DYNAMICS

An alternative algorithm is to use a technique called **molecular dynamics** to calculate the behavior of the system. The general idea in molecular dynamics is to use the solutions of Newton's equations of motion, like those you learned in freshman physics, to describe the behavior of molecules. Recall back to your first semester of freshman physics when they discussed Newton's equations of motion (i.e., $\vec{F} = m\vec{a}$). Newton's equations of motion describe how particles move in the presence of an external field, and when particles collide. For example, you can use Newton's equations of motion of describe how a cannonball moves under the influence of gravity. You can also use Newton's equation of motion to describe what happens when two billiard balls collide.

In the same way, one can use Newton's equations of motion to describe what happens when molecules collide. Consider a simple reaction where A collides with BC to produce products:

$$A + BC \longrightarrow AB + C \tag{6.82}$$

Atoms A, B, and C obey

$$\frac{d\vec{x}_A}{dt} = \vec{v}_A \tag{6.83}$$

$$\frac{d\vec{x}_B}{dt} = \vec{v}_B \tag{6.84}$$

$$\frac{d\vec{x}_C}{dt} = \vec{v}_C \tag{6.85}$$

$$m_A \frac{d\vec{v}_A}{dt} = \vec{F}_A \tag{6.86}$$

$$m_B \frac{d\vec{v}_B}{dt} = \vec{F}_B \tag{6.87}$$

$$m_C \frac{d\vec{v}_C}{dt} = \vec{F}_C \tag{6.88}$$

where $\vec{x}_A$, $\vec{x}_B$, and $\vec{x}_C$ are the position vectors for atoms A, B, and C, and $\vec{v}_A$, $\vec{v}_B$, and $\vec{v}_C$ are their velocities; $\vec{F}_A$, $\vec{F}_B$, and $\vec{F}_C$ are the forces on the molecules; and m_A, m_B, and m_C are the masses of atoms A, B, and C, respectively. Therefore, one can calculate how all of the atoms move as a function of time by numerically integrating equations (6.83)–(6.88).

The hard part is determining the forces. Generally, people use quantum mechanics to calculate $E(\vec{x}_A, \vec{x}_B, \vec{x}_C)$, the total energy of the system. The force on A is then given by

$$\vec{F}_A = -\frac{dE(\vec{x}_A, \vec{x}_B, \vec{x}_C)}{d\vec{x}_A} \qquad (6.89)$$

Note that the derivative in equation (6.89) is a gradient because $\vec{x}_A$ is a vector. If one knows E, one can calculate the position of the atoms as a function of time. We will show how to calculate E in Chapter 10.

In order to use molecular dynamics (MD) to calculate rates, you

1. Pick initial conditions for the reaction $A + BC \rightarrow AB + C$. The initials conditions include
   a) The velocity of A toward BC
   b) How closely A collides with BC
   c) The initial rotational state of BC, including where B is relative to C
   d) The initial vibrational state of BC, including how far apart B is from C.
2. Numerically integrate equations (6.83)–(6.88) to see what happens when A collides with BC.
3. Repeat the calculation thousands of millions of times starting with different initial conditions.

You can then average to calculate a reaction rate.

6.6 SOLVED EXAMPLES

Example 6.A   Calculation of the Equilibrium Constant for Adsorption of a Gas on a Finite Number of Sites   We will use statistical mechanics to calculate equilibrium constants later in this book. In this section, we will calculate the equilibrium constant for a simple case. The adsorption of a gas, assuming that the gas adsorbs onto a fixed array at sites, is

$$A + S \rightleftharpoons A_{ad} \qquad (6.A.1)$$

The array of sites can be a two-dimensional array of sites as on a surface, or a three-dimensional array of sites, as in a liquid or a solid. To simplify the analysis, we will assume that there are no interactions between molecules on adjacent sites.

Our approach will be to use statistical mechanics to calculate the chemical potential of the gas and the surface layer, and then set the two expressions equal to calculate the equilibrium constant. In practice, we will first calculate an expression for the canonical partition function for the adsorbed layer, use that expression to calculate the chemical potential of the adsorbed layer, and set the chemical potential on the surface equal to the chemical potential in the gas phase to calculate an equilibrium constant.

Solution

Step 1. Calculate the canonical partition function. Let's start with the canonical partition function. According to equation (6.72), $Q_{canon}^N = g_a q^N$.

    Step 1a. Calculate g_a. Let's calculate an expression for g_a, the number of equivalent surface arrangements. Consider N_a different (e.g., distinguishable) molecules adsorbing on S_0 sites. The first molecule can adsorb on S_0 sites, the second molecule can adsorb on $(S_0 - 1)$ sites, and so on. Therefore, the total number of arrangements is given by

$$g_a^D = (S_0)(S_0 - 1)(S_0 - 2) \cdots (S_0 - N_a + 1) = \frac{S_0!}{(S_0 - N_a)!} \quad (6.A.2)$$

If the N_a molecules are indistinguishable, several of these arrangements are equivalent. The number of equivalent arrangements can be generated by considering the N_a sites which hold molecules. If the first molecule is on any N_a of these sites, and the second molecule is on any $N_a - 1$ of those sites, and so forth, the arrangement will be equivalent. The number of equivalent arrangements is given by

$$N_a(N_a - 1)(N_a - 2) \cdots 1 = N_a! \quad (6.A.3)$$

Therefore, the total number of inequivalent arrangements will be given by

$$g_a = \frac{S_0!}{(S_0 - N_a)!N_a!} \quad (6.A.4)$$

Step 1b. Combine to calculate Q_{canon}^N. Combining equations (6.72) and (6.A.4) shows that for a lattice gas with no adsorbate–adsorbate interactions

$$Q_{canon}^N = \frac{S_0!}{(S_0 - N_a)!N_a!}(q_A)^{Na} \quad (6.A.5)$$

where q_A is the molecular partition function for an adsorbed molecule.

Step 2. Calculate the Helmholtz free energy. The Helmholtz free energy at the layer, A_s is given by

$$A_s = k_B T \ln(Q_{canon}^N) \quad (6.A.6)$$

Combining equations (6.A.5) and (6.A.6) yields

$$A_s = -k_B T\{N_a \ln(q_A) + \ln(S_0!) - \ln(N_a!) - \ln(S_0 - N_a)!\} \quad (6.A.7)$$

One can use Stirling's approximation to simplify equation (6.A.7). According to Stirling's approximation

$$\ln(X!) = X \ln X - X \quad (6.A.8)$$

for any X. If one uses equation (6.A.8) to evaluate the log terms in equation (6.A.7), one obtains

$$A_s = -k_B T\{N_a \ln(q_A) + S_0 \ln S_0 - N_a \ln N_a - (S_0 - N_a)\ln(S_0 - N_a)\} \quad (6.A.9)$$

Step 3. Calculate the chemical potential of the adsorbed layer. The chemical potential of the layer, μ_s is defined by

$$\mu_s = \left(\frac{\partial A_s}{\partial N_s}\right)_{S_0, T} \quad (6.A.10)$$

Substituting equation (6.A.9) into equation (6.A.10) yields

$$\mu_s = k_B T\{\ln(N_a) - \ln(S_0 - N_a) - \ln q_A\} \tag{6.A.11}$$

Step 4. Calculate the chemical potential for the gas. Next, let's calculate μ_s, the chemical potential for an ideal gas at some pressure, P. Let's consider putting N_g molecules of A in a cubic box that has longer L on a side. If the molecules are indistinguishable, we freeze all of the molecules in space. Then we can switch any two molecules, and nothing changes. There are $N_g!$ ways of arranging the N_g molecules. Therefore

$$g_a = \frac{1}{N_g!} \tag{6.A.12}$$

substituting equation (6.A.12) into equation (6.72) yields

$$Q^N_{canon} = \frac{(q_g)^{N_g}}{(N_g)!} \tag{6.A.13}$$

where A_g is the Helmholtz free energy in the gas phase and q_g is the partition function for the gas-phase molecules. Substituting equation (6.A.13) into equation (6.A.6), applying Stirling's approximation, and substituting the resulting equation into equation (6.A.10) yields

$$\mu_g = k_B T\{\ln(q_g) - \ln N_g\} \tag{6.A.14}$$

Step 5. Set $\mu_g = \mu_a$ to calculate how much adsorbs. At equilibrium

$$\mu_s = \mu_a \tag{6.A.15}$$

Substituting equation (6.A.11) and (6.A.14) into equation (6.A.15) and rearranging yields

$$\ln\left(\frac{N_a}{N_g(S_0 - N_a)}\right) = \ln\left(\frac{q_a}{q_g}\right) \tag{6.A.16}$$

Taking the exponential of both sides of equation (6.A.16); we obtain

$$\frac{N_a}{N_g(S_0 - N_a)} = \frac{q_a}{q_g} \tag{6.A.17}$$

Note that N_a is the number of molecules in the gas phase. N_a is the number of adsorbed molecules and $(S_0 - N_a)$ is the number of bare sites. Consequently, the left-hand side of equation (6.A.17) is equal to K_A, the equilibrium constant for the reaction

$$A_g + S \Longleftrightarrow A_{(ad)} \tag{6.A.18}$$

Consequently

$$K_A = \frac{q_a}{q_g} \tag{6.A.19}$$

Equation (6.A.19) is a key result. It says that the equilibrium constant for adsorption is equal to the ratio of the partition function in the adsorbed phase to the partition function in the gas phase. We derived equation (6.A.19) for adsorption, but actually the results are general. We can always write the equilibrium constant for any process as the ratio of the partition functions for two phases.

In the homework set we ask the reader to use (6.A.14) to show that K_{AB}, the equilibrium constant for the gas-phase reaction

$$A + B \rightleftharpoons C \tag{6.A.20}$$

is given by

$$K_{AB} = \frac{q_C}{q_A q_B} \tag{6.A.21}$$

Equation (6.A.21) is a key result because it says that one can calculate equilibrium constants as a ratio of partition functions.

Example 6.B   Calculation of Molecular Velocities.

(a) Derive an expression for the average velocity of a ideal gas molecule.

(b) Derive an expression for the average internal energy.

(c) Plug in numbers into your expression at temperature $T = 273$ K.

Solution

(a) Molecular velocities can be calculated using the classical partition function, equation (6.78). According to equation (6.80), one can calculate the expectation value of the molecular velocity, $\langle v \rangle$, from

$$\langle v \rangle = \frac{1}{Q_{classical}} \frac{1}{h^{3m}} \int \int \cdots \int v e^{-\beta U} d\vec{r}_1 d\vec{r}_2 \cdots d\vec{r}_m d\vec{p}_1 d\vec{p}_2 \cdots d\vec{p}_m \tag{6.B.1}$$

We will consider a single molecule whose energy is independent of position. Substituting $Q_{classical}$ from equation (6.79) into equation (6.A.20), noting that the momentum p is equation to the mass times the velocity, and canceling out all of the excess integrals yields

$$\langle v \rangle = \frac{\int \int \int v e^{-\beta U} d\vec{v}}{\int \int \int e^{-\beta U} d\vec{v}} \tag{6.B.2}$$

where $\vec{v}$ is the velocity vector. In freshman physics, you learned that the energy of a particle moving with a velocity v is

$$U = \tfrac{1}{2} m_g v^2 \tag{6.B.3}$$

From vector calculus you know that in spherical coordinates.

$$d\vec{v} = v^2 \, dv \sin\theta \, d\theta \, d\phi \tag{6.B.4}$$

Substituting (6.B.3) and (6.B.4) into equation (6.B.2) yields

$$\langle v \rangle = \frac{\int_0^{2\pi} \int_0^{\pi} \int_0^{\infty} v \exp\left(-\frac{\beta}{2} m_g v^2\right) v^2 \, dv \sin\theta \, d\theta \, d\phi}{\int_0^{2\pi} \int_0^{\pi} \int_0^{\infty} \exp\left(-\frac{\beta}{2} m_g v^2\right) v^2 \, dv \sin\theta \, d\theta \, d\phi}$$

(6.B.5)

Performing the θ and ϕ integrals yields

$$\langle v \rangle = \frac{\int_0^{\infty} v^3 \exp\left(-\frac{\beta}{2} m_g v^2\right) dv}{\int_0^{\infty} v^2 \exp\left(-\frac{\beta}{2} m_g v^2\right) dv}$$

(6.B.6)

Looking up the integrals in the CRC

$$\langle v \rangle = \frac{\dfrac{1}{2 \left(\dfrac{\beta m_g}{2}\right)^2}}{\left(\dfrac{1}{\left(\dfrac{\beta m_g}{2}\right)}\right) \sqrt{\dfrac{2\pi}{\beta m_g}}}$$

(6.B.7)

Performing the algebra noting $\beta = 1/k_B T$ yields

$$\langle v \rangle = \left(\frac{8 k_B T}{\pi m_g}\right)^{1/2}$$

(6.B.8)

(b) A similar derivation shows that the average translational energy, $\langle U_T \rangle$, is

$$\langle U_T \rangle = \frac{\iiint \frac{1}{2} m_g v^2 e^{-\beta U} \, d\vec{v}}{\iiint e^{-\beta U} \, d\vec{v}} = \frac{3}{2} k_B T$$

(6.B.9)

(c) Substituting numbers into equation (6.B.8) allows one to calculate the following table of molecular velocities of gases at 0°C:

Molecule	Average Velocity, $\langle v \rangle$, meters/second	Molecular Diameter
Hydrogen	1687	2.74
Helium	1197	2.18
Carbon monoxide	453	3.12
Nitrogen	453	2.74
Krypton	262	4.16
Xenon	209	4.85

The numbers in this table are calculated in O'Hanlon (1980).

Example 6.C   Calculation of the Partition Function for HBr at 300 K   Consider the following data:

Frequency, ν	2650 cm^{-1}
Bond length	1.414 Å
m_H	1 amu
m_{Br}	80 amu

Calculate the (a) translational, (b) rotational, and (c) vibrational partition functions for HBr.

Solution

(a) **The translational partition function per unit volume.** From Table 6.7

$$q_t^3 = \frac{(2\pi m_g k_B T)^{3/2}}{h_p^3} \tag{6.C.1}$$

where q_t is the translational partition function per unit volume m_g is the mass of the gas atom in atomic mass units (amu), k_B is Boltzmann's constant, T is the temperature, and h_p is Planck's constant. Equation (6.C.1) is not that convenient, so first I wish to derive a simple expression. One can rewrite equation (6.C.1) as

$$q_t^3 = \left(\frac{m_g}{1\ amu}\right)^{3/2} \left(\frac{T}{300\ K}\right)^{3/2} \frac{(2\pi \times 1\ amu \times k_B \times 300\ K)^{3/2}}{h_p^3} \tag{6.C.2}$$

Next, let us evaluate the third term on the right of equation (6.C.2)

$$\frac{(2\pi \times 1\ amu \times k_B \times 300\ K)^{3/2}}{h_p^3}$$

$$= \frac{\left(2\pi(1\ amu)\right)\left(\dfrac{1.66 \times 10^{-27}\ kg}{1\ amu}\right)\left(1.381 \times 10^{-23}\dfrac{kg \cdot m^2}{second^2 \cdot K}\right)(300\ K)\Big)^{3/2}}{\left(6.626 \times 10^{-34}\dfrac{kg \cdot m^2}{second}\right)^3 \left(\dfrac{10^{10}\ Å}{m}\right)^3}$$

$$= \frac{0.977}{Å^3} \tag{6.C.3}$$

Combining 6.C.2 and 6.C.3 yields

$$q_t^3 = \left(\frac{m_g}{1\ amu}\right)^{3/2} \left(\frac{T}{300\ K}\right)^{3/2} \frac{0.977}{Å^3} \tag{6.C.4}$$

Equation (6.C.4) is the equation we will actually use to evaluate the translational partition function. For our case $m_g = 81$ amu, T = 300 K. Substituting in the numbers, we obtain

$$q_t = \left(\frac{81\ amu}{1\ amu}\right)^{3/2} \left(\frac{300\ K}{300\ K}\right)^{3/2} \frac{0.977}{Å^3} = \frac{712}{Å^3} \tag{6.C.5}$$

(b) The rotational partition function. According to Table 6.6

$$q_r^2 = \frac{8\pi^2 I\, k_B T}{S_n h_p^2} \tag{6.C.6}$$

where q_r is the rotational partition function, I is the moment of inertia, k_B is the Boltzmann's constant, h_p is Planck's constant, T is the temperature, and S_n is a *symmetry number* (one for HBr). Again I will first derive a simplified equation and then use it for calculations. Rearranging equation (6.C.6) yields

$$q_r^2 = \left(\frac{8\pi^2 (1\text{ amu·Å}^2) k_B (300\text{ K})}{h_p^2} \right) \left(\frac{1}{S_n} \right) \left(\frac{T}{300\text{ K}} \right) \left(\frac{I}{1\text{ amu·Å}^2} \right) \tag{6.C.7}$$

Next, let's calculate the first term on the right of equation (6.C.7):

$$\frac{8\pi k_B (300\text{ K})(1\text{ amu·Å}^2)}{h_p^2}$$

$$= \frac{8\pi^2 \left(1.38 \times 10^{-23} \dfrac{\text{kg·meter}^2}{\text{second}^2 \text{·K}} \right) (300\text{ K})(1\text{ amu·Å}^2)}{\left(6.626 \times 10^{-34} \dfrac{\text{kg·meter}^2}{\text{second}} \right)^2 \left(\dfrac{10^{10}\text{Å}}{\text{meter}} \right)^2 \left(\dfrac{1\text{ amu}}{1.66 \times 10^{-27}\text{ kg}} \right)}$$

$$= 12.4 \tag{6.C.8}$$

Combining (6.C.7) and (6.C.8) yields

$$\boxed{q_r^2 = 12.4 \left(\frac{T}{300\text{ K}} \right) \left(\frac{I}{1\text{ amu·Å}^2} \right) \left(\frac{1}{S_n} \right)} \tag{6.C.9}$$

Equation 6.C.4 is the equation we will use to calculate rotational particle function. Next, I will find a value for I. From your physical chemistry book

$$I = \mu (r_{AB})^2 \tag{6.C.10}$$

where

$$\mu = \frac{m_H m_{Br}}{(m_H + m_{Br})} \tag{6.C.11}$$

where m_H and m_{Br} are the atomic masses at H and Br. Plugging values of m_H and m_{Br} into equation (6.C.11) yields

$$\mu = \frac{(1\text{ amu})(80\text{ amu})}{1\text{ amu} + 80\text{ amu}} = 0.988\text{ amu} \tag{6.C.12}$$

Plugging μ into equation (6.C.10), with $r_{AB} = 1.414$ Å, yields

$$I = (0.988\text{ amu})(1.414\text{ Å})^2 = 1.97\text{ amu·Å}^2 \tag{6.C.13}$$

Substituting in I from equation (6.C.13) and $S_n = 1$ into equation (6.C.9) yields

$$q_r^2 = 12.4 \left(\frac{300 \text{ K}}{300 \text{ K}}\right) \left(\frac{1.97 \text{ amu·Å}^2}{1 \text{ amu·Å}^2}\right) \left(\frac{1}{1}\right) = 24.4 \tag{6.C.14}$$

(c) The vibrational partition function. From Table 6.6

$$q_v = \frac{1}{1 - \exp(-h_p v / k_B T)} \tag{6.C.15}$$

where q_v is the vibrational partition function, h_p is Planck's constant, v is the vibrational frequency, k_B is Boltzmann's constant, and T is the temperature. First, it is useful to derive an expression for the term in the exponent. One can show

$$\frac{h_p v}{k_B T} = \left(\frac{300 \text{ K}}{T}\right) \left(\frac{v}{1 \text{ cm}^{-1}}\right) \frac{h_p(1 \text{ cm}^{-1})}{k_B(300 \text{ K})} \tag{6.C.16}$$

I found the conversion factor:

$$\frac{h_p(1 \text{ cm}^{-1})}{k_B(300 \text{ K})} = 4.78 \times 10^{-3} \tag{6.C.17}$$

Therefore

$$\frac{h_p v}{k_B T} = 4.78 \times 10^{-3} \left(\frac{v}{1 \text{ cm}^{-1}}\right) \left(\frac{300 \text{ K}}{T}\right) \tag{6.C.18}$$

Equation (6.C.18) is the expression we will use to evaluate $\dfrac{h_p v}{k_B T}$. For our case

$$\frac{hv}{k_B T} = (4.78 \times 10^{-3}) \left(\frac{2650 \text{ cm}^{-1}}{1 \text{ cm}^{-1}}\right) \left(\frac{300 \text{ K}}{300 \text{ K}}\right) = 12.7 \tag{6.C.19}$$

Substituting (6.C.19) into (6.C.15) yields

$$q_v = \frac{1}{1 - \exp(-12.7)} = 1.0 \tag{6.C.20}$$

Example 6.D   Derivation of the Ideal-Gas Law   Use the results in this chapter to derive the ideal-gas law. Consider loading N identical nonlinear molecules into a box of volume V.

Solution   According to equation (6.48):

$$P = k_B T \left(\frac{d \ln Q_{\text{canon}}^N}{dV}\right)_{T,N} \tag{6.D.1}$$

In order to use equation (6.D.1), we need an expression for Q_{canon}^N. According to equation (6.72).

$$Q_{\text{canon}}^N = g_a q^N \tag{6.D.2}$$

In order to use equation (6.D.2), we will need an expression for q. According to equation (6.77)

$$q = q_T^3 q_r^3 q_v^{3Na-6} g_e e^{-\beta U_0} \tag{6.D.3}$$

where q_T is the translational partition function for the purpose of derivation, we will also need an expression for q_T^3. From Table 6.6, we have

$$q_T^3 = \frac{(2\pi m_g k_B T)^{3/2}}{h_p^3} a_x a_y a_z \tag{6.D.4}$$

Note that $a_x a_y a_z$ is the volume per molecule. Consequently

$$a_x a_y a_z = \frac{V}{N} \tag{6.D.5}$$

Combining equations (6.D.2)–(6.D.5) yields

$$Q_{canon}^N = g_a q_r^{3N} q_v^{3N} \frac{(2\pi m_g k_B T)^{3N/2}}{h_p^{3N} N^N} V^N \tag{6.D.6}$$

Taking the log of both sides of equation (6.D.6) yields

$$\ln Q_{canon}^N = \ln\left(g_a q_r^{3N} q_v^{3N} \frac{(2\pi m_g k_B T)^{3N/2}}{h_p^{3N} N^N}\right) + N \ln V \tag{6.D.7}$$

Combining equations (6.D.7) and (6.42) yields

$$P = k_B T \left(\frac{\partial}{\partial V} \ln\left(g_a g_r^{3N} q_v^{3N} \frac{(2\pi m_g k_B T)^{3N/2}}{h_p^{3N} N^N}\right)\right)_{T,N} + N k_B T \frac{\partial \ln V}{\partial V} \tag{6.D.8}$$

The first term on the right side of equation (6.D.8) is zero since none of the terms depend on the volume. Therefore

$$P = N k_B T \frac{\partial \ln V}{\partial V} = \frac{N k_B T}{V} \tag{6.D.9}$$

Rearranging equation (6.D.9) yields

$$\boxed{PV = N k_B T} \tag{6.D.10}$$

Example 6.E   An Illustration of the Use of Equations (6.15) and (6.20) The objective of this problem is to illustrate the use of equations (6.15) and (6.20). Consider two atoms attached to a wall via springs. Calculate the canonical partition function for the vibration of the atoms, assuming that $k_B T = 1$ kcal/mol and that the atoms are distinguishable. Assume that each atom has three vibrational energy levels as follows:

Atom 1 (Top Atom)		Atom 2 (Bottom Atom)	
n_1	U_1, kcal/mol	n_2	U_2, kcal/mol
1	−4	1	−5
2	−3	2	−3
3	−0.5	3	−1

Solution 1   We can enumerate the states of the system and their energies in the table that follows. In the table, n_1 and n_2 indicate the states of atoms 1 and 2, n is an index of the energy levels of the system, U_1 and U_2 are the vibrational energies of atoms 1 and 2, and U_n is the total energy of the system. For future reference, we also tabulate values of g_n, the degeneracy of each of the energy levels. (All of the degeneracies are 1.)

n	n_1	n_2	U_1	U_2	U_n	g_n
1	1	1	−4	−5	−9	1
2	1	2	−4	−3	−7	1
3	1	3	−4	−1	−5	1
4	2	1	−3	−5	−8	1
5	2	2	−3	−3	−6	1
6	2	3	−3	−1	−4	1
7	3	1	−0.5	−5	−5.5	1
8	3	2	−0.5	−3	−3.5	1
9	3	3	−0.5	−1	−1.5	1

For $\beta = 1$, the canonical partition function becomes

$$Q_{canon}^N = \sum_{n=1}^{9} g_n e^{-\beta U_n} = e^9 + e^7 + e^5 + e^8 + e^6 + e^4 + e^{5.5} + e^{3.5} + e^{1.5} = 13{,}069$$

(6.E.1)

Solution 2   Next, I want to derive some important results. A key point for later work is that we do not have to sum over n. Rather, we can sum over any group of variables provided we sum each state in the system once and only once. For example, instead of summing over n, we can sum over n_1 and n_2:

$$Q_{canon}^N = \sum_{n_1=1}^{3} \sum_{n_2=1}^{3} g_n e^{-\beta U_n}$$

(6.E.2)

Note that there are still nine terms in the sum in equation (6.E.2). Substituting, $g_n = 1$ and $U_n = U_1 + U_2$ into equation (6.E.2) yields

$$Q_{canon}^N = \sum_{n_1=1}^{3} \sum_{n_2=1}^{3} e^{-\beta(U_1+U_2)}$$

(6.E.3)

Note that U_1 does not depend on n_2. One can rearrange the sum in equation (6.E.3) as follows:

$$Q_{canon}^N = \sum_{n_1=1}^{3} e^{-\beta U_1} \sum_{n_2=1}^{3} e^{-\beta U_2} \qquad (6.E.4)$$

or:

$$Q_{canon}^N = \left(\sum_{n_1=1}^{3} e^{-\beta U_1}\right)\left(\sum_{n_2=1}^{3} e^{-\beta U_2}\right) = q_1 q_2 \qquad (6.E.5)$$

Equation (6.E.5) says that the canonical partition function is equal to the product of two molecular partition functions, where

$$q_1 = \sum_{n_1=1}^{3} e^{-\beta U_1} \qquad (6.E.6)$$

$$q_2 = \sum_{n_2=1}^{3} e^{-\beta U_2} \qquad (6.E.7)$$

Plugging in the numbers, we obtain

$$Q_{canon}^N = (e^4 + e^3 + e^{0.5})(e^5 + e^3 + e^1) = (76.33)(171.2) = 13{,}069 \qquad (6.E.8)$$

Conclusion: For independent particles, we can sum over n, or sum over n_1 and n_2, and get the same answer.

The Interacting Case   So far, I have treated n as a scalar. However, I can also treat n as a vector, specifically, $n = (n_1, n_2)$. So the (1,3) state will have $n_1 = 1$, $n_2 = 3$; thus, the (1,3) state is the $n = 3$ state in the table in Solution 1 of Example 6.E. Note that everything will be the same regardless of whether I treat n as a scalar or as a vector.

Another key point is that there is nothing in the derivation that required the two atoms to be independent. For example, consider two interacting atoms, where the strength of the interaction is $U_{1,2}$:

n	n_1	n_2	U_1	U_2	$U_{1,2}$	U_n	g_n
1	1	1	−4	−5	0.5	−8.5	1
2	1	2	−4	−3	0	−7	1
3	1	3	−4	−1	0	−5	1
4	2	1	−3	−5	0	−8	1
5	2	2	−3	−3	0.5	−5.5	1
6	2	3	−3	−1	0	−4	1
7	3	1	−0.5	−5	0	−5.5	1
8	3	2	−0.5	−3	0	−3.5	1
9	3	3	−0.5	−1	0.5	−1.0	1

We can calculate the partition function as before:

$$Q_{canon}^N = \sum_{n=1}^{9} g_n e^{-\beta U_n} = e^{8.5} + e^7 + e^5 + e^8 + e^{5.5} + e^4 + e^{5.5} + e^{3.5} + e^1 = 9721$$

$$(6.E.9)$$

Again, instead of summing over n, we can sum over n_1 and n_2:

$$Q_{canon}^N = \sum_{n_1=1}^{3} e^{-\beta U_1} \sum_{n_2=1}^{3} g_n e^{-\beta(U_n-U_1)} \qquad (6.E.10)$$

Now the energy does not separate, so there is no advantage to doing things in this manner. However, the formulation still works.

The Role of Degeneracy   Next, I want to change topics and examine the role of degeneracy. Next, let's consider two equivalent particles and see how degeneracy enters the mix. First, I will assume that the atoms are distinguishable.

Atom 1 (Top Atom)		Atom 2 (Bottom Atom)	
n_1	U_1, kcal/mol	n_2	U_2, kcal/mol
1	−5	1	−5
2	−3	2	−3
3	−1	3	−1

We can enumerate the states of the system as follows:

n	n_1	n_2	U_1	U_2	U_n	g_n
1	1	1	−5	−5	−10	1
2	1	2	−5	−3	−8	1
3	1	3	−5	−1	−6	1
4	2	1	−3	−5	−8	1
5	2	2	−3	−3	−6	1
6	2	3	−3	−1	−4	1
7	3	1	−1	−5	−6	1
8	3	2	−1	−3	−4	1
9	3	3	−1	−1	−2	1

For $\beta = 1$, the canonical partition function becomes

$$Q_{canon}^N = \sum_{n=1}^{9} g_n e^{-\beta U_n} = e^{10} + e^8 + e^6 + e^8 + e^6 + e^4 + e^6 + e^4 + e^2 = 29{,}315 \quad (6.E.11)$$

Next, let's consider the two atoms to be indistinguishable. If we ignore quantum effects, we can treat that case in the same way as before:

Atom 1 (Top Atom)		Atom 2 (Bottom Atom)	
n_1	U_1, kcal/mol	n	U_2
1	−5	1	−5
2	−3	2	−3
3	−1	3	−1

We can enumerate the state of the system as follows:

n	n_1	n_2	U_1	U_2	U_n	g_n
1	1	1	−5	−5	−10	1
2	1	2	−5	−3	−8	1
3	1	3	−5	−1	−6	1
4	2	1	−3	−5	−8	1
5	2	2	−3	−3	−6	1
6	2	3	−3	−1	−4	1
7	3	1	−1	−5	−6	1
8	3	2	−1	−3	−4	1
9	3	3	−1	−1	−2	1

Another way to solve this example is to note that the (1,3) state is equivalent to the (3,1) state, and so on; thus, the (1,3) and (3,1) states are degenerate. Therefore, another way to calculate the partition function for the system is to consider the (1,3) and (3,1) state to be different representations of the same state.

n	(n_1, n_2)	The Greater of U_1 and U_2	The Lesser of U_1 and U_2	U_n	g_n
1	(1,1)	−5	−5	−10	1
2	(1,2) or (2,1)	−5	−3	−8	2
3	(1,3) or (3,1)	−5	−1	−6	2
4	(2,2)	−3	−3	−6	1
5	(2,3) or (3,2)	−3	−1	−4	2
6	(3,3)	−1	−1	−2	1

In this case, we need to put in the degeneracy because the $n = 2$ state has contributions from two independent modes the (1,2) or the (2,1) state. If I set $g_n = 2$, everything works out.

The canonical partition function becomes

$$Q_{canon}^N = \sum_{n=1}^{9} g_n e^{-\beta U_n} = e^{10} + 2e^8 + 2e^6 + e^6 + 2e^4 + e^2 = 29{,}315 \qquad (6.E.12)$$

Conclusion: Classically, the degeneracy depends on how we choose to compute the sum. We can compute it directly (with a degeneracy of 1), or we can lump some terms together (for a degeneracy of 2 or more). Quantum-mechanically, there are corrections to this idea because of something called "exchange". See Hill (1960) for details.

In this book, we will normally choose to work with systems with a degeneracy of 1. However, there are a few examples where we will be forced to use a degeneracy of something other than 1.

Next, let's compute the internal energy of the system using the results above:

$$\langle U \rangle = \frac{\sum\limits_{n=1}^{9} g_n U_n e^{-bU_n}}{Q_{canon}^N}$$

$$= \frac{(-10)e^{10} + (-8)e^8 + (-6)e^6 + (-8)e^8 + (-6)e^6 + (-4)e^4 + (-6)e^6 + (-4)e^4 + (-2)e^2}{29,315} = -9.32 \quad (6.E.13)$$

Example 6.F   Calculation of Properties from Experiment   Next, I want to use the results in this chapter to determine nonequilibrium properties. Assume that I am doing a laser experiment where I shine a laser into a vacuum system filled with some species A. I measure the state of 38,227 of the atoms in the system and obtain the results listed in Table 6.F.1. Calculate the entropy of the system.

Solution   From equation (6.39)

$$S = k_B \sum_n P_n \ln \left(\frac{P_n}{g_n} \right) \quad \text{(entropies in cal/mol-K)} \quad (6.F.1a)$$

$$S = R \sum_n P_n \ln \left(\frac{P_n}{g_n} \right) \quad \text{(entropies in cal/mol-K)} \quad (6.F.1b)$$

We will use the equation (6.F.1b) here. Where k_B is Boltzmann's constant, R is the gas law constant, P_n = the probability that system is in state n. We can calculate the probabilities from $P_n = \dfrac{\text{number of times system is in the state } n}{\text{total observations (i.e. 38227)}}$. The results are given in Table 6.F.2. Plugging into equation (6.F.1.b) yields.

$$S = R \sum_n P_n \ln \left(\frac{P_n}{g_n} \right)$$

$$= [1.98 \text{ cal/(mol·K)}][0.664 \ln(0.664) + 0.082 \ln(0.082) + 0.010 \ln(0.010)$$

Table 6.F.1   The state distribution for Example 6.F

n	n_1	n_2	U_1	U_2	U_n	Number of Times the System is in This State
1	1	1	−5	−5	−10	25,375
2	1	2	−5	−3	−8	3,120
3	1	3	−5	−1	−6	398
4	2	1	−3	−5	−8	2,875
5	2	2	−3	−3	−6	501
6	2	3	−3	−1	−4	2,150
7	3	1	−1	−5	−6	255
8	3	2	−1	−3	−4	3,550
9	3	3	−1	−1	−2	3
Total						38,227

Table 6.F.2 The probability distribution for the data in Table 6.F.1

n	n_1	n_2	U_n	Number of Times the System Is in This State	p_n
1	1	1	−10	25,375	$25,375/38,227 = 0.664$
2	1	2	−8	3,120	$3,120/38,227 = 0.082$
3	1	3	−6	398	$398/38,227 = 0.010$
4	2	1	−8	2,875	$2,875/38,227 = 0.075$
5	2	2	−6	501	$501/38,227 = 0.013$
6	2	3	−4	2,150	$2,150/38,227 = 0.056$
7	3	1	−6	255	$255/38,227 = 0.007$
8	3	2	−4	3,550	$3,550/38,227 = 0.093$
9	3	3	−2	3	$3/38,227 = 0.00008$
Total				38,227	1

$$+ 0.075 \ln(0.075) + 0.013 \ln(0.013) + 0.056 \ln(0.056) + 0.007 \ln(0.007)$$

$$+ 0.093 \ln(0.093) + 0.00008 \ln(0.00008)]$$

$$= 237 \text{ cal/(mol·K)} \tag{6.F.2}$$

Now, I want to go back and make a point. Let's repeat the discussion after equation (6.E.12). In the discussion before equation (6.E.12), we computed the internal energy of a system by actually summing the probabilities. We then computed the internal energy by summing over the states and dividing by the partition function:

$$
\langle U \rangle = \frac{\sum_{n=1}^{9} g_n U_n e^{-\beta U_n}}{Q_{canon}^N}
$$

$$
= \frac{(-10)e^{10} + (-8)e^8 + (-6)e^6 + (-8)e^8 + (-6)e^6 + (-4)e^4 + (-6)e^6 + (-4)e^4 + (-2)e^2}{29,315}
$$

$$
= -9.32 \text{ kcal/mol} \tag{6.F.3}
$$

In the 1930–1960s (i.e., when most statistical mechanics books were written), people actually did calculations this way. An alternate approach was to derive an equation for the partition function and calculate properties. For example

$$
\langle U \rangle = \frac{\sum_{n=1}^{9} g_n U_n e^{-\beta U_n}}{Q_{canon}^N} = \left(\frac{\partial \ln(Q_{canon}^N)}{\partial \beta} \right)_{N,T} \tag{6.F.4}
$$

However, that method works only for the simplest examples where analytic solutions are possible. In reality, you see calculations like that in textbooks, but you seldom see such calculations in the literature. What you really see most are measurements or calculations where you get data like that in Table 6.F.2.

If I can do an experiment or a calculation where I can measure the fraction of the time that a system stays in each state, I can calculate all of the properties of the system.

Example 6.G Monte Carlo Calculations Assume that you do not know the results in Table 6.F.1. Devise an algorithm to compute the probabilities.

Solution This is a tough problem. People took years to develop a suitable algorithm. Fortunately, there is a standard algorithm, called the *Metropolis algorithm*, which works.

The idea behind the algorithm is to step through the canonical ensemble and measure the state of the system at each stage. One then calculates properties as an ensemble average of the properties of the system, as described in Example 6.F.

Consider the system of oscillators described in Example 6.E. The canonical ensemble for this example is a set of systems identical to the first. We can represent that ensemble via a series of boxes. Each box is labeled with the number n, the quantum number for the system, where n goes from one to nine. Figure 6.G.1 shows part of the canonical ensemble for the system. There are a series of boxes, with the state of the system marked. Now consider constructing an algorithm to step through these boxes, by starting in the upper left corner of the figure, moving to the left and then looking back around. One can show that the results calculated in this way will eventually converge to the exact result in the limit of a large enough number of boxes.

Metropolis et al. (1953) showed that one can actually do the calculations as follows:

1. Start at some system in the ensemble.
2. Move to the next system in the ensemble. That system may have the same state or a different state. (See Figure 6.G.1.) Use the following procedure to calculate the next state of the system.

   Choose a second state either at random, or by incrementing or decrementing one of the quantum numbers, or by moving one of the atoms.

   If U goes down, move to that state.

   If U goes up go to the new state when $\exp(-\beta \Delta U) >$ Rand, where Rand is a random number between 0 and 1.

Repeat. Then calculate all properties as an ensemble average.

Metropolis et al. proved that after a long time, the distribution of states generated by such a procedure is equal to the exact distribution of states. This is called the **Monte Carlo method** to calculate properties.

The following is a C program that uses the Metropolis algorithm to calculate the average energy of a system of oscillators:

1	1	3	8	7	5	3	1	1	3	2
2	1	5	4	1	2	1	5	4	2	2

Figure 6.G.1 Part of the canonical ensemble for the example above. Each box represents a different snapshot of the system. The number represents the state of that snapshot.

```
float u[9]={-10.,-8.,-6.,-8.,-6.,-4.,-6.,-4.,-2.};
main()
{ int i,newi,kount;
  float u_avg, delta_u,beta,ran;
  beta=.3333;
  printf("\n enter a value for beta");scanf("%f", &beta);

i=rndf(9); /* rndf(x) returns a random number between 0
   and x */
for(kount=0;kount<100;kount++) /* 100 steps to randomize*/
{ newi=rndf(9);/* choose a new random state for the system */
      delta_u=u[newi]-u[i];
      if(exp(-beta*delta_u)>rndf(1.0))i=newi;

}
u_avg=0;
for(kount=0;kount<1000;kount++) /* 1000 steps to compute avg
energy*/
{  newi=rndf(9);/* choose a new random state for the system */
   delta_u=u[newi]-u[i];
   ran=rndf(1.0);
   if(exp(-beta*delta_u)>ran)i=newi;
   u_avg=u_avg+u[i];

}
   u_avg=u_avg/1000.;
   printf("\n Average Energy = %5.3f",u_avg);
 return 0;
}
```

A compiled version of the program is available from http://www.Wiley.com/chemicalkinetics

Example 6.H   Discussion Problem for Class   Assume that we have a system of nine states whose energies are given in Table 6.F.2. Let's do a Monte Carlo routine by hand using the list of random numbers in Table 6.H.1. The procedure is to

(a) Pick a random state to start the algorithm and start at 2500 K. ($k_B T = 5$ kcal/mol).
(b) Pick a random new state.
(c) Calculate ΔU.
(d) If the energy goes down, go to that state.
(e) If the energy goes up, go to the state if $\exp(-\Delta U/k_B T) >$ Rand, where Rand is the first random number.
(f) Draw a block for the new state.
(g) Repeat steps (b)–(f) 20 times.

What does the distribution of states look like at 2500 K?
Repeat for 500 K ($k_B T$) = 1 kcal/mol.
I ask students to choose random steps in class. We then do the calculations. Try it yourself.

Table 6.H.1   Random numbers for Example 6.H

Step	Rand	Step	Rand	Step	Rand
0	0.417	7	0.914	14	0.435
1	0.984	8	0.607	15	0.289
2	0.069	9	0.183	16	0.614
3	0.460	10	0.593	17	0.542
4	0.885	11	0.157	18	0.745
5	0.484	12	0.838	19	0.471
6	0.506	13	0.473	20	0.210

6.7 SUGGESTIONS FOR FURTHER READING

A good historical account of kinetic theory can be found in:

S. G. Brush, *The Kind of Motion We Call Heat*: *A History of the Kinetic Theory of Gases in the 19th Century*, New York, 1986.

Thermochemical Methods are discussed in:

S. Benson, *Thermochemical Kinetics*, Wiley, New York, 1976.

N. Cohen, Thermochemistry of organic free radicals. In Z.B. Alfassi, ed., *General Aspects of the Chemistry of Radicals*, Wiley, Chichester, 1999.

Applications on surfaces are discussed in:

R. I. Masel, *Principles of Adsorption and Reaction on Solid Surfaces*, Chapter 3, Wiley, New York, 1996.

E. Shustorovich, ed., *Metal-Surface Reaction Energetics*: *Theory and Applications to Heterogeneous Catalysis, Chemisorption, and Surface Diffusion*, VCH, New York, 1991.

General references in statistical mechanics include:

D. Chandler, *Introduction to Modern Statistical Mechanics*, Oxford University Press, New York, 1987.

T. L. Hill, *An Introduction to Statistical Thermodynamics*, Dover, New York, 1986.

D. A. McQuarrie, *Statistical Thermodynamics*, University Science Books, Hill Valley, CA, (2000).

C. E. Hecht, *Statistical Thermodynamics and Kinetic Theory*, Dover, New York, 1998.

T. Davis, *Statistical Mechanics of Phases, Interfaces, and Thin Films*, VCH, New York, 1995.

M. P. Allen, and D. J. Tildgey, *Computer Simulation in Chemical Physics*, Kluwer Academic Publishers, New York, 1993.

J. R. Beeler, *Radiation Effects Computer Experiments*, North-Holland, Publ., Amsterdam, 1983.

H. C. Gould, and J. Tobochnik, *An Introduction to Computer Simulation*, Addison-Wesley, Reading, MA, 1988.

J. M. Hailey, *Molecular Dynamic Simulation*: *Elementary Methods*, Wiley, New York, 1992.

T. L. Hill, *An Introduction to Statistical Thermodynamics*, Addison-Wesley, Reading, MA, 1960.

W. G. Hoover, *Computational Statistical Mechanics*, Elsevier, New York, 1991.

6.8 PROBLEMS

6.1 Define the following in three sentences or less:

(a) Statistical ensemble   (f) Degeneracy

(b) Grand canonical ensemble   (g) Monte Carlo method

(c) Canonical ensemble   (h) Metropolis algorithm

(d) Microcanonical ensemble   (i) Molecular dynamics

(e) Partition function

6.2 Use Benson's methods to estimate the heats of formation of

(a) Neopentane

(b) Diethylether

(c) $CH_3CH_2O\bullet$ adsorbed on platinum

6.3 Use Benson's method to calculate the heat of reaction for the following reactions:

(a) $H + C_2H_6 \rightarrow H_2 + C_2H_5$

(b) $C_2H_6 \rightarrow H_{(ad)} + C_2H_{5(ad)}$ on platinum

(c) n-Butane to 1-methyl-propane

6.4 Use Benson's methods to calculate the equilibrium constant for the following reactions:

(a) $H_2 + C_2H_5OH \leftrightharpoons C_2H_6 + H_2O$

(b) n-Pentane $\rightarrow$ 2 methyl butane

(c) 1-Pentyl radical $\rightarrow$ 2 pentyl radical

6.5 Earlier in this chapter, we discussed the ensemble average and the state average. How are they the same, and how are they different? Consider using each average to calculate the equation of state (i.e., PV as a function of N and T) of a jar filled with water vapor. Exactly what would you do to calculate the average in each case? Use the canonical ensemble to solve this example.

6.6 Describe the

(a) Similarities and differences between the canonical ensemble, the grand canonical ensemble, and the microcanonical ensemble.

(b) How do you decide which one to use for a given problem?

6.7 In this book, we discussed only three ensembles: the canonical, the microcanonical, and the grand canonical ensembles. However, one can easily define other ensembles. Consider a constant P, N, and T ensemble:

(a) Is there any reason why we could not define such an ensemble?

(b) Would this be a statistical ensemble?

(c) What would vary as you moved from one element in the ensemble to the next? That is, which of P, V, N, T, S, A, U, H, or G would vary and which would be constant?

6.8 (a) Give a physical interpretation of the partition function.

(b) What determines the size of the partition function?

(c) Consider an ideal gas at one atmosphere. Qualitatively, why is the vibrational partition function smaller than the rotational partition function, which, in turn, is smaller than the translational partition function?

6.9 (a) Give a physical interpretation of the statistical mechanical expression for the entropy.

(b) How is the entropy related to the partition function?

(c) What do the results in Problem 6.8 tell you about the relative contribution of translation, rotation, and vibration to the entropy of a system?

6.10 The purpose of this problem is to compare the properties of H_2, HD, and D_2. H_2, HD, and D_2 are very similar molecules. However, there are some important differences.

(a) There is a quantum effect in H_2 and D_2. Half of the rotational states in HD are missing in H_2 and D_2. How will that affect the rotational partition functions of the molecules?

(b) The moment of inertia is larger in D_2 than in HD, which, in turn, is larger than that in H_2. How will that affect the rotational partition function of the two molecules?

(c) The vibrational frequency of H_2 is larger than that in HD, which, in turn, is larger than that in D_2. How will that affect the vibrational frequency of the molecules?

(d) Calculate the equilibrium constant for the reaction $H_2 + D_2 \leftrightharpoons 2HD$, assuming that the partition functions of all of the molecules are the same.

(e) Now assume all the partition functions in the various molecules are different. How will your results in (d) change?

6.11 Estimate the equilibrium constant for the reaction $H_2 + C_2H_4 \leftrightharpoons C_2H_6$.

(a) Use Benson's methods to estimate the heat of reaction.

(b) Assume that the heat of reaction $(-\Delta H)$ is 32.7 kcal/mol at 25°C. Use estimates for the various partition functions rather than calculating ΔG.

6.12 Discuss molecular dynamics as a way of calculating thermodynamic properties.

(a) What is it?

(b) What do you actually do in the calculations?

(c) Where is MD useful? What are the limitations of MD?

(d) How would you use MD to model the behavior of a set of molecules in a box?

(e) What would you have to know to solve problem (d) by computer?

(f) How would your program in (d) be different if you had adiabatic walls or isothermal walls?

(g) How could you use MD to examine the kinetics of adsorption on surfaces?

(h) How could you use MD to examine the behavior of a polymer molecule in solution? What would you need to know to do the simulation?

(i) How could you use MD to examine the diffusion of a molecule across a membrane? What would you need to know to do the simulation?

(j) How could you use MD to examine the flow of gases in very fine pores? What would you need to know to run the simulations?

(k) How could you use MD to simulate the behavior of colloids (i.e., micrometer-range particles in solution)? What would you need to know to run the simulations?

(l) Be prepared for me to ask you to set up an MD calculation for a specific example.

6.13 Consider using MD to calculate the equation of state (i.e., PV as a function of N and T) of a jar filled with water vapor. Exactly what would you do to calculate the average in each case? Do not assume an ideal gas. Calculate the pressure as the force on a wall.

6.14 In 1874, Kundt and Warburg measured the viscosity of a number of gases in a concentric tube viscometer. They found that the viscosity was constant down to 10^{-4} atm, as expected from the kinetic gas law. However, at lower pressures, the measured viscosity was too low. Boltzmann argued that the nonideal behavior was caused by the breakdown of a nonslip condition at the walls.

(a) How would you set up an MD simulation to test Boltzmann's hypothesis?

(b) How would you put in the interaction with the walls?

(c) Should the walls be flat and adiabatic? Will you get any momentum exchange with flat adiabatic walls?

(d) How will you account for real (rough) walls?

(e) How will you account for the energy exchange at the walls?

6.15 You are used to using the Navier–Stokes equation to simulate flow in a pipe, but at low pressures, the continuum approximation breaks down and you need to use MD to simulate the flow pattern.

(a) How would you set up an MD simulation to calculate the flow down a pipe at low pressure?

(b) How would you put in the interactions with the walls?

(c) Should the walls be flat and adiabatic? Will you get any momentum exchange with flat adiabatic walls?

(d) How will you account for real (rough) walls?

(e) How will you account for the energy exchange at the walls?

6.16 MD has been used to simulate the flow of powders down a chute. Assume that you have spherical hard-sphere particles.

(a) What information would you need to do the MD simulation?

(b) Now consider constructing a phase diagram for the flow down the chute where you plot the steady-state velocity of the particles versus the driving force [i.e., g cos (chute angle)]. Show that in the dilute limit (i.e., particles not touching each other), the terminal velocity of the particles down the chute will vary linearly with the driving force.

(c) Now consider interactions between the particles. There are two possible flow patterns: one where particles move randomly, and one where particles move in tandem. Use the principle of microscopic reversibility to find a critical value of the particle density where the phase transition will take place.

6.17 Discuss the Monte Carlo (MC)/Metropolis algorithm as a way of calculating thermodynamic properties.

(a) What is it?

(b) What do you actually do in the calculations?

(c) Where is MC useful? What are the limitations of MC?

(d) How would you use MC to model the behavior of a set of molecules in a box?

(e) What would you have to know to simulate (b) on a computer?

(f) How could you use MC to examine the behavior of a polymer molecule in solution?

What would you need to know to do the simulation?

(g) How could you use MD to simulate the behavior of colloids (i.e., micrometer-range-particles in solution). What would you need to know to run the simulations?

(h) In the cases in (d), (f), and (g), what would MD tell you that MC could not?

(i) In the cases in (d), (f), and (g), what would MC tell you that MD could not?

(j) Be prepared for me to ask you to set up an MC calculation for a specific example.

6.18 Consider using MC to calculate the equation of state (i.e., PV as a function of N, and T) of a jar filled with water vapor. Exactly what would you do to calculate the average in each case? Use the canonical ensemble. Do not assume an ideal gas, and calculate the pressure as the force on a wall.

6.19 In this book, we use the Metropolis algorithm to calculate the state of a system by minimizing the free energy. However, in chemical process design, one can also use similar methods to optimize a process. Consider a generic process. Let's define $\$U_n$ as the profit you make when the process is in condition n.

(a) Define an ensemble for this system.

(b) Let's assume the probability that the system is in state n as given by a Boltzmann distribution, specifically, $p_n = \exp(+\$U_n/\$KT)/Q$, where, $\$KT$ is a constant. Derive an expression for the partition function in the system.

(c) How could one use the Metropolis algorithm to calculate the expectation value of the profit?

(d) Use the thermodynamic argument to show that the algorithm goes to a state that maximizes the profit as $\$KT$ goes to zero.

6.20 Another application of the Metropolis algorithm is in process control. Consider trying to control a distillation column. Assume that the column is in some state n, and that the overhead composition is O_n. Also assume that there are an infinite number of states of the system, each of which has a probability P_n. The Metropolis algorithm can be used to calculate the performance of the system for this case.

(a) Describe the ensemble that you are using.

(b) Provide a block diagram of the algorithm. What would you do to each step? What would you need to know to actually implement the algorithm?

6.21 Describe, in your own words, what a statistical ensemble is like, and why an ensemble average is equal to a time average. How is an ensemble average different from a state average? Why do we need to include the degeneracy when we compute the state average? Why not include the degeneracy in the ensemble average?

6.22 Describe the canonical, microcanonical, and grand canonical ensembles. How are they similar? How are they different?

6.23 In statistical mechanics, we use ensemble average properties to calculate time-averaged properties. However, the same idea can be used for many other systems. Consider calculating the average number of cars per linear mile on a highway. There are two ways we can calculate the average. One is to count the number of cars in a given mile of highway, and do a time average of that number. The second way is to divide up the highway into one-mile strips and average the number of cars over all of the strips.

(a) Show that the two averages are the same in the limit of an infinite highway, with no exits, for any arbitrary speed distribution for the cars, provided that all of the cars are moving with some nonzero velocity.

(b) How will stalled cars at various points on the highway affect your averages? First consider a single stalled car, and then a distribution of stalled cars. Consider the stalled cars in your one-mile sample, and stalled cars outside of your one-mile sample separately.

(c) Extend these ideas to statistical mechanics. How would a molecule that does not sample all of the phase space change your statistical averages?

6.24 Elsewhere in this book, we will use the equations in this chapter to calculate the equilibrium properties of molecules. However, equations (6.8), (6.20), and (6.40) are applicable to nonequilibrium situations. Eres et al. Gurnick, and McDonald, *J. Chem. Phys.* **81**, 5552, (1984) examined the photodissociation of cyanogen (C_2N_2) into CN radicals. They sampled the system several times, and determined the populations in a series of states. The following data were obtained:

n	Relative Number of Molecules in the State	Vibrational Energy, cm^{-1}	Rotational Energy, cm^{-1}	n	Relative Number of Molecules in the State	Vibrational Energy, cm^{-1}	Rotational Energy, cm^{-1}
1	7000	2526	0	15	2300	2515	0
2	6500	2526	200	16	2000	2515	200
3	6000	2526	400	17	1800	2515	400
4	5000	2526	600	18	1500	2515	600
5	4200	2526	800	19	1000	2515	800
6	4000	2526	1000	20	800	2515	1000
7	3800	2526	1200	21	600	2515	1200
8	3500	2526	1400	22	400	2515	1400
9	3300	2526	1600	23	200	2515	1600
10	3000	2526	1800	24	50	2515	1800
11	2500	2526	2000	25	20	2515	2000
12	2000	2526	2200	26	10	2515	2200
13	1000	2526	2400	27	3	2515	2400
14	900	2526	2600	28	2	2515	2600

(a) Use the relative number of molecules in the table above to calculate the probability that the system is in each of the states in the table. For the purposes of this problem, assume that these are the only states available to the molecules.

(b) Use the data in the table to calculate the average rotational, and vibrational energies of the photodissociated products.

(c) Calculate the entropy of the system assuming that the states in the table represent all possible states of the system, and that equation (6.40) applies.

6.25 In the section we noted that one can derive equation (6.31) from the thermodynamic relationship

$$\left(\frac{\partial \langle U \rangle}{\partial V}\right)_{N,S} = -P \tag{6.32}$$

The objective of this problem is for you to derive the equation.

(a) Substitute equation (6.20) into equation (6.32) to derive an expression for $\left(\frac{\partial \langle U \rangle}{\partial V}\right)_{N,S}$.

(b) Use equation (6.40) to show that $\left(\frac{\partial p_n}{\partial V}\right)_{N,S} = 0$.

(c) Derive equation (6.31).

7

INTRODUCTION TO REACTION RATE THEORY

PRÉCIS

So far in this book, we have been discussing what is called classical kinetics: rate laws for chemical reactions and how those rate laws arise from the mechanism of the reaction. In the next several chapters, we will be changing topics and starting to discuss how reactions happen on a molecular level. Our objective will be to formulate an expression for the rate of reaction in terms of the rates of all of the elementary processes that occur during reaction and the forces on all of the molecules. This is a large topic. There are two parts: predicting the preexponentials and predicting the activation barriers. The object of Chapter 7 is to provide a brief overview of the prediction of preexponentials with collision theory, transition state theory, and the RRKM model so that the reader can get an introduction to the material. Chapters 8–11 will fill in the details.

7.1 INTRODUCTION

In this chapter, we will be discussing what is called *reaction rate theory*. The objective of reaction rate theory is to relate the rate constants for reactions to the properties of the reactants, products, and transient intermediates.

The earliest work on reaction rate theory came from Arrhenius (1889). Arrhenius was interested in why activation barriers arose in chemical reactions. Arrhenius considered a simple reaction:

$$A \longrightarrow B \qquad (7.1)$$

and proposed that if one looked at a chemical system containing A and B, there were two kinds of A molecules in the system: **reactive** A molecules (i.e., A molecules that had the right properties to react), and **unreactive** A molecules (i.e., A molecules that did not have the right properties to react).

359

Arrhenius noted that there is a distribution of molecular velocities as indicated in Figure 7.1. Most of the molecules are too cold to react. However, there are a few molecules in the tail of the Boltzmann distribution that *can* react.

Arrhenius then derived an equation for the rate of reaction. The derivation assumed that equilibrium was maintained between the reactive and unreactive A molecules. We reproduce the arguments below.

Recall that at equilibrium, the concentration of molecules of A that have the right properties to react, $C_A^{\dagger}$, is related to the overall concentration of the unreactive A molecules by

$$C_A^{\dagger} = C_A^u e^{-(\Delta G^{\dagger}/k_B T)} \tag{7.2}$$

where C_A^u is the concentration of the unreactive A molecules and $\Delta G^{\dagger}$ is the free energy change (kJ/molecule) when one converts an unreactive A molecule into a reactive one. Normally, most of the A molecules will be unreactive. Consequently, $C_A^u \cong C_A$, where C_A is the total concentration of A molecules. If it is then assumed that the reactive molecules react via a first-order rate law, one can show

$$r_A = -K_0 C_A^{\dagger} = -K_0 C_A e^{-(\Delta G^{\dagger}/k_B T)} \tag{7.3}$$

where K_0 is the rate constant for the reaction of the reactive A molecules. According to equation (7.3), the rate constant for the overall reaction is given by

$$k_1 = K_0 e^{-(\Delta G^{\dagger}/k_B T)} \tag{7.4}$$

Note:

$$\Delta G^{\dagger} = \Delta H^{\dagger} - T\Delta S^{\dagger} \tag{7.5}$$

where $\Delta H^{\dagger}$ is the average enthalpy to convert an unreactive A molecule into a reactive one, and $\Delta S^{\dagger}$ is the entropy of the same process. Substituting equation (7.5) into equation (7.4) yields

$$k_1 = \left(K_0 e^{(\Delta S^{\dagger})/k_B}\right) e^{-[(\Delta H^{\dagger})/k_B T]} \tag{7.6}$$

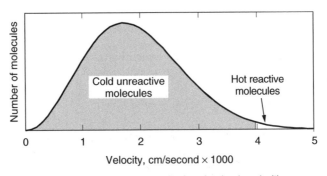

Figure 7.1 The Boltzmann distribution of molecular velocities.

Equation (7.6) is equivalent to

$$k_1 = k_0 e^{-(E_a/k_B T)} \tag{7.7}$$

with

$$k_0 = K_0 e^{(\Delta S^\dagger)/k_B} \tag{7.8}$$

In equation (7.6), k_1 is the rate constant for a reaction (7.1), k_0 is the preexponential for the reaction, E_a is the activation energy for the reaction, k_B is Boltzmann's constant, and T is temperature.

Equation (7.7) was Arrhenius' key result. At the time this work was done, there were many empirical rules to predict how rates vary with temperature. Arrhenius was the first person to derive a theoretical expression. When the expression was found to fit data, Arrhenius' expression, which was renamed *Arrhenius' law*, was universally adopted in kinetics.

Arrhenius's law was one of the early successes of physical chemistry. It applied to both the behavior of chemical reactions and the growth and reproduction of plants and simple animals as described in Chapter 2. Consequently, Arrhenius' law was an important advance.

7.2 COLLISION THEORY

Arrhenius never was able to provide a model for K_0 in equation (7.8). Fortunately, Trautz (1918) and Lewis (1918) independently proposed the collision theory of reactions. The idea in the original version of collision theory was that molecules need to collide if the molecules are going to react. Figure 7.2 shows a few trajectories that we have calculated using a program from Chapter 8. In the top case, A comes in to hit BC, then A flies away again. According to Trautz and Lewis, this would be an unreactive collision. In the second case, A comes in and hits B–C, and C flies away. That is a reactive collision. Trautz and Lewis proposed that one could calculate the reaction rate by looking at how many collisions occurred and multiplying by a reaction probability that is a function of energy. As a result, one can calculate the reaction rate from the number of times that the hot molecules collide.

Trautz and Lewis then used the kinetic theory of gases to calculate a collision rate. If one assumes that all hot molecules react when they collide, then one can show that the rate of the reaction

$$A + BC \longrightarrow AB + C \tag{7.9}$$

is

$$r_{A-BC} = Z_{ABC} P_{reaction} \tag{7.10}$$

where Z_{ABC} is the rate of collisions between A and BC in units of molecules/($\text{Å}^3 \cdot$second) and $P_{reaction}$ is the probability of reaction once the molecules collide.

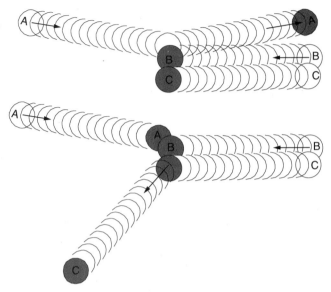

Figure 7.2 A collision between an A molecule and BC molecules.

Using Arrhenius' model in Section 7.1, we write

$$P_{\text{reaction}} = e^{-\Delta G^{\dagger}/k_B T} \tag{7.11}$$

Combining equations (7.10) and (7.11) shows

$$r_{A \to BC} = Z_{ABC} e^{-\Delta G^{\dagger}/k_B T} \tag{7.12}$$

Next, let's derive an equation for Z_{ABC}. First consider the reaction between the molecules A and BC. To simplify the analysis, we will treat the molecules as two billiard balls. Figure 7.3 shows some of the possible collisions between the billiard balls. In the figure, A is moving toward BC. A starts at the upper left, and the motion of A is indicated as a series of circles. In the top case, a hot A molecule approaches a hot BC but it misses. In the second case, the hot A just misses B–C. In the third case, the hot A collides with the hot BC. Trautz and Lewis assumed that in the first two cases, no reaction occurs, but in the third case, reaction happens with unity probability, provided the molecules have a minimum energy $E_T^{\ddagger}$.

Notice that any A molecule that gets within a distance b_{coll} of the BC molecule will collide with the BC molecule. Therefore the collision diameter becomes a distance. Any collisions with a distance less than b_{coll} react.

Next, we will derive an expression for the rate of molecular collisions. For the purposes of derivation, consider the motion of a given BC molecule toward an A molecule as shown in Figure 7.4. During the collision BC moves toward A as shown in this figure. We will define $v_{A \to BC}$ as the velocity of A toward BC.

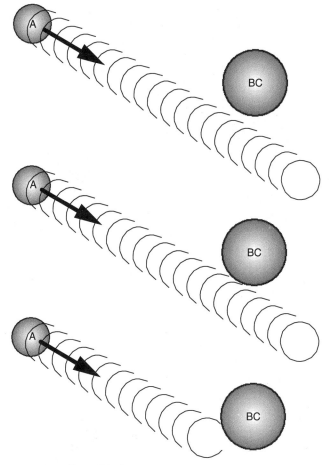

Figure 7.3   Some typical billiard ball collisions.

Now let us consider a small increment of time t_c. If we treat A as a fixed point, then during the collision BC will move a distance L_{ABC} given by

$$L_{ABC} = v_{A \to BC} t_c \tag{7.13}$$

Next, we want to note that if there is an A molecule within the cylinder in Figure 7.4, there will be a collision between A and BC sometime during the time slice t_c.

If there are two molecules within the cylinder, there will be two collisions. In general, the number of collisions will be

$$\begin{pmatrix} \text{Number of collisions} \\ \text{of the given} \\ \text{BC molecule} \end{pmatrix} = \begin{pmatrix} \text{volume of} \\ \text{cylinder} \end{pmatrix} C_A \tag{7.14}$$

where C_A is the concentration of the A molecules in the reacting mixture, measured in molecules/cm^3.

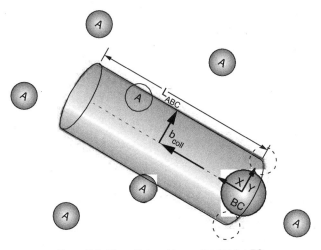

Figure 7.4 The collision of two molecules A + BC.

The volume of the cylinder is

$$\begin{pmatrix} \text{Volume of} \\ \text{cylinder} \end{pmatrix} = \pi(b_{coll})^2 L_{ABC} \tag{7.15}$$

For the purposes of derivation, it's useful to define a new variable by

$$\sigma^c_{A \to BC} = \pi(b_{coll})^2 \tag{7.16}$$

where b_{coll} is the collision diameter combining equations (7.13)–(7.16). Then

$$\frac{\begin{pmatrix} \text{Number of collisions} \\ \text{of the given BC} \\ \text{molecule} \end{pmatrix}}{t_c} = v_{A \to BC} C_A \sigma^c_{A \to BC} \tag{7.17}$$

Now consider a volume v of gas. The number of BC molecules in the gas is given by

$$\text{Number of BC molecules} = v C_{BC} \tag{7.18}$$

where C_{BC} is the concentration of BC in the reacting mixture, measured in molecules/cm^3. By definition

$$Z_{ABC} = \frac{\begin{pmatrix} \text{total number of} \\ \text{collisions at A with} \\ \text{all BCs} \end{pmatrix}}{(\text{volume})(\text{time})} = \frac{\begin{pmatrix} \text{total number of} \\ \text{collisions of A with} \\ \text{all BCs} \end{pmatrix}}{(v)t_c} \tag{7.19}$$

The total number of collisions is

$$
\begin{pmatrix}
\text{Total number of} \\
\text{A molecules with all} \\
\text{BC molecules}
\end{pmatrix}
=
\begin{pmatrix}
\text{number of collisions at} \\
\text{a given BC molecule}
\end{pmatrix}
\begin{pmatrix}
\text{number of BC} \\
\text{molecules in the} \\
\text{volume}
\end{pmatrix}
\tag{7.20}
$$

Combining equations (7.17)–(7.20) yields

$$
Z_{ABC} = C_A C_{BC} \sigma^c_{A \to B} v_{A \to BC}
\tag{7.21}
$$

Equation (7.21) gives us the collision rate if we know the velocity of A toward BC. In a real system, there will be a distribution of velocities, so we have to average equation (7.21) over the velocity distribution. The result is

$$
Z_{ABC} = \bar{v}_{A \to BC} C_A C_{BC} \sigma^c_{A \to BC}
\tag{7.22}
$$

where $\bar{v}_{A \to BC}$ is the average velocity of A toward BC. Equation (7.22) gives the total rate of collisions between hot A molecules and hot BC molecules.

Substituting equations (7.5) and (7.19) into equation into equation (7.12) yields

$$
r_{A \to BC} = (\bar{v}_{A \to BC} \sigma^c_{A \to BC} e^{\Delta S^{\dagger}/k_B})(e^{-\Delta H^{\dagger}/k_B T}) C_A C_{BC}
\tag{7.23}
$$

where $\bar{v}_{A \to BC}$ is the average velocity of A moving toward BC.

Substituting equation (7.16) into (7.23) and setting $\Delta H^{\dagger} = E_A$ yields

$$
r_{A \to BC} = \pi d^2_{coll} \bar{v}_{ABC} e^{-E_a/k_B T} (e^{\Delta S^{\dagger}/k_B}) C_A C_{BC}
\tag{7.24}
$$

where $r_{A \to BC}$ is the rate of reaction between A and BC, $\bar{v}_{ABC}$ is the average velocity of A toward BC, C_A and C_{BC} are the concentrations of A and BC, and d_{coll} is the collision diameter of A–BC as indicated in Figure 7.4. E_A is the average amount of energy needed to get the reaction to happen in kJ/molecule, k_B is Boltzmann's constant $[1.38 \times 10^{-23}$ J/(molecule·K)], T is the absolute temperature, and $\Delta S^{\ddagger}$ is the entropy of the molecules that react in kJ/(molecule·K).

Equation (7.23) is just a second-order rate law, with an activation barrier of $\Delta H^{\ddagger}$ and a preexponential, k_0, which is given by

$$
k_0 = \bar{v}_{A \to BC} \sigma^c_{A \to BC} e^{\Delta S^{\dagger}/k_B}
\tag{7.25}
$$

Neither Trautz nor Lewis had a way to calculate $\Delta S^{\dagger}$. Consequently, they set $\Delta S^{\dagger}$ to zero in equation (7.25) to obtain

$$
\boxed{k_0 = \bar{v}_{A \to BC} \sigma^c_{A \to BC}}
\tag{7.26}
$$

Equation (7.26) is the key result for simple collision theory.

Trautz and Lewis also asserted that the molecular velocity in equation (7.26) should be calculated, ignoring the fact that we are considering hot molecules. We will derive an equation for $\bar{v}_{A \to BC}$ is Section 8.16.4. The result is

$$\bar{v}_{A \to BC} = \left(\frac{8k_B T}{\pi \mu_{ABC}} \right)^{1/2} \tag{7.27}$$

where

$$\frac{1}{\mu_{ABC}} = \frac{1}{m_A} + \frac{1}{m_B + m_C} \tag{7.28}$$

and m_A, m_B, and m_C are respectively the masses of A, B, and C in atomic mass units $(1 \text{ amu} = 1.66 \times 10^{-24} \text{ grams})$.

I find it convenient to rewrite equation (7.27) as

$$\bar{v}_{ABC} = 2.52 \times 10^{13} \frac{\text{Å}}{\text{second}} \left(\frac{T}{300 \text{ K}} \right)^{1/2} \left(\frac{1 \text{ amu}}{\mu_{ABC}} \right)^{1/2} \tag{7.29}$$

For example, for the reaction $F + H_2 \to HF + H$ at 400 K

$$m_F = 19 \text{ amu}$$

$$m_{H_2} = 2 \text{ amu}$$

$$\frac{1}{\mu_{ABC}} = \frac{1}{19 \text{ amu}} + \frac{1}{2 \text{ amu}} = \frac{21}{38 \text{ amu}}$$

$$\mu_{ABC} = \frac{38}{21} \text{ amu}$$

$$\bar{v}_{ABC} = (2.52 \times 10^{13} \text{ Å/second}) \left(\frac{400 \text{ K}}{300 \text{ K}} \right)^{1/2} \left(\frac{1 \text{ amu}}{\frac{38}{21} \text{ amu}} \right)^{1/2}$$

$$= 2.2 \times 10^{13} \text{ Å/second}$$

There is one other detail. It is not clear whether you should use van der Waal's radii, covalent radii, or some other radii in the calculation. Generally, **covalent radii give better predictions**, although Lewis' original papers suggest van der Waal's radii instead. Trautz and Lewis picked numbers for d_{coll}, but did not have a theory to predict it, so this was a limitation of their model.

7.2.1 Predictions of Collision Theory

Equations (7.23) and (7.26) were big advances. Equation (7.23) correctly predicted that an elementary reaction between two molecules would show second-order kinetics. That was quite important to people deriving rate equations. Equation (7.26) gave a numerical value for the preexponential for the first time. Again, that was quite important. Further, the numerical value was close to the experimental values.

It is interesting to put some numerical values into equation (7.26) so that we can compare to experiments. Table 7.1 shows the molecular velocities of a number of molecules computed at 273 K. Notice that most of the molecular velocities in the table are between 2×10^{12} and 1×10^{13} Å/second. If we assume an average velocity of

Table 7.1 Molecular velocities and collision diameters for a number of molecules at 273 K

Molecule	Molecular Velocity, Å/second	Collision Diameter, Å
He	1.2×10^{13}	2.2
N_2	4.5×10^{12}	3.5
O_2	4.2×10^{12}	3.1
H_2O	5.6×10^{12}	3.7
C_2H_6	4.37×10^{12}	3.5
C_6H_6	2.7×10^{12}	5.3

4×10^{12} Å/second and an average collision diameter of 3 Å, substitute into equation (7.16), and substitute the resultant value of σ_A^c into equation (7.26), we predict a preexponential of about $(4 \times 10^{12}$ Å/second$) \times [\pi(3 \text{ Å})^2] = 1.1 \times 10^{14}$ Å^3/second.

It is useful to compare this result to the compilation of preexponentials in Table 7.2. Notice that most of the preexponentials in Table 7.2 are between 10^{12} and 2×10^{14} Å^3/(molecule·second). Equation (7.26) predicts 1×10^{14}. Therefore, it seems that equation (7.26) predicts preexponentials that are *usually* within an order of magnitude of the experimental value.

According to equation (7.26), preexponentials for reaction are determined by two factors: the size of the molecules, as given by the cross section; and the average velocity of the molecules, as given in Table 7.1. According to equation (7.26), if everything else is equal, faster-moving molecules will react more quickly than will slower-moving ones. These trends generally agree with experiments. Equation (7.26) also predicts that bigger molecules (i.e., those with larger cross sections) will react more quickly than will smaller molecules. Again, that agrees with experiment. Collision theory also gives preexponentials that are generally within the right order of magnitude. That is quite a significant result considering that collision theory was first proposed in 1918.

Table 7.3 shows a few additional values of the preexponentials calculated from equation (7.30). Details are given in the solved examples. Generally, **collision theory tends to overestimate the preexponentials for reaction**, although the errors are often less than a factor of 10. To put this in perspective, in 1918, one could not measure preexponentials to within a factor of 10, so this was as good an approximation as the data itself.

Table 7.2 A selection of the preexponentials reported by Westley (1980)

Reaction	Preexponential, Å^3/(molecule·second)	Reaction	Preexponential, Å^3/(molecule·second)
$H + C_2H_6 \rightarrow C_2H_5 + H_2$	1.6×10^{14}	$O + C_2H_6 \rightarrow OH + C_2H_5$	2.5×10^{13}
$H + CH \rightarrow H_2 + C$	1.1×10^{12}	$O + C_3H_8 \rightarrow (CH_3)_2CH + OH$	1.4×10^{10}
$H + CH_4 \rightarrow H_2 + CH_3$	1×10^{14}	$O_2 + H \rightarrow OH + O$	1.5×10^{14}
$O + H_2 \rightarrow OH + H$	1.8×10^{13}	$OH + OH \rightarrow H_2O + O$	1×10^{13}
$O + OH \rightarrow O_2 + H$	2.3×10^{13}	$OH + CH_4 \rightarrow H_2O + CH_3$	5×10^{13}
$O + CH_4 \rightarrow CH_3 + OH$	2.1×10^{13}	$OH + H_2CO \rightarrow H_2O + HCO$	5×10^{13}
$O + CH_3 \rightarrow H + CH_3O$	5×10^{13}	$OH + CH_3 \rightarrow H + CH_3O$	1×10^{13}
$O + HCO \rightarrow H + CO_2$	5×10^{12}	$OH + CH_3 \rightarrow H_2O + CH_2$	1×10^{13}

Table 7.3 **Preexponentials calculated from equation (7.30) for a number of reactions compared to experimental data**

Reaction	Calculated Preexponential Assuming b_{coll} = van der Waals Radius, Å³/(molecule·second)	Calculated Preexponential Assuming b_{coll} = Covalent Radius, Å³/(molecule·second)	Experimental Preexponential, Å³/(molecule·second)
$H + C_2H_6 \rightarrow C_2H_5H_2$	6.2×10^{14}	2.0×10^{14}	1.6×10^{14}
$H + CH \rightarrow H_2 + C$	4×10^{14}	2.0×10^{14}	1.1×10^{12}
$O + C_2H_6 \rightarrow OH + C_2H_5$	1.9×10^{14}	7.6×10^{13}	2.5×10^{13}
$OH + OH \rightarrow H_2O + O$	1.25×10^{14}	5.8×10^{13}	1×10^{13}
$H + O_2 \rightarrow OH + O$	4.0×10^{14}	2×10^{14}	1.5×10^{14}

7.2.2 Failure of Collision Theory

Unfortunately, there are a few examples where the preexponentials differ substantially from those predicted by collision theory. The reaction

$$CH_3CH_2CH_3 + O: \longrightarrow CH_3\overset{\bullet}{C}HCH_3 + \bullet OH \qquad (7.30)$$

has a preexponential of 1.4×10^{10} Å³/(molecule·second). That is over two orders of magnitude lower than predicted from collision theory [equation (7.26)]. In contrast, the reaction

$$2O_2 \longrightarrow 2O\bullet + O_2 \qquad (7.31)$$

has a preexponential of 5.8×10^{15} Å³/(molecule·second). That is about two orders of magnitude larger than predicted from collision theory. These are special cases. In most cases, collision theory gets about the right answer. Still, there are a number of examples where Trautz and Lewis' version of collision theory cannot explain the data.

The Trautz–Lewis model failed because it treated the collision between the reactants as a billiard ball collision. Every molecule of A that collided with BC was assumed to react with a fixed probability. In reality, the reaction probability varies with how the collision occurs. For example, consider the following two reactions:

$$CH_3CH_2CH_3 + O: \longrightarrow CH_3\overset{\bullet}{C}HCH_3 + \bullet OH \qquad (7.32a)$$

$$CH_3CH_2CH_3 + O: \longrightarrow \bullet CH_2CHCH_3 + \bullet OH \qquad (7.32b)$$

During reaction (7.32a), the oxygen reacts with the hydrogen on the propane's middle carbon. If the incoming oxygen hits the hydrogen on the middle carbon, then the reaction can occur. However, if the oxygen hits anywhere else, a $\bullet CH_2CH_2CH_3$ species will form. The Trautz–Lewis model ignores the fact that you need special collision geometry to allow the reaction to happen. As a result, their model tends to overestimate rates of reaction such as reaction (7.32a).

This example shows that, in some cases, one needs molecules to collide in the correct way for the desired reaction to occur. The Trautz–Lewis model ignores the need for a special collision geometry to get a desired reaction, so it does not always give a good prediction of the rate.

Polanyi noted that one can approximately account for these effects by considering the $\Delta S^{\ddagger}$ term in equation (7.25). Recall from Chapter 6 that the entropy is a measure of how many configurations are available in the system, specifically:

$$\text{Configurations} = \exp\left(\frac{S}{k_B}\right) \tag{7.33}$$

If, we look at reaction (7.32), there are many configurations of the reactants that do not lead to reaction. We can define $\Delta S^{\dagger}$ by

$$\exp\left(\frac{\Delta S^{\dagger}}{k_B}\right) = \frac{\text{configurations which lead to reaction}}{\text{average number of configurations of the reactants}} \tag{7.34}$$

$\Delta S^{\dagger}$ is called the **entropy of activation**. Generally, $\exp(\Delta S^{\dagger}/k_B)$ is less than 1.0, and it can be much less then 1 in a case such as reaction (7.32), where only special configurations lead to reaction. There are also a few cases where $\exp(\Delta S^{\dagger}/k_B)$ is greater than 1, because there are more accessible configurations in the transition state than in the reactants.

Polanyi also defined **loose transition states** and **tight transition states**. Loose transition states have $\Delta S^{\dagger}$ near zero, so any collision can lead to reaction. Tight transition states have negative values of $\Delta S^{\dagger}$. Tight transition states arise when a special configuration of the reactants is needed in order for a reaction to occur.

Unfortunately, in 1931, Polanyi was not able to get an expression of $\Delta S^{\ddagger}$, so he could not correct collision theory to account for this effect.

Another weakness of the model is that it does not explain activation barriers. Neither Arrhenius, nor Trautz, nor Lewis was able to explain why the reactants needed to be hot in order for reaction to occur. Trautz and Lewis just assumed — without explaining this assumption — that reactions had barriers.

7.3 THE MARCELLIN–WIGNER–POLANYI MODEL

Marcellin (1914–1920), Wigner (1932), and Polanyi (1931, 1932) provided the first model that explained why reactions have barriers. Their idea was that the reactants need to cross a mountain in a potential energy surface in order for the reaction to occur. See Figure 7.5.

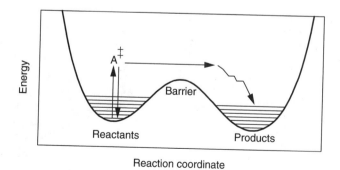

Figure 7.5 Polanyi's picture of excited molecules.

According to Marcellin, Wigner, and Polanyi, the activation barrier is associated with the energy to get over the barrier.

Figure 7.6 shows what the barrier really looks like for the reaction

$$H + CH_3OH \longrightarrow H_2 + CH_2OH \tag{7.35}$$

There is a two-dimensional barrier in Figure 7.6 rather than the one-dimensional barrier in Figure 7.5. The two-dimensional barrier is called a *potential energy surface*. The potential energy surface looks like a gorge through the mountains with two broad valleys and a saddle point in between. Point X in the figure corresponds to the minimum in the potential at the reactants, Y is the minimum at the products, and there is a rise corresponding to the saddle point. Figure 7.7 shows the saddle point in more detail. Notice that there is a hill. The hill is the barrier to the reaction.

People usually plot the potential energy surface as a contour plot. The contour plot is shown on the right of Figure 7.6. The lines are contours of constant energy, and are

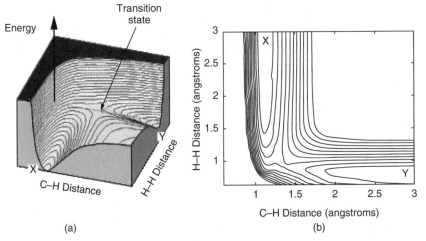

(a)   (b)

Figure 7.6 A potential energy surface for the reaction $H + CH_3OH \rightarrow H_2 + CH_2OH$ from the calculations of Blowers and Masel. The lines in the figure are contours of constant energy. The lines are spaced 5 kcal/mol apart.

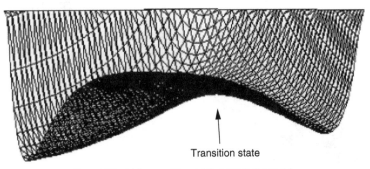

Figure 7.7 A blowup of the saddle point in Figure 7.6.

spaced 5 kcal/mol apart. Point X in the figure corresponds to the minimum in the potential at the reactants, Y is the minimum at the products, and there is a rise corresponding to the saddle point.

Examples 7.G and 8.A show more information about potential energy surfaces. Please look at these two examples before proceeding with this chapter.

Marcellin, Wigner, and Polanyi noted that molecules need to get over the rise in the potential energy surface in order for reaction to occur. Therefore, they proposed that activation barriers are associated with the rise in the potential energy surfaces. In the literature, this is called rise the **transition state**, or the **col** (saddle), in the potential energy surface. Generally, it is still believed that activation barriers are associated with the hills in the potential energy surface.

In the literature, people sometimes say that the barrier height is equal to the activation energy. In fact this is only an approximation. For example, Table 7.4 compares the activation barriers for a number of reactions to the energy of the saddle point on the potential energy surface. One sees qualitative agreement, that reactions with larger saddle point energies also have higher activation barriers. However, the activation barrier is not exactly equal to the saddle point energy.

7.3.1 Tunneling

Generally, when a hydrogen atom is being transferred during a reaction, the barrier is a little less than the saddle point because of a quantum-mechanical process called *tunneling*. In Chapter 8, we will show that quantum-mechanically an atom that does not have quite enough energy to go over a barrier still has some probability of going through that barrier. In these cases, the activation barrier is lower than the barrier in the potential energy surface because some atoms can get through the barrier.

Tunneling arises because of the uncertainty principle. Classically one might say that the reactants are just approaching the barrier. However, quantum-mechanically there is some uncertainty in the position of the molecule, and so when you think that the reactants are just approaching the barrier, the wavefunction for the molecules will have a component on the other side of the barrier as shown in Figure 7.8. This uncertainty in the atomic

Table 7.4 A comparison of the saddle point energy from accurate ab initio calculations and the activation barrier measured experimentally for a number of reactions

Reaction	Saddle Point Energy, kcal/mol	E_a, kcal/mol
$H + H_2 \rightarrow H_2 + H$	9.60	8.0
$D + H_2 \rightarrow DH + H$	9.60	7.78
$H + CH_4 \rightarrow H_2 + CH_3$	13.5	11.9
$H + CH_3CH_3 \rightarrow H_2 + C_2H_5$	11.8	9.7
$F + H_2 \rightarrow HF + H$	5.8	1.7
$H + CH_3CH_2CH_3 \rightarrow H_2 + C_3H_7$	10.4	8.2
$H + CH_3OH \rightarrow CH_2OH + H$	9.8	9.0
$CH_3 + CF_3I \rightarrow CH_3I + CF_3$	4.95	7.5

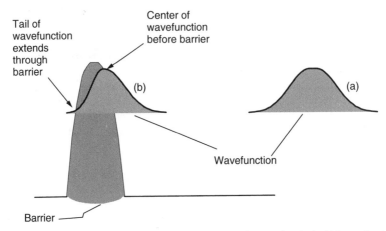

Figure 7.8 A diagram showing the extent of the wavefunction for a molecule. In (a) the molecule is by itself. In (b) the molecule is near a barrier. Notice that the wavefunction has a finite size (i.e., there is some uncertainty in the position of the molecule). As a result, when a molecule approaches a barrier, there is a component of the molecule on the other side of the barrier.

positions allows a molecule to transport through a barrier. We call the transport through the barrier tunneling.

Tunneling is most important for reactions in which a hydrogen atom is being exchanged. The effect is that activation energies for hydrogen transfer reactions are generally less than the energy of the barrier on the potential energy surface.

7.3.2 Dynamic Effect

Most ligand transfer reactions have activation energies that are slightly higher than the saddle point energies. Physically, when the reaction $A + BC \rightarrow AB + C$ occurs, the BC bond needs to accumulate enough energy to break. Generally, it is easier to accumulate enough energy if there is a little excess energy in the system. Consequently, the average energy of the reacting molecule is usually slightly higher than the saddle point energy.

We will discuss tunneling and dynamic effects in detail in Chapter 8. However, the thing to remember for now is that activation energies are usually slightly different from the energy of the saddle point in the potential energy surface.

7.4 TRANSITION STATE THEORY

Marcellin was killed in World War I, and neither Wigner nor Polanyi were able to derive a simple expression for the rate. However, Eyring (1937) took an equation that appeared in Tolman (1927), and modified it using Wigner's ideas to derive what is called *transition state theory*. We will reproduce the key arguments in Chapter 9. The idea builds on Arrhenius' model described in Section 7.1. First you use statistical mechanisms to calculate an expression for K_{equ}, the equilibrium concentration of hot molecules. Then you plug back into collision theory to get the rate constant. Consider the reaction

$$A + BC \longrightarrow AB + C \qquad (7.36)$$

According to equation (6.7), K_{equ}, the equilibrium constant for the production of the hot reactive molecules, is

$$K_{equ} = \frac{q_{ABC}^{\dagger}}{q_A q_{BC}} e^{-E_T^+/k_B T} \qquad (7.37)$$

where E^+ is the average energy needed to traverse the barrier, $q_{ABC}^{\dagger}$ is the partition function per unit volume for the hot molecules, q_A and q_{BC} are the partition functions per unit volume for A and BC, k_B is Boltzmann's constant, and T is the temperature. If we substitute equations into Arrhenius' model, assuming that the reaction probability is equal to the probability that a molecule is hot when the collision occurs, we obtain

$$k_{A \to BC} = K_0 \frac{q_{ABC}^{\dagger}}{q_A q_{BC}} e^{-E^+/k_B T} \qquad (7.38)$$

where $k_{A \to BC}$ is the rate constant for reaction (7.36), and K_0 is the rate constant for the reaction of hot molecules. Equation (7.38) is exact, but we will need an expression for K_0. We can get it from collision theory.

First, let us define a new partition coefficient q^+, by

$$q^+ = \frac{q_{ABC}^{\dagger}}{q_{A \to BC}^t} \qquad (7.39)$$

In equation (7.39) $q_{A \to BC}^t$ is the partition function for the translation of A toward BC and q^+ is the partition function for all of the other modes of the reacting A–B–C complex.

Combining equation (7.38) and (7.39) yields

$$k_{A \to BC} = (K_0 q_{A \to BC}^t) \frac{q^+}{q_A q_{BC}} e^{-E^\dagger/k_B T} \qquad (7.40)$$

Eyring (1935, 1937, 1944) proposed that one could replace q^+ in equation (7.5) with $q_T^{\ddagger}$, the partition function of the transition state, and E^+ the average energy of the molecules will react with $E^{\ddagger}$, the energy of the transition state. See Eyring (1944, 1980). The result is

$$k_{A \to BC} = (K_0 q_{A \to BC}^t) \frac{q_T^{\ddagger}}{q_A q_{BC}} e^{-E^{\ddagger}/k_B T} \qquad (7.41)$$

In Chapter 9 we will use collision theory to show

$$K_0 q_{A \to BC}^t = \left(\frac{k_B T}{h_P} \right) \qquad (7.42)$$

Substituting equation (7.42) into equation (7.41) yields

$$k_{A \to BC} = \left(\frac{k_B T}{h_P} \right) \frac{q_T^{\ddagger}}{q_A q_{BC}} e^{-E_T^{\ddagger}/k_B T} \qquad (7.43)$$

where k_B is Boltzmann's constant, T is the temperature, h_P is Planck's constant, $E_T^{\ddagger}$ is the energy of the saddle point in the potential energy surface, q_A and q_{BC} are the partition functions for the reactants, and $q_T^{\ddagger}$ is the partition function for all of the modes of transition state except the motion over the barrier.

Equation (7.43) is the key result for transition state theory. One should also memorize equation (7.43) before proceeding through the rest of this chapter.

7.4.1 Key Predictions of Transition State Theory

Transition state theory was a major advance. It provided a way to predict the magnitude of the preexponential. It also gave lots of parameters (i.e., the q values) that could be used to fit data.

It is useful to use transition state theory to get an order of magnitude for the rate constant. In Example 7.B, we show that for most first-, second-, and third-order reactions, $q_T^{\ddagger}/q_A q_{BC}$ is between 0.1 and 10 in units of molecules, angstroms, and seconds, while $k_B T/h_P \sim 10^{13}$/seconds. Consequently, transition state theory predicts

> Preexponentials for elementary reactions are usually between 10^{12} and 10^{14} in units of Å, molecules, and seconds independent of the order of the reaction or other details.

Other key predictions of transition state theory include

- Transition state theory identifies the barrier to reaction with the saddle point in the potential energy surface. Saddle point energies can be calculated exactly as described in Chapter 11, which then allows activation barriers to be calculated.
- Transition state theory allows you to calculate b_{coll} in equation (7.16) rather than guess it. In Chapter 9, we will show that for a simple collision b_{coll} is just the distance between the reactants at the transition state.
- Transition state theory suggests that $P_{reactions}$ can be less than $e^{-E_T^{\ddagger}/k_B T}$. For example, in the reaction

$$D + CH \longrightarrow C + HD$$

  the incoming deuterium has to hit the hydrogen atom for reaction to occur. Transition state theory accounts for that by saying that the transition state must be a linear CHD molecule.
- Transition state theory also predicts that small changes in the shape of the potential energy surface produce large changes in $q_T^{\ddagger}$, which, in turn, changes the rate.

We work out solutions in Examples 7.B–7.E. One should read those examples before proceeding with this chapter.

Transition state theory was a major advance. It gave an equation that could be used to make practical calculations before the days of computers. It gave an explanation of barriers to reactions that could be explained to college freshmen. The model contained lots of parameters (i.e., the q terms) that could be fit to data. When these ideas were first proposed, no one knew what potential energy surfaces for reactions were like. Transition state theory allowed people to fit their data to an empirical model, and so transition state theory explained a lot of data that had not been explained before.

Transition state theory has dominated our thinking about reactions since the early 1950s, so it was quite an important advance.

7.4.2 Relationship between Transition State Theory and Collision Theory

One of the details is that transition state theory is closely related to collision theory. Next, we will show that transition state theory and collision theory are identical for the collision of structureless particles. Let's assume that the transition state for the reaction is the hard-sphere collision in Figure 7.4. Next, we will obtain expressions for q_A, q_{BC}, and $q_T^{\ddagger}$.

A only has translational motion. According to Table 6.6

$$q_t^3 = \frac{(2\pi m_A k_B T)^{3/2}}{h_P^3} \tag{7.44}$$

where q_t is a translational partition function, m_A is the mass of A, k_B is Boltzmann's constant, T is the temperature, and h_P is Planck's constant.

Similarly, BC is a hard-sphere that can translate. There is also a BC vibration, but we will ignore rotation of the BC molecule. Therefore

$$q_{BC} = \frac{(2\pi(m_B + m_C)k_B T)^{3/2}}{h_P^3} q_{V,BC} \tag{7.45}$$

where m_B and m_C are the masses of B and C, respectively, and $q_{V,BC}$ is the partition function for the BC vibration.

The transition state can translate and rotate. There is also a partition function for the B–C vibration. Collision theory ignores changes in the BC vibration. Therefore

$$q_T^{\ddagger} = (q_t^{\ddagger})^3 (q_r^{\ddagger})^2 q_{V,BC} \tag{7.46}$$

where $q_t^{\ddagger}$ is the translational partition function of the transition state, $q_r^{\ddagger}$ is its rotational partition function, and $q_{V,BC}$ is the partition function for the BC vibration. From the above we have

$$(q_t^{\ddagger})^3 = \frac{(2\pi(m_A + m_B + m_C)k_B T)^{3/2}}{h_P^3} \tag{7.47}$$

According to the results in Chapter 6

$$(q_r^{\ddagger})^2 = \frac{8\pi^2 I k_B T}{h_P^2} \tag{7.48}$$

with

$$I = \mu_{ABC}(b_{coll})^2 \tag{7.49}$$

Substituting all of the q values into equation (7.43), we obtain

$$k_{A \to BC} = \frac{k_B T}{h_P} \frac{\dfrac{(2\pi(m_A + m_B + m_C)k_B T)^{3/2}}{h_P^3}}{\dfrac{(2\pi m_A k_B T)^{3/2}(2\pi(m_B + m_C)k_B T)^{3/2}}{h_P^6}}$$

$$\times \left(\frac{q_{V,ABC}}{q_{V,ABC}}\right) \frac{8\pi^2 \mu_{ABC}(b_{coll})^2 k_B T}{h_P^2} e^{(-E_T^{\ddagger}/k_B T)} \tag{7.50}$$

Crossing out like terms yields

$$k_{A \to BC} = (k_B T)^{1/2} \left(\frac{m_A + m_B + m_c}{m_A (m_B + m_C)} \right)^{3/2} \mu_{ABC} \times (b_{coll})^2 \left(\frac{8}{\pi} \right)^{1/2} e^{-E_T^{\ddagger}/k_B T} \qquad (7.51)$$

Note that according to equation (7.28)

$$\mu_{ABC} = \cfrac{1}{\cfrac{1}{m_A} + \cfrac{1}{m_B + m_C}} = \frac{(m_A)(m_B + m_C)}{m_A + m_B + m_C} \qquad (7.52)$$

Substituting equation (7.52) into (7.51) yields

$$k_{A \to BC} = (k_B T)^{1/2} \frac{1}{\mu_{ABC}^{1/2}} \left(\frac{8}{\pi} \right)^{1/2} \pi (b_{coll})^2 \exp^{-E_T^{\ddagger}/k_B T} \qquad (7.53)$$

Further rearrangements yields

$$k_{A \to BC} = \left(\frac{8 k_B T}{\pi \mu_{ABC}} \right)^{1/2} \pi (b_{coll})^2 e^{-E_T^{\ddagger}/k_B T} \qquad (7.54)$$

According to equation (7.27)

$$\left(\frac{8 k_B T}{\pi \mu_{ABC}} \right)^{1/2} = \bar{v}_{A \to BC} \qquad (7.55)$$

Substituting (7.55) into (7.54) yields

$$k_{A \to BC} = \bar{v}_{A \to BC} \pi (b_{coll})^2 e^{-E_T^{\ddagger}/k_B T} \qquad (7.56)$$

Notice that equation (7.56) is identical to equation (7.26) (collision theory). Therefore, transition state reduces to collision theory when the transition state geometry is a hard-sphere.

Transition state theory differs from collision theory in two key ways:

- One can actually make a calculation. With transition state theory the problem with collision theory is that one never knows b_{coll} in equation (7.16) and thus one cannot make any detailed calculations. Transition state theory replaces b_{coll} with d_{coll}. The distance between the reactants is calculated using transition state geometry. In Chapter 11 we will show that one can use ab initio methods to calculate the transition state geometry. Therefore d_{coll} can be calculated exactly.

- Transition state theory allows you to consider reactions like reaction (7.32a) where only special configurations lead to the desired products. In such a case the partition function for the transition state will be reduced since fewer configurations lead to reaction. See equation (7.34).

7.4.3 Limitations of Transition State Theory

Unfortunately, transition state theory also makes errors, so one has to be careful when using it. For example, Figure 7.9 gives a plot of the experimental values of the rate constant divided by k_BT/h_P for a number of reactions versus the value of the same quantity calculated from transition state theory. Notice that there is very little correlation between transition state theory and the experiment. Physically, Figure 7.9 tests whether the $q_T^{\ddagger}/q_A q_{BC}$ term in equation (7.43) is correctly modifying the rate as the shape of the potential energy surface changes. The figure shows that transition state theory is not quite getting the right answer.

Transition state theory does not work because transition state theory assumes that the reactants remain at a pseudoequilibrium as the reactants traverse the potential energy surface and that tunneling can be ignored. A typical $A + BC$ collision might last only 10^{-12} seconds. There is no time for all of the normal modes to reach thermal equilibrium. As a result, there are some dynamical corrections to the rate equation. When the reaction

$$A + BC \longrightarrow AB + C \qquad (7.57)$$

occurs, one has to get energy into the B–C bond to break the bond. One also has to transfer momentum into C to carry C away. It happens that the energy and momentum transfer rates play a significant role in reactions. Those influences are ignored in transition state theory.

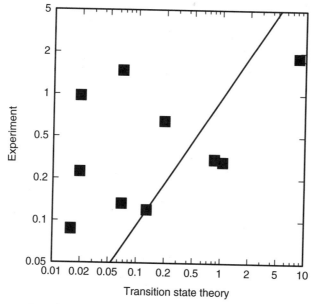

Figure 7.9 A comparison of the kh/k_BT measured experimentally for the reactions in Table 7.4 to the transition-state approximation of the same quantities.

There is a simple way to think about this process. Consider a hypothetical reaction such as that shown in Figure 7.2. During reaction, the molecules collide with one another as shown in Figure 7.2. A comes in and then collides with BC. Then C flies away. Marcellin showed that one can view the process as a trajectory on a potential energy surface, where R_{AB}, the distance from A to B, and R_{BC}, the distance from B to C, change with time. Figure 7.10 shows a typical trajectory. Notice that the trajectory needs to go around a sharp turn on the potential energy surface in order for reaction to happen. The sharp turn corresponds to converting momentum from the A–B motion to B–C motion. Consequently, one needs to have enough momentum transfer and sidewise motion to get a reaction to happen.

Sidewise motion comes from vibrational energy of the BC molecule. In Chapter 8, we will find that one has to carefully balance the translational energy from the incoming A molecule with the vibrational energy in the B–C molecule in order for the reaction to happen. If there is too little vibrational energy, the reactants never make it around the turn in the potential energy surface. If there is too little translational energy, the reactants never make it up the hill. All of these effects can be explored by computing the trajectories of molecules over potential energy surfaces, as described in Chapter 8.

At present, in general, there are no simple ways to predict a rate of reaction when energy transfer is important. People discuss the effect by rewriting equation (7.43) as

$$k_{A \to BC} = \left(\frac{k_B T}{h}\right) \kappa(T) \left(\frac{q_T^\ddagger}{q_A q_{BC}}\right) e^{-E_T^\ddagger / k_B T} \tag{7.58}$$

where $\kappa(T)$ is called the *transmission factor*. Physically, $\kappa(T)$ accounts for the fact that some collisions with enough total energy do not make it over the transition state. $\kappa(T)$ also accounts for the fact that some molecules can go through the barrier rather than over the barrier due to a tunneling process as discussed in Section 7.3.1. It is possible to do complex calculations to look at each state of the system and use an approximation to

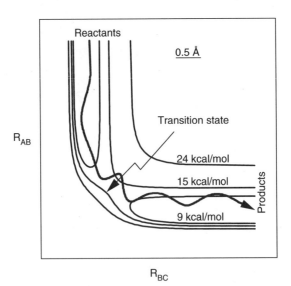

Figure 7.10 A trajectory on a potential energy surface.

estimate the probability that the molecules react. The calculations are complicated, but in principle one can calculate $\kappa(T)$.

In practice, κ is often treated as a "fudge" factor. It is temperature-dependent and can be greater than 1 when tunneling is important. Generally, it is hard to predict κ accurately, which limits the utility of equation (7.58). Accurate Methods to calculate κ will be discussed in Chapter 9.

7.5 RICE–RAMSPERGER–KASSEL–MARCUS (RRKM) THEORY

Next, we want to change topics and start to discuss unimolecular isomerization reactions. Unimolecular isomerizations are reactions in which a single molecule rearranges to form stable products. They can be written in general as

$$A \Longrightarrow B \tag{7.59}$$

Examples of unimolecular reactions include

$$\underset{H_2C-CH_2}{\overset{CH_2}{\diagup \diagdown}} \Longrightarrow CH_2{=}CH{-}CH_3 \tag{7.60}$$

$$CD_2CO + h_p\nu \longrightarrow CD_2CO^* \longrightarrow CD_2^* + CO^* \tag{7.61}$$

where $h_p\nu$ is a photon and the asterisks (*) represent excited species. Unimolecular reactions are different from ligand exchange reactions.

Recall that an elementary reaction

$$A \longrightarrow B \tag{7.62}$$

is impossible. Instead, we need a collision partner X to get the reaction to occur. In Chapter 5, we wrote the reaction with a collision partner as

$$A + X \longrightarrow B + X \tag{7.63}$$

Equation (7.63) is an approximation. It says that the collision partner collides with the reactant to immediately produce the product. Lindemann, (1922) however, found that the process does not occur immediately. Instead, it was better to think about the reaction as occurring in following two steps:

$$A + X \underset{2}{\overset{1}{\rightleftharpoons}} A^* + X$$

$$A^* \overset{3}{\longrightarrow} B \tag{7.64}$$

First the collision partner collides with the reactant A, to form A^*, an excited A molecule. Then the excited A decomposes to yield products.

Next it is useful to calculate a rate equation for reaction (7.64). The derivation follows that in Chapter 4. First one writes equations for the rate of formation of all the species.

Then one applies the steady-state approximation to A*. The result is the following approximation to the rate equation:

$$r_B = \frac{k_1 k_3 [A][X]}{k_3 + k_2 [X]} \tag{7.65}$$

Equation (7.65) is Lindemann's approximation to the rate of a unimolecular reaction.

In order to understand equation (7.65), we will consider the special case that the only collision partners are other reactants so that $[X] = [A]$. Equation (7.65) becomes

$$r_B = \frac{k_1 k_3 [A]^2}{k_3 + k_2 [A]} \tag{7.66}$$

According to equation (7.66), the reaction should be second-order at low pressure and first-order at high pressure (i.e., when $k_2[A] \gg k_3$). Figure 7.11 shows a log–log plot of the rate data for the reaction

$$CH_3NC \longrightarrow CH_3CN \tag{7.67}$$

and indeed the log–log plot is second-order at low pressure (i.e., slope $= 2$) and first-order at high pressures (slope $= 1$).

If one compares to the Lindemann model in detail, one finds that it fits very well at high and low pressures, but that there are some small deviations at intermediate pressures. Also, the value of k_3 that you need to fit the data is way too large. In Figure 7.9, we used $k_3 = 10^{17}$/second, while according to transition state, the k_3 should be 10^{14}/second, at most.

The reason why k_3 is so large is that the excited complex lasts for a long time in comparison to usual molecular collisions. The complex might last 10^{-7} seconds, compared to 10^{-13} seconds for a transition state. With a bimolecular collision, the hot molecules

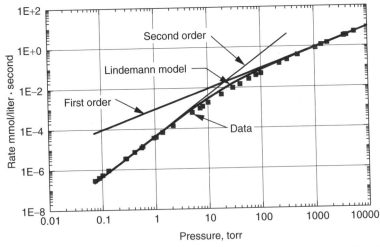

Figure 7.11 The rate of the reaction $CH_3NC \rightarrow CH_3CN$ as a function of pressure. (Data of Schneider and Rabinowicz (1962).

collide once, and if they do react during the collision, the reactants fly apart. In a unimolecular reaction, though, reactants cannot fly apart. Instead, the reacting atoms bump against each other multiple times. In effect, you get multiple collisions. That increases the rate.

Kassel (1928a,b,c), Rice, and Ramsperger (1927, 1928) worked out a classical equation for the rate of reaction for cases where the molecules stay hot for a long time. Transition states theory assumes that reaction occurs only when there is enough energy in the $A \rightarrow BC$ translation to carry the reactants over the transition state. However, Rice, Ramsperger, and Kassel assumed that the activation energy could be distributed over s vibrational modes and one would still get reaction. They then asserted that the rate would be enhanced by a factor of

$$\frac{\text{Number of ways to put an energy } E_a \text{ in s modes}}{\text{Number of ways to put an energy of } E_a \text{ in one mode}}$$

They then counted the modes, assuming that the spacing at the vibrational levels was much smaller than $k_B T$ to estimate the enhancement. After some algebra, they obtained

$$k_3 = \left(\frac{k_B T}{h_p}\right) \frac{1}{(s-1)!} \left(\frac{E_a}{k_B T}\right)^{s-1} \frac{q^{\ddagger}}{q_A q_{BC}} e^{-E_a/k_B T} \tag{7.68}$$

where s is the number of modes that can store energy. According to equation (2.34), $\frac{E_a}{k_B T}$ is on the order of 20. Methylisocyanide (CH_3NC) has 12 vibrational modes, and if we arbitrarily assume that six of them contribute to reaction, then according to equation (7.68), the rate will be enhanced by a factor of

$$\frac{1}{(6-1)!} (20)^{6-1} = 26,667 \tag{7.69}$$

Physically, if one can store up energy in all of the modes, the rate will be faster than if all of the energy needs to be deposited in the breaking bond during a single collision. As a result, the rate constant for a unimolecular reaction tends to be bigger than the rate constant for a bimolecular reaction.

If one tests equation (7.68) in detail, one finds that it explains why the k_3 in the Lindemann model is so large, but it does not explain the errors in the Lindemann model at medium pressures. The reasons for the errors are

- The rate constant varies according to how much energy is put in the molecule. If you use a laser to put considerable excess energy in the reactants, the reactants react more quickly than if there is only a small amount of excess energy.
- There are quantum effects, the assumption that the level spacing is less than $k_B T$ is not accurate for vibrations.

Marcus derived a modification of the Rice–Ramsperger–Kassel theory for that case. According to the model, the rate of reaction for an molecule excited with an energy E^* is given by

$$r_{A \rightarrow B} = \frac{q_M^{\ddagger}}{h_p q_A} \tag{7.70}$$

where h_P is Planck's constant, q_A is the partition function per unit volume of the reactant A, and $q_M^{\ddagger}$ is the partition function for the molecules that have accumulated enough energy to react. Note that $q_M^{\ddagger}$ is different from $q_T^{\ddagger}$. $q_T^{\ddagger}$ includes only states that have energy in the mode that translates over the saddle point in the potential energy surface. These are states where energy is localized in the bond that breaks during the reaction. $q_M^{\ddagger}$, on the other hand, includes all of the states of the system, which, added up together, have enough energy to cross the barrier independent of whether the energy is localized in any specific bond. The key assumption in equation (7.70) is that if the excited complex has enough energy to react, the excited complex will eventually decay into products.

Equation (7.70) is one version of the RRKM model. Figure 7.12 shows how well the model fits in the best case. It happens that in this case the RRKM model predicts that the rate should show a staircase pattern with increasing energy for reasons that are described in Chapter 9. The data seem to show the same staircase pattern. We will give other examples in Chapter 9 and will find that the theory does not always do as well as shown in Figure 7.12. Still, with unimolecular reactions, we can predict very accurate rate constants easily. One can also make accurate predictions with bimolecular reactions, but in those cases the calculations are much more elaborate.

There is one other important detail: the rate constants for unimolecular reactions change according to what collision partner is used. Table 7.5 gives examples of this. Generally, it is not uncommon for the rate constant for a unimolecular reaction to vary by more than an order of magnitude according to which collision partner is used. Unfortunately, there is not enough room here to describe all of the effects.

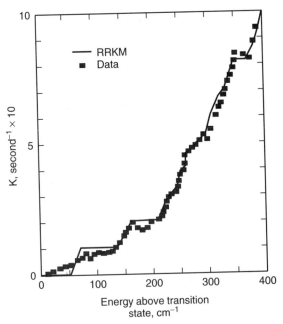

Figure 7.12 A comparison of the rate of the photolysis of ketene (CH_2CO) as a function of energy to the predictions of the RRKM model.

Table 7.5 The preexponentials [in $cm^6/(mol^2 \cdot second)$] for a series of unimolecular reactions, as you change the collision partner

Reaction	k_0 When x = Argon	k_0 When x = Water	k_0 When x = N_2
$NO_2 + X \rightarrow NO + O + X$	1.7×10^{14}	6.7×10^{15}	1.57×10^{15}
$H_2O + X \rightarrow OH + H + X$	2.1×10^{15}	3.5×10^{17}	5.1×10^{16}
$HO_2 + X \rightarrow O_2 + H + X$	1.5×10^{15}	3.2×10^{16}	2×10^{15}
$H_2 + X \rightarrow H + H + X$	6.4×10^{17}	2.6×10^{15}	
$O_2 + X \rightarrow 2O + X$	1.9×10^{13}		1.0×10^{14}

Source: Data of Westley (1980).

7.6 SUMMARY AND PLAN FOR THE NEXT TWO CHAPTERS

In summary, in this chapter we reviewed three key methods that can be used to estimate preexponentials for reaction: collision theory, transition state theory, and the RRKM model. All three of these models are based on Arrhenius' idea that a reaction happens when two hot molecules collide with the right configuration to react. Collision theory assumes that reaction happens whenever two hot molecules collide. The assumption is made that it does not matter how the two molecules come together. For example, in the reaction $A + BC \rightarrow AB + C$, collision theory assumes that the rate will be the same whether A first collides with B or A first collides with C. Transition state theory takes the opposite view and says that a reaction occurs only if the collision reaches the transition state. Transition state theory also assumes that equilibrium is manufactured throughout the collision process. The RRKM model goes back to the collision theory assumptions, and says that if you have a unimolecular reaction, a reaction will occur whenever the initial excitation of the reactant A puts enough energy into the molecule independent of where the initial energy is deposited. Each of these models are approximations. However, they do work in a wide variety of systems.

In the next two chapters, we will be expanding these ideas in detail. In particular, in Chapter 8, we will consider collision theory and try to understand what happens when atoms or molecules collide. We will see how one calculates the reaction probability and will provide software so that students can use the results to calculate rates. We will then use our findings to derive transition state theory, the RRKM model, and related results in Chapter 9. I divided up the discussion in this way because, when I teach the material to undergraduates, I find that I cannot get to catalysis unless I skip Chapter 8 and 9. However, I believe that collision theory is essential to our understanding of reactions. Therefore, I urge the readers to read the chapters and use the findings to learn about how molecules react.

7.7 SOLVED EXAMPLES

Example 7.A   A Collision Theory Calculation   Use collision theory to calculate the preexponential for the reaction

$$H + CH_3CH_3 \longrightarrow H_2 + CH_2CH_3 \qquad (7.A.1)$$

at 500 K.

Solution According to collision theory

$$k_0 = \pi d_{coll}^2 \bar{v}_{ABC} \tag{7.A.2}$$

First let us calculate $\bar{v}_{ABC}$. According to equation (7.26)

$$\bar{v}_{ABC} = 2.52 \times 10^{13} \text{ Å/second} \left(\frac{T}{300 \text{ K}}\right)^{1/2} \left(\frac{1 \text{ amu}}{\mu_{ABC}}\right)^{1/2} \tag{7.A.3}$$

with

$$\mu_{ABC} = \mu \frac{1}{\dfrac{1}{M_A} + \dfrac{1}{M_{BC}}} \tag{7.A.4}$$

For reaction (7.A.1)

$$\mu_{ABC} = \frac{1}{\dfrac{1}{1 \text{ amu}} + \dfrac{1}{30 \text{ amu}}} = 0.968 \text{ amu} \tag{7.A.5}$$

Plugging in the numbers shows that at 500 K:

$$\bar{v}_{ABC} = 2.52 \times 10^{13} \left(\frac{500 \text{ K}}{300 \text{ K}}\right)^{1/2} \left(\frac{1 \text{ amu}}{0.968 \text{ amu}}\right)^{1/2} = 3.31 \times 10^{13} \text{ Å/second} \tag{7.A.6}$$

There is some question about what values of d_{coll} to use in the calculation. Hydrogen has a van der Waals diameter of 1.5 Å, while ethane has a van der Waals diameter of 3.5 Å. One approximation to d_{coll} is

$$d_{coll} = \frac{1.5 \text{ Å} + 3.5 \text{ Å}}{2} = 2.5 \text{ Å} \tag{7.A.7}$$

Substituting (7.A.5) and (7.A.6) into equation (7.A.2) yields

$$k_0 = \pi \frac{(2.5 \text{ Å})^2}{\text{molecule}} (3.31 \times 10^{13} \text{ Å/second}) = 6.49 \times 10^{14} \frac{\text{Å}^3}{\text{molecule·second}} \tag{7.A.8}$$

Example 7.B Approximate Evaluation of $\dfrac{q^{\ddagger}}{q_A q_{BC}}$ Earlier in this chapter, we said that typically $\dfrac{q^{\ddagger}}{q_A q_{BC}}$ Is within an order of magnitude or two of unity. The object of this problem is to evaluate $\dfrac{q^{\ddagger}}{q_A q_{BC}}$ using the approximations given in Chapter 6 for a case where A, B, and C are atoms, not molecular ligands.

Solution Consider a second-order reaction $A + BC \rightarrow AB + C$. Recall from Chapter 6 that if a molecule contains n_a atoms, the molecule will have $3n_a$ total vibrational, translational, and rotational modes. All molecules can translate in three directions, so every molecule has three translational modes. A linear molecule can rotate in two directions, so it has two rotational modes. Such a molecule will have $3n_a - 3 - 2 = 3n_a - 5$ vibrational modes. A nonlinear molecule can rotate in three directions, so it will have three rotational modes. Such a molecule will have $3n_a - 3 - 3 = 3n_a - 6$ vibrational modes.

If A contains a single atom, then the atom can only translate, not vibrate. Therefore

$$q_A = (q_t^3)_A \qquad (7.B.1)$$

where q_t is a translational partition function and the parentheses designate that we are considering molecule A. The BC molecule can translate and rotate in two directions. It has six total modes. There are three translational modes and two rotational modes. Therefore, there must be one vibrational mode. Consequently

$$q_{BC} = (q_t^3 q_v q_r^2)_{BC} \qquad (7.B.2)$$

where q_t is a translational partition function per unit length, q_v is a vibrational partition function, and q_r is a rotational partition function and the parens are used to designate a BC molecule.

The A–BC transition state has three atoms, so it must have $3n_a = 9$ total modes. If we treat A–BC as linear, then the molecule will have three translational modes and two rotational modes. That leaves $9 - 3 - 2 = 4$ vibrational modes. Therefore

$$q^* = (q_t^3 q_v^4 q_r^2)_{TST} \qquad (7.B.3)$$

where q^* is the partition function for the transition state and the parentheses are used to indicate that this is for the transition state. The transition state approximation is to treat one of the vibrational modes as a translational mode:

$$q_A^* = (q_t^3 q_v^3 q_r^2 q_{A \rightarrow BC})_{TST} \qquad (7.B.4)$$

where $q_{A \rightarrow BC}$ is the partition function for the motion of A toward BC. In transition state theory, we want $q^\ddagger$, not q^*. By definition

$$q^\ddagger = \frac{q^*}{q_{A \rightarrow BC}} = (q_t^3 q_v^3 q_r^2)_{TST} \qquad (7.B.5)$$

Combining equations (7.B.1)–(7.B.5) yields

$$\frac{q^\ddagger}{q_A q_{BC}} = \frac{(q_v^3 q_r^a q_t^3)_{TST}}{(q_t^3)_A (q_t^3 q_v q_{r^2})_{BC}} \qquad (7.B.6)$$

Now we will make the approximation that

$$(q_t)_{TST} \approx (q_t)_A \approx (q_t)_{BC} = q_t \qquad (7.B.7)$$

$$(q_r)_{TST} \approx (q_r)_{BC} \qquad (7.B.8)$$

$$(q_v)_{TST} \approx (q_v)_{BC} = q_v \qquad (7.B.9)$$

Substituting (7.B.7)–(7.B.9) into equation (7.B.6) yields

$$\frac{q^\ddagger}{q_A q_B} = \frac{q_v^2}{q_t^3} \qquad (7.B.10)$$

According to Table 6.5, $q_V \approx 1$, $q_t \approx 3/\text{Å}$. Substituting into equation (7.B.10) yields

$$\frac{q^\ddagger}{q_A q_{BC}} = \left(\frac{1^2}{(3/\text{Å})^3}\right) = 0.037 \ \text{Å}^3 \qquad (7.B.11)$$

One can do the same analysis for a nonlinear transition state. The result is

$$\frac{q^\ddagger}{q_A q_{BC}} = \frac{q_v q_r}{q_t^3} \frac{1 \times 50}{(3/\text{Å})^3} = 1.85 \ \text{Å}^3 \qquad (7.B.12)$$

Equations (7.B.11) and (7.B.12) are estimates of the partition functions from transition state theory.

Notice that in this chapter we saw that $q^\ddagger/q_A q_{ABC}$ is between 0.1 and 10 but we got smaller values. This is an artifact that is created because we have ignored the fact that the transition state is bigger, than the reactants, so its rotational partition function is bigger, too. Also, vibrational partition functions in transition states tend to be bigger than in the reactants. Those two factors produce another factor of 5–10 in the partition function. Multiplying the results in (7.B.11) and (7.B.12) by 5 shows that $q^\ddagger/q_A q_B$ should be between 0.1 and 10 in angstrom units (Å).

Example 7.C   A True Transition State Theory Calculation   In Example 7.B, we used the approximation in equations (7.B.7)–(7.B.9) to calculate the partition functions. Here we want to do the calculation more carefully. The reaction

$$F + H_2 \longrightarrow HF + H \qquad (7.C.1)$$

is one of the most heavily studied reactions in the literature. It is important in chemical lasers. Stark and Werner [1996] calculated a potential energy surface using exact methods, while Stechleretal [1985] fit a potential energy surface to the reaction, and adjusted the potential energy surface so that transition state theory fits the experimental results.

Solution   In the solution here, we will first compute the rate using transition state theory with a potential energy surface that has been adjusted so that transition state theory works. Then, we will use the exact results and see how things change.

Table 7.C.1 shows the parameters used in the calculations. Generally, the adjusted potential energy surface was calculated by extending the H–F bond in the transition state by 30% and reducing the barrier by 3 kcal/mol. We also assumed a curvature of $310 \ \text{cm}^{-1}$ for the barrier, that is we assumed the barrier looks like an inverted parabola with a vibrational frequency of $310 \ \text{cm}^{-1}$. All of the other parameters do not matter to the calculation.

According to transition state theory:

$$k_{F \to H_2} = \left(\frac{k_B T}{h_P}\right) \frac{q_{F-H_2}^{\ddagger T}}{q_{H_2}} e^{-E_T^\ddagger / k_B T} \qquad (7.C.2)$$

It is useful to divide up the partition functions in equation (7.C.2) into the contributions from the translation, vibration, rotation, and electronic modes:

Table 7.C.1 Parameters used to calculate the transition state theory rate constant for $F + H_2 \rightarrow HF + H$ (the exact parameters are also shown for comparison)

	Transition State		Reactants	
	Exact	Used for Transition State Calculations	F	H_2
r_{HF}	1.34 Å	1.602 Å		
r_{HH}	0.801 Å	0.756 Å		0.7417 Å
ν_{H-H} stretch	~3750 cm^{-1}	4007 cm^{-1}		4395.2 cm^{-1}
ν_{FH_2} bend	?	397.9 cm^{-1}		
ν_{FH_2} Bend	?	397.9 cm^{-1}		
Curvature barrier	?	310 cm^{-1}		
$E^{\ddagger}$	5.6 kcal/mol	1.7 kcal/mol		
M	21 amu	21 amu	19 amu	2 amu
I	5.48 amu·Å^2	7.09 amu·Å^2		0.275 amu·Å^2
g_e	4	4	4	1

$$k_{F \rightarrow H_2} = \frac{k_B T}{h_P} 1^{\ddagger} \left(\frac{q^{\ddagger}}{q_{H_2} q_F} \right)_{trans} \left(\frac{q^{\ddagger}}{q_{H_2} q_F} \right)_{vibr} \left(\frac{q^{\ddagger}}{q_{H_2} q_F} \right)_{rot} \left(\frac{q^{\ddagger}}{q_{H_2} q_F} \right)_{elec} e^{-E_T^{\ddagger}/k_B T} \quad (7.C.3)$$

where $1^{\ddagger}$ is an extra factor of 2 that arises because there are two equivalent transition states: one with the fluorine attacking one hydrogen and the other with one fluorine attacking the other hydrogen.

Now it is useful to use the results in Chapter 6 to calculate the various terms in equation (7.C.3). According to Table 6.7

$$q_t = \left(\frac{2\pi m k_B T}{h_P^2} \right)^{1/2} \quad (7.C.4)$$

where q_t is the translational partition function for a single translational mode of a molecule, m is the mass of the molecule, k_B is Boltzmann's constant, T is the temperature, and h_P is Planck's constant. For our particular reaction, the fluorine can translate in three directions, the H_2 can translate in three directions, and the transition state can translate in three directions. Consequently

$$\left(\frac{q^{\ddagger}}{q_{H_2} q_F} \right)_{trans} = \frac{\left(\dfrac{2\pi m_{\ddagger} k_B T}{h_P^2} \right)^{3/2}}{\left(\dfrac{2\pi m_F k_B T}{h_P^2} \right)^{3/2} \left(\dfrac{2\pi m_{H_2} k_B T}{h_P^2} \right)^{3/2}} \quad (7.C.5)$$

where m_F, m_{H_2}, and $m_{\ddagger}$ are the masses of fluorine, H_2, and the transition state.

Performing the algebra; we obtain

$$\left(\frac{q^{\ddagger}}{q_F q_{H_2}} \right)_{trans} = \left(\frac{m_{\ddagger}}{m_F m_{H_2}} \right)^{3/2} \left(\frac{h_P^2}{2\pi k_B T} \right)^{3/2} \quad (7.C.6)$$

Let's calculate the last term in equation (7.C.6). Rearranging the last term shows

$$\left(\frac{h_P^2}{2\pi k_B T}\right)^{3/2} = \left(\frac{300 \text{ K}}{T}\right)^{3/2} \left(\frac{h_P^2}{2\pi k_B (300 \text{ K})}\right)^{3/2} \tag{7.C.7}$$

Substituting the numbers yields

$$\left(\frac{h_P^2}{2\pi k_B T}\right)^{3/2} = \left(\frac{300 \text{ K}}{T}\right)^{3/2}$$

$$\times \left(\frac{(6.626 \times 10^{-34} \text{ (kg·meter}^2/\text{second})^2 \left(\frac{10^{10} \text{ Å}}{\text{meter}}\right)^2 \left(\frac{\text{amu}}{1.66 \times 10^{-27} \text{ kg}}\right)}{2\pi (1.381 \times 10^{-23} \text{ kg·meter}^2)/(\text{second}^2 \cdot \text{K})(300 \text{ K})}\right)^{3/2}$$

$$\tag{7.C.8}$$

Doing the arithmetic yields

$$\left(\frac{h_P^2}{2\pi k_B T}\right)^{3/2} = \left(\frac{300 \text{ K}}{T}\right)^{3/2} 1.024 \text{ Å}^3 \cdot \text{amu}^{3/2} \tag{7.C.9}$$

Combining equations (7.C.6) and (7.C.9) yields

$$\left(\frac{q^\ddagger}{q_F q_{H_2}}\right)_{\text{trans}} = \left(\frac{M_\ddagger}{(M_F) M_{H_2}}\right)^{3/2} \left(\frac{300 \text{ K}}{T}\right)^{3/2} (1.024 \text{ Å}^3 \cdot \text{amu}^{3/2}) \tag{7.C.10}$$

Setting $T = 300$ K, $M_\ddagger = 21$ amu, $M_F = 19$ amu, $M_{H_2} = 2$ amu yields

$$\left(\frac{q^\ddagger}{q_F q_{H_2}}\right)_{\text{trans}} = \left(\frac{21 \text{ amu}}{(19 \text{ amu})2 \text{ amu}}\right)^{3/2} 1.024 \text{ Å}^3 \cdot \text{amu}^{3/2} = 0.42 \text{ Å}^3 \tag{7.C.11}$$

Next, let's calculate the ratio of the rotational partition functions. The fluorine atom does not rotate, so

$$\left(\frac{q^\ddagger}{q_{H_2} q_{F_2}}\right)_{\text{rot}} = \left(\frac{q^\ddagger}{q_{H_2}}\right)_{\text{rot}} \tag{7.C.12}$$

According to table (6.7)

$$q_r = \frac{8\pi k_B T I}{h_P^3} \tag{7.C.13}$$

where k_B is Boltzmann's constant, T is the temperature, h_P is Planck's constant, and I is the moment of inertia of the molecule. Combining (7.C.12) and (7.C.13) yields

$$\left(\frac{q^\ddagger}{q_{H_2}}\right)_{\text{rot}} = \left(\frac{8\pi k_B T I^\ddagger / h_P^3}{8\pi k_B T I_{H_2} / h_P^3}\right) = \frac{I^\ddagger}{I_{H_2}} \tag{7.C.14}$$

Substituting in the adjusted value of $I^\ddagger$ and I_{H_2} from Table 7.C.1 yields

$$\left(\frac{q^\ddagger}{q_{H_2}}\right)_{\text{rot}} = \frac{I^\ddagger}{I_{H_2}} = \frac{7.091 \text{ amu·Å}^2}{0.275 \text{ amu·Å}^2} = 25.8 \tag{7.C.15}$$

Next, let's calculate the vibrational partition functions. According to Table 6.5

$$q_v = \frac{1}{1 - \exp\left(-\dfrac{h_p v}{k_B T}\right)} \tag{7.C.16}$$

Let's first get an expression for the exponential term in equation (7.C.16). It is easy to show

$$\frac{h_p v}{k_B T} = \left(\frac{h_p(1 \text{ cm}^{-1})}{(k_B)(300 \text{ K})}\right) \left(\frac{300 \text{ K}}{T}\right) \left(\frac{v}{1 \text{ cm}^{-1}}\right) \tag{7.C.17}$$

Plugging in values of h_p and k_B from the end of the book yields.

$$\frac{h_p v}{k_B T} = \frac{(2.85 \times 10^{-3} \text{ kcal/(mol·cm}^{-1}))(1 \text{ cm}^{-1})}{(1.980 \times 10^{-3} \text{ kcal/(mol·K)})(300 \text{ K})} \left(\frac{300 \text{ K}}{T}\right) \left(\frac{v}{1 \text{ cm}^{-1}}\right) \tag{7.C.18}$$

Note we actually used $h_p \mathbb{C}/N_a$ and k_B/N_a in equation (7.C.16), and not h_p where N_a is Avogadro's number and $\mathbb{C}$ is the speed of light, to get the units right. Doing the arithmetic in equation (7.C.18) yields

$$\frac{h_p v}{k_B T} = (4.784 \times 10^{-3}) \left(\frac{300 \text{ K}}{T}\right) \left(\frac{v}{1 \text{ cm}^{-1}}\right) \tag{7.C.19}$$

Table 7.C.2 shows numerical values for various values of v. The vibrational partition function ratio equals

$$\left(\frac{q^{\ddagger}}{q_{H_2}}\right)_{vib} = \frac{q^{\ddagger}_{HH} q^{\ddagger}_{bend} q^{\ddagger}_{bend}}{(q_{H-H})_{H_2}} = \frac{(1)(1.19)(1.19)}{1} = 1.42 \tag{7.C.20}$$

Next, let's calculate the ratio of the partition functions for the electronic state. Let's consider only the ground electronic state:

$$\left(\frac{q^{\ddagger}}{q_{H_2} q_F}\right)_{elec} = \frac{g^{\ddagger}_e}{(g_e)_{H_2}(g_e)_F} = \frac{4}{1 \times 4} = 1 \tag{7.C.21}$$

Finally, let's calculate $k_B T / h_p$:

$$\frac{k_B T}{h_p} = \frac{(1.381 \times 10^{-23} \text{ (kg·meter}^2)/(\text{second}^2 \cdot \text{molecule·K})(300 \text{ K})}{6.626 \times 10^{-34} \text{ (kg·meter}^2)/\text{second}} \left(\frac{T}{300 \text{ K}}\right)$$

$$= 6.25 \times 10^{12}/\text{molecule·second} \left(\frac{T}{300 \text{ K}}\right) \tag{7.C.22}$$

Table 7.C.2 The vibrational partition function

Mode	v, cm^{-1}	$h_p v / k_B T$	q_v
$q^{\ddagger}_{HH}$	4395.2	21.0	1.0
$(q_{HH})_{H_2}$	4007	19.2	1.0
$q^{\ddagger}_{bend}$	379.9	1.82	1.19

Putting this all together allows one to calculate a preexponential:

$$k_0 = 1^{\ddagger} \left(\frac{k_B T}{h_P} \right) \left(\frac{q^{\ddagger}}{q_{H_2} q_F} \right)_{trans} \left(\frac{q^{\ddagger}}{q_{H_2} q_F} \right)_{rot} \left(\frac{q^{\ddagger}}{q_{H_2} q_F} \right)_{vib} \left(\frac{q^{\ddagger}}{q_{H_2} q_F} \right)_{elec} \qquad (7.C.23)$$

Plugging in the numbers yields

$$k_0 = 2(6.25 \times 10^{12}/(\text{molecule·second}))(0.42 \text{ Å}^3)(25.8)(1.42)(1)$$

$$= 1.92 \times 10^{14} \text{ Å}^3/(\text{molecule·second}) \qquad (7.C.24)$$

If one uses the actual transition state geometry, the only thing that changes significantly is the rotational term. One obtains

$$\left(\frac{q^{\ddagger}}{q_{H_2}} \right)_{rot} = \left(\frac{I^{\ddagger}}{I_{H_2}} \right)_{rot} = \frac{5.48 \ (\text{amu·Å}^2)}{0.275 \ (\text{amu·Å}^2)} = 19.9 \qquad (7.C.25)$$

Then k_0 becomes

$$k_0 = 2(6.25 \times 10^{12}/\text{molecule·second})(0.42 \text{ Å}^3)(19.9)(1.4)(1)$$

$$= 1.46 \times 10^{14} \text{ Å}^3/(\text{molecule·second}) \qquad (7.C.26)$$

One can also calculate the preexponential via old collision theory. In collision theory, one considers the translations and rotations, but not the vibrations:

$$k_0 = 1^{\ddagger} \left(\frac{k_B T}{h_P} \right) \left(\frac{q^{\ddagger}}{q_{H_2} q_F} \right)_{trans} \left(\frac{q^{\ddagger}}{q_{H_2} q_F} \right)_{rot} \qquad (7.C.27)$$

In equation (7.C.26), the rotational partition function should be calculated at the collision diameter, and not the transition state geometry. If we assume a collision diameter of 2.3 Å (i.e., the sum of the van der Waals radii), we obtain

$$I^{\ddagger} = (r_{F-H_2})^2 (\mu_{FH_2}) = (2.3 \text{ Å})^2 \left(\frac{(2 \text{ amu})(19 \text{ amu})}{21 \text{ amu}} \right) = 9.57 \text{ Å}^2 \cdot \text{amu} \qquad (7.C.28)$$

Substituting into equation (7.C.25) using the results given above yields

$$k_0 = 2(6.25 \times 10^{12}/(\text{mol·second}))(0.42 \text{ Å}^3) \left(\frac{9.57 \text{ Å}^2 \cdot \text{amu}}{0.275 \text{ Å}^2 \cdot \text{amu}} \right)$$

$$= 1.8 \times 10^{14} \text{ Å}^3/(\text{mol·second}) \qquad (7.C.29)$$

Table 7.C.3 summaries these results. In this particular example, collision theory did pretty well. Transition state theory does not improve the predictions if you use the real transition state geometry. You can fit the real data if you "adjust" the transition state geometry. Overall, transition state theory is almost the same as collision theory.

Of course, one should expect that from the calculations. Transition state theory makes two corrections to old collision theory:

Table 7.C.3   A comparison of the preexponential calculated by transition state theory and collision theory to the experimental value [in $\mathring{A}^3$/(mol·second)]

k_0 transition state theory with adjusted transition state geometry	1.92×10^{14}
k_0 transition state theory with exact transition state geometry	1.46×10^{14}
k_0 collision theory	1.8×10^{14}
k_0 experiment	2.3×10^{14}

1. Transition state theory uses the transition state diameter rather than the collision diameter in the calculation.
2. Transition state theory multiplies by two extra terms: (a) the ratio of the vibrational partition function, to the electronic partition function for the transition state and (b) the reactants.

Neither of these corrections is very big, and these corrections tend to cancel one another. The result in this example was that the prediction was less reliable than collision theory.

   This is a typical result. Transition state theory makes small adjustments to collision theory, but ignores other key terms such as the coupling between vibrations and translations. The net result is that for many examples, transition state theory is not more accurate than collision theory in estimating preexponentials for reactions.

Example 7.D   Activation Barriers for Reactions According to Transition State Theory   Use transition state theory to calculate the activation barrier for reaction 7.C.1.

Solution   According to transition state theory

$$E_a = E_T^{\ddagger} + E_0^{\ddagger} - E_0^{H_2} - E_0^{F} \qquad (7.D.1)$$

where $E^{\ddagger}$ is the energy of the transition state relative to the reactants and $E_0^{\ddagger}$, $E_0^{H_2}$, E_0^{F} are the zero-point energies of the various species. Recall that $E_0 = h_P v/2$, where h_P is Planck's constant. Therefore

$$E_a = E_T^{\ddagger} + \tfrac{1}{2} h_P (v_{HH}^{\ddagger} + v_{bend}^{\ddagger} + v_{bend}^{\ddagger} - v_{curvature}^{\ddagger} - v_{HH}^{H_2}) \qquad (7.D.2)$$

If we use the "adjusted" potential energy surface, we obtain

$$E_a = 1.7 \text{ kcal/mol} + \tfrac{1}{2}(2.859 \times 10^{-3} \text{ kcal/(mol·cm}^{-1}))$$
$$\times (4007 \text{ cm}^{-1} + 397.9 \text{ cm}^{-1} + 397.9 \text{ cm}^{-1} - 310 \text{ cm}^{-1} - 4392 \text{ cm}^{-1}) \quad (7.D.3)$$
$$= 1.7 \text{ kcal/mol} + 0.14 \text{ kcal/mol} = 1.84 \text{ kcal/mol}$$

One can also calculate the activation barrier based on the exact potential energy surface. Stark and Werner [1996] et al. did not report vibrational frequencies for the bending mode, so I will assume that they are the same as in the approximate potential energy surface. In

Table 7.D.1   A comparison of the activation barrier calculated from transition state theory to the experimental value (in kcal/mol)

E_a from the exact potential energy surface	5.38
E_a from the "adjusted" potential energy surface	1.84
E_a from experiment	1.58

that case

$$E_a = 5.6 \text{ kcal/mol} + \tfrac{1}{2}(2.859 \times 10^{-3} \text{ kcal/(mol·cm}^{-1}))$$

$$\times (3750 \text{ cm}^{-1} + 397.9 \text{ cm}^{-1} + 397.9 \text{ cm}^{-1} - 310 \text{ cm}^{-1} - 4392 \text{ cm}^{-1}) \quad (7.D.4)$$

$$= 5.6 \text{ kcal/mol} - 0.22 \text{ kcal/mol} = 5.38 \text{ kcal/mol}$$

Table 7.D.1 compares these calculations to the experimental results. Again, transition state theory gets you the right order of magnitude. However, it is not exact. In this particular case, tunneling lowers the barrier and tunneling is ignored in transition state theory. One can get exact results by adjusting the potential energy surface. An alternative is to do the calculation correctly including tunneling as described in Chapter 9.

Example 7.E   Temperature dependence of the preexponential   In Example 7.C, we calculated the preexponential term at 300 K to calculate k_0. How does the preexponential change with temperature?

Solution   The principal temperature dependence comes from the $k_B T/h_P$ term and the ratio of translational partition functions in equation (7.C.24). If one substitutes equation (7.C.10) into equation (7.C.24), one finds

$$k_0 = \left(\frac{k_B 300 \text{ K}}{h_P}\right)\left(\frac{300 \text{ K}}{T}\right)^{3/2}\left(\frac{q^\ddagger}{q_{H_2}q_{HF}}\right)_{\substack{trans \\ 300 \text{ K}}}\left(\frac{q^\ddagger}{q_{H_2}q_{HF}}\right)_{\substack{rot \\ 300 \text{ K}}}$$

$$\times \left(\frac{q^\ddagger}{q_{H_2}q_{HF}}\right)_{\substack{vib \\ 300 \text{ K}}}\left(\frac{q^\ddagger}{q_{H_2}q_{HF}}\right)_{\substack{elec \\ 300 \text{ K}}} \quad (7.E.1)$$

where $\left(\dfrac{q^\ddagger}{q_{H_2}q_{HF}}\right)_{\substack{trans \\ 300 \text{ K}}}$ is the translational partition function at 300 K. Comparing equations (7.D.1) and (7.C.22) shows

$$k_0 \approx k_0^{300 \text{ K}}\left(\frac{300 \text{ K}}{T}\right)^{1/2} \quad (7.E.2)$$

In the homework set, we ask the reader to look at a case where the number of rotational modes also changes. In that case, one predicts some extra temperature dependence due to the temperature dependence of the rotational partition function.

Example 7.F   Transition State Analogs   In the examples so far in this chapter, we considered exact transition state for a reaction. However, another approach is to find a stable molecule that is an analog of the transition state. The idea is to use the following expression:

$$k_0 = Z_{AB} e^{+\Delta S^{\ddagger}/k_B} \tag{7.F.1}$$

where k_0 is the preexponential, k_B is Boltzmann's constant, T is the temperature, h_P is Planck's constant, and $\Delta S^{\ddagger}$ is the entropy of formation of the transition state.

One then estimates $\Delta S^{\ddagger}$ by looking at a molecular analogy of the transition state. For example, consider the reaction

$$CF_3Br + CH_3 \longrightarrow CF_3 + BrCH_3 \tag{7.F.2}$$

The transition state for the reaction looks like

$$CF_3 \cdots Br \cdots CH_3 \tag{7.F.3}$$

Find an analogous molecule and use it to estimate $\Delta S^{\ddagger}$.

Solution Consider the stable molecule CF_3–S–CH_3. It has a structure like that of $CF_3 \cdots Br \cdots CH_3$. The bonds in CF_3–S–CH_3 are shorter than in the transition state, and the frustrated rotations have higher barriers, but CF_3–S–CH_3 is an analog of the transition state for reaction (7.F.2). Now consider the reaction

$$CF_3S + CH_3 \Longrightarrow CF_3SCH_3 \tag{7.F.4}$$

This reaction is similar to the reaction

$$CF_3Br + CH_3 \longrightarrow (CF_3 \text{---} Br \text{---} CH_3)_{TST} \tag{7.F.5}$$

where $(CF_3 \text{---} Br \text{---} CH_3)$ is the transition state for reaction (7.F.2), and it happens that I can look up (ΔS_A) the entropy change for reaction (7.F.4). According to the NIST tables, ΔS_A is -41.2 cal/(mol·K).

Years ago people often used the approximation

$$\Delta S^{\ddagger} = \Delta S_A \tag{7.F.6}$$

Combining equations (7.F.1) and (7.F.6) yields

$$k_0 \geq Z_{AB} e^{+\Delta S_A/k_B} \tag{7.F.7}$$

We put a "greater than or equal to" sign in equation (7.F.4) because ΔS_A is a lower bound to $\Delta S^{\ddagger}$ for the reaction. Generally, the analog has a smaller rotational partition function than the transition state, so the ΔS_A for the analog is always less than $\Delta S^{\ddagger}$. For example, $\Delta S^{\ddagger}$ for reaction (7.F.2) is -14 cal/(mol·K), while the reaction

$$CH_3S + CH_3 \Longrightarrow CF_3SCH_3 \tag{7.F.8}$$

has a ΔS of -41.2 cal/(mol·K). By comparison, collision theory gives an upper bound to the preexponential. The result is that one can get a range for the preexponential for the reaction.

Let us use the transition state analog to estimate k_0 for reaction (7.E.2) at 500 K. We will use the approximation

$$k_0 = Z_{AB}e^{-\Delta S_A/k_B} \tag{7.F.9}$$

First, we will estimate Z_{AB} following the methods in Example 7.A:

$$Z_{AB} = \pi(b_{coll})^2 \bar{v}_{ABC} \tag{7.F.10}$$

$$\mu_{ABC} = \cfrac{1}{\cfrac{1}{m_{CF_3Br}} + \cfrac{1}{m_{CH_3}}} = \cfrac{1}{\cfrac{1}{179 \text{ amu}} + \cfrac{1}{15 \text{ amu}}} = 13.8 \text{ amu} \tag{7.F.11}$$

At 500 K

$$\bar{v}_{ABC} = 2.52 \times 10^{13} \text{ Å/second} \left(\frac{500 \text{ K}}{300 \text{ K}}\right)^{1/2} \left(\frac{1 \text{ amu}}{13.8 \text{ amu}}\right)^{1/2} = 8.8 \times 10^{12} \text{ Å/second} \tag{7.F.12}$$

Assume a collision diameter of 2 Å:

$$Z_{AB} = \pi(2 \text{ Å})^2(8.8 \times 10^{12} \text{ Å/second}) = 1.1 \times 10^{14} \frac{\text{Å}^3}{\text{mol/second}} \tag{7.F.13}$$

Substituting into equation (7.F.9)

$$k_0 = (1.1 \times 10^{14} \text{ Å}^3/(\text{molecule·second})) \exp\left(\frac{-41.2 \text{ cal/(mol·K)}}{1.98 \text{ cal/(mol·K)}}\right) \tag{7.F.14}$$

Doing the arithmetic yields

$$k_0 = 1 \times 10^5 \text{ Å}^3/(\text{molecule·second}) \tag{7.F.15}$$

Benson (1986) also has one other good rule of thumb. If you have a transition state of the form CF_3–Br–CH_3, then the CF_3 and CH_3 bending and rocking models will have vibrational frequencies of about 70% of the vibrational frequency of the rocking modes in a stable molecule. That typically adds a factor of 3 to the rate. Therefore the value based on an analog is $k_n = 3 \times 10^5$ Å^3/(molecule·second). By composition, the experimental value is $10^{12\pm0.5}$ Å^3/molecule.

In my experience, the transition state analogs grossly underestimate preexponentials. However, they provide useful lower bounds, so if you measure a lower preexponential than you calculate for the analog, you know that something is wrong.

We will see later in this book that even though transition state analogs are not that useful in predicting $\Delta S^{\ddagger}$, the analogs are very useful in designing catalysts or enzymes. Catalysts and enzymes enhance rates by lowering the energy of the transition state for a reaction. One way to design a catalyst is to find a transition state analog (i.e., one with the correct ΔS_A) and then design a catalyst that strongly binds to the analog. If the catalyst binds strongly to the analog, the catalyst will usually also bind strongly to the transition state. The binding lowers $\Delta G^{\ddagger}$. Transition state analogs are very useful in designing catalysts even though they are not as useful for calculating rates.

There also is the converse, that if you want to stop a catalytic reaction, then a good way to do that is to add a transition state analog to block the transition state for the formation of an undesired product. For example, there is a class of AIDS drugs, called *protease inhibitors*, which work by blocking the transition state for one of the key steps in reproduction of the AIDS virus. People design the drugs by finding a transition state analog of the key step in the reproduction cycle of the virus, and then testing the transition state analog as a drug. The transition state analogs are very potent drugs. They really work!

Example 7.G   Understanding Contour Plots   Consider a potential v defined by:

$$v(r1, r2, r0, a, w, vp, wa, hr) = w*(Exp(-2*a*(r1 - r0))$$

$$-2*Exp(-a*(r1 - r0))) + (w + hr)*(Exp(-2*a*(r2 - r0))$$

$$-2*Exp(-a*(r2 - r0))) + vp*Exp(-a*(r1 + r2 - 2*r0)) +$$

$$w + wa*Exp(-4*a*a*((r1 - r0)^2 + (r2 - 3*r0)^2)) + \qquad (7.G.1)$$

$$wa*Exp(-4*a*a*(((r1 - 3*r0)^2) + ((r2 - r0)^2)))$$

$$If(v > 20 + abs(hr)) \; Then$$

$$v = 20 + abs(hr)$$

Make contour plots for the following function of r1 and r2 with parmeters r0, a, w, hr, vp, wa given by

(a) $w = 109$ kcal/mol, $wa = 0$, $hr = 0$, $vp = 258$ kcal/mol, $r0 = 0.74$ Å, $a = 1.2$ Å^{-1}

(b) $w = 109$ kcal/mol, $wa = 0$, $hr = 0$, $vp = 218$ kcal/mol, $r0 = 0.74$ Å, $a = 1.2$ Å^{-1}

(c) $w = 109$ kcal/mol, $wa = 0$, $hr = 0$, $vp = 158$ kcal/mol, $r0 = 0.74$ Å, $a = 1.2$ Å^{-1}

(d) $w = 109$ kcal/mol, $wa = -10$, $hr = 0$, $vp = 258$ kcal/mol, $r0 = 0.74$ Å, $a = 1.2$ Å^{-1}

(e) $w = 109$ kcal/mol, $wa = 20$, $hr = 0$, $vp = 178$ kcal/mol, $r0 = 0.74$ Å, $a = 1.2$ Å^{-1}

(f) $w = 109$ kcal/mol, $wa = 0$, $hr = -20$, $vp = 233$ kcal/mol, $r0 = 0.74$ Å, $a = 1.2$ Å^{-1}

Solution   I used microsoft excel to do the calculations. First I defined a microsoft excel module to calculate the v. My module is given in Table 7.G.1. The spreadsheet can be found in the instructors materials.

Then I set up a microsoft Excel spreadsheet to do the calculations. Table 7.G.2 shows the formulas in part of the spreadsheet. I named cell B1 w, cell B2 vp, cell D2, and cell

Table 7.G.1   The module used to calculate the function in equation (7.G.1)

```
Public Function v(r1,r2,r0,a,w,vp,wa,hr) As Variant
v = w*(Exp(-2*a*(r1 - r0)) - 2*Exp(-a*(r1 - r0)))
v = v + (w + hr)*(Exp(-2*a*(r2 - r0)) - 2*Exp(-a*(r2 - r0)))
v = v + vp*Exp(-a*(r1 + r2 - 2*r0))
v = v + w
v = v + wa*Exp(-4*a*a*((r1 - r0)^2 + (r2 - 3*r0)^2))
v = v + wa*Exp(-4*a*a*(((r1 - 3*r0)^2) + ((r2 - r0)^2)))
If (v > 20 + Abs(hr)) Then
v = 20 + Abs(hr)
End If
End Function
```

Table 7.G.2 The formulas in part of the spreadsheet[a] used to solve Example 7.G.2. A copy of this spreadsheet is available at http://www.wiley.com/chemicalkinetics

	A	B	C	D	E	F	G
01	w=	109.4	a=	1.2	wa=	0	hr=
02	vp=	258	r0=	0.74	sp=	0.05	0
03	r2\r1	0.5	=B3+sp	=C3+sp	=D3+sp	=E3+sp	=F3+sp
04	0.5	=v(B$3,$A4,r0, a,w,vp,wa,hr)	=v(C$3,$A4,r0, a,w,vp,wa,hr)	=v(D$3,$A4,r0, a,w,vp,wa,hr)	=v(E$3,$A4,r0, a,w,vp,wa,hr)	=v(F$3,$A4,r0, a,w,vp,wa,hr)	=v(G$3,$A4,r0, a,w,vp,wa,hr)
05	=A4+sp	=v(B$3,$A5,r0, a,w,vp,wa,hr)	=v(C$3,$A5,r0, a,w,vp,wa,hr)	=v(D$3,$A5,r0, a,w,vp,wa,hr)	=v(E$3,$A5,r0, a,w,vp,wa,hr)	=v(F$3,$A5,r0, a,w,vp,wa,hr)	=v(G$3,$A5,r0, a,w,vp,wa,hr)
06	=A5+sp	=v(B$3,$A6,r0, a,w,vp,wa,hr)	=v(C$3,$A6,r0, a,w,vp,wa,hr)	=v(D$3,$A6,r0, a,w,vp,wa,hr)	=v(E$3,$A6,r0, a,w,vp,wa,hr)	=v(F$3,$A6,r0, a,w,vp,wa,hr)	=v(G$3,$A6,r0, a,w,vp,wa,hr)
07	=A6+sp	=v(B$3,$A7,r0, a,w,vp,wa,hr)	=v(C$3,$A7,r0, a,w,vp,wa,hr)	=v(D$3,$A7,r0, a,w,vp,wa,hr)	=v(E$3,$A7,r0, a,w,vp,wa,hr)	=v(F$3,$A7,r0, a,w,vp,wa,hr)	=v(G$3,$A7,r0, a,w,vp,wa,hr)
08	=A7+sp	=v(B$3,$A8,r0, a,w,vp,wa,hr)	=v(C$3,$A8,r0, a,w,vp,wa,hr)	=v(D$3,$A8,r0, a,w,vp,wa,hr)	=v(E$3,$A8,r0, a,w,vp,wa,hr)	=v(F$3,$A8,r0, a,w,vp,wa,hr)	=v(G$3,$A8,r0, a,w,vp,wa,hr)
09	=A8+sp	=v(B$3,$A9,r0, a,w,vp,wa,hr)	=v(C$3,$A9,r0, a,w,vp,wa,hr)	=v(D$3,$A9,r0, a,w,vp,wa,hr)	=v(E$3,$A9,r0, a,w,vp,wa,hr)	=v(F$3,$A9,r0, a,w,vp,wa,hr)	=v(G$3,$A9,r0, a,w,vp,wa,hr)
10	=A9+sp	=v(B$3,$A10,r0, a,w,vp,wa,hr)	=v(C$3,$A10,r0, a,w,vp,wa,hr)	=v(D$3,$A10,r0, a,w,vp,wa,hr)	=v(E$3,$A10,r0, a,w,vp,wa,hr)	=v(F$3,$A10,r0, a,w,vp,wa,hr)	=v(G$3,$A10,r0, a,w,vp,wa,hr)
11	=A10+sp	=v(B$3,$A11,r0, a,w,vp,wa,hr)	=v(C$3,$A11,r0, a,w,vp,wa,hr)	=v(D$3,$A11,r0, a,w,vp,wa,hr)	=v(E$3,$A11,r0, a,w,vp,wa,hr)	=v(F$3,$A11,r0, a,w,vp,wa,hr)	=v(G$3,$A11,r0, a,w,vp,wa,hr)
12	=A11+sp	=v(B$3,$A12,r0, a,w,vp,wa,hr)	=v(C$3,$A12,r0, a,w,vp,wa,hr)	=v(D$3,$A12,r0, a,w,vp,wa,hr)	=v(E$3,$A12,r0, a,w,vp,wa,hr)	=v(F$3,$A12,r0, a,w,vp,wa,hr)	=v(G$3,$A12,r0, a,w,vp,wa,hr)
13	=A12+sp	=v(B$3,$A13,r0, a,w,vp,wa,hr)	=v(C$3,$A13,r0, a,w,vp,wa,hr)	=v(D$3,$A13,r0, a,w,vp,wa,hr)	=v(E$3,$A13,r0, a,w,vp,wa,hr)	=v(F$3,$A13,r0, a,w,vp,wa,hr)	=v(G$3,$A13,r0, a,w,vp,wa,hr)
14	=A13+sp	=v(B$3,$A14,r0, a,w,vp,wa,hr)	=v(C$3,$A14,r0, a,w,vp,wa,hr)	=v(D$3,$A14,r0, a,w,vp,wa,hr)	=v(E$3,$A14,r0, a,w,vp,wa,hr)	=v(F$3,$A14,r0, a,w,vp,wa,hr)	=v(G$3,$A14,r0, a,w,vp,wa,hr)
15	=A14+sp	=v(B$3,$A15,r0, a,w,vp,wa,hr)	=v(C$3,$A15,r0, a,w,vp,wa,hr)	=v(D$3,$A15,r0, a,w,vp,wa,hr)	=v(E$3,$A15,r0, a,w,vp,wa,hr)	=v(F$3,$A15,r0, a,w,vp,wa,hr)	=v(G$3,$A15,r0, a,w,vp,wa,hr)

[a]The actual spreadsheet extends to cell AF40.

D2 r0. Cell F1 was named wa, and cell f2 was named sp, where sp refers to the spacing of r1 and r2 points. I listed values of r1 in cells B3, C3, . . . , AF3. I listed values of r2 in cells A4, A5, . . . , A40. I then used the function defined by the module in Table 7.B.1 to calculate all of the numerical values.

I then used the three-dimensional (3D) plotting routines in Excel to produce a diagram of the potential. Excel has a 3D plotting routine. You use it by first choosing cells to be plotted, (A3 to AF40 in my example) and then choosing a 3D surface plot. I used the defaults for all of the values. I then went back and edited the vertical axis of the plot so that the major unit was 2.

Part A Figure 7.G.1 shows a diagram of the potential when w = 109 kcal/mol, wa = 0, vp = 258 kcal/mol, r0 = 0.74 Å, a = 1.74 Å^{-1}. The potential looks like a deep valley with a rise and then a fall. There is a saddlepoint in the middle of the potential.

Figure 7.G.2 shows a side view of the potential. Notice the rise in the potential near the saddle point.

Figure 7.G.3 shows another view of the potential in this case from above. There are a series of parabola-shaped contours leading up to the saddle point. Then there is a contour that looks somewhat like a flat body with two arms and two legs. Then there are a series of circular contours.

The parabolic contours are indicative of the potential rising in the direction of the center of the parabola. The contour with arms and legs is indicative of a saddle point. The circular contours are indicative of a potential that goes smoothly from reactants to products.

There are no connections from reactants to products below the saddle point. However, all of the contours above the saddle point connect from reactants to products.

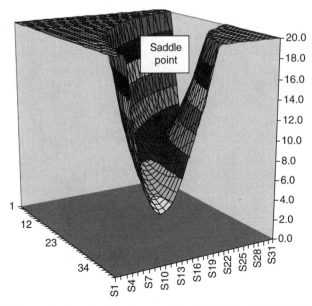

Figure 7.G.1 A diagram of the potential when w = 109 kcal/mol, wa = 0, hr = 0, vp = 258 kcal/mol, r0 = 0.74 Å, a = 1.2 Å^{-1} (elevation = 15, rotation = 60, perspective = 30, chart depth = 100).

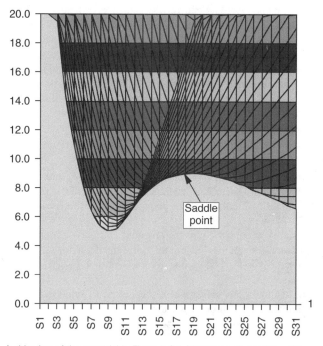

Figure 7.G.2 A side view of the potential in Figure 7.G.1 (elevation = 0, rotation = 90, perspective = 30, chart depth = 100).

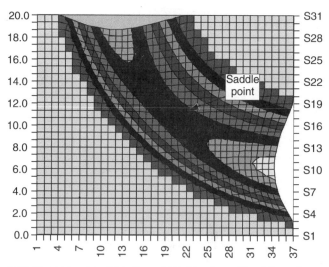

Figure 7.G.3 A top view of the potential in Figure 7.G.1 (elevation = 90, rotation = 0, perspective = 30, chart depth = 100).

It is very important that the reader be able to recognize the saddle points on these contour plots. Therefore the reader should examine these contour plots carefully before proceeding.

Part B   I solved part B using the same spreadsheet that I used for part A. I changed the value of Vp and then used the same plots as with part A.

Figure 7.G.4 shows a plot of the potential. In this case, there is a deep valley, but no hills between the reactants and products.

Figure 7.G.5 shows a top view of the potential. Notice that there are the hemispherical contours, but none of the parabolic contours, and none of the areas that look like a body with arms and legs.

Part C   Again this problem was solved using the spreadsheet in Table 7.G.1 and the same plotting commands. Figure 7.G.6 shows a diagram of the potential. In this case there is a valley as before, but now the potential falls, showing a well. A well corresponds to the formation of a stable molecule or molecular complex.

Figure 7.G.7 shows a top view of the potential. In this case there is a bull's-eye pattern of ellipses corresponding. The bull's-eye pattern corresponds to a stable complex or molecule.

Part D   Again I used the same spreadsheet as before to calculate the potential. In this case I set sp = 0.07 to get a better plot, and inputted the parameters accordingly.

Figure 7.G.8 shows a top view of the potential for this case. Notice that there are two elliptical areas corresponding to complexes or intermediates and a saddle point in between.

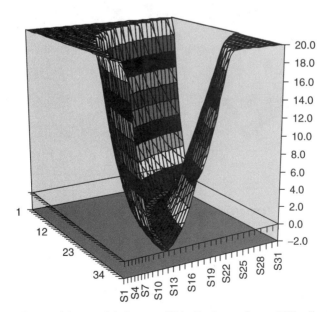

Figure 7.G.4   A diagram of the potential when w = 109 kcal/mol, wa = 0, vp = 218 kcal/mol, r0 = 0.74 Å, a = 1.2 Å^{-1} (elevation = 15, rotation = 60 perspective = 30, chart depth = 100).

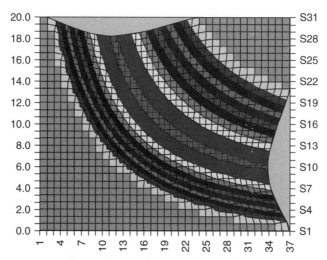

Figure 7.G.5 A top view of the potential in Figure 7.G.6 (elevation = 90, rotation = 0, perspective = 30, chart depth = 100).

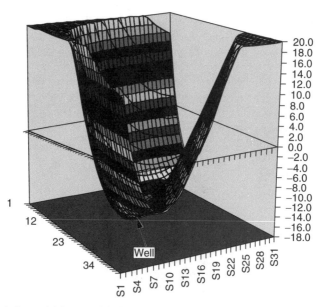

Figure 7.G.6 A diagram of the potential when w = 109 kcal/mol, wa = 0, vp = 158 kcal/mol, r0 = 0.74 Å, a = 1.2 Å^{-1} (elevation = 15, rotation = 60, perspective = 30, chart depth = 100).

Many reactions show potential energy surfaces like those in Figure 7.G.8, and so one should study the features carefully.

Part E I solved part E using the same spreadsheet as for part D with sp = 0.10. Figure 7.G.9 shows the plot. In this case, there are two saddle points and an intermediate or complex in between.

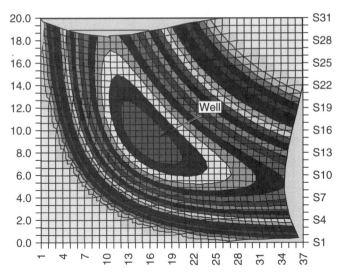

Figure 7.G.7 A top view of the potential in Figure 7.G.6 (elevation = 90, rotation = 0, perspective = 30, chart depth = 100).

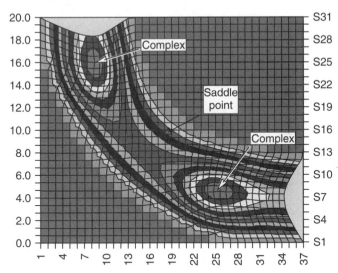

Figure 7.G.8 A top view of the potential for w = 109 kcal/mol, wa = −10, hr = 0, vp = 158 kcal/mol, r0 = 0.74 Å, a = 1.2 Å^{-1} (elevation = 90, rotation = 0, perspective = 30, chart depth = 100).

Part F I solved part F using the same spreadsheet as for part D with sp = 0.10. Figure 7.G.10 shows the plot. There are parabolic contours showing that the potential is rising, but there is no transition state between reactants and products.

Readers may want to reproduce these results themselves before proceeding to the next chapter.

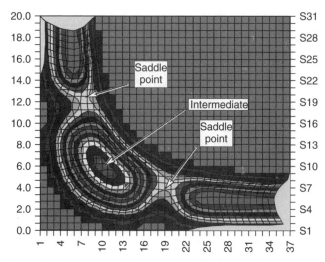

Figure 7.G.9 A top view of the potential for w = 109 kcal/mol, wa = 20, hr = 0, vp = 178 kcal/mol, r0 = 0.74 Å, a = 1.2 Å^{-1} (elevation = 90, rotation = 0 perspective = 30, chart depth = 100).

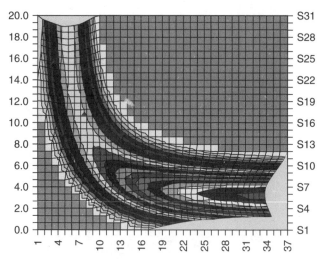

Figure 7.G.10 A top view of the potential for w = 109 kcal/mol, wa = 0, hr = −20, vp = 233 kcal/mol, r0 = 0.74 Å, a = 1.2 Å^{-1} (elevation = 90, rotation = 0, perspective = 30, chart depth = 100, sp = 0.1).

Example 7.H   Counting normal modes   Consider the reaction $H + CH_3CH_3 \rightarrow H_2 + CH_2CH_3$. How many modes should you include in $q^{\ddagger}$?

Solution   Consider the transition state $H \cdots H \cdots CH_2CH_3$. The transition state has 9 atoms so it must have $3 \times 9 = 27$ normal modes. The transition is nonlinear, so it has three translational modes, three rotational modes, so it must have $27 - 3 - 3 = 21$ vibrational modes.

One of the modes is the H–H–C asymmetric stretch. That is the mode that carries the incoming hydrogen along the reaction coordinate. That mode is already included in the $(k_B T/h_p)$ term in transition state theory so we should not include it in $q^{\ddagger}$. That leaves $21 - 1 = 20$ vibrational modes. Therefore, the transition state must have 3 translational modes, three rotational modes and 20 vibrational modes.

7.8 SUGGESTIONS FOR FURTHER READING

Other elementary discussions of transition state theory and collision theory can be found in:

K. J. Laidler, *Chemical Kinetics*, 3rd ed., Harper & Row, New York, 1987.

Additional suggestions for further reading are given in Chapters 8 and 9.

7.9 PROBLEMS

7.1 Define the following terms:

(a) Arrhenius' theory

(b) Collision theory

(c) Transition state theory

(d) RRKM model

(e) Lindemann model

(f) Transmission coefficient

(g) Wigner and Polanyi's model

(h) Tunneling

(i) Entropy of activation

(j) Saddle point energy

(k) Potential energy surface

(l) Reduced mass

(m) Collision partner

(n) Partition function

(o) Moment of inertia

(p) Transition state

(q) Col

(r) Loose transition state

(s) Tight transition state

(t) Transition state analog

7.2 In Section 7.4.1 we noted that transition state theory predicts that a bimolecular collision has a preexponential in the order of 10^{13} Å^3/(molecule·second). Calculate a typical value of the preexponential in units of **(a)** cm^3/(mol·second), **(b)** liter/(mol·second), **(c)** liter/(mol·hour), **(d)** dm^3/(mol·second).

7.3 Compare and contrast: collision theory, transition state theory, and the RRKM model. What are the key assumptions in each model? What are the key predictions of each model? Where do the models fail? How is transition state theory different from Wigner and Polanyi's model?

7.4 Describe in your own words why the preexponentials for unimolecular reactions are so large. Where does the large term come into the RRKM model? How large is the term?

7.5 What lead Arrhenius to propose that the rate varied exponentially with temperature? That is, what in Figure 7.1 and equation (7.2) leads to an exponential temperature dependence?

7.6 One of the ways that enzymes work is by binding to the transition state of a reaction. Assume that you have an enzyme that forms a 10-kcal/mol bond to the reactants (i.e., $\Delta G = 10$ kcal/mol), a 12-kcal/mol bond to the products, and a 28-kcal/mol bond to the transition state. By how much will the rate of reaction increase at 310 K in the presence of the enzyme?

7.7 Table 7.4 compares the saddle point energy to the activation barrier for a number of reactions.

(a) Plot the measured versus calculated activation barrier. Do you see any pattern?

(b) Why are the barriers lower than the saddle energy for hydrogen transfer reactions?

(c) How do the other barriers compare?

7.8 Figure 7.6a shows a potential energy surface for the reaction

$$H + CH_3OH \longrightarrow H_2 + CH_2OH$$

Make a copy of the figure. Start with the picture on the left of the figure and identify

(a) The minimum energy of the reactants when the reactants are far apart. (*Hint*: When the reactants are far apart, the H–H distance is large.)

(b) The minimum energy of the products.

(c) The vibrational well of the reactants.

(d) The vibrational well of the products.

(e) The transition state.

(f) What does the C–H potential look like when the H–H distance is large?

(g) What does the H–H potential look like when the C–H distance is large?

(h) What does the potential look like near the transition state?

(i) Repeat all of the questions on the contour plot on Figure 7.6b. Can you see the C–H and H–H potentials? Can you see the shape of the transition state?

7.9 In Section 7.4, we discussed the Lindemann model. We assumed that a unimolecular reaction $A \Rightarrow B$ followed

$$X + A \underset{2}{\overset{1}{\rightleftharpoons}} A^* + X$$

$$A^* \xrightarrow{3} B$$

and stated that

$$r_B = \frac{k_1[A][X]}{\dfrac{k_2}{k_3}[X] + 1} \qquad (P7.9.1)$$

(a) Use the steady-state approximation to derive equation (P7.9.1).

(b) Now let $[X] = [A]$. Convert equation (P7.9.1) to a linear form. Two linear forms are possible: one based on the Lineweaver–Burke transformation (see Example 3.A), and another based on the Eadié–Hofstee transformation. Derive an equation for both of them.

(c) Table P3.17 page 149 shows some data for the reaction $HNC \Rightarrow HCN$. How well do each of the lines fit?

(d) Why do the lines look so poor when Figure 7.10 shows that equation P7.9.1 fits the data quite well? Hint: look back to the discussion of errors in Example 3.A.

7.10 Use transition state theory to calculate the temperature dependence of the rate constant for

(a) The reaction $A + BC \rightarrow AB + C$, where A, B, and C are atoms. Assume a linear transition state.

(b) $A + BC \rightarrow AB + C$, where A, B, and C are atoms. Assume a nonlinear transition state.

(c) $AB + CD \rightarrow ABC + D$, where A, B, C, and D are atoms. Assume a linear transition state.

(d) $AB + CD \rightarrow ABC + D$, where A, B, C, and D are atoms. Assume a nonlinear transition state.

(e) Estimate the value of the preexponential at 500 K in all cases.

7.11 Example 7.G showed values of the potential given by equation (7.G.1).

(a) Set up the spreadsheet yourself and verify the findings. The original spreadsheet is posted at http://www.wiley.com/chemicalkinetics.

(b) Print out the numerical values of the spreadsheet for w = 100 kcal/mol, wa = 0, hr = 0, vp = 230 kcal/mol, r0 = 0.74 Å, a = 1.2 Å⁻¹.

(c) Consider the reaction $A + BC \rightarrow AB + C$, with r_1 = the AB distance and r_2 = the BC distance. Find the place corresponding to the reactants and the products. (*Hint*: Examine Example 8.A — this is similar except r_1 and r_2 are reversed.)

(d) Trace a path from the reactants to products as indicated in Example 8.A.

(e) Identify the saddle point on your plot.

(f) Show that your point is really a saddle point.

(g) What is the energy of the transition state?

7.12 Example 7.G showed values of the potential given by equation (7.G.1).

(a) Set up the spreadsheet yourself and verify the findings. The original spreadsheet is posted at http://www.wiley.com/chemicalkinetics.

(b) Make contour plots (i.e., top views) of the potential for the following parameters:

(1) w = 100 kcal/mol, wa = 0, hr = 0, vp = 218 kcal/mol, r0 = 0.74 Å, a = 1.2 Å⁻¹

(2) w = 100 kcal/mol, wa = 0, hr = −20, vp = 218 kcal/mol, r0 = 0.74 Å, a = 1.2 Å⁻¹

(3) w = 100 kcal/mol, wa = 0, hr = 20, vp = 218 kcal/mol, r0 = 0.74 Å, a = 1.2 Å⁻¹

(4) w = 100 kcal/mol, wa = 5, hr = 0, vp = 218 kcal/mol, r0 = 0.74 Å, a = 1.2 Å⁻¹

(5) w = 100 kcal/mol, wa = 0, hr = 0, vp = 188 kcal/mol, r0 = 0.74 Å, a = 1.2 Å⁻¹

(6) w = 100 kcal/mol, wa = −5, hr = 0, vp = 218 kcal/mol, r0 = 0.74 Å, a = 1.2 Å⁻¹

Use sp = 0.1 in all cases

(c) In each case label the reactants, products, transition states, or complexes on your plots following Problem 7.11(c).

(d) Which of the reactions are activated in the forward direction? [*Hint*: The reaction will be activated if (1) there is a saddle point with an energy above the reactants or (2) if the reaction is endothermic so that the system must go uphill to products.]

(e) Which of the reactions are activated in the reverse direction?

7.13   The reaction

$$\begin{array}{c} CH_2 \\ / \ \backslash \\ CH_2-CH_2 \end{array} \Longrightarrow CH_3CH=CH_2$$

has an activation barrier of 66 kcal/mol.

(a) Use the Kassel–Rice–Ramsperger model to estimate the preexponential for the reaction at 1000 K. (Assume that you are working in the high-pressure limit so that $k = k_3$.)

(b) Estimate the rate constant at 1000 K.

(c) Estimate the temperature dependence of the preexponential for the reaction.

7.14   D. R. Hershbach, et al. (1956), reported preexponentials for a number of reaction. Most were near 10^{13} Å/(molecule·second), but some were not. Table P7.14 gives a few examples where the preexponential differs substantially from 10^{13}. Go through each of the examples and tell us what you know about the transition state.

(a) Is it a loose (i.e., hard-sphere-like) or tight (i.e., have a very specific geometry)?

(b) Diagram a likely transition state from the data given.

(c) Estimate the preexponential for your transition state. Can you find a transition state with a small enough preexponential?

7.15   The preexponentials for surface reactions can be very different from 10^{13}. Consider the reaction

$$CO + S \longrightarrow CO_{ad} \qquad\qquad (P7.15.1)$$

Table P7.14   A selection of reactions that show preexponentials well below 10^{13}

Reaction	E_a, kcal/mol	$\log(k_0)$, Å³/(molecule·second)
$NO + O_3 \longrightarrow NO_2 + O_2$	2.6	11.1
$NO + O_3 \longrightarrow NO_3 + O$	7	12
$NO_2 + F_2 \longrightarrow NO_2F + F$	10.4	11.4
$NO_2 + CO \longrightarrow CO + CO_2$	31.5	12.3
$2NO_2 \Longrightarrow 2NO + O_2$	26.5	11.5
$NO + NO_2Cl \longrightarrow NOCl + N_2O$	6.4	11.1
$2NOCl \Longrightarrow 2NO + Cl_2$	20.3	12.2
$F_2 + ClO_2 \longrightarrow FClO_2 + F$	8.5	9.7
$2ClO \Longrightarrow Cl_2 + O_2$	0	10.0

(a) Assume that the transition state is a surface complex that does not translate, vibrate, or rotate. Estimate the preexponential for the reaction. Hint: Be careful with units. When you replace a translational with a vibrational partition function you replace q_T with q_r. However q_t (not q_T) appears in equation (7.B.3).

(b) Now assume that the preexponential can translate in two dimensions and rotate around one axis. How would the preeexponential change?

7.16 The objective of this problem is to calculate the rate of formation of NO from the air in your bedroom. Assume that the main reaction is

$$N_2 + O_2 \longrightarrow 2NO$$

with an activation barrier of 85 kcal/mol.

(a) What is the collision rate of N_2 with O_2 in air at 300 K and 1 atm?

(b) What fraction of the collisions have at least 85 kcal/mol?

(c) Assume that there are 2000 liters of air in your bedroom. Use collision theory to estimate the NO production rate in molecules/per hour.

(d) NO is poisonous. Will you ever produce enough NO to poison you? Be sure to consider the equilibrium concentration of NO.

7.17 Use transition state theory and collision theory and the data in Table P7.17 to estimate the rate of the reaction

$$H + HBr \longrightarrow H_2 + Br \qquad (P7.17.1)$$

at 300, 400, 500, 600, 700, 800, 900, and 1000 K.

(a) Estimate the preexponentials at each temperature

(b) Make an Arrhenius plot of the rate constant as a function of temperature to estimate a preexponential. How well does your preexponential fit agree with the results in (a)?

(c) How well does your activation barrier compare to $E^{\ddagger}$?

Table P7.17   Data for the transition state of reaction (P7.17.1)

	Linear Transition State	Reactants	
		H	HBr
r_{HH}	1.5 Å	—	1.414 Å
r_{HBr}	1.42 Å	—	—
v_{H-Br}	2340 cm^{-1}	—	2650 cm^{-1}
v_{HHBr} bend	460 cm^{-1}	—	—
v_{HHBr} bend	460 cm^{-1}	—	—
Curvature	320 cm^{-1}	—	—
I	10.4 amu·Å^2	—	1.99 amu·Å^2
$E^{\ddagger}$	1.21 kcal/mol	—	—
g_e	4	1	4

(d) How do the preexponentials compare to the experimental value of 4.2×10^{13} Å^3/(molecule·second)?

More Advanced Problems

7.18 The objective of this problem is to use collision theory and transition state theory to model adsorption. Consider a simple chemical reaction:

$$CO + S \longrightarrow CO_{ad} \qquad (P7.18.1)$$

where a CO molecule reacts with a surface site, S. Masel (1996) shows that one can model the rate via collision theory:

$$R_{ad} = (0.25v_{CO})\sigma_s C_{CO} C_s P_{stick} \qquad (P7.18.2)$$

where R_{ad} is the rate of adsorption in molecules/cm^2, v_{CO} is the average molecular velocity of the CO, C_{co} is the gas-phase CO concentration in molecules/cm^2, C_s is the concentration of surface sites, σ_s is the surface area taken up by one surface site, and P_{stick} is the probability that the CO sticks when it hits the surface. Notice that we use $0.25v_{CO}$ in equation (P7.18.2) rather than v_{CO}. If all the CO would impinge perpendicularly to the surface, then v_{CO} would have been the correct term. However, in reality, the COs impinge from all directions, and it works out that the correct velocity is $0.25v_{CO}$.

(a) Calculate the rate of adsorption of CO in molecules/(cm^2·second) at 300 K and 1 atm pressure. Assume that P_{stick} is 0.5, and that $C_s = 10^{15}$ sites/cm^2.

(b) How long will it take for all of the surface sites to be filled with gas?

(c) In reality, P_{stick} decreases as the surface coverage increases because gas cannot stick to sites that are filled with gas. Calculate how long it will take to fill 99% of the surface with gas assuming that $P_{stick} = 0.5(1 - \theta_{CO})$, where θ_{CO} is the fraction of the surface covered by CO.

(d) How low of a pressure do you need to reach to slow down the adsorption process, so that the adsorption process takes 100 seconds?

(e) The calculations so far in this problem assumed that the sticking probability was 0.5. Use transition state theory to get a "better" value. Assume that there is no barrier to adsorption (i.e., $E_a = 0$). Also assume that at the transition state, the CO is stuck to the surface, so it does not translate or rotate. Instead the transition state vibrates, with vibrational frequencies of 45, 49, 130, 133, and 2085 cm^{-1}.

7.19 In this chapter we have been examining chemical reactions, but the equations in the chapter are also useful for nuclear reactions. Consider the simple nuclear reaction:

$$^{207}Pb + n \longrightarrow {}^{208}Pb$$

where n is a neutron.

(a) Derive an equation for the rate of reaction as a function of the concentration of neutrons and lead atoms, the average velocity of the neutron and the lead atoms, and the crosssection for neutron capture. Note the crosssection $\sigma_x = \pi(b_{coll})^2$, where b_{coll} is the collision diameter.

(b) Assume that you have a nuclear reactor producing 1-eV neutrons. Calculate the rate constant of the reaction. Assume a crosssection of 0.37 barn/lead atom (1 barn = 10^{-20} $Å^2$). [*Hint*: First calculate the velocity of the neutrons from $E_n = \frac{1}{2}m_n(v_n)^2$, where E_n is the energy of the neutron, m_n is the mass of the neutron, and v_n is the velocity of the neutron].

(c) Show that the reaction is first-order in the neutron concentration.

(d) Now consider shooting neutrons through a flat plate. Show that if you start the neutrons moving through the plate at time, $t = 0$, the concentration of neutrons will obey

$$\frac{dC_n}{dt} = \sigma_x v_n C_n C_{Pb} \tag{P7.19.1}$$

where C_n is the neutron concentration, C_{Pb} is the concentration of lead atoms, v_n is the neutron's velocity, and σ_x is the cross section.

(e) Do a shell balance to show that the concentration of neutrons at any distance x from the front of the plate obeys

$$\frac{dC_n}{dx} = \sigma_x C_n C_{Pb} \tag{P7.19.2}$$

(f) Integrate equation (P7.19.2) to derive an equation for the fraction of the neutrons that are captured in passing through a lead plate as a function of the thickness of the plate. Assume that the plate is 22% ^{207}Pb and that the other lead isotopes have small neutron cross sections.

(g) How thick of a lead plate do you need to capture 99.999% of the neutrons?

7.20   Berry and Marshall (1918) studied the reaction

$$CH_3 + ClCF_3 \longrightarrow CH_3Cl + CF_3$$

They obtained the results in Table P7.20 (see also Figure P7.20). The objective of this problem is to use transition state theory to obtain a rate constant for the reaction.

(a) How many translational, vibrational, and rotational modes are there in the reactants, products, and transition state?

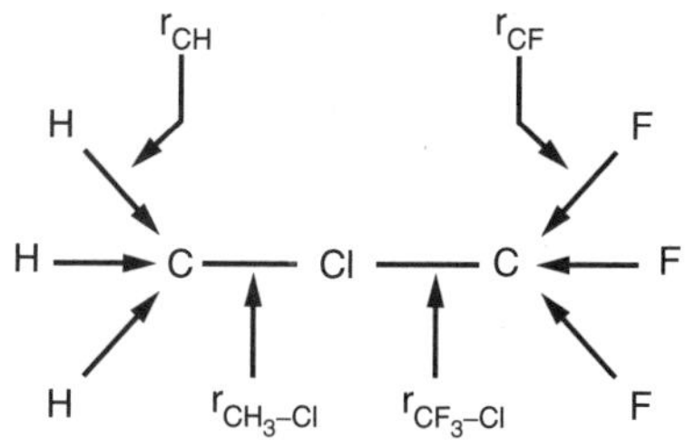

Figure P7.20   Transition state geometry for Problem 7.20.

Table P7.20 Data for Problem 7.20

	CH$_3$ (planar)	ClCF$_3$	TST
r$_{CH}$	1.078 Å		1.085 Å
r$_{CF}$		1.335 Å	1.331 Å
r$_{CF_3-Cl}$		1.749 Å	2.014 Å
r$_{CH_3-Cl}$			2.031 Å
Angle H–C–Cl			104.1°
Angle Cl–C–F		110.4°	109.2°
E‡			19.4 kcal/mol
Vibrations	384 cm^{-1}	337 cm^{-1}	14 cm^{-1}
	1407 cm^{-1}	332 cm^{-1}	79 cm^{-1}
	1477 cm^{-1}	402 cm^{-1}	79 cm^{-1}
	3061 cm^{-1}	531 cm^{-1}	242 cm^{-1}
	3240 cm^{-1}	531 cm^{-1}	242 cm^{-1}
	3240 cm^{-1}	752 cm^{-1}	274 cm^{-1}
		1095 cm^{-1}	483 cm^{-1}
		1220 cm^{-1}	483 cm^{-1}
		1220 cm^{-1}	647 cm^{-1}
			755 cm^{-1}
			755 cm^{-1}
			951 cm^{-1}
			1131 cm^{-1}
			1190 cm^{-1}
			1190 cm^{-1}
			1379 cm^{-1}
			1379 cm^{-1}
			2909 cm^{-1}
			3052 cm^{-1}
			3052 cm^{-1}
Curvature			101 cm^{-1}

(b) Next, you need to calculate the three moments of inertia for each of the species. Recall from freshman physics that

$$I_X = \sum_{atoms} m_i(X_i - X_{cm})^2$$

$$I_Y = \sum_{atoms} m_i(Y_i - Y_{cm})^2$$

$$I_Z = \sum_{atoms} m_i(Z_i - Z_{cm})^2$$

where X_i, Y_i, and Z_i are respectively the X, Y, and Z positions of atoms I; and X_{cm}, Y_{cm}, and Z_{cm} are the positions of the center of mass.

(c) Pick one of the carbon atoms as the origin, then calculate the X, Y, and Z positions of all of the other atoms in the reactants, products, and transition state.

(d) Calculate the position of the center of mass from

$$X_{cm} = \sum_i m_i X_i$$

$$Y_{cm} = \sum_i m_i Y_i$$

$$Z_{cm} = \sum_i m_i Z_i$$

(e) Next, calculate the three moments of inertia.

(f) Plug into transition state theory to calculate the rate constant for the reaction at 800 K.

7.21 One of the key steps in the reproduction cycle of the HIV-1 virus is for the virus to produce an enzyme protease that catalyzes the hydrolysis of a protein in your blood cells. The enzyme produces fragments that act as the building block for the production of more viruses. One class of AIDS drugs work by mimicking the transition state for the reaction and binding to the active site on the enzyme. *Erickson and Wlodawer*, (1993) describe the findings.

(a) Explain how the transition state analogs block the transition state.

(b) Look up the structure of Indinavir, Saquinavir, and Ritonavir. How are they similar or different?

(c) Look up how each of these drugs bind to the enzyme.

(d) What do these findings tell you about the transition state for the reaction on the protease?

8

REACTIONS AS COLLISIONS

PRÉCIS

In Chapter 7, we started to study collision theory and described how one can use the theory to estimate reaction rates. All of the analysis in Chapter 7 was for a hard-sphere collision. The objective of this chapter is to extend the ideas to more realistic collisions. In particular, we will show how one can use trajectory calculations to calculate rates, outline the forces involved, and provide qualitative information about reactive collisions. The material will draw heavily on simulations. A suitable simulation program is available from http://www.wiley.com/chemicalkinetics.

8.1 HISTORICAL INTRODUCTION

The idea that reactions were associated with collisions of molecules dates back to the start of kinetics. The objective of this section will be to provide an historic overview of the ideas so that we have a basis for further discussion.

Recall that in 1884 Van't Hoff had proposed that reactions would obey what is now called *Arrhenius law*:

$$k_1 = k_0 e^{-(E_a / k_B T)} \tag{8.1}$$

where k_1 is the rate constant for a reaction, k_0 is the preexponential for the reaction, E_a is the activation energy for the reaction, k_B is Boltzmann's constant, and T is the temperature. As noted in Chapter 2, this rate form was not initially accepted. However, equation (8.1) eventually was adopted by most investigators.

In Chapter 7, we noted that Arrhenius wrote a series of papers in which he tried to understand why reactions obey equation (8.1). Arrhenius (1889) considered an ideal

413

reaction:

$$A \longrightarrow B \tag{8.2}$$

Arrhenius proposed that if one looked at a chemical system containing A and B, there were two kinds of A molecules in the system: **reactive** A molecules (i.e., A molecules that had the right properties to react), and **unreactive** A molecules (i.e., A molecules that did not have the right properties to react). Arrhenius then assumed that equilibrium was maintained between the reactive and unreactive A molecules to obtain

$$k_1 = (K_0 e^{(\Delta S^{\ddagger})/k_B}) e^{-[(\Delta H^{\ddagger})/k_B T]} \tag{8.3}$$

Equation (8.3) is equivalent to equation (8.1) with

$$k_0 = K_0 e^{(\Delta S^{\ddagger})/k_B} \tag{8.4}$$

8.2 COLLISION THEORY

In 1899, Arrhenius did not have a model to estimate either K_0, $\Delta S^{\ddagger}$ or $\Delta H^{\ddagger}$. However, a few years after Arrhenius' work appeared, Trautz (1916, 1918) and Lewis (1916, 1918) independently proposed what is now called the *collision theory of reactions* in an attempt to get a value for K_0. We already discussed the findings in Chapter 7. The objective of this section is to review the key equations.

Recall that the objective of collision theory is to use a knowledge of molecular collisions to predict K_0 from equation (8.4). Trautz and Lewis proposed a model to do just that. The model builds on Arrhenius' concept that only hot molecules can react. The model assumes that the rate of reaction is equal to the rate of collisions of the molecules. Consequently, if one can calculate the collision rate, one will then get an expression for K_0. In the remainder of this section we will present an expression for K_0 following the derivation of Trautz and Lewis.

Trautz and Lewis considered ligand exchange reactions of the form

$$A + BC \longrightarrow AB + C \tag{8.5}$$

where a radical A abstracts an atom B from a molecule BC to yield products. Trautz and Lewis noted that the reaction occurs when a hot A molecule collides with a hot BC molecule. Under most circumstances, collisions should be rare events. Consequently, Trautz and Lewis assumed that the rate of reaction was equal to the rate of collisions of hot A molecules with hot BC molecules. Trautz and Lewis then used statistical mechanics to estimate a value for the collision rate, and thereby derived a value for K_0.

The derivation is given in Chapter 7. The result is

$$r_{A \to BC} = Z_{A \to BC} e^{-[(\Delta G^{\ddagger})/k_B T]} \tag{8.6}$$

where $Z_{A \to BC}$ is the rate of A—BC collisions. Equation (7.22) gives the total rate of collisions between hot A molecules and hot BC molecules.

$$Z_{A \to BC} = \bar{v}_{A \to BC} C_A C_{BC} \sigma_{A \to BC} \tag{8.7}$$

where $\bar{v}_{A \to BC}$ is the average velocity of A toward BC, C_A and C_{BC} are the concentrations of A and BC, $\sigma_{A \to BC}$ is the collision cross section, $\Delta G^{\ddagger}$ is the free energy of activation, k_B is Boltzmann's constant, and T is the temperature. Combining equations (8.6) and (8.7) yields

$$r_{A \to BC} = (\bar{v}_{A \to BC} \sigma^c_{A \to BC} e^{\Delta S^{\ddagger}/k_B})(e^{-\Delta H^{\ddagger}/k_B T}) C_A C_{BC} \tag{8.8}$$

where $\bar{v}_{A \to BC}$ is the average velocity of A moving toward BC.

Equation (8.8) is a second-order rate law, with an activation barrier of $\Delta H^{\ddagger}$ and a preexponential, k_0, given by

$$k_0 = \bar{v}_{A \to BC} \sigma^c_{A \to BC} e^{\Delta S^{\ddagger}/k_B} \tag{8.9}$$

Neither Trautz nor Lewis had a way to calculate $\Delta S^{\ddagger}$. Consequently, they set $\Delta S^{\ddagger}$ to zero in equation (8.9) to obtain

$$k_0 = \bar{v}_{A \to BC} \sigma^c_{A \to BC} \tag{8.10}$$

Trautz and Lewis also asserted that the molecular velocity in equation (8.10) should be calculated, ignoring that we are considering hot molecules. According to the results in Chapter 6, the average velocity of A moving toward BC is given by

$$\bar{v} = \left(\frac{8k_B T}{\pi \mu_{ABC}} \right)^{1/2} \tag{8.11}$$

where

$$\frac{1}{\mu_{ABC}} = \frac{1}{m_A} + \frac{1}{m_B + m_C} \tag{8.12}$$

and m_A, m_B, and m_C are the masses of A, B, and C, respectively. A derivation of equation (8.11) is given in Section 8.16.4.

Equation (8.10) is the key result for simple collision theory. One should memorize the equation before proceeding.

8.2.1 Predictions of Collision Theory

Next, it is useful to review the predictions of collision theory. We have already discussed the key predictions in Section 7.2. According to collision theory

- Preexponentials for reactions are all about 10^{13} or 10^{14} units of angstroms, molecules, and seconds.
- The preexponentials scale as the molecular diameter. Larger molecules are more reactive than smaller ones.
- It does not matter how the molecules collide, they will always find the right configuration to react.

The first two predictions have excellent agreement with experiment. Preexponentials are usually within a factor of 10 of those predicted by collision theory. Larger molecules are generally more reactive. The third prediction does not work as well. Molecules often do not find the right configuration to react. All these ideas are discussed in Section 7.2. Any reader who has not read the discussion of collision theory in Section 7.2 should do so before continuing with this chapter.

There are a few examples where the preexponentials differ substantially from those predicted by collision theory. The reaction

$$CH_3CH_2CH_3 + O \longrightarrow CH_3CHCH_3 + OH \tag{8.13}$$

has a preexponential of 1.4×10^{10} $\text{Å}^3/(\text{molecule·second})$. That is over two orders of magnitude lower than one would expect from [collision theory from equation (8.10)]. In contrast, the reaction

$$2O_2 \longrightarrow 2O + O_2 \tag{8.14}$$

has a preexponential of 5.8×10^{15} $\text{Å}^3/(\text{molecule·second})$. That is about two orders of magnitude larger than one would expect from equation (8.10). These are special cases. In most cases, collision theory predicts reasonable preexponential factors. Still, there are a number of examples where Trautz and Lewis' version of collision theory cannot explain the data.

The Trautz–Lewis model failed because it treated the collision between the reactants as a billiard ball collision. Every molecule of A that collided with BC was assumed to react with a fixed probability. In reality, the reaction probability varies with how the collision occurs. For example, consider a simple reaction:

$$(CH_3)_2CH_2 + O: \longrightarrow (CH_3)_2CH\bullet + \bullet OH \tag{8.15}$$

During reaction (8.15), the oxygen reacts with the hydrogen on the middle carbon of pentane. If the incoming oxygen hits the hydrogen on the middle carbon, the reaction can occur. However, if the oxygen hits anywhere else, a $\bullet CH_2CH_2CH_3$ will form. The Trautz–Lewis model ignores the fact that you need special collision geometry to allow the reaction to happen. As a result, their model tends to overestimate rates of reaction in reactions similar to (8.15).

This example shows that in some cases, one needs molecules to collide in the correct way for the desired reaction to occur. The Trautz–Lewis model ignores the need for a special collision geometry to get a desired reaction, so it does not always give a good prediction of the rate.

8.3 AN IMPROVED COLLISION THEORY

In the last section, we noted that the main weakness of the Trautz–Lewis version of collision theory is that it ignores the fact that one needs a special geometry in order for reaction to occur. In the remainder of this chapter, we will discuss some of the work that has been done to improve collision theory. The general approach will be to find a way to determine which collisions actually lead to reaction, and then find a way to use that information to predict a rate. Qualitatively, we will determine which of the collisions lead to reaction by doing a calculation to see how the reactants change during the collision, and using the calculation to determine whether reaction occurs.

In this section we will describe the qualitative effects and the key equations. The quantitative details will be discussed in Sections 8.4 and 8.5.

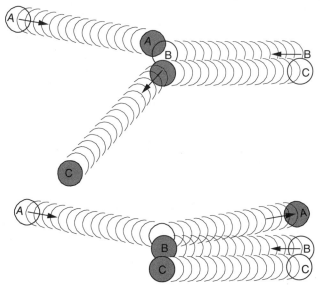

Figure 8.1 A typical collision between A and BC.

First, it is interesting to consider what a collision is like. Figure 8.1 shows two different collisions between A and BC. In the top case, A flew toward BC, A stuck, and C flew away. In this case, reaction happened. In the bottom case, A flew toward BC. A collided with BC but A left again. No reaction occurred.

Now imagine doing a thought experiment where one could watch thousands of molecules and see when collisions happen. If one could watch the collisions, one could calculate a reaction rate. In practice, one cannot watch the collisions of real molecules, but one can simulate the collisions in a computer.

In the next three sections, we will derive equations for the key collision process. First, we will derive an expression for the rate of reaction as a function of the probability that two molecules react when the molecules collide at a given angle and velocity. We will then show how to calculate the reaction probability by watching the dynamics of the particles. We will also need to discuss intermolecular forces. Some of the material is heavy going. We need to introduce several concepts over a short time. However, the reader should work through the materials in order to understand what is said later in the chapter, and in the chapters that follow.

8.3.1 The Reaction Probability

First, it is useful to use statistical mechanics to derive an expression for the rate as a function of the reaction probability. Recall that in statistical mechanics, one uses a statistical ensemble to calculate an average value of a function. Well, a rate is a function, so if one averages the rate over a statistical ensemble of all possible collisions, one can calculate an average rate of reaction. In this section, we will derive an equation for all the key quantities. The result looks formidable. However, in reality, the result is easier than it initially appears.

Let's define $P_{reaction}$ as the probability that a given A molecule reacts with a given BC molecule if the two molecules collide. One can show that the reaction probability varies with

- $v_{A \to BC}$, the velocity that the A molecule approaches the BC molecule
- E_{BC}, the internal state (i.e., vibrational–rotational energy) of the BC molecule before collision occurs
- The *impact parameter* $b_{A \to BC}$, which is a measure of how closely A collides with BC
- The *angle of approach*, where ϕ is a measure of the angle of the collision
- The initial position $\mathbb{R}_{BC}$ and velocity v_{BC} of B relative to C when collision occurs

Figure 8.2 page 420 shows how $b_{A \to BC}$ and ϕ are defined. The figure assumes that when atom A collides with the BC molecule, the A atom follows the trajectory shown in Figure 8.2. The impact parameter is defined as the between the point that atom A would intersect the plane going through the center of the BC bond and the center of mass of BC if the atom A would fly in a straight line and not be attracted or repelled by BC. The angle of approach is defined as the angle where the A impinges relative to the angle of the B–C bond.

Next we want to briefly discuss why the reaction probability varies with $v_{A \to BC}$, E_{BC}, $b_{A \to BC}$, ϕ, $\mathbb{R}_{BC}$, and v_{BC}. The velocity term is obvious. The faster the reactants are moving, the quicker they go over the activation barrier. The E_{BC} term is less obvious. Physically, though, if you put a lot of energy into the BC bond, the BC will stretch, and a bond that is stretching is easy to break. Consequently, the reaction probability generally rises as E_{BC} increases. The impact parameter is a third term. Physically, the reactants need to hit each other to react. Consequently, when $b_{A \to BC}$ is large, the reaction probably is negligible. While the reaction probability is significant when $b_{A \to BC}$ is small. The ϕ term is harder to understand. Consider a reaction such as

$$D + CH \longrightarrow C + HD \tag{8.16}$$

Notice that the reaction probability will be higher if the incoming deuterium hits the hydrogen than when the deuterium hits the carbon. The ϕ dependence accounts for that.

If you have stationary BC molecules, the first four are all the variables you need. However, if the reactants have internal motion, you need to consider that as well. For example, the CH could be rotating during reaction (8.16). The rotary motion could bring the hydrogen around to cause the hydrogen to collide with the deuterium. That will enhance the rate.

Generally, there are many variables that affect a reaction probability. One should memorize the list on the previous page before proceeding.

8.3.2 An Equation for the Rate as a Function of the Reaction Probability

In statistical mechanics, one calculates everything as an ensemble average. For example, let's assume that we have an expression for the reaction rate as a function of $v_{A \to BC}$, E_{BC}, ϕ, $b_{A \to BC}$, $\mathbb{R}_{BC}$, v_{BC}:

$$r_{A \to BC}(v_{A \to BC}, E_{BC}, \phi, b_{A \to B}, \mathbb{R}_{BC}, v_{BC}) \tag{8.17}$$

According to the results in Section 6.4.1, the overall rate, $r_{A \to BC}$, is calculated by averaging equation (8.17) over $v_{A \to BC}$, E_{BC}, ϕ, $b_{A \to BC}$, $\mathbb{R}_{BC}$, v_{BC}. In Chapter 6, we found that the correction expression is

$$r_{A \to BC} = \sum_{\substack{\text{states} \\ n}} r_{A \to BC}(v_{A \to BC}, E_{BC}, \phi, b_{A \to B}, \mathbb{R}_{BC}, v_{BC}) \mathbb{P}_n \qquad (8.18)$$

where $\mathbb{P}_n$ is the probability that a given state n is occupied.

For the work that follows, it is convenient to replace the probability in equation (8.18) by a distribution function, $D(v_{A \to BC}, E_{BC}, \phi, b_{A \to B}, \mathbb{R}_{BC}, v_{BC})$.

I hope that you remember the velocity distribution function $D(v_A)$ from physical chemistry. The velocity distribution function is the probability that the velocity is between v_A and $v_A + \delta(v_A)$. In the same way, $D(v_A \to BC, E_{BC}, \phi, b_{A \to B}, \mathbb{R}_{BC}, v_{BC})$ is the probability that $v_{A \to BC}$ is between $v_{A \to BC}$ and $v_{A \to BC} + \delta(v_{A \to BC})$, E_{BC} is between E_{BC} and $E_{BC} + \delta(E_{BC})$, and so on.

You use $D(v_{A \to BC}, E_{BC}, \phi, b_{A \to B}, \mathbb{R}_{BC}, v_{BC})$ in the same way that you use $D(v_A)$. So for example, recall from physical chemistry that if you wanted to calculate the velocity average of some function $F(v_A)$, you would calculate

$$\bar{F} = \int F(v_A) D(v_A) \, dv_A \qquad (8.19)$$

where $\bar{F}$ is the average value of F. In the same way the average rate becomes

$$r_{A \to BC} = \int\int\int\int\int\int r_{A \to BC}(v_{A \to BC}, E_{BC}, \phi, b_{A \to BC}, \mathbb{R}, v_{BC})$$

$$\times D(v_{A \to BC}, E_{BC}, \phi, b_{A \to BC}, \mathbb{R}_{BC}, v_{BC}) \, dv_{A \to BC} \, dE_{BC} \, d\phi \, db_{A \to BC} \, d\mathbb{R}_{BC} \, dv_{BC} \quad (8.20)$$

Equation (8.20) looks pretty formidable, but it is merely saying that one needs to average the rate over six variables. One should memorize this equation before proceeding.

8.3.3 Derivation of an Expression for $r_{A \to BC}(v_{A \to BC}, E_{BC}, \phi, b_{A \to BC}, \mathbb{R}, v_{BC})$

Next, we want to derive an expression for $r_{A \to BC}(v_{A \to BC}, E_{BC}, \phi, b_{A \to BC}, \mathbb{R}, v_{BC})$. In Section 7.2 we found that for a hard-sphere collision, the reaction rate is given by

$$r_{A \to BC} = \left(\begin{array}{c} \text{collision rate} \\ Z_{ABC} \end{array} \right) \left(\begin{array}{c} \text{probability of reaction} \\ \text{during a collision} \end{array} \right) \qquad (8.21)$$

where the collision rate is the same as given by equation (8.7). Combining equations (8.7) and (8.21) shows

$$r_{A \to BC} = C_A C_{BC} \bar{v}_{A \to BC} C_A C_{BC} \sigma_{A \to BC} P_{\text{reaction}} \qquad (8.22)$$

Equation (8.22) applies only if you have a hard-sphere collision. If you do not have a hard-sphere collision, then the reaction probability is going to vary with the distance between the reactions prior to collision $b_{A \to BC}$, the angle ϕ, and all the rest of the variables discussed in Section 8.3.2.

Equation (8.22) is not correct for this situation. However, one can derive a correct equation by considering the differential slice of area shown in Figure 8.2, calculating a

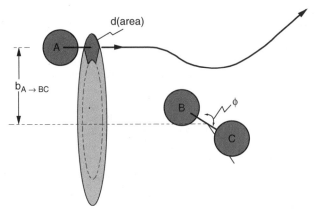

Figure 8.2 A typical trajectory for the collision of an A atom with a BC molecule as calculated by the methods in Section 8.3.2.

differential rate, and then integrating the differential rate $dr_{A \to BC}$ is given by

$$dr_{A \to BC} = C_A C_{BC} V_{A \to BC} P_{\text{reaction}} \, d(\text{area}) \tag{8.23}$$

where d(area) is the differential area, $(b_{A \to BC} db_{A \to BC} \, d\phi)$.

In reality, the A molecules have a distribution of velocities. Consequently, one has to average over the distribution of velocities. One also has to average over the distribution of internal energies of the B–C molecule since the reaction probability changes if the B–C molecules starts out hot. Combining equations (8.21)–(8.23) and integrating over all the variables yields

$$r_{A \to BC} = C_A C_{BC} \int\int\int\int\int\int P_{\text{reaction}}(v_{A \to BC}, E_{BC}, b_{A \to BC}, \phi, \mathbb{R}_{BC}, v_{BC}) v_{A \to BC}$$

$$\times D(v_{A \to BC}, E_{BC}, \mathbb{R}_{BC}, v_{BC}) \, dv_{A \to BC} \, dE_{BC} (b_{A \to BC} \, db_{A \to BC}) \, d\phi \, d\mathbb{R}_{BC} \, dv_{BC} \tag{8.24}$$

In equation (8.24) $D(v_{A \to BC}, E_{BC}, \mathbb{R}_{BC}, v_{BC})$ is the probability that a given pair of molecules will have a velocity, $v_{A \to BC}$; an internal energy, E_{AB}; an internal coordinate, $\mathbb{R}_{BC}$; and an internal velocity, v_{BC}: $b_{A \to BC}$ is the impact parameter: $P_{\text{reaction}}(v_{A \to BC}, E_{BC}, b_{A \to BC}, \mathbb{R}_{BC}, v_{BC})$ is the probability that the molecules will react when collision occurs, and $(b_{A \to BC} db_{A \to BC} \, d\phi)$ is the differential area. Note that the integral in equation (8.24) goes over only positive values of the velocity, since a collision occurs if A moves toward BC, but no collision occurs when A moves away from BC.

Again, we have chosen a sign convention so that if $v_{A \to BC}$ is a positive when atom A is moving toward BC, while if $v_{A \to BC}$ is negative, atom A is moving away from BC. We consider only positive velocities in equation (8.24), because if the velocity is negative, the A atom never collides with the BC molecule.

One also needs an expression for k_2 the rate constant for reaction (8.5). The result is

$$k_2 = \int\int\int\int\int\int P_{\text{reaction}}(v_{A \to BC}, E_{BC}, b_{A \to BC}, \phi, \mathbb{R}_{BC}, v_{BC}) v_{A \to BC}$$

$$\times D(v_{A \to BC}, E_{BC}, \mathbb{R}_{BC}, v_{BC}) \, dv_{A \to BC} \, dE_{BC} (b_{A \to BC} \, db_{A \to BC} \, d\phi) \, d\mathbb{R}_{BC}, \, dv_{BC} \tag{8.25}$$

It is also useful to define the cross section, σ_{BC} by

$$\sigma_{BC}(v_{A \to BC}, E_{BC}) = \int\int\int\int\int\int P_{reaction}(v_{A \to BC}, E_{BC}, b_{A \to BC}, \phi, \mathbb{R}_{BC}, v_{BC})$$

$$\times D_1(\mathbb{R}_{BC}, v_{BC}) b_{A \to BC} \, db_{A \to BC} \, d\phi \, d\mathbb{R}_{BC} \, dv_{BC} \qquad (8.26)$$

where $D_1(\mathbb{R}_{BC}, v_{BC})$ is the distribution function for $\mathbb{R}_{BC}$ and v_{BC}. Physically, the cross section is proportional to the rate constant for a set of molecules whose energy and velocity are fixed.

Equations (8.25) and (8.26) are the statistical mechanics expression for the reaction rate. Equation (8.25) looks very complicated, but actually it is not. The equation simply says that you need to average the reaction rate over the velocity distribution of A, the internal energy of BC, the impact parameter, and the angle of approach. Readers should convince themselves that Equation (8.25) computes a reaction rate as an average reaction probability before proceeding.

8.4 MOLECULAR DYNAMICS AS A WAY TO COMPUTE REACTION PROBABILITIES

Equation (8.25) is the key result from the last section. What it says is that one can compute a rate by picking values of the initial velocity, energy, impact parameter, and angle of approach, $\mathbb{R}_{BC}$, v_{BC}, and then compute the reaction probability. One repeats for a different angle or energy, and computes reaction probability again. The result is a series of plots like those in Figure 8.1. One then plugs into equation (8.25) to compute a rate. An example calculation is given in Example 8.C. This is an easy calculation with a computer. We ask the reader to do the calculation in the homework set, Problems 8.23–8.30.

The objective of this section is to describe how one computes the reaction probability. The general approach is to compute thousands of trajectories as A approaches BC. One then averages to calculate a rate.

Molecular dynamics (MD) is used to compute all of the trajectories. In MD, one takes an A and BC molecule, gives the atoms their initial velocities and positions, and uses a computer to calculate how all of the atoms move in the force field created by all of the other atoms in the system. For example, during the reaction

$$A + BC \longrightarrow AB + C \qquad (8.27)$$

you start A moving toward BC, as shown in Figure 8.1, and then compute what happens as the reaction proceeds. Generally, the calculations are done by solving Newton's equations of motion for all the atoms in the system.

That is not as difficult as it sounds. Consider a particle, P, that is moving in a force field. Think back to your days in freshman physics. In freshman physics, you probably solved a problem where you computed the trajectory of a cannon ball. The approach was to integrate Newton's equations of motion. If a cannon ball experiences a net force, $\vec{F}_P$,

---

[1] In the literature people sometimes integrate over the angular momentum vector. This is equivalent to the derivation here.

then the position and acceleration of the cannon ball $\vec{a}_P$ are given by

$$\vec{F}_P = m_P \vec{a}_P = m_P \frac{d^2 R_P}{dt^2} \tag{8.28}$$

where R_P is the position of the cannon ball, $\vec{a}_P$ is the acceleration, m_P is the cannon ball's mass, and t is time. Note that the force, position, and acceleration in equation (8.28) are vectors.[2] to a reasonable approximation atoms and molecules usually follow equation (8.28). Consequently, one can use equation (8.28) to determine what happens when atom A collides with molecule B–C.

Equation (8.28) is a simple second-order differential equation. If one knows the initial conditions for the variables in equation (8.28) (i.e., the initial position and initial velocity), one can use the numerical algorithms to the supplementary materials in Chapter 6 to numerically integrate equation (8.28) to calculate the position of particle P as a function of time. A computer program called ReactMD is available from Dr. Masel's Website and from http://www.wiley.com/chemicalkinetics. The reader should download that program before proceeding.

We call the position of a particle P as a function of time, the **trajectory** of the particle P. Trajectories are very important to the work in this chapter.

Now the next question is how to use equation (8.28) to calculate what happens during a reaction. Consider a molecule A reacting with a second molecule BC. Let's assume that we know the following: $v_{A \rightarrow BC}$, the velocity with which the A molecule approaches the B–C molecule; E_{BC}, the vibrational–rotational energy of the BC molecule before collision occurs; the initial velocity and position of BC; the "impact parameter," $b_{A \rightarrow BC}$; and the angle of approach, ϕ. Notice that if we know the forces in equation (8.28), we can calculate the trajectory of the A molecule as it approaches BC. We can also integrate equation (8.28) to obtain the change in the position of B and C. A typical result is given in Figure 8.1. Generally, either the reaction will occur or it will not. If A reacts with BC, the A atom will stay attached to B and the C atom will fly off. In contrast, if no reaction occurs, A will fly away by itself. One can use MD to calculate whether a reaction occurs at any set of initial conditions. Hence, one can get all of the data one needs to numerically integrate equation (8.25).

According to equation (8.25), the reaction rate is equal to the reaction probability (0 or 1 for these collisions) averaged over all of the initial conditions. One can compute the necessary integral by changing the value of the initial positions, and velocities and do the calculation again. If one repeats that process for a wide range of values of all of the parameters, one can plug into equation (8.25) to calculate the reaction rate.

8.4.1 Molecular Dynamics Simulation of Reactive Collisions

The hard part in the computation is the calculation of $P_{reaction}(v_{A \rightarrow BC}, E_{BC})$. In this section we will describe how one calculates $P_{reaction}(v_{A \rightarrow BC}, E_{BC})$.

Consider the interaction between three atoms A, B, and C. If A, B, and C are confined to a line, then one can replace $\vec{R}_A$, $\vec{R}_B$, and $\vec{R}_C$ — the position vectors for A, B, and C — by scalar quantities, R_A, R_B, and R_C. The classical equations of motion for A, B, and C are

$$m_A \frac{d^2 R_A}{dt^2} = F_A \tag{8.29}$$

---

[2] We use R to designate a position to distinguish R from a reaction rate.

$$m_B \frac{d^2 R_B}{dt^2} = F_B \qquad (8.30)$$

$$m_C \frac{d^2 R_C}{dt^2} = F_C \qquad (8.31)$$

where R_A, R_B, and R_C are the positions of atoms A, B and C; m_A, m_B, and m_C are the masses of A, B and C; t is time; and F_A, F_B, and F_C are the net forces on atoms A, B, and C.

If one knows the forces, one can numerically integrate equations (8.29)–(8.31) to calculate $P_{reaction}(v_{A \to BC}, E_{BC})$ as a function of the initial velocity and internal energy of the molecules.

8.4.2 How Do We Know the Forces?

In order to get any further on this topic, we will need to introduce considerable new material. First, we need to know all of the forces between the molecules. Then, we need to calculate how each of the molecules move. Sections 8.4.2–8.5 discuss the forces. Section 8.6 discusses how molecules move. Students generally find the materials in this section pretty difficult. However, I advise that you stay with it. It will get easier.

As noted above, in order to do any calculations, one needs to know the forces on all of the particles. For the purposes of discussion, it is useful to define a quantity called the *potential energy surface* for the system, $V_{Total}(R_A, R_B, R_C)$, which is the potential energy of the system as a function of R_A, R_B, and R_C, the position vectors of atoms A, B and C. Recall from freshman physics that if one knows the potential energy of a system of particles, one can calculate the forces on each particle from a very complicated looking equation:

$$\begin{aligned} \vec{F}_A &= -\nabla_A V_{total} \\ \vec{F}_B &= -\nabla_B V_{total} \\ \vec{F}_C &= -\nabla_C V_{total} \end{aligned} \qquad (8.32)$$

where $\vec{F}_A$, $\vec{F}_B$, $\vec{F}_C$ are the forces on atoms A, B, and C and

$$\begin{aligned} \nabla_A &= \left(\frac{\partial}{\partial x_A}, \frac{\partial}{\partial y_A}, \frac{\partial}{\partial z_A} \right) \\ \nabla_B &= \left(\frac{\partial}{\partial x_B}, \frac{\partial}{\partial y_B}, \frac{\partial}{\partial z_B} \right) \\ \nabla_C &= \left(\frac{\partial}{\partial x_C}, \frac{\partial}{\partial y_C}, \frac{\partial}{\partial z_C} \right) \end{aligned} \qquad (8.33)$$

where X_A, Y_A, and Z_A are the x, y, and z coordinates of A; X_B, Y_B, and Z_B are the x, y, and z coordinates of B; and X_C, Y_C, and Z_C are the x, y, and z of C. In actual practice, one never has to worry about equations (8.32) and (8.33). The computer does all the work automatically.

However, the key thing to notice is that if one knows the potential energy surface, V_{total}, for a reaction, one can calculate the trajectories of all of the atoms, which in turn can be used to calculate a reaction probability and hence a reaction rate.

Next, we need to discuss what forces between molecules are like. At this point, I need to discuss a little notation. When people talk about "intermolecular forces", they rarely plot the forces between molecules. Instead, they plot the intermolecular potentials, where as you may recall from above, the potential is related to the force by equation (8.32). The force is the derivative of the potential, or the potential is the integral of the force.

In the literature, it is common to see plots of the potential energy of the system as a function of the intermolecular distance. For example, Figure 8.3 shows a potential energy diagram for the interaction of two neon atoms with one another and the interaction between two fluorine atoms with one another. Notice that in both cases, the potential starts out at zero at long intermolecular distances, reaches a minimum at some intermediate distance, then rises again at short distances. In freshman physics you learned that the force is related to the potential by

$$F_{Ne} = -\frac{\partial V}{\partial R_{NeNe}} \tag{8.34}$$

At long distances, $\partial V/\partial R_{NeNe}$ is positive. According to equation (8.34), the force is negative. So the force pushes R_{NeNe} in the negative direction. R_{NeNe} shrinks, which means that the neons are pulled together. In contrast, at short distances, the $\partial V/\partial R_{NeNe}$ is negative according to equation (8.34), so the force is positive. The forces push R_{NeNe} in the positive direction. The neons move apart.

Another way to look at this result is to say that the neon–neon potential is attractive at long distances and repulsive at short distances. Physically, what is happening is that the neons attract at long distances because of dispersion (induced dipole–induced dipole) forces. In contrast, they repel at short distances because of a repulsion between the electron clouds and between the atomic cores.

There is a minimum in the potential of intermediate interatomic distances where all the forces balance. The minimum in the potential corresponds to the interatomic distance in neon dimers in the gas phase.

The neon–neon potential looks similar to the potential for the interactions of any two nonpolar molecules. For example, the potential for interaction between two methanes or ethane and a methane looks similar to that in Figure 8.3. There are some slight differences

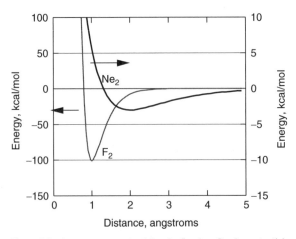

Figure 8.3   A neon–neon potential and a fluorine–fluorine potential.

with polar molecules (e.g., two waters) since the orientation of the dipole affects the new force. Still, one finds that the potential looks qualitatively like the neon–neon potential in Figure 8.3.

Figure 8.3 also shows a fluorine–fluorine potential. The fluorine–fluorine potential looks just like the neon–neon potential. The fluorines attract at long distances, so the potential goes down, but then there is repulsion at short distance as the atomic cores in the fluorine overlap. The well is deeper with two fluorines than with two neons because the fluorines can form a fluorine–fluorine bond. However, the fluorine–fluorine potential looks very similar to the neon–neon potential.

8.4.3 Intermolecular Forces

In order to proceed, we will need to understand why intermolecular potentials look the way they do. Most of you have already learned about intermolecular potentials in your physical chemistry course. Here we will review some of the key ideas to help you understand intermolecular potentials.

Four key forces come into play when atoms or molecules interact with one another:

- Dispersion forces
- Forces due to electron exchange and bonding
- Pauli repulsions
- Nuclear repulsions

8.4.3.1 *Dispersion Forces* Dispersion forces are weak, long-range physical interactions. Dispersion forces occur whenever there are two molecules in moderate proximity. Consider the two neon atoms shown in Figure 8.4. For the moment, we will treat the neon classically as a nucleus surrounded by electrons. Figure 8.4 shows the instantaneous position of the electrons in our idealized neon. If you start with a single isolated neon, then there will be a distribution of electrons around each neon. The nucleus and the electrons form an instantaneous dipole. The dipole is always fluctuating, and on average neon does not have any net dipole. But there is still an instantaneous dipole.

Now consider bringing the two neons together. Each neon will have an instantaneous dipole. The dipole on one of the neons can attract the dipole on the other neon, provided the two dipoles align. The dipole–dipole interaction lowers the energy of the system. That produces the lowering of the energy seen in Figure 8.3.

An important detail is that you get the lowering of the energy only when the dipoles are aligned. The electrons in the neons are each moving; in order to keep the dipoles aligned,

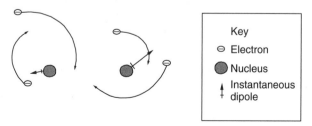

Figure 8.4 The interaction between two neon atoms.

when an electron in one of the neons moves, the electrons on the other neons need to move, too. If the electrons do not move together, the dipoles will go out of alignment. People call this simultaneous motion of the electrons on both neons *correlated motion* or *correlation*. Correlation allows the dipoles to stay aligned, so it produces the lowering of energy shown in Figure 8.3.

8.4.3.2 *Pauli Repulsions* The neon–neon potential is not attractive at all distances, however. Instead, the potentials are attractive at long distances, but repulsive at short distances as shown in Figure 8.3. It is useful to consider why the repulsions arise in terms of the interactions between the arbitals on each neon. One can understand why the repulsions arise with the aid of Figure 8.5. Figure 8.5 is a plot of the changes in the highest occupied molecular orbitals (HOMOs) of the pair of neons. Positive orbitals are lightly colored in the figure; negative orbitals are darkly colored.

The HOMOs on each neon start out as a p orbital on each neon. Recall that neon has the structure $1s^2 2s^2 2p^6$, so the p orbitals are the highest occupied orbitals of the system. The top part of Figure 8.5 shows what the p orbitals actually look like. In freshman chemistry you might have been told that the p orbitals look like a figure eight. However, when you actually calculate the shapes, you find that the p orbitals look more like two half-spheres with a space between.

Now consider moving the two neons together. The neons can come together with the two positive parts together or apart. Quantum-mechanically, orbitals of the same sign have a bonding attraction while orbitals of the different sign have an antibonding repulsion. The picture in the middle of Figure 8.5 shows the antibonding orbitals, while the picture on the bottom shows the bonding orbitals.

If you had just the bonding and antibonding orbitals, there would be little net repulsion. However, there is an additional effect—when you push the two neons together, the orbitals distort. I like to think of it as pushing two balloons together. The orbitals flatten at the point of intersection because the electrons in the orbitals repel as the result of a quantum effect associated with something called "electron exchange" (see Chapter 11). It

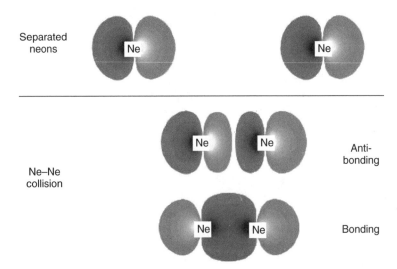

Figure 8.5   The changes in the highest occupied molecular orbital of Ne₂ as the neon atoms move together.

costs energy to distort the orbitals. As a result, when you push two neon atoms too close together, the neons repel. People call this interaction a **Pauli repulsion** because Pauli first showed how it would arise mathematically.

Think about pushing two balloons together. You get a repulsion as the balloon flattens. The same thing happens when you push two orbitals together. The orbitals distort, so they repel.

The net result is a long-range attraction and a short-range repulsion; therefore the neon–neon potential looks like the potential in Figure 8.3. The potential starts out at zero when the neon atoms are far away, reaches a minimum, and rises again. The general shape in Figure 8.3 is typical for the interaction of any molecule with any other; in other words one always sees an attraction potential at long distances and a repulsive potential at short distances.

Next, it is useful to ask how the interactions would change if we replace the two neons with two fluorines. Well, neon and fluorine are somewhat similar. Fluorine is next to neon in the periodic table, and both neon and fluorine are polarizable. However, there is a fundamental difference between the interaction of two neons and two fluorines. A neon atom has one more electron than a fluorine. The two fluorines can form a fluorine–fluorine sigma bond, and there is no antibonding interaction.

Figure 8.6 shows an orbital diagram of the other orbitals in the fluorines. For this case with fluorine, the antibonding orbital is empty, so there is no repulsion.

Of course, one can put only two electrons in a bond. If you try to put three electrons into a bond, you will get a Pauli repulsion. Consequently, there are Pauli repulsions in reactions such as

$$F + e^- + F \longrightarrow F_2^- \tag{8.35}$$

or

$$F + F^- \longrightarrow F_2^- \tag{8.36}$$

where e^- is an electron.

Now if one pushes the fluorines together, there is another effect, because the nonbonding orbitals in the fluorine begin to overlap. Again, that produces a Pauli repulsion, so the energy rises, and there is no exchange (bonding) interaction to lower the barrier. Detailed calculations show that the fluorine–fluorine potential looks very similar to the neon–neon potential in Figure 8.3. The only key difference is that the fluorine–fluorine potential is sharper and deeper than the neon–neon potential (see Figure 8.3).

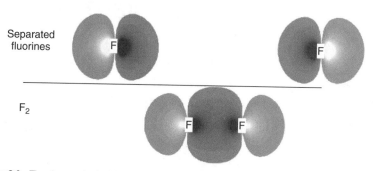

Figure 8.6   The changes in the highest occupied molecular orbital when two fluorines come together.

All atom–atom, atom–molecule, and molecule–molecule potentials look qualitatively like those in Figure 8.3. The potential is attractive at long range, and repulsive at short range, and there is a minimum in between.

8.4.4 Reactive Potentials

The discussion in Section 8.4.3 was for the interaction of two atoms, A and B. Next, we want to consider a generalized reaction:

$$A + BC \longrightarrow AB + C \tag{8.37}$$

Recall that reaction (8.37) is called an "exchange" reaction.

When an exchange reaction occurs, there are some extra forces to consider. For example, consider the reaction

$$Cl + F_2 \longrightarrow ClF + F \tag{8.38}$$

Here, the chlorine–fluorine and fluorine–fluorine potentials are both completely attractive. As a result, one might imagine that when a chlorine approaches an F_2, the chlorine would be attracted to both of the fluorines to form a stable ClF_2 complex. In fact, however, under most conditions, the chlorine atom reacts with the F_2 to form ClF and a fluorine atom.

8.5 POTENTIAL ENERGY SURFACES FOR REACTIONS

In this section, we will describe why you form ClF, and not a stable ClF_2 molecule, and we will describe the active potential. Let's consider reaction (8.38) in more detail.

Figure 8.7 shows an orbital diagram for the interaction of a chlorine atom with a F_2. The chlorine starts out with a p orbital, while the F_2 starts out with an intact sigma bond. The sigma bond is formed from two p orbitals, so there are nonbonding lobes on either side of the two fluorines corresponding to the nonbonding portions of the p orbitals.

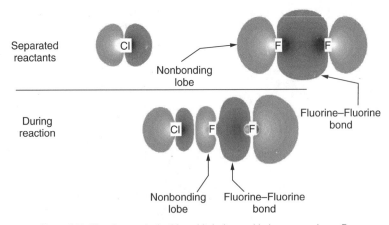

Figure 8.7 The changes in the $3A_{1g}$ orbital when a chlorine approaches a F_2.

When the chlorine approaches the fluorine, there is a Pauli repulsion between the orbital in the chlorine and the nonbonding lobe in the fluorine. The nonbonding lobe is pushed up against the F–F bond, so the F–F bond distorts. The F–F bond pushes up against the nonbonding orbital in the other fluorine. That pushes the second fluorine away. Eventually, the forces become so strong that the fluorine–fluorine bond breaks.

Most exchange reactions (i.e., reactions of the form A + BC → AB + C) show similar patterns. One orbital pushes into another. Eventually bonds break leading to products.

8.5.1   Description of Potential Energy Surfaces for Reactions

One can represent this information with a potential energy contour. Note that it is a multidimensional potential energy contour since the potential varies with the positions of the chlorine and both of the fluorines.

Next we will describe what the potential energy diagram is like. For the purposes of discussion, we will consider the reaction

$$Cl + {}^{18}F - {}^{19}F \longrightarrow Cl{}^{18}F + {}^{19}F \qquad (8.39)$$

and assume that the reaction occurs when the chlorine approaches the fluorine along the fluorine's bond in Figure 8.8.

There are three atoms in reaction (8.39). If one puts the origin on one of the atoms, then the other two atoms can each move in three directions. Therefore, in principle, one would need to consider a $2 \times 3 = $ six-dimensional potential energy surface to fully model the reaction. However, in fact, the key dimensions are the ones that involve breaking R_{ClF}, the distance from the incoming chlorine atom and the ${}^{18}F$; and R_{FF}, the length of the F–F bond. For future reference, it is important to note that R_{ClF} and R_{FF} are related to R_{Cl}, R_F^{18} and R_F^{19}, the positions of each of the atoms, by

$$R_{ClF} = |R_{Cl} - R_F^{18}|$$
$$R_{FF} = |R_F^{18} - R_F^{19}| \qquad (8.40)$$

Figure 8.9 shows several views of an appropriate potential energy surface for reaction (8.39). The top left figure is a three-dimensional (3D) plot of the potential energy surface. The plot is a plot of the energy of the system as a function of R_{FF} and R_{ClF}. The left side of the figure corresponds to the potential at large R_{ClF}. At this side of the plot, the chlorine atom is far away from the F_2, so the potential looks just like the F–F potential in Figure 8.3. There is a well and a barrier. Similarly, the right side of the figure corresponds to the large F–F distance. This corresponds to a stable ClF with a potential that is similar to the F–F potential. The back corner of the figure corresponds to all the atoms being

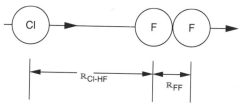

Figure 8.8   The geometry considered in the potential energy surface shown in Figure 8.9.

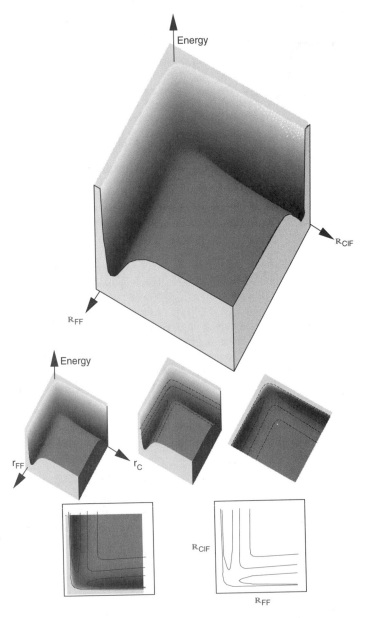

Figure 8.9 A potential energy surface for reaction (8.39).

close together. Notice that this region looks like a little hill in a miniature golf course. There is a saddle point on the top of the hill. In the literature, people call the top of the hill the **col** or the **transition state**.

Unfortunately, when you look in the literature, you find that you do not see potential plots like the one in the top left part of Figure 8.9. Instead, people plot contour plots like those in the bottom right of Figure 8.9. The contour plot represents the same information as the 3D plot. To get from the 3D plot to the contour plot, you first need to rotate the plot by 180°, and look straight down the plot as shown in the top right of Figure 8.9. Then you add contours as shown in the bottom left. Then you eliminate the shading to the plot on the bottom right of Figure 8.9.

The plot on the bottom right provides the same information as the plot on the upper left. The plot is just harder to see. The lines in the figure are contours of constant energy. The potential is zero in the upper right corner. The potential goes down, moving along the right side of the figure, reaching a minimum at point A. The potential then goes up again, moving to the lower right corner of the figure. Initially, the potential is almost constant, moving left from the lower right corner of the figure. Once one gets halfway across the figure, the potential starts to rise again. The potential eventually reaches a maximum at the lower left corner of the figure. Similarly, the potential starts out at zero at the upper right corner of the figure. The potential goes down moving left from there, reaching a minimum at point B. Then the potential rises again. Physically, the top left portion of the figure, R_{Cl-F} is large, while R_{FF} is small. Therefore, the top left portion of the figure corresponds to the initial state of the system where the chlorine atom is just beginning to approach the fluorine molecule. When chlorine is far away from the fluorine, there will be little interaction between the chlorine and the fluorine. As a result, the potential energy surface will look just like that in a fluorine molecule. The potential will be attractive at long F–F distances and repulsive at short F–F distances. The potential will not depend on R_{Cl-F}. In a similar way, in the bottom right portion of the figure, R_{ClF} is small, while R_{F-F} is large. Therefore, the bottom right portion of the figure corresponds to the final state of the reaction where the chlorine has reacted with the F_2 to form a stable ClF molecule and an isolated ^{19}F. In that case, there will be little interaction between the ClF molecule and the ^{19}F. As a result, the potential energy surface will look just like that in a ClF molecule. The potential will be attractive at long Cl–F distances and repulsive at short Cl–F distances. The potential will not depend on R_{F-F}.

Now, the lower left portion of the contour plot in Figure 8.9 is complex. The lower left portion of the curve corresponds to the case where the chlorine and the fluorine are close together. Obviously, if the chlorine and the F_2 get too close, the potential will be repulsive. However, at intermediate distances, there will be a metastable Cl–F bond and a metastable F–F bond. As a result, the potential energy of the system will be lower than in the case when the chlorine, ^{18}F, and ^{19}F are all far apart. Still, Figure 8.7 shows that there is a Pauli repulsion between the chlorine and the fluorine. The Pauli repulsion raises the energy of the system. The Pauli repulsion goes away when the ^{19}F moves away. The net effect is that the potential goes up slightly when the chlorine is close enough to the F_2 to react. The Pauli repulsion raises the energy, so it produces a barrier to the reaction.

Most potential energy surfaces for exchange reactions (i.e., reactions of the form $A + BC \rightarrow AB + C$) look just like the potential energy surface in Figure 8.9. There is usually a Pauli repulsion, so there is a small barrier to the reaction. The one exception is electron transfer reactions. For example, the reaction

$$Na + F_2 \Longrightarrow NaF + F \tag{8.41}$$

actually occurs via two steps. First an electron is transferred from the sodium to the F_2:

$$Na + F_2 \longrightarrow Na^+ + F_2^- \tag{8.42}$$

Then the ions react:

$$Na^+ + F_2^- \longrightarrow NaF + F \tag{8.43}$$

These reactions are very different from reaction (8.39) because the electron transfer eliminates most of the Pauli repulsions. A detailed description of these reactions will be given in Section 8.14. However, for now we will consider only reactions like reaction (8.39), where there is a significant barrier to reaction.

For future reference, we will call the saddle point in the potential in Figure 8.9 the **transition state** for the reaction. Later in this chapter, we will note that an understanding of the properties of the transition states is critical to the understanding of rates of chemical reactions.

Students often have reported some difficulty seeing the saddle point in the potential energy surface, and so I thought that a numerical example would be helpful. Table 8.A.4 (in Section 8.17, page 470) shows some numerical values of a potential energy surface calculated in Example 8.A. In the table, we have defined function $V(r_1, r_2)$ for the reaction $A + BC \rightarrow AB + C$. The potential is a function of r_1, the B–C distance, and of r_2, the A–B distance; the potential is also a function of a, r_0. We then used a spreadsheet to calculate numerical values at V as a function of r_1 and r_2. Table 8.A.2 shows the formulas in the spreadsheet. Column A gives the values of r_2, row B gives values of r_1, and the region from B4 to Q19 shows numerical values of the potential.

Now let's focus on the numerical values in Table 8.A.4. If we start out near the reactants, the potential is low. We started the table at a point where the energy is 3 kcal/mol. When the reactants come together, r_2 is reduced. Notice that the energy goes up with decreasing r_2.

Next, let us focus on the minimum-energy pathway from reactants to products in Table 8.1 (in Section 8.11). The shaded cells in Table 8.1 show the minimum-energy pathway going from the reactants to the products. Notice that as we move along the minimum-energy pathway from reactants to products, the energy goes up to 8.9 kcal/mol, and then back down to 3 kcal/mol.

The place where the potential is 8.9 kcal/mol is a saddle point in the potential energy surface. It is a local maximum if you approach along the shaded pathway in Table 8.A.4, but it is a local minimum if you go horizontally or vertically in the table. Thus the transition state is a saddle point in the potential energy surface.

We expand these ideas in Examples 7.G and 8.A. The reader should examine those examples before proceeding with this chapter.

8.6   REACTION AS MOTION ON POTENTIAL ENERGY SURFACES

Now that we have the preliminaries out of the way, we can talk about what reactions between molecules are like. The general scheme will be to use molecular dynamics to integrate equation (8.28) to calculate the trajectory of all of the atoms as a reaction proceeds.

In this section we will describe what the trajectories are like and how they lead to reaction. We will only consider reactions with activation barriers in this section. We will consider barrierless reactions (e.g., $Na + F_2 \rightarrow NaF + F$) in Section 8.14. For discussion purposes, we will consider a general reaction

$$A + BC \longrightarrow AB + C \tag{8.44}$$

During reaction (8.44), radical A approaches molecule BC. The A can react to extract atom B from BC or bounce without reacting. There are two key coordinates in the system: R_{AB}, the distance from A to B, and R_{BC}, the distance from B to C.

First, it is useful to note that there are three ways to visualize the reaction:

- As a trajectory in space where all the atoms move
- As a trajectory on the potential energy surface in Figure 8.9 where the bond lengths evolve
- As a plot of the motion of the atoms versus time

It is hard to explain what the motion of atoms is like without doing a simulation. A program called ReactMD is available from Dr. Masel's Website and http://www.wiley.com/chemicalkinetics, which simulates several cases with pictures and animations of the trajectories.

Figure 8.10 shows what the real trajectory is like. In this case, we plotted the positions of the atoms as a function of time. Generally, what happens is that an A atom comes up to the BC. The A atom stops because of the Pauli repulsion. If the A atom happens to

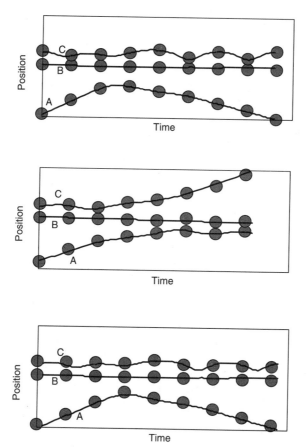

Figure 8.10 A series of trajectories during the reaction $A + BC \rightarrow AB + C$, with $M_A = M_C = 1$, $M_B = 19$, and various initial reactant configurations.

come up just when C is moving away, reaction happens. If not, A bounces off. In the top in the Figure 8.10 case, A is far from BC at the end of process, so no reaction has happened. In contrast, in the middle case, A and B are close together and C has flown away. In this case reaction has occurred.

Do not worry about understanding the details of Figure 8.10 now. We will explain it in further detail later in this chapter.

Marcellin (1912,1914a,b) showed that it is useful to plot the position–time data on a potential energy surface. Recall that when the A approaches the BC, R_{AB} and R_{BC} will change with time. One can plot that change as a trajectory (i.e., line) on the potential energy surface as shown in Figure 8.11. The solid dark line in Figure 8.11 starts at the reactants (i.e., the upper left) and ends at the products (i.e., the lower right). We will call that line a *reactive trajectory* or a *reactive collision*, since if the system follows that trajectory, the system reaches the products so reaction occurs. Similarly, the heavy dashed line starts out at the reactants (i.e., the upper left) and loops back to the upper left (i.e., back to the reactants). We will call such a trajectory a *nonreactive collision*, since if the system follows that line, the system never gets to products so no reaction occurs.

One can calculate a reaction rate from Figure 8.11 by starting with a distribution of molecules, computing a series of trajectories like those in Figure 8.11, and adding up the rate for the cases when the molecules react. Note that in the case shown in Figure 8.9, the potential goes up when the hydrogen comes close enough to the F_2 to react because of the Pauli repulsion between the hydrogen and the F_2. Therefore, it will take a certain amount of energy to get the hydrogen to react with the fluorine. This is a general result. There is always a Pauli repulsion during any exchange reaction. Consequently, one can conclude that:

> It usually costs energy to bring the reactants close enough that they can react. As a result, there is a barrier to most exchange reactions.

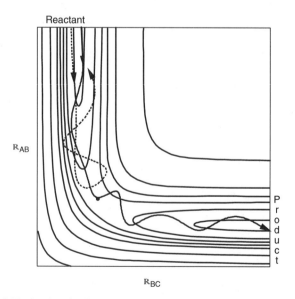

Figure 8.11 A series of typical trajectories for motion over a potential energy contour.

In Chapters 10 and 11, we will show that additional barriers arise because one needs to stretch or break bonds during chemical reactions.

8.7 QUALITATIVE FEATURES OF THE COLLISIONS: EXCHANGE REACTIONS

In the next several sections, we will discuss what reactions are like. We will consider three generic types of reaction: unimolecular reactions, bimolecular recombination reactions, and exchange reactions. The discussion in this section will consider exchange reactions. Other reaction types will be discussed starting at Section 8.10.

Exchange reactions are the most common elementary reactions. An exchange reaction is any reaction of the form

$$A + BC \longrightarrow AB + C \tag{8.45}$$

where a radical or atom A, collides with a molecule BC to transfer a ligand B and produce a new radical C. Examples of exchange reactions include

$$D\bullet + CH_3CH_3 \longrightarrow DH + \bullet CH_2CH_3 \tag{8.46}$$

$$F\bullet + CH_3Cl \longrightarrow FCH_3 + Cl\bullet \tag{8.47}$$

$$HO\bullet + CH_3CH_2CH_3 \longrightarrow HOCH_3 + \bullet CH_2CH_3 \tag{8.48}$$

In all cases, one is breaking one bond and forming another. The formation and destruction of bonds is one of the key features of ligand exchange reactions.

In the next several sections, we will describe what ligand exchange reactions are like on a molecular level. For most of the discussion, we will consider a special example: an exchange reaction $H + FH \rightarrow HF + H$. We will assume that the reaction follows the potential energy surface in Figure 8.12. The potential energy surface looks just like the

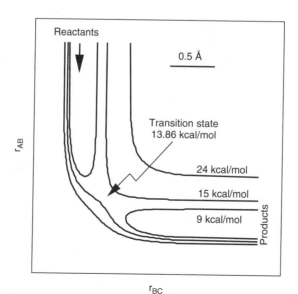

Figure 8.12   An idealized potential energy surface for the reaction $A + BC \rightarrow AB + C$.

ones we discussed earlier in the chapter with broad wells and a barrier in between. We have chosen parameters so that the reaction has a barrier of 13.86 kcal/mol, and the equilibrium bond lengths are 1.04 Å. A and C each have a mass of 1 amu. B has a mass of 19 amu. For discussion purposes, I have set the A–B and B–C bond energies to be 30 kcal/mol, not the 108 kcal/mol seen in HF. I have also artificially tripled the widths of the vibrational wells so that it will be easier to observe the vibrational motion.

Our general approach will be to start atoms moving, and calculate the positions as a function of time, to see whether reaction happens. A detail in the calculations is that one needs to specify the initial conditions to actually do the calculations. Generally one specifics the initial rotational and vibrational states of the BC molecule, the initial velocity of A toward BC, and the initial position of B and C as they vibrate in and out. In the work that follows, we usually fix the total energy and vary the other variables. In these calculations, the total energies is the sum of the rotational and vibrational energies of BC and the translational energy of A toward BC.

Figure 8.13 shows what typical trajectories are like. In this case, we put a total 18 kcal/mol into the incident molecules, and calculated the position of the atoms as a function of time. In both cases we put the same amount of energy in the molecule, but choose different values for the initial B–C position, so we got different results. In the top case (a), the atom A comes in and hits the BC molecule and reacts. Notice that A starts out far away from BC. However, as time evolves, A comes until it is in close proximity to BC. Then there is some complex behavior. Then after some time, C moves away, while A and B stay together. This is a reactive collision because we end up with A bound to B, and C far away.

The next case, (b), is a nonreactive trajectory. In this case, A + BC collides as before. However, A moves away at the end of the collision, so there is no reaction.

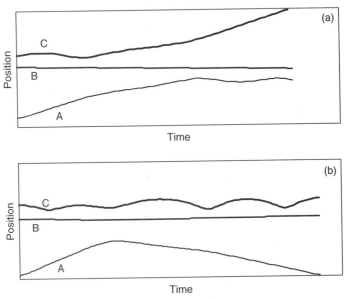

Figure 8.13 Two typical trajectories for the reaction $A + BC \Rightarrow AB + C$, with $M_A = M_C = 1$, $M_C = 19$, $E_{total} = 18$ kcal/mol ($E_{trans} = 14$ kcal/mol, $E_{tot} = 0$ kcal/mol, $E_{vib} = 4$ kcal/mol). Reaction occurs in the top trajectory but not in the bottom trajectory.

ReactMD generates thousands of trajectories like this. The reader should run some trajectories before proceeding.

Figure 8.14 shows a series of trajectories calculated by putting a total 13.8 kcal/mol into the reactants, and calculating the trajectories of the molecules. Notice that in all cases, no reaction occurs. To keep this in perspective, we choose a case where the incident energy was 13.8 kcal/mol, while the barrier was 13.86 kcal/mol. In all cases, no reaction happens.

One can understand this finding with the aid of Figure 8.15. Figure 8.15 is a replot of the data in Figure 8.14 on a potential energy surface. The trajectories all start out moving toward the transition state. However, there is not enough energy to cross the barrier. Consequently, no reaction happens.

> This is a general result. Classically, if the molecule does not have enough energy to cross the barrier, no reaction occurs. Quantum-mechanically, there is a small reaction rate due to a process call tunneling. See Sections 7.3.1, 9.3.1, and 9.5.

At higher energies, one does have some probability of reaction. Figure 8.16 shows a series of trajectories calculated by setting the energy of the incident molecules, and then adjusting all of the other parameters to optimize the probability of reaction. At an energy of 13.6 kcal/mol, the A atom comes in and makes it almost all the way to the transition state. However, the A does not have enough energy to make it over the hill, so the A

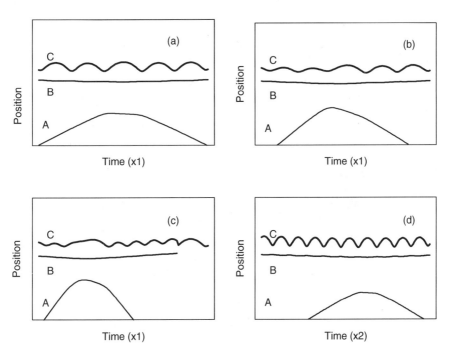

Figure 8.14 A sampling of the trajectories taken by putting a total of 13.8 kcal/mol into the reactants, choosing random initial positions for the atoms, and putting a random amount of rotational energy into the BC bond. We chose the initial velocity of A toward BC so that the total energy was 13.8 kcal/mol, and then we integrated the equations of motion to see whether reaction occurs. In this example, the barrier is 13.88 kcal/mol, so no reaction occurs.

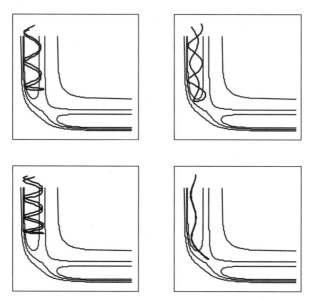

Figure 8.15 A replot of the data from Figure 8.14 on a potential energy surface.

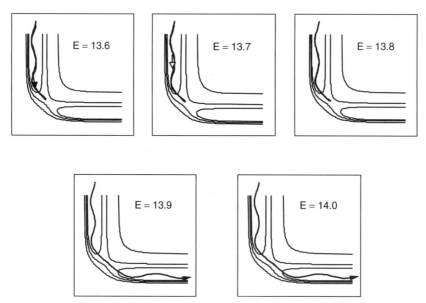

Figure 8.16 A series of trajectories calculated by fixing the total energy of the reactants and then optimizing all of the other parameters. The barrier is 13.88 kcal/mol for this example.

stops before it makes it over the barrier. In the particular case shown, the trajectory seems to double back on itself, so you see only one line in the figure. Actually, there are two lines, one on top of the other. The plots for E = 13.7 and 13.8 kcal/mol are similar to those for E = 13.6. There is no reaction. However, at E = 13.9, reaction occurs, so the trajectory (i.e, black line) leads from reactants to products. One also observes reaction at E = 14.0.

Notice that when the energy is below the saddle point energy, 13.86 kcal/mol, no reaction occurs. However, as soon as the energy goes above 13.86 kcal/mol, the reaction starts. Consequently, there is some probability of reaction whenever the incident molecules have enough energy to traverse the barrier.

This is not obvious from Figure 8.16, but, in fact, when the incoming molecules have barely enough energy to cross the barrier, the reaction probability is relatively small. One needs some excess energy to get reaction to happen at a reasonable rate.

Physically, the trajectory has to go through the pass of the top of the potential energy surface. Figure 8.17 shows a picture of the pass. A molecule with energy E can go through only the part of the pass where the energy is less than E. Figure 8.17 shows two different cases. At a low energy, the molecule can go through only the bottom of the pass, so the rate is low. However, as you raise the energy, the molecule can go through a wide portion of the pass. Consequently, the rate is higher.

Figure 8.18 shows some actual data. The figure shows how the cross section for the reaction $H + H_2 \rightarrow H_2 + H$ varies with the energy of the incoming H atom, where, as noted previously, the cross section is a quantity proportional to the rate constant for the reaction. Notice that the reaction probability is near zero until the molecules have enough energy to cross the barrier. Then the cross section increases rapidly. The cross section decreases again at high energy because the reactants fly past each other before the reaction can occur. Empirically, rates generally reach a maximum at energies 2–3 times the barrier and then decrease again.

Interestingly, there is a small probability of a reaction occurring when the energy is below the top of the barrier. This occurs because of a quantum effect called tunneling, which will be discussed in Chapter 9.

There are lots of other details that affect rates of reactions. When you do an MD calculation, you specify the initial positions and velocities of all of the atoms, and then numerically integrate equation (8.28) (Newton's equations of motion) to calculate how the system changes. It happens that when the energy is what is called "near threshold" (i.e., just above the barrier), small changes in the initial positions and velocities of the atoms can affect whether reaction actually happens.

Generally, reactions are rare events until you are several kcal/mol above the top of the barrier as shown in Figure 8.18. One key parameter is the relative motion of all of

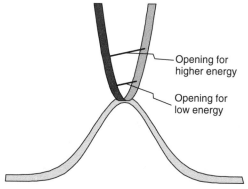

Figure 8.17   A blowup of the top of the barrier.

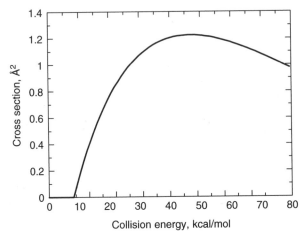

Figure 8.18 The cross section for reaction $H + H_2 \rightarrow H_2 + H$. [Adapted from Tsukiyama et al. (1988) and Levine and Bernstein (1987)].

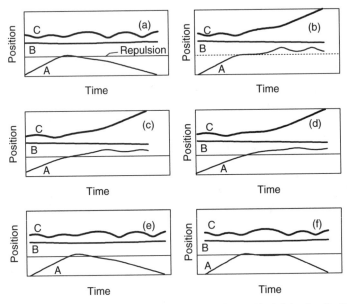

Figure 8.19 A series of cases calculated by fixing free energy at 18 kcal/mol, fixing the vibrational energy at 6 kcal/mol, and varying whether A hits when C is vibrating in toward B or out away from B.

the atoms at the moment of collision. Figure 8.19 shows a series of cases where we let B–C start vibrating, and then delayed the collision, so that A would hit B–C at different times along the B–C vibrational cycle. Sometimes A would hit when the B–C bond was stretched; at other times A would hit when the B–C bond is short. Notice that only about half of the trajectories lead to reaction even though the reactants have 18 kcal/mol, which is more than enough energy to cross the barrier. If you look carefully at the trajectories, you will notice that the critical variable is the motion of C when A crosses the dashed line in the figures. C vibrates up and back as the reaction proceeds. If A reaches the line in Figure 8.18 when C happens to be moving away, C continues to move away from B–C. So reaction occurs. If C is vibrating in toward B, no reaction occurs. Case (b) is hard to see, but actually C starts to vibrate away just when A hits the dashed line. This example illustrates that you need coordinated motion of all of the atoms in order for reaction to happen. If C is moving away from B anyway, C continues to move away. However, if C is moving in toward B, the incoming A molecule does not have enough momentum to push C away; consequently, no reaction occurs.

People often think about this effect in terms of what happens on a potential energy surface. Figure 8.20 shows a replot of the data in Figure 8.19 on a potential energy surface. Notice that the molecules need to go around a bend in the potential energy surface in order for reaction to happen. Well, the bend is like a bend in a racetrack, and the molecule needs to synchronize its turn in order to get around the track. If the molecule turns at the wrong time, the molecule does not make it around the track even though the molecule has more than enough energy to cross the barrier. Think about what happens at a curve in a racecourse. If the driver turns too fast, the car does not make it around the curve even through it has plenty of power to keep going. In the same way, the molecules need to coordinate their motion as the system reaches the transition state. In particular, the atoms need to be moving to the right as they approach the transition state in Figure 8.20. Motion to the right is equivalent to C moving away from B when A approaches BC.

A related effect is that the molecule needs to be moving in the right direction as it approaches the barrier. The molecules need to be moving up the hill as they approach the barrier in order for reaction to happen. Vibrations parallel to the hill are less effective in carrying molecules over the barrier.

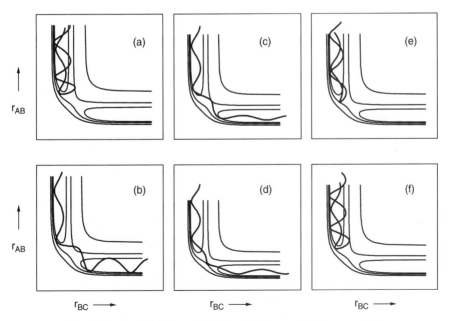

Figure 8.20 A replot of the results in Figure 8.19 on a potential energy surface.

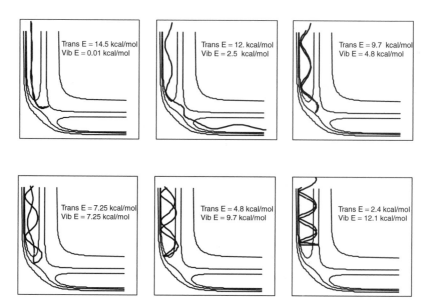

Figure 8.21 A series of cases where the molecules have enough energy to get over the barrier. Notice that some of the molecules do not make it, because the partitioning of the energy between translation and rotation is incorrect.

Figure 8.21 shows some examples where we varied the initial vibrational and translational energies on the molecules and tried to see if the molecules made it over the barrier. In these cases, there is more than enough energy to carry the molecules over the barrier, but sometimes, the molecules are moving in the wrong direction as they approach the barrier. There is no momentum to carry the reactants over the barrier, so no reaction happens.

One might think that the way to optimize the reaction is to put all of the energy in translation to give the molecules as much momentum as possible to go up the barrier. However, the results in Figure 8.21 show that does not really work. The reason is because the trajectory needs to make a left turn on the potential energy before the trajectory gets to the transition state. Consequently, it is necessary to balance the forward motion of the atoms and the turning motion to optimize the reaction probability. Generally, some of the energy needs to be in translation, but other energy needs to be in vibration to optimize the turning of the molecule. Again, there is a close analogy to a racetrack. If you ever watch a race, you will notice that the racecars go wide before they go around a curve. Well, the same thing happens with molecules. The trajectory needs to go wide on the potential energy surface to make it around the curve. The only way that molecules can go wide is if they accumulate some vibrational energy before they collide. Consequently, the optimal trajectories always have some vibrational energy. If the molecules have too much vibrational energy, though, the molecules do not have enough momentum to get over the barriers, so one has to carefully balance the vibrational and rotational energy to get reaction to happen. There also is a timing issue. A race driver who turns too early will never make it around the track. In the same way, when molecules vibrate too early, they never make it over the barrier. Perfect timing occurs when C is vibrating out when A comes in.

The trajectories in Figure 8.21 were plotted on a standard potential energy surface. There is some subtlety in looking at the motion on a potential energy surface in that trajectories can sometimes seem to turn around with no reason. This is an artifact associated with something subtle in the mathematics. Recall that the atoms are really moving on a nine-coordinate potential energy surface, where the nine coordinates are the X, Y, and Z coordinates of atom A; the X, Y, and Z coordinates of atom B; and the X, Y, and Z coordinates of atom C. When you collapse the nine-dimensional space onto a two-dimensional plot, you can get artifacts. In particular, atoms can seem to turn around for no reason.

It happens that if one uses some special coordinates, called *mass-weighted coordinates*, the artifacts vanish. The mass-weighted coordinates are defined by

$$R_1 = R_{AB} + \left[\frac{m_A m_C}{(m_A + m_B)(m_B + m_C)} \right]^{1/2} R_{BC} \tag{8.49}$$

$$R_2 = R_{BC} \left(\frac{m_B(m_A + m_B + m_C)}{(m_A + m_B)(m_B + m_C)} \right)^{1/2} \tag{8.50}$$

Figure 8.22 is a replot of the data in Figure 8.21 on mass-weighted coordinates. Notice that the potential energy surface is slightly skewed. However, the trajectories now look normal. In particular, the trajectories that do not make it stop because they are moving parallel to the hill, rather than up the hill.

Next it is interesting to consider how the shape of the potential energy surface affects the rate of reaction. The effect is substantial. If you have a potential energy that is banked like a racetrack, then the incoming molecule can ride around the curve. In such a case, transitional energy is easily transferred from the incoming A molecule to the C molecule. However, if you have an L-shaped potential energy, such as that in Figure 8.23, then you need vibrational energy to get around the curve. L-shaped potential energy surfaces are quite common when the reactants A, B, and C are atoms. They are less common when

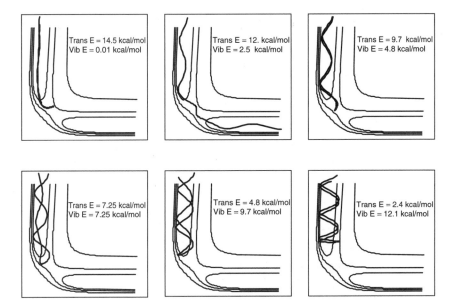

Figure 8.22 A replot of the data in Figure 8.21 in mass-weighted coordinates.

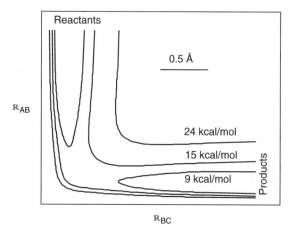

Figure 8.23   An L-shaped potential energy.

A and C are molecular ligands. Still, L-shaped potential energy surfaces do present some particular difficulties during reaction.

8.7.1   Polanyi Rules

Polanyi rules pose some special issues. Polanyi showed that with an L-shaped potential energy surface, the position of the transition state also has a large effect on the energy transfer barrier to reaction. Polanyi noted that one can characterize reactions according to whether they have an early, middle, or late transition state. Figure 8.24 shows an example of the potential energy surface with early, middle, and late transition states. An early transition state is one that arises when the reactants have hardly changed. A late transition state is a transition state where the species look similar to the products, and middle transition states are somewhere in between.

The reaction

$$CH_3OH + O \longrightarrow CH_3O + OH \qquad (8.51)$$

has an early transition state. The reaction

$$^{35}Cl + CH_3{}^{36}Cl \longrightarrow {}^{35}ClCH_3 + {}^{36}Cl \qquad (8.52)$$

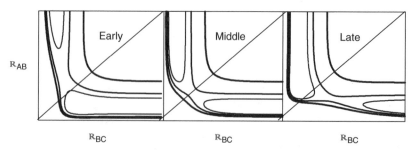

Figure 8.24   Potential energy surfaces with early, middle, and late transition states.

has a middle transition state. The reaction

$$D + CH_3CH_3 \longrightarrow DH + C_2H_5 \tag{8.53}$$

has a late transition state.

A reaction with an early transition state will reach the transition state before the reactants are significantly deformed. The B–C bond will not be stretched significantly. However, the A–B bond will be much larger than in a stable AB molecule. See Figure 8.24. In contrast, a reaction with a late transition state will reach the saddle point near the end of the reaction. The B–C bond will be almost broken, while the A–B bond will be about the same length as in a stable AB molecule. A reaction with a middle transition state is somewhere in between the two, where the bonds in both the reactants and products are stretched by similar amounts in the transition state.

In this section we will discuss some of the special effects which occur with L-shaped potential energy surfaces. The work will build on findings of Polanyi.

Polanyi showed that $P_{reaction}(v_{A \rightarrow BC}, E_{BC})$ changes according to whether the reaction has an early, middle, or late transition state. If the reaction has an early transition state, the reaction probability is higher when the translational energy is large than when the B–C vibrational energy is large. In contrast, if the reaction has a late transition state, the reaction probability is higher when the B–C vibrational energy is large than when the translational energy is large. Physically, when there is an early barrier to the reaction, most of the barrier is associated with getting A and BC in close proximity. In that case, the faster A moves toward BC, the more momentum A will carry into the collision. A's momentum allows it to get in close proximity to B–C. That increases the rate of reaction. In contrast, if the reaction has a late transition state, most of the barrier is associated with breaking the B–C bond. In that case, it is better to have energy in the B–C bond before the reaction, rather than having to transfer energy during the collision.

Polanyi showed that reactions generally retain energy during collision; thus, if you have an early transition state so that the A needs to be moving fast during the reaction, then the collisional energy generally stays with A. In that case, the A–B molecules come off hot (i.e., with a high vibrational energy) while C comes off cold (i.e., moving slowly). In contrast, if you have a late transition state, so that A starts out cold while B–C is excited, the A–B will end up cold while B and C will end up with a considerable amount of translational energy.

Polanyi showed that one can measure the vibrational and translational temperatures, by comparing the distribution of vibrational energy levels and velocities to those predicted by Boltzmann's equation. For example, Figure 8.25 shows how the vibrational energy is partitional in the DF produced during the reaction:

$$F + D_2 \longrightarrow DF + D \tag{8.54}$$

According to Boltzmann's distribution, the P_n, the probability that a given state, h, is occupied should follow

$$P_n = e^{-E_n/k_B T} \tag{8.55}$$

where E_n is the energy of the nth state.

Figure 8.25 shows how P_n actually varies. Notice that the ground state ($n = 0$) has a low probability of being occupied. The $n = 3$ vibrational level has a larger population than does the ground state. This is quite different from the scenario expected from the Boltzmann distribution. The average energy of the HF is 21.3 kcal/mol. Therefore, it

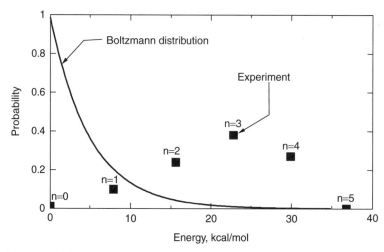

Figure 8.25   The distribution of vibrational energy produced during the reaction $F + H_2 \rightarrow HF + H$. [Results of Polanyi and Woodall (1972)].

seems that the HF is much hotter than the hydrogen atoms, which is as expected for an early transition state.

All of these trends can be explored with the ReactMD program. Any reader who has access to this program is advised to run the program and vary the parameters before proceeding with this chapter.

8.8   SUMMARY OF LINEAR COLLISIONS

At this point, it is useful to summarize what we have learned about linear $A + BC$ collisions.

- First, we found that to a reasonable approximation one can treat the collision of two molecules as a collision between two classical particles following Newton's equations of motion.
- The reactants must have enough total energy to get over the transition state (or col) in the potential energy surface in order for reaction to happen.
- It is not good enough for the molecules to just have enough energy. Rather, the energy needs to be correctly distributed between vibration and transition.
- Coordinated motions of the atoms are needed. In particular, it helps to have C moving away from B when A collides with BC.
- We also find that we need to localize energy and momentum into the B–C bond for reaction to happen.
- The detailed shape of the potential energy surface has a large influence on the rate.

These effects mean that the system has very complex behavior. Note, however, that the latter four effects cause perhaps a factor of only 10 or 100 in rate. There are always some trajectories that make it over the barrier, even though the molecules have barely enough energy to cross the barrier. If 1% of the trajectories make it, those trajectories will have an important effect on the rate.

8.9   THE NONLINEAR CASE

The discussion in the Sections 8.7 and 8.7.1 considered the dynamics of the reaction $A + BC \rightarrow AB + C$ where A, B, and C are confined to a line. With a linear collision, all of the A atoms that hit the BC molecule will react with a constant probability. However, in reality, A, B, and C are not confined to a line. Instead, the reaction occurs over a three-dimensional space. With a three-dimensional system, the reaction probability varies with how closely the molecules collide. Generally, reaction probabilities are higher when molecules collide head on than when the molecules miss. The object of the next few sections are to discuss these effects.

To start it is useful to note that one can quantify this effect by seeing how the reaction probability varies with the impact parameter $b_{A \rightarrow BC}$. Recall from Section 8.3.1 that the impact parameter is a measure of how close the reactants come to one another. Generally, the reaction probability is high if there is a head-on collision between the reactants. A head-on collision corresponds to an impact parameter of zero. In contrast, the reaction probability is near zero if the reactants miss. A large-impact parameter corresponds to the reactants missing. Generally, the reaction probability varies with the impact parameter. Consequently, the impact parameter has a large effect on the reaction.

Figure 8.26 shows how the reaction probability varies with the impact parameter at intermediate impact parameters. Generally, the reaction probability is highest with an impact parameter of 0. The reaction probability is relatively insensitive to the impact parameter when the impact parameter is small. However, there is a sharp cutoff at a larger impact parameter. As a result, the reaction rate is negligible unless atoms hit with a small impact parameter.

One can understand this behavior qualitatively with the aid of Figure 8.27. When the molecules collide with an impact parameter of zero, all of the momentum of atom A is directed toward BC. In contrast, when the impact parameter is nonzero, one can divide

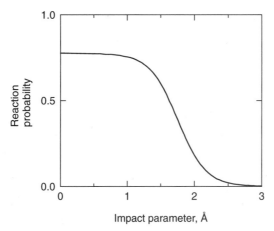

Figure 8.26   The variation in $P_{A \rightarrow BC}$ with changing impact parameter.

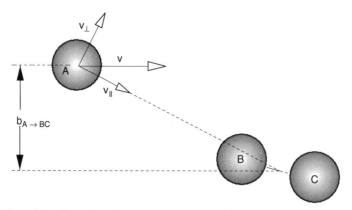

Figure 8.27   The typical trajectory for the collision of an A atom with a BC molecule.

A's momentum into two components as indicated in Figure 8.27. One component, $v_\parallel$, carries A toward BC, while the second component, $v_\perp$, carries A away from BC. $v_\perp$ increases as the impact parameter increases. As a result, the probability that atom A will move away from BC increases as the impact parameter increases.

In order to fully model this process for the reaction $A + BC \rightarrow AB + C$, one has to integrate the equations of motion [i.e., equations (8.29) to (8.32)] numerically. In general, that cannot be done analytically. However, one can get the structure of the solution by solving the equations of motion for a simpler reaction, $A + B \rightarrow AB$. A detailed derivation of the key equations will be given in Section 8.16.1 at the end of this chapter. The key result is that when A and B collide with a nonzero impact parameter, the distance from A to B, the R_{Ab}, will act as though A and B are moving in an "effective potential" given by

$$V_{eff} = \frac{(b_{A \rightarrow BC})^2 E_{kE}}{R_{AB}^2} + V_{AB}(R_{AB}) \tag{8.56}$$

where $V(R_{AB})$ is the A–B potential, E_{kE} is the kinetic energy of A moving toward B, $b_{A \rightarrow BC}$ is the impact parameter, and R_{AB} is the distance from A to B. Equation (8.56) arises because whenever $v_\perp$ is nonzero, there is a component of angular momentum that carries the reactants apart. One needs extra kinetic energy to overcome that component of angular momentum. The first term on the right of equation (8.56) is the extra kinetic energy you need to overcome the component of the angular momentum that carries the reactants apart. Therefore, we will call this barrier **the angular momentum barrier to reaction**. The extra angular momentum barrier grows as the impact parameter increases. Consequently, the extra angular momentum barrier keeps the molecules from reacting when the impact parameter is large.

Figure 8.28 shows a plot of the effective potential as a function of R_{AB} to see how that can work out. We choose a case where $V(R_{AB})$ is what is called a Lennard-Jones potential (i.e., a potential like that in Figure 8.30). When $b = 0$, there is no barrier to prevent A from approaching B. Note, however, that as $b_{A \rightarrow BC}$ increases, a barrier arises. This barrier prevents the reaction from occurring at large-impact parameters.

If one considers a more complex reaction, one finds that the arguments are more subtle. For example, with the reaction $A + BC \rightarrow AB + C$, one can transfer angular momentum from A to C, which slightly reduces the angular momentum barrier to reaction. Still,

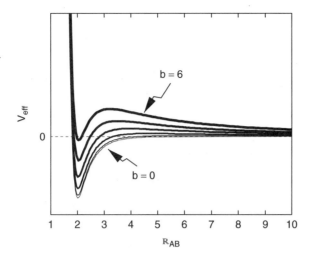

Figure 8.28 A plot of the effective potential as a function of R_{AB} for a modified Lennard-Jones Potential with $b_{A \to BC} = 0, 1, 2, 3, 4, 5, 6$.

angular momentum barriers arise in all gas-phase reactions:

The angular momentum barrier prevents reactions from occurring when molecules approach with large-impact parameters. As a result, no reaction occurs unless the reactants get close to each other.

8.9.1 Changes in the Potential Energy Due to Nonlinear Collisions

There is another effect that also changes the potential with nonlinear collisions. Recall that the potential energy surface in Figure 8.9 was calculated for the case where the chlorine atom approached the fluorine directly along the fluorine–fluorine bond. In that case, the potential energy surface is simple. However, if the chlorine approaches along a different angle, the potential energy surface changes.

First note that the incoming chlorine needs to form a bond with the p-orbital. When the chlorine approaches the F_2 at some finite angle away from the fluorine–fluorine bond, the overlap with the p orbital is reduced. That reduces the net attraction between the fluorine and the chlorine. Simultaneously, the overlap with the nonbonding orbitals in the F_2 increases. There is a Pauli repulsion between the nonbonding orbitals and the incoming chlorine. The result is that the barriers increase with a nonlinear collision. There are equations in the literature to quantify the increase, but when this book was being written, the equations did not seem to have good correlation with experiment. Still, the qualitative feature that the barriers to reaction increase with nonlinear collisions will be important to the analysis because the increase in barrier decreases the rate.

8.9.2 Influence on the Overall Rate

Now, it is useful to go back and see how the angular momentum barrier and the variation in the barriers to reaction will change the analysis in Section 8.2 (i.e., collision theory). Note that the analysis in Section 8.2 assumed that the reaction rate was equal to the collision rate times the probability that molecules that collided actually reacted. The probability is assumed to be constant for all molecules that hit each other. However, the results in the previous section show that the probability is not constant. Thus, some modification of the results in Section 8.2 are needed.

One can account for the variations in the reactive probability by integrating equation (8.24). It is useful to define a reaction cross section, $\sigma^r_{A \rightarrow BC}$, by integrating over the impact parameter, $b_{A \rightarrow BC}$, and ϕ_{ABC}, R_{ABC}, v_{ABC}:

$$\sigma_{A \rightarrow BC}(v_{A \rightarrow BC}, E_{BC}) = \int\int\int\int\int\int P_{reaction}(v_{A \rightarrow BC}, E_{BC}, b_{A \rightarrow BC}, \phi, R_{BC}, v_{BC})$$
$$\times D_1(R_{BC}, v_{BC})b_{A \rightarrow BC}\, db_{A \rightarrow BC}\, d\phi\, dR_{BC}\, dv_{BC} \qquad (8.57)$$

where $D_1(R_{BC}, v_{BC})$ is the distribution function for the internal states of the BC molecule, $b_{A \rightarrow BC}$ is the impact parameter, ϕ_{ABC} is the angle of approach, and R_{AB} and v_{BC} are the internal positions and velocities of the BC molecule as defined in Section 8.3.1.

For a spherically symmetric case:

$$\sigma^r_{A \rightarrow BC}(v_{A \rightarrow BC}, E_{BC}) = 2\pi \int P_{A \rightarrow BC}(b_{A \rightarrow BC}, v_{A \rightarrow BC}, E_{BC})b_{A \rightarrow BC}\, db_{A \rightarrow BC} \qquad (8.58)$$

where $P_{A \rightarrow BC}$ is the reaction probability as a function of b averaged over R_{BC}, v_{BC}, and ϕ.

The reaction probability is already in equation (8.58). Therefore, equation (8.25) for the rate constant becomes

$$k_{A \rightarrow BC} = \int\int\int v_{A \rightarrow BC}\sigma^r_{A \rightarrow BC}(v_{A \rightarrow BC}, E_{BC})D_2(v_{A \rightarrow BC}, E_{BC})dv_{A \rightarrow BC}\, dE_{BC} \qquad (8.59)$$

where D_2 is the distribution function for $v_{A \rightarrow BC}$ and E_{BC}.

Equations (8.58) and (8.59) can be used to calculate the rate constant for reaction if $P_{reaction}$ $(b_{A \rightarrow BC}, v_{A \rightarrow BC}, E_{BC})$ is known. Hence, they are quite useful. An example calculation is given in Example 8.D.

8.10 MORE COMPLEX REACTIONS

All of the simulations we have presented were for cases where A, B, and C are single atoms. However, one can generalize these effects to reactions where the individual species A, B, and C are molecular ligands rather than atoms. Generally, everything we have said so far in this chapter also applies to the case where A, B, and C are molecular ligands. A reaction involves concerted destruction and formation of bonds. Both bond formation and bond destruction processes must occur simultaneously. One can view this process as an energy transfer barrier to reaction. During the reaction $A + BC \rightarrow AB + C$, energy must accumulate in the B–C bond before any reaction occurs. No reaction occurs unless C leaves before A leaves, and A might stay around for only 10^{-14} seconds. As a result, it helps to have the atom B moving away from the rest of the molecule C when the attacking group A moves in toward B.

Another issue is that in order for reaction to occur, the B–C bond that needs to break (i.e., the B–C bond) must somehow acquire enough energy to break before the attacking group A leaves. Therefore, the rate at which energy accumulates in the B–C bond has an important influence on the reaction rate. Generally, one finds that when the system has barely enough energy to cross the activation barrier, reaction rarely occurs, because the collision energy seldom gets localized in the B–C bond. The reaction rate increases with increasing energy. Note, however, that if A comes in too fast, the A will leave before there is time for reaction to occur. In that case, the reaction rate will go down again.

Another interesting effect is that when you have bigger molecules, the collisions last longer, As a result, the reaction probability increases. It is hard to understand why that happens without detailed analysis. However, when you watch movies of collisions, you find that if you have a reaction of little molecules:

$$H + F_2 \longrightarrow HF + F \qquad (8.60)$$

the collision might last only 10^{-12} seconds. However, if two moderate-sized molecules react, such as

$$C_5H_{12}O + HC_5H_{12} \longrightarrow C_5H_{12}OH + C_5H_{12} \qquad (8.61)$$

the reactants might stick together for 10^{-10} seconds. There are more chances for reaction when the reactants stick together. As a result, the preexponentials for reactions tend to go up as the reactants get bigger.

Physically, collisions last longer with big molecules than with small molecules because of energy partitioning in the molecules. Recall that, in general, molecules have to be moving pretty fast (i.e., molecules need significant translational energy) in order to have enough energy to cross the barrier. When a collision occurs between two small molecules, the molecules continue to move quickly, and as a result, the collision lasts for only an instant. In contrast, with a big molecule, the translational energy can be converted into vibrational energy. It is like bouncing a bowling ball on a mattress. The mattress takes up energy, so the ball does not fly away. In the same way, a small molecule transfers energy to the bigger molecule. The molecules slow down, so the molecules stay in close proximity longer. The molecules can also get entangled with one another. As a result, the reactants stay in closer proximity longer with the big molecules than with small molecules.

8.10.1 State-Selected Chemistry

The other thing you see with a more complex molecule is *state-selected chemistry*. The objective of state-selected chemistry is to use a laser to excite certain vibrational or rotational modes in a molecule in a way that a desired reaction pathway is enhanced. For example, Crim [1996] examined the photodissociation of monodeuterated water (HOD). If you simply photolyze HOD, with a photon, v, you get two reactions:

$$HOD + v \longrightarrow H + OD \qquad (8.62)$$

$$HOD + v \longrightarrow HO + D \qquad (8.63)$$

Reaction (8.62) is very slightly favored over reaction (8.63) because of a quantum process called "tunneling." However, Crim (1996) found that if they excite the OD bond before photolysis, reaction (8.63) is favored. So far, the enhancements have been a factor of about 200. Unfortunately, you get that only with very expensive lasers, and it works only with small molecules. However, these effects are important in special reactions.

8.11 COMPLEXITIES: UNIMOLECULAR REACTIONS

So far, we have been discussing exchange reactions, where a ligand is exchanged from one species to the next. Those are the simplest reactions. However, there are other reaction types. In the next several sections, we will discuss these other reaction types.

In this section we will discuss unimolecular reactions. Unimolecular reactions are the second most common type of reaction. Unimolecular reactions are reactions of the form

$$BC \Longrightarrow products \qquad (8.64)$$

where a stable molecule BC is somehow converted into products. Examples of unimolecular reactions include

$$O_2 \Longrightarrow O + O \qquad (8.65)$$

$$CH_3CH_2CH_3 \Longrightarrow CH_3 + C_2H_5 \qquad (8.66)$$

$$\begin{array}{c} CH_2 \\ / \ \backslash \\ H_2C-CH_2 \end{array} \Rightarrow CH_2{=}CH{-}CH_3 \qquad (8.67)$$

In Chapter 4, we noted that unimolecular reactions are never elementary reactions. One can use molecular dynamics to see why that is so. If one starts with a stable molecule, the molecule is by definition stable. It never dissociates.

Consider trying to dissociate F_2. Figure 8.29 shows a plot of potential energy of F_2 as a function of the F–F distance. The plot is a duplicate of the plot in Figure 8.3. Now consider trying to dissociate an F_2 molecule with an initial state given by the line in the figure. If you start with the F_2 with a state on the line, the fluorines vibrate back and forth. However, the fluorines are trapped in the attractive well. In the state shown, the fluorine–fluorine bond has a total energy of 40 kcal/mol (i.e., the fluorines are held together by 40 kcal/mol). One needs to accumulate the 40 kcal/mol in order for the bond to break. If the F_2 is isolated from its surroundings, energy is considered, so there is no way to accumulate the 40 kcal/mol. As a result, the fluorine cannot fall apart.

> Simple unimolecular reactions are never elementary reactions.

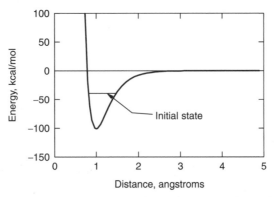

Figure 8.29 A plot of the fluorine–fluorine potential from Figure 8.3.

Of course, one does sometimes observe overall reactions like:

$$O_2 \Longrightarrow 2O \tag{8.68}$$

However, that is never an elementary reaction.

Instead, reactions like (8.65), (8.66), and (8.67) occur via bimolecular collision

$$X + O_2 \longrightarrow O + O + X \tag{8.69}$$

where X is a collision partner, as discussed below.

Physically, a hot X molecule collides with the O_2, and transfers energy to the O_2. The energy accumulates in the O–O bond. Eventually, after several collisions, the O–O bond has enough energy to break. Notice that one needs to accumulate energy in the O–O bond before reaction can happen, which is why a collision partner is needed. One can observe reactions of the form

$$AB + \nu \longrightarrow A + B \tag{8.70}$$

where ν is a photon, since the photon provides the energy (and momentum) needed to dissociate the molecules.

Another observation is that the rate constant varies according to the nature of the collision partner. For example, Table 8.1 shows how the rate of the reaction $O_2 + X \rightarrow 2O + X$ varies with the collision partner. The rate is zero in the absence of a collision partner. Further, one finds that the rate constant varies by a factor of almost 80 according to whether argon or oxygen is used as the collision partner.

The result of this discussion is that unimolecular reactions are very complicated, and one often needs detailed calculations to make useful predictions.

One of the key things you observe with a unimolecular reaction is that the reaction occurs in two stages. First, the collision partner collides with AB to form a hot AB molecule. Then, the hot molecule decomposes to form product. The hot molecule often lasts for 10^{-9} seconds. By comparison, ligand exchange reactions last only 10^{-13} seconds. Consequently, the preexponentials for unimolecular reactions tend to be much higher than the preexponentials for ligand exchange reactions.

At this point, we will defer our discussion of unimolecular reactions to the next chapter. The thing to remember for now is that unimolecular reactions are never elementary. Instead, one needs a collision partner for reaction to occur.

Table 8.1 Arrhenius parameters for the reaction $O_2 + X \rightarrow 2O + X$

Collision Partner	Preexponential, $\text{Å}^3/(\text{mol·second})$ at 4000 K	Activation Energy, kcal/mol
None	0	?
Ar	7.5×10^{14}	118.8
O_2	5.8×10^{16}	118.8

Source: Data from Westley (1980).

8.12 BIMOLECULAR RECOMBINATION REACTIONS

There is one other common type of reaction called a *bimolecular recombination reaction*. Bimolecular recombination reactions are any reaction of the form

$$B + C \longrightarrow BC \tag{8.71}$$

where two species bind together to form a stable molecule. Examples of bimolecular reactions include

$$F + F \longrightarrow F_2 \tag{8.72}$$

$$CH_3 + CH_2CH_3 \longrightarrow CH_3CH_2CH_3 \tag{8.73}$$

$$Na + F \longrightarrow NaF \tag{8.74}$$

One would think that reactions (8.72) and (8.74) would be quite rapid. After all, according to Figure 8.3, there is no barrier to bringing two fluorines together. In the same way, there is no barrier to bringing a sodium up to fluorines. Actually, however, the rate of reactions (8.72)–(8.74) and all other bimolecular recombination reactions is zero in the absence of a collision partner.

8.13 ENERGY TRANSFER BARRIER TO REACTIONS

Next, it is useful to discuss why reactions (8.72)–(8.74) do not occur in the absence of a collision partner. The difficulty is associated with an energy transfer barrier to reaction. According to data in the CRC, reaction (8.74) is 181 kcal/mol exothermic. If a sodium would collide with a fluorine, one might think that one could form a sodium fluoride molecule. However, it would be a very hot sodium fluoride molecule since the heat of reaction is 181 kcal/mol exothermic. In fact, heat balance shows that the sodium fluoride would heat up to 34,000 K! Well, at 34,000 K, the sodium fluoride has a high probability of dissociating. The result is that the sodium and fluorine never stay together very long. One can form stable sodium fluoride by cooling off the sodium fluoride before the sodium and the fluorine fly apart. However, one needs to transfer the heat in some way. The only way that heat can be transferred is by convection (i.e., collisions) or radiation (i.e., photons). Consequently, if the reaction does not produce a photon, then one would need a collision partner for reaction to happen.

Reaction (8.74) is a special case, because the reaction is so exothermic. However, next, we want to show that any reaction of the form $A + B \rightarrow AB$, where two reactants combine to form a single product, cannot occur in the absence of a collision partner or photon.

Consider the generalized reaction $A + B \rightarrow AB$. At the start of the reaction, A approaches B with some energy, E_{AB}. For the purpose of discussions, assume that the A–B potential, $V(R_{AB})$, looks like the potential in Figure 8.3. As A approaches B, A is first accelerated since $V(R_{AB})$ is attractive at long distances. However, A begins to slow down again as A and B become very close. One can calculate A's velocity from an energy balance:

$$\begin{pmatrix} \text{Total} \\ \text{energy} \end{pmatrix} = \begin{pmatrix} \text{kinetic} \\ \text{energy} \end{pmatrix} + \begin{pmatrix} \text{potential} \\ \text{energy} \end{pmatrix} \tag{8.75}$$

or

$$E_{AB} = \frac{1}{2} M_{AB} v_{AB}^2 + V(R_{AB}) \tag{8.76}$$

where v_{AB} is the velocity of A toward BC, M_{AB} is the effective mass of the AB pair, and $V(R_{AB})$ is the attractive potential between A and B.

It is useful to also calculate the net force on A from equation (8.29). Note that in order for the reaction $A + B \rightarrow AB$ to occur, A must fall into B's attractive well. A's velocity must be small (i.e., like a vibration). Further, the net force on A must be small, or else A will be pushed away from B. It works out, however, that when one actually calculates the velocity from equation (8.76) and the net force from equation (8.29), one finds that one can never get to the situation where both the force and the velocity are small at the same time.

Consider the case where an A atom collides with a B atom. Figure 8.30 shows a plot of A's velocity and the net force on A during a collision. A starts out moving slowly. However, it is accelerated by B's attractive well. The net result is that A's velocity is quite large when A is sitting over the minimum in the A–B potential. As a result, A flies past the potential well. A will eventually stop when the A–B bond is highly compressed. However, the net force on A is tremendous there, so A is pushed away. One can never get to the position where the net force on A is small, and where A is moving slowly enough that A stays bound to B. As a result, A cannot react with B.

The analysis above was for a Morse potential. However, detailed simulations with the program ReactMD show that the reaction $A + B \rightarrow AB$ never occurs. A always scatters (i.e., bounces away) from B independent of the form of A–B potential.

Physically, what is happening is that A is approaching B with some energy, E_{AB}. A has to lose that energy in order to fall into B's attractive well. However, there is no place

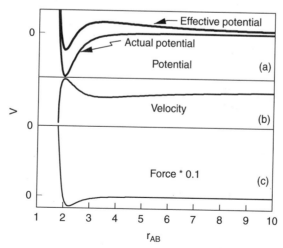

Figure 8.30   (a) A typical potential for the reaction $A + B \rightarrow AB$. (b) The velocity of A toward B. (c) The net force on A.

for the energy to go. As a result, A simply flies out again.

One can generalize this analysis to conclude that elementary reactions of the form

$$A + B \longrightarrow C$$

are impossible in the absence of collision partners or photons.

Reactions of the form $A + B \rightarrow AB + v$, where v is a photon, are possible because the photon can carry away energy and momentum. In practice, though, such reactions are rare.

Still, experimentally, one often observes reactions of the form $A + B \Rightarrow AB$. Note, however, that such reactions are never elementary reactions. Instead, one finds that the reaction actually follows a mechanism $A + B + X \rightarrow AB + X$, where X is again a **collision partner**. X can be another molecule in the system, or the walls (i.e., surfaces) of the reaction vessel. The function of the collision partner is to remove energy and momentum from the reactive complex so that A can fall into B's attractive well. Experimentally, the rate of a reaction of the form $A + B \Rightarrow AB$ goes to zero at low pressures because there are no collision partners. However, one can observe reactions of the form $A + B \Rightarrow AB$ at moderate pressures. At moderate pressures, collisions with other molecules in the system occur. Those collisions carry away energy and momentum from the AB complex. As a result, the reaction has a finite rate.

In the problem set, we ask the reader to show that the reaction looks elementary at high pressures. However, the reaction is not really elementary. In particular, the rate constants for the reaction vary with the environment since the collision partner plays an important role in the reaction.

For example, Table 8.2 shows the rate constant for the reaction $NO + O + X \rightarrow NO_2 + X$. Notice that the rate constant is about a factor of 60 lower when the reaction is run in argon than when the reaction is run in water vapor. Clearly, this is a significant effect. Therefore, it is incorrect to view reactions of the form $A + B \Rightarrow AB$ as elementary. Rather, it is better to view the reactions as needing a collision partner.

The same arguments do not apply to ligand exchange reactions. Ligand exchange reactions such as reactions (8.45)–(8.48) all have two species coming into the reaction

Table 8.2 Arrhenius parameters for the reaction $NO + O + X \rightarrow NO_2 + X$

Collision Partner	Preexponential, Activation $\text{Å}^3/(\text{mol}^2\cdot\text{second})$	Energy, kcal/mol
None	0	?
Ar	1.7×10^{14}	1.88
O_2	1.7×10^{15}	1.88
CO_2	3.84×10^{15}	1.88
H_2O	1.1×10^{16}	1.88

Source: Data from Westley (1980).

and two species leaving it. The energy to break bonds can come from the reactants, while the products can carry off any energy created in the process. As a result, the large energy transfer barriers to reaction seen with recombination reactions are not observed with ligand exchange reaction. Instead, one gets collisions, and those collisions lead to reaction.

This leads to the key point in this section:

All elementary chemical reactions must have two or more reactants and two or more products. Reactions of the form $A + B \rightarrow AB$, $AB \rightarrow A + B$, $A \rightarrow C$ are not elementary; collision partners or photons are needed before such reactions can occur.

The reader should memorize this box before proceeding.

8.14 ELECTRON TRANSFER REACTIONS

There is another class of reactions where angular momentum barriers play a key role: reactions that do not have a barrier. We will discuss them in this section.

Consider the reaction

$$Na + F_2 \Longrightarrow NaF + F \tag{8.77}$$

Reaction (8.77) looks very similar to reaction (8.38). The only difference is that we have replaced the chlorine by sodium. At first glance one might think that the potential energy surface for reaction (8.77) would look very similar to that for reaction (8.38). In fact, however, there is very little similarity. Reaction (8.77) actually occurs in two steps. First, the sodium atom approaches the F_2 and exchanges an electron.

$$Na + F_2 \longrightarrow Na^+ + (F_2)^- \tag{8.78}$$

Then the sodium ion reacts with the F_2 ion to yield products:

$$Na^+ + (F_2)^- \longrightarrow NaF + F \tag{8.79}$$

Experimentally, the electron transfer occurs when the reactants are 50 Å apart. Then the reaction proceeds quite quickly. We will describe the features of the electron transfer process in Chapter 11. However, a qualitative picture is given in Figure 8.31.

Figure 8.31 shows the energy of the neutral ($Na + F_2$) and ionized state of a NaF_2 complex as a function of the distance between the sodium and the F_2. When the molecules are far apart, the neutral species are stable. However, as the sodium approaches the F_2, the charged state has a lower energy. Quantum-mechanically, the electron always goes to the lowest energy state and electrons move at speeds close to the speed of light. Consequently, the electron jumps from the sodium to the F_2 as soon as the charged state has a lower energy than the neutral state. There are some details, based on something called an "avoided" crossing, which will be discussed in Chapter 10. However, the key thing to remember is that if the charged state has a lower energy than the neutral species, the charged species will form. Experimentally, the charged species form when the reactants are 50 Å apart (i.e., before there are significant Pauli repulsions).

Once the charged species form, the reaction happens very rapidly. Recall that Na^+ is a tiny ion. The Na^+ can move all the way up to the F^- without experiencing a significant

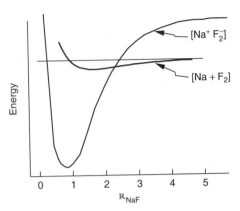

Figure 8.31   The energy of the neutral and charged states of NaF$_2$ as a function of R_{NaF} the sodium–fluorine distance.

Pauli repulsion. Further, the Na$^+$ is drawn in because the positive charge on the sodium is attracted to the negative charge on the F$_2^-$. As a result, the potential energy surface for the reaction does not show a barrier. Overall, reaction (8.79) is extremely rapid.

Still it is not instantaneous. Recall from Section 8.9 that every reaction has an angular momentum barrier as shown in Figure 8.28. The barrier grows as $b_{A \to BC}$ increases. This barrier prevents reaction (8.79) from occurring when $b_{A \to BC}$ is large. Consequently reaction (8.79) has a finite rate even though there is no barrier for reaction.

In a more general sense, angular momentum barriers to reaction are very important for any reaction with a negligible barrier. After all, the rate should be infinite without the angular momentum barrier to reaction. We will quantify these results in Chapter 9. The thing to remember for now is that reaction rates are finite for barriers reactions because of the angular momentum barrier to reaction.

8.15   SUMMARY

The three primary causes of reaction barriers are

- Hills in the potential energy surface
- Energy transfer barriers
- Angular momentum barriers to reaction

The hills arise because the electron clouds need to be compressed in order for the reactants to get close enough to react. If you do not have enough energy, no reaction can occur.

The energy transfer barriers arise because one needs to put energy into a bond to get a bond to break. If a molecule has enough energy to react, the molecule still might not react because the energy is in the wrong mode.

Similarly, one needs to remove energy to get a new bond to form. If there is no place for the energy to go, you could form only very hot molecules. Hot molecules are unstable. Angular momentum barriers are more subtle, because they are important only with large-impact parameters. Generally, if the impact parameter is large, the reactants will fly apart before they can react.

At this point we need to end our discussion. In this chapter, we have shown that there are three principal causes barriers to reaction. All three of these processes are important to reactions. One should study them in detail.

We have also found that one can explore all of the effects via trajectory calculations. Trajectory calculations are quite important to reactions. The reader should run some trajectories before proceeding further.

8.16   SUPPLEMENTAL MATERIAL

8.16.1   Derivation of the Angular Momentum Barrier to Reaction

As noted previously, there is an angular momentum barrier to reaction. The purpose of the next several pages will be to derive equation (8.56) for the barrier.

The derivation starts by looking at the classical equations of motion of A and B during a collision between A and B.

$$m_A \frac{d^2 R_A}{dt^2} = \vec{F}_A = -\nabla_A V(R_A, R_B) \tag{8.80}$$

$$m_B \frac{d^2 R_B}{dt^2} = \vec{F}_B = -\nabla_B V(R_A, R_B) \tag{8.81}$$

From Newton's second law, the force that A exerts on B must be equal and opposite to the force that B exerts on A:

$$\vec{F}_B = -\vec{F}_A \tag{8.82}$$

It is useful to change coordinates to what is called **center-of-mass coordinates**. Let's define a new vector $\vec{X}$ via

$$\vec{X} = \frac{m_A R_A + m_B R_B}{m_A + m_B} \tag{8.83}$$

$\vec{X}$ is the position of the center of mass of A and B. Taking the second derivative of equation (8.83), we obtain

$$\frac{d^2 \vec{X}}{dt^2} = \left(\frac{1}{m_A + m_B} \right) \left(m_a \frac{d^2 R_A}{dt^2} + m_B \frac{d^2 R_B}{dt^2} \right) \tag{8.84}$$

Substituting equations (8.80)–(8.82) into equation (8.84) yields

$$\frac{d^2 \vec{X}}{dt^2} = \frac{1}{m_A + m_B} (\vec{F}_A + \vec{F}_B) = 0 \tag{8.85}$$

Therefore, the center of mass of the system moves with a constant velocity throughout the collision. Now, let's develop an expression for R_{AB}, the distance from A to B. The definition of R_{AB} is

$$R_{AB} = R_B - R_A \tag{8.86}$$

Taking the second derivative of equation (8.86) yields

$$\frac{d^2 R_{AB}}{dt^2} = \frac{d^2 R_B}{dt^2} - \frac{d^2 R_A}{dt^2} \tag{8.87}$$

Substituting (8.80) and (8.81) into equation (8.87) yields

$$\frac{d^2 R_{AB}}{dt^2} = \frac{1}{m_B}\vec{F}_B - \frac{1}{m_A}\vec{F}_A \tag{8.88}$$

but $\vec{F}_A = -\vec{F}_B$. Therefore

$$\frac{d^2 R_{AB}}{dt^2} = \left(\frac{1}{m_A} + \frac{1}{m_B}\right)\vec{F}_B = \frac{1}{\mu_{AB}}\vec{F}_B \tag{8.89}$$

where μ_{AB} is defined by

$$\frac{1}{\mu_{AB}} = \frac{1}{m_B} + \frac{1}{m_A} = \frac{m_A + m_B}{m_A m_B} \tag{8.90}$$

The quantity μ_{AB} is called the **reduced mass** of the system. Rearranging equation (8.89) yields

$$\mu_{AB}\frac{d^2 \vec{R}_{AB}}{dt^2} = \vec{F}_B \tag{8.91}$$

$$\vec{F}_B = -\nabla_B V = -\left(\frac{d}{dX_B}, \frac{d}{dY_B}, \frac{d}{dZ_B}\right)V \tag{8.92}$$

but

$$\vec{R}_{AB} = \vec{R}_B - \vec{R}_A \tag{8.93}$$

$$\vec{F}_B = -\left(\frac{d}{dX_{AB}}, \frac{d}{dY_{AB}}, \frac{d}{dZ_{AB}}\right)V \tag{8.94}$$

where X_{AB}, Y_{AB}, and Z_{AB} are the x, y, and z coordinates of $\mathbb{R}_{AB}$.

It is useful to convert equation (8.91) to spherical coordinates; $\mathbb{R}_{AB}$, θ_{AB}, and ϕ_{AB}, where $\mathbb{R}_{AB}$, θ_{AB} and ϕ_{AB} are related to X_{AB}, Y_{AB}, and Z_{AB}, respectively, the X, Y, and Z components of $\mathbb{R}_{AB}$, by

$$X_{AB} = \mathbb{R}_{AB}\cos\theta_{AB}$$
$$Y_{AB} = \mathbb{R}_{AB}\sin\theta_{AB}\cos\phi_{AB} \tag{8.95}$$
$$Z_{AB} = \mathbb{R}_{AB}\sin\theta_{AB}\sin\phi_{AB}$$

For the moment, we will consider only $\phi_{AB} = 0$.

Let's define L_{AB} as the angular momentum of the AB pair by

$$L_{AB} = \mu_{AB}v_{A\to B}\mathbb{R}_{AB} = \mu_{AB}\left(\frac{d\theta_{AB}}{dt}\right)\mathbb{R}_{AB}^2 \tag{8.96}$$

where $v_{A\to B}$ is the instantaneous velocity of A toward B. Next, let's demonstrate that angular momentum is conserved during the collision. When $\phi_{AB} = 0$, it follows that

$$\theta_{AB} = \arctan\left(\frac{X_{AB}}{Y_{AB}}\right) \tag{8.97}$$

$$\frac{d\theta_{AB}}{dt} = \left(\frac{1}{1 + \left(\frac{Y_{AB}^2}{X_{AB}^2}\right)}\right) \frac{d}{dt}\left(\frac{Y_{AB}}{X_{AB}}\right)$$

$$= \left(\frac{1}{1 + Y_{AB}^2}\right)\left(\frac{1}{X_{AB}}\frac{\partial Y_{AB}}{\partial t} - \frac{Y_{AB}}{X_{AB}^2}\frac{\partial X_{AB}}{\partial t}\right) \tag{8.98}$$

Multiplying the top and bottom of equation (8.98) by X_{AB}^2 yields

$$\frac{d\theta_{AB}}{dt} = \frac{X_{AB}\dfrac{dY_{AB}}{dt} - Y\dfrac{dX_{AB}}{dt}}{X_{AB}^2 + Y_{AB}^2} = \frac{X_{AB}}{R_{AB}^2}\frac{dY_{AB}}{dt} - \frac{Y_{AB}}{R_{AB}^2}\frac{dX_{AB}}{dt} \tag{8.99}$$

Now consider:

$$\frac{dL_{AB}}{dt} = \frac{d}{dt}\left(\mu_{AB}R_{AB}^2\frac{d\theta_{AB}}{dt}\right) \tag{8.100}$$

Substituting equation (8.99) into equation (8.100) yields

$$\frac{dL_{AB}}{dt} = \frac{d}{dt}\left(\mu_{AB}\left(X_{AB}\frac{dY_{AB}}{dt} - Y_{AB}d\frac{X_{AB}}{dt}\right)\right) \tag{8.101}$$

Performing the algebra, we obtain

$$\frac{dL_{AB}}{dt} = \mu_{AB}\frac{dX_{AB}}{dt}\frac{dY_{AB}}{dt} + \mu_{AB}X_{AB}\frac{d^2Y_{AB}}{dt^2} - \mu_{AB}\frac{dX_{AB}}{dt}\frac{dY_{AB}}{dt} - \mu_{AB}Y_{AB}\frac{d^2X_{AB}}{dt^2} \tag{8.102}$$

where the first and third terms cancel. Therefore

$$\frac{dL_{AB}}{dt} = \mu_{AB}X_{AB}\frac{d^2Y_{AB}}{dt^2} - \mu_{AB}Y_{AB}\frac{d^2X_{AB}}{dt^2} \tag{8.103}$$

One can use equations (8.92) and (8.94) to show

$$\mu_{AB}\frac{d^2X_{AB}}{dt^2} = \cos\theta_{AB}\frac{\partial V(R_{AB})}{\partial R_{AB}} = -\frac{X_{AB}}{R_{AB}}\frac{\partial V(R_{AB})}{\partial R_{AB}} \tag{8.104}$$

$$\mu_{AB}\frac{d^2Y_{AB}}{dt^2} = \sin\theta_{AB}\frac{\partial V(R_{AB})}{\partial R_{AB}} = -\frac{Y_{AB}}{R_{AB}}\frac{\partial V(R_{AB})}{\partial R_{AB}} \tag{8.105}$$

Substituting equations (8.104) and (8.105) into equation (8.103) yields

$$\frac{dL_{AB}}{dt} = X_{AB}\left(-\frac{Y_{AB}}{R_{AB}}\frac{\partial V(R_{AB})}{\partial R_{AB}}\right) - Y_{AB}\left(-\frac{X_{AB}}{R_{AB}}\frac{\partial V(R_{AB})}{\partial R_{AB}}\right) \tag{8.106}$$

Note that all of the terms on the right of equation (8.106) cancel. Therefore

$$\frac{dL_{AB}}{dt} = 0 \tag{8.107}$$

Equation (8.106) is a key result because it shows that angular momentum is a conserved quantity. Next, we will show that the energy of the system, E_{AB}, is a conserved quantity where the energy is given by

$$E_{AB} = \tfrac{1}{2}\mu_{AB}v_{A\to B}^2 + V(R_{AB}) \tag{8.108}$$

where $v_{A\to B}$ is the instantaneous velocity of A toward B and $V(R_{AB})$ is the A–B potential. Note

$$(v_{A\to BC})^2 = \left(\frac{dX_{AB}}{dt}\right)^2 + \left(\frac{dY_{AB}}{dt}\right)^2 \tag{8.109}$$

Combining equations (8.108) and (8.109) yields

$$E_{AB} = \frac{1}{2}\mu_{AB}\left[\left(\frac{dX_{AB}}{dt}\right)^2 + \left(\frac{dY_{AB}}{dt}\right)^2\right] + V(R_{AB}) \tag{8.110}$$

Now consider the quantity $\dfrac{dE_{AB}}{dt}$. Differentiating (8.110), we obtain

$$\frac{dE_{AB}}{dt} = 2\mu_{AB}\left[\frac{dX_{AB}}{dt}\left(\frac{d^2X_{AB}}{dt^2}\right) + \frac{dY_{AB}}{dt}\left(\frac{d^2Y_{AB}}{dt^2}\right)\right] + \frac{dv(R_{AB})}{dR_{AB}}\frac{dR_{AB}}{dt} \tag{8.111}$$

Note

$$R_{AB} = \sqrt{X_{AB}^2 + Y_{AB}^2} \tag{8.112}$$

Therefore

$$\frac{dR_{AB}}{dt} = \frac{1}{\sqrt{X_{AB}^2 + Y_{AB}^2}}\left(X_{AB}\frac{dX_{AB}}{dt} + Y_{AB}\frac{dY_{AB}}{dt}\right) \tag{8.113}$$

Substituting equation (8.112) into the first term in (8.113) and rearranging yields

$$\frac{dR_{AB}}{dt} = \left[\frac{X_{AB}}{R_{AB}}\left(\frac{dX_{AB}}{dt}\right) + \frac{Y_{AB}}{R_{AB}}\left(\frac{dY_{AB}}{dt}\right)\right] \tag{8.114}$$

Substituting (8.114) into (8.111) yields

$$\frac{dE_{AB}}{dt} = \mu_{AB}\left[\left(\frac{dX_{AB}}{dt}\right)\left(\frac{d^2X_{AB}}{dt^2}\right) + \left(\frac{dY_{AB}}{dt}\right)\left(\frac{d^2Y_{AB}}{dt^2}\right)\right]$$
$$+ \frac{dV(R_{AB})}{dR_{AB}}\left[\frac{X_{AB}}{R_{AB}}\left(\frac{dX_{AB}}{dt}\right) + \frac{Y_{AB}}{R_{AB}}\left(\frac{dY_{AB}}{dt}\right)\right] \tag{8.115}$$

Rearranging equation (8.115) yields

$$\frac{dE_{AB}}{dt} = \left(\frac{dX_{AB}}{dt}\right)\left[\mu_{AB}\left(\frac{d^2X_{AB}}{dt^2}\right) + \frac{X_{AB}}{R_{AB}}\frac{dV(R_{AB})}{dR_{AB}}\right]$$
$$+ \left(\frac{dY_{AB}}{dt}\right)\left[\mu_{AB}\left(\frac{d^2Y_{AB}}{dt^2}\right) + \frac{Y_{AB}}{R_{AB}}\frac{dV(R_{AB})}{dR_{AB}}\right] \tag{8.116}$$

But according to equations (8.104) and (8.105)

$$\mu_{AB}\frac{dX_{AB}}{dt} + \frac{X_{AB}}{R_{AB}}\frac{dV(R_{AB})}{dR_{AB}} = 0 \tag{8.117}$$

$$\mu_{AB}\frac{dY_{AB}}{dt} + \frac{Y_{AB}}{R_{AB}}\frac{dV(R_{AB})}{dR_{AB}} = 0 \tag{8.118}$$

Therefore

$$\frac{dE_{AB}}{dt} = 0 \tag{8.119}$$

E_{AB} is a conserved quantity!

Next, we want to use conservation of energy to derive an equation for the angular momentum barrier for reaction. Equation (8.108) can be rewritten as

$$E_{AB} = \tfrac{1}{2}\mu_{AB}(v_{\parallel}^2 + v_{\perp}^2) + V(r_{AB}) \tag{8.120}$$

where $v_{\parallel}$ and $v_{\perp}$ are the two components of the molecule's velocity as defined in Figure 8.27. Next, it is useful to derive an expression for $v_{\perp}$ during the A–B collision. $v_{\perp}$ provides angular momentum to the A–B pair, so let's start by calculating the angular momentum of the system. Consider the situation long before a collision occurs. Atom A is moving toward atom B with a velocity, v_0. Combining equations (8.95) and (8.98) shows

$$L_{AB}^0 = \mu_{AB}R_{AB}^2\left(\frac{d\theta}{dt}\right) = X_{AB}\left(\frac{dY_{AB}}{dt}\right) - Y_{AB}\left(\frac{dX_{AB}}{dt}\right) \tag{8.121}$$

where L_{AB}^0 is the initial value of L_{AB}. Initially

$$Y_{AB} = b, \qquad \frac{dY_{AB}}{dt} = 0, \qquad \frac{dX_{AB}}{dt} = v_0 \tag{8.122}$$

where v_0 is the velocity of A toward B at the beginning of the reaction [i.e., when R_{AB} is large so that $V(R_{AB}) = 0$]. Substituting equation (8.122) into equation (8.121) shows

$$L_{AB}^0 = -b\mu_{AB}v_0 \tag{8.123}$$

Note that angular momentum is a conserved quantity. Therefore, L_{AB} is the value of the angular momentum at any point during the A–B collision must equal the initial value:

$$L_{AB} = L_{AB}^0 - b\mu_{AB}v_0 \tag{8.124}$$

For future reference, it is useful to note that when R_{AB} is large so that $V(R_{AB})$ is zero, equation (8.120) becomes

$$E_{AB} = \tfrac{1}{2}\mu_{AB}v_0^2 \tag{8.125}$$

Next, it is useful to derive an expression for $v_{\perp}$ in terms of the angular momentum for the collision.

$$v_{\perp} = R_{AB}\frac{d\theta_{AB}}{dt} \tag{8.126}$$

Substituting equation (8.96) into equation (8.126) yields

$$v_\perp = \frac{L_{AB}}{R_{AB}\mu} \tag{8.127}$$

Substituting equations (8.124) and (8.127) into equation (8.120) shows

$$E = \frac{1}{2}\mu_{AB}\left(\frac{dR_{AB}}{dt}\right)^2 + \frac{L_{AB}^2}{2\mu_{AB}R_{AB}^2} + V(R_{AB}) \tag{8.128}$$

Substituting equations (8.123) and (8.125) into equation (8.128) yields

$$E = \frac{1}{2}\mu_{AB}\left(\frac{dR_{AB}}{dt}\right)^2 + \left[\frac{E_{AB}b^2}{R_{AB}^2} + V(R_{AB})\right] \tag{8.129}$$

It is useful to define an effective potential, V_{eff}, by

$$V_{eff}(R_{AB}) = \frac{E_{AB}b^2}{R_{AB}^2} + V(R_{AB}) \tag{8.130}$$

Substituting (8.130) into equation (8.129) and rearranging yields

$$\frac{dR_{AB}}{dt} = \pm\sqrt{\frac{2}{\mu_{AB}}(E - V_{eff}(R_{AB}))} \tag{8.131}$$

Equations 8.130 and 8.131 are the key results cited earlier in the text.

8.16.2 Mass-Weighted Coordinates and the Affine Transformation

In the main body of this chapter, we found that we could represent the behavior of the system as a trajectory on a potential energy surface. If the reactants are confined to a plane, the general equations of motion for the reaction $A + BC \Rightarrow AB + C$ are

$$\frac{d^2R_A}{dt^2} = \frac{1}{m_A}F_A \tag{8.132}$$

$$\frac{d^2R_B}{dt^2} = \frac{1}{m_B}F_B \tag{8.133}$$

$$\frac{d^2R_C}{dt^2} = \frac{1}{m}F_C \tag{8.134}$$

where

$$F_A = -\frac{\partial V(R_{AB}, R_{BC})}{\partial R_A} = +\frac{\partial V(R_{AB}, R_{BC})}{\partial R_{AB}} \tag{8.135}$$

$$F_B = -\frac{\partial V(R_{AB}, R_{BC})}{\partial R_B} = +\frac{\partial V(R_{AB}, R_{BC})}{\partial R_{BC}} - \frac{\partial V(R_{AB}, R_{BC})}{\partial R_{AB}} \tag{8.136}$$

$$F_C = -\frac{\partial V(R_{AB}, R_{BC})}{\partial R_C} = -\frac{\partial V(R_{AB}, R_{BC})}{\partial R_{BC}} \tag{8.137}$$

where we have noted that

$$R_{AB} = R_B - R_A \tag{8.138}$$

$$R_{BC} = R_C - R_B \tag{8.139}$$

If you want to follow the trajectory, you need to see how R_{AB} and R_{BC} evolve in time. Combining equations (8.132)–(8.134) yields

$$\frac{d^2 R_{AB}}{dt^2} = \frac{d^2 R_B}{dt^2} - \frac{d^2 R_A}{dt^2} = \frac{1}{m_B} F_B - \frac{1}{m_A} F_A \tag{8.140}$$

$$\frac{d^2 R_{BC}}{dt^2} = \frac{d^2 R_C}{dt^2} - \frac{d^2 R_B}{dt^2} = \frac{1}{m_C} F_C - \frac{1}{m_B} F_B \tag{8.141}$$

Substituting equations (8.135)–(8.137) into equations (8.140) and (8.141) yields

$$\frac{d^2 R_{AB}}{dt^2} = -\left(\frac{1}{m_A} + \frac{1}{m_B}\right) \frac{\partial V(R_{AB}, R_{BC})}{\partial R_{AB}} + \frac{1}{m_B} \frac{\partial V(R_{AB}, R_{BC})}{\partial R_{BC}} \tag{8.142}$$

$$\frac{d^2 R_{BC}}{dt^2} = -\left(\frac{1}{m_B} + \frac{1}{m_C}\right) \frac{\partial V(R_{AB}, R_{BC})}{\partial R_{BC}} + \frac{1}{m_B} \frac{\partial V(R_{AB}, R_{BC})}{\partial R_{AB}} \tag{8.143}$$

By comparison, if R_{AB} and R_{BC} followed Newton's equation of motion, you would expect the R_{AB} to follow

$$\frac{dR_{AB}}{dt^2} = -\frac{1}{\mu_1} \frac{\partial V(R_{AB}, R_{BC})}{\partial R_{AB}} \tag{8.144}$$

where μ_1 is a constant.

Therefore, there is an extra term in equations (8.142) and (8.143) that one would not expect from Newton's equation of motion. In actual practice, that is not much of a problem, since people normally integrate equations (8.142) and (8.143) numerically. However, in the literature, people sometimes change the coordinates so that the system follows Newton's equation. So-called **mass-weighted coordinates** are used to eliminate the extra terms in equations (8.142) and (8.143). The mass-weighted coordinates, R_1 and R_2, are defined by

$$R_1 = R_{AB} + \left[\frac{m_A m_C}{(m_A + m_B)(m_B + m_C)}\right]^{1/2} R_{BC} \tag{8.145}$$

$$R_2 = R_{BC} \left(\frac{m_B(m_A + m_B + m_C)}{(m_A + m_B)(m_B + m_C)}\right)^{1/2} \tag{8.146}$$

Equations (8.145) and (8.146) are called the *affine transformation*. After considerable algebra, one can show that R_1 and R_2 obey

$$\mu_1 \frac{d^2 R_1}{dt^2} = -\frac{\partial V(R_{AB}, R_{BC})}{\partial R_1} \tag{8.147}$$

$$\mu_1 \frac{d^2 R_2}{dt^2} = -\frac{\partial V(R_{AB}, R_{BC})}{\partial R_2} \tag{8.148}$$

with

$$\frac{1}{\mu_1} = \frac{1}{m_A} + \frac{1}{m_B + m_C} \tag{8.149}$$

Figure 8.32 shows a diagram of the potential energy surface for the reaction $F + H_2 \rightarrow HF + H$ plotted in mass-weighted coordinates. One finds that the coordinate is inclined by an angle θ where

$$\sin\theta = \left[\frac{m_A m_C}{(m_A + m_B)(m_B + m_c)} \right]^{1/2} \tag{8.150}$$

The reaction $F + H_2 \rightarrow HF + H$ has an unusually large tilt because the fluorine is so much heavier than the hydrogens.

The advantage of mass-weighted coordinates is that if you think about the motion of the system as rolling a marble over the potential energy surface, then the marble will follow the correct trajectory in a mass-weighted coordinate for an impact parameter of zero. One still has an effective potential when the impact parameter is nonzero.

8.16.3 MD/MC Calculations of Reaction Rate Constants

One can, in principle, combine Monte Carlo (MC) and molecular dynamics (MD) calculations to estimate rate constants for each type of reaction discussed in the last two sections. The general procedure is to start with an ensemble of reactant molecules, and assume that the internal states of the reactant molecules are in equilibrium before reaction occurs. One then uses a Monte Carlo routine similar to that discussed in Chapter 6 to choose initial conditions for the molecules, while a molecular dynamics routine is used to calculate the reactive cross section. For example, if one were examining the reaction $A + BC \rightarrow AB + C$, one would use the Monte Carlo routine to pick initial values of $v_{A \rightarrow BC}$ and E_{BC}, where $v_{A \rightarrow BC}$ and E_{BC} are as defined at the beginning of Section 8.6. The calculations are easier if one modifies the Monte Carlo routine to pick only those values of $v_{A \rightarrow BC}$ and E_{BC} that meet the constraint that the total energy of the molecules be sufficient to traverse the activation barrier for reaction. One then picks several values of the impact parameter b and integrates equations (8.29)–(8.31) to calculate $P_{A \rightarrow BC}$ (b).

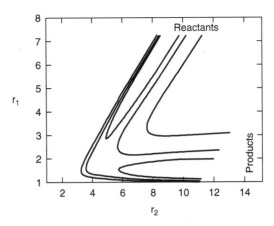

Figure 8.32 A potential energy surface for $F + H_2 \rightarrow HF + H$ in mass-weighted coordinates. After Bender, 1972.

One then plugs into equation (8.58) to calculate a reactive cross section. The process is repeated for thousands of values of $v_{A \to BC}$ and E_{BC}. One then takes an average using equation (8.59) to calculate a rate. Such calculations are easy to do provided one knows an accurate potential energy surface for the reaction. Generally, the results of the calculations agree quite well with experiment provided the potential energy surface is accurate and tunneling is negligible. As a result, such calculations are reasonably common, although not as common as you might imagine since the computations are quite time-consuming.

8.16.4 The Velocity Distribution in Center at Mass Coordinates

Last, we want to use the results in Section 8.16.1 to derive an expression for the $\bar{v}_{A \to BC}$, the average velocity of A toward BC.

According to the analysis in Section 6.4.1 and Example 6.B, the average velocity of any molecule is

$$\bar{v}_{A \to BC} = \frac{\iiint v_{A \to BC} \exp(-E/k_B T) \, d\vec{v}_{A \to BC}}{\iiint \exp(-E/k_B T) \, d\vec{v}_{A \to BC}} \tag{8.151}$$

According to equation (8.107), during a collision of A with BC, the total energy is given by

$$E = \tfrac{1}{2}\mu_{AB}(v_{A \to BC})^2 + V(R_{ABC}) \tag{8.152}$$

We will consider only the case where R_{ABC} is so large that $V(R_{ABC})$ is negligible. Equation (8.152) then becomes

$$E = \tfrac{1}{2}\mu_{ABC}(v_{A \to BC})^2 \tag{8.153}$$

Substituting equation (8.153) into equation (8.154) yields

$$v_{A \to BC} = \frac{\iiint v_{A \to BC} \exp(-\tfrac{1}{2}\mu_{ABC}(v_{A \to BC})^2) \, d\vec{v}_{A \to BC}}{\iiint \exp(-\tfrac{1}{2}\mu_{ABC}(v_{A \to BC})^2) \, d\vec{v}_{A \to BC}} \tag{8.154}$$

Equation (8.154) is identical to equation (6.106) except that the mass is μ_{AB}. We do the integral in Example 6.B. The results are

$$v_{A \to BC} = \left(\frac{8k_B T}{\pi \mu_{AB}}\right)^{1/2} \tag{8.155}$$

8.17 SOLVED EXAMPLES

Example 8.A   Properties of Potential Energy Surfaces   Assume that the potential energy surface for the reaction $D + H_2 \to HD + H$ is given by

$$V(r_1, r_2) = W(e^{-2a(r_1 - r_0)}) - 2e^{-a(r_1 - r_0)}$$

$$+ W(e^{-2a(r_2 - r_0)} - 2e^{-a(r_2 - r_0)}) + \tag{8.A.1}$$

$$+ V_P(e^{-a(r_1 + r_2 - 2r_0)}) + W$$

where $V(r_1, r_2)$ is the potential energy surface, r_1 is the H–H distance, r_2 is the H–D distance, $W = 109.4$ kcal/mol, $a = 1.95$ Å^{-1}, $r_0 = 0.74$ Å, and $V_P = 258$ kcal/mol.

(a) Use a spreadsheet to calculate values for the potential energy surface for r_1 and r_2 varying between 0.5 and 2 Å.

(b) Trace the minimum-energy pathway from reactants to products.

Solution   I solved this in Excel. A piece of my spreadsheet is shown in Tables 8.A.1 and 8.A.2. The original spreadsheet is available at http://www.wiley.com/chemicalkinetics.

First I defined my own function V using the macromodule. The macromodule allows you to define your own functions in Excel. You use it by first inserting a module worksheet in Excel, using the Insert-Macro-Module on the Excel menu. You then type in a program in a language called *Visual BASIC*.

Table 8.A.1 shows my module. The first line defined the function V as public (available to the rest of the system) and of type variant (general). I could have also declared it as single (single-precision real number) or double (double-precision real number) as described in Chapter 2. I then defined the function and included an end statement.

I then used the function in my spreadsheet. Table 8.A.2 shows the spreadsheet. I listed r_1 values in cells A4, A5, A6,... and r_2 values in cells B3, C3, D3,... . I named cell B1 = w, cell B2 = vp, cell D1 = a, and cell D2 = r0.

I then calculated a matrix of $V(r_1, r_2)$ values, so cell B4 contains $V(r_1, r_2)$ with $r_1 = b3$ and $r_2 = $a4$, cell B5 contains $V(r_1, r_2)$ with $r_1 = B3$ and $r_2 = $A5$, and cell C4 contains $V(r_1, r_2)$ with $r_1 = C3$ and $r_2 = $A4$.

Table 8.A.1   The module used to declare the function V in Excel

```
Public Function v(r1, r2, r0, a, w, vp) As Variant
v = w*(Exp(-2*a*(r1 - r0)) - 2*Exp(-a*(r1 - r0)))
v = v + w*(Exp(-2*a*(r2 - r0)) - 2*Exp(-a*(r2 - r0)))
v = v + vp*Exp(-a*(r1 + r2 - 2*r0))
v = v + w
End Function
```

Table 8.A.2   A part of the spreadsheet showing the formulas used to calculate the potential

	A	B	C	D
1	w=	109.4	a=	1.954
2	vp=	258	r0=	0.74
3	r2\r1	0.5	0.6	0.7
4	0.5	=v(B$3,$A4, r0,a,w,vp)	=v(C$3,$A4, r0,a,w,vp)	=v(D$3,$A4, r0,a,w,vp)
5	0.6	=v(B$3,$A5, r0,a,w,vp)	=v(C$3,$A5, r0,a,w,vp)	=v(D$3,$A5, r0,a,w,vp)
6	0.7	=v(B$3,$A6, r0,a,w,vp)	=v(C$3,$A6, r0,a,w,vp)	=v(D$3,$A6, r0,a,w,vp)
7	0.8	=v(B$3,$A7, r0,a,w,vp)	=v(C$3,$A7, r0,a,w,vp)	=v(D$3,$A7, r0,a,w,vp)

Table 8.A.3 Alternate spreadsheet to calculate the potential

	A	B	C	D	E
1	w=	109.4	a=	1.954	
2	Vp=	258	r0=	0.74	
3	r2\r1	0.5	0.6	0.7	0.8
4	0.5	=w*(EXP(-2*a*(B$3-r0))-2*EXP(-a*(B$3-r0)))+w*(EXP(-2*a*($A4-r0))-2*EXP(-a*($A4-r0)))+vp*EXP(-a*(B$3+$A4-2*r0))+w	=w*(EXP(-2*a*(C$3-r0))-2*EXP(-a*(C$3-r0)))+w*(EXP(-2*a*($A4-r0))-2*EXP(-a*($A4-r0)))+vp*EXP(-a*(C$3+$A4-2*r0))+w	=w*(EXP(-2*a*(D$3-r0))-2*EXP(-a*(D$3-r0)))+w*(EXP(-2*a*($A4-r0))-2*EXP(-a*($A4-r0)))+vp*EXP(-a*(D$3+$A4-2*r0))+w	=w*(EXP(-2*a*(E$3-r0))-2*EXP(-a*(E$3-r0)))+w*(EXP(-2*a*($A4-r0))-2*EXP(-a*($A4-r0)))+vp*EXP(-a*(E$3+$A4-2*r0))+w
5	0.6	=w*(EXP(-2*a*(B$3-r0))-2*EXP(-a*(B$3-r0)))+w*(EXP(-2*a*($A5-r0))-2*EXP(-a*($A5-r0)))+vp*EXP(-a*(B$3+$A5-2*r0))+w	=w*(EXP(-2*a*(C$3-r0))-2*EXP(-a*(C$3-r0)))+w*(EXP(-2*a*($A5-r0))-2*EXP(-a*($A5-r0)))+vp*EXP(-a*(C$3+$A5-2*r0))+w	=w*(EXP(-2*a*(D$3-r0))-2*EXP(-a*(D$3-r0)))+w*(EXP(-2*a*($A5-r0))-2*EXP(-a*($A5-r0)))+vp*EXP(-a*(D$3+$A5-2*r0))+w	=w*(EXP(-2*a*(E$3-r0))-2*EXP(-a*(E$3-r0)))+w*(EXP(-2*a*($A5-r0))-2*EXP(-a*($A5-r0)))+vp*EXP(-a*(E$3+$A5-2*r0))+w
6	0.7	=w*(EXP(-2*a*(B$3-r0))-2*EXP(-a*(B$3-r0)))+w*(EXP(-2*a*($A6-r0))-2*EXP(-a*($A6-r0)))+vp*EXP(-a*(B$3+$A6-2*r0))+w	=w*(EXP(-2*a*(C$3-r0))-2*EXP(-a*(C$3-r0)))+w*(EXP(-2*a*($A6-r0))-2*EXP(-a*($A6-r0)))+vp*EXP(-a*(C$3+$A6-2*r0))+w	=w*(EXP(-2*a*(D$3-r0))-2*EXP(-a*(D$3-r0)))+w*(EXP(-2*a*($A6-r0))-2*EXP(-a*($A6-r0)))+vp*EXP(-a*(D$3+$A6-2*r0))+w	=w*(EXP(-2*a*(E$3-r0))-2*EXP(-a*(E$3-r0)))+w*(EXP(-2*a*($A6-r0))-2*EXP(-a*($A6-r0)))+vp*EXP(-a*(E$3+$A6-2*r0))+w
7	0.8	=w*(EXP(-2*a*(B$3-r0))-2*EXP(-a*(B$3-r0)))+w*(EXP(-2*a*($A7-r0))-2*EXP(-a*($A7-r0)))+vp*EXP(-a*(B$3+$A7-2*r0))+w	=w*(EXP(-2*a*(C$3-r0))-2*EXP(-a*(C$3-r0)))+w*(EXP(-2*a*($A7-r0))-2*EXP(-a*($A7-r0)))+vp*EXP(-a*(C$3+$A7-2*r0))+w	=w*(EXP(-2*a*(D$3-r0))-2*EXP(-a*(D$3-r0)))+w*(EXP(-2*a*($A7-r0))-2*EXP(-a*($A7-r0)))+vp*EXP(-a*(D$3+$A7-2*r0))+w	=w*(EXP(-2*a*(E$3-r0))-2*EXP(-a*(E$3-r0)))+w*(EXP(-2*a*($A7-r0))-2*EXP(-a*($A7-r0)))+vp*EXP(-a*(E$3+$A7-2*r0))+w

The word *macromodule* does not work with all spreadsheets, so I have also included a spreadsheet with all of the formulas inserted directly. It is Table 8.A.3.

The numerical values in the spreadsheet are presented in Table 8.A.4. The reactants start out with r_2 large, so row 19 is close to the reactants. If we read across row 19, we find that the minimum comes in column E, so cell E19 is close to the reactants.

Similarly, the products have r_1 large, so column Q is close to the products. If we read down column Q, we find that the minimum comes in row 7, so cell Q7 is close to the products.

Table 8.A.4 Numerical values in the spreadsheet above

	A	B	C	D	E	F	G	H	I	J	K	L	M	N	O	P	Q	
1	w=	109	a=	1.95														
2	vp=	258	r0=	0.74														
3	r2\r1	0.5	0.6	0.7	0.8	0.9	1.0	1.1	1.2	1.3	1.4	1.5	1.6	1.7	1.8	1.9	2.0	
4	0.5	628.0	482.7	376.4	297.9	239.3	195.2	161.8	136.1	116.2	100.8	88.6	79.0	71.4	65.3	60.4	56.5	P
5	0.6	482.7	358.2	268.9	204.4	157.4	122.9	97.2	78.0	63.4	52.3	43.7	37.1	31.8	27.7	24.5	21.9	R
6	0.7	376.4	268.9	193.7	140.8	103.3	76.5	57.3	43.3	33.1	25.6	20.0	15.7	12.5	10.0	8.1	6.6	O
7	0.8	297.9	204.4	140.8	97.4	67.7	47.4	33.4	23.8	17.2	12.6	9.4	7.1	5.5	4.4	3.6	3.0	D
8	0.9	239.3	157.4	103.3	67.7	44.4	29.4	19.8	13.8	10.1	7.9	6.7	6.1	5.8	5.8	5.9	6.1	U
9	1	195.2	122.9	76.5	47.4	29.4	18.7	12.7	9.6	8.3	8.1	8.6	9.3	10.2	11.1	11.9	12.7	C
10	1.1	161.8	97.2	57.3	33.4	19.8	12.7	9.6	8.9	9.7	11.1	12.9	14.7	16.5	18.2	19.6	20.9	T
11	1.2	136.1	78.0	43.3	23.8	13.8	9.6	8.9	10.3	12.6	15.4	18.3	21.1	23.6	25.9	27.8	29.6	S
12	1.3	116.2	63.4	33.1	17.2	10.1	8.3	9.7	12.6	16.3	20.2	24.0	27.5	30.7	33.5	35.9	37.9	
13	1.4	100.8	52.3	25.6	12.6	7.9	8.1	11.1	15.4	20.2	25.1	29.6	33.7	37.4	40.6	43.3	45.6	
14	1.5	88.6	43.7	20.0	9.4	6.7	8.6	12.9	18.3	24.0	29.6	34.7	39.4	43.5	47.0	50.0	52.6	
15	1.6	79.0	37.1	15.7	7.1	6.1	9.3	14.7	21.1	27.5	33.7	39.4	44.4	48.8	52.7	55.9	58.7	
16	1.7	71.4	31.8	12.5	5.5	5.8	10.2	16.5	23.6	30.7	37.4	43.5	48.8	53.5	57.6	61.0	64.0	
16	1.8	65.3	27.7	10.0	4.4	5.8	11.1	18.2	25.9	33.5	40.6	47.0	52.7	57.6	61.8	65.4	68.5	
18	1.9	60.4	24.5	8.1	3.6	5.9	11.9	19.6	27.8	35.9	43.3	50.0	55.9	61.0	65.4	69.2	72.3	
19	2	56.5	21.9	6.6	3.0	6.1	12.7	20.9	29.6	37.9	45.6	52.6	58.7	64.0	68.5	72.3	75.6	
					reactants													

Next, we need to trace a path from reactants to products. We can take any path we want. However, the lowest energy path goes up then down.

The transition state lies between cells I10 and H11. Notice that the transition state is sort of in a local minimum, because most of the surrounding cells have a higher energy. However, the energy decreases along the shaded I10–H11 diagonal, so the transition state is really a saddle point.

We ask the reader to continue these results in the homework set.

Example 8.B   Numerical Integration of Newton's Equations of Motion   The objective of this example is to use the algorithms from Example 4.B (numerical integration) to numerically integrate Newton's equations of motion.

Integrate the equations of motion for the hypothetical reaction $A + B \Rightarrow AB$.

Solution   Consider two atoms, A and B, that satisfy

$$m_A \frac{d^2 R_A}{dt} = F_{AB} \tag{8.B.1}$$

$$m_B \frac{d^2 R_B}{dt} = -F_{AB} \tag{8.B.2}$$

where R_A and R_B are the positions of A and B, m_A and m_B are the masses of A and B, and F_{AB} is the force on A due to the presence of B. The negative sign in equation (8.B.2) arises because of Newton's second law (where any force is met by an equal and opposite force).

R_A, R_B, and F_{AB} in equations (8.B.1) and (8.B.2) are vectors. However, it is useful to convert them to scalar quantities. Let's define X_A, Y_A, and Z_A, and X_B, Y_B, and Z_B to be the X, Y, and Z coordinates of A and B; v_{XA}, v_{YA}, v_{ZA}, v_{XB}, v_{YB}, and v_{ZB} to be the X, Y, and Z components of the velocity of A and B; and F_X, F_Y, and F_Z to be the X, Y, and Z components of F_{AB}. Then equations (8.B.1) and (8.B.2) become

$$\frac{dX_A}{dt} = v_{XA} \tag{8.B.3}$$

$$\frac{dY_A}{dt} = v_{YA} \tag{8.B.4}$$

$$\frac{dZ_A}{dt} = v_{ZA} \tag{8.B.5}$$

$$m_A \frac{dv_{XA}}{dt} = F_X \tag{8.B.6}$$

$$m_A \frac{dv_{YA}}{dt} = F_Y \tag{8.B.7}$$

$$m_A \frac{dv_{ZA}}{dt} = F_Z \tag{8.B.8}$$

$$\frac{dX_B}{dt} = v_{XB} \tag{8.B.9}$$

$$\frac{dY_B}{dt} = v_{YB} \tag{8.B.10}$$

$$\frac{dZ_B}{dt} = v_{ZB} \tag{8.B.11}$$

$$m_B \frac{dv_{XB}}{dt} = -F_X \qquad (8.B.12)$$

$$m_B \frac{dv_{YB}}{dt} = -F_Y \qquad (8.B.13)$$

$$m_B \frac{dv_{ZB}}{dt} = -F_Z \qquad (8.B.14)$$

Therefore, one can calculate the motion of the two atoms by picking initial velocities and positions and numerically integrating equations (8.B.3)–(8.B.14).

We still need to get an expression for the sources. Let's define R_{AB} as the distance from A to B:

$$R_{AB} = \sqrt{(X_A - X_B)^2 + (Y_A - Y_B)^2 + (Z_B - Z_A)^2} \qquad (8.B.15)$$

According to equation (8.32):

$$F_X = -\frac{dV_{AB}}{dX_A}$$

where V_{AB} is the A–B potential.

Combining equations (8.B.14) and (8.B.15) yields

$$F_X = \frac{dV_{AB}}{dX_A} = -\frac{dR_{AB}}{dX_A} \frac{dV_{AB}}{dR_{AB}} = \frac{(X_B - X_A)}{R_{AB}} \frac{dV_{AB}}{dR_{AB}} \qquad (8.B.16)$$

Similarly

$$F_Y = \frac{(Y_B - Y_A)}{R_{AB}} \frac{dV_{AB}}{dR_{AB}} \qquad (8.B.17)$$

$$F_Z = \frac{(Z_B - Z_A)}{R_{AB}} \frac{dV_{AB}}{dR_{AB}} \qquad (8.B.18)$$

This gives us a system of equations that can be integrated numerically.

In order to do the numerical integration, one needs to convert the equations slightly. Let's define

$$W(1) = X_A, W(2) = Y_A, W(3) = Z_A, W(4) = X_B, W(5) = Y_B$$

$$W(6) = Z_B, W(7) = v_{XA}, W(8) = v_{YZ}, W(9) = v_{ZA} \qquad (8.B.19)$$

$$W(10) = v_{XB}, W(11) = v_{XB}, W(12) = v_{ZB}$$

The equations of motion become

$$\frac{dW(i)}{dt} = W(i+6) \qquad \text{for } i = 1\text{–}6 \qquad (8.B.20)$$

$$\frac{dW(i)}{dt} = \frac{W(i-3) - W(i-6)}{R_{AB} m_A} \frac{dV_{AB}}{dR_{AB}} \qquad \text{for } i = 7\text{–}9 \qquad (8.B.21)$$

$$\frac{dW(i)}{dt} = \frac{W(i-9) - W(i-6)}{R_{AB} m_A} \frac{dV_{AB}}{dR_{AB}} \qquad \text{for } i = 10\text{–}12 \qquad (8.B.22)$$

In order to do a numerical integration, you need to also know the Jacobian:

$$dF\,dW(i, j) = \frac{d}{dW(j)} \left(\frac{dWi}{dt} \right) \qquad (8.B.23)$$

The following computer program does an integration. Shaded lines in the programs are continuation lines.

```
PROGRAM integratemd
C  Solve diffequ for MD problem
   INTEGER KMAXX,NMAX,NVARS
   PARAMETER (KMAXX=200,NMAX=50,NVARS=12)
   INTEGER kmax,kount,nbad,nok,I
   REAL dxsav,eps,hstart,time1,time2,w(NVARS)
    COMMON /path/kmax,kount,dxsav,xp(KMAXX),yp(NMAX,KMAXX)
     REAL ma,mb
     REAL r0, well, range,JAB, rab,QAB
     common /potent/r0, well, range, JAB, rAB, QAB

     COMMON /masses/ma,mb
   EXTERNAL stiff,derivs,rkqs
     ma=1.
     mb=35.
     kmax=0
     hstart=1.e-3
     eps=1.e-6
     r0=2.
     well=5.
     range=1
c
time2=0.
c
   w(1)=-10
   w(2)=0.
   w(3)=0.
      w(4)=0.
      w(5)=0
      w(6)=0.
      w(7)=1.0
      w(8)=0.
      w(9)=0.
      w(10)=0.
      w(11)=0.
      w(12)=0.

c  next open data file and write out initial
conditions
      open(unit=4,file= traject.txt ,
      action= write ,status= replace )

      write(4,100)vab,time2, (w(i),i=1,12)
      write(6,100)vab,time2,w(1),w(7)

c next start solving the equation
```

```
      do while ((time2.lt. (20.))
         1 .and. (w(1).gt.(-10.1))
         2 .and. (w(1).lt.(10.1)))
         time1=time2
         time2=time1+0.1
c
c odeint integrates the rate equation from
ctime1 to time2
c
         call odeint(w,NVARS,time1,time2,
         eps,hstart,0,nok,nbad,derivs,
             rkqs)
100      format(1x,f5.2,12(   ,2x,f7.3))
      write(4,100)time2, (w(i),i=1,12)
      write(6,100)time2,w(1),w(7)
10       end do
      write(*, (/1x,a,t30,i3) )
 Successful steps: ,nok
      write(*, (1x,a,t30,i3) )  Bad steps: ,nbad
999   write(*,*)  NORMAL COMPLETION
   STOP
   END
```

```
SUBROUTINE derivs(time,w,f)
c
c  subroutine to calculate the
C  derivatives
c
   REAL time,w(12),f(12)
      REAL ma,mb
      integer I
      REAL r0, well, range,JAB, rab,QAB
      common /potent/r0, well, range, JAB,
      rAB,QAB
      COMMON /masses/ma,mb
C Assume a morse potential, i.e.
C    VAB= well*(exp(-2.*(rAB-r0)/range)
C    1 -2*exp(-(rAB-r0)/range))

      rAB=sqrt((w(1)-w(4))**2+

      (w(2)-w(5))**2+

      (w(3)-w(6))**2)

      QAB= 2*well/range*(exp(-(rAB-
r0)/range)
         -exp(-2*(rAB-r0)/range))/rab
      do 10 i=1,6
      f(i)=w(i+6)
10    enddo
   do 20 i=7,9
      f(i)= QAB*(w(i-3)-w(i-6))/ma
```

```
20    enddo
      do 30 i=10,12
      f(i)= QAB*(w(i-9)-w(i-6))/mb
30    continue
    return
    END

SUBROUTINE jacobn(time,w,dfdt,dfdw,n,nmax)
   INTEGER n,nmax,I
   REAL time,w(*),dfdt(*),dfdw(nmax,nmax)
    REAL r0, well, range,JAB, rab,QAB
common /potent/r0, well, range, JAB,
rAB,QAB
      do 15 i=1,n
    dfdt(i)=0.
      do 11 j=1,n
      dfdw(i,j)=0.0
11 enddo
15    enddo
rAB=( (w(1)-w(4))**2 + (w(2)-w(5))**2 +
      (w(3)-w(6))**2)
      rAb=sqrt(rab)
QAB= 2.*well/range*(exp(-(rAB-r0)/range)
      -exp(-2*(rAB-r0)/range))/rAB
      JAB= 2.*well/range/range*(
      -exp(-(rAB-r0)/range)
      +2.*exp(-2*(rAB-r0)/range)
    /rAB/rAB-QAB/Rab
20    do    i=1,6
      dfdw(i,i+6)=1.
20    enddo
      do 30 i=7,9
      do 29 j=1,3
      dfdw(i,j)= JAB*(w(i-3)-w(i-6))/ma
      *(w(j)-w(j+3))
      dfdw(i,j+3)= JAB*(w(i-3)-w(i-6))/ma
      *(w(j+3)-w(j))
29    enddo
      dfdw(i,i-3)=dfdw(i,i-3)+QAB/ma
      dfdw(i,i-6)=dfdw(i,i-6) -QAB/ma
30    enddo
      do 40 i=10,12
      do 39 j=1,3
      dfdw(i,j)= JAB*(w(i-9)-w(i-6))/ma
      *(w(j)-w(j+3))
      dfdw(i,j+3)= JAB*(w(i-9)-w(i-6))/ma
      *(w(j+3)-w(j))
39    enddo
      dfdw(i,i-9)=dfdw(i,i-9)+QAB/mb
      dfdw(i,i-6)=dfdw(i,i-6)-QAB/mb
40    enddo
    return
    END
```

Example 8.C Numerical Integration of Equation (8.135) The advantage of the computer program is that it is easily generalized to three dimensions. However, a simpler result can be obtained by integrating equations (8.134) and (8.135). The following program uses the IMSL Runge–Kutta routine to do that. There is some subtlety in the program since equation (8.135) has two roots, and you can get stuck when the velocity equals zero. However, the following equation takes care of those subtleties.

```fortran
PROGRAM radialinte
C  Solve diffequ for MD problem
      USE numerical_libraries
   INTEGER NVARS
   PARAMETER (NVARS=1)
   INTEGER i,ido
   REAL time1,time2,w(NVARS)
      Real parm(50)
      REAL ma,mb,mu
      Logical flag
      REAL r0, well, range,E,b
      common /potent/r0, well,range,E,flag,b
      COMMON /masses/ma,mb,mu
   EXTERNAL derivs
      ma=1.
      mb=35.
      mu=1./(1/ma+1/mb)
      parm(1)=1.e-3
      eps=1.e-6
      r0=2.
      well=5.
      range=1
      IDO=1
      flag=.false.
      b=0.
c
c The k's re the rate constants
   time2=0.
   w(1)=10
   E=0.5*mu*(1.**2)
c next open data file and write out initial conditions
open(unit=4,file='traject.txt',action='write'
,status='replace')
   write(4,100)time2, w(1)
   write(6,100)time2,w(1)
c next start solving the equation
   do while ((time2.lt. (20.))
      1 .and. (w(1).gt.(-10.1))
      2 .and. (w(1).lt.(10.1)))
      time1=time2
      time2=time1+0.1
   c
c IVPRK integrates the rate equation
   c
         CALL IVPRK (IDO, NVARS, derivs,
time1,time2,EPS,PARAM,W)
```

```
100     format(1x,f5.2,12(    , 2x,f7.3))
   write(4,100)time2, w(1)
   write(6,100)time2,w(1)
10         end do
999 write(*,*)    NORMAL COMPLETION
   STOP
   END
     subroutine derivs (N,t,w,wprime)
     integer N
     logical flag
     real t,w(*),wprime(*),Veff,VAB,rAB
     REAL ma,mb,mu
     REAL r0, well, range,E,b,Eeff
     common /potent/r0,well,range,E,flag,b
     COMMON /masses/ma,mb,mu
     rAB=w(1)
     VAB= well*(
     exp(-2.*(rAB-r0)/range)
     -2*exp(-(rAB-r0)/range))
     Veff=VAB+E*b*b/rAB/rAB
     Eeff= E-Veff
     if(Eeff.gt.0)then
     wprime(1)= -sqrt(2*mu*Eeff)
     if(flag)wprime(1) =-wprime(1)
     else
     wprime(1)= 10.
     Endif
     if(abs(wprime(1)).lt.0.01) flag=.true.
   return
   end
```

Example 8.D   Calculating the Rate Constant Using Equation (8.60)

Figure 8.17 shows some data for the cross section for the reaction $H + H_2 \rightarrow H_2 + H$ as a function of E_T, the translational energy of H approaching H_2. The energy is measured in center-of-mass coordinates as described in Section 9.9.

Assume that the cross section follows:

$$\sigma_{A \to BC}^r = \begin{cases} 0 & \text{For} \quad E_T \leq 0.35 \text{ eV} \\ 12 \text{ Å}^2 \left(\dfrac{1 - e^{0.35 \text{ eV} - E_T}}{5.08 \text{ eV}} \right) \times e^{-E_T/2 \text{ eV}} & \text{For} \quad E_T \geq 0.35 \text{ eV} \end{cases} \qquad (8.D.1)$$

where E_T is the translational energy. Calculate the rate constant for the reaction at 300 K.

Solution

According to equation (8.60), if there is no E_{BC} dependence, then

$$k_{A \to BC} = \int_0^\infty v_{A \to BC} \sigma_{A \to B}^r D(v_{A \to BC}) \, dv_{A \to BC} \qquad (8.D.2)$$

According to results in Example 6.D:

$$D(v_{A \to BC}) = \frac{(v_{A \to BC})^2 \exp\left(-\frac{\mu_{ABC}}{2k_BT}(v_{A \to BC})^2\right)}{\int_0^\infty (v_{A \to BC})^2 \exp\left(-\frac{\mu_{ABC}}{2}(v_{A \to BC})^2\right)} \tag{8.D.3}$$

where μ_{ABC} is the reduced mass of ABC, k_B is Boltzmann's constant, and T is the temperature. Looking up the integral in the CRC yields

$$D(v_{A \to BC}) = \frac{4}{\sqrt{\pi}} \left(\frac{\mu_{ABC}}{2k_BT}\right)^{3/2} (v_{A \to BC})^2 \exp\left(-\frac{\mu_{ABC}}{2k_BT}(v_{A \to BC})^2\right) \tag{8.D.4}$$

Combining equations (8.D.2) and (8.D.4) and substituting $E_T = \frac{1}{2}\mu_{ABC}(v_{A \to BC})^2$ yields

$$k_{A \to BC} = \sqrt{\frac{8k_BT}{\pi\mu_{AB}}} \int_0^\infty \left(\frac{E_T}{k_BT}\right) \sigma^r_{A \to BC} e^{-E_T/k_BT} \, d\left(\frac{E_T}{k_BT}\right) \tag{8.D.5}$$

Note

$$\sqrt{\frac{8k_BT}{\pi\mu_{AB}}} = \bar{v}_{A \to BC} \tag{8.D.6}$$

Therefore

$$k_{A \to BC} = \bar{v}_{A \to BC} \int_0^\infty \left(\frac{E_T}{k_BT}\right) \sigma^r_{A \to BC} e^{-E_T/k_BT} \, d\left(\frac{E_T}{k_BT}\right) \tag{8.D.7}$$

For future reference, it is useful to define an average cross section, $I_{A \to BC}$, by

$$I_{A \to BC} = \int_0^\infty \left(\frac{E_T}{k_BT}\right) \sigma^r_{A \to BC} e^{-E_T/k_BT} \, d\left(\frac{E_T}{k_BT}\right) \tag{8.D.8}$$

Equation (8.D.7) then becomes

$$k_{A \to BC} = \bar{v}_{A \to BC} I_{A \to BC} \tag{8.D.9}$$

Let's define a new variable, W, by

$$W = \frac{E_T - 0.35 \text{ eV}}{k_BT} \tag{8.D.10}$$

Substituting equation (8.D.5) into equation (8.D.4) yields

$$k_{A \to BC} = \bar{v}_{A \to BC} e^{-0.35 \text{ eV}/k_BT} \times \int_{-0.35 \text{ eV}/k_BT}^\infty$$

$$\times \left(\frac{Wk_BT + 0.35 \text{ eV}}{k_BT}\right) \sigma^r_{A \to BC} e^{(-W)} \, dW \tag{8.D.11}$$

For future reference, it is useful to note

$$k_B T = 0.6 \text{ kcal/mol} = 0.026 \text{ eV/molecule} \tag{8.D.12}$$

$$\frac{1}{\mu_{ABC}} = \frac{1}{2 \text{ amu/mol}} + \frac{1}{1 \text{ amu/mol}} = \frac{1.5}{\text{amu}} \tag{8.D.13}$$

According to equation (7.29)

$$\begin{aligned}
\bar{v}_{A \to BC} &= (2.52 \times 10^{13} \text{ Å/second}) \left(\frac{T}{300 \text{ K}}\right)^{1/2} \left(\frac{1 \text{ amu}}{\mu}\right)^{1/2} \\
&= (2.52 \times 10^{13} \text{ Å/second}) \left(\frac{300 \text{ K}}{300 \text{ K}}\right)^{1/2} \left(\frac{1 \text{ amu}}{1 \text{ amu}/1.5}\right)^{1/2} \qquad (8.D.14) \\
&= 3.08 \times 10^{13} \text{ Å/second}
\end{aligned}$$

Substituting equation (8.D.9) into (8.D.7) and adding the appropriate conversion factors yields

$$\begin{aligned}
k_{A \to BC} = (2.94 \times 10^{13} \text{ Å/second}) \, e^{-0.35 \text{ eV}/k_B T} \\
\times \int_{-0.35 \text{ eV}/k_B T}^{\infty} \left(\frac{W + 0.35 \text{ eV}}{k_B T}\right) \sigma_{A \to BC}^r \exp(-W) \, dW
\end{aligned} \tag{8.D.15}$$

Note: $\sigma = 0$ for $E_T < 0.35$ eV. Therefore

$$\begin{aligned}
k_{A \to BC} = \frac{2.94 \times 10^{13} \text{ Å}}{\text{molecule·second}} e^{-0.35 \text{ eV}/k_B T} \\
\times \int_0^{\infty} (W + 0.35 \text{ eV}/k_B T) \sigma_{A \to BC}^r e^{(-W)} \, dW
\end{aligned} \tag{8.D.16}$$

Let's define $I_{A \to BC}$ and $F(W)$ by

$$I_{A \to BC} = \int_0^{\infty} \left(\frac{W + 0.35 \text{ eV}}{k_B T}\right) \sigma_{A \to BC}^r e^{(-W)} \, dW \tag{8.D.17}$$

$$F(W) = \left(\frac{W + 0.35 \text{ eV}}{k_B T}\right) \sigma_{A \to BC}^r \tag{8.D.18}$$

Combining equations (8.D.15) and (8.D.16) yields

$$I_{A \to BC} = \int_0^{\infty} F(W) \, e^{(-W)} \, dW \tag{8.D.19}$$

One can conveniently integrate equation (8.D.17) using the Laguere integration formula:

$$\int_0^\infty F(W)e^{(-W)}\,dW = \sum_i B_i F(W_i) \qquad (8.D.20)$$

where the B_i and W_i values are as given in the following spreadsheet for the calculations:

	A	B	C	D	E	F
02	kbT=	=D₂*0.00198 /23.05	T= 2	300		
03					I=	=SUM (F5:F10)
04	W	Et	S	F(w)	B$_i$	term in sum
05	0.22285	=A5*kbT +0.35	=12*(1−EXP((0.35- $B5)/5.08))* EXP(−$B5/2)	=$B5*$C5 /kbt	0.458964	=D5*E5
06	1.118893	=A6*kbT +0.35	=12*(1−EXP((0.35- $B6)/5.08))* EXP(−$B6/2)	=$B6*$C6 /kbt	0.417	=D6*E6
07	2.99273	=A7*kbT +0.35	=12*(1−EXP((0.35- $B7)/5.08))* EXP(−$B7/2)	=$B7*$C7 /kbt	0.113373	=D7*E7
08	5.77514	=A8*kbT +0.35	=12*(1−EXP((0.35- $B8)/5.08))* EXP(−$B8/2)	=$B8*$C8 /kbt	0.0103991	=D8*E8
09	9.83747	=A9*kbT +0.35	=12*(1−EXP((0.35- $B9)/5.08))* EXP(−$B9/2)	=$B9*$C9 /kbt	0.000261017	=D9*E9
10	15.98287	=A10*kbT +0.35	=12*(1−EXP((0.35- $B10)/5.08))* EXP(−$B10/2)	=$B10*$C10 /kbt	8.98547e-7	=D10*E10

Here are the results:

	A	B	C	D	E	F
02	kbT=	0.02577	T=	300		
03					I=	0.747826
04	W	Et	S	F(w)	Bi	term in sum
05	0.22285	0.355743	0.011349	0.156665	0.458964	0.071904
06	1.118893	0.378834	0.056199	0.826153	0.417	0.34506
07	2.99273	0.427123	0.146036	2.420454	0.113373	0.274414
08	5.77514	0.498826	0.26998	5.225937	0.010399	0.054345
09	9.83747	0.603512	0.43199	10.11683	2.61E-04	0.002641
10	15.98287	0.76188	0.6385	18.87694	8.99E-07	1.7E-5

Therefore

$$I_{A \to BC} = 0.748 \text{ Å}^2 \tag{8.D.21}$$

$$k_{A \to BC} = 3.08 \times 10^{13} \frac{\text{Å}}{\text{second}} \times (0.748 \text{ Å}^2) e^{-0.35 \text{ eV}/k_B T} \tag{8.D.22}$$

$$k_{A \to BC} = 2.3 \times 10^{13} \frac{\text{Å}^3}{\text{molecule·second}} e^{-0.35 \text{ eV}/k_B T} \tag{8.D.23}$$

Notice that the activation barrier is about 0.35 eV (i.e., the minimum energy to get reaction) even though the reaction probability is small below 0.5 eV. It is not exactly 0.35 eV, though. In the problem set, we ask the reader to calculate the rate constant at other temperatures. Both $\bar{v}_{A \to BC}$ and $I_{A \to BC}$ are temperature-dependent. If you make an Arrhenius plot of the data, you find that the activation barrier is but close to but not exactly 0.35 eV, even though the reaction probability is negligible at E = 0.35 eV.

Example 8.E Calculating the Cross Section A program called ReactMD is available from Dr. Masel's website. Assume that you used the program to calculate the reaction probability as a function of impact parameter, and the data in Table 8.E.1 were obtained. Calculate the cross section for the reaction.

Solution The cross section is given by

$$\sigma = 2\pi \int_0^\infty P(b) b \, db \tag{8.E.1}$$

We can integrate using the trapezoid rule.
Here is a spreadsheet to do the calculations:

	A	B	C
03	b	P(b)	b P(b)
04	0	0.84	=A4*B4
05	0.2	0.83	=A5*B5
06	0.4	0.85	=A6*B6
07	0.6	0.78	=A7*B7
08	0.8	0.8	=A8*B8
09	1	0.75	=A9*B9
10	1.2	0.8	=A10*B10
11	1.4	0.83	=A11*B11
12	1.6	0.72	=A12*B12
13	1.8	0.21	=A13*B13
14	2	0.1	=A14*B14
15	2.2	0	=A15*B15
16			
17		integral=	=0.5*(C4+C15)+SUM(C5 : C14)
18		σ=	=2*PI()*C17

Table 8.E.1   Results from the calculation of reaction probability as a function of impact parameters

$b_{a \to BC}$, Å	P(b)	$b_{A \to BC}$, Å	P(b)
0	0.84	1.2	0.80
0.2	0.83	1.4	0.83
0.4	0.85	1.6	0.72
0.6	0.78	1.8	0.21
0.8	0.80	2.0	0.10
1.0	0.75	2.2	0

Here are the results:

	A	B	C
03	b	P(b)	
04	0	0.84	0
05	0.2	0.83	0.166
06	0.4	0.85	0.34
07	0.6	0.78	0.468
08	0.8	0.8	0.64
07	1	0.75	0.75
10	1.2	0.8	0.96
11	1.4	0.83	1.162
12	1.6	0.72	1.152
13	1.8	0.21	0.378
14	2	0.1	0.2
15	2.2	0	0
16			
17		integral=	6.216
18		σ=	39.05628

Example 8.F   Calculation of the Cross Section for a Hard-Sphere Collision.
Assume that you have two molecules colliding as shown in Figure 8.F.1, and assume that the reaction occurs whenever

$$\tfrac{1}{2}\mu_{AB}(v_\perp)^2 < E^{\ddagger} \tag{8.F.1}$$

Derive an expression for the cross section.

Solution   From Figure 8.F.1, we obtain

$$v_\perp = v_{AB} \sin\theta \tag{8.F.2}$$

where v_{AB} is the velocity of the molecule.

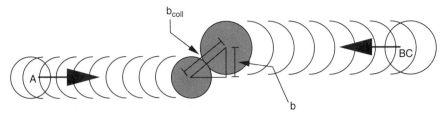

Figure 8.F.1 A hard-sphere collision.

From geometry, we have

$$\sin\theta = \sqrt{1 - \left(\frac{b}{b_{coll}}\right)^2} \tag{8.F.3}$$

Therefore

$$\frac{1}{2}\mu_{AB}(v_\perp)^2 = \frac{1}{2}\mu_{AB}(v_{AB})^2\left[1 - \left(\frac{b}{b_{coll}}\right)^2\right] = E_T\left[1 - \left(\frac{b}{b_{coll}}\right)^2\right] \tag{8.F.4}$$

where E_T is the translational energy.
   Therefore, reaction occurs whenever

$$E_T\left[1 - \left(\frac{b}{b_{coll}}\right)^2\right] > E^\ddagger \tag{8.F.5}$$

Let us define b_{coll} by

$$E_T\left[\left(1 - \left(\frac{b}{b_{coll}}\right)^2\right)\right] = E^\ddagger \tag{8.F.6}$$

or

$$(b_{crit})^2 = (b_{coll})^2\left(\frac{E_T - E^\ddagger}{E_T}\right) \tag{8.F.7}$$

By definition

$$P_{A\to BC} = \begin{cases} 1 & \text{if} \quad b \le b_{crit} \\ 0 & \text{if} \quad b > b_{crit} \end{cases} \tag{8.F.8}$$

where $P_{A\to BC}$ is the reaction probability.
   From equation (8.59)

$$\sigma_x^r = \pi\int_0^b bP_{A\to BC}(b)\,db \tag{8.F.9}$$

Combining equations (8.F.8) and (8.F.9) and integrating yields

$$\sigma_x^r = \pi(b_{crit})^2 \tag{8.F.10}$$

Combining equations (8.F.7) and (8.F.9) yields

$$\sigma_x^r = \begin{cases} \pi(b_{coll})^2 \left(\dfrac{E_T - E^{\ddagger}}{E_T} \right) & \text{for} \quad E_T > E^{\ddagger} \\ 0 & \text{elsewhere} \end{cases} \tag{8.F.11}$$

Equation (8.F.10) is called the *hard-sphere cross section*.

In the homework set, we ask the reader to show that if one plugs equation (8.F.11) into equation (8.D.7), one obtains

$$k_{A \to BC} = \pi(b_{coll})^2 e^{-E^{\ddagger}/k_B T} \tag{8.F.12}$$

8.18 SUGGESTIONS FOR FURTHER READING

A general discussion of the influence of dynamics on reactions can be found in:

A. J. Alexander and R. N. Zare, Anatomy of elementary chemical reactions, *J. Chem. Educ.* **75**, 1105 (1998).

R. D. Levine and R. B. Bernstein, *Molecular Reaction Dynamics and Chemical Reactivity*, Oxford University Press, Oxford, UK, 1987.

R. Schinke, *Photodissociation Dynamics*, Cambridge University Press, Cambridge, UK, 1993.

W. H. Miller, Quantum and semiclassical theory of chemical reaction rates, *Faraday Discuss.* **110**, 1 (1998).

G. C. Schatz, Quantum effects in gas phase bimolecular reactions, In W. Hase, ed., *Advances in Classical Trajectory Methods*, JAI Press, Stamford, 1998, p. 205.

D. M. Hirst, Potential energy surfaces and Reaction dynamics, Taylor & Francis, London (1985).

8.19 PROBLEMS

8.1   Define the following terms:

(a) Collision theory
(b) Molecular dynamics
(c) Newton's equations of motion
(d) Impact parameter
(e) Angle of approach
(f) Distribution function
(g) Trajectory
(h) Cross section
(i) Bimolecular exchange reaction
(j) Recombination reaction
(k) Barrierless reaction
(l) L-shaped potential surface
(m) Polanyi rules
(n) Collision partner
(o) Reaction collision

(p) Unreactive collision
(q) Reduced mass
(r) Reaction probability
(s) Classical equations of motion
(t) Effective potential
(u) Potential energy surface
(v) Dispersion force
(w) Electron exchange and bonding
(x) Pauli repulsion
(y) Transition state
(z) Mass-weighed coordinates
(aa) Early transition state
(bb) Middle transition state
(cc) Late transition state

8.2   Equation (8.25) has a giant integral.

    (a) Why do we need to integrate over all of these variables?

    (b) What does the distribution function do in the equation?

8.3   Explain in your own words why the potential looks the way it does in Figure 8.3. Why is there a well? Why does the potential rise again at short distances?

8.4   Explain in your own words why Figure 8.9 looks the way it does.

    (a) Why is there a well in the F–F and Cl–F potentials?

    (b) Why is there a repulsion at short F–F and Cl–F distances?

    (c) Why is there a barrier to reaction?

    (d) Identify the reactants, products, and transition state on the figure.

    (e) How does the contour plot correspond to the two-dimensional original?

8.5   The objective of this problem is to review some of the results in this chapter.

    (a) Why did we say that the top trajectory in Figure 8.13 lead to reaction while the bottom trajectory did not?

    (b) Which trajectories lead to reaction in Figure 8.15?

    (c) Why were there no reactions in Figure 8.15?

    (d) Which trajectories lead to reaction in Figure 8.16?

    (e) Why did some of the trajectories in Figure 8.16 lead to reaction while others did not?

    (f) Which trajectories lead to reaction in Figure 8.19?

    (g) What is different about the trajectories that lead to reaction in Figure 8.19 and those that do not?

    (h) Repeat (f) and (g) for Figure 8.20.

    (i) Repeat (f) and (g) for Figure 8.21.

8.6   Explain in your own words why Figure 8.18 looks the way it does.

    (a) Why is there negligible reaction when $E_T < 9$ kcal/mol? (*Note*: $E^{\ddagger} = 9.3$ kcal/mol.)

    (b) Why does the cross section rise from 10 to 40 kcal/mol?

    (c) Why does the cross section drop again at high energies?

    (d) If you work at a temperature of 500 K, what is the probability that a state at 9 kcal/mol, 12 kcal/mol, 25 kcal/mol, 50 kcal/mol, 200 kcal/mol will be occupied?

8.7   Explain in your own words why the trends summarized in Section 8.8 are observed.

8.8   Explain in your own words why the reaction probability decreases as the impact parameter increases.

8.9   Explain in your own words why unimolecular reactions are impossible in the absence of a collision partner.

8.10 Explain in your own words

    (a) How the angular momentum barrier to reaction arises? When is it important? How does it affect reaction?

(b) How the energy transfer barrier to reaction arises? When is it important? How does it affect reactions?

8.11 Explain how the energy and angular momentum barriers to reaction could affect the following reactions:

(a) $CH_3NC + X \rightarrow CH_3CN + X$ (d) $Na + Cl \rightarrow NaCl$

(b) $F + H_2 \rightarrow HF + H_2$ (e) $CH_3 + CH_3CH_3 \rightarrow CH_3CH_2CH_3 + H$

(c) $Cl + F_2 \rightarrow ClF + F$ (f) $H' + C_3H_8 \rightarrow H'C_3H_7 + H$

8.12 The reaction $CH_3CH_3 \Rightarrow H_2C{=}CH_2 + H_2$ proceeds by the following approximate mechanism:

$$CH_3CH_3 \xrightarrow{\ 1\ } 2CH_3$$

$$CH_3 + CH_3CH_3 \xrightarrow{\ 2\ } CH_4 + CH_2CH_3$$

$$CH_2CH_3 \xrightarrow{\ 3\ } H_2C{=}CH_2 + H$$

$$H + CH_3CH_3 \xrightarrow{\ 4\ } H_2 + CH_2CH_3$$

$$2CH_2CH_3 \xrightarrow{\ 5\ } CH_3CH_2CH_2CH_3$$

$$2H \xrightarrow{\ 6\ } H_2$$

$$2CH_3 \xrightarrow{\ 7\ } CH_3CH_3$$

On the basis of the results in this chapter:

(a) Which of the reactions need collision partners? If a reaction needs a collision partner, explain what the collision partner will do. If the reaction does not need a collision partner, explain why not.

(b) How do energy transfer barriers to reaction affect each of the reactions above?

(c) How do momentum transfer barriers to reaction affect each of the reactions above?

8.13 In Section 8.2.1 we noted that the reaction

$$CH_3CH_2CH_2 + O \longrightarrow CH_3CHCH_3 + OH \qquad (P8.13.1)$$

has an unusually low preexponential [1.4×10^{10} Å^3/(molecule·second)], while the reaction

$$CH_3CH_2CH_3 + O \longrightarrow CH_2CH_2CH_3 + OH \qquad (P.8.13.2)$$

has a normal preexponential [7×10^{12} Å^3/(molecule·second)].

(a) Make a diagram of the reactants. Based on the geometry alone, how much larger will the cross section for reaction (P8.13.2) be than for equation (P8.13.1)?

(b) Now put in the van der Waals radii. How much will the CH_3 groups constrain reaction (P8.13.1)? (*Hint*: Look at the fraction of the solid angle that is blocked).

(c) Are these effects large enough to explain the supposed factor of 500 difference in rate?

(d) Do you have any other ideas what could cause the preexponential or equation (P8.13.1)? (*Hint*: Qualitatively, how will rotational motion of the $CH_3CH_2CH_3$ affect the rate of both reactions?)

8.14 In Section 8.2 we noted that the reaction

$$2O_2 \longrightarrow 2O + O_2$$

has an unusually large preexponential [5.8×10^{15} Å^3/(molecule·second)].

(a) Use the preexponential to estimate the average cross section for the reaction, namely, $\sigma_{A \rightarrow BC}$ in equation (8.10).

(b) How can the cross section be that big? Would the attractive forces between two oxygen atoms be large enough to give this large of a cross section?

(c) What if a complex formed? Could that account for the results?

(d) What if the key reaction were:

$$O_2 + O_2^* \longrightarrow 2O + O_2 \qquad\qquad (P8.14.1)$$

where O_2^* is an excited O_2 molecule? How large would the O_2 need to be to account for the preexponential? That is, how large of an oxygen–oxygen bond would be needed to account for the preexponential?

(e) Now consider the reaction

$$2O \longrightarrow O_2 \qquad\qquad (8.14.2)$$

Assume that the potential follows

$$V_{AB}(R_{AB}) = 119 \text{ kcal/mol} \left[\left[e^{\frac{1.21 \text{ Å} - R_{AB}}{2.58 \text{ Å}}} - 1 \right]^2 - 1 \right]$$

In Figure 8.28, we found that when b is large there is a barrier to reaction. Calculate the barrier height as a function of b. [*Hint*: Use equation (8.56) to calculate V_{eff} as a function of R_{AB}. Use the solver function on Excel to calculate the barrier in the effective potential.]

(f) What value of b gives a barrier of $E = 2$ kcal/mol (i.e., $2k_BT$ at 500 K)?

(g) Assume that all molecules react when b is such that the barrier is less than E and no reaction occurs. Estimate the cross section for reaction (P8.14.2.)

(h) How does the cross section in (g) compare to the value estimated in a)?

8.15 How does the computer program in Example 8.B work?

(a) Show that the program is correctly plotting a trajectory.

(b) Where are the intermolecular forces calculated?

(c) Where is the actual integration done?

(d) Why are there 12 variables with only two atoms?

(e) What would you need to do to the program to enable it to handle the reaction

$$A + BC \longrightarrow AB + C?$$

(f) What extra equations would you need in (e)?

(g) What extra forces would you need? Be sure to consider the effects of orbital compression like that in Figure 8.7.

8.16 Describe in your own words the key features in the bottom picture in Figure 8.7. Where are orbitals being compressed? How do the forces arise on the rightmost fluorine atom? How is the fluorine–fluorine bond distorted?

8.17 Refer to Table P8.17. How does the preexponential calculated in Example 8.C compare to that estimated by:

(a) Collision theory (Chapter 7)

(b) Transition state theory (assume a linear transition state)

8.18 In Example 8.D, we derived an expression for the rate of the reaction $H + H_2 \rightarrow H_2 + H$. As part of the derivation, we calculated a messy integral. The integral is temperature-dependent.

(a) Calculate the preexponential and the rate of reaction at 200, 300, 500, 700 and 1000 K. The spreadsheet is available in the instructions materials.

(b) Make an Arrhenius plot of the reaction rate. How close is the activation barrier to the 0.35 eV expected from equation (8.D.23)?

(c) How close is the preexponential inferred from your Arrhenius plot to the preexponentials you estimated in (a)?

(d) Use transition state theory to see how the preexponential should vary with temperature.

(e) How well does the actual temperature dependence compare to that expected from transition state theory? (*Hint*: Make a log–log plot of the preexponential vs. T. How does your slope compare to what is expected?)

8.19 Assume that the cross section for a reaction follows

$$\sigma_x = \begin{cases} 0 & \text{for } E < E^\ddagger \\ \sigma_0 \left(1 - \dfrac{E^\ddagger}{E}\right) & \text{for } E \geq E^\ddagger \end{cases}$$

Table P8.17   Data for Problem 8.17

r	H	H_2	Linear TST
Mass	1 amu	2 amu	3 amu
R_{H-H}	—	0.7417 Å	0.87 Å
Vibrations	—	4395 cm^{-1}	2012 cm^{-1}
	—	—	965 cm^{-1}
	—	—	965 cm^{-1}
$E^\ddagger$	—	—	8.5 kcal/mol

(a) Assume $\sigma_o = 1.5$ Å^2 and $E^{\ddagger} = 13.86$ kcal/mol. Numerically integrate equation (8.D.7) to calculate the preexponential rate constant for the reaction to calculate the rate constant at 300 K.

(b) Integrate the expression analytically.

(c) How does your result compare to collision theory (Section 8.2)?

8.20 We discussed chemical reactions in this chapter, but the concepts are useful for nuclear reactions as well. Consider the reaction

$$n + {}^{207}Pb \longrightarrow {}^{208}Pb \qquad (8.20.1)$$

where n is a neutron.

(a) Assume that you have a nuclear reactor that is producing neutrons with the following energy distribution:

$$D(E) = \left(\frac{E}{E_O}\right)^{1/2} \frac{e^{-E/E_O}}{E_O} \qquad (8.20.2)$$

with $E_O = 1.5$ keV. Figure P8.20 shows data for the cross section for the reaction. Estimate the rate constant.

(b) How thick of a piece of lead will be needed to stop 99.99% of the neutrons? Lead is 22% ^{207}Pb. Assume the other isotopes of lead do not react with neutrons.

(c) Calculate the conversion of neutrons with an energy of 20 keV. Are they all trapped?

(d) In actual practice, the neutrons undergo something called "inelastic collisions", where they give up energy to the lattice. How will the inelastic collisions affect your answer in (c)?

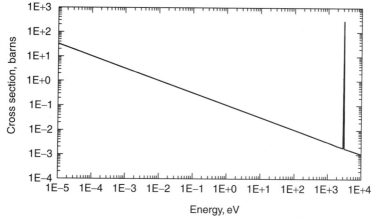

Figure P8.20 The cross section for the reaction $n + {}^{207}Pb \rightarrow {}^{208}Pb$ as a function of the energy of the neutron.

(e) Assume that the cross section for the inelastic collisions is 1000 barns independent of energy. How important will the inelastic collisions be in slowing down the 20-keV neutrons so that the neutrons are more easily trapped?

8.21   Salazart et al. (1997) measured the cross section for the reaction:

$$Ca + HBr \longrightarrow CaBr + H \qquad (P8.21.1)$$

and the following data in Table P8.21 were obtained.

(a) Use the procedure in Example 8.D to estimate the activation barrier and preexponential for the reaction at 200, 300, 400, 500, 600, and 700 K.

(b) How does a plot of the cross section versus energy compare to that in Figure 8.29?

(c) On the basis of the results, what is the barrier to the reaction?

(d) Make an Arrhenius plot of the rate constant versus 1/temperature to estimate the activation barrier for the reaction.

(e) Why didn't the wiggles in the plot of σ versus E affect the activation barrier or preexponential?

8.22   Knott, Proch and Kompa, J. Chem. Phys.. **108**, 527 (1998) measured the cross-section for the reaction

$$CO + H_2^{2-} \longrightarrow HCO^- + H^- \qquad (8.23.1)$$

For energies between 10^{-2} and 5 eV, the cross section followed

$$\sigma = \frac{18.2 \text{ Å}^2 \text{ (eV)}^{0.37}}{(E_T)^{0.37}} \qquad (P8.22.2)$$

(a) Make a plot of the cross section versus energy.

(b) On the basis of these results, what is the barrier to the reaction?

(c) Use the procedure in Example 8.D to integrate the rate equation to calculate the preexponential and activation energy at 100, 200, 300, 400, and 500 K.

(d) Why is the preexponential so big?

Table P8.21   The rate of reaction (P8.21.1) estimated from the results of Salazar (1997)

Energy, kJ/mol	σ	Energy kJ/mol	σ	Energy kJ/mol	σ
$E_T < 19.5$	0	23.4	0.50	26.7	0.45
20.5	0.39	23.7	0.47	27.0	0.44
20.7	0.81	24.0	0.48	27.3	0.44
21.0	1.00	24.3	0.50	27.6	0.43
21.3	0.78	24.6	0.50	27.9	0.43
21.6	0.51	24.9	0.48	28.2	0.43
21.9	0.42	25.2	0.47	28.5	0.42
22.2	0.48	25.5	0.46	28.8	0.42
22.5	0.59	25.8	0.46	29.1	0.41
22.8	0.61	26.1	0.46	29.4	0.41
23.1	0.56	26.4	0.45	29.7	0.41

(e) Are the reactants large enough to account for the large cross section?

(f) Are there strong enough attractions between the reactants to pull the hydrogen in so that the cross section is huge? (answer yes!)

(g) Make an Arrhenius plot of the data.

(h) Why does the Arrhenius plot show a barrier when the reaction is barrierless?

8.23 The program ReactMD is available from Dr. Masel's Website and in the instructions materials. Choose the reaction $A + BC \rightarrow AB + C$ and the case "choose random trajectories".

(a) What does the potential energy surface look like? How does it compare to Figure 8.9?

(b) Use the print command at the end of the trajectories. Make a plot of the potential energy surface. Identify the reactants, products, and transition state.

(c) On your plot, label the trajectories that lead to reaction.

(d) Calculate the reaction probability for the trajectories on your plot.

(e) Now rerun the cases again, resizing the plots so that you can see the animation and the motion on a potential energy surface side by side. Find a trajectory where C is moving away form B when A hits. What does that trajectory look like on the potential energy surface? What does that trajectory look like on a plot of position versus time?

(f) Use your results in (e) to explain why we say that "in order for a reaction to happen, C must be moving away from B when A hits."

(g) Now compare the plots on a simple potential energy surface and on mass-weighted coordinates. How are the trajectories the same or different?

8.24 This problem also uses ReactMD available from Dr. Masel's Website. Choose the reaction $A + BC \rightarrow AB + C$ and the case "choose random trajectories".

(a) Run trajectories, record the energy of the trajectory, and determine whether the trajectory lead to reaction.

(b) What is the highest energy you used where no reaction happened? What is the lowest energy? Where did you see a reaction?

(c) How do you rationalize your results given that the barrier is only 13.80 kcal/mol?

(d) Use your data to calculate the reaction probability versus energy. Divide up the energy scale into 0.5 kcal/mol increments (e.g., 13–13.5 kcal/mol, 13.5–14.0 kcal/mol) and calculate the reaction probability versus energy.

(e) An additional file, Does it react. TXT, is available from Dr. Masel's Website and in the instructions materials. Repeat your results in (d) using the data in this file.

(f) The calculations use a one-dimensional potential. Assume that you can convert to three-dimensions using $\sigma_x^r = 3 \text{ Å}^2 P_{reaction}(E_T)$.

(g) Make a plot of σ_X^r versus E_T. How does it compare to Figure 8.26?

(h) Fit your data to Equation 8.D.1.

(i) Follow the procedure in Example 8.D to calculate the preexponential and activation barrier at 500 K. (*Note*: You may have to interpolate to estimate a cross section.)

(j) Repeat (g) for t > 300, 700, 900, and 1100 K. Make an Arrhenius plot of your data. How do the preexponential and the activation barrier estimated from the slope of the plot compare to your result?

8.25 This example also uses ReactMD from Dr. Masel's Website. Pick the reaction $A + BC \rightarrow AB + C$.

(a) Choose the case "vary incident energy, optimize other parameters", and run trajectories. What is the minimum energy that leads to reaction? How does that compare to the barrier height? Plot out your results you will need them later in the problem.

(b) Now choose the case "vary incident energy, do not optimize." How do your results differ?

(c) Look carefully at the trajectories to see how the reactive trajectories differ from the unreactive trajectories. First determine how much the trajectory vibrates. How much vibrational energy is there in the "optimal" trajectories? How much vibrational energy is there in the trajectories that do not react?

(d) Next, look at when the trajectory turns on the potential energy surface. When does the optimal trajectory turn? When do the nonreactive trajectories turn?

(e) Relate the turning motion to coordinated motion. Look carefully at the coordinated motion by cascading and resizing the windows so that you can see the animation at the same time that you see the trajectory on the potential energy surface.

8.26 This example also uses ReactMD from Dr. Masel's Website. Choose the reaction $A + BC \rightarrow AB + C$ and the case "vary the timing of the collision."

(a) Run the case and see how the reactive trajectories differ from the unreactive trajectories. Look particularly at when the trajectory turns on the potential energy surface. When do the reactive trajectories turn? When do the unreactive trajectories turn?

(b) Now vary the position of the transition state from an early transition state to a late transition state. How do your results change?

(c) Repeat (a) and (b) for a late transition state.

8.27 This example also uses ReactMD from Dr. Masel's Website. Choose the reaction $A + BC \rightarrow AB + C$ and the case "vary the partitioning between translation and vibration."

(a) Run the case and look at the trajectories. What happens if you have no vibrational energy? What happens if you have too much vibrational energy and too little translational energy?

(b) Now change to an early transition state. How does the answer in (a) change?

(c) Now change to a late transition state. How does the answer in (a) change?

(d) How do your results compare to the Polanyi rules?

8.28 This problem uses ReactMD, available from Dr. Masel's Website, to calculate a rate constant for the reaction $A + BC \rightarrow AB + C$.

(a) First, in order to learn how to use the "fix the energy vary the initial conditions" option, the program calculates the reaction probability of an energy of 14.0 kcal/mol.

(b) Now assume that you want to calculate a rate constant following the procedure in Example 8.D. Set up an expression for $I_{A \to BC}$ following the procedure in Example 8.D. Assume that the cross section is zero for E < 13.88 kcal/mol.

(c) In order to use Laguere integration, you will need the cross section at a series of energies. Follow the procedure in Example 8.D to calculate those energies. (*Hint*: If you calculated things correctly, the first energy point should be 14.0 kcal/mol.)

(d) Use ReactMD to calculate the reaction probability at each energy. You will need at least 2000 trajectories of each energy to get an accurate answer.

(e) An additional file, reactionprob.TXT, is available from Dr. Masel's Website. Repeat your results in (d) using the data in this file.

(f) Assume $\sigma_x^r = (1 \text{ Å})^2 \times P_{reaction}$. Integrate equation (8.D.7) to calculate a rate constant assuming $\bar{v}_{ABC} = 1 \times 10^{13} \text{ Å/second}$.

8.29 Problems 8.24–8.28 assumed a collision where the reactants were confined to a line. Now consider a nonlinear collision.

(a) Use the "vary the impact parameter" option in ReactMD to see how the reaction probability varies with impact parameter. What do you find?

(b) Look carefully at the animation. How do the nonlinear collisions differ from the linear collisions? Do you see any "weird" trajectories, where the incoming A molecule inserts directly into the B–C bond?

(c) What do the "weird" trajectories look like when plotted on a potential energy surface?

(d) What does a trajectory with a large-impact parameter look like on a potential energy surface?

(e) Can you see evidence that atoms are hitting the angular momentum barrier to reaction? Describe what you see.

(f) Now try varying the angle of approach. How do the trajectories change?

8.30 The objective of this problem is to use ReactMD to calculate the collisional cross section. The procedure will closely follow Example 8.E. First you will use ReactMD to calculate the reaction probability. Then you will integrate equation 8.E.1 to calculate the cross section.

(a) Use ReactMD to calculate the reaction probability at several values of b for an energy of 10 kcal/mol.

(b) Integrate equation (8.D.4) to calculate the cross section.

(c) What would you need to do to calculate a rate constant from these results?

8.31 The objective of this problem is to understand how trajectories change as you change properties. Consider the reaction $A + BC \to AB + C$ confined to a plane.

(a) Assume that A starts with an impact parameter at 3 Å, and moves past BC. For the purposes of the problem we will assume

$$X_A = 3 \text{ Å}$$
$$Y_A = (10^{12} \text{ Å/second})t - 10 \text{ Å}$$
$$X_B = X_c = 0$$

$$Y_B = 1 \text{ Å} + 0.1 \text{ Å} \sin 10^{13}/\text{second t}$$

$$Y_C = 1 \text{ Å} - 0.1 \text{ Å} \sin 10^{13}/\text{second t}$$

where $t =$ time and X_A, X_B, X_C, Y_A, Y_B, and Y_C are the X and Y positions of A, B, and C.

(b) Show that these equations satisfy Newton's equations of motion, with B–C a harmonic oscillator. Assume that A does not interact with BC.

(c) Use a spreadsheet to plot X_A, Y_A, X_B, Y_B, and X_C,Y_C versus time.

(d) Next, use the spreadsheet to plot R_{AB} versus. R_{BC}.

(e) Why does R_{AB} decrease then increase again given that there is no interaction between A and BC.?

(f) Why are there all of the wiggles in the trajectories?

More Advanced Problems

8.32 The objective of this problem is to use collision theory to make some predictions about film growth. Assume that you are designing an evaporator (see Figure P8.32) to deposit copper. The evaporator works by heating copper in a crucible. The copper sublimes, filling the crucible with copper vapor.

(a) Use collision theory to estimate how many copper molecules will escape from a 1-cm^2 hole at the top of the crucible. Assume that the copper pressure inside the crucible is 10^{-9} Atm, and the temperature is 1500 K.

(b) Assume that the copper is being deposited on a 10-inch wafer and that all of the copper that leaves the crucible makes it onto the wafer. How long will it take to deposit a 10-μm thick layer of copper?

8.33 Example 8.A used the routine ODEINT to integrate the trajectories. However, ODEINT may not be available at your university. Convert the program so that it can use the IMSL subroutine IVPRK. What do the trajectories look like using I PAK?

8.34 In the examples in this chapter, we used trajectory calculations to model the reactions of two simple molecules. The objective of this problem is to decide what would change if the molecules are big, such as proteins.

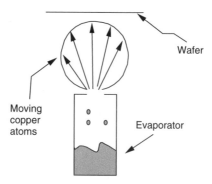

Figure P8.32   A rough diagram of the evaporator.

(a) How would the equations of motion change? Would the same equations of motion apply?

(b) How would the forces be different? Would the forces still be determined by path repulsions, or would other forces be important?

(c) How would the collisions be different? Would molecules just bounce, or would they stick together for a long time? Could you use the same computer codes to simulate the reactions?

8.35 The Materials Processing Simulation Center at Cal Tech (California Institute of Technology, Pasadena) has posted a number of simulations of the changes in the arbital shapes during Diels–Alder reactions. The URL [universal (also uniform) resource locator] address is http://www.wag.caltech.edu/gallery/gallery_quantum.html.

(a) Download the simulations and look at them.

(b) How do the Orbital distartions compare to those described in this chapter?

(c) What will the potential energy surface for a Diels–Alder reaction be like?

(d) What will you use as coordinates for a plot of the potential energy surface?

8.36 Get a copy of the following articles and write a three page description of the findings.

(a) Aoiz, F. J., Friedrich, B., Herrero, V. J., Rabanos, V. S., and Verdasco, J. E. Effect of pendular orientation on the reactivity of $H + DCl$ — A quasiclassical trajectory study, *Chem. Phys. Lett.* **289**, 132 (1998).

(b) Ansari, W. H., and Sathyamurthy, N. Classical mechanical investigation of collinear H-+H2-+H- dynamics, *Chem. Phys. Lett.* **289**, 487 (1998).

(c) Bolton, K., Hase, W. L., Schlegel, H. B., and Song, K. A direct dynamics study of the $F + C_2H_4 \rightarrow C_2H_3F + H$ product energy distributions, *Chem. Phys. Lett.* **288**, 621 (1998).

(d) Bujnowski, A. M., and Pitt, W. G. Water structure around enkephalin near a PE surface — A molecular dynamics study, *J. Colloid Interface Sci.* **203**, 47 (1998).

(e) Doubleday, C., Bolton, K., and Hase, W. L. Direct dynamics quasiclassical trajectory study of the thermal stereomutations of cyclopropane, *J. Phys. Chem.* **102**, 3648 (1998).

(f) Kobayashi, H., Takayanagi, T., and Tsunashima, S. Studies of the $N(D_2) + H_2$ reaction on revised potential energy surfaces, *Chem. Phys. Lett.* **277**, 20 (1997).

(g) Kumar, S., Kapoor, H., and Sathyamurthy, N. Dynamics of the Reaction $He + H_2^+ \rightarrow HeH^+ + H$ on the Aguado-Paniagua surface, *Chem. Phys. Lett.* **289**, 361 (1998).

(h) Lightstone, F. C., Zheng, Y. J., and Bruice, T. C. Molecular dynamics simulations of ground and transition states for the S(N)2 displacement of Cl- from 1,2-dichloroethane at the active site of *Xanthobacter autotrophicus* haloalkane dehalogenase, *J. Am. Chem. Soc.* **120**, 5611 (1998).

(i) Simka, H., Willis, B. G., Lengyel, I., and Jensen, K. F. Computational chemistry predictions of reaction processes in organometallic vapor phase epitaxy, *Prog. Crys. Growth Charact. Mater.* **35**, 117 (1997).

(j) Wang, H. B., Peslherbe, G. H., and Hase, W. L. Trajectory studies of S(N)2 nucleophilic substitution.4. Intramolecular and unimolecular dynamics of the $Cl-\ldots CH_3Br$ and $ClCH_3\ldots Br$-Complexes, *J. Am. Chem. Soc.* **116**, 9644 (1994).

(k) Pecina, O., and Schmickler, W. On the dynamics of electrochemical ion-transfer reactions, *J. Electroanal. Chem.* **450**, 303 (1998).

(l) Meijer, E. J., and Sprik, M. Ab initio molecular dynamics study of the reaction of water with formaldehyde in sulfuric acid solution, *J. Am. Chem. Soc.* **120**, 6345 (1998).

(m) Xie, J. Q., and Feng, J. Y. Molecular-dynamics simulation of silicon film growth from cluster beams, *Nucl. Instrum. Methods Phys. Res. Sec. B* **142**, 77 (1998).

(n) Valkealahti, S., and Manninen, M. Diffusion on aluminum-cluster surfaces and the cluster growth, *Phys. Rev. B* **57**, 15533 (1998).

(o) Bhatti, Q. A., and Matthai, C. C. Computer simulation of adatom dynamics on single stepped SiC(001) surfaces, *Thin Solid Films* **318**, 46 (1998).

(p) Munger, E. P., Chirita, V., Sundgren, J. E., and Greene, J. E. Destabilization and diffusion of two-dimensional close-packed Pt clusters on Pt(111) during film growth from the vapor phase, *Thin Solid Films* **318**, 57 (1998).

(q) Jager, H. U., and Weiler, M. Molecular dynamics studies of A-C-H film growth by energetic hydrocarbon molecule impact, *Diamond Relat. Mater.* **7**, 858 (1998).

(r) Yamahara, K., and Okazaki, K. Molecular dynamics simulation of the structural development in sol-gel process for silica systems, *Fluid Phase Equilib.* **144**, 449 (1998).

(s) Levine, S. W., Engstrom, J. R., and Clancy, P. A kinetic monte carlo study of the growth of Si on Si(100) at varying angles of incident deposition, *Surf. Sci.* **401**, 112 (1998).

(t) Chatfield, D. C., Eurenius, K. P., and Brooks, B. R. HIV-1 protease cleavage mechanism — A theoretical investigation based on classical MD simulation and reaction path calculations using a hybrid QM/MM potential, THEOCHEM — *J. Mol. Struct.* **423**, 79 (1998).

TRANSITION STATE THEORY, THE RRKM MODEL, AND RELATED RESULTS

PRÉCIS

In Chapter 8, we showed how one could calculate reaction rates exactly by keeping track of how molecules move. Often, however, it is too expensive to calculate a reaction rate exactly. Instead, one uses an approximation to estimate the rate of reaction. In this chapter, we will discuss three approximations: transition state theory, the Rice–Ramsperger–Kassel–Marcus (RRKM) model, and phase space theory. We will expand on the findings in Chapter 7 and show how these approximations can be used to estimate preexponentials for reaction to reasonable accuracy.

9.1 ARRHENIUS' MODEL AND TOLMAN'S EQUATION

The subject that we now call reaction rate theory got its start when Arrhenius wrote a series of papers showing how an activation barrier arose in chemical reactions. We presented Arrhenius' original derivation in Section 8.1. However, now it is useful to extend the arguments to the generalized reaction:

$$A + BC \longrightarrow AB + C \tag{9.1}$$

In the remainder of this section, we will provide a derivation of Tolman's equation for the rate constant for reaction (9.1):

$$k_{A \to BC} = \left(\frac{k_B T}{h_P} \right) \frac{q^+}{q_A q_{BC}} \exp \left(-\frac{E^+}{k_B T} \right) \tag{9.2}$$

where $k_{A \to BC}$ is the rate constant for reaction (9.1), k_B is Boltzmann's constant, T is the absolute temperature, h_P is Planck's constant, q_A is the microcanonical partition function per unit volume of the reactant A, q_{BC} is the microcanonical partition function per unit volume for the reactant BC, E^+ is the average energy of the molecules that react, and q^+ is the average partition function per unit volume of the molecules that react, divided by the partition function for the translation of A toward BC.

The derivation does require some patience. One can skip to equation (9.25) without loss of continuity.

The idea of derivation is to assume that the reaction occurs in two steps:

- First, hot molecules are created in the right configuration to react.
- Then the hot molecules react by coming together, moving over the barrier in the potential energy surface and on to products.

When we say that we are creating hot molecules in the right configuration, we are saying that we are producing molecules that are moving together with enough energy to react and in a configuration that will eventually lead to a reaction. In order to properly account for the motion, we will consider the entire trajectory and not just the time the molecules spend in the region of the transition state. We will calculate the rate as the concentration of hot molecules divided by the rate that the hot molecules react, that is the amount of time it takes to go from hot reactants over the transition state on to products. The final equation will resemble the equation for transition state theory, but the individual terms in the equation will be different.

Let's start with a system containing a mixture of A atoms and BC molecules. For the moment, let's treat each A–BC pair of molecules as one giant extended molecule. Arrhenius (1889) pointed out that at any instant, there are two groups of molecule pairs: those molecule pairs that are close enough and in the right configuration to react, and those molecule pairs that are not in the right configuration to react.

Following Arrhenius, let's call K_0, the rate constant for the pairs of molecules that are in the right configuration, to react. The rate of reaction, $r_{A \to BC}$, is equal to

$$r_{A \to BC} = K_0 N^\dagger \qquad (9.3)$$

where $N^\dagger$ is the number of molecule pairs in the right configuration to react. Equation (9.3) is a key equation for the derivation here.

In the next several sections we will derive equations for $N^\dagger$ and K_0. $N^\dagger$ will be calculated from equilibrium, and K_0 will be calculated from the time it takes the reactants to come together, cross the transition state, and move on to products.

First, let us use an equilibrium relationship to estimate $N^\dagger$. At equilibrium, the number of pairs of molecules in the right configuration to react is given by

$$N_{eq}^\dagger = K^\dagger [A][BC] \qquad (9.4)$$

where $K^\dagger$ is the equilibrium constant for production of molecule pairs in the right configuration to react and $N_{eq}^\dagger$ is the concentration of molecule pairs at equilibrium. There are two ways to produce the molecule pairs at equilibrium: either as a forward reaction from reactants or as a back reaction from products. If one wants to consider the forward reaction, one wants to consider only the hot molecules that were formed from the reactants. It works out that at equilibrium only half of the $N_{eq}^\dagger$ molecules were formed from reactants. The others were formed from the back reaction from the products. Consequently, $N^\dagger$, the concentration of pairs formed from reactants, is given by

$$N^\dagger = \tfrac{1}{2} N_{equ}^\dagger \qquad (9.5)$$

Combining equations (9.3)–(9.5) yields Arrhenius' expression for the rate of reaction (9.1):

$$r_{A \to BC} = K_0 \frac{K^\dagger}{2} [A][BC] \qquad (9.6)$$

In Arrhenius' approximation, $k_{A \to BC}$, the rate constant for reaction (9.1) is given by

$$k_{A \to BC} = K_0 \frac{K^\dagger}{2} \qquad (9.7)$$

Next it is useful to substitute the statistical mechanics for the equilibrium constant into equation (9.7). According to equation (6.7), the equilibrium constant is given by

$$K^\dagger = \frac{q^*}{q_A q_{BC}} \qquad (9.8)$$

In equation (9.8) q_A and q_{BC} are the microcanonical partition function per unit volume of the reactants A and BC and q^* is the microcanonical partition function per unit volume for the hot molecules relative to the reactants. Substituting equation (9.8) into equation (9.7) yields

$$k_{A \to BC} = \frac{K_0}{2} \frac{q^*}{q_A q_{BC}} \qquad (9.9)$$

Next, it is useful to define a new variable q^{hot} by

$$q^{hot} = q^* \exp\left(-\frac{E^+}{kT}\right) \qquad (9.10)$$

where E^+ is the average energy of the molecules that react, minus the energy of the reactants; k is Boltzmann's constant; and T is temperature. Physically q^{hot} is the rotational, translational, and vibrational partition function for the hot molecules calculated[1] relative to a molecule with energy E^+.

Substituting equation (9.10) into equation (9.9) yields

$$k_{A \to BC} = \frac{K_0}{2} \frac{q^{hot}}{q_A q_{BC}} \exp\left(-\frac{E^+}{k_B T}\right) \qquad (9.11)$$

Equation (9.11) is Tolman's version of Arrhenius' expression for the rate constant.

Next, we will follow the work of Herzfeld (1919) and use collision theory to get an expression for K_0. The idea in the derivation will be to calculate the time it takes for the reactants to come together and react.

Consider the collision between molecule A and molecule BC (see Figure 9.1), where the molecules start a distance d_{ABC} away, and d_{ABC}, is in the order of 3×10^5 Å (i.e., the

---

[1] We did not discuss it in Chapter 6, but recall from thermodynamics, that all thermodynamic quantities are always calculated with respect to a standard state. For example, heats of formation of species are usually calculated with respect to pure elements at 300 K. Free energies of species are calculated relative to pure molecules at 1 atm 300 K. In the same way, q^* is the partition function for the hot molecules calculated relative to the reactants, while q^{hot} is the partition functions calculated are with respect to a molecule with energy E^+.

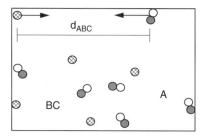

Figure 9.1   A generalized collision between A and BC.

mean free path in the gas phase). It will take a given time, t_{ABC}, for the molecules to get close enough to react:

$$t_{ABC} = \frac{d_{ABC}}{\bar{v}_{A \to BC}} \tag{9.12}$$

where $\bar{v}_{A \to BC}$ is the average velocity of A toward BC. In Chapter 6 we found that molecular velocities are typically in the order of 1000 meters/second = 10^{13} Å/second. At a typical mean free path of 3×10^5 Å, it will take 3×10^{-8} seconds for the reactants to come together and collide. By comparison, in Chapter 8 we found that molecules take about 10^{-13} seconds to cross the transition state. That is a very small time compared to the time it takes for the molecules to come together collide.

In the derivation here we will assume that the 10^{-13} seconds that the molecules take to cross the transition state is negligible compared to the 3×10^{-8} seconds it takes the molecules to come together. Therefore, the rate of reaction will be equal to the rate at which molecules come together in the right configuration to react.

Under this assumption, the rate constant for reaction is as follows:

$$k_0 = \frac{1}{t_{ABC}} = \frac{\bar{v}_{A \to BC}}{d_{ABC}} \tag{9.13}$$

According to the analysis in Example 6.C and the supplemental material for Chapter 8, for an ideal gas

$$\bar{v}_{A \to BC} = \left(\frac{8k_B T}{\pi \mu_{ABC}} \right)^{1/2} \tag{9.14}$$

where

$$\mu_{ABC} = \frac{m_A m_{BC}}{m_A + m_{BC}} \tag{9.15}$$

where m_A and m_{BC} are the masses of A and BC. We are examining hot molecules, so the average velocity could be slightly different from that in equation (9.14). But we will ignore that difference for the derivation here.

Combining equations (9.11), (9.13), and (9.14) yields.

$$k_{A \to BC} = \frac{1}{2\,d_{ABC}} \left(\frac{8kT}{\pi \mu_{ABC}} \right)^{1/2} \frac{q^{hot}}{q_A q_{BC}} \exp \left(-\frac{E^+}{k_B T} \right) \tag{9.16}$$

Next, we need to define some more partition functions. Recall from the analysis in Section 6.4.7 that the partition function for the hot molecules can be written as a product of the translational partition function for the molecules, the vibrational partition function for the molecules, and the rotational partition function for the molecules. Let's define

$$q_{A \to BC} = \text{the partition function for the translation of A toward BC} \qquad (9.17)$$

$$q^+ = q^{hot}_{notA \to BC \text{ translation}} = \text{the partition function for all of the normal}$$
$$\text{modes of the hot molecules except the A} \to \text{BC translation} \qquad (9.18)$$

The analysis in Section 6.4.7 shows

$$q^{hot} = q_{A \to BC}(q^{hot}_{not \ AB \to C \text{ translation}}) \qquad (9.19)$$

Combining equations (9.16) and (9.19) yields.

$$k_{A \to BC} = \left\{ \frac{q_{A \to BC}}{2 \, d_{ABC}} \left(\frac{8 k_B T}{\pi \mu_{ABC}} \right)^{0.5} \right\} \left(\frac{q^{hot}_{notA \to BC \text{ translation}}}{q_A q_{BC}} \right) \exp\left(-\frac{E^+}{k_B T} \right) \qquad (9.20)$$

Table 9.1 summarizes all of the partitions functions we have defined so far. One should refer to it as the derivation proceeds.

Next, we need an expression for $q_{A \to BC}$, the partition function for the motion of A toward BC. Table 7.5 gives an expression for q_t^{ideal}, the partition function for the motion of an ideal gas A *toward and away* from BC:

$$q_t^{ideal} = \frac{(2 \pi \mu_{ABC} k_B T)^{1/2} \, d_{ABC}}{h_P} \qquad (9.21)$$

where h_p is Planck's constant, μ_{ABC} is the reduced mass of A toward BC, and d_{ABC} is the distance average between A and BC. Strictly speaking, equation (9.21) might not apply to the hot molecules, but we will ignore that effort here. The term $q_{A \to BC}$ contains only the partition function for the motion of A toward BC. In an ideal gas, 50% of the A molecules will be moving toward BC and 50% will be moving away from BC. Therefore

$$q_{A \to BC} = \tfrac{1}{2} q_t \qquad (9.22)$$

Table 9.1 Definitions of the partition functions used in the derivation of Tolman's equation

q_A	The partition function of the reactant A
q_{BC}	The partition function of the reactant BC
q^*	The partition function of the hot molecules when energies are measured relative to the reactants
q^{hot}	The partition function of the hot molecules when energies are measured relative to a molecule with energy E^+
$q_{A \to BC}$	The partition function for the motion of A toward BC
$q^{hot}_{not \ A \to BC \text{ translation}}$	The partition function for all of the normal modes of the hot molecules except the A $\to$ BC translation [see equation (9.19)]
q^+	Another notation for $q^{hot}_{not \ A \to BC \text{ translation}}$

Combining equations (9.16) and (9.20)–(9.22) yields

$$k_{A \to BC} = \left\{ \frac{1}{2d_{ABC}} \left(\frac{8k_B T}{\pi \mu_{ABC}} \right)^{1/2} \frac{(2\pi \mu_{ABC} k_B T)^{1/2}}{2h_P} d_{ABC} \right\}$$
$$\times \frac{q_{not\ A-BC\ translation}^{hot}}{q_A q_{BC}} \exp \left(-\frac{E^+}{k_B T} \right) \qquad (9.23)$$

Performing the algebra, we obtain

$$k_{A \to BC} = \left(\frac{k_B T}{h_P} \right) \left(\frac{q_{not A \to BC\ translation}^{hot}}{q_A q_{BC}} \right) \exp \left(-\frac{E^+}{k_B T} \right) \qquad (9.24)$$

People often rewrite equation (9.24) as

$$k_{A \to BC} = \left(\frac{k_B T}{h_P} \right) \frac{q^+}{q_A q_{BC}} \exp \left(-\frac{E^+}{k_B T} \right) \qquad (9.25)$$

In equation (9.25) k_B is Boltzmann's constant, T is the absolute temperature, h_P is Planck's constant, q_A is the partition function per unit volume of the reactant A, q_{BC} is the partition function per unit volume for BC; E^+ is the average energy of the molecules that react, and q^+ is another notation for $q_{not\ A-BC\ translation}^{hot}$. Equation (9.25) is called *Tolman's equation*.

Equation (9.25) is a key result that we will be using throughout this chapter. The reader should memorize the equation before proceeding.

Table 9.2 compares transition state theory to Tolman's equation. Tolman's equation, equation (9.25), looks like the transition state theory result, but it is not the same. Equation (9.25) contains q^+, the partition function for the molecules that are posed to react, and E^+ the average energy of the molecules that react. Transition state theory replaces E^+ by $E_T^{\ddagger}$ and q^+ by $q_T^{\ddagger}$, where $E_T^{\ddagger}$ is the energy of the transition state and $q_T^{\ddagger}$ is the partition function for the transition state. Experimentally, E^+ is often slightly different from $E_T^{\ddagger}$. Similarly, q^+, is slightly different from $q_T^{\ddagger}$. As a result, equation (9.25) differs from transition state theory, despite the close apparent similarities between the two.

Equation (9.25) is very general. We made only a few assumptions in deriving equation (9.25): that the reactants are ideal gases, that the reacting mixture is in thermal equilibrium with the reaction vessel, and that the partition functions for the hot molecules is given by the ideal gas partition functions. All of these assumptions are excellent for gases at moderate pressures, provided the heat generated by the reaction is small enough that thermal equilibrium is maintained. Therefore, equation (9.25) is essentially exact.

Tolman (1920) used a version of equation (9.25) to derive **'Tolman's theorem'**, which says that

The activation barrier for a reaction is equal to the average energy of the molecules that react minus the average energy of the reactants.

Tolman's theorem is still thought to be correct.

In the next several sections, we will obtain some approximate expression for q^+. That will allow us to calculate rate constants from equation (9.25).

Table 9.2   A Comparison of Tolman's equation and transition state theory

Equation		Definitions
Tolman's	$k_{A \rightarrow BC} = \left(\dfrac{k_B T}{h_P} \right) \left(\dfrac{q^+}{q_A q_{BC}} \right) \exp \left(-\dfrac{E^+}{k_B T} \right)$	$E^+ =$ average energy of the hot molecules before they react
		$q^+ =$ partition function for the hot molecules before they react the partition function includes all of the normal modes of the AB–C complex except the normal mode carrying the species together
TST[a]	$k_{A \rightarrow BC} = \left(\dfrac{k_B T}{h_P} \right) \left(\dfrac{q_T^\ddagger}{q_A q_{BC}} \right) \exp \left(-\dfrac{E^\ddagger}{k_B T} \right)$	$E^\ddagger =$ Saddle point energy
		$q_T^\ddagger =$ partition function for molecules at the saddle point in the potential energy surface; the partition function includes all of the normal modes of the AB–C complex except the normal mode carrying the species over the barrier

[a]Transition state theory.

9.1.1   Relation to Collision Theory

As an example of the use of equation (9.25), it is useful to use equation (9.25) to derive collision theory. We already discussed collision theory in Section 7.2. We showed that in simple collision theory approximation, the rate constant for the reaction $A + BC \rightarrow AB + C$ is given by

$$k_{A \rightarrow BC} = \bar{v}_{A \rightarrow BC} \sigma_{A \rightarrow BC} \exp \left(-\frac{E^\ddagger}{k_B T} \right) \qquad (9.26)$$

where $\bar{v}_{A \rightarrow BC}$ is the average velocity of A toward BC, $\sigma_{A \rightarrow BC}$ is the cross section, and $E^\ddagger$ is the minimum internal energy needed for the reaction to occur.

In this section, we will show that equation (9.25) reduces to equation (9.26) provided the critical configuration is the hard-sphere collision in Figure 9.2.

Recall from Chapter 8 that in collision theory, one assumes that the reaction occurs when an A molecule hits a BC molecule as shown in Figure 9.2. The cross section for

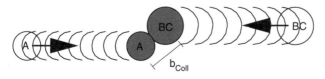

Figure 9.2   A hard-sphere collision between A and BC.

the collision is given by

$$\sigma_{A \to BC} = \pi (b_{coll})^2 \qquad (9.27)$$

where b_{coll} is the distance from A to BC at the point of collision. According to equation (9.14).

$$\bar{v}_{A \to BC} = \left(\frac{8k_B T}{\pi \mu_{A \to BC}} \right)^{1/2} \qquad (9.28)$$

The rate constant becomes

$$k_{A \to BC} = \pi \left(\frac{8k_B T}{\pi \mu_{A \to BC}} \right)^{1/2} (b_{max})^2 \qquad (9.29)$$

Next, let's use equation (9.25) to calculate the rate constant for the reaction. We will assume that, a reaction occurs whenever the reactants hit each other as shown in Figure 9.2. If we assume that the BC molecule is structureless, then the reacting pair has two rotational modes and three translation modes (plus a mode for the approach of A toward BC, which is already accounted for by the model). The modes separate if we have a hard-sphere collision. Consequently, q^{hot} is given by

$$q^{hot} = q_r^2 q_t^3 q_{A \to BC} \qquad (9.30)$$

where q_r is the rotational partition function and q_t is a translational partition function for the ABC complex and $q_{A \to BC}$ is the partition function of A moving toward BC. Comparing equations (9.19) and (9.30) and noting $q^{hot}_{not\ AB-C\ translation} = q^+$ shows

$$q^+ = q_r^2 q_t^3 \qquad (9.31)$$

Substituting the expressions for q_r and q_t in Table 6.6 into equation (9.31) with a symmetry number of 1.0 yields

$$q^+ = \frac{(2\pi(m_A + m_{BC})k_B T)^{3/2}}{h_P^3} \left(\frac{8\pi^2 I k_B T}{h_P^2} \right) \qquad (9.32)$$

where I is given by

$$I = \mu_{ABC}(b_{coll})^2 \qquad (9.33)$$

Note that we use the collision diameter, not d_{abc}, the initial distance from A to BC, in equation (9.33), because b_{coll}, not d_{abc}, appears in equation (8.113).

We also need expressions for q_A and q_{BC}. For structureless particles

$$q_A = \frac{(2\pi m_A k_B T)^{3/2}}{h_P^3} \qquad (9.34)$$

$$q_{BC} = \frac{(2\pi m_{BC} k_B T)^{3/2}}{h_P^3} \qquad (9.35)$$

Substituting equations (9.32)–(9.35) into equation (9.25) and doing pages of algebra yields

$$k_{A \to BC} = \pi (b_{coll})^2 \left(\frac{8 k_B T}{\pi \mu_{ABC}} \right)^{1/2} \tag{9.36}$$

Notice that equation (9.36) is identical to equation (9.29). Therefore, collision theory is equivalent to equation (9.25) provided one assumes that a reaction occurs whenever an A molecule collides with a hot BC molecule, and that the BC motion toward The saddle point is separable from the rest of the motion of the molecule, so that q^+ is given by equation (9.32).

This example illustrates that equation (9.25) can be used to derive useful rate equations provided we can calculate q^+.

9.2 CONVENTIONAL TRANSITION STATE THEORY

Next, we want to show how transition state theory arises from equation (9.25). We already described transition state theory in Chapter 7. The key assumption in transition state theory is that one can replace E^+ by $E^{\ddagger}$ and q^+ by $q^{\ddagger}$ in equation (9.25). Here, E^+ is the average energy of the molecules that react, $E^{\ddagger}$ is the transition state energy, q^+ is the partition function for molecules that react, and $q^{\ddagger}$ is the partition function at the transition state as given in Table 9.2. Equation (9.25) becomes

$$k_{A \to BC} = \left(\frac{k_B T}{h_p} \right) \left\{ \frac{q_T^{\ddagger}}{q_A q_{BC}} \left[\exp \left(-\frac{E^{\ddagger}}{k_B T} \right) \right] \right\} \tag{9.37}$$

When we replace E^+ by $E^{\ddagger}$, we are saying that the average energy of the molecules that react is equal to the transition state energy. In most cases that is a fair assumption. Physically, in the absence to tunneling, only molecules that have enough energy to cross the barrier can react, so E^+ must be at least $E^{\ddagger}$. E^+ can be smaller than $E^{\ddagger}$ if there is tunneling. We will show later in this chapter that E^+ can be a little larger than $E^{\ddagger}$ if hotter molecules are more reactive. Still, most molecules that react have an energy close to $E^{\ddagger}$, which makes the replacement of E^+ by $E^{\ddagger}$ a reasonable assumption.

The replacement of q^+ by $q^{\ddagger}$ is harder to justify. Physically, when you replace q^+ by $q^{\ddagger}$, you are saying that the partition function is not changing during reaction. That would be a good assumption if there were no changes in the translational or vibrational modes of the molecules during the reaction. However, in Chapter 8 we found that molecules need some vibrational energy to get around the bend in the potential energy surface. When you go around the bend, you convert vibration to translation over the barrier. Consequently, the vibrational and translational modes must be changing during reaction. One of the key assumptions in transition state theory is to ignore those changes. That introduces errors.

9.2.1 Derivation of Equation (9.37), the Conventional Transitional State Theory Expression for the Rate of Reaction

Next, it is useful to derive equation 9.42 more precisely. In particular, we want to show that if one assumes that the reaction probability [i.e, P_n in equation (9.9)] is zero for all of the states that do not have enough energy to cross the barrier and 1.0 for all of the states that do have enough energy to cross the barrier, that there are no changes in the

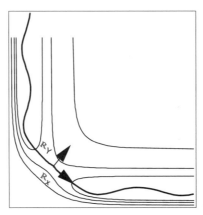

Figure 9.3 The minimum-energy pathway for motion over the barrier as determined by the trajectory calculations in Chapter 8 (see also Figure 8.21).

translational or vibrational modes of the molecules during the reaction, and that there are no quantum effects, one can derive equation (9.37).

The derivation of equation (9.37) is pretty straightforward. Let's consider a particle moving along the trajectory shown as the solid line in Figure 9.3, where the solid line is the minimum-energy trajectory that makes it over the barrier. One can define a new coordinate system, R_x and R_y as shown in the figure. R_x is motion along the trajectory, and R_y is motion perpendicular to the trajectory. One can then do all of the derivations in the new coordinate system using what is called a reaction path Hamiltonian to calculate the positions.

Next let's calculate q^{hot}, the partition function for the molecules that are in the right configuration using this new coordinate system. Note that the coordinate system separates in X,Y,Z coordinates. Therefore, we can write the partition function as

$$q^{hot} = q_X q_Y q_{everything\_else} \tag{9.38}$$

In equation (9.38), q_X is the partition function for motion in the X direction, q_Y is the partition function for motion in the Y direction, and $q_{everything\_else}$ is the partition function for all of the other modes of the molecule. Substituting equation (9.38) into equation (9.20) yields

$$k_{A \to BC} = \left\{ \frac{q_x}{2d_{ABC}} \left(\frac{8k_B T}{\pi \mu_{ABC}} \right)^{0.5} \right\} \left(\frac{q_y q_{everything\_else}}{q_A q_{BC}} \right) \exp \left(-\frac{E^+}{k_B T} \right) \tag{9.39}$$

Next we want to relate q_X, q_Y, and $q_{everything\_else}$ to properties of the transition state. Recall that q^{hot} is the partition function per unit volume for the molecules that are hot enough and in the right configuration to react. According to the discussion in Section 6.4.1, the partition functions can be written as a sum over all states of the A−BC molecular complex:

$$q^{hot} = \frac{\sum_N P_N^{hot} g_N \exp \left(-\frac{E_N - E^+}{k_B T} \right)}{V} \tag{9.40}$$

In equation (9.40), N is an index of all of the states of the A–BC molecular complex, g_N is the degeneracy of that state, P_N^{hot} is the probability that a molecule in the Nth state is hot enough and in the right configuration to react, E_N is the energy of the state, E^+ is the energy of the reference state, and V is the molar volume.

Similarly, the partition function for the transition state is given by

$$q^{TST} = \frac{\sum_N P_N^{TST} g_N \exp\left(-\frac{E_N - E^+}{k_B T}\right)}{V} \tag{9.41}$$

In equation (9.41), N is an index of all of the states of the A–BC molecular complex, g_N is the degeneracy of that state, P_N^{TST} is the probability that a molecule in the Nth state is present in the transition state, E_N is the energy of the state, E^+ is the energy of the reference state, and V is the molar volume.

By definition, the transition state includes only molecules with an energy greater than $E^{\ddagger}$, so P_N^{TST} equals zero for states with energy below the barrier and unity for molecules with energy above the barrier.

Notice that if $P_N^{TST} = P_N^{hot}$, then

$$q^{hot} = q^{TST} \exp\left(-\frac{E^+ - E^{\ddagger}}{k_B T}\right) \tag{9.42}$$

Therefore, if $P_N^{hot} = P_N^{TST}$, equation (9.42) becomes

$$k_{A \to BC} = \left\{\frac{q_x^{\ddagger}}{2\,d_{ABC}}\left(\frac{8k_B T}{\pi\mu_{ABC}}\right)^{0.5}\right\}\left(\frac{q_y^{\ddagger}q_{everything\_else}^{\ddagger}}{q_A q_{BC}}\right)\exp\left(-\frac{E^{\ddagger}}{k_B T}\right) \tag{9.43}$$

Table 9.3 summarizes the definitions of the partition functions in equations (9.39) and (9.43).

Table 9.3   Definitions of the partition functions used in the derivation of transition state theory

q_A	The partition function of the reactant A
q_{BC}	The partition function of the reactant BC
q^{hot}	The partition function of the hot molecules when energies are measured relative to a molecule with energy E^+
q^{TST}	The partition function for the transition state
q_X	The partition function for the motion of the HOT MOLECULES in the X direction in Figure 9.3
q_Y	The partition function for the motion of the HOT MOLECULES in the Y direction in Figure 9.3
$q_{everything\_else}$	The partition function for all other motions of the HOT MOLECULES
$q_x^{\ddagger}$	The partition function for the motion of the TRANSITION STATE in the X direction in Figure 9.3
$q_y^{\ddagger}$	The partition function for the motion of the TRANSITION STATE in the Y direction in Figure 9.3
$q_{everything\_else}^{\ddagger}$	The partition function for all the other motions of the TRANSITION STATE
$q^{\ddagger}$	$q_y^{\ddagger}q_{everything\_else}^{\ddagger}$

At this point, we have only assumed that $P_N^{Hot} = P_N^{TST}$.

In order to get to conventional state theory, we have to make the additional assumptions:

1. All the P_N^{hot} in equation (9.40) are 0 for states with energies below the barrier and unity for states with energy above the barrier so that $P_N^{hot} = P_N^{TST}$.
2. Motion along the R_x direction is pure translation.
3. Motion perpendicular to the R_x direction is pure vibration.
4. The P_N^{hot} are determined only by the properties of the transition state; this assumption ignores the findings from Chapter 8 that the reaction probability is higher when the potential energy surface is smoothly banked than when there is a sharp turn in the potential energy surface. It also ignores the fact that the vibrational levels of the reactants might not match the vibrational levels of the transition state.

If we make these four assumptions, then equation (9.43) becomes

$$k_{A \to BC} = \left\{ \frac{q_{trans}^\ddagger}{2\,d_{ABC}} \left(\frac{8k_B T}{\pi \mu_{ABC}} \right)^{0.5} \right\} \left(\frac{q^\ddagger}{q_A q_{BC}} \right) \exp \left(-\frac{E^\ddagger}{k_B T} \right) \qquad (9.44)$$

where $q^\ddagger = q_y^\ddagger q_{everything\_else}^\ddagger$. Substituting in the translational partition function from equations (9.21) and (9.22) and doing algebra yields the following equation (9.45).

$$k_{A \to BC} = \left(\frac{k_B T}{h_p} \right) \left[\left[\frac{q_T^\ddagger}{q_A q_{BC}} \right] \exp \left(-\frac{E^\ddagger}{k_B T} \right) \right] \qquad (9.45)$$

Equation (9.45) is the key result from transition state theory: When I have taught this course, students have had some trouble understanding what I am talking about when I discuss modes as being pure translation or vibration. Therefore, it is useful to consider a simple example, the isomerization of hydrogen isocyanide to hydrogen cyanide:

$$HNC \longrightarrow HCN \qquad (9.46)$$

In Chapter 8, we found that reaction (9.46) is not an elementary reaction, because we need a collision partner. However, for the purposes here, it is useful to treat reaction (9.46) as an elementary reaction so that we can illustrate the ideas. We will also assume that the transition state for the reaction is complex with a bridging hydrogen:

$$\begin{array}{c} H \\ / \; \backslash \\ C = N \end{array} \qquad (9.47)$$

First, let's look at the modes of hydrogen isocyanide. Hydrogen isocyanide has three atoms. According to the material in Chapter 6, a molecule with three atoms will have $3 \times 3 = 9$ total modes. It will have three transitional modes, so six modes are left. HNC is linear so the molecule will have two rotational modes; $6 - 2 = 4$ so HNC must have four vibrational modes.

For the purposes of discussion, we will label the vibrations as a symmetric and asymmetric H–N–C stretch and two H–NC bending modes. Following the analysis in Chapter 6 we obtain

$$q_{HNC} = q_t^3 q_r^2 (q_{sym})(q_{asym})(q_{bend\ A})(q_{bend\ B}) \tag{9.48}$$

where q_t is translational partition function for HNC per unit volume; q_r is a partition function for the rotation of HNC; and q_{syn}, q_{asyn}, $q_{bend\ A}$, $q_{bend\ B}$ are the partition functions for the symmetric and asymmetric vibrations and the two bending modes.

When the reaction occurs, the HNC bond bends until it breaks. One can write the partition function for the transition state as before:

$$q_{HNC}^{\ddagger} = (q_t^{\ddagger})^3 (q_r^{\ddagger})^2 q_{sym}^{\ddagger} q_{asym}^{\ddagger} q_{bend\ A}^{\ddagger} q_{bend\ B}^{\ddagger} \tag{9.49}$$

where we have put a double dagger on all of the modes of the molecule to indicate that we are looking at the transition state.

Our derivation of transition state theory makes the assumption that the bending mode of the molecule that carries the reaction over the transition state can be approximated as a translational rather than an excited vibrational partition function. Our derivation transition state theory also assumes that q_{sym} and q_{assym} are simple vibrational partition functions so that

$$q_T^{\ddagger} = (q_t^{\ddagger})^3 (q_r^{\ddagger})^2 q_{sym}^{\ddagger} q_{asym}^{\ddagger} q_{bend\ B}^{\ddagger} \tag{9.50}$$

Note that we have lost the $q_{bend\ A}$ mode in equation (9.50). The $q_{bend\ A}$ mode is q_x in equation (9.43).

Laidler (1987) provides several alternate derivations of transition state theory as summarized in Table 9.4. His derivations are similar to those here except that he treats $q_x^{\ddagger}$ differently. In one case he treats $q_x^{\ddagger}$ as a vibration in an inverted harmonic barrier, and

Table 9.4   Alternate assumptions that lead to transition state theory

Reference	Assumptions
This derivation	$P_N^{hot} = P_N^{TST} = 0$ for states below barrier, 1 for states above the barrier independent of the curvature of the barrier
	$q_x^{\ddagger}$ = translation with no curvature
	$q_y^{\ddagger}$ = vibration with no curvature
Laidler (1987)	$P_N^{hot} = P_N^{TST} = 0$ for states below barrier, 1 for states above barrier independent of curvature of the barrier
	$q_x^{\ddagger}$ = vibration in an inverted barrier with a frequency near zero and no curvature
	$q_y^{\ddagger}$ = vibration with no curvature
Laidler (1987)	$P_N^{hot} = P_N^{TST} = 0$ for states below the barrier, 1 for states above the barrier independent of the curvature of the barrier
	$q_x^{\ddagger}$ = classical translation over the barrier with no quantum effects
	$q_y^{\ddagger}$ = vibration with no curvature

then uses an algebraic approximation that applies only in the limit that the vibrational frequency is zero. In an alternate derivation he treats the motion in the x direction as a free-particle translation. Both derivations give the same final equation as the one given here, so they are equivalent.

If we apply Laidler's first approximations to reaction (9.46), we would say that the bending mode is a vibration in an inverted parabola with a near-zero frequency. In the second case we would say that the vibration is not quantized. In either case we would ignore the fact that the hydrogen moves in a curved pathway, and that the inverted barrier has a nonzero frequency.

9.3 SUCCESSES OF CONVENTIONAL TRANSITION STATE THEORY

There are four key successes of conventional transition state theory:

- Transition state theory generally gives preexponentials of the correct order of magnitude.
- Transition state theory is able to relate barriers to the saddle point energy in the potential energy surface.
- Transition state theory is able to consider isotope effects.
- Transition state theory is able to make useful prediction in parallel reactions such as reactions (7.32a) and (7.32b).

The first key prediction of transition state theory is that preexponentials should be about 10^{13} in units of molecule-angstroms and seconds independent of the order of the reaction. Tables 7.2 and 7.3 give a selection of some preexponentials. Generally, preexponentials are between 10^{10} and 10^{16}, which is qualitatively what one would expect from transition state theory.

The second key prediction of conventional transition state theory is associated with the energy to go over the barrier in the potential energy surface. Experimentally, that prediction is usually correct. For example, Table 7.4 shows the saddle point energy for a number of reactions and the activation barrier for those reactants. Notice that the activation energy is always within ± 5 kcal/mol of the saddle point energy. Generally, hydrogen transfer reactions have slightly lower barriers than expected from transition state theory, while most other reactions show barriers slightly higher than those expected from transition state theory.

The idea that the activation barrier is associated with the saddle point energy is quite useful in that it allows one to manipulate the activation energy for reactions.

> Experimentally, anything one does to lower the energy of the transition state for a reaction lowers the activation energy for the reaction.

One does not want to overstate this conclusion because in molecular dynamics (MD) simulations, the activation energy for a reaction is generally different from the energy of the transition state. Physically, one needs to localize energy in a specific bond in the reactive complex to get the reaction to go. Typically, there is a barrier to intermolecular energy transfer, so unless the reactive complex has some excess energy, it is unlikely for reaction to occur. The net result is that molecules that cross the saddle point and go on to form products are on the average hotter than one would expect from equilibrium. There can also be interesting tunneling corrections. As a result, activation barriers for reactions are generally somewhat different from what one would expect from equation (9.37).

The third key accomplishment of transition state theory is that it partially explains isotope effects. Consider a simple reaction:

$$H + H_2 \longrightarrow H_2 + H \tag{9.51}$$

One might think that the reaction rate would not change when one replaces the H_2 by D_2 in reaction (9.51). However, experimentally, the rate goes down by about a factor of 4 at 300 K (see Table 9.5). The rate changes because a H–D bond has a frequency different from that an H–H bond.

Recall that vibrations have zero-point energies, so molecules do not really fall to the bottoms of attractive wells. Rather, the actual energy of a molecule is above the base of the well. Similarly, the transition state has zero-point energy.

Consequently, according to Tolman's theorem, the activation barrier will be shifted by the zero-point energies of the molecules. Table 9.6 gives numerical values of the changes see also Figure 9.4 for activation barriers for these reactions. Table 9.6 also shows that there are changes in the frequency of the vibrational modes of the transition state, which changes $q_T^{\ddagger}$. Overall, conventional transitional state theory predicts about a factor of 12 variation in rate of the reactions in Table 9.5. By comparison, experimentally, there is a factor of 6 variation in rate. The data in Table 9.5 show that conventional transition state theory predicts the correct qualitative trends, on isotopic substitution, but not the quantitative details.

The fourth key success of transition state theory is that it allows one to predict something about the relative rates of parallel reaction such as reactions

$$CH_3CH_2CH_3 + OH \longrightarrow CH_3CHCH_3 + H_2O \tag{9.52}$$

Table 9.5 The relative rates of a series of reactions at 300 K

Reaction	Experimental Relative Rate	CTST Relative Rate
$D + H_2 \rightarrow DH + H$	6	12.3
$H + H_2 \rightarrow H_2 + H$	4	6.7
$D + D_2 \rightarrow D_2 + D$	2	1.7
$H + D_2 \rightarrow HD + D$	1	1

Table 9.6 Various contributions to the relative rate of the reactions in Table 9.5 at 300 K

Reaction	Symmetric Stretching Frequency, cm^{-1}	Bending Frequency, cm^{-1}	Δ Zero-Point Energy, kJ/mol	Relative Rate at 300 K
$D + H_2 \rightarrow DH + H$	1732	924	4.79	12
$H + H_2 \rightarrow H_2 + H$	2012	965	6.96	6.7
$D + D_2 \rightarrow D_2 + D$	1423	683	7.60	1.7
$H + D_2 \rightarrow HD + D$	1730	737	10.16	1

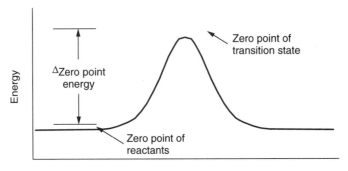

Figure 9.4   The activation barriers for the reactions in Table 9.6.

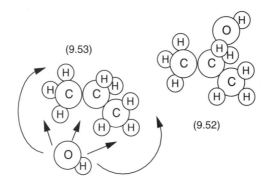

Figure 9.5   Possible transition states for reactions (9.52) and (9.53).

and

$$CH_3CH_2CH_3 + OH \longrightarrow CH_3CH_2CH_2 + H_2O \qquad (9.53)$$

Recall that reaction (9.53) can occur whenever the reactants collide, while reaction (9.52) can occur only when the reactants collide in a special configuration. If we translate that idea into the language of statistical mechanism, we would say that all of the states of the transition state contribute to reaction (9.53), while only special states contribute to equation (9.52). According to equation (9.41), $q_t^{\ddagger}$ will be smaller when fewer states contribute. Therefore, according to conventional transition state theory, the rate of reaction (9.52) will be smaller than the rate of Reaction (9.53) see Figure 9.5).

In actual practice, conventional transition state theory predicts that the rate of reaction (9.52) will be only 50 times smaller than rate of reaction (9.53) while experimentally, the rate is 500 times smaller. Therefore, transition state theory does predict the right qualitative trends, even if it does not give a quantitatively correct result. Still, there is the qualitative trend that as $q_t^{\ddagger}$ gets smaller, the rate gets smaller, too.

9.3.1   Limitations of Conventional Transition State Theory

One does not want to extrapolate the results too far, because there are many examples where changes in $q^{\ddagger}$ do not produce the results expected from transition state theory. One key prediction of conventional state theory is that rates will vary according to whether

$$\left[\frac{q_T^{\ddagger}}{q_A q_B}\right] = \exp\left(-\frac{-\Delta S^{\ddagger}}{k_B}\right) \tag{9.54}$$

is large or small, where $\Delta S^{\ddagger}$ is the entropy change in approaching the transition state and k_B is Boltzmann's constant. Physically, the term in square brackets in equation (9.54) represents the effects of the transition state entropy on the reaction. The term is small when there is a very narrow transition state (i.e., a sharp valley) and large when the transition state is wide. More precisely, the term in brackets in equation (9.54) is greater than one when the system gains degrees of freedom approaching the transition state. In contrast, the term is less than one when the system loses degrees of freedom when approaching the transition state. Years ago, people thought that when $q_T^{\ddagger}/q_A q_B$ increased, the rate would also increase.

Since late 1990s however, people have been able to calculate the transition state geometry and $q_T^{\ddagger}$ exactly using ab initio methods described in Chapter 11. These calculations have allowed people to test the idea that the rate constant increases as $[q_T^{\ddagger}/q_A q_{BC}]$ increases. Generally, it is found that there is **little correlation** between $[q_T^{\ddagger}/q_A q_{BC}]$ and the preexponential for reactions.

For example, Table 9.7 and Figure 9.6 show data for the quantity $h_p k_{A \rightarrow BC}^0 / k_B T$ for a number of reactions. According to equation (9.51)

$$\frac{h_p k_{A \rightarrow BC}^0}{k_B T} = \frac{q^{\ddagger}}{q_A q_{BC}} \tag{9.55}$$

Table 9.7 and Figure 9.6 show the data. Notice that there is no correlation between $h_p k_{A \rightarrow BC}^0 / k_B T$ and $q_T^{\ddagger}/q_A q_{BC}$. If equation (9.54) had worked, then all of the data in Figure 9.6 should have scattered around the line in the figure. However, clearly the data

Table 9.7   A comparison of experimental values of $(h_p k_{A \rightarrow BC}^0)/k_B T$ to those calculated from transition state theory

Reaction	Experiment $\dfrac{h_p k_{A \rightarrow BC}^0}{k_B T}$	Transition State Theory $\dfrac{q_T^{\ddagger}}{q_A q_{BC}}$
$NO + O_3 \rightarrow NO_2 + O_2$	0.15	.07
$NO + O_3 \rightarrow NO_3 + O$	1.05	0.02
$NO_2 + F_2 \rightarrow NO_2F + F$	0.26	0.02
$NO_2 + CO \rightarrow NO + CO_2$	0.32	1.05
$2NO_2 \rightarrow 2NO + O_2$	0.33	0.83
$NO + NO_2Cl \rightarrow NOCl + NO_2$	0.13	0.13
$2NOCl \rightarrow 2NO + Cl_2$	1.66	0.07
$NO + Cl_2 \rightarrow NOCl + Cl$	0.66	0.21
$H + H_2 \rightarrow H_2 + H$	2.34	8.33
$H + CH_4 \rightarrow H_2 + CH_3$	10.47	3.31
$CH_3 + C_2H_6 \rightarrow CH_4 + C_2H_5$	0.093	0.167

[a] The transition state calculations are from Laidler (1987). The experimental results are from Westley (1980) when available and, if not, from Laidler (1987)

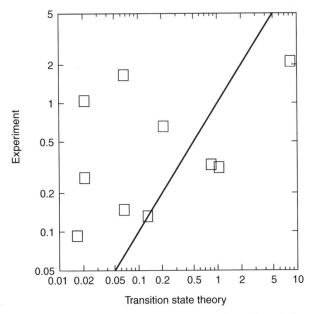

Figure 9.6 A plot of the results in Table 9.7. If the simple formation of transition state theory with constant transmission coefficients had worked, the results should have scattered around the line in the figure.

do not scatter around the line. In fact, for the examples shown, there is almost a negative correlation between the transition state theory predictions and the experiment.

To put the data in Table 9.7 in perspective, these were the first cases that people looked at because they were simple, and there were reasons to suspect that conventional transition state theory would work. As a result, the fact that the experimental values of $h_p k^0_{A \to BC}/k_B T$ show almost no correlation with those predicted by transition state theory suggests that the corrections associated with changes in $q^{\ddagger}_T/q_A q_{BC}$ do not have any correlation with variations in reaction rate.

The fact that conventional transition state makes sizable errors is quite important. From a practical standpoint, if one takes a simple approximation for k_0, the preexponential for a reaction

$$k_0 \approx \frac{k_B T}{h_P} \tag{9.56}$$

one would predict preexponentials within one or two orders of magnitude of the correct result. If one instead used

$$k_0 \approx \frac{k_B T}{h_P} \frac{q_T}{q_A q_B} \tag{9.57}$$

one would improve the quality of the predictions in some cases and decrease the quality of the prediction in other cases. On average, the predictions do not improve. We have included an example of a calculation of $q^{\ddagger}$ in Example 9.B. However, from a practical standpoint, conventional transition state theory does **not** predict more accurate rates of reaction than does collision theory except in special cases where a special reactant configuration is needed in order for reaction to occur.

Transition state theory is still rather useful in comparing the rates of two very similar reactions. We show below that the key assumption in transition state theory is that the errors in $q_x^{\ddagger}$ and $q_y^{\ddagger}$ exactly cancel. As long as the errors are constant, one can make useful comparisons even if the absolute rates are incorrect.

If one does isotope substitution, the errors are almost constant, except for the tunneling corrections, so transition state theory will give useful results. Transition state theory will also work well if one is looking at a case where the height of the barrier changes but the shape of the barrier does not. In Chapter 12, we will find that catalysts of solvents can sometimes change the height of the barrier without changing the shape of the barrier. However, it fails when we change the shape of the potential energy surface. Consequently, while transition is useful in comparing similar reactions, it is less useful for calculating absolute rates.

At this point it is useful to go back and ask which of the assumptions used to derive equation (9.45) caused the failures illustrated in Figure 9.6. Assumption 1 immediately preceding equation (9.44) is the critical one, and it is the one that made transition state theory famous. Transition state theory says that all of the trajectories that make it to transition state go on to make products, while collisions, which do not make it to the transition state, do not result in reaction. One can test that model using the methods in Chapter 8. When one does the calculations, one finds that one occasionally observes a trajectory like that in Figure 9.7, where the reaction crosses the transition state and goes back again. However, those are rare events, and do not produce significant error.

There is a related quantum effect that does cause some error. Recall from our discussion in Section 8.6 that according to classical trajectory calculations, if the reactant energy is below the top of the barrier, no reaction occurs. If the reactants have just enough energy to make it to the top of the barrier, the reactants stop at the transition state, and there is no reaction. In contrast, according to classical mechanics, if the reactants have enough extra energy to get over the barrier, the forward momentum of the reactions will carry them over the barrier, so the reaction probability will be unity.

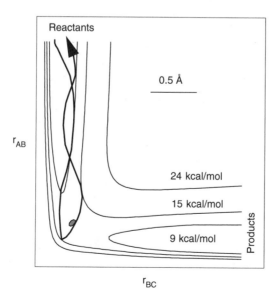

Figure 9.7 A recrossing trajectory.

Physically, classical mechanics views the reaction as motion of a ball over a hill. If you roll a ball up a hill and do not have enough energy to get over the hill, the ball will roll back down the hill. However, as soon as the ball has enough energy to traverse the hill, the ball will go over the hill and on to products.

The quantum result is different, however. Recall that there is some uncertainty in the momentum of reactants. If the reactants have an energy above the top of the barrier, then, on average, their momentum will carry them over the barrier. However, there will be a distribution of momentum including some movement in the negative direction. The result is that quantum-mechanically, the probability does not suddenly jump to unity as soon as you get above the top of the barrier. Instead, there is a gradual increase in the reaction probability, in the first 0.1–0.5 kcal over the top of the barrier. The result is that transition state theory slightly overestimates the rate of reaction. Transition state theory assumes that the reaction probabilities are unity when they are in fact less than unity.

A greater source of error occurs because of a quantum-mechanical effect called *tunneling*. If you have a reaction

$$A + BC \longrightarrow AB + C \qquad (9.58)$$

then quantum-mechanically, a molecule in a state just below the barrier has some probability of tunneling through the barrier as indicated in Figure 9.8. Physically, according to quantum mechanics there is some uncertainty in the position of the reactants. Classically one might say that the reactants are just approaching the barrier. However, quantum-mechanically there is some uncertainty in the position of the molecule, and so when you think that the reactants are just approaching the barrier, the wavefunction for the molecules will have a component on the other side of the barrier as shown in Figure 9.9. This uncertainty in the atomic positions allows a molecule to transport through a barrier. We call the transport through the barrier *tunneling*.

We will derive equations for the tunneling rate in Section 9.5. However, the thing to remember for now is that states just below the top of the barrier have some probability of tunneling through that barrier. Transition state theory ignores tunneling, so it can make a significant error when tunneling is important.

In practical situations, tunneling causes 30% errors in rate at normal temperatures, although it can cause larger effects at low temperatures or when hydrogen is involved.

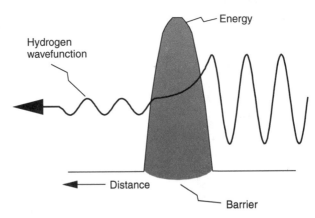

Figure 9.8 Tunneling through a barrier.

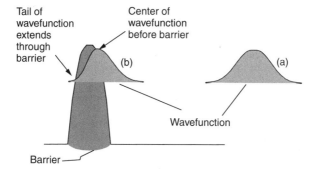

Tail of wavefunction extends through barrier

Center of wavefunction before barrier

(b)

(a)

Wavefunction

Barrier

Figure 9.9 A diagram showing the extent of the wavefunction for a molecule (a) by itself and (b) near a barrier. Notice that the wavefunction has a finite size (i.e., there is some uncertainty in the position of the molecule). As a result, when a molecule approaches a barrier, there is a component of the molecule on the other side of the barrier.

Another significant source of error arises because we are treating the motion parallel to the transition state as pure translation, and the motion perpendicular to the transition state as pure vibration.

In Section 9.2.1, we discussed these two assumptions in terms of the following example:

$$HNC \longrightarrow HCN \tag{9.59}$$

In this example, transtion state theory assumes that the bending of the HCN toward the transtion state can be treated as pure translation and the HCN bond stretching can be considered pure vibration. I think it is obvious that it is not such a good assumption to treat the HNC bending mode as a translational mode or the bond stretching as pure vibration. As a result, transition state theory clearly has problems for reaction (9.59).

The errors are less obvious if one has a reaction of the following form:

$$A + BC \longrightarrow AB + C \tag{9.60}$$

In reaction (9.60), A is moving toward BC, so at first sight, it would seem reasonable to treat the motion of A toward BC and over the transition state as translation. In practice, however, this is not such a good assumption.

Recall from Chapter 8 (see also Figure 9.10), the reactants need to turn on the potential energy surfaces as they approach the transition state, (i.e., vibrational energy is needed to

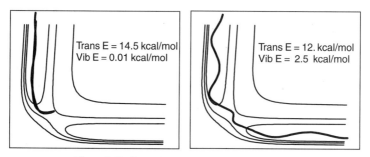

Trans E = 14.5 kcal/mol
Vib E = 0.01 kcal/mol

Trans E = 12. kcal/mol
Vib E = 2.5 kcal/mol

Figure 9.10 Two of the trajectories from Figure 8.21.

carry the reactants over the barrier). If you start with pure translational energy, as shown on the left of Figure 8.7, the system does not make it over the barrier. Instead, the reactants bounce apart as shown on the left of Figure 8.7. In contrast, if you put vibrational energy in, the reaction happens. In Chapter 8, we found that one observes reaction only if one gives the molecules the right amount of vibrational energy to carry the molecules over the barrier. The third key assumption in transition state theory is that the motion over the barrier is pure translation, and does not contain a vibrational component. The results in Chapter 8 show that this assumption is in error.

The fourth key assumption in conventional transition state theory is similar. According to the results in Chapter 8, during collision, some of the translational energy of the incident molecules is converted to vibration. The fourth key assumption in transition state theory is to ignore that coupling and assume that the motion is pure vibration. Again, the results in Chapter 8 show that this is a serious error. This assumption ignores the finding from Chapter 8 that it helps for the potential energy surface to be slowly banked so that molecules are directed into the transition state.

The discussion in the last few paragraphs would make it seem that transition state theory is a terrible approximation. However, it works out that there is often some cancellation of errors, so that transition state theory is a much better approximation than it would initially appear. Our third assumption is that the partition function for motion over the barrier is a translational — not a vibrational — partition function. Translational partition functions are in the order of 10^5, while vibrational partition functions are in the order of unity. If the real partition function were 60% translation and 40% vibration and tunneling raised the rate by 1.5, then $q_x^{\ddagger}$ would be given by

$$q_x^{\ddagger} = (q_t)^{0.6}(q_v)^{0.4}(1.5) = 1.5 \times 10^3 \tag{9.61}$$

where q_t is a translational partition function and q_v is a vibrational partition function. By comparison, transition state theory predicts q_x equals 10^5. This result implies that the $k_B T/h_p$ term in equation (9.45) is about a factor of 60 too large.

In contrast, if we assume that $q_y^{\ddagger}$ is 20% translation and 80% vibration, then $q_y^{\ddagger}$ will be given by

$$q_y^{\ddagger} = (q_v)^{0.8}(q_T)^{0.3} = 10 \tag{9.62}$$

by comparison, transition state theory predicts values of 1. Therefore, transition state theory under predicts $q_y^{\ddagger}$ by a factor of 10.

This brings up the cancellation of errors. According to transition state theory, the rate is proportional to the product of $q_x^{\ddagger}$ and $q_y^{\ddagger}$. If transition state theory makes $q_x^{\ddagger}$ a factor of 60 too big and $q_y^{\ddagger}$ a factor of 10 too small, then the product of $q_x^{\ddagger}$ and $q_y^{\ddagger}$ will be incorrect by a factor of 60×0.1 or a factor of 6. In Section 9.3.1, we will find that transition state theory is about a factor of 10 of the right answer and predicts many useful trends.

> In my view the key assumption in transition state theory is that the errors in $q_x^{\ddagger}$ and $q_y^{\ddagger}$ exactly cancel.

That is not an exact assumption, but it is a pretty good one. The assumption works best if you are trying to compare the rates of two very similar reactions so that the errors in one reaction will be almost the same as the errors in the second reaction. Transition state theory works less well in cases where you are trying to calculate absolute rate constants, so the errors matter.

There is also a related quantum effect that can be important. Transition state theory assumes that the vibrational motion of the reactants is easily converted into vibrational motion of the transition state. Sometimes, there is a mismatch in the vibrational frequencies. In that case the vibrational energy is not easily conversed, so one gets a reduction in the rate.

9.4   THE ROLE OF THE TRANSMISSION COEFFICIENT

Years ago, people tried to account for the errors by defining a **transmission coefficient** κ_T where the transmission coefficient is defined as the fraction of the reactive complexes that decay to products. In this formulation, the rate becomes

$$k_{A \to BC} = \left(\frac{kT}{h} \right) \kappa_T \frac{q^{\ddagger}}{q_A q_{BC}} \exp \left(\frac{-E^{\ddagger}}{k_B T} \right) \tag{9.63}$$

The transmission coefficient is usually less than unity in the absence of tunneling, but it is temperature-dependent. Generally, molecules with some energy in excess of that needed to cross the activation barrier have a higher probability of reacting to form products than molecules with barely enough energy to cross the barrier (see Figure 8.18). One gets more hot molecules with increasing temperature, so κ_T generally increases with increasing temperature. If one takes data over a wide range of temperatures, one finds that κ_T shows complex behavior with temperature. Still, one can fit κ_T to an Arrhenius form over a limited range of temperature;

$$\kappa_T = \kappa_T^0 \exp \left(\frac{E_T}{k_B T} \right) \tag{9.64}$$

Combining equations (9.63) and (9.64) shows that the apparent preexponential for reaction $k_{A \to BC}^0$ is given by

$$k_{A \to BC}^0 = \left(\frac{k_B T}{h_p} \right) \kappa_T^0 \left[\frac{q^{\ddagger}}{q_A q_{BC}} \right] \tag{9.65}$$

In the absence of tunneling κ_T is usually less than one, but κ_T^0 can be greater than one. Generally, κ_T^0 varies from 10^{-3} to 10^3. A κ_T^0 greater than one corresponds to data above the line in Figure 9.6, so clearly values of κ_T^0 greater than one are quite common. Further, Figure 9.6 shows that the variations in κ_T^0 are comparable to the variations in the terms in square brackets in equation (9.65). Physically, one changes κ_T by changing the shape of the potential energy. The highest values of κ_T come when the potential energy surface is shaped like a racetrack, with gently banked turns, and no sharp corners, so one can easily transfer energy into the bonds that need to get broken. There is some subtlety to the arguments because changes to the potential energy surface generally change both κ_T^0 and the term in brackets in equation (9.65). Often the variations in κ_T^0 are larger than the variations in the term in square brackets in equation (9.65). As a result, one does not necessarily speed up a reaction by increasing the terms in square brackets in equation (9.65) (i.e., when one gains degrees of freedom when approaching the transition state), nor does one necessarily slow down a reaction by reducing the terms in square brackets in equation (9.65).

9.5 TUNNELING

One reason for the errors in transition state theory is that transition state theory ignores tunneling. We briefly mentioned tunneling in Section 9.3. The idea in tunneling is that molecules that do not have quite enough energy to cross the activation barrier can still tunnel through the barrier and react as indicated in Figure 9.8.

Tunneling arises because of the uncertainty principle. Classically, the center of mass of the atom is a point with a fixed position. However, quantum-mechanically, the position of the atom can be represented by a wavefunction with a finite width as indicated in Figure 9.9. The wavefunction for a molecule does not end abruptly when the molecule hits a barrier. Instead, the wavefunction falls off exponentially. If the barrier is thin, there is some probability of the wavefunction transversing the barrier.

Tunneling corrections are most important for electron transfer reactions such as

$$Fe^{3+} + Fe^{2+} \longrightarrow Fe^{2+} + Fe^{3+} \tag{9.66}$$

because electrons tunnel easily. It is also important in proton transfer reactions:

$$OH + C_2H_6 \longrightarrow H_2O + C_2H_5 \tag{9.67}$$

However, tunneling is *usually* less important with any atom heavier than hydrogen, except at low temperatures.

9.5.1 An Equation for Tunneling Rates

Next, it is interesting to solve a simple example to illustrate how tunneling arises. Consider scattering of an atom through the square-wave potential barrier shown in Figure 9.11. For the purposes of derivation, it will be assumed that the barrier is given by

$$V(X_A) = \begin{cases} 0 & \text{for} \quad X_A > 0 \\ V_B & \text{for} \quad 0 > X_A > -d_b \\ 0 & \text{for} \quad X_A < -d_b \end{cases} \tag{9.68}$$

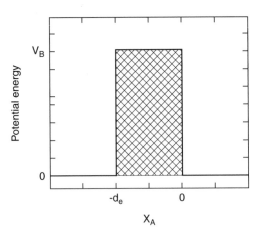

Figure 9.11 A plot of the square-well barrier.

Consider the collision of a beam of molecules onto the barrier. Assume that molecules start out at $X_A = \infty$ moving toward the barrier. The molecules will hit the barrier and then either reflect from the barrier or be transmitted through it. When $X_A > 0$, the Schroedinger equation is

$$\frac{\hbar^2}{2m_i}\frac{d^2}{dX_A^2}\psi_i = E\psi_i \tag{9.69}$$

There are two solutions of equation (9.69). A solution consisting of an atom directed toward the barrier

$$\psi_I = e^{-iK_iX_A} \quad \text{for} \quad X_A > 0 \tag{9.70}$$

and a reflected atom

$$\psi_r = c_i \exp(iK_iX_A) \tag{9.71}$$

where c_i is a constant and K_i is related to the incident energy, E_i, and mass, m_i, of the molecule by

$$K_i = \sqrt{\frac{2m_iE_i}{\hbar^2}} \tag{9.72}$$

where $\hbar$ is Planck's constant over 2π.

In the region where the potential is V_B, the Schroedinger equation becomes

$$\left(\frac{\hbar^2}{2m_i}\frac{d^2}{dX_A^2} + V_B\right)\psi_i = E\psi_i \tag{9.73}$$

The solution of the Schroedinger equation is

$$\psi_i = c_1 e^{-ip_iX_A} \tag{9.74}$$

where c_1 is a constant and

$$p_i = \sqrt{\frac{2m_i(E_i - V_B)}{\hbar^2}} \tag{9.75}$$

One can solve for c_1 by noting that the wavefunction and its derivatives must be continuous at $X_A = 0$. The result for when d_e is large is

$$\psi_i = e^{-iK_iX_A} + \left(\frac{K_i - p_i}{K_i + p_i}\right)e^{-iK_iX_A} \quad \text{for} \quad X_A \geq 0 \tag{9.76}$$

$$\psi_i(X_A) = \left(\frac{2K_i}{K_i + p_i}\right)e^{-ip_iX_A} \quad \text{for} \quad X_A \leq 0 \tag{9.77}$$

People often rewrite equations (9.76) and (9.77) as

$$\varphi_i = \begin{cases} \varphi_I + \varphi_r & \text{for} \quad X_Z > 0 \\ \varphi_T & \text{for} \quad X_A \leq 0 \end{cases} \tag{9.78}$$

where φ_I = incident wavefunction
φ_r = reflected wavefunction
φ_T = transmitted wavefunction

Analysis of equation (9.76) shows $\varphi_I = e^{-K_i X_i}$, which is the same as first term in equation (9.76):

$$\psi_r = \left(\frac{K_i - p_i}{K_i + p_i}\right) e^{iK_i X_A} \tag{9.79}$$

Equation (9.85) predicts $\varphi_r = [(K_i - p_i)/(K_i + p_i)]e^{iK_i X_A}$; physically, φ_r is the wavefunction for the molecules that reflect from the barrier without reacting, while φ_T represents the wavefunction for the molecules that tunnel into the barrier and so can react.

The surprising thing, however, is that even when E_i is less than V_B, the wavefunction is nonzero for $X_A < 0$ (i.e., within the barrier). As a result, quantum-mechanically, molecules penetrate into the barrier, even though classically, no molecules should penetrate into the barrier (i.e., to $X_A < 0$).

Figure 9.12 shows a plot of the wavefunctions when $E_i < V_B$. Notice that the wavefunction shows an exponential decay into the barrier. Thus, a molecule within the barrier probably decays as the barrier thickens. More precisely, $P(d_e)$, the probability of finding a molecule some distance, $-d_e$, in the barrier, is

$$P(-d_e) = \psi^*(-d_e)\psi(-d_e) \tag{9.80}$$

Substituting (9.79) into (9.80) yields

$$P(-d_e) = \left|\left(\frac{2K_i}{K_i + p_i}\right) \exp\left(-2d_e\sqrt{\frac{2m_i(V_B - E_i)}{\hbar^2}}\right)\right| \tag{9.81}$$

If the step in the potential ends at d_e, then there will be some probability of a molecule moving through the barrier. If one assumes a barrier of the form in equation (9.68), there is an artifact associated with the quantum reflection at the sharp drop in potential at $X_A = -d_e$. However, the key result is that the tunneling rate can be approximated by equation (9.81). Note that the tunneling rate decreases exponentially as the square root of the mass, and as $(V_B - E_i)$. Consequently, tunneling rates are highest if the barriers are thin (i.e., d_e is small), and the species that tunnels has a low mass.

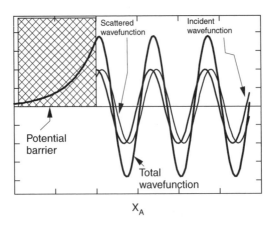

Figure 9.12 The real part of the incident (φ_i), scattered (φ_r), and total wavefunction (φ_T) for the square-well barrier.

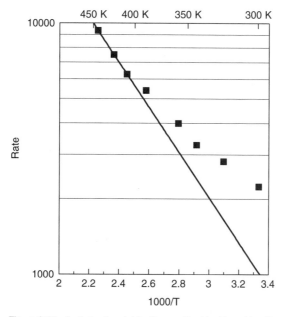

Figure 9.13 An Arrhenius plot for the reaction $H + H_2 \rightarrow H_2 + H$.

Generally, tunneling is very important for electron transfer reactions such as reaction (9.66) because electrons have low masses. Tunneling can also be seen in proton transfer reactions such as

$$D + H_2 \longrightarrow HD + H \qquad (9.82)$$

$$O + CH_4 \longrightarrow OH + CH_3 \qquad (9.83)$$

Tunneling rates are much higher with hydrogen than with deuterium, so one can easily detect them by looking for a large isotope effect. Generally, one can detect tunneling only at low temperatures. At higher temperatures, the molecules traversing the barrier swamp those that tunnel.

You can look for tunneling experimentally, by looking for deviations from Arrhenius' law. Molecules can react even though there is insufficient energy for them to cross the barrier. As a result, you get extra reaction at low temperatures. The extra reaction causes the Arrhenius plots to curve up at low temperatures. For example, Figure 9.13 shows some data for the following reaction:

$$H + H_2 \longrightarrow H_2 + H \qquad (9.84)$$

Note that the rate of 300 K is a factor of five higher than you would expect from Arrhenius' law. The factor of 5 increase is thought to be associated with tunneling.

9.5.2 Ekhart Approximation for Tunneling

Next we want to present an equation for tunneling. Equation (9.81) is not actually used in practice, because real potentials are not square wells. Instead, the potentials look more like hills. People use different approximations to account for the fact that the potential is more like a hill than a square wave.

The simplest approximation is to assume that the barrier is what is called an *Eckart potential*:

$$V(X_A) = \frac{A \exp(\textrm{R}/L)}{1 + \exp(\textrm{R}/L)} + \frac{B \exp(\textrm{R}/L)}{(1 + \exp(\textrm{R}/L))^2} \tag{9.85}$$

where A, B, and L are constants.

Figure 9.14 shows a plot of the Eckart potential; it is a simple hill with a banner of V_0 and a heat of reaction of $V_0 - V_1$, where

$$A = V_0 - V_1$$

$$B = ((V_0)^{1/2} - (V_1)^{1/2})^2$$

$$L = \frac{1}{2\pi v_C} \sqrt{\frac{2}{\mu_{AB}}} \left(\frac{V_0 V_1}{(V_0)^{0.5} + (V_1)^{0.5}} \right) \tag{9.86}$$

where v_c is the curvature of the barrier and μ_{AB} is the reduced mass.

One can solve the Schroedinger equation to obtain an exact expression for the tunneling rate through an Eckart barrier. The solution requires a knowledge of hypergeometric functions. A detailed derivation is given in Landau and Lifshitz [1965] or.

$$P(E^\ddagger - E) = \frac{\sin h(a) \sin h(b)}{\sin h^2((a + b)/2) + \cos h^2(c)} \tag{9.87}$$

where $P(E^\ddagger - E)$ is the tunneling probability, $E^\ddagger$ is the transition state energy, E is the energy of the particle, and

$$a = \frac{4\pi}{hpv_t} \sqrt{E^\ddagger - E + V_0} \left(\frac{V_0^{1/2}}{1 + (V_0 V_1)^{1/2}} \right)$$

$$b = \frac{4\pi}{hpv_t} \sqrt{E^\ddagger - E + V_1} \left(\frac{V_0^{1/2}}{1 + (V_0 V_1)^{1/2}} \right)$$

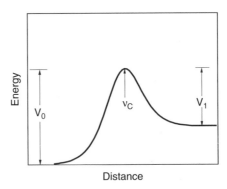

Figure 9.14 The Eckart potential.

$$c = 2\pi\sqrt{\frac{V_0V_1}{h_P\nu_i} - \frac{1}{16}} \tag{9.88}$$

and h_P is Planck's constant.

One then needs to integrate equation (9.87) over all energies. After considerable algebra and some simplification, the result is that the transmission coefficient for tunneling can be approximated by

$$\kappa(T) = 1 + \frac{1}{24}\left(\frac{h_P\nu_C}{k_BT}\right)^2\left(1 + \frac{k_BT}{E^{\ddagger}}\right) \tag{9.89}$$

where $\kappa(T)$ is the transmission coefficient, ν_i is the curvature of the barrier, k_B is Boltzmann's constant, T is temperature, and h_P is Planck's constant. Equation (9.89) is called *Wigner's approximation to the barrier*. Note that the curvature of the barrier is the frequency of vibrational mode that you would have if you turned the barrier upside down to form a potential well. Equation (9.89) is the tunneling approximation used most often in the literature. Example 9.A shows how it is used. One should read that example before proceeding.

9.5.3 Semiclassical Approximations for Tunneling

The alternative approach is to use a semiclassical approximation for tunneling. The semiclassical approximation actually arises from what is called a *path integral formation*. Details are given by Miller (1976, 1993, 1998). Here we will derive a simple equation to get the structure of the solutions.

According to equation (9.81), when a particle tunnels through a barrier of differential thickness dx, the wavefunction will be reduced by

$$\frac{\psi(x+dx)}{\psi(x)} = \exp\left(-dx\sqrt{\frac{2m_i(V_B - E_i)}{\hbar^2}}\right) \tag{9.90}$$

taking the log of both sides and rearranging, we obtain

$$\frac{\ln(\psi(x+dx)) - \ln(\psi(x))}{dx} = -\sqrt{\frac{2m_i(V_B - E_i)}{\hbar^2}} \tag{9.91}$$

in the limit that dx = 0:

$$\frac{d\ln(\psi(x))}{dx} = -\sqrt{\frac{2m_i(V_B - E_i)}{\hbar^2}} \tag{9.92}$$

Now let's integrate equation (9.92) for motion through a barrier. We will assume that $x = x_0$ is the point where the molecule hits the barrier.

Integrating equation (9.92), we

$$\ln(\psi(x)) - \ln(\psi(x_0)) = \left(-\int_{x_0}^{x}\sqrt{\frac{2m_i(V_B - E_i)}{\hbar^2}}dx\right) \tag{9.93}$$

Note that when you travel through a barrier, V_B varies with x as shown in Figure 9.14. Consequently, the square root cannot be taken out of the integral in equation (9.93).

Taking the exp( ) of equation (9.93) and rearranging yields

$$\psi(x) = \psi(x_0)\exp\left(-\int_{x_0}^x \sqrt{\frac{2m_i(V_B - E_i)}{\hbar^2}}\,dx\right) \tag{9.94}$$

The probability that the molecule gets through the barrier is $\psi^*\psi$. Therefore

$$P(x) = \exp\left(-2\int_{x_0}^x \sqrt{\frac{2m_i(V_B - E_i)}{\hbar^2}}\,dx\right) \tag{9.95}$$

Equation (9.95) is the semiclassical approximation for tunneling through a one-dimensional barrier.

An analogous equation is derived in Miller (1998) for the three-dimensional case. The result is

$$P(S) = \exp\left(-2\int_{S_0}^S \sqrt{\frac{2m_i(V_B - E_i)}{\hbar^2}}\,dS\right) \tag{9.96}$$

where S is a coordinate that moves along the trajectory that the molecule uses to traverse the barrier.

In order to use equation (9.96) quantitatively, one needs to make an assumption about the path that the molecule makes as it traverses the barrier. Figure 9.15 shows some of the paths that people use. Experimentally, the Marcus–Coltrin path seems to be a particularly good one, although other approximations often are used.

In order to use equation (9.96), one needs to pick an energy and do the integration over an assumed path. One then uses Laguere integration (see Example 9.8D) to estimate

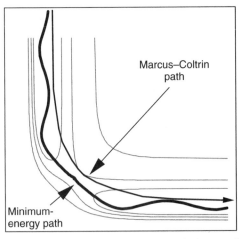

Figure 9.15 Some of the paths people assume to calculate tunneling rates.

a rate. The calculations are tedious, but accurate. Transition state theory with tunneling corrections usually can predict rate constraints accurate to a factor of approximately 1.5.

9.6 VARIATIONAL TRANSITION STATE THEORY

Over the years, there have been many attempts to improve transition state theory. One of the first approximations is called *variational transition state theory* (VTST). According to conventional transition state theory, the rate constant for the reaction is given by

$$k_{A \to BC} = \frac{k_B T}{h_P} \frac{q_T^{\ddagger}}{q_A q_{BC}} \exp[-E^{\ddagger}(k_B T)] \tag{9.97}$$

where $q_T^{\ddagger}$ is evaluated at the saddle point in the potential energy surface. Recall, from Section 9.4.1, that the actual barrier to the reaction is shifted by the zero-point energy and other effects. The zero-point energy can change during the course of the reaction. Consequently, the actual barrier is different from the energy of the transition state, and the maximum may occur at a point slightly different from the saddle point in the potential energy surface.

This effect is most important for a potential energy surface such as that in Figure 9.16, where the potential energy surface narrows significantly on the way to products. Recall that the zero-point energy increases as the potential narrows. As a result, the zero-point energy increases in Figure 9.16 moving to products. Consequently, the maximum barrier is not quite at the saddle point in Figure 9.16.

One can quantify this effect by defining ΔG_T at each point in the potential energy surface via

$$\frac{q_T}{q_A q_{BC}} \exp\left(-\frac{E}{k_B T}\right) = \exp\left(-\frac{\Delta G_T}{k_B T}\right) \tag{9.98}$$

According to variational transition state theory, the transition state for the reaction is the saddle point in a plot of ΔG_T rather than the saddle point in the total energy.

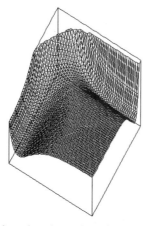

Figure 9.16 A potential energy surface where the maximum free energy occurs at a point different than the saddle point. Notice that the reactant vibrational well is much wider than the product vibrational well.

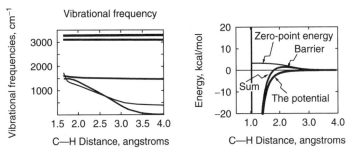

Figure 9.17 The vibrational frequency, zero-point energy potential energy, and the sum of potential and zero-point energies for reaction (9.51) as a function of the C–H bond length. [After Hase et al (1987)].

In actual practice, this effect is not very important except in barrierless reactions. For example, the reaction

$$X + H + CH_3 \longrightarrow HCH_3 + X \tag{9.99}$$

is barrierless. The reaction has two radicals, and there are no Pauli repulsions to keep the reactants apart. Consequently, there is no transition state. However, the H–C bond becomes increasingly localized as the reaction proceeds.

Figure 9.17 shows how the vibrational and zero-point energy changes as the reaction proceeds. Notice that the vibrational frequencies increase as the C–H bond shortens, which raises the zero-point energy. If everything is right, one can get a barrier under those circumstances.

Transition state theory does not make any predictions for reaction (9.99), because there is no transition state. However, variational transition state theory predicts a second-order rate constant of about 2.8×10^{14} Å^3/(molecule·second at 300 K. By comparison, collision theory gives a rate constant of 5×10^{14} Å^3/(molecule·second and the experimental value is 4.6×10^{14} Å^3/(molecule·second).

Truhlar et al (1985, 1996) review 33 cases where variational transition state was tested in detail. Reaction (9.99) is by far the best example of variational transition state within a factor of 1.7 of the right answer at 300 K. Unfortunately, variational transition state theory is a factor of 15 off at 1000 K. All of the other cases considered by Truhlar were worse. Of the 33 cases, 16 were off by a factor of 5 or more at 300 K, and 5 cases were off by a factor of 10 or more. None of the examples was within a factor of 10 at 1000 K. As a result, variational transition state theory without tunneling does not improve predictions very much.

Still, variational transition state theory, without tunneling, is discussed at length in the literature, so one should be familiar with it.

9.7 MULTICONFIGURATIONAL TRANSITION STATE THEORY

Fortunately, there is a version of transition state theory that does much better than collision theory. This theory is called *multiconfigurational transition state theory* (MCTST).

In my view, MCTST is not transition state at all; rather, it goes back to the exact result, equation (9.25):

$$k_{A \to BC} = \frac{k_B T}{h} \frac{q^+}{q_A q_{BC}} \exp[-E^+ k_B T] \tag{9.100}$$

with q^+ as given by equation (9.40):

$$q^+ = \sum_N g_N P_N^{hot} \frac{\exp[-(E_N - E^+)/k_B T]}{V} \tag{9.101}$$

where g_N is the degeneracy of state N of the reactants, E_N is the energy of the state, P_N^{hot} is the reaction probability when the reactants are in state N; E_N is the energy of system; $E^{\ddagger}$ is the saddle point energy; k_B is Boltzmann's constant, V is the volume per molecule, and T is the absolute temperature.

The idea in multiconfiguration transition state theory is to find a way to estimate the P_N^{hot} in equation (9.101) without making the assumptions in transition state theory. To set the stage for the work, Figure 9.18 is a replot of results in Figure 8.11, where some of the trajectories make it to products and some do not. Notice that a few of the trajectories that get past the transition state do not make it to products. However, all of the trajectories that make it to the right edge of the dashed box in Figure 9.18 do make it to products. Consequently, if one could find a way to calculate the probability that a given state m makes it from the top edge to the right edge of the box in Figure 9.18, one could obtain an accurate value for the P_m terms.

It happens that one can get an excellent approximation to the P_N^{hot} using the results in Section 6.3.1. In Section 6.3.1, we presented the approximation

$$P_N^{hot}(S) = \exp\left(-2\int_{S_0}^{S}\sqrt{\frac{2m_i(V_0 - E_i)}{\hbar^2}}\,dS\right) \tag{9.102}$$

for tunneling. People often use a closely related approximation for the P_N^{hot}:

$$P^{hot} = \frac{1}{1 + \exp\left(2\int_{S_0}^{S}\sqrt{\frac{2m_i(V_0 - E_i)}{\hbar^2}}\,dS\right)} \tag{9.103}$$

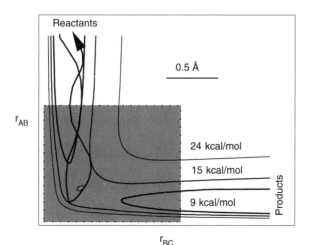

Figure 9.18 Some trajectories from the reactants to products.

where P^{hot} is an approximation to P_N^{hot} Note that equation (9.103) also applies to states with enough energy to cross the barrier. In order to use equation (9.103), one has to do a trajectory calculation. Still, the results are usually within 30% of the exact results, which is as good as you can usually measure anything in kinetics. An alternative is to use equation (9.87) to calculate P_N^{hot} that produces errors in the order of 2.

In actual practice, one calculates values of κ_N where

$$\kappa_N = P_N \exp\left(-\frac{(E_N - E^\ddagger)}{k_B T}\right) \qquad (9.104)$$

κ_N varies with the width of the barrier, the width of the opening over the barrier, and how much the potential energy surface curves in moving toward the barrier.

Truhlar et al (1996, 1984) compare the results of the MCTST calculations to the variational TST results for the 33 cases where the reaction rate is known exactly. In all cases, MCTST was within 30% of the exact result, while variational transition state theory (VTST) is often off by a factor of 10. This example illustrates the importance of tunneling in determining rates of biomolecular reactions.

9.8   UNIMOLECULAR REACTIONS

At this point we will be changing topics. So far we have been discussing bimolecular reactions. Next, we want to discuss unimolecular reactions.

A *unimolecular reaction* is a reaction that starts with a single molecule, AB, and the molecule either isomerizes to form a new species, C, or splits apart to form two or more products. Examples of unimolecular reactions include the isomerization of cyclopropane:

$$\begin{array}{c} CH_2 \\ / \ \backslash \\ H_2C - CH_2 \end{array} \Longrightarrow H_2C = CH - CH_3 \qquad (9.105)$$

and the scission of H_2 into atoms:

$$H_2 \longrightarrow 2H \qquad (9.106)$$

We discussed unimolecular reactions in Chapter 8. Unimolecular reactions are fundamentally different from bimolecular reactions. If we start with a unimolecular reaction, the reactants are inherently stable. The reactants must collide with a collision partner or be excited with a laser before reaction can occur. The fact that the reactant molecules must be excited before reaction happens is a fundamental feature of unimolecular reactions, which is not seen in bimolecular reactions.

There are three ways to initiate a unimolecular reaction:

- Via a laser or photon activation
- Via thermal activation
- Via chemical activation

Laser activation is simple. You shine a laser on a reactant molecule and put the molecule in an excited state, and then the molecule can fall apart. Laser methods are the most accurate way to study unimolecular reactions. *Thermal activation* is a much more complex process. The reactant molecule collides with a collision partner to produce a hot species that then decomposes. *Chemical activation* is even more complicated. A radical combines with a stable species to yield an excited molecule that then decomposes.

In the next several sections, we will discuss each process separately, and derive equations for the rate of the reaction.

9.9 LASER PHOTOLYSIS

Let's start by considering a laser photolysis experiment. In a laser photolysis experiment, one excites a molecule with a laser, and then the molecule breaks into products. For example, when deuterated ketene (CD_2CO) is excited with a UV laser, one forms an excited form of ketene, (CD_2CO^*). The (CD_2CO^*) then decomposes into products:

$$CD_2CO + h\nu \xrightarrow{\ 1\ } CD_2CO^* \xrightarrow{\ 1\ } CD_2 + CO^* \qquad (9.107)$$

Next, we will want to derive an expression for k_2, the rate at which excited molecules decay into products. Our derivation will be to first use Arrhenius' model to calculate an expression for k_{-2}, the rate of the reverse reaction:

$$CD_2^* + CO^* \xrightarrow{\ -2\ } CD_2CO^* \qquad (9.108)$$

and then use the equilibrium expression

$$\frac{k_2}{k_{-2}} = K_2^{eq} = \frac{q_{CD_2} q_{CO}}{q^*} \qquad (9.109)$$

where k_2 and k_{-2} are the rate constants for reactions 2 and -2, K_2^{eq} is the equilibrium constant; q^* is the partition function for an excited ketene molecule, and q_{CD_2} and q_{CO} are the partition functions for CO and H_2.

First, we want to note that equation (9.25) does not apply directly to reaction (9.108). Equation (9.25) was derived for the case where the reactant molecules were in thermal equilibrium with a heat bath. When you excite a molecule with a laser, you deposit a specific quantum of energy into the molecule. Therefore, the system is not in thermal equilibrium. Instead, the product molecule has a specific translational energy, E_T, and a specific vibrational–rotational energy, E_{VR}. When we run the reaction in reverse, we need to start with reactants with a specific translational and vibrational–rotational energy. The reactants are not in thermal equilibrium. Consequently, equation (9.25) does not apply.

Next, we will derive an equivalent to equation (9.25) for a case where molecules have a specific E_T and E_{VR}. Most of the derivation is the same as in Section 9.1. In particular, equations (9.11), (9.13), and (9.19) apply, except that we have to use different values of the partition functions, since the system is not in thermal equilibrium with a heat bath, and the $E^{\ddagger}/k_B T$ term in all of the expressions goes away since we do not have a thermal distribution of energies. Combining equations (9.11), (9.13), and (9.19) for the case where the reactions have a specific energy yields

$$k_{-2} = \frac{1}{2d_{ABC}} \bar{v}_{ABC}(d_{ABC} q_{A \to BC}) \frac{q^{\ddagger}}{q_{CD_2} q_{CO}} \qquad (9.110)$$

Next, we note that the distribution of states with a laser differs from the distribution in a thermal experiment. As a result, one needs new expressions for $\bar{v}_{ABC}$ and $q_{A \to BC}$. First, let's get an expression for $\bar{v}_{ABC}$. If we assume that the molecules have a translational energy of E_T, then the average velocity is related to the energy by

$$\tfrac{1}{2}\mu_{ABC}(\bar{v}_{ABC})^2 = E_T \tag{9.111}$$

Solving equation (9.111) for $\bar{v}_{ABC}$ yields

$$\bar{v}_{ABC} = \left(\frac{2E_T}{\mu_{ABC}}\right)^{1/2} \tag{9.112}$$

Equation (9.112) gives the velocity of molecules with a translational energy of E_T.

Next, we will derive an expression for $q_{A \to BC}$. Recall that the partition function for translation comes from the solution of the Schroedinger equation for a particle in a box. The energy levels for a particle in a box are given by

$$E_n = \frac{n^2 h_P^2}{8\mu_{ABC}d_{ABC}^2} \tag{9.113}$$

where n is the index of the states in the box, h_P is Planck's constant, μ_{ABC} is the effective mass, and d_{ABC} is the size of the box. In a laser photolysis experiment, E_T is fixed between E_T and $E_T + dE_T$. Consequently, $q_{A \to BC}^* d_{AB}$ equals the number of states between E_T and $E_T\, dE_T$:

$$q_{A \to BC}d_{ABC} = \frac{dn}{dE_n} \tag{9.114}$$

Solving equation (9.113) for n and plugging into equation (9.114) yields

$$q_{A \to BC}d_{ABC} = \frac{d_{ABC}}{h_P}\left(\frac{\mu_{ABC}}{E_T}\right)^{1/2} \tag{9.115}$$

Substituting equations (9.112) and (9.115) into (9.110) yields

$$k_{-2} = \frac{1}{2d_{ABC}}\left(\frac{2E_T}{\mu_{ABC}}\right)^{1/2}\left(\frac{d_{ABC}}{h_P}\right)\left(\frac{2\mu}{E_T}\right)^{1/2}\frac{q^\ddagger}{q_{CD_2}q_{CO}} \tag{9.116}$$

Canceling out terms in equation (9.116) yields

$$k_{-2} = \frac{1}{h_P}\frac{q^\ddagger}{q_{CD_2}q_{CO}} \tag{9.117}$$

Substituting equation (9.117) into equation (9.109) yields

$$k_2 = \frac{1}{h_P}\frac{q^\ddagger}{q^*} \tag{9.118}$$

where $q^{\ddagger}$ is the partition function for molecules that react, q_{A*} is the partition function for the reactants after excitation by the laser, and h_P is Planck's constant. Equation (9.118) is called the *microcanonical expression for a rate constant*. Equation (9.118) is the fundamental equation that one uses to analyze laser photolysis.

9.9.1 Simplifications of the Microcanonical Rate Equations

Next, we need to get expressions for the partition functions in equation (9.118). For the purposes of derivation, we will assume that the molecule is excited with an energy E^* and use E^* as the energy zero for equation (9.40). Equation (9.40) becomes

$$q^{\ddagger} = \sum_m P_m g_m \qquad (9.119)$$

where P_m is the probability that a reaction occurs when the system is in state m and g_m is the degeneracy in state m. Note that one should consider states only with a total energy of E^* in equation (9.119).

There is some subtlety in equation (9.119). When you initially excite ketene with a laser, you put all of the energy of the laser into vibrations, rotations, and electronic excitations. However, when the ketene dissociates, energy is deposited into translation, rotation, vibration, or electronic excitations. Let us call E_T, the energy that is put into translation, and E_{VR} the energy put in rotations, vibrations, and electronic excitations. Then the sum in equation (9.119) should go over all states that satisfy

$$E_T + E_{VR} = E^*. \qquad (9.120)$$

Note that whenever

$$E_{VR} \leq E^* \qquad (9.121)$$

you can always pick a value of E_T and satisfy equation (9.120). Therefore, the sum in equation (9.119) should go over all states that satisfy equation (9.121).

The analysis is different for q_{A*}. By definition

$$q_{A*} = \sum_n g_n \qquad (9.122)$$

where the sum in equation (9.122) goes over states where $E_{VB} = E^*$.

Next we want to put equations (9.118) to (9.122) into standard notation. In the literature, people often write equation (9.118) differently. Recall that q^* is the number of total states of the system with an energy between E^* and $E^* + dE_T$. In the literature, people call that $N(E^*)$ rather than q^* to make it clear that the energy is fixed. $N(E^*)$ is defined by

$$N(E^*) = q^* \qquad (9.123)$$

Similarly, $q^{\ddagger}$ is just the average value of g_n when E^* is fixed and E_{VR} is allowed to vary from 0 to E^*. People often define

$$G(E^*) = \sum_m P_m g_m \qquad (9.124)$$

to emphasize the dependence on E^*.

Substituting the definitions of $G(E^*)$ and $N(E^*)$ into equation (9.118) yields

$$k_2(E^*) = \frac{1}{h_P} \frac{G(E^*)}{N(E^*)} \tag{9.125}$$

where we have noted that k_2 is a function of E^*. Equation (9.125) is the exact result for the rate constant for decomposition of a hot molecule with energy E^*.

9.9.2 Microcanonical Transition State Theory

Unfortunately, equation (9.125) is not that useful. Notice that in order to use the equation, one needs an expression for the P_m terms, which are the probability that a molecule in state m reacts. In general, one will not know the P_m value but can, however, estimate them from transition state theory. In transition state theory, $P_m = 0$ if $E_{VR} < E^{\ddagger}$, $P_m = 1$ if $E_{VR} \geq E^{\ddagger}$, where E_{VR} is equal to the vibrational rotation of the molecules and $E^{\ddagger}$ is the total energy at the transition state. Let's define a new quantity, $G^T(E^*)$, by

$$G^T(E^*) = \sum_s g_s \tag{9.126}$$

The sum in equation (9.126) goes over the states with $E_{VR} > E^{\ddagger}$ at the transition. $G^T(E^*)$ is the transition state approximation to $G(E)$.

Combining equations (9.125) and (9.126) yields

$$\boxed{k_2(E^*) = \frac{1}{h_P} \frac{G^T(E^*)}{N(E^*)}} \tag{9.127}$$

where $N(E^*)$ is the number of reactant states with an energy between E^* and $E^* + \delta E^*$, $G^T(E^*)$ is the total number of states with E_{VR} from $E^{\ddagger}$ to E^*, and h_P is Planck's constant. People call equation (9.127) the **microcanonical transition state approximation** to the rate. The use of the approximation in (9.126) is called **microcanonical transition state theory** (MTST). This equation will be very important for our discussion in the rest of this chapter.

Equation (9.127) is the key equation used to analyze laser photolysis. One should memorize the equation before proceeding.

9.10 THE RRKM MODEL

There is one subtlety in the definition of $G^T(E^*)$; one needs to consider which state to include in the sum in equation (9.126). All of the states of the system are considered in the exact result, equation (9.124), but the P_m values could be small for some of the states. Those states do not contribute to the reaction. For example, in reaction (9.107), the out-of-plane bending of the CD_2 might not be very important to the reaction. Well, in the exact result, that is not much of an issue since P_m is small for a state that does not contribute to the reaction. However, one assumes that the P_m values are 1.0 when one computes the sum in equation (9.126). Therefore, one needs to be careful to only include states that actually contribute to the reaction in the sum in equation (9.126)

Now it is not obvious which states contribute to the reaction. With a bimolecular reaction such as reaction (9.1), one would include only one group of states: the states of the A–BC molecular complex at the saddle point in the potential energy surface. However, that does not work for laser photolysis. Instead, one has to include many more states on the sum.

The reason is subtle. When you excite a molecule with a laser, the molecule goes into a metastable state. There is no place for energy to go, and so the molecule stays excited until something happens to deexcite the molecule. Molecules stay hot for perhaps 10^{-8} seconds. By comparison, a bimolecular collision lasts only 10^{-12} seconds. Consequently, there is much more time for energy to get localized in a bond when you excite a molecule with a laser than in a bimolecular collision. In a bimolecular collision, you need to initially put all of the energy into the bond that breaks in order to get reaction to happen. However, in laser photolysis, bonds can break even if you do not start with energy in the correct bonds.

Rice and Ramsperger (1927,1928) and Kassel (1928 a,b,c, 1935) suggested a model to quantify this idea. The original model does not work well for laser photolysis because it ignored that there are specific energy levels in a molecule. However, Marcus (1952) suggested a simple modification to the model that can explain the data. The model is now called the **RRKM model**.

The idea in the RRKM model is that when you excite a molecule with a laser, it does not matter so much which modes of the molecule are excited initially. There is rapid energy flow within a molecule, so even if you put energy into the wrong vibrational mode, the energy will eventually get over to the right mode and the bonds will break. More specifically, RRKM theory assumes that

- $N(E^*)\delta E^*$ is the number of vibrational modes of the reactants with a vibrational energy between E^* and $E^* + \delta E^*$.
- $G^T(E^*)$ is the number of vibrational modes of the transition state with a vibrational energy between $E^{\ddagger}$ and E^* independent of whether the mode directly couples to bond scission.

One then treats the rotational modes separately using

$$k_2(E^*) = \frac{1}{h_P} \left(\frac{q_R^{\ddagger}}{q_{A^*}} \right) \frac{G_V^T(E*)}{N_V(E*)} \tag{9.128}$$

where $G_V^T(E*)$ is the number of vibrational states at the transition state, with an energy between $E^{\ddagger}$ and E^*; $N_V(E^*)\delta E^*$ is the number of vibrational states of the reactants with an energy between E^* and $(E^* + \delta E^*)$; $q_R^{\ddagger}$ is the rotational partition function for the transition state; and q_{A^*} is the rotational partition function for the excited products.

Quantitatively, the RRKM model assumes that equation (9.124) works and that $N(E^*)$ is the total number of vibrational modes of the reactant with a vibrational energy between E^* and $E^* + \delta E^*$, and that $G^T(E^*)$ is the total number of vibrational modes of the reactants with a vibrational energy between $E^{\ddagger}$ and E^*.

It is easy to show that

$$G^T(E^*) = \int_{E^{\ddagger}}^{E^*} N^T(E^*) \, dE^* \tag{9.129}$$

where $N^T(E^*)\delta E^*$ is the number of modes of the transition state with energy between E^* and $(E^* + dE^*)$. $N^T(E^*)$ has dimensions of reciprocal molecules and, $G^{\ddagger}(E^*)$ has dimensions of energy per/molecule.

Solved Example 9.B is an RRKM calculation. One should read that example before proceeding with the chapter.

9.10.1   Qualitative Results

The RRKM model is one of the key results in the literature. Therefore, we want to describe it in detail. First, we will work out the structure of the solution predicted by equation (9.127), where $G^T(E^*)$ contains all of the modes of the molecule. For the purpose of discussion, it is useful to define $G^O(E^*)$:

$$G^O(E^*) = \sum_n g_n \tag{9.130}$$

where the sum goes over all the states of the reactants, not just the states at the transition state.

Just to give the flavor, Figures 9.19 and 9.20 show a plot of the vibrational levels of chloroform and $G^O(E^*)$ for chloroform and some other molecules. Chloroform has a series of distinct levels, so $G^O(E^*)$ shows a series of steps. The step structure is a key feature of $G^O(E^*)$ and $G^T(E^*)$. The step structure occurs only near zero energy. At high energies, $G^T(E^*)$ appears to be smoothly varying and $N(E^*)$ is constant. When you look at unimolecular reactions, you usually probe the energy region near threshold (i.e., where E^* is only slightly greater than $E^{\ddagger}$). In that case, one expects a step structure in $G^T(E^*)$ because the number of states about threshold is small. In contrast, the reactant energy is large compared to the zero point of the reactants. In such a case, $N(E^*)$ is smoothly varying.

For example, Figure 9.21 shows a plot of $G^T(E^*)$ and $k(E^*)$ for reaction (9.107). Again, one observes a series of steps in the $G^T(E^*)$ versus energy. Figure 9.21 also shows the

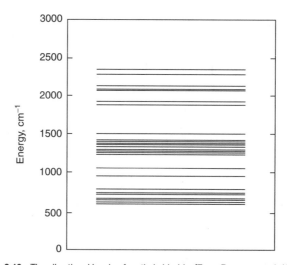

Figure 9.19 The vibrational levels of methyl chloride. [From Pearson, et al. (1965).]

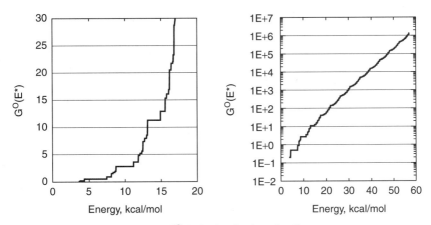

Figure 9.20 $G^O(E^*)$ for the vibrations of methane.

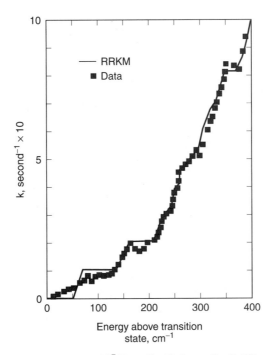

Figure 9.21 A plot of $G^T(E^*)$ and $k_2(E^*)$ for reaction (9.107).

rate constant for reaction (9.107) as a function of energy. Under the conditions of these experiments, $N(E^*)$ is almost constant. Notice, that the rate constant does not fit exactly $G^T(E^*)$ for the first 300 cm^{-1}. However, the rate constant tracks $G^T(E^*)$ almost exactly after that.

Figure 9.22 shows a different case, the isomerization of ketene:

$$H_2^{13}C^{12}CO \longrightarrow H_2^{12}C^{13}CO \qquad (9.131)$$

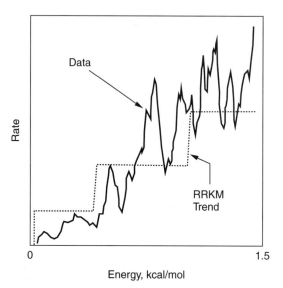

Figure 9.22 A plot of $k_2(E^*)$ for reaction (9.131).

This reaction shows a series of peaks and valleys in the rate constant. Notice that RRKM does not fit the data very well. When you examine the same reaction at higher energies, equation (9.127) fits exactly. Experimentally, (9.127) fits most unimolecular decomposition reactions, except within the first $300-800$ cm^{-1} of what is called the threshold energy, where the threshold energy is the minimum energy needed to traverse the barrier. To keep this in perspective, $k_B T = 208$ cm^{-1} at 300 K. Therefore, the implications of Figures 9.21 and 9.22 is that the RRKM works quite well, except at energies within a few k_B T of the transition state.

The RRKM model is not exact because it assumes that the P_m values in equation (9.124) are 0.0 below threshold and 1.0 above threshold. In reality, one can get some contributions from states below threshold due to tunneling. Further, P_m is not 1.0 for states just above threshold, since energy needs to be localized in a specific bond for a reaction to occur. There is some probability that the energy never becames localized. That produces errors just at the threshold. The approximation usually works great once you are 1000 cm^{-1} or so above threshold, however, because enough excess energy is deposited in the specific bonds to allow the bonds to break.

Equation (9.130) is the key equation that people use to analyze data from laser photolysis experiments. Consequently, the equation is quite important.

Generally, RRKM calculations are easy. We include a sample calculation in Example 9.B. It happens that if you know the vibrational frequencies of the molecules, you can calculate $G^O(E^*)$ and $N(E^*)$ exactly using what is called the *Beyer–Swinehart algorithm* given in Example 9.B. If you know the vibrational levels of a molecule, and assume that vibrational levels are the same in the hot molecules as in the ground state, you can do the entire calculation in about 2 min on a Pentium PC. And the neat thing is that the calculations work. Again, there is some subtlety because people seldom consider all of the modes in the molecule the modes that are presumed to couple into products only. Still, equation (9.130) has proved quite accurate for most laser photolysis reactions with well-defined transition states. The one exception is the selective chemistry example mentioned in Section 9.10.

9.10.2   Non-RRKM Behavior

The RRKM model has its strengths and weaknesses. In actual practice, the model works most of the time for unimolecular reactions with well-defined transition states, but there are some special examples where the model does not work. For example, in Chapter 8, we noted that Crim (1996) found that if he excited the O–D bond in HOD, he could get the OD bond to break preferentially.

This example is special because the O–D stretching mode is largely uncoupled to the other modes of the molecule.

Figure 9.23 shows another example: the isomerization of stilbene, where RRKM is off by about an order of magnitude.

This presents the general finding that

> Non-RRKM behavior arises whenever intramolecular energy transfer is slow compared to vibrational times, so that energy is not quickly localized in the bonds being broken.

For some reason, cases like this are often ignored in the literature. People instead assume that the coupling is strong enough that the reaction rate of a unimolecular reaction is relatively insensitive to where you put the initial energy excitation. Experimentally, the coupling can be slow. Consequently, one needs to consider energy transfer within a molecule to get an accurate rate constant. Intramolecular energy transfer is difficult to model, because quantum effects associated with your having individual vibrational modes, rather than individual levels, play a key role (see. Leitner and Wolynes, 1996, for details). Still, it is important to consider it.

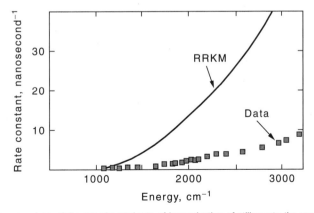

Figure 9.23   A comparison of the experimental rate of isomerization of stilbene to the predictions of the RRKM model.

9.11   BARRIERLESS REACTIONS

Another weakness of RRKM is that the model does not apply to reactions that do not have a well-defined transition state. If the system does not have a well-defined transition state, then you cannot calculate $q_R^{\ddagger}$ or $G^T(E^{\ddagger})$. Consequently, you cannot do RRKM calculations.

People use several models to get around that. In order to understand the details, we will consider the following reaction:

$$O + O + X \longrightarrow O_2 + X \qquad (9.132)$$

Figure 9.24 shows the oxygen–oxygen potential. Notice that the potential is completely attractive. There is no barrier. As a result, when reaction (9.132) occurs, there is no transition state for the reaction.

However, recall from Section 8.9 that there is still an angular momentum barrier to reaction. According to the analysis in Section 8.9, the oxygen atoms move in an effective potential given by

$$V_{eff} = \frac{(b_{A \to BC})^2 E_{KE}}{R_{AB}^2} + V_{AB}(R_{AB}) \qquad (9.133)$$

where V_{eff} is the effective potential $V_{AB}(R_{AB})$ is the actual potential, E_{KE} is the translational energy for the collision between the two oxygen atoms, and R_{AB} is the distance between the two oxygen atoms.

Figure 9.25 shows a plot of V_{eff} as a function of R_{AB} for various values of $b_{A \to BC}$. The figure was calculated for $E_{KE} = 1$ kcal/mol. Notice that there is a barrier in the effective potential when $b_{A \to BC} > 0$ even though the actual reaction is barrierless.

Let's define b_{crit} such that the height of the barrier is equal to E_{KE} when $b_{A \to BC} = b_{crit}$. Notice that the height of the barrier increases as $b_{A \to BC}$ increases. The barrier equals E_{KE} when $b_{A \to BC} = b_{crit}$. Consequently, when $b_{A \to BC} > b_{crit}$, the barrier height will be larger than E_{KE}. If the barrier height is larger than E_{KE}, no reaction can occur in the absence of tunneling.

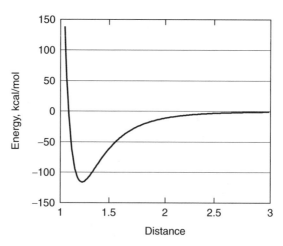

Figure 9.24 An oxygen–oxygen potential.

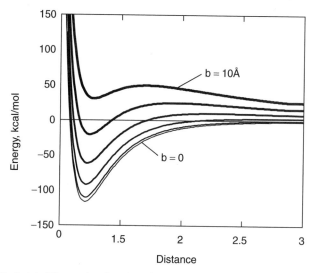

Figure 9.25 A plot of V_{eff} as a function of R_{AB} for various values of $b_{A \to BC}$, and $E_{KE} = 1$ kcal/mol.

On the other hand, when $b_{A \to BC} \geq b_{crit}$, the reaction can occur easily. Therefore, there is a barrier to reaction (9.132) at $b_{A \to BC} = b_{crit}$.

9.12   COLLISION THEORY FOR BARRIERLESS REACTIONS

Next, let us calculate an approximation to the rate of reaction using collision theory. According to collision theory, the cross section for reaction, σ_x, is given by

$$\sigma_x = 2\pi \int_0^\infty P_{reaction}(b_{A \to BC})b_{A \to BC}\, db_{A \to BC} \tag{9.134}$$

where $P_{reaction}(b_{A \to BC})$ is the reaction probability and $b_{A \to BC}$ is the impact parameter. For the purposes here, we will assume that

$$P_{reaction} = \begin{cases} 0 & \text{for} \quad b_{A \to BC} > b_{crit} \\ 1 & \text{for} \quad b_{A \to BC} \leq b_{crit} \end{cases} \tag{9.135}$$

Combining equations (9.134) and (9.135) yields

$$\sigma_x = 2\pi(b_{crit})^2 \tag{9.136}$$

Substituting into equation (9.21) yields

$$k_{A \to BC} = \bar{v}_{ABC}(2\pi(b_{crit})^2) \tag{9.137}$$

where $k_{A \to BC}$ is the rate constant for the reaction and $\bar{v}_{ABC}$ is the average velocity of A toward BC in reduced-mass coordinates. We use equation (9.137) to calculate the rate constants in Example 9.D. The reader may want to refer to that example.

9.12.1   Collision Theory for Barrierless Dissociation Reactions

Next, let us consider the reverse of reaction (9.132):

$$O_2 + h_P\nu \longrightarrow 2O \tag{9.138}$$

where a photon reacts with O_2 to yield two oxygen atoms with E_{KE}.

According to microscopic reversibility, k_{138}, the rate constant for reaction (9.138), is related to k_{132}, the rate constant for reaction (9.132), by

$$k_{138} = k_{132}K_{equ} \tag{9.139}$$

where K_{equ} is the equilibrium constant for the reaction:

$$O_2 + h_P\nu \rightleftharpoons 2O \tag{9.140}$$

Next we will make the approximation that the equilibrium constant for reaction (9.140) is unity. Substituting equation (9.137) into equation (9.139) and setting $K_{equ} = 1$ yields

$$k_{138} = 2\pi(b_{crit})^2 \tag{9.141}$$

Equation (9.141) is the collision theory approximation to the rate constant for reaction (9.138).

9.12.2   Phase Space Theory

The equations in Sections 9.10.2 and 9.10.3 were classical, but people often use quantum-mechanical methods to develop similar results. The quantum-mechanical version of the classical theory of barrierless reactions is called **phase space theory**.

The quantum-mechanical methods go back to equation (9.127):

$$k_{A \to BC} = \frac{1}{h_P} \frac{G^T(E^*)}{N(E^*)} \tag{9.142}$$

where $k_{A \to BC}$ is the rate constant, $G^T(E^*)$ is the number of states with a rotational–vibrational energy between E^* and $E^\ddagger$, and $N(E^*)$ is the number of reactant states with a rotational–vibrational energy between E^* and $(E^* + d(E^*))$. RRKM theory suggests that you should calculate $G^T(E^*)$ and consider only states of the transition state. Phase space theory instead assumes that

- $G^T(E^*)$ should be evaluated of the products.
- Rotational modes should be included in the calculation on $G^T(E^*)$ provided $E_{VR} < E_{KE}$.

The inclusion of rotational states with large impact parameters makes $G^T(E^*)$ much larger in phase space theory than in RRKM.

Therefore, one calculates $G^T(E^*)$ by summing all of the vibrational and rotational levels of the products subject to the constraint that

$$E_{VR} \leq E^* \tag{9.143}$$

and angular momentum is conserved.

There is another version of the theory called **orbiting phase space theory**. Orbiting phase space theory adds the additional constraint that the angular momentum is less than the angular momentum when $b_{A\to BC} = b_{crit}$, so that the reactants can get over the angular momentum barrier to reaction.

Phase space theory calculations end up to be very similar to RRKM calculations, except that you directly include the rotational levels in the sums and look at the properties of the products rather than the transition state.

The major result of phase space calculations is that preexponentials for barrierless reactions can be huge. For example, a reaction such as

$$X + O_2 \longrightarrow 2O + X \qquad (9.144)$$

with $X = O_2$ has a preexponential of 5.8×10^{15} Å^3/(molecule·second). Barrierless ionic reactions often have preexponentials above 10^{16} Å^3/(molecule·second). Physically, b_{crit} is often in the order of 30 Å. Consequently, cross sections for barrierless reactions can be gigantic.

9.13 THERMAL UNIMOLECULAR REACTIONS

Now we will be changing topics. The analysis in the Sections 9.9–9.11 considered unimolecular reactions initiated with a laser. Next we will consider unimolecular reactions initiated thermally. For example, if you heat up cyclopropane, it will isomerize to propylene, via a single concerted process:

$$\begin{array}{c} CH_2 \\ / \ \backslash \\ H_2C-CH_2 \end{array} \Longrightarrow CH_2 = CH-CH_3 \qquad (9.145)$$

Other examples of thermal unimolecular reactions are given in Section 8.10. In Chapter 5, we noted that true unimolecular reactions are rare. Most simple isomerization reactions proceed via an initiation propagation mechanism. However, there are some special cases where one heats up the reactants, and gets a unimolecular isomerization reaction. Those cases will be discussed in this section.

Unimolecular reactions are much more complex than bimolecular reactions. Recall, from Chapter 8, that it is impossible for a unimolecular reaction $A \Rightarrow B$ to be an elementary reaction because there is no way to conserve energy in the process. Instead, one needs a collision partner to allow the reaction to occur.

In 1922, Lindemann proposed that one could treat the unimolecular reaction as a series of bimolecular collisions, and then use an early version of transition state theory to calculate the rate. Basically, Lindemann proposed that all unimolecular reaction proceed via what is now called the **Lindemann mechanism**. Consider the following overall reaction:

$$A \Longrightarrow B \qquad (9.146)$$

According to the Lindemann mechanism, the first step in the process is the interaction of A with a collision partner to yield A*, an excited A molecule:

$$X + A \xrightarrow{\ 1\ } A^* + X \qquad (9.147)$$

The A^* can be deexcited via collisions:

$$X + A^* \xrightarrow{\ 2\ } A + X \tag{9.148}$$

Or the A^* can react to form products:

$$A^* \xrightarrow{\ 3\ } B \tag{9.149}$$

The Lindemann mechanism forms the basis for the discussion of all unimolecular reactions, so it is quite important.

Lindemann (1922) proposed that one could treat reactions (9.147), (9.148), and (9.149) as standard elementary steps. He then used the steady-state approximation to derive a rate equation for the Lindemann mechanism. We derived the result in Chapter 4. According to Lindemann, r_B, the rate of any unimolecular reaction, is given by

$$r_B = \frac{\left(\dfrac{k_1 k_3}{k_2}\right)[A]}{1 + \dfrac{k_3}{k_2[X]}} \tag{9.150}$$

There are two key limits in this equation. The high-pressure limit is where $(k_3/k_2[X])$ is negligible, and the low-pressure limit is where $(k_3/k_2[X])$ is large. In the high-pressure limit, equation (9.150) becomes

$$r_B = \frac{k_1 k_3}{k_2}[A] \tag{9.151}$$

The rate of reaction should be first-order in A and zero-order in X. The generally experiments reproduce those trends. In contrast, in the low-pressure limit, equation (9.150) becomes

$$r_B = k_1[A][X] \tag{9.152}$$

So the rate should be first-order in A and X. Experimentally, the rate is first-order in [A] but not in [X]. In the literature, people seldom plot the rate versus [X]. Instead, they define a new quantity, $k_1[X]$ and use it in the equations.

Over the years, people have tried to fit a considerable amount of data to equation (9.150). Generally, one can get a reasonable fit to real data, but there are small errors. In order to see that, it is useful to define a new quantity:

$$k_L = \frac{\dfrac{k_1 k_3}{k_2}}{1 + \dfrac{k_3}{k_2[X]}} \tag{9.153}$$

Note that if one substitutes equation (9.153) into equation (9.150), one finds that the rate of reaction is given by

$$r_B = k_L[A] \tag{9.154}$$

So k_L is a pseudo-first-order rate constant for the reaction $A \Rightarrow B$.

Figure 9.26 shows a plot of k_L versus [X] calculated for the isomerization of isocyanide. The Lindemann model predicts that k_L should be constant at high pressures and

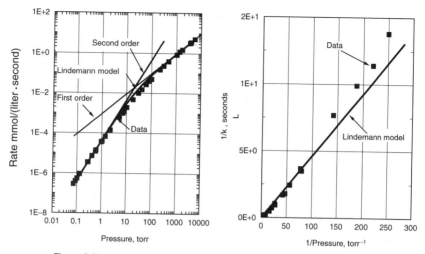

Figure 9.26 A plot of k_1 versus [X] calculated by the Lindemann mechanism.

decrease at low pressures. The data show a similar trend, although the results are different in detail. Generally, the Lindemann model fits the qualitative but not the quantitative trends in the data.

9.13.1 Hinshelwood's Theory

There is one key point that one has to watch out for in Figure 9.26. Notice that in order to fit the data, one had to assume that $k_2 = 10^{21}$ Å^3/(molecule·second) By comparison, collision theory predicts a k_2 of 10^{14} Å^3/(molecule·second) Consequently, in order to fit the data on the basis of Lindemann's model, one has to assume that k_2 is a factor of 10^7 larger than you would expect from transition state theory.

Hinshelwood (1927) offered the first explanation of why k_2 could be unusually large. His idea was that energy could be stored in many modes of the molecule rather than just the one mode that contributes to moving over the barrier. Then energy would be transferred, allowing reaction to happen with many modes. In such a case k_1 would be much larger than you would expect if only one mode contributed to reaction. Hinshelwood assumed that there were s_H modes in the molecule, which contributes to the reaction. Hinshelwood also made the simplifying assumption that all the vibrational modes have the same vibrational frequency, and then used nonquantum statistical mechanics to show

$$\frac{dk_1(E^*)}{k_2} = (s_H)! \left(\frac{E^*}{k_B T}\right)^{s_H} \exp\left(\frac{-E^*}{k_B T}\right) dE^* \tag{9.155}$$

where s_H is the number of modes that can store energy for the reaction, $E^* + \delta E^*$ is the energy of the molecule, k_B is Boltzmann's constant, T is the absolute temperature, and $dk_1(E^*)$ is the contribution to molecules with an energy between $E^{\ddagger}$ and $E^* + \delta E^*$.

The system needs to have an energy of at least $E^{\ddagger}$ to get over the barrier in the potential energy surface, where $E^{\ddagger}$ is the saddle point energy. Integrating equation (9.155) from $E^{\ddagger}$

to infinity yields

$$\frac{k_1}{k_2} = ((s_H - 1)!) \left(\frac{E^{\ddagger}}{k_B T}\right)^{s_H - 1} \exp\left(\frac{-E^{\ddagger}}{k_B T}\right) \tag{9.156}$$

Hinshelwood used collision theory to calculate k to obtain a result for k_1. Recall from Chapter 2 that for a typical reaction $E^{\ddagger}/k_B T$ is in the order of $5-10$. So, if s_H is 10, one should observe a factor of $10^7 - 10^{10}$ change in k_2.

Figure 9.26 compares the results of Hinshelwood's analysis to the data. Generally, Hinshelwood's calculations were within an order of magnitude of the correct result but not quantitative.

9.13.2   RRKM Model

The problem with the Hinshelwood model is that it explicitly considers the variation of k_1 with E^*, but does not explicitly consider the variation of k_3 with E^* where $k_3{}^2$ is the rate constant for decomposition of molecules with energy E^*. In Section 9.8 and 9.9, we noted that k_3 is a strong function of E^*. Consequently, one needs to consider the E^* dependence explicitly.

According to the Lindemann model, $dk_L(E^*)$, the contribution to k_L from molecules with energy between E^* and $E^* + \delta E^*$, is

$$dk_L = \frac{\dfrac{dk_1(E^*)k_3(E^*)}{k_2}}{1 + \dfrac{k_3(E^{\ddagger})}{k_2[X]}} \tag{9.157}$$

From equilibrium

$$\frac{dk_1(E^*)}{k_2} = \frac{q^*}{q_A} \exp\left(\frac{-E^*}{k_B T}\right) dE^* \tag{9.158}$$

where q^* is the partition function for A molecules, with an energy between E^* and $E^* + \delta E^*$ as described in Section 9.9, and q_A is the partition function for the reactants. In Section 9.9, we also noted

$$q^* = N(E^*) \tag{9.159}$$

Similarly, according to equation (9.130), we obtain

$$k_3(E^*) = \frac{G(E^*)}{h_P N(E^*)} \tag{9.160}$$

Combining equations (9.157) and (9.160) and integrating yields

$$k_L = \int_{E^{\ddagger}}^{\infty} \frac{\dfrac{G(E^*)}{h_P q_A}}{1 + \dfrac{G(E^*)}{h_P N(E^{\ddagger})k_2[X]}} \exp\left(\frac{-E^*}{k_B T}\right) dE^* \tag{9.161}$$

where $E^{\ddagger}$ is the minimum energy necessary for the reaction to occur.

---

[2] In his paper, Hinshelwood actually calculated k_3 from k_1. However, the expression did not fit the data.

Equation (9.161) is the exact result of the RRKM model, but it is rarely presented in this form in the literature. Recall that $G(E^*)$ includes all states of the molecules. Often, rotational levels of a molecule cannot carry the molecules over the transition state. In that case, it is useful to separate the rotations from the vibrations in equation (9.37). Let's define $G_V(E^*)$ and $N_V(E^{\ddagger})$ by

$$G_V(E^*) = \frac{G(E^*)}{q_r^{\ddagger} l^{\ddagger}} \tag{9.162}$$

$$N_V(E^*) = \frac{N(E^*)}{q_r^{\ddagger} l^{\ddagger}} \tag{9.163}$$

where $q_r^{\ddagger}$ is the rotational partition function of the transition state and $l^{\ddagger}$ is the *symmetry number*. The symmetry number accounts for the fact that there may be multiple equivalent bonds in a molecule to break. So, for example, if you look at the reaction

$$C_2H_6 + X \longrightarrow C_2H_5 + H + X \tag{9.164}$$

you will notice that $l^{\ddagger}$ equals 6.

Substituting equations (9.162) and (9.163) into equation (9.161) yields

$$k_L = l^{\ddagger} \frac{q_r^{\ddagger}}{h_P q_{Ar} q_{Av}} \int_{E^{\ddagger}}^{\infty} \frac{G_V(E^*)}{1 + \dfrac{G_V(E^*)}{h_p k_2 N_V(E^*)[X]}} \frac{\exp(-E^* k_B T)}{E^*} \tag{9.165}$$

Equation (9.165) is the form of the RRKM equation that appears most often in the literature.

Figure 9.27 shows how well the RRKM model fits the data. Notice that the model fits the pressure dependence of the rate equation about as accurately as the rate data have been measured. Holbrook et al. showed that one can fit the data even better by adjusting

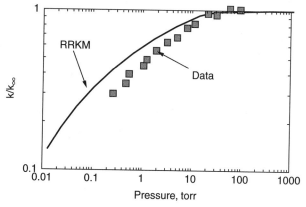

Figure 9.27 A comparison of the RRKM model to experimental data for the reaction between 1,2-dichlorocyclopropane and 2,3-dichloropropene. [Data of Hase et al. (1987). Calculations of Holbrook et al. (1996)].

Table 9.8 Preexponentials [in cm^6/(mol^2·second)] for a series of unimolecular reactions on changing of the collision partner

Reaction	k_0 When X = Argon	k_0 When X = Water	k_0 When X = N$_2$
$NO_2 + X \longrightarrow OH + H + X$	1.7×10^{14}	6.7×10^{15}	1.57×10^{15}
$H_2O + X \longrightarrow OH + H + X$	2.1×10^{15}	3.5×10^{17}	5.1×10^{16}
$HO_2 + X \longrightarrow O_2 + H + X$	1.5×10^{15}	3.2×10^{16}	2×10^{15}
$H_2 + X \longrightarrow H + H + X$	6.4×10^{17}	2.6×10^{15}	
$O_2 + X \longrightarrow 2O + X$	1.9×10^{13}		1.0×10^{14}

Source: Data of Westley (1980).

the parameters in the calculation. However, this fit was done with no adjustment of parameters.

The one thing that is not considered in the RRKM model is that the reaction rate changes substantially as one changes the collision partner. Table 8.4 shows one example of that, while Table 9.8 shows some additional data. Notice that the rate of reaction changes by about a factor of 20–60 as you change the collision partner. Well, that effect is not accounted for in the RRKM model.

Physically, what is happening is that gas-phase collisions do not necessarily excite molecules. Instead, one has to adjust k_1 and k_2 by the probability that a collision leads to excitation of the A molecules. Experimentally, argon is a worse collision partner than water or N$_2$. Physically, argon interacts so weakly with molecules that argon is not an effective collision partner. In contrast, water, which has a large dipole moment, interacts much more strongly. Empirically, one can account for this effect by multiplying k_1 and k_2 in equation (9.165) by a factor β_c to account for ineffective energy transfer. Qualitatively, this seems to work. One does need to have a way to estimate β_c to do the calculation. Holbrook et al. (1996) provide a good description of the models, but unfortunately, I do not have room to discuss the findings here.

The key point to remember, though, is that the rate of unimolecular reactions varies strongly with the collision partner, so one needs to be careful in predicting rates.

9.14 SUMMARY

In summary, then, in this chapter we reviewed transition state theory, the RRKM model, and the phase space model in detail.

All three start with the exact results that $k_{A \rightarrow BC}$, the rate constant for a molecule with an energy E^*:

$$k_{A \rightarrow BC} = \frac{1}{h_P} \frac{q^{\ddagger}}{q_r} \tag{9.166}$$

while for a thermal distribution of energies:

$$k_{A \rightarrow BC} = \left(\frac{k_B T}{h_p} \right) \frac{q^+}{q_r} \exp \left(-\frac{E^+}{k_B T} \right) \tag{9.167}$$

where q^+ is a modified partition function for the molecules that react, q_r is the partition function for the reactants, h_p is Planck's constant, k_B is Boltzmann's constant, T is temperature, and E^+ is the average energy of the molecules which react. The models then make various approximations for q^+ and q_r.

Transition state theory makes the assumption that q^+ is the partition function evaluated at the transition state, ignoring motion along the reaction coordinate. At each energy, this approximation includes only states with energy of E^*.

The RRKM model is similar to transition state theory in that it evaluates q^+ at the transition state. However, it assumes that all of the states that have enough vibrational energy to carry the reactants over the transition state are included in the evaluation of the partition function, not just the states moving over the transition state.

The phase space model goes one step further, and assumes that the transition state lies either at the barrier in the effective potential, or alternatively at infinity so that all angular momentum states of the products are included in the evaluation of the partition function.

All of these models have the strengths and all fail in cases such as

$$\text{HOD} + h\nu \longrightarrow \text{OH} + \text{D} \tag{9.168}$$

where the detailed dynamics matter. Still, the methods work in most situations, so they are of considerable interest. In the literature, people often treat the RRKM model and transition state theory as though they are exact. Although they are not exact, they certainly are very good approximations in practice.

9.15 SOLVED EXAMPLES

Example 9.A   Tunneling Corrections Using the Eckart Barrier   In Example 7.C we calculated the preexponential for the following reaction:

$$\text{F} + \text{H}_2 \longrightarrow \text{HF} + \text{H} \tag{9.A.1}$$

To what extent will the preexponential change at 300 K if we consider tunneling?

Solution   From equation (9.63),

$$k_{A \to BC} = \kappa(T) \left[\left(\frac{k_B T}{h_p} \right) \frac{q^{\ddagger}}{q_A q_{BC}} \exp\left(\frac{-E^{\ddagger}}{k_B T} \right) \right] \tag{9.A.2}$$

We already evaluated the term in brackets in equation (9.A.2) in Example 7.C. Substituting that result, we obtain

$$k_{A \to BC} = \kappa(T)[2.05 \times 10^{14} \text{ Å}^3/(\text{molecule·second})] \tag{9.A.3}$$

According to equation (9.89)

$$\kappa(T) = 1 + \frac{1}{24} \left(\frac{h_p \nu_C}{k_B T} \right)^2 \left(1 + \frac{k_B T}{E^{\ddagger}} \right) \tag{9.A.4}$$

where, according to Table 7.C.1, $\nu_C = 310$ cm^{-1}. Equation (7.C.17) says

$$\frac{h_p \nu_C}{k_B T} = \nu(4.784 \times 10^{-3} \text{ cm}) \left(\frac{300 \text{ K}}{T} \right) \tag{9.A.5}$$

Substituting $\nu_i = 310$ cm^{-1} at $T = 300$ K into equation (9.A.5) yields

$$\frac{h_p \nu_C}{k_B T} = (310 \text{ cm}^{-1})(4.784 \times 10^{-3} \text{ cm}) \left(\frac{300 \text{ K}}{300 \text{ K}} \right) = 1.48. \tag{9.A.6}$$

Substituting into equation (9.A.4) yields

$$\kappa(T) = 1 + \frac{1}{24}(1.48)^2 \left(1 + \frac{0.00198 \text{ kcal/(mol·K) at 300 K}}{5.6 \text{ kcal/mol}} \right) = 1.10 \tag{9.A.7}$$

Therefore the rate will go up by only 10% at 300 K. At 100 K

$$\frac{h_p \nu_i}{\kappa T} = (310 \text{ cm}^{-1})(4.784 \times 10^{-3} \text{ cm}) \left(\frac{300 \text{ K}}{100 \text{ K}} \right) = 4.45 \tag{9.A.8}$$

Substituting into equation (9.A.4) yields

$$\kappa(T) = 1 + \frac{1}{24}(4.45)^2 \left(1 + \frac{0.00198 \times 100}{5.6} \right) = 1.85 \tag{9.A.9}$$

The tunneling correction in this example is smaller than normal, because the barrier has such a small curvature. A typical number would be 1000 cm^{-1}. Still it illustrates the point that tunneling becomes more important as the temperature drops.

Finally, it is interesting to note that if we replace the hydrogen with a deuterium in the reaction

$$F + DH \longrightarrow FD + H \tag{9.A.10}$$

the curvature drops to 215 cm^{-1}. In that case

$$\frac{h_p \nu_i}{k_B T} = (215 \text{ cm}^{-1})(4.784 \times 10^{-3} \text{ cm}) \left(\frac{300 \text{ K}}{100 \text{ K}} \right) = 3.08 \tag{9.A.11}$$

$$\kappa(T) = 1 + \frac{1}{24} \left(\frac{h_p \nu_i}{k_B T} \right)^2 \left(1 + \frac{k_B T}{E^\ddagger} \right) = 1.41 \tag{9.A.12}$$

so the tunneling correction for deuterium is half that of hydrogen.

Example 9.B   An RRKM Calculation   In Section 9.9 we noted that RRKM calculations are often used to model photolysis reactions. In this example we will calculate $k_{A \to BC}$ for the reaction

$$C_2H_5 \longrightarrow C_2H_4 + H \tag{9.B.1}$$

Table 9.B.1   Data for the reaction $C_2H_5 \rightarrow C_2H_4 + H$

	Reactants	Transition State
Vibrational	111	409
frequencies cm^{-1}	409	432
	851	863
	1007	895
	1099	996
	1295	1026
	1527	1305
	1589	1328
	1618	1583
	1625	1682
	3123	3278
	3193	3365
	3229	3302
	3268	3392
	3373	
Rotational	0.713 (1,2)	0.769 (1,2)
modes cm^{-1}	3.44 (1,1)	2.68 (1,1)
$E^{\ddagger}$		13,046 cm^{-1}
		(37.3 kcal/mol)

[a]The notation 0.769 (1,2) means that the rotational mode has a degeneracy of 2 and a σ of 1 and a frequency of 0.769 cm^{-1}.

Source: Data from Gilbert and smith (1990).

for E^* between 13,046 cm^{-1} (37.3 kcal/mol) and 15,046 cm^{-1} (43 kcal/mol) (data is given in Table 9.B.1).

Solution   According to equation (9.129)

$$k_{A \rightarrow BC} = \frac{1 G_V^{\ddagger}(E^*)}{h_p N_V(E^*)} \frac{q_r^{\ddagger}}{q_{rA}} \qquad (9.B.2)$$

There are two common ways to calculate $G_V^{\ddagger}(E^*)$ and $N_V(E^*)$ the Beyer–Swinehart (1973) direct count algorithm, and the semiclassical Marcus–Rice (1951) and Whitten–Rabinovich approximations (1964). The following is a computer program to do the direct count:

```
      Program Beyer_Swinehart
C! density of vibrational states by
C! Beyer-Swinehart algorithm
      implicit none
      integer(2), parameter :: MODES=15
      integer(2), parameter :: points=5000
      integer(2):: vibr_freq(MODES)
      integer(2):: vibr_degen(MODES)
      integer i, j
      integer(2):: start_frequency=0
      real(8) n(0:points)
      real(8) g(0:points),x,y
      real :: energy_scale=2.
```

```
c!energy_scale equals spacing for energy bins IN cm-1
      data vibr_freq/111,409,851,1067,1099,
      1 1295,1527,1589,1618,1625,3123,
      2 3193,3229,3268,3373/
      data vibr_degen/ 15*1/
      do 5 i=1,MODES
      vibr_freq(i)=vibr_freq(i)/energy_scale
 5    enddo
      start_frequency=start_frequency/energy_scale
C! next initialize arrays
      do 2 i=1,points
      n(i)=0
      g(i)=1
 2    enddo
      n(0)=1
      g(0)=1
c! count the number of modes
      do 10j=1,MODES
      do 9i=vibr_freq(j),points
      n(i)=n(i)+n(i-vibr_freq(j))*vibr_degen(j)
      g(i)=g(i)+g(i-vibr_freq(j))*vibr_degen(j)
      if(mod(i,500).eq.0)write(*,*)i,n(i)
 9    enddo
 10   enddo
      n(0)=0.
c! next write data in format for microsoft Excel, lotus
      open(unit=8,file= statedens.csv ,status="replace",action="write"
      write(8,101)
      write(8,102)
 101    format("E', 'E','N(E)','G(E)'")
 102    format("cm-1/molecule','kcal/mole','/cm-1','dimensionless'")
      do 20 I=start_frequency,points,100
      x=I*energy_scale
      y=x*2.859e-3
      n(i)=n(I)/energy_scale
      g(i)=g(I)-1.0
      write(8,100)x,y,n(i),g(i)
 20   enddo
 100    format(f9.1,',',f9.3,',',e15.7,',',e15.7)
      stop
      end
```

This program calculates n(E*) by dividing the energy scale into a series of cells and counting the number of vibrational bands in each cell. The algorithm goes through with the lowest frequency and puts a 1 in each cell where the frequency or its overtone arises. It then goes through with the second frequency and adds one to the cell if the cell is an overtone, or if some combination of the previous frequencies are in the cell. The effect is a complete count of the number of frequencies contributing to the cell. The algorithm takes only 15 s on a 1995 Pentium PC, so it is quite robust. I recommend that you use the direct-count method for RRKM calculations.

Still, I want to note that there are several analytic approximations for $n(E^*)$ and $G(E^*)$ available in the literature. The simplest approximation comes from classical mechanics:

$$G(E^*) = \frac{(E^*)^S}{(S)! \prod h_p \nu_i} \tag{9.B.3}$$

$$N(E^*) = \frac{dG(E^*)}{dE^*} = \frac{(E^*)^{S-1}}{(S-1)! \prod h_p \nu_i} \tag{9.B.4}$$

Marcus and Rice (1951) showed that a better approximation is

$$G(E^*) = \frac{(E^* + \alpha E_Z)^S}{S! \prod h_p \nu_i} \tag{9.B.5}$$

where α is a constant between 0 and 1, and E_Z is the zero-point energy

$$E_Z = \frac{h_p}{2} \sum \nu_i \tag{9.B.6}$$

Marcus and Rice assume that α is a constant: typically 1.0. Whitten and Rabinovich (1964) proposed different approximations where it is assumed that α varies via the following formula:

$$\alpha = 1 - \beta W\left(\frac{E^*}{E_Z}\right)$$

$$\beta = (S-1)\frac{\displaystyle\sum_i \nu_i^2}{\left(\displaystyle\sum_i \nu_i\right)^2} \tag{9.B.7}$$

$$W\left(\frac{E^*}{E_Z}\right) = \begin{cases} \dfrac{1}{5\left(\dfrac{E^*}{E_Z}\right) + 2.73\left(\dfrac{E^*}{E_Z}\right)0.75 + 3.51} & \text{for } \dfrac{E^*}{E_Z} \le 1 \\[4mm] \exp\left(-2.4191\left(\dfrac{E^*}{E_Z}\right)\right) & \text{for } \dfrac{E^*}{E_Z} > 1 \end{cases} \tag{9.B.8}$$

I do not like any of these approximations, so I always use the direct-count method. Getting back to Example 9.B, first I ran the program and picked out the values of $n(E^*)$ for energies between 13,046 and 15,046 cm^{-1}. Then I reran the program, put in the vibrational modes of the transition state, and looked at the region between 0 and 2000 cm^{-1}. (*Note*: The barrier is 13,046 cm^{-1}, so if the total energy is 14,046 cm^{-1}, the transition state will have only 1000 cm^{-1} of vibrational energy. Table 9.B.2 shows some of these results.)

Table 9.B.2   Calculated values of n(E*) and G(E*)

E*, cm^{-1}	N$_V$(E), cm^{-1}	E* − E‡, cm^{-1}	G$_V^T$(E* − E‡)	k$_2$, second^{-1}
13,046	3,818	0	0	0
13,246	4,286	200	0	0
13,446	4,807	400	0	0
13,646	5,388	600	2	1.17 × 10^7
13,846	6,044	800	2	1.04 × 10^7
14,046	6,758	1,000	8	3.73 × 10^7
14,246	7,485	1,200	9	3.79 × 10^7
14,476	8,443	1,400	19	7.10 × 10^7
14,696	9,439	1,600	24	8.02 × 10^7
14,846	10,513	1,800	43	1.29 × 10^8
15,046	11,715	2,000	55	1.48 × 10^8

Next, let's calculate k. First, let's calculate

$$q_r = \frac{\pi^{1/2}}{\sigma_A \sigma_B \sigma_C} \left(\frac{8\pi I_A k_B T}{h_p^2}\right)^{1/2} \left(\frac{8\pi I_B k_B T}{h_p^2}\right)^{1/2} \left(\frac{8\pi I_C k_B T}{h_p^2}\right)^{1/2} \tag{9.B.9}$$

where I_A, I_B, and I_C are the three moments of inertia; σ_A, σ_B, and σ_C are the symmetry factors; h_p is Planck's constant; k_B is Boltzmann's constant; and T is temperature.

The rotational frequency ω_r satisfies

$$\omega_r = \frac{h_p}{8\pi I_A} \tag{9.B.10}$$

so

$$q_r = \frac{\pi^{1/2}}{\sigma_A \sigma_B \sigma_C} \left(\frac{(k_B T)^3}{h_p^3 \omega_r^A \omega_r^B \omega_r^C}\right)^{1/2} \tag{9.B.11}$$

where ω_r^A, ω_r^B and ω_r^C are the rotational frequencies of the reactants. Plugging equation (9.B.11) and an equivalent equation for $q_r^{\ddagger}$ into equation (9.B.3) yields

$$k_2 = \frac{G_V^T(E* - E^{\ddagger})}{h_p N_V(E*)} \left(\frac{\sigma_A \sigma_B \sigma_C}{\sigma_A^{\ddagger} \sigma_B^{\ddagger} \sigma_C^{\ddagger}}\right) \left(\frac{\omega_A \omega_B \omega_C}{\omega_A^{\ddagger} \omega_B^{\ddagger} \omega_C^{\ddagger}}\right)^{1/2} \tag{9.B.12}$$

Next, let's substitute numbers in equation (9.B.12) for $E* = 15,046$ cm^{-1}. Taking numbers from Table 9.B.2 and noting $h_p = 3.33 \times 10^{-11}$ cm^{-1}/second yields

$$k_2 = \frac{55}{(11,715 \text{ cm}^{-1})(3.33 \times 10^{-11} \text{ cm}^{-1}/\text{second})} \left(\frac{(344)(0.713)(0.713)}{(268)(0.769)(0.769)}\right)^{1/2} \tag{9.B.13}$$

Doing the arithmetic shows that $k_2 = 1.48 \times 10^8$ second.

Example 9.C   Calculation of k_{uni}   Next, let's use the results in Example 9.B to calculate k_{uni}, the high-pressure limit of k_2.

Solution   According to equation (9.165)

$$k_2 = \ell^{\ddagger} \left(\frac{q_r^{\ddagger}}{q_{Ar}} \right) \frac{1}{h_p q_{Av}} \times \int_{E^{\ddagger}}^{\infty} \left(1 + \frac{G_V^{\ddagger}}{\frac{G_V^{\ddagger}(E^*)}{h_p k_2 N_V(E^*)[X]}} \right) \exp \left(\frac{-E^*}{k_B T} \right) dE^* \quad (9.C.1)$$

k_{uni} is the limit of k_2 for large [X]. Taking the limit of equation (9.C.1) as [X] goes to infinity yields

$$k_{uni} = \ell^{\ddagger} \left(\frac{q_r^{\ddagger}}{q_{Ar}} \right) \frac{1}{h_p q_{Av}} \int_{E^{\ddagger}}^{\infty} G_V(E^*) \exp \left(\frac{E^*}{k_B T} \right) dE^* \quad (9.C.2)$$

Substituting equation (9.B.12) into equation (9.C.1) and rearranging yields:

$$k_{uni} = \left(\frac{\sigma_A \sigma_B \sigma_C}{\sigma_A^{\ddagger} \sigma_B^{\ddagger} \sigma_C^{\ddagger}} \right) \left(\frac{\omega_A \omega_B \omega_C}{\omega_A^{\ddagger} \omega_B^{\ddagger} \omega_C^{\ddagger}} \right)^{1/2} \frac{1}{h_p q_{Av}} \exp \left(\frac{-E^{\ddagger}}{k_B T} \right)$$
$$\times \int_0^{\infty} G_V(E^* - E^{\ddagger}) \exp \left(\frac{E^{\ddagger} - E^*}{k_B T} \right) d(E^* - E^{\ddagger}) \quad (9.C.3)$$

Table 9.C.1 is a spreadsheet for the calculation of the integral in (9.C.3).

Table 9.C.1   Spreadsheet used to calculate the results in Example 9.C

	A	B	C
01		T=	700
02		integral=	=SUM(C5:C55)*100
03	'E'	'G(E)'	g(e)exp()
04	'cm-1/molecule'	'dimensionless'	
05	0	0	=B5*EXP(-0.004784*A5*(300/T))
06	200	0	=B6*EXP(-0.004784*A6*(300/T))
07	400	0	=B7*EXP(-0.004784*A7*(300/T))
08	600	2	=B8*EXP(-0.004784*A8*(300/T))
09	800	2	=B9*EXP(-0.004784*A9*(300/T))
10	1000	8	=B10*EXP(-0.004784*A10*(300/T))
11	1200	9	=B11*EXP(-0.004784*A11*(300/T))
12	1400	19	=B12*EXP(-0.004784*A12*(300/T))
13	1600	24	=B13*EXP(-0.004784*A13*(300/T))
14	1800	43	=B14*EXP(-0.004784*A14*(300/T))
15	2000	55	=B15*EXP(-0.004784*A15*(300/T))
	Spreadsheet Continues Down To Row 55		

Table 9.C.2   Numerical values in the spreadsheet used to calculate the results in Example 9.C

	A	B	C
01		T=	700
02		integral=	1407.563
03	'E'	'G(E)'	g(e)exp()
04	'cm-1/molecule'	'dimensionless'	
05	0	0	0
06	200	0	0
07	400	0	0
08	600	2	0.584485
09	800	2	0.387871
10	1000	8	1.029585
11	1200	9	0.768651
12	1400	19	1.076849
13	1600	24	0.902665
14	1800	43	1.073244
15	2000	55	0.910976
	Spreadsheet Continues Down To Row 55		

Table 9.C.2 shows the numerical values of the spreadsheet in Table 9.C.1. In the spreadsheet, we used equation (7.C.7) to evaluate $E^* - E^{\ddagger}$:

$$\exp\left(-\left[\frac{E^* - E^{\ddagger}}{k_B T}\right]\right) = \exp\left(\frac{-h_p \nu}{k_B T}\right) = \exp\left[(-4.784 \times 10^{-3} \text{ cm}) \nu \left(\frac{300 \text{ K}}{T}\right)\right] \tag{9.C.4}$$

The integral works out to be 1407 cm^{-1}. We also have to calculate q_{Av}. It is important to be careful with the integration since we have not kept track of zero-point energy. One can show

$$q_{Av} = q_v \exp\left(\frac{-E_Z}{k_B T}\right) \tag{9.C.5}$$

where q_v is the vibrational partition function and E_Z is the zero-point energy.

I find it useful to compute q_{Av} via numerical integration:

$$q_{Av} = \int_0^\infty N(E^*) \exp\left(\frac{-E^{\ddagger}}{k_B T}\right) dE^* \tag{9.C.6}$$

Numerical integration yields $q_{Av} = 15.9$ using 2-cm^{-1} steps for the numerical integration. Substituting into equation (9.C.3) yields

$$k_{uni} = \left(\frac{1 \times 1 \times 1}{1 \times 1 \times 1}\right) \left(\frac{(344)(0.713)(0.713)}{(268)(0.769)(0.769)}\right)^{1/2} \frac{1407 \text{ cm}^{-1}}{(15.9)3.33 \times 10^{-11} \text{ cm}^{-1}} \times \exp\left(\frac{-E^{\ddagger}}{k_B T}\right) \tag{9.C.7}$$

where $h_p = 3.33 \times 10^{-11}$ cm^{-1}/second, doing the arithmetic

$$k_{uni} = 2.79 \times 10^{12}/\text{second} \left[\exp \left(\frac{-E^{\ddagger}}{k_B T} \right) \right] \tag{9.C.8}$$

at 700 K. Notice that the RRKM preexponentials are not substantially larger than those from transition state theory. Physically, one needs a large reaction cross section to get a large preexponential, and the transition state is not large for reaction (9.B.1).

Example 9.D A Classical Calculation for the Preexponential of a Barrierless Reaction   Estimate the cross section for the reaction

$$O + O + X \longrightarrow O_2 + X \tag{9.D.1}$$

with $X = O_2$ at a kinetic energy of 2 kcal/mol. Assume that $V_{OO}(R_{OO})$, the oxygen–oxygen potential, follows:

$$V_{OO} = (116 \text{ kcal/mol}) \left(\left(\frac{1.21 \text{ Å}}{R_{OO}} \right)^{12} - 2 \left(\frac{1.21 \text{ Å}}{R_{OO}} \right)^{6} \right) \tag{9.D.2}$$

Solution   According to equation (9.136)

$$\sigma = 2\pi (b_{crit})^2 \tag{9.D.3}$$

where b_{crit} is the value of $b_{A \to BC}$, where the barrier in the effective potential V_{eff} is equal to E_{KE}, the kinetic energy. According to equation (9.133)

$$V_{eff} = \frac{(b_{A \to BC})^2 E_{KE}}{R_{OO}^2} + V_{OO}(R_{OO}) \tag{9.D.4}$$

Next, let us derive an approximation for b_{crit}. For the purposes of derivation we will first ignore the first term in parentheses in equation (9.D.2) and instead assume

$$V_{OO} = -\frac{C}{R_{OO}^6} \tag{9.D.5}$$

where C is a constant.
  For our example

$$C = 2(1.21 \text{ Å})^6 (116 \text{ kcal/mol}) \tag{9.D.6}$$

Combining equations (9.D.4) and (9.D.5) yields

$$V_{eff} = \frac{(b_{A \to BC})^2 E_{KE}}{R_{OO}^2} - \frac{C}{R_{OO}^6} \tag{9.D.7}$$

The maximum occurs when

$$\frac{dV_{eff}}{dR_{OO}} = 0 \tag{9.D.8}$$

Substituting equation (9.D.7) in (9.D.8) and performing the algebra yields

$$0 = -\frac{2(b_{A\to BC})^2 E_{KE}}{R_{OO}^3} + \frac{6C}{R_{OO}^7} \tag{9.D.9}$$

Solving equation (9.D.9) for R_{OO} yields

$$R_{OO} = \left(\frac{3C}{(b_{A\to BC})^2 E_{KE}}\right)^{1/4} \tag{9.D.10}$$

Substituting equation (9.D.10) into equation (9.D.7) yields

$$V_{eff}^{max} = \frac{\frac{2}{3}(b_{A\to BC})^3 (E_{KE})^{1.5}}{\sqrt{3C}} \tag{9.D.11}$$

where we have noted that this is the maximum value of V_{eff}. When $b_{A\to BC} = b_{crit}$, then $V_{eff}^{max} = E_{KE}$. Therefore,

$$E_{KE} = \frac{\frac{2}{3}(b_{crit})^3 (E_{KE})^{1.5}}{\sqrt{3C}} \tag{9.D.12}$$

Solving equation (9.D.12) for b_{crit} yields

$$b_{crit} = \left(\frac{3}{2}\right)^{1/3} (3)^{1/6} \left(\frac{C}{E_{KE}}\right)^{1/6} \tag{9.D.13}$$

Plugging in the numbers yields

$$b_{crit} = (1.5)^{1/3} (3)^{1/6} \left(\frac{2(1.21\ \text{Å})^6 (116\ \text{kcal/mol})}{2\ \text{kcal/mol}}\right)^{1/6} = 3.67\ \text{Å} \tag{9.D.14}$$

Plugging into equation (9.D.3) yields

$$\sigma_x = 2\pi(3.67\ \text{Å})^2 = 84\ \text{Å} \tag{9.D.15}$$

where σ_x is the cross section

Next, we will go back to the real potential, and use the solver function in Excel to calculate a better value of b_{crit}. Our approach will be to guess $b_{crit} = 3.6$ or 3.7 Å, use the solver function to maximize V_{eff}, by varying V_{OO}, and then interpolate to get a better value of b_{crit}. The spreadsheet in Table 9.D.1 does the calculations (see also Table 9.D.2). The result is that $b_{crit} = 3.67$ Å as before.

Example 9.E   The Classical Approximation for $k_{A\to BC}$ Next we will calculate a value of the rate constant for the reaction

$$2O + X \longrightarrow O_2 + X \quad \text{with} \quad X_2 = O_2 \tag{9.E.1}$$

Table 9.D.1   The spreadsheet used to calculate the results in Example 9.D

	A	B	C
04	Guess a value of b_{crit}	R_{00} to maximize V_{eff} for this guess of b_{crit} (The values in this column werecalculated using the solver function to maximize V_{eff} by varying R_{00})	Calculate maximum value of V_{eff} for this guess of b_{crit}
05	3.6	3.02675698	=2*(A5/B5)^2+116*((1.21/B5)^12 −2*(1.21/B5)^6)
06	3.7	2.98531074	=2*(A6/B6)^2+116*((1.21/B6)^12 −2*(1.21/B6)^6)
07	=A5+(2− C5)/(C6− C5)*(A6-A5)	2.99690597	=2*(A7/B7)^2+116*((1.21/B7)^12 −2*(1.21/B7)^6)

Table 9.D.2   Numerical values in the spreadsheet used to calculate the results in Example 9.D

	A	B	C
04	Guess a value of b_{crit}	R_{00} to maximize V_{eff} for this guess of b_{crit}	Calculate maximum value of V_{eff}
05	3.6	3.02675698	1.884271
06	3.7	2.98531074	2.045876
07	3.671612	2.99690597	1.999095

Solution   Combining equations (9.D.3) and (9.D.13) yields

$$\sigma_X = 2\pi \left(\frac{3}{2}\right)^{2/3} (3)^{1/3} \left(\frac{C}{E_{KE}}\right)^{1/3} \tag{9.E.2}$$

According to equation (9.C.8)

$$k_{A \to BC} = \bar{v}_{A \to BC} \int_0^\infty (E_{KE}/k_B T)\sigma_x \exp(E_{KE}/k_B T)\, d(E_{KE}) \tag{9.E.3}$$

As in Example 9.C, we will use a spreadsheet (Table 9.E.1; see also Table 9.E.2) to calculate the integral.

Therefore, the integral is 86 $Å^2$.

Table 9.E.1 Spreadsheet used to calculate the integral in equation (9.E.3) using Laguere integration

	A	B	C	D	E	F
01						
02	kbT=	=D2*0.00198	T=	300		
03			Cc=	=116*2* γ1.21γ 6	Integral=	=SUM (F5:F10)
04	W	E$_T$	σ	F(w)	B$_i$	term in sum
05	0.22285	=A5*kbT	=2*PI()*(3^(1/3))*((Cc/B5)^(1/3))	=$B5*$C5/kbT	0.458964	=D5*E5
06	1.118893	=A6*kbT	=2*PI()*(3^(1/3))*((Cc/B6)^(1/3))	=$B6*$C6/kbT	0.417	=D6*E6
07	2.99273	=A7*kbT	=2*PI()*(3^(1/3))*((Cc/B7)^(1/3))	=$B7*$C7/kbT	0.113373	=D7*E7
09	5.77514	=A8*kbT	=2*PI()*(3^(1/3))*((Cc/B8)^(1/3))	=$B8*$C8/kbT	0.0103991	=D8*E8
09	9.83747	=A9*kbT	=2*PI()*(3^(1/3))*((Cc/B9)^(1/3))	=$B9*$C9/kbT	0.000261017	=D9*E9
10	15.98287	=A10*kbT	=2*PI()*(3^(1/3))*((Cc/B10)^(1/3))	=$B10*$C10/kbT	0.000000898547	=D10*E10

Table 9.E.2 Numerical values in the spreadsheet used to calculate the integral in equation (9.E.3) using Laguere integration

	A	B	C	D	E	F
01						
02	kbT=	0.594	T=	300		
03			Cc=	728.1154	Integral=	86.14471
04	W	E$_T$	σ	F(w)	B$_i$	term in sum
05	0.22285	0.132373	159.9634	35.64785	0.458964	16.36108
06	1.118893	0.664622	93.4176	104.5243	0.417	43.58663
07	2.99273	1.777682	67.29803	201.4048	0.113373	22.83387
09	5.77514	3.430433	54.05519	312.1763	0.0103991	3.246352
09	9.83747	5.843457	45.26165	445.2601	0.000261017	0.11622
10	15.98287	9.493825	38.50111	615.3583	0.00000089854	0.000553

Next we need to calculate μ for $X = O_2$ (where amu = atomic mass units):

$$\mu = \cfrac{1}{\cfrac{1}{32 \text{ amu}} + \cfrac{1}{32 \text{ amu}}} = 16 \text{ amu} \qquad (9.E.4)$$

From equation (7.A.4)

$$\bar{v}_{A \to BC} = (2.52 \times 10^{13} \text{ Å/s}) \left(\frac{T}{300 \text{ K}} \right)^{1/2} \left(\frac{1 \text{ amu}}{\mu} \right)^{1/2} \qquad (9.E.5)$$

Plugging in the numbers at 300 K yields

$$\bar{v}_{A \to BC} = (2.52 \times 10^{13} \text{ Å/second}) \left(\frac{300 \text{ K}}{300 \text{ K}} \right)^{1/2} \left(\frac{1}{16} \right)^{1/2} = 0.63 \times 10^{13} \text{ Å/second}$$

$$k_{A \to BC} = \bar{v}_{A \to BC} \times \text{integral} = (0.63 \times 10^{13} \text{ Å/second})(86.1 \text{ Å}^2)$$
$$= 5.42 \times 10^{14} \text{ Å}^3/\text{second} \qquad (9.E.6)$$

By comparison, the experimental value is 5.8×10^{15} Å^3/second. The experimental value is larger than the computed value because the potential in equation (9.D.3) is not quite right.

Example 9.F   Unimolecular Decomposition   Calculate the preexponential for the reaction

$$X + O_2 \longrightarrow 2O + X \qquad (9.F.1)$$

Solution   From equilibrium

$$k_{(9.F.1)} = k_{(9.E.1)} \rho_{stp} \exp \left(+\frac{\Delta S}{k_B T} - \frac{\Delta H}{k_B T} \right) \qquad (9.F.2)$$

where $K_{(9.F.1)}$ and $K_{(9.E.1)}$ are the rate constant for reactions (9.F.1) and (9.E.1), ΔS and ΔH are the energy and enthalpy for the reaction, and ρ_{stp} is the molecular density of stp (standard temperature and pressure) (i.e., where ΔS and ΔH were measured).

According to the *NIST Chemistry Webbook* (*http://www.webbook.nist.gov*), at stp

$$\Delta H = 119.1 \text{ kcal/mol}$$

$$\Delta S = 27.92 \text{ cal/mol} \qquad (9.F.3)$$

In principle, we should slightly modify these numbers, since we are working at 700 K. However, I will ignore those corrections here. Plugging equation $K_{(9.E.1)}$ and equation (9.F.3) into equation (9.F.2) yields

$$k_{(9.F.1)} = \left(5.42 \times 10^{14} \frac{\text{Å}^3}{\text{mol·second}} \right) \left(\frac{6.0 \times 10^{23} \text{ molecules}}{22.4 \text{ liters}} \right) \left(\frac{1}{10^{27}} \frac{1}{\text{Å}^3} \right)$$
$$\times \exp \left(\frac{+27.92}{1.98} \right) \exp \left(\frac{-119 \text{ kcal/mol}}{k_B T} \right) \qquad (9.F.4)$$

$$k_{(9.F.1)} = 1.9 \times 10^{16}/\text{second} \exp\left(\frac{-119 \text{ kcal/mol}}{k_B T}\right) \tag{9.F.5}$$

It is interesting to compare the results in Examples 9.B and 9.F. Notice that in Example 9.B, there is a well-defined transition state, and K is about $10^{12}/\text{s}^{-1}$. In Examples 9.E and 9.F, there is no transition state. Instead, the barrier in the effective potential limits the rate. The preexponentials are larger. Physically, the rate-determining step during reaction (9.F.1) is

$$O_2^* + X \longrightarrow 2O + X \tag{9.F.6}$$

where O_2^* is an excited O_2 molecule with a greatly extended bond. The fact that the bond is highly extended in the activated complex makes the preexponential unusually large. One seldom observes large preexponentials unless the bonds in the activated complex are unusually large.

Example 9.G   Benson's Rules   The examples so far assume that we know accurate values for the properties of the transition state. The purpose of this example is to estimate the vibrational properties, when one does *not* know accurate values of the vibrational frequencies.

Solution   The steps are (1) counting the number of the modes in the molecule and transition state, (2) assigning the modes to specific vibrations, and (3) guessing the frequencies of the vibrations.

Let's consider the example, the elimination of HCl from chloroethane:

$$CH_3CH_2Cl \longrightarrow CH_2CH_2 + HCl \tag{9.G.1}$$

The original molecule has 8 atoms or $8 \times 3 = 24$ total modes. There will be three translations and three rotations. Therefore, there will be $24 - 6 = 18$ vibrational modes. We can look up the frequencies of those modes: 3014, 2986, 2977, 2946, 2887, 1470, 1448(2), 1385, 1289, 1251, 1081, 1031, 964, 786, 677, 336, and 250 cm^{-1}. The notation 1448(2) means that the vibrational mode has a degeneracy of 2.

We can also look up the rotational constants ($v = 0.174[1, 2], 1.03 [1,1]$) Now what do we do about the transition state? Let's assume that the transition state looks like

$$\begin{array}{c} \text{H--Cl} \\ | \quad | \\ \text{H} \diagdown \underset{/}{\text{C}} \! == \! \underset{\diagdown}{\text{C}} \! - \text{H} \\ \text{H} \qquad \text{H} \end{array} \tag{9.G.2}$$

There are still 8 atoms, so there will be a total of 24 modes. Three of those are rotations, three of those are translations, and one mode carries the reactants over the transition state. That leaves 17 vibrational modes. Next, we will use Benson's rules to guess what all the vibrational modes are. Benson proposes that if one does not know the vibrational frequencies, one can use the results in Table 9.G.1 to guess frequencies. In the table dotted bonds indicate forming or breaking; C⋯C is a reaction where you take a single bond and make it into a double bond (or the reverse).

Gilbert and Smith also suggest that you treat R groups that are not directly involved in the transition state as free rotors rather than bends that are not included in $G_V^{\ddagger}(E^*)$. They give the list of rotational constants shown in Table 9.G.2.

Table 9.G.1 **Approximate vibrational frequencies (in cm^{-1}) for a number of stretches and bonds**

		Stretches in Stable Bonds			
Group	Frequency	Group	Frequency	Group	Frequency
C_S–H	3000	C–C	1000	C–O	1150
C_D–H	3100	C=C	1650	S–H	2600
C_T–H	3200	C≡C	2050	N–H	3300
C=O	1700	O–H	3700	C–F	1100
C–CL	650	C–Cl	650	C–Br	550
		Stretches in Partial Bonds in Transition States			
C$\cdots$C	675	C$\doteq$O	1150	C$\cdots$Cl	490
C$\doteq$C	1300	C$\doteq$C	1850	C$\cdots$Br	420
C$\cdots$H	2200	C$\cdots$F	820	C$\cdots$	375
H$\cdots$F	3000	H$\cdots$Cl	2200	H$\cdots$B$_r$	1900
		Bends in Stable Bonds			
≡C–H	700	=C–H	1100	–C–H	1000
H–C–H	1450	C–C=C	300	H–C–C	1150
H–O–C	1200	In plane H–C=C	1150	C–C=C	420
O–C=O	420	C=C=C	850	H–C–C rock	700
C–C–Cl	400	Out of plane H–C=C	700	C–C–Br	360
C–C–C	420	C–O–C	400	O–C–O	400
		Bends in Partial Bonds in Transition States			
H$\cdots$C–H	1000	H$\cdots$C–C	800	H$\cdots$O–C	840
H–C$\doteq$C	1150	C$\cdots$C=C	290	C–C$\doteq$C	420
O$\doteq$C$\doteq$O	420	C=C$\doteq$C	420	HC$\doteq$C rock	700
C–C$\cdots$Cl	250	C–C$\cdots$B$_r$	250	C–C$\cdots$I	220
C–C$\cdots$C	300	C–O$\cdots$C	280	C–C$\cdots$O	280

Source: After Benson (1976) and Gilbert and Smith (1990).

Table 9.G.2 **Moments of inertia of some groups approximated by free rotors**

R Group	I amu·A^2	σ	R Group	I amu·A^2	σ
–CH$_3$(sp^3)	3.2	3	CH$_3$(sp^3) (planar)	3.6	3
–CH$_2$D	4.2	1	C$_2$H$_5$	34.7	1
Isopropyl	73	1	tert-butyl	110	3
–CH$_2$F	28	1	–CH$_2$Cl	70	1
–CH$_2$Br	185	1	–CH$_2$I	265	1

Source: After Gilbert and Smith (1990).

Table 9.G.3 The vibrational frequencies of the transition state following Benson's rules

$4 \times$ –C–H	3150(4)
C$\cdots$H	2000
C$\doteq$C	1300
C$\cdots$Cl	490 (stretch)
Cl$\cdots$C$\doteq$C	280 (bend)
H–C–H	1450
$2 \times$ H–C–C bend	1150(2)
H–C–C rock	700
$2 \times$ H–C–C bend	1150(2)
H–C–C rock	700
H–C–H bend	1450
H$\cdots$Cl	2200
H–C$\doteq$C rock	700

To use Table 9.G.1, one needs to decide which 17 modes to pick. This is an issue because if the transition state were a stable molecule, it would have 18 modes, and one would want to pick 17.

Gilbert and Smith give the following rules for selecting the correct frequencies:

1. Choose all stretches; for a noncyclic molecule, there are $N - 1$ of these, while there are N a cyclic molecule.
2. Then choose all bends involving a heavy atom (e.g., Cl).
3. Then for the first CH_2 group: one H–C–H twist/wags, and one H–C–C rock.
4. Then for each CH_3 group: two H–C–H twist/wags, three H–C–H bends, and one free internal rotation.
5. Then for any additional CH_2 group: add one H–C–H bend, two H–C–H twist/wags, and one H–C–C rock, as in (2), plus a C–C–C bend and an appropriate free internal rotor (e.g., $-C_2H_5$ would be the additional rotor when one goes from chloroethane to chloropropane).

Following these rules, we obtain the frequencies listed in Table 9.G.3.

In Table 9.G.3, notice that there are still 18 modes, so we need to eliminate one. One could eliminate either the C$\cdots$Cl or C$\cdots$H since they carry the complex over the transition state. Gilbert and Smith recommend eliminating the C$\cdots$Cl since it is the lowest frequency and therefore the bond that stretches the most. The solution to this problem is then Table 9.G.3 with the C$\cdots$Cl stretch eliminated.

9.16 SUGGESTIONS FOR FURTHER READING

A good alternative derivation of transition state theory can be found in:

K. J. Laidler, *Chemical Kinetics*, 3rd ed., Harper & Row, New York, 1987.

An review of modern versions of transition state theory can be found in:

D. G. Truhlar, B. C. Garrett, and S. J. Klippenstein, Current status of transition-state theory, *J. Phys. Chem.* **100**, 12771–12800 (1996).

W. H. Miller, Quantum and semiclassical theory of chemical reaction rates, *Faraday Discuss.* **110**, 1 (1998).

Good books and articles on unimolecular reactions include:

T. Baer, and W. L. Hase, *Unimolecular Reaction Dynamics*, Oxford University Press, Oxford, UK, 1996.

K. Holbrook, M. J. Pilling, and S. H. Robinson, *Unimolecular Reactions*, Wiley, New York, 1996.

D. L. Thompson, Practical methods for calculating rates of unimolecular reactions, *Int. Rev. Phys. Chem.* **17**, 547 (1998).

9.17 PROBLEMS

9.1 Define the following terms:

(a) Transition state theory

(b) Variational transition state theory

(c) Tolman's theorem

(d) Tunneling

(e) Eckart barrier

(f) Semiclassical tunneling approximation

(g) Transmission coefficient

(h) Microcanonical rate equation

(i) Microcanonical transition state theory

(j) Density of states

(k) RRKM model

(l) Non-RRKM model

(m) Barrierless reaction

(n) Effective potential

(o) Phase space theory

(p) Lindemann mechanism

(q) Hinshelwood's theory

9.2 Review transition state theory, RRKM, and variational transition state theory. What are the key assumptions in each model? When will the models work? When will they fail?

9.3 In Chapter 7 we noted that preexponentials for unimolecular reactions are often on the order of 10^{15}/second. Look at the preexponentials calculated in the solved problems. When do you get a large preexponential? When is the preexponential in the order of 10^{13}/second?

9.4 One of the key assumptions in transition state theory is that as the transition state widens, more molecules can get through.

(a) How well does that assumption work?

(b) Draw an analogy to pouring milk back into a plastic jug. How does the size of the opening affect your ability to pour milk back into the jug? Is there a limit to how much increase in the opening of the jug affects your ability to pour milk into the jug?

(c) How does a funnel affect the ease of getting milk into the jug?

(d) How well are your findings in (b) and (c) incorporated into transition state theory? Does transition state consider increasing the size of the opening? Does transition state theory consider the idea that if the opening is already big, making it even bigger does not help much because there is a bottleneck elsewhere? Does transition state theory consider the role of the potential energy surface in funneling molecules into the transition state and back on to products?

(e) How do your answers in (d) change if you consider variational transition state theory?

(f) How do your answers in (d) change if you consider multiconfigurational transition state theory?

9.5 Explain in your own words.

(a) Why is the preexponential for reaction (7.32b) larger than that for reaction (7.32a)?

(b) How are the effects incorporated into transition state theory?

(c) How would you incorporate the multiple pathways for reaction (7.32) into transition state theory.

9.6 Figure 9.6 shows a plot of the rate constant for the reactions in Table 9.7 versus the value estimated from transition state theory.

(a) How would the curves change if you considered tunneling? Would the points move to the left or to the right?

(b) Which points would move the most?

(c) Would a consideration of tunneling improve the predictions?

9.7 Figure 9.13 and Table P9.7 show some data for the reaction $H + H_2 \rightarrow H_2 + H$.

(a) There is a line in the figure showing the best fit through the high-temperature portion of the data, but the line is not the best fit through the data. Use a least-squares procedure to fit the data to an Arrhenius plot to estimate a preexponential and activation barrier for the reaction.

(b) Also, fit an Arrhenius plot to the three highest temperature points to estimate a high-temperature activation barrier and preexponential.

(c) In which case will tunneling be more important: high temperature or low temperature?

(d) Compare your activation barriers. How has tunneling affected the barrier to reaction?

(e) Repeat (d) for preexponentials.

9.8 In Problem 7.17 you were asked to calculate the rate of the reaction

$$H + HBr \longrightarrow H_2 + Br$$

at several temperatures.

Table P9.7   Rate data for the reaction $H + H_2 \rightarrow H_2 + H$

Temperature, K	Rate Constant, cm^3/(mol·second)	Temperature, K	Rate Constant, cm^3/(mol·second)
300	2240	388	5400
323	2820	408	6280
343	3280	423	7510
358	4000	443	9360

(a) Repeat the calculations, including the effect of tunneling.

(b) Answer all of the other questions in Problem 7.17.

9.9 Describe, in your own words, the things that cause the activation energy to differ from the height of the barrier in the potential energy surface. Be sure to consider the effects of

(a) The zero-point energy of the reactants and the transition state

(b) Tunneling

(c) Temperature dependence of the partition functions

(d) Dynamic corrections due to the need to direct atoms into the transition state

In each case, be sure to indicate whether the effect raised or lowers the barrier. Look through the solved problems, and your solutions to Problems 9.3 and 9.7 to estimate the order of magnitude of each of the effects.

9.10 Describe everything you know about tunneling.

(a) What is it?

(b) When is it important?

(c) What factors determine the tunneling rules?

9.11 In Table 9.B.2 we calculated the $N_V(E^*)$ and $G^{\ddagger}(E^*)$ via a direct count. Compare the values in the table to (a) the classical approximation; (b) the Marcus–Rice approximation with $\alpha = 0.7$; (c) the Whitten–Rabinovich formula. How do the values compare?

9.12 In Example 9.B. we calculated the rate constant for reaction $C_2H_5 \rightarrow C_2H_4 + H$ for $E^* - E^{\ddagger}$ up to 2000 cm^{-1}.

(a) Continue the calculations up to 20,000 cm^{-1}. Does the rate constant continue to rise?

(b) Construct a log–log plot of the rate constant versus $(E^* - E^{\ddagger} = 4090$ cm$^{-1})$. Do the data follow the Hinshelwood formula $k(E^*) = \alpha(E^* - E^{\ddagger})^s$ where S is the number of modes and α is a constant that couple to the reaction?

(c) How does your value of S compare to the value you used in your calculations. (*Hint:* Did you exclude any of the normal modes of the reactants in the RRKM calculation?)

9.13 Consider the reaction $CH_3CH_2Cl \rightarrow CH_2{=}CH_2 + HCl$.

(a) Use the results in Example 9.G and the RRKM method to estimate $k(E^*)$ for the reaction. Assume $q_r^{\ddagger} = q_{rA}$.

(b) Calculate the rate constant for the reaction at 300 K. Assume $E^{\ddagger} = 54.5$ kcal/mol.

9.14 Repeat Problem 9.13 for the reaction $CH_3CH_2CH_2Cl \rightarrow CH_3CH{=}CH_2 + HCl$ using the methods in Example 9.G to estimate the frequencies of the $CH_3CH_2CH_2Cl$.

9.15 Estimate the cross section for the reaction $O + O + X \rightarrow O_2 + X$.

(a) Assume that the O–O potential satisfies

$$V_{OO} = (116 \text{ kcal/mol}) \left[\left(\exp \left(-\frac{R_{OO} - R_1}{R_2} \right) - 1 \right)^2 - 1 \right] \qquad (P9.15.1)$$

with $R_1 = 1.21$ Å, $R_2 = 2.58$ Å, $E_{KE} = 2$ kcal/mol.

(b) Notice that the cross section came out much differently than in Example 9.D. Where is the difference in the calculation?

(c) Compare the potential in equation (P9.15) and that in equation (9.D.2). Plot the two potentials. How do they differ?

(d) What do your results tell you about the influence of the potential on the cross section?

9.16 Repeat Problem 9.15 for the reaction $O_2 + X \rightarrow 2O + X$.

9.17 Explain in your own words how isotopic substitution affects reaction rate.

(a) How do changes in the zero-point energy change activation barriers?

(b) How do changes in the rotational frequencies of the species change the preexponential?

(c) How do changes in the vibrational frequencies change the preexponential?

(d) Is (c) a significant effect?

(e) How do changes in tunneling rates change the preexponential and activation barrier?

(f) How will the larger momentum during the collision (i.e., heavier molecules transfer more momentum) affect the rate?

In all cases, indicate whether a change from hydrogen to deuterium will increase or decrease the rate.

9.18 Figure P9.18 shows a potential energy surface for a series of reactions. They all have a barrier of 14 kcal/mol.

(a) Which ones will be effective in funneling molecules into the transition state?

(b) Which have the largest openings?

(c) On the basis of your analysis of the dynamics of the collision, which transition states are easiest to get over?

(d) From your results in (a)–(c), guess which reactions will have the largest preexponential. Which will have the smallest?

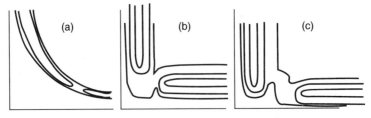

Figure P9.18   Some sample potential energy surfaces.

(e) Repeat (d) assuming transition state theory is valid.

(f) How do your results in (d) and (e) compare?

More Advanced Problems

9.19 At long distances, the potential for an ion–molecule collision follows

$$V(R) = -\frac{1}{2}\frac{\alpha_\pi e^2}{R^4}[Z_i]$$

Where e is the electronic charge (4.8×10^{-10} electrostatic units), α_π is the polarizability of the molecule, and Z_i is the charge on the ion, and R is the distance from the ion to the molecule.

(a) Derive an expression for the reactive cross section as a function of ion energy following the solution in Example 9.D.

(b) Integrate your expression to derive an expression for the rate constant.

(c) Your result in (b) is called the *Langevin model* for ion–molecule reactions. Scott et al., (1997) *Journal of Chemical Physics*, **106**, (1997), 3982 measured the rate constants for several reactions of the type $X^+ + H_2 \to HX^+ + H$ and $X^+ + H \to$ products, where $X = CO_2$, CO, SO_2, NO_2, CS_2, CN, $C_2N_{2+}C_2H_3$. How well do their results compare to your predictions?

(d) Fairley et al. (1998) Scott, Milligan, McLagan, McEwan, *Int. J. Mass Spectrometry*, **172**, (1998), 79, specifically examined the reaction

$$SO_2 + H_2{}^+ \longrightarrow HSO_2{}^+ + H$$

What did they find that was different about this reaction?

(e) Are their arguments for a barrier reasonable?

10

WHY DO REACTIONS HAVE ACTIVATION BARRIERS?

PRÉCIS

A knowledge of activation barriers plays a critical role in chemical kinetics. In Chapter 5, we showed that if one knows how to estimate activation barriers for a series of elementary reactions, one can predict the mechanisms of the overall reaction. In Chapters 7–9, we showed that if one knows the activation barriers and potential energy surfaces, one can predict the rate constants for each of the elementary steps in the mechanism. So far, we have assumed that we knew the potential, and have used that knowledge to make predictions.

In the next two chapters, we will concentrate on finding ways to predict activation barriers for reaction. The main focus of this chapter (10) will be to provide an overview of why chemical reactions have activation barriers. We will focus on the general models that people have proposed to estimate activation barriers for elementary chemical reactions. Some of the ideas are old. However, some of the ideas are more modern and not completely established. Most of the discussion in Chapter 10 will be qualitative. We will quantify the ideas in Chapter 11.

10.1 INTRODUCTION

The idea that reactions are activated dates back to Faraday's work in 1834. Faraday proposed that chemical reactions are not instantaneous because there was an electrical barrier to reaction. Faraday postulated that molecules were held apart by an electrical force and that one needs to overcome that force in order for reaction to happen. Faraday's ideas were proposed 75 years before people could measure an activation barrier. Consequently, Faraday never quantified his ideas. However, Faraday's ideas still were very important in the development of kinetics.

Not much progress was made in quantifying the activation barriers until Arrhenius' work appeared in 1889. Arrhenius (1889) proposed that reactions would follow

Arrhenius' law

$$\text{Rate} = \text{rate}_o \times e^{-E_a/k_B T} \qquad (10.1)$$

where rate is the rate of reaction, E_a is the activation barrier, k_B is Boltzmann's constant, T is the temperature, and rate_o is a constant. Arrhenius also derived equation (10.1) by assuming that only hot (i.e., excited) molecules could react. According to Arrhenius, the activation barrier was associated with the energy the molecules needed to get them hot enough to react.

Later Bodenstein expanded on these ideas. Bodenstein noted that overall reactions occur via a series of elementary steps where bonds break and form. Bodenstein showed that Arrhenius' law is applicable only to elementary reactions. Overall reactions often show deviations from Arrhenius' law.

10.1.1 Models for Activation Barriers for Elementary Reactions

Over the years many models were proposed to understand why elementary chemical reactions are activated.

In 1935, Polanyi and Evans discussed the idea that bonds stretch during elementary reactions. The stretching causes a barrier. See Evans and Polanyi (1935, 1936, 1937, 1938). Bonds also break. The bond scission also causes a barrier.

More recent studies indicate that activation barriers are subtle. There are many different reasons why activation barriers arise during elementary chemical reactions, the main causes of these effects are listed in Table 10.1. Also

- Bonds need to stretch or distort during reaction. It costs energy to stretch or distort bonds. Bond stretching and distortion is one of the major causes of barriers to reaction.
- In order to get molecules close enough to react, the molecules need to overcome Pauli repulsions (i.e., electron–electron repulsions) and other steric effects. The Pauli repulsions are another major cause of barriers to reaction.
- In certain special reactions, there are quantum effects that prevent the bonds in the reactants from converting smoothly from the reactants to the products. Quantum effects can produce extra barriers to reaction.
- There are also a few special cases where the reactants need to be promoted into an excited state before a reaction can occur. The excitation energy provides an additional barrier to reaction.

For example, consider the following elementary reaction:

$$H + CH_3CH_2 \longrightarrow H_2 + C_2H_4 \qquad (10.2)$$

Table 10.1 Principal causes of barriers to chemical reactions

Bond stretching and distortion
Orbital distortion due to Pauli repulsions
Quantum effects
Special reactivity of excited states

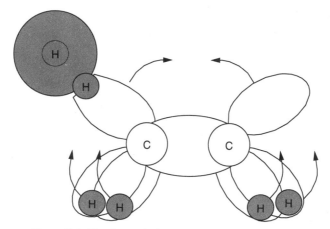

Figure 10.1 The changes in the geometry during reaction (10.2).

Figure 10.1 shows the changes in the geometry during the reaction. First the hydrogen needs to come in and get close enough to react. Then the hydrogen is transferred from a carbon atom to the hydrogen. Simultaneously, the other hydrogens flip up, the carbon–carbon bond shortens, and an orbital on each carbon rehybridizes to form a π (pi) bond. There are barriers to getting the incoming hydrogen close enough to the methyl group to react. There are barriers to the stretching of the carbon–hydrogen bond. There is a Pauli repulsion associated with the shortening of the carbon–carbon bond. There is also a quantum effect associated with the formation of the double bond. All of these effects occur simultaneously, and so it is difficult to estimate a barrier for a reaction like reaction (10.2).

One can use quantum-mechanical methods, as described in Chapter 11, to calculate the strength of the barrier. However, you do not necessarily know why the barrier arises from such calculations. Therefore, in this chapter, we will break up the problem and look at one effect at a time.

I want to say at the beginning that, while there are many theories about why barriers arise in chemical reactions, the models have not yet been completely quantified except in special cases. Consequently, while we do have many models, one still does not always know which model applies in each situation.

10.2 BARRIERS ASSOCIATED WITH BOND EXTENSIONS

In the remainder of this chapter we will discuss each of the effects in Table 10.1. To start off, we will discuss the barriers that arise because bonds need to be distorted during an elementary reaction.

The idea that bond stretching could cause barriers to reaction was first discussed by Evans and Polanyi (1935, 1936, 1937, 1938). Polanyi argued that bonds are extended during reaction, and that this phenomenon caused barriers. Polanyi had previously found empirically that barriers to reaction for similar reaction could be fit to what is now called the *Polanyi relationship*:

$$E_a = E_a^0 + \gamma_P \Delta H_r \tag{10.3}$$

where E_a^0 is called the *intrinsic activation barrier* and γ_P is called the *transfer coefficient*. Polanyi and Evans derived equation (10.3). The derivation is given in Section 10.3.

Later Marcus (1955, 1964, 1968, 1969) showed that a better approximation is given by

$$E_a = \left(1 + \frac{\Delta H_r}{4E_a^0}\right)^2 E_a^0 \qquad (10.4)$$

Equations (10.3) and (10.4) are the key equations that people use to determine how bond stretching affects barriers to reaction. In the next several sections we will show how equations (10.3) and (10.4) arise, and describe when the equations work and when they fail.

10.3   DERIVATION OF THE POLANYI RELATIONSHIP

First, let us derive the Polanyi relationship, equation (10.3). The Polanyi relationship was first derived by Evans and Polanyi in 1935. At the time, there was a lot of interest in acid-catalyzed reactions. During an acid-catalyzed reaction an acid reacts with a stable species to yield products. For example, consider the acid-catalyzed reaction

$$CH_3CH_2OH \Longrightarrow C_2H_4 + H_2O \qquad (10.5)$$

and assume that the reaction is taking place in aqueous solution, so that the reaction is catalyzed by hydronium ions, $[H_3O]^+$. Reaction (10.5) has been examined at great depth in the literature. The first step is a proton transfer from the hydronium to the ethanol:

$$CH_3CH_2OH + [H_3O]^+ \longrightarrow [CH_3CH_2OH_2]^+ + H_2O \qquad (10.6)$$

Then the protonated ethanol loses water:

$$[CH_3CH_2OH_2]^+ + X \longrightarrow [CH_3CH_2]^+ + H_2O + X \qquad (10.7)$$

Then another water reacts with the ethyl group to regenerate the hydronium:

$$H_2O + [CH_3CH_2]^+ \longrightarrow [H_3O]^+ + CH_2CH_2 \qquad (10.8)$$

Evans and Polanyi assumed that in the rate-determining step of an acid-catalyzed reaction, is proton transfer from the acid to the reactant.

Next, we will derive an equation for the activation energy for the proton transfer process. We will consider the reaction $B - H + R \rightarrow B^- + HR^+$, where R is an proton acceptor, B is a conjugate base, and H is a hydrogen atom. For the purposes of derivation we will assume that one can separate the proton transfer process into two parts: the scission of the bond between the proton and its conjugate base (i.e., the negative ion produced when the proton dissociates), and the formation of a bond between the proton and the reactant to derive equation (10.3).

Figure 10.2 will be used in the derivation. Consider the transfer of a proton from an acid BH to a reactant R. We can define two quantities: r_{BH}, the length of the B–H bond; and r_{HR}, the length of the R–H bond as indicated in Figure 10.2. Following Evans and

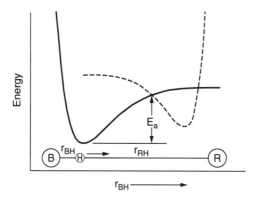

Figure 10.2 The energy changes that occur when a proton H is transferred between a conjugate base B and a reactant R. The solid line is the energy of the B–H bond; the dotted line is the energy of the H–R bond.

Polyani, we will assume that during the reaction the distance between B and R does not change. This assumption will be relaxed later in this chapter.

During the reaction, the proton–conjugate base bond breaks: r_{BH} increases while r_{RH} decreases. Figure 10.2 shows how the energy of the B–H and R–H bonds change as the reaction proceeds. The energy of a typical B–H bond looks like the solid line in Figure 10.2, where the energy of the proton increases as the proton moves from its equilibrium position. We have drawn a Morse potential, but the same arguments could be made for other potentials. The energy of the R–H bond looks similar to the potential for the B–H level. However, when we plot the data onto Figure 10.2, we need to be careful because the x axis in Figure 10.2 is the length of the B–H bond. When r_{BH} increases, r_{RH} decreases. As a result, when we plot the potential energy function of the R–H bond as a function of r_{BH}, the curve will be backwards as indicated by the dotted line in Figure 10.2.

Now consider what happens during the reaction. Initially, the B–H bond breaks, so the potential will ride up the solid contour shown in Figure 10.2. However, once the R–H bond begins to form, the potential can go back down the dotted line in Figure 10.2. During the reaction, the scission of the B–H bond occurs simultaneously with the formation of the R–H bond. Thus, in principle, the total energy should be the sum of the two curves. However, Evans and Polanyi suggested that one can obtain a reasonable approximation by assuming that the potential energy of the system went up the B–H curve as the reaction began, and then went down the R–H curve as the reaction went to completion. Evans and Polanyi also identified E_a in Figure 10.2 as the activation energy of the reaction.

Evans and Polanyi's model, although not quantitative, was quite important. It shows that one can consider a reaction as a sum of two processes: one involving bond scission and the other involving bond formation. One can then define the potential energy surface for each of the bond scission processes, and by considering the curve crossing between the two potential energy surfaces, one can get some information about rates.

The Evans–Polanyi model has had wide applicability in the physical–organic chemistry literature. As an example of the utility of the Evans–Polanyi model, consider how Figure 10.2 changes when one changes the strength of the acid by replacing one of the ligands on the acid with a different ligand. For the moment, assume that the acid dissociates more easily. By definition, when the acid strength increases, the B–H bond becomes easier to break in solution. This corresponds to an increase in the free energy of the B–H bond. The free energy could increase if either the position or the shape of the

B–H curve in Figure 10.2 changes. However, in the examples considered by Evans and Polanyi, the changes in the shape of the potential did not make enough of a difference in the free energy to make a significant change in the acid strength. Therefore, Polanyi and Evans proposed that one can get a qualitative picture of how a change in acid strength affected the reaction by simply displacing the solid curve in Figure 10.2 up or down.

Figure 10.2 shows the result of an upward displacement. Notice that as the solid curve is displaced upward, the activation energy for proton transfer goes down. For small displacements, the activation energy varies linearly with the change in the energy of the reaction.

On a more fundamental level, the Evans–Polanyi model provides an explanation of why activation energies arise in reactions. The idea is that when a reaction occurs, one bond breaks and another one forms. One can distinguish between two cases: the case in Figure 10.3, where the process is activated; and the case in Figure 10.4, where the curves cross at the minimum in the BH potential so the activation energy is zero. Fundamentally, the two cases are similar. However, when the reaction starts in Figure 10.3, the energy of the system goes up as the B–H bond is stretched, and that increase is larger than the lowering in energy due to formation of the R–H bond. As a result, the total energy of the system goes up as the reaction proceeds. That causes the reaction to be activated. In contrast, in the case in Figure 10.4, the lowering of the total energy of the system due to bond formation is larger than the energy increase due to bond scission. As a result, there is no activation barrier to reaction.

It happens that most real cases look closer to the case in Figure 10.3 than the case in Figure 10.4. Consider reaction $A + BC \rightarrow AB + C$. In the initial part of the reaction, A is moving in toward B–C. At long range, the A–BC potential is attractive, but as A gets closer, the A–BC potential is usually repulsive. The electron cloud of A overlaps with the electron cloud of B–C. That produces an electron–electron repulsion (i.e., a Pauli repulsion) of the type discussed in Chapter 9. The Pauli repulsions tend to keep the reactants from getting closer than their van der Waals radii. Now, as the reaction proceeds, the B–C bond begins to break and the A–B bond begins to form. However, the van der Waals radii of the molecules are usually larger than the covalent radii of the atoms. As a result, when atom B begins to be transferred to A, atom A is still pretty far away from B, as illustrated in Figure 10.5. As a result, the B–C bond needs to stretch before it breaks. Generally one observes about 25% bond extensions. The B–C bond loses energy as the

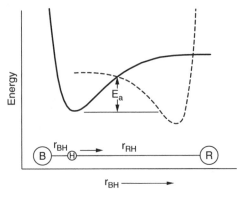

Figure 10.3 A diagram illustrating how an upward displacement of the B–H curve affects the activation energy when the B–R distance is fixed.

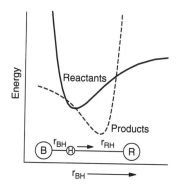

Figure 10.4 A diagram illustrating a case where the activation energy is zero.

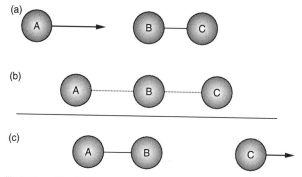

Figure 10.5 An illustration of the A–BC geometry: (a) A approaching B; (b) the transition state, (c) near the end of the reaction.

B–C bond is stretched. However, the A–B bond is still very long, so the net bonding between A and B is relatively small. This produces a situation that is analogous to the situation in Figure 10.4, where the B–C bond starts to break but the A–B bond barely starts to form. Figure 10.3 shows that in such a case, there is a finite barrier to reaction. Figures 10.3 and 10.5 apply to most reactions. As a result, most reactions are activated.

A more careful analysis indicates that activation energies can also arise because of bond distortions. However, the basic implication of the Evans–Polanyi model is that activation barriers arise because of the Pauli repulsions and because of the bond scissions and bond distortions that occur during reaction. The activation is zero only when no bonds break or distort during reaction, or when the reaction is so exothermic that bond formation dominates over bond scission. Both cases are rare. As a result, most reactions are activated.

10.4 THE POLANYI RELATIONSHIP

The analysis in the previous section was all qualitative. However, Polanyi and Evans showed that one can use the model to derive equation (10.3). In the material that follows, we will reproduce Polanyi and Evans' derivation to see how the ideas arise.

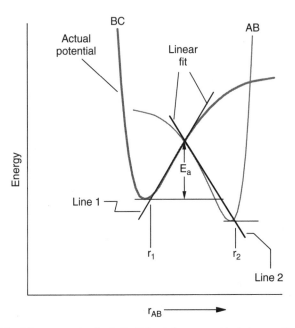

Figure 10.6   A linear approximation to the Polanyi diagram used to derive equation (10.11).

Consider the reactions $A + BC \rightarrow AB + C$. One can model the reaction as motion along the potential energy in Figure 10.12. The reactants come together at a constant value of r_{BC}, then atom B is transferred, and the reactants move away along a line of constant, r_{AB}. It is useful to consider a slice of the potential energy along the dashed line in Figure 10.12. Figure 10.6 shows how the energy varies along the slice. The horizontal axis (abscissa; x axis) in the figure is the distance along the dashed line, measured from the upper left corner of Figure 10.12. The vertical axis (ordinate; y axis) is the energy measured relative to the reactants. Notice that the potential goes up, reaches a maximum and then goes down.

Polanyi suggested that one can view the potential as being composed of a reactant potential E_1 and a product potential E_2. The two potentials are assumed to cross at the transition state.

For the purposes of derivation, it is useful to fit lines to the potential energy contour for the reaction as shown in Figure 10.6, where the lines are chosen to fit the tangent of the AB and BC potentials in the transition state. For the purposes of the derivation, it will be useful to assume that E_1 and E_2 are given by

$$E_1 = E_{reactant} + |Sl_1|(r_{ABC} - r_1) \quad \text{(reactants)} \tag{10.9}$$

$$E_2 = E_{product} + |Sl_2|(r_2 - r_{ABC}) \quad \text{(products)} \tag{10.10}$$

where $E_{reactant}$ is the reactant energy; $E_{product}$ is the product energy; E_1 is the line moving up from the reactants; E_2 is the line moving from the products; r_{ABC} is the distance along a coordinate, moving from left to right along the dashed line as in Figure 10.12; and Sl_1 and Sl_2 are the slopes of the two potential energy curves at the transition state.

Notice from Figure 10.6, E_a is equal to the energy at which line 1 intersects line 2. Solving equations (10.9) and (10.10) simultaneously for $E_a = E_1 = E_2$ and noting that $E_{product} - E_{reactant} = \Delta H_r$ yields

$$E_a = \left(\frac{|Sl_1||Sl_2|}{|Sl_1| + |Sl_2|} \right) (r_2 - r_1) + \frac{|Sl_1|}{|Sl_1| + |Sl_2|} \Delta H_r \qquad (10.11)$$

Substituting

$$E_a^0 = \left(\frac{|Sl_1||Sl_2|}{|Sl_1| + |Sl_2|} \right) (r_2 - r_1) \qquad (10.12)$$

$$\gamma_P = \frac{|Sl_1|}{|Sl_1| + |Sl_2|} \qquad (10.13)$$

into equation (10.11) yields

$$\boxed{E_a = E_a^0 + \gamma_P \Delta H_r} \qquad (10.14)$$

Equation (10.11) is called the *Polanyi relationship*. Equations (10.11) and (10.14) are the key Evans–Polanyi results. The implication of equation (10.11) is that if one had some way to change either the enthalpy or the free energy of a reaction, the activation energy for the reaction would also change. The activation barrier would also change if $|Sl_1|$ or $|Sl_2|$ were to change, or if $(r_2 - r_1)$ changed.

10.4.1 Key Predictions of the Polanyi Relationship

Physically, one can change the enthalpy of a reaction by adding a substituent group to one of the reactants. For example, consider the reaction between acetic acid and ethanol to yield ethylacetate:

$$CH_3COOH + CH_3OH \longrightarrow CH_3COOCH_3 + H_2O \qquad (10.15)$$

One can vary the acid strength of the acetic acid by fluorinating the methyl group. That makes the reaction more exothermic, which in turn leads to a lowering of the activation barrier for the reaction.

Experimentally, one often finds that the activation barrier for a series of closely related reactions varies linearly with the heat of reaction. Figure 10.7 shows how the activation barriers for a number of hydrogen transfer reactions vary with the heat of reaction. Notice that in the case of hydrocarbons, there is an approximately linear relationship between the heat of reaction and the activation barrier for the reaction. Consequently, the Polanyi relationship has proved to be quite a useful way to correlate data.

Another key prediction of the Polanyi relationship is that in general strong bonds will be harder to break than weaker bonds. To see that, it is useful to compare the potentials for forming a strong bond and a weaker one. Figure 10.8 shows the plot of the potential for breaking a strong bond and breaking a weaker bond. Notice that as you increase the strength of a bond, you increase Sl_1. According to equation (10.12), if everything else is equal, that should increase the barrier to reaction.

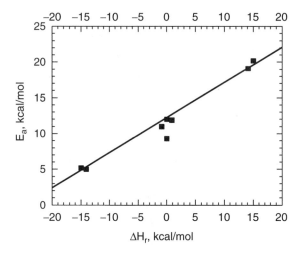

Figure 10.7 A plot of the activation barriers for the reaction $R + HR' \rightarrow RH + R'$ with R, R = H, CH_3, OH plotted as a function of the heat of reaction ΔH_r.

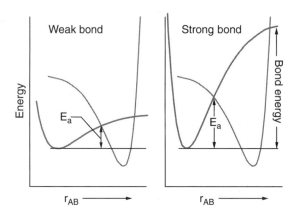

Figure 10.8 A schematic of the curve crossing during the destruction of a weak bond and a strong one for the reaction $AB + C \rightarrow A + BC$.

Another important effect is that, if everything else is equal, as you increase $(r_2 - r_1)$, the barriers to reaction increase. One calls reactions with small values of $(r_2 - r_1)$ reactions with **tight transition states**. If $(r_2 - r_1)$ is small, one does not have to stretch the bonds very much for reaction to happen. Consequently, if everything else is equal and there are no other effects, the barriers to reaction will be small.

One calls reactions with large values of $(r_2 - r_1)$, reactions with **loose transition states**. If $(r_2 - r_1)$ is large, one has to stretch the bonds a lot in order to get reaction to happen. Consequently, if everything else is equal and there are no other effects, the barriers to reaction will be large.

In actual practice, everything is rarely equal. Stronger bonds are not necessarily harder to break than weaker bonds. Reactions with looser transition states do not necessarily have

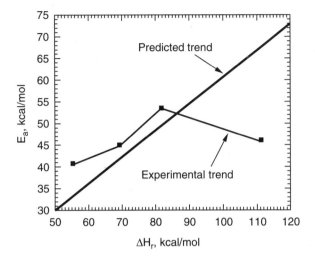

Figure 10.9 The activation barrier for the reaction $X + CH_3X \rightarrow XCH_3 + X$. The numbers are from the calculations of Glukhoutsev, et al. (1995).

larger barriers to reaction. For example, Figure 10.9 shows some data for the reaction

$$X^- + CH_3X \longrightarrow XCH_3 + X^- \quad (10.16)$$

for $X = F^-, Cl^-, Br^-, I^-$. Notice that the activation barrier hardly changes as the strength of the CH_3X bond increases. At first, one might think that this result violates the Polanyi relationship, but actually it does not. What is happening physically is that as you change the strength of the CH_3X bond, you increase Sl_1 and Sl_2 but you decrease $(r_2 - r_1)$. The net effect is that the barrier to reaction is hardly changed.

Another limitation is that one needs to be careful when using the Polanyi relationship with polar and nonpolar molecules. For example, Figure 10.10 shows a plot of the activation barrier for a number of hydrogen transfer reactions of the form

$$RH + R' \longrightarrow R + HR' \quad (10.17)$$

with R, R' = t-butyl, phenyl, CCl_3, CF_3, CH_3CH_2, H, H_2N, OH, CH_3, CH_3O. Notice that although there is some correlation to ΔH, the correlation is poor. Roberts and Steele report a correlation coefficient of 0.544. This example shows the difficulty with using the Polanyi relationship for molecules with widely varying properties. In particular, polar molecules and nonpolar molecules do not belong on the same plot.

This brings up an important point — activation barriers are subtle. If everything else is equal, increasing the heat of reaction (i.e., making the reaction less exothermic) will increase the barriers to reaction. If everything else is equal, increasing the strength of the bonds that are being formed and destroyed will increase the barriers to reaction. If everything else is equal, increasing the amount of bond distortion in the transition state will increase the barrier to reaction. If everything else is equal, pulling electrons out of the transition state, which lowers the Pauli repulsions, will lower the barriers to reaction. However, everything else is rarely equal. A substituent may increase the bond strength

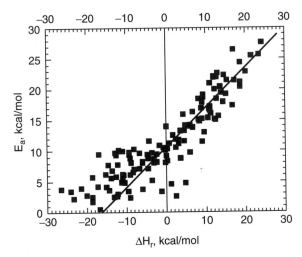

Figure 10.10   A Polanyi relationship for a series of reactions of the form $RH + R' \rightarrow R + HR'$. [Data from Roberts and Steele (1994).]

and decrease the amount of bond distortion. It is very difficult to know how the activation barrier will change in such a situation.

10.4.2   Limitations of the Polanyi Relationship

Another difficulty is that the Polanyi relationship does not work all of the time. According to the Polanyi relationship, the activation barrier for a reaction should vary linearly with the heat of reaction. Well, the linearity works, only over a limited range of ΔH_r. If one goes to a large range of ΔH_r, one finds that the Polanyi plots are generally nonlinear. Figure 10.11 shows a typical plot that one might observe in the literature. In this case one plots $\log(k)$ versus ΔH_r. Note that $\log(k)$ is proportional to $-E_a$, so according to the Polanyi relationship, the plot should be linear. However, the plot is nonlinear. The nonlinearity is typical. Consequently, the Polanyi relationship is limited to small ranges of ΔH.

Another difficulty is that the Polanyi relationships gives unphysical results for very exothermic reactions. Consider a system following the Polanyi equation:

$$E_a = 12 \text{ kcal/mol} + 0.5(\Delta H_r) \tag{10.18}$$

Notice that according to equation (10.18), a reaction with a ΔH_r more negative than -24 kcal/mol will have a negative activation barrier. That is physically impossible. Thus, the Polanyi relationship fails in the limit of very exothermic reactions.

Another physical impossibility occurs when $\Delta H_r > 24$ kcal/mol. For example, if we had a reaction with a ΔH_r of $+30$ kcal/mol, equation (10.18) would predict a barrier of 27 kcal/mol. Yet, the system must go up 30 kcal/mol to get to the products. Consequently, the activation barrier to reaction must be more than 30 kcal/mol.

One can generalize this result. In Chapter 4 we found that the activation barrier of an elementary reaction must be greater than the heat of reaction. Equation (10.18) predicts that when ΔH_r is greater that $+24$ kcal/mol, the activation barrier will be less than the heat

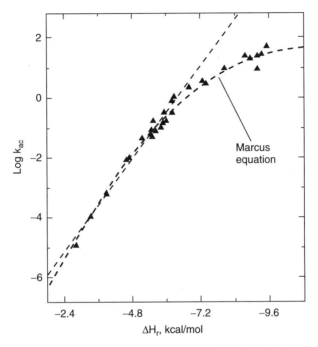

Figure 10.11   A Polanyi plot for the enolization of $NO_2(C_6H_4)O(CH_2)_2COCH_3$. Note that $\ln(k_{ac})$ is proportional to E_a. [Data of Hupke and Wu (1977).]

of reaction. Consequently, the Polanyi relationship fails in the limits of very endothermic reactions.

Generally, the Polanyi relationship is a useful approximation for reactions that are neither too endothermic nor too exothermic. However, the Polanyi relationship fails if the reactions are very endothermic or very exothermic.

10.5   THE MARCUS EQUATION

Marcus (1955, 1968) proposed an extension of the Polanyi relationship in order to overcome these difficulties with very endothermic and very exothermic reactions. Marcus considered a general reaction of the form

$$A + BC \longrightarrow AB + C \qquad (10.19)$$

and derived an equation for E_a using a modification of Polanyi's methods.

In the remainder of this section, we will derive a formula for E_a using Marcus' work as a guide. Our derivation will assume that the reaction follows the solid trajectory of Figure 10.12, where the reactants come together, a reaction occurs, and then the products fly apart. We will divide the trajectory into three parts: a part where the reactants come together without being significantly distorted, a part where atom B is transferred, and a part where the reactants fly away. In this approximation, the free energy of activation is the work it takes to bring the reactants to point X, w_r^1, plus the free energy it takes to

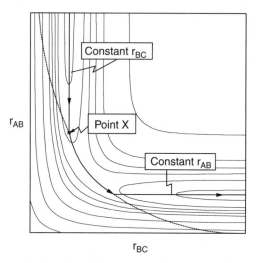

Figure 10.12   An approximation of the minimum-energy trajectory for the reaction A + BC → AB + C.

transfer atom B, E_a^1:

$$E_a = E_a^1 + w_r^1 \tag{10.20}$$

Marcus postulated with w_r^1 would be nearly constant for a group of closely related reactions, and that E_a^1 could be calculated using a modification of Polanyi's derivation described earlier in this section. For the purposes of derivation, we will define a quantity, r_X, as the distance along the dashed line in Figure 10.12 with $r_X = 0$ at point X. We will then assume that the energy of the A–B and B–C bonds follows the dotted lines in Figure 10.13 as a function of r_X where the energies show a Lennard-Jones dependence of r_X. To simplify the analysis, we will fit each of the potential contours in Figure 10.3 with a parabolic function near the transition state as indicated in Figure 10.13.

In the next few pages, we will derive Marcus' equation. For the purpose of derivation, we will assume that $E_{left}(r_X)$ and $E_{right}(r_X)$, the energies of the left and right parabolas in Figure 10.13, are given by

$$E_{left}(r_X) = SS_1(r_X - r_1)^2 + E_1 \tag{10.21}$$

$$E_{right}(r_X) = SS_2(r_X - r_2)^2 + \Delta H_r + E_2 \tag{10.22}$$

where SS_1, SS_2, r_1, and r_2 are fitting parameters. In equations (10.21) and (10.22) we have noted that E_{left} and E_{right} are functions of r_X. We will note in the material to follow that SS_1 and SS_2 are related to the vibrational frequency of the atom B as it is being transferred.

Note that since we are measuring energies from the horizontal line at the bottom of the A–B curve in Figure 10.13, E_a^1 is equal to the energy where E_{left} equals E_{right}.

For the purpose of derivation, it is useful to define $r^\ddagger$ as the value of r_X where $E_{left} = E_{right}$.

From equations (10.21) and (10.22)

$$SS_1(r^\ddagger - r_1)^2 + E_1 = SS_2(r^\ddagger - r_2)^2 + E_2 + \Delta H_r \tag{10.23}$$

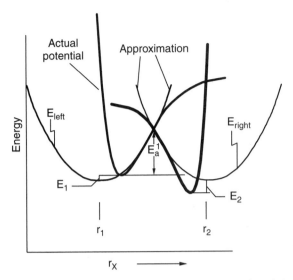

Figure 10.13 An approximation of the change in the potential energy surface that occurs when ΔH_r changes.

One can simplify equation (10.23) by assuming

$$SS_1 = SS_2 \tag{10.24}$$

$$E_2 = E_1 \tag{10.25}$$

Substituting SS_2 from equation (10.24) into equation (10.23), solving for $r^{\ddagger}$, and then substituting E_2 from equation (10.25) yields

$$r^{\ddagger} = \frac{(r_1 + r_2)}{2} + \frac{E_2 - E_1 + \Delta H_r}{2SS_1(r_2 - r_1)} = \frac{r_1 + r_2}{2} + \frac{\Delta H_r}{2SS_1(r_2 - r_1)} \tag{10.26}$$

Substituting $r^{\ddagger}$ into equation (10.21) yields

$$E_a^1 = E_{\text{left}}(r^{\ddagger}) = \left(\frac{r_2 + r_1}{2} - r_1 + \frac{\Delta H_r}{2SS_1(r_2 - r_1)} \right)^2 SS_1 + E_1 \tag{10.27}$$

Rearranging equation (10.27), we obtain

$$E_a^1 = \left(1 + \frac{\Delta H_r}{SS_1(r_2 - r_1)^2} \right)^2 \frac{(r_2 - r_1)^2}{4} SS_1 + E_1 \tag{10.28}$$

This equation can be put into a standard form by defining a quantity E_a^0 by

$$E_a^0 = \frac{(r_2 - r_1)^2}{4} SS_1 \tag{10.29}$$

Substituting equation (10.29) into equation (10.28) yields

$$E_a^1 = \left(1 + \frac{\Delta H_r}{4E_a^0}\right)^2 E_a^0 + E_1 \tag{10.30}$$

Substituting equation (10.30) into equation (10.20) yields

$$E_a = \left(1 + \frac{\Delta H_r}{4E_a^0}\right)^2 E_a^0 + w_r \tag{10.31}$$

with

$$w_r = w_r^1 + E_1 \tag{10.32}$$

People usually assume $w_r = 0$ to simplify the analysis. If $w_r = 0$, then

$$\boxed{E_a = \left(1 + \frac{\Delta H_r}{4E_a^0}\right)^2 E_a^0} \tag{10.33}$$

Equation (10.33) is called the *Marcus equation*. Marcus was awarded the 1993 Nobel Prize in Chemistry for his work on the Marcus equation and other contributions.

10.5.1 Qualitative Features of the Marcus Equation

It is useful to consider the qualitative features of the Marcus equation. Figure 10.14 shows a plot of the variation in E_a with varying ΔH_r calculated from equation (10.33) for a typical set of parameters. E_a actually varies parabolically with ΔH_r. Still, if we examine a limited range of ΔH_r, we see that E_a varies approximately linearly with ΔH_r. In contrast, when we work over an extended range of ΔH_r nonlinearities are seen. Therefore, according to the Marcus equation, one would expect reactions to obey linear free-energy relationships over a reasonable range of free energy. However, one would expect there to be some deviations from linearity if data are taken over a wide range of free energy.

Experimentally, one often finds that E_a varies linearly with ΔH_r in data taken over a limited range of ΔH_r. However, one usually observes some curvature if one takes data over a wide range of free energy. For example, consider the data in Figure 10.11. Notice that the data follow a curved line that is similar to the curved lines shown in Figure 10.14. Hupke and Wu (1977) fit the data in Figure 10.11 to equation (10.33); the results are given in Figure 10.11. Notice that the fit is good. Hence, we can explain the curvature in Figure 10.11 using the Marcus equation.

Marcus has shown that one can predict the fitting parameters in his equation from first principles, and fit real data. Further, it has been found that all of the examples in the literature that show curved free-energy plots can be explained via the Marcus formalism except those that show a change in mechanism with changing ΔH_r. Hence, the Marcus equation is quite useful in explaining data.

There is one unusual prediction of the Marcus equation. Notice that according to equation (10.33), the rate of reaction should reach a maximum (i.e., E_a should reach a

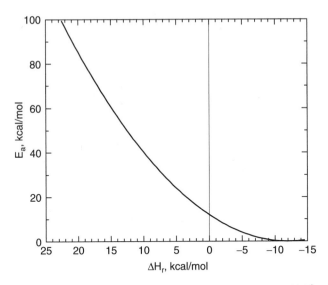

Figure 10.14 A plot of the activation barrier predicted from the Marcus equation with $E_a^0 = 12$ kcal/mol.

minimum) when $\Delta H_r = \Delta H_r^{max}$ where ΔH_r^{max} satisfies

$$1 + \frac{\Delta H_r^{max}}{8E_a^0} = 0 \qquad (10.34)$$

The rate should decrease with increasing $-\Delta H_r$ when $-\Delta H_r \geq E_a^0$.

Figure 10.15 illustrates why the maximum occurs. Most reactions show Polanyi diagrams like those in Figure 10.15a, where a reaction needs to go up a potential energy

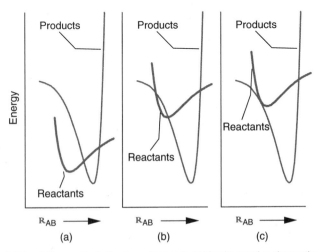

Figure 10.15 A Polanyi diagram illustrating how changes in ΔH for the reaction change the barriers to the reaction $AB + C \rightarrow AB + C$: (a) normal case; (b) saturation; (c) Marcus inverted region.

contour and back down again to occur. However, if the reaction is sufficiently exothermic, one can get to a situation where the energy curves for the reactants and products cross at the equilibrium point for the reactants as illustrated in Figure 10.15b. Notice that at this particular potential, the activation barrier is zero. Now consider what happens when the reaction becomes more exothermic as illustrated in Figure 10.15c. Note that as we continue to increase the potential, the intersection of the energy curves for the reactants and products moves to the left, causing the activation energy for the reaction to increase.

We call the region where the rate decreases with increasing driving force the **Marcus inverted region**. Up until 1980, no one had observed Marcus inverted behavior. However, a number of electron transfer reactions that show inverted behavior have been discovered since 1980. See W. H. Miller (1991) for details. There is a complication in electron transfer reactions in that the rate of the reaction is moderated by something called a *Frank–Condon factor*, which can decrease with increasing $-\Delta H_r$. We will discuss this effect later in this chapter. At present, the best available information is that Frank–Condon factors are quite important in the **Marcus inverted region**. As a result, one cannot explain the data with the Marcus equation alone. Still, the Marcus equation gives the correct qualitative behavior even in the inverted region, and so it is quite powerful.

10.5.2 Strengths of the Marcus Formulation

One of the major strengths of the Marcus formulation is that it eliminates the difficulties of the Polanyi relationship for very exothermic and very endothermic reactions. Figure 10.16 is a plot of the height of the barrier as a function of the heat of reaction. A plot of $E_a = \Delta H_r$ is also shown in the figure. Notice that the barriers are never negative. Further, the barriers are always greater than the heat of reaction. Consequently, unlike the Polanyi relationship, the Marcus equation never gives a physically impossible result.

Another strength of the Marcus equation is that it allows the activation barrier to vary nonlinearly with the heat of reaction. For example, Figure 10.11 showed some data where the activation barrier varies nonlinearly with the heat of reaction. Figure 10.11 shows a fit of the data to the Marcus equation. Notice that the equation fits very well.

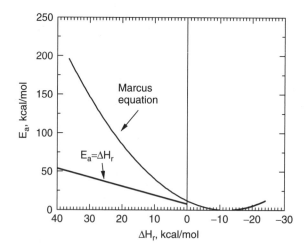

Figure 10.16 A plot of the activation barrier calculated from the Marcus equation as a function of the heat of reaction for a reaction with an intrinsic barrier of 12 kcal/mol.

One subtlety in all of this is that Cohen and Marcus (1968) have shown that E_a^0 is not constant. Rather, E_A^0 can vary over a series of reactions. That can produce complex behavior.

Fortunately, Marcus has shown that one can often estimate E_a^0 from something called 'identity reactions'. A simple calculation is shown in Example 10.A. Generally the Marcus equation allows one to estimate activation barriers from first principles. Again an example calculation is shown in Example 10.A.

10.5.3 Weaknesses of the Marcus Equation

Unfortunately, there are many examples where the Marcus equation does not fit as well (see Figures 10.17 and 10.18). Generally failures occur when you consider data taken over a wide range of ΔH_r. Figure 10.18 shows some data for some electron transfer reactions

$$R^* + CH_3CN \longrightarrow R^+ + CH_3CN^- \tag{10.35}$$

where R^*, an excited species, is deexcited via electron transfer to acetonitrile. Figure 10.17 shows data for intramolecular electron transfer processes across a spacer molecule:

$$R^- - Sp - R^+ \longrightarrow R^+Sp - R^- \tag{10.36}$$

In the first case the activation barrier reaches a maximum and then levels off. In the second case the activation barrier reaches a maximum and then decreases again. However, the decrease is much smaller than that predicted from the Marcus equation. Both of these cases show that while the Marcus equation eliminates the difficulties with the Polanyi

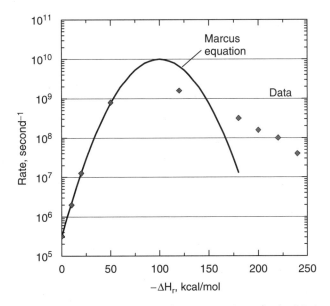

Figure 10.17 The rate of intramolecular electron transfer across a spacer molecule plotted as a function of the heat of reaction. [Data of J. R. Miller et al., (1984)].

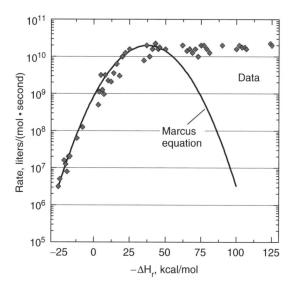

Figure 10.18   The rate of florescence quenching of a series of molecules in acetonitrite plotted as a function of the heat of reaction. [Data of Rehm and Weller (1970).]

relationship for very exothermic reactions, the Marcus equation does not entirely eliminate the difficulties.

Another weakness of the Marcus equation is that one cannot easily explain the trends in Figures 10.9 and 10.10 with the Marcus equation unless one assumes that E_a^0 varies over the series of reactions. The Marcus equation is basically a curve-crossing model. As one increases the strength of the bonds, one increases the slopes of the all of the lines, so that the barriers to reaction should increase. However, in Figure 10.9 we find that the activation barrier is not increasing substantially when one makes some substantial changes in the bond strength. People have fit the data in Figures 10.9 and 10.10 by assuming that E_a^0 varies over the series of reactions. However, if E_a^0 is treated as a variable, the Marcus equation loses its predictive power. Further, the variations in E_a^0 have not been physically reasonable.

The Marcus equation does not give the correct result for the data in Figures 10.17 and 10.18 because those figures assume that E_a^0 is constant in the series of reactions. According to the derivation of the Marcus equation in Section 10.5, E_a^0 is constant only when $(r_2 - r_1)$ in equation (10.29) is constant. In practice, $(r_2 - r_1)$ can change during reaction. Notice that according to equation (10.29), if we change $(r_2 - r_1)$, we change the barrier to reaction; when $(r_2 - r_1)$ increases, the barriers to reaction increase.

The Marcus equation gives a different result. According to the Marcus equation, when we decrease $(r_2 - r_1)$, we first decrease the barriers to reaction, and then increase them again.

The reason why the data in Figure 10.9 do not show an increase in the barrier to reaction is that $(r_2 - r_1)$ is increasing as slopes Sl_1 and Sl_2 are decreasing. The net result is that the barriers do not change.

The data in Figure 10.17 is more complicated. The Marcus equation works in the normal region for the data in Figure 10.17; however, the method fails in the Marcus inverted region. Physically, the situation is as illustrated in Figure 10.19. If one assumes

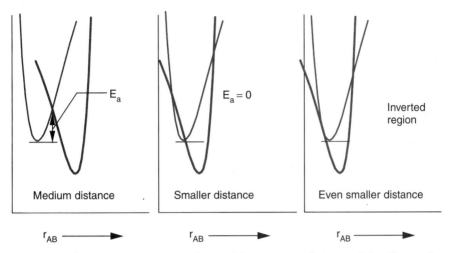

Figure 10.19 Changes in the curve-crossing model as $(r_2 - r_1)$ decreases during the reaction $AB + C \rightarrow A + BC$.

that the $(r_2 - r_1)$ is constant during the reaction, then the situation will be as seen on the right of Figure 10.19. The curves will cross in the inverted region, so the barriers to reaction should increase as ΔH becomes more negative. Note, however, that during a bimolecular reaction, $(r_2 - r_1)$ decreases as the reactants come together. Before the system gets to the situation on the right of Figure 10.19, the system will get to the situation in the middle of Figure 10.19, where the barrier to reaction is zero. Thus the barrier to reaction will be zero even though there will be a barrier at smaller values of $(r_2 - r_1)$. The derivation of the Marcus equation ignores this possibility, so it does not fit the data in Figure 10.17.

The situation in the data in Figure 10.18 is even more complex. This is a special reaction where one is transferring an electron across a molecule. $(r_2 - r_1)$ is constrained, so one cannot reach the situation in the center of Figure 10.19. As a result, one does observe an inverted region. One does not follow the Marcus equation exactly, however, because $(r_2 - r_1)$ still varies to compensate for the inverted region.

Generally one finds that the Marcus equation works well when $(r_2 - r_1)$ is constrained. If one considers a unimolecular reaction such as:

$$
\begin{array}{cccc}
\text{H} & \text{H} & \text{R} & + \\
\text{R}'-\text{C}-\text{C}-\text{C}-\text{C}-\text{H} \\
\text{H} & \text{H} & \text{H} & \text{H}
\end{array}
\longrightarrow
\begin{array}{cccc}
\text{H} & \text{H} & + & \text{R} \\
\text{R}'-\text{C}-\text{C}-\text{C}-\text{C}-\text{H} \\
\text{H} & \text{H} & \text{H} & \text{H}
\end{array}
\qquad (10.37)
$$

$(r_2 - r_1)$ represents the carbon–carbon bond length. The carbon–carbon bond length does not change significantly during reaction, so it is reasonable to assume that $(r_2 - r_1)$ is constant in such a situation. However, if one looks at a bimolecular reaction such as:

$$ \text{H} + \text{CH}_3\text{CH}_2\text{R} \longrightarrow \text{CH}_4 + \text{CH}_2\text{R} \qquad (10.38) $$

$(r_2 - r_1)$ changes as one changes the heat of reaction. The Marcus equation does less well in this situation.

10.6   PAULI REPULSIONS

In order to develop a better model for the energy at the transition state, one needs to be able to predict $(r_2 - r_1)$ at the transition state. Recall from Chapter 8 that during a bimolecular reaction there are repulsions that push the reactants apart. These are called *Pauli repulsions*. As the Pauli repulsions get stronger, the reactants are pushed apart so that $(r_2 - r_1)$ increases. If the Pauli repulsions weaken, the reactants can get closer together so that $(r_2 - r_1)$ decreases. As a result, the Pauli repulsions play a key role in determining $(r_2 - r_1)$. The bond energies also play a key role. If you are forming a strong bond during the reaction, the strong bond tends to pull the reactants together. As a result, the transition state gets tighter; specifically, $(r_2 - r_1)$ decreases. In contrast, weaker bonds increase $(r_2 - r_1)$. In general, $(r_2 - r_1)$ is determined by a balance between the bond energies and the Pauli repulsions.

10.6.1   The Origin of Pauli Repulsions

In order to model $(r_2 - r_1)$, one needs to understand Pauli repulsions. Pauli repulsions are basically electron–electron repulsions. Recall from freshman physics that electrons repel each others via a Coulomb force. The Coulomb force occurs with electrons on molecules in the same way that the Coulomb force occurs when there are two isolated electrons. As a result, when two reactant molecules come together, the electrons on one reactant molecule repel the electrons on the other reactant molecule. That can produce a barrier to reaction.

The actual interaction is more subtle than it would appear from the previous paragraph. Recall that when two hydrogen atoms come together, the electrons do not repel. Instead, the electrons on the two hydrogen atoms pair up to form a bond. In Chapter 11 we will show that, because of a quantum effect called *exchange*, the electron–electron repulsion will go to zero when two electrons pair up into a bond. As a result, the electron–electron repulsion is subtler that it would first appear.

Now consider the reaction

$$D + H_2 \longrightarrow DH + H \tag{10.39}$$

The incoming deuterium can form a bond with the hydrogen, so there is an attractive interaction. Yet, the H_3 complex is not stable; that is, there is no stable H_3 molecule. Consequently, there must also be a repulsive interaction to drive the H_3 apart. This repulsion is called a **Pauli repulsion**. Note that the Pauli repulsion is more complex than a simple Coulomb repulsion because it goes to zero when the electrons in a molecule pair up to form a bond.

10.6.2   A Physical Picture of the Pauli Repulsions

One needs to use quantum mechanics to derive an exact expression for the Pauli repulsions. However, it is easy to get a graphical picture of the interactions. In the text that follows, we will describe a qualitative method that one can use to understand the interactions. The ideas will come out of molecular orbital (MO) theory. When I teach this material, I usually review MO theory. The notes from my review are in Chapter 11. Consider the reaction

$$D + H_2 \longrightarrow DH + H \tag{10.40}$$

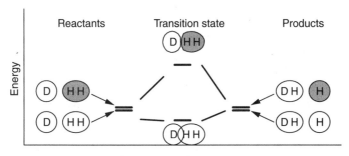

Figure 10.20   A diagram of the key molecular orbitals for the reaction $D + H_2 \rightarrow DH + H$.

The deuterium starts out with an electron in a 1s orbital while the H_2 starts out with two electrons in a σ orbital. The 1s orbital is spherical as indicated in Figure 10.20. The σ orbital starts out almost spherical, but I have drawn it more elliptical to make the diagram clearer.

When the 1s orbital interacts with the σ orbital, there are two possible molecular orbitals MOs of the system: a bonding MO where the wavefunction on the deuterium has the same sign as the wavefunction on the H_2, and an antibonding MO where the wavefunction on the deuterium has the opposite sign as the wavefunction on the H_2. In Figure 10.20 I have arbitrarily assigned a positive sign to the orbital on the deuterium and have then let the orbital on the H_2 have either a positive or a negative sign. In order to make the figure easy to see, I have made the positive orbitals light-colored and the negative orbitals dark-colored.

A key thing to remember from diagrams like this is that the light-colored orbitals attract other light-colored orbitals, dark-colored orbitals attract dark-colored orbitals, but dark-colored orbitals repel light-colored orbitals provided the orbitals are occupied.

Now consider what happens as the deuterium approaches the H_2. In the bonding state of the transition state (i.e., the bottom state in Figure 10.20), the orbitals are all of the same sign. Orbitals of the same sign attract. Therefore there is an attractive interaction between the deuterium and the H_2. In contrast, in the top state, the orbitals are of different signs. Orbitals of different signs repel. The repulsion causes the orbitals of different signs to be distorted. The distortion produces a barrier to reaction.

Another way to view this is that there is a bonding (attractive) interaction when two orbitals come together with the same sign and an antibonding interaction when two orbitals come together with different signs. For future reference we will call the former case the **bonding state** of the system, and the latter case the **antibonding state** of the system.

A subtlety is that there is always an antibonding state in the system even when there is no barrier to reaction. Physically, one gets a barrier only when the antibonding state is occupied. If all of the antibonding states are empty, there will be no electron–electron repulsions, and therefore no barriers.

We need to keep track of the electron count to see if the antibonding state is occupied. There are three electrons in the transition state for reaction (10.40). We can put two electrons in the bonding molecular orbital, so we have a two-electron attraction. However, quantum-mechanically, one can put at most two electrons into a molecular orbital. The third electron must go into the antibonding orbital. That produces a one-electron repulsion.

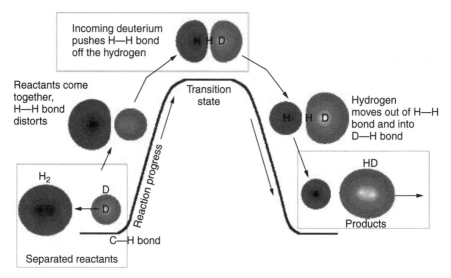

Figure 10.21   A diagram of the orbital distortions during the reaction $D + H_2 \rightarrow DH + H$.

Detailed calculations show that the attraction is 93 kcal/mol while the repulsion is 103 kcal/mol. That produces a net barrier of 10 kcal/mol.

Figure 10.21 shows a detailed picture of the orbital distortions. Notice that the orbitals distort during reaction. I like to think of the orbitals as balloons. When the deuterium comes in, it pushes the hydrogen–hydrogen bond out of the way. That allows the hydrogen atom on the left of the H_2 to be transferred to the deuterium. In this particular case, the left hydrogen is at the node between the two orbitals at the transition state for the reaction. In other reactions, we have found that the transition state can be slightly earlier or later than the point at which the hydrogen is transferred. However, one usually sees orbitals pushing into one another during reactions.

Quantum-mechanically, exothermic reactions have finite activation barriers only when filled orbitals of different signs push into one another, causing orbital distortions. If there are no orbital distortions, an exothermic reaction will not be activated. An endothermic reaction will still be activated because the system has to go uphill to get to products.

Next, let's consider a slightly different reaction:

$$D + [H_2]^+ \longrightarrow [DH_2]^+ \tag{10.41}$$

The MO diagram for reaction (10.41) is the same as that for reaction (10.40). However, there is an important difference. There are only two electrons in the transition state for reaction (10.41). One can put both electrons into the bonding state. The antibonding state is empty. Consequently, there is no barrier to reaction (10.41). There is a barrier to the following reaction, however

$$D + [H_2]^+ \longrightarrow [DH_2]^+ \longrightarrow [DH]^+ + H \tag{10.42}$$

There is no barrier to forming $[DH_2]^+$, but $[DH_2]^+$ is a stable molecule. It costs energy to break the $[DH_2]^+$ apart. Consequently, there is a barrier to reaction (10.42).

One can generalize these results to any reaction. Pauli repulsions cause barriers to reactions. The barriers arise when orbitals of different signs push up against one another. If all of the atomic orbitals have the same sign, there will be little or no barrier. Molecules would rather form MOs where all of the individual atomic orbitals have the same sign. Unfortunately, the MO with all of the atomic orbitals of the same sign can hold only two electrons. If there are three or more electrons in the system, there will be an antibonding interaction. Consequently, there will be a barrier to reaction.

10.6.3 Qualitative Picture of the Role of Pauli Repulsions on the Barriers to Reaction

Next, we want to discuss how to quantify these effects. To start off, it is important to note that at present it is very difficult to predict the strength of the Pauli repulsions exactly. Generally, one needs to do quantum-mechanical calculations, and then do analysis to estimate the strength of the Pauli repulsions. So far such calculations have been done for only a few special cases, so there is not much to say. Still, we do have pictures. I think that the pictures are important enough that I want to include them.

Figure 10.22 shows a diagram of the orbital distortions during the reaction

$$H + CH_3CH_3 \longrightarrow H_2 + CH_2CH_3 \tag{10.43}$$

This case is very much like the case in Figure 10.21. The hydrogen comes in and pushes the orbital off one of the methyl hydrogens. Again, the orbitals look like balloons pushing into one another. Lee and Masel (1996,1997) showed that the orbital distortion causes there to be a barrier to reaction.

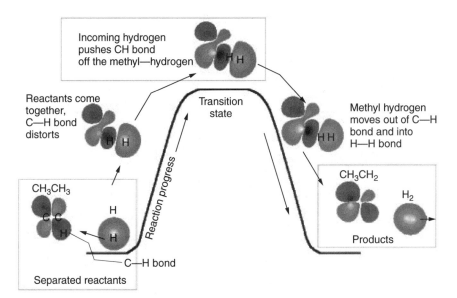

Figure 10.22 A diagram of the orbital distortions during the reaction $H + CH_3CH_3 \rightarrow H_2 + CH_2CH_3$. The diagram shows only the interaction with the E state of ethane (the C–H bond). Other MOs of the ethane also distort.

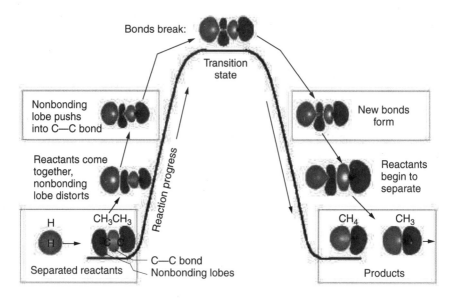

Figure 10.23   A diagram of the orbital distortions during the reaction $H + CH_3CH_3 \rightarrow CH_4 + CH_3$. Only the $3A_{1g}$ state of ethane (the state for the carbon–carbon bond) is shown.

Figure 10.23 shows a diagram of the orbital distortions during the reaction

$$H + CH_3CH_3 \longrightarrow CH_4 + CH_3 \qquad (10.44)$$

This case is more complicated. The hydrogen starts out with a spherical orbital, while the ethane starts out with a carbon–carbon bond, and two nonbonding lobes. When the hydrogen collides with the ethane, the hydrogen orbital pushes into the nonbonding orbital on the methyl group. The nonbonding orbital on the methyl group then pushes into the carbon–carbon bond. The net effect is that the carbon–carbon bond breaks. Again, one can view the process as balloons pushing into one another, causing bonds to break.

Notice that the orbital distortions are much larger for the reaction in Figure 10.23 than for the reaction in Figure 10.22. The result is that the barriers to reaction are larger for the reaction in Figure 10.23 than for the reaction in Figure 10.22. Lee and Masel estimate a barrier of 10 kcal/mol for the reaction in Figure 10.22 and a barrier of 42 kcal/mol for the reaction in Figure 10.23.

One of the questions in the literature is whether the large barrier to the reaction in Figure 10.23 is caused by the Pauli repulsions or by the fact that bonds need to be distorted during reaction. Notice that you are moving hydrogens out of the way during the reaction in Figure 10.23. Those displacements could be responsible for the barriers to reaction.

In order to separate the effects of bond distortion and orbital displacements, Blowers and Masel (1999a,b) calculated the activation barriers for a number of reactions of the form

$$CH_3OH + H \longrightarrow \text{products} \qquad (10.45)$$

Figure 10.24 compares the activation barriers to the bond distortion energies and the orbital distortion energies. Notice that the activation barriers increase as the orbital

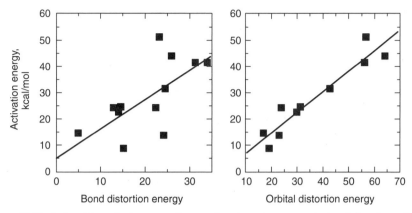

Figure 10.24 A plot of the activation energy for a number of reactions versus the bond distortion and the orbital distortion energy results of Blowers and Masel (1999a,b).

distortion energies increase. There is an excellent correlation between the barriers and the orbital distortion energies. In contrast, there is only a poor correlation between the activation energies and the bond distortion energies. Therefore, it seems that orbital distortions are more important than bond distortions in determining the barriers to reaction.

One should not infer, from the last paragraph, that bond displacements do not produce barriers. That is not true. Instead, bond distortions and orbital distortions are intimately connected. You get orbital distortions whenever you get bond distortions so that the orbital distortion energy contains a component due to the bond distortion energy. However, the orbital distortion energy has other components; one can get orbital distortions without bending or stretching any bonds. Those orbital distortions are important: more important, in fact, than the bond distortions. Consequently, the orbital distortion energy has a larger influence on the barriers than does the bond distortion energy.

In your organic chemistry class, you learned about orbital distortions in terms of **steric effects**. If you have a large group blocking a site, the reaction rate at that site will be reduced. Steric effects are important to a wide range of reactions, as you learned in organic chemistry.

Still, it is important to realize that steric effects are really manifestations of Pauli repulsions and apply only to situations where Pauli repulsions are important. Consider the following reaction:

$$[(CH_3)_2CH(NO_2)] + H \longrightarrow [(CH_3)_2C(NO_2)] + H_2 \qquad (10.46)$$

The carbon atom in reaction (10.46) is blocked by a methyl group, and so the reaction rate is slow. Kresge (1974) estimates a barrier of 19 kcal/mol. However, if you change the reaction slightly

$$[(CH_3)_2CH(NO_2)] + H^+ \longrightarrow [(CH_3)_2C(NO_2)]^+ + H_2 \qquad (10.47)$$

the reaction is unactivated. Physically, during reaction (10.46) the incoming hydrogen needs to overcome a large Pauli repulsion (i.e., electron–electron repulsion) in order to get the reaction to occur. In contrast, in reaction (10.47), a proton, rather than a hydrogen, comes in. The proton does not contain any electrons, and you need electrons to get Pauli repulsions. Consequently, there are no Pauli repulsions in reaction (10.47). The absence of Pauli repulsions makes reaction (10.47) unactivated even though reaction (10.46) has a barrier of 19 kcal/mol.

There is one other case where Pauli repulsions disappear: a case where an electron is transferred at long distances. Consider the reaction

$$H + F_2 \longrightarrow HF + F \tag{10.48}$$

One might think that there is a Pauli repulsion in reaction (10.48). After all, there are electrons on the hydrogen and the F_2 that can repel. However, in practice, the hydrogen transfers an electron to the fluorine at long distances, producing H^+ and $[F_2]^-$. The electron transfer eliminates the Pauli repulsions. Consequently, reaction (10.48) is unactivated.

10.6.4   Empirical Extensions of the Polanyi Relation

In the literature, it has been common to account for the effects discussed in Sections 10.6.2–10.6.4 by empirically extending the Polanyi relationship. For example, Roberts and Steele examined a series of reaction of the form

$$BH + R \longrightarrow B + HR \tag{10.49}$$

where B is a conjugate base and R is a different conjugate base. Roberts and Steele found that the activation barriers for exothermic reactions of the form in equation (10.49) could be fit by the equation

$$E_a = E_a^O \left(\frac{H_{BH}H_{RH}}{(H_{HH})^2} \right) + \gamma_P \Delta H + \gamma_X (\chi_B - \chi_R) + \gamma_s (s_B + s_r) \tag{10.50}$$

where H_{BH} and H_{RH} are the strengths of the BH and RH bonds, H_{HH} is a normalization factor (the strength of an HH bond), γ_p is the transfer coefficient, χ_B and χ_p are the Mullikin electronegativities of B and R, s_B and s_r are structure factors fit to the data, and E_a^O, γ_X and γ_s are constants.

Physically, the first term on the right of equation (10.50) is the intrinsic barrier multiplied by a correction factor to account for the fact that stronger bonds are harder to break than are weaker bonds. The second term is the standard transfer coefficient times heat of reactions as discussed in Section 10.3. The third term is a constant times the difference in electronegativity of the two species. Physically, when the two species have very different electronegativities, there is partial charge transfer, which lowers the Pauli repulsions. The last term in equation (10.50) accounts for the fact that larger groups can have larger Pauli repulsions.

At this point, equations like (10.50) have not been derived theoretically. In fact, up until the late 1990s, people did not know why they worked. As a result, the equations are not highly respected in the literature. Still, they work very well.

One other unexplained feature at present is that the γ_p values are different for exothermic and endothermic reactions. γ_p is usually about 0.3 for exothermic reactions and about 0.7 for endothermic reactions. That has not yet been explained.

10.7   A MODEL FOR THE ROLE OF PAULI REPULSIONS ON THE BARRIERS TO REACTION

One can get some insight into equation (10.50), by developing a model to estimate $(r_2 - r_1)$ and then plugging that result into the Marcus equation. Blowers and Masel derived an equation to estimate $(r_2 - r_1)$. The idea in the Blowers–Masel model is to calculate E_a by explicitly considering how Pauli repulsions affect $(r_2 - r_1)$. Let's consider the reaction

$$H + CH_3CH_3 \longrightarrow CH_4 + CH_3 \tag{10.51}$$

We discussed the qualitative features of potential energy surfaces for reactions in Sections 8.4.2, 8.4.3, and 8.5. During reaction (10.51), a carbon–carbon bond breaks and a carbon–hydrogen bond forms. According to the analysis in Sections 8.4.2, 8.4.3, and 8.5, respectively, one can approximate the potential energy surface for the system as the sum of three terms: the energy to stretch the C–C bond, the energy to form a new C–H bond, and the energy associated with the orbital distortions when the reactants come together:

$$V(\rho_{CC}, \rho_{CH}) = E_{CC}(\rho_{CC}) + E_{CH}(\rho_{CH}) + V_{Pauli} \tag{10.52}$$

where $V(\rho_{CC}, \rho_{CH})$ is the potential energy surface for the system, ρ_{CH} and ρ_{CC} are the lengths of the C–H and C–C bonds, E_{CC} is the energy to stretch the carbon–carbon bond as a function of the bond length, E_{CH} is the energy that one gains in forming a carbon–hydrogen bond, and V_{Pauli} is the energy associated with the orbital distortions.

Blowers and Masel approximated E_{CC} and E_{CH} by what are called *Morse potentials*:

$$E_{CC}(\rho_F) = w_{CC}([\exp(-\alpha_{CC}(\rho_{CC} - \rho_{CC}^e)) - 1]^2 - 1) \tag{10.53}$$

$$E_{CH}(\rho_{CH}) = w_{CH}([\exp(-\alpha_{CH}(\rho_{CH} - \rho_{CH}^e)) - 1]^2 - 1) \tag{10.54}$$

where ρ_{CC} and ρ_{CH} are the lengths of the C–C and C–H bonds; ρ_{CC}^e and ρ_{CH}^e are the equilibrium bond lengths; w_{CC} and w_{CH} are the bond dissociation energies for the carbon–carbon and carbon–hydrogen bonds in kcal/mol, which are available from the CRC (the CRC calls them *bond strengths*); and α_{CC} and α_{CH} are constants.

Figure 10.25 shows a plot of the Morse potential as a function of the bond length. The potential starts out high (i.e., very repulsive) and at short bond distances. The potential decreases as the bond length increases, reaching a minimum when $\rho_{CC} = \rho_{cc}^e$. The potential then goes to zero at long bond distances. If we change α_{CC}, the width of the minimum changes but the depth does not change. Real atomic potentials look very similar to this, which is why the Morse potential has proved so useful.

For the purposes of derivation, we will also assume that the Pauli repulsion can be approximated via a simple exponential:

$$V_{Pauli} = V_0 \exp(-\beta_{CC}\rho_{CC} - \beta_{CH}\rho_{CH}) \tag{10.55}$$

where β_{CC}, β_{CH}, and V_0 are constants.

Figure 10.26 shows a plot of a Pauli potential calculated from equation (10.55). The potential is small unless both the carbon–carbon and carbon–hydrogen bond lengths are small, which corresponds to the case where the reactants are close enough together that significant orbital distortions are seen. The potential stiffens as β increases.

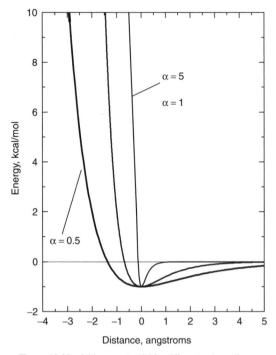

Figure 10.25   A Morse potential for different values of α_{CC}.

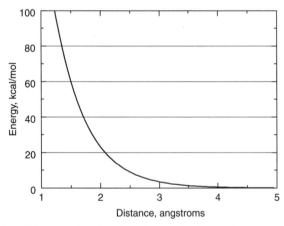

Figure 10.26   A plot of V_{Pauli} as a function of the interatomic distance for some typical values of the parameters.

Figure 10.27 compares some data for some helium–helium interactions to equation (10.55). Notice the good fit. Therefore, it seems that equation (10.55) is a good approximation.

Next we will derive an equation that allows us to calculate E_a from the Blowers–Masel model. Combining equations (10.52)–(10.55) yields what looks like a complex expression

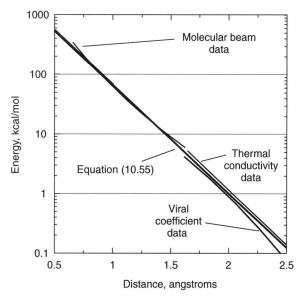

Figure 10.27 A comparison of Pauli repulsion determined experimentally for helium–helium collisions to that calculated from equation (10.55). [Data from Bernstein and Muckerman (1967)].

for the potential energy:

$$V(\rho_{CC}, \rho_{CH}) = w_{CC}([\exp(-\alpha_{CC}(\rho_{CC} - \rho_{CC}^e)) - 1]^2 - 1) - w_{CC}$$
$$+ w_{CH}([\exp(-\alpha_{CH}(\rho_{CH} - \rho_{CH}^e)) - 1]^2 - 1) \qquad (10.56)$$
$$+ V_0 \exp(-\beta_{CC}\rho_{CC} - \beta_{CH}\rho_{CH})$$

Figure 10.28 is a plot of the potential energy surface calculated from equation (10.59) for reaction (10.51). Blowers and Masel showed that the potential energy surface is virtually identical to the ab initio potential energy surface for the same reaction. Therefore, the model does reproduce the qualitative trends in the potential energy surfaces.

Next, we will simplify equation (10.56). First it is useful to define n_{CC} and n_{CH}, the Pauling bond order of the carbon–carbon and carbon–hydrogen bond, via

$$n_{CC} = \exp(-\alpha_{CC}(\rho_{CC} - \rho_{CC}^e)) \qquad (10.57)$$

$$n_{CH} = \exp(-\alpha_{CH}(\rho_{CH} - \rho_{CH}^e)) \qquad (10.58)$$

Pauling showed that with this definition, n_{CC} is 1.0 for a first-order bond and 2.0 for a second-order bond. During reaction (10.51), the $n_{CC} = 1.0$ and $n_{CH} = 0.0$ at the start of the reaction. As the reaction proceeds, n_{CC} decreases while n_{CH} increases, until at the end of the reaction $n_{CC} = 0.0$ while $n_{CH} = 1.0$.

Substituting equations (10.57) and (10.58) into equation (10.56) yields

$$V(n_{CC}, n_{CH}) = w_{CC}([n_{CC} - 1]^2 - 1) - W_{CC}$$
$$+ w_{CH}([n_{CH} - 1]^2 - 1) + V_P(n_{CC})^{q_{CC}}(n_{CH})^{q_{CH}} \qquad (10.59)$$

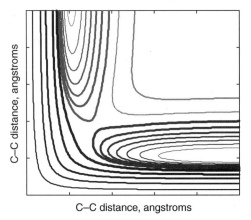

C–C distance, angstroms

Figure 10.28 A potential energy surface calculated from equation (10.59) with $w_{CC} = 95$ kcal/mol, $w_{CH} = 104$ kcal/mol, $V_P = 300$ kcal/mol, $q_{CC} = 0.7$, $q_{CH} = 0.5$.

where

$$q_{CC} = \frac{\beta_{CC}}{\alpha_{CC}}, \qquad q_{CH} = \frac{\beta_{CH}}{\alpha_{CH}} \qquad (10.60)$$

$$V_P = V_0 \exp(+\beta_{CC}\rho^e_{CC} + \beta_{CH}\rho^e_{CH}) \qquad (10.61)$$

At the transition state

$$\frac{\partial V}{\partial n_{CC}} = 0, \qquad \frac{\partial V}{\partial n_{CH}} = 0 \qquad (10.62)$$

Substituting equation (10.59) into equation (10.62), solving for n_{CC} and n_{CH}, and then substituting back into equation (10.59) allows one to derive an expression for the energy of the transition state. For special case that $q_{CC} = q_{CH} = 1$, Blowers and Masel find

$$E_a = \begin{cases} 0 & \text{when } \dfrac{\Delta H_r}{4E^0_a} < -1 \\[2ex] \dfrac{(w_0 + 0.5\Delta H_r)(V_P - 2w_0 + \Delta H_r)^2}{(V_P)^2 - 4(w_0)^2 + (\Delta H_r)^2} & \text{when } -1 \le \dfrac{\Delta H_r}{4E^0_a} \le 1 \\[2ex] \Delta H_r & \text{when } \dfrac{\Delta H_r}{4E^0_a} > 1 \end{cases} \qquad (10.63)$$

where E_a is the activation barrier and

$$w_0 = \frac{w_{CC} + w_{CH}}{2} \qquad (10.64)$$

It works out that V_P is related to the intrinsic barrier, E^0_a, by

$$V_P = 2w_0 \left(\frac{w_0 + E^0_a}{w_0 - E^0_a} \right) \qquad (10.65)$$

In order to use equation (10.63), one first chooses E_a^0 and w_0, calculates V_P and then plugs into equation (10.65). The results are virtually independent of w_0. Consequently, one can use an average value of w_0 (i.e. 100 kcal/mol) in all of the calculations with little error.

10.7.1 Qualitative Features of the Model

The Blowers–Masel model makes four key predictions:

- The activation barrier varies nonlinearly with the heat of reaction, approaching zero for very exothermic reactions and ΔH_r for very endothermic reactions.
- There is no inverted region.
- The results are similar to the Marcus equation when $-1 \leq \Delta H_r/4E_a^0 \leq 1$.
- Increases in the bond energies can increase or decrease the intrinsic barriers to reaction.

Figure 10.29 compares the activation barriers computed from the Blowers–Masel model to the activation barrier to barriers computed from the Marcus equation. Both models give virtually identical results when -36 kcal/mol $\leq \Delta H \leq 36$ kcal/mol. However, there are derivations at higher energies. The barriers vary nonlinearly with the heat of reaction in qualitative agreement with Figure 10.11.

Still, the Marcus equation predicts and inverted region where E_a^0 grows to infinity for very exothermal reactions; no inverted behavior is seen with the Blowers–Masel model. The Blowers–Masel model also predicts that E_a approaches ΔH_r at large ΔH_r. While the Marcus equation predicts that E_a diverges from ΔH_r at large ΔH_r.

Figure 10.29 compares both models to data. Notice that the data seem to fit the Blowers–Masel model much better than does the Marcus equation. Physically, the

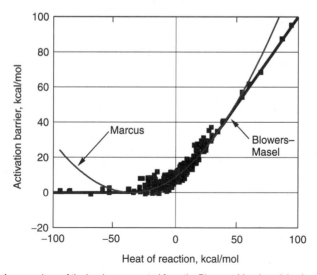

Figure 10.29 A comparison of the barriers computed from the Blowers–Masel model to barriers computed from the Marcus equation and to data for a series of reactions of the form $R + HR^1 \rightarrow RH + R^1$ with $w_0 = 100$ kcal/mol and $E_a^0 = 10$ kcal/mol.

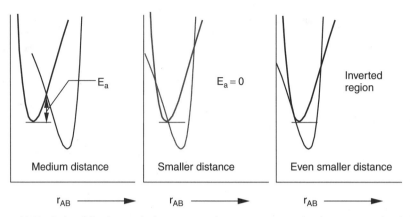

Figure 10.30   A plot of the changes in the curve crossing energy as two molecules come together during the reaction $AB + C \rightarrow A + BC$.

Blowers–Masel model allows the size of the transition state to vary. Generally the reactants come together; specifically, $(r_2 - r_1)$ changes as the reaction occurs. One can imagine the potential energy surface as being constructed as a series of curve-crossing (see Figure 10.30) models. When the reactants are far away, there is a large barrier to reaction. However, as the distance decreases, the barrier decreases. You always get to the situation that the barrier is zero before you get into the inverted region. The reactants react when the barrier is zero, so the system never reaches the inverted region. As a result, the Blowers–Masel model does not allow Marcus inverted behavior to occur.

The other major difference between the Blowers–Masel model and the Marcus equation is that the Marcus equation predicts that increases in the bond energy in the reactants will always increase the intrinsic activation barrier. However, the Blowers–Masel model allows for nonmonotonic behavior. For example, Figure 10.31 shows the variation in the intrinsic barriers with changing bond strength, w_0, computed from the Blowers–Masel model with $V_P = 400$ kcal/mol. Notice that the intrinsic activation barrier increases and then decreases with increasing bond strength. Figure 10.31 also shows some ab initio results. The data show some similar trends, although the data do not fit equation (10.63) exactly. Physically, equation (10.63) does not fit exactly because of the assumption $q_{CC} = q_{CH} = 1$; if you adjust q_{CC} and q_{CH}, you can fit the data in Figure 10.31 exactly. The final equation $q_{CC} \neq 1$ is more complex than equation (10.63).

Physically, increases in the bond strength in the reactants make the bonds harder to break, which raises the intrinsic barrier to reaction. However, the increases in the bond strength also pull the reactants together. That decreases the size of the transition state, $(r_2 - r_1)$, which, in turn, lowers the intrinsic barrier to reaction. As a result, the intrinsic barriers to reaction can increase or decrease with increasing w_0, as shown in Figure 10.31.

One of the main advantages of the Blowers–Masel model is that it allows one to fit a wider data set than does the Marcus equation. For example, the data in Figure 10.17 are easily fit to the Blowers–Masel model, whereas they do not fit the Marcus equation. The Blowers–Masel model also allows one to fit data as shown in Figure 10.9. Thus, the model is useful.

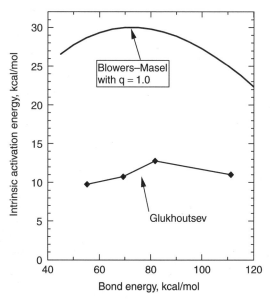

Figure 10.31 The variation in the intrinsic barrier with changing bond strength, w_0, computed from the Blowers–Masel model with $V_P = 400$ kcal/mol. Glukhoutsev's (1995) ab initio calculations of the activation barrier for the reaction $X + CH_3X \rightarrow XCH_3 + X$ are included for comparison.

10.7.2 Comparison of the models

Next, it is useful to compare the three models described so far in this chapter. Figure 10.32 shows a plot of the Marcus equation, the Polanyi relationship, and the Blowers–Masel approximation for identical sets of parameters. Notice that all three models are very similar. The data in Figure 10.29 follow the Blowers–Masel approximation most closely,

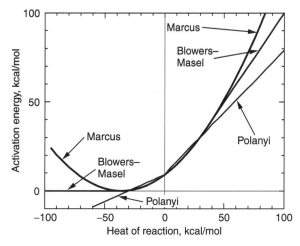

Figure 10.32 A comparison of the Marcus equation, the Polanyi relationship, and the Blowers–Masel approximation for $E_a^0 = 9$ kcal/mol and $w_0 = 120$ kcal/mol.

but the other two models look very similar to the Blowers–Masel approximation over the range of conditions where most data are measured (i.e., -50 kcal/mol $< \Delta H_r <$ 50 kcal/mol). The Marcus equation deviates from the Blowers–Masel approximation only for very endothermic and very exothermic reactions. In order to fit data with the Polanyi relationship, one needs to draw several lines. However, if one does that, one still fits data quite well. In practice, there are some small differences between the models; in particular, the Blowers–Masel model fits the data over the widest range. However, in practice, all three models give very similar predictions except for very endothermic or very exothermic reactions.

10.8 LIMITATIONS OF THE MODELS: QUANTUM EFFECTS

The big limitations of the models that we have discussed so far is that the models ignore quantum effects. One might initially suppose that quantum effects are not important to reactions. After all, bonds break, and most bonds look fairly classical. Still, there is one important class of reactions where the quantum effects are quite important: four-center reactions. Consider the following reaction:

$$H_2 + I_2 \longrightarrow 2HI \tag{10.66}$$

Years ago, people thought that reaction (10.66) proceeded by a four-centered reaction. Two bonds form during the reaction, and two are destroyed. Therefore, one could draw a stick diagram and show two bonds forming and two bonds breaking. However, if you work out the wavefunctions, it turns out that during the reaction, you end up either trying to put four electrons in a single molecular orbital, or having an electron with a spinup and a spindown simultaneously. Quantum-mechanically, one can put at most two electrons into an orbital, one with spinup and another with spindown. It is impossible for an electron to have two different spins at the same time. Therefore, it ends up that the bonding in the four-centered transition state is not possible.

In a more general way, reaction (10.66) cannot occur via a four-centered transition state because it is impossible to continuously transform the reactant wavefunctions into product wavefunctions. Instead, one must break some bonds before the reaction can proceed. This effect is not limited to four-centered reactions. There is a whole series of reactions called *symmetry-forbidden reactions* that have high barriers due to quantum effects.

10.9 THE CONFIGURATION MIXING MODEL

We are going to need to do a considerable amount of algebra before we can derive the key equations that we want to use to see whether the bonds are transformed smoothly during the course of a reaction. Pross (1985) summarized the key ideas from several groups in what he calls the **configuration mixing model**. The idea is to treat a reaction as a process where the reactant wavefunctions are continuously transformed into product wavefunctions.

The analysis builds on some of the concepts of MO theory, which we will describe in detail in Chapter 11. Recall that in MO theory, one can write the wavefunction, ψ, for any MO of a system as an antisymmetrized product of the wavefunctions, ϕ_a, of all of the atoms in the system. For example, ψ_σ, the wavefunction for the σ bond in a diatomic molecule AB, is given by

$$\psi_\sigma = \frac{1}{\sqrt{2}}(\phi_A + \phi_B)(\overline{\phi_B} + \overline{\phi_A}) \tag{10.67}$$

where ϕ_A is the wavefunction for an electron in the bonding orbital of A with the spin pointing up, $\overline{\phi_A}$ is the wavefunction for an electron in the bonding orbital of A with the spin pointing down, ϕ_B is the wavefunction for an electron in the bonding orbital of B with the spin pointing up, $\overline{\phi_B}$ is the wavefunction for an electron in the bonding orbital of B with the spin pointing down, and ψ is the antisymmetrizer. There are three important MOs for excited states of this system: ψ_σ^*, the wavefunction for the σ^* orbital of the A–B molecule:

$$\psi_{\sigma^*} = \frac{1}{\sqrt{2}}(\phi_A - \phi_B)(\overline{\phi_A} - \overline{\phi_B}) \tag{10.68}$$

$\Psi_{A:^-B^+}$, the wavefunctions for both electrons on A:

$$\Psi_{A:^-B^+} = \phi_A\overline{\phi_A} \tag{10.69}$$

and $\Psi_{A^+B:^-}$, the wavefunctions for both electrons on B:

$$\Psi_{A^+B:^-} = \phi_B\overline{\phi_B} \tag{10.70}$$

According to the configuration interaction (CI) model described in Section 11.2, one can write the wavefunction for any state of the system as a sum of the Slater determinates for these MOs, plus the Slater determinates for additional excited states. For example, if A and B form a bond with a significant dipole moment, then the wavefunction for the bonding state of the system, Ψ_{A-B}, can be written as

$$\Psi_{A-B} = c_\sigma\psi_\sigma + c_{A:^-B^+}\Psi_{A:^-B^+} + c_{A^+B:^-}\Psi_{A^+B:^-} + c_{\sigma^*}\psi_{\sigma^*} \tag{10.71}$$

where we used Ψ rather than ψ in equation (10.71) to indicate that Ψ_{A-B} is a wavefunction that has been computed with the CI model. Similarly, $\Psi_{A\bullet B\bullet}$, the wavefunction for one electron being on A and one being on B with no interaction between the two, is

$$\Psi_{A\bullet B\bullet} = \frac{1}{\sqrt{2}}(\psi_\sigma + \psi_\sigma^*) \tag{10.72}$$

For the discussion that follows, we will designate the wavefunction in equation (10.72) as the [A•B•] state of the system, and the wavefunction in equation (10.71) as the [A–B] state of the system. In some cases, it will also be useful to discuss the [A$^+$B$^-$] state of the system. The [A$^+$B$^-$] state of the system is just the [A–B] state in the limit $|c_{A^+B:^-}| \gg |c_\sigma|$, $|c_{A:^-B^+}|$, $|c_{\sigma^*}|$.

In the material that follows, we will consider how the electronic configuration of the A–B molecule changes as a reaction proceeds and use that information to predict the activation barrier for the reaction between A and B.

The derivation will start with a simple example where a sodium atom approaches a chlorine atom and reacts to form sodium chloride:

$$Na• + Cl• \longrightarrow Na^+Cl^- \tag{10.73}$$

When the two atoms are far apart, the [Na•Cl•] configuration is the ground state of the system while [Na⁺Cl⁻] configuration is the first excited state. In contrast, in sodium chloride, [Na⁺Cl⁻] state is the ground state and the [Na•Cl•] state is the first excited state.

Now consider doing a thought experiment where a sodium atom approaches a chlorine atom but the sodium atoms and the chlorine atom are prevented from exchanging electrons. According to calculations, when the sodium atom approaches the chlorine atom, the energy of both the [Na•Cl•] and [Na⁺Cl⁻] configurations go down as indicated in Figure 10.33. However, the [Na⁺Cl⁻] configuration is stabilized much more than is the [Na•Cl•] configuration. That is why the equilibrium structure is [Na⁺Cl⁻].

Now, let's do the same process but allow the sodium atom and chlorine atom to exchange electrons. When the sodium atom and the chlorine atom initially approach one another, they start out in the [Na•Cl•] configuration. However, during the reaction, there is an exchange of electrons between the sodium and the chlorine to yield the [Na⁺Cl⁻] species. The exchange of electrons occurs gradually during the course of the reaction. At any point along the reaction coordinate, one can represent the electronic structure of the system as a mixture of the [Na•Cl•] and [Na⁺Cl⁻] configurations.

For future reference, we will indicate the reaction pathway as the lower dashed line in Figure 10.33. The reactants start out in the [Na•Cl•] configuration, but, as the reaction

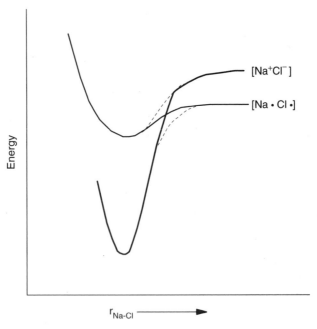

Figure 10.33 A schematic showing how the energy of the [Na•Cl•] and [Na⁺Cl⁻] configurations of NaCl change as a sodium atom approaches a chlorine atom.

proceeds, there is a continuous change in the configuration of the system from the [Na•Cl•] state into the [Na$^+$Cl$^-$] state.

Note that if you did not get mixing of the configurations, the sodium and the chlorine would stay in the [Na•Cl•] configuration. The atoms could not get into the [Na$^+$Cl$^-$] configuration. Hence, one would not get any reaction. As a result, it is the mixing of configurations that allows the sodium and chlorine reaction to form (metastable) sodium chloride.[1]

This example illustrates the idea that reactions can be represented as a conversion of the electronic structure of the reactants from one configuration to another. In the following discussion, we will model the changes in electronic structures, and use that information to make prediction about reactions.

We will start by noting that the conversion of one state into another can be modeled as something called an **avoided crossing**. Notice that if there were no interactions between the [Na•Cl•] and [Na$^+$Cl$^-$] configurations of sodium chloride, the curves for the energies of the two states in Figure 10.33 would cross. In reality, however, there are strong interactions between the two configurations. In Section 10.9.1, we will show that when two states interact, they mix to form an upper state whose energy is greater than that of either of the two individual states, and a lower state whose energy is lower than that of either of the two individual states. The energy of the two states are represented by the dashed line in Figure 10.33. Notice that because of the interactions between the [Na•Cl•] and [Na$^+$Cl$^-$] configurations, the energies of the eigenstates of the system never cross; hence the name **avoided crossing**. One does not always get avoided crossings when energy curves for reactions cross. Later in this chapter, we will provide several examples where the curves actually cross. However, the significance of the avoided crossing is that it provides a mechanism to allow the system to change from one configuration to another. When avoided crossings happen, one can continuously transfer an electron from one state of the system to another. That allows bonds to rearrange so that a reaction can take place. If there is no avoided crossing, there is no change in the configuration of the reactants and hence no reaction. As a result, avoided crossings are very important to the theory of reactions.

Woodward and Hoffmann (1970) show that one can model most reactions as a change in the electronic configuration of the reactants and an avoided crossing. (The main exception is in photochemical reactions where the change in configuration can be driven by a photon rather than an avoided crossing.) Hence, the idea of describing a reaction as a change in the electronic configuration of the reactants with an avoided crossing is quite useful.

In the materials to follow, it will be useful to draw what is called a *configuration mixing diagram for a reaction*, where the configuration mixing diagram is a plot showing the energy of each of the configurations of the system change as a function of the reaction coordinate (e.g., bond order). The diagram also shows how the states mix.

Figure 10.34 shows a configuration mixing diagram for the reaction of a sodium atom with a chlorine atom. The configuration mixing diagram for the reaction between a sodium atom and a chlorine atom contains the same information as was given in Figure 10.33. There are two solid lines in the coordination mixing diagram: one starting at the [Na•Cl•] state of the reactants and going to the [Na•Cl•] state in the products, and the other starting at the [Na$^+$Cl$^-$] state in the reactants and going to the [Na$^+$Cl$^-$] state in the products. There are also dashed lines indicating the avoided crossing. The implication

---

[1] As noted in Chapter 8, when an isolated sodium atom reacts with an isolated chlorine atom, a considerable amount of energy is released. The energy needs to be dissipated before a stable NaCl crystal can form. As a result, initially reaction (10.73) produces a hot metastable complex not crystalline sodium chloride.

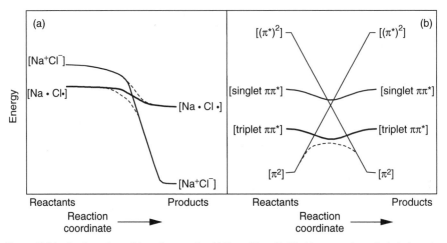

Figure 10.34 Configuration mixing diagram for (a) Na• + Cl• → NaCl, (b) conversion of *cis*-butene to *trans*-2-butene.

of this configuration mixing diagram is that during the reaction, the [Na•Cl•] and the [Na$^+$Cl$^-$] states mix continuously to convert the reactants to products.

Figure 10.34 also shows the configuration mixing diagram for the conversion of *cis*-butene, into *trans*-butene.

$$
\begin{array}{c}
H_3C \qquad CH_3 \qquad\quad H_3C \qquad H\\
\diagdown\;\; \diagup \qquad\qquad\; \diagdown\;\; \diagup\\
C{=}C \qquad\longrightarrow\qquad C{=}C\\
\diagup\;\; \diagdown \qquad\qquad\;\; \diagup\;\; \diagdown\\
H \qquad\;\; H \qquad\qquad\; H \qquad CH_3
\end{array}
\qquad (10.74)
$$

The system starts outs with two p electrons in a π^2 orbital. For future reference, we note that there are several excited states of the system: a triplet $\pi\pi^*$ state, a singlet $\pi\pi^*$ state, and a singlet $(\pi^*)^2$ state. Now consider what happens when one rotates the :CHCH$_3$ group in the right half of *cis*-2-butene around the carbon–carbon bond, leaving the CH$_3$HC: group on the left fixed. Note that the rotation reverses the sign of all of the p orbitals on the right :CHCH$_3$ group. We start out with an electron in a π orbital with a wavefunction, ψ_π, given by

$$
\psi_\pi = \frac{1}{\sqrt{2}}(\phi_{p\cdots left} + \phi_{p\cdots right}) \qquad (10.75)
$$

where $\phi_{p\cdots left}$ and $\phi_{p\cdots right}$ are the wavefunctions for the p orbitals on the left and right :CHCH$_3$ group. Notice that when we rotate the right :CH$_2$CH$_3$ group $\phi_{p\cdots right}$ is converted into $-\phi_{p\cdots right}$. As a result, the π orbital is converted into a π^* orbital with

$$
\psi_\pi = \frac{1}{\sqrt{2}}(\phi_{p\cdots left} - \phi_{p\cdots right}) \qquad (10.76)
$$

Similarly, the π^* orbitals are converted to bonding π orbitals. As a result, the original $(\pi^*)^2$ state is converted into the ground state of the system while the original π^2 state is converted into a $(\pi^*)^2$ state.

Now consider what happens as the reaction proceeds. We start within the π^2 state. However, as we begin to twist the molecule, we begin to mix in some of the excited states. First, the triplet $\pi\pi^*$ state begins to interact with the π^2 state. Later the original $(\pi^*)^2$ state plays a role. This diagram illustrates two important conclusions from the configuration mixing model: (1) the mixing of excited states with the ground state plays an important role in determining the potential energy surface for a reaction, and (2) often many different excited states play a role in the reaction.

The idea that the mixing of the ground state with the excited states of the system determines potential energy surfaces provides an alternative view of why reactions are activated.

Recall that when people first started to apply reaction rate theory, it was not obvious why activation barriers arose during reactions. The reaction of hydrogen atoms with a deuterium tritium molecules illustrates the difficulty:

$$H + DT \longrightarrow HD + T \qquad (10.77)$$

Note that as this reaction proceeds, a deuterium–tritium bond breaks and a hydrogen–deuterium bond forms. Initially, there is a single D–T bond. Half way through the reaction, there is half a H–D bond and half a D–T bond. At the end of the reaction, there is a single H–D bond. Hence, the total bond order is conserved during reaction.

Up to the 1990s, people thought that reactions were activated because bonds broke during reaction. We now know that the barrier for reaction (10.77) occurs because orbitals are distorted. In fact, when one does a quantum calculation of the barriers, one writes the wavefunction for the system ψ as a sum of the wavefunctions for all of the states in the system ϕ_m:

$$\psi = \sum_{states,m} C_m\phi_m \qquad (10.78)$$

In equation (10.78) the C_m terms are a series of coefficients and ϕ_m is the wavefunction for the mth state of the system. When orbitals distort, one finds that one gets contributions to ψ from the excited states of the system. As a result, one can say that one is mixing excited states into the ground state as the reaction proceeds.

The configuration mixing model explains in part why activation barriers arise during reaction. The general idea is that as a reaction proceeds, excited states are mixed into the ground-state wavefunction. The mixing of excited states raises the energy of the system and hence produces an activation barrier. Physically, when excited states get mixed into the ground-state wavefunction, the wavefunctions are distorted. Hence, one might want to think about a barrier as being associated with distortion of orbitals in the reactants as a reaction proceeds.

It is useful to consider how the Marcus–Polanyi relationship arises from the configuration mixing model. In the next few pages, we will show that the Marcus equation arises naturally from the configuration mixing model. Our approach will be to start with a two-state model where one of the states represents the reactant configuration while the other state represents the product configuration. We will then consider how a substituent affects the energy of each states. An analysis similar to that in Section 11.5 will be used to derive an equation analogous to the Marcus equation.

Consider the following reaction:

$$A\bullet + B\bullet \longrightarrow A^+ + B^- \qquad (10.79)$$

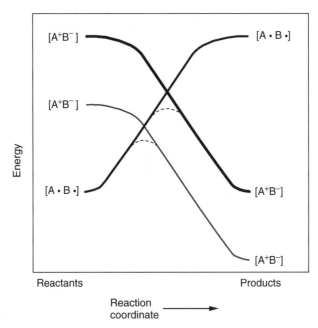

Figure 10.35   Configuration mixing model for the reaction $A\bullet + B\bullet \rightarrow A^+ + B^-$ showing how changes in the energy of the [AB] configuration affect the reaction.

A configuration mixing diagram for the reaction is shown as the solid line in Figure 10.35. The [A•B•] configuration is the ground state of the reactants and an excited state of the products, while the [A^+B^-] configuration is an excited state of the reactants and the ground state of the products. During the reaction, the configuration of the system moves up the [A•B•] curve and then down the [A^+B^-] curve.

Now consider making a change that displaces the curve for the [A^+B^-] state up or down. Notice that such a displacement will cause the activation barrier to change. One can model the change by examining how the interaction between the curves shift up or down as the reaction proceeds.

Notice the close correspondence between the configuration mixing diagram in Figure 10.35 and the Polanyi diagram in Figure 10.13. In both cases, the activation barrier is given as a curve crossing on a reaction coordinate diagram. Hence, the analysis in Section 11.5 also applies to the configuration mixing diagram Figure 10.35. In particular, if we fit the curves in Figure 10.35 to lines, we can derive the Polanyi relationship. If we fit the curves to parabolas, we will derive the Marcus equation. Hence, the Polanyi relationship and Marcus equation apply as well to the situation depicted in Figure 10.35 as to the situation described in Section 10.5.

Pross and Shalk showed that one could use the configuration mixing model to calculate the intrinsic barrier for reaction. The idea is that in the configuration mixing model, intrinsic barriers are associated with the energy to move electrons from one MO to another. If one models the interaction between the system as a curve crossing, then one can get some information about the barriers from Figure 10.36.

Figure 10.36 shows four cases. The first two cases show how changes in the gap between the reactions affects the barriers to reaction. Notice that as you increase the gap,

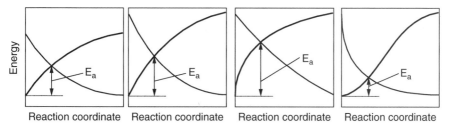

Figure 10.36   A diagram showing how changes in the configuration mixing model affects the intrinsic barrier to reaction.

it costs more energy to put an electron into an excited state. Consequently, it costs more energy to mix in the excited state. That raises the barriers to reaction.

Another way to change the barriers is to change the slopes in Figure 10.36. Notice that as the slopes decrease, the barriers decrease. Physically, the slopes are a measure of how much of the excited state is mixed into the ground state during the initial stages of reaction. If the orbitals in the reactions are distorted more, the initial contribution from the excited states will grow. That will raise the barriers to reaction. Generally, the configuration mixing model makes predictions that are similar to those of the Blowers–Masel model. Further, the configuration mixing model allows you to quantify the barriers to reaction; the details will be discussed in Chapter 11.

Unfortunately, the configuration mixing model does not work in detail. The key assumption in the configuration mixing model is that one can write the wavefunction for the transition state as a sum of terms such as that in equation 10.71, where all the terms in the sum are wavefunctions for states of the reactants. Quantum-mechanically, that is not correct. Physically, when a reaction occurs, bonds are extended. The extended bonds contain contributions from orbitals that are bigger than the orbitals in the reactants. In particular, big orbitals called "diffuse functions" play a key role. The big orbitals are missing from the sum in equation (10.71). As a result, the configuration mixing model does not work in detail.

10.10   SYMMETRY-FORBIDDEN REACTIONS

Another source of derivations from the configuration mixing model occurs during something called a **symmetry-forbidden reaction**. Throughout the last section, we assumed that when the energy contours for two states of the system cross, the states will mix to provide a pathway for a reaction. However, in 1965, Woodward and Hoffmann (1965a,b,c) noted that sometimes states cannot mix. When states do not mix, one does not have a convenient pathway to convert the reactants into products. As a result, the reaction rate is negligible. Hoffmann showed that the states will not mix when the symmetries of the states are wrong. Hence, he called reactions that show negligible rates, because the reactant and product configurations do not mix, **symmetry-forbidden reactions**.

Woodward and Hoffmann (1970) wrote a famous book, *The Conservation of Orbital Symmetry*, which describes, in detail, the role of symmetry in determining rates of reactions. In the materials that follow, we will summarize the key ideas from Woodward and Hoffmann's analysis. We will also review some other interpretations of the ideas due to Pearson (1976) and Fukui (1975). One should refer to Woodward/Hoffmann (1970), Pearson (1976), or Fukui (1952,1957,1975) for further details.

10.10.1   Forbidden Crossings

In the next several sections, we will be discussing symmetry-forbidden reactions, namely, reactions that cannot occur because symmetry prevents the key states from mixing. In this section, we will show that one can determine when states can mix by examining a quantity $\beta_{10}(\mathbf{X})$ defined by

$$\beta_{10}(\mathbf{X}) = \iiint \Psi_1^*(\mathbf{X})\mathcal{H}_1(\mathbf{X})\Psi_0(\mathbf{X})\,d\bar{r} \tag{10.80}$$

where $\Psi_0(\mathbf{X})$ is the wavefunction for the initial state of the system, $\Psi_1^*(\mathbf{X})$ is the wavefunction for the final state of the system, and $\mathcal{H}$ is called the *transition Hamiltonian*. States can mix whenever β_{10} is nonzero. No mixing occurs when β_{10} is zero. The derivation is complex and could be skipped without difficulty. It is useful to consider when states can mix. Consider a reaction where molecules A and B come together and react. When the two molecules are far apart, we can describe the electronic structure of the system by a Hamiltonian, $\mathcal{H}_0^0$, with a ground-state wavefunction Ψ_0, and a first excited-state wavefunction Ψ_1^0, where Ψ_0^0 and Ψ_1^0 satisfy

$$\mathcal{H}_0^0 \Psi_0^0 = E_0^0 \Psi_0^0 \tag{10.81}$$

$$H_0^0 \Psi_1^0 = E_1^0 \Psi_1^0 \tag{10.82}$$

with $E_0^0 < E_1^0$.

Now consider doing a thought experiment where we move the two molecules together but do something to prevent the two molecules from reacting. Let's define a Hamiltonian for the nonreactive case as $\mathcal{H}_0(\mathbf{X})$, where we have noted that $\mathcal{H}_0$ is a function of $\mathbf{X}$, where $\mathbf{X}$ is a generalized reaction coordinate. We will assume that when $\mathbf{X} = 0$, $\mathcal{H}_0(\mathbf{X})$ will equal the reactant Hamiltonian and when $\mathbf{X} = 1$, $\mathcal{H}_0(\mathbf{X})$ will equal the product Hamiltonian. We will also define $\Psi_0(\mathbf{X})$ and $\Psi_1(\mathbf{X})$ to be the eigenstates of the Hamiltonian that need to mix for the reaction to occur, and we will define $E_0(\mathbf{X})$ and $E_1(\mathbf{X})$ to be the corresponding eigenvalues of $\mathcal{H}_0(\mathbf{X})$.

Figure 10.37 shows a configuration mixing diagram for the reaction between A and B. The solid lines in the figure are plots of $E_0(\mathbf{X})$ and $E_1(\mathbf{X})$ as a function of the reaction coordinate. We have set up the system so that during the reaction Ψ_0 changes from the ground state to the first excited state while Ψ_1 changes from the first excited state to the ground state.

Now consider moving the molecules together and trying to get the reaction to occur. When reaction occurs, the Hamiltonian of the system will be different from $\mathcal{H}_0$, so let's call the new Hamiltonian $\mathcal{H}_1(\mathbf{X})$. For the derivation that follows, it is useful to define a constant $(\mathcal{H})$ by

$$\mathcal{H}_1(\mathbf{X}) = \mathcal{H}_0(\mathbf{X}) + \mathcal{H}(\mathbf{X}) \tag{10.83}$$

The question that we want to address is whether the two states can mix when the atoms move together, so let's assume that the new ground state of the system has a wavefunction $\Psi'(\mathbf{X})$ that is a mixture of the wavefunctions for the two original configurations:

$$\Psi'(\mathbf{X}) = C_0(\mathbf{X})\Psi_0(\mathbf{X}) + C_1(\mathbf{X})\Psi_1(\mathbf{X}) \tag{10.84}$$

We want the new wavefunction to satisfy

$$\mathcal{H}_1(\mathbf{X})\Psi'(\mathbf{X}) = E'(\mathbf{X})\Psi'(\mathbf{X}) \tag{10.85}$$

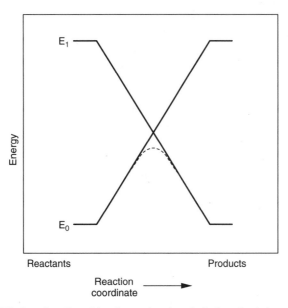

Figure 10.37   A configuration mixing diagram for a hypothetical reaction between A and B.

A detailed derivation is given in Masel (1996). Masel shows that there are two energy levels of the system with energies given by

$$E'_-(\boldsymbol{\mathcal{X}}) = \frac{E'_0(\boldsymbol{\mathcal{X}}) + E'_1(\boldsymbol{\mathcal{X}})}{2} - \frac{\sqrt{(E'_0(\boldsymbol{\mathcal{X}}) - E'_1(\boldsymbol{\mathcal{X}}))^2 + 4\beta_{01}(\boldsymbol{\mathcal{X}})\beta_{10}(\boldsymbol{\mathcal{X}})}}{2} \qquad (10.86)$$

$$E'_+(\boldsymbol{\mathcal{X}}) = \frac{E'_0(\boldsymbol{\mathcal{X}}) + E'_1(\boldsymbol{\mathcal{X}})}{2} + \frac{\sqrt{(E'_0(\boldsymbol{\mathcal{X}}) - E'_1(\boldsymbol{\mathcal{X}}))^2 + 4\beta_{01}(\boldsymbol{\mathcal{X}})\beta_{10}(\boldsymbol{\mathcal{X}})}}{2} \qquad (10.87)$$

and wavefunctions given by

$$\Psi_-(\boldsymbol{\mathcal{X}}) = \begin{cases} \Psi_a(\boldsymbol{\mathcal{X}}) & \text{when} \quad E_0(\boldsymbol{\mathcal{X}}) \le E_1(\boldsymbol{\mathcal{X}}) \\ \Psi_b(\boldsymbol{\mathcal{X}}) & \text{otherwise} \end{cases} \qquad (10.88)$$

$$\Psi_+(\boldsymbol{\mathcal{X}}) = \begin{cases} \Psi_a(\boldsymbol{\mathcal{X}}) & \text{when} \quad E_0(\boldsymbol{\mathcal{X}}) \le E_1(\boldsymbol{\mathcal{X}}) \\ \Psi_b(\boldsymbol{\mathcal{X}}) & \text{otherwise} \end{cases} \qquad (10.89)$$

with

$$\Psi_a(\boldsymbol{\mathcal{X}}) = \frac{\Psi_0(\boldsymbol{\mathcal{X}}) + \left(\dfrac{\beta_{10}(\boldsymbol{\mathcal{X}})}{E'_-(\boldsymbol{\mathcal{X}}) - E'_1(\boldsymbol{\mathcal{X}})}\right)\Psi_1(\boldsymbol{\mathcal{X}})}{\sqrt{1 + \left(\dfrac{\beta_{10}(\boldsymbol{\mathcal{X}})}{E'_-(\boldsymbol{\mathcal{X}}) - E'_1(\boldsymbol{\mathcal{X}})}\right)^2}} \qquad (10.90)$$

$$\Psi_b(\boldsymbol{\mathcal{X}}) = \frac{\Psi_1(\boldsymbol{\mathcal{X}}) + \left(\dfrac{\beta_{01}(\boldsymbol{\mathcal{X}})}{E'_+(\boldsymbol{\mathcal{X}}) - E'_0(\boldsymbol{\mathcal{X}})}\right)\Psi_0(\boldsymbol{\mathcal{X}})}{\sqrt{1 + \left(\dfrac{\beta_{01}(\boldsymbol{\mathcal{X}})}{E'_+(\boldsymbol{\mathcal{X}}) - E'_0(\boldsymbol{\mathcal{X}})}\right)^2}} \qquad (10.91)$$

10.11 QUALITATIVE RESULTS

Now, let's consider what happens when a reaction occurs. At the start of the reaction $\beta_{10}(\mathcal{X}) = 0$, so $\Psi_-(\mathcal{X}) = \Psi_0(\mathcal{X})$. Now, let's assume that when the molecules come together, $\beta_{10}(\mathcal{X})$ is nonzero. Note that according to equation (10.90), when $\beta_{10}(\mathcal{X})$ is nonzero, we get mixing of the states. If $\beta_{10}(\mathcal{X})$ is nonzero, the final wavefunction will approach $\Psi_1(\mathcal{X})$. Therefore, a reaction can occur whenever $\beta_{10}(\mathcal{X})$ is nonzero when the molecules collide.

The dashed line in Figure 10.37 is a plot of E as a function of $\mathcal{X}$ for a typical value of $\beta_{10}(\mathcal{X})$. Note that when $\beta_{10}(\mathcal{X})$ is nonzero, the energy of the lower eigenstate goes up over a hill and eventually ends up at E_0. This is typical for a reaction.

10.11.1 Reactions with Negligible Coupling

Note, however, that there are some cases where $\beta_{10}(\mathcal{X})$ is zero when two molecules collide. When $\beta_{10}(\mathcal{X})$ is zero throughout the collision process, $\Psi_-(\mathcal{X})$ is equal to Ψ_0 as the reactants come together. Hence, there would be no coupling of the state. Therefore, we conclude that the motion of the nuclei will convert one configuration of a system into another only when $\beta_{10}(\mathcal{X})$ is nonzero during the collision.

For future reference, we will call $\beta_{10}(\mathcal{X})$ the **coupling constant**, and note that reactions occur only when the coupling constant is nonzero.

In Nobel Prize–winning work, Woodward and Hoffmann (1970) provided several examples where $\beta_{10}(\mathcal{X})$ was zero during a reaction. In this section, we will consider a simple example that shows how $\beta_{10}(\mathcal{X})$ can be zero. One should refer to Woodward and Hoffmann (1970) or Pearson (1976) for many other examples.

Consider the following simple reaction:

$$H_2 + D_2 \longrightarrow 2HD \tag{10.92}$$

One can imagine that the reaction can go by either the chain propagation mechanism shown in Figure 10.38 or the four-centered reaction shown in Figure 10.39. Wright (1970, 1975) has also proposed a concerted six-centered reaction (e.g., $2H_2 + D_2 \rightarrow H_2 + 2HD$).

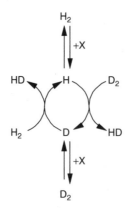

Figure 10.38 The chain propagation mechanism for H_2/D exchange.

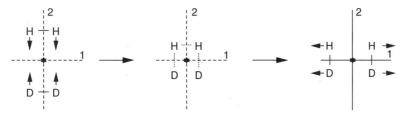

Figure 10.39 A hypothetical four-centered mechanism for H_2/D_2 exchange. The dotted lines in the figure denote mirror planes that are preserved during the reaction (see the text). This reaction is symmetry-forbidden.

At first sight one might think that the four-centered reaction would dominate. The six-centered reaction requires three molecules to come together simultaneously in the gas phase which is highly improbable. The first step in the chain propagation mechanism (Figure 10.38) is hydrogen–hydrogen bond scission. Hydrogen–hydrogen bond scission is highly activated. In contrast, superficially, during the four-centered reaction one is simply exchanging bonds within the molecule. As we have drawn the picture, there is no change in bond order anywhere in the reaction. Hence, from a superficial analysis, one might conclude that the four-centered reaction is possible.

In fact, however, the four-centered reaction has never been observed. At high temperatures, reaction (10.92) goes via the chain propagation mechanism. There has been some discussion about the mechanism at low temperatures where the six-centered reaction may predominate. However, there is no evidence that the reaction ever occurs at a measurable rate via a four-centered reaction. Calculations of Conroy and Malli (1969) indicate that the four-centered reaction has an activation energy of 515 kJ/mol! By comparison, the H–H bond strength is only 435 kJ/mol.

In the next few paragraphs, we will show that the reason why the reaction does not occur via the four-centered reactions is that the β_{10} values for the transformation of some of the reactant configurations in reaction (10.92) into production configurations are zero. As a result, the reaction in Figure 10.39 cannot occur via a simple gas-phase collision.

In order to derive the key results, it will be useful to view the reaction by sitting on a point half way between the hydrogen and the deuterium as indicated by the dot in Figures 10.39 and 10.40. One can imagine a transition state where there are both H–H and H–D bonds. If we model this transition state as one big molecule, then we can use molecular orbital (MO) theory to tell us how the configuration of the molecule changes as the reaction proceeds. There are four key MOs in the system. They are depicted in

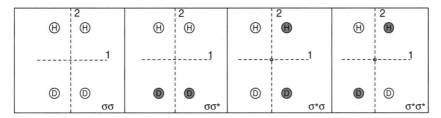

Figure 10.40 A schematic of the key molecular orbitals for the transition state of reaction (10.92). Positive atomic orbitals are depicted as open circles; negative orbitals are depicted as shaded circles.

Figure 10.40. For further reference, we will label the MO in terms of the bonding of one of the hydrogen atoms. The $\sigma\sigma$ MO will be an MO where one of the hydrogens is bound to the other hydrogen via a sigma bond between the hydrogens, and a σ^* bond between the hydrogens and the deuteriums. Figure 10.40 shows a schematic of the wavefunctions for the various states. All of the orbitals in the wavefunction for the $\sigma\sigma$ state, $\psi_{\sigma\sigma}$, have the same sign. In $\psi_{\sigma\sigma^*}$, the orbitals on both hydrogens have the same sign, but the orbitals on the two deuteriums have opposite signs. In $\psi_{\sigma^*\sigma}$, the orbitals on each adjacent H–D have the same sign but the orbitals on the H–H and D–D have opposite signs. In $\psi_{\sigma^*\sigma^*}$, the wavefunction for the orbitals has alternating signs.

Now consider what happens when we move the H_2 and D_2 together. We start out with no net interaction between the H_2 and D_2, so the wavefunction for the system $\Psi_{\text{reactants}}$ can be approximated by

$$\Psi_{\text{reactants}} = \psi_{\sigma\sigma}\psi_{\sigma\sigma^*} \tag{10.93}$$

At the end of the reaction, there are H–D bonds, but there are no net interactions between the two HD. Therefore, the wavefunction for the products Ψ_{products} can be approximated by $\psi_{\text{products}} = \psi_{\sigma\sigma}\psi_{\sigma^*\sigma}$. As a result, during the reaction electrons are moved from the $\sigma\sigma^*$ to the $\sigma\sigma^*$ MO.

Woodward and Hoffmann (1970) show that it is convenient to represent the transfer of electrons by a correlation diagram where the correlation diagram is very similar to a configuration mixing diagram. We keep track of what happens to each individual MO (i.e., the ψs) while in the configuration mixing diagram, and we keep track of what happens to the total wavefunction for the system (i.e., the Ψs).

Figure 10.41 shows a correction diagram for the four-centered reaction depicted in Figure 10.39. The reaction starts out with $\sigma\sigma$ and $\sigma\sigma^*$ MOs occupied and the $\sigma^*\sigma$ and $\sigma^*\sigma$ MOs empty. However, during the reaction, electrons are moved from the $\sigma\sigma^*$ to the $\sigma^*\sigma$ MO.

Now the questions is, whether the reaction can occur as depicted in Figure 10.39. According to the results in Section 10.11, the motion of the H_2 and D_2 toward one

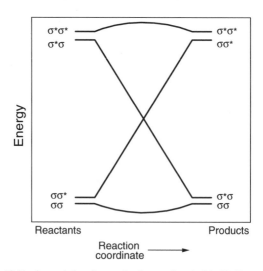

Figure 10.41 A correlation diagram for the reaction depicted in Figure 10.39.

another in the configuration shown in Figure 10.39 will allow the electrons in the $\sigma\sigma^*$ MO to be moved into the $\sigma^*\sigma$ MO whenever β is nonzero. Therefore, we can tell if reaction can occur by computing β. For further discussion, we will note that there are two mirror planes in Figure 10.39: one that lies half way between the H and the D_2, and, one that goes between the midpoints of the H–H and D–D bonds. We will designate those two mirror planes as planes 1 and 2.

According to equation (10.80), the coupling constant for conversion of the $\sigma\sigma^*$ state into the $\sigma^*\sigma$ state $\beta_{(\sigma\sigma^*)(\sigma^*\sigma)}$ is given by

$$\beta_{(\sigma\sigma^*)(\sigma^*\sigma)} = \iiint \psi^*_{\sigma\sigma^*} \mathcal{H} \psi_{\sigma^*\sigma} \, d\bar{r} \tag{10.94}$$

Note that $\mathcal{H}$ is completely symmetric. Assume that the interaction of the two mirror planes is the origin for Figure 10.39. If we move up, we move in the y direction. If we move to the right, we move in the positive x direction. Note that from symmetry

$$\mathcal{H}(x, y) = \mathcal{H}(-x, y) = \mathcal{H}(x, -y) = \mathcal{H}(-x, -y) \tag{10.95}$$

It is useful to rewrite equation (10.94) as

$$\beta_{(\sigma\sigma^*)(\sigma^*\sigma)} = \int_0^\infty \int_0^\infty \psi_{\sigma\sigma^*} \mathcal{H} \psi_{\sigma^*\sigma} \, dx \, dy + \int_0^\infty \int_{-\infty}^0 \psi^*_{\sigma\sigma^*} \mathcal{H} \psi_{\sigma^*\sigma} \, dx \, dy$$
$$+ \int_0^\infty \int_{-\infty}^0 \psi^*_{\sigma\sigma^*} \mathcal{H} \psi_{\sigma^*\sigma} \, dx \, dy + \int_{-\infty}^0 \int_{-\infty}^0 \psi^*_{\sigma\sigma^*} \mathcal{H} \psi_{\sigma^*\sigma} \, dx \, dy \tag{10.96}$$

For the purposes of derivation, it is useful to define a quantity "INT" by

$$\mathrm{INT} = \int_0^\infty \int_0^\infty \psi_{\sigma\sigma^*} \mathcal{H} \psi_{\sigma^*\sigma} \, dx \, dy \tag{10.97}$$

Now consider a quantity INT' given by

$$\mathrm{INT}' = \int_0^\infty \int_{-\infty}^0 \psi^*_{\sigma\sigma^*} \mathcal{H} \psi_{\sigma^*\sigma} \, dx \, dy \tag{10.98}$$

INT is related to INT' by a transformation that switches y with $-y$. Notice that when we switch y to $-y$, $\psi_{\sigma^*\sigma}$ does not change while $\psi_{\sigma\sigma^*}$ is converted into $-\psi_{\sigma\sigma^*}$. Therefore

$$\mathrm{INT}' = -\mathrm{INT} \tag{10.99}$$

A similar argument shows

$$\int_0^\infty \int_{-\infty}^0 \psi^*_{\sigma\sigma^*} \psi_{\sigma^*\sigma} \, dx \, dy = -\mathrm{INT} \tag{10.100}$$

$$\int_{-\infty}^0 \int_{-\infty}^0 \psi^*_{\sigma\sigma^*} \psi_{\sigma^*\sigma} \, dx \, dy = \mathrm{INT} \tag{10.101}$$

Substituting equations (10.97)–(10.101) into equation (10.96) yields

$$\beta_{(\sigma\sigma^*)(\sigma^*\sigma)} = 0 \qquad (10.102)$$

Therefore, if we move an H_2 molecule toward a D_2 in the configuration shown in Figure 10.39, there will be no coupling between the $\sigma\sigma^*$ and the $\sigma^*\sigma$ MOs. As a result, no reaction will occur. One can get some small amount of coupling if one distorts the geometry (e.g., by twisting the H_2 relative to the D_2). However, $\beta_{(\sigma\sigma^*)(\sigma^*\sigma)}$ is always small. As a result, one does not get a significant reaction.

Now it is useful to go back and consider why the picture of the transition state in Figure 10.37 was wrong. In Figure 10.37, we assumed that it will be possible to simultaneously break a H–H bond and form a H–D bond. However, if we multiply out all of the wavefunctions, we will find that during the reaction, we need to transform wavefunctions that look like $\phi_1\bar{\phi}_2\phi_3\bar{\phi}_4$ into wavefunctions that look like $\phi_1\bar{\phi}_2\bar{\phi}_3\phi_4$, where ϕ_1, ϕ_2, ϕ_3, and ϕ_4 are the atomic orbitals on atoms 1, 2, 3 and 4 (e.g., the two hydrogen atoms and the two deuterium atoms, respectively) and the bars represent spins. Physically, when we try to share an electron between the two states, as we would do if there were partial H–H bond and a partial H–D bond, we would find that we need a single electron to spinup and down at the same time. The electron cannot be both spinup and spindown simultaneously. Therefore, it is impossible to simultaneously break the H–H bond and form a H–D bond. That explains why Conroy and Malli (1969) found that the four-centered reaction has an activation energy of 515 kJ/mol.

Another way to get the same result is to go back to the stick model of the four-centered reaction in Figure 10.41. In the stick model, one is saying that in the transition state for the reaction, there are four σ bonds in the system with four total electrons. According to MO theory, the only way for the four-centered bonding to be stable would be for the four-centered bond to be one of eigenstates of the system. However, the eigenstates of the system are the four states shown in Figure 10.41. Of those, only the $\sigma\sigma$ state is bonding in both directions. One can put two electrons in the $\sigma\sigma$ state, but the other two electrons must be put into other molecular orbitals. These other MO are antibonding orbitals. As a result, the four-centered bonding in Figure 10.41 is not an eigenstate of the system. Therefore, quantum-mechanically, the four-centered bonding cannot occur.

A more subtle analysis shows that the reaction

$$H_2{}^+ + D_2{}^+ \longrightarrow \begin{bmatrix} H{-}H \\ |\quad| \\ D{-}D \end{bmatrix}^+ \longrightarrow 2HD^+ \qquad (10.103)$$

where $H_2{}^+$ and $D_2{}^+$ are positively charged molecules, is allowed. However, in reaction (10.103) there are only two electrons in the transition state. One can put both electrons in the SS state so that the bonding is possible.

This example illustrates an important feature of Woodward and Hoffmann's analysis: There are many simple reactions that look plausible on paper but cannot occur because they involve bonding schemes that are not possible quantum-mechanically. We can draw a stick diagram like that in Figure 10.39 for many proposed transition states. However, there is no guarantee that the bonding shown in the stick diagram is possible. If it is not, the proposed reaction will have a very high activation barrier. Hence, one has to do some analysis before one assumes that one can get a proposed reaction to occur. A generalization of the analysis above shows that four-center reactions of species with σ

bonds are usually symmetry-forbidden. (There are exceptions where a nonbonding orbital can change the symmetry of the system.) Therefore, one would rarely expect to see four-centered reactions experimentally.

10.12   CONSERVATION OF ORBITAL SYMMETRY

One can imagine repeating the analysis in the last section for a series of reactions. One could simply work out all of the integrals. However, Woodward and Hoffmann found a trick that simplifies the analysis. The trick comes from group theory. Recall from Chapter 2 that we can classify surface structures based on the symmetry elements in the surface. In the same way, we can classify reaction paths in terms of their symmetry elements. For example, if we assume that hydrogen and deuterium are equivalent in the reaction in Figure 10.39 then the reaction coordinate has three mirror planes: the two lines indicated in the figure and the plane of the paper. There also are three 2-fold axes: two at the plane of the paper at the mirror planes and one perpendicular to the plane of the paper at the dot. The dot is also an inversion center. In a similar way, one can classify the symmetry of all of the MO of the system. Woodward and Hoffmann (1970) show that the symmetry of the MO determines whether a reaction can occur. If orbital symmetry is conserved, the coupling constants will be nonzero as the reaction can occur. However, if the orbital symmetry is not conserved, no reaction can occur.

10.12.1   Some Results from GroupTheory

We need to review some group theory to see how this idea arises. Recall that in group theory, one classifies the symmetry of molecules and the reaction pathways in terms of their point group. In Chapter 2, we noted that there are only 13 three-dimensional Bravis lattices and three space groups. In the same way, there are only 47 important point groups. There are more point groups than space groups because one can have 5-, 7-, 8-,..., fold axis in molecules, but not in a repeating three-dimensional (3D) structure. Cotton (1971) provides an excellent overview of the application of group theory to molecules and reactions, and there is no room to discuss all of the important ideas here. However, the key feature of the analysis is that if one knows the symmetry elements in a molecule or reaction path, one can use the tables in Cotton (1971) to find the point group.

The most important tables in Cotton's book are the character tables near the end of the book. The character tables include a listing of all of the important point groups and their symmetry elements. For example, Table 10.2 is a character table for the D_{2h} point group. The D_{2h} point group has three perpendicular 2-fold axes, three mirror planes, an inversion center, and an identity element (i.e., an element that does not change anything). Note that the transition state in Figure 10.39 has all of these symmetry elements and no extras. Therefore, we say that the reaction in Figure 10.39 follows the D_{2h} point group.

The character table also lists something called the **irreducible representations** of the point group. The idea of a group representation is more subtle than there is room to discuss here. However, basically, if we solve the Schroedinger equation for a molecule (or transition state) that has a given symmetry, then we can classify the independent solutions as being a member of one of several irreducible representations. (The irreducible representations are just a way to classify the solution of the Schroedinger equation in terms of their symmetry properties.) Each independent MO of a system will belong to a single irreducible representation. Hence, one can classify the symmetry of MO in terms of their irreducible representations.

Table 10.2   A character table for the D$_{2h}$ point group

Irreducible Representation	Identity	z 2-Fold Axis	y 2-Fold Axis	x 2-Fold Axis	Inversion Center	xy Mirror Plane	xz Mirror Plane	yz Mirror Plane	Rotations and Orbitals
A_g	1	1	1	1	1	1	1	1	x^2, y^2, z^2
B_{1g}	1	1	-1	-1	1	1	-1	-1	r_z, xy
B_{2g}	1	-1	1	-1	1	-1	1	-1	r_y, xz
B_{3g}	1	-1	-1	1	1	-1	-1	1	r_x, yz
A_u	1	1	1	1	-1	-1	-1	-1	
B_{1u}	1	1	-1	-1	-1	-1	1	1	z
B_{2u}	1	-1	1	-1	-1	1	-1	1	y
B_{3u}	1	-1	-1	1	-1	1	1	-1	x

Source: Adapted from Cotton (1971).

The character table makes it easy to find the irreducible representation for each MO. Notice all the ones and minus ones in the character table (Table 10.2). The ones and minus ones tell what happens when each of the symmetry operations of the point group acts on a given MO. For example, the xy mirror plane converts y to $-y$. If we look down the xy mirror plane column in Table 10.2, we find that B_{2u} irreducible representation has a character of -1. Therefore, $\Psi_{B_{2u}}$, the wavefunction of an MO of B_{2u} symmetry in the D$_{2h}$ point group, will be converted to $-\Psi_{B_{2u}}$ when y is converted to y. One can therefore determine the irreducible representation of a given MO by examining how the symmetry elements in the system affect the given MO.

For example, consider $\Psi_{\sigma^*\sigma}$ in Figure 10.40. $\Psi_{\sigma^*\sigma}$ does not change when it is reflected around the xy or yz mirror plane (i.e., the plane of the paper and plane 2 in Figure 10.40, respectively). However, it switches signs when it is reflected around the xz mirror plane (i.e., plane 1 in Figure 10.40). Therefore, the irreducible representation of $\Psi_{\sigma^*\sigma}$ should have a 1 in the column for the xy or yz mirror planes and a -1 in the column for the xz mirror plane. Table 10.2 shows that only the B_{2u} irreducible representation has the correct symmetry. Therefore, we conclude that $\Psi_{\sigma^*\sigma}$ has B_{2u} symmetry. The reader may want to show that $\Psi_{\sigma\sigma}$ has A_g symmetry, $\Psi_{\sigma\sigma^*}$ has B_{3u} symmetry, and $\Psi_{\sigma^*\sigma}$ has B_{1g} symmetry.

Now it is useful to go back to our analysis of equation (10.96). Note that we used the symmetry of the various wavefunctions to show that $\beta_{(\sigma\sigma^*)(\sigma^*\sigma)}$ is zero. Note that we could have done the analysis directly from the character table. For example, if we start with INT from equation (10.97) and let y go to $-y$ (i.e., reflect around the xz mirror plane), $\psi_{\sigma^*\sigma}$ does not change while $\psi_{\sigma\sigma^*}$ is converted into $-\psi_{\sigma\sigma^*}$, which is completely symmetric (i.e., a member of the A_g irreducible representation), so $\psi_{\sigma\sigma^*}$ does not change either. Therefore

$$INT' = (-1)(1)(1)INT \qquad (10.104)$$

As a result, one could have arrived at the results in Section 10.11.1 directly from the character table without doing detailed analysis.

One can generalize this result to show that the coupling coefficients are always zero unless orbital symmetry is conserved. Consider transferring an electron from $\psi_{\sigma\sigma^*}$ to some other wavefunction ψ_x. The transfer coefficient for the transfer, β_x, is given by

$$\beta_x = \int \psi_{\sigma\sigma^*}^* \psi_x \, d\vec{r} \qquad (10.105)$$

If we start with a wavefunction of B_{2u} symmetry such as $\psi_{\sigma\sigma^*}$ and multiply by $\mathcal{H}$, we will still end up with a function of B_{2u} symmetry, since $\mathcal{H}$ is totally symmetric. Now, if we multiply by ψ_x we will get a new function that we will call Ξ. One can show from something called the **great orthogonality theorem** that if Ξ changes sign under any of the group operations when we integrate over all space, half the integral will be positive and half will be negative. As a result, β_x will be zero. Notice that we can tell whether Ξ changes sign directly under all of the group operations from the character table. If the irreducible representation of ψ_x has ones in any of the places where the irreducible representation of $\psi_{\sigma\sigma^*}$ has minus ones, or if the irreducible representation of ψ_x has minus ones in any of the places where the irreducible representation of $\psi_{\sigma\sigma^*}$ has ones, Ξ the integral in equation (10.97) will switch signs under at least one of the symmetry operations of the D_{2h} space group. As a result, β_x will be zero. The only instance where β_x will not be zero is when the irreducible representations of ψ_x and $\psi_{\sigma\sigma^*}$ have ones and minus ones in just the same places in the character table. That happens only when ψ_x and $\psi_{\sigma\sigma^*}$ have the same orbital symmetry (i.e., are members of the same irreducible representation).

The conclusion for the analysis in the last paragraph is that during the reaction, we can convert one molecular orbital into another only when both orbitals are of the same symmetry (i.e., members of the same irreducible representation). As a result, orbital symmetry is conserved during concerted chemical reactions.

10.13 EXAMPLES OF SYMMETRY-ALLOWED AND SYMMETRY-FORBIDDEN REACTIONS

It is useful to use these ideas to make some predictions about chemical reactions. Clearly, the first example is the four-centered reaction in Figure 10.39. Notice that we start with electrons in A_g and B_{2u} molecular orbitals and end up with electrons in the A_g and B_{3u} molecular orbitals. The orbital symmetry changes during the reaction. Hence, the reaction is forbidden.

Now consider the reaction

$$H + D_2 \longrightarrow HD + D \qquad (10.106)$$

where a hydrogen atom reacts with D_2. For the purpose of analysis, let's assume that the reaction occurs with the hydrogen atoms and the two deuterium atoms lying in a line.

Figures 10.20 and 10.21 show MO diagrams for the reaction. One starts with a symmetric orbital and an antisymmetric orbital at the start of the reaction, and ends up with orbitals of the same symmetry. Consequently, orbital symmetry is conserved during reaction (10.22).

Calculations indicate that the reaction in Figure 10.12 has an activation energy of 34 kJ/mol while the reaction in Figure 10.39 has an activation energy of 515 kJ/mol.

This example illustrates an important point: We need to look carefully at a reaction pathway before we decide whether a reaction is allowed or forbidden. Many reactions that look feasible on paper do not occur because of symmetry constraints. In particular, symmetric four-centered reactions are usually forbidden, so they are rarely seen. Physically, they have large barriers to reaction.

10.14 SUMMARY

In summary, then, the results in this chapter showed that activation barriers are associated with

- Bond stretching
- Orbital distortion
- Quantum effects

We have provided some qualitative models to explain the variations. We will quantify the effects in Chapter 11.

10.15 SOLVED EXAMPLES

Example 10.A   Using the Marcus Equation to Estimate Barriers for Reactions   One of the more intriguing applications of the Marcus equation is in predicting the barriers for reactions. The idea is to estimate E_a^0 for an identity reaction, and then use that information to predict barriers.

An identity reaction is a reaction where a group is exchanged on a carbon center:

$$X + R - X \longrightarrow X - R + X \tag{10.A.1}$$

where X could be a neutral species or a charged ligand. Notice that ΔH_r is zero for an identity reaction according to equation (10.23). Consequently, $E_{a,xx}^0$, the intrinsic barrier

$$E_{a,xx}^0 = E_{a,xx} \tag{10.A.2}$$

where $E_{a,xx}$ is the activation barrier for reaction (10.A.1).

Marcus supposed that when you have a nonidentity reaction

$$X + R - Y \longrightarrow X - R + Y \tag{10.A.3}$$

One can estimate $E_{a,xy}^0$, the intrinsic barrier for reaction (10.A.3), from

$$E_{a,xy}^0 = \tfrac{1}{2}(E_{a,xx}^0 + E_{a,yy}^0) \tag{10.A.4}$$

where $E_{a,yy}^0$ is the intrinsic barrier for reaction (10.A.1) when X is replaced by Y. To see how this works, let us try to estimate the activation barrier for the reaction

$$H' + CH_2 = CHCH_2OH \longrightarrow CH_2 = CHCH_2H' + OH \tag{10.A.5}$$

Equation (10.A.5) is equivalent to equation (10.A.3) with $X = H'$ and $Y = OH''$. Lee et al. (1995a) calculated the activation energy for the reactions

$$OH' + CH_2CHCH_2OH \longrightarrow CH_2CHCH_2OH' + OH \tag{10.A.6}$$

$$H' + CH_2 = CHCH_3 \longrightarrow CH_2 = CHCH_2H' + H \tag{10.A.7}$$

If we measure energies relative to the bound complexes, reaction (10.A.6) has an activation barrier of 27.9 kcal/mol, while reaction (10.A.7) has an activation barrier of 55.1 kcal/mol.

According to equation (10.A.4)

$$E_{a,xy}^0 = \tfrac{1}{2}(27.9 + 55.1) = 41.5 \text{ kcal/mol} \tag{10.A.8}$$

According to data in Lee et al, reaction (10.A.5) is 47 kcal/mol exothermic (i.e., $\Delta H_r = -47$). According to equation (10.33)

$$E_a = E_a^0 \left(1 + \frac{\Delta H}{4E_a^0}\right)^2 = 41.5 \left(1 + \frac{-47}{(4)(41.5)}\right)^2 = 21.3 \text{ kcal/mol} \qquad (10.A.9)$$

By comparison the actual value is 17.3 kcal/mol.

10.16 SUGGESTIONS FOR FURTHER READING

The connections between thermodynamics and kinetics are discussed in:

A. Pross, *Theoretical and Physical Principles of Organic Reactivity*, Wiley, New York, 1995.

S. S. Shaik, H. B. Schlegel, and S. Wolfe, *Theoretical Aspects of Physical Organic Chemistry*, Wiley, New York, 1992.

The connection between orbital pictures and molecular forces is discussed in:

R. F. W. Bader, *Atoms in Molecules: A Quantum Theory*, Oxford University Press, Oxford, UK, 1994.

10.17 PROBLEMS

10.1 Define the following terms:

(a) Polanyi relationship

(b) Marcus equation

(c) Intrinsic activation barrier

(d) Transfer coefficient

(e) Marcus inverted region

(f) Pauli repulsion

(g) Configuration mixing diagram

(h) Configuration mixing model

(i) Correlation diagram

(j) Symmetry-forbidden reaction

(k) Avoided crossing

(l) Tight transition state

(m) Loose transition state

10.2 Why do activation barriers arise during chemical reactions? What are the key factors in determining whether a reaction is activated? Why are activation energies reduced for fairly exothermic reactions? What happens with very exothermic reactions? Why?

10.3 How do Pauli repulsions affect barriers to reactions? Would there be a barrier in the absence of Pauli repulsions? Explain.

10.4 Describe in your own words how orbitals change as reactions proceed. How do atom transfer processes occur? What extra forces occur when a methyl group is transferred?

10.5 Describe the Polanyi equation. How does it arise? What are the key assumptions in the derivation? When is the equation useful? When does it fail?

10.6 Describe the Marcus equation. How does it arise? What are the key assumptions in the derivation of the Marcus equation? When is the equation useful? When does it fail? What are the advantages of the Marcus equation over the Polanyi relationship? When is the Marcus equation appropriate? When should you use the Blowers–Masel approximation instead?

10.7 Compare and contrast the Marcus equation, the Polanyi relationship, and the Blowers–Masel approximation. Be sure to consider the key features of the derivation and the predictions. Why does the Marcus equation predict parabolic behavior for very exothermic and endothermic reactions while the Blowers–Masel approximation predicts linear behavior? What are the key different assumptions of the two models?

10.8 The data in Figure 10.11 show nonlinear behavior with ΔH_r.

   (a) The plot actually shows $\ln(k)$ versus ΔH_r. Show that if data in the figure followed the Polanyi relationship and the preexponential were constant, a plot of $\ln(k)$ versus ΔH_r should be linear.

   (b) Use the Marcus equation to explain why the curve is nonlinear.

   (c) Does the Blowers–Masel approximation predict any different trends?

10.9 What are the key physical forces that are represented by the intrinsic barrier?

   (a) According to the Marcus equation, what variables affect the intrinsic barrier?

   (b) According to the Blowers–Masel equation, what variables affect the intrinsic barrier?

   (c) How would charge transfer prior to reaction affect the intrinsic barrier?

   (d) How would steric repulsions affect the intrinsic barrier?

10.10 Explain the key trends you see in the data in Figure 10.29. Why is there so much variability in the points? (*Hint*: look at the Blowers–Masel approximation and equation (10.11)—what varies as you change reactions?)

10.11 Compare the behaviors in Figures 10.17 and 10.18. Why do you see inverted behavior in Figure 10.17 but not in Figure 10.18?

10.12 In your organic chemistry class you learned about steric effects in reactions.

   (a) How do the steric interactions affect the Blowers–Masel approximation?

   (b) How does the tightness of the transition state change as the steric interactions change?

   (c) How will your results in (b) affect the predictions of the Polanyi relationship?

   (d) How will your results in (b) affect the predictions of the Marcus equation?

10.13 The objective of this problem is to use what you learned in organic chemistry class to estimate the intrinsic barriers for a series of reactions of the form

$$HO + CHR_3 \longrightarrow H_2O + CR_3$$

   (a) How would the C–H bond energy change as you change the R group from H to CH_3 to $C(CH_3)_3$.

   (b) How will the Pauli repulsions change? (Hint: How does the size of the electron clouds change?)

   (c) Will the barrier go up or down?

   (d) How would an electron-withdrawing group affect the intrinsic barrier?

Table P10.14 **Preexponentials and activation barriers for the reaction NH + H–R → NH$_2$ + R**

R	E_a, kcal/mol	k_0, cm^3(mol·second)	R	E_a kcal/mol	k_0 cm^3(mol·second)
CH$_3$	20	9×10^{13}	CH(CH$_3$)$_2$	14.5	6×10^{13}
CH$_2$CH$_3$	16.5	7×10^{13}	(CH$_3$)$_2$CO	11.5	5×10^{13}

Source: Data of Rhorig and Wagner, Berichte Bunsen Gesellschaft Physik. Chem. 98, 858 (1994)

10.14 Rhorig and Wagner, Berichte Bunsen Gesellschaft Physik. Chem. **98**, 858 (1994) examined a number of hydrogen abstraction reactions of the form

$$NH + H\text{–}R \longrightarrow NH_2 + R$$

Their data are given in Table P10.14.

(a) Use data in the CRC and the NIST Webbook to estimate the heat of reaction for each of the reactions in the table.

(b) Fit the data to a Polanyi plot. How well do they fit?

(c) Try fitting the data to the Marcus equation. Assume that the intrinsic barrier is constant. How well do the data fit?

(d) Does the assumption that the intrinsic barrier is constant make sense, or should the intrinsic barrier vary? (*Hint*: Look at what determines an intrinsic barrier.)

(e) Examine the variations in the preexponential. According to transition state theory, is the tightness of the transition state changing?

(f) Use the data in Table P10.14 to estimate the role of the tightening of the transition state in raising or lowering the intrinsic barrier. Should the intrinsic barrier go up or down?

(g) Try fitting the data to the Blowers–Masel equation. Assume that the V_P is constant. How well do the data fit?

(h) Does the assumption that the V_P is constant make sense, or should V_P vary and instead the intrinsic barrier be constant?

10.15 Maslack, Vallombroso, Chapman, and Narvaez, Angew. Chem. 33 (1994) examined the rate constant for a number of bond cleavage reactions. Their data are given in Table P10.15.

(a) Maslack et al. showed that $\ln(k)$ versus ΔH_r is nonlinear. Show that if data followed the Polanyi relationship and the preexponential were constant, a plot of $\ln(k)$ versus ΔH_r should be linear.

(b) Use the Marcus equation to explain why the curve is nonlinear.

(c) How well does the Marcus equation fit the data?

(d) Does the Blowers–Masel approximation predict any different trends?

(e) How well does the Blowers–Masel approximation fit the data?

(f) Try adjusting the temperature in the fit; do you get a better fit?

(g) What do you conclude from the fact that you need to adjust the temperature?

Table P10.15   Rate constants at 300 K and heats of reaction for the series of reactions considered by Maslack, Vallombroso, Chapman, and Narvaez, Angew. Chem. 33 (1994)

Reaction number	$\ln(k)$, $cm^3/(mol\cdot second)$	H_r, kcal/mol	Reaction Number	$\ln(k)$, $cm^3/(mol\cdot second)$	H_r, kcal/mol
(1c)	−7.4	24.8	(3a)	5.9	5.7
(1d)	−7.0	23.9	(3b)	7.0	2.8
(1e)	−4.7	18.7	(3c)	8.6	−1
(1f)	−3.2	18.2	(3d)	9.6	−12.3
(1g)	−2.5	16.3	(3e)	10.5	−17.6
(1h)	−2.7	15.9	(4a)	8.8	4.6
(2a)	−0.2	15.2	(4b)	9.1	4
(2b)	0.7	14.5	(4c)	9.6	3
(2c)	1.3	13.2	(4d)	6.4	3.9
(2d)	1.6	13.2	(4e)	7.4	4.8
(2e)	1.8	10.7	(5)	0.2	15.6
(2f)	5.2	7.8	—	—	—
(2g)	5.7	6.6	—	—	—
(2h)	5.6	6.2	—	—	—

10.16   Lee, Kim and Lee, J. Computational Chem. 16 (1995b) examined the activation energy for a number of allyl transfer reactions:

$$X^- + CH_2{=}CHCH_2Y \longrightarrow CH_2{=}CHCH_2X + Y^-$$

Table P10.16 shows a selection of their results

(a) Fit the data to a Polanyi plot. How well do they fit?

(b) Try fitting the data to the Marcus equation. Assume that the intrinsic barrier is constant. How well do the data fit?

(c) Does the assumption that the intrinsic barrier is constant make sense, or should the intrinsic barrier vary? (*Hint*: Use the additivity assumption to estimate the barriers.)

Table P10.16   Activation energy for a number of allyl transfer reactions

X	Y	ΔH_r, kcal/mol	E_a, kcal/mol	X	Y	ΔH_r, kcal/mol	E_a, kcal/mol
H	H	0	55.1	F	H	+67.2	74.4
NH_2	NH_2	0	43.3	OH	H	+47.1	64.3
OH	OH	0	27.9	NH_2	H	+26.1	59.2
F	F	0	15.0	Cl	H	+87.5	92.2
PH_2	PH_2	0	33.3	SH	H	+67.3	77.6
SH	SH	0	24.9	PH_2	H	+47.2	67.4
Cl	Cl	0	18.8	OH	F	−16.1	13.1
OH	Cl	−38.9	5.6	NH_2	F	−40.7	5.8
NH_2	Cl	−60.1	3.5	Cl	F	+16.9	32.7
SH	Cl	−19.8	13.45	SH	F	+0.4	22.0
PH_2	Cl	−41.1	7.28	PH_2	F	−20.4	15.0

Source: Results of Lee, Kim and Lee, *J. Computational Chem.* **16** (1995) 1045.

(d) Try fitting the data to the Blowers–Masel equation. Assume that the V_P is constant. How well do the data fit?

(e) Does the assumption that the V_P is constant make sense, or should V_P vary and instead E_A^0 be constant?

(f) Use the methods in Example 10.A to estimate the barriers to all of the reactions. How well does the method work?

10.17   Bernasconi and Ni (1994) describe two different methods to estimate the intrinsic barriers for a series of reactions of the form

$$X^- + HX \longrightarrow XH + X^- \qquad \text{(P10.17.1)}$$

They were not able to measure the intrinsic barriers directly, but they were able to get an indirect measure by examining relative rates of reactions. The results are given in Table P10.17. The question is how you can do an experiment to see which measure of the intrinsic barrier is accurate. By way of background, nonidentity proton exchange experiments

$$X^- + HY \longrightarrow XH + Y^- \qquad \text{(P10.17.2)}$$

can easily be done, so the object is to find a way to use an exchange reaction to distinguish between the two measures of intrinsic barriers.

(a) How many nonidentity reactions can you form from the X groups in Table P10.17?

(b) Use Marcus' additivity postulate to estimate the intrinsic barriers for each of these nonidentity reactions.

(c) Which nonidentity reaction will show the largest differences?

(d) Look in Bernasconi and Ni's paper and find data for that reaction.

(e) Which measure of the intrinsic barrier fits best?

10.18   Consider the following series of reactions:

$$HO + CH_4 \longrightarrow H_2O + CH_3$$

$$HO + CH_3CH_3 \longrightarrow H_2O + CH_2CH_3$$

Table P10.17   Intrinsic barriers for a number of reactions of the form $X^- + HX \to XH + X^-$

X	E_a^0, kcal/mol Estimated Using an Amine Reference, kcal/mol	E_a^0, kcal/mol Estimated Using a 9-Cyanofluorene Reference, kcal/mol
$[(CH_3)NH_2]^+$	6.80	9.37
Benzylmaloniumnitrite	11.52	13.87
9-Cyanofluorene	15.40	13.05
1,3-Indandione	16.72	14.37
4-Nitrophenylacetonitrite	17.62	15.17
(3-Nitrophenyl)nitromethane	21.74	19.39
(4-Nitrophenyl)nitromethane	21.96	19.61
Phenylnitromethane	22.64	20.17

$$HO + CH_2(CH_3)_2 \longrightarrow H_2O + CH(CH_3)_2$$
$$HO + CH(CH_3)_3 \longrightarrow H_2O + C(CH_3)_3$$

(a) Use Benson's approximation from Chapter 6 to estimate the heat of reaction for each of the reactions and the energies of the bonds that form and break.

(b) Estimate the strength of the Pauli repulsions, assuming that the hydrogen–methyl potential is given by

$$V_r = (1.5 \text{ kcal/mol}) \exp \left(\frac{4.5 \text{ Å} - R_{O-H}}{0.75 \text{ Å}} \right)$$

Be sure to consider that in the reaction $H + CH_2(CH_3)_2 \rightarrow H_2 + CH(CH_3)_2$ the incoming hydrogen is in close proximity to three methyl groups.

(c) Use the Blowers–Masel approximation to estimate the barriers for the reaction.

(d) Make a Polanyi plot of the data and estimate the intrinsic activation barrier and transfer coefficient.

(e) Why is the transfer coefficient negative?

(f) Does the derivation of the Polanyi relationship allow the transfer coefficient to be negative?

(g) Does the Marcus equation allow the transfer coefficient to be negative?

10.19 Why are there orbital symmetry barriers to reaction? Where do they arise, and where are they important? How does the orbital symmetry barrier affect the reaction $H_2 + D_2 \rightarrow 2HD$? What would the orbital diagrams look like for a symmetry-forbidden reaction?

10.20 (a) Show mathematically that the four-centered reaction $A_2 + B_2 \rightarrow 2AB$ is symmetry-forbidden.

(b) What is the physical significance of this result? (*Hint*: Show that in the four-centered configuration, it is impossible to simultaneously have an A–A, B–B, and two A–B bonds.)

10.21 Lopes et al. (1999) examined the role of acyloxymethyl as a way to protect antibiotics from degradation (hydrolysis) by bacteria.

(a) Read the paper and report on the findings

(b) The Brønsted plot is curved. How do Lopes et al explain the curvature?

(c) Could the curvature be due the a shift in the position of the transition state as suggested from the Marcus formulation?

10.22 Consider the reaction $D + H_2 \rightarrow HD + H$. Figures 10.20 and 10.21 show diagrams of the key orbitals during the reaction.

(a) The pictures only show states that would be expected to be occupied. What will the empty states look like? (*Hint*: In the empty states, bonding orbitals are converted to antibonding orbitals.)

(b) Draw a configuration mixing diagram during the reaction. (*Hint*: The configuration mixing diagram links up MOs in the ground state of reactants with MOs with the same pattern of plusses and minuses in the excited states of the products and vice versa.)

(c) Do the orbitals really follow the configuration mixing diagram? Specifically can you see the excited state being mixed into the ground state?

(d) Are there any extra interactions not considered in the configuration mixing diagram?

(e) Draw a correlation diagram for the system. (*Hint*: The correlation diagram links up MOs in the ground state of the reactants with MOs with the same pattern of plusses and minuses in the products.)

(f) Do the orbitals really follow the configuration mixing diagram? In other words, can you see a continuous transition from reactants to products?

(g) Are there any extra interactions not considered in the configuration mixing diagram?

More Advanced Problems

10.23 Roberts and Steele J. Chem. Soc., Perkin Trans. 2, (1994), fit the activation barriers for a number of atom transfer reactions to a complex expression.

(a) Read Roberts and Steele's paper and explain why all of the factors arise.

(b) Relate the findings to the Blowers–Masel model. How are the variations in the Pauli repulsions considered in the Roberts–Steele model?

(c) Are there any factors in the Roberts–Steele model that are ignored in Blowers–Masel's model?

10.24 Zavitsas J. Chem. Soc., Perkin Trans. 3, (1998), refit the activation barriers for a number of atom transfer reactions.

(a) Read Zavitsas paper and explain why all of the factors arise.

(b) How does Zavitsas model compare to that of Roberts and Steele J. Chem. Soc., Perkin Trans. 2, (1994)?

(c) Relate the findings to the Blowers–Masel model. How are the variations in the Pauli repulsions considered in Zavitsas model?

(d) Are there any factors in Zavitsas model that are ignored in the Blowers–Masel model?

11

MORE ABOUT ACTIVATION ENERGIES

PRÉCIS

An understanding of potential energy surfaces is critical to an understanding of reaction rates. In Chapter 10, we showed why activation barriers arise, and provided a qualitative picture of them. In this chapter we will quantify the results. First we will show how one can use the results from Chapter 10 to estimate barriers. We will then show how one can calculate the barriers from first principles. There are books on computational methods and activation barriers, and we cannot cover everything. Here we will review the ideas so that the reader knows where to start.

11.1 INTRODUCTION

In Chapter 10 we reviewed some of the models that are used to understand activation barriers for reactions. We showed why barriers arise, and briefly mentioned how one can start to predict activation barriers.

In this chapter we will expand the discussion and ask how one can predict activation barriers for reactions either from data or from first principles. First, we will show how the methods from Chapter 10 can be used to correlate data for a wide range of reactions. We will find that if one knows data for a given series of reactions, it is often possible to estimate the activation barriers for another reaction in the series. In general, the more information one has about a series, the more accurate predictions one can make. So if we know the activation barriers for 50 reactions in a series, it will often be possible to accurately estimate the activation barrier for a 51st reaction.

Unfortunately, one seldom knows the barriers for a series of reactions. In those cases, one can still use an approximation based on the Polanyi, Marcus and Blowers–Masel equations from Chapter 10. Generally, the Marcus equation or the Blowers–Masel extension can be used to predict modestly accurate barriers with very little information. The methods are good enough if you need barriers accurate to only ± 3–5 kcal/mol.

The only way to get more accurate activation barriers is to either measure them or calculate them using quantum-mechanical methods. In this chapter, we will briefly review how quantum-mechanical methods can be used to estimate barriers. It was hard to know how much quantum-mechanics to include in this chapter. Quantum-mechanical methods are the way the field is moving. Yet, there is not enough room here to cover even a small fraction of the literature. In this chapter quantum-mechanical methods are briefly reviewed. Then there are references at the end of the chapter if the reader wishes a more in-depth coverage.

11.2   EMPIRICAL CORRELATIONS FOR REACTION RATES: THE POLANYI RELATIONSHIP AND BRØNSTED CATALYSIS LAW

First let us talk about correlations based on the Polanyi relationship, correlations started in (1914). At the time, the Polanyi relationship had not yet been derived, but people did know *Arrhenius' law:*

$$k = k_0 \exp\left(-\frac{E_a}{k_B T}\right) \qquad (11.1)$$

In equation (11.1), k is the rate constant for a reaction, k_0 is the preexponential, E_a, is the activation energy, k_B is Boltzmann's constant, and T is the absolute temperature. People began to wonder if there were any correlations between activation energies and other properties that they could measure. The work was limited because in 1900 people did not know as much as people know today. Still, the findings provide a basis from which to correlate a wide range of rate data.

The earliest successful attempt to correlate rate data that I know about is the work of Taylor (1914). Taylor proposed that there was a correlation between the rate of a series of acid-catalyzed reactions and the strength of the acid. Taylor's original correlation did not prove to have wide applicability. However, Brønsted and Pederson (1924) expanded on Taylor's work and were able to arrive at some correlations that have wide applicability.

Brønsted and Pederson (1924) were examining the acid-catalyzed decomposition of nitramide:

$$NH_2NO_2 \Longrightarrow N_2O + H_2O \qquad (11.2)$$

They found that when they plotted the log of the rate constant for the decomposition process, k_{na}, against the log of the equilibrium constant for the dissociation of the acid catalyst, K_{ac}, the data fell onto a straight line (see Figure 11.1). Brønsted (1928) quantified these data and proposed that, in general, the rate constant of an acid-catalyzed reaction, k_{ac}, was related to the equilibrium constant for the dissociation of the acid, K_{ac}, by

$$\ln[k_{ac}] = \gamma_{Brøn} \ln(K_{ac}) + \ln(\beta_{Brøn}) \qquad (11.3)$$

where $\gamma_{Brøn}$ and $\beta_{Brøn}$ are constants. Similar correlations have since been proposed by many other investigators. They are now called *Brønsted relationships* in the literature, to acknowledge the pioneering contributions of Johannes N. Brønsted.

Over the years, a number of other reactions have been examined. See Brønsted (1928), Pederson (1934), and Bell (1941, 1973) for details. It has been found that there often is a general correlation between the log of the rate constant of an acid-catalyzed reaction and the acid strength of the catalyst. Typical data are shown in Figures 11.2 and 11.3. Most

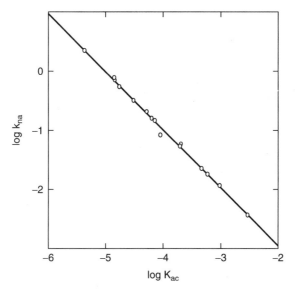

Figure 11.1 The rate constant for the acid-catalyzed dissociation of nitramide, k_{na}, as a function of the dissociation constant for the acid K_{ac}. [Data from Brønsted and Pederson (1924).]

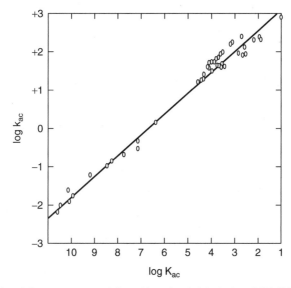

Figure 11.2 A plot of the rate constant of the acid-catalyzed dehydration of $CH_3CH(OH)_2$ in acetone. [Replot of the data of Bell and Higginson (1949).]

Brønsted plots are linear over a limited range of K_{ac} where we note that K_{ac} is related to the pH of the acid. Bell and Higginson (1949) have found one case, the dehydration of $CH_3CH(OH)_2$, which shows a linear Brønsted plot over a wide range of K_{ac}. This case has been widely reproduced in the literature and so we include it in Figure 11.2. However, most of the examples that have been examined in the literature show behavior like that

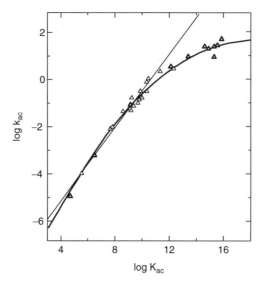

Figure 11.3   A Brønsted plot of the base-catalyzed enolization of $NO_2(C_6H_4)O(CH_2)_2COCH_3$. The solid line is a fit of the data by the Marcus equation. [Data from Hupke and Wu (1977).]

illustrated in Figure 11.3, where the Brønsted plot is nonlinear but can be approximated by a linear relationship provided that the data are taken over a limited range of K_{ac}. There are a few examples such as those in Figures 9.23 and 9.24 where the Brønsted plot does not work at all. However, the Brønsted relationship has proved to be quite useful. See Bell (1973) for details.

Houriti and Polanyi (1935) proposed an extension of the Brønsted relationship. They noted that the acid-catalyzed reactions studied by Brønsted all followed the same general scheme where one extracted a proton from the acid and the proton catalyzed the reaction. From thermodynamics, K_{ac} is given by

$$\ln[K_{ac}] = -\frac{\Delta G_{ac}}{k_B T} \tag{11.4}$$

where ΔG_{ac} is the free-energy change that occurs when the acid dissociates, k_B is Boltzmann's constant, and T is the absolute temperature. Substituting k_{ac} from equation (11.1) and K_{ac} from equation (11.3) into equation (11.4) and taking the log of both sides yields

$$E_a = \gamma_{Brøn}(\Delta G_{ac}) + kT \ln\left[\frac{k_0^{ac}}{\beta_{Brøn}}\right] \tag{11.5}$$

Therefore, the implication of the Brønsted relationship is that there is a linear relationship between the activation energy for the reaction and the free-energy it takes to remove a proton from the acid. Houriti and Polanyi (1935) generalized this idea to propose that for any homologous series of reactions, the activation energy for the reaction, E_a, is related to the free-energy change during the reaction $(-\Delta G_r)$ by

$$E_a = \gamma_P(\Delta G_r) + E_a^0 \tag{11.6}$$

where γ_P and E_a^0 are constants. Various references call γ_P the *reaction coefficient* or the *transfer coefficient*. E_a^0 is called the *intrinsic barrier to the reaction*. Equation (11.6) is called the **Polanyi relationship**. Later Evans and Polanyi were able to derive a related equation

$$E_a = \gamma_P(\Delta H_r) + E_a^0 \tag{11.7}$$

where ΔH_r is the heat of reaction. In the literature equations (11.6) and (11.7) are both called the Polanyi relationship. One does need to be careful, however, because the two equations are not identical.

In the literature, the Polanyi relationship is also sometimes written as

$$k_B T \ln\left[\frac{k_B T}{k_{ac} h_P}\right] = \Delta G_r^\ddagger = \gamma_P(\Delta G_r) + E_a^0 \tag{11.8}$$

where $\Delta G_r^\ddagger$ is the free-energy of activation and h_P is Planck's constant. Equation (11.8) is derived by starting with the transition state theory expression for the rate constant:

$$k_{ac} = \left(\frac{k_B T}{h_P}\right) \exp\left(-\frac{\Delta G_r^\ddagger}{k_B T}\right) \tag{11.9}$$

solving for $\Delta G_r^\ddagger$, substituting $\Delta G_r^\ddagger$ for E_a in equation (11.6), and then rearranging.

Equation (11.6) implies that if you make a change that changes the free-energy of the reaction, the free-energy of activation also undergoes a corresponding change. Physically, equation (11.6) arises because the transition state has properties of both the reactants and the products. As a result, when you make a change in either the reactants or the products, the change will be reflected in the properties of the transition state.

11.2.1 Application of Polanyi Relationships in Electrochemistry

Equation (11.7) has been widely applied in the electrochemical literature. In an electrochemical process, one inserts two electrodes into a reaction mixture, applies a voltage across the electrodes, and uses the voltage to drive the reactions.

One of the questions in electrochemistry is how the rate varies with the applied potential. Recall from freshman chemistry that the free-energy of formation of an ion on an electrode is the half-cell potential of the ion (i.e., the free-energy of formation of the ion) plus the applied potential. As a result, the free-energy of the ion changes by one volt when one changes the potential on the electrode by one volt. Consequently, according to equation (11.6), the activation energy for a reaction should vary linearly with the applied potential.

We can derive an equation for the change in rate as a function of potential. Consider a simple electron transfer reaction, where a reactant A receives n_e electrons to produce a species B:

$$A + n_e e^- \longrightarrow B^{n_e^-} \tag{11.10}$$

According to Faraday's law, the free-energy of the reaction in the cell, ΔG_r, varies with the applied potential, η_v, according to

$$\Delta G_r = \Delta G_r^0 + n_e \mathcal{F} \eta_v \tag{11.11}$$

where $\mathcal{F}$ is Faraday's constant and ΔG_r^0 is the free-energy change at zero applied potential.

Combining equations (11.8) and (11.11) yields

$$\ln\left(\frac{k_r}{\left(\frac{k_B T}{h}\right)}\right) = \left(\frac{E_a^0 + \gamma_P \Delta G_r^0}{k_B T}\right) + \left(\frac{n_e \mathcal{F} \gamma_P}{k_B T}\right) \eta_v \tag{11.12}$$

According to equation (11.12), a plot of the log of k_r versus the applied potential should be linear with a slope, Sl, given by

$$Sl = \left(\frac{n_e \mathcal{F} \gamma_P}{k_B T}\right) \tag{11.13}$$

Note $\dfrac{k_B T}{\mathcal{F}} = 26$ meV at 300 K. Hence, according to equation (11.12), a 120-meV change in potential will produce about an order of magnitude change in rate when $n_e = 1$, and $\gamma_P = 0.5$.

In an electrochemical experiment, one seldom measures the rate of a reaction directly. Rather, one builds an electrochemical cell with an anode and cathode and measures the current produced by the cell as a function of the applied voltage. We will show below that if we measure potentials relative to the equilibrium point of the cell and work at potentials that satisfy

$$|\eta_v| > 5 \frac{k_B T}{\mathcal{F} \gamma_P} \tag{11.14}$$

the current will be proportional to the rate of the reaction. Hence, if the electrochemical reaction follows the Polanyi relationship, a semilog plot of the current versus voltage should be a straight line for potentials that satisfy equation (11.14).

Experimentally, Tafel (1905) showed that the current was linear for a number of oxidation reactions. For example, Figure 11.4 shows how a rate of hydrolysis of water varies with potential under a variety of different conditions. Notice that the data are linear over an extended region of potentials. There are deviations at high and low potentials and the exact magnitude of the current varies with conditions. However, the Polanyi relationship works over 11 orders of magnitude in rate. Tafel (1905), Houriti and Polanyi (1935), Conway (1952), and Vetter (1967) review results for hundreds of electrochemical reactions, and all of them show the same general trends seen in Figure 11.4. The current is nonlinear at low potentials but shows near-linear behavior over an extended range of potentials. Further, the transfer coefficient, γ_H, is between 0.4 and 0.6 for all except one of the examples cited in these reviews. As a result, it appears that in electrochemical systems, the Polanyi relationship works over an extended range of potentials and the transfer coefficient is nearly constant for all the examples. Electrochemical systems that follow the Polanyi relationship are said to obey **Tafel kinetics**. Most electrochemical systems obey Tafel kinetics. The main exceptions so far are systems that show a change in mechanism with changing potential.

There is a deviation from linearity at very low potentials in Figure 11.4. Physically, the deviations at low potential occur because at low potential the reactions are reversible. The forward reaction follows a linear free-energy relationship. The backward reaction follows one also. The net current is the difference between the two. One can go through a derivation by keeping track of the two terms. A suitable derivation appears in Bard and Faulkner (1980), Conway (1985), or Masel (1996). If one scales the potential so that when

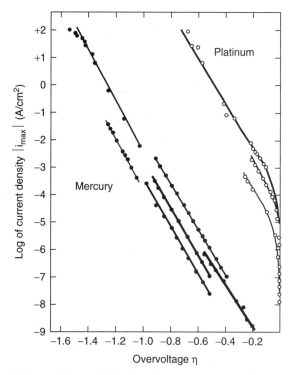

Figure 11.4 The current produced during the electrolysis of water as a function of the applied potential over mercury and platinum electrodes. The various lines are data taken at different conditions. [Data from Vetter (1967).]

$\eta_V = 0$, the net rate of reaction is zero, i_{net}, the net current (i.e., the difference between the current for the forward and reverse reaction) becomes

$$i_{net} = i_0 \exp\left[-\frac{n_e \mathcal{F} \gamma_a}{k_B T} \eta_V\right] - i_0 \exp\left[-\frac{n_e \mathcal{F} \gamma_c}{k_B T} \eta_V\right] \qquad (11.15)$$

where γ_a and γ_c are the transfer coefficients for the forward and reverse reactions, respectively.

Equation (11.15) is called the *Butler–Volmer* equation, since it was first derived by Butler (1932) on the basis of work by Volmer and Erdey-Gruz (1930). Figure 11.5 is a plot of the current predicted by the Butler–Volmer equation as a function of the applied potential. The figure shows that if one made a semilog plot of the current versus potential, one would expect to observe nonlinearties at low potential due to the reverse reaction on the cathode. The shapes are similar to the experimental curves in Figure 11.4. Hence, one can use equation (11.15) to qualitatively explain the deviations from linearity seen in Figure 11.4. Vetter (1967) shows that, in fact, the fit is quantitative. As a result, he concludes that the data in Figure 11.4 follow the Polanyi relationship even at low potentials even though the plot of current versus potential is nonlinear.

The situation is different at high potentials, however. Notice that the mercury data in Figure 11.4 begin to turn over at high potentials. This saturation effect is often attributed to a mass transfer limitation. However, calculations described below indicate that the rate

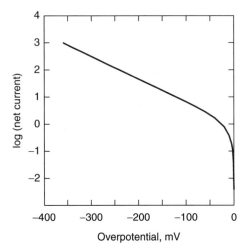

Figure 11.5 A plot of the current predicted by the Butler–Volmer equation as a function of the applied potential

will saturate even in the absence of a mass transfer limitation. It is now thought that there are real deviations from linear free-energy behavior in the high regime.

Physically, these deviations occur because the Polanyi diagrams show nonlinearities when data are taken over a wide region of potential as described by the Marcus equation or the Blowers–Masel equation.

In reality, the Marcus equation is an oversimplification of what is happening in an electrochemical reaction. There are some other effects as described by Dogandze (1971). However, these effects are small. Usually, the data show linear behavior over an extended range of potential. The Polanyi relationship usually works in electrochemical systems except under extreme conditions.

11.3 LINEAR FREE-ENERGY RELATIONSHIPS

The Polanyi relationship, itself, rarely appears in the organic chemistry literature. However, a variation of the Polanyi relationship called a **linear free-energy relationship** or (**LFER**) is often used to correlate data in the physical–organic chemistry literature.

The use of linear free-energy relationships arose from the work of Hammett. In 1924, Hammett speculated that the Brønsted catalysis law might be extended to the reactions of organic molecules. At the time, no one had suitable thermodynamic data for decomposition of organic molecules. However, Hammett realized that one can get a measure of a molecule's free-energy by looking at the molecule's acidity. Therefore, he supposed that one might be able to create Brønsted plots for organic reactions using acidity as a free-energy scale.

It took several years to develop this idea. However, in 1933, Hammett and Pfluger examined the reaction of trimethylamine with a series of substituted methylbenzoates:

$$X-\bigcirc\!\!\!\!-COOCH_3 \;\xrightarrow{\;H^+\;}\; X-\bigcirc\!\!\!\!-COO^- \qquad (11.16)$$
$$+\,N(CH_3)_3 \qquad\qquad\qquad +\,[N(CH_3)_4]^+$$

They found that there was a general linear correlation between the log of k_X, the rate constant for the hydrolysis of X–methylbenzoate, and the log of K_X, the dissociate constant of the corresponding benzoic acid. Hammett expanded this work to include a wide variety of substituted benzoic acids. He found that there was usually a simple linear relationship between the rate constant for reactions of meta- and para-substituted benzoic acids and the dissociation constant of the acid. For example, Figure 11.6 shows a plot of the log of the rate constant for the reaction

$$X-\underset{}{\bigcirc}\text{COOCH}_2\text{CH}_3 \xrightarrow{\ \text{OH}^-\ } X-\underset{}{\bigcirc}\text{COOH} \qquad (11.17)$$
$$+ \text{H}_2\text{O} \qquad\qquad\qquad + \text{CH}_3\text{CH}_2\text{OH}$$

and the log of the equilibrium constant for the dissociation of the corresponding substituted benzoic acid:

$$X-\underset{}{\bigcirc}\text{COOH} + \text{H}_2\text{O} \longrightarrow X-\underset{}{\bigcirc}\text{COO}^- + \text{H}_3\text{O}^+ \qquad (11.18)$$

The data in Figure 11.6 have been normalized to the values for X = hydrogen, that is, the rate constant for the hydrolysis of ethylbenzoate, k_H; and the dissociation constant for benzoic acid, K_H. Notice that all the points for the meta and para groups lie close to the same line. There are significant deviations from the line for ortho substitutions. We will show below that one can understand the linear relationship in Figure 11.6 by looking at how the substituent affects the free-energy of the reaction. We will also find

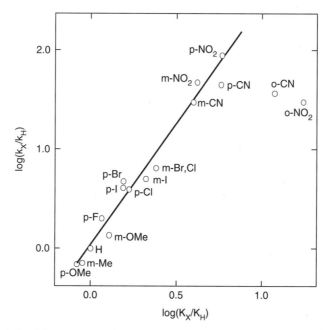

Figure 11.6 A plot of the rate constant for the hydrolysis of a series of substituted ethyl benzoates versus the dissociation constant of the corresponding benzoic acid. [Data of Ingold and Nathan (1936) and Evans et al. (1937).]

that the deviations in the ortho position imply that there is an additional effect of an ortho-substituted group. In the example in Figure 11.6, there is a steric hindrance; a group bound to the for ortho position gets in the way of the reaction.

Most linear free-energy plots look like Figure 11.6. One observes a near-linear relationship between the rate of the reaction and the free-energy of dissociation of the corresponding benzoic acid for meta- and para-substituted species, but not ortho-substituted species.

Hammett fit this data to a line of the form

$$\ln\left(\frac{k_X}{k_H}\right) = \gamma_H \ln\left(\frac{K_X}{K_H}\right) = \gamma_H \sigma_1^0 \tag{11.19}$$

where γ_H is a constant, and

$$\sigma_1^0 = \ln\left(\frac{K_X}{K_H}\right) \tag{11.20}$$

σ_1^0 is now called the Hammett constant.

For the work that follows, it is interesting to relate equation (11.19) to equation (11.6). Let's first consider the left side of the equation. Substituting expressions for k_X and k_H from equation (11.1) into $\ln(k_X/k_H)$ yields

$$k_B T \ln\left(\frac{k_X}{k_H}\right) = E_a^H - E_a^X + k_B T \ln\left(\frac{k_X^0}{k_H^0}\right) \tag{11.21}$$

where E_a^H and E_a^X are the activation energies for the hydrolysis of ethylbenzoate and X-substituted ethylbenzoate and k_H^0 and k_X^0 are the preexponentials for the same reactions. Similarly, substituting expressions for K_X and K_H from equation (11.4) into $\ln(K_X/K_H)$ shows

$$k_B T \ln\left(\frac{K_X}{K_H}\right) = \Delta G_{ac}^H - \Delta G_{ac}^X \tag{11.22}$$

where ΔG_{ac}^H and ΔG_{ac}^X are the free energies of dissociation of benzoic acid and X-substituted benzoic acid, respectively.

Substituting equations (11.21) and (11.22) into equation (11.19) shows

$$E_a^X = \gamma_H(\Delta G_{ac}^X - \Delta G_{ac}^H) + E_a^H + k_B T \ln\left(\frac{k_X^0}{k_H^0}\right) \tag{11.23}$$

Therefore, equation (11.19) is almost a linear free-energy relationship with

$$E_a^0 = E_a^H = \gamma_H \Delta G_{ac}^H + k_B T \ln\left(\frac{k_X^0}{k_H^0}\right) \tag{11.24}$$

Of course, in principle, the last term in equation (11.24) can vary with the substituent, X. Therefore, equation (11.19) is not quite a linear free-energy relationship. It is easy to show, however, that if one substitutes

$$\sigma_1^0 = \ln\left(\frac{K_X}{K_H}\right) + \frac{1}{\gamma_H} \ln\left(\frac{k_X^0}{k_H^0}\right) \tag{11.25}$$

for the last term in equation (11.24), the extra term cancels, and so one does get an exact linear free-energy relationship. The implication of equation (11.25) is that one should adjust σ_1^0 slightly to account for variations in preexponential factors.

People in the physical–organic chemistry community have adjusted the values of σ_1^0 to account for these effects. Many of the adjustments were done empirically. Reviews of this work include Chapman and Shorter (1972, 1978) and Wells (1968). The corrected values are labeled σ^0 in Table 11.2. Figure 11.7 shows a replot of the data from Figure 11.6 versus σ^0 to illustrate the kind of fit that is obtained. Generally, one can fit a wide range of data provided that one adjusts the constants appropriately.

Plots of $\ln(k)$ versus σ^0 are called **Hammett plots** in the literature. Generally, Hammett plots have proved to be very useful in physical–organic chemistry.

In the literature people sometimes call Hammett plots *linear free-energy relationships*. Of course, there is a very important difference between a plot of the rate as suggested by equation (11.23) and a true linear free-energy plot, as given by equation (11.6). ΔG_r in equation (11.6) is the overall free-energy change for the reaction, while ΔG_{ac}^X is the free-energy of dissociation of benzoic acid. Therefore, equation (11.23) is a true linear free-energy relationship only when ΔG_r is linearly dependent on ΔG_{ac}.

There is some important physics in this assumption. The hydrolysis of ethylbenzoate occurs via electrophilic attack of a hydroxyl group onto the α carbon:

$$X-\!\!\underset{}{\bigcirc}\!\!-\overset{\overset{O}{\|}}{C}-OCH_2CH_3 \qquad (11.26)$$
$$\nwarrow OH^-$$

The transition state is a charged complex:

$$X-\!\!\underset{}{\bigcirc}\!\!-\overset{\overset{O}{\|}^{\ominus}}{\underset{\underset{H}{O}}{C}}-OCH_2CH_3 \qquad (11.27)$$

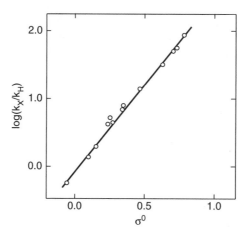

Figure 11.7 A replot of the meta and para data from Figure 11.6 using the modified values of σ^0 from the tabulation of Exner (1978)

Notice that if the substituent X withdraws electrons, it will stabilize the transition state and thereby enhance the rate. If the species donates electrons, it will destabilize the transition state and thereby decrease the rate.

Now consider the effect of an electron-donating or electron-withdrawing group on the dissociation of benzoic acid:

$$X-\bigcirc-COOH + H_2O \longrightarrow X-\bigcirc-COO^- + H_3O^+ \qquad (11.28)$$

Notice that the dissociation process produces a charged species. Hence, an electron-withdrawing substituent will enhance the dissociation of benzoic acid, while an electron-donating substituent will reduce the dissociation of benzoic acid. As a result, the influence of a given substituent group in stabilizing the transition state shown in (11.27) is similar to the influence of the substituent group in enhancing the dissociation of benzoic acid. As a result, one would expect there to be a correlation between the effect of a given substituent on the acidity of benzoic acid, and the effect of the substituent on rate of the dissociation of ethylbenzoate, as seen in Figure 11.6.

One can imagine many generalizations of this idea. For example, if a given reaction has a negatively charged transition state similar to the one in (11.27), then a given substituent may have a similar effect on the rate of the given reaction as it does on the rate of the dissociation of benzoic acid. In such a case, one would expect there to be a simple relationship between the rate constant for the reaction and the dissociation constant of benzoic acid.

As an example, consider the reaction between hydrogen peroxide and a series of substituted benzenesulfinic acids:

$$X-\bigcirc-SO_2^- + H_2O_2 \longrightarrow X-\bigcirc-SO_3^- + H_2O \qquad (11.29)$$

This reaction also has a negatively charged transition state, so one might expect it to follow a correlation similar to those described earlier in this section. Figure 11.8 shows

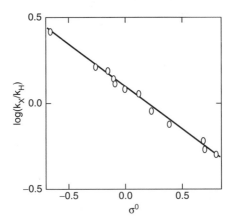

Figure 11.8 A plot of the log of the rate constant for the reaction of a series of substituted benzenesulfinic acids with hydrogen peroxide against σ^0 values determined for substituted benzoic acids. [Data of Lindberg (1966).]

a plot of Lindberg's (1966) measurements of the rate constant for the reaction against σ^o values determined for substituted benzoic acids. Notice that one still gets a linear fit even though there is no benzoic acid in the reaction. The slope of the plot is negative because the benzene ring is less charged in the transition state than at the beginning of the reaction; the more an electron-withdrawing group destabilizes the transition state, the more it destabilizes the reactants.

Experiments have shown that there are many reactions that follow the Hammett correlation with σ^o's for substituted benzoic acids even though there is no benzoic acid in the system. For example, Table 11.1 lists a series of acids whose dissociation constants correlate with σ^o for benzoic acid. The constant γ_{K_a} is the slope of a log–log plot of the dissociation constant of the acid versus the dissociation constant of benzoic acid. Notice that many of the acids are different from benzoic acid. However, a given substituent changes the dissociation constant of the acid in the same way as the substituent changes the dissociation constant of benzoic acid. The implication of Table 11.1 is that if we want to construct a linear free-energy plot for a reaction of any of the acids listed in the table, we could simply correlate the data to σ^o values calculated for benzoic acid.

One should not push this idea too far, however. In a standard Hammett plot, one is assuming that a substituent affects the transition state in the same way that it affects benzoic acid. If molecules are different enough from benzoic acid, the substituents can have quite different effects on the molecules than on benzoic acid. Hence corrections would be needed.

For example, Figure 11.9 shows a Hammett plot for the reaction:

$$(XAr)_3C\text{–}Cl \rightleftharpoons (XAr)_3C^+ + Cl^- \tag{11.30}$$

Notice that the measured rate is often one or two orders of magnitude different from the rate predicted by the linear free-energy curve. Hence, one could not use the line in Figure 11.9 to make useful predictions.

The deviations from the line in Figure 11.9 occur because the transition state for reaction (11.30) is a positively charged species. The positively charged species has little in common with the negatively charged transition state seen during benzoic acid dissociation.

Table 11.1   A series of acids whose dissociation constants correlate with the dissociation constants of substituted benzoic acids at 25°C

Acid	Solvent	γ_{K_a}
	H_2O	1.00
$XC_6H_4COOH \rightleftharpoons XC_6H_4COO^- + H^+$	50% aqueous C_2H_5OH	1.60
	C_2H_5OH	1.96
$XC_6H_4CH_2COOH \rightleftharpoons XC_6H_4CH_2COO^- + H^+$	H_2O	0.49
$XC_6H_4CH_2CH_2COOH \rightleftharpoons XC_6H_4CH_2CH_2COO^- + H^+$	H_2O	0.21
$XC_6H_4CH=CHCOOH \rightleftharpoons XC_6H_4CH=CHOO^- + H^+$	H_2O	0.47
$XC_6H_4(NH_3)^+ \rightleftharpoons XC_6H_4NH_2 + H^+$	H_2O	2.77
	30% aqueous C_2H_5OH	3.44
$XC_6H_4OH \rightleftharpoons XC_6H_4O^- + H^+$	H_2O	2.11
$XC_6H_4PO(OH)_2 \rightleftharpoons XC_6H_4PO.OH.O^-$	H_2O	0.76
	50% aqueous C_2H_5OH	0.99

Source: Johnson (1973).

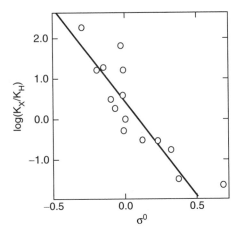

Figure 11.9 A Hammett plot of the equilibrium constant for the ionization of $(XAr)_3C-Cl$ in liquid SO_2. [The data are replotted from data in Isaacs (1987).]

Table 11.2 A selection of the Hammett constants recommended by Exner (1978), Johnson (1973), and Hansch et al. (1991)

Substituent	σ_m^0	σ_p^0	σ_p^+	σ_p^-	σ_m^+	σ_I	σ_R
H	0.0	0.0	0.0	0.0	0.0	0.0	0.0
$-CH_3$	−0.07	−0.17	−0.31	−0.17	−0.07	−0.04	−0.11
$-C_2H_5$	−0.07	−0.15	−0.30	−0.19	—	−0.02	−0.10
$-(CH_2)_3CH_3$	−0.08	−0.16	−0.29	−0.12	—	+0.04	−0.12
$-C(CH_3)_3$	−0.10	−0.20	−0.26	−0.13	−0.06	−0.03	−0.12
$-OCH_3$	+0.12	−0.27	−0.78	−0.26	+0.05	+0.29	−0.45
$-OCH_2CH_3$	+0.34	−0.14	−0.81	−0.28	—	+0.26	−0.44
$-F$	+0.34	+0.15	−0.07	−0.03	+0.35	+0.50	−0.34
$-Cl$	+0.37	+0.23	+0.11	+0.19	+0.40	+0.47	−0.22
$-Br$	+0.39	+0.23	+0.15	+0.25	+0.41	+0.47	−0.20
$-I$	+0.34	+0.06	+0.14	+0.27	+0.36	+0.41	−0.16
$-NO_2$	+0.71	+0.78	+0.79	+1.27	+0.67	+0.72	−0.15
$-CN$	+0.56	+0.66	+0.66	+1.00	+0.56	+0.57	−0.13
$-NC$	+0.48	+0.49	+0.60	+0.90	—	—	—
$-N(CH_3)_2$	−0.16	−0.83	−1.70	−0.12	—	+0.07	−0.33
$N(CH_3)_3^+Cl^-$	+0.88	+0.82	+0.41	+0.77	+0.36	+0.73	+0.15
$-NH_2$	−0.16	−0.66	−1.30	−0.15	−0.16	+0.12	−0.48
$-COOH$	+0.37	+0.45	+0.43	+0.77	—	+0.28	+0.16
$-COOR$	+0.35	+0.44	+0.48	+0.64	+0.37	+0.32	+0.14
$-CF_3$	+0.43	+0.54	+0.61	+0.65	—	+0.42	+0.08
$-OH$	+0.12	−0.37	−0.92	−0.37	—	+0.25	−0.40

In order to account from these differences, workers in the physical–organic chemistry community have determined different values of σ for different classes of reactions as indicated in Table 11.2. σ_m^0 and σ_p^0 are the adjusted Hammett constants for meta- and para-substituted benzoic acid. σ_m^- and σ_p^- are used to correlate data for nucleophilic reactions (i.e., reactions with negative ions). σ_m^+ and σ_p^+ are values of the Hammett constants, which have been adjusted for aryl chlorides. They are usually used to correlate data for reactions with electrophiles, or reactions with a positively charged transition state. σ_I and σ_R are constants used to correlate data from alkanes. People have found that reactions of para-substituted amines with neucleophiles do not follow any of these plots because of the presence of the lone pair on the amine. Therefore, they have defined yet another Hammett constant, σ_p^- for this situation. Note that we list values of σ_p^- only where the values deviate significantly from σ_p^0.

Figure 11.10 shows how an example of well-corrected values of σ fit actual data. A comparison of Figure 11.9 and Figure 11.10 shows that, indeed, one does fit data better if one uses modified values of σ. Hence, there is some advantage in using a σ-modified scale. The disadvantage of this approach is that if we need to define a new σ scale for each set of reactions, we have in effect limited the generality of the methods. Therefore, people generally limit themselves to about four σ scales. There is one big advantage of having several σ scales, however. One can use the different σ scales to learn something about transition states for reactions. Reactions that follow the σ^0 scale have transition states similar to those in the dissociation of benzoic acid. The carbon center can have a small residual positive or negative charge. However, the charges are usually small. In contrast, reactions that correlate with the σ^- scale have a partial lone pair (i.e., a large negative charge) in the transition state. Reactions that correlate to the σ^+ have an empty orbital (e.g., a large positive charge) in the transition state. As a result, if we know that a reaction follows a specific σ scale, we can learn some important information about the transition state for the reaction. This is the key result in this section.

There are two other key σ scales in Table 11.2: the σ_R scale and the σ_I scale. The σ_I scale is called the *inductive scale*. It measures the ability of substituents to withdraw electrons to a reaction center. A species that has a strong interaction with the negative charge on the carbon center in the transition state in reaction (11.27) would have a large value of σ_I. The σ_R scale is called the *resonant scale*. It measures the ability of a substituent to distribute the charge in a π bond or ring. A species that has a large value

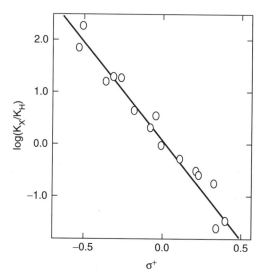

Figure 11.10 A replot of the data from Figure 11.9 on the σ^+ scale

of σ_R would increase the rate of reaction (11.26) when the species adsorbs in the para position. Much smaller interactions will be seen in the meta position. See Isaacs (1987) for a good description of the resonant interaction.

11.3.1   Applications of Linear Free-Energy Relationships in Chemistry

Hammet plots are widely used in the physical–organic chemistry literature. The plots are also called linear free-energy relationships (LFERs). It is quite common to create a plot of the log of the rate constant for a reaction versus the Hammett constant for the reaction. If you have a large series of reactions, you can fit the data to a Hammett plot. If the plot is linear, people say that the reaction has a single rate-determining step. Sometimes they also draw conclusions about the nature of the transition state. Reactions that follow the σ^0 scale have transition states similar to those in the dissociation of benzoic acid. The carbon center can have a small residual positive or negative charge. Generally, however the charges are small. In contrast, reactions that correlate with the σ^- scale have a partial lone pair (i.e., a large negative charge) in the transition state. Reactions that correlate to the σ^+ have an empty orbital (i.e., a large positive charge) in the transition state. Hence, if we know that a reaction follows a specific σ scale, we can learn some important information about the transition state for the reaction.

In industry people also use Hammett plots to interpolate data. If you measure the rate constants for a series of reactions and find that they follow a linear Hammett plot, it is often safe to assume that other similar reactions also follow a similar Hammett plot. As a result, one can estimate the rate constants for a series of closely related reactions without having data for each reaction. We include an example calculation in solved Example 11.A. The reader should examine that example before proceeding.

The weakness in using Hammett plots is that one needs considerable data before one can do anything useful. Generally, there is no way to predict the slope or the intercept of the Hammet plot. One often does not know which σ scale to use without taking extensive data. As a result, while LFERs are convenient ways to correlate data, they have relatively little predictive value.

11.4   THE POLANYI RELATIONSHIP

Fortunately, one can start to make predictions using the Polanyi relationship. Recall from Chapter 10 that Polanyi and Evans showed that E_a, the activation energy for an elementary reaction, is related to ΔH_r, the heat of reaction, by

$$E_a = E_a^0 + \gamma_P(\Delta H_r) \qquad (11.31)$$

where E_a^0 is the intrinsic activation barrier and γ_P is a constant. People sometimes rewrite the Polanyi relationship as

$$E_a = E_a^0 + \gamma_P(\Delta G_r) \qquad (11.32)$$

Notice that if one can guess E_a^0 and γ_P one can predict activation barriers for elementary reactions. One can use that knowledge to predict the mechanism of the reaction as described in Chapter 5. One can then use the steady-state approximation to predict the rate of an overall reaction.

In the remainder of this chapter we will discuss the ideas that have surfaced to estimate the barriers for elementary gas-phase reactions. We do only gas-phase reactions, because the methods have not yet been extensively tested in other situations.

To start off, it is useful to note that some years ago Seminov considered a small series of atom transfer reactions and found that they satisfied.

$$E_a = 12 \text{ kcal/mol} + 0.3(\Delta H_r) \qquad (11.33)$$

if the reaction was exothermic and

$$E_a = 12 \text{ kcal/mol} + 0.7(\Delta H_r) \qquad (11.34)$$

if the reaction was endothermic. Equations (11.33) and (11.34) are called the **Seminov relationships**. (Seminov, 1935).

Figures 11.11 and 11.12 show a comparison of the activation energies of a number of reactions of the form

$$R-H + X \longrightarrow R + H-X \qquad (11.35)$$

to the predictions of the Seminov (1935) relationships. The data are from Westley (1980). Notice that there is a reasonable fit provided the heat of reaction is between +40 and −40 kcal/mol. All the activation barriers are within ±9 kcal/mol. The root-mean-square (RMS) error is 4 kcal/mol. By comparison, the experimental error is ±3 kcal/mol.

The weakness of the Seminov relationship and the related Polanyi does not do well when the heat of reaction is more than 40 kcal/mol or less than −40 kcal/mol.

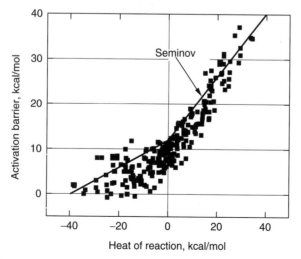

Figure 11.11   A comparison of the activation energies of a number of hydrogen transfer reactions to those predicted by the Seminov relationships [equations (11.33) and (11.34)].

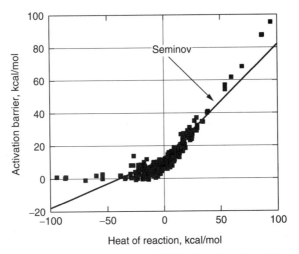

Figure 11.12 A comparison of the activation energies of 482 hydrogen transfer reactions to those predicted by the Seminov relationships over a wider range of energies.

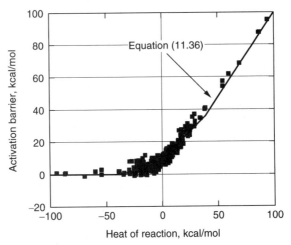

Figure 11.13 A comparison of the activation energies of 482 hydrogen transfer reactions to those predicted by equation (11.36).

One can get a better fit with the equation

$$
E_a = \begin{cases}
\Delta H_r & \text{when } \Delta H_r > 36 \\
9 \text{ kcal/mol} + 0.75\Delta H_r & \text{when } 0 < \Delta H_r < 36 \\
9 \text{ kcal/mol} + 0.25\Delta H_r & \text{when } 0 > \Delta H_r > -36 \\
0 & \text{when } \Delta H_r < -36
\end{cases} \quad (11.36)
$$

Data are shown in Figure 11.13. In this case all the activation barriers are within 7 kcal/mol. The RMS error goes down to 2.6 kcal/mol.

The Blowers–Masel approximation discussed in Chapter 10 also does quite well in fitting the data. Recall that according to the Blowers–Masel approximation, the activation barrier, E_a, is given by

$$E_a = (w_0 + 0.5\Delta H_r)\frac{(V_P - 2w_0 + \Delta H_r)^2}{V_P^2 - 4w_0^2 + (\Delta H_r)^2} \qquad (11.37)$$

where w_0 is the average of the bond energies of the bond that forms and the bond that breaks, ΔH_r is the heat of reaction, and V_P is related to the intrinsic barrier, E_a^0, by

$$V_P = 2w_0 \left(\frac{w_0 + E_a^0}{w_0 - E_a^0}\right) \qquad (11.38)$$

Figure 11.14 shows the fit that is obtained if one assumes that the average bond energy is 105 kcal/mol (the average of all the real bond energies). In this case the RMS error is 2.7 kcal/mol. There is one point, out of the 482 reactions that has an error of 8.4 kcal/mol, but the rest of the barriers are within 6.5 kcal/mol.

One can extend these ideas to many different reactions. Table 5.4 shows the Polanyi relationships for a number of reactions. One can use this expression to fit a number of different reactions with and an RMS error of 5 kcal/mol.

One can also estimate values of the activation barrier using the Blowers–Masel approximation. Table 11.3 shows the values of the intrinsic barrier that one should use to do the calculations. Generally the activation energies for atom transfer reactions can be predicted to about ±3–5 kcal/mol, that is, about the same accuracy as the measurements. There is much less work on ligand transfer reactions, but the few examples that had been examined when this book was written still do follow the equation with an error of ±3–5 kcal/mol.

The one major weakness of the method is that experimentally, activation barriers are different with primary, secondary, and tertiary carbons. Other steric effects also play an important role, in principle, because E_a^0 is varying. However, no knows how to predict the variation, so the results are uncertain.

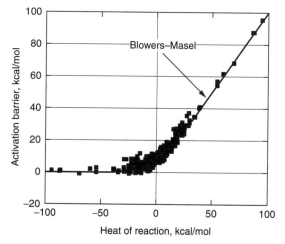

Figure 11.14 A comparison of the activation energies of 482 hydrogen transfer reactions to those predicted by the Blowers–Masel approximation with $E_a^0 = 9.5$ kcal/mol and $w_0 = 105$ kcal/mol.

Table 11.3 Suggested values of the intrinsic barriers for it elementary gas-phase reactions[a]

Reaction	Intrinsic Barrier Range, kcal/mol	Best Estimate of the Intrinsic Barrier, kcal/mol
Atom transfer reactions (e.g. $RH + X \rightarrow R + HX$)	7–15	10
Ligand transfer reactions to hydrogen (i.e, $H + R–X \rightarrow HR + X$)	40–50	45
Other ligand transfer reactions	Usually 40–60	50

[a]These values are used to estimate the activation barrier using the Blowers–Masel approximation.

The advantage of these equations is that they work on a wide range of reactions, and you can make useful predictions with pencil-and-paper calculations. The one disadvantage is that the methods are limited to types of molecules that have been examined in the past. If you were working with ions, none of the correlations would work unless you had a correlation for the specific type of ion that you are examining.

Still, industrially, people use these methods to estimate activation barriers for applications like reactor design. Most companies have their own set of correlations; for example, an oil company might have 500 different correlations for all the different types of reactions that occur in a refinery. People do not find the large number of correlations to be particularly satisfying. After all, if you have a new reaction, which correlation should you use? Still, industrially, people use data or correlations to estimate kinetics most of the time.

11.5 QUANTUM-MECHANICAL METHODS

People are starting to switch to ab initio quantum-mechanical methods, however. The idea in the ab initio methods is to solve the Schroedinger equation

$$\mathcal{H}(\hat{r}, \hat{R})\Psi(\hat{r}, \hat{R}) = E\Psi(\hat{r}, \hat{R}) \tag{11.39}$$

to calculate the energy of a single molecule or a group of molecules as a function of the atomic positions. In equation (11.39), $\mathcal{H}(\hat{r}, \hat{R})$ is the Hamiltonian for the system. The Hamiltonian depends on $\hat{R}$, the position vector for all the atoms in the system, and $\hat{r}$, the position vector for all the electrons, $\Psi(\hat{r}, \hat{R})$ is the wavefunction, and E is the energy. $\hat{R}$ is a giant vector that has the x, y, and z positions of all the atoms in the system. Similarly, $\hat{r}$ is a giant vector that has the x, y, and z positions of all the electrons in the system.

In actual practice, one fixes the atomic positions and then solves the Schroedinger equation for a specific value of the atomic coordinates. One then repeats the process several times to get the energy of the system as a function of the atomic positions. This energy is the potential energy of the system, since we are not allowing the atoms to move, so the atoms have no kinetic energy. The potential energy surface is simply a plot of the energy as a function of atomic position calculated from equation (11.39). One can then find the saddle point(s) in the potential energy surface and use it (them), to estimate the activation barrier as described in Chapter 9.

There is commercial software available to do all the calculations. Professor Masel happens to use a program called GAUSSIAN. Other common software is called CERIUS, GAMESS, HONDO, CADPAC, GRADSCF, JAGUAR, and Q-Chem. The software allows one to easily compute energies, using various approximations. It is also possible to hone in on the saddle point in the potential energy surface, and thereby get an approximation to the activation energy.

The advantages of the ab initio methods are that they have a sound basis in theory and, in principle, they can be done to arbitrary accuracy. In practice, one gets only reasonable accuracy if one does a very large calculation. Consequently, in practice the methods are limited to small molecules. However, ab initio methods have the aura of being exact. As a result, they are very well respected in the literature even though in 1998 activation barriers calculated by way of ab initio methods were often less accurate than activation barriers calculated by the methods described in Section 11.4.

In the next few sections, we will describe how one uses ab initio quantum-mechanical methods to calculate potential energy surfaces. There is insufficient room in this book to describe in detail how one actually solves the Schroedinger equation. Good reviews include the texts of Levine (1991, 1994), Foresman and Frisch (1996), Hehre (1986), Dykstra (1993), Szabo and Ostiund (1996) or Bauschlicher (1989). Still, we wanted to give the reader a brief summary of the methods so that the reader could understand the key ideas. One is referred to Levine (1991), Foresman and Frisch (1996), Hehre(1986), Dykstra (1988), Beder (1994), Grant and Grahm (1995), Szabo (1996), or Bauschlicher (1989) for a more comprehensive treatment.

First, it is useful to define all the language people use when they do this type of calculation. To start, we want to note that the idea that one can compute a potential energy surface, specifically, the total energy of the system as a function of the atomic coordinates, is an approximation. When the atoms in the system are moving fast enough, the electrons will not be able to keep up. As a result, in principle, the potential energy surface should be a function of the atomic positions and velocities. However, the velocity effect is important only when the atomic kinetic energy is large enough to cause electronic excitations. This is not the usual situation. In most cases, one can approximate the potential energy as being a function of the positions but not the velocities of all the atoms in the system. In the literature people call that approximation the **adiabatic approximation** or the **Born–Openheimer approximation**. You often read papers where they say that activation energies are computed within the adiabatic approximation or the Born–Oppenheimer approximation. Such calculations assume that the electron states are not affected by the atomic motions. This is a very good approximation for ground states.

11.5.1 The Hartee–Fock Approximation

Next we want to note that in practice one never solves the Schroedinger equation exactly. Instead, there are a series of approximations that one uses to estimate the properties of the molecules. One common, although inaccurate, approximation is called the **Hartree–Fock** method. The key assumption in the Hartree–Fock method is that we can write the wavefunction for a molecule or a transition state as "an antisymmetrized product of one-electron wavefunctions." I will explain this phrase words in a moment, but first I want to write down an equation for the Hartree–Fock approximation to the wavefunction:

$$\Psi_{HF}(\hat{r}, \hat{R}) = \forall \psi_1(\vec{r}_1)\psi_2(\vec{r}_2)\psi_3(\vec{r}_3)\psi_4(\vec{r}_4)\cdots\psi_N(\vec{r}_N) \tag{11.40}$$

In equation (11.40) $\Psi_{HF}(\hat{r}, \hat{R})$ is the Hartree–Fock approximation to the total wavefunction for the molecule; $\psi_1, \psi_2, \psi_3, \ldots$ are the wavefunctions for the various molecular orbitals of the system, where each spin state is considered to be an individual molecular orbital;

$\vec{r}_1, \vec{r}_2, \ldots$ are the positions of each of the electrons in the system; N is the number of electrons; and $\forall$ is something called the *antisymmetrizing operator*.

Let us break apart equation (11.40). First note that the wavefunction is constructed as a product of the molecular orbitals of the system. Recall from freshman chemistry or organic chemistry that the states of a molecule are represented by molecular orbitals (MOs) and not atomic orbitals. Generally the molecular orbitals extend over the entire molecule and not just one bond.

Next, it is important to note that according to equation (11.40), the wavefunction is a product, not a sum, of the molecular orbitals. This is how the Schroedinger equation normally works. Szabo (1996) shows that if one has two completely independent states of the system, then the wavefunction for the system should be the product of the wavefunctions for the individual states. The fact that the wavefunction is written as a product of states in equation (11.40) implies that the wavefunctions of the individual electrons are in some sense independent. This idea will be discussed in more detail later in this section.

Finally, equation (11.40) contains the antisymmetrizer. The antisymmetrizer is needed to force the wavefunction to obey the Pauli exclusion principle and to properly account for a quantum-mechanical process called **electron exchange**. Electron exchanges arise because all the electrons in a given molecule are equivalent and indistinguishable. Quantum-mechanically, when one has indistinguishable particles, the particles can "exchange", that is, switch places. Electron I moves from molecular orbital i to molecular orbital j. Simultaneously, electron J moves from molecular orbital j to molecular orbital i. The energy of the system does not change because both electrons are equivalent during such a process. However, electrons are *fermions* (i.e., particles with a spin of $\frac{1}{2}$). Pauli showed that in order for the energy to come out correctly, the wavefunction must change sign during exchange. The antisymmetrizer in equation (11.40) is used to assure that electrons change sign during electron exchange. Basically, the antisymmetrizer takes a wavefunction and converts it into what is called a *Slater determinant*:

$$\forall \Psi_{HF} = \frac{1}{\sqrt{N!}} \begin{vmatrix} \psi_1(\vec{r}_1) & \psi_2(\vec{r}_1) & \psi_3(\vec{r}_1) & \cdots & \psi_N(\vec{r}_1) \\ \psi_1(\vec{r}_2) & \psi_2(\vec{r}_2) & \psi_3(\vec{r}_2) & \cdots & \psi_N(\vec{r}_2) \\ & & \cdots & & \\ & & \cdots & & \\ & & \cdots & & \\ \psi_1(\vec{r}_N) & \psi_2(\vec{r}_N) & \psi_3(\vec{r}_N) & \cdots & \psi_N(\vec{r}_1) \end{vmatrix} \tag{11.41}$$

where $\psi_i(\vec{r}_j)$ is the wavefunction for the jth electron when it is in the ith molecular orbital and N is the total number of electrons.

It is useful to consider a simple example to see how this all works out. Consider a helium atom, and let's assume that a wavefunction can be written as an antisymmetrized product of a 1s wavefunction with spin α and a 1s wavefunction with spin β:

$$\Psi_e^1(\hat{r}) = \forall(\psi_{1s}^\alpha(\vec{r}_1)\psi_{1s}^\beta(\vec{r}_2)) \tag{11.42}$$

Following equation (11.41)

$$\sqrt{2}\Psi_e^1(\hat{r}) = \begin{vmatrix} \Psi_{1s}^\alpha(\vec{r}_1) & \Psi_{1s}^\beta(\vec{r}_1) \\ \Psi_{1s}^\alpha(\vec{r}_2) & \Psi_{1s}^\beta(\vec{r}_2) \end{vmatrix} = \Psi_{1s}^\alpha(\vec{r}_1)\Psi_{1s}^\beta(\vec{r}_2) - \Psi_{1s}^\alpha(\vec{r}_2)\Psi_{1s}^\beta(\vec{r}_1) \tag{11.43}$$

Now consider a different possible wavefunction where two electrons are in the same orbital:

$$\Psi_e^2(\hat{r}) = \forall(\psi_{1s}^\alpha(\vec{r}_1)\psi_{1s}^\alpha(\vec{r}_2)) \qquad (11.44)$$

Substituting into equation (11.41) yields

$$\sqrt{2}\Psi_e^2(\hat{r}) = \begin{vmatrix} \Psi_{1s}^\alpha(\vec{r}_1) & \Psi_{1s}^\alpha(\vec{r}_1) \\ \Psi_{1s}^\alpha(\vec{r}_2) & \Psi_{1s}^\alpha(\vec{r}_2) \end{vmatrix} = \Psi_{1s}^\alpha\Psi_{1s}^\alpha - \Psi_{1s}^\alpha\Psi_{1s}^\alpha = 0 \qquad (11.45)$$

Now consider a third case that is the sum of the previous two:

$$\Psi_e^3(\hat{r}) = \forall\Psi_{1s}^\alpha(\vec{r}_1)\Psi_{1s}^\beta(\vec{r}_2) + \forall\Psi_{1s}^\alpha(\vec{r}_1)\Psi_{1s}^\alpha(\vec{r}_2) \qquad (11.46)$$

Substituting equation (11.46) into equation (11.43) and doing the algebra yields

$$\Psi_e^3(\hat{r}) = \forall\Psi_{1s}^\alpha(\vec{r}_1)\Psi_{1s}^\beta(\vec{r}_2) \qquad (11.47)$$

Therefore, one effect of the antisymmetrizer is to cancel out any terms in the wavefunction where there are two electrons in the same orbital with the same spin. The antisymmetrizer prevents two electrons from being in the same orbital with the same spin, so it forces the wavefunction to satisfy the Pauli exclusion principle. Two electrons can be in the same orbital with a different spin, but there are only two possible spins. Consequently, there can be two, but no more than two, electrons in any orbital on any atom or molecule.

There is another key effect of the antisymmetrizer: It lowers the energy of the system when two electrons pair up into a bond. To see that, let's go back to the helium case again and first calculate the energy assuming that there is no antisymmetrizer. Then we will add in the antisymmetrizer.

Recall from quantum mechanics that one can calculate the energy of the system from

$$\langle\Psi_e|\mathcal{H}|\Psi_e\rangle = E \qquad (11.48)$$

where Ψ_e is the wavefunction, $\mathcal{H}$ is the Hamiltonian, and $\langle\Psi_e|\mathcal{H}|\Psi_e\rangle$ is a notation for a giant integral:

$$\langle\Psi_e|\mathcal{H}|\Psi_e\rangle = \int\int\int\int\int\int \Psi_e^* H\Psi_e \, d\vec{r}_1 \, d\vec{r}_2 \qquad (11.49)$$

In equation (11.49) $\vec{r}_1$ is the position of electron 1 and $\vec{r}_2$ is the position of electron 2. Let us assume that the Hamiltonian can be divided into a kinetic energy term, a potential energy term for interaction between the two electrons, and a potential energy term for interaction with the nucleus:

$$\mathcal{H} = \mathcal{T} + \mathcal{V}_{ee} + \mathcal{V}_{eN} \qquad (11.50)$$

where $\mathcal{T}$ is the kinetic energy, $\mathcal{V}_{ee}$ is the potential energy for the interaction between the two electrons, and $\mathcal{V}_{eN}$ is the energy for the interaction between the electrons and the nucleus.

Let's first assume that the wavefunction is given by

$$\Psi_e^{bad}(\hat{r}) = \psi_{1s}^\alpha(\vec{r}_1)\psi_{1s}^\beta(\vec{r}_2) \qquad (11.51)$$

Substituting equations (11.50) and (11.51) into equation (11.48) yields

$$E_{bad} = \langle \Psi_e^{bad} | \mathcal{T} | \Psi_e^{bad} \rangle + \langle \Psi_e^{bad} | \mathcal{V}_{ee} | \Psi_e^{bad} \rangle + \langle \Psi_e^{bad} | \mathcal{V}_{eN} | \Psi_e^{bad} \rangle \tag{11.52}$$

where E_{bad} is the energy of the state whose wavefunction is given by equation (11.51). Let's focus on the middle term on the right side of equation (11.52). Recall that $\mathcal{V}_{ee}$ is a Coulomb potential:

$$\mathcal{V}_{ee} = \frac{e^2}{|\vec{r}_1 - \vec{r}_2|} \tag{11.53}$$

where e is the charge on an electron. Substituting equations (11.51) and (11.53) into equation (11.49) yields

$$\langle \Psi_e^{bad} | \mathcal{V}_{ee} | \Psi_e^{bad} \rangle = \iiint \iiint \psi_{1s}^{\alpha}(\vec{r}_1) * \psi_{1s}^{\beta}(\vec{r}_2) * \frac{e^2}{|\vec{r}_1 - \vec{r}_2|} \psi_{1s}^{\alpha}(\vec{r}_1) \psi_{1s}^{\beta}(\vec{r}_2) \, d\vec{r}_1 \, d\vec{r}_2 \tag{11.54}$$

Equation (11.54) is positive, which means that the second term in equation (11.52) raises the energy of the system. Physically, the electrons repel each other. Repulsions raise the energy of the system.

Next, we will derive the same result using the wavefunction in equation (11.43). Substituting equations (11.43) and (11.53) into equation (11.49) yields

$$\langle \Psi_e^1 | \mathcal{V}_{ee} | \Psi_e^1 \rangle = \left(\frac{1}{2} \right) \iiint \iiint \psi_{1s}^{\alpha}(\vec{r}_1) * \psi_{1s}^{\beta}(\vec{r}_2) * \frac{e^2}{|\vec{r}_1 - \vec{r}_2|} \psi_{1s}^{\alpha}(\vec{r}_1) \psi_{1s}^{\beta}(\vec{r}_2) \, d\vec{r}_1 \, d\vec{r}_2$$
$$- \left(\frac{1}{2} \right) \iiint \iiint \psi_{1s}^{\alpha}(\vec{r}_2) * \psi_{1s}^{\beta}(\vec{r}_1) * \frac{e^2}{|\vec{r}_1 - \vec{r}_2|} \psi_{1s}^{\alpha}(\vec{r}_2) \psi_{1s}^{\beta}(\vec{r}_1) \, d\vec{r}_1 \, d\vec{r}_2 \tag{11.55}$$

It is useful to compare equations (11.54) and (11.55). Notice that you get an extra term in equation (11.55). The extra term is negative, which implies that it lowers the energy of the system. Therefore, the net effect of exchange is to lower the repulsions between the electrons when the electrons pair up into bonds.

The first term on the right side of equation (11.55) is called the **Coulomb energy**; the second term is called the **exchange energy**. The Coulomb energy is the energy that you would have if there were no exchange. The exchange energy is the extra energy one gets because of the bonding. The exchange energy always has a negative sign, which means that it lowers the energy of the system. A more detailed analysis shows that exchange reduces the repulsions only between two electrons that are paired up in a bond. If the two electrons are not paired up, or if there are more than two electrons, there will be a strong repulsion.

We will be using the definitions of the Coulomb repulsion and exchange energy extensively later in this chapter. The reader should be thoroughly familiar with the definitions before proceeding.

11.5.2   The Hartree–Fock–Roothan method

Next, we want to briefly discuss how to actually solve the Schroedinger equation using the Hartree–Fock approximation. The first step is usually to the expand each ψ_I in equation

(11.40) as a sum of atomic basis functions:

$$\psi_i = \sum_j C_{i,j} \phi_j \tag{11.56}$$

where the ϕ_j are a series of atomic functions (s orbital, p orbitals, etc), and the $C_{i,j}$ are a series of coefficients yet to be determined. One then plugs equations (11.40) and (11.56) back into the Schroedinger equation and tries to calculate the $C_{i,j}$ values.

The calculations are difficult. The electrons repel and the repulsions change as you change the coefficients. As a result, it is often difficult to do the calculations. Roothan found a feasible way to do the calculations. The method is now called the **Hartree–Fock-Roothan** method.

In actual practice, one does what is called a **self-consistent field** (SCF) calculation. One first guesses a series of coefficients, and then plugs into equation (11.55) to calculate the strength of the Coulomb and exchange terms. One can then use those numbers to calculate the average potential that any one electron feels, as a result of the average interaction with all the other electrons in the system. Once one knows the potential, one then goes back and solves the Schroedinger equation for the potential. One then uses the potential to calculate the wavefunction for each of the states in the system and the coefficients in equation (11.56). Once one knows the coefficients, one can get a better guess for the potentials, so one can repeat the calculation and gain further accuracy. In actual practice, one repeats the calculations hundreds of times until everything is internally consistent. The details are available in Levine (1999). The key thing to remember for now is that we are calculating each of the orbital coefficients by repetitively solving the Schroedinger equation until everything is internally consistent.

The **variational principle** is often used in part of the calculation. The variational principle holds that if you have an approximate wavefunction with some undetermined coefficients, then the best value of the coefficients can be determined by varying the coefficients to minimize the total energy of the system.

11.5.3 Practical Hartree–Fock Calculations

In actual practice, one does not have to worry about the details. Commercial codes can do all the calculations. In actual practice, one specifies the geometry of the molecule and the basis set that you want to use, and the code does the rest. We will describe some of the details later in this chapter. The thing to remember for now is that Hartree–Fock calculations can be done with very little effort.

Unfortunately, the problem is that Hartree–Fock calculations are not accurate enough to make useful predictions. The problem is that there are small errors in the Hartree–Fock approximation, and the errors matter to the chemistry. Recall that in the Hartree–Fock approximation, one is calculating the total energy of a molecule. The total energy is a big number compared to an activation barrier. For example, a propane molecule has a total energy of about 74,000 kcal/mol. A 1% error in the total energy will still be 740 kcal/mol. By comparison, the activation barrier for hydrogen transfer from propane is 13 kcal/mol. If we wanted to calculate an activation barrier accurate to within 10% (i.e., 1.3 kcal/mol), then we would need a calculation that is accurate to 1.3/74,000 or about 1 part in 57,000. That means that small errors in the Hartree–Fock approximation will have a significant effect.

11.5.4 Correlation

One key error in the Hartree–Fock approximation is to assume that the wavefunction can be written as a product of one-electron wavefunctions. The one-electron wavefunctions are found by solving the Schroedinger equation for the motion of an electron in the average field created by all the other electrons in the system. Recall that the electron actually sees the instantaneous field created by all the other electrons, not the average field. If one electron moves into a region of space, then, on average, all the other electrons should move out of the way. One can view this process as a coordinated motion where one electron moves into a region of space and a second electron moves out of the way.

Note that the Hartree–Fock approximation assumes that the wavefunction can be written as a product of *one*-electron wavefunctions. It is impossible to represent coordinated motion of electrons with a one-electron wavefunction. One would need two-electron wavefunctions to represent the coordinated motion of two electrons, and three-electron wavefunctions to represent the coordinated motion of three electrons. The Hartree–Fock approximation does not allow for two- or three-electron wavefunctions. That creates a 0.5% error in the total energy.

Again, we need to keep track of how big a 0.5% error is. Recall that propane has a total energy of about 74,000 kcal/mol, so a 0.5% error is about 370 kcal/mol. By comparison, we are going to try to predict an activation barrier of 12 kcal/mol. Clearly, we will not be able to tolerate a 370-kcal/mol error when we are trying to calculate a property accurate to 12 kcal/mol.

The way people handle the coordinated motion of the electrons is to add an energy called the **correlation energy** to the Hartree–Fock energy to get accurate properties. The correlation energy is a small correction to the total energy. Still it is about 0.5–1% of the total energy and varies from one system to the next, so it needs to be considered in any calculation.

There are three main approaches to estimating the size of the correlation energy:

- The Configuration interaction (**CI**) method
- Perturbation theory
- Density functional theory

In the CI method, one simply goes back and puts some two-electron wavefunctions into the approximation, then adds in three-electron wavefunctions, four-electron wavefunctions, and so on until everything converges. The hard part is to figure out how to do that systematically. The approach that is usually used is to write the wavefunction as a sum of Hartree–Fock wavefunctions

$$\Psi(\hat{r}) = \sum_i c_i \Psi_{HF}^i(\hat{r}) \qquad (11.57)$$

where $\Psi(\hat{r})$ is the true wavefunction for the system; $\Psi_{HF}^0(\hat{r})$ is the Hartree–Fock wavefunction for the ground state; $\Psi_{HF}^1(\hat{r})$, $\Psi_{HF}^2(\hat{r})$, and so on are the Hartree–Fock wavefunctions for the excited states of the system; and the c_i are a series of coefficients that are yet to be determined. It was not immediately obvious that this function allows one to account for the coordinated motion of the electrons. However, by summing, you are allowing dependent motions. Therefore, in principle, one can account for the coordinated motions exactly using the sum in equation (11.57).

In actual practice, one limits the sum to a few thousand states. It is common to include all the states with one or two excited electrons, and perhaps a few states with three excited electrons. The calculations end up to be huge. However, they do have the advantage that they will eventually converge to the exact result.

In the literature there are several names for this method. CI is the general name, and then people add extra terms such as CIS, CISD, CISD(T). CIS is a CI method, where the sum in equation (11.57) includes all the excited states where a single electron is put into an excited state, CISD includes excitations where up to two electrons are put in excited states. CISDT includes triple excitations. There is another version of the algorithm called QCISD(T) or CCSD(T). CCSD(T) treats single, and double, as in CISD, and then uses an expansion to get some approximation to the higher order excitations. QCISD is similar except that it uses a slightly less accurate expansion. Table 11.4 lists several other methods. We will discuss them in the next few sections.

There is one important detail: about 80% of the correlation energy is associated with the double excitations. Consequently, if you do not have the double excitations in the calculations, the energies are not accurate enough for chemical purposes.

11.5.5 Moller–Plesset Perturbation Theory

The alternative is to choose an approximation for the correlation energy. The idea is to calculate the energy as in equation (11.55) and then add in an extra term for the correlation energy:

$$E = \langle \Psi_e | \mathcal{T} | \Psi_e \rangle + \langle \Psi_e | \mathcal{V}_{eN} | \Psi_e \rangle + \langle \Psi_e | \mathcal{V}_{Coul} | \Psi_e \rangle + \langle \Psi_e | \mathcal{V}_{ex} | \Psi_e \rangle + \mathcal{V}_{corr} \qquad (11.58)$$

In equation (11.58), $\mathcal{T}$ is the kinetic energy operator, $\mathcal{V}_{eN}$ is the potential for the interaction of the electrons with the nuclei, $\mathcal{V}_{Coul}$ is the potential for the Coulomb interaction in the absence of exchange, $\mathcal{V}_{ex}$ is the exchange energy, and $\mathcal{V}_{corr}$ is the extra energy associated with the correlation. One then uses an approximation to calculate the correlation energy.

The most common approximation is called *Moller–Plesset perturbation theory*, abbreviated MP2, MP3, or MP4. The idea of the MPn is to start with the Hartree–Fock wavefunction and use something called *perturbation theory* to estimate the correlation energy.

The idea in the Moller–Plesset method is to use an analytic approximation for the coefficients in (11.57), and then plug back into equation (11.49) to calculate the energy change. MP2 uses the approximation

$$c_i = \frac{\langle \Psi^i_{HF}(\hat{r}) | \mathcal{H}_1 | \Psi^0_{HF}(\hat{r}) \rangle}{E_i - E_0} \qquad (11.59)$$

to estimate the coefficients. Equation (11.59) comes from a Taylor series expansion of the energy, where the Taylor series is truncated to second order. A derivation of the equation is given in Levine (1991). In equation (11.59), $\Psi^1_{HF}(\hat{r})$ is the Hartree–Fock wavefunction for the ith excited state, $\Psi^0_{HF}(\hat{r})$ is the ground-state wavefunction, E_i is the Hartree–Fock energy of the ith excited state, E_0 is the Hartree–Fock wavefunction for the ground state, and $\mathcal{H}_1$ is the error in the Coulomb potential associated with the Hartree–Fock approximation.

It happens that according to equation (11.59), all the c_i values are zero for the excited states of the closed shells with a single excited electron. Consequently, the main contribution to the MP2 energy comes from the doubly excited states.

Generally, MP2 gets about 80% of the correlation energy. For propane, the absolute energy is still off by 74 kcal/mol. Usually, one is using MP2 to calculate energy differences, and there is some cancellation of errors. Still, energy differences might be off by 20 kcal/mol. Clearly, one cannot calculate chemically accurate energies using MP2. MP2, however, can be used to calculate fairly accurate geometries and vibrational frequencies. For details, see Pople et al. (1993), Scott and Radom (1996) or Bytheway and Wong (1998).

There are higher-order MPn methods as well (MP3, MP4, etc.). They include more terms in the Taylor series, which sometimes improves the energies. Unfortunately, the extra terms sometimes decrease the accuracy of the calculation. Consequently, MP3 and MP4 are used much less than MP2.

The big advantage of the MP2 methods is that they are much faster than CI methods. You can calculate very accurate transition state geometries or the shape of the potential energy surface with MP2. One does need to go back and calculate the absolute energies or barriers using a more accurate method. However, MP2 does give good geometries and vibrational frequencies, so it is useful for many purposes.

11.5.6   Composite Methods

If one wants very accurate energies, one usually uses a CCSD(T) method with a large basis set (see Section 11.5.7) or what is called a **composite method.** The CCSD(T) method is also used in Section 11.5.4. Here we will describe the composite method.

A composite method starts with a CI calculation and then adds corrections to extrapolate the errors to zero. You usually do a small CCSD(T) type calculation and then add in MP2 or MP4 corrections to account for the effects missing in the small CCST(T) calculations. All the methods are pretty empirical. However, they do give chemically useful results. Table 11.4 gives a selection of the methods. Generally all the very accurate calculations of energies are done with the composite methods, so they are quite important. One should read the table before proceeding.

11.5.7   Density Functional Theory

The alternative approach in the literature is to go more empirical. The idea in the semiempirical method is to use an empirically derived approximation for some of the energies in equation (11.58). One then calculates the other terms self-consistently. If you look at the literature from 1960 to 1999, you find that the literature has gone through a cycle. In 1960 people used only empirical methods. Then in the 1970 and 1980s the CI-type methods came into their own, as computers were developed that could handle the computations. Now the semiempirical methods are making a comeback.

One very common set of empirical methods is based on something called *density functional theory* (DFT). Some years ago, Kohn and Sham (1965) showed that, in principle, you do not have to know the wavefunction to calculate the exchange and correlation energies in equation (11.58); all that you need to know is the electron density over all space. Therefore, if you know the electron densities everywhere, you can, in principle, calculate the exchange and correlation energy exactly.

Density functional theory builds on that idea. First, it notes that the exchange and correlation integrals in equation (11.58) can rigorously be replaced by functionals of the electron density. A *functional* is an operator that transforms one function, the electron density, into another function, the energy. Density functional theory is derived by replacing

Table 11.4 Some common methods used to solve the Schroedinger equation for molecules[a]

Method	Description
HF	One-electron wavefunctions, no correlation energy as described in Section 11.5.2
CI	One of a number of methods where the configuration interaction is used to estimate the correlation energy as described in Section 11.5.4. In the literature the CI keyword is sometimes erroneously used to denote the CIS method
GVB	A minimal CI calculation where the sum in equation (11.57) includes 2 configurations per bond. *The configurations are to improve bond dissociation energies.*
MCSCF, CASSCF	A CI calculation where the sum in equation (11.57) includes all the single excitations of the "active orbitals" and ignores excitations of the "inactive" orbitals. In the limit of a large number of configurations, MCSCF gives the exact result. However, often people just use a few configurations and still call the calculation MCSCF.
CIS	A CI calculation where the sum in equation (11.57) includes all the single excitations. This method tends to somewhat be inaccurate.
CID	A CI calculation where the sum in equation (11.57) includes all the double excitations.
CISD	A CI calculation where the sum in equation (11.57) includes all the single and double excitations.
CISDT	A CI calculation where the sum in equation (11.57) includes all the single, double, and triple excitations.
CCSD, QCISD	An improved version of a CISD calculation, where the single and double excitations are included exactly, and an approximation is used to estimate the coefficients for the higher-order excitations.
CCSD(T) QCISD(T)	An improved version of a CCSD calculation, where a better approximation for the triple excitations is used.
MP2, MP3, MP4	A calculation where Moller–Plesset perturbation theory is used to estimate the correlation energy as described in Section 11.5.5. The various numbers refer to the level of perturbation theory used in the calculation.
G2, G3	A combined method where you compute the energy as a weighted sum of CCSD(T) with different basis sets, an MP4 calculation with a large basis set, plus other corrections.
G2(MP2)	A combined method where you substitute a MP2 calculation for the MP4 calculation in the G2 calculation
CBS	A different combined method, where you use a series of intermediate calculations to extrapolate the CCSD(T) results to infinite basis-set size.

[a] All the methods in this table are based on the Hartree–Fock method plus corrections.

the exact exchange and correlation functional with an approximate functional fit to data a calculation. A further assumption is that the function is dependent only on the local value of the electron density and perhaps the first derivatives of the electron density. Nonlocal effects and, second derivatives are ignored.

Recall that fundamentally, the exchange and correlation energies are related to an integral of the electron density. One key assumption in density functional theory is that the exchange and correlation energy are dependent only on the local value of the electron density and not on the integral (i.e. nonlocal) properties.

A second key assumption in the current implementations of density functional theory is that the exact functional is replaced by an approximate functional with parameters. The parameters are fit to exact results in one class of systems and then used in another class of systems. For example, the approximation for the exchange and correlation functions are fit to data for metals and are then used for molecules and ions. If you had an exact functional, this procedure would work. However, no one knows its accuracy when an approximate functional is used.

A third key assumption in density functional theory is that something called *the self-interaction correction* can be replaced by a scaling rule. This is a complicated concept, but the idea is that rigorous DFT gives too large of an exchange/correlation energy, because it does not properly account for the fact that an electron cannot exchange with itself. People correct for the excess exchange by multiplying the rigorous DFT exchange energy by an empirical scaling factor. As soon as one includes an empirical factor, one creates errors.

In practice, people have calculated the exchange and correlation scaling factors for some simple system, for example, a uniform electron gas, and then fit the results to functions of the electron density and their gradients. One then assumes that the exchange and correlation scaling factors in a molecule are the same as the exchange and correlation scaling factors in the model system where the functions were derived. There are many variations of the approach as summarized in Table 11.5. You see many DFT calculations in the literature. Some show good agreement with experiment. Others do not. When this book was written, it was unclear which DFT method works best.

In my experience, DFT works much better for the static properties of molecules than for activation energies. When this book was being written, all the functionals that people were using were derived for a uniform electron gas or an electron gas with a simple

Table 11.5 Some common density functional methods used to solve the Schroedinger equation for molecules[a]

Method	Description
	Exchange approximations
Slater	A simple exchange approximation where the exchange energy is approximated as being $\frac{2}{3}$ times the integral of the electron density to the $\frac{4}{3}$ power. This is exact for a uniform electron gas.
Becke	A modification of the Slater approximation, where corrections are included for changes in the exchange energy due to gradients in the electron density
Perdew–Wang	A modification of the Becke approximation. Perdew and Wang (1992) fit the exchange energy for an electron gas as $V_{ex} = F_1[\rho_e] + F_2 \left[\dfrac{d\rho_e}{dx} \right]$, where V_{ex} is the exchange energy, ρ_e, is the electron density, and F_1 and F_2 are functions that were fit to calculations for electron gases.
Modified Perdew–Wang	Modifications of the Perdew–Wang method where different functionals are used.
	Correlation Approximation
VWN	A correlation approximation due to Vosko, Wilk, and Nusair (1980), that assumes that the correlation is only a function of the local electron density. The function is calculated assuming that you have a uniform electron gas and the given density. This method is sometimes called the *local spin density method*.

Table 11.5 (*continued*)

Method	Description
Becke	A modified version of the VWN approximation, where an extra correction is added to account for variations in the correlation energy due to the gradients in the electron density. The Becke gradient approximation is optimized for metals.
LYP	A different modification of the VWN approximation, where an extra correction is added to account for variations in the correlation energy due to the gradients in the electron density. The LYP gradient approximation is optimized for molecules
Perdew–Wang	A modification of the Becke approximation. Perdew and Wang fit the exchange energy for an electron gas as $V_{cor} = F_3[\rho_e] + F_4\left[\dfrac{d\rho_e}{dx}\right]$, where V_{cor} is the exchange energy, ρ_e is the electron density, and F_3 and F_4 are functions that were fit to calculations for electron gases.
Modified Perdew–Wang	There are a number of modifications of Perdew–Wang in the literature. Generally, people use different functions to tailor F_3 and F_4 for a specific application. I find it difficult to use the methods, because one never knows whether they are going to work. However, they are common in the literature.

Mixed Methods

Method	Description
X-alpha	An approximation that uses the Slater approximation for the exchange integral and ignores the correlation. It is common in X-alpha to also change the coefficient in the Slater approximation from $\frac{2}{3}$ to 0.7.
LDA, LDSA	A method that used the Slater approximation for the exchange and the VWN approximation for the correlation.
B3LYP	A hybrid method where the exchange energy is approximated as a weighted average of the Hartree–Fock exchange energy and the Slater–Becke exchange energy, while the correlation is calculated via the LYP approximation.
AM1, MINDO, CNDO, Hückel	These are several semiempirical methods. These methods are similar, in spirit, to the DFT methods, but one also approximates the coulomb repulsions with empirical functions.

[a] All the methods in this table are based on density functional theory (DFT). Generally, in DFT, one combines approximations for the exchange integral with approximations for the correlation function.

gradient in electron density. In my experience, these functionals are very accurate in systems where the gradients in the electron density are small. For example, the atoms in a bulk metal are surrounded by many free-electrons. The local environment looks like a free-electron gas. Consequently, DFT works quite well for bulk metals. DFT also works well for paraffins. The electron density around an sp^3-hybridized carbon is fairly uniform, so DFT works well.

Unfortunately, DFT does not work as well for transition states or molecules with unusual bonding. Transition states are, by nature, unsymmetric. As a result, if one chooses a functional that is derived for a symmetric environment, one does not get good energies with DFT.

For example, a DFT method called Local Spin Density Approximation (LSDA) or Slater-Vosko-Wilk-Nusair, with Nusau SVWN predicts that the reaction $H + H_2 \rightarrow H_2 + H$ has an activation barrier of -2.81 (i.e., the transition state is 2.81 kcal/mol more stable than the reactions). The experimental result is that the reaction has a barrier of $+9.7$ kcal/mol.

Table 11.6   The energy of the transition state of the reaction H + H$_2$ → H$_2$ + H calculated by a number of methods[a]

Exchange Approximation	Correlation Approximation	Transition State Energy, kcal/mol
	DFT Methods	
Slater	VWN	−2.81
Slater	LYP	−3.45
Slater	Perdew–Wang	−3.58
Becke	VWN	+3.65
Becke	LYP	+2.86
Becke	Perdew–Wang	+2.84
Perdew–Wang	VWN	+2.75
Perdew–Wang	LYP	+1.98
Perdew–Wang	Perdew–Wang	+1.95
	Non-DFT Methods	
	MP2	+13.21
	CCSD(T)	+9.91
	G2	+9.8
	Experiment	+9.7

[a] All of the calculations used a 6−311G++(3df, 3pd) basis set.
Source: Results of Johnson et al. (1994).

Table 11.6 gives several other results. Notice that none of the DFT methods predict reasonable activation barriers for the H + H$_2$ reaction. In contrast, MP2 and CCSD(T) do give estimates that are of the correct order of magnitude.

In principle, one could do better if one had the correct functionals. Johnson et al. (1994) showed that if adds a "self-interaction correction" (i.e., a functional), one can get reasonable barrier heights for the H + H → H$_2$ + H reaction with DFT. Unfortunately, the self-interaction correction does not work with hydrocarbons. When this book was written, the proper functionals did not exist for reactions of hydrocarbons. As a result, in my view DFT was not useful for kinetics. Nevertheless, is was being widely used in the literature.

11.5.8   Other Semiempirical Methods

Table 11.5 also lists several semiempirical methods: AM1, MINDO, CNDO, and Hückel. These are methods that are similar in spirit to DFT methods in that one replaces the exact energy with an approximate functional. However, the functionals are not as accurate as those for DFT. Generally, the functionals were fit to data for molecules in the 1960s and never improved. Also, one generally approximates the Coulomb energy as well as the exchange and correlation energy. The net effect is that semiempirical calculations are not respected in the literature. I cannot say that they are "bad", because their accuracy is similar to that of most DFT methods, which are well respected. However, you rarely see semiempirical methods anymore, because people think that they are much worse than DFT.

11.5.9 Basis-Set Issues

There is one other detail in the ab initio calculations. The accuracy of the calculation is determined by the accuracy of the computational method and the accuracy of the basis set that is used. Recall that the first step in any quantum-mechanical calculation is to expand the wavefunctions for the MOs of the system as a sum of atomic basis functions [see equation (11.56)]. Consequently, the accuracy of the results of the calculation will depend on the accuracy of the basis set.

Table 11.7 gives a list of some common basis sets. The list is long, and there is no good way to describe them all. However, I wanted to cover some key features so the reader could get some idea what the basis sets are like. All of these basis sets are constructed as a sum of Gaussians,

$$\phi_j = \sum_n C_{j,n} g_n \tag{11.60}$$

where the $C_{j,n}$ terms are a series of constants and g_n is a function of the form

$$g_n = I_n x^i y^j z^k \exp\left(-\alpha_n \left(x^2 + y^2 + z^2\right)\right) \tag{11.61}$$

In equation (11.61) x, y, and z are the x, y, and z distances from a given atom, and I_n, x, y, z, and α_n are constants. Functions of the form in equation (11.61) are called **Gaussians**. When $i = j = k = 0$, the function is called an *s orbital*. When $i + j + k = 1$, the orbital is called a *p orbital*. When $i + j + k = 2$, the orbital is called a *d orbital*. People generally fit the constants in equation (11.61) to atomic orbitals to "derive" a basis set.

To understand how a basis set is defined, it is useful to consider a basis set for oxygen. Huzinaga did a Hartree–Fock calculation for a single-oxygen atom and found that he could fit his resultant 1s and 2s wavefunctions using equation (11.60), with $i = j = k = 0$. The other parameters are listed in Table 11.8.

Table 11.7   Some of the basis sets commonly used for ab initio calculations

STO–3G, 3–21G, 3–21++G, 3–21G*, 3–21++G*, 3–21GSP, 4-31G, 4-22GSP, 6–31G, 6–31++G, 6–31G*, 6–31G**, 6–31+G*, 6–31++G*, 6–31++G**, 6–31G(3df,3pd), 6–311G, 6–311G*, 6–311G**, 6–311+G*, 6–311++G**, 6–311++G(2d,2p), 6–311G(2df,2pd), 6–311++G(3df,3pd), MINI, MIDI, SVP, SVP + Diffuse, DZ, DZP, DZP + Diffuse, TZ, cc-pVDZ, cc-pVTZ, cc-pVQZ, cc-pV5Z, cc-pV6Z, pV6Z, cc-pVDZ(seg-opt), cc-pCVDZ, cc-pCVTZ, cc-pCVQZ, cc-pCV5Z, aug-cc-pVDZ, aug-cc-pVTZ, aug-cc-pVQZ, aug-cc-pV5Z, aug-cc-pV6Z, aug-cc-pCVDZ, aug-cc-pCVTZ, aug-cc-pCVQZ, aug-cc-pCV5Z, LANL2DZ ECP, SBKJC VDZ ECP, CRENBL ECP, CRENBS ECP, DZVP, DZVP2, TZVP

Table 11.8   Parameters for Huzinaga's calculations for oxygen

Gaussian	g_1	g_2	g_3	g_4	g_5	g_6	g_7	g_8	g_9
α_n, angstroms^{-1}	14,771	2222	516	153	51.4	18	6.44	1.76	0.538
$C_{j,n}$ for the 1s orbital	0.0012	0.009	0.043	0.144	0.356	0.461	0.140	−0.006	0.001
$C_{j,n}$ for the 2s orbital	−0.0003	−0.002	−0.010	−0.036	−0.095	−0.196	−0.037	0.596	0.526

Notice that there are nine Gaussians in the calculation. One could imagine including all nine Gaussians in the calculation. However, when one does so, one finds that the calculations take too long. So Huzinaga proposed that one could eliminate some of the terms to simplify things. For example, in the DZ basis set Huzinaga proposed that he would include g_1 through g_6 on the 1s orbitals and g_6 through g_8 on the 2s orbitals. The g_9 contribution was ignored. People call this the DZ basis set in the literature. An alternative name is 6-2G.

People call the process where you start with nine functions and grouping terms a **contraction**. People call the resultant basis sets a **contracted** basis set.

Poble devised the notation 6-2G is a designation telling how many Gaussians are in each component of the wavefunction. The number before the dash refers to the number of Gaussians used to represent the core electrons. A 6 implies a sum of 6 gaussians. There is one number that implies that there is only one function. The number 2 refers to the number of Gaussians on the valence electrons. In this case there is one function on the valence electron, made of two Gaussians.

Notice that the 6-2G basis set has left off some key components in Table 11.8. Consequently, there are some errors.

There are a number of other basis sets in Table 11.7. The idea of the other basis sets is to add more functions to get a better representation. For example, the 6-2G basis set uses only one basis function for each electron. That may be an acceptable representation for an H_2 but not for a big molecule. In a big molecule, the atomic orbitals are distorted. One needs to allow the orbitals to distort to get good energies.

People allow for orbital distortions by placing two or more basis functions on each of the valence electrons. For example, in the 6-31G basis set, the core electrons are still represented by a single function with six Gaussians, but there are two different basis functions on the valence electrons. One of the basis functions is a sum of three Gaussians; the other is a single Gaussian. The 6-311G basis set includes three basis functions for each valence electron, one with three Gaussians, and two with one Gaussian each. In Table 11.7 some of the wavefunctions also have designations like 6-311+G(3df, 2p). The plus sign indicates that along with all the core and valence electrons, you include a **diffuse function** in the basis set. A diffuse function is a wavefunction with a small value of α_n. g_9 in Table 11.8 is a diffuse function. A diffuse function is bigger than a normal atomic wavefunction. The diffuse functions are not that important for molecules, but you need them when there are extended bonds, as in a transition state. The (3df,2p) designation indicates that one includes extra p, d, and f orbitals in the calculation. The notation (3df,2p) implies that one adds two extra p orbitals to each hydrogen and three extra d orbitals and one extra f orbital to all the other atoms in the system. These additional orbitals are very important for ionic reactions, because ions polarize bonds. They can also be important in, for example, isomerization reactions, where the transition state has a character different from that of the reactants or products.

Many other basis sets are listed in Table 11.7. Each basis set is optimized for some specific purpose, so, for example, LANL2DZ is optimized for transition metal atoms; 6-311+G(3df,2p) is optimized to give the best energies for hydrocarbons. 6-31G* often gives excellent vibrational frequencies and geometries for the same molecules.

I cannot tell you how to choose basis sets appropriately. Each is special for a given set of properties; a basis set that gives good energies does not necessarily give good vibrational frequencies. Thus, you need to be careful to choose the correct basis set for your problem and test it on systems where the key information is known. Another key thing is to avoid mixing basis sets. If you mix a basis set, you create errors that are difficult to correct. One can calculate accurate transition state energies, geometries, and vibrational frequencies using ab initio methods. As a result, the methods are quite useful. However, one has to be careful, or else one will get poor results.

11.5.10 Combining the Basis Sets with the Methods

Once one chooses a basis set and a method (eight possible methods are listed in Table 11.9), one can start to do calculations. People report the calculation by the method and then the method and then the basis set, so a MP2/6-31G* calculation is a MP2 calculation done with a 6-31G* basis set. MP2/6-31G* calculations usually give excellent geometries and vibrational frequencies, but not good energies. One can obtain reasonable energies using one of the combined methods G2, G2+, or CBS. B3LYP also gives reasonable energies for stable hydrocarbons, but not transition states, species with strain, or unusual bonds.

11.5.11 Calculating a Potential Energy Surface Using ab initio Methods

Next, I want to describe how one goes about calculating the properties of the transition state. Generally, the approach is to guess at the geometry of the transition state, solve the Schroedinger equation at that geometry, use a search algorithm to optimize to the saddle point, and then calculate properties.

In practice, the hardest part is getting a good initial estimate to the geometry of the bonds that break and form. There is nothing about this in the literature. However, Professor Masel's group has been doing these types of calculations for gas-phase reactions of hydrocarbons and found some rules that you can use to guess a reasonable geometry for the transition state:

- The active bonds (i.e., the bonds that break and form) are usually about 25% longer than in a stable molecule. For example, consider the following reaction:

$$H + H_3CCH_3 \longrightarrow H\cdots H_3C\cdots CH_3 \longrightarrow CH_4 + CH_3 \qquad (11.62)$$

At the transition state the C–H bond length will be about 25% longer than in methane and the C–C bond length will be about 25% longer than in ethane.

Table 11.9 Average accuracy of a number of methods for the bond energy of stable hydrocarbons

Method	Average Error, kcal/mol
CBS-Q	1.0
G2	1.2
G2(MP-2)	1.5
CBS-4	2.0
B3LYP/6-311+G(3df,2p)	2.7
B3LYP/6-311+G(2d,p)	3.9
MP2/6-311+G(2d,p)	8.9
HF/6-311+G(2d,p)	46.1

Source: Foresman and Frisch (1996).

- The bond lengths for all the inactive bonds will be half way between the bond lengths in the reactants and the products. Thus, the C–H bonds in the terminal CH_3 group will be half way between the bond lengths in an ethane and a methyl radical.
- All the bond angles in the transition state will be half way between the bond angles in the reactants and the products. This means that C–H bonds in the terminal CH_3 group will be half way between the bond angles in an ethane and those a methyl radical.

That still leaves one angle to specify. For example, is it better for the incoming hydrogen to approach the ethane along the C–C bond axis or at a tetrahedral angle, with the hydrogen approaching between the two hydrogens. In my experience, both are possible, so I try both and see which converges quicker.

The rules listed above generally give you reasonable initial estimates of the transition state geometry for ligand exchange reactions of hydrocarbons.

The next step is to optimize the transition state geometry. Shegel has developed a great algorithm called *synchronous transit* to do that. The algorithm tries to search for a saddle point in the potential energy surface. The algorithm is part of a commercial code called GAUSSIAN, so in principle, one can find transitions states fairly easily. In practice, the algorithm does have some convergence problems, so there is a little bit of art to getting it to work. However, one can often calculate the transition state geometry fairly accurately.

In my experience, accurate transition state geometries and vibrational properties can be calculated with MP2/6-31+G* or MP2/AUG-CC-PVDZ calculations. People also use B3LYP, but I have not found B3LYP to give as accurate transition state geometries as MP2. Activation barriers are not accurately predicted with MP2 calculations. However, accurate transition state energies can be calculated using any of the combined methods, for example, CBS or CCSD(T).

Density functional methods such as B3LYP seem to give excellent properties for stable molecules. However, they do not do well for transition states because of limitations of the functionals described in Section 11.5.7.

One can also use these methods to calculate many other properties (see Levine for details). In practice these calculations can easily be done for small molecules. Larger molecules are a problem because of computer time limitations, but they are theoretically possible. The advantage of the ab initio methods is that if you do them properly, you know that you are getting the right answer. The disadvantage is that they still require a considerable amount of computer time.

11.6 SUMMARY

At this point I need to close the discussion. I have talked about a variety of methods that one can use to estimate activation barriers. Empirical methods are widely used in industry to model a wide variety of reactions. Ab initio calculations are the wave of the future; people do them now for small molecules, but in the future large molecules will be examined as well.

11.7 SOLVED EXAMPLES

Example 11.A   Sample Ab Initio Calculations   Calculate the equilibrium geometry of ethane at the MP2/6-31+G(d) level.

Solution   I decided to solve this problem using GAUSSIAN98. GAUSSIAN calculations are simple. You specify (1) the geometry of the molecule and (2) the computational method. The GAUSSIAN98 program does the rest.

Table 11.A.1 shows my input file. The first line tells the computer that I want to use MP2/6-31+G(d) to solve the problem. I added an opt keyword to say that I want to optimize the geometry. I also added a test keyword. This prevents huge output files from being stored on disk.

The second line is blank by convention, while the third line lists the name of the job, in this case "ethane molecule."

The fourth line is again blank.

The fifth line contains two numbers; the charge and the spin multiplicity of the molecule. The charge is the charge on the molecule. A neutral molecule will have a charge of zero. An ion with a +1 charge will have a charge of +1.

The spin multiplicity is the degeneracy of the electronic configuration. Recall that each electron has a spin of $+\frac{1}{2}$ (up) or $(-\frac{1}{2})$ down. It is useful to define a quantity "net spin" by

$$\text{Net spin} = \sum_{\text{all electrons}} (\text{spin of each electron})$$

If all of the electrons are paired into bonds, the net spin will be zero. If there is one unpaired electron, the net spin will be 0.5. If there are two unpaired electrons, there are two possibilities: (1) if both electrons have the same spin, the net spin will be $\frac{1}{2} + \frac{1}{2} = 1$; (2) if both electrons have the opposite spin, the net spin will be $\frac{1}{2} - \frac{1}{2} = 0$.

The spin multiplicity is defined by

$$\text{Spin multiplicity} = 2(\text{net spin}) + 1$$

The ground states of most stable molecules have a multiplicity of 1 because all of the electrons are paired up into bonds. Radicals, on the other hand, usually have a multiplicity

Table 11.A.1   The input file needed to calculate the equilibrium geometry of ethane using GAUSSIAN 98

```
#MP2/6-31+G(d) opt test
Ethane molecule

0 1
C
C 1 RCC
H 1 RCH 2 ACH
H 1 RCH 2 ACH 3 DHH
H 1 RCH 2 ACH 4 DHH
H 2 RCH 1 ACH 3 DCT
H 2 RCH 1 ACH 6 DHH
H 2 RCH 1 ACH 7 DHH

RCC 1.87
RCH 1.09
ACH 111.
DHH 120.
DCT 60.
```

of 2 because there is always a net spin. A diradical such as •CH$_2$CH$_2$CH$_2$• or an oxygen atom can have a multiplicity of 3.

Physically, the spin multiplicity is the degeneracy of the spin state. A stable molecule will usually have a single degenerate spin state. A radical has a doubly degenerate spin state: one with spin up, one with spin down. For our case, the charge is zero because we are choosing a neutral molecule. The multiplicity is one because we are looking at the ground state of a neutral molecule and all the electrons are paired.

Next comes the geometric description. There are two ways to enter geometries in GAUSSIAN. One can either list the X, Y, and Z coordinates for each atom or list the bond lengths and bond angles. I chose the latter.

People call the list of bond lengths and bond angles the "Z matrix." I create Z matrices by first listing all of the atoms and then adding coordinates. There are two carbons and six hydrogens in ethane, so I made a list of eight atoms: two carbons and six hydrogens. The first atom is atom 1, the second is atom 2, and so on as shown in Figure 11.A.1.

Then I need to put in bond lengths. To simplify things, use the following definitions:

RCC = the carbon–carbon bond length

RCH = the carbon-hydrogen bond length

ACH = the CH bond angle

DCH = the rotation angle between one hydrogen atom and the adjacent carbon atom

DCT = the rotation of the methyl group on one side of the molecule relative to the carbon atom on the other side of the molecule

Now let's focus on the input file.

The first line puts the first atom at the origin.

The second line puts atom 2 at a distance RCC away from atom 1.

The third line puts atom 3 at a distance RCH from atom 1 and an angle of ACH from atom 2.

The fourth line puts atom 4 at a distance of RCH from atom 1 and at an angle of ACH from atom 2 and rotated by DCH from atom 3.

The fifth line puts atom 5 at a distance of RCH from atom 1 and at an angle of ACH from atom 2 and rotated by DCH from atom 4.

The sixth line puts atom 6 at a distance of RCH from atom 2 and at an angle of ACH from atom 1 and rotated by DCT from atom 3.

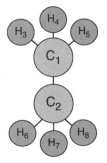

Figure 11.A.1 A diagram of ethane. The numbers represent the order of the atoms in the z-matrix.

The seventh line puts atom 7 at a distance of RCH from atom 2 and at an angle of ACH from atom 1 and rotated by DCH from atom 6.

The eighth line puts atom 8 at a distance of RCH from atom 2 and at an angle of ACH from atom 1 and rotated by DCH from atom 7.

The remaining lines in the input file give initial guesses for the input variables. For example, I guessed RCC = 1.87 Å, RCH = 1.09 Å, ACH = 111°, DCH = 120° clockwise, and DCT = 60°. Most of the numbers come from typical bond lengths and bond angles from the CRC. I deliberately picked a bond length that was too long for the carbon–carbon bond, because I wanted to show that GAUSSIAN calculates the correct bond length if I guess a bad one.

I ran GAUSSIAN with the input file shown in Table 11.A.1. Table 11.A.2 shows part of the output from the run. The GAUSSIAN job actually produces 1100 lines of output. These are the last 29 lines. From the output RCC = 1.528 Å, RCH = 1.094 Å, ACH = 111.1°, DHH = 120°, and DCT = 60°. Notice that GAUSSIAN optimized the structure. The carbon–carbon bond length was reduced from 1.87 Å to 1.5277 Å. All of the other variables were changed by a smaller amount. The run took about 90 seconds on

Table 11.A.2 Part of the output from GAUSSIAN for the input shown

Final structure in terms of initial Z-matrix:

```
C
C,1,RCC
H,1,RCH,2,ACH
H,1,RCH,2,ACH,3,DHH,0
H,1,RCH,2,ACH,4,DHH,0
H,2,RCH,1,ACH,3,DCT,0
H,2,RCH,1,ACH,6,DHH,0
H,2,RCH,1,ACH,7,DHH,0
Variables:
RCC = 1.52770877
RCH = 1.09407761
ACH = 111.11640213
DHH = 120.
DCT = 60.

1\1\ NATIONAL CENTER FOR SUPERCOMPUTING APPLICATIONS-
BILLIE\FOpt\RMP2-FC\6-31+G(d)\C2H6\RMASEL\30-Jun-1999\0\\#MP2/6-31+G(D)
OPT TEST\\ Ethane molecule\ \ 0,1\C,0.,0.,-.7638543853\C,0.,0.,
0.7638543853\H,1.0206107823,0.,-1.1580110214\H,- 0.5103053912,
-0.8838748649,-1.1580110214\H, -0.5103053912,0.8838748649,
-1.1580110214\H,0.5103053912,0.8838748649,1.1580110214\H,-.0206107823,
0.,1.1580110214\H,0.5103053912,-0.8838748649,1.1580110214\Version=HP-
PARisc-HPUX-G98RevA.6\HF=-79.2291737\MP2=-79.497602\RMSD=3.599e-09
\RMSF=1.313e-04\Dipole=0.,0.,0.\PG=D03D [C3(C1.C1),3SGD(H2)]\\@

LOVE IS BLIND, THAT'S WHY ALL THE WORLD LOVES A LOUVER.
Job cpu time: 0 days 0 hours 1 minutes 34.6 seconds.
File lengths (MBytes): RWF= 14 Int= 0 D2E= 0 Chk= 6 Scr= 1
Normal termination of Gaussian 98.
```

the NCSA (National Center for Supercomputing Application) computer, which is typical for a molecule of this size.

Example 11.B   Energy Calculations   Use the results in Table 11.A.2 to estimate the total energy and the correlation energy of ethane.

Solution   GAUSSIAN lists the energies in the output block at the end of the program. The HF = is the Hartree–Fock energy in hartrees; MP2 = is the MP2 energy in hartrees, where 1 hartree = 627.5095 kcal/mol.

From the output MP2 = −79.497602 hartrees. Therefore the total energy is given by

Total energy = −79.497602 hartrees × 627.5095 (kcal·mol)/hartree = −49885.5 kcal/mol

Again from the output HF = −79.2291737

HF = −79.2291737 hartrees × 627.5095(kcal·mol)/hartree = −49717.05 kcal/mol

The correlation energy is the difference between the HF and the MP2 energy:

Correlation energy = HF-MP2 = −49885.5 − (−49717.05) = −168.45 kcal/mol

The correlation energy is only $168.45/49885.5 = 0.3\%$ of the total energy. However, the correlation energy is still −168 kcal/mol so it cannot be ignored.

Example 11.C   Ab Initio Calculation of Heats of Reaction   Calculate the ΔU (i.e. energy) of reaction for the reaction

$$H + CH_3CH_3 \longrightarrow H_2 + CH_2CH_3 \qquad (11.C.1)$$

at the MP2/6-31+G(d) level. Ignore the zero-point correction (defined below).

Solution   I will use GAUSSIAN to calculate the total energy of the reactants and products. The heat of reaction ΔU will be estimated as a difference between the total energy of the reactants and that of the products.

We already have the total energy of ethane from Example 11.B. Tables 11.C.1–11.C.3 show the input file needed to calculate the energy of hydrogen atoms, hydrogen molecules, and ethyl radicals.

Table 11.C.1 shows the input file for a hydrogen molecule. The input file is similar to the input file in Table 11.A.1. The first line is the method. The second line is blank.

Table 11.C.1   Input file to calculate the energy of a hydrogen molecule

```
#mp2/6-31+G(d) opt test

hydrogen molecule

0 1
H 0., 0., 0.
H 0.8,0., 0.
```

Table 11.C.2 Input file to calculate the energy of an ethyl radical

```
#mp2/6-31+G(d) opt test

Ethyl radical

0 2
C
C 1 RCC
H 1 RCH1 2 ACH1
H 1 RCH2 2 ACH2 3 DHH1
H 2 RCH3 1 ACH3 4 DCT
H 2 RCH4 1 ACH4 5 DHH2
H 2 RCH5 1 ACH5 6 DHH3

RCC 1.52
RCH1 1.09
RCH2 1.09
RCH3 1.09
RCH4 1.09
RCH5 1.0
ACH1 111.
ACH2 111.
ACH3 111.
ACH4 111.
ACH5 111.
DHH1 120.
DHH2 120.
DHH3 120.
DCT 60.
```

Table 11.C.3 Input file to calculate the energy of a hydrogen atom

```
#hf/6-31+G(d) opt test

hydrogen atom

0 2
H
```

The third line is the title. The fourth line is blank. The fifth line shows the charge and multiplicity. The sixth and seventh lines say that I have two hydrogen atoms.

There is one special thing in the file. I decided to specify the X, Y and Z positions for each atom, so the first hydrogen atom is at 0.,0.,0. and the second hydrogen atom is at 0.8, 0.0, 0.0. I used 0.8 Å as a guess of the bond length. That is a typical bond length. GAUSSIAN98 will calculate a better value for me.

I submitted the input file to GAUSSIAN98 and got 1000 lines of output. Table 11.C.4 shows the last 15 lines of the output file. I have highlighted two places in the file. These are the final X, Y and Z positions of each hydrogen.

Table 11.C.4   The last 15 lines of output of the GAUSSIAN run for a hydrogen molecule

```
Test job not archived.
1\1\ NATIONAL CENTER FOR SUPERCOMPUTING APPLICATIONS-
BILLIE\FOpt\RMP2-FC\6 31+G(d)\H2\RMASEL\01-Jul-1999\0\\
#MP2/6 31+G(D)OPT TEST\\hydrogen molecule\\0,1\H,
.3687587189,0.,0.\H,0.3687587189,0.,0.\\
Version=HP-PARisc-HPUX-G98RevA.6\State=1 SGG\HF= 1.1267865
\MP2=-1.1441408\RMSD=2.175e 12\RMSF=1.631e-06\Dipole=0.,0.,0.
\PG=D*H[C*(H1.H1)]\\@

GARLIC THEN HAVE POWER TO SAVE FROM DEATH

BEAR WITH IT THOUGH IT MAKETH UNSAVORY BREATH,

AND SCORN NOT GARLIC LIKE SOME THAT THINK

IT ONLY MAKETH MEN WINK AND DRINK AND STINK.

 SIR JOHN HARRINGTON,

  THE ENGLISHMAN S DOCTOR   ,1609
Job cpu time:0 days 0 hours 0 minutes 27.3 seconds.
```

GAUSSIAN shifts the molecules so that the x,y,z origin is at the center of mass of the molecule. There is one atom at $+0.3687587$ Å, 0, 0, and another atom at -0.3687587 Å. The bond length is equal to

$$\text{Bond length} = +0.3687587 \text{ Å} - (-0.3687587 \text{ Å}) = 0.7375 \text{ Å}$$

The MP2 energy is MP2 $= -1.1441408$ hartrees.

Table 11.C.2 shows the input file for the ethyl radical. I created this file by starting with the input file for an ethane molecule in Table 11.A.1, and then deleted atom 5. I had to renumber the rest of the atoms accordingly. I made three other changes. First I changed the C–C bond length to the value from Example 11.A. Then I changed the multiplicity to 2, since I have a radical with two spin states: one with spinup the other with spindown. Finally, I let all of the bond lengths and bond angles be different, because I was not sure how the bond lengths would change.

I submitted the job and got almost 2000 lines of output. Table 11.C.5 shows the last 50 lines. Notice that the geometry is very similar to than of ethane. There are two MP2 energies in the output: one labeled MP2=, the other labeled PMP2=. PMP2 is a version of MP2 that corrects for an error called *spin contamination*. Generally PMP2 energies are more accurate than MP2 energies. The PMP2 energy is -78.8423993 hartrees. Note that the number wraps onto the next line.

Table 11.C.3 shows the input file for a hydrogen atom. It is similar to the others except that there is only one atom.

There are two special parts to the input file: (1) the charge and multiplicity (the charge is zero as expected) and the multiplicity. The multiplicity is 2 because a hydrogen atom is a radical, and there are two possible spin states, one with spinup and another with spindown.

The second unusual thing in the input file is that the method is HF/6-31+G*, not MP2/6-31+G*. It happens that GAUSSIAN will not do a MP2 calculation on a hydrogen atom. Recall, that hydrogen atoms have only one-electron, and so the correlation energy is zero. GAUSSIAN98 creates an error message when the correlation energy is zero.

Table 11.C.5   The last 50 lines of output of the GAUSSIAN run for a ethyl radical

```
Final structure in terms of initial Z-matrix:
C
C,1,RCC
H,1,RCH1,2,ACH1
H,1,RCH2,2,ACH2,3,DHH1,0
H,2,RCH3,1,ACH3,4,DCT,0
H,2,RCH4,1,ACH4,5,DHH2,0
H,2,RCH5,1,ACH5,6,DHH3,0
 Variables:
RCC=1.49164467
RCH1=1.08276737
RCH2=1.08276737
RCH3=1.10052188
RCH4=1.09413499
RCH5=1.09413499
ACH1=120.7158591
ACH2=120.7158591
ACH3=111.75803712
ACH4=111.36638532
ACH5=111.36638532
ACH2=120.7158591
ACH3=111.75803712
ACH4=111.36638532
ACH5=111.36638532
DHH1=167.91010516
DHH2=119.56434422
DHH3=120.87131156
DCT=83.95505258
Test job not archived.
1\1\ NATIONAL CENTER FOR SUPERCOMPUTING APPLICATIONS-
BILLIE\FOpt\UMP2-FC\6-31+G(d)\C2H5(2)\RMASEL\01-Jul-1999\0\\#MP2/6-
31+G(D) OPT TEST\\Ethyl radical\\0,2\C,0.0435464704,-0.0754246992,-
0.7908402212\C, -0.0497426233,0.0861567509,0.689089647\H,0.9284363327,
0.2432829634,- 1.3272920193\H,-0.674907393,-0.6824079682,-
1.3272920193\H, 0.4317890897,-0.7478806415,1.2216844785\H,
0.4434632206,1.0044425007,1.0217015027\H,-1.0916043325,0.1181708357,
1.0217015027\\Version=HP-PARisc-HPUX-G98RevA.6\State=2-A'\HF=-
78.5988922\MP2=-78.8403239\PUHF=-78.6021196\\**PMP2= -78.8423993**
\S2=0.762683\S2-1=0.753582\S2A=0.75011\RMSD=5.287e-09\RMSF=4.764e-05
\Dipole=0.0316089,-0.0547482,0.1081351\PG=CS [SG(C2H1),X(H4)]\\@

TO BEHOLD HARMS OF OUR OWN HANDS DOING,
WHERE NONE BESIDE US WROUGHT,
CAUSES SHARP RUING...

                        SOPHOCLES
Job cpu time: 0 days 0 hours 4 minutes .4 seconds.
File lengths (MBytes): RWF= 14 Int= 0 D2E= 0 Chk= 6 Scr= 1
Normal termination of Gaussian 98.
```

Fortunately, when the correlation energy is zero, the MP2 energy is the same as the HF energy. GAUSSIAN will do an HF calculation on a hydrogen. Consequently I ended up using a HF calculation for this problem.

Again, I submitted the job to GAUSSIAN and got 800 lines of output. Table 11.C.6 shows four of the last 15 lines. The energy is highlighted HF = −0.4982329 hartrees.

Table 11.C.7 summarizes these results. The ethane data are from Table 11.A.2. The hydrogen atom, hydrogen molecule, and ethyl radical data are from Tables 11.C.4, 11.C.5, and 11.C.6, respectively.

The heat of reaction ΔU can then be approximated by

$$\Delta U = \text{(ethyl energy)} + \text{(H}_2 \text{ energy)} - \text{(ethane energy)} - \text{(hydrogen atom energy)}$$

Plugging in the numbers yields

$$\Delta U = (-78.8423993 \text{ hartrees}) + (-1.1441408 \text{ hartrees}) - (-79.497602 \text{ hartrees})$$

$$- (-0.4982329 \text{ hartrees}) = 0.0092948 \text{ hartrees};$$

$$0.0092948 \text{ hartrees} \times 627.5095 (\text{kcal·mol})/\text{hartree} = +5.83 \text{ kcal/mol}$$

This compares to an experimental value of −2 kcal/mol.

Physically, the heat of reaction is a little off for two reasons:

1. Limitations of MP2/6-31+G(d)
2. Errors associated with ignoring the energy associated with the zero point motion.

The bigger effect is the limitations of MP2/6-31+G(d). MP2/6-31+G(d) gives total energies within 1% of the exact result, but still, the total energy could be off by 0.7 hartrees = 439 kcal/mol. You cancel some of those errors when you take the difference. However, the errors are in the order of 10 kcal/mol.

If you want better numbers, you need to use a bigger basis set and a more exact computational procedure. For example, the G2 energy is within 0.5 kcal/mol for reaction (11.C.1).

There is a smaller error associated with ignoring zero-point energies. Recall from physical chemistry that the vibrational levels of a molecule have some zero point energy. I have ignored that in the calculation above.

Table 11.C.6   The four last lines of output of the GAUSSIAN run for a hydrogen molecule

```
1\1\ NATIONAL CENTER FOR SUPERCOMPUTING APPLICATIONS-BILLIE\SP\UHF\
6 31+G(d)\H1(2)\RMASEL\30-Jun-1999\0\\#HF/6 31+G(D)\\hydrogenatom
\\0,2\H\\Version=HP-PARisc-HPUX-G98RevA.6\HF= 0.4982329\S2=0.75\S2 1=0.
\S2A=0.75\RMSD=1.039e 16\Dipole=0.,0.,0.\PG=KH\\@
```

Table 11.C.7   Summary of the MP2 and PMP2 energies from the GAUSSIAN runs

Molecule	Hydrogen Atom	Hydrogen Molecule	Ethyl Radical	Ethane
Total energy, hartrees	−0.4982329	−1.1441408	−78.8423993	−79.497602

Example 11.D   Finding a Transition State Geometry   Find the MP2/6-31+G* transition state for the reaction

$$H + CH_3CH_3 \longrightarrow H_2 + CH_2CH_3$$

Solution   Again I used GAUSSIAN98. I assumed that the incoming hydrogen was removing atom 3 from the ethane. Table 11.D.1 shows my input file (see also Table 11.D.2). I started with the input file for ethane from Example 11.A and added an extra hydrogen atom, atom 9. As a first guess, I assumed that the hydrogen approached the ethane along the C–H bond axis. I then made separate bond lengths and angles for each of the atoms because I did not know whether they would all be the same.

Table 11.D.1   The input file used to calculate the transition state geometry with GAUSSIAN

```
#mp2/6-31+G(d) fopt=(ts,calcFC) test

Ethane + H TST

0 2
C
C 1 RCC
H 1 RCH1 2 ACH1
H 1 RCH2 2 ACH2 3 DHH1
H 1 RCH3 2 ACH3 4 DHH2
H 2 RCH4 1 ACH4 3 DCT
H 2 RCH5 1 ACH5 6 DHH3
H 2 RCH6 1 ACH6 7 DHH4
H 3 RHH 1 AHH 4 DDD

RCC 1.5277
RCH1 1.368
RCH2 1.094
RCH3 1.094
RCH4 1.094
RCH5 1.094
RCH6 1.094
ACH1 111.116
ACH2 111.116
ACH3 111.116
ACH4 111.116
ACH5 111.116
ACH6 111.116
DHH1 120.
DHH2 120.
DHH3 120.
DHH4 120.
DCT 60.
RHH 0.923
AHH 179.99
DDD 30.
```

Table 11.D.2   The last 50 lines of the GAUSSIAN output for the input file in Table 11.D.1

```
Final structure in terms of initial Z-matrix:
C
C,1,RCC
H,1,RCH1,2,ACH1
H,1,RCH2,2,ACH2,3,DHH1,0
H,1,RCH3,2,ACH3,4,DHH2,0
H,2,RCH4,1,ACH4,3,DCT,0
H,2,RCH5,1,ACH5,6,DHH3,0
H,2,RCH6,1,ACH6,7,DHH4,0
H,3,RHH,1,AHH,4,DDD,0
RCC=1.50649999
RCH1=1.40503065
RCH2=1.0897126
RCH3=1.0897126
RCH4=1.09400415
RCH5=1.09726808
RCH6=1.09400414
ACH1=105.71279353
ACH2=116.07822611
ACH3=116.07822856
ACH4=111.14856702
ACH5=111.18949181
ACH6=111.14856823
DHH1=112.04403856
DHH2=135.91190293
DHH3=119.79675132
DHH4=119.79675236
DCT=60.20325946
RHH=0.89156416
AHH=177.55185199
DDD=-121.71173834
Test job not archived.
1\1\ NATIONAL CENTER FOR SUPERCOMPUTING APPLICATIONS-BILLIE\FTS\UMP2-F
C\6-31+G(d)\C2H7(2)\RMASEL\01-Jul-1999\0\\#MP2/6-31+G(D)FOPT=(TS,CALC
FC) TEST\\Ethane+H TST\\0,2\C,-0.191417856,-0.0000003522,-0.66397428
64\C,-0.0958224513,0.0000045617,0.8394896316\H,1.1342378662,-0.0000175
905,-1.1295363549\H,-0.5884380138,-0.9072190105,-1.118730105\H,-0.5884
172807,0.907223721,-1.118737394\H,0.4352345185,0.8854278664,1.20122039
09\H,-1.0916754189,0.0000177636,1.300221767\H,0.4352143245,-0.88542794
92,1.2012274817\H,1.987285848,-0.0000300703,- 1.3887578568\\Version=HP-
PARisC-HPUX-G98RevA.6\HF=-79.6912708\MP2=-79.9591999\PUHF=-79.698677\
PMP2-0=-79.9642596\S2=0.788293\S2-1=0.763863\S2A=0.750436\RMSD=2.302e-0
9\RMSF=4.930e-05\Dipole=-0.317008,0.0000045, 0.165609\PG=C01 [X(C2H7)]\
\@

HONESTY IN A LAWYER IS LIKE A HEN'S HIND LEGS.
-- MAGNUS OLESON, LAKE WOBEGON PATRIARCH, C.1875
Job cpu time: 0 days 0 hours 10 minutes 7.1 seconds.
File lengths (MBytes): RWF= 148 Int= 0 D2E= 0 Chk= 6 Scr= 1
Normal termination of Gaussian 98.
```

In order to run the job, I needed an initial guess of all of the bond lengths. I assumed that most of the bond lengths and bond angles were the same as those in ethane. The one exception was the bond that broke. In Example 11.A we found that CH bonds in ethane are 1.09 Å long, but if the C–H bond breaks, it must extend first. I guessed that it stretched by 25%, since Blowers and Masel (1999) indicated that most bonds stretch by 25% before breaking, so I used

$$RCH1 = (1.25)^*(1.09 \text{ Å}) = 1.368 \text{ Å}.$$

Similarly, the H–H bond in the transition state should be 25% longer than the H–H bond in H_2. In Example 11.C we found that the H–H bond was 0.7375 Å long, so I guessed

$$RHH = (1.25)^*0.7375 \text{ Å} = 0.923 \text{ Å}$$

There are two other important statements in the input file. First, I used the command opt = (ts,CalcFc). That tells GAUSSIAN to optimize to a transition state. Second, the multiplicity was 2. I am starting with a radical (i.e., a H atom) with an unpaired electron, and since electrons are conserved, I will always have at least one unpaired electron in the system. Therefore, the multiplicity is 2.

Table 11.D.7 shows the output file. It contains the geometry.

Example 11.E   Calculating an Activation Energy Using Ab Initio Methods   Calculate an activation energy for the reaction

$$H + CH_3CH_3 \Longrightarrow H_2 + CH_2CH_3$$

Solution   If we ignore the zero-point correction, the height of the barrier is

$$E_a = \text{(transition state energy)} - \text{(reactant energy)}$$

$$= \text{(transition state energy)} - \text{(hydrogen energy)} - \text{(ethane energy)}$$

One can read the PMP2 energies from Tables 11.C.7 and 11.D.2:

$$E_a = (-79.9642596 \text{ hartrees}) - (-79.497602 \text{ hartrees}) - (-0.4982329 \text{ hartrees})$$

$$= 0.031575 \text{ hartrees}; \quad 0.031575 \text{ hartrees} \times 627.5095 \text{ (kcal·mol)/hartree}$$

$$= 19.8 \text{ kcal/mol}.$$

By comparison, the experimental value in Westley is 9.6 kcal/mol.

Example 11.F   Activation Barriers via the Blowers–Masel Approximation   Use (a) the Blowers–Masel approximation and (b) the Marcus equation to estimate the activation barrier for the reaction

$$H + CH_3CH_3 \longrightarrow H_2 + CH_2CH_3$$

Solution

(a) According to the Blowers–Masel approximation, the activation barrier is as given by equation (11.37):

$$E_a = (w_0 + 0.5^* \Delta H_r) \left(\frac{(V_p - 2w_0 + \Delta H_r)^2}{(V_p)^2 - 4(w_0)^2 + (\Delta H_r)^2} \right) \tag{11.F.1}$$

where w_0 is the average bond energy of the bonds that break and form, ΔH_r is the heat of reaction, and V_P is given by

$$V_P = 2w_0 \left(\frac{w_0 + E_a^0}{w_0 - E_a^0} \right) \tag{11.F.2}$$

Where E_a^0 is the intrinsic barrier given in Table 11.3. According to Table 11.3, $E_a^0 = 10$ kcal/mol. One could calculate a value of w_0 from data in the CRC. However, I decided to assume a typical value for a C–H bond, specifically, 100 kcal/mol. I chose $\Delta H_r = -2$ kcal/mol from Example 11.C. Plugging into equation (11.F.2) shows

$$V_P = 2(100 \text{ kcal/mol}) \left(\frac{100 \text{ kcal/mol} + 10 \text{ kcal/mol}}{100 \text{ kcal/mol} - 10 \text{ kcal/mol}} \right) = 244.4 \text{ kcal/mol} \tag{11.F.3}$$

Plugging into equation (11.F.1) shows

$$E_a = (100 \text{ kcal/mol} + 0.5^*(-2 \text{ kcal/mol}))$$

$$\times \left(\frac{(244.4 \text{ kcal/mol} - 2(100 \text{ kcal/mol}) + (-2 \text{ kcal/mol}))^2}{(V_p)^2 - 4(w_0)^2 + (\Delta H_r)^2} \right) = 9.0 \text{ kcal/mol}$$

By comparison, the experimental value from Westley is 9.6 kcal/mol.

(b) According to the Marcus equation

$$E_a = E_a^0 \left(1 + \frac{\Delta H}{4E_a^0} \right)^2 \tag{11.F.4}$$

From the data above, $E_a^0 = 10$ kcal/mol, $\Delta H_r = -2$ kcal/mol. Plugging into equation (11.F.4), we obtain

$$E_a = 10 \text{ kcal/mol} \left(1 + \frac{-2 \text{ kcal/mol}}{4(10 \text{ kcal/mol})} \right)^2 = 9.0 \text{ kcal/mol}$$

which is the same as the Blowers–Masel approximation.

11.8 SUGGESTIONS FOR FURTHER READING

Good books on quantum chemistry include:

I. N. Levine, *Quantum Chemistry*, 4th ed., Prentice-Hall, Englewood Cliffs, NJ, 1991.

A. Szabo and N. S. Ostlund, *Modern Quantum Chemistry: Introduction to Advanced Electronic Structure Theory*, Dover, New York, 1996.

A good introduction to practical ab initio methods is:

J. B. Foresman and A. E. Frisch, *Exploring Chemistry with Electronic Structure Methods*, 2nd ed., Gaussian, Pittsburgh, 1996.

G. Grant and W. Graham Richards, *Computational Chemistry*, Clarendon Press, Oxford, UK, 1995.

F. Jensen, *Introduction to Computational Chemistry*, Wiley, New York, 1999.

11.9 PROBLEMS

11.1 Define the following terms:

(a) Brønsted catalysis law

(b) Polanyi relationship

(c) Linear free-energy relationship

(d) Transfer coefficient

(e) Tafel plot

(f) Hammett constant

(g) Antisymmetrizer

(h) Coulomb energy

(i) SCF method

(j) Variational principle

(k) Correlation energy

(l) Exchange

(m) Exchange energy

(n) Hartree–Fock method

(o) Correlation interaction

(p) MP2

(q) Density functional theory

(r) Basis set

(s) Contracted basis set

(t) Diffuse function

(u) Polarization function

(v) 6-311+G(2df,2p)

11.2 Explain the importance of electron exchange in everyday life.

(a) If there were no electron exchange, would electrons pair up to form bonds? (*Hint:* What would the interaction energy be in the absence of exchange?)

(b) How does electron exchange allow chemical bonds to form?

(c) Given your results in (a) and (b), would you be alive if there were no electron exchange?

11.3 The reaction

$$CH_3OH + 4OH^- \longrightarrow CO + 4H_2O + 4 \text{ electrons}$$

is being run in electrochemical solution.

(a) Estimate the percentage increase in rate when you increase the potential from 0.2 volt to 1 volt.

(b) Estimate the percentage decrease in rate when you increase the potential from 0.2 volt to 0.02 volt. Be sure to consider the fact that the reaction is reversible.

Assume a transfer coefficient of 0.5 in all cases.

11.4 Agusti et al. Journal of Physical Chemistry. A 102(1998)10723 measured the rate of iron-catalyzed oxidation of a series of substituted benzenes. Data are given in Table P11.4.

(a) How well do the data follow a linear free-energy relationship?

(b) Which σ scale works best?

(c) What do your findings tell you about the transition state for the reaction?

(d) On the basis of your findings, predict the rate constant for R = NH_2.

11.5   Cencione et al. Journal of the Chemical Society-Faraday Transactions. 94(1998) 2933 measured the rate of the S_N2 reaction between a phosphate group and a series of substituted benzenes. Data are given in Table P11.5.

(a) How well do the data follow a linear free-energy relationship?

(b) Which σ scale works best?

(c) What do your findings tell you about the transition state for the reaction?

(d) On the basis of your findings, predict the rate constant for R = NH_2.

11.6   Myeloperoxidase is an enzyme used to oxidize various toxins, and to produce natural disinfectants. Burner et al. Journal of Biological Chemistry. 274(1999) 9494 measured the rate of myeloperoxidase-catalyzed oxidation of a series of substituted benzoic acid hydrazides. Data are given in Table P11.6

(a) How well do the data follow a linear free-energy relationship? Assume that the substituent constants for benzoic acid hydrazines are the same as for Benzoic acid.

(b) Which σ scale works best?

(c) What do your findings tell you about the transition state for the reaction?

(d) On the basis of your findings, predict the rate constant for R = NH_2.

11.7   Estimate the activation barriers for the following gas-phase reactions using (1) the Polanyi relationship, (2) the Seminov relationship, (3) the Blowers–Masel approximation, (4) the Marcus equation, heat of formation data are given in Table P11.7.

(a) $OH + CH_3CH_2CH_3 \rightarrow CH_2CH_2CH_3 + H_2O$

(b) $O + O_3 \rightarrow 2O_2$

Table P11.4   The rate of iron-catalyzed oxidation of a series of substituted benzenes (RC_6H_5)

Substituent	Cl	Br	H	CH_3	OCH_3	OH
k, minutes^{-1}	8.20	7.40	5.30	3.10	2.70	1.80

Source: Data of Agusti et al. Journal of Physical Chemistry A. 102(1998)10723

Table P11.5   The rate of the S_N2 reaction between a phosphate and a series of substituted benzenes (RC_6H_5).

Substituent	OH	F	Cl	Br	I	NC	t-Butyl	OCH_3
k, liter/(mol·second)	5.3×10^8	7×10^7	6.9×10^6	4×10^6	1.2×10^7	6.5×10^6	1×10^8	4.6×10^7

Source: Data of Cencione et al. Journal of the Chemical Society-Faraday Transactions. 94(1998) 2933

Table P11.6 The rate of myeloperoxidase-catalyzed oxidation of a series of substituted benzoic acid hydrazides $R(C_6H_3)(NH_2)(COOH)$

Substituent	p-NH_3^+	p-OH	p-OCH_3	H	p-Cl	m-OCH_3	m-NO_2	p-NO_2
k, 10^{-6} liter/(mol·second)	645.8	149.35	27.1	17.4	9.51	1.18	1.80	1.27

Note: m = meta, p = para.
Source: Data of Burner et al. Journal of Biological Chemistry. 274(1999), 9494

Table P11.7 Standard heats of formation of some species

Molecule	Heat of Formation, kcal/mol	Molecule	Heat of Formation, kcal/mol	Molecule	Heat of Formation, kcal/mol
O	+59.55	CH_3	+34.82	$CH_3CH_2CH_3R$	−25.02
O_3	+34.10	C_2H_4	+12.54	$CH_2CH_2CH_3$	+23.9
OH	+9.32	C_2H_5	+28.4	CH_3CHCH_3	+22.0
H	+52.10	C_2H_6	−20.04	CH_4	−17.89
CH_3OOCH_3	−30.0	CH_3O	+4.1	CH_2OH	−2.0

11.8 Castro et al. Journal of Organic Chemistry. 64(1999),2310 examined the reaction between 4-nitro-phenyl-O-ethyl thiolcarbonate and a series of substituted phenoxy anions. Data are given in Table P11.8.

(a) How well do these data follow the Brønsted relationship?

(b) How well do these data fit a Hammett plot? Assume that the substituent constants for pentafluoro are 5 times those of parafluoro.

(c) On the basis of these data, estimate the rate constant for the reaction with $R = p$-OH.

11.9 In 1934, Rice and Herzfeld proposed that the pyrolysis of ethane follows the following mechanism:

$$C_2H_6 \xrightarrow{\ 1\ } 2CH_3$$

$$CH_3 + C_2H_6 \xrightarrow{\ 2\ } CH_4 + C_2H_5$$

$$C_2H_5 \xrightarrow{\ 3\ } C_2H_4 + H$$

$$H + C_2H_6 \xrightarrow{\ 4\ } H_2 + C_2H_5$$

Table P11.8 Rate constant for the reaction 4-nitro-phenyl-O-ethyl thiolcarbonate and a series of substituted phenoxy anions ($R-C_6H_4COO^-$)

R	p-CH_3O	p-H	p-Cl	p-CN	Pentafluoro
k, mol/liter·second	1060	135	121	4.8	0.94
$pK_a = ln(K_a)$	10.3	9.9	9.4	7.8	5.3

$$2CH_3 \xrightarrow{5} C_2H_6$$

$$2H \xrightarrow{6} H_2$$

$$CH_3 + C_2H_5 \xrightarrow{7} C_3H_8$$

(a) Which steps need collision partners?

(b) Estimate the heat of reaction for each of the steps using the data in Table P11.7.

(c) Use the Polanyi relationship to estimate the activation barriers for each of the reactions in the mechanism.

(d) Use the Blowers–Masel approximation to estimate the activation barriers for each of the reactions in the mechanism.

(e) Use the Marcus equation to estimate the activation barriers for each of the reactions in the mechanism.

(f) On the basis of your results, would any step be rate-determining?

(g) Estimate a rate constant for each step at 1000 K, assuming a typical value of the preexponential.

11.10   Consider the gas-phase reaction

$$CH_3OH \Longrightarrow CO + 2H_2 + \text{other products}$$

(a) Propose a feasible mechanism for the reaction.

(b) Estimate the heat of reaction for each of the steps using the data in Table P11.7.

(c) Use the Polanyi relationship to estimate the activation barriers for each of the reactions in the mechanism.

(d) Use the Blowers–Masel approximation to estimate the activation barriers for each of the reactions in the mechanism.

(e) Use the Marcus equation to estimate the activation barriers for each of the reactions in the mechanism.

(f) Estimate a rate constant for each step at 1000 K assuming a typical value of the preexponential.

11.11   The free-radical polymerization of ethylene is thought to go by the following mechanism:

$$CH_3OOCH_3 \xrightarrow{1} 2CH_3O$$

$$CH_3O + C_2H_4 \xrightarrow{2} CH_3OC_2H_4$$

$$CH_3OC_2H_4 + C_2H_4 \xrightarrow{3} CH_3O(C_2H_4)_2$$

$$CH_3O(C_2H_4)_n + C_2H_4 \xrightarrow{4} CH_3O(C_2H_4)_{n+1}$$

$$CH_3O(C_2H_4)_n + CH_3O(C_2H_4)_m \xrightarrow{5} CH_3O(C_2H_4)_{n+m}OCH_3$$

$$CH_3O(C_2H_4)_n + CH_3O(C_2H_4)_mCH_2CH_2(C_2H_4)_p \xrightarrow{6}$$

$$CH_3O(C_2H_4)_nH + CH_3O(C_2H_4)_mCHCH_2(C_2H_4)_p$$

(a) Estimate approximate heats of reaction for each of the steps using the data in Table P11.7. Guess any unknown values on the basis of data on analogous molecules.

(b) Use the Polanyi relationship to estimate the activation barriers for each of the reactions in the mechanism.

(c) Use the Blowers–Masel approximation to estimate the activation barriers for each of the reactions in the mechanism.

(d) Use the Marcus equation to estimate the activation barriers for each of the reactions in the mechanism.

(e) On the basis of your results, would any step be rate-determining?

11.12 Zhang and Brauman Journal of the American Chemical Society. 121(1999)2508 examined the intrinsic barriers for the exchange of chlorine in a series of substituted benzoyl chlorides (RC_6H_4COCl). Table P11.12 shows some of their results.

(a) How well do these results fit a Hammett plot?

(b) What is the significance of the changes in the intrinsic barrier?

(c) Zhang and Brauman also did benzyl chlorides. There the barrier was constant. Speculate why the intrinsic barrier is constant with the benzyl chlorides but not the benzoyl chlorides.

(d) How close are the barriers to those expected from the Blowers–Masel analysis?

(e) Are the barriers on average larger or smaller than those expected from the Blowers–Masel analysis?

(f) Speculate why Zhang and Brauman's results are different than those expected from the Blowers–Masel analysis. (*Hint:* Zhang and Brauman considered negative ions while Blowers and Masel considered radicals. Which terms would change in Blowers and Masel's analysis if ions were considered?)

11.13 In this chapter we suggested that you use a constant value of E_a^0 in the Blowers–Masel and Marcus analysis. In fact, E_a^0 varies. Describe the effects of the following factors in changing E_a^0:

Table P11.12   The intrinsic barriers for the exchange of chloride ions in a series of substituted benzoyl chlorides

R	p-CH$_3$	H	m-CH$_3$	m-OCH$_3$	m-F	m-CF$_3$
E_a^0, kcal/mol	12.0	11.0	11.3	12.6	9.9	9.3

Source: Data of Zhang and Brauman Journal of the American Chemical Society. 121(1999)2508

(a) Steric hindrances

(b) Orbital symmetry limitations

(c) Bond distortions

(d) Substituents (see Problem 11.12)

(e) Changes in the splitting between the HOMO and the LUMO (the highest occupied and lowest unoccupied molecular orbitals).

11.14 Chizanowski and Wiekowski, Langmuir 14 (1998) 1967 examined the reaction

$$CH_3OH + 4OH^- \longrightarrow CO + 4H_2O + 4 \text{ electrons}$$

in a methanol fuel cell. They found that the reaction was an order of magnitude faster on a "Pt(111)" electrode than on a "Pt(110)" electrode.

(a) According to the equations in this chapter, how is it possible for the rate to be different on two such similar catalysts? What terms could change in the equations?

(b) What experiments could you do to tell if the increase in rate was due to a change in the preexponential or a change in the activation barrier?

11.15 The objective of this problem is to use your favorite quantum mechanical code to calculate the energy of methane

(a) Calculate the energy using HF/6-31G**

(b) Calculate the energy MP2/6-31G**

(c) Calculate the energy using CCSD/6-31G**

(d) How large is the correlation energy in case (c)?

(e) How good is the MP2 estimate of the correlation energy?

(f) How does the correlation energy compare to the total energy on a percentage basis?

(g) How large is the correlation energy on an absolute basis (e.g., kcal/mol)?

11.16 Consider the reaction

$$OH + CH_4 \longrightarrow H_2O + CH_3$$

(a) Estimate the heat of reaction using experimental heats of formation in Table P11.7.

(b) Estimate the heat of reaction using Benson's methods (Chapter 6).

(c) Use your favorite quantum-mechanical code to calculate the heat of reaction at the PMP2/6-31G** level.

(d) Use your favorite quantum-mechanical code to calculate the heat of reaction at the CCSD(T)/6-31G** level.

(e) Use your favorite quantum-mechanical code to calculate the heat of reaction at the G3 level. If the G3 level is not implemented on your computer try G2 or G4 or some other similar composite method.

(f) Compare your answers. What do you conclude?

11.17 Consider the reaction

$$OH + CH_4 \longrightarrow H_2O + CH_3$$

(a) Use the approximation in Section 11.5 to guess a transition state geometry.

(b) Use your favorite quantum-mechanical code to calculate the energy of the transition state at the PMP2/6-31G* level.

(c) Try optimizing the geometry to see if your energy changes.

(d) Calculate the energy of the reactants at the PMP2/6-31G* level.

(e) Calculate the activation barrier for the reaction.

(f) How does your calculation compare to the experimental value of 5.0 kcal/mol?

(g) Try using the Blowers–Masel approximation to estimate the activation barrier. Calculate the heat of reaction using the data in Table P11.7.

11.18 Koh et al. Journal of Organic Chemistry. 63(1998) 9834 studied the reaction between a series of substituted phenyl chloroformates and a series of substituted pyridines.

(a) Read the paper and report on the findings.

(b) What is the significance of the linear Brønsted plot?

(c) What is the significance of the linear Hammett plot?

(d) What is the significance of the fact that Hammett plot is linear on the σ_p^- scale?

11.19 Use the material in this chapter to calculate an equation for

(a) The properly antisymmetrized orbitals of H_2. To simplify the problem, consider only a 1s orbital on each hydrogen.

(b) The properly antisymmetrized orbitals of F_2. To simplify the problem, consider only the 1s, 2s and 2p orbitals on each fluorine.

11.20 Abraham et al. Archives Of Toxicology, 72 (1998) 227, J. Chemical Soc, Perkin Trans II 11(1998) 2405, examine the "nasal pungency" of a number of organic molecules. The nasal pungency is the minimum amount that you can smell. Abraham et al. note that smelling is a process where the organic molecules dissolve and then react in your nose. The rate of the reaction should follow a linear free-energy relationship.

(a) Read the papers. What do the authors say about the form of the free-energy relationship.

(b) How well do the data fit the Hammett relationship?

(c) From your results in (b), what can you say about your relative ability to smell organic molecules? Is the process solubility-limited or reaction-limited?

12

INTRODUCTION TO CATALYSIS

PRÉCIS

In the last few chapters we have been discussing principles of reactions, with a particular emphasis on reactions in the gas phase. However, now we will be changing topics and start to discuss how one can change the rate of a reaction by changing the chemical environment where the reaction occurs. In particular, we will discuss how one can use a catalyst or solvent to promote a desired reaction. Our approach will be a little bit nonstandard in that we start by asking what we can do to increase rates. We will develop the ideas into predictive tools later in this book.

12.1 INTRODUCTION

So far we have been discussing how reactions occur when the reactants are isolated from their environment. However, in most cases one does not run reactions that way. Instead, one runs the reaction in solution and/or in the presence of a catalyst. The solvent and catalyst are active participants in the reaction. If you run the reaction in the wrong solvent, you will not get as much desired product. Usually the solvent and catalysts are carefully chosen to maximize the rate of production of some desired product or to eliminate some side reaction. In the remainder of this chapter, we will give an overview of how catalysts or solvents work. Additional details will be given in later chapters.

Let's start with catalysts.

> Ostwald defined a catalyst as a substance that changes the rate of reaction but that is not itself consumed in the process.

When students read Ostwald's definition, students often think that the catalysts are not active participants in the reaction. However, in reality, the catalysts are active participants in the reaction. The catalyst usually reacts with the reactants to form a stable complex:

$$\text{Reactants} + \text{catalyst} \Longrightarrow \text{complex}$$

Then the complex rearranges to yield products and regenerate the catalyst:

$$\text{Complex} \Longrightarrow \text{products} + \text{catalyst}$$

Notice that the catalyst is regenerated at the end of the reaction so there is no net consumption of catalyst.

One way to think about a catalytic reaction is to consider the catalyst to be analogous to a printing press. A printing press takes in reactants, paper and ink, goes through a cycle of steps, and produces a product: a printed page. The printing press is not changed during the process. In the same way a catalyst takes in reactants goes through a series of steps and produces products. The catalyst is not changed in the process.

People often talk about catalytic reactions in terms of a catalytic cycle. The catalytic cycle is the sequence of steps that the system goes through, starting with reactants and ending with products. For example, Figure 12.1 shows a catalytic cycle for the formation of acetic acid via the Monsanto process:

$$\text{CH}_3\text{OH} + \text{CO} \Longrightarrow \text{CH}_3\text{COOH} \qquad (12.1)$$

There are two catalysts: HI and $[\text{Rh(CO)}_2\text{I}_2]^-$. Both catalysts are active participants in the reaction. The HI reacts with the alcohol in the first step of the reaction. The $[\text{Rh(CO}_2)\text{I}_2]^-$ reacts with the product of the HI reaction. Still, both catalysts cycle through the reaction and are regenerated. The catalysts are not consumed even though the catalysts are active participants in the reaction.

Most catalytic reactions look like the reaction in Figure 12.1. The catalyst participates in the reaction, but the catalyst is regenerated, so the catalyst is not consumed in the reaction.

Table 12.1 lists the catalysts used to promote some common reactions. The rate enhancement seen under typical conditions is also included in the table. Rates typically increase by factors of $10^3 - 10^9$ with gas-phase catalysts. Solid- or liquid-phase catalysts can often increase rates by factors of $10^{10} - 10^{20}$, and there are cases where rate enhancements as large as 10^{40} are seen. Clearly, these effects are large enough that they need to be understood.

Table 12.1   The rate enhancement of a number of reactions in the presence of a catalyst

Reaction	Catalyst	Rate Enhancement	Temperature, K
Ortho $\text{H}_2 \Longrightarrow$ para H_2	Pt (solid)	10^{40}	300
$2\text{NH}_3 \Longrightarrow \text{N}_2 + 3\text{H}_2$	Mo (solid)	10^{20}	600
$\text{C}_2\text{H}_4 + \text{H}_2 \Longrightarrow \text{C}_2\text{H}_6$	Pt (solid)	10^{42}	300
$\text{H}_2 + \text{Br}_2 \Longrightarrow 2\text{HBr}$	Pt (solid)	1×10^8	300
$2\text{NO} + 2\text{H}_2 \Longrightarrow \text{N}_2 + 2\text{H}_2\text{O}$	Ru (solid)	3×10^{16}	500
$\text{CH}_3\text{COH} \Longrightarrow \text{CH}_4 + \text{CO}$	I_2 (gas)	4×10^6	500
$\text{CH}_3\text{CH}_3 \Longrightarrow \text{C}_2\text{H}_4 + \text{H}_2$	NO_2 (gas)	1×10^9	750
$(\text{CH}_3)_3\text{COH} \Longrightarrow (\text{CH}_3)_2\text{CH}_2\text{CH}_2 + \text{H}_2\text{O}$	HBr (gas)	3×10^8	750

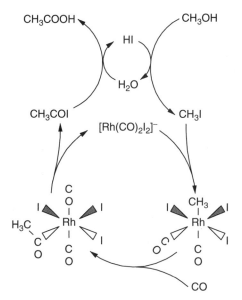

Figure 12.1 A schematic of the catalytic cycle for acetic acid production via the Monsanto process. The arrows in this figure are as described in Chapter 5.

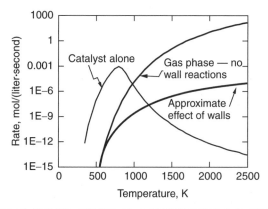

Figure 12.2 The rate of hydrogen oxidation on a platinum-coated pore calculated with (a) only heterogeneous (catalytic) reactions, (b) only radical reactions, and (c) combined radical–homogeneous reactions.

Another key experimental observation is that catalysts do not work over a wide range of conditions. For example, Figure 12.2 shows some rates of reaction calculated for hydrogen oxidation in a small pore. Notice that the rate of reaction reaches a maximum at an intermediate temperature, and then declines. The catalyst is ineffective at high temperature.

Interestingly, at very high temperatures, the gas-phase reactions are much faster than the catalytic reactions. At the highest temperatures, the catalyst slows down the reaction by promoting termination reactions.

12.1.1 Classes of Catalysts

Next we want to change topics and discuss the various types of catalysts. There are two broad classes of catalysts: **homogeneous catalysts** and **heterogeneous catalysts**. Homogeneous catalysts are substances you add to a reacting phase to speed up a given reaction, while heterogeneous catalysts act at the boundary of a phase to promote reactions.

In the next few sections we will list some of the key catalyst systems. The sections are written to provide a general overview of the kinds of materials that are catalytically active and not to teach people why the catalysts work. When I teach the course, my students often find the breadth of the material overwhelming. However, I usually tell students to just memorize the material, particularly the material in the tables, and then reread the chapter.

12.2 OVERVIEW OF HOMOGENEOUS CATALYSTS

Let's start with homogeneous catalysts. Homogeneous catalysts are substances that are added to a reacting phase to speed up some desired reaction. Examples of homogeneous catalysts include.

- Acids or bases
- Metal salts
- Enzymes
- Radical initiators
- Solvents

In the next several sections we will list several examples of acid catalysts, metal catalysts, and enzyme and radical initiators. The lists are not meant to be complete. However, we wanted to give the reader a picture of the wide variety of materials that show catalytic properties.

12.2.1 Acids and Bases

Table 12.2 lists a number of reactions commonly catalyzed by acids or bases. Acid catalysis includes acids such as HF or H_2SO_4 and organic acids such as acetic acid or triflouroacetic acid. Basic catalysts include compounds such as sodium hydroxide. Generally any strong acid or strong base can be used to catalyze reactions.

Acids and bases act by interacting with various hydrocarbon species to form **carbocations**. Recall that a carbocation is a hydrocarbon ion with a positive charge. Carbocations are very reactive species. The production of the very reactive species allows the reaction to occur at an enhanced rate. For example, during the acid-catalyzed alkylation reaction

$$\text{Benzene} + \text{ethylene} \Longrightarrow \text{ethylbenzene} \qquad (12.2)$$

a proton reacts with the ethylene to form an ethyl ion:

$$H^+ + CH_2CH_2 \longrightarrow [CH_3CH_2]^+ \qquad (12.3)$$

Table 12.2 Some reactions commonly catalyzed by acids and bases

Reaction	Example	Typical Application
Isomerization (rearranging the structure of a molecule)	$CH_2=CHCH_2CH_3 \Rightarrow$ $CH_3CH=CHCH_3$	Octane enhancement, monomer production, paraxylene production,
Alkylation (making very small molecules into a bigger molecule)	$CH_3CH=CHCH_3 +$ $CH_3CH_2CH_2CH_3 \Rightarrow$ $(CH_3CH_2)CH(CH_3)(C_4H_9)$	Pharmaceutical production, monomer production Fine chemicals; butane + olefin $\Rightarrow$ octane
Cracking (taking a big molecule and breaking it into two smaller ones)	$C_{12}H_{24} \Rightarrow C_7H_{14} + C_5H_{10}$	Crude-oil conversion, digestion
Esterfication (attaching an acid to a base eliminating water)	$CH_3CH_2OH + CH_3COOH \Rightarrow$ $CH_3COOCH_2CH_3 + H_2O$	Soap production, fragrance production
Aldol condensation reactions (combining two aldehydes by eliminating water)	$2CH_3CH_2CH_2CHO \Rightarrow$ $CH_3CH_2CH_2CH =$ $C(CHO)CH_2CH_3 + H_2O$	Fine chemicals pharmaceutical production
Alcohol dehydration (removing a hydrogen and an OH from an alcohol, producing a double bond)	$CH_3CH_2OH \Rightarrow CH_2 =$ $CH_2 + H_2O$	Alternative fuels
Cationic polymerization	Propylene $\Rightarrow$ polypropylene	Polymer production

The ethyl ion reacts with benzene to yield an ethylbenzene ion:

$$[CH_3CH_2]^+ + C_6H_6 \longrightarrow [CH_3CH_2C_6H_6]^+ \tag{12.4}$$

Then the ethylbenzene ion loses a proton:

$$[CH_3CH_2C_6H_6]^+ \longrightarrow CH_3CH_2C_6H_5 + H^+ \tag{12.5}$$

Reaction (12.2) has a significant Pauli repulsion, but reactions (12.3)–(12.5) do not. The absence of Pauli repulsions makes acid-catalyzed reactions much quicker than the corresponding uncatalyzed reactions.

We discussed some of the reactions of carbocations in Chapter 5. Most industrial carbocation reactions are catalyzed by acids or bases.

12.2.2 Metal Atoms

Metal atoms can also be used as homogeneous catalysts. Table 12.3 shows the most important examples. Generally the transition metal catalysts work by binding to the intermediates of a reaction, thereby increasing the concentration of the key intermediates. For example, the Wilkinson catalyst, $Rh(P(C_6H_5)_3)_3Cl$, catalyzes the hydrogenation of ethylene. The reaction occurs via the following steps:

$$H_2 + 2S \longrightarrow 2H(ad) \tag{12.6}$$

$$C_2H_4 + S \longrightarrow C_2H_4(ad) \tag{12.7}$$

$$C_2H_4(ad) + H(ad) \longrightarrow C_2H_5(ad) + S \tag{12.8}$$

$$C_2H_5(ad) + H(ad) \longrightarrow C_2H_6 + 2S \tag{12.9}$$

where S is an empty site on the rhodium cluster, and the (ad)s denote species attached to the cluster. In this case the cluster is stabilizing hydrogen atoms and ethyl groups. In Chapter 4, we showed that the rate of a cyclic reaction is proportional to the concentration of the intermediates of the reaction. In fact, the Wilkinson catalyst stabilizes

Table 12.3   Examples of reactions catalyzed by homogeneous transition metal catalysts

Reaction	Catalyst
Olefin polymerization	$[TiCl_2(C_5H_5)_2]^{2+}$ or $TiCl_2/Al(C_2H_5)_3$ (Ziegler–Natta catalyst)
Olefin hydrogenation	$Rh(P(C_6H_5)_3)_3Cl$ (Wilkinson catalyst)
$C_2H_4 + H_2O \rightarrow$ acetaldehyde (Wacker process)	$PdCl_2(OH)_2$
$C_2H_4 + H_2 + CO \rightarrow$ propylaldehyde (hydroformylation)	$HCo(CO)_4$
$CH_3OH + CO \rightarrow CH_3COOH$ (monsanto carbonylation process)	$[Rh(CO)_2I_2]^{1-}$
$H_2O_2 + CH_3CH_2OH \rightarrow CH_3CHO + 2H_2O$	Fe^{2+}

Table 12.4   Some examples of enzymes listed in the Brookhaven National Labs' protein

Oxidoreductases[a]		Transferases[b]		Hydrolases[c]	
NADH peroxidase (oxidizes NADH with peroxides)	NADH + $H_2O_2 \Rightarrow$ NAD(+) + $2H_2O$	Dimethylallyl-*cis*-transferase (transfers dimethylallyl groups)	Dimethylallyl diphosphate + isopentenyl diphosphate $\Rightarrow$ diphosphate + dimethylallyl-*cis*-isopentenyldiphosphate	Carboxylesterase (promotes hydrolysis of ester linkages)	A carboxylic ester + H_2O $\Rightarrow$ an alcohol + a carboxylic anion
Ferroxidase (oxidizes Iron)	$4Fe^{2+} + 4H^+ + O_2 \Rightarrow$ $4Fe^{3+} + 2H_2O$	Glycoaldehyde transferase (transfers glucoaldeydes); also called transketolase	Sedoheptulose 7-phosphate + D-glyceraldehyde 3-phosphate $\Rightarrow$ D-ribose 5-phosphate + D-xylulose 5-phosphate	1,4-D-Glucan glucanohydrolase (also called-amylase)	Hydrolysis of 1,4-glucosidic linkages in oligosaccharides and polyasaccharides
Glucose oxidase (oxidizes glucose)	β-D-Glucose + $O_2 \Rightarrow$ D-glucono-1,5-lactone + H_2O_2	Alanine aminotransferase (transfer amino groups from alanine)	L-Alanine + 2-oxoglutarate $\Rightarrow$ pyruvate + L-glutamate	Interleukin 1-β-converting enzyme	Release of interleukin 1-β by specific hydrolysis at 116-Asp-|-Ala-117 and 27-Asp-|-Gly-28 bonds

[a] Promote oxidation reduction reactions.
[b] Promote transfer of functional groups.
[c] Promote hydrolysis/cleavage reactions.
[d] Promote addition of CO_2, H_2O and NH_3 to double bonds or formations of double bonds via elimination of CO_2, H_2O or NH_3.
[e] Promote isomerization reactions.
[f] Promote bond formation; generally used to catalyze entothermic reactions requiring ATP.

the intermediate. That raises the concentration of the intermediates. As a result, the rate of reaction is tremendously enhanced.

The mechanism of metal catalysis will be discussed in detail in Chapter 14. However, the thing to remember for now is that metal atoms stabilize radical intermediates. The stabilization speeds up rates.

Industrially, hydroformulation is the largest application of homogeneous catalysts. The Monsanto process is also the main route that people use to produce acetic acid. Ziegler–Natta catalysts are used to produce polyethylene. There are many other examples of reactions catalyzed by transition metal compounds in solution; however, Table 12.3 lists the main reactions used commercially.

12.2.3 Enzymes

Enzymes are another very important class of catalysts. Enzymes are complicated proteins. They work principally by four different routes:

database

Lyases[d]		Isomerases[e]		Ligases[f]	
Carbonate dehydratase (dehydrates carbonates)	$H_2CO_3 \leftrightarrow CO_2 + H_2O$	Maleate isomerase (promotes *cis-trans* isomerization of Maleate)	Maleate $\Rightarrow$ fumarate	Leucine–t-RNA ligase	ATP + L-leucine + t-RNA(leu) $\Rightarrow$ AMP + diphosphate + L-leucyl-t-RNA(leu).
Citrate dehydratase	Citrate $\leftrightarrow$ *cis*-aconitate + H_2O	Cholestenol δ•-isomerase	5-α-Cholest-7-en-3-β-ol $\Rightarrow$ 5-α-cholest-8-en-3-β-ol	Pyruvate carboxylase	ATP + pyruvate + (HCO_3) $\Rightarrow$ ADP + phosphate + oxaloacetate
Pyruvate decarboxylase	A 2-oxo acid $\leftrightarrow$ an aldehyde + CO_2	Mannose isomerase	D-Mannose $\Rightarrow$ D-fructose	Aspartate–ammonia ligase	ATP + L-aspartate + $NH_3 \Rightarrow$ AMP + diphosphate + L-asparagine

- Enzymes bind to the reactants in such a way that key bonds in the reactants are stretched. That makes bonds easier to break.
- Enzymes lower the energy of the transition state. The analysis in Chapter 9 shows that lowering the transition state energy enhances the rate of reaction.
- Enzymes work like metals to stabilize key intermediates.
- When you have a bond-forming reaction, the enzymes push the reactants together. That also promotes reaction.

Table 12.4 lists several different enzymes. Basically, enzymes are classified according to what they do rather than how they are constructed. For example, NADH peroxidase is an enzyme used to react NADH (nicotinamide adenine dinucleotide) with hydrogen peroxide and superoxide ($H_2O_2^-$) where NADH is a key molecule in the energy cycle of cells. In Table 12.4 we list "NADH peroxidase" as though NADH peroxidase were a single enzyme. In fact, NADH peroxidase represents a whole family of enzymes. For example, horseradish NADH peroxidase is slightly different from mammalian NADH peroxidase.

Another important idea is that enzymes often work with "cofactors", which are additional substances needed to get the enzyme to work. So, for example, FAD (flavine adenine dinucleotide) is a cofactor for NADH peroxidase.

Table 12.4 lists six key classes of enzymes:

- Oxidoreductases
- Transferases
- Hydrolases
- Lyases
- Isomerases
- Ligases

Oxidoreductases are enzymes that promote oxidation–reduction reactions. Transferases are enzymes that promote transfer of functional groups from a donor molecule to an acceptor molecule. Hydrolases are enzymes that promote hydrolysis reactions (i.e, scission of bonds with additions of water). Lyases are enzymes that promote elimination of CO_2, H_2O, and NH_3 from organic molecules leading to the formation of double bonds, and the reverse reactions where water, CO_2, NH_3, are added to double bonds. Isomerases are enzymes that promote unimolecular isomerization reactions. Ligases are enzymes that promote endothermic bond-formation processes, consuming ATP in the process.

Enzymes are discussed in detail in elsewhere. See Faber (2000), Jencks (1987), Fersht (1999), or Sinnot (1997). The thing to remember for now is that there are many different enzymes, and they each perform a specific function.

Another key point is that if you want to find an enzyme to do a specific reaction, you can look in a standard table and find out whether a suitable enzyme exists. I usually look things up in the protein database at which in 2000 was at http://www.rcsb.org/pdb/ or http://www.chem.qmw.ac.uk/iubmb/enzyme/

12.2.4 Radical Initiators

Radical initiators represent a fourth important class of catalysts. Recall that in many reactions, the slow step is the initiation step where radicals are formed. If one can find a way to form the radicals more easily, the rate of reaction will be enhanced. Radical initiators are molecules that decompose into radicals very easily, possibly in the presence of light. The radicals then initiate the reaction.

For example, in Chapters 5 and 10 we considered the reaction

$$C_2H_6 \Longrightarrow C_2H_4 + H_2 \tag{12.10}$$

and found that the slow step was the initiation step where the C–C bond broke. Well, an iodine–iodine bond is much easier to break than a carbon–carbon bond. Consequently, the iodine can decompose at modest temperatures:

$$X + I_2 \longrightarrow 2I + X \tag{12.11}$$

Then the iodine can react with ethane to start the reaction:

$$I + CH_3CH_3 \longrightarrow HI + CH_2CH_3 \tag{12.12}$$

There are many variations of this. For example, the carbon–carbon bonds in acetaldehyde are much weaker than the carbon–carbon bonds in ethane. Consequently, acetaldehyde is a useful initiator for ethane dehydration.

Free-radical initiators are most important for polymerization reactions. Molecules like ethane are hard to decompose into radicals, and you need radicals or ions to start free-radical polymerization. Consequently one adds a molecule that is easy to decompose, like benzoyl peroxide. The benzoyl peroxide decomposes into radicals:

$$R{-}O{-}O{-}R \longrightarrow 2\,RO\bullet \tag{12.13}$$

Then the radical reacts with the ethylene to start the polymerization process:

$$RO\bullet + CH_2CH_2 \longrightarrow ROCH_2CH_2\bullet \tag{12.14}$$

Free-radical initiators tend to give polymers with varying properties because of chain transfer reactions, so they are less useful than the transition metal polymerization catalysts. Still, the catalysts are used when one wants a soft material.

Free-radical processes are also important to atmospheric chemistry. Chlorine atoms produced from photolysis of chlorocarbons catalyze the destruction of ozone via the process:

$$Cl + O_3 \longrightarrow ClO + O_2 \tag{12.15}$$

The ClO can then react via a number of processes to reduce the ozone layer. One particular reaction is

$$ClO + O_3 \Longrightarrow Cl + 2O_2 \tag{12.16}$$

Others reactions also occur.

Table 12.5 gives several other examples of free-radical initiators. They are used extensively.

Table 12.5   Some examples of reactions initiated or catalyzed by free radicals and similar species

Reaction	Initiator	Reaction	Catalyst
Olefin polymerization	Peroxides, $(Ph)_3CC(Ph)_3$	$2SO_2 + O_2 \Rightarrow SO_3$ (lead chamber process)	NO/NO_2
Hydrocarbon dehydrogenation	Iodine, NO_2, chlorine atoms	Ozone depletion	Cl
Hydrocarbon oxidations	$[(CH_3CH_2)_4N][I]$, $[(CH_3CH_2)_4N]$ $[C_6H_5COO]$	—	—

Table 12.6   The rate of reaction (12.17) in several solvents[a]

Solvent	Rate constant, liter/(mol·second)
Gas phase	$\approx 10^{-45}$
Water	3.5×10^{-5}
Methol	3×10^{-6}
Methyl cyanide	0.13
DMF	2.5

[a] All measurements have been extrapolated to 25°C.

12.2.5   Solvents

There is one other class of homogeneous catalysts that is usually discussed separately from all other types of homogeneous catalysts: solvents. Solvents can act just like catalysts. They speed up reactions and change selectivities. For example, Table 12.6 shows some data for the rate of the reaction

$$CH_3I + NaCl \longrightarrow CH_3Cl + NaI \qquad (12.17)$$

Notice that the rate in DMF is a factor of 10^6 larger than that in water, and 10^{45} larger than that in the gas phase. Clearly DMF is acting like a catalyst.

In the literature, people seldom discuss solvents as catalysts, but solvents can act just like catalysts. Generally solvents speed up rates of ionic reactions by factors of 10^{30} or more. Increases in rates of nonionic reactions are more unusual. Solvents are less able to selectively catalyze reactions than are the other types of homogeneous catalysis. Still, many solvents show considerable catalytic activity. Therefore I consider solvents to be a class of homogeneous catalysts.

12.3   INTRODUCTION TO HETEROGENEOUS CATALYSTS

At this point, we will be changing topics. So far we have been surveying homogeneous catalysts; however, now we move on and start to discuss heterogeneous catalysts.

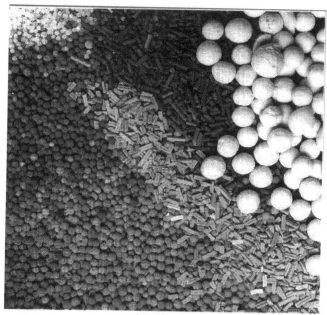

Figure 12.3 Pictures of some heterogenous catalysts. [From Wijngaarden and Westerterp (1991) With permission.]

Figure 12.3 shows pictures of some heterogeneous catalysts. Heterogeneous catalysts can be pellets, powders or other solids.

Heterogeneous catalysts are quite similar chemically to homogeneous catalysts. They are substances that are added to a reacting mixture to speed up a reaction. However, while homogeneous catalysts are generally substances that dissolve into the reacting mixture, heterogeneous catalysts are solids that do not dissolve. Instead, reaction occurs at or near the interface between the solid and the reactant mixture.

The fact that the reaction occurs at an interface means that reactants have to diffuse to the interface before the reaction can occur. Consequently, heterogeneously catalyzed reactions are often a little slower than homogeneously catalyzed ones.

Still, industrially, heterogeneous catalysts are almost always preferred over homogeneous catalysts. Recall that with a homogeneous catalyst, the catalyst is added to the reacting phase. When the reaction is completed, either the catalyst must be separated from the products, or you end up losing the catalyst. Many catalysts contain expensive metals, such as rhodium. You would not want to throw rhodium away. Other catalysts contain acids; the acids are disposal problems. In contrast, a heterogeneous catalyst is a solid. The solid is easily separated from the reacting mixture. As a result, it is generally much easier to use a heterogeneous catalyst rather than a homogeneous catalyst in a chemical process.

Examples of heterogeneous catalysts include

- Supported metals
- Transition metal oxides and sulfides
- Solid acids and bases
- Immobilized enzymes and other polymer-bound species

Generally, heterogeneous catalysts are just like homogeneous catalysts except that the active species are attached to a solid. The solid allows the catalyst to be separated from the reaction mixture quite easily. The easy separation generally makes heterogeneous catalysts preferred over homogeneous catalysts in industrial practice.

In the next several sections we will list several examples of catalysts in each of these classes. The lists are not meant to be complete. However, we wanted to give the reader a picture of the wide variety of materials that show catalytic properties so that the reader will have a general idea about what is available.

12.3.1 Supported Metal Catalysts

The simplest heterogeneous catalyst is a supported metal catalyst. Figure 12.4 shows a diagram of a supported metal catalyst. The catalyst consists of a series of small platinum particles on an inert silica (SiO_2) support. Generally, the platinum provides most of the catalytic activity. However, the silica is there to both add mechanical strength to the catalyst, and to spread out the platinum over more surface area.

Supported metals are fundamentally similar to the homogeneous transition metal catalysts discussed in Section 12.2.2. The catalyst contains an active metal that does chemistry similar to that discussed in Section 12.2. However, the metal is attached to the support. The support spreads out the metal and holds the metal in place, so the catalyst is easy to separate from the reacting mixture. Supports tend to be high-surface-area materials: aluminum oxide, silicon dioxide, and activated carbon. Generally, most supported metal catalysts contain group VIII or group Ib metals, although other transition metals and rare earths also show catalytic activity.

Table 12.7 shows a selection of the reactions commonly run on supported metal catalysts. Generally metals are used to hydrogenate and dehydrogenate hydrocarbons, to oxidize substances, and to generally catalyze hydrocarbon conversions. Supported metal catalysts are used in the catalytic converter in your car and your wood stove. People

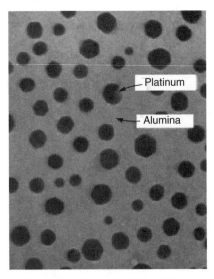

Figure 12.4 A supported metal catalyst.

Table 12.7   A selection of the reactions catalyzed by supported metals

Reaction	Catalyst	Reaction	Catalyst
Hydrocarbon hydrogenation, Dehydrogenation	Pt, Pd, Ni	$CO + H_2 \Rightarrow$ hydrocarbons (Fischer–Tropsch)	Fe, Rh
CO oxidation, total oxidation of hydrocarbons	Pt, Pd, Cu, Ni, Fe, Rh, Ru	Steam reforming for production of hydrogen	Ni plus additives
$CO + 2H_2 \Rightarrow CH_3OH$	Cu/ZnO	Reforming (isomerization of oil)	Pt/Re/Al2O3
$2CO + 2NO \Rightarrow 2CO_2 + N_2$	Pt, Rh, Ru (catalytic converter)	$2NH_3 + O_2 \Rightarrow N_2O_5 + 3H_2O$	Pt
$N_2 + 3H_2 \Rightarrow 2NH_3$	Fe, Ru, Rh	Alcohols $+ O_2 \Rightarrow$ aldehydes $+ H_2O$, e.g., $2CH_3OH + O_2 \Rightarrow 2H_2CO + H_2O$	Ag, Cu
$2\ C_2H_4 + O_2 \Rightarrow$ 2-ethylene oxide	Ag, Cu	$R{-}R' + H_2 \Rightarrow RH + HR'$(hydrogenolysis)	Ni, Co, Rh, Ru

are starting to sell supported metal catalysts to clean the air in your home. Most major industrial chemicals are made with a supported metal or a metal oxide catalyst. Chapter 14 will be devoted to a detailed discussion of supported metal catalysts. However, the thing to remember for now is that virtually all reactions where you add or subtract an atom from a molecule are run on a supported metal catalysts or a metal oxide catalyst.

Supported metal catalysts work just like homogeneous transition metal catalysts. For example, platinum catalyses the hydrogenation of ethylene via the same series of steps as the Wilkinson catalyst:

$$H_2 + 2S \longrightarrow 2H(ad) \tag{12.18}$$

$$C_2H_4 + S \longrightarrow C_2H_4(ad) \tag{12.19}$$

$$C_2H_4(ad) + H(ad) \longrightarrow C_2H_5(ad) + S \tag{12.20}$$

$$C_2H_5(ad) + H(ad) \longrightarrow C_2H_6 + 2S \tag{12.21}$$

where S is an empty site on the surface of the platinum and the (ad) denote species attached to the surface of the platinum. In this case the platinum is stabilizing hydrogen atoms and ethyl groups. The stabilization tremendously speeds up the rate of reaction as discussed in Section 1.2.

12.3.2 Special Reactions on Supported Metal

Reactions (12.18)–(12.21) can occur on either a cluster or a supported metal catalyst. However, there are a few reactions that occur only on supported metal catalysts.

For example, the conversion of acetylene to benzene occurs readily on a palladium catalyst, but at the time this book was written, the reaction had not yet been seen on a cluster compound.

The overall reaction is

$$3C_2H_2 \Longrightarrow C_6H_6 \tag{12.22}$$

Physically, in order for reaction (12.22) to occur, the catalyst needs to simultaneously coordinate three acetylene molecules. That is easy on a metal surface because there are many atoms on the surface to hold the acetylene. In contrast, the reaction is difficult on a single transition metal atom in solution. I expect that as people begin to make larger cluster compounds, they will observe reaction (12.22). However, at present, metal cluster compounds with multiple metal atoms are difficult to make. In contrast, it is easy to get a cluster of atoms on a supported metal catalyst.

The mechanism of metal catalysis will be discussed in detail in Chapter 14. However, the thing to remember for now is that like metal atoms, metal surfaces stabilize radical intermediates. In Chapter 4 we found that an increase in the intermediate can tremendously enhance the rate of a reaction.

12.3.3   Transition Metal Oxides, Nitrides, Sulfides, and Carbides

In Table 12.7, we listed reactions that are commonly run on pure metals. However, there are a number of reactions that are instead run on transition metal oxides or sulfides they are listed in Table 12.8. I usually think of a transition metal oxide, nitride, sulfide, or carbide catalyst as being a transition metal catalyst that has been poisoned (i.e., slowed down)

Table 12.8   A selection of the reactions catalyzed by transition metal oxides, nitrides, and sulfides

Reaction	Catalyst	Reaction	Catalyst
$2\,SO_2 + O_2 \Rightarrow 2\,SO_3$	V_2O_5	$CO + H_2O \Rightarrow CO_2 + H_2$ (water–gas shift)	FeO, CuO, ZnO
Hydrodesulfurization	CoS, MoS, WS	$2(CH_3)_3COH \Rightarrow$ $(CH_3)_3COC(CH_3)_3 +$ H_2O	TiO_2
$CH_3CH{=}CH_2 + O_2 \Rightarrow$ $CH_2{=}CHCHO +$ H_2O	$(Bi_2O_3)_x(MoO_3)_y$ (bismuth molybate) uranium antimonate	$2\,CH_3CH =$ $CH_2 + 3O_2 + 2NH_3 \Rightarrow$ $2CH_2{=}CHC{\equiv}N +$ $6H_2O$ (aminoxidation)	$(FeO)_x(Sb_2O_3)_y$
$4NH_3 + 4NO + O_2 \Rightarrow$ $4N_2 + 6H_2O$ (selective catalytic reduction)	V_2O_5, TiO_2	Benzene $+ O_2 \Rightarrow$ maleic anhydride $+$ water naphthylene $+$ $O_2 \Rightarrow$ phthalic anhydride $+$ water	$(V_2O_5)_x(PO_4)_y$
$CH_3CH_2(C_6H_5) +$ $O_2 \Rightarrow$ $CH_2{=}CH(C_6H_5) +$ H_2O (styrene production)	FeO	Selective oxidation of hydrocarbons	NiO, Fe_2O_3, V_2O_5, TiO_2, CuO, Co_3, O_4, MnO_2
Aromatiztion, e.g. heptane $\Rightarrow$ toluene H_2 or H_2O	Cr_2O_3/Al_2O_3	Hydrodenitrogenation	NiS, MoS

via addition of oxygen, nitrogen, carbon, or sulfur. The same types of chemistry occur on the metal–metal oxide, nitride, sulfide, or carbide catalyst as on the pure metal. However, the reaction is slower on the metal–metal oxide, nitride, sulfide, or carbide catalyst. In particular, side reactions are suppressed. As a result, metal oxide, nitride, sulfide, and carbide catalysts can sometimes be used to promote very selective transformations.

For example, let's assume that you want to design a system to selectively oxidize ethane to ethylene. You can oxidize ethane to ethylene over a platinum mesh catalyst provided you run the reaction at high temperatures (1000°C), and keep the residence time short (10^{-3} seconds or less). However, if you run the reaction longer, the ethylene is further oxidized to eventually produce CO_2 and water.

One can avoid the need for a short residence time by running the process on a metal oxide catalyst. Commercially, one normally runs the reaction on an iron oxide catalyst. The metal oxide is less reactive than platinum. The primary reaction to produce acetylene is slowed. However, the secondary reaction to produce CO_2 and H_2O is slowed to a greater extent. As a result, the iron catalyst can selectively oxidize the ethane to ethylene and not produce much CO_2.

Generally, the mechanism of the oxidation of ethylene is similar to the mechanism of the hydrogenation of ethylene [reactions (12.19), (12.20), and (12.21)], except that the reaction occurs in reverse. First the ethane dissociatively adsorbs, then the ethylene is dehydrogenated and then the product desorbs:

$$C_2H_6 + 2S \longrightarrow C_2H_5(ad) + H(ad) \tag{12.23}$$

$$C_2H_{5(ad)} + S \longrightarrow C_2H_{4(ad)} + H_{(ad)} \tag{12.24}$$

$$C_2H_{4(ad)} \longrightarrow C_2H_4 + S \tag{12.25}$$

Oxygen then reacts with the hydrogen to produce

$$2H_{ad} + O_{ad} \longrightarrow H_2O \tag{12.26}$$

All of these steps are the reverse of the steps mentioned in Section 12.3.1.

In the literature, it is unusual to find the mechanism written this way. The reason is that in a metal oxide catalyst, there are two types of sites: (1) metal sites, which I will call S_M sites; and (2) oxygen sites, which I will call S_{Ox} sites. Ethylene normally binds to a metal site while hydrogen prefers to bind to an oxide site. As a result one can rewrite the mechanism as:

$$C_2H_6 + S_M + S_{Ox} \longrightarrow C_2H_{5(ad,M)} + H_{(ad,Ox)} \tag{12.27}$$

$$C_2H_{5(ad,M)} + S_{Ox} \longrightarrow C_2H_{4(ad,M)} + H_{(ad,Ox)} \tag{12.28}$$

$$C_2H_{4(ad,M)} \longrightarrow S_M + C_2H_4 \tag{12.29}$$

where the notation$_{(ad,M)}$ is used to denote a species attached to a metal atom while, $H_{(ad,Ox)}$ denotes a hydrogen attached to a lattice oxygen. Notice that the $H_{(ad,Ox)}$ is basically a hydroxyl group. In Chapter 5, we found that hydroxyls can react with hydrocarbons like ethane via the reaction

$$OH + C_2H_6 \longrightarrow C_2H_5 + H_2O \tag{12.30}$$

The $H_{(ad,Ox)}$ can react in the same way:

$$H_{(ad,Ox)} + C_2H_6 \longrightarrow C_2H_5 + H_2O + \square \qquad (12.31)$$

where the $\square$ represents a lattice site that has been depleted of oxygen.

Reaction (12.31) depletes lattice oxygen, which is replaced via O_2 adsorption:

$$O_2 + 2\,\square \longrightarrow S_{Ox} \qquad (12.32)$$

This example illustrates the one major difference between reactions on transition metals and on transition metal oxides. On transition metals, all of the reacting species are adsorbed on the metal. With metal oxides, the lattice oxygen is a major player in the reaction.

People often think of reactions on metal oxides as oxidation reduction reactions, where metal is reduced, then reoxidized. Generally, there is more going on than just oxidation–reduction. However, oxidation–reduction is a key part of the catalytic cycle.

12.3.4   Solid Acids And Bases

In Section 12.3.2, we discussed transition metal oxides. Transition metal oxides have weak metal–oxygen bonds, and so oxygen can be easily exchanged. By comparison, the metal–oxygen bond in alumina is so strong that the bond is not easily broken. As a result, alumina is not a good oxidation catalyst. Still, there is some important chemistry that occurs on metal oxides with strong metal–oxygen bonds. The objective of this section is to give an overview of the key chemistry.

Generally, metal oxides with strong metal–oxygen bonds are acids or bases. For example, you know that H_2SO_4 is a strong acid. Well, H_2AlO_3 is a strong acid, too. $AlCl_3$ and $FeCl_3$ are strong Lewis acids. There are strong Lewis acid sites on Al_2O_3. Silica (SiO_2) can stabilize the acid sites, so silica aluminas are quite effective solid acid catalysts.

There are also special acids, like (SbF_5) and sulfated zirconia, which are much stronger acids than H_2SO_4. These compounds are also effective acid catalysts.

Na_2O and MgO are strong bases. They are used to catalyze basic reactions.

Table (12.9) shows several other examples. All of the materials in Table (12.9) are strong acids or bases.

Industrially, most acid-catalyzed reactions are run on a special class of solid acids called **zeolites**. Zeolites are highly porous silica/aluminas. The pore structure is very uniform. Generally, the structures consist of cages that are stacked one on top of another. Figure 12.5 shows one of the cages in a material called *faugasite*. The uniform pores

Table 12.9   Some common solid acids and bases

Material	Type	Material	Type
Silica/alumina	Solid acid	Mordenite	Zeolite
Alumina	Solid acid	ZSM-5	Zeolite
Y-zeolite faugasite	Zeolite	VFI	Large-pore zeolite
Sodalite	Zeolite	Offretite	Zeolite
HF–SbF$_5$	Superacid	HSO$_3$F	Superacid
$H_2[Ti_6O_4(SO_4)_4(OEt)_{10}]$	Superacid	Sulfated zirconia	Superacid
MgO	Solid base	Na$_2$O	Base

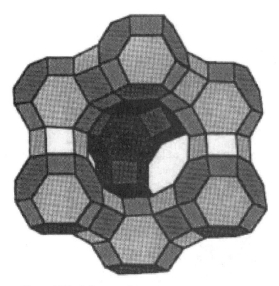

Figure 12.5 A diagram of the pore structure in faugasite.

allow people to design zeolites with special mass transfer properties that enhance the selectivities of reactions. This idea will be discussed in greater detail later in this chapter. Industrially, most of the acid-catalyzed reactions listed in Table 12.2 are actually run on solid acids. The one major exception is butane alkylation, which is still commonly catalyzed with hydrofluoric acid.

12.3.5 Polymer-Bound Catalysts

All of the heterogeneous catalysts we have discussed so far in this chapter were heterogeneous by their nature. The catalysts were solids. Solids have the advantage that they are easy to separate at the end of a reaction, so they are generally cheaper to use than are homogeneous catalysts. In the 1970s and 1980s there was a lot of effort devoted toward starting with a homogeneous catalyst, and anchoring the catalyst to a polymer to produce a material that is catalytically active and easy to separate from the products. For example, immobilized enzymes, that is, enzymes bound to polymer beads, were tried as reusable enzyme catalysts. People also tried bonding transition metal salts to polymers and using the resultant materials as catalysts. Acid groups were also bound to polymers. Acid-bound materials are currently being used as catalysts. When this book was being written, none of the other materials had made it into industrial practice. Still, people continue to discuss polymer bound catalysts in the literature, and so I decided to mention them.

12.3.6 Photocatalysts

There is one other class of catalysts that I did not have a place for anywhere else, so I decided to list separately: photocatalysts. Photocatalysts are catalysts that are inactive under normal conditions, but when the catalysts are irradiated with UV light, the catalysts become activated. At present, there are only three important photocatalysts:

- TiO_2 and $TiSrO_3$
- Fe_2O_3
- Mercury vapor

TiO_2 is commonly used as a catalyst for destruction of hydrocarbon wastes in aqueous environments. The main mode of catalytic action is for water to react with the O^{2-} in the titania surface to yield

$$O^{-2} + H_2O \longrightarrow 2OH^- \tag{12.33}$$

TiO_2 is a semiconductor. When light hits the semiconductor, the light interacts with the semiconductor to produce free electrons and holes (positive charges). The charges can promote reaction. The holes can then react with the OH^-s to yield OH radicals:

$$h^+ + OH^- \longrightarrow OH\bullet \tag{12.34}$$

where h^+ is a hole.

The OH radicals then oxidize hydrocarbons:

$$OH\bullet + C_6H_6 \longrightarrow C_6H_5 + H_2O \tag{12.35}$$

Electrons react with O_2 to regenerate the O^{2-} lattice:

$$4e^- + O_2 \longrightarrow 2O^{-2} \tag{12.36}$$

which can also oxidize species.

Fe_2O_3 works similarly to TiO_2 except that Fe_2O_3 is much less catalytically active than TiO_2.

Mercury vapor works via a different mechanism. Mercury is relatively unreactive in its ground state, but once the mercury is promoted into the $^3P_1$ state, the mercury becomes quite reactive. For example, the excited mercury can extract hydrogen from a hydrocarbon:

$$Hg(^3P_1) + H - R \longrightarrow HgH + R\bullet \tag{12.37}$$

Then the mercury falls back to its ground ($^1S_0$) state, releasing the hydrogen:

$$HgH \longrightarrow Hg(^1S_0) + H \tag{12.38}$$

The result is that mercury vapor can initiate gas-phase radical chemistry and act much like the radical catalysts described earlier in this chapter.

Industrially, people rarely use mercury anymore because mercury is so toxic. However, there are many papers on the topic.

12.3.7 Summary

At this point, then, we have listed many different types of catalysts and briefly mentioned how they work. When I teach this course I find that students are often overwhelmed by all of the variations; and in reality there are reasons to say that the variations are overwhelming. In fact, that has been the main point of the sections so far in this chapter: there are many variations. Many substances can act as catalysts.

As I said earlier, I recommend that the reader memorize everything in Tables 12.2–12.9. The material in these tables will be necessary background for the things that follow.

12.4   GENERAL OVERVIEW OF CATALYTIC ACTION

At this point, we will be changing topics. So far we have been discussing what catalysts are like. Now, we will start to discuss how catalysts work.

Catalysts work by changing the local environment around the reactants. The change in the local environment stabilizes intermediates and modifies the forces between the reactants. These changes can be used to promote a desired reaction.

The idea that rates of reactions can vary with the local environment around the reactants goes back to the early days of kinetics. In 1817, sir Humphry Davy found that he could prevent explosions in coal mines if he surrounded the candles used to illuminate the mines with a platinum shield. A few years later, Michael Faraday examined the fundamental processes in Davy's lamp. In 1834 Faraday proposed that the platinum was catalyzing the termination reactions in the flame by acting in two key ways. The platinum was holding the reactants in close proximity so that they could react. The platinum was also modifying the forces between the reactants to stimulate the reaction.

A few years later, Jöns Berzelius (1836a,b) did extensive studies on catalysts. At that time people did not know about molecules. However, Berzelius proposed that the catalyst changed the rate by modifying the forces between the reactants to stimulate reactions. Another idea in the literature came from the work of Jean-Baptiste Perrin (1919), who suggested that catalysts were able to transfer energy into the reactants and thereby overcome the energy requirements needed to activate molecules. Finally, Paul Sabatier (1913) suggested that the catalysts stabilized intermediates, thereby promoting a reaction.

Over the years there have been many arguments about whether Faraday, Berzelius, Sabatier, or Perrin were correct. However, we now know that they were all correct. Catalysts *can* work in a variety of different ways, and if you look at different reactions, you can find many different modes of catalyst action. When I teach my catalysis course, I give the following list of ways a catalyst can be designed to work:

- Catalysts can be designed to help initiate reactions.
- Catalysts can be designed to stabilize the intermediates of a reaction.
- Catalysts can be designed to hold the reactants in close proximity.
- Catalysts can be designed to hold the reactants in the right configuration to react.
- Catalysts can be designed to block side reactions.
- Catalysts can be designed to sequentially stretch bonds and otherwise make bonds easier to break.
- Catalysts can be designed to donate and accept electrons.
- Catalysts can be designed to act as efficient means for energy transfer.

It is also important to realize that

- One needs a catalytic cycle to get reactions to happen.
- Mass transfer limitations are more important when a catalyst is present.

There are many ideas here so it is useful to consider an example. Consider the reaction

$$H_2 + Br_2 \Longrightarrow 2\,HBr \tag{12.39}$$

In chapter 5 we found that the reaction goes via the following mechanism:

$$Br_2 \longrightarrow 2Br$$

$$Br + H_2 \longrightarrow HBr + H$$

$$H + Br_2 \longrightarrow HBr + Br$$

$$2\,Br \longrightarrow Br_2 \tag{12.40}$$

In actual practice, the reaction goes through a catalytic cycle, as will be discussed later in this chapter. However, for now we will assume that the reaction goes only once, and try to see how a catalyst can work.

Figure 12.6 shows how the free energy of the system changes during the reaction in the gas phase. The system goes uphill during the initiation step, and more uphill during the first reaction step. Then everything is downhill to products.

The first way a catalyst can work is to help initiate the reaction by reducing the initial steep rise in energy as shown in Figure 12.6b. According to the Polanyi relationship, when you reduce the initial rise, you will decrease the barriers to reaction.

The next way that a catalyst works is to stabilize the intermediates as shown in Figure 12.6c. The idea is to bind to the reactants just enough that the reaction never has to go uphill. That way the overall barrier for reaction is reduced.

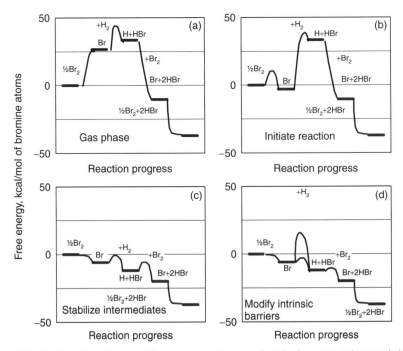

Figure 12.6 An illustration of some of the ways a catalyst can affect the free-energy changes during a reaction.

The third main way a catalyst works is to adjust the intrinsic barriers at each stage of the reaction as indicated in Figure 12.6d. Generally you want to lower the intrinsic barriers for desired reactions and raise the intrinsic barriers for undesirable reactions. One can lower the intrinsic barriers by changing either the entropy or the enthalpy of activation. You can lower the entropy of activation by holding the reactants close together and in the right configuration to react. You can lower the enthalpy of activation by either stretching bonds, thereby making them easier to break, or by adjusting the charge to moderate the Pauli repulsions between the reactants.

These ideas will be discussed in detail in the next several sections. It would be unusual for a catalyst to do all of these things. The very best catalysts might help initiate the reaction, stabilize the reactants, hold the reactants in close proximity and in the right configuration to react, and possibly block side reactions. However, few catalysts work that efficiently. I view the list above as a list of things that you would ideally want the catalyst to do. However, few catalysts do *all* of those things. In the next several sections we will choose examples to illustrate each of these effects.

Another point is that much of the material from here on will be qualitative. People know in a qualitative way how catalytic reactions work. However, very few of the ideas have been quantified enough that one can use them to make practical calculations.

I used the word "catalyst" in the last few paragraphs, but I do not want to restrict the discussion to conventional catalysts. People now know, for example, that a solvent can act like a catalyst. Solvents can modify the forces between reactants, stabilize intermediates, hold the reactants in close proximity, and act as efficient media for energy transfer. Other condensed phases can also act as catalysts. In the literature it has become common to treat the effects of solvents separately from the effects of catalysts. However, we now know that the principles of catalytic action and the principles of solvation are almost the same. Therefore, I find it convenient to treat catalysts and solvents as having fundamentally the same types of influence on rates.

In the next several sections we will discuss how a solvent or a catalyst can modify the rate of a reaction. We will concentrate on the general principles that apply to a large number of systems. Specific information about different catalyst systems will be given in Chapters 13 and 14.

12.5 CATALYSTS CAN BE DESIGNED TO INITIATE REACTIONS

The basis for all of our analysis will be a simple observation, discussed in Masel (1996): Catalysts seldom change the mechanisms of reactions; catalysts change only the initiation processes and the rates and selectivities of reactions. For example, McKenney et al. (1963) found that at 200 K the rate of the reaction

$$C_2H_6 \Longrightarrow C_2H_4 + H_2 \qquad (12.41)$$

goes up by a factor of about 10^9 if one adds NO_2 to the reacting mixture. The mechanism of reaction (12.41) does not change substantially in the presence of the catalyst. The only major change is in the initiation step.

Recall from Chapter 5 that reaction (12.41) goes via the following mechanism:

Initiation

$$C_2H_6 \longrightarrow 2CH_3 \qquad (12.42)$$

Transfer

$$CH_3 + C_2H_6 \longrightarrow C_2H_5 + CH_4 \qquad (12.43)$$

Propagation

$$C_2H_5 \longrightarrow C_2H_4 + H \qquad (12.44)$$

$$H + C_2H_6 \longrightarrow C_2H_5 + H_2 \qquad (12.45)$$

Termination

$$2CH_3 \longrightarrow C_2H_6 \qquad (12.46)$$

$$2C_2H_5 \longrightarrow C_4H_{10} \qquad (12.47)$$

$$CH_3 + C_2H_5 \longrightarrow C_3H_8 \qquad (12.48)$$

If fact, NO_2 changes the reaction in a very simple way. Recall that NO_2 is paramagnetic. It has an unpaired electron. The presence of the unpaired electron makes NO_2 a "stable radical". NO_2 can do radical chemistry even though NO_2 is a stable species.

Now, let's use the material in Chapter 5 to guess how the presence of NO_2 would affect the rate. Let's consider the reaction

$$NO_2 + C_2H_6 \Longrightarrow \text{products} \qquad (12.49)$$

According to the analysis in Chapter 5, when an NO_2 molecule collides with ethane, the main reaction will be a hydrogen transfer process:

$$NO_2 + C_2H_6 \longrightarrow C_2H_5 + HNO_2 \qquad (12.50)$$

Once one forms the C_2H_5 the C_2H_5 can react via reactions (12.44) and (12.45). Reaction (12.50) is about 20 kcal/mol endothermic, and according to the analysis in Chapter 5, it should have an intrinsic barrier of 14 kcal/mol. If one plugs the intrinsic barrier into the Polanyi relationship, one finds that reaction (12.50) should have an activation barrier of about 25 kcal/mol. By comparison, reaction (12.42) has an activation barrier of about 90 kcal/mol. The result is that reaction (12.50) is usually much faster than reaction (12.42). The increase in the initiation process means that reaction (12.41) is much faster in the presence of NO_2 than in the absence of NO_2.

This example illustrates a key point:

Catalysts can initiate reactions. The mechanisms are similar to the mechanisms without a catalyst, but the initiation process is much faster with the catalyst.

The increase in the initiation rate produces a tremendous increase in the overall rate of reaction. One can quantify the effect using analysis similar to that in Section 12.7. In the case above, the presence of NO_2 increases the rate by a factor of 10^9 at 750 K.

Table 12.10 shows several other examples where catalysts initiate reactions. Notice that a wide range of reactions can be initiated by catalysts. Not all catalysts work this way, of course. However, catalysts can be designed to initiate chemical reactions.

Table 12.10   Some examples of reactions initiated by catalysts

Reaction	Catalyst	Mechanism of Initiation
$CH_3\ CH_3 \Rightarrow C_2H_4 + H_2$	NO_2	$NO_2 + CH_3CH_3 \rightarrow HNO_2 + CH_3CH_2\bullet$
$CH_3COH \Rightarrow CH_4 + CO$	I_2	$X + I_2 \rightarrow 2I + X$
		$I + CH_3COH \rightarrow HI + CH_3CO\bullet$
Ethylene $\Rightarrow$ polyethylene	$\bullet ROO R$	$ROO R \rightarrow 2 RO\bullet\ RO\bullet + CH_2=$
		$C_6H_5OCH_2CH_2\bullet + nCH_2CH_2 \Rightarrow$
		polyethylene
		$CH_2 \rightarrow ROCH_2CH_2\bullet$
$H_2 + Br_2 \Rightarrow 2HBr$	Metallic platinum	$Br_2 + 2S \rightarrow 2Br\bullet_{ad}$
Propylene $\Rightarrow$ polypropylene	Ti^+	$T_1^+ + propylene \rightarrow CH_3CHTiCH_2^+$
		$CH_3CHT_1CH_2^+ + nC_3H_6 \Rightarrow$
		polypropylene
$C_2H_5OH \Rightarrow C_2H_4 + H_2O$	H^+	$C_2H_5OH + H^+ \rightarrow [C_2H_5OH_2]^+$
		$[C_2H_5OH_3]^+ \rightarrow [C_2H_5]^+ + H_2O$
		$[C_2H_5]^+ \rightarrow C_2H_4 + H^+$
$2O_3 \rightarrow 3O_2$	Cl	$O_3 + Cl \rightarrow O_2 + ClO$

12.6   CATALYSTS CAN BE DESIGNED TO STABILIZE INTERMEDIATES

Another way that catalysts can change rates of reaction is by stabilizing intermediates. In Chapter 2, we noted that at 200°C the rate of the reaction

$$H_2 + Br_2 \Longrightarrow 2HBr \qquad (12.51)$$

goes up by a factor of 10^8 in the presence of a platinum catalyst. Well, the mechanism of reaction (12.51) does not change dramatically in the gas phase and on the catalyst. It is just that the rates of many of the elementary reactions are much slower in the gas phase than on the catalyst.

Platinum is able to catalyze reaction (12.51) mainly because the platinum stabilizes the intermediates of the reaction. In Chapter 5 we found that in the gas phase reaction (12.51) goes by the following mechanism:

$$X + Br_2 \xrightarrow{\ 1\ } 2Br + X$$

$$Br + H_2 \xrightarrow{\ 2\ } HBr + H$$

$$H + Br_2 \xrightarrow{\ 3\ } HBr + Br$$

$$X + 2Br \xrightarrow{\ 4\ } Br_2 + X$$

$$H + HBr \xrightarrow{\ 5\ } H_2 + Br$$

$$X + Br + H \xrightarrow{\ 6\ } HBr + X \qquad (12.52)$$

In Chapter 5, we found that during reaction (12.52), most HBr is being produced in a catalytic cycle. First a bromine atom reacts with a H_2 to produce an HBr and an H atom. The hydrogen atom then reacts with the Br_2 to generate more HBr and regenerate the bromine atom. Note that in Section 4.3 we found that the rate of reaction (12.52) is directly proportional to the bromine concentration in the gas phase, so if we could find a way to increase the bromine atom concentration, we would increase the rate.

Now, it should not take much to enhance the rate. One can show that at 200°C, the bromine atom concentration should be about 10^{-17} mol/liter. Bromine atoms are the active centers during reaction (12.52), and with only 10^{-17} mol/liter of bromine atoms, the rate is limited by a shortage of bromine atoms.

One could imagine trying to find a way to increase the concentration of bromine atoms using a solvent. After all, we know that many different molecules will dissociate in solution. If we could find a solvent that dissociated the Br_2 to yield bromine atoms, we would increase the rate of reaction.

I do not know of any simple solvent that will dissociate Br_2. However, Br_2 does dissociate when Br_2 adsorbs on platinum to yield chemisorbed bromine atoms. Physically, the platinum is acting just like a solvent when it dissociates the Br_2. The free electrons in the platinum form a cage around the bromine atoms in exactly the same way that a solvent forms a cage around an ion in solution. The solvent cage is different in a metal than in a liquid because electrons behave differently than solvent molecules. Still, the process is fundamentally just like solvation. The electrons in the platinum solvate the bromine atoms, so that bromine atoms are stable on the platinum surface.

Just to quantify things, at saturation, the bromine atom concentration in the adsorbed layer is about 50 mol/liter. By comparison under similar conditions, the bromine atom concentration is about 10^{-17} mol/liter in the gas phase.

This is a general finding. The first key role of any catalyst is to increase the concentration of the active species. Increases of 20 orders of magnitude in the intermediate concentration are typical. That has a tremendous influence on the rate.

In the literature it is common to view the stabilization of intermediates in terms of the changes in the enthalpy of the system. To orient the reader, Figure 12.7 shows the enthalpy changes during the gas-phase reaction $H_2 + Br_2 \Rightarrow 2HBr$. The curve is very similar to that in Figure 12.6 except that the vertical axis is enthalpy, not free energy.

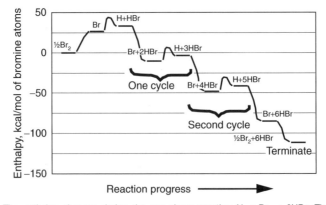

Figure 12.7 The enthalpy changes during the gas-phase reaction $H_2 + Br_2 \Rightarrow 2HBr$. The enthalpy is measured as the enthalpy per mole of bromine atoms produced in the initiation step.

Focusing on the figure, we see that first the Br_2 dissociates to form bromine atoms. That costs 26.74 kcal/mol. Then the bromine atoms react with H_2 to produce hydrogen atoms and HBr. That reaction is another 6.8 kcal/mol endothermic. Then there is a cycle where a Br_2 reacts with a hydrogen atom to produce HBr and a bromine atom. That reaction is 44 kcal/mol exothermic. Then the bromine atom reacts with H_2 to yield HBr and the hydrogen atom. That reaction is 6.8 kcal/mol endothermic.

The cycle repeats several times. Each time the system goes through the cycle, two molecules of HBr are produced. Consequently. the enthalpy of the system goes down by twice the heat of formation of HBr or 37.16 kcal/mol. We indicated two cycles in the figure, but the system actually goes through the cycle hundreds of times. Eventually, the reaction terminates.

In the literature people often redraw Figure 12.7, assuming that the reaction terminates after completing one cycle. The diagram assuming that the reaction terminates after completing one cycle is shown in Figure 12.8. The key feature of Figure 12.8 is that the reaction needs to go uphill twice for the reaction to proceed. That makes the reaction rather slow.

Figure 12.9 shows how the enthalpy changes that occur during the reaction vary if you run the reaction on a platinum surface. The platinum binds to the intermediates of the reaction. The enthalpy of formation of the intermediate is lowered by the strength of the adsorbate–surface bond. This stabilization of intermediates is a key process on platinum.

Notice that on platinum all of the steps in the mechanism except the termination step lower the enthalpy of the system. There is only one uphill step, and its barrier is small. That makes the reaction $H_2 + Br_2 \Rightarrow 2HBr$ very rapid on Pt(111).

Most catalysts speed up reactions by binding to some key intermediate and thereby stabilizing the intermediate. That increases the intermediate concentration, and thereby

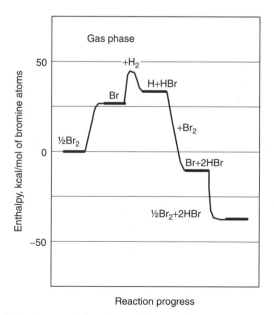

Reaction progress

Figure 12.8 The enthalpy changes during the gas-phase reaction $H_2 + Br_2 \Rightarrow 2HBr$ assuming that the reaction terminates after one cycle.

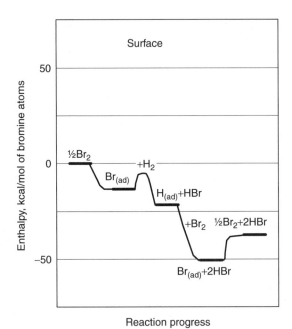

Figure 12.9 The enthalpy changes during the Rideal–Eley surface reaction $H_2 + Br_2 \Rightarrow 2HBr$ on Pt(111) assuming that the reaction terminates after one cycle.

increases the rate. One has to design the catalyst to selectively bind to the desired intermediate, and that is not always possible. However, most catalysts do increase the concentration of key intermediates.

12.6.1 Stabilization of Multiply Bound Intermediates

An alternative is for the catalyst to open up a new reaction pathway that would not be seen in the gas phase. In Section 12.6, we stated that catalysts seldom change the mechanism of a reaction. However, there are exceptions. The exceptions occur because the primary mechanisms of a reaction can vary with conditions. Catalysts can also stabilize intermediates that would be present only at very low concentrations in the gas phase. As a result, a minor reaction pathway in the gas phase can become a major reaction pathway in the presence of a catalyst.

Generally in the gas phase, one sees only monoradicals, that is, radicals with one dangling bond. However, on a solid catalyst one can see di- or triradicals, namely, species with multiple bonds to the catalyst. The ability to form multiple bonds allows reactions to occur that would not occur at reasonable rates in the gas phase.

For example, consider a simple reaction:

$$N_2 + 3H_2 \rightleftharpoons 2NH_3 \tag{12.53}$$

Reaction (12.53) is extremely slow in the gas phase. According to the analysis in Chapter 5, in the gas phase, reaction (12.53) should occur via an initiation–propagation mechanism:

Initiation

$$X + H_2 \longrightarrow 2H + X \qquad (12.54)$$

Propagation

$$H + N_2 \longrightarrow NH + N \qquad (12.55)$$

$$N + H_2 \longrightarrow NH + H \qquad (12.56)$$

$$NH + H_2 \longrightarrow NH_2 \qquad (12.57)$$

$$H_2 + NH_2 \longrightarrow NH_3 + H \qquad (12.57a)$$

Termination

$$X + 2H \longrightarrow H_2 + X \qquad (12.58)$$

However, according to data in the CRC, reaction (12.55) is about 150 kcal/mol endothermic. Consequently, in the gas phase, reaction (12.55) is extremely slow at any reasonable conditions.

Everything changes in the presence of an iron catalyst. On iron, the nitrogen readily dissociates via the reaction

$$N_2 + 6S \longrightarrow 2N(ad) \qquad (12.59)$$

One can then hydrogenate the adsorbed nitrogen via the following reactions:

$$N(ad) + H(ad) \longrightarrow NH(ad) + S \qquad (12.60)$$

$$NH(ad) + H(ad) \longrightarrow NH_2(ad) + S \qquad (12.61)$$

$$NH_2(ad) + H(ad) \longrightarrow NH_3(ad) + S \qquad (12.62)$$

Notice that when reaction (12.59) occurs, six nitrogen–surface bonds form. According to Table 6.5, on iron, ΔH_r for forming a single nitrogen surface bond is -14 kcal/mol. Therefore, the heat of reaction of reaction (12.59) is $(-14) \times 6 = -84$ kcal/mol (exothermic). Consequently, there is no thermodynamic barrier to reaction (12.59). Consequently, reaction (12.59) can occur on an iron catalyst even though reaction (12.59) occurs negligibly slowly in the gas phase.

One can use the data in Table 6.4 to show that reactions (12.60)–(12.62) are each 21 kcal/mol endothermic on iron. If we substitute the 21 kcal/mol into the Polanyi relationship, we find that each reaction should have an activation barrier of about 28 kcal/mol. Such reactions are quite feasible at reasonable temperatures, which is why one uses an iron catalyst.

One does not see a similar reaction in the gas phase since according to data in the CRC handbook, in the absence of a catalyst, reaction (12.59) is 225 kcal/mol endothermic.

This example illustrates an important point. Nitrogen radicals with three unpaired bonds are fairly unstable. However, when nitrogen adsorbs onto an iron catalyst, the iron can form multiple bonds to the nitrogen. The possibility of forming multiple bonds allows reaction (12.53) to occur.

One can generalize these results to many other situations. One can design a catalyst that stabilizes covalently bonded di- and triradicals. That allows the catalytic reactions to go via a reaction pathway that is not available in the gas phase.

> Generally, the ability to form multiply bound species allows a catalyst to produce species that would not be produced at reasonable rates in the gas phase.

12.6.2 Stabilization of Ionic Intermediates

Another way that a catalyst can work is to stabilize ionic intermediates, thereby allowing ionic pathways to occur. Consider a simple isomerization reaction:

$$RHC=CRH \Longrightarrow RRC=CHH \qquad (12.63)$$

It is hard to find a feasible mechanism for reaction (12.63) in the gas phase. One possibility is for the R group to break off:

$$X + RHC=CRH \longrightarrow RHC=CH + R\bullet + X \qquad (12.64)$$

Then there could be an addition reaction:

$$R\bullet + RHC=CRH + X \longrightarrow R_2HCCRH\bullet + X \qquad (12.65)$$

Then a β-hydrogen elimination:

$$R_2HCCH\bullet + X \longrightarrow R_2C=CH + H\bullet + X \qquad (12.66)$$

Reaction (12.64) is 120 kcal/mol endothermic. Further, this mechanism does not have a catalytic cycle. Consequently, according to the results in Section 5.4, the mechanism will have a very slow rate.

On the other hand, if one starts with a proton, H^+ in an acid catalyst, it is easy to find a catalytic cycle. First the proton reacts with the olefin to yield a carbocation:

$$RHC=CRH + H^+ \longrightarrow [RHC=CRHH]^+ \qquad (12.67)$$

Then the carbocation isomerizes:

$$[RHC=CRHH]^+ \longrightarrow [RRHC=CHH]^+ \qquad (12.68)$$

Then the new ion loses a proton to form the products:

$$[RRHC=CHH]^+ \longrightarrow RRC=CHH + H^+ \qquad (12.69)$$

Reactions (12.67)–(12.69) have barriers of 20 kcal/mol or less. Notice that the addition of the acid catalyst has allowed the reaction to occur via an ionic species. None of the reactions have large barriers. The result is that the reaction is quite facile (speedy).

This example shows that an acid catalyst can facilitate reactions by helping to create ionic intermediates.

Again, one does have to design the catalyst properly to stabilize the ionic intermediates. Not all catalysts will work. Still, one can design solid acid catalysts that stabilize ionic intermediates and thereby catalyze reactions like reaction (12.63).

12.6.3   The Effect of Intermediate Stabilization on Rates — Can We Have Too Much of a Good Thing?

There is a subtle point in all of this — it is possible to stabilize an intermediate so much that the intermediate becomes unreactive. Therefore, one needs to be careful to choose the catalyst carefully.

In this section, we quantify how much rates go up in the presence of a catalyst and show that one can overstabilize an intermediate and thereby decrease the rate. We will go back to the example in Section 12.6, namely

$$H_2 + Br_2 \Longrightarrow 2HBr \tag{12.70}$$

and derive an expression for the reaction. To put the discussion in perspective, in Section 12.6, we found that the concentration of bromine atoms increases by a factor of 10^{17} in the presence of a platinum catalyst. Yet, the data in Table 12.1 show that the rate increases by a factor of only 10^8. Other catalysts show much smaller rate enhancements even though the other catalysts increase the bromine atom by a factor of more than 10^{13}. Therefore, it appears that there is something else going on that prevents a 10^{17} increase in the intermediate concentration from producing a 10^{17} increase in rate.

In the remainder of this section, and in the next section, we will try to explain why the rate does not show a 10^{17} increase . To start off, it is useful to recall that a steady-state treatment of mechanism (12.52) shows that the rate of HBr formation via mechanism (12.52) is

$$r_{HBr} = 2k_2[H_2][Br] \tag{12.71}$$

where r_{HBr} is the rate of HBr formation; $[H_2]$ and $[Br]$ are the concentrations of hydrogen molecules and bromine atoms, respectively; and k_2 is the rate constant for the second reaction in mechanism (12.52). Therefore, if everything were equal, and k_2 did not change, one would expect the rate to increase by a factor of 10^{17} when the bromine atom concentration goes up by a factor of 10^{17}. However, experimentally, the rate goes up by a factor of only 10^8.

There are several reasons why the rate changes by a factor of 10^8. There are some mass transfer limitations, and there is a second reaction pathway involving a direct reaction between adsorbed hydrogen atoms and adsorbed bromine's. However, the key effect is that when you stabilize the bromine atoms on the platinum surface, the bromine atoms become less reactive so k_2 goes down.

This is also a general effect. Radicals are very reactive in the gas phase. You can stabilize the radicals by solvating them on a metal surface. However, then the radical becomes less reactive.

In the next section, we will quantify this effect. Our main tool will be the Polanyi relationship, which we derived in Chapter 10:

$$E_a = E_a^0 + \gamma_P \Delta H_r \tag{12.72}$$

where E_a is the activation barrier, E_a^0 is the intrinsic barrier to reaction, ΔH_r is the heat of reaction, and γ_P is the transfer coefficient. According to equation (12.72), when you change the heat of reaction, you also change the activation barrier of the reaction. Next, let's work out the implications of the Polanyi relationship for the reaction:

$$Br + H_2 \xrightarrow{\ 2\ } HBr + H \tag{12.73}$$

When you stabilize the bromine atoms, the free energy of the bromine atoms goes down. The enthalpy of the bromine atoms goes down, too, which means that the ΔH_r of reaction (12.73) becomes more positive. According to equation (12.72), when ΔH_r becomes more positive, the activation barrier for reaction (12.73) will increase. Consequently, k_2 will go down. Therefore, the implication of equation (12.72) is that when we stabilize the bromine atoms, the rate constants for reaction (12.73) will always go down.

Again, this is also a general effect. Whenever you stabilize a reactive intermediate, you increase the concentration of the intermediate. That generally increases the rate. However, there is always the secondary effect that the intermediate becomes less reactive. The loss of reactivity of the intermediate tends to partially counteract the effects of the increased stability of the intermediate, so the rate does not increase as much as the intermediate concentration increases.

Interestingly, there are some cases where one stabilizes the intermediate so much that the reaction slows down. For example, Sachtler and Fahrenfort (1958) and Fahrenfort et al. (1960) examined the decomposition of formic acid on a number of catalysts. Additional data is in Sachtler (1960). They found that the main reaction is

$$HCOOH \Longrightarrow H_2 + CO_2 \qquad (12.74)$$

Sachtler and Fahrenfort (1958) suggested that reaction (12.74) occurred via a very simple mechanism: The formic acid first adsorbs to form a formate intermediate. Then the formate intermediate decomposes to yield CO_2 and H_2.

$$HCOOH \longrightarrow HCOO^-_{(ad)} + H_{ad}$$
$$H_{(ad)} + HCOO^-_{ad} \longrightarrow CO_2 + H_2 \qquad (12.75)$$

Figure 12.10 shows how the rate of formic acid decomposition changes as one changes the binding energy of the formate intermediate. Notice that the rate increases, reaches a maximum, and then declines. People call plots with a maximum like those in Figure 12.10 **volcano plots**. Most catalytic systems follow volcano plots.

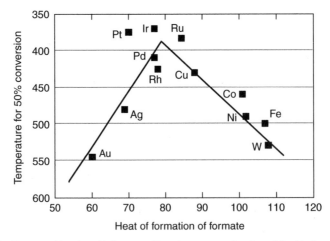

Figure 12.10 The rate of formic acid decomposition changes as a function of the binding energy of the formate intermediate.

The idea that rates reach a maximum was discovered by Sabatier (1913). Sabatier suggested that all reactions will follow a volcano plot. After all, an increase in the intermediate concentration will usually increase the rate. Still, Sabatier noted that you do not want the intermediate to be so stable that you produce the intermediate and not the product. Therefore, Sabatier proposed that volcano plots would be a universal phenomenon.

Sabatier also proposed what is now called **the principle of Sabatier**:

> The best catalysts are substances that bind the reactants strongly, but not too strongly.

If you have a weakly bound intermediate, increasing the binding energy will increase the intermediate concentration, thus increasing the rate. On the other hand, if you have a strongly bound intermediate, the intermediate concentration is already high. In such a case, increasing the binding energy of the intermediate does not increase the intermediate concentration substantially. However, k_2 decreases. According to equation (12.71), if k_2 decreases, and the intermediate concentration does not increase by a comparable amount, the rate should decrease.

In actual practice, however, Sabatier's principle is not quite a universal phenomenon. There are few reactions where one observes a leveling off rather than a decline of the rate when you stabilize the intermediates too much. However, this is the exception. Most catalytic reactions follow Sabatier's principle.

12.7 DERIVATION OF SABATIER'S PRINCIPLE

Next, we want to show how the maxima arises in detail. Our approach will be to work out an equation for the rate, and use the equation to see if there is a maximum. Let's go back to reaction (12.51), HBr formation, and assume that we are running the reaction in an adsorbed layer. There are two reaction pathways for HBr formation: a Rideal–Eley mechanism

$$Br_2 + 2S \longrightarrow 2Br_{ad}$$
$$Br_{ad} + H_2 \longrightarrow HBr + H_{ad}$$
$$H_{ad} + Br_2 \longrightarrow HBr + Br \qquad (12.76)$$

and a Langmuir–Hinshelwood mechanism

$$H_2 + 2S \longrightarrow 2H_{ad}$$
$$Br_2 + 2S \longrightarrow 2Br_{ad}$$
$$H_{ad} + Br_{ad} \longrightarrow HBr \qquad (12.77)$$

Where the names Rideal–Eley and Langmuir–Hinshel were discussed in Chapter 5. The Rideal–Eley mechanism is very similar to the gas-phase reaction, while the Langmuir–Hinshelwood reaction usually occurs only on a surface. For the purposes of this section we will ignore the Langmuir–Hinshelwood reaction, (12.77), even though on a real catalyst, the Langmuir Hinshelwood reaction is faster than the Rideal–Eley reaction at moderate pressures. The Rideal–Eley reaction dominates at high pressures, however.

Following the discussion in Chapter 5, we will assume that the adsorption of bromine goes via the reaction

$$\tfrac{1}{2}Br_2 + S \rightleftharpoons Br_{ad} \tag{12.78}$$

where S is a site on the surface that is available to hold gas as described in Chapter 5. We will also assume that reaction (12.78) is in equilibrium. At equilibrium, the bromine concentration is given by the following equation:

$$K_{Br} = \frac{[Br_{ad}]}{[S][Br_2]^{1/2}} \tag{12.79}$$

In equation (12.79), K_{Br} is the equilibrium constant for the adsorption process; $[Br_2]$, $[Br_{ad}]$, and $[S]$ are the concentrations of Br_2, adsorbed bromine atoms, and bare surface sites.

Solving equation (12.79) for $[Br_{ad}]$, and substituting the result into equation (12.71) yields

$$r_{HBr} = 2k_2 K_{Br}[H_2][S][Br_2]^{1/2} \tag{12.80}$$

Next, we will attempt to consider how changes in the binding energy of the bromine atoms changes the rate of HBr formation. Well, when the binding energy of the bromine changes, K_{Br} goes up and k_2 goes down. In order to quantify the effect, we will assume that K_{Br} is given by

$$K_{Br} = \exp\left(\frac{-(\Delta G_{ad})}{k_B T}\right) = \exp\left(\frac{-(\Delta H_{ad} - T\Delta S_{ad})}{k_B T}\right) \tag{12.81}$$

where ΔG_{ad} is the free energy of adsorption, ΔH_{ad} is the heat of adsorption, ΔS_{ad} is the entropy of adsorption, k_B is Boltzmann's constant, and T is the absolute temperature.

According to Arrhenius' law, k_2 is given by

$$k_2 = k_2^0 \exp\left(\frac{-(E_{a,2})}{k_B T}\right) \tag{12.82}$$

where k_2^0 is the preexponential and $E_{a,2}$ is the activation barrier for reaction 2.

Substituting equation (12.72) into equation (12.82) yields

$$k_2 = k_2^0 \exp\left(\frac{-(E_{a,2}^0 + \gamma_{P,2}\Delta H_{r,2})}{k_B T}\right) \tag{12.83}$$

where $E_{a,2}^0$ is the intrinsic barrier for reaction 2 in mechanism (12.76), $\Delta H_{r,2}$ is the heat of reaction, and $\gamma_{P,2}$ is the transfer reaction.

Note that as we change the heat of adsorption, ΔH_r changes, too. Let's define ΔH_0 as the heat of reaction when the heat of adsorption is zero. It is easy to show that when the heat of adsorption is nonzero, ΔH_r, the heat of reaction, is given by

$$\Delta H_{r,2} = \Delta H_0 - \Delta H_{ad} \tag{12.84}$$

Substituting equations (12.81) and (12.83) into equation (12.80) and then substituting in equation (12.84) yields

$$r_{HBr} = 2k_{Br} \exp\left(\frac{(1 - \gamma_{p,2})\Delta H_{ad}}{k_B T}\right)[H_2][S][Br_2]^{1/2} \tag{12.85}$$

with

$$k_{Br} = 2k_2^0 \exp\left(\frac{-(E_{a,2}^0 - \gamma_{p,2}\Delta H_0 + T\Delta S_{ad})}{k_B T}\right) \tag{12.86}$$

There are two interesting cases to consider: (1) the case where [S] is constant, so that there is no limit to the adsorbed bromine concentration; and (2) a case where [S] decreases as bromine adsorbs, so that the surface can hold only a finite amount of bromine.

Figure 12.11 shows a plot of the rate of HBr formation as calculated from equation (12.85), assuming that [S] is constant. Notice that the rate of HBr formation increases monotonically as the binding energy of the Br increases. Physically, if we assume that [S], the number of bare sites, is constant, we are in effect assuming that there is nothing to slow down the adsorption of bromine. In such a case, the bromine concentration continues to increase as the binding energy of the bromine increases, and so the reaction rate increases monotonically with coverage.

Quite a different effect occurs if one assumes that the reaction follows a Langmuir adsorption isotherm. According to the analysis later in this chapter, if one adsorbs hydrogen and bromine onto the platinum catalyst, the hydrogen and bromine take up sites. [S], the number of free sites, decreases as hydrogen and bromine adsorb. One can quantify this effect using an equation that we will derive in Section 12.17:

$$[S] = \frac{S_0}{1 + K_{Br_2}\sqrt{P_{Br_2}} + K_{H_2}\sqrt{P_{H_2}}} \tag{12.87}$$

Figure 12.12 shows a plot of the rate of HBr formation calculated from equation (12.85), with [S] from equation (12.87) and $\gamma_p = 0.5$. Notice that Figure 12.12 shows a volcano plot similar to that in Figure 12.10, where the rate reaches a maximum a heat of adsorption (i.e., heats of formation of the intermediate) and then declines. Physically, when ΔH_a

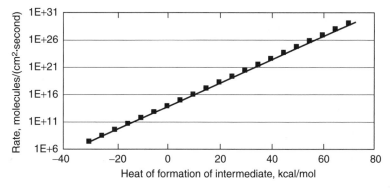

Figure 12.11 The rate of HBr formation as calculated from equation (12.85), with [S] $= 1 \times 10^{14}$ cm^2 and $\gamma_p = 0.5$, T $= 500$ K, $P_{H_2} = P_{Br_2} = 1$ atm.

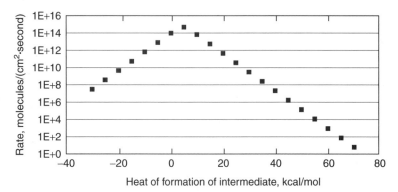

Figure 12.12 The rate of HBr formation calculated from equation (12.85), with [S] from equation (12.87) and $\gamma_p = 0.5$, $T = 500$ K, $P_{H_2} = P_{Br_2} = 1$ atm.

is small, increases in ΔH_a increase the intermediate concentration substantially. The increases in ΔH_a also decrease the reactivity of the intermediate since the intermediate is being stabilized, but that is a smaller effect. The net result is that the rate increases.

In contrast, once the surface fills up with reactants, further increases in ΔH_a do not produce substantial increases in the intermediate concentration. However, the increases in ΔH_a still decrease the reactivity of the intermediate. The net result is that the rate decreases. If one works through the numbers, one finds that there is an maximum rate at intermediate values of ΔH_a.

One also finds that for a given reaction on a given catalyst, there is an optimal temperature where the rate is maximized. That is why there is a maximum in Figures 2.16 and 12.2.

In actual practice, one almost always observes a maximum in the rate of reaction with increasing intermediate bond strength. Physically, with a heterogeneous catalyst, there are always a finite number of sites on the catalysts to hold reactants. Once all of the sites are filled, further increases in the bond strength of the intermediates mainly decreases the reactivity. With a homogeneous catalyst, there are only a finite number of complexes that you can make, before the catalysts are saturated. Once all the catalyst molecules are attached to reactants, further increases in bond strength decrease the activity.

In practice, Sabatier's principle works for most real catalysts. The one exception is in polymerization catalysts, such as a free-radical initiator. For example, a peroxide can initiate a free-radical polymerization via the mechanism

$$ROOR \longrightarrow 2\,RO\bullet$$

$$RO + CH_2{=}CH_2 \longrightarrow ROCH_2CH_2\bullet$$

$$ROCH_2CH_2\bullet + CH_2{=}CH_2 \longrightarrow ROCH_2CH_2CH_2CH_2\bullet \qquad (12.88)$$

In this case, one wants the initiator, $RO\bullet$, to have as strong of a bond with the reactants as possible. Physically, the reactivity of the radical does not decline with increasing bond strength, so a high bond strength is helpful.

There is one other detail of note. One does not need to know the heat of adsorption of the key intermediates to construct a plot like Figure 12.12. Instead, one can use any measure of the bond strength as the x axis in Figure 12.12, and one will still get a similar curve. This is very important because it often is quite difficult to know the heat of formation of a reactive intermediate. However, if one can find some other measure of the bond strength that is proportional to the heat of formation of the reactive intermediate, one can still construct a volcano plot.

In the literature, people often plot volcano plots as a function of the heat of formation of the bulk oxide per mole of oxygen. Figure 12.13 shows a plot of the heat of adsorption of a number of adsorbed intermediates as a function of the heat of formation of the bulk oxide per mole of oxygen. Notice that the heat of formation of the bulk oxide per mole of oxygen is proportional to the heat of adsorption of the reactive intermediate. A surface that binds some intermediates strongly tends to bind most adsorbates strongly as well. Therefore, one can construct a volcano plot using the heat of formation of the bulk oxides as an x axis (abscissa).

The advantage of this approach is that one can construct the volcano plot from data that are readily available, and so one can construct volcano plots for a wide number of reactions. Table 12.11 lists the heats of formation of a number of oxides. One can use the data in the table to construct volcano plots.

There is one subtlety in all of this. Table 12.11 gives the total heat of formation per mole of oxide. We actually want to scale the heat of formation to moles of adsorbate. Well, in the literature, there are two approaches to go to moles of adsorbate. One idea is to assume that the heat of formation of the intermediate is proportional to the heat of formation of the oxide *per mole of oxygen*. The second idea is to assume that the heat of formation of the intermediate is proportional to the heat of formation of the oxide *per mole of metal*. One also customarily assumes that if there are multiple oxides, the strongest oxide is the most important. If one makes these assumptions, one can plot the rate of a catalytic reaction versus the heat of formation of the oxide, and expect to observe a volcanolike curve.

There are two key types of plots in the literature:

- Plots of the log of the reaction rate versus the heat of formation of the oxide *per mole of oxygen*

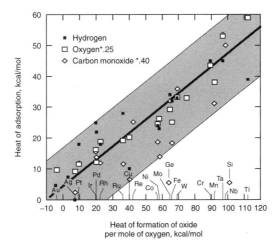

Figure 12.13 The heat of formation of adsorbed oxygen, hydrogen, and carbon monoxide on a number of metals as a function of the heat of formation of the corresponding oxide per mole of oxygen.

Table 12.11 **The heat of formation of several metal oxides**

Oxide	Heat of Formation, kcal/mol	Oxide	Heat of Formation, kcal/mol	Oxide	Heat of Formation, kcal/mol
AgO	−7.3	IrO_2	−40.1	PtO	−9.7
AgO_2	−6.3	FeO	−63.7	Re_2O_7	−297.5
As_2O_5	−218.6	Fe_2O_3	−197.5	RhO	−21.7
Au_2O_3	+19.3	PbO	−52.4	RuO_2	−52.5
Bi_2O_3	−137.9	MnO_2	−125.5	SeO_2	−55
CdO	−60.9	Mn_2O_3	−232.1	SnO	−68.4
CoO	−57.2	HgO	−21.7	TiO_2	−218
Cr_2O_3	−269.7	MoO_2	−130	V_2O_5	−373
CuO	−37.1	MoO_3	−180.3	V_2O_2	−200
Cu_2O	−39.8	NiO	−58.4	WO_2	−136.3
Ga_2O_3	−258	PbO	−52.4	W_2O_5	−337.9
In_2O_3	−222.5	PdO	−20.4	ZnO	−83.17

Source: Data from the CRC or Landolt-Bornstein.

- Plots of the log of the reaction rate versus the heat of formation of the oxide *per mole of metal*

Plots of the log of the reaction rate versus the heat of formation of the oxide *per mole of oxygen* are called **Sachtler–Fahrenfort** plots (Sachtler and Fahrenfort, 1958). Plots of the log of the reaction rate versus the heat of formation of the oxide *per mole of metal* are called **Tanaka–Tamaru** plots. (Tanaka and Tamaru, 1963)

Figure 12.14 shows a Sachtler–Fahrenfort plot and a Tanaka–Tamaru plot for the hydrogenation of ethylene. Both plots show some correlation to the data, although clearly there are lots of variations. (Without the line, the figures look like scatterplots.) Generally, Sachtler–Fahrenfort plots fit better than do Tamaru–Tanaka plots.

One has to be careful not to extrapolate these plots too far. Selenium, mercury, and lead are inactive for ethylene hydrogenation even though they have the heat of formation of the oxides is similar to those for the transition metal. Still, the Sachtler–Fahrenfort plots are a useful way to correlate data, even though they do not fit exactly.

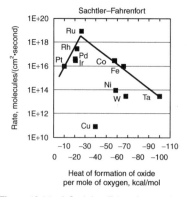

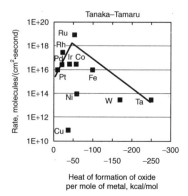

Figure 12.14 A Sachtler–Fahrenfort and Tanaka–Tamaru plots for the hydrogenation of ethylene.

In summary, then, the results in this section show that if a substance binds the intermediates of a reaction strongly, but not too strongly, the substance will show some catalytic activity. However, one does not want to overdo the bond strength, or else the substance will be catalytically inactive.

12.8 CATALYSTS CAN BE DESIGNED TO HOLD THE REACTANTS IN CLOSE PROXIMITY

Another way that a catalyst can work is to hold the reactants in close proximity to each other. For example, in Section 12.1 we noted that in 1817, Davy discovered that if one surrounded a candle with a platinum gauze, the platinum would prevent the flame from causing an explosion. In this case, the platinum is acting to catalyze the termination reactions in the flame so the candle does not cause an explosion.

In 1834, Faraday proposed that the main role of the catalyst is to hold the reactants in close proximity so they can react.

The reactions in a flame are pretty complex. There are many catalytic cycles. Still, the most important intermediates are the hydroxyls and the hydrogen atoms. Figure 12.15 shows a catalytic cycle for the hydroxyls.

As noted above, Davy found that platinum can catalyze the termination reactions in a flame. In the gas phase, the main quenching reactions are

$$2H \longrightarrow H_2 \tag{12.89}$$

$$H + OH \longrightarrow H_2O \tag{12.90}$$

Both reactions are slow in the gas phase because the concentration of intermediates is low. When you add a platinum catalyst, the catalyst concentrates the intermediates. Hydrogens adsorb on the catalyst and wait for another hydrogen atom or hydroxyl to hit the catalyst and react with the hydrogen. The catalyst in effect concentrates the reactants. That speeds up the termination reaction.

There are some subtleties here because the platinum also speeds up the initiation reactions. For example, if you put a platinum wire into a hydrogen/oxygen mixture, the platinum speeds up the initiation process more than the platinum speeds up the termination process. In that case, the flame is initiated by platinum. On the other hand, with a candle

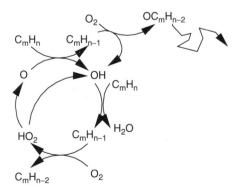

Figure 12.15 A simplified version of the reactions in a flame.

(i.e., a paraffin flame), the platinum speeds up the termination reactions more than the initiation reactions. As a result, the platinum quenches the flame.

In both cases, the platinum is concentrating the intermediates of the reaction so that the reactants are in close proximity to react. It is just that in one case the platinum speeds the initiation reactions more than the termination reactions while in the other case the opposite occurs.

12.9   CATALYSTS CAN BE DESIGNED TO HOLD THE REACTANTS IN THE CORRECT CONFIGURATION TO REACT

In the previous section, we noted that catalysts hold the reactants in close proximity and thereby speed up rates of reaction. Still, it is interesting to note that one could get even higher increases in rate if the catalyst could also hold the reactants in just the right configuration to react. Pushing the reactants together would also produce an additional increase in rate. It is, in fact, possible to design a catalyst that holds the reactants in just the right configuration to react. Very few catalysts do this, but the ones that do are especially efficient.

There is one specific example that illustrates the effects particularly well: the conversion of acetylene to benzene:

$$3C_2H_2 \Longrightarrow C_6H_6 \tag{12.91}$$

Reaction (12.91) is very rapid on a palladium catalyst. Figure 12.16 shows what is called the *active site* on the catalyst. The active site is defined as the arrangement of surface atoms where the reaction occurs. It happens that when acetylene adsorbs on palladium, the acetylene binds in what is called a "bridge bound state," where the acetylene is held off center of what is called a "threefold hollow" on the palladium surface. The bonding position is shown in Figure 12.16. It happens that if you put three acetylenes onto the palladium, the three acetylenes form a hexagonal structure similar to the hexagonal structure in benzene. The bond lengths are also about right for benzene. That promotes benzene formation.

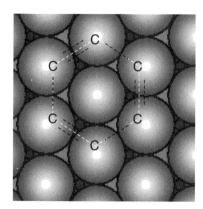

Figure 12.16   The active site for reaction (12.91) on a palladium catalyst.

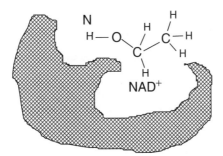

Figure 12.17 A cartoon of the reaction of ethanol and NAD^+ on the active site of liver alcohol dehydrogenase. [Adapted from Oppenheimer and Handlon (1992).]

If one changes the geometry by, for example, going to a square surface, the reaction is not observed. This example clearly illustrates the idea that if a catalyst holds the reaction in the right configuration to react, the reaction will occur at an unusually high rate.

This process commonly happens in enzyme-catalyzed reactions. For example, the first step in the destruction of alcohol by your liver is a hydrogen transfer from the alcohol to an ion called NAD^+ [nicotinamide adenine dinucleotide (oxidized form)]:

$$NAD^+ + CH_3CH_2OH \Longrightarrow NADH + [CH_3CHOH]^+ \qquad (12.92)$$

Reaction (12.92) occurs on an enzyme called *liver alcohol dehydrogenase*. Figure 12.17 shows how the reaction occurs in liver alcohol dehydrogenase. The diagram is adapted from Oppenheimer and Handlon (1992). The NAD^+ fits into a pocket in the enzyme. The alcohol sits over the NAD^+ in a bent configuration, with a hydrogen in the alcohol pushing into the NAD^+. The close proximity of the alcohol and the NAD^+ facilitates the reaction.

12.10 CATALYSTS STRETCH BONDS AND OTHERWISE MAKE BONDS EASIER TO BREAK

Some enzymes work in another way, too. They stretch bonds and otherwise make bonds easier to break. For example, Figure 12.18 shows a diagram of an enzyme called *Lysozyme 161L*. The name Lysozyme means that the molecule is part of a wide class of enzymes made in a part of a cell called a lysozome. The number 161L is the listing in the protein database. (The listings go in order.) Generally, lysozymes are enzymes that animals use to kill bacteria. For example, there are lysozymes in your tears that kill bacteria in your eyes.

Microbiologists say that lysozymes work by catalyzing the hydrolysis of polysaccharides in the cell walls of the bacteria, causing the cell walls to rupture. Let me translate. Polysaccharides are carbohydrates (sugars). In this case the polysaccharides consist of NAM (*N*-acetylmuramic acid) and NAG (*N*-acetylglucosamine) units. Both molecules are shown in Figure 12.19. Lysozyme catalyzes the hydrolysis of the bond between the NAG and NAM units:

$$-NAG-NAM-NAG-NAM-NAG-NAM- + H_2O \longrightarrow$$

$$-NAG-NAM-NAG-NAM-OH + H-NAG-NAM- \qquad (12.93)$$

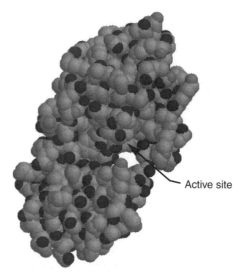

Figure 12.18 A picture of Lysozyme 161L. This figure was generated using a program called RASMOL, using data in the protein database from an x-ray-diffraction spectrum generated by Weaver and Matthews (1987).

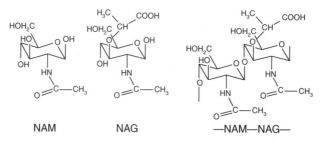

Figure 12.19 The structure of NAM, NAG, and a NAM–NAG repeat unit.

The active site for the hydrolysis reaction is indicated by the arrow in Figure 12.18. Generally a six-sugar –NAG–NAM–NAG–NAM–NAG–NAM–unit binds to the active site. The bonds in the NAM that are being hydrolyzed are stretched, and the NAM is distorted to a near-planar configuration. Then a proton is donated from the enzyme, breaking the polysaccharide bond. Then water is added to the resultant fragment.

The first step is for the enzyme to stretch. This makes it easy for the proton to come in. In addition, the lysozyme has a breathing mode where the opening (cleft) in the lysozyme opens and closes like a Pac-Man. When the Pac-Man closes, the proton is pushed into the stretched bond in the sugar, which also facilitates reaction. Note that both of these interactions are modifying the forces between the reactants, to facilitate the reaction.

The enzyme is also stabilizing the ionic intermediate formed when the proton reacts with the polysaccharide, cleaving the NAG.

This example shows that enzymes can stretch bonds. Bond stretching makes bond scission very likely to occur.

12.11   STABILIZATION OF TRANSITION STATES

Another mode of enzymatic action is to stabilize the transition state for a reaction. When you stabilize the transition state, you lower the intrinsic barrier to reaction and thereby speed up the reaction. Recall from equation 9.2 that anything one does to increase $q^{\ddagger}$, the partition function for the transition state, will increase the rate of reaction. One way to increase $q^{\ddagger}$ is to stabilize the transition state.

The lysozyme example illustrates this effect as well. Note that in the discussion above, we said that the NAM is stretched into a planar configuration when the polysaccharide binds to the enzyme. In the transition state for reaction (12.93), NAM has a planar configuration. The enzyme stabilizes that planar configuration. As a result, the binding of the polysaccharide to the enzyme has stabilized the transition state for reaction (12.93).

Most enzymes are thought to stabilize the transition states for reactions. That is part of the reason why enzymes are so catalytically active.

12.11.1   Catalytic Antibodies

There is a separate branch of chemistry called *catalytic antibodies* that tries to exploit these findings. The idea is to build an antibody that selectively binds to the transition state for a reaction. Note that by definition, when an antibody binds to a specific molecule, that molecule will be stabilized. Well, when the antibody binds to a transition state, the transition state will be stabilized. That increases $q^{\ddagger}$ and thereby speeds up reactions.

For example, Gouverneur, et al. (1993) synthesized a series of catalytic antibodies for the Diels–Alder reaction:

$$\text{(12.94)}$$

The antibody was designed to bind to the transition state for the reaction. That lowered the energy of the transition state.

Experimentally, the catalytic antibody was quite a good catalyst for the reaction. This case is particularly interesting since the catalytic antibody does chemistry that is not found in nature, and unlike a standard metal catalyst, it produces enantiomerically pure products. The result is that these compounds look quite interesting.

There is a new branch of chemistry developing on the basis of these findings. Generally, people first synthesize a *hapten*, which has a structure similar to that of the transition state for a reaction. They then feed the hapten to mice or bacteria. The mice or bacteria treat the hapten as an invading molecule, and build antibodies to the hapten. Those antibodies are separated, and some of them show catalytic activity.

These results show that materials that stabilize transition states can enhance reaction rates.

12.11.2   Transition Metal Catalysts

Transition metal catalysts (e.g., platinum) can occasionally work similarly to catalytic antibodies in stabilizing the transition state for a reaction. Note that during a reaction one is mixing excited states into the ground state of a system. If one can stabilize those antibonding orbitals, one will also stabilize the transition state. As a result, one can lower the energy of the transition state and thereby facilitate a reaction. Generally d electrons are able to interact with antibonding orbitals to stabilize them. S and p electrons are less able to facilitate reactions.

For example, consider the dissociation of hydrogen on platinum:

$$H_2 + 2S \longrightarrow 2H_{ad} \tag{12.95}$$

Reaction (12.95) is a classic four-centered symmetry-forbidden reaction. So, on the basis of the analysis in Section 10.9, one would expect reaction (12.95) to have a large intrinsic barrier. Well, on aluminum the reaction does have a large barrier. People observe little dissociation of energies up to 50 kcal/mol even though the reaction is 60 kcal/mol exothermic. If one plugs these measurements into the Polanyi relationship with a γ_P of 0.5, one finds that the intrinsic barrier is at least 80 kcal/mol. An intrinsic barrier of 80 kcal/mol is consistent with our expectations for a symmetry forbidden reaction.

On platinum, however, the reaction is unactivated. The heat of adsorption is 13 kcal/mol, and so if we plug into the Polanyi relationship, we find that the intrinsic barrier is less than 7 kcal/mol. Clearly, H_2 dissociation on platinum does not have the large intrinsic barrier one expects for a symmetry-forbidden reaction.

This occurs because the reaction is no longer symmetry-forbidden on platinum. Figure 12.20 shows a correlation diagram for reaction (12.95). In the absence of the platinum, the correlation diagram is identical to the diagram for reaction (10.11). During the reaction the $\sigma\sigma^*$ state is lost and a $\sigma^*\sigma$ state forms. This is a standard symmetry-forbidden reaction.

When the platinum is present, everything changes. The d bands in the platinum add some extra states, which I have labeled A and E in the diagram, both of which are partially filled with electrons. In the material that follows we will show that during the reaction, the electrons from the $\sigma\sigma^*$ state can flow into the A band, while electrons from the E band can flow into the $\sigma^*\sigma$ state. That allows the reaction to occur with a minimal barrier.

A good way to understand this interaction is to look at what happens when electrons flow into the $\sigma^*\sigma$ state. The $\sigma^*\sigma$ state consists of an antibonding state in the H_2 plus a corresponding state in the metal. Figure 12.21 shows a diagram of the antibonding state.

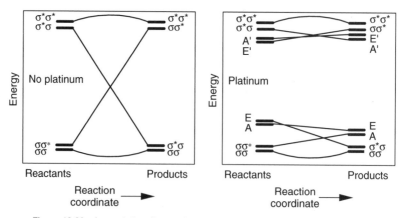

Figure 12.20 A correlation diagram for reaction (12.95) with and without platinum.

Notice that there is a sign change in the antibonding state. Consequently, according to the material in Section 10.11, if one has a state in the metal with a similar sign change, and one can take electrons out of that state, then one can get the reaction to occur smoothly as shown in Figure 10.35.

Figure 12.21 shows a diagram of the key orbitals during the reaction. There are two kinds of orbitals in platinum: s orbitals and d orbitals. The s orbitals spread out over the whole surface. There are no big sign changes, so the s band cannot interact with the antibonding orbitals on the H_2. In contrast, the E states in the d bands have the same sign changes as in the antibonding orbitals in the H_2. Consequently, one can transfer electrons from the d bands to the antibonding orbitals in the H_2 with minimal difficulty.

In fact, there is some subtlety to the arguments, because one wants to break the H–H bond, not simply stabilize the antibonding orbitals. Well, the d bands do that, too. Recall that orbitals of the same sign attract while orbitals of different sign repel. If one looks at the s bands, one finds that there is an attractive interaction on one side of the hydrogen but a repulsive interaction on the other side. The net effect is that the s bands do not attract the antibonding orbitals. On the other hand, if one orients the hydrogen as shown in the lower diagram in Figure 12.21, both sides of the antibonding orbitals on the hydrogen will be attracted to the d bands on the platinum. The overlap increases as the hydrogen dissociates. The net effect is that the antibonding orbitals in the hydrogen are pulled toward the platinum atoms. That rips the hydrogen apart.

The net effect is that the intrinsic barrier for the reaction has been reduced by over 70 kcal/mol. Generally, the 70 kcal/mol or more reduction in intrinsic barriers occurs only for symmetry-forbidden reactions: dissociation of diatomic molecules into two atoms, or scission of double or triple bonds. Most other reactions show similar intrinsic barriers on both transition metals and nontransition metals.

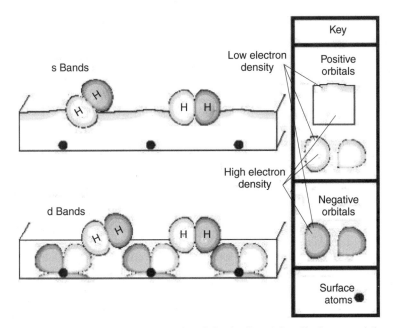

Figure 12.21   A diagram of the key interactions during the dissociation of hydrogen on platinum.

12.11.3 Acid Catalysts

Another way that catalysts can change the intrinsic barriers is to change the net charges on the reacting species. The idea is simple. An electronegative catalyst can remove charge from the reactants. That can produce changes in the barriers to reaction.

In order to see how that works, consider the reaction discussed in Section 12.6.2:

$$RHC{=}CRH \Longrightarrow RRC{=}CHH \tag{12.96}$$

In Section 12.6.2 we noted that in acid solution the proton reacts with the olefin to yield a carbocation:

$$RHC{=}CRH + H^+ \longrightarrow [RHC{=}CRHH]^+ \tag{12.97}$$

Then the carbocation isomerizes:

$$[RHC{=}CRHH]^+ \longrightarrow [RRHC{=}CHH]^+ \tag{12.98}$$

Then the new ion loses a proton to form the products:

$$[RRHC = CHH]^+ \longrightarrow RRC{=}CHH + H^+ \tag{12.99}$$

In this case the acid is doing several things. The acid is initiating the reaction. However, the acid solution is also stabilizing the charged species. The stabilization of the charges is important to the reaction. In the material that follows we will discuss how the charges affect the barriers to reaction.

In order to determine how the charge on the molecule affects the barriers to reaction, it is useful to consider a reaction where a neutral hydrogen atom reacts with the olefin to yield a radical:

$$X + RHC{=}CRH + H\bullet \longrightarrow RH\overset{\bullet}{C}{-}\overset{R}{C}H_2 + X \tag{12.100}$$

Then the radical isomerizes:

$$X + RH\overset{\bullet}{C}{-}\overset{R}{C}H_2 \longrightarrow RH\overset{R}{C}{-}\overset{\bullet}{C}H_2 + X \tag{12.101}$$

Then there is loss of a hydrogen atom:

$$X + RH\overset{R}{C}{-}\overset{\bullet}{C}H_2 \longrightarrow R\overset{R}{C} = CH_2 + H\bullet + X \tag{12.102}$$

First, let us consider reaction (12.100).

Figure 12.22 shows some of the key orbitals during reaction (12.100). The reactant, $RHC{=}CRH$, starts with a π-bond. Recall that a π-bond forms when two p orbitals line up and bind together. The π bond looks like an extended P orbital as shown in Figure 12.22. The hydrogen starts out with a spherical orbital as we saw in Chapter 10.

Now consider what happens when reaction (12.100) starts. When reaction (12.100) starts, the hydrogen approaches the ethylene. In Chapter 10, we found that there will be a bonding and antibonding interaction. Figure 12.22 shows the bonding and antibonding molecular orbitals. The two MOs look almost the same, except that in the bonding MO

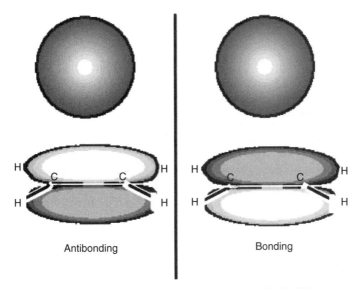

Antibonding   Bonding

Figure 12.22   A diagram of the key MOs during reaction(12.100).

the top lobe on the ethylene has the same sign as the s orbital while in the antibonding MO, the lobe has a sign different from that of the s orbital.

Now consider what happens when the orbitals come together. This case is very similar to the cases we discussed in Section 10.6.3. Notice that in the antibonding state, a positive orbital is pushing up against a negative orbital. Physically, we are pushing one balloon of electrons into another, so there is a repulsion exactly like the repulsions discussed in Chapter 10. The repulsive interaction pushes up the energy of the system. Consequently the reaction is activated.

A more detailed analysis shows that the repulsion occurs because there are three electrons in the system. If there were only two electrons, we could put both electrons into the bonding state, and so we would not have to put electrons into orbitals of different signs. Consequently, one could eliminate the repulsive antibonding interaction. However, the bonding MO can hold only two electrons. If one has three electrons, one of the electrons must be put into the antibonding MO. That produces the repulsive interaction.

Notice that if one starts with H^+, there will be only two electrons in the system, so there will be no repulsion. Consequently, the fact that one has protons or hydroniums, rather than neutral hydrogen atoms, in acid solution eliminates most of the barriers to reaction (12.100).

One can also run reaction (12.100) on a solid. Solid catalysts are great because one can modify the charges on the protons. Recall that in a covalently bonded molecule such as water, the hydrogen has a small net positive charge. That is why water has a net dipole. In a solid, one can increase the charge. On certain solids, called **superacid catalysts** the charge is nearly $+1$. That promotes easy proton transfer. Consequently, superacid catalysts are able to lower the intrinsic barriers to reaction (12.100)

The superacid catalysts also promote reaction (12.101). Reaction (12.101) is more complex than reaction (12.100). Reaction (12.101) starts with a radical. There is a half-filled P orbital on one carbon, and a C–R bond on the other carbon. Figure 12.23 shows a very approximate diagram of the key orbitals during reaction (12.101).

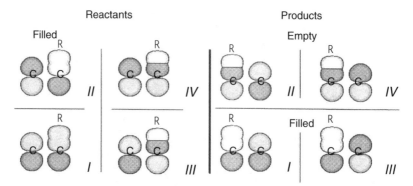

Figure 12.23 A rough diagram of the key MOs during reaction (12.101).

During reaction (12.101) the system starts out with the orbitals in the left of Figure 12.23. There are four MOs labeled I, II, III, and IV. In the diagram, we have arbitrarily assigned the lobe on the R group to have a positive sign, and then considered all possible signs on the p hybrids on the carbon. In MO I, both the p's are positive; in MO II, the left p is negative, while the right p is positive. In the MO III the right p is negative and the left p is positive, while in MO IV, both p's are negative. In MOs I and II, the π group has a bonding interaction with the carbon, while in III and IV, there is a sign change in moving from R to C and so the interaction is antibonding. As a result, orbitals I and II are bonding orbitals while orbitals III and IV are antibonding orbitals. In contrast, at the end of the reaction, the R group migrates from the right to the left of the molecule. In this case, orbitals I and III are bonding while orbitals II and IV are antibonding.

Now consider moving the R group. Notice that the R group needs to move across the molecule for reaction to occur. However, in order to move the positive orbital on the R group in orbital II, it will need to displace the negative nonbonding orbital on the carbon. In Chapter 10 we found that such orbital displacements have large barriers. The net effect is that 1,2 displacements have large barriers with neutral radicals.

Notice that the repulsion occurs only because the electrons in the R group are pushing up against the nonbonding orbital in the molecule. If one modifies the charges on the molecule, one can remove the electrons from the nonbonding orbital. If one puts a $+1$ charge on the molecule, there will be no repulsions.

Well, again on **superacid catalysts** the charge is nearly $+1$. That promotes easy isomerization. The net result is that the superacid catalyst is able to promote reaction (12.101).

These results show that catalysts can modify the intrinsic barriers to reaction. These modifications allow very selective reactions to occur.

These results show that catalysts can be designed to modify the changes on the reactants in a way that facilitates reaction.

12.12 CATALYSTS CAN BE DESIGNED TO BLOCK SIDE REACTIONS

Another thing that catalysts can do is to block side reactions. The idea is simple. You design the catalyst so that it is shaped in such a way that the reactants can get together only when the reactants are in the correct configuration to react via the desired pathway. The result is that you get desired reaction and not an undesired one.

One specific example where this effect is very important is in the polymerization of propylene:

$$n\ H_2C=CHCH_3\ \longrightarrow\ \left[-C-\underset{\underset{CH_3}{|}}{C}-\right]_n \tag{12.103}$$

Polypropylene undergoes free-radical or cationic polymerization as discussed in Chapter 5. When the free-radical reaction occurs, the methyl group can go on the top or the bottom of the molecule. If the methyl groups are distributed randomly, the polymer has poor mechanical properties. As a result, the polymer cannot be used in many applications. In contrast, if one can control the positions of the methyl groups, one can produce polymers with much better mechanical properties. Such polymers are very valuable.

There are two key forms of oriented polypropylene: isotactic polypropylene, where all of the methyl groups are on one side

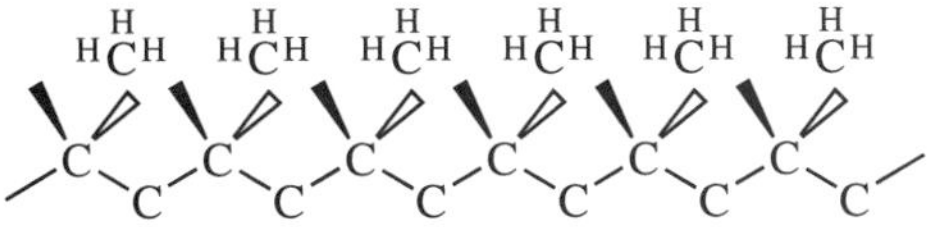

$$(12.104)$$

and syndiotactic polypropylene, where the methyl groups alternate from side to side:

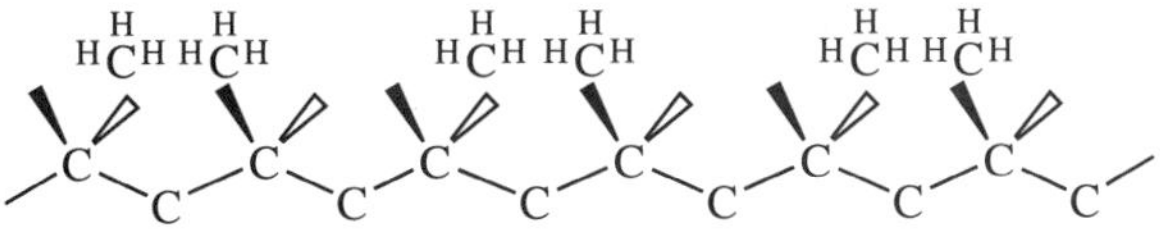

$$(12.105)$$

Consider making isotatic polyethylene. Figure 12.24 shows a diagram of a step during the production of isotactic polypropylene where a single propylene unit is added to a growing polymer chain. Notice that one can add the propylene, with a methyl group facing in the correct direction or in the wrong direction. If one would add the propylene with the methyl group facing the wrong way, polymerization would still occur, but one would not end up with isotactic polypropylene.

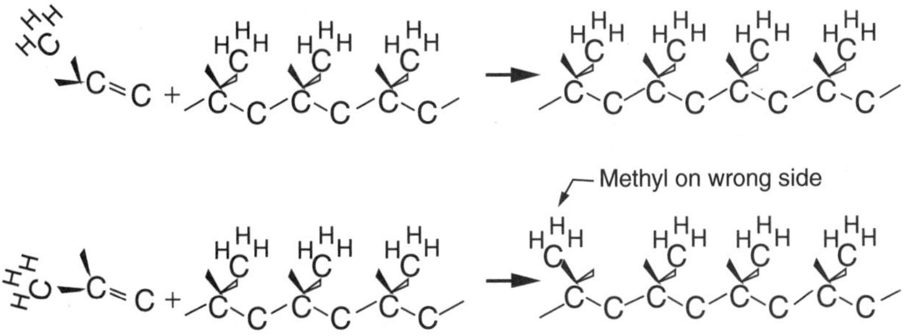

Figure 12.24   A rough diagram of one step during the production of isotatic polypropylene.

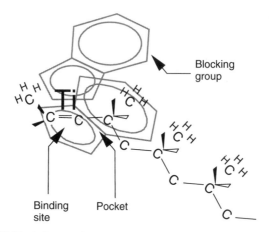

Figure 12.25 A diagram of propylene polymerization in a Ziegler–Natta catalyst.

When one runs these reactions industrially, one makes these polymers with what is called a *Ziegler–Natta catalyst*. The Ziegler–Natta catalyst is designed to produce only isotactic propylene. Figure 12.25 shows a diagram of the active site in a Ziegler–Natta catalyst. The titanium atom actually catalyzes the reaction. However, the titanium atom is surrounded by ligands that enclose the site. There are two pockets in the ligands. During the reaction, propylene squeezes into the pockets and reacts. The ligands prevent the propylene from twisting. That prevents the side reaction. The net result is that all of the methyl groups line up on the same side.

This example shows that there can be an advantage to designing a catalyst with blocking groups to prevent side reactions.

12.13 CATALYSTS CAN BE DESIGNED TO DONATE AND ACCEPT ELECTRONS

Another mode of catalytic action is for the catalyst to donate or accept electrons. Recall that in Section 2.3.2 we found that certain metal oxides are effective oxidation catalysts because they can store and give up lattice oxygen. Well, these catalysts work by a redox process where the metal is oxidized and reduced. For example, consider CO oxidation over a copper substrate. There are two stable copper oxides: Cu^{1+} oxide (i.e., Cu_2O) and Cu^{2+} oxide (i.e., CuO). During the reaction, the Cu^{1+} oxide is first oxidized to a Cu^{2+} oxide:

$$Cu_2O + \tfrac{1}{2}O_2 \longrightarrow 2CuO \tag{12.106}$$

then the Cu^{2+} oxide reacts with CO to yield a Cu^{1+} oxide:

$$2CuO + CO \longrightarrow Cu_2O + CO_2 \tag{12.107}$$

Notice that in step (12.106) the copper donates an electron to the oxygen, while in step (12.107), the oxygen donates electrons back to the copper. The donation and release of electrons makes copper a good catalyst for CO oxidation.

There are several examples where the electron transfer process is more direct. For example, the electron transfer reaction

$$Fe^{2+} + V^{4+} \longrightarrow Fe^{3+} + V^{3+} \tag{12.108}$$

is slow because the $+2$ charge on the iron repels the $+3$ charge on the vanadium. Further, the solvation shells on the iron and vanadium get in the way of each other. The repulsion is less with Cu^{1+} because Cu^{1+} has only a $+1$ charge. Further copper has a much smaller solvation shell than either iron or nickel. As a result, copper ions can facilitate reaction (12.108).

The main mechanism of the reaction is electron exchange. First Cu^{1+} reacts with V^{4+} donating an electron:

$$Cu^{1+} + V^{4+} \longrightarrow Cu^{2+} + V^{3+} \tag{12.109}$$

Then the Cu^{1+} is regenerated by reaction with Fe^{2+}:

$$Fe^{2+} + Cu^{2+} \longrightarrow Fe^{3+} + Cu^{1+} \tag{12.110}$$

Notice that the copper is donating electrons in reaction (12.109) and accepting the electron back again in reaction (12.110). This is another example where the catalyst works by donating and accepting charge.

Another application of this technique is in fuel cells. Fuel cells are commonly used to generate electricity. Figure 12.26 shows a diagram of a simple hydrogen fuel cell. Generally, hydrogen reacts on a platinum catalyst on the anode of the fuel cell to produce protons and electrons:

$$H_2 \Longrightarrow 2H^+ + 2e^- \tag{12.111}$$

where e^- is an electron. The protons are then transmitted through a polymer membrane in the fuel cell, where they react on the cathode:

$$O_2 + 4H^+ + 4e^- \Longrightarrow 2H_2O \tag{12.112}$$

Fuel cells can produce enough power to run an automobile. Notice that the catalyst on the anode is accepting electrons, while the catalyst on the cathode is giving up electrons. This is a further example showing that catalysts can work by accepting and donating electrons.

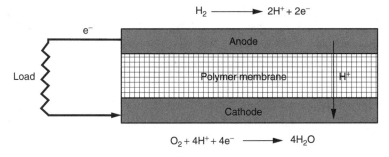

Figure 12.26 A diagram of a polymer fuel cell.

12.14 CATALYSTS CAN ACT AS AN EFFICIENT MEANS FOR ENERGY TRANSFER

There is one other mode of catalytic action that people do not talk about much in the literature. Catalysts can be designed to act as an efficient means for energy transfer. Recall from Chapter 9 that the rate of most unimolecular reactions and bimolecular recombination reactions is limited by the rate at which energy enters or leaves the molecule. For example, the reaction

$$Na + Cl \longrightarrow NaCl \tag{12.113}$$

has a zero rate in the absence of a collision partner even though the reaction is 98 kcal/mol exothermic. If you add a collision partner, the reaction goes very quickly. In a sense, the collision partner is acting like a catalyst to carry heat away from the reactants.

It happens that solid catalysts are often much better collision partners than are gaseous species. For example, when Polanyi (1932) first examined reaction (12.113), he found that most of the reaction occurred on the walls of his vessel. The walls were acting as an efficient means of heat transfer, and thereby allowed reaction to occur. All solid catalysts or solvents have similar effects. The catalysts provide very efficient means for heat transfer. That allows reaction to occur. In practice, most unimolecular and bimolecular reactants occur much more rapidly in the presence of a liquid or a solid. Therefore, the reaction is being catalyzed even though the catalyst never binds strongly to the reactants.

12.15 MASS TRANSFER EFFECTS ARE MORE IMPORTANT WHEN A CATALYST IS PRESENT

Finally, it is important to point out that mass transfer effects become more important when catalysts are present. Recall that, according to collision theory, the rate of reaction is determined by the rate that molecules collide, multiplied by a factor that accounts for the fact that not all molecules that collide react. In a solution, the collision rate is determined by mass transfer effects, while the fraction that reacts is determined by what happens after the reactants collide. A catalyst can modify the fraction of the molecules that react once the molecules collide. However, the catalyst cannot make the reactants collide more quickly. As a result, as you speed up a reaction with a catalyst, the rate of reaction will eventually be limited by the rate of collisions.

The rate of collisions is determined by the rate at which the reactants diffuse together. The diffusion rate is, in turn, determined by the mass transfer rate. The net effect is that mass transfer became more important as you improve catalysts, so mass transfer controls the rate of reaction on the very best catalysts.

When students first hear that mass transfer is controlling the rate, they think that the catalyst is not doing its job. However, catalysts are supposed to speed up rates of reactions, and if one speeds up a reaction enough, one always finds that diffusion starts to play a role. The best catalysts speed up a reaction so much that the reaction rate is instantaneous once the reactants collide. The result is that the mass transfer effects control the rate of reaction.

People usually quantify the role of mass transfer on the basis of two parameters: the **Thiele parameter**, Φ_p; and what I call a *mass transfer factor*, η. The Thiele parameter is defined by

$$\Phi_p = \frac{\text{reaction rate}}{\text{diffusion rate}} \tag{12.114}$$

The mass transfer factor is defined by

$$\eta_e = \frac{\text{actual reaction rate}}{\text{reaction rate if mass transfer were instantaneous}} \qquad (12.115)$$

In the literature people call the mass transfer factor an **effectiveness factor**. We derive equations for Φ_P and η_e in the supplemental material. The result for a first-order reaction occurring in a spherical catalyst pellet is

$$\eta_e = \frac{1}{\Phi_P}\left[\frac{1}{\tanh(3\Phi_P)} - \frac{1}{3\Phi_P}\right] \qquad (12.116)$$

$$\Phi_P = y_P\sqrt{\frac{k_1}{D_e}} \qquad (12.117)$$

where y_P is the radius of the pellet, k_1 is the rate constant for the reaction, and D_e is the diffusivity of the reactant in the pellet.

Figure 12.27 shows a plot of the mass transfer factor versus the Thiele parameter for diffusion into a solid catalyst pellet. Basically, the mass transfer factor is unity when the Thiele parameter is much less than one, and then tails off linearly when the Thiele parameter is greater than one. One can show that when $\phi_p > 10$, the term in brackets in equation (12.116) is between 0.9 and 1, under these circumstances:

$$\eta_e \approx \frac{0.95}{\phi_p} \qquad (12.118)$$

To keep Figure 12.27 in perspective, there is some confusion created by calling η_e an effectiveness factor. The mass transfer factor does not measure the effectiveness of a catalyst. *Good catalysts can have low mass transfer factors, while bad catalysts can have high mass transfer factors.* Generally, if a catalyst is not speeding up the reaction very much, the Thiele parameter will be small, which means that, according to Figure 12.27, the mass transfer factor will be close to unity. Thus, a bad catalyst can have a large mass transfer factor.

In contrast, if the catalyst speeds up a reaction by a considerable amount Φ_p will be large. Figure 12.27 shows that the mass transfer factor is reduced under such circumstances.

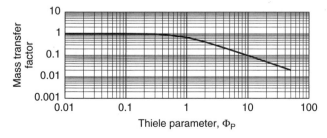

Figure 12.27 A plot of the mass transfer factor versus the Thiele parameter for diffusion in a porous catalyst pellet.

All of this is a balance because you can also get a large Φ_p if you have designed the catalyst pellet wrong so that D_e is too small. However, that is rarely an issue with modern support materials. Usually, a small mass transfer factor is associated with a very active catalyst. Consequently, *the mass transfer factor does not measure the effectiveness of a catalyst even though the mass transfer factor is often called an effectiveness factor in the literature.*

I believe it is important to think of η_e as a **mass transfer** factor, not an effectiveness factor. Generally, when η_e is small, you are near the mass transfer limit, so there is little that you can do to improve the catalyst except speedup the mass transfer rate. On the other hand, when η_e is large, the reaction rate is slow compared to the diffusion rate so that you can do to speed up the reaction. When I teach this material, I call η_e a mass transfer factor. However, η_e is still called an "effectiveness factor" in the literature.

12.15.1 Shape-Selective Catalysis

The other thing to recognize is that mass transfer limitations are not necessarily bad. In industrial practice, mass transfer limitations are used as design variables to improve the efficiency of catalysts.

One of the problems in designing a catalyst is that most catalytic materials speed up a wide variety of reactions. The rate of the desired reaction is enhanced. However, often the rates of some undesirable reactions are also enhanced. Fortunately, there is a design strategy to get around this difficulty:

- Use a solid catalyst.
- Design the solid in such a way that only the desired product can get out of the catalyst.
- Add enough catalytic components so that even if a side product is formed, the side product will be converted into the desired product.

An example of this strategy comes in the production of *para*-xylene. Xylene can be made via the alkylation of toluene over an acid catalyst

$$CH_3C_6H_5 + CH_3OH \Longrightarrow CH_3C_6H_4CH_3 + H_2O \qquad (12.119)$$

The alkylation reaction produces *meta*- and *ortho*-xylene in addition to *para*-xylene. People run the reaction in a catalyst that has acid sites in an interconnecting pore structure like that in Figure 12.28. The *o*-, *m*-, and *p*-xylene are formed at the acid sites, and then need to diffuse out of the pores before the xylene can leave the reactor.

Note that if you align *m*- and *o*-xylene standing with the methyl facing up, the *m*- and *o*-xylene are wider than the para-xylene. As a result, if one designs a structure like that in Figure 12.28 with pores that just fit para-xylene, *o*- and *m*-xylene will not be able to diffuse down the pores, so the *o*- and *m*-xylene get trapped in the catalyst. One then adds an occasional isomerization site to the catalyst. The isomerization site converts the *o*- and *m*-xylene to *p*-xylene. The net result is that the catalyst produces mainly *p*-xylene, even though the alkylation reaction, reaction (12.119), produces similar amounts of *o*-, *m*-, and *p*-xylene.

People call this type of catalyst a **shape-selective catalyst**, because you are able to select molecules according to their shape and how easily they diffuse. Shape-selective catalysis is very important, because one can design very selective catalysts. Still, it is not

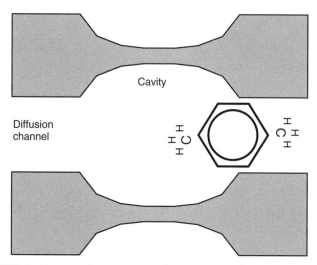

Figure 12.28 An interconnecting pore structure that is selective for the formation of paraxylene.

so easy to make a catalyst with just the right size pores, so this is an area where there is more art than science.

Table 12.12 lists several different materials that are used as shape-selective catalysts. Generally, shape-selective catalysts are made out of materials called **zeolites**. Zeolites are silica aluminas, with very uniform pores. Figure 12.5 shows a single unit cell in a zeolite. Generally, zeolites consist of small cages stacked on top of each other. There are small interconnecting pores between the cages. Table 12.12 shows the pore sizes in some zeolites, while Table 12.13 shows the dimensions of some typical molecules. Notice that you can choose the pore sizes so that only certain molecules can squeeze through the pores. Thus zeolite can be used to selectively produce a designed product, while not producing a larger side product.

Zeolites can be used with a wide variety of catalyst materials, and can be used to promote selectivity in a wide number of reactions. Therefore, they are very useful for catalyst design.

Table 12.12 Pore sizes in some zeolites

Zeolite	Size of Diffusion Channel, Å	Size of Cavity, Å
Chabazite	3.6 × 3.7	5
Zeolite A	4.1 × 4.1	6.5
Erondite	3.6 × 5.2	11.6
Ferrierite	4.3 × 5.5	6.5
ZSM-5	5.5 × 5.6	10.5
Offretite	6.4 × 6.4	6.5
Mordenite	6.7 × 7.0	10.5
Faugasite	7.4 × 7.4	11.9
VFI	13 × 13	—

Table 12.13 Minimum diameters of some molecules

Molecule	Minimum Diameter, Å
Linear alkane	4
Isoalkane	5.5
Benzene	5.1
Paraxylene	5.1
Orthoxylene	5.7
2-Methyl alkenes	5.1
Naphthalene	7.3

12.16 SUMMARY OF POTENTIAL CATALYST FUNCTIONS

In summary, then, in the last few sections, we showed that catalysts *can* change rates in eight key ways:

- Catalysts can be designed to help initiate reactions.
- Catalysts can be designed to stabilize the intermediates of a reaction.
- Catalysts can be designed to hold the reactants in close proximity.
- Catalysts can be designed to hold the reactants in the right configuration to react.
- Catalysts can be designed to block side reactions.
- Catalysts can be designed to sequentially stretch bonds and otherwise make bonds easier to break.
- Catalysts can be designed to donate and accept electrons.
- Catalysts can be designed to act as an efficient means for energy transfer.

It is also important to realize that

- One needs a catalytic cycle to get reactions to happen.
- Mass transfer limitations are more important when a catalyst is present.

One of the questions I get when I teach this material is "Here are all of these effects, how do I know which effect will be the most important in a given catalyst system?" Unfortunately, at present we rarely know the answer to that question. As noted in Section 12.1, one would like the catalyst to do all of these things. For example, platinum is a wonderful catalyst for olefin hydrogenation. The platinum speeds up the initiation step during the hydrogenation process. The platinum stabilizes the intermediates of the reaction. The platinum holds the reactants in close proximity. The platinum also lowers the intrinsic barrier to hydrogenation.

Platinum does everything so it is a great catalyst. It is not clear how to know that a priori. Generally, the most important effect of a catalyst is to stabilize intermediates and initiate reactions. However, all of the other effects also occur to some extent with any catalyst. Consequently, catalytic reactions are less well understood than, for example, gas-phase reactions.

12.17 KINETICS OF CATALYTIC REACTIONS

Next, we will be changing topics to discuss the kinetics of catalytic reactions. So far in this chapter, we learned that most catalytic reactions are the same. The reactants bind to the catalyst. Then, the system goes through a catalytic cycle where intermediates are formed and destroyed. Finally, the products leave the catalyst and the catalyst is regenerated. In the next two sections, we will derive equations for the rates of catalytic reactions. We discussed many of the details in Section 5.10, but it is useful to repeat some of the findings here:

- First, all reactions go by a catalytic cycle. For example, Figure 12.29 shows two different catalytic cycles for the production of water. The reactants adsorb, then react.
- There are three kinds of mechanisms as shown in Figure 12.30: a Langmuir–Hinshelwood mechanism where all of the reactants are adsorbed; a Rideal–Eley mechanism, where a gas-phase species collides with an adsorbed species and reacts; and a precursor mechanism, where there is one strongly bound reactant bound reactant and one reactant in a weakly bound precursor state.

There are several examples of these mechanisms in Section 5.10. The reader should review Section 5.10 before proceeding.

Next, we note that it is important to think about the rate of a catalytic reaction in terms of a **turnover number**, T_N. The turnover number is also called the **turnover frequency**.

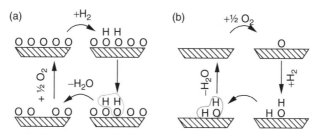

Figure 12.29 Catalytic cycles for the production of water via (a) disproportion of OH groups, and (b) the reaction $OH_{(ad)} + H_{(ad)} \rightarrow H_2O$.

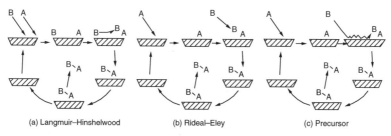

Figure 12.30 Schematic of (a) Langmuir–Hinshelwood, (b) Rideal–Eley, (c) precursor mechanism for the reaction $A + B \Rightarrow AB$ and $AB \Rightarrow A + B$.

In Chapter 2 we defined the turnover number as

$$T_N = \frac{R_A}{N_S} \tag{12.120}$$

where R_A is the rate of reaction per unit area and N_S is the number of metal atoms per unit area. Physically, the turnover number is the rate that the catalytic cycle occurs on each metal atom, measured in number of catalytic cycles per second.

Figure 12.31 shows some typical turnover numbers for catalytic reactions. The fastest reactions occur at rates of 100/second, while the slowest occur at 10^{-4}/second. There are a few examples of enzyme-catalyzed reactions that occur at 10^6/second. In my experience, in industrial reactions, turnover numbers are most commonly between 0.1 and 10/second. Faster reactions are usually mass-transfer-limited. Slower reactions are usually indicative of a catalyst needing improvement.

12.17.1 Langmuir Rate Laws

As mentioned at the beginning of this section, next we want to discuss the kinetics of catalytic reactions. We briefly reviewed the kinetics of catalytic reactions in Section 2.7. Recall that we found that catalytic reactions show complex kinetics. Rates do not vary linearly with concentration. Arrhenius plots are curved. For example, Figures 12.32 and 12.33 show data for the reaction $CO + \frac{1}{2}O_2 \Rightarrow CO_2$. Notice that as we increase the CO partial pressure, the rate goes up, reaches a maximum, and then declines. In the remainder of this section we will derive an equation for the rate of a catalytic reaction to try to understand why the rates show such complex behavior.

The derivation will be based on Langmuir's model of adsorption. Langmuir proposed that gases bind to fixed sites on a catalyst's surface. During reaction, reactants bind to the surface sites as illustrated in Figure 12.34. Some sites will be empty, and some sites will be covered by reactants A or B. Langmuir proposed that adsorption will occur only when a gas-phase A molecule reacts with an empty site. Consequently, the rate of adsorption will be proportional to S, the number of empty sites on the catalyst surface.

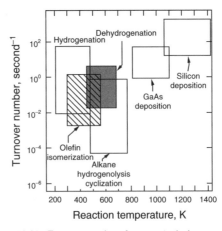

Figure 12.31 Turnover numbers for some typical processes.

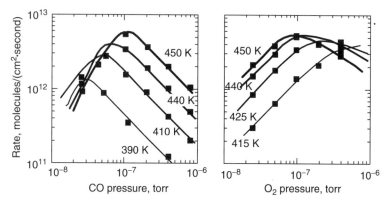

Figure 12.32   The influence of the CO pressure on the rate of CO oxidation on Rh(111). [Data of Schwartz et al. (1986).]

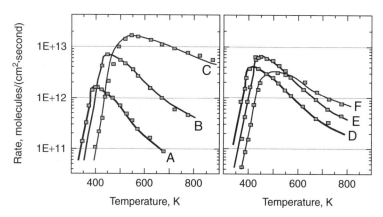

Figure 12.33   The rate of the reaction $CO + \frac{1}{2}O_2 \Rightarrow CO_2$ on Rh(111). (a) $P_{CO} = 2.5 \times 10^{-8}$ torr, $P_{O_2} = 2.5 \times 10^{-8}$ torr; (b) $P_{CO} = 1 \times 10^{-7}$ torr, $P_{O_2} = 2.5 \times 10^{-8}$ torr; (c) $P_{CO} = 8 \times 10^{-7}$ torr; $P_{O_2} = 2.5 \times 10^{-8}$ torr; (d) $P_{CO} = 2 \times 10^{-7}$ torr, $P_{O_2} = 4 \times 10^{-7}$ torr; (e) $P_{CO} = 2 \times 10^{-7}$ torr, $P_{O_2} = 2.5 \times 10^{-8}$ torr; (f) $P_{CO} = 2.5 \times 10^{-8}$ torr, $P_{O_2} = 2.5 \times 10^{-8}$ torr. [Data of Schwartz et al. (1986).]

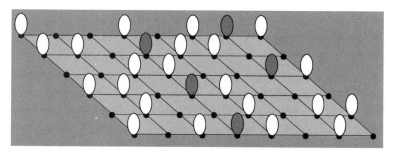

Figure 12.34   Langmuir's model for the adsorption of gas on a solid catalyst. The light gray area represents the surface of the catalyst. The black dots are the sites on the catalyst are sites that are available to adsorb gas. The white ovals are the adsorbed A molecules. The dark ovals are adsorbed B molecules.

Consider the reaction

$$A \Longrightarrow C \qquad (12.121)$$

and assume that the reaction occurs via the following mechanism:

$$S + A \underset{2}{\overset{1}{\rightleftharpoons}} A_{ad}$$

$$A_{Ad} \underset{4}{\overset{3}{\rightleftharpoons}} C_{ad} \qquad (12.122)$$

$$C_{ad} \underset{6}{\overset{5}{\rightleftharpoons}} C + S$$

In reaction (12.121), A is a gas-phase molecule, S is an *empty* site, A_{ad} is an adsorbed A molecule, C_{ad} is an adsorbed C molecule, and C is a gas-phase C molecule.

We will assume that the reaction is being run in the presence of a species B, which adsorbs on the catalyst:

$$S + B \underset{8}{\overset{7}{\rightleftharpoons}} B_{ad} \qquad (12.123)$$

but does not participate in the reaction.

During reaction (12.122), A reacts with a bare site on the catalyst, S, to yield an adsorbed complex. Then the adsorbed complex rearranges to form a new species, C, and then the C desorbs, regenerating a bare site.

For purposes of derivation, it is useful to define

$S_0 =$ the total concentration of sites available to adsorb gas (i.e., the total number of black dots in Figure 12.33), N/cm^2

$[S] =$ the concentration of empty sites, N/cm^2 (i.e., the number of black dots that do not contain A or B)

$[A_{ad}] =$ the concentration of adsorbed A molecules, N/cm^2

$[B_{ad}] =$ the concentration of adsorbed B molecules, N/cm^2

$[C_{ad}] =$ the concentration of adsorbed C molecules, N/cm^2

$P_A =$ the partial pressure of A

$P_B =$ the partial pressure of B

$P_C =$ the partial pressure of C

Following the analysis in Section 4.3, r_B, the net rate of formation of B, can be measured anywhere along the reaction path. If we consider reactions 3 and 4, we find that r_C is given by:

$$r_C = k_3[A_{ad}] - k_4[C_{ad}] \qquad (12.124)$$

where k_3 and k_4 are the rate constants for reactions 3 and 4, and $[A_{ad}]$ and $[C_{ad}]$ are the concentrations of adsorbed A and C. $[A_{ad}]$ and $[C_{ad}]$ have units of molecules/cm^2 (or mol/cm^2) for a solid catalyst, molecules/cluster for a metal cluster catalyst, or molecules/enzyme for an enzyme catalyst.

One can calculate $[A_{ad}]$ and $[C_{ad}]$ from the steady-state approximation:

$$0 = r_{A_{ad}} = k_1 P_A[S] - k_2[A_{ad}] - k_3[A_{ad}] + k_4[C_{ad}] \tag{12.125}$$

$$0 = r_{C_{ad}} = k_6 P_C[S] - k_5[C_{ad}] - k_4[C_{ad}] + k_3[A_{ad}] \tag{12.126}$$

where $r_{A_{ad}}$ and $r_{C_{ad}}$ are the rates of formation of adsorbed A and B, P_A and P_C are the partial pressures of A and C over the reactive surface, and $[S]$ is the concentration of empty surface sites.

In the literature, it is common to assume that reactions 3 and 4 are rate-determining. Under such circumstances, k_3 will be much smaller than k_2 and k_6, while k_4 will be much smaller than k_5 and k_1. Therefore the last two terms in equations (12.125) and (12.126) will be negligible. Under such circumstances

$$[A_{ad}] = \left(\frac{k_1}{k_2}\right) P_A[S] \tag{12.127}$$

$$[C_{ad}] = \left(\frac{k_6}{k_5}\right) P_C[S] \tag{12.128}$$

Similarly for B

$$[B_{ad}] = \frac{k_8}{k_7} P_B[S] \tag{12.129}$$

Rearranging equations (12.127)–(12.129) yields

$$\frac{[A_{ad}]}{P_A[S]} = \left(\frac{k_1}{k_2}\right) \tag{12.130}$$

$$\frac{[B_{ad}]}{P_B[S]} = \left(\frac{k_8}{k_7}\right) \tag{12.131}$$

$$\frac{[C_{ad}]}{P_c[S]} = \left(\frac{k_6}{k_5}\right) \tag{12.132}$$

Equations (12.130) and (12.131) imply that there is an equilibrium in the following reactions:

$$A + S \underset{2}{\overset{1}{\rightleftharpoons}} A_{ad}$$

$$B + S \underset{7}{\overset{8}{\rightleftharpoons}} B_{ad} \tag{12.133}$$

$$C + S \underset{5}{\overset{6}{\rightleftharpoons}} C_{ad}$$

One needs an expression for $[S]$ to complete the analysis. One can get an expression for $[S]$ by assuming that on any catalyst, there are a finite number of sites to hold the reactants. Each site can be bare, or it can be covered by A, B or C as indicated in Figure 12.33. If we define S_0 as the total number of sites in the catalyst, we can show

$$S_0 = [S] + [A_{ad}] + [B_{ad}] + [C_{ad}] \tag{12.134}$$

Substituting equations (12.127)–(12.129) into equation (12.134) and then solving for [S] yields

$$[S] = \frac{S_0}{1 + \dfrac{k_1}{k_2}P_A + \dfrac{k_8}{k_7}P_B + \dfrac{k_6}{k_5}P_C} \tag{12.135}$$

Substituting equation (12.135) into equations (12.127) and (12.128) yields

$$[A_{ad}] = \frac{\left(\dfrac{k_1}{k_2}\right)P_A S_0}{1 + \dfrac{k_1}{k_2}P_A + \dfrac{k_8}{k_7}P_B + \dfrac{k_6}{k_5}P_C} \tag{12.136}$$

$$[C_{ad}] = \frac{\left(\dfrac{k_6}{k_5}\right)P_C S_0}{1 + \dfrac{k_1}{k_2}P_A + \dfrac{k_8}{k_7}P_B + \dfrac{k_6}{k_5}P_C} \tag{12.137}$$

According to the analysis in Section 4.3, the equilibrium constant for the adsorption of A is given by

$$K_A = \frac{k_1}{k_2} \tag{12.138}$$

Similarly, the equilibrium constants for the adsorption of B, and C are given by

$$K_B = \frac{k_8}{k_7} \tag{12.139}$$

$$K_C = \frac{k_6}{k_5} \tag{12.140}$$

Substituting equations (12.138)–(12.140) into equations (12.136) and (12.137) yields

$$[A_{ad}] = \frac{K_A P_A S_0}{1 + K_A P_A + K_B P_B + K_C P_C} \tag{12.141}$$

$$[C_{ad}] = \frac{K_C P_C S_0}{1 + K_A P_A + K_B P_B + K_C P_C} \tag{12.142}$$

Substituting equations (12.141) and (12.142) into equation (12.124) yields

$$r = \frac{k_3 K_A P_A S_0 - k_4 K_C P_C S_0}{1 + K_A P_A + K_B P_B + K_C P_C} \tag{12.143}$$

In the catalysis literature, equation (12.143) is called the **Langmuir–Hinshelwood** expression for the rate of the reaction A $\Rightarrow$ C, since we assumed that the reaction obeyed a Langmuir–Hinshelwood mechanism and the equation was first derived for surface reaction by Langmuir (1915). The same equation is called the *Michaelis–Menten equation* in the enzyme literature since it was also derived by Michaelis and Menten (1913).

12.17.2 Langmuir–Hinshelwood–Hougan–Watson Rate Laws

Next, I want to talk about a trick. In the homework set, we ask the reader to derive several Langmuir–Hinshelwood–Michaelis–Menten rate equations. One could follow the derivation in Section 12.17. However, Hougan and Watson (1943) found a trick that makes it easier to do the derivation. In this section, we will discuss the trick.

Let's go back to the example in Section 12.17. Notice that k_3 is much smaller than k_2 while k_4 is much smaller than k_5 in equations (12.125) and (12.126). Consequently, the last two terms in equations (12.125) and (12.126) are negligible. Consider equation (12.125). If the last two terms in equation (12.125) are negligible, then

$$k_1 P_a [S] \cong k_2 [A_{ad}] \qquad (12.144)$$

Notice that the left side of equation (12.144) is the rate of reaction 1, while the right side of equation (12.144) is the rate of reaction 2. Therefore, the implication of equation (12.144) is that when steps 3 and 4 are rate-determining, the rate of reaction 1 will almost equal the rate of reaction 2. A similar analysis shows that when reactions 3 and 4 are rate-determining, the rate of reaction 5 must almost equal the rate of reaction 6.

Note, however, that this is only an approximation. The rate of reaction 5 cannot be exactly equal to the rate of reaction 6, because then there would be no net production of products. Similarly, if the rate of reaction 1 were exactly equal to the rate of reaction 2, there would be no net consumption of reactants. However, the implication of the preceeding derivation is that one can calculate an accurate value of the surface concentration of each of the species attached to the catalyst by assuming that the rate of reaction 1 approximately equals the rate of reaction 2 and the rate of reaction 5 approximately equals the rate of reaction 6, even though in reality, the rates are not exactly equal.

One can extend these ideas to other situations. Consider, for the moment, how the arguments would change if reactions 5 and 6 were rate-determining in reaction (12.122). If 5 and 6 were rate-determining, then a steady-state approximation on C_{ad} would give

$$0 = r_{B_{ad}} = k_3 [A_{ad}] - k_4 [C_{ad}] - k_5 [C_{ad}] \qquad (12.145)$$

As before, the last term in equation (12.145) is negligible:

$$k_3 [A_{ad}] \cong k_4 [C_{ad}] \qquad (12.146)$$

The left side of equation (12.146) is the rate of reaction 3, while the right side is the rate of reaction 4. Therefore, the implication of equation (12.146) is that one can calculate the concentrations of adsorbed A and C by assuming that the rate of reaction 3 is equal to the rate of reaction 4. Now consider reactions 1 and 2. The steady-state approximation for A_{ad} implies

$$0 = r_{A_{ad}} = k_1 P_A [S] - k_2 [A_{ad}] + k_4 [C_{ad}] - k_3 [A_{ad}] \qquad (12.147)$$

Note, however, that according to equation (12.146), the last two terms in equation (12.147) are approximately equal when reactions 5 and 6 are rate-determining. Therefore, when reactions 5 and 6 are rate-determining, one obtains

$$k_1 P_A [S] \cong k_2 [A_{ad}] \qquad (12.148)$$

The left side of equation (12.148) is the rate of reaction 1 and the right side is the rate of reaction 2. Therefore, the implication of equation (12.148) is that when reactions 5 and 6 are rate determining, the rate of reaction 1 is approximately equal to the rate of reaction 2. We note again that the rates are not exactly equal. However, the key result is that one can calculate accurate surface concentrations by assuming that the rates are almost equal even though the rates are not exactly equal.

One can generalize these results to say that when there is a simple rate-determining step in a reaction, one can calculate an accurate rate equation by assuming that

- All the steps before the rate-determining step are in equilibrium with the reactants. Consequently, the concentration of all of the species before the rate-determining step can be calculated via an equilibrium expression with the reactants.
- All the steps after the rate-determining step are in equilibrium with the products. Consequently, the concentration of all of the species after the rate-determining step can be calculated via an equilibrium expression with the products.
- Sites are conserved, so one can calculate the concentration of bare sites via a site balance.

For example, if we have the reaction $A + B \Rightarrow C$ following the mechanisms

$$A + S \rightleftharpoons A_{ad} \tag{12.149}$$

$$B_2 + 2S \rightleftharpoons 2B_{ad} \tag{12.150}$$

$$A_{ad} + B_{ad} \longrightarrow C \tag{12.151}$$

then the rate of production of C is given by

$$r_c = k_{151}[A_{ad}][B_{ad}] \tag{12.152}$$

If step (12.151) is rate-determining, then it is okay to assume that steps (12.149) and (12.150) are in equilibrium:

$$\frac{[A_{ad}]}{P_A[S]} = K_A \tag{12.153}$$

$$\frac{[B_{ad}]^2}{P_B[S]^2} = K_B \tag{12.154}$$

In equations (12.152), (12.153), and (12.154), [S] is concentration of empty sites, P_A and P_B are the partial pressures of A and B_2, $[A_{ad}]$ and $[B_{ad}]$ are the concentrations of adsorbed A and B, K_A and K_B are the equilibrium constants for the adsorption of A and B, and k_{150} is the rate constant for reaction (12.151). Some of the terms in equation (12.154) are squared because, according to (12.150), when B_2 adsorbs, it produces two adsorbed B atoms.

Solving equations (12.153) and (12.154) for $[A_{ad}]$ and $[B_{ad}]$ and substituting that result into equation (12.152) yields

$$r_c = k_{151}K_A(K_B)^{0.5}P_A(P_B)^{0.5}[S]^2 \tag{12.155}$$

Next, we need an expression for [S]. We get it from a site balance let's define S_0 as the total number of sites available to adsorb gas (i.e., black dots in Figure 12.33). Notice that the black dots can be empty or they can be covered by A or B. Therefore

$$S_0 = [S] + [A_{ad}] + [B_{ad}] \tag{12.156}$$

Substituting $[A_{ad}]$ and $[B_{ad}]$ from equations (12.153) and (12.154) into (12.156) yields

$$S_0 = [S] + K_A P_A [S] + (K_B)^{0.5} (P_B)^{0.5} [S] \tag{12.157}$$

Solving equation (12.157) for [S] yields

$$[S] = \frac{S_0}{1 + K_A P_A + (K_B)^{0.5} (P_B)^{0.5}} \tag{12.158}$$

Substituting (12.158) into (12.155) yields the Langmuir–Hinshelwood–Hougan–Watson rate equation:

$$r_c = \frac{k_{150} K_A (K_B)^{0.5} (S_0)^2 P_A (P_B)^{0.5}}{(1 + K_A P_A + (K_B)^{0.5} (P_B)^{0.5})^2} \tag{12.159}$$

We work out more examples in solved Examples 12.B and 12.C. The reader should examine those examples before proceeding.

12.17.3 Qualitative Features

Next, we want to discuss the predictions of the model. Figure 12.35 shows a plot of the rate of the A $\Rightarrow$ C reaction as a function of the reactant pressure calculated from equation (12.143) with $k_4 = 0$. Notice that the rate of reaction goes up and then levels off. Furthermore, as the partial pressure of B rises, the rate goes down even though we are considering the reaction A $\Rightarrow$ C, so B is not participating in the reaction.

Notice that the trends in Figure 12.35 are quite different from the trends seen with gas-phase reactions. In the gas phase, the rate would increase with increasing A concentration and never level off. A species such as B, which is not participating in the reaction, would have no effect on the rate.

It is useful to compare the results in Figure 12.35 to the experimental results in Figure 2.14. Notice that the experimental results follow the trends expected from equation (12.143). In fact, the lines in Figure 2.14 were calculated from equation (12.143). Therefore, it seems that the experiments do follow the trends in equation (12.143). Clearly, the kinetics of catalytic reactions look quite different from the kinetics of gas-phase reactions even though the mechanisms of the two reactions look quite similar.

The reason why the rate shows this weird behavior is that the concentration of A on the catalyst does not have a simple relationship to the partial pressure of A. Figure 12.36 is a plot of $[A_{ad}]$, the concentration of A on the catalyst, calculated from the equation (12.141). Notice that the concentration of A on the catalyst increases as the partial pressure of A increases, and then levels off. As we increase the partial pressure of B, the surface concentration of A on the catalyst decreases, even though there is no direct reaction between A and B.

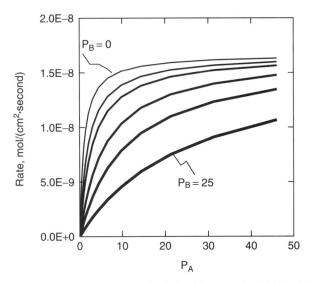

Figure 12.35 A plot of the rate of the reaction A ⇒ C calculated from equation (12.143) with $k_4 = 0$, $P_B = 0$, 1, 2, 5, 10 and 25, $K_A = K_B = 1$.

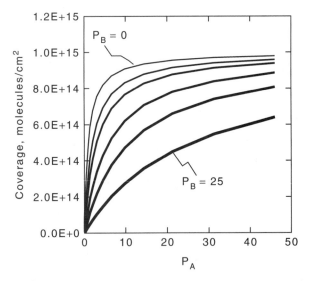

Figure 12.36 A plot of $[A_{ad}]$ calculated from the Langmuir adsorption isotherm, equation (12.141), with $K_A = 1$, $K_B = 1$, $P_B = 0, 1, 2, 5, 10, 25$.

This happens because there are only a finite number of sites on the catalyst to hold gas and A and B are competing for those sites. At low pressures, there are plenty of bare sites, so A simply adsorbs. However, at higher A pressures, the surface will start to fill up with gas. In that case, when one increases the partial pressure of A in the gas phase, one does not put that much more A onto the surface.

I find it useful to consider another quantity, θ_A, defined by

$$\theta_A = \frac{[A_{ad}]}{S_0} \tag{12.160}$$

where θ_A is the fraction of the sites on the catalyst that are covered with A. Substituting equation (12.141) into equation (12.160) shows

$$\theta_A = \frac{K_A P_A}{1 + K_A P_A + K_B P_B} \tag{12.161}$$

A similar derivation for B shows

$$\theta_B = \frac{K_B P_B}{1 + K_A P_A + K_B P_B} \tag{12.162}$$

Equation (12.161) is called the *Langmuir adsorption isotherm*. It was first derived by Langmuir (1912).

Figure 12.37 is a plot of θ_A as a function of P_A for various values of P_B. Notice that as θ_A approaches 1.0, a change in the partial pressure of A does not change the surface concentration of A significantly, because one cannot squeeze more A molecules onto the surface. Further, when P_B increases at constant P_A, the surface concentration of A decreases!

Molecule B takes up some of the surface sites and thereby partially blocks the adsorption of A. Consequently, as the partial pressure of B rises, the surface concentration of A decreases.

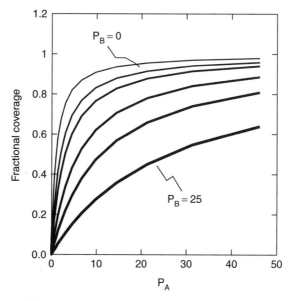

Figure 12.37   A plot of the Langmuir adsorption isotherm.

Next, it is useful to go back and consider the implication of these results for the reaction:

$$A \Longrightarrow C \tag{12.163}$$

According to equation (12.124), when $K_4 = 0$

$$r_c = k_3 [A_{ad}] \tag{12.164}$$

Therefore, the rate is proportional to the concentration of A on the catalyst. Figure 12.36 shows a plot of the concentration of A on the surface. Notice that at high A pressures, $[A_{ad}]$ is almost independent of P_A. Consequently, according to (12.163), increases in P_A do not significantly enhance the rate of reaction (12.163). Similarly, the presence of B reduces $[A_{ad}]$. Consequently, the rate of reaction (12.163) goes down. The result is that the presence of B decreases the rate of the reaction even though B does not directly participate in the reaction.

Notice, however, that according to equation (12.162), the fractional coverage of B goes to zero when P_A is large. Therefore, B will not affect the rate at high enough A partial pressures.

Experimentally, one usually observes such trends, which is why equation (12.143) has prove to be so useful.

12.17.4 Alternative Isotherms

There is one other detail. When you fit real adsorption data, you sometimes have to use an equation for θ other than the Langmuir adsorption isotherm. In the literature, there are a number of adsorption isotherms, where an *adsorption isotherm* is defined as an equation or plot of $[A_{ad}]$ versus P_A. Masel (1996) lists several isotherms. The most important are the Toth equation

$$\theta_A = \left(\frac{K_A P_A^{C_{Toth}}}{1 + K_A P_A^{C_{Toth}}} \right)^{1/C_{Toth}} \tag{12.165}$$

and the multisite model:

$$\theta_A = \sum_{i=1}^{n} \left(\frac{X_i K_A^i P_A}{1 + K_A^i P_A} \right) \tag{12.166}$$

In equations (12.165) and (12.166) i is a sum over different sites in a sample and C_{toth}, X_i, K_a^i are parameters that are adjusted to fit data as described in Chapter 4 of Masel (1996). The Toth equation arises because of attractive or repulsive interactions between molecules. The multisite model works because there can be different sites (i.e., places where reactants can adsorb) on a catalyst with different activities. For condensable substances (e.g., H_2O), people sometimes use what is called the Brunauer–Emmett–Teller (BET) equation:

$$[A_{ad}] = \frac{S_0 c_B \left(\dfrac{P_A}{P_A^S} \right)}{\left(1 - \dfrac{P_A}{P_A^S} \right) \left[1 + (c_B - 1) \dfrac{P_A}{P_A^S} \right]} \tag{12.167}$$

where $[A_{ad}]$ equals the concentration of A on the catalyst, S_0 is the concentration of sites on the catalyst, P_A is the partial pressure of A, and P_A^S is the pressure you would have to go to to condense toward A. equation (12.167) arises because, with a condensable gas, you can get what is called a *multilayer*, where additional layers of gas condense on top of the first layer of gas.

Experimentally, one usually gets better kinetics with the Langmuir adsorption isotherm than with any other models. However, there are exceptions. See Masel (1996) for details.

12.17.5   Extension to More Complex Reactions

So far, we have discussed only the reaction A $\Rightarrow$ C, but it is helpful to consider a slightly more complicated case:

$$A + B \Longrightarrow C \tag{12.168}$$

This case was first considered by Langmuir [1919]. Langmuir assumed that the reaction followed a Langmuir-Hinshelwood mechanism:

$$
\begin{aligned}
S + A &\Longrightarrow A_{ad} & (1) \\
S + B &\Longrightarrow B_{ad} & (2) \\
A_{ad} + B_{ad} &\Longrightarrow C + 2S & (3)
\end{aligned}
\tag{12.169}
$$

Langmuir also assumed that step (3) in reaction (12.169) was much slower than the rest, so the surface concentrations of A and B maintained a dynamic equilibrium. If reaction 3 is elementary, then r_C, the rate of formation of C, is equal to

$$r_C = k_3[A_{ad}][B_{ad}] \tag{12.170}$$

where $[A_{ad}]$ and $[B_{ad}]$ are the concentrations of A and B on the surface of the catalyst in molecules/cm^2 and k_3 is the rate constant for reaction 3. If it is assumed that there are a fixed number of sites, S_0, to hold gas then equation (12.168) can be rewritten

$$r = k_{AB}\theta_A\theta_B \tag{12.171}$$

where θ_A and θ_B are the fractional coverage of A and B, that is

$$\theta_A = \frac{[A_{ad}]}{S_0}, \qquad \theta_B = \frac{[B_{ad}]}{S_0} \tag{12.172}$$

and $k_3(S_0)^2 = k_{AB}$. Langmuir then assumed that the adsorption of A and B would follow a Langmuir adsorption isotherm from equations (12.161) and (12.162) to obtain

$$r_c = \frac{k_{AB}K_AK_BP_AP_B}{(1 + K_AP_A + K_BP_B)^2} \tag{12.173}$$

Equation (12.173) is the key result in this section. One should memorize it before proceeding.

Figure 12.38 shows a plot of the r_C/k_{AB} as a function of P_A calculated from equation (12.173) with $K_A = K_B = 1$. Notice that the rate goes up with increasing P_A, reaches a maximum, and then declines. The results are in qualitative agreement with the data in

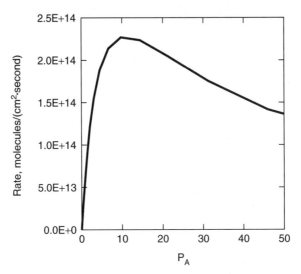

Figure 12.38   A plot of the rate calculated from equation (12.173) with $K_B P_B = 10$.

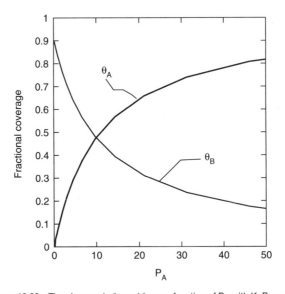

Figure 12.39   The changes in θ_A and θ_B as a function of P_A with $K_B P_B = 10$.

Figure 2.17. In fact, the fit is quantitative. The lines in Figure 2.17 are actual fits to the data via equation (5.127).

Just to keep this plot in perspective, note that for a gas-phase reaction, an increase in the partial pressure of the reactants always produces an increase in the reaction rate. However, the data in Figure 2.17 and the plot in Figure 12.38 show that with a catalytic reaction, an increase in the partial pressure of the reactants may decrease, not increase, the rate.

Physically, what is happening is that according to equation (12.161), when P_A increases, more A will adsorb. As a result, θ_A will increase. However, according to equation (12.162), when P_A increases, θ_B decreases because there are fewer vacant sites to hold B (see Figure 12.39). Now consider the product of θ_A and θ_B in equation (12.171). Note that since θ_A goes up and θ_B goes down, the product of the two can increase or decrease with increasing pressure of A. If the fractional decrease in θ_B is smaller than the fractional increase in θ_A, the net rate will increase. However, if the fractional decrease in θ_B exceeds the fractional increase in θ_A, the net rate will decrease. Consequently, an increase in the partial pressure of A will not necessarily increase the rate of the reaction between A and B, and it may decrease it.

12.17.6 Further Extensions

One can extend these ideas to a wide variety of reactions. For example, consider the case where reaction (12.168) is run in the presence of a species D_2 that dissociatively adsorbs on the catalyst and blocks sites but does not otherwise participate in the reaction. One can write equilibrium constants for A and B as before:

$$\frac{[A_{ad}]}{P_A[S]} = K_A \tag{12.174}$$

$$\frac{[B_{ad}]}{P_B[S]} = K_B \tag{12.175}$$

Now let's assume that D_2 dissociatively adsorbs to form two D atoms, that is, that during the adsorption process, D follows

$$D_2 + 2S \rightleftharpoons 2D_{ad} \tag{12.176}$$

At equilibrium:

$$\frac{[D_{ad}]^2}{P_{D_2}(S_0)^2} = K_D \tag{12.177}$$

where $[D_{ad}]$ is the concentration of adsorbed D atoms, P_{D_2} is the partial pressure of D_2, S_0 is the number of sites in the catalyst, and K_D is the equilibrium constant for the adsorption of D. The site balance becomes

$$S_0 = [S] + [A_{ad}] + [B_{ad}] + [D_{ad}] \tag{12.178}$$

Substituting equations (12.174), (12.175), and (12.177) into equation (12.178) and rearranging yields

$$[S] = \frac{S_0}{1 + K_A P_A + K_B P_B + (K_D P_{D_2})^{1/2}} \tag{12.179}$$

Substituting equation (12.179) into Equations (12.174) and (12.175), and then substituting the result into equation (12.170), yields

$$R_C = \frac{k_{AB} K_A K_B P_A P_B}{\left(1 + K_A P_A + K_B P_B + (K_D P_{D_2})^{1/2}\right)^2} \tag{12.180}$$

Equation (12.180) is the rate equation for reaction (12.168) in the presence of an inert species D_2 that dissociates.

Masel (1996, Chapter 7) provides many other examples of these methods. In the problem set, we also ask the readers to do some similar examples. The derivations are simple but amazingly powerful. One can fit kinetic data for most catalytic reactions, most enzyme reactions, and many reactions in electronic materials production using equation (12.180).

12.18   SUMMARY

In summary, then, in this chapter, we described many different types of catalysts:

Homogeneous catalysts:

- Acids or bases
- Metal salts
- Enzymes
- Radical initiators
- Solvents

Heterogeneous catalysts:

- Supported metals
- Metal oxides, carbides, and sulfides
- Solid acids and bases
- Immobilized enzymes and other polymer-bound species
- Photocatalysts

We found that all catalysts work in basically the same way:

- Catalysts can be used to help initiate reactions.
- Catalysts can be used to stabilize intermediates.
- Catalysts can be used to hold the reactants in close proximity.
- Catalysts can be used to hold the reactants in the right configuration to react.
- Catalysts can be used to block side reactions.
- Catalysts can be used to modify the forces between the reactants, which changes the intrinsic barriers to reactions.
- Catalysts can be used to act as efficient means for energy transfer.

It is also important to realize that

- One needs a catalytic cycle to get a reaction to happen.
- Mass transfer limitations are more important when a catalyst is present.
- Most catalytic reactions also follow Langmuir–Hinshelwood rate laws.

In the next few chapters, we will be examining each type of catalyst in detail. However, we recommend that the reader carefully study the materials in this chapter before proceeding to Chapters 13 and 14.

12.19 SUPPLEMENTAL MATERIAL

12.19.1 Derivation of equation 12.116

In section 12.15, we provided the following approximation for the effectiveness factor for the reaction $A \Rightarrow B$ occurring in a spherical pellet of radius y_p:

$$\eta_e = \frac{1}{\Phi_P}\left(\frac{1}{\tanh(3\Phi_P)} - \frac{1}{\Phi_P}\right) \tag{12.181}$$

with Φ_P given by equation (12.117), or equivalently (for first order)

$$\Phi_P^2 = \frac{y_P^2\left(-r_A^0\right)}{C_A^0 D_e} \tag{12.182}$$

where y_p is the diameter of the pellet $\left(-r_A^0\right)$ is the reaction rate (in the absence of mass transfer), D_e is the effective diffusing into the pellet, and C_A^0 is the concentration of the reactant outside the pellet.

Equation (12.181) predicts that the actual rate of reaction is less than the rate in the absence of mass transfer resistances. The rate is reduced because the reactants are used up as they diffuse into the pellet, as shown in Figure 12.40. The average concentration of reactants in the pellet is less than the concentration in the gas phase. As a result, the rate is less than the rate one would have if the concentrations were equal to the gas-phase concentration.

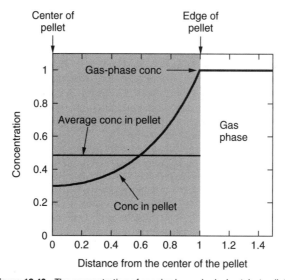

Figure 12.40 The concentration of species in a spherical catalyst pellet.

The purpose of this section is to derive equation (12.181). The effectiveness factor η_e is defined by

$$\eta_e = \frac{\text{actual rate for a catalyst pellet with mass transfer limitations}}{\text{rate in the absence of mass transfer limitations}} \qquad (12.183)$$

Let's consider the reaction $A \Rightarrow B$ occurring in the spherical pellet shown in Figure 12.40.

In this section, we derive an equation for the effectiveness factor assuming that all of the mass transfer resistance is associated with diffusion of reactants into the pellet. We will also assume that the diffusion process follows Fick's law with an *effective diffusiity*, D_e.

First let us derive an expression of the concentration of A diffusing into the pellet. A mass balance on A shows

$$\left(4\pi y^2 D_e \left(-\frac{dC_A}{dy}\right)\right)_{y+\Delta y} - \left(4\pi y^2 D_e \left(-\frac{dC_S}{dy}\right)\right)_y = 4\pi y^2 \Delta y \,(-r_A) \qquad (12.184)$$

where y is the distance from the center of the pellet, C_A is the concentration of A at a distance y, Δy is the thickness of the slice, D_e is the effective diffusivity, and r_A is the rate of the $A \Rightarrow B$ reaction. The first term in equation (12.184) is the amount that diffuses into the slice. The second term in equation (12.184) is the amount that diffuses out. The last term is the amount that is used up. The negative sign in the first term expression arises because A is diffusing in the negative direction.

Taking the limit as Δy goes to zero yields

$$\frac{d^2 C_A}{dy^2} + \frac{2}{y}\frac{dC_A}{dy} + \frac{r_A}{D_e} = 0 \qquad (12.185)$$

The boundary conditions on equation (12.185) are

$$\frac{dC_A}{dy} = 0 \quad \text{at} \quad y = 0 \qquad (12.186)$$

$$C_A = C_A^0 \quad \text{at} \quad y = y_p \qquad (12.187)$$

Note that equation (12.186) arises because the concentration profile must be symmetric around $y = 0$, or else the second term in equation (12.185) will grow to infinity.

In order to solve equation (12.187), it is useful to define a new variable, $\mathbb{C}_A$, by

$$\mathbb{C}_A = \frac{yC_A}{y_p C_A^0} \qquad (12.188)$$

where C_A^0 is the concentration of A in the fluid and y_p is the radius of the pellet.

One can show

$$\frac{C_A^0 y_p}{y}\frac{d^2\mathbb{C}_A}{dy^2} = \frac{d^2 C_A}{dy^2} + \frac{2}{y}\frac{dC_A}{dy} \qquad (12.189)$$

Equation (12.185) becomes

$$\left(\frac{C_A^0 y_p}{y}\right)\frac{d^2\mathbb{C}_A}{dy^2} + \frac{r_A}{D_e} = 0 \qquad (12.190)$$

For simplicity, we will assume

$$r_A = -k_A C_A \tag{12.191}$$

Substituting equations (12.188) and (12.191) into equation (12.190) yields

$$\frac{d^2 C_A}{dy^2} - \frac{k_A}{D_e} C_A = 0 \tag{12.192}$$

The solution of equation (12.192) is

$$C_A = \frac{\sinh\left(\sqrt{\frac{k_A}{D_e}} y^4\right)}{\sinh\left(\sqrt{\frac{k_A}{D_e}} y_P\right)} \tag{12.193}$$

Converting equation (12.193) to standard form yields

$$C_A = C_A^0 \frac{y_P}{y} \frac{\sinh\left(3\Phi y/y_P\right)}{\sinh\left(3\Phi_P\right)} \tag{12.194}$$

where Φ_P is called the Thiele modulus defined by

$$\Phi_P = \frac{y_P}{3} \sqrt{\frac{k_A}{D_e}} \tag{12.195}$$

Next, we want to compute how much product is consumed by the pellet. There are two ways to do that. One could integrate the rate over the volume of the pellet. However, we can also calculate how much A is diffusing into the pellet and note that at steady state the amount of A consumed by the pellet equals the amount of A that diffuses into the pellet. The algebra is simpler if we choose the latter derivation. From the definition of the diffusivity, we have

$$(\text{A consumed}) = -4\pi y_P^2 D_e \frac{dC_A}{dy} \tag{12.196}$$

Next, let's define r_A^{mr}, the average rate of reaction per unit volume of pellet. From a mass balance

$$(\text{A consumed}) = \frac{4\pi}{3} y_P^3 r_A^{mr} \tag{12.197}$$

Combining equations (12.196) and (12.197) yields

$$r_A^{mr} = -\frac{3D_e}{y_P} \left(\frac{dC_A}{dy}\right)_{y=y_P} \tag{12.198}$$

Substituting equation (12.194) into equation (12.198) yields

$$r_A^{mr} = -\frac{3D_e C_A^0}{y_P} \left[\frac{3\phi_P}{y_P} \frac{\cosh\left(3\Phi_P\right)}{\sinh\left(3\Phi_P\right)} - \left(\frac{y_P}{(y_P)^2}\right) \frac{\sinh\left(3\Phi_P\right)}{\sinh\left(3\Phi_P\right)}\right] \tag{12.199}$$

Factoring out $3\Phi_P/y_p$ from the big bracket on the left of equation (12.199) yields

$$r_A^{mr} = \frac{3D_e C_A^0}{y_P}\left(\frac{3\Phi_P}{y_P}\right)\left(\frac{1}{\tanh(3\Phi_P)} - \frac{1}{3\Phi_P}\right) \qquad (12.200)$$

Further rearranging yields

$$r_A^{mr} = -\left(\frac{9D_e}{y_P^2}\right)(C_A^0)(\Phi_P)\left(\frac{1}{\tanh(3\Phi_P)} - \frac{1}{3\Phi_P}\right) \qquad (12.201)$$

Note that from equation (12.195)

$$\frac{1}{\Phi_P^2} = \left(\frac{9D_e}{y_P^2}\right)\left(\frac{1}{k_A}\right) \qquad (12.202)$$

Substituting equation (12.202) into equation (12.201) yields

$$r_A^{mr} = -\left(\frac{k_A}{\Phi_P^2}\right)C_A^0\Phi_P\left(\frac{1}{\tanh(3\Phi_P)} - \frac{1}{3\Phi_P}\right) \qquad (12.203)$$

The effectiveness factor η_e is defined by

$$\eta_e = \frac{r_A^{mr}}{r_A(\text{no mass transfer limitation})} \qquad (12.204)$$

where from equation (12.191)

$$r_A(\text{no mass transfer limitation}) = -k_A C_A^0 \qquad (12.205)$$

Substituting equation (12.203) and (12.205) into equation (12.204) yields

$$\eta_e = \frac{1}{\Phi_P}\left(\frac{1}{\tanh 3\Phi_P} - \frac{1}{3\Phi_P}\right) \qquad (12.206)$$

Equation (12.206) was cited in Section 12.15.

In the homework set, we ask the reader to do a similar derivation for diffusion into a flat plate of thickness 2 L. The result is

$$\eta_e = \frac{\tanh \Phi_P}{\Phi_P} \qquad (12.207)$$

where

$$\Phi_P = L\sqrt{\frac{k_a}{D_e}} \qquad (12.208)$$

Similarly, one can integrate equation (12.185) for other rate forms (see also Figure 12.41).

$$r_A = kC_A^n \qquad (12.209)$$

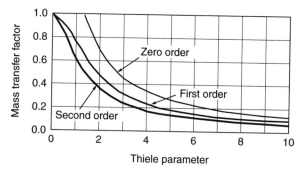

Figure 12.41 A plot of the effectiveness factor for reaction in a flat plate where the rate follows equations (12.207) and (12.208).

and

$$r_A = \frac{k_2 C_A}{(1 + k_2 C_A)^m}$$ (12.210)

In all cases equation (12.206) is an excellent approximation whenever the factor is greater than about 2.

12.20   SOLVED EXAMPLES

Example 12.A   Fitting Data to Langmuir–Hinshelwood and Michaelis–Menten Rate Forms   Steger and Masel (1998) examined the etching of copper with hexafluoropentanedione ($CF_3COCH_2COCF_3$). The main reaction is

$$4(CF_3COCH_2COCF_3) + 2Cu + O_2 \Longrightarrow 2Cu(CF_3COCHCOCF_3)_2 + 2H_2O \quad (12.A.1)$$

Table 12.A.1 shows some data for the rate of copper etching via the reaction. Fit the data to the the Langmuir–Hinshelwood expression:

$$R_e = \frac{k_1 P_{O_2}^{0.5}}{1 + K_2 P_{O_2}^{0.5}}$$ (12.A.2)

where R_e is the etch rate, P_{O_2} is the partial pressure of oxygen, and k_1 and K_2 are constants.

Table 12.A.1   The rate of copper etching measured by Steger and Masel

R_e μm/minute	P_{O_2}, torr	R_e μm/minute	P_{O_2}. torr	R_e μm/minute	P_{O_2}, torr
0.3	0.1	0.8	1.5	1.2	7.5
0.5	0.3	0.9	2.0	1.3	10.0
0.6	0.5	1.0	3.0	1.4	15.0
0.8	1.0	1.1	5.0	1.5	20.0

Solution   Generally people try to solve problems of this type by transforming equation (12.A.2) to a linear form and then using a least-squares approach to fit the data. There are two key transformations: the Lineweaver–Burke transformation and the Eadie–Hofstee transformation as discussed in example 3.A. Both transformations were developed by biochemists to fit data to fit data to the Michaelis–Menten equation and then were adopted by catalytic chemists.

The Lineweaver–Burke transformation rearranges equation (12.A.2) by taking one over both sides:

$$\frac{1}{R_e} = \frac{1}{k_1 P_{O_2}^{0.5}} + \frac{K_2}{k_1} \tag{12.A.3}$$

Therefore a plot of $1/R_e$ versus $P_{O_2}^{-0.5}$ should be linear.

The Eadie–Hofstee transformation can be derived by multiplying equation (12.A.2) by $(1 + P_{O_2}^{0.5})$:

$$R_e + R_e K_2 P_{O_2}^{0.5} = k_1 P_{O_2}^{0.5} \tag{12.A.4}$$

Dividing equation (12.A.4) by $P_{O_2}^{0.5}$ yields

$$\frac{R_e}{P_{O_2}^{0.5}} = k_1 - R_e K_2 \tag{12.A.5}$$

Therefore the plot of $R_e/P_{O_2}^{0.5}$ versus R_e should be linear.

One can also rearrange the Lineweaver–Burke transformation as

$$\frac{P_{O_2}^{0.5}}{R_e} = \frac{1}{k_1} + \frac{K_2}{k_1} P_{O_2}^{0.5}$$

Therefore the plot of $P_{O_2}^{0.5}/R_e$ versus $P_{O_2}^{0.5}$ should be linear.

If one had perfect data measured to several significant figures, one would get the same result, independent of whether one plotted the data as shown in equation (12.A.3), (12.A.5), or (12.A.6). However, in actual practice the three methods give slightly different results. Let's illustrate that point using the data in Table 12.A.1. Figure 12.A.1 shows a plot of the data using three methods. Notice that all the plots look linear, which suggests that all three methods work.

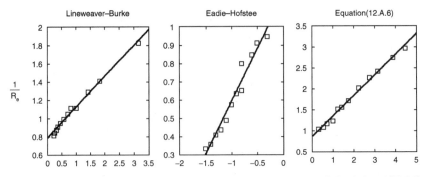

Figure 12.A.1   A plot of the data in Table 12.A.1 according to equations (12.A.3), (12.A.5), and (12.A.6).

The plots are not completely revealing, though, so one needs numbers to compare the results.

I have analyzed the data with the help of the spreadsheet in Table 12.A.2 . In the spreadsheet, column A is the pressure, column B is the rate, column C is one over the square root of the pressure, column D is one over the rate, column E is minus the rate, column F is the rate over the square root of the pressure, column G is the square root of the pressure, and column H is the square root of the pressure divided by the rate.

Table 12.A.3 shows the numerical values in the spreadsheet.

In the Lineweaver–Burke method, one fits the data to equation (12.A.3). Column C should be linear with column D. In fact, it is, and the regression output (cells D16 and D17) show

$$\frac{1}{R_e} = \frac{0.879}{P_{O_2}^{0.5}} + 0.477 \tag{12.A.7}$$

A comparison of equations (12.A.3) and (12.A.7) yields

$$\frac{1}{k_1} = 0.879 \tag{12.A.8}$$

or

$$k_1 = \frac{1}{0.879} = 1.137 \tag{12.A.9}$$

Similarly, comparison of equations (12.A.3) and (12.A.7) shows

$$\frac{K_2}{k_1} = 0.477 \tag{12.A.10}$$

or

$$K_2 = 0.477 \times k_1 = 0.477 \times 1.137 = 0.544 \tag{12.A.11}$$

In the Eadie–Hofstee method, one fits the data to equation (12.A.5). Column E should be linear as column F. The regression output 14 (cells F16 and F17) show.

$$\frac{R_e}{P_{O_2}^{0.5}} = 1.166 - 0.577 \times R_e \tag{12.A.12}$$

Therefore

$$k_1 = 1.166 \tag{12.A.13}$$

$$K_2 = 0.577 \tag{12.A.14}$$

Similarly, one can fit the data to equation (12.A.6). Column G should be linear with column H. The regression out put (cells H16 and H17) shows.

$$\frac{P_{O_2}^{0.5}}{R_e} = 0.869 + 0.490 \times P_{O_2}^{0.5} \tag{12.A.15}$$

Therefore

$$\frac{1}{k_1} = 0.869 \tag{12.A.16}$$

Table 12.A.2 The formulas used to fit the rate data by the various methods

	A	B	C	D	E	F	G	H
			Lineweaver–Burke		Eadie–Hoffstee		Equation (12.A.6)	
01								
02	pressure	rate	1/p^0.5	1/rate	"–rate	"rate/p^0.5	p^.5	p^0.5/rate
03	0.1	0.3	=1/A3^0.5	=1/B3	=–B3	=+B3/(A3^0.5)	=+A3^0.5	=+G3/B3
04	0.3	0.5	=1/A4^0.5	=1/B4	=–B4	=+B4/(A4^0.5)	=+A4^0.5	=+G4/B4
05	0.5	0.6	=1/A5^0.5	=1/B5s	=–B5	=+B5/(A5^0.5)	=+A5^0.5	=+G5/B5
06	1	0.8	=1/A6^0.5	=1/B6	=–B6	=+B6/(A6^0.5)	=+A6^0.5	=+G6/B6
07	1.5	0.8	=1/A7^0.5	=1/B7	=–B7	=+B7/(A7^0.5)	=+A7^0.5	=+G7/B7
08	2	0.9	=1/A8^0.5	=1/B8	=–B8	=+B8/(A8^0.5)	=+A8^0.5	=+G8/B8
09	3	1	=1/A9^0.5	=1/B9	=–B9	=+B9/(A9^0.5)	=+A9^0.5	=+G9/B9
10	5	1.1	=1/A10^0.5	=1/B10	=–B10	=+B10/(A10^0.5)	=+A10^0.5	=+G10/B10
11	7.5	1.2	=1/A11^0.5	=1/B11	=–B11	=+B11/(A11^0.5)	=+A11^0.5	=+G11/B11
12	10	1.3	=1/A12^0.5	=1/B12	=–B12	=+B12/(A12^0.5)	=+A12^0.5	=+G12/B12
13	15	1.4	=1/A13^0.5	=1/B13	=–B13	=+B13/(A13^0.5)	=+A13^0.5	=+G13/B13
14	20	1.5	=1/A14^0.5	=1/B14	=–B14	=+B14/(A14^0.5)	=+A14^0.5	=+G14/B14
15								
16			slope	=SLOPE(D3:D14,C3:C14)	slope	=SLOPE(F3:F14,E3:E14)	slope	=SLOPE(H3:H14,G3:G14)
17			intercept	=INTERCEPT(D3:D14,C3:C14)	intercept	=INTERCEPT(F3:F14,E3:E14)	intercept	=INTERCEPT(H3:H14,G3:G14)
18			r2	=RSQ(D3:D14,C3:C14)	r2	=RSQ(F3:F14,E3:E14)	r2	=RSQ(H3:H14,G3:G14)

Table 12.A.3   The numerical values in the spreadsheet used to fit the rate data by the various methods

	A	B	C	D	E	F	G	H
01			Lineweaver–Burke		Eadie–Hoffstee		Equation (12.A.6)	
02	pressure	rate	1/p^0.5	1/rate	"-rate	"rate/p^0.5	p^.5	p^0.5/rate
03	0.1	0.3	3.162278	3.333333	-0.3	0.948683	0.316228	1.054093
04	0.3	0.5	1.825742	2	-0.5	0.912871	0.547723	1.095445
05	0.5	0.6	1.414214	1.666667	-0.6	0.848528	0.707107	1.178511
06	1	0.8	1	1.25	-0.8	0.8	1	1.25
07	1.5	0.8	0.816497	1.25	-0.8	0.653197	1.224745	1.530931
08	2	0.9	0.707107	1.111111	-0.9	0.636396	1.414214	1.571348
09	3	1	0.57735	1	-1	0.57735	1.732051	1.732051
10	5	1.1	0.447214	0.909091	-1.1	0.491935	2.236068	2.032789
11	7.5	1.2	0.365148	0.833333	-1.2	0.438178	2.738613	2.282177
12	10	1.3	0.316228	0.769231	-1.3	0.411096	3.162278	2.432521
13	15	1.4	0.258199	0.714286	-1.4	0.361478	3.872983	2.766417
14	20	1.5	0.223607	0.666667	-1.5	0.33541	4.472136	2.981424
15								
16			slope	0.879058	slope	0.576817	slope	0.489952
17			intercept	0.477853	intercept	1.165903	intercept	0.869251
18			r2	0.994711	r2	0.961547	r2	0.992587

or

$$k_1 = 1.150 \qquad\qquad (12.A.17)$$

Similarly

$$\frac{K_2}{k_1} = 0.490 \qquad\qquad (12.A.18)$$

or

$$K_2 = 0.490 \times k_1 = 0.490 \times 1.150 = 0.564 \qquad\qquad (12.A.19)$$

Table 12.A.4 summarizes these results. Notice that the three methods give different values of the coefficients although the differences are not large.

Figure 12.A.2 shows a replot of the data on a semilog scale. In this particular case there is almost no difference between the methods.

In the homework set we will ask the reader to show that the Lineweaver–Burke plot is sensitive to errors. Physically, when one uses the Lineweaver–Burke method, one is fitting to 1/Re rather than Re. 1/Re is very sensitive to noise at low concentrations. As a result, Lineweaver–Burke plots sometimes fail.

If you try to derive your own method, be careful to compare errors on Re not 1/Re, or else you will do a poor job of fitting real data.

In my experience, it is better to avoid the linear methods and use nonlinear least-squares method instead. The nonlinear least-squares method can be done using the solver function in Microsoft Excel or Lotus 1-2-3.

Table 12.A.4 Fits of the data to equations (12.A.3), (12.A.5), and (12.A.6)

	Lineweaver–Burke	Eadie–Hofstee	Equation (12.A.6)
Slope	0.879	0.577	0.490
Intercept	0.477	1.166	0.869
Regression coefficient	0.9994711	0.961547	0.992587
k_1	1.137	1.166	1.150
K_2	0.544	0.577	0.564

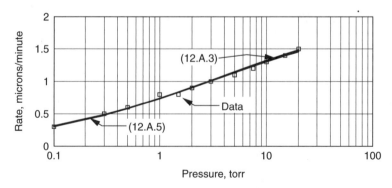

Figure 12.A.2 A plot of the data in Table 12.A.1 fit by using the Lineweaver–Burke method and the Eadie–Hofstee method.

Here is the spreadsheet I set up to do the calculations:

	A	B	C	D	E
5	pressure	rate	rate calculated from equation(12.A.2)	error^2	abs(error)
6	0.1	0.3	=C$19*($A6^0.5)/(1+C$20*($A6^0.5))	=(C6-$B6)^2	=ABS(C6-$B6)
7	0.3	0.5	=C$19($A7^0.5)/(1+C$20*($A7^0.5))	=(C7-$B7)^2	=ABS(C7-$B7)
8	0.5	0.6	=C$19*($A8^0.5)/(1+C$20*($A8^0.5))	=(C8-$B8)^2	=ABS(C8-$B8)
9	1	0.8	=C$19*($A9^0.5)/(1+C$20*($A9^0.5))	=(C9-$B9)^2	=ABS(C9-$B9)
10	1.5	0.8	=C$19*($A10^0.5)/(1+C$20*($A10^0.5))	=(C10-$B10)^2	=ABS(C10-$B10)
11	2	0.9	=C$19*($A11^0.5)/(1+C$20*($A11^0.5))	=(C11-$B11)^2	=ABS(C11-$B11)
12	3	1	=C$19*($A12^0.5)/(1+C$20*($A12^0.5))	=(C12-$B12)^2	=ABS(C12-$B12)
13	5	1.1	=C$19*($A13^0.5)/(1+C$20*($A13^0.5))	=(C13-$B13)^2	=ABS(C13-$B13)
14	7.5	1.2	=C$19*($A14^0.5)/(1+C$20*($A14^0.5))	=(C14-$B14)^2	=ABS(C14-$B14)
15	10	1.3	=C$19*($A15^0.5)/(1+C$20*($A15^0.5))	=(C15-$B15)^2	=ABS(C15-$B15)
16	15	1.4	=C$19*($A16^0.5/)/(1+C$20*($A16^0.5))	=(C16-$B16)^2	=ABS(C16-$B16)
17	20	1.5	=C$19*($A17^0.5)/(1+C$20*($A17^0.5))	=(C17-$B17)^2	=ABS(C17-$B17)
18					
19		k_1	1.15667844795306 (Calculated via solver)	=SUM(D6:D17)	=SUM(E6:E17)
20		K_2	0.570642862949246 (Calculated via solver)		

In the spreadsheet, $c19 = k_1$, $c20 = K_2$, column C, contains the rates were calculated from equation (12.A.2). Column D contains the individual errors squared. Cell d19 is the sum of the squares of the errors. When I set up the spreadsheet, I guessed $k_1 = 1$ and $K_2 = 0.5$ and used the solver function in Excel to minimize the d19 by varying c19 and c20. The result is that Excel calculated that the best k_1 is 1.157 and the best K_2 is 0.571.

There is one subtlety in the calculation. Notice that I calculated the error squared in column D of the spreadsheet. If you suspect that you have bad points in the calculation, it is better to instead minimize the absolute value of the error. Column E in the spreadsheet shows the absolute value of the error. Again, one can use the solver function in Excel to minimize the d19 by varying c19 and c20. The result is that Excel calculated that the best k_1 is 1.131 and the best K_2 is 0.554. Additional details are in Problem 3.A.

Example 12.B   Simple Langmuir Derivation   The reaction $A + B \Rightarrow C$ obeys

$$NH + H_2 + X \longrightarrow NH_3 + X$$

Derive an equation for the rate of formation of C as a function of the partial pressures of A and B. Assume that reaction 3 is rate-determining.

Solution

$$r_C = k_3[A_{ad}][B_{ad}] \tag{12.B.1}$$

Assume that reaction 1 is in equilibrium:

$$\frac{[A_{ad}]}{SP_A} = K_1 \tag{12.B.2}$$

Similarly for reaction 2

$$\frac{[B_{ad}]}{SP_B} = K_2 \tag{12.B.3}$$

Combining (12.B.1)–(12.B.3) yields

$$r_C = K_1 K_2 k_3 P_A P_B S^2 \tag{12.B.4}$$

I need S to complete solution; I get it from a site balance:

$$S_0 = S + [A_{ad}] + [B_{ad}] \tag{12.B.5}$$

Combining (12.B.2), (12.B.3), and (12.B.5) yields

$$S_0 = S + SK_1 P_A + K_2 P_B \tag{12.B.6}$$

Solving (12.B.6) for S, we obtain

$$S = \frac{S_0}{1 + K_1 P_A + K_2 P_B} \tag{12.B.7}$$

Combining equations (12.B.4) and (12.B.7) yields

$$r_C = \frac{K_1 K_2 k_3 P_A P_B (S_0)^2}{(1 + K_1 P_A + K_2 P_B)^2} \tag{12.B.8}$$

Example 12.C   Derivation of a Langmuir–Hinshelwood rate equation   Tanaka (1960) proposed that the decomposition of silane, SiH_4, on a silicon wafer followed the following mechanism:

$$SiH_4 + S \rightleftharpoons SiH_{4ad} \tag{12.C.1}$$

$$SiH_{4ad} + S \longrightarrow SiH_{3ad} + H_{ad} \tag{12.C.2}$$

$$SiH_{3(ad)} \rightleftharpoons Si + S + \tfrac{3}{2}H_2 \tag{12.C.3}$$

$$2H_{(ad)} \rightleftharpoons H_2 + 2S \tag{12.C.4}$$

Derive a Langmuir–Hinshelwood–Hougan–Watson rate equation for the reaction, assuming that reaction (12.C.2) is rate-determining, and irreversible.

Solution If reaction (12.C.2) is rate-determining, then the rate of silicon deposition, r_{Si}, is given by

$$r_{Si} = k_2 \left[SiH_{4(ad)} \right] [S] \qquad (12.C.5)$$

Following Hougan and Watson (1943), we will assume that reactions (12.C.1), (12.C.3), and (12.C.4) are in equilibrium.

$$\frac{\left[SiH_{4(ad)} \right]}{[S] P_{SiH_4}} = K_{SiH_4} \qquad (12.C.6)$$

$$\frac{\left[SiH_{3(ad)} \right]}{[S] P_{H_2}^{3/2} [Si]} = K_{SiH_3} \qquad (12.C.7)$$

$$\frac{\left[H_{(ad)} \right]}{[S] P_{H_2}^{1/2}} = K_H \qquad (12.C.8)$$

where K_{SiH_4}, K_{SiH_3} and K_H are the equilibrium constants for reactions (12.C.1), (12.C.3) and (12.C.4), respectively.

Combining equations (12.C.5) and (12.C.6) yields

$$r_{Si} = k_2 K_{SiH_4} P_{SiH_4} [S]^2 \qquad (12.C.9)$$

We get [S] from a site balance:

$$S_0 = [S] + \left[SiH_{4(ad)} \right] + \left[SiH_{3(ad)} \right] + \left[H_{(ad)} \right] \qquad (12.C.10)$$

where S_0 is the total number of surface sites. Note that if you have a growing water, then, after you deposit a silicon atom, the silicon atom is a site for further adsorption. Adsorbed silicon does not use up sites. Therefore, there is no [Si$_{ad}$] term in equation (12.C.10). Instead [Si] = S_0. Substituting equations (12.C.6)–(12.C.8) into equation (12.C.10) yields

$$S_0 = [S] + K_{SiH_4} [S] P_{SiH_4} + K_{SiH_3} [S] P_{H_2}^{3/2} [Si] + [K_H] [S] P_{H_2}^{1/2} \qquad (12.C.11)$$

Noting that $[S_i] = S_0$ and solving for [S] yields

$$[S] = \frac{S_0}{1 + K_{SiH_4} P_{SiH_4} + K_{SiH_3} P_{H_2}^{3/2} S_0 + K_H P_{H_2}^{1/2}} \qquad (12.C.12)$$

Substituting equation (12.C.12) into equation (12.C.9) yields

$$r_{S_i} = \frac{k_2 K_{SiH_4} P_{SiH_4} S_0^2}{\left(1 + K_{SiH_4} P_{SiH_4} + K_{SiH_3} P_{H_2}^{3/2} S_0 + K_H P_{H_2}^{1/2} \right)^2} \qquad (12.C.13)$$

Equation (12.C.13) is the Langmuir–Hinshelwood–Hougan–Watson rate equation for silicon deposition.

Example 12.D   Distinguishing between Mechanisms   Assume that the decomposition of silane instead obeys

$$SiH_4 + 4S \longrightarrow Si + 4H_{(ad)} \tag{12.D.1}$$

$$2H_{(ad)} \longrightarrow H_2 + 2S \tag{12.D.2}$$

Reaction (12.D.2) is rate-determining. How would you tell the difference between this mechanism and the mechanism in Example 12.C: (a) spectroscopically or (b) on the basis of kinetic analysis.

Solutions   There are two different solutions: a spectroscopic solution and a kinetic solution.

(a) The spectroscopic solution is to do spectroscopy of the surface, and see what species are present. For example, if reaction (12.C.2) is rate-determining, then there should be SiH_4 on the surface; if reaction (12.D.2) is rate-determining, adsorbed hydrogen atoms should be seen. There are some uncertainties because the concentrations could be low. However, one can often find conditions where the reactants in the rate-determining step are on the surface of the water. One can then use IR (infrared) to see which species are present. If silane is present, reaction (12.C.2) is rate-determining. If hydrogen atoms are present, then (12.D.2) is rate-determining. If both are present, no reaction is rate-determining.

(b) You can try to determine which mechanism is better by seeing whether equation (12.C.13) or the equivalent expression, reactions (12.D.1) and (12.D.2), work better.

First let us derive an equation for the rate, assuming that reaction (12.D.2) is rate-determining. According to the analysis in Chapter 4, $r_{12.C.2}$, the rate of reaction (12.D.2), is given by

$$r_{12.C.2} = k_{2C}[H_{(ad)}]^2 \tag{12.D.3}$$

If reaction 12.D.1 is in equilibrium, then

$$\frac{[H_{(ad)}]^4}{[S]^4 P_{SiH_4}} = K_{12.C.1} \tag{12.D.4}$$

where $K_{12.C.1}$ is the equilibrium constant for reaction (12.D.1).

Rearranging equation (12.D.4) shows

$$[H_{(ad)}] = [S](K_{12.C.1})^{0.25}(P_{SiH_4})^{0.25} \tag{12.D.5}$$

Substituting (12.D.5) into (12.D.3) yields

$$r_{12.C.2} = k_{2C}[S]^2(K_{12.C.1})^{0.5}P_{SiH_4}^{0.5} \tag{12.D.6}$$

Again, we calculate [S] from the balance:

$$S_0 = [S] + [H_{(ad)}] \tag{12.D.7}$$

where S_0 is the total number of sites on the surface.

Substituting (12.D.5) into (12.D.7) and rearranging shows

$$[S] = \frac{S_0}{1 + (K_{12.C.1})^{0.25}(P_{SiH_4})^{0.25}}$$ (12.D.8)

Combining equations (12.D.6) and (12.D.8) yields

$$r_{12.C.2} = \frac{k_{2C}S_0^2(K_{12.C.1})^{1/2}(P_{SiH_4})^{0.5}}{(1 + (K_{12.C.1})^{0.25}(P_{SiH_4})^{0.25})^2}$$ (12.D.9)

Now the question is how to do experiments to distinguish between equations (12.C.13) and (12.D.9). The easier way to find out is to put both rate equations in a spreadsheet and look for differences.

For the purposes here, it is useful to consider a case where P_{H_2} is negligible in the denominator of equation (12.C.13). In that case (12.C.13) becomes

$$r_{Si} = \frac{k_2 K_{SiH_4} P_{SiH_4} S_0^2}{(1 + K_{SiH_4} + P_{SiH_4})^2}$$ (12.D.10)

Table 12.D.1 shows a spreadsheet that I used to illustrate the difference between equations (12.C.14) and (12.D.9). In the spreadsheet I set column A to the pressure, column B to the rate calculated from equation (12.C.14), and column C to be the rate from equation (12.D.9). I picked $S_0 = 1$, $k_2 = 1$ $K_{SiH_4} = 1$. I then calculated an error in column D, and used the solver function of the spreadsheet to find values of k_{2C} and $K_{12.C.1}$ to minimize cell D$4, the total difference between the values of the rate was calculated from equations (12.D.9) and (12.C.14).

Figure 12.D.1 shows a plot of the data. Notice that the two curves look similar at low pressure, but there are significant differences at high pressure. Equation (12.D.9) rises over the range of pressure shown. One can show that the rate eventually saturates at high pressures. In contrast, equation (12.C.14) reaches a maximum and then declines.

Figure 12.D.1 tells us how to run experiments to distinguish between the mechanisms. One runs the reaction over a wide range of pressure, and looks to see how the rate varies. If the rate reaches a maximum and declines, then equation (12.C.14) will be better. If the

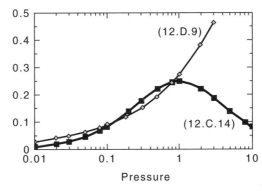

Figure 12.D.1 A plot of the rate calculated via equations (12.C.14) and (12.D.9).

Table 12.D.1 The formulas in the spreadsheet used to create Figure 12.D.1

	A	B	C	D
01	KSiH4=	1		
02	k2s0=	1		
03	K12.c.1=	6.295E - 08		
04	k2s0C=	1115.97		=SUM(D6:D16)
05	pressure	12.B.14	12.C.9	error
06	0.01	=k2s0*KSiH4*$A6 / (1+KSiH4*$A6)^2	=k2s0C*(ABS(K12.c.1)*$A6)^0.5 / (1+(ABS(K12.c.1)*$A6)^0.25)^2	=ABS(B6-C6)^2
07	0.02	=k2s0*KSiH4*$A7 / (1+KSiH4*$A7)^2	=k2s0C*(ABS(K12.c.1)*$A7)^0.5 / (1+(ABS(K12.c.1)*$A7)^0.25)^2	=ABS(B7-C7)^2
08	0.03	=k2s0 KSiH4*$A8 / (1+KSiH4*$A8)^2	=k2s0C*(ABS(K12.c.1)*$A8)^0.5 / (1+(ABS(K12.c.1)*$A8)^0.25)^2	=ABS(B8-C8)^2
09	0.05	=k2s0 KSiH4*$A9 / (1+KSiH4*$A9)^2	=k2s0C*(ABS(K12.c.1)*$A9)^0.5 / (1+(ABS(K12.c.1)*$A9)^0.25)^2	=ABS(B9-C9)^2
10	0.08	=k2s0 KSiH4*$A10 / (1+KSiH4*$A10)^2	=k2s0C*(ABS(K12.c.1)*$A10)^0.5 / (1+(ABS(K12.c.1)*$A10)^0.25)^2	=ABS(B10-C10)^2
11	=10*A6	=k2s0*KSiH4*$A11 / (1+KSiH4*$A11)^2	=k2s0C*(ABS(K12.c.1)*$A11)^0.5 / (1+(ABS(K12.c.1)*$A11)^0.25)^2	=ABS(B11-C11)^2
12	=10*A7	=k2s0*KSiH4*$A12 / (1+KSiH4*$A12)^2	=k2s0C*(ABS(K12.c.1)*$A12)^0.5 / (1+(ABS(K12.c.1)*$A12)^0.25)^2	=ABS(B12-C12)^2
13	=10*A8	=k2s0*KSiH4*$A13 / (1+KSiH4*$A13)^2	=k2s0C*(ABS(K12.c.1)*$A13)^0.5 / (1+(ABS(K12.c.1)*$A13)^0.25)^2	=ABS(B13-C13)^2
14	=10*A9	=k2s0*KSiH4*$A14 / (1+KSiH4*$A14)^2	=k2s0C*(ABS(K12.c.1)*$A14)^0.5 / (1+(ABS(K12.c.1)*$A14)^0.25)^2	=ABS(B14-C14)^2
15	=10*A10	=k2s0*KSiH4*$A15 / (1+KSiH4*$A15)^2	=k2s0C*(ABS(K12.c.1)*$A15)^0.5 / (1+(ABS(K12.c.1)*$A15)^0.25)^2	=ABS(B15-C15)^2
16	=10*A11	=k2s0*KSiH4*$A16 / (1+KSiH4*$A16)^2	=k2s0C*(ABS(K12.c.1)*$A16)^0.5 / (1+(ABS(K12.c.1)*$A16)^0.25)^2	=ABS(B16-C16)^2
17	=10*A12	=k2s0*KSiH4*$A17 / (1+KSiH4*$A17)^2	=k2s0C*(ABS(K12.c.1) $A17)^0.5 / (1+(ABS(K12.c.1)*$A17)^0.25)^2	=ABS(B17-C17)^2
18	=10*A13	=k2s0*KSiH4*$A18 / (1+KSiH4*$A18)^2	=k2s0C*(ABS(K12.c.1)*$A18)^0.5 / (1+(ABS(K12.c.1)*$A18)^0.25)^2	=ABS(B18-C18)^2
19	=10*A14	=k2s0*KSiH4*$A19 / (1+KSiH4*$A19)^2	=k2s0C*(ABS(K12.c.1)*$A19)^0.5 / (1+(ABS(K12.c.1)*$A19)^0.25)^2	=ABS(B19-C19)^2
20	=10*A15	=k2s0*KSiH4*$A20 / (1+KSiH4*$A20)^2	=k2s0C*(ABS(K12.c.1)*$A20)^0.5 / (1+(ABS(K12.c.1)*$A20)^0.25)^2	=ABS(B20-C20)^2
21	=10*A16	=k2s0*KSiH4*$A21 / (1+KSiH4*$A21)^2	=k2s0C*(ABS(K12.c.1)*$A21)^0.5 / (1+(ABS(K12.c.1)*$A21)^0.25)^2	=ABS(B21-C21)^2

rate goes up and saturates, then equation (12.D.9) will be better. If the rate merely rises, then one cannot simply distinguish between the rate equations.

On occasion one can still use the F test in Example 3.C to distinguish between the mechanisms. However, without a difference in trend, it is often difficult to reach a definite conclusion.

Example 12.E Constructing Sachtler–Fahrenfort and Tanaka–Tamaru Plots
Table 12.E.1 contains some data for the rate of ethylene hydrogenation on a number of metals at 0 C. Next, let's construct a Sachtler–Fahrenfort plot of the data.

Table 12.E.1   The spreadsheet for Example 12.E

	A	B	C	D	E	F	G
4	Metal	rate	ΔH_f Of Oxide	Oxygens in oxide	Metal atoms in oxide	ΔH_f per mole of oxide	ΔH_f per mole of metal
5	Pt	1.0E+16	−9.7	1	1	=$C5/D5	=$C5/E5
6	Pd	3.0E+16	−20.4	1	1	=$C6/D6	=$C6/E6
7	Ir	3.0E+16	−40.1	2	1	=$C7/D7	=$C7/E7
8	Rh	3.0E+17	−21.7	1	1	=$C8/D8	=$C8/E8
9	Ru	9.0E+18	−52.5	2	1	=$C9/D9	=$C9/E9
10	Cu	8.3E+10	−39.8	1	2	=$C10/D10	
11	Cu	8.3E+10	−37.1	1	1		=$C11/E11
12	Co	3.0E+16	−57.2	1	1	=$C12/D12	=$C12/E12
13	Ni	1.0E+14	−58.4	1	1	=$C13/D13	=$C13/E13
14	Fe	1.0E+16	−197.5	3	2	=$C14/D14	=$C14/E14
15	Fe	1.0E+16	−63.7	1	1		
16	W	3.0E+13	−136	2	1	=$C16/D16	=$C16/E16
17	Ta	3.0E+13	−499.9	5	2	=$C17/D17	=$C17/E17

Solutions   There are two types of plots: Sachtler–Fahrenfort plots and Tanaka–Tamaru plots. Sachtler–Fahrenfort plots are graphs of the rate of reaction on a series of metal catalysts as a function of the heat of formation of the metal per mole of oxygen, while Tanaka–Tamaru plots are graphs of the rate of reaction on a series of metal catalysts as a function of the heat of formation of the metal per mole of metal.

If we wanted to construct a Sachtler–Fahrenfort plot of the data, we would calculate the heat of formation of the oxide per mole of oxygen. A spreadsheet to do the calculations is given in Table 12.E.1. Column A lists the metal; column B lists the rate. Column C in the heat of formation of the oxide per mole from Table 12.11, while column D is the number of oxygens in the corresponding oxide. The heat of formation per mole of oxygen is given in column F. Similarly, column E gives the number of metal atoms in the oxide, while column G gives the heat of formation of the oxide per mole of metal.

The one subtlety in the analysis comes when there are two oxides of the same material. For example, copper is available as both Cu_2O and CuO. The heat of formation of Cu_2O per mole of oxygen is greater that the heat of formation of CuO per mole of oxygen. Consequently, one uses the heat of formation of Cu_2O when constructing a Sachtler–Fahrenfort plot. In contrast, the heat of formation of CuO per mole of copper is larger than the heat of formation of Cu_2O per mole of copper. Therefore, one uses the heat of formation of CuO when constructing the Tamaru–Tanaka plot.

The Sachtler–Fahrenfort plot is a graph of the rate data (i.e., column A) versus column F in the spreadsheet. The Tanaka–Tamaru plot is a graph of the rate data (i.e., column A) versus column G in the spreadsheet. The actual plot is given in Figure 12.14 in the chapter.

12.21   SUGGESTIONS FOR FURTHER READING

Introductory books on catalysis include:

M. Boudart, *Kinetics of Heterogeneous Catalytic Reactions*, Princeton University Press, Princeton, NJ, 1984.

G. C. Bond, *Heterogeneous Catalysis, Principles and Applications*, Clarendon Press, Oxford, UK 1987.

B. C. Gates, *Catalytic Chemistry*, Wiley, New York, 1992.

More advanced treatments include:

H. Pines, *The Chemistry of Catalytic Hydrocarbon Conversions*, Academic Press, San Diego, 1991.

G. V. Smith and F. Notheisz, *Heterogeneous Catalysis in Organic Chemistry*, Academic Press, San Diego, CA, 1999.

J. M. Thomas and W. J. Thomas, *Principles and Practice of Heterogeneous Catalysis*, Wiley, New York, 1996.

J. R. Anderson and M. Boudart eds., *Catalysis:Science and Technology*, Elsevier, Amsterdam, 1984–1998.

Mass transfer effects are well described in:

R. I. Wijngaarden and K. R. Westerterp, *Industrial Catalysts*, New York, VCH, 1998.

Enzyme catalysis is covered in

K. Faber, *Biotransformations in Organic Chemistry*, 4th ed, Springer, NY 2000.

A. Fersht, *Structure and Mechanism in Protein Science*, 3rd ed, W. H. Freeman 1999.

12.22   PROBLEMS

12.1   Define the following terms:

(a) Homogeneous catalyst

(b) Heterogeneous catalyst

(c) Volcano plot

(d) Isomerization

(e) Alkylation

(f) Cracking

(g) Aldol condensation

(h) Ziegler–Natta catalyst

(i) Wilkinson catalyst

(j) Wacker process

(k) Hydroformylation

(l) Monsanto carbonylation process

(m) Oxidoreductases

(n) NADH peroxidase

(o) Hydrolases

(p) Lyases

(q) Isomerase

(r) Langmuir adsorption isotherm

(y) Fischer–Tropsch synthesis

(z) Hydrogenolysis

(aa) Hydrodesulfurization

(bb) Selective catalytic reduction

(cc) Aminoxidation

(dd) Water–gas shift

(ee) Faugasite

(ff) zeolite

(gg) Superacid

(hh) ZSM-5

(ii) Mordenite

(jj) Sodalite

(kk) Photocatalysts

(ll) Langmuir–Hinshelwood kinetics

(mm) Sabatier's principle

(nn) Adsorption

(oo) Adsorbate

(s) Effectiveness factor

(t) Michaelis–Menten Kinetics

(u) Transferases

(v) Ligases

(w) Menschutkin reactions

(x) Supported metals

(pp) Molecular adsorption

(qq) Dissociative adsorption

(rr) Langmuir–Hinshelwood mechanism

(ss) Rideal–Eley mechanism

(tt) Precursor mechanism

(uu) Surface sites

12.2 How do catalysts change rates of reactions? List the key ways that catalysts change reactions and give an example of each of the effects.

12.3 Describe the key differences between Langmuir–Hinshelwood, Rideal–Eley, and precursor mechanisms. How would you distinguish between them experimentally.

12.4 Describe in your own words why the kinetics of surface reactions are so different from the kinetics of gas-phase reactions. What makes the kinetics different?

12.5 Summarize the various types of catalysts described in the first half of this chapter. Provide an illustration of each type of catalyst and an explanation of how the catalyst works.

12.6 Visit the Websites of some of the large catalyst manufacturers:

(a) Johnson Matthey (*http://www.matthey.com/divisions/catalytic.htm*), Engel-hard (*http://www.engelhard.com/*), W. R. Grace (*http://www.gracedavison.com/*), United Catalysts Süd-chemie (*http://www.unitedcatalysts.com/*). Find a list of catalysts that they offer and describe what the catalysts do. Then go to the Shell and Exxon sites, search the word catalyst, and see what you find.

(b) Visit the Websites of the major enzyme manufacturers: Novozymes (*http://www.novozymes.dk/*), Genencor (*http://www.genencor.com/*), Royal Gist Brochades (*http://www.gist-brocades.com/*). What do you find?

12.7 Describe in your own words when Sabatier's principle works.

(a) When do you get a maximum as in Figure 12.12?

(b) When does the rate continue to increase as in Figure 12.11?

(c) How would changes in the intrinsic barriers to reaction affect a volcano plot?

12.8 Find five examples of catalytic reactions in your home or your car. Be sure to consider reactions in your oven, your washing machine, your stomach, your intestines, and so on.

12.9 Diagram the catalytic cycle for

(a) Ethylene hydrogenation catalyzed by the Wilkinson catalyst.

(b) Platinum-catalyzed ethylene hydrogenation

(c) The copper-catalyzed reaction: $Fe^{2+} + V^{4+} \Rightarrow Fe^{3+} + V^{3+}$

(d) The dehydration of ethanol

(e) Hydrogen oxidation on a fuel cell

(f) Free-radical polymerization of styrene

(g) Photocatalysis of CO oxidation

12.10 Your stomach contains a solution of hydrochloric acid that is buffered to a pH of 2.0. On the basis of the results in this chapter

    (a) What would you expect the acid to do to?

        (1) Proteins

        (2) Fats

        (3) Sugars and starches

        (4) Invading bacteria

    (b) Your saliva contains the enzyme lingual amylase. Look in the protein database or a biochemistry book and find out what each enzyme does to the substances listed in items (1)–(4) (above). What is the main mode of enzyme action?

    (c) Your intestines contain α-amylase, trypsin, chymotrypsin, and elastase. Look in the protein database and find out what each enzyme does to the substances in items (1)–(4). What is the main mode of enzyme action?

12.11 Enzymes are widely used in detergents. The enzymes catalyze the hydrolysis of stains.

    (a) Look in the protein database and find enzymes that would be active for the hydrolysis of

        (1) Proteins

        (2) Fats

        (3) Sugars and starches

        (4) Cellulose

    (b) Detergents have a pH of about 8.5. Would any of the compounds you found be stable at a pH of 8.5?

    (c) Look at some of the Websites for the commercial enzyme manufacturers: Novozymes (*http://www.novozymes.dk/*), Genencor (*http://www.genencor. com/*). What do they tell you about their enzymes?

    (d) The following is a noninclusive list of some of the major commercial detergent enzymes: Alcalase, Esperase, Maxatase, Optimase, Durazym, Kazusase, Opticlean, Savinase, Celluzyme, Duramyl. Look up each of the enzymes and see what they do.

12.12 In Example 12.A we used a variety of methods to fit Steger and Masel's data to rate equations.

    (a) Set up the spreadsheets yourself and reproduce the findings

    (b) Assume that Steger and Masel misrecorded the data so the rate was 0.03 at a pressure of 0.1 torr. Reanalyze the data. How well does each of the methods do?

    (c) What do the results in this problem tell you about the sensitivity of the various methods to errors?

12.13 Ethylene hydrogenation was one of the first catalytic reactions to be studied in detail. Langmuir and Hinshelwood (1918) proposed that the reaction went via a Langmuir–Hinshelwood reaction:

$$H_2 + 2S \rightleftharpoons 2H_{ad} \qquad\qquad (P12.13.1)$$

$$H_2C=CH_2 + S \rightleftharpoons H_2C=CH_{2(ad)} \qquad \text{(P12.13.2)}$$

$$H_2C=CH_{2(ad)} + 2H_{ad} \longrightarrow C_2H_6 \qquad \text{(P12.13.3)}$$

Rideal and Eley proposed a Rideal–Eley mechanism:

$$H_2 + 2S \rightleftharpoons 2H_{ad} \qquad \text{(P12.13.4)}$$

$$H_2C=CH_2 + H_{ad} \rightleftharpoons CH_3CH_{2(ad)} \qquad \text{(P12.13.5)}$$

$$CH_3CH_{2(ad)} + H_{ad} \longrightarrow C_2H_6 \qquad \text{(P12.13.6)}$$

Houriti and Polanyi proposed a third mechanism:

$$H_2 + 2S \rightleftharpoons 2H_{ad} \qquad \text{(P12.13.7)}$$

$$H_2C=CH_2 + S \rightleftharpoons H_2C=CH_{2(ad)} \qquad \text{(P12.13.8)}$$

$$H_2C=CH_2 + H_{ad} \rightleftharpoons CH_3CH_{2(ad)} \qquad \text{(P12.13.9)}$$

$$CH_3CH_{2(ad)} + H_{ad} \longrightarrow C_2H_6 \qquad \text{(P12.13.10)}$$

(a) Derive a rate equation for each mechanism.

(b) What experiments would you do to distinguish between the mechanisms? For the purpose of this problem, assume that you are working in 1920 and do not know how to measure the surface concentration of species.

(c) Repeat part (b) assuming that you can measure surface concentrations

12.14 The synthesis of ethylene (C_2H_4) is one of the largest chemical processes in the United States.

(a) Find a reasonable mechanism for the conversion of ethane (CH_3CH_3) into ethylene and hydrogen.

(b) What other products would you expect?

(c) Use the steady-state approximation to derive an expression for the rate of the overall reaction.

(d) Estimate the preexponentials and activation barriers for each of the reactions.

(e) How could you catalyze the reaction? Pick one homogeneous catalyst and one heterogeneous catalyst and explain how they would work. Do not pick the catalysts from part (f) or (g) (in this problem) or Problem 12.6.

(f) In industrial practice, people use water and a zeolite (a solid catalyst) as catalysts. The water dissociates on the zeolite to yield OH and H. The water also reacts with carbon deposited on the pores of the zeolite. Explain how the water/zeolite mixture would catalyze the reaction.

(g) In Section 12.1.1 we noted that I_2 can catalyze the decomposition of ethane. How would that work?

(h) Find a reasonable mechanism for the decomposition of ethane in the presence of iodine.

(i) Estimate all of the preexponentials and rate constants for the reaction in part (h).

(j) Use the steady-state approximation to derive an equation for the rate of reaction for the mechanism in part (h).

(k) Show that the rate is higher than in the case in part (a).

12.15 Repeat Problem 12.14, parts (g)–(k) using acetaldehyde as a catalyst. Could ethanol also be used as a catalyst. (*Hint*: What products are formed when ethanol decomposes?)

12.16 What type of catalyst would you use for the following reaction:

(a) Hydrogenation of corn oil to form margarine

(b) Oxidation of carbon monoxide in your home furnace

(c) Hydrogenation of furfurylamine

(d) Hydrolysis of protein stains on clothes

(e) Production of methylamine from methanol and ammonia

(f) Transformation of 2,4-pentanedione into 3-methyl-2-cyclopentanone

(g) Partial oxidation of C_3–C_6 olefins

(h) Softening of a pair of jeans by hydrolyzing the celluose.

12.17 Jennings (1991) summarized the available data for ammonia synthesis on a number of catalysts. Table P12.17 summarizes the data.

(a) How well do these data follow a Sachtler–Fahrenfort plot?

(b) On the basis of your findings, what other materials would be good catalysts for ammonia synthesis?

(c) In reality, lead and other substances you would guess from Table 6.5 are not good catalysts for the reaction. What does that tell you?

12.18 Sinfelt and co-workers did a considerable amount of data on ethane hydrogenolysis. Table P12.18 gives a selection of their results.

(a) How well do these data follow a Sachtler–Fahrenfort plot and Tanaka–Tamaru plot?

(b) Are there any points way off of the curve? What do you make of that?

(c) On the basis of your plots and the data in Table 12.11, what other substances would you expect to be active for this reaction?

Table P12.17   The relative rate of ammonia synthesis over a number of catalysts

Metal	Relative Rate	Heat of Adsorption of Nitrogen, kcal/mol of nitrogen atoms
Ru	1.0	−30
Fe	0.01	−42
Rh	0.005	−27
Re	1×10^{-5}	−99
Pt	1×10^{-7}	−71

Note: Heat of adsorption of $N_2 = 3^*$ energy in Table 6.5.

Source: Data from Jennings (1991).

Table P12.18   The rate of ethane hydrogenolysis at 523 K and 10 atm over an number of supported metal catalysts

Substance	Rate Molecules/cm^2/sec	Substance	Rate Molecules/(cm^2·second)
Co	9×10^{12}	Rh	2×10^{14}
Ni	4×10^{14}	Pd	2×10^{9}
Cu	$>1 \times 10^{6}$	Os	2×10^{16}
Ru	6×10^{14}	Ir	5×10^{13}
Re	2×10^{13}	Pt	2×10^{9}

Source: Data of J. H. Sinfelt, Adv Catalysis, **23**, 91 (1973a,b), J. Catalysis, **29**, 308 (1973).

  (d) Experimentally, none of these other metals actually shows significant catalytic activity. What do you conclude from that?

12.19 The reaction $2CH_3CH_2OH \Rightarrow CH_3CH_2CH=CH_2 + 2H_2O$ is being run in an acid ˎcatalyst.

  (a) Find a feasible mechanism for the reaction.
  (b) What side products do you expect?
  (c) If you run the reaction in a zeolite, could you eliminate some of the side products?
  (d) What zeolite would you choose?

12.20 Imbert, Gnep, and Guisset, (1997) *J. Catalysis.* **172**, 307 examined the isomerization of creosol, $CH_3(C_6H_4)OH$.

  (a) Find a feasible mechanism for the reaction
  (b) What side products do you expect?
  (c) If you run the reaction in a zeolite, could you eliminate some of the side products?
  (d) What zeolite would you choose?
  (e) Imbert et al. actually ran the reaction in ZSM-5. They found that initially they got mainly *para*-creosol. However, after they ran the reaction for a long time so that there was plenty of time for all species to diffuse in and out of the catalyst, *meta*-creosol was the major product. How can you account for these observations?

12.21 Ilao, Yamamoto, Segawa (1996). *Journal of Catalysis.* **161**, 20 examined the reaction between methanol and ammonia to yield methylamine.

  (a) What general class of reaction would the addition of a methyl group to ammonia fall under?
  (b) What type of catalyst would you use?
  (c) What side products would you expect?
  (d) What would you do to avoid the formation of dimethyl and trimethyl amine?
  (e) Look at the lists of catalysts in this chapter. Which one is properly sized to avoid the formation of dimethyl and trimethyl amine?
  (f) How do your predictions in a), b), c) compare to Ilao, Yamamoto, Segawa's findings?

12.22 Haupfear, Olson and Schmidt (1994). *Journal of the Electrochemical Society.* **141**, 1943 examined the kinetics of SiO_2 deposition from tetraethylorthosilicate. $(CH_3CH_2O)_4Si$ also called TEOS. The main reaction is:

$$(CH_3CH_2O)_4Si \Longrightarrow SiO_2 + 4CH_2CH_2 + 2H_2O$$

(a) Derive a Langmuir–Hinshelwood rate equation for the reaction, assuming that the mechanism of the reaction is $TEOS \rightleftharpoons TEOS_{(ad)} \rightarrow$ products.

(b) Haupfear, et al. found that the rate of decomposition is reduced when ethylene is fed into the reactor even though the reaction is irreversible. Show that your rate equation predicts that the ethylene will slow down the reaction.

(c) Would the ethylene affect the activation barrier for the reaction. In other words, would a plot of log rate versus 1/T change? Try to be quantitative.

(d) What effect would you expect acetylene to have on the reaction rate? How about ammonia?

12.23 In Table 2.9 we said that the hydrogenation of ethylene over a nickel catalyst obeys

$$r_{C_2H\cdot} = \frac{k_1 S_0 K_2 K_3 P_{H_2} P_{C_2H_4}}{(1 + K_2 P_{H_2} + K_3 P_{C_2H_4})^2} \qquad (P12.23.1)$$

The objective of this problem is to work out the behavior of equation (P12.23.1). Assume

$$k_1 = 10^{13}/second \times exp(-18 \ kcal/mol)/k_B T,$$

$$K_2 = 10^{-11}/atm \times exp(16 \ kcal/mol)k_B T,$$

$$K_3 = 10^{-10}/atm \times exp(14 \ kcal/mol)k_B T), \ S_0 = 10^{15} \ molecules/cm^2$$

(a) Plot the rate as a function of the C_2H_4 pressure at 300 K. Assume an H_2 pressure of 1 atm and consider C_2H_4 pressures from 0.5 to 10 atm.

(b) Why does the rate decline at high C_2H_4 pressures?

(c) Plot the rate as a function of temperature for temperatures between 200 and 500 K. Assume a C_2H_4 pressure of 1 atm and a H_2 pressure of 1 atm.

(d) Why does the rate decline at high pressures?

12.24 The decomposition of urea in the presence of an enzyme called *urease* follows the following rate law:

$$r_{urea} = -\frac{k_1 K_2 [urea]}{1 + K_2 [urea]}$$

Table P12.24 shows rate data for the reaction.

(a) Fit the data to the rate law using a Lineweaver–Burke plot, the Eadie–Hofstee plot, and a nonlinear fit to the data. (*Hint*: Look back to Examples 3.A and 12.A.)

(b) Why do the different methods give you such different numbers?

12.25 Vannice and Poondi (1997). *J. Catalysis.* **169**, 166 examined benzyl alcohol hydrogenation over on a number of supported platinum catalysts. They found that the main reaction was hydrogenation of the benzyl alcohol to toluene.

Table P12.24   Rate data for Problem 12.24

Urea Concentration, mol/liter	Rate, mol/(liter·minute)	Urea Concentration, mol/liter	Rate, mol/(liter·minute)
0.0001	200	0.03	10,400
0.0003	700	0.05	10,800
0.0005	1200	0.1	11,000
0.001	2000	0.3	11,500
0.003	4500	0.5	12,000
0.005	6300	1	12,100
0.01	8300	2	12,100

(a) Write down a balanced overall reaction.

(b) Propose a feasible mechanism for the reaction.

(c) Derive a Langmuir–Hinshelwood rate equation for the reaction assuming that the rate-determining step is the addition of the second hydrogen to benzaldehyde.

(d) How well does your expression fit the rate data in Table P12.25?

(e) Next, derive an expression for the production of side products. Assume that the rate-determining step is the addition of another hydrogen.

(f) Vannice et al. found that when they poisoned the catalyst with TiO_2 they were able to enhance the rate. Explain this result.

12.26 Sauer and Ollis, (1994). *J. Catalysis.* **149**, 81 examined the photooxidation of acetone on a TiO_2 catalyst. Under the conditions used by these investigators, the main reaction was

$$4O_2 + CH_3COCH_3 \Longrightarrow 3CO_2 + 3H_2O$$

Assume that the reaction occurs via the following mechanisms:

$$CH_3COCH_3 + S \Longleftrightarrow CH_3COCH_{3(ad)} \qquad (P12.26.1)$$

$$CH_3COCH_{3(ad)} + H^+ \Longleftrightarrow [CH_3COHCH_3]_{(ad)}^+ \qquad (P12.26.2)$$

$$[CH_3COHCH_3]_{(ad)}^+ + 4O_2 \longrightarrow 3CO_2 + 3H_2O + H^+ \qquad (P12.26.3)$$

where S is an active site on the catalyst and h^+ is a hole generated by a photon.

Table P12.25   Rate data for problem 12.25

Benzyl Alcohol Pressure, atm	Hydrogen Pressure, atm	Rate μmol/second	Benzyl Alcohol Pressure, atm	Hydrogen Pressure, atm	Rate μmol/second
0.004	1.00	0.66	0.004	0.80	0.64
0.008	1.00	0.65	0.004	0.67	0.62
0.008	1.00	0.64	0.004	0.67	0.58
0.004	1.00	0.78	0.004	0.53	0.55
0.004	0.85	0.61	0.004	0.53	0.48
0.004	0.80	0.72	0.004	0.38	0.48
0.004	0.80	0.77	0.004	0.38	0.42

(a) Derive equations for the rate of reaction assuming that reaction (P12.26.1) is rate-determining.

(b) Derive equations for the rate of reaction assuming that reaction (P12.26.2) is rate-determining.

(c) Derive equations for the rate of reaction assuming that reaction (P12.26.3) is rate-determining.

(d) Table P12.26 shows some rate data extrapolated from Sauer and Olis' results. Use an F test as described in Example 3.B to determine which rate equation fits best.

12.27 In Section 12.13 we presented the mechanism of hydrogen oxidation in a fuel cell. However, that mechanism works with only one kind of fuel cell: a fuel sell with a polymer electrolyte. The electrolyte is active at 60°C.

$$H_2 \Longrightarrow 2H^+ + 2e^-$$

$$O_2 + 4H^+ + 4e^- \Longrightarrow 2H_2O$$

However, under other conditions, alternative mechanisms occur:

With an alkaline electrolyte at 120°C

$$H_2 + 2OH^- \Longrightarrow 2H_2O + 2e^-$$

$$O_2 + 2H_2O + 4e^- \Longrightarrow 4OH^-$$

With a molten carbonate at 600°C

$$H_2 + 2CO_3{}^{2-} \Longrightarrow H_2O + CO_2 + 2e^-$$

$$O_2 + 2CO_2 + 4e^- \Longrightarrow 2CO_3{}^{2-}$$

With a molten salt at 900°C

$$H_2 + O^{2-} \Longrightarrow H_2O + 2e^-$$

$$O_2 + 4e^- \Longrightarrow 2O^{2-}$$

Table P12.26   Rate data for Problem 12.26

Acetone Concentration, mg/m^3	Rate, $mg/(cm^3 \cdot minute)$	Acetone Concentration, mg/m^3	Rate, $mg/(cm^3 \cdot minute)$	Acetone Concentration, mg/m^3	Rate, $mg/(cm^3 \cdot minute)$
0	0	25	0.08	150	0.25
3	0.01	30	0.10	170	0.26
4	0.02	40	0.12	200	0.27
5	0.02	50	0.14	250	0.29
7	0.03	70	0.17	300	0.3
10	0.04	100	0.21	350	0.032
15	0.06	130	0.23	400	0.32

(a) Guess a feasable mechanism for each of the reactions on a platinum catalyst. (*Hint*: The reactions above are not elementary reactions. Find elementary reactions that occur on platinum. There should be a different mechanism on the anode and cathode. See Figure 5.10 for ideas.)

(b) Diagram the catalytic cycle for each reaction. Be sure to consider the production and destruction of the catalytic site. (*Hint*: In each case there is one catalytic cycle on the anode and one on the cathode.)

(c) How is electricity generated?

(d) Look up the half-cell potential for each of these reactions. How much voltage will you produce?

12.28 The production of indendiol is one of the key steps in the manufacture of an AIDS drug called Trixovan (see Figure P12.28). The objective of this project is for you to propose ways to make indendiol.

(a) One scheme to make the product is to oxidize indene with hydrogen peroxide. Search the protein database to find an enzyme that would catalyze the simultaneous addition of two OH groups.

(b) An alternative scheme is to add iodine across the double bond in indene to form a indene diiodate and then hydrolyze the product (replace iodine with water). What kind of catalyst would you use to catalyze the addition of iodine?

(c) What type of catalyst would you use to catalyze the hydrolysis reaction? (*Hint*: What type of catalyst works for the reverse reaction, i.e., dehydration?)

(d) Repeat (b) and (c) when the first step is to phosphorylating the indene.

12.29 The reaction $CH_3CH_2OH + NAD^+ \Rightarrow CH_3CHO + NADH + H^+$ is critical to the destruction of alcohol in the human body.

(a) Look up NAD and NADH in a biochemistry book. What are they?

(b) Look in the protein database and find an enzyme that would catalyze the reaction in your liver.

(c) Assume that the reaction follows the mechanism:

$$NAD^+ + S \rightleftharpoons NAD^+_{(ad)} \qquad (P12.29.1)$$

$$NAD^+_{(ad)} + CH_3CH_2OH \rightleftharpoons [CH_3CH_2OH{:}NAD]^+_{(ad)} \qquad (P12.29.2)$$

$$[CH_3CH_2OH{:}NAD]^+_{(ad)} \longrightarrow NADH + [CH_3CH_2O]^+_{(ad)} \qquad (P12.29.3)$$

$$[CH_3CH_2O]^+_{(ad)} \rightleftharpoons CH_3CHO + H^+ + S \qquad (P12.29.4)$$

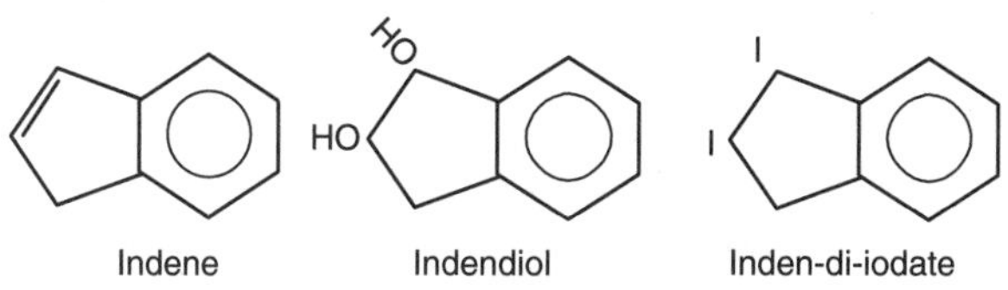

| Indene | Indendiol | Inden-di-iodate |

Figure P12.28 Structures of some of the intermediates used in the production of Trixovan.

where $[CH_3CH_2OH:NAD]^+_{(ad)}$ is a NAD, ethanol complex and S is a site on the enzyme. Derive an equation for the rate of the reaction assuming that reaction (P12.29.3) is rate-determining.

(d) How do your kinetics differ from Langmuir–Hinshelwood kinetics?

(e) Look up a biochemistry book, and find out how much of the NAD^+ is available in a typical human being?

(f) Does the amount of NAD^+ vary with the weight of the individual?

(g) Use your findings to suggest a possible hangover cure.

(h) Would a vitamin work?

12.30 *The US Chemical Industry Statistical Handbook for 1997* gives a chart showing the feedstocks used to produce all of the major chemicals in the United States. Figure P12.30 shows part of the benzene chain, specifically, the chemicals produced starting with benzene. The chart includes the major organic chemicals used in the process, but not hydrogen, oxygen, CO, or water.

(a) Look at each step (i.e., arrow) and write a balanced reaction for the step, adding hydrogen oxygen, water, or CO as needed.

(b) Use your knowledge from Chapter 5 to suggest a possible mechanism for each reaction.

(c) Suggest a possible catalyst or series of catalysts for each step.

(d) Look in the chemical processes handbook. How do your guesses for catalysts compare to the ones actually used?

12.31 In the human body, cells use the reaction

$$Glucose + ATP \Longrightarrow glucose\text{-}6\text{-}phosphate + ADP$$

to trap glucose in cells. The kinetics of the reaction varies between various tissues. Your heart and brain use an enzyme called *hexokinase*. Your liver uses a different enzyme called *glucokinase*.

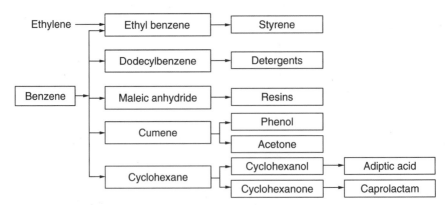

Figure P12.30   A part of the benzene chain for the production of industrial chemicals. [Adapted from *The U.S. Chemical Industry Statistical Handbook for 1997* and *Dr D. L. Burdick and Dr W. L. Leffler Petrochemical Chart.*]

Table P12.31   The reactivity of hexokinase and glucokinase for phosphorylating glucose

Blood Glucose, mmol/liter	Rate with Hexokinase, mmol/hour	Rate with Glucokinase, mmol/hour	Blood Glucose, mmol/liter	Rate with Hexokinase, mmol/hour	Rate with Glucokinase, mmol/hour
0.05	0.3	0.02	3.0	0.97	0.7
0.1	0.5	0.03	4.0	1.0	0.9
0.3	0.75	0.10	5.0	1.0	1.0
0.5	0.8	0.15	6.0	1.0	1.1
1.0	0.9	0.3	7.0	1.0	1.2
2.0	0.95	0.5	10.0	1.0	1.4

(a) Fit the data in Table P12.31 to Michaelis–Menten kinetics.

(b) How do the results above help your brain and heart to survive famine? What happens as the blood glucose level drops? Where does the glucose go? To your heart or to your liver?

(c) Now assume that there is an excess of glucose in your bloodstream. What happens? (*Note*: 5 mmol/liter is a normal glucose level in humans.)

12.32 In Section 12.5 we found that NO_2 can catalyze the decomposition of ethane. NO_2 catalysis occurs at low temperature, where there are not many radicals in the system. However, at higher temperatures, NO_2 can instead inhibit the decomposition of ethane. The main mechanism is for the NO_2 to trap the radicals in the system (i.e., NO_2 reacts with the radicals to form stable species). The objective of this problem is to develop a criterion to see whether NO_2 will enhance or inhibit ethane decomposition.

(a) Write a complete mechanism for ethane decomposition including the fact that NO_2 can initiate the mechanism as discussed in Section 12.5 and that NO_2 can trap the hydrogen atoms and ethyl radicals.

(b) Use the steady-state approximation to develop a rate equation for the reaction.

(c) Use the results in Chapters 5, 7, and 11 to guess a preexponential and an activation barrier for each step.

(d) Use your results to calculate the rate of ethane decomposition at 500, 800, and 1200 K, assuming that you have a reactor filled with 1 atm of ethane and 0.01 atm of NO_2.

(e) How do your results compare to the case in Figure 12.9?

12.33 Blaser, Jalett, Garland, Studer, Thies, and Wirthtijani (1998). *J. Catalysis*. **173**, 282 examined the kinetics of the hydrogenation of ethyl pyruvate ($CH_3COCOOC_2H_5$) to ethyl lactate ($CH_3CHOHCOOC_2H_5$) on a 5% Pt/Al_2O_3 catalyst in toluene in both presence and absence of a chiral modifier.

(a) What is the stoichiometry of the reaction? Provide a balanced overall reaction.

(b) Propose a feasible mechanism for the reaction.

(c) Blaser et al. propose that in the absence of the chiral modifier, the rate-determining step in the reaction is the addition of an adsorbed hydrogen to an adsorbed ethyl pyruvate. Derive a Langmuir–Hinshelwood rate equation for the reaction, assuming that this mechanism is correct.

(d) Blaser et al. also propose that in the presence of the chiral modifier, the addition of the first hydrogen to the ethyl pyruvate rate is rapid, while the addition of a second hydrogen to the adsorbed ethyl pyruvate is rate-determining. Derive a Langmuir–Hinshelwood rate equation for the reaction assuming that this mechanism is correct.

(e) What types of experiments would you do to distinguish between your rate equations in c) and d)?

(f) Which equation fits the data in Table P12.33 better?

12.34 Schürch, et al. (1998). *J. Catalysis.* **173**, 187 examined the role of certain amines in modifying the stereospecificity of ethyl pyruvate ($CH_3COCOOC_2H_5$) hydrogenation to ethyl lactate ($CH_3CHOHCOOC_2H_5$) on a 5% Pt/Al_2O_3 catalyst.

(a) If you were trying to modify the stereospecificity of a reaction, what would you do? (*Hint*: Draw an analogy to the production of isotactic polyethylene.)

(b) Amines bind moderately strongly to platinum. How could the amine work?

(c) If you were designing the modifiers, what other species would you choose? Look specifically at thiols since they bind even more strongly to platinum.

(d) Would you anticipate side reactions? (*Hint*: What other reactions would the platinum do? What other reactions would the alumina do?)

12.35 Keane, (1997). *Journal of the Chemical Society-Faraday Transactions.* **93**, 2001 examined the hydrogenation of methylacetoacetate (MAA) ($CH_3COCH_2COOCH_3$) to methyl 3-hydroxybutyrate ($CH_3CHOHCH_2COOCH_3$) over a Ni/SiO_2 catalyst.

(a) What is the reaction? Provide a balanced equation.

(b) Derive a Langmuir–Hinshelwood rate equation for the reaction assuming that the addition of the first hydrogen is rate-determining

(c) Derive a Langmuir–Hinshelwood rate equation for the reaction assuming that the addition of the second hydrogen is rate-determining.

(d) Which equation fits the data in Table P12.35 better?

(e) Why is there a maximum rate at intermediate MAA concentrations?

12.36 Many different catalysts can decompose H_2O_2: catalyse from potatoes, MnO_2, iron, and iron oxide. Put a 30% solution of H_2O_2 into a Petri dish, than add

Table P12.33   Rate data for problem 12.33 - rates/gram of catalyst

Rate,mmol/ (gm·hour)	H_2 Pressure, bar	Ethyl Pyruvate Concentration, mol/liter	Rate mmol/ (gm·hour)	H_2 Pressure, bar	Ethyl Pyruvate Concentration, mol/liter
0	21	0	17	10	3
20	21	0.3	24	20	3
20.5	21	1	30	30	3
25	21	2.5	34	40	3
24	21	3	42	60	3
23	21	4.2	48	80	3
18	21	6.8	54	100	3
15	21	9.0	68	160	3

Table P12.35   Rate data for Problem 12.34

Rate, mmol/ (liter·minute)	MAA Concentraion, mmol/liter	H_2 pressure, atm	Rate, mmol/ (liter·minute)	MAA Concentration, mmol/liter	H_2 pressure, atm
2.1	1	1.0	2.0	1	0.5
7.9	5	1.0	7.2	5	0.5
11.5	10	1.0	10.0	10	0.5
13.5	30	1.0	11.0	30	0.5
12.8	40	1.0	10.2	40	0.5
20	50	1.0	9.4	50	0.5
10.4	70	1.0	8.0	70	0.5
8.5	100	1.0	6.4	100	0.5

Note: The 0.5-atm data were made up for this problem. Keane actually measured at only 1 atm.

a slice of potato, some MnO_2, a clean nail, and a rusty nail. Which substance decomposes the H_2O_2 the quickest?

12.37 Consider a first-order reaction occurring in a 5-mm-diameter spherical catalyst pellet.

(a) Calculate the effectiveness factor for the reaction. Data: $k_1 = 8$/second, $D_e = 0.8$ cm²/second.

(b) How would the effectiveness factor change if you changed to 2-mm-diameter spherical catalyst pellets?

(c) How would the effectiveness factor change for 4-mm-diameter pellets with a k_e of 1/min?

12.38 As noted in Problem 12.11, enzymes are commonly added to detergents to eliminate stains. The detergents work best in cold water. The object of this problem is to quantify the idea that the enzymes work best in cold water.

(a) Consider removing a food stain from your shirt. Food stains contain proteins that take a week to set. Guess at a kinetic expression for the setting of the stain assuming the activation barrier to be 26 kcal/mol, that is, the value from Figure 2.8, assuming that the half-life is a week.

(b) Now consider the reaction between the stain and the enzyme. Develop a kinetic expression for the reaction between the enzyme and the stain. Assume that

(1) The reaction between the enzyme is catalyzing the hydrolysis of the protein in the stain,

(2) The reaction follows a Rideal–Eley mechanism; that is, the enzyme binds to the stain, then water is added without binding to the enzyme,

(3) The binding energy (i.e., ΔG_{ad}) between the protein and the enzyme is 25 kcal/mol,

(4) The reaction has a preexponential of 10^{13}/second and an activation barrier of 18 kcal/mol.

(5) $S_0 = 1$ enzyme site/1000 linkages in the stain.

(c) At this point, you have two parallel reactions. Integrate the rate equations to calculate the concentration of the stain, the portion of the stain that is set, and the portion of the stain that is hydrolyzed as a function of time.

(d) Plot your results as a function of time at T = 273, 290, 300, 340 K. What fraction of the stain is set in each case?

12.39 In the supplemental material, we derived an equation for the effectiveness factor for diffusion into a flat plate. The objective of this problem is to derive an expression for the diffusion into a flat slab of thickness 2L. Assume that the top and bottom of the slab are surrounded by gas and that the slab is infinite in the x and y directions.

(a) Use a shell balance to show that the slab obeys

$$D_e \frac{d^2 C_A}{dz^2} + r_A = 0 \qquad (P12.39.1)$$

where z is the distance from the center of the slab, C_a is the concentration of the reactant, and r_A is the rate of formation of the reactant A.

(b) Substitute in a first-order rate law to derive a differential equation for the concentration in the slab.

(c) What are the boundary conditions for your rate equation?

(d) Show that the solution of the differential equation is

$$C_A = C_A^0 \left(\frac{\cosh(\Phi_P z/L)}{\cosh(\Phi_P)} \right) \qquad (P12.39.2)$$

with

$$\Phi_P = L \sqrt{\frac{k_A}{D_e}} \qquad (P12.39.3)$$

(e) Now use your results to derive an equation for the effectiveness factor.

(f) Make a plot of the effectiveness factor as a function of the Thiele modulus. How do your results compare to those for a spherical catalyst pellet?

12.40 Chen, Lu Pradier, Paul and Flodstrom (1997). *J. Catalysis*. **177**, 3, examined the effect of SO_2 on the reduction of NO by butene over a platinum catalyst. They found that small additions of SO_2 enhanced the rate of reaction, while large amounts of SO_2 stopped the reaction.

(a) Use a Langmuir–Hinshelwood rate expression to show how large amounts of SO_2 could stop the reaction.

(b) Now consider small amounts of SO_2. Small amounts of SO_2 act as a poison to weaken the bonds between the reactants and the platinum. How could that increase the rate of reaction?

(c) SO_2 also creates acid sites. How would that affect the reaction?

More Advanced Problems

12.41 Gouverneur, Houk, Pascual-Teresa, Beno, Janda, and Lerner (1993). *Science* **262**, 204 produced catalytic antibodies to control the percentage of cis and trans isomers in a series of Diels–Alder reactions. Their technique is to calculate the transition state geometry, find a "hapten" that has a geometry that closely mimics the transition state, and then create antibodies to the hapten. The antibody strongly binds the transition state and so acts like a catalyst for the reaction.

(a) In the paper they use RHF/3-21G methods to calculate the transition state geometry for the reaction. Recalculate the transition state geometry at the MP2/6-31G* level. To simplify things, replace the part of the diene past the nitrogen with a hydrogen. How much does the geometry change? See Longcharich, Brown and Houk (1989), *J. Organic Chem*, **54**, 1129 for more details of the transition state calculation.

(b) How would the changes in the geometry of the transition state affect the conclusions in the paper? Is the real transition state geometry closer or farther away from the hapten geometry?

12.42   Korre, Klein, and Quann (1997). *Industrial & Engineering Chemistry Research*. **36**, 2041 say that they used "inhibition studies" to examine "the kinetics of naphthalene and phenanthrene hydrocracking over a presulfided NiW/USY catalyst in an 1-L batch autoclave at $P_{H2} = 68.1$ atm and $T = 350$ degrees C."

(a) What is the overall reaction? Write a balanced equation for the reactants and major products.

(b) Describe the catalyst. What do you know about the structure of the catalyst from the properties given in the paper?

(c) How was selective inhibition used to help determine kinetics?

(d) Korre, et al. conclude that the reaction occurs on two sites. What is the evidence for this conclusion? Do you agree with the conclusion?

12.43   Prins, Jian, and Flechsenhar (1997a,b). *Polyhedron*. **16**, 3235, (1997). *Journal of Catalysis*. **168**, 491 examined the kinetics and mechanism of aniline hydrodenitrogenation. They find that the data are not easily fit by a Langmuir–Hinshelwood rate expression, but a multisite model works.

(a) Use the multisite model to derive an expression for the hydrodenitrogenation of analine.

(b) Read Prins' paper. What evidence does he provide for the multisite model? Is the evidence convincing?

12.44   HJ. Bart W. Kaltenbrunner and H. Landschutzer, examined the kinetics of the esterification of acetic acid with propyl alcohol catalyzed by the ion exchange resin Dowex Monosphere at 650°C. The mixture is very nonideal.

(a) How do Bart et al. account for the nonideal mixing properties in their kinetic model? Do the changes make sense? In other words, does it make sense to modify the kinetic equations in this way?

12.45   Read the following papers and report on the findings; and then devise a possible homework problem based on the results.

(a) M. C. Ilao, H. Yamamoto, and K. Segawa, Shape-selective methylamine synthesis over small-pore zeolite catalysts. *J. Catal.* **161**(1); 20–30 (1996).

(b) H. U. Blaser, H. P. Jalett, Garland, M., Studer, M., Thies, H., and A. Wirthtijan, Kinetic studies of the enantioselective hydrogenation of ethyl pyruvate Catalyzed by a cinchona modified PT/AL$_2$O$_3$ catalyst. SO *J. Catal.* **173**(2); 282–294 (1998).

(c) Megalofonos, S. K., and Papayannakos, N. G. Kinetics of catalytic reaction of methan and hydrogen sulphide over MOS_2. *Appl. Catal. A* **165**(1-2); 249–258 (1997).

(d) Mahajani, S. M., and Sharma, M. M. Reaction of glyoxal with aliphatic alcohols using cationic exchange resins as catalysts. *Org. Process Res. Dev.* **1**(2); 97–105 (1997).

(e) Olsbye, U., Wurzel, T., and Mleczko, L. Kinetics and reaction engineering studies of dry reforming of methane over a $NI/LA/AL_2O_3$ catalyst. *Ind. Eng. Chem. Res.* **36**(12); 5180–5188 (1997).

(f) Mark, M. F., Mark, F. and Maier, W. F., Reaction kinetics of the CO_2 reforming of methane. *Chem. Eng. Technol.* **20**(6); 361–370 (1997).

(g) Lechert, H. The mechanism of faujasite growth studied by crystallization kinetics. *Zeolites* **17**(5-6); 473–482 (1996).

(h) Bond, G. C., Hooper, A. D., Slaa, J. C., and Taylor, A. O. Kinetics of metal-catalyzed reactions of alkanes and the compensation effect. *J. Catal.* **163**(2); 319–327 (1996).

(i) Yu, Y. S., Bailey, G. W., and Jin, X. C. Application of a lumped, nonlinear kinetics model to metal sorption on humic substances. *J. Environ. Qual.* **25**(3); 552–561 (1996).

(j) Loumagne, F., Langlais, F., and Naslain, R. Reactional mechanisms of the chemical vapour deposition of sic-based ceramics from CH_3SICL_3/H-2 gas precursor. *J. Crys. Growth* **155**(3-4); 205–213 (1995).

(k) Kim, D. S., and Lee, Y. H., Growth mechanism of room-temperature deposited A-SIC-H films by ion-assisted RF glow discharge. *J. Electrochem. Soc.* **142**(10); 3493–3504 (1995).

(l) Holm, D. R., Hill, C. G., and Conner, A. H., Kinetics of the liquid phase hydrogenation of furan amines. *Ind. Eng. Chem. Res.* **34**(10); 3392–3398 (1995).

(m) Kim, E. J., and Gill, W. N. Low pressure chemical vapor deposition of silicon dioxide films by thermal decomposition of tetra-alkoxysilanes. *J. Electrochem. Soc.* **142**(2); 676–682 (1995).

(n) Sauer, M. L., and Ollis, D. F. Acetone oxidation in a photocatalytic monolith reactor. *J. Catal.* **149**(1); 81–91 (1994).

(o) Creighton, J. R. The surface chemistry and kinetics of tungsten chemical vapor deposition and selectivity loss. *Thin Solid Films* **241**(1-2); 310–317 (1994).

(p) Edwards, M. E., Villa, C. M., Hill, C. G., Jr., and Chapman, T.W. Effectiveness factors for photocatalytic reactions occurring in planar membranes. *Ind. Eng. Chem. Res.* **35**(3), 712–720 (1996).

(q) Haupfear, E. A., and Schmidt, L. D. Kinetics of boron deposition from BBr_3 plus H_2. *Chem. Eng. Sci.* **49**(15), 2467 (1994).

(r) Kuhne, H., How hydrogen influences axial growth rate distribution during silicon deposition from silane. *Semicond. Sci. Technol.* **8**(11) 2018–2022 (1993).

(s) Garant, H., and Lynd, L. Applicablity of competitive and noncompetitive kinetics to the reductive dechlorination of chlorinated ethenes. *Biotechnol. Bioeng.* **57**, 751 (1998).

(t) Ishikawa, T., Yoshimura, K., Sakai, K., Tamoi, M., Takeda, T., and Shigeoka, S. Molecular characterization and physiological role of a glyoxysome-bound ascorbate peroxidase from spinach. *Plant Cell Physiol.* **39**(1); 23–34 (1998).

(u) Leaf, T. A., Srienc, F. Metabolic modeling of polyhydrozybutyrate biosynthesis. *Biotechnol. Bioeng.* **57**(5); 557–570 (1998).

(v) Zorko, M., Pooga, M., Saar, K., Rezaei, K., and Langel, U. Differential regulation of gtphase activity by mastoparan and galparan. *Arch. Biochem. Biophys.* **349**(2); 321–328 (1998).

(w) Hoh, C. Y., and Cordruwisch, R. Experimental evidence for the need of thermodynamic considerations in modeling of anaerobic environmental bioprocesses. *Water Sci. Technol.* **36**(10); 109–115 (1997).

(x) Fadiloglu, S., and Soylemez, Z. Kinetics of lipase-catalyzed hydrolysis of olive oil. *Food Res. Int.* **30**(3-4); 171–175 (1997).

(y) Jones, A. W., Jonsson, K. A., and Kechagias, S. Effect of high-fat, high-protein, and high-carbohydrate meals on the pharmacokinetics of a small dose of ethanol. *Br. J. Clin. Pharmaco.* **44**(6); 521–526 (1997).

(z) Iwuoha, E. I., Devillaverde, D. S., Garcia, N. P., Smyth, M. R., Pingarron, J. M., Reactivities of organic phase biosensors. 2. The amperometric behaviour of horseradish peroxide immobilized on a platinum electrode modified with an electrosynthetic polyaniline film. *Biosensors Biolectron.* **12**(8); 749–761 (1997).

(aa) Egger, D., Wehtje, E., and Adlercreutz, P. Characterization and optimization of phospholipase A(2) catalyzed synthesis of phosphatidylcholine. *Biochim. Biophys. Acta* **1343**, 76–84 (1997).

(bb) Aoyama, T., Yamamoto, K., Kotaki, H., Sawada, Y., and Igo, T. Pharmacodynamic modeling for change of locomotor activity by methylphenidate in rats. *Pharm. Res.* **14**(11); 1601–1606 (1997).

(cc) Davies, R.R., and Distefano, M.D., A semisynthetic metalloenzyme based on a protein cavity that catalyzes the enantioselective hydrolysis of ester and aminde substrates. *J. Am. Chem. Soc.* **119**(48); 11643–11652 (1997).

(dd) Franssen, O., Vanooijen, R.D., Deboer, D., Maes, R.A.A., Herron, J.N., and Hennink, W.E. Enzymatic degradation of methacrylated dextrans. *Macromolecules* **30**(24); 7408–7413 (1997).

(ee) Turhan, M., and Mutly, M. Kinetics of kappa-casein/chymosin hydrolysis. *Milchwissenschaft* **52**(10); 559–563 (1997).

(ff) Deka, R.C., and Vetrivel, R., Adsorption sites and diffusion mechanism of alkylbenzenes in large pore zeolite catalysts as predicted by molecular modeling techniques. *J. Catal* **174**(1); 88–97 (1998).

12.46 Look at the following journals: *Journal of Catalysis, Applied Catalysis A, Applied Catalysis B, Biotechnology & Bioengineering.* Find an interesting article and report on its findings.

13

SOLVENTS AS CATALYSTS

PRÉCIS

The objective of this chapter is to discuss solvents as catalysts. In the literature, people seldom think of solvents as catalysts. However, solvents have effects similar to those of catalysts. Solvents stabilize intermediates. Solvents modify intrinsic barriers. Solvents change rates. Mass transfer effects are more important in solvents than in the gas phase. There also are some interesting energy transfer effects.

In this chapter we will outline the key effects. We show how a solvent acts like a catalyst and describe some of the key equations. We cannot review everything, but Reichardt (1988) provides a comprehensive review of the subject, while Tapia and Bertran (1996) provides an up-to-date treatment of the theory. Connors' (1990) treatment is also very good.

13.1 INTRODUCTION

The objective of this chapter is to describe reactions in liquid solvents. If you lookup the descriptions of reactions in liquids in most of the kinetics books written before 2000, you will find that they all treat reactions in liquids as a special topic that is separate from reactions in the gas phase, and separate from reactions on solid catalysts. However, my view is that reactions in liquids are very similar to reactions on catalysts. Generally, a solvent has many of the same effects as a catalyst. In particular

- Solvents can initiate reactions.
- Solvents stabilize intermediates.
- Solvents stabilize transition states and thereby modify the intrinsic barriers to reactions.
- Solvents act as efficient means for energy transfer.

- Solvents can donate or accept electrons.
- Mass transfer limitations are more important when solvents are present.

In Chapter 12 we illustrated these ideas by considering the reaction [see equation(12.39)]

$$H_2 + Br_2 \Longrightarrow 2HBr$$

In Chapter 5 we found that the reaction goes via the following mechanism [see equation (12.40)]:

$$Br_2 \longrightarrow 2Br$$

$$Br + H_2 \longrightarrow HBr + H$$

$$H + Br_2 \longrightarrow HBr + Br$$

$$2Br \longrightarrow Br_2$$

Figure 13.1a shows how the free energy of the system changes as the reaction proceeds in the gas phase. Generally, during the initial stages of the reaction, the energy rises. But then the free energy of the system falls again as the reaction proceeds.

One of the key roles of a solvent is to avoid the initial rise in energy associated with initiating the reaction. Physically, the solvent lowers the energy of the bromine atoms.

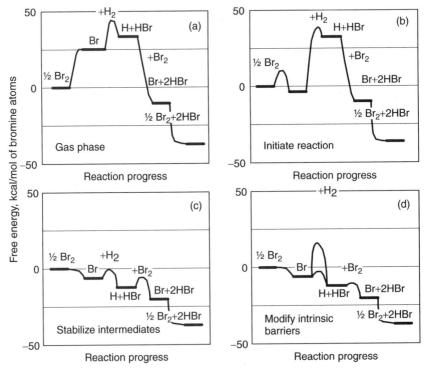

Figure 13.1   An illustration of some of the ways a solvent can affect the free energy changes during the reaction. $H_2 + Br_2 \Rightarrow 2HBr$.

That lowers the energy of the bromine state, so you avoid the first initial barrier as shown in Figure 13.1b.

The solvent can also stabilize the other species in the reaction as shown in Figure 13.1c. That eliminates the need for the system to ever go uphill. As a result, the activation barrier for the reaction is reduced.

Generally, solvents raise the intrinsic barriers for all reactions, as illustrated in Figure 13.1d. However, some species are more affected than others. That allows you to control selectivities.

The last of the effects listed on pages 795 and 796 are almost the same as those listed in Chapter 12. Solvents rarely orient the reactants in the right configuration to react. Solvents may encourage rather than inhibit side reactions. Still, solvents do many of the same things as catalysts and follow many of the same equations. Therefore, I think that it is useful to regard solvents as catalysts. However, solvents are seldom as selective as conventional catalysts.

In this chapter we will review all of the effects so that the reader can have a consistent view of them. In Section 13.2–13.4 we provide a qualitative overview of the role of solvents on rates. In Section 13.5 we will review solvation. The influence of solvation on rates will be discussed in Sections 13.6 and 13.7. Section 13.4 will consider the influence of mass transfer on the rate processes and other dynamic effects. The effects of solvents is a large topic, and we include only an overview here. Additional information can be found in Reichardt(1988) or Tapia and Bertran (1996).

13.2 EXAMPLES OF THE ROLE OF SOLVENTS IN CHANGING RATES OF REACTIONS

As a start on the discussion, it is useful to give a qualitative overview of the effects of solvents on rates of reactions. There are several key effects:

- Solvents can stabilize ionic intermediates.
- Solvents can stabilize ionic transition states.
- Solvents can also stabilize radical intermediates.
- Solvents moderate the collisions between molecules.
- Solvents are efficient media for energy transfer.
- The collision frequency is lower in solution than in the gas phase because solvent molecules get in the way, but the net effect is that the presence of solvents often increases the rate of reaction.

Generally solvents have a much larger effect on ionic reactions than on radical reactions. However, the solvent affects both ionic and radical reactions. We will summarize the key effects in the next few sections to provide the reader with a qualitative picture of the influence of solvents on rates.

First let us consider some data. Tables 13.1–13.3 show data on four different reactions in a number of solvents. Table 13.1 shows data for a Diels–Alder reaction. Notice that in this case, the solvent has only a small effect on the rate. The rate constant is almost the same in solution as in the gas phase. Varying the solvent makes at most a factor of 4 change in rate.

In contrast, Table 13.2 shows data for an S_N2 reaction. There the rate is 10^{45} higher in DMF than in the gas phase. The rate is a factor of 10^6 less in water than DMF.

Table 13.1   The rate of the Diels–Alder reaction (2 cyclopentadiene → cyclopentadiene dimer) at 300 K

Solvent	Rate Constant, liter/(mol·second)	Solvent	Rate Constant, liter/(mol·second)
Gas phase	6.9×10^{-7}	Carbon disulfide	9.3×10^{-7}
Ethanol	19×10^{-7}	Tetrachloromethane	7.9×10^{-7}
Nitrobenzene	13×10^{-7}	Benzene	6.6×10^{-7}
Paraffin oil	9.8×10^{-7}	Neat liquid	5.2×10^{-7}

Source: Data of Wasserman (1952).

Table 13.2   The rate of the S_N2 reaction NaCl + CH_3I → CH_3Cl + NaI at 350 K

Solvent	Rate constant, liter/(mol·second)	Solvent	Rate constant, liter/(mol·second)
gas phase	$\sim 10^{-45}$	—	—
Water	3.5×10^{-6}	Methylcyanide	0.13
Methanol	3.1×10^{-6}	Dimethylformamide	2.5

Note: The gas-phase rate in the table is estimated from the ab initio calculations of Glukhoutsev et al. (1995).
Source: Data from Parker (1969)

Table 13.3   The rate of some association reactions in various solvents

Solvent	Dielectric Constant	Rate of the Reaction $(C_2H_5)_3N + CH_3I \rightarrow [(C_2H_5)_3NCH_3]^+ + [I]^-$ at 100°C, liter/(mol·second)	Rate of the Reaction $CH_3COOCOCH_3 + 2C_2H_5OH \rightarrow 2CH_3COOC_2H_5 + H_2O$ at 50°C, liter/mol·second
Hexane	1.89	0.5×10^{-5}	0.0119
Benzene	2.28	39×10^{-5}	0.0053
Chlorobenzene	5.62	160×10^{-5}	0.0046
Methoxybenzene	9	400×10^{-5}	0.0029
Acetone	20.7	265×10^{-5}	—
Nitrobenzene	35	1383×10^{-5}	0.0024

Source: Data from, Menschutkin (1890) and Laidler (1965).

Table 13.3 shows two intermediate cases. In the case on the left, the solvent changes the rate by a factor of 10^3, while in the case on the right the rate changes by a factor of 4.

One can speculate why the solvents play such a large role in Table 13.2, but not in Table 13.1. Notice that the reaction in Table 13.1 involves only neutral species. Generally, solvents have only a small effect on reactions of neutral species.

In contrast, the solvent has a tremendous effect on reaction rates of ions. For example, Table 13.2 shows data for an S_N2. The rate of reaction is tiny in the gas phase, but substantial in solution. The rate of ionic reactions tend to be substantially modified in the presence of solvents. Further, the rate varies with the solvent.

Association reactions tend to be in the middle as shown in Table 13.3. The rate of the trans esterification reaction $CH_3COOCOCH_3 + 2C_2H_5OH \rightarrow 2CH_3COOC_2H_5 + H_2O$ changes by a factor of only 4 with changing solvent. The rate is similar to the rate in the

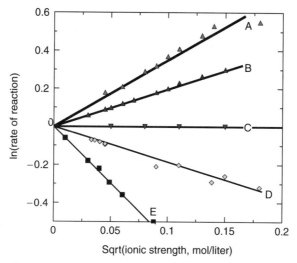

Figure 13.2 The effect of ion strength on a number of reactions after Laidler (1987). The points are data. The lines are predictions of equation (13.39). (a) $Co(NH_3)_5Br^{2+} + Hg^{2+} \rightarrow$; (b) $S_2O_8^{2-} + I^- \rightarrow$; (c) $[(Cr(urea)_6]^{3+} + H_2O \rightarrow$; (d) $Co(NH_3)_5Br^{2+} + OH^- \rightarrow$ products; (e) $Fe^{2+} + Co(C_2O_4)_3^{3-} \rightarrow$ products.

gas phase. On the other hand, the reaction $(C_2H_5)_3N + CH_3I \rightarrow [(C_2H_5)_3NI] - [CH_3]^+$ changes by a factor of 2000 with changing solvent. The rate is much higher in solution than in the gas phase.

In fact, the reaction $(C_2H_5)_3N + CH_3I \rightarrow [(C_2H_5)_3NI]^-[CH_3]^+$ is a special case. You start with a neutral amine and end up with an ionic species. In 1890, Menschutkin discovered a series of reactions where ions form from neutral species. They are now called **Menschutkin reactions** in the literature. Menschutkin reactions show important solvent effects. However, most other reactions of neutral species show only small variations between solvents and the gas phase.

The other second key effect you see during ion reactions in solution is that the presence of other ions can have an effect on rates. Figure 13.2 shows some of the data. Notice that the addition of a small amount of sodium chloride can change the rate by a factor of $3-10$ even though sodium chloride does not directly participate in the reaction. Of course, the effects shown in Figure 13.2 are much smaller than the effects listed in Table 13.2. The salt is making only a minor perturbation to a giant enhancement due to the presence of the solvent. Still, people discuss this effect in great detail in the literature, so we have to mention it, too.

Finally, solvents greatly affect the mass transfer rates during reactions. We will work out all of the details in Section 13.8. However, the key feature is that the solvent slows down the diffusion rate of the reactants, which affects the collision rate. In addition, the solvent tends to trap the reactants in a solvent cage. That has an important effect on the rate.

13.2.1 Summary of the Qualitative Trends

Now it is useful to step back and summarize the key ideas that we have covered so far. We see that solvents can have tremendous affects on rates of reactions. The effects are huge; the reaction $NaCl + CH_3I \rightarrow CH_3Cl + NaI$ is a factor of 10^{45} faster in DMF than in the gas phase. Generally solvents make a huge difference to the rate of an initiation reaction. They also have a major effect if the reactants are ions. Propagation reactions of neutral species have smaller effects, but still the effects are important in special examples.

13.3   WHY DO SOLVENTS CHANGE RATES?

Next, we want to ask why solvents change rates. I want to show that solvents change rates for the same reasons that catalysts change rates:

- Solvents stabilize intermediates.
- Solvents can initiate reactions.
- Solvents stabilize transition states and thereby modify the intrinsic barriers to reactions.
- Solvents act as efficient means for energy transfer.
- Solvents can donate or accept electrons.
- Mass transfer limitations are more important when solvents are present.

The next several sections will review these ideas to show qualitatively how solvents change rates of reaction. In particular, we will show that the key role of a solvent is to solvate the intermediates and transition states for a reaction as illustrated in Figure 13.1, thereby promoting the reaction.

13.3.1   Solvents Solvate Species

To start off, it is important to note that the most important role of the solvent is to solvate the reactants and intermediates of a reaction. This may seem obvious, but in order for a reaction to occur in a solvent, the reactants and intermediates must first dissolve in the solvent. If the reactants are not soluble, no reaction can occur.

If one of the intermediates is not soluble, the intermediate will precipitate out, and the reaction will stop at the intermediate. The products of a reaction do not have to be soluble in the solvent. Nothing bad happens if the products of a reaction precipitate out. Still, the solvent must solvate the reactants and intermediates in order for the reaction to occur. Solvation, then, becomes a key way that solvents affects rates of reactions.

13.3.2   Solvation of Intermediates

One of the key ways that solvents enhance reactions is to solvate the key intermediates of a reaction and thereby enhance the rate of reaction. For example, consider the reaction

$$CH_3I + NaCl \longrightarrow CH_3I + NaI \tag{13.1}$$

Reaction (13.1) is slow in the gas phase. However, if you dissolve the reactants in dimethylformamide (DMF), the reaction is extremely rapid. First the sodium chloride dissociates in DMF to form sodium ions and chloride ions. Then the chloride ions can undergo a S_N2 reaction with methyl iodide to form the methylchloride:

$$Cl^- + CH_3I \longrightarrow ClCH_3 + I^- \tag{13.2}$$

There is also an S_N1 reaction in the gas phase, where the methyliodide dissociates to yield $[CH_3]^+$ and I^-. The $[CH_3]^+$ then reacts with Cl^-. At 350 K, the chloride ion concentration is a factor of 10^{55} larger in DMF or water solution than in the gas phase. That tremendously increases the rate.

Experimentally, reaction (13.1) can be rapid in solution. In contrast, little reaction can be seen in the gas phase or by directing a beam of methyliodide at a sodium chloride sample.

Notice that the DMF has acted just like a catalyst. The presence of the solvent has increased the concentration of a key intermediate by a factor of 10^{55}. The increase in the intermediate concentration produces a tremendous increase in rate.

The rate does not go up by a factor of 10^{55}, though. Recall that when you attach a radical to a catalyst, the radical becomes less reactive. In the same way, when you dissolve the chloride ions in DMF, the ions become less reactive. Calculations indicate that the overall rate goes up by a factor of only 10^{45} even though the concentration of chloride ions goes up be a factor of 10^{55}. That is because the chloride ions are a factor of 10^{10} less reactive in DMF than in the gas phase.

It is useful to put these results onto a potential energy plot similar to those mentioned in Chapter 12. Figure 13.3 shows some thermodyamic data for the reaction

$$Cl^- + CH_3Br \longrightarrow ClCH_3 + Br^- \tag{13.3}$$

If you run the reaction in the gas phase, you need to first create a chloride ion. That takes 88 kcal/mol, compared to the solution.

Once you create a chloride ion, the ion is very reactive. Chloride ions form stable ion–molecule complexes with CH_3Br. The reaction to form a $ClCH_3$ and a bromide ion only has a barrier of 2.6 kcal/mol.

By comparison, the chloride ion is stable in aqueous solution. One can get chlorides to form at room temperature, but then the chloride ions are less reactive than in the gas

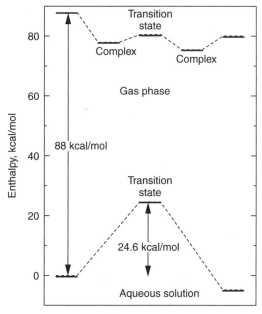

Figure 13.3 A comparison of the energy changes during the reaction $Cl^- + CH_3Br \to ClCH_3 + Br^-$. [Data from Reichardt (1988) p. 31].

phase. Reaction (13.3) has an activation barrier of 24.6 kcal/mol compared to 2.6 kcal/mol in the gas phase.

This example shows that a solvent can act just like a catalyst. The solvent stabilizes the intermediates of a reaction. The stabilized intermediates are less reactive than in the gas phase. However, the concentration of intermediates is substantially enhanced. The net effect is that ionic reactions show much higher overall rates in polar solvents than in the gas phase.

For some reason, in the literature, people often present these results differently. They ignore the 88 kcal/mol to form the ion and look only at the relative rates of reaction after they have formed the necessary ions. Table 13.4 shows the data. Notice that the chloride ion is a factor of 10^{16} more reactive in the gas phase than in aqueous solution. Clearly, this is a substantial effect.

Another important feature in Tables 13.2 and 13.4 is that the rate of reaction can vary by a factor of 10^6 with changing solvent. The data in Tables 13.2 and 13.4 indicate that water is less effective than DMF for S_N2 reactions of anions (negatively charged ions).

Experimentally the best solvent varies with the reaction. Table 13.5 shows data for S_N1 reaction. Water is one of the best solvents for the S_N1 reaction in Table 13.5 and one of the worst solvents for the S_N2 reaction in Table 13.4.

One can generalize these findings to many reactions. Some solvents favor one reaction while other solvents favor other reactions. Generally, what are called **protic** solvents favor S_N1 reactions of ions. **Polar aprotic** solvents favor S_N2 reactions of ions. Nonpolar solvents favor radical reactions.

Table 13.6 gives examples of each of these types of solvents. Polar protic solvents are solvents that form hydrogen bonds. Generally, the solvent molecule contains at least one hydrogen atom that is bonded to either an oxygen or a nitrogen. The hydrogen or

Table 13.4   The rate of the S_N2 reaction $Cl^- + CH_3Br \rightarrow ClCH_3 + Br^-$ at 298 K

Solvent	Rate Constant, liter/(mol·second)	Solvent	Rate Constant, liter/(mol·second)
Gas phase	2.1×10^{-11}	—	—
Acetone	5.5×10^{-21}	Methanol	1.0×10^{-26}
DMF	9.3×10^{-22}	Water	8.3×10^{-27}

Source: Data from Reichardt (1988).

Table 13.5   The rate constant a prototypical S_N1 reaction: the solvolysis of p-methoxyneophyl-tolunesulfonate in a number of solvents at 75°C

Solvent	k, minute^{-1}	Solvent	k, minute^{-1}
Formic acid	7.1	Methylcyanide	3.6×10^{-3}
Water	4.0	DMF	3.0×10^{-3}
Acetic acid	0.19	Acetic anhydride	2.0×10^{-3}
Methanol	0.1	Pyridine	1.3×10^{-3}
Ethanol	3.8×10^{-2}	Acetone	5.1×10^{-4}
$C_7H_{15}COOH$	4.4×10^{-3}	Ethylacetate	6.8×10^{-5}
Dimethyl Sulfoxide	1.1×10^{-2}	Dioxane	5.1×10^{-5}
Nitromethane	7.2×10^{-3}	Diethylether	$\sim 3 \times 10^{-6}$

Source: Data of Smith et al. (1961).

Table 13.6   Some examples of protic and aprotic solvents

Protic solvents: good for S_N1 reactions of anions
  Water, ethanol, methanol, acetic acid, formic acid, ammonia, ethane thiol

Polar aprotic solvents: good for S_N2 reactions of anions
  Acetone, dimethylsulfoxide (DMSO) [$(CH_3)_2S{=}O$], dichloromethane, ethers,
  Dimethylformamide (DMF) [$(CH_3)_2NCHO$], cyclohexanone, acetaldehyde

Nonpolar aprotic solvents: good for radical reactions
  Ethylene, benzene

oxygen allows a hydrogen bond, to form. Generally, polar protic (i.e., hydrogen-bonding) solvents speed up S_N1 reactions.

Polar aprotic solvents are molecules with a dipole moment that do not form a hydrogen bond. Generally, they are polar molecules with no hydrogen atoms bonded to an oxygen or nitrogen. Polar aprotic solvents favor S_N2 reactions of anions.

Nonpolar aprotic solvents are simple nonpolar molecules. Ions are rarely not soluble in nonpolar solvents, so nucleophilic reactions are unfavorable in nonpolar aprotic solvents. Radicals are soluble in nonpolar solvents, however. As a result, you see radical reactions mainly in nonpolar aprotic solvents.

13.3.3   Solvents Can Initiate Reactions

A related effect is that solvents can help initiate reactions. For example, benzoylperoxide ($Ph_3COOCPh_3$) is a stable molecule. (*Note:* Ph is a phenyl group.) However, when you dissolve the benzyolperoxide in a polarizable organic solvent such as ethylene, the benzoylperoxide dissociates into two Ph_3CO radicals. People usually discuss benzoylperoxide as being a catalyst for ethylene polymerization, since the benzyol peroxide dissociates into radicals that start the polymerization. However, the ethylene is acting like a catalyst, too! When benzoyl peroxide dissolves in ethylene, the benzoyl peroxide dissociates. The peroxide radical is more soluble in the ethylene than the neutral molecule. As a result, one gets a higher concentration of the radical in ethylene solution than in, for example, argon solution.

Generally this effect is much smaller than the effects discussed in Sections 13.3.1 and 13.3.2. Radical reactions can be enhanced by a factor of 10–100 in the presence of a solvent, while enhancements of 10^{45} are seen with ionic reactions. Still there are some important examples where the solvent stabilizes a radical intermediate.

13.4   SOLVENTS STABILIZE TRANSITION STATES AND THEREBY MODIFY THE INTRINSIC BARRIERS TO REACTIONS

The effects discussed in Sections 13.3.2 and 13.3.3 were associated with stabilizing the reactants and intermediates during reactions. Another key effect of solvent is to modify the intrinsic barriers to a reaction by solvating transition states. In this section we will briefly review those effects.

Recall from transition state theory that to some approximation the rate constant for reaction is given by

$$k_1 = \left(\frac{k_B T}{h_p}\right) \exp\left(\frac{-\Delta G^{\ddagger}}{k_B T}\right) \tag{13.4}$$

where k_1 is the rate constant for a reaction, k_B is Boltzmann's constant, T is the absolute temperature, h_p is Planck's constant, and $\Delta G^{\ddagger}$ is the free energy of activation.

Consequently, anything you do to decrease $\Delta G^{\ddagger}$ will enhance the rate. A solvent can change $\Delta G^{\ddagger}$. Thus, the solvent can also change the rate.

For future reference I want to note that the ratio of the rate constant in a solvent, S, to the rate constant in a reference solvent is given by

$$\ln\left(\frac{k_S}{k_R}\right) = \left(\frac{\Delta G_R^{\ddagger} - \Delta G_S^{\ddagger}}{k_B T}\right) \tag{13.5}$$

where $\Delta G_S^{\ddagger}$ is the free energy of activation in the solvent S and $\Delta G_R^{\ddagger}$ is the free energy of activation in the reference solvent.

Next, I want to derive an equation to relate the change in rate to the change in the free energy of solvation of the reactants and the transition state. Let's consider an ideal reaction $A \rightarrow C$. One can compute the free energy of reaction in solution by the thermodynamic path in Figure 13.4. First we will define $\Delta G_g^{\ddagger}$ as the free energy of activation in the gas phase. Then the free energy of reaction in solution will be

$$\Delta G_S^{\ddagger} = \Delta G_g^{\ddagger} + (\Delta G_{g \rightarrow s}^{\ddagger} - \Delta G_{g \rightarrow s}^{A}) \tag{13.6}$$

where $\Delta G_s^{\ddagger}$ is the free energy of activation in the solution, $\Delta G_g^{\ddagger}$ is the free energy of activation in the gas phase, $\Delta G_{g \rightarrow s}^{A}$ is the free-energy change taking the reactants from the gas phase into solution, and $\Delta G_{g \rightarrow s}^{\ddagger}$ is the free-energy change taking the transition state from the gas phase into solution. $\Delta G_{g \rightarrow s}^{A}$ is called the *free energy of solvation of the reactants* and $\Delta G_{g \rightarrow s}^{\ddagger}$ is called the *free energy of solvation of the transition state*. Equation (13.6) says that the free energy of activation of a reaction in the solution is equal to the free energy of activation in the gas phase, plus a correction equal to the difference between the free energy of solvation of the transition state and the free energy of solvation of the reactants.

From thermodynamics

$$\Delta G_{g \rightarrow s}^{\ddagger} = k_B T \ln(\phi^{\ddagger}) \tag{13.7}$$

$$\Delta G_{g \rightarrow s}^{A} = k_B T \ln(\phi^{A}) \tag{13.8}$$

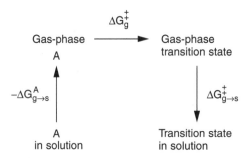

Figure 13.4   A thermodynamic path used to derive equation (13.6).

where ϕ^A and $\phi^{\ddagger}$ are the fugacity coefficients of the reactants and the transition state, respectively. Combining equations (13.5)–(13.8) yields

$$k_{sol} = k_g \left(\frac{\phi^A}{\phi^{\ddagger}} \right) \qquad (13.9)$$

where k_{sol} is the rate constant in solution, k_g is the rate constant in the gas phase, and ϕ^A and $\phi^{\ddagger}$ are the fugacity coefficients of the reactants and the transition state, respectively. One can also develop an expression for the rate constant in solution compared to the rate constant in an ideal solution. The derivation is the same as above. The result is

$$k_{sol} = k_{ideal} \left(\frac{\gamma^A}{\gamma^{\ddagger}} \right) \qquad (13.10)$$

where k_{sol} is the rate constant in solution, k_{ideal} is the rate constant in ideal solution, and γ^A and $\gamma^{\ddagger}$ are the activity coefficients of the reactants and the transition state, respectively.

Equations (13.6), (13.9), and (13.10) are the key results for the remainder of this chapter. The reader should memorize these equations before proceeding

13.4.1 Qualitative Predictions of Equation (13.6)

Next, we want to briefly discuss the qualitative predictions of equation (13.6). Recall from thermodynamics that the solubility of species A in solution is related to $-\Delta G^A_{g \rightarrow s}$. The larger the value of $-\Delta G^A_{g \rightarrow s}$ the more soluble the species will be in solution.

Now consider computing the size of the difference $(\Delta G^{\ddagger}_{g \rightarrow s} - \Delta G^A_{g \rightarrow s})$. If the reactants are more soluble in the solution than the transition state, then $(\Delta G^{\ddagger}_{g \rightarrow s} - \Delta G^A_{g \rightarrow s})$ will be positive. According to equation (13.6), the activation energy for the reaction will be larger in solution than in the gas phase. In contrast, if the transition state is more soluble than the reactants, then $(\Delta G^{\ddagger}_{g \rightarrow s} - \Delta G^A_{g \rightarrow s})$ will be negative, so the activation barrier will be smaller in solution than in the reactants.

That has important implications for reactions. For example, consider the reaction

$$(C_2H_5)_3N + CH_3I \longrightarrow [(C_2H_5)_3NCH_3]^+[I^-] \qquad (13.11)$$

Data for this reaction are given in Table 13.3. The reaction starts with neutral species but you end up with ions. The transition state is half way in between the reactants and products, and therefore you would expect the transition state to be partially ionic. According to equation (13.6), the rate of reaction (13.11) should be enhanced when the solubility of the ionic transition state is larger than the solubility of the reactants, while the rate should be reduced when the transition state is less soluble than the reactants.

Recall that ions are more soluble in polar solutions, whereas neutral species are more soluble in nonpolar solutions. The transition state for reaction (13.11) is ionic, while the reactants are neutral. Therefore, according to equation (13.6), the rate of reaction (13.9) should be enhanced in a polar solvent such as nitrobenzene and inhibited in a nonpolar solvent such as hexane. Experimentally, the rate of reaction (13.11) is a factor of 2700 higher in nitrobenzene than in hexane.

These results demonstrate that solvents can make a substantial change in the rate of a reaction by stabilizing the transition state of the reaction.

13.5 INTRODUCTION TO SOLVATION

So far all of our discussion has been qualitative but in the next several sections we will try to develop some equations for the influence of solvents on rates of reactions. Our approach will be to first review the theory of solvation and then apply the theory to rates of reaction. In practice, I have never found the equations to be very useful because they do not fit data very well. However, the qualitative insights are useful.

In this section, we will review the theory of solvation. I am assuming that the reader has already had a detailed introduction to solvation in a thermodynamics or physical chemistry course. Here we will briefly review the ideas. One should refer to a thermodynamics book for more information.

The word **solvation** is defined as *the process of forming a solution with a dissolved species.* The dissolved species is called the **solute**. The species that does the dissolving is called the **solvent**.

The thermodynamics of solvation are controlled by the free-energy changes that occur when a solution is formed. Let's consider dissolving a solid in a liquid. In order for a solute to dissolve in a solvent, three things must occur:

- Solute molecules must separate from each other and move into the solvent.
- Some solvent molecules must separate to make room for the solute.
- The solute and the solvent must mix together.

One can express the free-energy change during this process as

$$\Delta G_{solvation} = \Delta G_{solute} + \Delta G_{solvent} + \Delta G_{mix} \qquad (13.12)$$

where $\Delta G_{solvation}$ is the free-energy change during solvation, ΔG_{solute} is the free-energy change associated with the solute molecules separating from each other, $\Delta G_{solvent}$ is the free-energy associated with the solvent molecules moving apart to make room for the solute, and ΔG_{mix} is the energy you get back when the solute mixes into the solvent. Generally, $\Delta G_{solvent}$ and ΔG_{solute} are positive and ΔG_{mix} is negative. A solution forms only when the sum $(\Delta G_{solute} + \Delta G_{solvent} + \Delta G_{mix})$ is negative. Two phases form if $(\Delta G_{solute} + \Delta G_{solvent} + \Delta G_{mix})$ is positive.

Equation (13.12) is a key result, because it says that you need to consider the solute, the solvent and the solution in order to understand the free-energy of solvation.

In order to get the flavor of equation (13.12), let us consider dissolving sodium chloride in water. When sodium chloride dissolves in water, the sodium and chloride ions need to be removed from the salt crystal, the water molecules need to move apart to make spaces for the ions, and the ions need to move into the spaces. ΔG_{solute} is large and positive because we are breaking the ionic bonds (i.e. the electrostatic interactions) between the sodium ions and the chloride ions, $\Delta G_{solvent}$ is large and positive because we are breaking the hydrogen bonds in the water, while ΔG_{mix} is large and negative because the ions strongly interact with the water. It is not immediately obvious that $\Delta G_{solute} + \Delta G_{solvent} + \Delta G_{mix}$ is negative, but in fact it is, which makes sodium chloride very soluble in water.

Table 13.7 shows data for the solubility of a number of other alkyl halides. Lithium fluoride is relatively insoluble because the electrostatic interaction between the lithium and the fluoride is huge in LiF. The lattice energy (ΔH_{solute}) is 247 kcal/mol. ΔG_{solute} is so positive that the solubility is small. Cesium iodide has a modest solubility because

Table 13.7   Solubility of some alkyl halides in water

Salt	Solubility, mols/liter	Lattice Energy (ΔH_{solute}), kcal/mol
LiF	0.10	247
NaCl	6.11	188
KBr	4.49	167
CsI	1.69	144
LiI	12.32	174
CsF	24.16	177

the cesium and the iodine are so large. In order for the cesium and iodide to dissolve in water, many hydrogen bonds must break. $\Delta G_{solvent}$ limits the solubility. Sodium chloride and potassium bromide are intermediate cases. The ions are not so large that you break water–water bonds. Yet they are not so small that the electrostatic interactions predominate. Interestingly, lithium iodide and cesium flouride are more soluble than any of the other salts in Table 13.7. In both cases one ion is large while the other is small. The large size reduces the lattice energy, compared to LiF, while the fact that there is at least one small ion allows some of the solvation to occur without disrupting the water too much.

These results show that, in order to understand solubility, one needs to consider all three terms on the rightside of equation (13.12). If one considers only the direct interactions between the solute and the solvent, one will not get the right answer.

In thermodynamics, people discuss solvation of an ionic salt in terms of

- The **hydrophobic interaction**
- The **electrostatic interaction**
- **Solvation forces**

The hydrophobic interaction is associated with the energy to move the solvent molecules apart to make spaces for the solute. It is large in water, and smaller in non-hydrogen-bonding fluids. Generally, the hydrophobic interaction is proportional to the heat of vaporization of the solvent per unit volume. We will provide equations for the hydrophobic interaction in Section 13.7.4.

The electrostatic interaction is associated with the change in the lattice energy when the solute dissolves. In freshman physics you learned that the force between two ions A and B is given by Coulomb's law:

$$V = \frac{Z_A Z_B e^2}{\varepsilon r_{AB}} \tag{13.13}$$

where V is the interaction potential, Z_A and Z_B are the charges on A and B, e is the charge on an electron, ε is the dielectric constant of the medium, and r_{AB} is the distance between the ions. According to equation (13.7), small ions have larger electrostatic attractions than do bigger ions. Solvents moderate the electrostatic interaction. Notice that the ε is 1.0 in vacuum or in a salt crystal, but 78 in water. As a result, the electrostatic interaction between ions will be much smaller in water than in the gas phase. Solvents with a large dielectric constant (e.g., water) are more effective in moderating the electrostatic interaction than a solvent with a low dielectric constant (e.g., pentane) would be.

Of course, equation (13.13) works only at low ion concentrations. At higher ion concentrations, the ions arrange themselves to lower the net attraction between a given pair of ions. Positive charges surround negative ions and vice versa. As a result, the net force between the ions is reduced. Debye and Hückel (1923) showed that the interaction energy can be approximated by

$$V = \left(\frac{Z_A Z_B e^2}{\varepsilon r_{AB}}\right) - 2 Z_A Z_B (k_B T) Q_D \sqrt{I} \tag{13.14}$$

where V is the interaction potential, Z_A and Z_B are the charges on A and B, e is the charge on an electron, ε is the dielectric constant of the medium, r_{AB} is the distance between the ions, Q_D is a property of the fluid called the Debye–Hückel constant [0.51 (liters/mol)$^{0.5}$ for water at 300 K], k_B is Boltzmann's constant, T is the temperature, and the I is called the *ionic strength*. I is calculated from

$$I = \frac{1}{2} \sum_{ions} (Z_i)^2 C_i \tag{13.15}$$

where Z_i is the charge of ion i in solution, and C_i is the ion concentration. The sum in equation (13.15) goes over all the ions in solution. According to equation (13.15) a 1 molar (1 M) sodium chloride solution has an ionicity of 1 mol/liter, while a 1 molar $MgCl_2$ solution has an ionicity of 3 mol/liter.

Equation (13.12) and (13.13) consider the solvent moderates as the electrostatic interactions between ions. The solvent has many other effects. Examples include

- Chemical bonds between the solvent and the solute
- Ion–solvent interactions
- Dipole–solvent interactions
- Quadrupole–solvent interactions
- Pauli repulsions

In the literature, people often lump all of these other effects under the title *solvation forces*. Solvation forces can also have major effects.

One key effect is that the ions interact directly with the solvent. Kirkwood (1934) found that $\Delta U_{g \to s}$, the energy to moving an ion from the gas phase into a medium with a dielectric constant, ε, can be approximated by

$$\Delta U_{g \to s} = -\left(\frac{\varepsilon - 1}{2\varepsilon + 1}\right) \frac{(Z_A)^2 e^2}{r_A} \tag{13.16}$$

In equation (13.16) Z_A is the charge on the ion, r_A is the radius of the ion, and e is the charge on an electron. Kirkwood also found that the energy to move a dipole into a medium is given by

$$\Delta G_{g \to s} = -\left(\frac{\varepsilon - 1}{2\varepsilon + 1}\right) \left(\frac{(\mu_A)^2}{r_A}\right) \tag{13.17}$$

where μ_A is the dipole moment of the molecule, r_A is the radius of the molecule, and ε is the dielectric constant of the medium. These equations are useful to estimate $-\Delta G_{g \to s}^A$ when electrostatic interactions predominate.[1]

---

[1] Equations (13.16) and (13.17) consider only the effects of the electrostatic interactions with the medium. They ignore the hydrophobic interaction and other effects.

Covalent effects are controlled by the donor capability of the solvent. To understand why the donor capability is important, note that the solubility of HF in water is much higher than the solubility of LiF in water. In both cases the electrostatic and hydrophobic interactions are similar. However, HF can react with water via the reaction

$$HF + H_2O \longrightarrow H_3O^+ + F^- \tag{13.18}$$

while no similar reaction is seen with LiF. The extra reaction enhances the solubility of HF.

Next, it is useful to quantify how a direct reaction between the solute and water would affect the solubility. When a reaction occurs, the free-energy of mixing, ΔG_{mix}, will be unusually large (negative). According to equation (13.12), if ΔG_{mix} is negative, $\Delta G_{solvation}$ will be negative, too. That tremendously increases the solubility.

People use two common measures to identify the binding power of solvents. Aprotic solvents are often characterized via the donor number (DN) of a solvent. The donor number is the energy of the bond formed with BF_3 (a strong lewis base). Table 13.8 lists the donor values for a series of solvents. The energies vary from 2 to 30 kcal/mol.

There are no good measures of the bonding power of protic solvents. Kamlet and Taft have proposed a three parameter fit where Π_T is a measure of the ability of the solvent to form polar bonds, α_T is a measure of the proton donor ability of a solvent, and β_T is

Table 13.8 Properties of some common solvents

Solvent	ε	δ, calories/cm^3	DN, kcal/mol	Π_T	α_T	β_T
Hexane	1.89	7.3	—	−0.08	0	0
Dioxane	2.21	10.0	14.8	0.55	0	0.37
Carbon tetrachloride	2.23	8.6	—	0.28	0	0
Benzene	2.28	9.2	—	0.59	0	0.10
Toluene	2.38	8.9	—	0.54	0	0.11
Diethylether	4.34	7.4	19.2	0.27	0	0.47
Chloroform	4.70	9.3	—	0.58	0.4	0
Chlorobenzene	5.62	9.9	—	0.71	0.0	0.07
Ethylacetate	6.02	9.5	17.1	0.55	0.0	0.45
Acetic acid	6.19	10.1	—	0.64	1.12	—
THF^a	7.32	9.1	20	0.58	0.0	0.55
Dichloromethane	8.9	9.7	2.4	0.82	0.30	0.00
t-Butyl alcohol	10.9	10.6	—	0.41	0.68	1.01
Pyridine	12.3	10.7	33.1	0.87	0	0.64
n-Butyl alcohol	17.7	10.8	—	0.47	0.79	0.88
Acetone	20.7	9.9	17.0	0.71	0.08	0.48
Acetic anhydride	21	10.3	10.5	0.76	0	—
Ethanol	24.3	12.7	—	0.54	0.83	0.77
Methanol	32.6	14.5	—	0.60	0.93	0.62
Nitrobenzene	35	10.0	8.1	1.01	0	0.39
DMF	36.7	12.1	26.6	0.88	0	0.69
DMSO	49	12.0	29.8	1.00	0	0.76
Water	78.5	23.4	—	1.09	1.17	0.18
Formamide	110	19.2	—	0.97	0.71	—

^{a}Tetrahydropuran.

a measure of the proton acceptor ability of a solvent. Kamlet and Taft then took a large set of kinetic data and fit it to the equation

$$\ln\left(\frac{k_S}{k_R}\right) = \frac{\Delta G_S^{\ddagger}}{k_a T} = s\prod_T + a\alpha_T + b\beta_2 \qquad (13.19)$$

where k_S is the rate constant of a given reaction in a solvent s; k_R is the rate constant of the given reaction in hexane; s, a, and b are constants that depend only on the reaction; and Π_T, α_T, and β_T are constants that depend only on the solvent. They then did a multiparameter regression to calculate all of the parameters. Table 13.8 shows the values.

I have never been able to use these numbers quantitatively because I do not know how to estimate s, a, and b. However, the scale gives a useful qualitative measure of the ability of a solvent to solubilize species. For example, water and DMSO are both polar molecules; however, water is a better proton donor and DMSO is a better proton acceptor.

The values in Table 13.8 would also make it seem that water is a better proton donor than a proton acceptor, but that is an artifact of the way the parameterization is done. Because α_T and β_T are dimensionless parameters that are scaled artificially, the fact that α_T is bigger than β_T for water is not significant.

13.6 QUALITATIVE PREDICTIONS OF THE EFFECTS OF SOLVENTS ON RATES

Next, I want to demonstrate that when you run a reaction in a solvent, the changes in the electrostatic forces and the hydrophobic forces during solvation are usually much larger than the changes in the solvation forces unless there is a direct chemical bond between the reactants and the solvent.

First, let us consider how the solvent will affect the electrostatic interaction. Consider the interaction between a sodium ion and a chloride ion. In the gas phase, $\varepsilon = 1.0$. The sodium has a radius of 0.97 Å; the chloride has a radius of 1.81 Å. One can calculate the electrostatic energy in the gas phase from equation (13.11).

$$V = \frac{(1)(-1)(1 \text{ esu})^2}{1(0.97 \text{ Å} + 1.81 \text{ Å})} \times \left(\frac{332 \text{ kcal/(mol·Å)}}{(1 \text{ esu})^2}\right) = -119 \text{ kcal/mol}$$

where esu = electrostatic unit.

If we run the same reaction in water ($\varepsilon = 78.5$), we obtain

$$V = \frac{(1)(-1)(1 \text{ esu})^2}{78.5(0.97 \text{ Å} + 1.81 \text{ Å})} \times \left(\frac{332 \text{ kcal/(mol·Å)}}{(1 \text{ esu})^2}\right) = -1.5 \text{ kcal/mol}$$

Notice that when we put the sodium chloride in water, the electrostatic interaction was reduced by over 117 kcal/mol. Consequently, shielding of the electrostatic ions can produce changes in the order of 100 kcal/mol.

The hydrophobic effect is smaller, but similar. If you put a big ion in water, you might break four to six hydrogen bonds. If we assume that each hydrogen bond is in the order of 10 kcal/mol, then the hydrophobic effect is in the order of 40–60 kcal/mol per ion.

By comparison, you learned in thermodynamics that dipole (van der Waals) forces are usually 5 kcal/mol or less. This 5 kcal/mol is a much smaller effect than the 100 kcal/mol due to the electrostatic and hydrophobic effects. Consequently, we can conclude that the electrostatic and hydrophobic effects are much more important than dipole effects in solvation.

Now consider how a solvent will affect a reaction. Notice that if you have an ion, the solvent will have a huge effect, because the solvent can change the electrostatic interaction. On the other hand, the solvent will have much less influence when the species are all neutral. A solvent with a large dielectric constant will tend to stabilize charge separation in the transition state. That will have a large influence on the rate.

The data in Table 13.3 illustrate this effect. There is charge separation in the transition state for the reaction $(C_2H_5)_3N + CH_3I \rightarrow [(C_2H_5)_3NCH_3]^+ + [I]^-$. A solvent with a high dielectric constant would be expected to stabilize the transition state, which should enhance the rate. The data in Table 13.3 are basically in agreement with the qualitative predictions, although the methoxybenzene does not quite follow the expected trend.

In contrast, the reaction $CH_3COOCOCH_3 + 2C_2H_5OH \rightarrow 2CH_3COOC_2H_5 + H_2O$ has no charges in the transition state. One would expect the solvent to make much less difference there.

Similar arguments apply to the data in Tables 13.1 and 13.2. There are no charges involved in the reaction in Table 13.1. The rate is almost the same as in the gas phase. On the other hand, there are important charges in the reaction in Table 13.2. There the solvent makes a much larger difference.

13.6.1 Hughes–Ingold Rules

Next, we want to see if we can get some information about the sign of the effect of solvents on rates. The analysis will build on the work of Hughes and Ingold, who worked out a qualitative scheme to understand the role of the solvent in a number of ionic reactions [see Hughes et al. (1933, 1990); Ingold (1939)]. Their key findings are

- If the transition state for a reaction has a larger charge than the reactants do, then the rate of reaction will increase as the polarity of the solvent increases.
- If the transition state for a reaction has a smaller charge than the reactants, then the rate of reaction will decrease as the polarity of the solvent increases.
- If the net charge remains the same, but the charge is dispersed, then there will be a small decrease in rate as the polarity of the solvent increases.
- If the net charge remains the same, but the charge is localized, then there will be a small increase in rate as the polarity of the solvent increases.

It is interesting to apply these rules to S_N1 and S_N2 reactions in solution. Recall that the notation S_N refers to a reaction where a nucleophile replaces another nucleophile on an organic molecule. The overall reaction for both S_N1 and S_N2 reactions is

$$Y^- + R_3C\text{–}X \Longrightarrow YCR_3 + X^- \tag{13.20}$$

However, the mechanisms are different. In an S_N1 reaction, the organic molecule ionizes in solution producing a carbocation and a nucleophile. The carbocation then reacts with another nucleophile to yield a new stable molecule:

$$R_3C\text{–}X \longrightarrow R_3C^+ + X^- \tag{13.21}$$

$$R_3C^+ + Y^- \longrightarrow R_3CY \tag{13.22}$$

Table 13.9 Implications of the Hughes–Ingold rules for S$_N$-type reactions

Mechanism	Reactants	Transition state	Change in Charge in the Transition state	Effect of Increase in Solvent Polarity on Rate
S$_N$2	Y$^-$ + RX	$^{\delta-}$Y$\cdots$R$\cdots$X$^{\delta-}$	Dispersed	Small decrease
S$_N$2	Y + RX	$^{\delta+}$Y$\cdots$R$\cdots$X$^{\delta-}$	Increased	Large increase
S$_N$2	Y$^-$ + RX$^+$	$^{\delta-}$Y$\cdots$R$\cdots$X$^{\delta+}$	Reduced	Large decrease
S$_N$2	Y + RX$^+$	$^{\delta+}$Y$\cdots$R$\cdots$X$^{\delta-}$	Dispersed	Small decrease
S$_N$1	RX	$^{\delta+}$R$\cdots$X$^{\delta-}$	Increased	Large increase
S$_N$1	RX$^+$	$^{\delta+}$R$\cdots$X$^{\delta+}$	Dispersed	Small decrease

where R_3C^+ is a carbocation and X^- and Y^- are nucleophiles. In contrast to a S$_N$2 reaction, the reactant does not ionize. Instead, there is a direct replacement where the Y^- collides with the R_3C–X and bond formation and bond destruction occur in a single step.

$$Y^- + R_3C\text{-}X \longrightarrow YCR_3 + X^- \tag{13.23}$$

Generally, S$_N$1 reactions are favored when there are large R groups around the reaction center, while S$_N$2 reactions are favored when R = H.

Hughes and Ingold considered the effect of the charge stabilization on the rate of reaction. Table 13.9 summarizes their predictions. Generally, if the transition state has more charge than the reactants do the rate will be increased as you increase the "polarity" of the solvent. The opposite is true if the transition state has less charge than do the reactants. Spreading the charge produces a small decrease in the rate. Table 13.9 provides many examples. One should examine the table before proceeding.

13.6.2 Hydrophobic Effects

Another affect of the solvent is associated with the hydrophobic effect. In Section 13.5 we noted that when a small ion dissolves in a protic (i.e., hydrogen-bonding) solvent, the solvent is barely disrupted. On the other hand, when a large ion dissolves, the hydrogen bonds are disrupted. As a result, protic solvents favor small ions and small transition states.

In contrast, with a polar aprotic solvent, the hydrophobic effect is much smaller, because there are no hydrogen bonds to break. As a result, large ions are more soluble in aprotic solvents than in protic solvents.

Now consider the implication of these results to S$_N$1 and S$_N$2 reactions. During an S$_N$2 reaction of an ion, the transition state is always larger than the individual reactants. Consequently, S$_N$2 reactions are faster in aprotic solvents than in protic solvents. The effects are large. The data in Table 13.2 show a factor of 10^6 difference between the aprotic solvent, DMF; and the protic solvent, water.

The reverse is true in an S$_N$1 reaction. With an S$_N$1 reaction, one starts with a large species and makes two smaller ions. As a result, S$_N$1 reactions tend to be favored in protic solvents. Table 13.8 shows some data. Notice that the rate is a factor of 10^6 higher in a protic solvent, formic acid than in a polar aprotic solvent, diethylether.

Generally, S$_N$1 reactions are fastest in protic solvents, while S$_N$2 reactions are fastest in polar aprotic solvents.

Of course, there are exceptions. For example, the reaction

$$Br^- + (CH_3)_3CP(C_6H_5)_3 \Longrightarrow BrC(CH_3)_3 + [P(C_6H_5)_3]^- \tag{13.24}$$

should occur via a S_N1 reaction since the CH_3 groups block the S_N2 reaction. However, the ions are large, so the hydrophobic effect does not allow a S_N1 reaction to occur. In practice, reaction (13.24) has a tiny rate that does not vary much between protic and aprotic solvents.

Most reactions follow the general rules above, although reaction (13.24) is an exception.

13.7 SEMIQUANTITATIVE PREDICTIONS

The discussion in Section 13.6 was qualitative, but in the literature, there have been many attempts to quantify the ideas. There are four key semiquantitative models:

- The double-sphere model
- The single-sphere model
- Debye–Hückel theory
- Regular solution theory

We will discuss these models in the next several sections.

13.7.1 The Double-Sphere Model

The double-sphere model was proposed by Scatchard in 1931 and 1932. It attempts to understand how solvents affect the rate of simple ionic reactions. In the next section, we will derive the model, and then see how well it works.

Consider the second step in a S_N1 reaction where two ions come together to form the product:

$$R_3C^+ + Y^- \longrightarrow R_3CY \tag{13.25}$$

Let's first run the reaction in a hypothetical solvent that has all of the same properties of the real solvent except that the dielectric constant is infinite. Let's call $\Delta G^{\ddagger}_{\infty}$ the free-energy of activation in the hypothetical solvent. Then, from equation (13.6), the free-energy of activation in a real solvent is given by

$$\Delta G^{\ddagger}_S = \Delta G^{\ddagger}_{\infty} + (\Delta G^{\ddagger}_{\infty \to s} - \Delta G^R_{\infty \to s}) \tag{13.26}$$

where $\Delta G^R_{\infty \to s}$ is the free-energy change in taking the reactants from the hypothetical solvent to the real solvent and $\Delta G^{\ddagger}_{\infty \to s}$ is the free-energy change in taking the transition state from the hypothetical solvent to the real solvent.

The idea in the double-sphere model is to assume that the main contribution to the free-energy change is the electrostatic interaction, and the hydrophobic interactions in forming the transition state are negligible. You further assume that there is no charge transfer in the

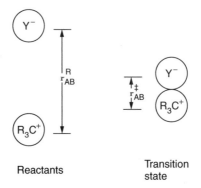

Figure 13.5   The double-sphere model of a reaction.

transition state, so the two ions are still ions at the transition state. Under this assumption, the free-energy changes can be approximated by the potential in equation (13.13):

$$\Delta G^R_{\infty \to s} = \frac{Z_A Z_B e^2}{\varepsilon r^R_{AB}} \tag{13.27}$$

$$\Delta G^\ddagger_{\infty \to s} = \frac{Z_A Z_B e^2}{\varepsilon r^\ddagger_{AB}} \tag{13.28}$$

where Z_A and Z_B are the charges on the reactants, ε is the dielectric constant of the solvent, e is the charge on an electron, and r^R_{AB} and $r^\ddagger_{AB}$ are the average distance between the reactants in solution and at the transition state as indicated in Figure 13.5. The double-sphere model makes the further assumption that $r^R_{AB} \gg r^\ddagger_{AB}$, so that $\Delta G^\ddagger_{\infty \to s} \gg \Delta G^R_{\infty \to s}$. Combining equations (13.5) and (13.20)–(13.22) yields

$$k_B T \ln \left(\frac{k_S}{k_\infty} \right) = \Delta G^\ddagger_\infty + \frac{Z_A Z_B e^2}{\varepsilon r^\ddagger_{AB}} \tag{13.29}$$

Equation (13.29) is the key result from the double-sphere model. In equation (13.29) k_S is the rate constant in the solvent S, k_∞ is the rate constant in the hypothetical solvent, k_B is Boltzmann's constant, T is the temperature, $\Delta G^\ddagger_S$ is the free energy of activation in the solvent S, $\Delta G^\ddagger_\infty$ is the free energy of activation in the hypothetical solvent, Z_A and Z_B are the charges on the reactants, ε is the dielectric coefficient of solvent S, $r^\ddagger_{AB}$ is the distance between the charges in the transition state, and e is the charge on an electron.

13.7.2   The Single-Sphere Model

The single-sphere model is similar to the double-sphere model. The ideas are similar, except that the Kirkwood formula, equation (13.16), is used to estimate the free energies:

$$\Delta G^\ddagger_{g \to s} = - \left(\frac{\varepsilon - 1}{2\varepsilon + 1} \right) \frac{(Z_A)^2 e^2}{r_A} \tag{13.30}$$

$$\Delta G^B_{g\to s} = -\left(\frac{\varepsilon - 1}{2\varepsilon + 1}\right)\frac{(Z_B)^2 e^2}{r_B} \tag{13.31}$$

$$\Delta G^{\ddagger}_{g\to s} = -\left(\frac{\varepsilon - 1}{2\varepsilon + 1}\right)\frac{(Z_{\ddagger})^2 e^2}{r_{\ddagger}} \tag{13.32}$$

where $\Delta G^A_{g\to s}$, $\Delta G^B_{g\to s}$, $\Delta G^{\ddagger}_{g\to s}$ are respectively the free-energy in taking the reactants, products and transition state into solution; Z_A, Z_B, and $Z_{\ddagger}$ are the charges on the reactants, product, and transition state, respectively; r_A, r_B, and $r_{\ddagger}$ are the radii of the reactants, products and transition state; and ε is the dielectric constant of the solvent. Plugging equations (13.30)–(13.32) into equation (13.6) yields

$$k_B T \ln\left(\frac{k_S}{k_g}\right) = \Delta G^{\ddagger}_S = \Delta G^{\ddagger}_g + \left(\frac{\varepsilon - 1}{2\varepsilon + 1}\right)\left(\frac{(Z_A)^2 e^2}{r_A} + \frac{(Z_B)^2 e^2}{r_B} - \frac{(Z_{\ddagger})^2 e^2}{r_{\ddagger}}\right) \tag{13.33}$$

where k_S is the rate constant in solvent S, k_g is the rate constant in the gas phase, k_B is Boltzmann's constant, T is the temperature, $\Delta G^{\ddagger}_S$ is the free energy of activation in solvent S, $\Delta G^{\ddagger}_g$ is the free energy of activation in the gas phase. Also Z_A, Z_B, and $Z_{\ddagger}$ are the charges on the reactants, product and transition state; r_A, r_B, and $r_{\ddagger}$ are the radii of the reactants, products, and the transition state; and ε is the dielectric constant of the solvent.

One can obtain a similar result for a molecule with dipoles rather than charges:

$$k_B T \ln\left(\frac{k_S}{k_g}\right) = \Delta G^{\ddagger}_S = \Delta G^{\ddagger}_g + \left(\frac{\varepsilon - 1}{2\varepsilon + 1}\right)\left(\frac{(\mu_A)^2}{(r_A)^3} + \frac{(\mu_B)^2}{(r_B)^3} + \frac{(\mu_A)^2}{(r_{\ddagger})^3}\right) \tag{13.34}$$

where r_A, r_B and $r_{\ddagger}$ are the radii of the reactants, products and transition state; μ_A, μ_B, and $\mu_{\ddagger}$ are the corresponding dipole moments; and ε is the dielectric constant of the solvent. Similarly, a reaction that starts with neutral species and ends up with two charges Z_A and Z_B should follow:

$$k_B T \ln\left(\frac{k_S}{k_g}\right) = \Delta G^{\ddagger}_S = \Delta G^{\ddagger}_g + \left(\frac{\varepsilon - 1}{2\varepsilon + 1}\right)\frac{Z_A Z_B e^2}{r^{\ddagger}_{AB}} \tag{13.35}$$

where $r^{\ddagger}_{AB}$ is the separation between the two charges in the transition state.

Equation (13.29) predicts that the rate will vary linearly with $1/\varepsilon$, while equation (13.35) predicts that the rate will vary linearly with $(\varepsilon - 1)/(2\varepsilon + 1)$. However, both equations give very similar results numerically.

For future reference, we will call $(\varepsilon - 1)/(2\varepsilon + 1)$ the **Kirkwood constant**. The Kirkwood constant is a measure of the ability of solvents to stabilize charges. Generally the larger the Kirkwood constant, the more charges will be stabilized.

Unfortunately, the quantitative agreement between theory and experiment is mixed. Figure 13.6 compares equation (13.35) to data for the reaction

$$(CH_3CH_2)_3N + CH_3CH_2I \longrightarrow [(CH_3CH_2)_3NI]^- + [CH_3CH_2]^+ \tag{13.36}$$

in a series of acetone–dioxane mixtures. One finds that the data follow a straight line when plotted against one over the dielectric constant (i.e., $1/\varepsilon$) or the Kirkwood constant [i.e., $(\varepsilon - 1)/(2\varepsilon + 1)$].

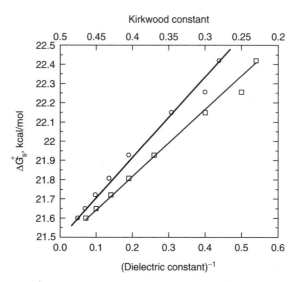

Figure 13.6 A plot of $\Delta G_s^{\ddagger}$ for reaction (13.36) in a series of dioxane–acetone mixtures. (o) data plotted against $1/\varepsilon$; (□) data plotted against $(\varepsilon - 1)/(2\varepsilon + 1)$].

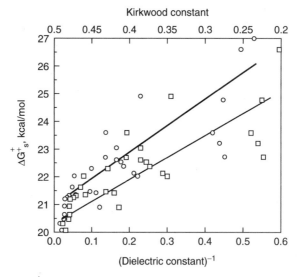

Figure 13.7 A plot of $\Delta G_s^{\ddagger}$ for reaction (13.36) in a series of different solvents. [key: (o) data plotted versus $1/\varepsilon$; (□) data plotted versus $(\varepsilon - 1)/(2\varepsilon + 1)$].

Clearly the theory works. On the other hand, Figure 13.7 shows a plot of the rate of reaction (13.36) in many solvents. One no longer sees a good agreement between theory and experiment.

One might wonder why the data agreed so well with theory in Figure 13.6 and so poorly in Figure 13.7. Well, note that the hydrophobic interaction was ignored in the derivation of equations (13.32)–(13.35). One can redo the derivation, putting in the hydrophobic interaction, and it works out that equation (13.33) should be modified to

$$\Delta G_S^{\ddagger} = \Delta G_r^{\ddagger} + \left(\frac{\varepsilon - 1}{2\varepsilon + 1} \right) \left(\frac{(Z_A)^2 e^2}{r_A} + \frac{(Z_B)^2 e^2}{r_B} - \frac{(Z_{\ddagger})^2 e^2}{r_{\ddagger}} \right) \qquad (13.37)$$

with

$$\Delta G_r^{\ddagger} = \Delta G_g^{\ddagger} + \Delta G_h^A + \Delta G_h^B - \Delta G_h^{\ddagger} \qquad (13.38)$$

where ΔG_h^A, ΔG_h^B, and $\Delta G_h^{\ddagger}$ are the hydrophobic interactions in the reactants, products, and transition state respectively. If ΔG_h^A, ΔG_h^B, and $\Delta G_h^{\ddagger}$ are constant, then $\Delta G_r^{\ddagger}$ will be constant, too. In that case a plot of $\Delta G_S^{\ddagger}$ versus $(\varepsilon - 1)/(2\varepsilon + 1)$ will be linear. However, if ΔG_h^A, ΔG_h^B, and $\Delta G_h^{\ddagger}$ vary from one solvent to the next, scatter will be seen.

We will see later in this chapter that dioxane and acetone have almost the same hydrophobic interaction. As a result, the hydrophobic interaction is almost constant in Figure 13.6. So, the theory works. In contrast, Figure 13.7 shows data for many different solvents. The hydrophobic interactions vary from one solvent to the next; in that case, the theory shows considerable variations.

In my experience, equations (13.33)–(13.36) work qualitatively. According to equation (13.33), if the transition state has a larger charge than the reactants, the rate of reaction will increase as the Kirkwood constant of the solvent increases. The opposite is true if the charge in the transition state is smaller than the charge in the reactants. If charge is more diffuse in the transition state (i.e., $r_{\ddagger}$ is larger than r_A and r_B), then increases in the Kirkwood constant will decrease the rate while the reverse is true if the charge in the transition state becomes more localized. These predictions agree qualitatively with experiment and with the Hughes–Ingold rules, which is why the theory is well respected. However, the predictions do not fit the data exactly, since they ignore variations in ΔG_h^A, ΔG_h^B, and $\Delta G_h^{\ddagger}$.

13.7.3 Debye–Hückel Theory

In the literature there have been many attempts to find ways to study reactions under conditions where the hydrophobic interaction is constant but the effective dielectric constant of the fluid varies. One common approach is to use salts (e.g., sodium chloride) to vary the effective dielectric constant of a fluid. In Section 13.6.1 we found that when you run a reaction in solution, the electrostatic interactions are reduced.

Plugging equations (13.5) and (13.14) into equation (13.26) yields

$$\ln \left(\frac{k_S}{k_\infty} \right) = \left(\frac{G_{S \to \infty}^{\ddagger} - G_{S \to \infty}^R}{k_B T} \right) = \left(\frac{Z_A Z_B e^2}{\varepsilon r_{AB}} \right) - 2 Z_A Z_B Q_D \sqrt{I} \qquad (13.39)$$

where k_S is the rate constant in a solvent S with ionicity I, k_∞ is the rate constant in a hypothetical solvent with an infinite dielectric constant, $G_{S \to \infty}^R$ and $G_{S \to \infty}^{\ddagger}$ are the free energies in taking the reactants and the transition state from the solvent S to a hypothetical solvent with an infinite dielectric constant, T is the temperature, k_B is Boltzmann's constant, Z_A and Z_B are the charges on the reactants, e is the charge on an electron, ε is the dielectric constant of the solvent, r_{AB} is the distance between the ions in the transition state, and Q_D is the Debye–Hückel constant of the solution.

Equation (13.39) implies that the log of the rate constant will vary linearly with the square root of the ionicity. Figure 13.2 compares this prediction to experiment. Notice that this approximation is quite good, which is why the models are well respected.

We want to say again that the key assumption in equations (13.36) and (13.39) is that the hydrophobic interactions are constant. This is a useful approximation if you are starting with a solvent and adding ions. However, it does not give accurate results if you change solvents.

In my experience, the equations in this section do not work quantitatively, because they ignore the hydrophobic interaction.

13.7.4   Regular Solution Theory

At present, there are no accurate ways to consider the influence of the hydrophobic interaction on rates of reactions. There is a rough approximation, called regular solution theory which puts in some of the interactions. Then, there are Monte Carlo and molecular dynamics methods, which put in all of the interactions, but are computationally intractable. At present, no simple model works, and even the computationally demanding calculations show only modest agreement with theory.

In this section we will review regular solution theory. Monte carlo methods are reviewed in Tapia and Bertran (1996). The idea in regular solution theory is to assume that the hydrophobic interactions control the variations between various solvents. One then makes a series of assumptions to estimate rates.

In this section, we will derive the key equations. Our derivation will leave off some details that are included in more exact treatments. However, the derivation gives the essence of the arguments.

Consider moving an isotopically labeled A molecule from a pure A phase into a solvent phase, S. In order for the reaction to occur, you need to create a cavity in the S phase to hold the A molecule. Let's call V_A the volume of the cavity. Regular solution theory approximates the energy to create the cavity by

$$E_{cavity} = V_A \times (CED_S) \tag{13.40}$$

where E_{cavity} is the energy needed to create the cavity; V_A is the volume of the cavity; and CED_S is the cohesive energy density of fluid, S, that is, the energy per unit volume needed to remove A molecules from the fluid S to create a hole. Equation (13.40) assumes that the entire energy needed to create the cavity is associated with the hydrophobic interaction. Electrostatic interactions are ignored.

Next, let us calculate the energy required to move an isotopically labeled A molecule from the A phase to the S phase. We will assume that there are no volume changes in mixing, so that the partial molar volume of molecule A is the same in the A phase as in the S phase. We will also assume that there are no AA interactions in the S phase. That is true only in the dilute limit. Under these assumptions the energy change in mixing ΔE_{mix} is given by

$$\Delta E_{mix} = V_A \times (-(CED_A) - (CED_S) + 2(CED_{AS})) \tag{13.41}$$

where (CED_A) is the cohesive energy of fluid A, (CED_S) is the cohesive energy of fluid S, and (CED_{AS}) is the cohesive energy of an AS pair of molecules. The constant 2 in equation (13.41) is needed to make ΔE_{mix} zero when A and S are the same chemicals.

Equation (13.39) is reasonably accurate. However, there is an unknown term (CED_{AS}). Schatchard (1931) and Hildebrand and Wood (1933) proposed that one could evaluate (CED_{AS}) by making the second assumption:

$$(CED_{AS}) = [(CED_A)(CED_S)]^{1/2} \tag{13.42}$$

Equation (13.42) is a gross approximation. It assumes that the AS interactions are a simple geometric average of the A–A and S–S interactions and ignores the electrostatic interactions between the solvent and the solute.

Combining equations (13.41) and (13.42) yields:

$$\Delta E_{mix} = V_A \times ((CED_A)^{1/2} - (CED_S)^{1/2})^2 \tag{13.43}$$

We note again that equation (13.43) was derived under the assumption that there were no volume changes in mixing and that the mixture was dilute. Other situations are covered in the standard chemical thermodynamics textbooks.

Next, it is useful to define δ_A and δ_S, the solubility parameter for the species A and S, by

$$\delta_A = (CED_A)^{1/2} \tag{13.44}$$

$$\delta_S = (CED_S)^{1/2} \tag{13.45}$$

Combining equations (13.40)–(13.42) yields

$$\Delta E_{mix} = V_A \times (\delta_A - \delta_S)^2 \tag{13.46}$$

Finally we will make the assumption that the entropy changes in mixing are negligible, so that the free-energy change during mixing is equal to ΔE_{mix}:

$$\Delta G^A_{A \to s} = \Delta E_{mix} \tag{13.47}$$

where $\Delta G^A_{A \to s}$ is the free-energy needed to take a molecule A from an ideal A phase to an S phase. Combining equations (13.44) and (13.45) yields

$$\Delta G^A_{A \to s} = V_A \times (\delta_A - \delta_S)^2 \tag{13.48}$$

Equation (13.46) is the key result from regular solution theory.

Next, we want to use regular solution theory to predict the effect of the solvent on the rate of reaction $A + B \to TST \to$ products. Plugging equation (13.48) into equation (13.26) yields

$$\Delta G^{\ddagger}_S = \Delta G^{\ddagger}_{ideal} + (V_{\ddagger}(\delta_{\ddagger} - \delta_S)^2 - V_A(\delta_A - \delta_S)^2 - V_B(\delta_B - \delta_S)^2) \tag{13.49}$$

where $\Delta G^{\ddagger}_S$ is the free energy of activation in a solvent, S; $\Delta G^{\ddagger}_{ideal}$ is the free energy of reaction in a hypothetical reference solution where all of the free energies of solvation (i.e., $\Delta G^A_{A \to s}$'s values) are zero; V_A, V_B and $V_{\ddagger}$ are respectively the partial molar volumes

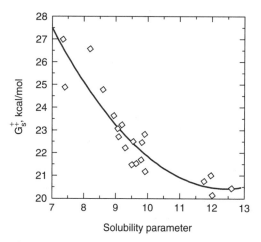

Figure 13.8 A plot of $\Delta G_s^\ddagger$ for reaction (13.36) in a series of different solvents as a function of the solubility parameter of the solvent.

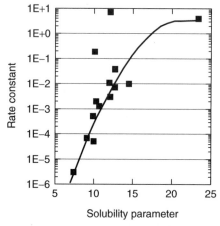

Figure 13.9 A plot of the rate constant for the reaction in Table 13.5 in a series of different solvents as a function of the solubility parameter of the solvent.

on A, B, and the transition state in the S solution; and $\delta_A, \delta_B, \delta_\ddagger$ and δ_S are the solubility parameters for A, B, the transition state, and the solvent, respectively.

Equation (13.49) implies that the best solvents for a reaction are solvents where $\delta_\ddagger = \delta_S$. It also implies that the free-energy of activation should vary parabolically with the solubility parameter. In a larger way, the equation suggests that one can use the solubility parameter as a rough indication of the strength of the hydrophobic interactions between the reactant molecules. We will show later in this section that this is a good approximation.

Unfortunately, regular solution theory does not work quantitatively. Figure 13.8 is a plot of $\Delta G_s^\ddagger$ for reaction (13.36) in a series of different solvents as a function of the solubility parameter of the solvent. One finds reasonable parabolic behavior, with a few significant deviations. The model has three fitting parameters: — $\Delta G_{ideal}^\ddagger$, $\delta_\ddagger$, and $V_\ddagger$ — so we cannot use the model to extrapolate data. However, the model still fits better than the single-sphere or double-sphere model.

Figure 13.9 shows a similar plot for the S_N1 reaction in Table 13.5. One finds that again there is qualitative agreement, but some significant deviations.

In my experience, regular solution theory works modestly well. However, it ignores the electrostatic interactions and other effects, so its utility is limited.

13.7.5 Quantitative Models

In the literature, there have been many attempts to make quantitative predictions. Tapia and Bertran (1996) have presented an excellent review of the methods. Generally one needs an elaborate calculation to get any useful quantitative information. Monte Carlo or molecular dynamics calculations of the type described in Chapters 6 and 8 give some useful insights. Quantum-mechanical calculations of the type described in Chapter 11 can be modified to include the effect of a uniform dielectric medium. These calculations also can give useful insights. Unfortunately, the computations are huge, and just starting to appear. As a result, I decided to not review the material here. Tapia and Bertran's (1996) book is excellent. One should refer to that text for an in-depth treatment.

13.7.6 Summary

In summary, then, in this section (13.7) we reviewed some of the models that are used to explain how reaction rates change as you change solvents. Generally, we found that solubility is often dominated by two effects: the electrostatic effect and the hydrophobic effect. The electrostatic effect is quantified by the single-sphere or double-sphere model. These models explain why the Hughes–Ingold rules arise, and predicts that the Kirkwood constant would be a useful way to rank solvents. Experimentally this model works only modestly well, but it is well respected. The hydrophobic effect is quantified by regular solution theory. Regular solution theory suggests that larger species are harder to solvate than smaller species. According to the model the difficulty in solvation scales as the cohesive energy density of the solvent. Again this model works modestly well, but it fails to capture key physics.

Unfortunately, at present, no one has combined the two models. One can consider either the electrostatic interaction or the hydrophobic interaction, but not both. The net result is that the models are only modestly accurate. One can predict qualitative trends, but not quantitative details.

13.8 DYNAMIC CORRECTIONS

There is one other important detail about reactions in solutions. There is some question as to whether transition state theory, equation (13.4), works in solution. There are two key issues:

- Mass transfer limitations change the nature of the collision process between molecules in solution. The correct first term in equation (13.4) might not be the gas-phase value $k_B T/h_p$.
- There are dynamic corrections in solution that are different from those in the gas phase. Those dynamic corrections are not properly accounted for in equation (13.4).

In order to understand the effects, let us start by reviewing what collisions are like in the gas phase. Figure 8.1 shows what a collision is like during the reaction $A + BC \rightarrow AB + C$ in the gas phase. Reactant A comes in from hundreds of angstroms away, and collides with BC. Then A either reacts or does not react. Then the particles separate. The entire collision process lasts perhaps 10^{-12} seconds.

One could never get a collision process like that in Figure 8.1 to occur in solution. There is solvent in solution. Reactant A cannot come in from hundreds of angstroms away and directly collide with BC. Instead, A must diffuse through the solvent until it encounters BC. The diffusion process in solution is slower in a solvent that is the corresponding process in the gas phase. As a result, the reaction rate in solution will be reduced.

There is a compensating effect, however. Generally A and BC will be held in a solvent cage. When collision occurs, the two solvent cages merge. Then A and BC will be trapped in a big solvent cage until one of them diffuses away. The hydrophobic effect will tend to trap A and BC in the solvent cage, and so once A and BC get together, they tend to stay together for much longer than they would in the gas phase. As a result, A and BC have many more chances to react than in the gas phase.

The net result is that the collision rate is lower in solution than in the gas phase, but once the collisions occur, there is a much higher chance that reaction will occur in solution than in the gas phase. That will produce a change in the rate.

There is a third effect as well. The solvent is always present, so when reaction occurs, some solvent molecules might need to be displaced. That can also affect the rate.

In the next several sections we will review some of the attempts to model these dynamics. First we will discuss the role of diffusion on rates, then we will mention the effect of the solvent cage. Finally we will discuss the role of solvent displacements on rates.

13.8.1 Diffusion of the Reactants through the Solvent

First let us derive an equation for the rate of collisions of molecules. Consider the diffusion of an ion A toward an ion BC in solution (see Figure 13.10). From transport or physical chemistry you know that the flux of A toward BC is given by

$$J_{A \rightarrow BC} = -(D_A + D_{BC}) \left[\left(\frac{\partial C_A}{\partial r} \right) + \left(\frac{C_A}{k_B T} \right) \left(\frac{\partial V}{\partial r} \right) \right] \qquad (13.50)$$

where $J_{A \rightarrow BC}$ is the flux of A toward BC at a radius r from A, D_A and D_{BC} are the diffusivities of A and BC in solution, C_A is the concentration of A at a distance r away from BC, $V(r)$ is the attractive potential between A and BC, k_B is Boltzmann's constant, and T is the absolute temperature. There are two diffusivities in equation (13.50) because A and BC are both diffusing.

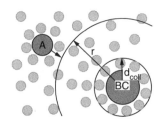

Figure 13.10 The diffusion of an ion A toward BC.

In the supplemental material we integrate equation (13.50) to obtain the rate of reaction of A with BC, allowing A and BC to diffuse together and getting stuck in the solvent cage. The result is

$$r_A = -k_D[A][BC] \tag{13.51}$$

where r_A is the reaction rate, [A] and [BC] are the concentrations of A and BC, and k_D is a diffusion limited rate constant given by

$$k_D = \frac{4\pi(D_A + D_{BC})I_{ABC}N_A}{1 + \dfrac{4\pi(D_A + D_{BC}I_{ABC}N_A}{k_1(\exp(-V(d_{coll})/k_BT))}} \tag{13.52}$$

where k_1 is the actual rate constant, d_{coll} is the collision diameter, D_A and D_{BC} are the diffusivities of A and BC in solution, $V(d_{coll})$ is the attractive potential between A and BC at the collision diameter, k_B is Boltzmann's constant, T is the absolute temperature, N_A is Avogadro's number, and I_{ABC} is given by

$$I_{ABC} = \left(\int_{d_{coll}}^{\infty} \frac{\exp(V(r)/k_BT)}{r^2} \, dr \right)^{-1} \tag{13.53}$$

Detailed derivations of equations (13.51)–(13.53) are given in the supplemental material.

There are two key limits of equation (13.52): the fast diffusion limit and the slow diffusion limit.

In the fast diffusion limit, the second term in the denominator of equation (13.52) is much greater than 1; in the slow diffusion limit, the second term in the denominator of equation (13.52) is much less than 1.

Let's consider the fast diffusion case first. In the fast diffusion limit, equation (13.52) reduces to

$$k_D = k_1 \exp\left(\frac{-V(R)}{k_BT} \right) \tag{13.54}$$

with $R = d_{coll}$. If $V(R)$ is given by a modification of equation (13.13), then

$$V(R) = \frac{Z_A Z_{BC} e^2}{\varepsilon R} \tag{13.55}$$

Combining equations (13.5), (13.54) and (13.55) yields

$$-k_BT \ln\left(\frac{k_1}{k_1^0} \right) = \Delta G_S^{\ddagger} = \Delta G_{\infty}^{\ddagger} + \frac{Z_A Z_{BC} e^2}{\varepsilon R} \tag{13.56}$$

Equation (13.56) is equivalent to equation (13.29), so in the fast diffusion limit, diffusion does not change the rate constant.

On the other hand, in the slow diffusion limit, equation (13.52) reduces to

$$k_D = 4\pi(D_A + D_{BC})I_{ABC}N_A \tag{13.57}$$

If the reactants are not charged, then it is usually a good approximation to ignore the attraction between the reactants (i.e., $V(R) \approx 0$). In that case $I_{ABC} = d_{coll}$. Here, d_{coll} is the collision diameter is given by

$$k_D = 4\pi(D_A + D_{BC})R\, d_{coll} N_A \tag{13.58}$$

Equation (13.58) is the key equation for diffusion-limited reactions.

Next, it is useful to calculate the order of magnitude for the rate constant in equation (13.58). According to data on pages 3–258 and 10–24 in Perry's sixth edition, diffusivities of small molecules in aqueous solution are usually in the order of 10^{-5} cm²/second. Substituting into equation (13.58) with a 2-Å collision diameter yields

$$k_D = 4\pi(10^{-5}\text{cm}^2/\text{second})\left(\frac{2\text{ Å}}{\text{molecule}}\right)\left(\frac{10^8\text{Å}}{\text{cm}}\right)^2$$

$$= 2 \times 10^{12}\text{Å}^3/(\text{second·molecule}) \tag{13.59}$$

By comparison, data in Chapter 7 show that preexponentials for gas-phase reactions are in the order of 2×10^{13} under similar conditions. Therefore the diffusion limit has reduced the preexponential by about an order of magnitude.

In practice, most real data are taken in the fast diffusion regime. The criterion for fast diffusion is

$$1 \ll \frac{4\pi(D_A + D_{BC})I_{ABC} N_A}{k_1\left(\exp\left(-V(R)/k_B T\right)\right)} \tag{13.60}$$

Next, let us calculate the order of magnitude of the term on the right of equation (13.60). If the there are no attractions or repulsions between the molecules, then $I_{ABC} = d_{coll}$ and $V(R) = 0$. The Right hand side (RHS), of equation (13.60) reduces to the following:

$$\text{RHS} = \frac{4\pi(D_A + D_{BC})I_{ABC}}{k_1\left(\exp\left(-V(R)/k_B T\right)\right)} = \frac{4\pi(D_A + D_{BC})\, d_{coll}}{k_1} \tag{13.61}$$

Let us assume

$$k_1 = (2 \times 10^{13}\text{Å}^3/\text{second})\exp\left(\frac{-E_a}{k_B T}\right) \tag{13.62}$$

Substituting equation (13.62) into equation (13.61) and choosing $(D_A + D_{BC}) = 10^{-5}$cm²/second and $d_{coll} = 2$ Å/molecule shows

$$\text{RHS} = \frac{4\pi(D_A + D_{BC})\, d_{coll}}{k_1} = \frac{4\pi(10^{-5}\text{cm}^2/\text{second})(2\text{Å})/\text{molecule}}{2 \times 10^{13}\text{ Å}^3/\text{second}(\exp(-E_a/k_B T))}\left(\frac{10^8\text{ Å}}{\text{cm}}\right)^2$$

$$= 0.1 \times \exp\left(\frac{E_a}{k_B T}\right) \tag{13.63}$$

Table 13.10 shows values of RHS for several values at the activation barrier. To put the table in perspective, note that in Chapter 2 we found that most real reactions have activation barriers near 20 kcal/mol. The RHS of equation (13.60) will be in the order of 10^{13}, which is much greater than 1. The RHS of equation (13.60) is less than 1.0 only when the activation barrier is less than 2 kcal/mol. In that case the reaction will be virtually instantaneous.

In the homework set we ask the readers to consider other examples. There are some examples in biological systems where diffusion limitations have a significant effect on the

Table 13.10 The value of the RHS of equation (13.60) for some typical sets of parameters for diffusion of small molecules in water

E_a, kcal/mol	T, K	RHS	E_a	T, K	RHS
0	300	0.1	5	300	453
1	300	0.54	10	300	2×10^6
2	300	2.9	15	300	1×10^{10}
3	300	15.6	20	300	4×10^{13}
4	300	85	25	300	2×10^{17}

rate of a homogeneous reaction in a solvent. Reactions in solids are often diffusion-limited. However, the vast majority of practical examples of homogeneous reactions in liquids are taken in the fast diffusion regime.

There are some laboratory examples that are not in the fast diffusion regime, however. Consider the photolysis reaction discussed in Sections 9.8 through 9.10:

$$CH_2CO + h\nu \longrightarrow CH_2 + CO \tag{13.64}$$

In Chapter 9 we found that the ketene (CH_2CO) is excited by a photon. Then in the gas phase, the products form over the next 10^{-10} seconds. Now consider what would happen in a viscous solution. We would still excite the ketene, but the products might not diffuse away. That could lead to a diffusion limitation.

In my experience, diffusion-limited rate constants are hardly ever important in homogeneous reactions, although they are important in heterogeneous reactions, or reactions of biological molecules.

13.8.2 Dynamic Effects: Kramers' Theory

There is another effect that can, on occasion, be important — solvents can change the dynamics of a reactive collision. Recall that in solution the reactants are in close contact with the solvent. The interactions with the solvent change according to how the reactants move.

Consider trying to have a reactive collision where a hot A molecule collides with a BC molecule. Note that the hot molecule will be in close contact with the solvent, so unless the reaction occurs very quickly, the hot A molecule will be deexcited by interaction with the solvent.

Then there is an opposing effect. On average, some of the solvent molecules can be hot. Those hot molecules can transfer energy to the reactants, thereby promoting reaction.

The net effect is usually to change the rate constant by one or two orders of magnitude. That is enough to be important, but not so large that you cannot handle the effect theoretically.

Hendrik Kramers (1940) worked out the theory years ago. The idea behind the theory is to look at two processes: (1) the excitation of molecules due to the interactions with hot solvent molecules and (2) the loss of molecules due to energy transfer to the fluid. One can then solve for the change in the distribution of molecules from the Folker–Planck equation

$$\frac{\partial D(r, v, t)}{\partial t} + v \frac{\partial D(r, v, t)}{\partial r} + \frac{1}{m} \frac{\partial V}{\partial r} \frac{\partial D(r, v, t)}{\partial v}$$

$$= \beta v \frac{\partial D(r, v, t)}{\partial v} + \beta D(r, v, t) + \beta \frac{k_B T}{m} \frac{\partial^2 D(r, v, t)}{\partial v^2} \tag{13.65}$$

where $D(r_1 v_1 t)$ is the velocity distribution of the molecules at position r and time t, v is the velocity, m is the mass of the molecule, k_B is Boltzmann's constant, T is the temperature, and β is a coefficient of friction between the molecules and the solvent.

I know that most of my readers have not seen the Folker–Planck equation before. It can be derived by writing a mass-and-energy balance on the reacting molecules, and then forcing the ratio of the rate of excitation of molecules to the rate of deexcitation to the point where thermal equilibrium is maintained. Kramers (1940), modeled the deexcitation process as classical friction. Recall from freshman physics that when a particle is moving through a fluid, the particle is a frictional force equal to a friction coefficient, f_e, times the velocity of the particles. Kramers proposed that the same frictional force would occur on the molecular level to derive equation (13.65).

Kramers actually solved the equation for a unimolecular reaction [equation (13.65)]. The key findings from the analysis was that there are three regimes:

- A low-friction regime where the reactants are not strongly coupled to the solvent. In this case the rate of reaction looks like that expected from RRKM theory or the equivalent.
- A medium-friction regime where the coupling with the solvent is stronger. In this case the solvent is an effective collision partner for the reactants, so the rate looks like that from transition state theory.
- A high-friction regime where molecules have so many collisions with the solvent that hot molecules are deexcited before they have a chance to go over the barrier in the potential energy surface. In this case, the rate of reaction is less than that expected from transition state theory.

In 1940. Kramers actually derived an equation for the rate at which molecules cross the barrier. I do not want to include the equations here because it does not work. However, the equation is discussed in many other texts.

There is one key point from all of this analysis: A solvent is a wonderful collision partner. Under normal conditions a reaction in a solvent is in the medium-friction regime. There are many collisions, so transition state theory usually works pretty well. Of course, it is possible to have too much of a good thing. In that case hot molecules are deexcited so quickly that no reaction can occur. Detailed calculations show that collective excitations of the solvent can also have an important influence on the rate. Talkner and Hanggi (1995) have a good description of the effects. One should refer to their book for further details.

13.9   SUMMARY

In summary, then, in this chapter we discussed how solvents affect rates of reactions. Generally, we found that solvents act just like catalysts:

- Solvents stabilize intermediates.
- Solvents stabilize transition states.
- Solvents act as efficient means for energy transfer.
- Mass transfer limitations are more important when solvents are present.

The effects are huge. Rates in solution can be a factor of 10^{40} higher than in the gas phase, and can vary by a factor of 10^6 from one solvent to the next.

The key qualitative findings are that aprotic solvents are best for S_N1 reactions, polar protic solvents are best for S_N2 reactions, while nonpolar solvents are best for radical reactions. There are also the Hughes–Ingold rules to see how solvent polarity affects rates.

Unfortunately, though, when this book was being written, people did not have good models to understand the variations. In this chapter we mentioned the single-sphere model, the double-sphere model, and regular solution theory. All three models could explain the qualitative variations in rate with changing solvent. However, the quantitative agreement is not as good. Errors as large as a factor of 100 in rate are seen. There are no better models at present, although molecular dynamics calculations are beginning to give useful insights.

13.10 SUPPLEMENTAL MATERIAL: DERIVATION OF EQUATION (13.52)

The objective of this section is to derive equation (13.52), the rate constant for reaction in solution for the case where the rate constant of the reaction is mass transfer limited. We will consider two cases, a case where there are no forces between the molecules, and a case where there is an interaction potential between the molecules.

13.10.1 Mass-Transfer-Limited Reaction Rates in the Absence of an Attractive Potential between A and BC

First we will derive an expression for the rate, ignoring the effects of any interactions between A and BC.

Consider the reaction $A + BC \rightarrow$ products and assume that A reacts with some rate constant k_{coll} whenever A gets to within a collision diameter d_{coll} of BC. During the reaction A and B must diffuse together as shown in Figure 13.10. Next, we want to derive a differential equation for the rate that A collides with B via diffusion.

We will start by writing a mass balance around the sphere of radius r around the BC molecule in Figure 13.10. Let's call $J_{A \rightarrow BC}$ *the flux of A toward BC*. At steady state, the rate that A diffuses into the sphere must equal the rate at which A is consumed by reaction:

$$\begin{pmatrix} \text{Flux of a} \\ \text{into the sphere} \end{pmatrix} \begin{pmatrix} \text{surface area} \\ \text{of sphere} \end{pmatrix} = \begin{pmatrix} \text{rate at which A} \\ \text{is consumed} \end{pmatrix} \qquad (13.66)$$

The flux is given by $J_{A \rightarrow BC}$. The area is $4\pi r^2$. We will assume that the rate of reaction is given by

$$\mathbb{R}_A = -k_{coll}[A]_o \qquad (13.67)$$

where $\mathbb{R}_A$ is the rate of formation of A in molecules/second/(molecule of BC), k_{coll} is the rate constant when A and BC collide, and $[A]_o$ is the average concentration of A in molecules/cm^3 at a distance d_{coll} from BC.

Combining equations (13.66) and (13.67) and noting that the rate of consumption of A is equal to minus the rate of formation of A yields

$$(J_{A \rightarrow BC})(4\pi r^2) = k_{coll}[A]_o \qquad (13.68)$$

where r is the distance from BC, $J_{A \rightarrow BC}$ is the flux at that distance, k_{coll} is the rate constant when A and BC collide, $[A]_o$ is the average concentration of A at a distance

d_{coll} from BC, and [BC] is the BC concentration in solution. Note k_{coll} has units of $(cm)^3/(molecule \cdot second)$.

First let's consider the case where there is no attraction or repulsion between the reactants. From mass transfer theory you know that

$$J_{A \to BC} = (D_A + D_{BC})\left[\left(\frac{\partial C_A}{\partial r}\right)\right] \tag{13.69}$$

where D_A and D_{BC} are the diffusivities of A and BC in solution. Equation (13.69) contains the sum of D_A and D_{BC} since both A and BC are diffusing.

Combining equations (13.68) and (13.69) and rearranging yields

$$\left(\frac{k_{coll}[A]_o}{4\pi r^2}\right) - (D_A + D_{BC})\left[\left(\frac{\partial C_A}{\partial r}\right)\right] = 0 \tag{13.70}$$

with the boundary condition

$$C_A = [A] \quad \text{at} \quad r = \infty \tag{13.71}$$

where [A] is the average concentration of A in the bulk of the solution measured in molecules/cm^3. Equation (13.70) is a general equation for the diffusion of A toward BC, ignoring any effects of the interactions between A and BC.

Next, we want to derive an expression for $[A]_o$ the average concentration of A at a distance of one collision diameter from BC. The solution of equation (13.70) is

$$C_A = [A] - \frac{k_{coll}[A]_o}{4\pi(D_A + D_{BC})r} \tag{13.72}$$

At the collision diameter

$$r = d_{coll} \quad \text{and} \quad C_A = [A]_o \tag{13.73}$$

Combining equations (13.72) and (13.73) yields

$$[A]_o = [A] - \frac{k_{coll}[A]_o}{4\pi(D_A + D_{BC})d_{coll}} \tag{13.74}$$

solving equation (13.74) for $[A]_o$ yields

$$[A]_o = \left(\frac{[A]}{1 + \dfrac{k_{coll}}{4\pi(D_A + D_{BC})d_{coll}}}\right) \tag{13.75}$$

Equation (13.75) gives the average concentration of A, at a distance d_{coll} from BC.

Next, we will obtain an expression for the rate. Substituting equation (13.75) into equation (13.67) yields

$$\mathbb{R}_A = -\left(\frac{k_{coll}[A]}{1 + \dfrac{k_{coll}}{4\pi(D_A + D_{BC})d_{coll}}}\right) \tag{13.76}$$

Equation (13.76) gives the rate of formation of A per B molecule. One can convert that to a rate per unit volume by multiplying by the number of B molecules per unit volume:

$$r_A(\text{molecules}) = -\left(\frac{k_{coll}[A][B]}{1 + \dfrac{k_{coll}}{4\pi(D_A + D_{BC})d_{coll}}} \right) \tag{13.77}$$

where $r_A(\text{molecules})$ is the rate of formation of A in molecules/(cm^3·second), and [B] is the concentration of B in molecules/cm^3.

One can convert this to moles by multiplying by Avogadro's number, N_A:

$$r_A N_A = -\left(\frac{k_{coll}(C_A N_A)(C_B N_A)}{1 + \dfrac{k_{coll}}{4\pi(D_A + D_{BC})d_{coll}}} \right) \tag{13.78}$$

In equation (13.78), r_A is the rate of formation of A in mol/(cm^3·second), C_A is the concentration of A in mol/cm^3, C_B is the concentration of B in mol/cm^3, k_{coll} is the rate constant when A and BC collide, d_{coll} is the collision diameter in cm/molecule, N_A is Avogadro's number, and D_A and D_{BC} are the diffusivities of A and BC in solution.

Rearranging equation(13.78) yields

$$r_A = -\left(\frac{k_1 C_A C_B}{1 + \dfrac{k_1}{4\pi(D_A + D_{BC})N_A d_{coll}}} \right) \tag{13.79}$$

with

$$k_1 = N_A k_{coll} \tag{13.80}$$

Equation 13.79 can be put into standard form by multiplying the top and bottom by $\dfrac{4\pi(D_A + D_{BC})N_A d_{coll}}{k_1}$. The result is

$$r_A = -\left(\frac{4\pi(D_A + D_{BC})N_A d_{coll} C_A C_B}{1 + \dfrac{4\pi(D_A + D_{BC})N_A d_{coll}}{k_1}} \right) \tag{13.81}$$

Equation (13.81) is the key equation for the rate of diffusion-limited reactions, for the case where the interaction between A and BC has a negligible effect on the rate of collisions between A and BC.

13.10.2 Mass-Transfer-Limited Reaction Rates in the Presence of an Interaction Potential between A and BC

Next, we will consider the effect of the A + BC interaction on the collision rate. Physically, if A and B have an opposite charge, they will attract each other in solution. That will increase the rate of collisions, which will, in turn, increase the rate of reaction.

We will derive an equation for the change in rate in the next few pages. The derivation will follow the derivation in Section 13.10.1, except that we will add an extra term to account for the influence of the electric field.

First we need a suitable diffusion equation. Recall that an ion in solution will move in response to an electric field. Opposite charges attract. Similar charges repel. In Chapter 8 we noted that when a molecule moves in a potental $V(r)$, the molecule will experience a force, F, given by

$$F = -\frac{\partial V(r)}{\partial r} \qquad (13.82)$$

where F is the force and $V(r)$ is the potential at point r. In Chapter 8, we used equation (13.82) to derive expressions for the trajectories of molecules in the gas phase. The same equation also applies the motion of a molecule in solution.

The electrical force causes ions to diffuse. Einstein showed that the flux of ions through a stationary solution is given by

$$J_A = (D_A)\left[\left(\frac{\partial C_A}{\partial r}\right) + \left(\frac{C_A}{k_B T}\right)\left(\frac{\partial V(r)}{\partial r}\right)\right] \qquad (13.83)$$

In equation (13.83), J_A is the flux of A through the solution, D_A is the diffusivity of A, C_A is the concentration of A at a point r in solution, $V(r)$ is the electric potential at point r, k_B is Boltzmann's constant, and T is the temperature.

We want the flux of A toward BC to be such that A and BC can both move in solution. It is given by

$$J_{A\to BC} = (D_A + D_{BC})\left[\left(\frac{\partial C_A}{\partial r}\right) + \left(\frac{C_A}{k_B T}\right)\left(\frac{\partial V(r)}{\partial r}\right)\right] \qquad (13.84)$$

In equation (13.84), $J_{A\to BC}$ is the flux of A toward BC, D_A is the diffusivity of A in solution, D_{BC} is the diffusivity of BC in solution, C_A is the concentration of A at a point r from a given BC molecule, $V(r)$ is the electric potential at point r, k_B is Boltzmann's constant, and T is the temperature.

Combining equations (13.68) and (13.84) yields

$$\left(\frac{k_{coll}[A]_o}{4\pi r^2}\right) = (D_A + D_{BC})\left[\left(\frac{\partial C_A}{\partial r}\right) + \left(\frac{C_A}{k_B T}\right)\left(\frac{\partial V(r)}{\partial r}\right)\right] \qquad (13.85)$$

with the boundary condition

$$C_A = [A] \quad \text{at} \quad r = \infty \qquad (13.86)$$

where [A] is the average A concentration in the bulk of the solution measured in molecules/cm^3.

Equation (13.85) can be integrated using an integrating factor. Consider the quantity

$$\exp\left(-\frac{V(r)}{k_B T}\right)\frac{\partial}{\partial r}\left[C_A \exp\left(\frac{V(r)}{k_B T}\right)\right]$$

From calculus

$$\exp\left(-\frac{V(r)}{k_B T}\right)\frac{\partial}{\partial r}\left[C_A \exp\left(\frac{V(r)}{k_B T}\right)\right] = \left[\left(\frac{\partial C_A}{\partial r}\right) + \left(\frac{C_A}{k_B T}\right)\left(\frac{\partial V(r)}{\partial r}\right)\right] \qquad (13.87)$$

Substituting equation (13.85) into equation (13.87) and multiplying by the exponential yields

$$\frac{\partial}{\partial r}\left[C_A \exp\left(\frac{V(r)}{k_B T}\right)\right] = \left(\frac{k_{coll}[A]_o}{4\pi(D_A + D_{BC})}\right)\left(\frac{\exp\left(\frac{V(r)}{k_B T}\right)}{r^2}\right) \tag{13.88}$$

Integrating from infinity to r assuming that $V(r)$ goes to zero at infinity shows

$$[A] - C_A\left(\exp\left(\frac{V(r)}{k_B T}\right)\right) = \left(\frac{k_{coll}[A]_o}{4\pi(D_A + D_{BC})}\right)\int_r^\infty \left(\frac{\exp\left(\frac{V(r)}{k_B T}\right)}{r^2}\right) dr \tag{13.89}$$

In equation (13.89), C_A is the concentration of A at some distance r from BC, $[A]$ is the average concentration of A in the solution, $[A]_o$ is the concentration at the collision diameter, D_A is the diffusivity of A in solution, D_{BC} is the diffusivity of BC in solution, $V(r)$ is the electric potential at point r, k_B is Boltzmann's constant, and T is the temperature. Equation (13.89) is a general expression for the concentration at any given point in solution.

Next, we want to derive an expression for $[A]_o$ the concentration of A at the collision diameter. At the collision diameter

$$r = d_{coll} \quad \text{and} \quad C_A = [A]_o \tag{13.90}$$

Substituting equation (13.90) into equation (13.89) yields

$$[A] - [A]_o\left(\exp\left(\frac{V(d_{coll})}{k_B T}\right)\right) = \left(\frac{k_{coll}[A]_o}{4\pi(D_A + D_{BC})}\right)\int_{d_{coll}}^\infty \left(\frac{\exp\left(\frac{V(r)}{k_B T}\right)}{r^2}\right) dr \tag{13.91}$$

Next, it is useful to define a quantity I_{ABC} by

$$\frac{1}{I_{ABC}} = \int_{d_{coll}}^\infty \left(\frac{\exp\left(\frac{V(r)}{k_B T}\right)}{r^2}\right) dr \tag{13.92}$$

Substituting equation (13.92) into equation (13.91) yields

$$[A] - [A]_o\left(\exp\left(\frac{V(d_{coll})}{k_B T}\right)\right) = \left(\frac{k_{coll}[A]_o}{4\pi(D_A + D_{BC})I_{ABC}}\right) \tag{13.93}$$

Solving equation (13.93) for $[A]_o$ yields

$$[A]_o = \frac{[A]}{\left(\exp\left(\frac{V(d_{coll})}{k_B T}\right)\right) + \left(\frac{k_{coll}}{4\pi(D_A + D_{BC})I_{ABC}}\right)} \tag{13.94}$$

Next, we want an expression for the rate of reaction. Substituting equation (13.94) into equation (13.67) shows

$$
\mathbb{R}_A = \frac{k_{coll}[A]}{\left(\exp\left(\dfrac{V(d_{coll})}{k_B T}\right)\right) + \left(\dfrac{k_{coll}}{4\pi(D_A + D_{BC})I_{ABC}}\right)} \tag{13.95}
$$

where r_A is the rate of a formation in molecules/second/(molecules of B).

One can convert this to a rate per unit volume by multiplying by the B concentration and avogadro's number following the derivation in Section 13.10.1. The result is

$$
r_A = \frac{k_1 C_A C_{BC}}{\left(\exp\left(\dfrac{V(d_{coll})}{k_B T}\right)\right) + \left(\dfrac{k_1}{4\pi(D_A + D_{BC})I_{ABC}N_A}\right)} \tag{13.96}
$$

In equation (13.96), r_A is the rate of formation of A in mol/(liter·second), k_1 is the rate constant in liters/(mol·second), C_A and C_{BC} are the concentrations in mol/liter, D_A is the diffusivity of A in solution, D_{BC} is the diffusivity of BC in solution, $V(d_{coll})$ is the electric potential at the collision diameter k_B is Boltzmann's constant, T is the temperature, N_A is Avogadro's number, and I_{ABC} is the integral given by equation (13.92).

Equation (13.96) can be put into standard form by multiplying the top and bottom by $\left(\dfrac{4\pi(D_A + D_{BC})I_{ABC}N_A}{k_1}\right)$. The result is

$$
r_A = \frac{4\pi(D_A + D_{BC})I_{ABC}N_A C_A C_{BC}}{\left(\dfrac{4\pi(D_A + D_{BC})I_{ABC}N_A}{k_1 \exp(-V(d_{coll})k_B T)}\right) + 1} \tag{13.97}
$$

Equation (13.97) is a simple second-order rate law with a rate constant, k_D, given by

$$
\boxed{k_D = \frac{4\pi(D_A + D_{BC})I_{ABC}N_A}{\left(\dfrac{4\pi(D_A + D_{BC})I_{ABC}N_A}{k_1 \exp(-V(d_{coll})/k_B T)}\right)} + 1} \tag{13.98}
$$

Equation (13.98) is equivalent to equation (13.52).

13.11   SUGGESTIONS FOR FURTHER READING

Good books on the effects of solvents on rates of reactions include:

C. Reichardt, *Solvents and Solvent Effects in Organic Chemistry*, 2nd ed, VCH, New York, 1988.

A. Connors, *Chemical Kinetics: The Study of Reaction Rates in Solution*, VCH, New York, 1990.

O. Tapia and J. Bertrán, *Solvent Effects and Chemical Reactivity*, Kluwer Academic Publishers, New York, 1996.

13.12   PROBLEMS

13.1   Define the following terms:

(a) Solute

(b) Solvent

(k) Hydrophobic interaction

(l) Electrostatic interaction

(c) Solvation

(d) S_N1 reaction

(e) S_N2 reaction

(f) Protic solvent

(g) Aprotic solvent

(h) Polar aprotic solvent

(i) Nonpolar aprotic solvent

(j) Menschutkin reaction

(m) Solvation forces

(n) Kirkwood constant

(o) Donor number

(p) Cohesive energy density

(q) Single-sphere model

(r) Double-sphere model

(s) Regular solution theory

13.2 Describe in your own words how solvents affect rates of reactions.

(a) What are the key effects?

(b) What reactions will be most affected?

(c) What measures of solvent properties do you use to quantify the effects?

13.3 Describe the various terms in equation (13.12). Why does each of the terms arise? What are the key forces that contribute to each term? How would the terms be different in protic solvents, polar aprotic solvents, and nonpolar aprotic solvents?

13.4 Table 13.7 lists the solubility of several species in water. Explain the trends in the table, such as why LiF is insoluble and CsF is very soluble.

13.5 Explain in your own words why salt affects the rates of the various reactions in Figure 13.2.

(a) How does the salt affect the forces between the reactants?

(b) According to the double sphere model, how would the change in force change the reaction rate?

(c) Be quantitative with your predictions in (b). What types of reaction should be enhanced by the additions of salts? What reactions should be decreased by the additions of salts?

(d) Are the trends in Figure 13.2 as you would expect from the single-sphere model? Specifically, does salt decrease the rate when the reactants attract?

13.6 Consider the following reactions:

$$CH_2ClCOO^- + OH^- \longrightarrow CH_2OHCOO^- + Cl^-$$

$$CH_2ClCOOK + KOH \Longrightarrow CH_2OHCOOK + KCl$$

$$C_5H_5N + C_2H_5I \longrightarrow [C_5H_5(C_5H_5N)]^+I^-$$

$$C_2H_5O^- + C_2H_5I \longrightarrow C_2H_5OC_2H_5 + I^-$$

$$S_2O_8^{2-} + 2I^- \longrightarrow I_2 + 2SO_4^{2-}$$

(a) How do changes in ionic strength affect the rate of reaction?

(b) How do changes in the dielectric constant of the solution affect the rates of reaction?

(c) How do changes in the Kirkwood constant of the solution affect the rate of reaction?

(d) How do changes in the cohesive energy density of the solvent affect the rate of reaction?

13.7   Table 13.9 indicates how increases in the "polarity" of the solvent affect the rate of a reaction.

(a) Use the ideas later in this chapter to indicate how changes in the Kirkwood constant would affect each of the classes of reaction in the table.

(b) Consider the series of solvents: acetone, DMF, water, methanol. Rank the solvents for the reaction in Table 13.4 according to Kirkwood's theory and the single-sphere model.

(c) How well do the predicted trends compare to the data in Table 13.4?

(d) Repeat for the double-sphere model. Does that do any better?

(e) Why did the theory fail so badly? What key forces are missing in the models? How can you account for the difference between the experimental results and the models?

(f) Does regular solution theory do any better? Do you get the right trends?

(g) Go back to the data in Tables 13.1–13.3. In which cases does the theory give the correct trends?

13.8   Figures 13.6 and 13.7 show the rate of reaction (13.36) in (a) a series of acetone–dioxane mixtures and (b) a variety of other solvents?

(a) Why does the double-sphere model work in the case in Figure 13.6 but not in the case in Figure 13.7?

(b) Consider the other solvent mixtures

(1) Dioxane–nitrobenzene

(2) Dioxane–acetic acid

(3) Dioxane–methanol

In which cases will the data fit the double-sphere model? (Hint: Be sure to consider hydrogen bonding.)

(c) Repeat (b) for the reaction in Table 13.4. Would the theory work?

13.9   Table 13.3 shows data for two different association reactions.

(a) How well do the data fit the trends expected from the single-sphere model?

(b) How well do the data fit the trends expected from the double-sphere model?

(c) How well do the data fit the trends expected from regular solution theory?

(d) Account for any significant differences in terms of the forces on the molecules and the Hughes–Ingold rules.

13.10   Table 13.5 shows data for the rate of an S_N1 reaction in several solvents.

(a) How well do the data fit the trends expected from the single-sphere model?

(b) How well do the data fit the trends expected from the double-sphere model?

(c) How well do the data fit the trends expected from the regular solution theory?

(d) Account for any significant differences in terms of the forces on the molecules and the Hughes–Ingold rules.

13.11   Kim and Russell (1999) examined the kinetics of the cyclization of 2,3-diethynlquinoxaline. Their data are given in Table P13.11.

(a) Fit the data to the double-sphere model and account for any discrepancies.

Table P13.11   The rate constant for the cyclization of 2,3-diethynlquinoxaline in various solvents

Solvent	k, second^{-1}	ε	Solvent	k, second^{-1}	ε
Acetonitrile	3.3×10^{-5}	36.64	Dioxane	1.9×10^{-4}	2.21
Methanol	3.7×10^{-5}	33	CCl$_4$	2.6×10^{-4}	2.23
Benzene	1.2×10^{-4}	2.38	THF	7.1×10^{-4}	7.52

Source: Data of Kim and Russell, *Tetrahedron Letters*, **40** 3835 (1999).

Table P13.12   The rate constant for the reaction between 4-nitrophenylacetate and imidazole in various solvents

Solvent	k, leter/ (mol·second)	ε	Solvent	k, leter/ (mol·second)	ε
Toluene	0.75×10^{-3}	2.39	Propylene Carbamate	4.45×10^{-3}	64.95
THF	2.28×10^{-3}	7.43	DMF	9.9×10^{-3}	36.71
Methanol	3.22×10^{-3}	32.63	DMA	12.5×10^{-3}	37.78
Acetone	2.62×10^{-3}	20.56	water	57.8×10^{-3}	78.39
Acetonitrile	1.9×10^{-3}	35.95	diethylether	0.7×10^{-3}	4.23

Source: Data of Schmeer, Six, and Steinkirchner, *Solution Chemistry* **28**, 211 (1999).

(b) Fit the data to the single-sphere model and account for any discrepancies.

(c) Fit the data to the regular solution theory and account for any discrepancies.

(d) Try fitting the data to the Taft–Kamlet relationship. What do you find?

13.12   Schmeer et al (1999) examined the reaction between 4-nitrophenylacetate and imidazole. Their data are given in Table P13.12.

(a) Fit the data to the double-sphere model and account for any discrepancies.

(b) Fit the data to the single-sphere model and account for any discrepancies.

(c) Fit the data to the regular solution theory and account for any discrepancies.

(d) Try fitting the data to the Taft–Kamlet relationship. What do you find?

13.13   Verify that

(a) C_A from equation (13.72) satisfies equation (13.70).

(b) C_A from equation (13.89) satisfies equation (13.85).

(c) Explain where the various terms in equation (13.98) arise. Where did they come into the derivation? How large is each of term?

13.14   The object of this problem is to see when diffusion-limited rate constants will be important. Consider a reaction $A + BC \rightarrow$ products with a preexponetial of 10^{13} Å^3/(molecule·second, and an activation energy of 20 kcal/mol.

(a) Show that in the limit that $V(r) = 0$, the criterion for fast diffusion is

$$1 \ll \left(\frac{4\pi(D_A + D_{BC})d_{coll}}{k_1} \right)$$

(b) Calculate the value of $\left(\dfrac{4\pi(D_A + D_{BC})d_{coll}}{k_1}\right)$ for the reaction at 300 K. Assume $d_{coll} = 2$ Å, $(D_A + D_{BC}) = 10^{-5} cm^2/second$.

(c) According to the Stokes–Einstein relationship, the diffusivity of a molecule is given by

$$D_A = \frac{k_B T}{2\pi d_M \mu_s}$$

where D_A is the diffusivity, k_B is Boltzmann's constant, T is the temperature, d_M is the molecular diameter, and μ_s is the viscosity of the solvent. How large of a molecule would be needed before the diffusion limitations become important? Assume that the d_{coll}, the collision diameter of the reactive group on the molecule remains 2 Å.

(d) How do your results in (c) compare to the size of (1) proteins, (2) growing polymer chains, and (3) micrometer-sized silica particles coagulating in solution?

13.15 In the supplemental material we derived an equation for the diffusion-limited rate of a reaction of the form $A + BC \rightarrow$ products. Redo the analysis for the reaction $2A \rightarrow$ products. Assume $V(r) = 0$.

(a) How does the differential equation for the diffusion of the reactants change?

(b) How do the boundary conditions change?

(c) How does the solution of the differential equation change?

(d) Can you still calculate $[A]_o$ analytically?

(d) How does your final expression change?

13.16 The reaction $2I \rightarrow I_2$ has been reported to be mass-transfer-limited in carbon tetrachloride.

(a) Estimate the rate constant for the reaction. Assume $D_I = 1.5 \times 10^{-5} cm^2/second$ and a collision diameter of 2.5 Å.

(b) Use material from Chapter 3 to estimate the half-life of the reaction assuming an initial iodine atom concentration of 10^{-3} molar.

(c) What method would you use to study the rate of the reaction. Hint: Look back to Table 3.1.

14

CATALYSIS BY METALS

PRÉCIS

The next major class of catalysts are metal catalysts. In this chapter we will discuss how metals act like catalysts. First, we will give an overview of the structure of metal catalysts. Then we will review what reactions are like on metals. Next, we will briefly review the mechanisms of some common reactions on metals. Finally, we will provide some information about catalyst selection. Metal catalysis is a large subject, and I cannot cover everything. However, there is a list of excellent references at the end of this chapter.

14.1 INTRODUCTION

Metal catalysts are quite important. Industrially, metal and metal oxide catalysts are used to catalyze most large-scale chemical processes. Almost all large-scale chemicals, petroleum products, and polymers are made using catalysts. Consequently, the metal catalysts are quite important to the world economy.

In this chapter we will give an overview of metal catalysts. First we will note that there are two key types of catalysts:

- Transition metal cluster compounds
- Supported metal catalysts

In this chapter we will briefly describe what each type of catalyst is like. We will particularly consider the structure of the catalyst, and briefly mention how the structure allows the catalyst to function.

Next, we will discuss mechanisms of reaction on surfaces. We will find that all metal-catalyzed reactions occur after completing the same general cycle:

- First, the reactants adsorb onto a site on the surface of the metal.

- Next, there are a series of steps where the molecules attached to the surface atoms rearrange, one ligand at a time, to produce products.
- Finally, the products desorb, regenerating the bare site.

All catalytic reactions follow this general cycle. We give several examples of mechanisms later in this chapter. Our list is by no means complete but we wanted to include a few key practical examples for the reader. We will find that the special thing about metal catalysts is that you rarely see elementary where a molecule attached to the surface rearranges. Instead, you find processes where reactants form one bond to the surface and then break a different reactant-surface bond. The surface is involved at every step in the process. This is different from the pattern in acid catalysis; in an acid catalyst, ligands can rearrange without forming or breaking any bonds to the surface.

Once we finish our discussion of mechanism, we will briefly discuss how metal catalysts work. We will find that the metal catalysts work by all of the mechanisms described in Chapter 12.

- Metal catalysts can help initiate reactions.
- Metal catalysts can stabilize the intermediates of a reaction.
- Metal catalysts can hold the reactants in close proximity and in the right configuration to react.
- Metal catalysts can be designed to block side reactions.
- Metal catalysts can stretch bonds and otherwise make bonds easier to break.
- Metal catalysts can donate and accept electrons.
- Metal catalysts can act as efficient means for energy transfer.

We give examples of each of these effects to give the reader a good molecule-level picture of how the catalyst functions.

Once we explain how catalysts work, we discuss trends over the periodic table, and try to explain why transition metals are such good catalysts. This is an evolving area, so our understanding is incomplete. However, I wanted to give readers a broad overview of the types of metals used as catalysts.

Finally, I provide a list of catalysts for some common reactions. This list is by no means complete. But I wanted to give the reader an overview of the kinds of metals that are used.

14.1.1   Transition Metal Clusters

To start, I want to describe what metal catalysts are like. I will begin with transition metal cluster compounds. Figure 14.1 shows the structure of some transition metal cluster compounds that are commonly used as catalysts. A transition metal cluster is basically a transition metal atom surrounded by ligands. All of the transition metals can be used as catalysts, although some are more active than others.

You run catalytic reactions by putting the clusters in solvents and then adding reactants, and allowing the reactants to react.

One of the key issues in designing catalytic clusters is to find a way to create bare sites on the cluster by removing ligands. There is a simple rule called the **18-electron rule**, which tells us how many ligands a transition cluster needs in order to be stable.

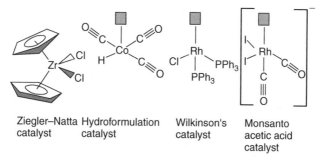

Ziegler–Natta   Hydroformulation   Wilkinson's   Monsanto
catalyst   catalyst   catalyst   acetic acid
                                                    catalyst

Figure 14.1 The structure of some transition metal clusters commonly used as catalysts. (Ph = a phenyl group). The square boxes in the figure represent empty sites.

> A metal atom in an organometallic cluster would like to have a filled shell with 18 electrons. If the metal atom has 18 electrons, the atom is comfortable. If not, the metal tries to bond to other electron donors until the metal atom is surrounded by 18 electrons.

For example, a nickel atom has 9 d electrons and 1 s electron, or a total of 10 electrons. It needs to add 8 more electrons to form a stable compound. A CO ligand can donate 2 electrons. Consequently, $Ni(CO)_4$ is a stable compound. $[Ni(CO)_5]^{2+}$ is also a stable compound where the superscript notation 2+ indicates that the complex has a +2 charge. Table 14.1 gives the electron counts for some typical ligands. One can use it to predict the stable configuration of metal cluster compounds. Additional examples are given in solved Example 14.B. The reader might want to read that example before proceeding.

The 18-electron rule is not an absolute requirement. There are many stable inorganic commands, such as $NiCl_2$, which do not satisfy the 18-electron rule. However, all of the catalytic clusters that I am aware of do satisfy the 18-electron rule.

Generally, active transition metal catalysts consist of transition metal clusters with weakly bound ligands such as PPh_3 (triphenylphosphine) groups. The PPh_3 groups are removed during reaction to produce an active site that is available for reaction. That

Table 14.1   The electron count for some common ligands

Ligand	Electron Count	Ligand	Electron Count
Alkyl (i.e., methyl, ethyl)	1	CO	2
H	1	Olefins	2
OH, Cl, Br, I, F, CN	1	Ethers, ketones	2
NO (bent)	1	PR_3	2
NO (linear)	3	RCN	2
Alkylidyne (M≡C–R)	3	Carbenes (M=CR$_2$)	2
Allyl	3	Amines (e.g., NH$_3$)	2
Linear amindo (M=N–R)	4	Bent Amindo (M=N–R)	2
Dienes (e.g., cyclobutadiene)	4	Cyclopentadyenyl	5
Diols	4	Arenes (e.g., benzene)	6
Cycloheptatriene	7	Cyclooctatetraene	7

produces a metal atom with only 16 electrons and a space to hold adsorbate. Reactants can bind to the space, and be activated for reaction.

Details of the reactions will be given later in this chapter. The thing to remember for now is that transition metal clusters surrounded by weakly bound ligands can often catalyze simple reactions. Generally, catalytic clusters are surrounded by 16–17 electrons. Clusters surrounded by 18 electrons have no place for reaction to occur, 15-electron clusters bind the products too strongly to allow reaction to occur.

14.1.2 Supported Metal Catalysts

The other key class of metal catalysts consist of supported metal catalysts. Figure 12.3 shows a picture of a typical support metal catalyst. The catalyst consists of a series of small metal particles on a metal oxide support. The metals usually do most of the chemistry. However, the support is crucial to the operation of the catalyst.

Table 14.2 lists some of the common support materials. The support is generally a high-surface-area metal oxide. Support particles can come in many different shapes as indicated in Figure 14.2. The oxide consists of a highly porous structure as indicated in Figure 14.3.

The support provides a framework to hold the metal. Recall that metal catalysts work by binding intermediates to the surface of the catalyst. In Chapter 12 we found that

Table 14.2 Some common support materials

γ-Alumina (γ-Al_2O_3) (a Highly Porous Alumina)	High-Surface-Area Silica	Zeolites (Table 12.12)	High-Surface-Area ZnO
High-surface-area MgO	High-surface-area Zr_2O_3	Activated carbon	High-surface-area Ti_2O_3

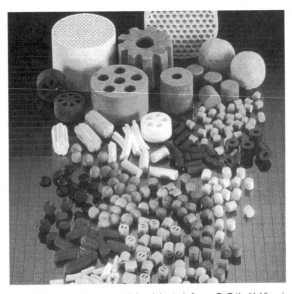

Figure 14.2 Pictures of some support materials. Adapted from G. Ertl, H. Vinozinger, T. Weiikamp, Preparation of catalysis, © 1998, J. Wiley, with permission.

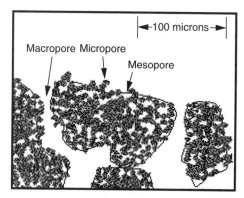

Figure 14.3 A cross-sectional diagram of a typical catalyst support.

increasing the surface area of the catalyst leads to higher reactivity. One way to get a high surface area is to grind up the metal into tiny particles. Tiny metal powders are not stable because if the particles touch, they stick to one another. However, one can get little particles to stay little if one puts them on a support. The oxide support holds the metal particles to prevent them from touching each other. As a result, one can maintain a high surface area and still keep the metal stable.

Key features of a support are that it be a very high surface area material and that the structure be very porous so that reactants can get in and out very easily. Surface areas as high as 500 square meters (m^2) of surface per gram of support are possible. The structures are often 60% open area. There is also often a complex pore structure to facilitate mass transfer.

For example, a high-surface-area silica (SiO_2) can be made by reacting tetraethoxysilicon, $Si(OCH_2CH_3)_4$, also called TEOS, with water. When TEOS reacts with water, it forms what is called a *hydrogel*, which is a polymer of silicon dioxide and silicon hydroxide, where each silicon dioxide and hydroxide unit is surrounded by water. The polymer molecules ball up into small particles about 100 Å across.

Initially the hydrogel is soluble in water. However, as you add more TEOS to produce more polymer molecules, the polymer molecules agglomerate into what is called a *floc* (the short term for floccule). The flocs are lumps of particles, a few hundred angstroms across. The flocs are the primary building blocks of the material.

Generally, after you make the flocs you remove some of the water. The flocs then combine to form the porous structure shown in Figure 14.3.

The hydrogel floc does not have that much surface area. However, if you continue to dry the floc, you can pull the water out of the 100-Å polymer particles. That produces tiny pores or cracks within the silicon dioxide polymer called **micropores**. The micropores are typically 5–20 Å across. Then there are **mesopores**. The mesopores are the spaces between the particles. The mesopores are typically 50–100 Å. Then there are the **macropores**. The macropores are the spaces between the flocs. The macropores are typically several micrometers across.

This highly porous structure has a tremendous surface area. It is possible to create a silica with a surface area of 600 m^2 per gram of material. That corresponds to taking all of the surface area of a football field and rolling it up into a volume of 5 cm^3. This provides a tremendous surface area to hold metal.

The oxide support can also be used to block sites and limit the production of undesired products as described in Section 12.15.1. Supports also do acid catalysis as described in Section 12.3.4.

An important concept in catalysis is that only special high-surface-area forms of a material are suitable as a support. For example, quartz is also made from silicon dioxide. However, quartz does not have as high of a surface area as does silica made as described above.

14.1.3   Metal Particles

The metal particles in a supported metal catalysts are generally crystalline. Figure 14.4 shows a diagram of the structure of a typical metal particle in a supported metal catalyst. The metal particles are the other key components in a supported metal catalyst. Recall from freshman chemistry that crystalline materials generally have regular arrangements of atoms that repeat throughout the material. Notice the hexagon across the top of the cluster in Figure 14.4. The hexagon repeats itself across the top of the cluster. We also see other structures repeating on the sides of the particle. Still, one does not have a regular, diamondlike shape. Instead, the surface is jagged. The jagged nature of clusters is an important feature of supported metal catalysis.

People talk about the jagged structures in terms of steps, kinks, and terraces. Figure 14.5 shows pictures of extended surfaces with steps, kinks, and terraces. Terraces are big flat regions on the surface; steps are regions that look like a step or staircase. Kinks are places where the step edge is jagged. The reader might want to look back at Figure 14.4 and find steps and kinks before proceeding with this chapter.

14.1.4   Introduction to Crystallography

Next, I want to give a short introduction to the crystallography of metals. When I was writing this book, I was not sure whether to include a discussion of crystallography, because you can understand catalysis without understanding crystallography. However, many of the papers in catalysis use crystallography notation. Consequently, I decided to review crystallography before proceeding. Of course, the reader can skip Section 14.2 without loss of continuity.

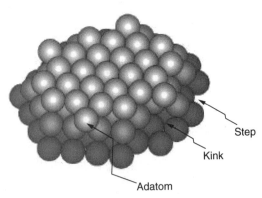

Figure 14.4   A diagram of the structure of a "typical" metal particle on a supported platinum catalyst.

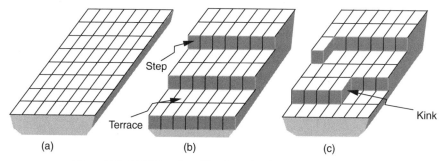

Figure 14.5 (a) A flat surface; (b) a stepped surface; (c) a kinked surface.

Crystallography is the study of crystals. Most metals and other solids form regular arrangements of atoms called *crystals*. Key features of crystals are that the atoms repeat in a regular pattern and the crystals can be macroscopic (i.e., like a diamond).

Different metals have different crystal structures. The most common crystal structures are the **simple cubic (SC)**, **body-centered cubic (BCC)**, **face-centered cubic (FCC)**, and **hexagonal close-packed (HCP)**. I am assuming that most of the readers of the book know what it means for a metal to be crystalline with a given crystal structure, so I will not repeat the definition here.

One of the key things you should remember from freshman chemistry is that every crystal structure has a repeat unit called a **unit cell**, and one can classify the crystal structures in terms of the **Bravis lattice,** where for example, BCC is one type of Bravis lattice and FCC is another type of Bravis lattice.

Figure 14.6 shows how the Bravis lattice varies over the periodic table. Platinum, palladium, nickel, gold, copper, silver, iridium, and rhodium show an FCC structure. Cobalt, ruthenium, osmium, and rhenium show an HCP structure. Iron shows a BCC structure at room temperature, although it goes to an FCC structure at high temperature.

Figure 14.7 shows a diagram of the basic repeat unit of a simple cubic, face-centered cubic, body centered cubic, and hexagonal close-packed unit cell. The simple cubic is a cube with atoms on each corner. The BCC lattice has an atom in the center of the lattice

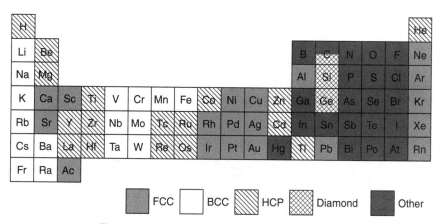

Figure 14.6 The Bravis lattice for crystals of the elements.

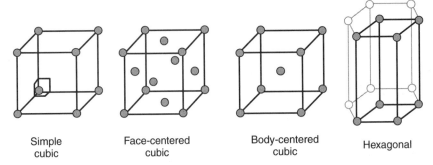

Simple cubic	Face-centered cubic	Body-centered cubic	Hexagonal

Figure 14.7 The basic repeat unit for the SC BCC, FCC, and HCP lattices. Only the nuclear positions are shown.

while the FCC lattice has atoms in the center of the faces. The hexagonal lattice looks entirely different.

I find the pictures in Figure 14.7 deceiving in that differences between the various structures appear to be unrealistically exaggerated. Figure 14.8 shows another view of the three lattices, in this case with space-filling atoms. I have included lines in the figure to enable the reader to identify the unit cell.

Let us start with the FCC lattice. The FCC lattice has a unit cell with four atoms in the corners and one atom in the center of each face. The BCC structure has four atoms in the corners and one atom in the center of the cell. Now focus on the BCC face that is most exposed in Figure 14.8. Notice that if we start with the top middle atom, we can draw a square through the middle atom in the center of each edge of the cube. There is an atom in the center of that cube as well. Therefore, at first sight, it is hard to distinguish between a BCC lattice and an FCC lattice. There is a difference. All of the atoms in the FCC lattice touch each other, while the atoms do not touch each other in the BCC lattice. However, this is a small difference, and in fact at first sight the BCC and FCC lattices look almost the same.

Now concentrate on the sides of the HCP lattice. Notice that the side of the HCP lattice also has a squarelike structure with an atom in the middle. The structure is actually a diamond, not a square, and the spacings are different. Still, there is considerable similarity between the FCC, BCC, and HCP structures.

The other key point is that all three lattices show similar close-packed structures. The top plane in the HCP structure is a hexagon, as is the lightly colored plane in the FCC

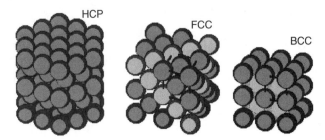

Figure 14.8 Another view of the HCP, FCC, and BCC lattices.

structure. The plane across the diagonal in the BCC lattice is not obviously hexagonal. However, if you look at a larger picture of the plane, it looks almost hexagonal.

The conclusion from this is that there is no really significant difference between the FCC, BCC, and HCP structures. Some metals assume a FCC structure. Others assume a BCC or HCP structure. However, all three structures look almost the same on an atomic level. The structures give much different x-ray diffraction patterns, however.

All transition metals assume a BCC, FCC, or HCP structure, so if you know the BCC, FCC, and HCP lattices you will know the bulk surface of all of the transition metals.

14.1.5 The Structure of Metal Surfaces

At this point we will be changing topics and starting to discuss the structure of metal surfaces. To start, let's look back and focus on the arrangement of the atoms in the surface of the crystal in Figure 14.4. Notice that some faces look relatively flat while other faces look rather bumpy. The flat surfaces can be quite different chemically from bumpy surfaces.

The flat terraces are called **close-packed planes**, while the bumpy surfaces are called **stepped surfaces**. In the literature, people often talk about **steps, kinks**, and **terraces**. These terms are defined in Figure 14.5.

Notice that if we look at the plane on the right side of the cluster in Figure 14.4, it has a kink and a step. There also is an **adatom** on the front surface of the cluster. An adatom is an extra atom that is not surrounded by other atoms. People discuss steps, kinks, terraces, and adatoms in the literature, so you should know what they are talking about.

14.1.6 Miller Indices

Next, I want to mention the standard notation that one uses to describe surface structure. Surface crystallography is an old subject. Réné Häuy did important work in 1806. In 1839 William Miller defined some notation to describe the structure of surfaces called the **Miller indices**. Today, most people use Miller indices to designate surface structure.

The Miller indices are a series of three numbers used to designate the arrangement of the planes in a metal. Generally one indicates the atomic arrangement by M(ijk), where M is the metal and i, j, and k are the Miller indices of the given plane. For example, Pt(111) will refer to the "one–one–one" face of platinum. The structure of the (111) face is given in Figure 14.9.

I have to tell you that I am not a fan of Miller's notation. One of the difficulties is that the (111) face of an FCC material looks nothing like the (111) face of a BCC or HCP material, so Pt(111) looks nothing like W(111) (platinum is FCC; tungsten is BCC). It happens that Pt(111) looks just like W(110). I will describe some other complications below. Still, Miller's notation has been universally adopted in the literature, so we are forced to use it.

The best way to remember the relationship between the Miller indices and the surface structure is to memorize the structures. Figure 14.9 shows a diagram that I use to remember the structures. The diagram is called a **stereographic triangle**. The stereographic shows all of the surface structures possible with an ideal FCC lattice. The figure is in the shape of a triangle because it came from field ion microscopy of field emission tips. You can actually see the atomic arrangements on metal surfaces in a field ion image, and the atomic arrangements happen to follow a triangular pattern in the image. The position of a given plane in the stereographic triangle is the place where the planes

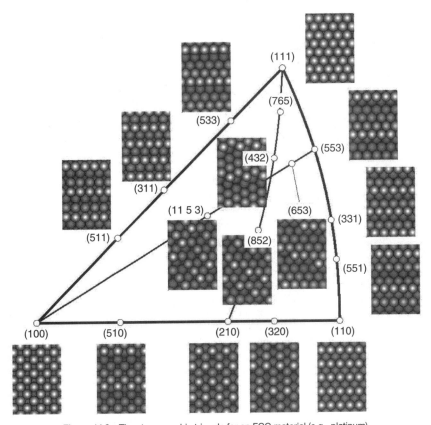

Figure 14.9 The stereographic triangle for an FCC material (e.g., platinum).

appeared in a triangular image, while the structure is as it would be if no atoms moved when the surface formed.

Figure 14.9 is pretty complicated, so it is useful to look at the corners and then move into the center. The plane at the lower left corner of the figure is called (100). The (100) plane is a square surface. There is a square consisting of an arrangement of atoms. The reader should find the square before proceeding. The plane at the upper corner of the triangle is called (111). (111) is a hexagonal surface. The plane at the lower right corner of the stereographic triangle is called (110). (110) has a washboard structure.

Next, let us examine the line between (100) and (111). Planes along the line have the last two indices the same. Surfaces along the line have a surface structure somewhere between that of (100) and (111). For example Pt(511) has terraces with a (100) orientation and steps with a (111) orientation.

Figure 14.9 applies only to a FCC structure. If you have a BCC plane (ijk), then the BCC plane will look like the FCC plane with indices (i (j + k) (j − k)), so a W(110) face will look like Pt(111) or a W(111) plane will look like a Pt(120).

HCP structures are more complex. The (001) face of an HCP material is hexagonal. The other faces with a zero in the indices (e.g., 100, 101, 201) are stepped. All of the other surfaces are kinked. People sometimes use a four-digit code for the HCP structures. The third digit can be ignored; for example, the Re(0001) face is the Re(001) face.

I have a few little tricks that help me to keep track of everything:

- The order of the indices does not matter for BCC or FCC lattices. Pt(210), Pt(012), Pt(120), and Pt(201) all look the same. Unfortunately, the order of the indices does matter for HCP lattices.
- Pt(111) is hexagonal; Pt(100) is square. All the other faces are stepped or kinked. In an FCC material, if one of the indices is zero or if two indices are the same, the surface is a stepped surface. If not it is a kinked surface.
- If you instead have a BCC material, you can use the rules that a BCC plane with indicies (i j k) will look very similar to the FCC plane with indices (i (j + k) (j − k)). You use the formula to calculate the equivalent structure in an FCC material. You then figure out its structure.

I sometimes forget what a surface looks like, and in that case I use vector arithmetic to work out the surface structure.

There are three key structures in the FCC lattice: the (100) structure, the (111) structure, and the (110) structure. The idea in vector arithmetic is to relate the structure of the (i j k) plane to the structure of the (111), (110), and (100) planes.

The steps are

1. Reorder the indices so that the largest index is first; for instance, Pt(15l) would be changed to Pt(511). Note that Pt(511) and Pt(151) have an identical surface structure since platinum is FCC.
2. Use vector arithmetic to expand the plane as a sum of (111), (110), and (100):

$$Pt(511) = 4(100) + (111)$$

3. Notice that the (100) has the highest coefficient (4). Therefore, the Pt(511) will have (100) terraces.
4. The other coefficients give the indices of the steps.

For example, $Pt(711) = 6Pt(100) + Pt(111)$.

Therefore, you would expect the structure of Pt(711) to have Pt(100) and Pt(111) components with a longer (100) structure than a (111) structure. We call the structure (100), *terraces* and (111), steps.

You have to be careful with the vector arithmetic because Miller indices were developed in 1839, and they *do not quite work like modern-day vectors*. An example is (311) = 2(100) + (111), which would suggest (100) terraces and (111) steps. However, in reality Pt(311) has a structure half way between that of Pt(100) and Pt(111). Therefore, while one can use the vector arithmetic to get an idea of what a plane looks like, the *vector arithmetic will not give you an exact picture*. Still, the vector arithmetic allows you to get an approximate mental picture of a plane, even if it is not exact.

Next, let us use the vector arithmetic to work out what the Pt(410) plane looks like. Note that (410) = 3(100) + (110), so you would expect Pt(410) to have (100) terraces and (110) steps as is observed. Next, let us consider Pt(653). Note (653) = 3(111) + 2(110) + (100). Consequently, one would expect Pt(653) to have a complicated structure with more (111) than anything else. People talked about Pt(653) as a **kinked surface** because there are kinks in the surface plane. Pt(i j k) will be a kinked surface if all three indices were nonzero and all three indices were different.

I do not have room here to include more details about Miller indices. The thing to remember for now is that people use Miller indices to denote surface structure. If you see the notation Pt(531), then you know that it is a platinum surface that is FCC and can figure out its structure using the rules above. [Pt(531) is kinked.] There is a long discussion of Miller indices in Chapter 2 of Masel (1996). Professor Masel also distributes a program from his Web page to calculate the geometry of a plane, given its Miller indices. The reader is referred to those sources for further information.

14.2   ELEMENTARY REACTIONS ON METALS

At this point I want to change topics and review something that is very important to catalysis: the mechanisms of reactions on metals. First, I will review some general ideas about metal reactions; then I will discuss what types of elementary reactions occur on metals. Finally I will give a series of examples.

We already started to discuss the mechanisms of reactions on metals in Chapter 5. In Chapter 5 we noted that reactions on metals usually look like radical reactions in the gas phase. Usually, there is an initiation step, where a bare site is created on the metal atom. Then there is an **adsorption** step, where the reactants attach themselves to the catalyst. Then there are a series of transformation processes, where bonds in the reactants are broken or formed one at a time to yield products. Then there is a **desorption** step, where the products are removed from the catalyst.

> The adsorption–reaction–desorption cycle occurs in all catalytic reactions.

For example, Figure 12.1 showed the catalytic cycle for the Monsanto process for acetic acid production. Notice that first the reactant adsorbs on the catalyst. Then bonds are broken and formed one step at a time to yield adsorbed products. Then the product desorbs.

We will describe the mechanisms of reaction on surfaces in detail in Section 14.3. Look for the adsorption–reaction–desorption cycle in all of the cases.

At this preliminary stage in the discussion, we will first consider the kinds of elementary reactions that occur on metals. Six key classes of elementary reactions occur on metals:

- Simple molecular adsorption reactions
- Dissociative adsorption reactions
- Bond scission reactions
- Addition reactions
- Recombination reactions
- Desorption reactions

In the next three sections we will describe each of them in detail, to give the reader a better idea of the reactions that do and do not occur on metals.

14.2.1   Adsorption

Let's start with adsorption. Adsorption is usually the first step in a catalytic cycle. During adsorption a molecule comes into contact with the metal and sticks.

In Chapter 5 we noted that there are two key kinds of adsorption:

- Molecular adsorption
- Dissociative adsorption

In **molecular adsorption** an incoming molecule attaches itself to the metal and no bonds in the molecule are broken. For example, a CO molecule can come down to a platinum to form an adsorbed CO

$$CO + S \longrightarrow CO_{(ad)} \tag{14.1}$$

or an olefin can react with a $HCo(CO)_3$ cluster to form $HCo(CO)_3(H_2C=CHR)$.

$$HCo(CO)_3 + H_2C=CHR \longrightarrow HCo(CO)_3(H_2C=CHR) \tag{14.2}$$

The interaction is usually a Lewis acid/Lewis base type of interaction. Recall from Section 14.1.1 that a metal atom would like to be surrounded by 18 electrons. However, the cobalt atom in $HCo(CO)_3$ is surrounded by only 16 electrons. As a result, the metal atom would like to bond to any ready source of electrons such as the olefin and is held in what is called a *molecularly adsorbed state*. Then the olefin forms a weak Lewis acid/Lewis base bond and is held in this molecularly adsorbed state.

A molecular adsorption process usually follows an equation such as the Langmuir adsorption isotherm discussed in Section 12.17. One of the key features for the discussion here is that the bonds are very weak: 10–25 kcal/mol, compared to 60–100 kcal/mol for a molecular bond. As a result, molecularly bound species can be displaced, or can react while on the surface.

In reaction 14.1, we indicate that the reaction occurs via a reaction with a bare site. The bare site is less obvious in reaction (14.2), but if we look at the molecule in detail, there is indeed an empty site as indicated in Figure 14.10.

There is a second kind of adsorption process called **dissociative adsorption**. Dissociative adsorption is a process where an adsorbate sticks by breaking a bond. Examples of dissociative addition reactions include

$$H_2 + 2S \longrightarrow 2H_{ad}$$
$$HCo(CO)_3 + H_2 \longrightarrow H_3Co(CO)_3 \tag{14.3}$$

Dissociative adsorption is fundamentally different from molecular adsorption in that bonds do break. Therefore, electrons need to rearrange in the adsorbate and the metal to accommodate the new bonding. Inorganic chemists call the process an **oxidative addition**

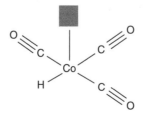

Figure 14.10   A diagram of $HCo(CO)_3$. The box denotes a bare site.

reaction since the adsorbate is formally being oxidized. The metal center is donating electrons to form an adsorbate–metal bond.

In reality, dissociative adsorption usually occurs in two steps. First the adsorbate adsorbs molecularly. Then the molecule dissociates. However, there is no activation barrier to the dissociation process other than an energy transfer barrier, so people sometimes think of the reaction as a single concerted process.

It happens that the rate of dissociative adsorption varies over the periodic table, as discussed in Section 5.11. In particular, metals near the center of the d series (e.g., tungsten) tend to dissociate simple molecules more readily than do metals on either side. One should reread Section 5.11 and examine Figure 5.12 again, before proceeding with this chapter.

Dissociation adsorption occurs on many metals. Molecular adsorption often occurs as well. A key point for the discussion later in this chapter is that one sees only molecular or dissociative adsorption on a metal. On acid catalysts one can observe a process in which an incoming reactant molecule directly attaches itself to another adsorbed molecule. However, one rarely sees direct attachment of adsorbants in metal catalysis. Instead, the reactants always attach to the metal. The direct participation of the metal is one feature that makes metal catalysts unique.

14.2.2   Reactions of Adsorbed Species

Next, we want to discuss what happens once a molecule adsorbs onto the metal and is activated to react. In particular, we will discuss the kinds of reactions that occur after species are adsorbed onto the metal. We will concentrate on four key types of reaction:

- Bond dissociation reactions
- Fragmentations
- Association reactions
- Single-atom recombinations

All of these reactions involve transfers of atoms or ligands to or from the surface. We will find that it is unusual to observe direct transfer of atoms from one adsorbed species to the next during reactions on metals.

First, let us discuss bond dissociation reactions. Bond dissociation reactions are another kind of oxidative addition reaction. In a bond dissociation reaction a bond in an adsorbate breaks and an atom or other ligand gets transferred to the metal atom. Examples of bond scission reactions include all of the reactions discussed in Section 5.12. For example, when ethanol adsorbs on platinum, the ethanol sequentially dehydrogenates via the sequence of reactions:

$$CH_3CH_2OH_{(ad)} + S \longrightarrow CH_3CH_2O_{(ad)} + H_{(ad)}$$

$$CH_3CH_2O_{(ad)} + S \longrightarrow CH_3CHO_{(ad)} + H_{(ad)}$$

$$CH_3CHO_{(ad)} + S \longrightarrow CH_3CO_{(ad)} + H_{(ad)} \tag{14.4}$$

$$CH_3CO_{(ad)} + S \longrightarrow CO_{(ad)} + CH_{3(ad)}$$

Notice that atoms or molecular ligands are transferred one at a time to produce products. The single-atom transfers to and from the metal are very characteristic of reactions on

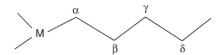

Figure 14.11 The α, β, γ, and δ positions from a metal center.

metals. They are seen in most metal-catalyzed reactions. Ligand transfer reactions are less characteristic, but they can occur.

There are several types of bond scission processes. They are called α scissions, β scissions, γ scissions, and δ scissions. During an α scission a hydrogen atom or an R group bonded to the carbon closest to the surface is transferred to the surface. During a β scission a hydrogen atom or R group bonded to the β carbon is transferred to the surface. During a γ and δ scissions atoms or R groups attached to the γ and δ carbon are transferred to the metal (the γ and δ carbons are defined in Figure 14.11). Experimentally, α and β scissions generally occur more readily than do γ and δ scissions. However, α, β, γ, and δ scissions can all be observed.

In Chapter 5 we noted that during a β-scission process one forms a double bond so the process is thermodynamically favorable. γ and δ scissions are less favorable although they are seen occasionally.

One can also break the C–C bonds rather than C–H bonds. For example, when propanol adsorbs on platinum, the propanol dehydrogenates to form a $CH_3CH_2CO_{(ad)}$ intermediate. The $CH_3CH_2CO_{(ad)}$ intermediate reacts via

$$CH_3CH_2CO_{(ad)} + S \longrightarrow CH_3CH_{2(ad)} + CO_{(ad)} \qquad (14.5)$$

Reactions where the C–C bond breaks and ligands are transferred to the metal are called **fragmentation reactions**. Fragmentation reactions have larger intrinsic barriers than do hydrogen transfers. However, fragmentation reactions are often seen when there are no alpha hydrogens.

The opposite of a fragmentation reaction is an **association reaction**. An association reaction is a reaction where two adsorbed species come together to form a bond. For example, during a process called hydroformulation, one might get a reaction like

$$CH_3CH_{2(ad)} + CO_{(ad)} \longrightarrow CH_3CH_2CO_{(ad)} + S \qquad (14.6)$$

A key feature of association reactions is that you are starting with two adsorbed species, but end up with a single species at the end of the reaction.

In practice, one often sees **combined displacement–association reactions** such as

$$CO + CH_3CH_{2(ad)} + CO_{(ad)} \longrightarrow CO_{(ad)} + CH_3CH_2CO_{(ad)} \qquad (14.7)$$

During reaction (14.7), the incoming CO pushes the ethyl group off of the adsorbed site to yield products. You start with a CH_3CH_2–metal bond and a CO–metal bond, and end up with a CH_3CH_2CO–metal bond, a CO–metal bond, and a CH_3CH_2–CO bond. The net bonding to the metal is conserved and you have a new C–CO bond. Reaction (14.7) is about 64 kcal/mol exothermic while reaction (14.6) is endothermic. As a result, reaction (14.7) is much more rapid than reaction (14.6).

During reaction (14.7) one is transferring a molecular ligand. However, in Chapter 5 we found that it is often easier to transfer an individual atom. Single-atom transfers can occur without another molecule immediately adsorbing. For example, during ethylene hydrogenation, one key step is

$$CH_2CH_{2(ad)} + H_{(ad)} \longrightarrow CH_3CH_{2(ad)} + S \tag{14.8}$$

During reaction (14.8) an adsorbed hydrogen reacts with an adsorbed ethylene to yield an adsorbed ethyl group. Reaction (14.8) is called a **hydrogen migration** reaction in the inorganic chemistry literature. One of the interesting things is that one can get what are called 1,1 and 1,2 migrations. During a 1,1 migration the hydrogen ends up on the α carbon in the ethyl group while in 1,2 migrations the hydrogen ends up on the β carbon. One can see both possibilities on metal cluster compounds. However, on supported metals, the 1,2 migrations seem to dominate.

One can have all kinds of other migration reactions. Halogens (fluorine, chlorine, bromine) follow many of the same pathways as hydrogen. Alkyl groups generally add at the α position. People also discuss **acetyl** $(-\overset{R}{C}=O)$ **migrations**. They are seen on cluster compounds, but rarely on metal surfaces. Generally, single-atom migrations predominate in most metal-catalyzed reactions. However, there are important examples of migrations of alkyl groups and other ligands.

Generally, most catalytic reactions have a series of migration reactions where the desired products are formed on the surface of the catalyst.

There is one other key idea about molecular rearrangement on metals—the metal is involved in every step of the process. Molecular ligands are transferred to the metal, and surface species recombine. All of the steps involve breaking or forming of a bond to the metal. One seldom observes direct transfer of ligands between adsorbed species. Instead, one *usually* observes only transfers to or from the metal.

14.2.3   Desorption

So far, we have been discussing only transfers that occur on the surface. However, one other type of transfer reaction is important: desorption of the products off of the catalyst. In this section we will review the types of processes that occur during desorption of species from metal catalysis. We will find that there are four key kinds of desorption processes:

- Simple molecular desorptions
- Recombinative desorptions
- Displacement reactions
- β Scissions

All four types occur quite regularly in metal clusters and supported metal catalysts. In this section we will describe each of these reactions in enough detail so that the reader can recognize each of them.

Let us start by describing simple molecular desorptions. During a simple molecular desorption process, the adsorbate detaches itself from the surface to yield gas-phase products:

$$CO_{(ad)} \longrightarrow CO + S \tag{14.9}$$

In all cases the molecule leaves the catalyst without undergoing significant rearrangement. There are some special things that need to happen to allow molecular desorption to occur. In particular

- We usually have to have a stable molecule already formed on the surface, although direct desorption of radicals can occasionally be observed.
- The adsorbate cannot be bound too strongly to the surface; a strongly bound molecule will fragment rather than desorb.

We rarely have both of these conditions satisfied. In which case, some other desorption process is needed.

The second most common desorption process is a **recombinative desorption** process. In a recombinative desorption process, two adsorbed radicals combine together to form a stable species that then leaves the surface. For example, ethyl groups are strongly bound to platinum. They cannot simply desorb. However, ethyl groups can react with hydrogen to form a stable species; ethane. The main reaction is

$$CH_3CH_{2(ad)} + H_{(ad)} \longrightarrow CH_3CH_3 + 2S \qquad (14.10)$$

Hydrogen atoms are also strongly bound to platinum, so hydrogen atoms cannot simply desorb. However, hydrogen can combine to yield a stable species, H_2:

$$2H(ad) \longrightarrow H_2 + 2S \qquad (14.11)$$

That is what actually occurs. Notice that during reactions (14.10) and (14.11) two species that are strongly bound to the surface react to form a stable molecule that leaves the surface.

Recombinative desorption processes are the most common type of desorption process on a supported metal catalyst.

In the inorganic literature, people call recombinative desorption processes **reductive eliminations**. Reductive eliminations do occur on clusters, but they are rare. Notice that reactions (14.10) and (14.11) form two bare sites. Recall that a metal atom would like to be surrounded by 18 electrons. Well, when you form a bare site on a bulk metal, the metal atoms can pull electrons out of the bulk. As a result, the metal atom can still be satisfied. However, there is no ready source of electrons on a metal cluster. Instead, when a reductive elimination occurs, the metal atom goes from an 18-electron environment to a 16-electron environment. That costs energy. As a result, reductive eliminations do not occur as readily on a metal cluster as on a supported metal catalyst.

On clusters, it is more common to observe a **displacement reaction**. In a displacement reaction a gas-phase molecule or a molecule in solution comes in and displaces a ligand from the surface. Examples include

$$CH_3CH_{2(ad)} + H_2 \longrightarrow CH_3CH_3 + H_{(ad)} \qquad (14.12)$$

$$CO + 2H_{(ad)} \longrightarrow H_2 + CO_{(ad)} \qquad (14.13)$$

$$CO + CH_2CH_{3(ad)} + H_{(ad)} \longrightarrow CH_3CH_3 + CO_{(ad)} \qquad (14.14)$$

During reaction (14.12), an H_2 comes in and then there is a concerted process where one of the hydrogen atoms interacts with the ethyl group to form ethane and at the same time the other hydrogen atom forms a bond to the cluster. Reaction (14.12) occurs mainly on clusters. It is rarely seen on bulk metals, although it has been observed on metals at low temperature.

Reactions (14.13) and (14.14) are similar. During reaction (14.13) a CO comes and pushes the two hydrogens together. The two hydrogens combine and desorb. During reaction

(14.14) a CO comes and pushes the hydrogens toward the ethyl group. The hydrogen and the ethyl group combine and desorb. Reactions (14.13) and (14.14) both preserve the electron count; according to Table 14.1, CO donates two electrons while a hydrogen and an ethyl group each donate one. As a result, both reactions can occur easily on a supported cluster. Reactions (14.12)–(14.14) can also occur on metal surfaces, although they are less common on surfaces because of the high intrinsic barriers to the reactants.

There is another reaction, a β–scission, which is very characteristic of a reaction on a cluster. In a β–scission process, the species leaves the surface, and the β hydrogen stays behind. For example, if one starts with a deuterated ligand, then

$$R_2CDCH_{2(ad)} \longrightarrow R_2C{=}CH_2 + D_{(ad)} \qquad (14.15)$$

β–scissions are thermodynamically favored under most circumstances. They occur readily on transition metal clusters, but less readily on supported metal catalysts. On supported metals, simple desorptions and recombinations dominate under most conditions, although β–scissions and displacement reactions can occasionally be seen.

14.2.4 Summary of Elementary Reactions on Metals

At this point we have covered all of the principal elementary reactions that occur during reactions on metals. I want to reiterate that a key feature of reactions on metal catalysts is that atoms or ligands are transferred to or from the metal in every step of the reaction. One seldom observes direct transfer of atoms or ligands between one adsorbed species and another, although transfer can occur during a displacement reaction. Single-atom transfers occur more easily than the transfer of molecular ligands. In the previous section we discussed the reactions that do occur. However, it is also important to recognize that the direct transfer of a ligand from one molecule to another or within a molecule rarely occurs on metal catalysts. Such transitions will occur on an acid catalyst, however.

14.3 OVERALL MECHANISMS OF METAL-CATALYZED REACTIONS

At this point we will be changing topics. In the previous three sections we discussed the types of elementary reactions that occur on metal catalysts. Next we want to combine the results in Sections 14.2.1–14.2.3 to see how the elementary reactions fit together into a mechanism.

First it is important to note a few general rules about reactions on catalysts:

- There must be bare sites on the catalyst to start the reaction.
- Then at least one of the reactants must adsorb on the bare sites.
- Then there are a series of bond dissociation reactions, fragmentations, association reactions, and single-atom recombinations which convert the adsorbed reactants into products
- Then the products desorb.

Catalytic reactions always go through a catalytic cycle:

- Adsorption
- Reaction
- Desorption

You need have sites to adsorb the reactants. The reactants need to stick to those bare sites. Then you need to get reactions that convert the reactants to products to occur. Then you need to get the products off the catalyst.

In the following sections, we will examine the mechanisms of a few industrially important reactants to give the reader a picture of what overall catalytic reactions are like. Catalysts are used for thousands of reactions, so there are many combinations. I choose the following examples to illustrate the general principles:

- Olefin hydrogenation
- Paraffin dehydrogenation
- Metal-catalyzed isomerization
- CO oxidation
- Partial oxidation of ethylene to ethylene oxide
- Hydroformylation

There are hundreds of other examples. The works by Bowker (1988), King and Woodruff (1981), Pines (1991), Ertl (1997, 1999a,b), Andersona and Boudart (1984–1988), Gault (1981), Emmett (1954), and Hoffmann (1988) provide hundreds of other examples. One should refer to this other work for more details.

14.3.1   Olefin Hydrogenation

Let us start with ethylene hydrogenation. Figures 14.12 and 14.13 show the mechanism of ethylene hydrogenation on a supported platinum catalyst and on a metal cluster. The

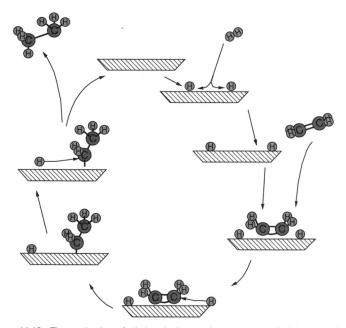

Figure 14.12   The mechanism of ethylene hydrogenation on supported platinum catalysts.

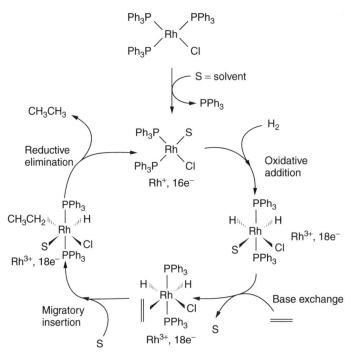

Figure 14.13   The mechanism of ethylene hydrogenation on a RhCl(PPh₃)₃ cluster. (Wilkinson's catalyst).

overall reaction is

$$CH_2{=}CH_2 + H_2 \Longrightarrow CH_3CH_3 \qquad (14.16)$$

In the cases in Figures 14.12 and 14.13 the hydrogen reversibly adsorbs on the catalyst.

$$H_2 + 2S \Longrightarrow 2H_{(ad)} \qquad (14.17)$$

Then ethylene adsorbs:

$$CH_2{=}CH_2 + S \longrightarrow CH_2{=}CH_{2(ad)} \qquad (14.18)$$

Then a migration reaction occurs where a hydrogen is transferred to the ethyl group:

$$CH_2CH_{2(ad)} + H_{(ad)} \longrightarrow CH_3CH_{2(ad)} + S \qquad (14.19)$$

Then there is a recombinative desorption regenerating the bare site:

$$CH_3CH_{2(ad)} + H_{(ad)} \longrightarrow CH_3CH_3 + 2S \qquad (14.20)$$

In the diagrams, the cycles look identical on a cluster and on a supported metal catalyst. These examples are unusual, however. In most cluster compounds, the reductive elimination and oxidative addition steps are replaced by a single displacement reaction:

$$CH_3CH_{2(ad)} + H_2 \longrightarrow CH_3CH_3 + H_{(ad)} \qquad (14.21)$$

The displacement reaction keeps 18 electrons around the metal center at every step in the hydrogenation process. As a result, reaction (14.21) is favored in most cluster compounds.

In summary, then, olefin hydrogenation goes by a mechanism where the reactants adsorb. Then there is a migration step, where a hydrogen atom is transferred to an ethyl group. Then the products desorb. The reactions are very similar on clusters and supported catalysts. However, there is an important difference in that the desorption processes are often different on a cluster than on a supported metal catalyst.

Hydrogenation reactions are the most important reactions on metals. One uses them at some stage in the production of most industrial chemicals. Therefore, one should memorize the mechanism in detail.

14.3.2 Paraffin Dehydrogenation

Next, we want to discuss the mechanism of the dehydrogenation of paraffins. The dehydrogenation of paraffins to olefins is also quite important industrially. The main reaction is

$$RCH_2CH_3 \Longrightarrow RCH{=}CH_2 + H_2 \tag{14.22}$$

Reaction (14.22) is the opposite of reaction (14.16). It generally follows the mechanism in Figure 14.12, except that all of the reactions go in reverse.

$$RCH_2CH_3 + 2S \longrightarrow RCH_2CH_{2(ad)} + H_{(ad)} \tag{14.23}$$

$$RCH_2CH_{2(ad)} + S \longrightarrow RCH{=}CH_{2(ad)} + H_{(ad)} \tag{14.24}$$

$$RCH{=}CH_{2(ad)} \longrightarrow RCH{=}CH_2 + S \tag{14.25}$$

$$2H_{(ad)} + S \longrightarrow H_{2(ad)} \tag{14.26}$$

Generally reaction (14.23) is rate-determining. Reaction (14.23) seldom occurs on a cluster. Reaction (14.23) has a significant intrinsic barrier because of the proximity effect discussed in Section 5.12.1. Reaction (14.23) usually occurs only at a significant rate at high temperature. Transition metal clusters are unstable at the conditions where reaction (14.23) occurs. As a result, no one has found a cluster with significant activity for reaction (14.22).

14.3.3 Isomerization

Hydrocarbons can also isomerize in metal surfaces. Isomerization reactions are not characteristic of reactions on metals. For example, neopentane isomerizes only on platinum and iridium. Larger molecules can isomerize, though. I wanted to briefly review the mechanism of the isomerization process.

Figure 14.14 shows the mechanism of 3-methylhexane isomerization on platinum or nickel. During the reaction, the 3-methylhexane adsorbs. Then the adsorbed species transfers three hydrogens to the metal, yielding a triply bound species. Next, there is a reductive elimination to yield a ring structure. The ring then undergoes an oxidative addition. Finally, there is a recombination, and the resultant product desorbs.

The process depicted in Figure 14.14 is called a *five-centered isomerization*, because one needs at least five carbon atoms to form the ring; six or more carbon atoms also work. One does not see these kinds of five-centered mechanisms in smaller molecules because

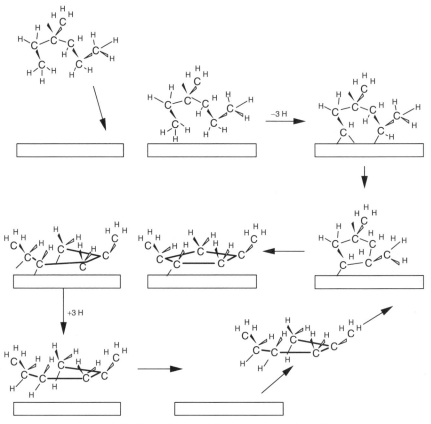

Figure 14.14   The mechanism of the 3-methylhexane isomerization.

there is too much ring strain. Five-centered isomerization processes are very common when moderate-size hydrocarbons adsorb on metal surfaces.

That is not to say that smaller molecules cannot isomerize. It is just that the isomerization processes are much slower, and generally the isomerization process can be observed only on a few metals, usually only platinum, iridium, and alloys containing platinum and iridium. For example, neopentane will isomerize on platinum and iridium, but negligible isomerization is observed on all of the other metals in the periodic table. Figure 14.15 shows one of the proposed mechanisms of neopentane isomerization. The neopentane adsorbs, and loses hydrogens. Then there is a 1,2 shift, where a methyl group migrates from one carbon center to the next. Next, there are a series of recombination steps where the adsorbed species gain hydrogens. Finally, the product desorbs.

In the literature, there is still some uncertainty as to whether the mechanism in Figure 14.15 actually occurs. After all, the 1,2 shift is characteristic of a reaction of an ion, but is not a typical process for a reaction on a metal.

Blowers and Masel have shown that small platinum clusters might be electronegative enough to stabilize carbocations. Once the carbocation forms, the 1,2 shift can occur. This is new (at the time of writing) work, however. It remains to be seen whether this theoretical prediction will be borne out by experiment.

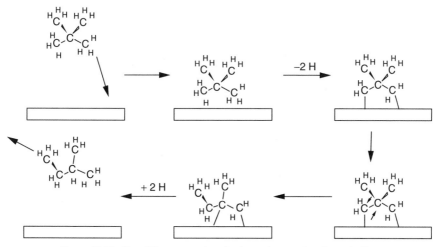

Figure 14.15   One of the proposed mechanisms of neopentane isomerization.

14.3.4   CO Oxidation

Next we want to consider CO oxidation. CO oxidation is an important reaction for automotive pollution control. During the reaction, CO reacts with oxygen to produce CO_2.

$$CO + \tfrac{1}{2}O_2 \Longrightarrow CO_2 \tag{14.27}$$

CO is poisonous, so reaction (14.27) is quite important. Figure 14.16 shows a catalytic cycle for the reaction. The cycle looks much like the cycles described earlier in this chapter. First the CO and the oxygen adsorb

$$O_2 + 2S \longrightarrow 2O_{(ad)} \tag{14.28}$$

$$CO + S \longrightarrow CO_{(ad)} \tag{14.29}$$

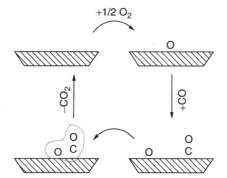

Figure 14.16   The catalytic cycle for CO oxidation.

Then there is a recombinative desorption, yielding the CO product and regenerating the bare site.

$$CO_{(ad)} + O_{(ad)} \longrightarrow CO_2 + 2S \qquad (14.30)$$

In the literature, people also discuss a precursor mechanism for CO oxidation, where the CO reacts before the CO attaches to a surface site. In my view, however, the evidence for such a mechanism is rather weak (at least it was rather weak when this book was being written). Still, the alternative mechanism is widely discussed in the literature.

14.3.5 Partial Oxidation of Ethylene to Ethylene Oxide

Another key reaction is the partial oxidation of ethylene to ethylene oxide, which runs on a silver catalyst. The mechanism is analogous to that for reactions (14.16) and (14.27). First the reactants adsorb:

$$O_2 + 2S \longrightarrow 2O_{(ad)} \qquad (14.31)$$

$$CH_2{=}CH_2 + S \longrightarrow CH_2{=}CH_{2(ad)} \qquad (14.32)$$

Then there is a migration reaction:

$$CH_2{=}CH_{2(ad)} + O_{(ad)} \longrightarrow \underset{CH_2-CH_{2(ad)}}{\overset{O}{\triangle}} \qquad (14.33)$$

Then the product desorbs:

$$\underset{CH_2-CH_{2(ad)}}{\overset{O}{\triangle}} \longrightarrow \underset{CH_2-CH_2}{\overset{O}{\triangle}} \qquad (14.34)$$

These steps are analogous to those in the previous section.

There is one important detail, though. The reaction occurs on silver, but not on platinum. On platinum one instead observes complete oxidation of the complex to CO_2, and H_2O:

$$CH_2CH_2 + 3O_2 \Longrightarrow 2CO_2 + 2H_2O \qquad (14.35)$$

The reason is that ethylene dissociates before ethylene oxide desorbs from a platinum surface.

14.3.6 Other Partial Oxidation Reactions of Hydrocarbons

Catalysts are used to catalyze a large series of partial oxidation reactions of hydrocarbons. Most other partial oxidation reactions follow the mechanism described in Section 14.3.7. First, oxygen adsorbs. Then the species being oxidized reacts with the oxygen. Then products desorb.

There is another detail which is important. Most partial oxidation reactions are run on metal oxides, not pure metals. In a metal oxide, you have what is called **lattice oxygen**: an oxygen that is part of the bulk oxide. During reaction, the lattice oxygen reacts with the hydrocarbon to form a new partially oxidized product. The lattice is strongly bound to the metal. Consequently, according to Sabatier's principle, the lattice oxygen is less reactive than is chemisorbed oxygen. That is good for partial oxidation because the less reactive reaction reacts less; it only partially oxidizes the hydrocarbon.

Another key idea in partial oxidation chemistry is that you do not want the reactant to bind too strongly to the catalyst. If things bind too strongly, the partially oxidized species cannot desorb. In that case, one usually observes total oxidation of the hydrocarbons.

14.3.7 Hydroformylation

There is one other key reaction that I want to discuss: a reaction called **hydroformylation**. During hydroformylation you make aldehydes from olefins and CO via the reaction

$$CO + RCH=CH_2 + H_2 \Longrightarrow RCH_2CH_2CHO \tag{14.36}$$

Figure 14.17 shows a picture of the catalytic cycle for the hydroformylation.

The cycle starts with a rhodium hydride catalyst with an empty site. The olefin adsorbs onto the catalyst:

$$RCH=CH_2 + S \longrightarrow RCH=CH_{2(ad)} \tag{14.37}$$

Then there is a hydride migration driven by adsorption of CO:

$$CO + RCH=CH_{2(ad)} + H_{(ad)} \longrightarrow RCH_2CH_{2(ad)} + CO_{(ad)} \tag{14.38}$$

Next, the alkyl group migrates to the CO:

$$RCH_2CH_{2(ad)} + CO_{(ad)} \longrightarrow RCH_2CH_2CO_{(ad)} + S \tag{14.39}$$

Then a H_2 comes in and displaces the $RCH_2CH_2CO_{(ad)}$ group:

$$RCH_2CH_2CO_{(ad)} + H_2 \longrightarrow RCH_2CH_2COH + H_{(ad)} \tag{14.40}$$

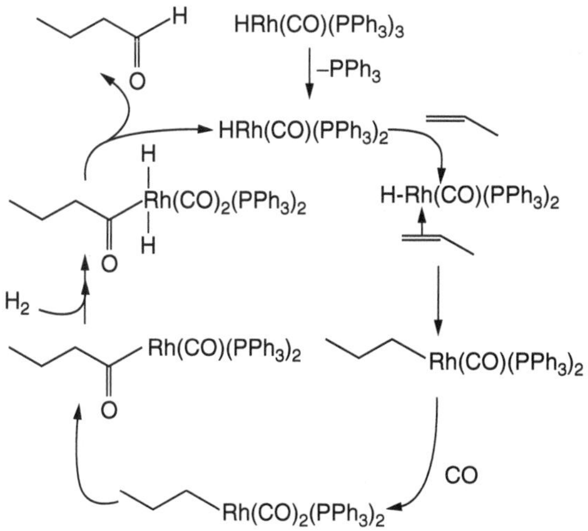

Figure 14.17   The catalytic cycle for hydroformylation over a rhodium hydride cluster.

In the literature, people say that reaction (14.40) is not a true displacement reaction because the hydrogen ends up on a different site than the original $RCH_2CH_2CO_{(ad)}$ group.

14.3.8 Other Examples

There are many other examples in the literature. I decided not to discuss NO reduction $(2NO + 2CO \Rightarrow N_2 + 2CO_2)$ and ammonia synthesis $(N_2 + 3H_2 \Rightarrow 2\ NH_3)$ because I wanted to use them in the problem sets. Basically all of the mechanisms look the same. Reactants approach the surface and adsorb, the adsorbed species react, and then the products desorb. Hydrogen, oxygen, nitrogen, and paraffins dissociatively adsorb, while CO and most unsaturated hydrocarbons adsorb molecularly. Then there are a series of rearrangements where atoms are transferred to or from the surface one at a time to yield products. Then the products desorb. All of the mechanisms look basically the same. As I said previously, the catalytic cycle consists of adsorption–reaction–desorption. A key feature of metal catalysis is that the metal usually directly participates in every step of the reaction. The participation of the metal makes metal catalysis unique.

14.4 PRINCIPLES OF CATALYTIC ACTION

At this point, I again want to change topics. So far in this chapter we have discussed reactions on metals, but have not explained how metals catalyze reactions. In this section we will discuss why transition metals are effective catalysts.

The role of metals in catalyzing reactions is a vast subject, and I cannot mention everything that is known. Still, I want to summarize some of the key ideas. Specifically, we will discuss how

- Metals can help initiate reactions
- Metals can stabilize the intermediates of a reaction
- Metals can hold the reactants in close proximity and in the right configuration to react
- Metals can stretch bonds and otherwise make bonds easier to break
- Metals can donate and accept electrons

Metals also can be designed to block side reactions and to act as an efficient means for energy transfer.

In the next several sections we will go over each of these effects one at a time and try to provide examples of how catalysts work.

14.4.1 Metals Initiate Reactions

First let us start by reviewing how metals help to initiate reactions. Let us consider the following example:

$$CH_2{=}CH_2 + H_2 \Longrightarrow CH_3CH_3 \tag{14.41}$$

Imagine running reaction (14.41) in the gas phase. Following the discussion in Chapter 5, we note that reaction (14.41) could occur via an initiation–propagation reaction. The initiation step will be

$$H_2 \longrightarrow 2H \tag{14.42}$$

The propagation steps are

$$CH_2CH_2 + H \longrightarrow CH_3CH_2 \qquad (14.43)$$

$$CH_3CH_2 + H_2 \longrightarrow CH_3CH_3 + H \qquad (14.44)$$

The termination steps is

$$2H \longrightarrow H_2 \qquad (14.45)$$

Notice that reaction (14.41) is unlikely to occur in the gas phase. Reaction (14.41) is 104 kcal/mol endothermic; by the time one gets to a high enough temperature for reaction (14.41) to occur, the ethylene polymerizes or pyrolyzes.

On a metal, though, the set of reactions

$$H_2 + 2S \rightleftharpoons 2H_{(ad)} \qquad (14.46)$$

$$CH_2=CH_2 + S \longrightarrow CH_2=CH_{2(ad)} \qquad (14.47)$$

$$CH_2CH_{2(ad)} + H_{(ad)} \longrightarrow CH_3CH_{2(ad)} + S \qquad (14.48)$$

$$CH_3CH_{2(ad)} + H_2 \longrightarrow CH_3CH_3 + H_{(ad)} \qquad (14.49)$$

is quite feasible. For example, on platinum, reaction (14.46) is 13 kcal/mol exothermic. The intrinsic barrier is 10 kcal/mol. As a result, while we need to go to high temperatures to get reaction (14.42) to occur, reaction (14.46) occurs at a reasonable rate. Experimentally reaction (14.46) occurs readily on platinum at 100 K.

This brings up one key reason that metals speed up reactions. In the gas phase, initiation reactions are very slow. The rate of reaction is limited to the rate at which radicals form via the initiation process. The slow formation of radicals limits the rate of gas-phase reactions. On a metal, though, this limitation does not apply. In Section 5.12 we noted that gases such as H_2, and O_2 will rapidly dissociate on most transition metals. N_2 will dissociate on metals near the middle of the periodic Table (e.g., iron, rhodium). The ability of metals to rapidly dissociate molecules allows the initiation reactions to occur rapidly. In Chapter 10 we showed that when you have rapid initiation reaction, the rate of the overall reaction is also usually greatly enhanced.

One key reason why metals are such effective catalysts is that they initiate reactions that would not occur at reasonable temperatures in the gas phase.

14.4.2 Metals Stabilize Intermediates

Next, we want to discuss the role of metals in stabilizing radical intermolecules. These ideas were discussed in Chapter 12. Recall that in the gas phase, radicals are rather unstable species. Consequently, the radical concentration in the gas phase is always low. In contrast, radicals can bind to a surface. It is not unusual for the radical concentration to increase by a factor of 10^{20}, thus producing a tremendous increase in rate.

For example, during reaction (14.42) one produces hydrogen atoms. Hydrogen atoms are rather unstable species. They have a heat of formation of 52 kcal/mol. Therefore, the reaction is slow. In contrast, on platinum reaction (14.46) is 13 kcal/mol exothermic [i.e., the heat of formation of H_{ad} is (3 kcal/mol)/2 = 6.5 kcal/mol]. Hydrogen atoms are 58.5 kcal/mol more stable on the surface than in the gas phase. As a result, at 300 K reaction (14.41) is a factor of 10^{16} faster on the surface than in the gas phase. Clearly, this is a significant effect.

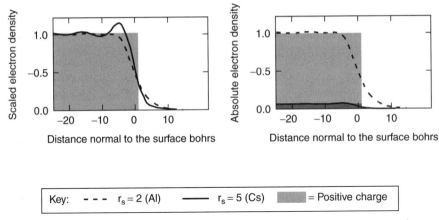

Key: - - - $r_s = 2$ (Al) ——— $r_s = 5$ (Cs) ▓▓▓ = Positive charge

Figure 14.18 The electron density extending out from a metal surface. (Note 1 bohr = 0.52 Å).

Now, to really understand reactions on metals you have to recognize that generally, metal surfaces stabilize radical intermediates and not ionic intermediates. Recall that metals conduct electricity because there are lots of free electrons in the metal. In bulk, the electrons are confined to the metal. However, on a surface, the electrons can spill out into the gas phase (see Figure 14.18). The result is that there is a ready source of electrons available above a surface.

Those electrons can stabilize radicals. Years ago people talked about the surface forming a local covalent bond to the radicals. Although we know that such a picture is too simplistic, there is a weak bond between the radical and the metal. The bond lowers the energy of the radical so that the radical is stable on the surface.

Metal surfaces can also occasionally form a bond to an ionic intermediate. However, ionic intermediates are much less common on metals than are radical intermediates.

Sections 12.6 and 12.7 discussed how stabilization of intermediates leads to enhanced reaction. Generally, rates of simple reactions are proportional to the concentration of the intermediates; as a result, increases in the intermediate concentration increases the rate of reaction. Of course, you can get too much of a good thing. In Section 12.7 we showed that if you stabilize the intermediates too much, the intermediates will become unreactive. Still, there is the general effect that increases in intermediate concentrations lead to increases in rate. The fact that metals can increase the intermediate concentrations can lead to enhanced rates.

14.4.3 The Role of d Bands in Lowering the Intrinsic Barriers to Reaction

The stabilization of intermediates is not sufficient to explain why platinum is such a good catalyst and, for example, aluminum is not. Platinum and aluminum will both stabilize radical intermediates. However, aluminum is not a good catalyst for hydrocarbon reactions even though platinum is a wonderful catalyst.

The reason why platinum is a better catalyst than aluminum is that platinum can dissociate H_2 or O_2 while aluminum cannot. In Section 14.4.1 we noted that the reaction

$$H_2 + 2S \longrightarrow 2H_{(ad)} \tag{14.50}$$

occurs readily at 100 K on platinum. Surprisingly, though, reaction (14.50) does not occur at a reasonable rate on aluminum or silicon at 800 K, even though the reaction is 54 kcal/mol exothermic on silicon and 74 kcal/mol exothermic on aluminum. No reaction is seen on potassium and magnesium, even though the reaction is more than 70 kcal/mol exothermic. Yet, reaction (14.50) occurs readily at 100 K on platinum even though the reaction is only 13 kcal/mol exothermic.

Masel (1996) estimated the intrinsic barriers on all the surfaces and found that the intrinsic barrier for reaction (14.50) is at least 80 kcal/mol on silicon and more than 90 kcal/mol on potassium and aluminum. By comparison, the intrinsic barrier is only 13 kcal/mol on platinum. Clearly, something is different on platinum than on silicon, potassium, and aluminum that allows reaction (14.50) to occur.

In order to understand this effect, we first have to consider why reaction (14.50) has a high activation barrier on silicon. Recall that in Chapter 10 we found that four-centered reactions are symmetry-forbidden. The dissociative adsorption of H_2 can be written

$$H_2 + 2S \longrightarrow \begin{array}{c} H\text{-}H \\ | \quad | \\ S\text{-}S \end{array} \longrightarrow \begin{array}{c} H \quad H \\ | \quad | \\ S\text{-}S \end{array} \qquad (14.51)$$

Notice that reaction (14.51) is analogous to the four-center reactions discussed in Section 10.12.1. In Section 10.12.1 we noted that in most cases in the gas phase a four-center reaction is symmetry-forbidden.

A similar effect occurs on silicon. Reaction (14.50) is symmetry-forbidden on silicon, so the reaction does not occur at a reasonable rate even though the reaction is 54 kcal/mol exothermic on silicon.

The arguments are more subtle on metals because the electrons in metals are held in bands. However, Masel (1996) has shown that H_2 dissociation is symmetry-forbidden on those metals that interact weakly with the antibonding orbitals in H_2 and symmetry-allowed on metals that interact strongly with the antibonding orbitals in H_2. The results in Section 10.11 imply that a symmetry-allowed reaction will have a much lower intrinsic barrier than will a symmetry forbidden reaction. Thus, the interactions with the antibonding orbitals play a key role in determining the intrinsic activation barriers for a reaction.

The first issue we want to address is why we care about antibonding orbitals during H_2 dissociation. Well, quantum-mechanically, antibonding orbitals are critical to bond scission. Recall that quantum-mechanically, a bond in a molecule can break in one of two ways. Either bonding electrons can be removed from the molecule, or electrons can be shared with the antibonding orbitals into the molecule. If one would remove all of the bonding orbitals in a H_2 molecule, one would end up with two positively charged ions (i.e., two H^+ atoms). That is not thermodynamically possible. As a result, one cannot break the H–H bond by simply removing all of the electrons from the H_2. Rather, the H atoms will still have electrons even after the H_2 dissociates. If all of those electrons are in the H_2 bond, the H_2 will still have a partial bond. It is only when one puts electrons into antibonding orbitals that the bond order will go to zero. Quantum-mechanically, the bond order is the number of bonding electrons minus the number of antibonding electrons. Therefore, if one can put electrons into antibonding orbitals, one can break the H_2 bond. Consequently, the transfer of electrons into antibonding orbitals is a key step in H_2 bond scission. No H_2 bond scission can occur unless electrons are transferred into the antibonding orbitals in the H_2.

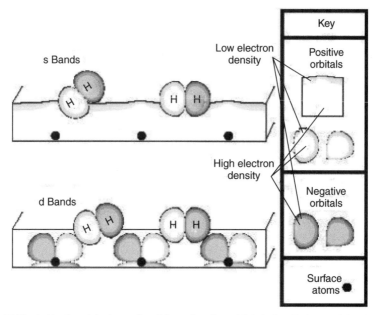

Figure 14.19   A side view of the interaction of the antibonding orbitals in H_2 with the s and -bands at the Γ point in Pt(100).

There are two kinds of orbitals in metals: s orbitals and d orbitals. Both orbitals join together to form bands. Figure 14.19 shows a side view of the orbital. In Chapter 8 and 10 we found that orbitals of the same sign have an attractive interaction while orbitals of different signs have a repulsive interaction. Well, the s band in platinum has a strong interaction with the σ^* orbital in the hydrogen when the H_2 is standing up on the surface. However, when the hydrogen bends over, a repulsive interaction comes in, which prevents dissociation. As a result, hydrogen dissociation is symmetry-forbidden on what are called *s-band metals*: metals without any d electrons available for bonding.

Everything changes when d electrons are present. Notice that the d lobes have an attractive interaction with the σ^* orbital in the hydrogen. That promotes easy bond scission.

One also has to consider the correlation diagram during the reaction. Figure 12.20 shows a correlation diagram for reaction (14.51). Notice that the d electrons change a symmetry-forbidden reaction into a symmetry-allowed process. The results in Section 10.11 show that rates are tremendously enhanced under those circumstances.

A similar mechanism occurs during most bond scission reactions. Generally d bands in metals are properly configured to interact with the antibonding orbitals in adsorbates. That weakens the bonds, and breaks the symmetry constraints. H_2, O_2, and other similar molecules dissociate readily on transition metals but do not dissociate on, for example, alumina or potassium. That makes transition metals good catalysts.

The transition metals also help to catalyze bond formation reactions. That is a little more subtle, but recall that at the transition state in a bond formation reaction, bonds in the reactants are extended. When you stabilize antibonding orbitals, you make it easier to stretch bonds. That lowers the energy of the extended bond, which, in turn, lowers the barrier of the bond scission reaction. Overall, d electrons facilitate bond scission, which is why transition metals are such great catalysts.

14.4.4 The Role of the s Bands in Lowering the Intrinsic Barriers to Reaction

Interactions with the s bands also affect the catalytic activity of metals. Recall that according to the results in Chapter 7, Pauli repulsions play a key role in chemical reactions. When you have a molecule, the Pauli repulsions tend to be large because there is no place for the electrons to go. However, in the presence of a metal, the electrons can in principle flow into the metal. As a result, in principle, the Pauli repulsions should be reduced.

At present it is unclear when this is a significant effect. Measurements indicate that the intrinsic barriers for many hydrogen transfer reactions on platinum are similar to those in the gas phase. Therefore, the electron transfer process does not appear to be important on platinum. No one knows about other metals, however.

14.4.5 Redox Chemistry

There is a related effect called *redox chemistry*. In redox (reduction–oxidation) chemistry the charge on the metal changes during reaction. For example, consider the oxidation of toluene to benzyl aldehyde:

$$C_6H_5CH_3 + O_2 \Longrightarrow C_6H_5\overset{H}{C}{=}O + H_2O \tag{14.52}$$

The reaction can be catalyzed by cobalt ions. It is thought that the reaction proceeds via the following series of steps:

$$Co^{3+} + C_6H_5CH_3 \longrightarrow Co^{2+} + H^+ + C_6H_5\overset{H}{\underset{H}{C}}\bullet \tag{14.53}$$

$$C_6H_5\overset{H}{\underset{H}{C}}\bullet + O_2 \longrightarrow C_6H_5\overset{H}{\underset{H}{C}}OO\bullet \tag{14.54}$$

$$C_6H_5\overset{H}{\underset{H}{C}}OO\bullet + Co^{2+} \longrightarrow C_6H_5\overset{H}{C}{=}O + Co^{3+} + OH^- \tag{14.55}$$

$$OH^- + H^+ \longrightarrow H_2O \tag{14.56}$$

The key thing about redox chemistry is that the metal can have two different oxidation states, the metal atom can take up an electron and thereby eliminate part of the Pauli repulsions, the metal can then give back the electron to yield the products. The transfer of electrons is also very characteristic of metal-catalyzed reactions, although many reactions occur without transforming electrons.

14.4.6 Metals Hold the Reactants in Close Proximity

Another way that metals catalyze reactions is to hold the reactants in close proximity and in the right configuration to react. The key ideas were discussed in Sections 12.8 and 12.9, but I did want to mention the ideas again because they are important.

In Chapter 12 we noted that one of the key ways that metals work is to hold reactants in the correct configuration to react. For example, consider the following reaction:

$$3HC \equiv CH \Longrightarrow C_6H_6 \qquad (14.58)$$

In order to form benzene, you need to first form a hexagonal intermediate. If you adsorb acetylene onto a hexagonal surface of palladium, you will get hexagonal intermediates. Reaction (14.58) is fast on the hexagonal faces of palladium. In contrast, it is much harder to form the hexagonal intermediates on the square faces of palladium. Much less reaction is seen on the square faces. This example shows that holding the reactants in the right configuration to react is quite important to catalysis.

Amazingly, in 1929 Balandin (1929a,b) proposed a model to understand how the surface configuration affects rates of the reaction. Balandin's idea is that surface reactions involve a series of bond scission and bond production processes. Balandin proposed that at each stage in the reaction, the surface must be configured to hold each of the reactive fragments. As a result, rates of reaction will be highest when the arrangement of atoms is correct.

Balandin (1958, 1969) defined what he called a **multiplet** for a reaction. The multiplet is a group of surface atoms that are properly configured to hold the reactive fragments formed during the reaction. For example, on oxide catalysts, ethanol can react via two reaction pathways: a dehydrogenation pathway to form acetaldehyde

$$CH_3CH_2OH \Longrightarrow CH_3CHO + H_2 \qquad (14.59)$$

and a dehydration pathway to form ethylene:

$$CH_3CH_2OH \Longrightarrow CH_2CH_2 + H_2O \qquad (14.60)$$

Figure 14.20 shows a diagram of the multiplet for these two reactions from Balandin (1929a,b). Balandin (1929a,b) proposed that reaction (14.59) requires one to have two catalytic centers, one to couple the two hydrogens and one to hold the aldehyde fragments. Similarly, reaction (14.60) requires that there to be two different reaction centers: one to couple the water and one to hold the olefin products. Balandin noted that one might need quite a different site to form water than to couple two hydrogens. Olefins would bind

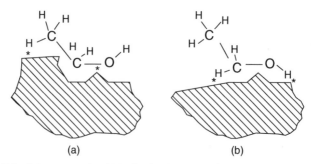

(a)   (b)

Figure 14.20 Balandin's suggested multiplet for the decomposition of ethanol (a) to form ethylene and (b) to form acetaldehyde. The asterisks in the figures represent places on the surface where reaction can occur.

differently than aldehydes. Thus, one would need a different site to produce ethylene than to produce acetaldehyde.

Balandin expanded his model by assuming that all of the elementary steps in a catalytic reaction occur on the same site. As a result, he suggested that only a site that is configured to do all of the steps in a catalytic reaction would be catalytically active. Balandin noted that, in particular, bond lengths are critical. Thus, one might need a special configuration of atoms to get a given reaction to occur.

Since Balandin's initial work, it has become clear that although his ideas have some qualitative significance, quantitatively they seldom work so well. For example, it is now known that the dehydrogenation of ethanol follows the mechanism in Figure 5.15, where the molecule sequentially loses hydrogens. Hence, the active site does not have to perform a number of tasks simultaneously. However, the idea that, in order for a metal catalyst to promote a given elementary reaction, the catalyst must contain sites that are properly configured to hold the reactants and products of the reaction does apply to most catalytic reactions.

14.4.7 Structure-Sensitive Reactions

The multiplet model is a subset of a larger field called **structure-sensitive reactions**. The idea in structure-sensitive reactions is that the rate of a given elementary surface reaction can depend on the structure of the surface. The catalyst must contain sites that are properly configured to hold the reactants and products of the reaction. The catalyst must also be configured to properly stabilize the transition state of the reaction. Structure-sensitive reaction arise because different faces of transition metals have different capabilities for binding the reactants, products, and transition state. The effects can be quite substantial, although they are not yet well understood.

The idea that reaction rates were structure-sensitive goes back to the 1925 measurements of Pease and Stewart. They found that if they adsorbed a small amount of mercury onto a copper catalyst, the rate of adsorption of hydrogen went down by a factor of 200 even though the equilibrium constant for the adsorption changed by only one order of magnitude. If the adsorption process were occurring uniformly over the surface of the copper, the decrease in rate would be the same as the decrease in the amount adsorbed. However, it was not. As a result, Pease and Stewart (1925) proposed that adsorption was not occurring uniformly over the solid surface. Rather, it was occurring on a series of distinct sites on the surface of the copper.

Taylor (1925), (1948) expanded these ideas substantially. Taylor's idea was that if one needed special sites for adsorption, one would also need special sites for reaction. Taylor showed experimentally that it was possible to poison a reaction without poisoning the adsorption process. As a result, Taylor suggested that most surface reactions occur only on special sites that he termed **active sites**.

The idea that there are active sites implies that reaction rates vary with surface structure. If the rate of a reaction were independent of structure, then, by definition no one site would be more able to catalyze a reaction than any other. As a result, when one says that there are active sites, one is also saying that the rate of reaction varies with surface structure.

In 1969 Boudart proposed that one could classify reactions into two general classes: **structure-sensitive reactions** and **structure-insensitive reactions**. In Boudart's definitions, structure-sensitive reactions are those whose rates vary with surface structure and structure-insensitive reactions are reactions whose rates do not depend on surface structure.

I am not aware of any reaction that is structure-insensitive under all conditions. Some reactions (e.g., ammonia synthesis from nitrogen and hydrogen) are structure-sensitive under most conditions. Other reactions (e.g., ethylene hydrogenation) are structure-sensitive only under special conditions. However, I am not aware of any truly structure-insensitive reaction. Hence, I do not believe that it is that useful to classify reactions as being completely structure-sensitive or structure-insensitive.

A much better classification is to note that some reactions display structure sensitivity under a wider range of conditions than do others. Under this classification, those reactions that show structure sensitivity under a wide range of conditions are called structure-sensitive.

One of the earliest ways to see if a reaction is structure-sensitive was to run the reaction on a supported catalyst, and see if the reaction rate varied with a particle size. Experimentally, the distribution of sites in a supported catalyst varies with particle size. Hence, if one observes a variation in rate with particle size, one knows that a reaction is structure-sensitive.

For example, Figure 14.21 shows how the rate of the reaction

$$N_2 + 3H_2 \longrightarrow 2NH_3 \tag{14.61}$$

over an iron catalyst varies with particle size. Note that the specific rate (i.e., the rate per unit surface area) varies by four orders of magnitude with particle size. As a result, Boudart et al. (1975) concluded that reaction (14.61) was structure-sensitive.

There is a danger with this type of measurement, however. Although one might not directly observe a variation in rate with particle size, one cannot state unequivocally that the reaction is structure-insensitive. The distribution of sites in a supported catalyst varies

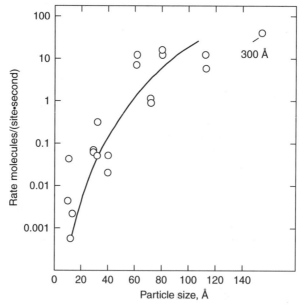

Figure 14.21 The rate of the reaction $N_2 + 3H_2 \rightarrow 2NH_3$ over an iron catalyst as a function of size of the iron particles in the catalyst. [Data of Boudart et al. (1975).]

with both the size of the catalyst particles and how the catalyst is processed. As a result, it is not obvious that if one simply varies the particle size, one samples all possible surface structures.

Another effect is that when one varies the particle size, one does not vary the distribution of all of the sites in the same way. For example, the concentration of kinks may change significantly. However, the concentration of (111) sites may change by only a small amount.

The effect of these uncertainties is that different investigators often come to different conclusions about whether a reaction is structure-sensitive. For example, Bond (1956) and Sinfelt and Lucchesi (1963) found that the rate of ethylene hydrogenation varied with particle size while Dorling et al. (1969) and Schlatter and Boudart (1972) found that it did not. For years it was thought that Bond's and Sinfelt's measurements were incorrect. However, another possibility was that Bond's catalysts and Sinfelt and Lucchesi's catalysts had some surface structures different from those of the catalysts used by Dorling et al. Recently, Backman and Masel (1988) found that the rate of ethylene hydrogenation is a factor of about 8 higher on Pt(111) than on (5×20)Pt(100). Hence, it is quite possible that the different investigators came to different conclusions about the structure sensitivity of ethylene hydrogenation because the catalysts used by these various investigators had a different distribution of sites.

Another way to look for structure sensitivity is to run reactions on a variety of faces of single crystals and look for a variation in rate with crystal face. For example, Figure 14.22 shows how the rate of nitric oxide decomposition varies as a function of crystal face for NO dissociation on the faces of platinum along the principal zone axes of the stereographic triangle. This reaction shows the largest variation in rate with crystal face observed so far. On Pt(410) Park et al. (1985) found that NO dissociatively adsorbed at temperatures above 150 K, and molecularly adsorbed below that level. On Pt(310) Sugai (1993) found that the NO dissociated at room temperature (his minimum temperature). In contrast, negligible dissociation is detected on Pt(100) below 400 K, while NO dissociation does not start on Pt(111) until the sample is heated above 1000 K. Analysis of this data indicates that the

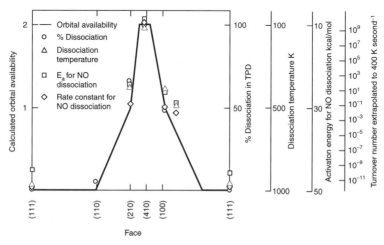

Figure 14.22 The rate of nitric oxide dissociation on several of the faces of platinum along the principle zone axes of the stereographic triangle. [Adapted from Masel (1986).]

activation energy for N–O dissociation varies from 6 kcal/mol on Pt(410) to 45 kcal/mol on Pt(111). This is equivalent to a 10^{21} variation in the rate with crystal face at 450 K.

The NO example shows an especially large variation in rate with crystal face. However, several other examples show two or three orders of magnitude variation in rate with crystal face. Key reactions in ammonia synthesis, ethane hydrogenolysis, C–O bond scission in methanol, and C–O bond scission in carbon monoxide have each been observed to show several orders of magnitude variation in rate with crystal face. Hence, the NO results are not an isolated case. Rather, structure sensitive reactions are fairly common.

The one major problem with doing measurements like those in Figure 14.22 is that unless one is lucky or has a theoretical model, it is easy to miss some important chemistry. There are many examples that so far have shown only less than half an order of magnitude variation in rate with crystal face. However, it is unclear whether these examples really do show small variations in rate with crystal face or whether the measurements have not yet been run on the correct surface geometries. Note, for example, that NO decomposition shows much higher rates on Pt(310) and Pt(410) than on any other face. Pt(310) and Pt(410) are not special in any other way. Hence, it would have been easy to miss the unusual activity of Pt(310) and Pt(410).

Another effect is that it is unclear that one can access all possible surface geometries with single crystals. Work of Sachdev et al. (1992) has shown that small particles of metals have surface sites that are not present in bulk materials. Thomas (1976) has shown that irregular structures such as dislocations can have an important influence on the reactivity of a surface. Hence, it is unclear whether one can necessarily produce the particular sites that were especially active for reactions by cutting single crystals. Therefore, if one observes a large variation in rate with crystal, face on a single crystal, one knows that the reaction is structure-sensitive. However, one cannot say the opposite. A reaction that does not show large variation in rate with crystal face may or may not be structure-insensitive.

I find it useful to classify reactions not by whether they are structure sensitive or insensitive, but rather to note that some reactions show more structure sensitivity than others. Table 14.3 shows some typical results. Generally reactions where carbon–hydrogen bond production or scission is rate-determining (i.e., the slow step in the reaction) are less structure-sensitive than are reactions where the scission of carbon–carbon, carbon–oxygen, or carbon–nitrogen bonds is rate-determining. The rate of scission of double or triple bonds is more structure-sensitive than that of single-bond scission. Oxidation reactions show inconsistent trends. For example, Engle and Ertl (1979) found that at 400 K the rate of CO oxidation on palladium varies by a factor of 6 with crystal face.

Table 14.3 The structure sensitivity of a series of reactions

Reaction	Largest Variation in Rate with Geometry Observed Prior to 1999
$2CO + O_2 \rightarrow 2CO_2$	6
$C_2H_4 + H_2 \rightarrow C_2H_6$	12
$CH_3OH \rightarrow CH_{2(ad)} + H_2O$	>100
$C_2H_6 + H_2 \rightarrow 2CH_4$	10^4
$N_2 + 3H_2 \rightarrow 2NH_3$	10^5
$2NO + 2H_2 \rightarrow N_2 + 2H_2O$	$\sim 10^{21}$

However, at 500 K the rate of the reaction was almost independent of crystal face. In later data, Sander et al. (1992) found that under the same conditions, the reaction rate oscillates on Pt(100), Pt(110), and Pt(210) but not on Pt(111).

Part of the reason for the inconsistent trends seen with oxidation reactions is that most oxidation reactions can go via either a precursor mechanism or a Langmuir–Hinshelwood mechanism. Generally, precursor mechanisms and Rideal–Eley mechanisms tend to be much less structure-sensitive than Langmuir–Hinshelwood mechanisms. As a result, if one runs a reaction under conditions where the precursor mechanism predominates, one will see little structure sensitivity even though the same reaction shows structure sensitivity under conditions where the Langmuir–Hinshelwood reaction prevails.

There is another complication, as well; reactions may be structure-sensitive for reasons that have nothing to do with the main reaction pathway. For example, a few paragraphs ago we discussed ethylene hydrogenation and noted that Backman and Masel (1988) found that ethylene hydrogenation showed a higher rate on Pt(111) than on (5×20)Pt(100). Note, however, that during ethylene hydrogenation a carbonaceous deposit builds up on the platinum surface. This carbonaceous deposit blocks the surface to further reaction. On Pt(111) the carbonaceous overlayer has a fairly open structure. There are many empty metal sites available to catalyze the reaction. However, on (5×20)Pt(100) the surface is more nearly covered by the carbonaceous fragments. As a result, one reason why (5×20)Pt(100) surface is less active than the Pt(111) surface is that under reaction conditions there are fewer metal sites exposed on the (5×20)Pt(100) surface than on the Pt(111) surface. Backman and Masel found that they could account for all of their measured structure sensitivity by keeping track of how many sites were exposed. Thus, even though Backman and Masel found that the overall rate of ethylene hydrogenation was structure-sensitive, they did not find that the individual steps in the main reaction pathway were structure-sensitive. Rather, they concluded that the reaction was structure-sensitive because during the reaction the number of exposed metal sites was different on Pt(111) and (5×20)Pt(110).

This example illustrates an important point. Surface reactions are complicated. Sometimes one can find a reaction that would not be structure-sensitive under idealized conditions. However, the reaction can still appear to be structure-sensitive because of some secondary processes in the reaction mechanism leading to site blockage. Manogue and Katzer (1974) call this **secondary structure sensitivity**. Burwell (1967) and Butt and Onal (1987) have documented many examples of secondary structure sensitivity.

Structure sensitivity is discussed in much greater detail in Chapters 6 and 10 in Masel (1996). One should refer to that book for further details.

14.4.8 Summary of Modes of Catalytic Action

In summary, then, in this section we found that

- Metals can help initiate reactions — in particular, they facilitate bond scission processes.
- Metals can stabilize the intermediates of a reaction, particularly radical intermediates,
- Metals can lower the intrinsic barriers to bond scission. The d electrons promote bond scission and bond formation. The s electrons promote redox chemistry.
- Metals can hold the reactants in close proximity and in the right configuration to react.

• Metals can donate and accept electrons and thereby modify the intrinsic barriers for reaction.

We did not discuss the following points, but

• Metals can be designed to block side reactions.
• Metals can act as an efficient means for energy transfer.

In Chapter 12 we noted that rates of reactions can go up tremendously in the presence of a metal because of these effects.

14.5   TRENDS OVER THE PERIODIC TABLE

At this point we will again change topics. In the last several sections we have discussed why rates of reactions are enhanced by metals. So far the discussion would make it seem that all metals behave the same. All transition metals stabilize radical intermediates. All transition metals lower intrinsic barriers. All transition metals can do redox chemistry.

In fact, the differences between the various metals are rather small. All of the transition metals catalyze ethylene hydrogenation and CO oxidation. The mechanisms are all the same. There are some details that are different. However, the big picture is that most group VIII metals have similar catalytic properties.

Still, the metals do not have identical properties. For example, in Section 14.4.1 we noted that the rate of ethylene hydrogenation goes up by a factor of 10^{16} on platinum. It goes up by a factor of 10^{17} on iridium but a factor of only 10^{15} on palladium. These may seem to be small differences, but a factor of 10 in rate is very important industrially.

In the next few sections we will review those differences. We want to say at the start that the absorbed differences between the transition metals are small. However, the differences are certainly significant industrially.

14.5.1   Copper, Silver, and Gold

We will start with copper, silver, and gold. Copper, silver, and gold catalysts are used for partial oxidation and occasionally for partial hydrogenation.

From a surface chemistry standpoint, the surfaces of copper, silver, and gold are fairly inert. Hydrocarbons, alcohols, amines, or H_2 interact weakly with copper, silver, and gold. Generally, one does not observe any reactions of these species on clean copper, silver, or gold surfaces. Instead, the species adsorb and then desorb again. In contrast, most alcohols, amines, and hydrocarbons dehydrogenate when they adsorb on platinum and other group VIII metals. Thus, copper, silver, and gold can be thought of as being fairly inert surfaces, at least for dehydrogenation.

Copper does react with carboxylic acids. For example, formic acid (HCOOH) will decompose to CO_2 and H_2 on Ag(110). Copper and silver will react with alkyliodides. For example, methyliodide reacts to form methyl groups and iodides. Thiols also react. For example, *tert*-butylthiol $((CH_3)_3CSH)$ reacts to form *tert*-butylthiolate $(CH_3CH_2S-_{ad})$ on Cu(110), Ag(110), and Au(110).

At present, no one has any general guidelines to predict what species will react on copper, silver, and gold. However, quantitatively, copper, silver, and gold will react with species that are easy to decompose. Generally, bonds with dissociation energies less than

70 kcal/mol can often react. However, clean gold, silver, and copper do not react with species with strong bonds (i.e., bond dissociation energies greater than 70 kcal/mol).

O_2 is an exception to these general rules. O_2 will dissociate on Ag(110), Cu(110), and Cu(111) and on gold nano-particles even though the oxygen–oxygen bond strength in O_2 is 119 kcal/mol. Thus, there are no hard-and-fast rules about the species that react on copper and silver.

The fact that O_2 does dissociate makes copper and silver active for partial oxidation. For example, Wachs and Madix (1978) found that when methanol adsorbs on Cu(110), the methanol desorbs again. However, when methanol and oxygen coadsorb on Cu(110), the oxygen can extract hydrogens from the methanol to yield formaldehyde

$$CH_3OH_{(ad)} + O_{(ad)} \longrightarrow H_2CO + H_2O \tag{14.62}$$

The ability of copper and silver to oxidize species leads to some interesting selectivities. For example, methanol can be oxidized to yield adsorbed formaldehyde on platinum. However, formaldehyde decomposes on platinum. As a result, very little formaldehyde desorbs. Copper and silver are unusual because they are able to selectively oxidize methanol to formaldehyde and not significantly decompose the formaldehyde product.

The result of the discussion in the last paragraph is that copper and silver are excellent partial oxidation catalysts. For example, methanol oxidation to formaldehyde is commonly run on a copper or silver catalyst, while the oxidation of ethylene to ethylene oxide is commonly run on silver. Copper supported on ZnO or Cr_2O_3 also shows mild hydrogenation activity, although the activity is absent in copper metal.

From the literature, it is not clear why copper, silver, and gold are so inert. Hammer and Norskov (1995) discuss the issues. Part of the reason is that copper, silver, and gold bind species weakly. For example, Benziger (1991) reports that hydrogen has a binding energy of 10–13 kcal/mol on the various faces of copper, while H_2 barely binds to silver and gold. Copper, silver, and gold do not stabilize the intermediates for hydrogenation and dehydrogenation reactions. They are not effective catalysts for such reactions.

That is not the whole story, however, because more recent data for coadsorption of alkyl ligands and atomic hydrogen suggest that copper and silver are relatively inactive for hydrogenation processes, even when the hydrogenation process is thermodynamically favorable. For example, consider the hydrogenation of a methyl group:

$$H_3C_{(ad)} + H_{(ad)} \longrightarrow CH_4 \tag{14.63}$$

One can use the data in Table 6.5 to estimate the heat of reaction for reaction (14.63) as described in Chapter 6. When one does that, one finds that reaction (14.63) is about 8 kcal/mol exothermic on Cu(111) and 35 kcal/mol exothermic on Ag(111) and only 2 kcal/mol exothermic on Pt(111).

Therefore, on the basis of the Marcus equation, one would expect reaction (14.63) to be much more rapid on copper and silver than on platinum. Surprisingly, though, Zhou and White (1993) and Liu et al. (1992) found that reaction (14.63) occurs rapidly at 200 K on Pt(111). Yet, Zhou et al. (1992) and Chiang and Bent (1992) found that when they coadsorbed methyl groups and hydrogen on copper or silver, the hydrogenation rate was low. The main pathway is coupling methyl groups to yield ethane and ethylene. There is a small amount of CH_4 production, but that is not the main reaction pathway. Xi and Bent (1992) suggest that Rideal–Eley mechanisms predominate. In contrast, Wang and Masel (1991) find that reaction (14.63) is rapid on Pt(110). These results show that copper and silver are ineffective for hydrogenation reaction even when the thermodynamics of the hydrogenation process is extremely favorable. Thus, thermodynamics alone (i.e., the Polanyi relationship) does not explain the relative inactivity of copper and silver for hydrogenation.

At present, it is unclear whether the inability of pure copper, silver, and gold to promote hydrogenation is associated with an orbital symmetry limitation or some other factor. Certainly, the absence of unpaired d electrons partially explains the lack of reactivity. Copper, silver, and gold do not have d electrons available for bonding, so many reactions will be at least partially symmetry-forbidden. However, Rodriguez et al. (1992) found that a copper monolayer on Re(0001) was active for hydrogenation–dehydrogenation, even though the d bands were still filled. There is something in addition to the d bands that makes copper, silver, and gold especially inert to hydrogenation. Presently, we do not know why copper, silver, and gold are so inert although there is the 1995 paper by Hammer and Norskov.

14.5.2 Platinum, Palladium, and Nickel

At this point, we will leave our discussion of gold, silver, and copper, and move on to platinum, palladium, and nickel. Now, when one moves from gold, silver, and copper to platinum, palladium, and nickel, one greatly changes the reactivity. Although gold and silver are virtually inert for hydrogenation–dehydrogenation, platinum, palladium, and nickel are reasonably active for hydrogenation–dehydrogenation. Platinum, palladium, and nickel are used extensively as hydrogenation–dehydrogenation catalysts. They also are useful for total oxidation.

Experimentally, platinum and palladium are the best hydrogenation catalysts in ultrahigh vacuum (UHV). For example, when one coadsorbs ethylene and hydrogen on Pt(111), Pt(110), or Pt(210) at low temperatures, ethane forms on heating. In contrast, when ethylene and hydrogen are coadsorbed on Ni(111) or Rh(111), the hydrogen desorbs before any hydrogenation occurs. Thus, in UHV platinum and palladium are the best hydrogenation catalysts. At higher pressures, nickel and rhodium are very active hydrogenation catalysts, because the hydrogen remains on the surface at higher temperatures.

Platinum, palladium, and nickel are also active dehydrogenation catalysts. For example, ethanol decomposes via the mechanism shown in Figure 5.15, whereas methanol decomposes via the mechanisms shown in Figure 5.14. In all cases one observes sequential dehydrogenation of the adsorbed species. Alkanes, however, are fairly inert on platinum, palladium, and nickel, in UHV, because of the proximity effect discussed in Chapters 5 and 10. Alkanes will dehydrogenate at higher pressures or in molecular beams.

Finally, platinum, palladium, and nickel are all effective total oxidation catalysts. Oxygen dissociates between 50 and 300 K on all three metals. The adsorbed oxygen is a strong nucleophile that reacts with hydrocarbons, alcohols, and other compounds to yield partially oxygenated species on the surface. However, unlike the gold, silver, and copper cases discussed above, partially oxidized species rarely desorb from the surfaces of platinum, palladium, or nickel. Instead, partially oxygenated species further react (and decompose) to eventually yield CO, CO_2, and water.

In the literature there has been little discussion of why platinum, palladium, and nickel are so much more reactive than gold, silver, and copper. People often cite thermodynamic effects. However, in fact, the data in Table 6.5 show that the strengths of the metal–carbon, metal–nitrogen, metal–oxygen and metal–hydrogen bonds are very similar on platinum, palladium, and copper (and rhodium and iridium). Yet, copper is much less reactive than platinum or palladium for hydrogenation and dehydrogenation. Thus, thermodynamic effects cannot explain why copper is so much less active than platinum for hydrogenation–dehydrogenation.

In my view, the main difference between platinum, palladium, and copper is associated with the d bands. The d-bands have electrons of the right symmetry to stabilize the transition state for C–H bond scission and bond formation processes. As a result, one would expect platinum and palladium to be much more active than copper for C–H bond scission and bond formation processes as is observed. One can account for the differences between platinum and copper based just on interactions with the d bands. Therefore, I believe that the differences in the d bands are causing copper to be much less active than platinum for C–H bond scission. Still, this view is not yet universally accepted in the literature, so the reader needs to treat it with caution.

In the discussion above we have treated platinum, palladium, and nickel as though they were identical. However, experimentally there are some differences in reactivity. The data in Table 3.5 show that oxygen binds more weakly to platinum and palladium than to nickel. According to the Polanyi relationship, weakly bound species are more reactive (i.e., less stable) than strongly bound species. Therefore, according to the Polanyi relationship, one would expect the oxygen adsorbed on platinum or palladium to be more reactive than oxygen adsorbed on nickel.

Experimentally, the weakly bound oxygen on platinum and palladium is much more reactive than the more strongly bound oxygen on nickel. For example, in Chapter 6 we noted that platinum is an excellent catalyst for the reaction of hydrogen and oxygen to yield water. A stochiometric (i.e., 2 : 1) mixture of hydrogen/oxygen gas is stable at 300 K and 1 atm. However, if one puts a platinum wire into the mixture, the mixture will explode. No explosion is seen with nickel. In fact, Lapuloulade and Neil (1973) found that hydrogen and oxygen do not react on a 300-K nickel surface in UHV, while Anton and Cadogan (1991) found that the reaction is almost instantaneous on platinum or palladium under similar conditions. Thus, oxygen on platinum and oxygen on palladium are much more reactive than oxygen on nickel, in agreement with the predictions of the Polanyi relationship.

The data in Table 6.5 also show that nickel binds hydrogen and carbon more strongly than does platinum or palladium. According to the Polanyi relationship, one would therefore expect nickel to be more reactive than platinum or palladium for carbon–hydrogen bond scission processes. That prediction agrees with experiment. For example, on platinum adsorbed ethylene can react to form ethylidyne, $Pt\equiv C-CH_3$, but no ethylidyne is observed on nickel. Such a result is consistant with the thermodynamics of the reactions.

All of these differences between the behavior of platinum and nickel are well understood. However, there is one difference between platinum and nickel that is not well understood, While nickel is an excellent hydrogenation–hydrogenolysis catalyst, platinum also shows activity for isomerization of small hydrocarbons. For example, Table 14.4 shows some data for the reaction of n-pentane and hydrogen over commercial platinum and nickel catalysts. On nickel, methane is the main product. However, on platinum, the pentane isomerizes to produce isopentane and cyclopentane.

At present, we do not know why platinum is such a better isomerization catalyst than nickel; however, the difference appears to be a difference in the acidity of the surface. Recall from Chapter 5 that isomerization reactions are faster on acid surfaces than in other systems. Experimentally, a hydrogen-saturated platinum surface is more

Table 14.4 The products of n-pentane hydrogenolysis over nickel and platinum catalysts

Catalyst	T, °C	pH$_2$, atm	P$_{pent}$, atm	Molar Percentage						S$_{iso}$ (%)	S$_{cycl}$ (%)
				C$_1$	C$_2$	C$_3$	C$_4$	Iso-C$_5$	c-C$_5$		
Pt/SiO$_2$ (16 wt%)	312	0.9	0.1	5	17	15	3	52	6	67	8
	346	0.9	0.1	6	20	18	4	43	7	59	9
Ni/SiO$_2$ (9 wt%)	350	2.5	0.5	85.9	6.6	4.7	8.1	0	0	0	0
	350	5.0	0.5	77.0	8.5	8.3	5.6	0	0	0	0
	350	5.0	2.0	52.0	0.5	3.0	4.4	0	0	0	0

key: iso-C$_5$ = isopentane; c-C$_5$ = cyclopentane; S$_{iso}$ + S$_{cycl}$ + C$_{cracking}$ = 1 (cracking = hydro-genolysis); (experiments performed in an open-flow apparatus under the conditions indicated, by respective authors.)

Source: Table adapted from ponec (1983).

acidic than a hydrogen-saturated nickel or copper surface. The experiments show clearly that platinum is a much better isomerization catalyst than nickel, which we will find to be characteristic of an acidic surface. At present, it is unclear that the acidity is the only factor, however.

14.5.3 Iridum, Rhodium, Cobalt, Osmium, Ruthenium, Iron, Rhenium, Tungsten, Molybdenum, and Chromium

In the next part of our discussion, we will move left from platinum, palladium, and nickel in the periodic table. We will start with iridium, rhodium, cobalt, osmium, ruthenium, iron, rhenium, tungsten, molybdenum and chromium (i.e., the other group VIII, VIIa, and VIa metals).

Generally, one uses iridium, rhodium, cobalt, osmium, ruthenium, iron, rhenium, tungsten, molybdenum, and chromium when one needs to produce a multiply bound intermediate to get a reaction to occur. For example, ammonia synthesis is commonly run on iron or rhenium catalysts. The main reaction is

$$N_2 + 3H_2 \Longrightarrow 2NH_3 \tag{14.64}$$

In Chapter 5 we noted that the reaction occurs via a triply bound nitrogen species; the rate-determining step in reaction (14.64) is the scission of a nitrogen–nitrogen bond to form the triply bound species. Experimentally, such a process is much more rapid on iron than on, for example, nickel. As a result, iron is the best catalyst for the reaction. Similarly, one commonly adds rhenium to an automotive catalyst to promote NO reduction:

$$2NO + 2H_2 \Longrightarrow N_2 + 2H_2O \tag{14.65}$$

$$2NO + 2CO \Longrightarrow N_2 + 2CO_2 \tag{14.66}$$

Again one needs to produce a multiply bound species. Generally, one uses metals in the center of the periodic table to form multiply bound intermediates. Metals near the center of the periodic table have many free electrons. As a result, they are more effective in stabilizing multiply bound species (i.e., adsorbed radicals with two or three dangling bonds).

14.5.4   Other Metals

There have been far fewer studies of the metals to the left of tungsten in the periodic table. The oxides of titanium, vanadium, zirconium lanthanum, and cerium, are commonly used as redox catalysts. The pure metals, however, do not show significant catalytic activity. Thorium and uranium oxide also show redox activity, although the radioactivity of thorium and uranium precludes their large-scale use.

14.6   MODELS FOR TRENDS OVER THE PERIODIC TABLE

In the literature, people have proposed several models to explain the variations in rate over the periodic table. The models have, at best, only qualitative agreement with data. Still, the models are discussed at length in the literature, so I wanted to present them.

One idea in the literature is that one can get some insight into the variations in rate with changing metal via a volcano plot. Recall from Section 12.6.2 that theoretically, there should be a correlation between the reactivity of a catalyst and the heat of adsorption of key intermediates.

One can represent that variation via a volcano plot, namely, a plot of the activation barrier versus the heat of adsorption of the intermediate. When people first proposed volcano plots, they did not know the heat of adsorption of the key intermediates, so they assumed that the heat of adsorption was proportional to the heat of formation of the bulk oxide. They also assumed that the intrinsic barrier did not change as one changes the metal, and that the transfer coefficient γ_P is constant. Those assumptions end up limiting the utility of volcano plots.

For example, Figure 14.23 examines the rate of ethylene hydrogenation on a series of metals considered by Sachtler and Fahrenfort. One observes what has been interpreted as a volcano plot for these metals. I guess that one can see the volcano, although it is hard to see it looking at the raw data.

Figure 14.24 shows a volcano plot for ethylene hydrogenation over a wider range of metals. Notice that there is no obvious correlation between the rate and the heat of adsorption. This is typical behavior. Theoretically, the data should follow a volcano plot,

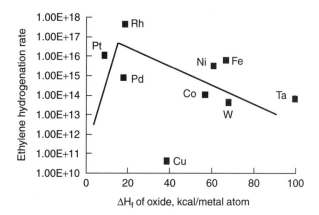

Figure 14.23   A volcano plot for the variations in the rate of ethylene hydrogenation over a subset of the transition metals.

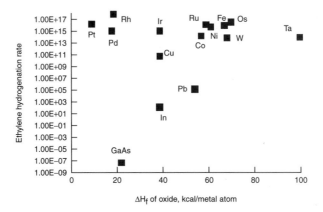

Figure 14.24   A repeat of Figure 14.23 with a larger data set.

if you plot the rate versus $E_a^0 + \gamma_P \Delta H_{Ad}$ where E_a^0 is the intrinsic activity barrier, ΔH_{ad} is the heat of adsorption, and γ_p is the transfer coefficient. However, when you plot the data against the heat of formation of oxygen, you ignore variations in E_a^0 and γ_P. As a result, the volcano plots have considerable scatter.

In my experience, volcano plots work if you are comparing very similar metals. However, they do not work when you look over a wide range of the periodic table.

In the literature, many people recognized that volcano plots often fail. In those cases they have tried other variables to try to get correlations. One variable that people use is something called the **%d character** of the metals. The %d character is a number that Pauling and Yost defined in 1932 to account for the fact that the heats of formation of transition metal oxides were greater than expected by bonds only with the s electrons. The %d character is an empirical number that has very little theoretical significance. Nevertheless, the %d character is used as a measure of the availability of d electrons to contribute to reactions. It is an imperfect measure, to say the least. After all, it was measured in 1932. However, it is the only measure available at present.

Figure 14.25 shows how the rate of ethylene hydrogenolysis varies for the 3d (period 4), 4d (period 5), and 5d (period 6) metals. Notice that in the data shown there is a

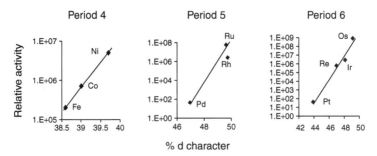

Figure 14.25   A plot of the rate of the ethylene hydrogenolysis ($H_2 + C_2H_6 \Rightarrow 2CH_4$) as a function of the %d character of the metal.

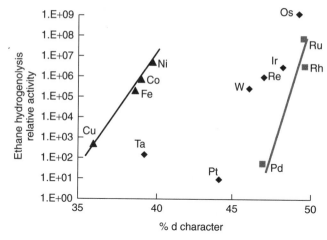

Figure 14.26   A replot of the results in Figure 14.25 on a wider data set.

reasonable fit. This particular correlation has been widely interpreted as indicating that the d electrons play a key role in the hydrogenolysis process.

Unfortunately, again, as we consider more data, the plots do not fit as well. For example, Figure 14.26 shows how rates vary for a wider range of metals. Notice that the tantalum point is way off the curve. Clearly, the correlation is not nearly as good as in Figure 14.25.

Unfortunately at present — as this book is being written — there is nothing better. People use these old models because that is the only thing that exists. There is a real need to find better correlations for reactions in catalysis.

14.7   METAL CATALYSTS FOR SPECIFIC ORGANIC TRANSFORMATIONS

At this point I wanted to change topics once again, and provide a compendium of catalysts used for specific reactions. Table 14.5 summarizes the results. When I was writing the book, I realized that students are going to view this as a table of reactions, with no content. Still, the table is very useful, because if you have a specific reaction that you want to catalyze, you can look in the table and determine which catalyst to use. I will discuss the idea in the solved problems. However, for now I just wanted to include a table of reactions and catalysis for students to use.

14.7.1   Promoters

There is one other detail of note. Industrially one rarely uses pure metals as catalysts. Instead, one adds a series of alloying agents called **promoters** to modify the catalyst's activity. The promoters change the activity of the catalyst by suppressing undesirable side reactions and slightly increasing the rate of the primary reaction. No one know how the promoters work in detail. However, promoters are very important to industrial catalysts.

Table 14.5 Supported metal catalysts for some typical reactions (R is an alkyl group, X is a halogen)

Reaction	Metal	Support	Temperature, °C	Pressure, bar
Hydrogenations				
$R\text{–}CH=CH\text{–}R + H_2 \rightarrow RCH_2CH_2R$	Pd > Pt > Rh=Ru = Ir=Ni	C > Al$_2$O$_3$=BaSO$_4$=CaCO$_3$	5–100	3–10
$R\text{–}C\equiv C\text{–}R + 2H_2 \rightarrow RCH_2CH_2R$	Pd	C > Al$_2$O$_3$	20–200	1–50
$R\text{–}C\equiv C\text{–}R + H_2 \rightarrow RCH=CHR$	Pd	CaCO$_3$ > C > BaSO$_4$	5–50	1–3
$R\text{–}CH=CH\text{–}R + H_2 \rightarrow RCH_2CH_2R$	Pt > Rh > Pd = Rh	C > Al$_2$O$_3$	5–100	1–10
$R\text{–}CH=CH\text{–}CH=CH\text{–}R + H_2 \rightarrow RCH_2CH_2\text{–}CH=CH\text{–}R$	Pd > Ru > Pt = Rh=Ru	Al$_2$O$_3$ > C > CaCO$_3$	5–100	1–3
$R\text{–}CH=CH\text{–}CH=CH\text{–}COOH + H_2 \rightarrow RCH_2CH_2\text{–}CH=CH\text{–}COOH$ (edible oil upgrading)	Ni > Cu/Cr$_2$O$_3$	SiO$_2$ > Al$_2$O$_3$	150–200	1–3
Cyclohexene + H$_2$ → cyclohexane	Pt > Pd = Rh	C > Al$_2$O$_3$	5–150	3–10
$R\text{–}NO_2 + 3H_2 \rightarrow R\text{–}NH_2 + 2H_2O$	Pd = Pt > Rh	C	50–150	3–50
$C_6H_5NO_2 + 2H_2 \rightarrow C_6H_5NHOH + H_2O$	Pt ≫ Pd = Ir	CaCO$_3$ > BaSO$_4$ > Al$_2$O$_3$	5–100	1–5
$R_2CO + H_2 \rightarrow R_2CHOH$	Ru > Rh > Pt	C > Al$_2$O$_3$	5–150	1–50
$C_6H_5CN + H_2 + H_2O \rightarrow C_6H_5CHO$	Pd	C	5–100	1–10
$RCN + 2H_2 \rightarrow RCH_2NH_2$	Rh	C ≫ Al$_2$O$_3$	5–100	1–10
$N_2 + 3H_2 \rightarrow 2NH_3$	Ru, Rh > Fe	C > unsupported	200–550	150–500
$CO + 2H_2 \rightarrow CH_3OH$	Cu	ZnO	200–300	50–100
$CO + 3H_2 \rightarrow CH_4 + H_2O$	NI	Al$_2$O$_3$	250–350	20–40
Oxidations				
$4NH_3 + 7O_2 \rightarrow 4NO_2 + 6H_2O$ (used to make nitric acid)	Pt	Gauze	900–1000	8
$CH_3CH=CH_2 + O_2 \rightarrow CH_2=CH\text{–}CHO + H_2O$	Bi$_2$O$_3$–MoO$_3$	SiO$_2$	400–500	1–3
$2CH_3CH=CH_2 + 2NH_3 + 3O_2 \rightarrow 2CH_2=CH\text{–}CN + 6H_2O$ (aminoxidation)	Bi$_2$O$_3$–MoO$_3$	SiO$_2$	400–500	1–3

Reaction	Catalyst	Support	Temperature	Pressure
$2CH_2{=}CH_2 + O_2 \rightarrow$ 2-ethylene oxide	Ag	α-Al_2O_3	200–300	10–20
$2CH_2{=}CH_2 + O_2 \rightarrow 2CH_3CHO$	Pd	Al_2O_3	150	1–2
$2CH_2{=}CH_2 + O_2 \rightarrow 2CH_3CHO$ (Wacker process)	Palladium cluster	Homogeneous catalyst	100–150	5–20
$CH_2{=}CHCH_2CH_3 + \frac{1}{2}O_2 \rightarrow CH_2{=}CH{-}CH{=}CH_2 + H_2O$	MgO:FeO	Al_2O_3	300–500	1–2
$ortho$-$CH_3C_6H_4(CH_3) + 3O_2 \rightarrow HOOCC_6H_4COOH + 2H_2O$	$V_2O_5 > TiO_2 > ZrO_2$	Al_2O_3	350–400	1
$2SO_2 + SO_3$	V_2O_5	S_1O_2	350–450	2–7
$CH_3OH + \frac{1}{2}O_2 \rightarrow H_2CO + H_2O$	Silver	Granules	500–600	1–2

Pollution Control Reactions

Reaction	Catalyst	Support	Temperature	Pressure
$CO + \frac{1}{2}O_2 \rightarrow CO_2$	$Ru > Rh > Pt = Pd$	Al_2O_3	≥ 150	1
$2NO + 2CO \rightarrow N_2 + 2CO_2$	$Ru > Rh > Pd > Pt$	Al_2O_3	≥ 200	1
$6NO + 4NH_3 \rightarrow 5N_2 + 6H_2O$	$V_2O_5 > TiO_2$	Al_2O_3	300	1

Other Important Reactions

Reaction	Catalyst	Support	Temperature	Pressure
$CH_4 + 2H_2O \rightarrow CO_2 + 4H_2$ (steam reforming)	Ni/C_aO	Al_2O_3	500–850	20–40
$CH_3CH{=}CH_2 + H_2 + CO \rightarrow CH_3CH_2CH_2CHO$ (hydroformulation)	$Rh(CO)PPh_3 > HCo(CO)_8$	Homogeneous catalyst	100–200	150–250
Hydrodesulfurization $C_4H_4S + 4H_2 \Rightarrow C_4H_{10} + H_2S$	$COO/Mo_2O_3 > NiO/Mo_2O_3$	Al_2O_3	250–400	5–100
High-temperature water–gas shift $CO + H_2O \Rightarrow CO_2 + H_2$	Fe_3O_4	Cr_2O_3	350–450	~30
Low-temperature water–gas shift $CO + H_2O \Rightarrow CO_2 + H_2$	Cu/ZnO	Al_2O_3	200–250	~30
Reforming of crude oil	$Pt/Sn > Pt/Re$	Acidic alumina	450–525	5–50
$2C_2H_4 + 4HCl + O_2 \rightarrow 2C_2H_4Cl_2 + 2H_2O$	$CuCl_2$	Al_2O_3	250	1–10

14.8 THE BLOWERS–MASEL EQUATION AS A QUANTITATIVE TOOL FOR CATALYST SELECTION

So far all of discussion in this chapter has been qualitative. Recently (at the time of writing), though, Masel's group has started to do some work to quantify the ideas in this chapter. Although the ideas are new, I decided to present them anyway, because they are related to the work discussed in this chapter and are useful.

In this section we will discuss how one can use the Blowers–Masel equation as a quantitative tool for catalyst selection. The idea is to formulate a feasible mechanism and then use the Blowers–Masel equation to find a suitable catalyst for that mechanism. Generally the steps are

- Guess at a mechanism using the rules in Chapter 5 and the rules earlier in this chapter.
- Postulate a series of feasible catalysts with the aid of Figure 5.12. Usually, you should only consider group Ib, group VIII, and group IIb metals.
- Use Blowers–Masel equation and the data in Table 6.5 to estimate the barrier for each of the steps in the mechanism on each of the metals.

Most metal–catalyzed reactions run at turnover numbers of 1/second and temperatures between 100 and 300°C. According to equation (2.15), a reaction with a rate constant of 1/second at 300°C (573 K) will have an activation barrier of about 573 K × 0.06 kcal/(mol·K) = 34 kcal/mol. Therefore, all of the steps in the catalytic mechanism must have an activation barrier less than 34 kcal/mol. You usually want the activation barriers for all of the desired steps to be 25 kcal/mol or less so that the catalyst can run at 200°C or less.

Often one also considers possible side reactions. Generally, one wants the activation barriers for all of the undesirable side reactions to be at least 10 kcal/mol larger than the activation barriers for all of the desired reactions.

This is new work, and it is not yet generally accepted in the literature. However, it is my work so I decided to include it in this chapter.

First, it is useful to review the key features of a feasible mechanism:

1. The mechanism must lead from reactants to products
2. The reaction must have a complete catalytic cycle — you need to start with bare active sites, and regenerate the sites at the end of the reaction.
3. The first step is usually the adsorption of one of the reactants. Usually, one of the reactants dissociatively adsorbs onto an active site to start the reaction going.
4. Then there are a series of steps of the kind described in Section 5.2, 5.2.1, 5.2.2, 5.2.3 leading to products. One cannot postulate any other types of reactions; they seldom occur at a high rate on a metal.
5. All of the proposed intermediates must be feasible and have an analog in inorganic chemistry. You cannot, for example, postulate species with nitrogen–nitrogen double bonds, if nitrogen–nitrogen double bonds have never been observed in metal cluster compounds.
6. There must be a desorption step where the products desorb and new sites are created to hold more reactants.

An illustration of postulating a mechanism is given in Example 14.A. Finding a feasible mechanism is the first step to the Blowers–Masel method of finding a suitable catalyst.

Once one has a mechanism, one needs to find a catalyst to catalyze the reaction. One can simply assume that the reaction will occur in all of the group VIII, Ib, and IIb metals and use the Blowers–Masel equation as described below to see if the catalyst can work. However, one can save some work by eliminating some of the metals. I often use Figure 5.12 to narrow my choice of metals. Recall that the initial step in most reactions on metals is the dissociation adsorption of one of the reactants. Well, if the reactant will not adsorb dissociatively, no reaction is possible. Figure 5.12 shows which metals will dissociate different molecules. There are three key categories: molecular adsorption, activated dissociative adsorption, and unactivated dissociative adsorption. In my experience, the most active catalysts usually lie along the boundary between dissociation and activated dissociation of a key reactant, so I usually concentrate on these metals in my analysis.

Once I have identified a metal, I then use the Blowers–Masel equation to estimate the activation barrier for each step in the mechanism. Recall that according to the Blowers–Masel equation, the activation barrier for a given reaction can be estimated from

$$E_a = \begin{cases} 0 & \text{when} \quad \Delta H_r < -4E_a^0 \\ \Delta H_r & \text{when} \quad \Delta H_r > 4E_a^0 \\ (w_0 + 0.5\Delta H_r)\left(\dfrac{(V_p - 2w_0 + \Delta H_r)^2}{(V_p)^2 - 4(w_0)^2 + (\Delta H_r)^2}\right) & \text{otherwise} \end{cases} \quad (14.67)$$

In equation (14.67), E_a is the activation barrier, w_0 is the average bond energy for the bonds that break and form during the reaction, ΔH_r is the heat of reaction, and V_p is a constant that is related to the intrinsic barrier for the reaction by

$$V_p = 2w_0\left(\frac{w_0 + E_a^0}{w_0 - E_a^0}\right) \quad (14.68)$$

In practice, one needs a value of E_a^0 to use in the calculations. Table 14.6 gives some approximate values. At this point, we have only approximations to the intrinsic barriers. The numbers in the table have an uncertainty of $\pm 20\%$. Still, the results are good enough that one can find possible catalysts.

An illustrative calculation is given in Example 14.A. In my experience, one does not want to overvalue the analysis. One can find candidate catalysts. However, one still has to

Table 14.6   Approximate values[a] of E_a^0 for reactions on transition metal surfaces

Reaction type	E_a^0, kcal/mol
Hydrogen transfer	10
Scission of C–C, C–O single bonds	40
Double-bond scission	50

[a]These values apply only to reactions on transition metals. Intrinsic barriers are a barrier of 2 higher or do not apply to transition metals.

Note: Structure sensitive

test the catalysts experimentally to see if they are active, and if there is some unanticipated side reaction.

Still, the methods do allow one to find possible catalysts. They also allow you to understand the trends described earlier in this chapter. One should read Example 14.A before proceeding.

14.9   KINETICS OF REACTIONS ON METALS

Next, I want to change topics. In the last several sections we were discussing trends over the periodic table. In this section I want to briefly review reactions on metals. The kinetics of reactions on metals generally follow the Langmuir–Hinshelwood-type rate laws that were discussed at the beginning of Section 12.17 and in Sections 12.17.1–12.17.5. Rates can be first-order, second-order or even negative order in the reactant concentrations as discussed in Sections 12.17.2 and 12.17.4.

Rates are usually expressed as a turnover number, T_N where the turnover number is the rate in molecules of product/surface site/sec. Typical turnover numbers are in the order of 1/second for active catalysts as discussed in Section 12.17. A typical metal has 10^{15} sites/cm^2 so at a turnover number of 1, 1 cm^2 of catalyst will produce only 10^{15} product molecules/second. Consequently, one needs a tremendous surface area to produce a significant amount of catalyst.

Fortunately, supported catalysts do have a tremendous surface area. It is easy to prepare a catalyst with 1 m^2 of active metal per cm^3 of catalyst. Catalyst loadings as high as 10 m^2 of active metal per cm^3 of catalyst can be obtained. One liter of catalyst with a surface area of (1 m^2 of active metal per cm^3 of catalyst will produce about 100 mol/hour of product at a turnover number of 1.

One key feature is that the rate of reaction is proportional to the surface area of the metal. If one increases the metal surface area by adding more metal or spreading the metal out on the support, one can generally increase the rate of reaction.

There are some special forms of the kinetic rate laws that apply only to reactions on metals. Mars–Van Krevelan kinetics apply to oxidation reactions on reducible metal oxides. In Section 14.3.6 we noted that when an oxidation reaction of a hydrocarbon occurs on a reducible metal oxide, the lattice oxygen from the metal oxide reacts with the hydrocarbon to produce products. Then oxygen adsorbs and reoxidizes the metal oxide. The lattice oxygen is different from oxygen bound to the surface, so the Langmuir–Hinshelwood–Hougan–Watson rate expression fails.

There is another expression, called a *Mars–Van Krevelan rate law*, which applies to this situation:

$$r = \frac{k}{\left(\dfrac{1}{K_{HC}P_{HC}}\right)^m + \left(\dfrac{1}{K_{O_2}P_{O_2}}\right)^n} \tag{14.69}$$

In equation (14.69) P_{HC} is the hydrocarbon partial pressure; P_{O_2} is the oxygen partial pressure; and k, K_{HC}, K_{O_2}, m, and n are constants. Equation (14.69) is the Mars–Van Krevelan rate expression for hydrocarbon oxidation.

People also have a rate equation for ammonia synthesis called the *Tempkin–Pyzhev equation*:

$$r = k_1[N_2]\left(\frac{[H_2]^3}{[NH_3]^2}\right)^m - k_2\left(\frac{[NH_3]^2}{[H_2]^3}\right)^{1-m} \tag{14.70}$$

In equation (14.70) $[N_2]$, $[H_2]$, and $[NH_2]$ are the partial pressures of nitrogen, hydrogen, and ammonia, and k_1, k_2, and m are constants. Industrially, one usually runs the ammonia synthesis reaction under conditions where the reaction is reversible. The second term on the right of the expression accounts for the fact that there is a reverse reaction. The factor of $\left(\dfrac{[H_2]^3}{[NH_3]^2} \right)$ in the expression arises because the surface concentration of hydrogen is reduced by the presence of ammonia and vice versa.

There are Langmuir-type rate equations for each of these reactions, and in my experience you cannot detect the difference experimentally. However, people often discuss these mechanisms in the literature, so I wanted to mention them before I close.

14.10 SUMMARY

In summary, then in this chapter we provided a very brief outline of catalysis by metals. We found that metals are very active for bond scission reactions and less active for isomerization reactions. Metals catalyze reactions by stabilizing intermediates, by acting as reaction initiators, by lowering intrinsic barriers, by doing redox chemistry, by holding the reactants in the right configuration to react, and by other effects. The d bands in metals are critical to reactions because the d electrons lower the intrinsic barriers to bond scission. We also reviewed how the reactivity of metals varies over the periodic table, although we were not able to explain all of the effects.

I work in metal catalysis, and when I reread this Chapter, I found more than 100 additional topics that I could have included. However, I am limited on what I can include in the space available. Other excellent bibliographic references are included at the end of the chapter.

14.11 SOLVED EXAMPLES

Example 14.A   Using the Blowers–Masel Approximation to find Feasible Catalysts
The production of ammonia from nitrogen and hydrogen is one of the largest chemical processes worldwide. The overall reaction is $N_2 + 3H_2 \Rightarrow 2NH_3$.

a) Find a feasible mechanism for the reaction on a metal catalyst.

b) Look carefully at Figure 5.12 and decide which metals are likely catalysts. Assume that all of the steps in the mechanism have to either be inactivated or have a small activation barrier. Which metals will be active for the key dissociation processes during the mechanism?

c) Now think about the recombination steps. If a recombination step is too endothermic, it will not be feasible. Use the data in Table 6.5 to estimate the heat of reaction for each of the steps.

d) From your results choose a possible metal catalyst.

Solution
   a) The approach is to

1) Postulate a feasible mechanism.
2) Use Table 6.1 and 6.5 to estimate ΔH_r for each of the steps in the mechanism.

3) Use the Blowers–Masel approximation to estimate E_a for the reaction.

4) E_a must be less than 34 kcal/mol, and preferably below 30 kcal/mol for every step in the mechanism.

We want a mechanism that leads from reactants to products, using the pathways earlier in this chapter. We require a complete catalytic cycle and all of the species be stable. The initial step is likely to be the dissociative adsorption of one of the reactants. Then atoms would be added or subtracted to get to products.

There are two possible initial reactions: one where N_2 dissociatively adsorbs, and one where H_2 dissociatively adsorbs. For the analysis here, we will consider both possibilities and see which one produces a viable catalytic cycle.

Let us assume that the dissociation of nitrogen is the initial step. When the nitrogen adsorbs, it needs six sites because each nitrogen has three bonds to the surface. Once the nitrogen dissociatively adsorbs, we will have nitrogen atoms on the surface. In order to get to an ammonia product, we have to hydrogenate the nitrogen atoms. There are two ways to do the hydrogenation: either we can dissociate the hydrogen into atoms and add hydrogen atoms one at a time, or we can add molecular hydrogen directly. That gives two possible mechanisms which I have labeled Mechanism I and Mechanism II below.

Mechanism I (Langmuir Hinshelwood)

$$N_2 + 6S \longrightarrow 2N_{(ad)}$$

$$H_2 + 2S \longrightarrow 2H_{(ad)}$$

$$N_{(ad)} + H_{(ad)} \longrightarrow NH_{(ad)} + 2S$$

$$NH_{(ad)} + H_{(ad)} \longrightarrow NH_{2(ad)} + 2S$$

$$NH_{2(ad)} + H_{(ad)} \longrightarrow NH_3 + 2S$$

Mechanism II (Eley Rideal)

$$N_2 + 6S \longrightarrow 2N_{(ad)}$$

$$N_{(ad)} + H_2 \longrightarrow NH_{(ad)} + H_{ad}$$

$$NH_{(ad)} + H_2 \longrightarrow NH_{2(ad)} + H_{ad}$$

$$NH_{2(ad)} + H_2 \longrightarrow NH_3 + H_{ad}$$

$$2H_{(ad)} \longrightarrow H_2 + 2S$$

The last step in Mechanism II might seem unnecessary, but recall that every catalytic reaction must go through a complete catalytic cycle and regenerate active sites. The last step generates active sites.

In mechanism I nitrogen adsorbs dissociatively, the hydrogen dissociates too. Then there is a series of steps leading to products. All of the steps in Mechanism I are of the type expected for reactions on metals, and we have a complete cycle regenerating active sites. Therefore, Mechanism I is a feasible mechanism on a metal surface. In fact, one can show that Mechanism I is a classic Langmuir-Hinshelwood mechanism of the type we discussed in Chapters 5 and 12.

Mechanism II is very similar to mechanism I except that the hydrogen is added via a displacement reaction. All of the steps are again feasible on metals and the process leads from reactants to products and regenerates an active site. Therefore Mechanism II is also a feasible mechanism.

Displacement reactions are very characteristic of reactions on metal clusters, and are much rarer on metal surfaces. Therefore, it seems that Mechanism I is more likely on bulk metals while Mechanism II is more likely on cluster compounds.

Mechanisms I and II assumed that the initial step was dissociation of nitrogen. One can also imagine a mechanism where the initiation step is the dissociative adsorption of hydrogen. In that case one would have to add nitrogen. If the nitrogen dissociates, the mechanism will be the same as in Mechanism I above. If instead the nitrogen does not dissociate, one could imagine a series of steps like those give in Mechanism III

Mechanism III (not feasible)

$$H_2 + 2S \longrightarrow 2H_{(ad)}$$

$$N_2 + H_{(ad)} \longrightarrow HN_{2(ad)}$$

$$HN_{2(ad)} + H_{(ad)} \longrightarrow HN_2H_{(ad)}$$

etc.

In each case the reactant hydrogens dissociatively adsorb and then atoms are added one step at a time to produce products so initially the reaction looks feasible. Still the $HN_{2(ad)}$ intermediate contains a nitrogen–nitrogen double. I am not aware that nitrogen–nitrogen double bonds have ever been observed. Therefore, Mechanism III is very unlikely.

The conclusion from all of this is that Mechanism I is the most likely mechanism on metal surfaces, while Mechanism II is the most likely mechanism on metal clusters. Experiments agree with these conclusions.

b) Next we want to see which metals are likely to be active for the reaction. Notice that the first step in the mechanism is dissociation of nitrogen. According to the data in Figure 5.12, nitrogen dissociates readily on Fe, Mn, Cr, Mo and W, and has a small barrier on Co, Ru, Rh, Re and Os. These metals are the most likely candidates as catalysts. (In fact they all show some catalytic activity).

One can use the Blowers–Masel equation to check that these metals are feasible for N–N dissociation. The idea is to

1. Estimate the heat of reaction using Table 6.5.
2. Calculate V_p using eq. 14.68.
3. Estimate the activation barrier using the Blowers–Masel equation, eq. 14.67.

In order to see how that works, consider the reaction

$$N_2 + 6S \longrightarrow 2N_{(ad)}$$

on an iron catalyst. Assume that the reaction breaks a N–N triple bond and forms 6 metal-nitrogen single bonds.

First let us estimate ΔH_r from Table 6.5. According to Table 6.5, a single nitrogen-surface bond contributes -14 kcal/mol ΔH_r. The reaction forms 6 bonds, (three to each nitrogen).

$$\Delta H_r = 6 \times (-14 \text{ kcal/mol}) = -84 \text{ kcal/mol}$$

We can estimate V_P from equation 14.68.

$$V_p = 2w_0 \left(\frac{w_0 + E_a^0}{w_0 - E_a^0} \right)$$

First we need a value for w_0, the average energy of the bonds which break and form during the reaction. According to data on the NIST webbook, N_2 has a bond energy of 225.94 kcal/mol. The reaction is 84 kcal/mol exothermic. Therefore the total bond energy in the products is $84 + 225.94 = 309.94$ kcal/mol

From equation 10.64,

$$w_0 = (225.94 + 309.94)/2 = 268 \text{ kcal/mol}$$

Next we will calculate V_p from equation 14.68 using $E_a^0 = 50$ kcal/mol from Table 14.6

$$V_p = 2 \times (268 \text{ kcal/mol}) \left(\frac{268 \text{ kcal/mol} + 50 \text{ kcal/mol}}{268 \text{ kcal/mol} - 50 \text{ kcal/mol}} \right) = 782 \text{ kcal/mol}$$

Next we can estimate E_a from equation 14.67

$$E_a = (w_0 + 0.5\Delta H_r) \left(\frac{(V_p - 2w_0 + \Delta H_r)^2}{(V_p)^2 - 4(w_0)^2 + (\Delta H_r)^2} \right)$$

Plugging in the numbers shows

$$E_a = (268 + 0.5(-84)) \left(\frac{(782 - 2 \times 268 - 84)^2}{(782)^2 - 4(268)^2 + (-84)^2} \right) = 17.8 \text{ kcal/mol}$$

According to the discussion in section 14.8 this is a reasonable activation barrier. Consequently, iron should be able to dissociate N_2 at room temperature and therefore catalyze the first step in Mechanism I.

The same calculations show that E_a is about 36 kcal/mol on platinum. According to the discussion in section 14.8, the activation of all of the steps must be below 34 kcal/mol, and should be around 25 kcal/mol. E_a is too large on platinum, which means that platinum should not be as good of a catalyst for the ammonia synthesis reaction as iron.

Similar calculations can also be done with, Mn, Cr, Mo, and W. All should be able to dissociate N_2 at room temperature. Ru, Rh, CO, N_1 and O_5 should be able to dissociate the N_2 at 200°C.

c) To find out which catalyst is best, we need to examine the hydrogenation reactions. For example consider the reaction

$$N_{(ad)} + H_{(ad)} \longrightarrow NH_{(ad)} + 2S \tag{14.A.1}$$

During this reaction, you break a nitrogen-surface bond and a hydrogen-surface bond, and form a NH bond.

Let us estimate the heat of reaction of iron. According to Table 6.1 forming a nitrogen–hydrogen bond contributes -2.1 kcal/mol to ΔH_r. According to Table 6.5, forming an H-surface bond on iron contributes $-(11 \text{ kcal/mol})$ while forming a surface N bond contributes -14 kcal/mol. Breaking the bonds contribute the opposite, that is $\Delta H_1 = +11$ kcal, and $\Delta H_2 = +14$ kcal/mol.

Therefore one can estimate ΔH_r for Reaction 14.A.1 as

$$\Delta H_r = -2.1 + 11 + 14 = 22.9 \text{ kcal/mol} \tag{14.A.2}$$

The reaction is endothermic but not so highly endothermic that it is impossible.

One can also estimate the activation barrier for reaction 14.A.1 from the Blowers-Masel equation as was discussed in Chapter 11 assuming an intrinsic barrier of 10 kcal/mol as suggested in Table 11.3. The calculation is the same as that in Example 11.F. First we calculate V_p given by equation 11.F.2

$$V_p = 2w_0 \left(\frac{w_0 + E_a^0}{w_0 - E_a^0} \right) \tag{14.A.3}$$

where w_0 is the average energy of the bond which breaks and forms, and E_a^0 is the intrinsic barrier. According to the CRC, the N–H bonds in ammonia have a bond energy of 110 kcal/mol, so I assumed $w_0 = 110$ kcal/mol. Plugging in the numbers shows:

$$V_p = 2(110 \text{ kcal/mol}) \left(\frac{110 \text{ kcal/mol} + 10 \text{ kcal/mol}}{110 \text{ kcal/mol} - 10 \text{ kcal/mol}} \right) = 264 \text{ kcal/mol} \tag{14.A.4}$$

We then estimate the activation energy by plugging into Equation (11.F.1)

$$E_a = (w_0 + 0.5\Delta H_r) \left(\frac{(V_p - 2w_0 + \Delta H_r)^2}{(V_p)^2 - 4(w_0)^2 + (\Delta H_r)^2} \right) \tag{14.A.5}$$

Plugging $w_0 = 110$ kcal/mol, $V_p = 264$ kcal/mol, $\Delta H_r = 22.9$ kcal/mol yields

$$E_a = (110 \text{ kcal/mol} + 0.5 \times 22.9 \text{ kcal/mol})$$
$$\times \left(\frac{(264 \text{ kcal/mol} - 2 \times 110 \text{ kcal/mol} + 22.9 \text{ kcal/mol})^2}{(264 \text{ kcal/mol})^2 - 4(110 \text{ kcal/mol})^2 + (22.9 \text{ kcal/mol})^2} \right) \tag{14.A.6}$$

Doing the arithmetic shows

$$E_a = 24.9 \text{ kcal/mol} \tag{14.A.7}$$

According to Figure 2.15, this is a reasonable barrier for a reaction at about 100°C. Therefore iron would be an effective catalyst for Mechanism I.

Let us repeat the same analysis on chromium. According to Table 6.6 a chromium–nitrogen bond contributes −44 kcal/mol to ΔH_r so breaking a chromium metal bond will contribute +44 kcal/mol. Similarly, breaking a chromium hydrogen bond will contribute 14 kcal/mol. Therefore on chromium:

$$\Delta H_r = -2.1 + 14 + 44 = 55.9 \text{ kcal/mol} \tag{14.A.8}$$

which is very high. One could calculate an activation energy as above, but note that Equation 11.F.1 only applies to cases where $\Delta H_r < 4E_a^0$. In our case $\Delta H_r > 4E_a^0$ so according to Equation 14.67

$$E_a = \Delta H_r = 55.9 \text{ kcal/mol} \tag{14.A.9}$$

In Chapter 2 we found that activation barriers for reactions near room temperature are closer to 20 kcal/mol so 55.9 kcal/mol does not seem feasible. A more detailed analysis using eq. 2.32 shows that a reaction with a barrier of 55.9 kcal/mol would have to be run at a temperature, T given by

$$T = 55.9 \text{ kcal/mol}/(0.06 \text{ kcal/mol-}°K) = 931 \text{ K} \qquad (14.A.10)$$

That is very hot for a typical chemical process. Therefore, we conclude that chromium will not be expected to be an effective catalyst for ammonia synthesis. In contrast iron is feasible.

d) A similar analysis shows that ruthenium and rhodium would also be expected to be feasible catalysts. Molybdenum and tungsten have higher E_a's but still they would be in the acceptable range.

Commercially, iron is the most popular catalyst. Rhodium and ruthenium are also used, although they are more expensive. Tungsten and molybdenum are used less often although they do show activity.

Example 14.B   Review of the 18 Electron Rule   Consider the following molecules:

$$Fe(CO)_5, Fe(CO)_4, Fe(CO)_4H, Fe(CO)_4H_2, Mn(CO)_5, [Mn(CO)_5]\text{-}, Fe(cp)_2,$$

$$Co(cp)_2, Ti(cp)_2Cl_2, Ti(cp)_2Cl_2(C_2H_4), Rh(PPh_3Rh(PPh_3)_4Cl, Ni(al)_2.$$

Note cp = cyclopentadienyl, al = allyl, Ph = phenyl

a) Which of the molecules obeys the 18 electron rule?

b) Which would be useful catalysts?

Solution

a) We use the electron counting rule:

$Fe(CO)_5$	Fe = 8 CO = 2	total = $8 + 5 \times 2 = 18$
$Fe(CO)_4$	Fe = 8 CO = 2	total = $8 + 4 \times 2 = 16$
$Fe(CO)_4H$	Fe = 8 CO = 2 H = 1	total = $8 + 4 \times 2 + 1 = 17$
$Fe(CO)_4H_2$	Fe = 8 CO = 2 H = 1	total = $8 + 4 \times 2 + 1 \times 2 = 18$
$Mn(CO)_5$	Mn = 7 CO = 2	total = $7 + 5 \times 2 = 17$
$[Mn(CO)_5]^-$	Mn = 7 CO = 2 e⁻ = 1	total = $7 + 5 \times 2 + 1 = 18$
$Fe(cp)_2$	Fe = 8 cp = 5	total = $8 + 2 \times 5 = 18$
$Co(cp)_2$	Co = 9 cp = 5	total = $9 + 2 \times 5 = 19$
$Ti(cp)_2Cl_2$	Ti = 4 cp = 5 Cl = 1	total = $4 + 2 \times 5 + 2 \times 1 = 16$
$Ti(cp)_2Cl_2(C_2H_4)$	Ti = 4 cp = 5 Cl = 1 C_2H_4 = 2	total = $4 + 2 \times 5 + 2 \times 1 + 2 = 18$
$Rh(PPh_3)_3Cl$	Rh = 9 PPh_3 = 2 Cl = 1	total = $9 + 3 \times 2 + 1 = 16$
$Rh(PPh_3)_4Cl$	Rh = 9 PPh_3 = 2 Cl = 1	total = $9 + 4 \times 2 + 1 = 18$
$Ni(al)_2$	Ni = 10 al = 3	total = $10 + 2 \times 3 = 16$

Therefore $Fe(CO)_5$, $Fe(CO)_4H_2$, $[Mn(CO)_5]^-$, $Fe(cp)_2$, $Ti(cp)_2Cl_2(C_2H_4)$ and $Rh(PPh_3)_4Cl$ obey the 18 electron rule.

b) Usually catalysts contain 16 or 17 electrons. From the analysis above $Fe(CO)_4$, $Fe(CO)_4H$, $Mn(CO)_5$, $Ti(cp)_2Cl_2$ $Rh(PPh_3)_3Cl$, and $Ni(al)_2$ could all be considered catalysts. One does need to do a calculation like that in Example 14.A to see is they are active catalysts.

People also sometimes say that 18 electron molecules with easy leaving groups (phosphines or amines) are catalysts. In this classification, $Rh(PPh_3)_4Cl$ could be considered a catalyst. Note I prefer to think of $Rh(PPh_3)_4Cl$ as a catalyst precursor and $Rh(PPh_3)_3Cl$ as the catalyst since only $Rh(PPh_3)_3Cl$ directly participates in a catalytic cycle.

14.12 SUGGESTIONS FOR FURTHER READING

Additional references for catalysis by metals include:

V. Ponec and G. Bond, *Catalysis by Metals*, Elsevier, Amsterdam, 1995.

R. I. Masel, *Principles of Adsorption and Reaction on Solid Surfaces*, Wiley, New York, 1996.

J. R. Anderson and M. Boudart (eds.,) *Catalysis: Science and Technology*, 10 vol., Elsevier, Amsterdam, 1984–1998.

Introductory books on catalysis include:

M. Boudart, G. Djego-Mirajossov, *Kinetics of Heterogeneous Catalytic Reactions*, Princeton University Press, Princeton, NJ, 1984.

G. C. Bond, *Heterogeneous Catalysis, Principles and Applications*, Clarendon Press, Oxford, 1987.

B. C. Gates, *Catalytic Chemistry*, Wiley, New York, 1992.

H. F. Rase, *Handbook of Commercial Catalysts*, CRC Press, Boca Raton, FL, 2000.

Thomas, J. M. and Thomas, W.J. *Principles and Practice of Heterogeneous Catalysis*, Wiley, NY 1996.

14.13 PROBLEMS

14.1 Define the following terms:

(a) Transition metal

(b) Macropore

(c) Micropore

(d) Mesopore

(e) Floc

(f) Terrace*

(g) Step*

(h) Kink*

(i) Stepped surface

(j) Kinked surface BCC

(k) FCC

(l) HCP

(m) Unit cell

(n) Bravis lattice

(o) Stereographic triangle

(p) Adsorption

(q) Molecular adsorption

(r) Dissociative adsorption

(s) Oxidative addition

(t) Bond dissociation reaction

(u) Fragmentation reaction

(v) Promoter

(w) Association reaction

(x) Single-atom recombination

(y) α scission

(z) β scission

(aa) γ scission

(bb) δ scission

(cc) Displacement reaction

(dd) Displacement–association reaction

(ee) Hydrogen migration

(ff) Acetyl migration

(**gg**) Molecular desorption

(**hh**) Recombinative desorption

(**ii**) Reductive elimination

(**jj**) Hydroformulation

(**kk**) Wilkinson catalyst

(**ll**) Structure–sensitive reaction

(**mm**) Multiplet

(**nn**) Active site

(**oo**) Secondary Structure sensitivity

(**pp**) Volcano plot

(**qq**) %d character

*Define these terms in the way they are discussed in this chapter.

14.2 Earlier in this chapter we noted that transition metals are active catalysts.

(**a**) Why are transition metals better hydrogenation catalysts than are nontransition metals?

(**b**) Are non–transition metals able to bind intermediates?

(**c**) What do the d bands do to promote catalytic activity?

(**d**) How are the intrinsic barriers different on transition and non–transition-metal catalysts?

14.3 Describe in your own words how catalytic activity varies over the periodic table.

(**a**) Why are some metals better catalysts than others?

(**b**) What measures exist to correlate trends in reactivity over the periodic table?

(**c**) How well do those measures correlate data?

14.4 Review how surface structure affects rates of catalytic reactions.

(**a**) In a general way, what effects are seen?

(**b**) How large are the effects?

(**c**) What types of reaction are most affected?

(**d**) Why does the degree of structure sensitivity vary from one reaction to the next?

14.5 Imagine that you are running ethylene hydrogenation on a palladium catalyst a 10-cm^3 well-mixed reactor. You feed 2 mol/hour of ethylene and hydrogen into the reactor, and get 1.8 mol/hour of ethane product.

(**a**) Calculate the rate in $\text{mol/(cm}^3 \cdot \text{hour)}$.

(**b**) Calculate the rate in mol/hour per gram of catalyst. Assume that the catalyst has an average density of 1.2 grams/cm^3.

(**c**) Calculate the rate in mol/hour per cm^2 of catalyst surface. Assume that the catalyst has a surface area of $100 \text{ m}^2\text{/gram}$ and that 2% of the surface is covered by palladium (i.e. the palladium area is $2 \text{ m}^2\text{/gm}$).

(**d**) Calculate the turnover number for the reaction. Assume that the palladium has 10^{15} sites/cm^2.

(**e**) How would the rate, in $\text{mol/(cm}^3 \cdot \text{hour)}$, change if you doubled the amount of palladium on the catalyst? Assume that the palladium surface area doubles.

14.6 Why are metal catalysts attached to a support?

(**a**) What does the support do to promote catalytic activity?

(**b**) What does the support do to enhance selectivity?

(c) What are some desirable features of catalyst supports?

(d) What are some undesirable features of catalyst supports?

14.7 Earlier in this chapter we said that a support might have a surface area of 100 m^2/gram. Assume that the support consisted of small spherical alumina particles.

(a) Calculate how small the spheres would need to be to get a surface area of 100 m^2/gram.

(b) How many Al_2O_3 units would be in each sphere?

(c) Now think about forming the spheres into pellets, and loading the pellets into a reactor 30 meters tall, 1 meter across. Would spheres be able to support all of this weight? (*Hint*: Think about a building; would one beam be sufficient to support a building? How about multiple beams?)

14.8 People design catalysts with macropores, mesopores, and micropores as indicated in Figure 14.3.

(a) Why do you need macropores? How would the mass transfer into the catalyst change if you had only mesopores and micropores?

(b) Why do you need mesopores? How would the mass transfer into the catalyst change if you had only macropores and micropores?

(c) Why do you need micropores? How would the surface area of the catalyst change if you had only macropores and mesopores?

14.9 Compare and contrast platinum and silver as catalysts.

(a) What types of reaction are catalyzed by platinum?

(b) What types of reactions are catalyzed by silver?

(c) Explain the difference in chemistry on the basis of what you know about the properties of the two metals.

14.10 Review the various ways that metals catalyze reactions.

(a) List the major ways that transition metals speed up reactions.

(b) Give an example of each effect.

(c) How would the reaction differ if you ran them on a lead surface?

(d) How would the reaction differ if you ran them on an aluminum surface?

14.11 One of the key operational problems is avoiding "poisoning." A molecule is called a "poison" if the molecule adsorbs strongly on the catalyst and blocks active sites. In Section 12.17.4, we derived a Langmuir–Hinshelwood expression for the rate of the reaction $A + B \Rightarrow C$.

(a) Derive an expression for the rate of the reaction in the presence of a poison X, which strongly adsorbs on the active sites, but does not otherwise participate in the reaction. Your expression should express the rate as a function of the variables in Section 12.17.4: K_{poison}, the equilibrium constant for adsorption of the poison; and P_{poison}, the partial pressure of the poison.

(b) Assume that you are running a reaction with 10 atm of each reactant, and that the adsorption equilibrium constant for adsorption of each reactant is about 1/atm. Find a value of K_{poison} such that if the partial pressure of the poison were 10^{-5} atm, the rate would be reduced by 50%.

14.12   Figure 14.9 shows the structure of several planes of FCC materials.

(a) Make a photocopy of the figure. Label all of the steps on the photocopy.

(b) Label all of the kinks on a second photocopy.

(c) Make a list of all of the planes in the figure and indicate the orientation (i.e., Miller indices) of the terraces and steps.

14.13   Describe in your own words the structure of Pt(111), Pt(110), Pt(100), Pt(510), Pt(511), Pt(551), Pt(789), W(111), W(110), W(100), W(510), W(511), W(551). W(789) (*Hint*: Platinum is FCC and tungsten is BCC.)

14.14   Make a photocopy of Figure 14.4.

(a) Label all of the steps and kinks visible on the figure.

(b) What is the orientation (i.e., Miller indices) of the steps and terraces?

14.15   Consider the following list of reactions: (1) ethylene hydrogenation, (2) butane dehydrogenation, (3) three-centered isomerization of 3-methylhexane, (4) five-centered isomerization of 3-methylhexane, (5) CO oxidation, (6) ammonia synthesis, (7) ethylene oxidation to ethylene oxide.

(a) Diagram the catalytic cycle for each reaction.

(b) Suggest a catalyst for each reaction.

14.16   Consider the following molecules: $Mo(C_6H_6)_2$, $Zr(cp)_2(CH_3)_2$, $Mo(PPh_3)_4H_4$, $Pt(CO)_3$, $(al)NiCl_2Ni(al)$

(a) Which of them obey the 18-electron rule?

(b) Which of them are likely catalysts?

(c) If you oxidized (i.e., removed an electron from) the cluster, would you expect the catalytic activity to increase or decrease?

14.17   In Example 14.A we showed that iron is likely to be an effective catalyst for ammonia synthesis.

(a) Reproduce the findings yourself and verify that you get the same results.

(b) Consider the following other metals: Pd, Rh, Os, Ir. Which of them would be expected to be active for the reaction?

(c) Consider the Eley–Rideal mechanism (Mechanism II). Which metals would be expected to be active for it?

14.18   For the reaction $2NO + 2CO \Rightarrow N_2 + 2CO_2$

(a) Propose a reasonable mechanism for the reaction.

(b) Consider the following other metals: Pd, Rh, Os, Ir. Use an analysis like that in Example 14.A to indicate which of the metals would be expected to be active for the reaction.

14.19   Reforming catalysts are made by depositing a catalytically active metal on a catalytically active support. Consider the following reactions. Which are likely to occur on a metal such as palladium? Which are likely to occur on the support? *Hint:* the support is a strong acid. How would your conclusions change on platinum?

(a) $RCH_2CH_2R' \rightarrow RCH{=}CHR' + H_2$.

(b) $RCH{=}CHR' + H^+ \rightarrow [RCHCH_2R']^+$

(c) $[RR'CHCH_2]^+ \rightarrow [R_2CHCH_2]^+$

(d) $[R_2CHCH_2]^+ \rightarrow R_2C{=}CH_2 + H^+$

(e) $R_2C{=}CH_2 + H_2 \rightarrow R_2HCCH_3$

14.20 McEwen et al., discovered that platinum was still able to catalyze ethylene hydrogenation even when the platinum was covered by a monolayer of silica.

(a) How is this possible? (*Hint:* Look at Figure 14.18.)

(b) What does this tell you about the role of the metal itself, and about free electrons on rates of catalytic reactions?

(c) Given the results above, why doesn't aluminum or lead catalyze ethylene hydrogenation?

14.21 Somorjai (1994) summarizes data for propane hydrogenolysis on a number of catalysts. The data are summarized in Table P14.21.

(a) How well do these data fit a Sachtler–Fahrenfort plot?

(b) Account for the deviations you observe in part (a). (*Hint:* What key parameter is not considered in the Sachtler–Fahrenfort plot?)

14.22 The decomposition of methylamine has been examined on a number of platinum faces.

(a) Provide a feasible pathway for methylamine decomposition on Pt(111). Be sure to justify your answer.

(b) What products do you expect to see?

(c) How could you form cyanogen ($N{\equiv}C{-}C{\equiv}N$)?

(d) How would the reaction differ on an acid catalyst?

14.23 The hydrogenation of ethylene, $C_2H_4 + H_2 \longrightarrow C_2H_6$, can be run on either a platinum or a nickel catalyst.

(a) Provide a feasible mechanism for the reaction. Be sure to justify your answer.

(b) Guess what the rate-determining step will be. Be sure to justify your answer.

(c) Use the material from Section 12.17 to derive a rate expression for the reaction.

(d) Compare the reactivity of platinum and nickel. Which metal do you expect to be a more active catalyst? What is the advantage of nickel?

Table P14.21   The Rate of Propane Hydrogenolysis[a]

Catalyst	Rate molecule/(cm²·second)	Catalyst	Rate molecule/(cm²·second)
6% Co/SiO$_2$	5×10^{13}	Fe/SiO$_2$	2×10^{10}
7% Co/SiO$_2$	3×10^{14}	W film	2×10^{14}
Ni/SiO$_2$	3×10^{12}	Ru/Al$_2$O$_3$	7×10^{10}
Ni/SiC	1×10^{13}	Pt/Al$_2$O$_3$	1×10^{11}
Ni/Al$_2$O$_3$	2×10^{15}	Pt film	2×10^{11}
Ni powder	4×10^{9}	Pt powder	1.6×10^{9}

[a] All rate data have been extrapolated to 250°C and 10 atm pressure.

(e) Why do people use a platinum or nickel catalyst? What are the advantages of platinum and nickel over (1) iron, (2) copper, (3) tungsten? Consider the catalytic activity, the cost of the catalyst, and the resistance of the catalyst to poisoning by CO.

14.24 The hydrogenation of benzene to cyclohexane is a key step in the production of nylon. The reaction is usually run on either a palladium or a nickel catalyst.

(a) Provide a feasible mechanism for the reaction. Be sure to justify your answer.

(b) Guess what the rate-determining step will be. Be sure to justify your answer.

(c) The catalyst is often pre–sulfided to eliminate hydrogenolysis reactions. How would presulfiding change the selectivity of the reaction? (*Hint:* Consider the ensemble size and the changes in electronic structure during sulfiding—assume that formation of a sulfide is similar to formation of an oxide.)

(d) An alternate approach is to add copper to eliminate hydrogenolysis reactions. How would modifications with copper differ from modifications by sulfur?

(e) Why not instead run the reaction on platinum, copper, rhodium, or cobalt? What would be the advantages and disadvantages of each of these other catalysts? Consider the catalytic activity, the cost of the catalyst, and the resistance of the catalyst to poisoning by CO.

14.25 Hydrofinishing is a process used to reduce the olefin and aromatic concentration in industrial solvents by hydrogenating undesirable species. Generally, one runs the reactions over a nickel catalyst, pretreated treated with copper to reduce the hydrogenolysis activity.

(a) Why is nickel a good catalyst?

(b) What does the copper do?

14.26 People often use palladium catalysts to hydrogenate acetylenes to olefins. Consider the reaction $C_2H_2 + H_2 \rightarrow C_2H_4$:

(a) Provide a feasible mechanism for the reaction. Be sure to justify your answer.

(b) Experimentally, one can run under conditions where the ethylene is not appreciably further hydrogenated to ethane. How is that possible? (*Hint:* Consider the difference in the heat of adsorption of acetylene and ethylene on a palladium catalyst. Which binds more strongly?)

(c) How would the reaction differ on nickel or platinum? (*Hint:* On the basis of the results in Table 6.5, how would the heat of adsorption change?)

(d) Why not run the reaction on a copper catalyst instead? How are the binding energies of various species different on copper and palladium? What other differences are there between the reactivity of copper and palladium?

14.27 One often runs methanol synthesis (i.e., $CO + 2H_2 \rightarrow CH_3OH$) on a Cu/ZnO catalyst. However, during the reaction $Fe(CO)_5$ is formed on the walls of the reaction vessel. The $Fe(CO)_5$ moves onto the catalyst, where it is reduced back to iron metal.

(a) What new chemistry would you expect the iron to do?

(b) What new reaction products could form?

14.28 Copper chromite is often used as a partial hydrogenation catalyst. The copper chromite hydrogenates olefins easily. However, benzene rings are not as easily hydrogenated.

(a) What is it about benzene rings that might make them harder to hydrogenate?

(b) Why would copper chromite be a less active hydrogenation catalyst than platinum?

(c) According to the principle of Sabatier, what do you need to do to get a selective hydrogenation reaction to work?

(d) Explain why platinum usually does total hydrogenation, while copper chromite does selective hydrogenation. (*Hint*: Why does copper do selective oxidation?)

(e) What would happen with increasing temperature and hydrogen pressure? Could you do selective hydrogenation with platinum if the contact time were short enough?

REFERENCES

Abraham, M. H., Kumarsingh, R., Cometto-muñiz, J. E., Cain, W. S. (1998a). *Arch. Toxicol.* **72**, 227.

Abraham, M. H., Kumarsingh, R., Cometto-muñiz, J. E., Cain, W. S., Rosés, M.,Bosch, E., and Díaz, M. L. (1998b). *J. Chem. Soc., Perkin Trans. 2* **11**, 2405.

Agusti et al. (1998b). *J. Phys. Chem. A* **102**, 10723.

Alexander, A. J., and Zare R. N. (1998). *J. Chem. Educ.* **75**, 1105.

Alfassi, Z. B., ed. (1999). *General Aspects of the Chemistry of Radicals*, Wiley, Chichester.

Allen, M. P., and Tildgey, D. J. (1993). *Computer Simulation in Chemical Physics*, Kluwer Academic Publishers, New York.

Allsopp, P. G. (1987). *Agric. and Forest Meteoro.* **41**, 165.

Anderson, J. R., and Boudart, M., eds. (1984–1998). *Catalysis: Science and Technology*, 10 vol. Elsevier, Amsterdam.

Anton, A. B., and Cadogan, D. C. (1991). *J. Vac. Sci Technol. A* **9**, 1890.

Aranda, A., Daële, D., Lebras, G., Poulet, G. (1998). *Int. J. Chem. Kinet.* **30**, 249.

Arrhenius, S. A. (1899). *Z. Phy. Chem.* **4**, 226.

Avogádro, A. (1811). *J. Physi.* **73**, 58.

Avogádro, A. (1814). *J. Physi.* **78**, 131.

Avogádro, A. (1814). *J. Physi.* **778**, 131.

Backman, A. L., and Masel, R. I. (1988). *J. Vac. Sci. Technol.* **6**, 1137.

Backman, A. L., and Masel, R. I. (1992). *J. Vac. Sci. Technol. A* **9**, 1789.

Bader Richard, F. W. (1994). *Atoms in Molecules: A Quantum Theory*, Oxford University Press, Oxford, UK.

Baer, T., and Hase, W. L. (1996). *Unimolecular Reaction Dynamics*, Oxford University Press, Oxford, UK.

Bailey, J. E., and Ollis, D. F. (1977). *Biochemical Engineering Fundamentals*, McGraw-Hill, New York.

Balandin, A. A. (1929a). *Z. Phys. Chem.* **32**, 289.

Balandin, A. A. (1929b). *Z. Phys. Chem.* **33**, 167.

Balandin, A. A. (1958). *Adv. Catal.* **10**, 96.

Balandin, A. A., (1969). *Adv. Catal.* **19**, 1.

Bamford, C., and Tipper, C. F. (1969–1999). ed. *Comprehensive Chemical Kinetics*. Elesvier, NY.

Bard, A. J., and Faulkner, L. R. (1980). *Electrochemical Methods*, Wiley, New York.

Bauschlicher, C. W. (1989). *Recent Advances in Electronic Structure Theory and Their Influence on the Accuracy of ab initio*, National Aeronautics and Space Administration.

Becker, K. H., Geiger, H., and Wiesen, P. (1996). *Int. J. Chem. Kinet.* **28**, 15.

Bell, R. P. (1941). *Acid Base Catalysis*, Oxford University Press, Oxford, UK.

Bell, R. P. (1973). *The Proton in Chemistry*, Chapman & Hall, London.

Bell, R. P., and Higginson, W. C. E. (1949). *Proc. R. Soc. London, Ser. A* **197**, 141.

Bender, C. F., O'Neil, S. V., Pearson, P. K., and Schaefer, H. F. (1972). *Science* **176**, 1412.

Benson, D. S., and Scheeline, A. (1996). *J. Phys. Chem.* **100**, 18911.

Benson, S. W. (1971). *Thermochemical Kinetics*, Wiley, New York.

Benson, S. W. (1976). *Thermochemical Kinetics*, 2nd ed., Wiley, New York.

Benson, S. W., and Buss, J. H. (1958). *J. Chem. Phys.* **29**, 546.

Benson, S. W., and Srinivansan, R. (1955). *J. Chem. Phys.* **23**, 200.

Benziger, J. B. (1991). in E. Shustorovich, ed., *Metal Surface Reaction Energetics*, VCH, New York.

Bernasconi, C. F., and Ni, O. X. (1994). *J. Org. Chem.* **59**, 4910.

Bernoulli, B., (1738). *Hydrodynamica*, Johannes Benhold, Dusekeri, 1738, p. 200. A translation is given in S. G. Brush, *Kinetic Theory*, Pergamon, Oxford, 1966.

Bernstein, R. B., and Muckerman, J. T., (1967). *Adv. Chem. Phys.* **12**, 389.

Berthelot, M. (1862). *Poggendorf's Ann.* **66**, 110.

Bertholl, A. (1912). *J. Chem. Phys.* **10**, 573.

Berzelius, J. J. (1836). *Fort. Physic. Wissenshaft Tubingen*, **243**.

Berzelius, J. J. (1836). *Edinberg New Philos J.* **21**, 223.

Bevington, P. R., and Robinson, D. K. (1992). *Data Reduction and Error Analysis for the Physical Sciences*, McGraw-Hill, New York.

Beyer, T., and Swinehart, E. F. (1973). *Comm. Assoc. Comput. Machin.* **16**, 372.

Blaser, H. U., Jalett, H. P., Garland, M., Studer, M., Thies, H., and Wirthfijani, A. (1998). *J. Catal.* **173**, 282.

Blowers, P., Ford, L., and Masel, R.I. (1998). *J. Phys. Chem.* **102**, 9267–9277.

Blowers, P., and Masel, R. I. (1999a). *AIChE J.* **45**, 1794–1801.

Blowers, P., and Masel, R. I. (1999b). *J. Phy. Chem. A* **103**, 7047–7054.

Bodenstein, M., and Lund, S. C. (1907). *J. Phys.* **49**, 168.

Bodenstein, M., and Lund, S. C. (1907). *Z. Phys. Chem.* **57**, 108.

Boltzmann, L. (1872). *Wein. Ber.* **66**, 275.

Boltzmann, L. (1877). *Wein. Ber.* **74**, 553.

Boltzmann, L. (1877a). *Ann. Phy.* **160**, 175.

Boltzmann, L. (1877). *Ann. Phys.* **160**, 175.

Boltzmann, L. (1877b). *Wein. Ber.* **74**, 553.

Bond, G. C. (1956). *Trans. Faraday Soc.* **52**, 1235.

Bond, G. C. (1987). *Heterogeneous Catalysis: Principles and Applications*, Clarendon Press, Oxford.

Boni, A. A., and Penner, R. C. (1976). *Combust. Sci. Technol.* **15**, 99.

Boudart, M. (1969). *Adv. Catal.* **20**, 153.

Boudart, M., Delboville, A., Dumesic, J. A., Khammouma, S., and Topsoe, H. (1975). *J. Catal.* **37**, 486.

Bowers M. J., (1977). *Gas Phase Ion Chemistry*, Academic Press, New York.

Bowker, M. (1998). *The Basis and Applications of Heterogeneous Catalysis*, Oxford University Press.

Boyle, R. (1660). *New Experiments Physico-Mechanical Touching the Spring of Air* (letter). Reprint given in *Boyle's Works*, Thomas Birch, London, 1772 and S. G. Brush, *Kinetic Theory*, Pergamon, Oxford, 1966.

Boyle, R. (1662) *A Defense of the Doctrine Touching the Spring and Weight of Air*, Oxford University Press, Oxford. Reprint given in *Boyle's Works*, Thomas Birch, London, 1772, and S. G. Brush, *Kinetic Theory*, Pergamon, Oxford, 1966.

Brauman, J. I., and Blair, L. V. (1971). *J. Am. Chem. Soc.* **93**, 3911.

Broadbelt, L. J., Stark, S. M., and Klein, M. T. (1994). *Ind. Eng. Chem. Res.* **33**, 790.

Brønsted, J. N. (1928). *Chem. Rev.* **5**, 231.

Brønsted, J. N., and Pederson, K. (1924). *Z. Phy. Chem.* **108**, 185.

Brovard, M. (1988). *Reaction Dynamics*, Oxford University Press, New York.

Brown, N. F., and Barteau, M. A. (1996). *Journal of Physical Chemistry*, **100**(6), 2269–2278.

Buchanan, J. (1870). *Dtsch. Chem. Ges.* **3**, 485.

Buchanan, J. I. (1871). *Br. Assoc. Rep.* **81**, 67.

Burdick, D. L., and Leffler, W. L. (1989). *Petrochemical Chart.* Pemwell books, Tulsa, UK.

Burner, U., Obinger, C., Paumam, M., Furtmüller, P. G., and Treltle, A. J. (1999). *J. Biol Chem.* **274**, 9494.

Burwell, R. L. (1967). *Catal. Rev.* **57**, 1895.

Butler, T. A. (1932). *Trans. Faraday Soc.* **28**, 379.

Butt, J. B., and Onal, I. (1987). *Trans. Faraday Soc.* **78**, 1982.

Bytheway, I., and Wong, M. W. (1998). *Chem. Phys. Lett.* **282**(3–4), 219–226.

Carneiro, J. W. D., Schleyer, P. V., Saunders, M., Remington, R., Schaefer, H. F., Rauk, A., and Sorensen, T. S. (1994). *J. Am Chem. Soc.* **116**, 3483.

Castro, E. A., Pavef, P., and Santos, J. G. (1999). *J. Org. Chem.* **64**, 2310.

Censione, S. S., Gonzalez, M. C., and Mártife, D. G. (1998). *J. Chem. Soc., Faraday Trans.* **94**, 2933.

Chandler, D. (1987). *Introduction to Modern Statistical Mechanics*, Oxford, New York.

Chang, J. S., and Hong, J. (1995). *J. Biotechnol.* **42**, 189.

Chaplain, M. A. J., and Byrne, H. M. (1996). *Wounds* **8**, 42.

Chapman, D. L., and Underhill K. (1913). *Trans. Chem Soc.* **99**, 498.

Chapman, N. B., and Shorter, J. (1972). *Advances in Linear Free Energy Relationships*, Plenum, New York.

Chapman, N. B., and Shorter, J. (1978). *Correlation Analysis in Chemistry*, Plenum, New York.

Charles, J. A. (1787). Unpublished; cited in *Dictionary of Scientific Biography* **3**, 409 (1805).

Charles, J. A. (1787). Unpublished; cited in *Dictionary of Scientific Biography* **3**, 207 (1971).

Chen, W. H. Lu Pradier, Paul, J., and Flodstrom, A. (1997). *J. Catal.* **177**, 3.

Chiang, C. M., and Bent, B. E. (1992). *Surf. Sci.* **279**, 79.

Chizanowski, W., and Wiekowski, A. (1998). *Langmuir* **14**, 1967.

Chlebicki, S., Shiman, L. Y., Guskou, A. K., and Makarov, M. T. (1997). *Int. J. Chem. Kinet.* **29**, 73.

Christiansen, J. A. (1922). *Z. Phys. Chem.* **103**, 99.

Chuchani, G., and Martin, I. (1997). *J. Phys. Org. Chem.* **10**, 121.

Chung, T. C., and Lu, H. L. (1998). *J. Polym. Sci. A* **36**, 1017.

Clausen, H. (1890). *Jahr. Landwirtsch.* **19**, 893.

Clausius, R. *Ann. Phys.* (Leipzig) **100**, 353 (1857).

Cohen, A. O., and Marcus, R. A. (1968). *J. Phys. Chem.* **72**, 4249.

Cohen, N. (1996). *J. Phys., Chem. Ref. Data* **25**, 1411.

Cohen, N. (1999). In Z. B. Alfassi, ed., *General Aspects of the Chemistry of Radicals*, Wiley Chichester. p. 297.

Cong, Y. U., and Masel, R. I. (1998). *Surface Science*, **396**, 1–15.

Connors, K. A., (1990). *Chemical Kinetics: The Study of Reaction Rates in Solution*, VCH, New York.

Conroy, H., and Malli, G. (1969). *J. Phys. Chem.* **50**, 5049.

Conway, B. E. (1952). *Electrochemical Data*, Elsevier, London.

Conway, B. E., (1985). *Mod. Aspects Electrochem.* **16**, 103.

Cornish-Bowden, A. (1995). *Analysis of Enzyme Kinetic Data*, Oxford University Press, Oxford, UK.

Costa M. M. S., Boldrini, J. L., and Bassanezi, R. C. (1995). *Math. Biosci.* **125**, 191.

Cotton, F. A. (1971). *Chemical Applications of Group Theory*, Wiley, New York.

Crim, F. F. (1996). *J. Chem. Phys.* **100**, 12725.

Dalton, J. (1804) *Philos. Mag.* **19**, 79.

Dalton, J. (1805) *Gilbert's Ann. Phy.*, **21**, 409.

Dalton, J. (1805). *Memoirs of the Literary and Philosophical Society of Manchester* **1**, 244–258.

Davis, T., (1995). *Statistical Mechanics of Phases, Interfaces, and Thin Films*, VCH, New York.

Davy, H. (1817). *Philos. Trans. R. Soc. London* **107**, 77.

Debye, P. J. W., and Hühckel, E. (1923). *Z. Phys.* **24**, 305.

de Levie R. (1986). *J. Chem. Educ.* **63**, 10.

Dobereiner M. (1829). *Ann. Phys. Leipzig* **15**, 301.

Dogandze, R. R. (1971). In N. S. Hush, ed., *Reactions of Molecules at Electrodes*, Wiley, New York, p. 135.

Dorling, T. A., Eastlake, M. J., and Moss, R. L. (1969). *J. Catal.* **14**, 23.

Dunbar, R., and McMahon, T. (1998). *Science* **279**, 194–197.

Dykstra, C. E. (1993). *Introduction to Quantum Chemistry*, Prentice-Hall, Englewood Cliffs, NJ.

Eberhard, J., and Howard, C. J. (1996). *Int. J. of Chem. Kinet.* **28**, 31.

Eley, D. D., and Rideal, E. K. (1940). *Nature* **146**, 401.

Eley, D. D., and Rideal, E. K. (1941). *Proc. Roy Soc London A* **178**, 429.

Emmett, P. H. (1954). *Catalysis*, Academic Press, NY.

Engle, T., and Ertl, G. (1978). *J. Chem. Phys.* **69**, 1267.

Engle, T., and Ertl, G. (1979). *Adv. Catal.* **28**, 1.

Eres, D. Gurnick, M., and McDonald, J. D. (1984). *J. Chem. Phys.* **81**, 5552.

Ertl, G., Knozinger, H., and Weitkamp, J. (1999). *Environmental Catalysis*, John Wiley & Sons, NY.

Ertl, G., Neuman, M., and Streit, K. (1977). *Surf. Sci.* **64**, 393.

Ertl, G., Knoezinger, H., Weitkamp, J., and Knozinger, H. (1997). *Handbook of Heterogeneous Catalysis*, John Wiley & Sons, NY.

Ertl, G., Knoezinger, H., Weitkamp, J., and Knozinger, (1999). *J. Preparation of Solid Catalysis*, John Wiley & Sons, NY.

Espenson, J. H. (1995). *Chemical Kinetics and Reactions Mechanisms*, McGraw-Hill, New York.

Essen, W., and Harcourt, A. V. (1865). *Proc. R. Soc.* **14**, 470.

Estenfelder, M., Lintz, H. G., Stein, B., and Gaube, J. (1998). *Chem. Eng. Process.* **37**, 109.

Evans, M. G., and Polanyi, M. (1935). *Trans Faraday Soc.* **31**, 875.

Evans, M. G., and Polayni, M. (1936). *Trans. Faraday Soc.* **32**, 1333.

Evans, M. G., and Polayni, M. (1937). *Trans. Faraday. Soc.* **33**, 448.

Evans, M. G., and Polayni, M. (1938). *Trans. Faraday. Soc.* **34**, 11.

Evans, D. P., Gordon, J. J., and Watson, H. B. (1937). *J. Chem. Soc.* p. 1430.

Exner, O. (1978). In N.B. Chapman and J. Shorter, eds., *Correlation Analysis in Chemistry*, Plenum, New York, p. 439.

Eyring, H. (1935). *J. Chem. Phys.* **3**, 107.

Eyring H. (1937). *J. Chem Phys.* **3**, 492.

Eyring, H., and Lin, S. M. (1980). *Chemical Kinetics*, Wiley, New York.

Eyring, H., Walter, J. L., and Kimball, G. E. (1944). *Quantum Chemistry*, Wiley, New York, p. 307.

Faber, K. (2000). *Biotransformations in Organic Chemistry*, 4th ed, Springer, NY.

Fahrenfort, J., van Riegen, L. L., and Sachtler, W. H. M. (1960). *Z. Electrochem.* **64**, 216.

Fairley, D. A., Scott, G. B. I., Milligan, D. B., MacLagan, R. G. A. R., and McEwan M. S. (1998). *Int. J. Mass Sprectram.* **172**, 79.

Fajans, K. (1920). *Berichte. Deutsch. Chem. Gesell.* **53**, 643.

Faraday, M. (1819). *Q. J. Sci. Arts* **7**, 106.

Faraday, M. (1834). *Philos. Trans. R. Soc. London* **124**, 55.

Felder, R. M., and Rousseau, R. W. (1978). *Elementary principles of chemical processes*, Wiley.

Fersht, A. (1999). *Structure and Mechanism in Protein Science*, W. H. Freeman, NY.

Fogler, S. (1998). *Elements of Chemical Reaction Engineering*, 3rd ed., Prentice Hall, Upper Saddle River, NJ.

Foresman, J. B., and Frisch, A. E. (1996). *Exploring Chemistry with Electronic Structure Methods*, 2nd ed., Gaussian, Pittsburgh.

Fossey, J., Lefort, D., and Sorba, J. (1995). *Free Radicals in Organic Chemistry*, Wiley, New York.

Frey, F.E. (1934). *Ind. Eng. Chem.* **26**, 198.

Fukuda, K., Nagashima, S., Noto, Y., Onishi, T., and Tamary, K. (1968). *Trans. Far. Soc.* **64**, 522.

Fukui, K. (1975). *The Theory of Orientation and Stereospecificity*, Springer-Verlag, Berlin.

Fukui, K., Yonezawa, T., and Shingo, H. (1952). *J. Phys. Chem.* **20**, 722.

Fukui, K., Yonezawa, T., and Shingo, H. (1957). *J. Phys. Chem.* **26**, 831.

Gates, B. C. (1992). *Catalytic Chemistry*, Wiley, New York.

Gault, F. G. (1981). *Adv. Catalysis* **30**, 1.

Gay-Lussac, J. L. (1802). *Ann. Chi.* **43**, 137.

Gibbs, J. W. (1902). *Statistical Mechanics*, Yale University Press, New Haven, CT.

Gilbert, R. G., and Smith, S. C. (1990). *The Theory of Unimolecular and Recombination Reactions*, Blackwell, Oxford.

Glukhoutsev, M. N., Pross, A., and Radom, L. (1995). *J. Am. Chem. Soc.* **117**, 9012.

Goldfinger, P., Letort, M., and Niclause, N. (1945). *Contributio a l'étude de la structure molecular*, Desuer, Liege, p. 283.

Gordon, R. T., Herm, R. R., and Hershbach, D. R. (1968). *J. Chem. Phys.* **49**, 2084.

Gould, H. C., and Tobochnik, J. (1960). *An Introduction to Statistical Thermodynamics*, Addison-Wesley, Reading, MA.

Gouverneur, V. E., Houk, K. N., de Pascual-Teresa, B., Beno, B., Janda, K. D., and Lerner, R. M. (1993). *Science* **262**, 204.

Gradshteyn, I. S., and Ryzhik, I. M. (1965). *Tables of Integrals, Series and Products.* Academic Press, NY

Grant, G., and Graham, R. W. (1995). *Computational Chemistry*, Clarendon Press, Oxford.

Gray, P., and Scott, S. K. (1990). *Chemical Oscillations and Instabilities*, Clarendon Press, Oxford.

Groebe, K., and Muellerhieser, W. (1996). *Int. J. Radiat. Oncol.* **34**, 395.

Grossman, R. B. (1998). *The Art of Writing Reasonable Organic Reaction Mechanisms*, Springer-Verlag, Berlin.

Guldberg, G. M., and Waage, P. (1869). *Forh. Vidensk. Selsk. Krist.* **35**, 92.

Hammer, B., and Norskov, J. K. (1995). *Nature* **376**(6537), 238–240.

Hammett, L. P. (1937). *J. Am. Chem. Soc.* **59**, 96.

Hammett, L. P., and Pfluger, H. L. (1933). *J. Am. Chem. Soc.* **55**, 4079.

Hansch, C., Leo, A., and Taft, R. W. (1991). *Chem. Rev.* **91**, 165.

Harcourt, A. V., and Essen, W. (1865). *Proc. R. Soc. London* **14**, 470.

Harcourt, A. V., and Essen, W. (1866). *Philos. Trans. R. Soc. London* **156**, 193.

Harcourt, A. V., and Essen, W. (1867). *Philos. Trans. R. Soc. London* **157**, 117.

Harris, S. J. (1990). *Appl. Phys. Lett.* **56**, 2298.

Harris, S. J., and Goodwin, D. G. (1993). *J. Phys. Chem.* **97**, 23.

Hase, W. L., Mondro, S. L., Duchovic, R. J., and Hirst, D. M. (1987). *JACS*, **109**, 2916.

Hasenclever, D., Loeffler, M., and Diehl, U. (1996). *Ann Oncol.* **7**, 95.

Haupfear, E. A. Olson, E. C., and Schmidt, L. D. (1994). *J. Electrochem. Soc.* **141**, 1943.

Häuy, R. (1806). *Traite de Minérologic*, Paris.

Hecht, C. E. (1998). *Statistical Thermodynamics and Kinetic Theory*, Dover, New York.

Hehre, W. J., Radom, L., Schleyer, P. V. R., and Pople, J. A. (1986). *Ab initio Molecular Orbital Theory*, Wiley, NY.

Heinrich, A. (1993). *The Hot-Blooded Insects*. Harvard University Press, Cambridge, MA.

Hershbach, D. R. (1966). *Adv. Chem. Phys.* **10**, 1.

Hershbach, D. R., Johnson, H. S., Pitzer, K. S., and Powell, R. E. (1956). *J. Chem. Phys.* **25**, 736.

Herzfeld, K. F. (1919). *Ann. Phys. (Leipzig)* **59** 735.

Herzfeld, K. F. (1922). *Phys A* **23**, 95.

Higginson, W. C., and Wright, R. H. (1955). *J. Chem. Soc.*, pp. 955, 1551.

Hildebrand, J. H., and Wood, S. E. (1933). *J. Chem. Phys.* **1**, 817.

Hill, T. L. (1986). *An Introduction to Statistical Thermodynamics*, Dover.

Hill, T. L. (1960). *An Introduction to Statistical Thermodynamics*, Addison-Wesley.

Hinshelwood, C. N. (1927). *Proc. R. Soc. London, Ser. A* **113**, 230.

Hinshelwood, C. N. (1946). *Proc. R. Soc. London, Ser. A* **1888**, 1.

Hirst, D. M. (1985). *Potential Energy Surfaces: Molecular Structure and Reaction Dynamics*, Taylor & Francis, London.

Hoffmann, R. (1988). *Solids and Surfaces: A Chemist's View of Bonding in Extended Structures*, VCH Publishers.

Hoffman, R., and Woodward, R. B. (1932). *The Conservation of Orbital Symmetry*, Academic Press, New York.

Holbrook, K., Pilling, M. J., and Robinson, S. H. (1996). *Unimolecular Reactions*, Wiley New York.

Hoover, W. G. (1991). *Computational Statistical Mechanics*, Elsevier, New York.

Hougan, A. O., and Watson, K. M. (1943). *Chemical Process Principles*, Wiley, NY.

Houriti, J., and Polanyi, M. (1933). *Acta Physicochim. URSS* **2**, 505.

Houriti, J. and Polanyi, M. (1934). *Trans. Farady. Soc.* **30**, 1164.

Huang, T., Seebauer, E. G., and Schmidt, L.D. (1987). *Surf. Sci.* **188**, 21.

Hughes, E. D., Ingold, C. K., and Patel, C. S. (1993). *J. Chem. Soc. London*, 526.

Hughes, E. D., Ingold, C. K., and Bateman, L. C. (1940). *J. Chem. Soc. London*, 1017.

Hupke, D. J., and Wu, D. (1977). *J. Am. Chem. Soc.* **99**, 7653.

Hyser, E. S. (1970). *Free Radical Chain Reactions*, Wiley, New York.

Ilao, M. C. Yamamoto, H., and Segawa, K. (1996). *J. Catal.* **161**, 20.

Imbert, F. E. Gnep, N., and Guisnet, M. (1997). *J. Catal.* **172**, 307.

Ingold, C. K. (1969). *Structure and Mechanism in Organic Chemistry*, Cornell University Press, Ithica.

Ingold, C. K., and Nathan, W. S. (1936). *J. Chem. Soc.* **58**, 222.

Isaacs, N. S. (1987). *Physical Organic Chemistry*, Longman, Essex.

Jacobs, P. A. (1982). *Catal. Rev.* **24**, 415.

Jacobs, P. A. (1984). In Delanney, ed., *Characterization of Heterogeneous Catalysts*, Dekker, New York, p. 364.

Jaffe, R. L., Henry, V., and Anderson, J. B. (1973). *J. Chem. Phys.* **59**, 112 8.

Jencks, W. P. (1987). *Catalysis in Chemistry and Enzymology*, Dover, NY.

Jennings, J. R. (1991). *Catalytic Ammonia Synthesis: Fundamentals and Practice*, Plenum NY.

Jenson, F. (1999). *Introduction to Computational Chemistry*, Wiley, New York.

Johnson, B. J., Gonzales, C. A., Gill, P. M. W., and Poble, J. A. (1994). *Chem. Phys. Lett.* **221**, 100.

Johnson, C. D. (1973). *The Hammett Equation*, Cambridge University Press, Cambridge, UK.

Johnston, H. S., and Parr, C. (1963). *J. Am. Chem. Soc.* **85**, 2544.

Juang, D. Y., Lee, J. S., and Wang, N. S. (1995). *Int. J. Chem. Kineti.* **27**, 1111.

Kassel, L. S. (1928a). *J. Chem. Phys.* **32**, 229.

Kassel, L. S. (1928b). *J. Phys. Chem.* **32**, 255.

Kassel, L. S. (1928c). *J. Phys. Chem.* **32**, 1065.

Kassel, S. L. (1932). *The Kinetics of Homogeneous Gas Phase Reactions*, Am. Chem. Soc., New York.

Kassel, L. S. (1935). *Kinetics of Homogeneous Gas Reactions*, Am. Chem. Soc., New York.

Keane, M. A. (1997). *J. Chem. Soc., Faraday Trans.* **93**, 2001.

Kelvin, W. T. (1870). *Nature (London)* **1**, 551.

Kim, E. J., and Gill, W. N. (1995). *J. Electrochem. Soc.* **142**, 676.

Kim, C. S., and Russell, K. C. (1999). *Tetraderon Lett.* **40**, 3835.

Kirkwood, J. G. *J. Chem. Phys.* **2**, 134 (351).

King, D. A. (1981). *The Chemical Physics of Solid Surfaces and Heterogeneous Catalysis*, Elsevier Scientific Pub Co.

Knott, W. J. Proch, D., and Kompa, K. L. (1998). *J. Chem. Phys.* **108**, 527.

Koh, H. J., Han, K. L., Lee, H. W., and Lee, I. (1998). *J. Org. Chem.* **63**, 9834.

Kohn, W., and Sham, L. J. (1965). *Physical Review*, **140**, A1133–A1138.

Kooij, D. M. (1893). *Z. Phys. Chem.* **12**, 155.

Korre, S. C. Klein, M. T., and Quann, R. J. (1997). *Ind. Eng. Chem. Res.* **36**, 2041.

Kramers, H. A. (1940). *Physica (The Hague)* **7**, 284.

Kresge, A. J. (1974). *Chem. Soc. Rev.* **2**, 475.

Kresge, A. J., Caldin, E., and Gold, U. (1975). *Proton Transfer Reactions*, Chapman & Hall, London.

Laidler, K. J. (1965). *Chemical Kinetics*, Harper and Row, New York.

Laidler, K. J. (1987). *Chemical Kinetics*, 3rd ed. Harper & Row, New York.

Laidler, K. J., and Wojciechowski, B. W. (1961). *Proc. R. Soc. London. Ser. A* **260**

Landau, L. D., and Lifshitz, E. M. (1965). Quantum Mechanics: Nonrelativistic theory, Pergamon, NY, p. 79.

Landau, L. D., and Lifshitz, E. M. (1965). *Quantum Mechanics*, 2nd ed, Pergamon Press, Oxford, p. 75.

Langmuir, I. (1912). *J. Am. Chem. Soc.* **34**, 1310.

Langmuir, I. (1913). *J. Am. Chem. Soc.* **35**, 105.

Langmuir, I. (1915). *J. Am. Chem. Soc.* **37**, 1139.

Langmuir, I. (1918). *J. Am. Chem. Soc.* **40**, 1361.

Langmuir, I. (1920). *J. Am. Chem. Soc.* **44**, 2190.

Lapuloulade, J., and Neil, K. S. (1973). *Surf. Sci.* **35**, 288.

Lavosier, A. (1789). *Traiteé Elémentaire de Chimie*, présené dans un order nouveau et d'après les découvertes récentes, avec Figures, Paris, Cuchet.

Lavoisier, A., Marveau, L. B., and Berthollet, C. L. (1787). *Methode de Nomenclature Chimique*, Paris.

Lee, I., Kim, C. K., and Lee, B. S. (1995a). *J. Phys. Org. Chem*, **8**, 473.

Lee, I., Kim, C. K., and Lee, B. S. (1995b). *J. Comput. Chem.* **16**, 1045.

Lee, W. T., and Masel, R. I. (1996). *J. Phys. Chem.* **100**, 10945.

Lee, W. T., and Masel, R. I. (1997). *J. Catal.* **165**, 80.

Lee, Y. T., McDonald, J. D., Lebruton, P. R., and Hershbach, D. R. (1968). *J. Chem. Phys.* **49**, 2447.

Leffler, J. E. (1993). *An Introduction to Free Radicals*, Wiley, New York.

Leitner, D. M., and Wolynes, P. G. (1996). *J. Chem. Phys.* **105**(24), 11226.

Leitner, D. M., and Wolynes, P. G. (1997). *Chem. Phys. Lett.* **280**, 411.

Levine, I. N. (1991). *Quantum Chemistry*, 4th ed., Prentice-Hall, Englewood Cliffs, NJ.

Levine, I. N. (1999). *Quantum Chemistry*, 5th ed., Prentice-Hall, Upper Saddle River, NJ.

Levine, R. D. (1999). *Quantum Mechanics of Molecular Rate Processes*, Dover, New York.

Levine, R. D., and Bernstein, R. B. (1987). *Molecular Reaction Dynamics and Chemical Reactivity*, Oxford University Press, New York.

Lewis, G. N. (1923a). *Thermodynamics and the Free Energy of Chemical Substances*, McGraw-Hill, New York.

Lewis, G. N. (1923b). *Valence and the Structure of Atoms and Molecules*, Am. Chem. Soc., New York.

Lewis, W. C. M. (1916). *J. Chem. Soc.* **109**, 796.

Lewis, W. C. M. (1918). *J. Chem. Soc.* **113**, 471.

Li, I., Yeh, J. M., Price, and Lee, Y. T. (1989). *J. Am. Chem. Soc.* **111**, 5597.

Lindberg, B. J. (1966). *Acta Chem. Scand.* **20**, 1843.

Lindemann, F. A. (1922). *Trans. Faraday Soc.* **17**, 598.

Liu, Z. M., Zhou, X. L., and White, J. M. (1992). *Chem. Phys. Lett.* **198**, 615.

Loffler, D. G., and Schmidt, L. D. (1976a). *J. Catal.* **41**, 440.

Loffler, D. G., and Schmidt, L. D. (1976b). *Surf. Sci.* **59**, 195.

Logan, S. R. (1996). *Fundamentals of Chemical Kinetics*, Longmans, Essex.

Longcharich, R. J., Brown, F. K., and Houk, K. N. (1989). *J. Org. Chem.* **54**, 1129.

Lopes, F., Moreira, R., and Lley, J. (1999). *J. Chem. Soc., Perkin Trans. 2*, p. 431.

Lorenzini, R., and Passoni, L. (1999). *Computer Physics Communications.* **117**(3), 241.

Marshall, R. J., and Marshall, P. (1998). *Int. J. Chem. Kinet.* **30**, 179.

Manogue, W. H., and Katzer, J. R. (1974). *J. Catal.* **32**, 166.

Marcellin, R. (1912). *C. R. Hebd. Seances Acad. Sci.* **151**, 1052.

Marcellin, R. (1914a). *J Chem. Phys.* **12**, 451.

Marcellin, R. (1914b). *C. R. Hebd. Seances Acad. Sci.* **158**, 116.

Marcellin, R. (1920). *C. R. Hebd. Seances Acad. Sci.* **161**, 1052.

March, J. (1992). *Advanced Organic Chemistry: Reactions, Mechanisms, and Structure*, 4th ed., Wiley, New York.

Marcus, R. A. (1952). *J. Chem. Phys.* **20**, 359.

Marcus, R. A. (1955). *J. Chem. Phys.* **24**, 966.

Marcus, R. A. (1964). *Annu. Rev. Phys. Chem.* **15**, 155.

Marcus, R. A. (1968). *J. Chem. Phys.* **72**, 891.

Marcus, R. A. (1969). *J. Chem. Phys.* **91**, 7224.

Marcus, R. A., and Calvin, M. E. (1977). *J. Chem. Phys.* **67**, 2609.

Marcus, R. A., and Rice, O. K. (1951). *J. Phys. Colloid. Chem.* **55**, 894.

Marcus, Y. (1998). *The Properties of Solvents*, Wiley, Chichester.

Martens, J. A., and Jacobs, R. A. (1990). In J. B. Moffit, ed., *Theoretical Aspects of Heterogeneous Catalysis*, Van Nostrand-Reinhold, New York.

Masel, R. I. (1986). *Catal. Rev.* **28**, 335.

Masel, R. I. (1996). *Principles of Adsorption and Reaction on Solid Surfaces*, Wiley, New York.

Maslack, P. Vallombroso, T. M. Chapman, W. H., and Narvaez, J. N. (1994). *Anger. Chem.* **33**, 73.

Maxwell, J. C. (1860). *Philos. Mag.* **19**, 19.

Maxwell, J. C. (1860). *Philos. Mag.* **20**, 19.

Maxwell, J. C. (1860). *Philos. Mag.* **20**, 21.

McCarty, J., Falconer, J., and Madix, R. J. (1973). *J. Catalysis* **30**, 235.

McEwen, A. B., Maier, W. F., Flemins, R. H., and Bauran, S. M. (1987). *Nature (London)* **329**, 531.

McKenney, D. J., Wojciechowski, B. W., and Laidler, K. J. (1963). *Can. J. Chem.* **41**, 1954.

McLewis, W. C. (1918). *J. Chem. Soc.* **113**, 471.

McQuarrie, D. A. (2000). *Statistical Thermodynamics, University Science Books*, Mill Valley CA (note: new edition due in 2000).

Menschutkin, N. (1890). *Z. Phys. Chem.* **6**, 41.

Metropolis, N., Rosenbluth, A. W., Rosenbluth, M. N., Teller, A. H., and Teller, A. E. (1953). *J. Chem. Phys.* **21**, 1087.

Meyer, V., and Raum, W. W. (1895). *Chem. Ges. Ber.* **28**, 2804.

Meyers, H. (1927). *Exp. Zool.* **49**, 1.

Mezaki, R., and Inoue, H. (1991). *Rate Equations of Solid-Catalyzed Reactions*, University of Tokyo Press, Tokyo.

Michaelis, L., and Menten, M. L. (1913). *Biochem Z.* **49**, 333.

Miller, A. (1999). *Writing Mechanisms in Organic Chemistry*, Academic Press, San Diego, CA.

Miller, J. R., Calcetewa, L. T., and Closs, G. L. (1984). *J. Am. Chem. Soc.* **106**, 3047.

Miller, W. H. (1839). *A. Treatise on Crystallography*, J. T. Deighton, Cambridge, UK.

Miller, W. H. (1976). *Acc. Chem. Res.* **9**, 306.

Miller, W. H. (1991). In J. R. Bolton, N. Magataga, and G. McLendon, eds., *Electron Transfer in Organic, Inorganic and Biological Systems*, Adv. Chem. Ser., Vol. 228, Am. Chem. Soc., Washington, DC.

Miller, W. H. (1993). *Acc. Chem. Res.* **26**, 174.

Miller, W. H. (1998). *Faraday Discuss.* **110**, 1.

Mondre, S. L., Duchovic, R. J., Hurst, D. M., and Hase, W. U. (1987). *J. Am. Chem. Soc.* **109**, 2916.

Monod, J. (1942). *Recherches sur la Croissance des Cultures Bactériénnes.* Hermann, Paris.

Moore, J. W., and Pearson, R. G. (1981). *Kinetics and Mechanism*, Wiley, New York.

Moore, J. W., and Pearson, R. G. (1986). *Kinetics and Mechanism*, 2nd ed., Wiley, New York.

Mortier, W. J. in J. Moffat, ed. (1991). *Theoretical Aspects of Heterogeneous Catalysis*, Van Nsotrand Reinhold, NY, p. 135.

Newton, I. (1687). *Philosophiae Naturalis Principia Mathematica*, London. A translation is given in S. G. Brush, *Kinetic Theory*, Pergamon, Oxford, 1966.

O'Hanlon, J. F. (1980). *A User's Guide to Vacuum Technology*, Wiley, New York.

Olah, G., Burrichier, A., Rasell, G., Gnamn, R., Chrisie, K. O., and Prakagh, G. K. S. (1997). *J. Am. Chem. Soc.* **119**, 8035.

Olmstead, W. N., and Brauman, J. L. (1977). *J. Am. Chem. Soc.* **99**, 4219.

Oppenheimer, and Handlon, (1992). *Enzyme* **20**, 453.

Ostwald, F. W. (1884). *Prak. J. Chem.* **28**, 385.

Ostwald, W. H., (1885). *Lehrbuch der Allgemeinen Chemie*, Liepzig.

Ostwald, W. H. (1902). *Grundiss der Allgemeinen Chemie*, Liepzig.

Oxley, J. C., Smith, J. L., Zheng, W., Rogers, E., and Coburn, M. D. (1997). *J. Phys. Chem. A* **101**, 5646.

Park, Y. O., Banholzer, W. F., and Masel, R. I. (1985). *Surface Sci*, **155**, 341.

Park, Y. O., Banholzer, W. F., and Masel, R. I. (1985). *Adv. Surface Sci.* **19**, 145.

Parker, A. J. (1969). *Chem. Rev.* **69**, 1.

Pauling, L. (1931a). *J. Am. Chem. Soc.* **53**, 1367.

Pauling, L. (1931b). *J. Am. Chem. Soc.* **53**, 3225.

Pauling, L., and Yost, D. M. (1932). *Proc. Natl. Acad. Sci. U.S.A.* **18**, 414.

Pauling, L. (1939). *The Nature of the Chemical Bond*, Cornell University Press, Ithaca, NY.

Pauling, L. (1960). *The Nature of the Chemical Bond*, 3rd ed., Cornell University Press, Ithaca, NY.

Pearson, M. J., Rabonowitz, B. A., and Whitten, G. Z. (1965). *J. Chem. Phys.* **42**, 2470.

Pearson, R. G. (1969). *J. Am. Chem. Soc.* **91**, 4947.

Pearson, R. G. (1976). *Symmetry Rules for Chemical Reactions*, Wiley, New York.

Pease, R. N. (1928). *J. Am. Chem. Soc.* **50**, 1179.

Pease, R. N., and Stewart, R. J. (1925). *J. Am. Chem. Soc.* **47**, 1235.

Pederson, K. J. (1934). *J. Phys. Chem.* **38**, 581.

Perdew, J. P., and Wang, Y. (1992). *Phys. Rev. B* **45**, 13244.

Perkins, J. M. (1994). *Radical Chemistry*, Ellis Horwood, New York.

Perrin, J. (1919). *Ann. Phys. (Leipzig)* **11**, 1.

Pines, H. (1991). *The Chemistry of Catalytic Hydrocarbon Conversion*, Academic Press, San Diego, CA.

Pines, H., and Stalick, W. M. (1977). *Base Catalyzed Reactions of Hydrocarbons*, Academic Press, New York.

Polanyi, J. C., and Woodall, K. J. (1972). *J. Chem. Phys.* **57**, 1574.

Polanyi, M. (1931). *Phys. Chem. B* **12**, 279.

Polanyi, M. (1932). *Atomic Reactions*, Williams-Norgate, London.

Polanyi, M. (1935). *Trans. Faraday Soc.* **31**, 875.

Ponec, V. (1983). *Adv. Catal.* **32**, 149.

Ponec, V., and Bond, G. (1995). *Catalysis by Metals*, Elsevier Amsterdam.

Pople, J. A., Anthony, P. S., Wong, M. W., and Radom, L. (1993). *Israel Journal of Chemistry*, **33**, 345.

Press, William, H. ed. Saul A. Teukolsky, Michael Metcalf, (1996). *Numerical Recipes in Fortran 90* Oxford University Press.

Priestley, J. (1778). *Experiments on Different Kinds of Air*, J. Johnson, London.

Priestley, J. (1790). *Experiments on Different Kinds of Air*, 2nd ed., J. Johnson, London.

Prins, R., Jian, M., and Flechsenhar, M. (1997a). *Polyhedron* **16**, 3235.

Prins, R., Jian, M., and Flechsenhar, M. (1997b). *J. Catal.* **168**, 491.

Pross, A. (1985). *Adv. Inorg. Chem.* **21**, 99.

Pross, A. (1995). *Theoretical and Physical Principles of Organic Reactivity*, Wiley, New York.

Ranley, J. R., Rust, F. F., and Vaughn, W. E. (1948). *J. Am. Chem. Soc.* **70**, 88.

Rehm, D. R., and Weller, A. (1970). *Is. J. Chem.* **8**, 259.

Reichardt, C. (1988). *Solvents and Solvent Effects in Organic Chemistry*, 2nd ed., VCH, New York.

Rhodin, T. N., and Ertl, G. (1988). *The Nature of the Surface Chemical Bond*, North-Holland Pub. Co.

Rhorig, M., and Wagner, H. G. (1994). *Ber. Bunsenges. Phys. Chem.* **98**, 858.

Rice, F. O. (1932). *The Aliphatic Free Radicals*, Baltimore, MD.

Rice, F. O., and Herzfeld, K. F. (1934). *J. Am. Chem. Soc.* **56**, 284.

Rice, O. K., and Ramsperger, H. C. (1927). *J. Am. Chem. Soc.* **49**, 1616.

Rice, O. K., and Ramsperger, H. C. (1928). *J. Am. Chem. Soc.* **50**, 617.

Roberts, P. P., and Steele, A. J. (1994). *J. Chem. Soc., Perkin Trans. 2*, p. 2155.

Rodriguez, J. A., Campbell, R. A., and Goodman, D. W. (1992). *J. Vac. Sci. Technol., A* **10**, 2540.

Rootsaert, W. J. M., and Farcus, W. H. M. (1935). *Orthohydrogen, Parahydrogen and Heavy Hydrogen*, Cambridge University Press, Cambrige, UK.

Roseveare, W. E. (1931). *J. Am. Chem. Soc.* **53**, 1651.

Sabatier, (1913). *Catalysis in Organic Chemistry*, Paris; reprinted in English, Van Nostrand, New York, 1923.

Sachdev, A., Masel, R. I., and Adams, R. B. (1992). *J. Catal.* **136**, 320.

Sachtler, (1960). *Z. Phys. Chem.* **26**, 16.

Sachtler, W. H. M., and Fahrenfort, J. (1958). *Proc. Int. Cong. Catal., 1st.* 1958.

Salazar, M. G., Urena, A. G., and Roberts, G. (1997). *Isr. J. Chem.* **37**, 353.

Sander, M., Imbihl, R., and Ertl, G. (1992). *J. Chem. Phys.,* **97**, 5193.

Sauer, M. L., and Ollis, D. F. (1994). *J. Catal.* **149**, 81.

Schatchard, G. (1931). *Chem. Rev.* **8**, 321.

Schatchard, G. (1932). *Chem. Rev.* **10**, 229.

Schatz, G. C. (1998). In W. Hase, ed., *Advances in Classical Trajectory Methods*, JAI Press, Stamford, CT, p. 205.

Scheele, C. W. (1777). *Chemische Ubhandung von der Luft und dem Fever*, Germann Press, Upsulla. An English translation appears in L. Dobbin, *The Collected Works of Carl Willhelm Scheele*, Bell, London, 1931.

Schlatter, J. C., and Boudart, M. (1972). *J. Catal.* **17**, 482.

Schmeer, G. Six, C., and Stemkirchner, J. (1999). *Journal of Solution Chem.* **28**, 211.

Schneider, F. W., and Rabinovitž, B. S. (1962). *J. Am. Chem. Soc.* **84**, 4215.

Schürch, M., Heinz, T., Aeschimann, R., Mallat, T., Pfaltz, A., and Baiker, A. (1998). *J. Catal.* **173**, 187.

Schwab, G. G. (1883). *Chem. Central Bl.* **14**, 403, 417.

Schwab, G. M., and Schwab-Agallidis, E. (1943). *Berichte. Deutsch. Chem. Gesell.* **76**, 1228.

Schwartz, S. B., Schmidt, L. D., and Fisher, G. B. (1986). *J. Phys. Chem.* **90**, 6194.

Scoles, G. (1988). *Atomic and Molecular Beams*, Oxford University Press, Oxford, UK.

Scott, A. P., and Radom, L. (1996). *Journal of Physical Chemistry*, **100**(41), 16502–16513.

Scott, G. B., Fairley, D. A., Freeman, C. G., McEwan, M. J., Spanel, P., and Smith, D. (1997). *J. Chem. Phys.* **106**, 3982.

Semenov, N. N. (1935). *Chemical Kinetics and Chain Reactions*, Oxford University Press, Oxford, UK.

Shaik, S. S., Schlegel, H. B., and Wolfe, S. (1992). *Theoretical Aspects of Physical Organic Chemistry*, Wiley, New York.

Shapley, H. (1920). *Proc. Nat. Acad. Sci. U.S.A.* **6**, 241.

Shapley, H. (1924). *Proc. Natl. Acad. Sci. U.S.A.* **10**, 436.

Shustorovich, E. (1986). *Sur. Sci. Rep.* **6**, 1.

Shustorovich, E., ed. (1991). *Metal-Surface Reaction Energetics: Theory and Applications to Heterogeneous Catalysis, Chemisorption, and Surface Diffusion*, VCH, New York.

Sinfelt, J. H. (1973a). *Adv. Catal.* **23**, 91.

Sinfelt, J. H. (1973b). *J. Catal.* **29**, 308.

Sinfelt, J. H., Carter, H. L., and Yates, D. J. (1972). *J. Catal.* **24**, 283.

Sinfelt, J. H., Bent, R. E., Kao, C. T., and Somorjai, G. A. (1988a). *Surf. Sci.* **202**, 388.

Sinfelt, J. H., Bent, R. E., Kao, C. T., and Somorjai, G. A. (1988b). *Surf. Sci.* **206**, 124.

Sinfelt, J. H., and Lucchesi, P. J. (1963). *J. Am Chem. Soc.* **85**, 3365.

Sinnott, M. (1997). *Comprehensive Biological Catalysis*, Academic, Pr NY.

Smith, A. M. (1992). *Environ. Entomol.* **21**, 314.

Smith, S. G., Fainberg, A. A., and Winstein, S. (1961). *J. Am. Chem. Soc.* **83**, 618.

Somorjai, G. A. (1994). *Introduction to Surface Chemistry and Catalysis*, Wiley, NY.

Spouge, J. L. (Shrager, R. I., Dimitrov, D. S.) (1996). *Math. Biosci.* **138**, 1.

Stark, K., and Werner, H. J. (1996). *J. Chem. Phys.* **104**, 6515.

Steckler, R., Truhlar, D. G., and Garrett, B. C. (1985). *J. Chem. Phys.* **82**, 5494.

Steger, R., and Masel, R. I. (1998). *Thin Solid Films* **336**, 1–9.

Steger, R., Cadwell, L., and Masel, R. I. (1994). *Proc. AIChE Top. Con., 'Synth. Process. Electron. Mater.* (1994), p. 142.

Steinfeld, J. I., Fransisco, J. S., and Hase, W. L. (1989). *Chemical Kinetics and Dynamics*, Prentice-Hall, Englewood Cliffs, NJ.

Steinfeld, J. I., Fransisco, J. S., and Hase, W. L. (1998). *Chemical Kinetics and Dynamics*, 2nd ed., Prentice Hall, Upper Saddle Rever, NJ.

Sugai, S., Takeuchi, K., Ban, T., Miki, H., Kawasaki, K., and Kioka, T. (1993). *Sur. Sci.* **282**, 67.

Swihart, M. T., and Carr, R. W. (1998). *J. Phys Chem.* **102**, 1542.

Szabo, A., and Ostlund, N. S. (1996). *Modern Quantum Chemistry: Introduction to Advanced Electronic Structure Theory*, Dover, New York.

Tafel, J. (1905). *Z. Phys. Chem.* **50**, 641.

Talkner, P., and Hanggi, P. (1995). *New Trends in Kramer's Reaction Rate Theory*, Kluwer Academic Publishers, Boston.

Tamaru, K., Boudart, M., and Taylor, H. S. (1955). *J. Chem. Phys.* **59**, 801.

Tanaka, K., and Tamaru, K. (1963). *J. Catal.* **2**, 366.

Tapia, O., and Bertran, J. (1996). *Solvent Effects and Chemical Reactivity*, Kluwer Academic Publishers, New York.

Taylor, H. S. (1914). *Z. Elektrochem.* **20**, 201.

Taylor, H. S. (1925). *Proc. R. Soc. London, Ser. A* **108**, 105.

Taylor, H. S. (1948). *Adv. Catal.* **1**, 1.

Thenard, J. (1818). *Ann. Chem. Phys.* **9**, 314.

Thomas, J. A. (1976). *Adv. Catal.* **19**, 108.

Thomas, J. M., and Thomas, W. J. (1996). *Principles and Practice of Heterogeneous Catalysis*, Wiley, New York.

Thon, N. (1926). *Z. Phys. Chem.* **124**, 327.

Thompson, D. L. (1998). *Int. Rev. Phy. Chem.* **17**, 547.

Tolman, R. C. (1920). *J. Am. Chem. Soc.* **42**, 2506.

Tolman, R. C. (1927). *Statistical Mechanics with Applications to Chemistry and Physics*, Am. Chem. Soc., New York.

Tolman, R. C. (1938). *The Principles of Statistical Mechanics*, Clarendon Press, Oxford.

Trautz, M. (1916). *Z. Anorg. Chem.* **96**, 1.

Trautz, M. A. (1918). *Z. Anorg. Chem.* **102**, 81.

Trulhar, D. G., and Garrett, B. C. (1984). *Annu. Rev. Phys. Chem.* **35**, 159.

Truhlar, D. G., Hase, W. L., and Hanes, J. T. (1983). *J. Chem. Phys.* **87**, 2264.

Truhlar, D. G., Isaacson, A. D., and Garier, B. C. (1985). In M. Baer, ed., *The Theory of Chemical Reactions*, Vol. 6, CRC Press, Boca Raton, FL, p. 65.

Truhlar, D. G., Garrett, B. C., and Klippenstein, S. J. (1996). *J. Phys. Chem.* **100**(31), 12771–12800.

Tsukiyama, K., Katz, B., and Bersohn, R. (1988). *J. Chem. Phys.* **84**, 1934.

U.S. Chemical Industry. (1997). *Statistical Handbook.* Chemical Manufacturers Association, Arlington, Virginia.

Vannice, M. A., and Poondi, D. (1997). *J. Catal.* **169**, 166.

Van't Hoff, J. H. (1878). *Ansichren Uber die Organishe Chemie*, Braunsweig Press.

Van't Hoff, J. H. (1883). *Studies in Chemical Dynamics*, Edward Arnold, London.

Van't Hoff, J. H. (1884). *Etudes de Dynamique Chemie*, Muller, Amsterdam.

Van't Hoff, J. H. (1886). *Etudes de Dynamique Chemie*, Edward Arnold, London.

Van't Hoff, J. H. (1896). *Studies in Chemical Dynamics*, Edward Arnold, London.

Vetter, K. J. (1967). *Electrochemical Kinetics*, Academic Press, New York.

Vogel, P. (1985). *Carbocation Chemistry*, Elsevier, New York.

Volmer, M., and Erdey-Gruz, T. (1930). *Z. Phys. Chem.* **150**, A203.

Von Steiger, A. L. (1920). *Berichte. Deutsch. Chem. Gesell.* **53**, 666.

Vosko, S. H., Wilk, L., and Nusair, M. (1980). *Canadian J. Phys.* **58**, 1200.

Wachs, I. E., and Madix, R. J. (1978). *J. Catal.* **53**, 208.

Walker, T. J. (1962). *Evolution (Lawrence, Kans.)* **16**, 407.

Wang, J., and Masel, R. I. (1991). *J. Am. Chem. Soc.* **113**, 5850.

Wang, J., and Masel, R. I. (1991). *Surf. Sci.* **243**, 199.

Wlodawer, A., Erickson, J. W. (1993). *Annu. Rev Biochem.* **62**, 543–585.

Wasserman, A. (1952). *Monatsh. Chem.* **83**, 543.

Wasserman, R., Achayai, R., Sibatai, C., Shm, K. H. (1996). *Math. Biosci.* **136**, 111.

Weaver, L. H., and Matthews, B. W. (1987). *J. Mol. Biol.* **193**, 189.

Weinberg, W. H. (1992). *J. Vac. Sci. Technol., A* **10**, 2271.

Welling, R. G., Lyons, L. L., Elliot, R., Amidon, G. L. (1977). *J. Clin. Pharmacol.* **17**, 199.

Wells, R. P. (1968). *Linear Free Energy Relationships*, Academic Press, London.

Westley, F. (1980). *Table of Recommended Rate Constants for Chemical Reactions Occurring in Combustion*, NBS, Washington, DC.

Whitten, G. Z., and Rabinovitz, B. S. (1964). *J. Chem. Phys.* **41**, 1883.

Wigner, E. (1932). *Z. Phys. Chem. B* **15**, 203.

Wigner, E., and Pelzer, H. (1932). *Z. Phys. Chem. B* **15**, 445.

Wijngaarden, R. I., and Westerterp, K. R. (1998). *Industrial Catalysts*, Wiley–VCH, New York.

Wilhelmy, R. L. (1850). *Poggendorf's Ann.* **81**, 413.

Woodward, R. B., and Hoffmann, R. (1965a). *J. Am. Chem. Soc.* **87**, 395.

Woodward, R. B., and Hoffmann, R. (1965b). *J. Am. Chem. Soc.* **87**, 2046.

Woodward, R. B., and Hoffmann, R. (1965c). *J. Am. Chem. Soc.* **87**, 2511.

Woodward, R. B., and Hoffmann, R. (1970). *The Conservation of Orbital Symmetry*, Verlag Chemie, Deerfield Beach, FL, (1992).

Wright, J. S. (1975). *Can J. Chem.* **53**, 549.

Wright, J. S. (1970). *Chem. Phys. Lett.* **6**, 476.

Xi, M., and Bent, B. E. (1994). *J. Vac. Sci. Technol., B* **10**, 2440.

Yagasaki, E., and Masel, R. (1994). *Catal. Spec. Rep., R. Soc. Chem.* **11**, 165.

Juang, D. Y., Lee, J. S., and Wang, N. S. (1995). *Int. J. Chem. Kinet.* **27**, 111.

Young, R. J., and Lovell, P. (1991). *Introduction to Polymers*, Chapter 2, Chapman & Hall, London.

Zavitsas, A. A. (1998). *J. Chem. Soc., Perkin Trans. II*, p. 499.

Zhong, M., and Brauman, J. L. (1999). *J. Am. Chem. Soc.* **121**, 2508.

Zhou, X. L., and White, J. M. (1993). *J. Vac. Sci. Technol., A* **11**, 2210.

Zhou, X. L., Blass, P. M., Koel, B. E., and White, J. M. (1992). *Sur. Sci.* **271**, 427, 452.

INDEX